HARMSWORTH
NATURAL HISTORY

WITH PHOTOGRAPHS AND SKETCHES FROM LIFE

CHIEF CONTRIBUTORS

RICHARD LYDEKKER, F.R.S.
SIR HARRY JOHNSTON
PROFESSOR J. R. AINSWORTH-DAVIS, M.A.

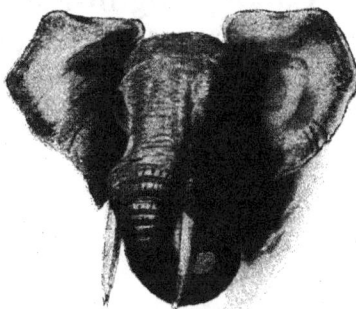

THIRD VOLUME

British Library Cataloguing-in-Publication Data
A catalogue record for this book is available from
the British Library

CONTENTS OF VOLUME III

VERTEBRATES—*continued*

INVERTEBRATES, OR BACKBONELESS ANIMALS

LIST OF SPECIAL COLOUR PLATES

ILLUSTRATIONS IN TEXT

ILLUSTRATIONS IN TEXT—*continued*

iv

GULLS

"The bird on the left is the great black-headed gull; to the right of it, in the foreground, is the lesser black-backed gull, on the right of which are glaucous gulls. Behind these, the laughing gull is flying, and a pair

HERRING-GULLS

"'Herring-gulls are in the habit of following the shoals of the fish from which they take their name, and may often be seen hovering above the fry, preparatory to taking a plunge among them in the water.'"

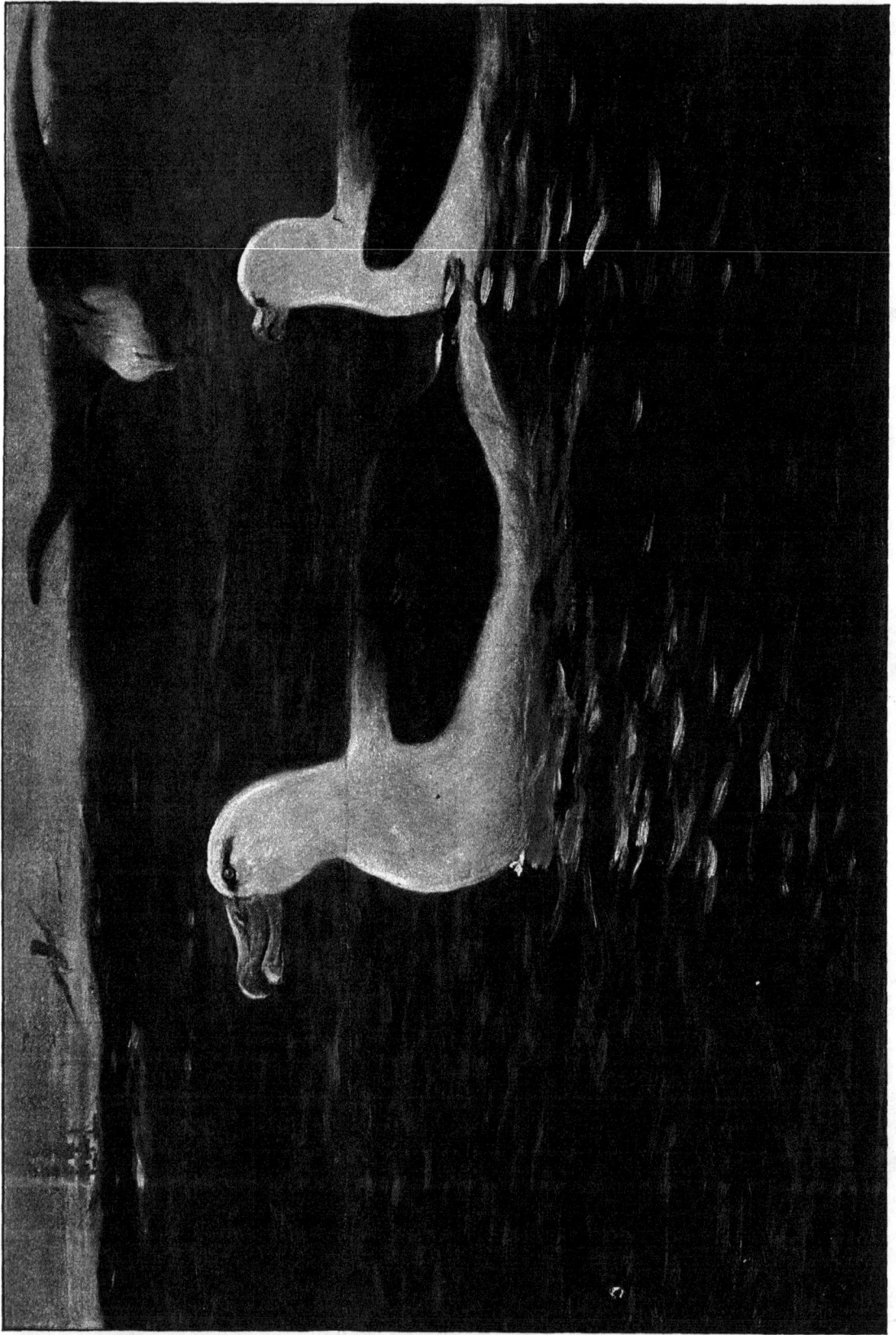

ALBATROSSES

"Although during calm or moderate weather this splendid bird sometimes rests on the surface of the water, it is almost constantly on the wing, and is equally at ease while passing over the glassy surface during the stillest calm, or flying with meteor-like swiftness before the most furious gale."

GUILLEMOT

PUFFIN

STORM-PETREL.

LITTLE GREBE

"The storm-petrel, except during the prevalence of severe storms and in the breeding season, is but seldom seen in the neighbourhood of land. Essentially a child of the ocean, it is frequently met with far out at sea, where it will follow vessels for considerable distances."

HARMSWORTH NATURAL HISTORY

THIRD VOLUME

BIRDS. ORDER XV. GAVIÆ

CHARACTERISTICS OF GULLS

FORMERLY associated with the petrels, the terns, skimmers, gulls, and skuas (order Gaviæ) are now regarded as nearly allied to the Limicolæ, with which they agree in the arrangement of their plumage (page 1332). Externally these birds are characterised by the prevalence of pure grey and white in the adult plumage and the complete webbing of the three front toes, as well as by their long wings, in which the fifth secondary quill is wanting. Their skulls differ from those of the typical Limicolæ in the absence of basi-pterygoid processes on the inferior surface of the rostrum, while the hind extremity of the lower half of the beak is abruptly truncated at its articulation with the skull ; and in the wing the flat bone corresponding to the first joint of the human forefinger has two circular perforations — a feature distinguishing the skeleton from that of any member of the plover tribe and their allies. Throughout the group there are deep grooves on the upper surface of the skull for glands, the development of these being very variable among the Limicolæ.

Except in the skimmers, the beak is simple, and may be either straight or hooked. In the wing there are ten large primaries, and one minute and concealed quill ; the whole plumage is remarkably compact, the contour feathers having after-shafts ; there are twelve tail feathers; the spinal feather-tract is well defined by bare lateral areas on the neck, and forked on the upper part of the back; and the oil-gland is tufted. In their down-clad and active young these birds resemble plovers, but the down is of a more complex type.

The first toe is raised above the level of the others, with which it is not connected by membrane, and the nasal apertures in the skull are of the long schizorhinal type, while the external nostrils are elongated, and placed rather low down on the sides of the base of the beak. In the general structure of the palate, as well as in the presence of a process on the outer side of the humerus, or upper wing-bone, gulls resemble plovers. Rarely, if ever, exceeding three in number,

the eggs are spotted or scrolled with dark markings on a light-coloured ground.

Gregarious and noisy in their habits, gulls and terns are chiefly frequenters of the coasts, although many of them may be found on inland waters, while all may be driven inland by stress of weather. Even in England gulls may be seen following the plough in search of worms, and in parts of Argentina, at a distance of some two hundred miles from the sea, they appear to dwell permanently inland, nesting in the lagunas. In Britain, while some species breed on coast cliffs, others nest on islands and inland lakes, grassy downs, and peat mosses.

All the members of the group are birds of powerful and sustained flight, and capable of floating in the air with scarce a movement of their wings, while they are equally at home on the surface of the water, where their webbed feet enable them to swim with facility. Terns are, however, more essentially aerial and aquatic birds than gulls, their short legs not being well adapted for walking on land. In the neighbourhood of the sea the food of all these birds consists mainly of fish and refuse, but when inland they consume worms, insects, the eggs and young of other birds, and offal.

Although the group as a whole has a cosmopolitan distribution, it is noteworthy that in

HERRING-GULLS

the great area lying between South America and the neighbourhood of Australia and New Zealand not a single gull is to be met with, although terns are abundant. Along the southern shores of Australia, and also in New Zealand, a large, dark-mantled gull (*Larus pacificus*) makes its appearance as an isolated form. Geologically, the group is an old one, remains of an extinct genus (*Halcyornis*) occurring in the London Clay, a formation belonging to the Eocene, or lowest, division of the Tertiary period ; while other forms, which have been assigned to the existing genus *Larus*, occur in beds pertaining to the lower portion of the Miocene period. The order may be divided into two families, the first of which is again split up into three subfamilies.

CLASSIFICATION OF GAVIÆ MENTIONED IN THIS WORK

TERNS, SKIMMERS, AND GULLS

INCLUDING the three groups terns, skimmers, and gulls, each of which represents a subfamily, the present family (*Laridæ*) is characterised as follows. The beak has no cere at its base; there are two notches on each side of the hind margin of the breast-bone; the toes may be either partially or fully webbed; and the claws are feeble or of moderate length.

MARSH-TERNS

The terns, of which eleven genera are here recognised, constitute a subfamily (*Sterninæ*) characterised by the straight and rather slender beak, in which the two halves are of nearly equal length, by the feather-tracts resembling those of the plovers in their arrangement, and by the slight or distinct forking of the tail. In Britain the group is represented by two genera, and it is to these that attention is here mainly confined. The marsh-terns form a genus (*Hydrochelidon*) represented by four species, three of which are British. Belonging to a group of genera in which the tail may be either nearly square or deeply forked, and the head devoid of elongated plumes at the gape of the mouth, marsh-terns are specially distinguished by the rounded or slightly pointed tail feathers, the short tail, which is less than half the length of the wing, the small beak, with the culmen less than twice the length of the metatarsus, and the feeble feet, in which the webs are considerably indented between the toes. All these terns nest in marshes, either on tussocks of grass or among floating vegetation.

The best known of the British species is the black tern (*H. nigra*), in which the under parts in the adult summer plumage are dark leaden grey, the upper tail coverts and tail being grey, the beak black, the chin and sides of the face like the under parts, the crown nearly black, and the under wing coverts pale grey. On the other hand, the white-whiskered tern (*H. hybrida*), which is but an occasional straggler to Britain from the south, has the beak blood-red, the chin

and sides of the face pure white, the throat and breast grey, passing into a blackish grey on the abdomen, and the under wing coverts white. Abundant in Southern Europe, this tern ranges over a large part of Africa and most of Asia.

The beautiful white-winged black tern (*H. leucoptera*), which is likewise a rare visitor to the British Isles, differs from both the preceding in that the upper tail coverts and tail are white in the adult summer plumage, the under parts being black as far as the vent, and the under wing coverts of the same hue, while the beak is dull red. Rare in Northern Europe, this species is more common in the south and east, whence its range extends over the greater portion of Europe. Mr. H. Saunders writes that "the black tern breeds in colonies, the nest being situated in marshes, and formed of decayed pieces of equisetum and other plants, or heaps of wrack, which rise and fall with the tide; sometimes they are placed on the firmer hummocks of bog in the middle of

BLACK TERN AND YOUNG

THE TERN

shallow parts. The eggs are three in number, of various shades of ochreous clay, olive brown, or olive green, blotched with dark brown, especially at the larger end. The food of this tern consists chiefly of beetles and dragon-flies, with some small fish ; it is also very partial to leeches."

TYPICAL TERNS

The true terns, of which the common tern (*Sterna fluviatilis*) is the typical form, differ from the marsh-terns and their allies by the distinctly pointed outer tail feathers ; and they are further characterised by the shortness of the metatarsus, the moderately elongated tail, and the compressed and slender beak. With two exceptions, these terns have the crown of the head black ; while, as a rule, the under parts are white or grey, although in the eastern black-bellied tern (*S. melanogaster*) they are black. The common species, as well as the Arctic (*S. macrura*), roseate (*S. dougalli*), and Sandwich tern (*S. cantiaca*), are large-sized birds belonging to a group of the genus in which the forehead is black to the culmen of the beak ; whereas the sooty tern (*S. fuliginosa*) is the representative of another group in which the front of the forehead is white in the adult plumage.

From all the above the lesser tern (*S. minuta*), together with several other species, may be distinguished by its inferior dimensions, the length of the wing being less than 8 inches, whereas in the other groups it varies from 9½ to 12 inches. The broad-billed tern (*S. eurygnatha*) is an inhabitant of the Atlantic coast of America, from South Brazil to the island of Trinidad, and is represented by a closely allied species on the Pacific coast of the New World.

There is a prevalent idea that shore-nesting birds make no nest, but lay their eggs indiscriminately among the shingle. This, however, is a complete misconception, so far at least as the lesser tern is concerned. As a matter of fact, this bird excavates a conical pit in the sand about two inches deep. Immediately round the " crater " a narrow zone of sand is cleared from shingle ; and, when completed and containing its full clutch of two or three eggs, the deepest part of the nest is filled with broken shells, into which the eggs are wedged with their points downwards. As the eggs are disproportionately large in relation to the bird, it is manifest that the position in which they are placed renders them most easily covered by the brooding bird. It has been assumed that the " crater " is excavated by the female pressing down the sand in the manner that sparrows dust themselves by the roadside, but Mr. Patten, who has recorded these observations, is of opinion that the work is done with the beak.

NODDIES

As an essentially tropical genus of the subfamily, brief mention must be made of the noddies, typically represented by *Anous stolidus*. These birds belong to a group of the subfamily differing from the one including the two last genera by the graduated tail, in which the feathers are pointed, and the outermost shorter than the next pair. As a genus, noddies are characterised by the short middle toe, the strong decurved beak, and by the fourth pair of tail feathers, counting from the outer side, exceeding all the others in length. The common noddy appears to be generally distributed throughout the tropics, one of its best-known breeding haunts being the Tortuga Islands, off Florida, where it nests in immense colonies. Its general colour is dark ; but, like the allied species, it has a light grey patch on the crown of the head and forehead.

SKIMMERS

The remarkable birds known as skimmers, or scissor-bills, constitute a subfamily (*Rhynchopinæ*) distinguished not only from the terns (which they otherwise resemble), but likewise from all other birds, by the peculiar structure of the beak, which is elongated and compressed to a knife-like form, with the lower half considerably longer than the upper one ; the latter being freely movable. The single genus of the subfamily is represented by three species, of which the black skimmer (*Rhynchops nigra*), distinguished by its dark beak, is North American, while the yellow-beaked skimmer (*R. albicollis*) is Indian, the third species inhabiting the Nile and Red Sea littoral. The American species has been observed flying close to the water, with the lower half of the beak immersed beneath the surface, doubtless searching for food.

FORK-TAILED GULLS

The gulls proper, as distinguished from the other members of the order, form the third subfamily

BLACK SKIMMER

BLACK-HEADED GULLS

Formerly rare in collections, this gull (*R. rossi*) has more recently been obtained abundantly off Point Barron; but its true Polar haunts appear to be as yet undiscovered, although it has been seen in summer in Boothia Felix and Franz-Josef Land.

TYPICAL GULLS

Represented by nearly half a hundred species, the typical gulls of the genus *Larus* differ from both the preceding by the squared tail, and are further characterised by the full development of the first, or hind, toe. The genus includes both the largest and the smallest representatives of the subfamily; and while some species assume a dark head in the breeding season, others lose all trace of dark tints in this region when adult.

LITTLE GULL

The smallest of the dark-headed species visiting the British Isles is the little gull (*L. minutus*), easily recognised by its diminutive size, and, when in flight, by the slaty black under surface of the wings. A straggler to Southern Norway, this gull is common in the Mediterranean countries, ranging eastwards to the Caucasus, and thence northwards across Siberia to the Lena. Nesting in colonies in the vicinity of Lake Ladoga, together with the common tern, the female generally lays three or four eggs, which may be distinguished from those of the former by the orange red, instead of yellow, colour of their yolks.

BLACK-HEADED GULLS

Whereas the species just mentioned is only a very occasional visitor, the black-headed gull (*L. ridibundus*) is a common denizen of the British coasts, showing an especial partiality for flat shores during winter, but in spring seeking marshes for the purpose of breeding.

(*Larinæ*) of the typical family; the great majority of them belonging to the genus *Larus*, although the kittiwake and an allied species from the North Pacific are separated as *Rissa*, while Sabine's gull and a kindred but very rare species from the Galapagos Islands constitute the genus *Xema*; and Ross's gull (*Rhodostethia rossi*) and the ivory gull (*Pagophila eburnea*) respectively represent distinct generic types.

As a subfamily, gulls are characterised by the upper half of the beak being longer than the lower one, over which its tip is bent down; while the tail is usually squared, although in one genus forked, and in a second wedge-shaped. Some of the smaller gulls, like so many of the terns, assume a dark head and neck in the summer plumage.

Sabine's gull (*Xema sabinei*), together with the Galapagos fork-tailed gull (*X. furcatum*), may be at once distinguished by the forking of the tail—a character in which these species agree with the great majority of terns, as they also do in the assumption of a dark head during the breeding season. Not a very uncommon straggler—especially in the immature state—to the British Isles, Sabine's gull breeds in Arctic America and Siberia, generally in company with the Arctic tern; two eggs being laid by the female on the bare ground. Of the second and larger species little is known, only a few examples having found their way into European collections.

ROSS'S GULL

As the forked tail serves to distinguish the members of the last genus, so the single representative of the genus *Rhodostethia* is equally well demarcated by the wedge-like contour of the same appendage. It is likewise characterised by its small dove-like beak; while the delicate pink hue of its plumage is also a striking feature, although one shared by some other members of the subfamily.

LAUGHING GULLS

GULLS ON THE THAMES EMBANKMENT IN WINTER

In this species the head and upper part of the neck are dark brown, and the beak is lake red in the summer dress; but in the Mediterranean black-headed gull (*L. melanocephalus*) the head is jet-black, and the beak coral-red, with a dark band in front of the angle, while the primary quills, in fully adult examples, are white, instead of parti-coloured.

In America the group is represented by the laughing gull (*L. atricilla*), distinguished from the British species by its larger dimensions, and characterised by having the first three outer primaries black, with minute white tips. The largest member of the group is the great black-headed gull (*L. ichthyaëlus*), ranging eastwards from the Levant to China, and northwards to Mongolia. In length the male may measure as much as 27 inches (against 16 inches in the black-headed gull); and in the breeding plumage the adult has the head and upper part of the neck jet-black, the beak orange red, passing into red at the angle of the lower half, and the first primary mainly white, with a black streak along the outer web; while on the second, third, and fourth quills of the same series the black forms a bar, followed by a broad white tip (page 1351).

COMMON GULL

The somewhat misnamed common gull (*L. canus*) is the first of a group of mostly large species, characterised by the pure white head and neck in the summer plumage of the adult. Measuring 18 inches in length, the adult in summer has the head and neck pure white, the back grey, the primaries mostly black, with white spots or tips, the beak yellow at the tip and greenish yellow at the base, and the total number of flight-feathers thirty-one. Breeding throughout Northern Europe and Asia, the common gull is now only a winter visitor to England, although it still nests in Ireland and Scotland. Its nests may be either scattered singly along the shore or aggregated into larger or smaller colonies. On the two sides of North America this gull is severally represented by an allied species.

HERRING-GULL

Another British species demanding notice is the herring-gull (*L. argentatus*), which considerably exceeds

the last in size, measuring upwards of 23 inches in length. In the adult summer plumage the head is white, the mantle pale pearl-grey, the beak wholly yellow, a ring round the eye yellow, and the legs flesh-coloured; and there are thirty-four flight-feathers (page 1352). The range of this species includes Northern Europe, the islands of the Atlantic, such as the Azores, where it breeds, and North America. In winter it visits the north of Africa.

Of this species in America, where it is often called the harbour-gull, Miss M. O. Wright observes that " the eggs are very interesting, because no two are of the same colour, being of every shade of blue and grey, from the colour of summer sky and sand to the tint of the many-coloured, water-soaked rocks themselves. The markings vary also in shape and size, and are in every shade of brown, through lilac and purple, to black. The parents are very devoted to their nests, and take turns in sitting. When the young are first hatched, though covered with down, they are very weak in the leg and helpless; but in the course of a few hours the little gulls are strong enough to walk, and the instinct to hide at the approach of anything strange comes to them very suddenly, so that a gull only three or four hours old will slip out of the nest, and either hide beneath a few grass-blades or flatten itself in the sand, where, owing to its spotted, colour-protective down, it is almost invisible, so well does Nature care for her children—provided that man does not interfere. When a gull nests in a tree, however, the little birds, not feeling the same necessity for hiding, do not try to leave the nest until the growth of their wings will let them fly."

In Southern Europe, as well as in Siberia and various parts of America, the herring-gull is replaced by several very closely allied birds, mainly distinguished by the darker or lighter hue of the mantle, the pattern of the quill-feathers, and the colour of the legs and of the ring round the eye. Herring-gulls are in the habit of following the shoals of the fish from which they take their name, and may often be seen hovering above the fry, preparatory to taking a plunge among them in the water. Their chief food consists, however, of various marine animals thrown up by the tide; although during the spring, and after rough weather, they frequently wander far inland.

BLACK-BACKED GULLS

Somewhat superior in size to the common gull (its length being about 21 inches), the lesser black-backed gull (*L. fuscus*) in the adult summer plumage has the head white, the primaries blackish with white tips, the back blackish, and the legs bright yellow. This species is resident in Britain, and ranges eastwards to the Caspian, while southwards it extends into Africa, and westwards to the Canaries.

Of the great black-backed gull (*L. marinus*), which is larger than all the preceding, a well-known naturalist, who writes under the nom-de-plume of "A Son of the Marshes," observes that these birds "are not particular as to the nature of their food, so long as there is enough of it; a rat or a bird, a fish or a snail, or bread and milk will suit them equally well. Tradition said that in the early days of our oldest inhabitants the great black-backed gull bred on some of the wild flats of the Kentish coast and in a portion of the lonely salt marshes of Essex."

In attacking young lambs, these gulls invariably begin by pecking out the eyes of their victims; and so many as nine of these marauders have been captured during a single evening by setting a number of traps round a dead lamb. In length this gull measures upwards of 28 inches; and in the adult breeding plumage the head is white, the back blackish, and the legs flesh colour; the number of flight feathers being thirty-four. Essentially an oceanic species, the great black-backed gull is mainly an inhabitant of both sides of the North Atlantic, although it has been procured on the Pacific side of North America, and in winter it ranges so far south as the Canaries. In the Southern Hemisphere it is replaced by the southern black-backed gull (*L. dominicanus*), characterised by its stout beak, brownish black mantle, and olive-coloured legs.

GLAUCOUS GULL

Largest of all the British species, the greater white-winged, or "glaucous," gull (*L. hyperboreus*, or *glaucus*), in which the males may measure fully 32 inches, is readily distinguished by the adult summer plumage being nearly white throughout, as well as by the comparative shortness of the wings and feet.

Essentially an Old World Arctic bird, this gull only wanders in winter to temperate and tropical Europe; but in the North Pacific it is represented by the allied *L. glaucescens*, ranging from America to Kamchatka, and distinguished by the faint grey mottlings on the wings.

Another occasional wanderer to the British Isles from the North is the Iceland gull (*L. leucopterus*), which may be distinguished from the last by its length not exceeding 22 inches, and likewise by the proportionately much longer wings and legs. Bonaparte's gull (*L. philadelphia*), a small species with a greyish black head and upper neck, is remarkable for its habit of breeding in tall trees.

KITTIWAKES

Represented only by the common circumpolar kittiwake (*Rissa tridactyla*) and an allied North Pacific species (*R. brevirostris*) from the region lying between Alaska and Kamchatka, these gulls are distinguished by the shortness of the metatarsus and the absence or rudimentary condition of the first, or hind, toe. It is not a little curious that while in most districts examples of the common kittiwake in which the latter toe persists are but rarely met with, in Bering Sea this condition is much more common. Measuring 15 inches in length, the kittiwake, in the summer plumage of the adult, has the upper parts white and grey, the tail white, the first to the fifth primaries tipped with black, the under parts white, the beak yellow, and the legs brownish black.

The kittiwake is a resident in the British Isles, where it breeds in numbers of rocky cliffs, and feeds chiefly on surface-swimming fry of fishes and marine invertebrates. The nests, which are usually placed close together on narrow ledges of rock, are built of seaweed, and generally contain three eggs. The Pacific species, which exhibits a similar varia-

KITTIWAKES NESTING

tion with regard to the first toe, may be distinguished by its orange-red legs.

IVORY GULL

Conspicuous on account of its uniform delicate white plumage, faintly suffused with a rosy tint, in marked contrast to which stand out the jet-black legs and greenish yellow beak, the lovely ivory gull (*Pagophila eburnea*) alone represents a genus characterised by the shortness of the beak, the long and slightly graduated tail, and the connection of the first toe (of which the claw is unusually long) with the metatarsus by means of a distinct web. A circumpolar inhabitant of the Arctic seas, this gull wanders into temperate regions during the winter; its breeding places being in Spitzbergen and other regions in the Far North. In contrast to the snowy white of the adult, the young of the ivory gull are conspicuously spotted with black. This is, perhaps, the most exquisitely beautiful member of a beautiful group of birds.

SKUAS

CLOSELY allied to the gulls, the skuas form a family (*Stercorariidæ*) by themselves, characterised by the following features. The beak has a cere at the base, and the tip of its upper half is hooked, the breast-bone has a single notch on each side, the blind appendages (cæca) of the intestine are larger than in the preceding family, and the completely webbed toes are furnished with strong, sharp, hooked claws.

Represented by seven species, the skuas may all be included in a single genus; and while four of them breed only in the colder regions of the Northern Hemisphere, there are three southern species, one of which (*Stercorarius chilensis*) is found on the western coast of South America so far as the Straits of Magellan, and thence northwards to Rio de Janeiro, the other ranging from Tierra del Fuego to the Cape, New Zealand, and the Indian Ocean. In the last edition of "Yarrell's British Birds" it is stated that "the skuas may be considered as forming a conspicuous portion of the predaceous division among the swimming birds, as indicated by their powerful and hooked beak and claws. Their food is fish, but they devour also the smaller water-birds and their eggs, the flesh of whales, as well as other carrion, and are observed to tear their prey in pieces while holding it under their crooked talons. They rarely take the trouble to fish for themselves, but, watching the smaller gulls and terns while thus employed, they no sooner observe one to have been successful than they immediately give chase, pursuing it with fury; and having obliged it from fright to disgorge the recently swallowed fish they descend to catch it, being frequently so rapid and certain in their movements and aim as to seize their prize before it reaches the water."

Of the two members of the group breeding within the limits of the British Isles, the largest is the great skua (*S. catarrhactes*), its only resort within those limits being the Shetlands.

Measuring about 24 inches in length, the great skua has the two middle tail feathers less than an inch longer than the others, and may be further distinguished by the white bases to the flight feathers; the general colour being dark brown. It nests in a hole of about a foot in diameter, laying one or two eggs on a lining of moss and heather. It is to this species that the three southern birds alluded to above are allied.

Taking the other species in their order of size, the long-tailed skua (*S. parasiticus*), which measures 22 inches in length, has the two central tail feathers about 9 inches longer than the rest. Essentially an Arctic species, this bird is but a very occasional visitor to the British Isles. Temminck's, or the "pomatorhine," skua (*S. pomatorhinus*), on the other hand, is a regular winter visitor to the last-named area, and may be distinguished by the two middle tail feathers being twisted upwards and exceeding the others in length by 4 inches, the total length of the bird being 21 inches. Lastly, we have Richardson's skua (*S. crepidatus*), measuring an inch less than the last, and distinguished by the two middle tail feathers being only 3 inches longer than the others. Circumpolar and subarctic in its breeding range, this species is much more abundant in Britain than either of the others, nesting not only in the Hebrides, Orkneys, and Shetlands, but likewise on the mainland in the counties of Caithness and Sutherland.

R. LYDEKKER

TEMMINCK'S SKUA

RICHARDSON'S SKUA

From photographs by Mr. A. J. R. Roberts

GREAT SKUA

ORDER XVI. ALCÆ
GENERAL CHARACTERISTICS OF THE AUK TRIBE

ALTHOUGH at one time grouped with the divers and grebes, to which they approximate in the backward position of the legs the auks (order Alcæ) are now regarded as relatives of the plover and gull tribes, with which they are sometimes brigaded under the title of Charadriiformes. In the circumstance that the vertebræ of the back articulate by means of cup-and-ball joints, as well as in the characters of the skull, these birds are indeed essentially gull-like, although they differ by the absence of a process to the lower end of the humerus, as well as in the upward prolongation of the crest of the tibia. The latter feature is, however, probably merely an adaptation for the purpose of increasing the swimming and diving powers of these birds. The members of the order here mentioned are the following :

ORDER	Pigeon-guillemot ....U. columba	Knob-billed auk ....S. pusillus
Auk Tribe—Alcæ	GENUS 3	Parrot-auk ....S. psittaculus
FAMILY	Short-beaked Guillemots—Brachyrhamphus	GENUS 6
Auks—Alcidæ	SPECIES	Cerorhyncha
GENUS 1	Marbled guillemot..Brachyrhamphus marmoratus	SPECIES
Typical Auks—Alca	Black-throated guillemot ....B. antiquus	Rhinoceros-auk ....Cerorhyncha monocerata
SPECIES	GENUS 4	GENUS 7
Great auk....................Alca impennis	Little Auk—Alle	Puffins—Fratercula
Razorbill .... A. torda	SPECIES	SPECIES
GENUS 2	Little auk .... Alle nigricans	Arctic puffin ....Fratercula arctica
Guillemots—Uria	GENUS 5	Horned puffin ....F. corniculata
SPECIES	Pacific Pigmy Auks—Simorhynchus	Whiskered puffin ....F. cirrhata
Guillemot ....Uria troile	SPECIES	GENUS 8
Pallas's guillemot .... U. arra	Tufted auk .... Simorhynchus cristatellus	Tufted Puffin—Lunda
Brünnich's guillemot .... U. bruennichi		
Black guillemot ....U. grylle		

AUKS

ALL the members of the group are included in the single family *Alcidæ*, and all have heavy and compact bodies, with the legs placed far back, the plumage close and elastic, and the head relatively large. The first, or hind, toe is absent or rudimentary, and the short metatarsus, which is not laterally compressed, is covered with small scales, although generally with a row of larger plates in front. In the wing there are eleven primaries, and from fifteen to nineteen secondaries ; and, except in the great auk (where there are eighteen), there are twelve feathers in the short tail, which may be either rounded or graduated. The long front toes are fully webbed, and furnished with sharp, claw-like nails. The beak is subject to great variation, and in some cases is very short, deep, and laterally compressed. The oil-gland is tufted.

Dr. R. W. Shufeldt, after reviewing the various arrangements prepared by other writers, considers that these birds should form an order, the Alcæ, which is connected, on the one hand, with the plover group through the gulls and their allies, and, on the other, through the petrels, with the penguins, loons, grebes, and their extinct toothed ally *Hesperornis*.

The members of the order are mainly confined to the colder regions of the Northern Hemisphere, where they are specially numerous in the North Pacific. The breeding range extends in America so far as California and Maine, and in the Old World to Japan and the Tagus ; but some of the species wander still farther south, although none are denizens of the Southern Hemisphere. All the species are inhabitants of the shore or the ocean.

THE GREAT AUK

The typical members (*Alca*) of the family are characterised by the large size of the compressed beak, marked in front by oblique grooves, and feathered at its base close up to the slit-like nostrils, which are almost concealed by a dense, velvety feathering, completely filling the pits in which they are situated. The wings are more or less short, and the tail is graduated, with its component feathers pointed.

On account of its extinction during the nineteenth century, as well as from being the largest representative of the family, and the only bird in the Northern Hemisphere incapable of flight, the great auk, or garefowl (*A. impennis*), is a species of great interest. In common with many other northern sea-birds, it was formerly known as the penguin—a name now transferred to the well-known birds of the Southern Hemisphere ;

GREAT AUKS, OR GAREFOWL

A PAIR OF YOUNG RAZORBILLS   Photo, L. Medland

and in size may be roughly compared to a goose, its total length being about 32 inches. It is especially characterised by the rudimentary condition of the wings, which, owing to the reduction in the length of the ulna and bones of the digits, were quite useless in flight; while it is further distinguished by the beak being equal in length to the head, and furnished with numerous grooves on its lower as well as on its upper half. In colour the plumage of the head, neck, and back is black, but the under parts, as well as a characteristic spot in front of the eye, are white. As already mentioned, the tail is peculiar in carrying sixteen feathers.

Confined to the North Atlantic, and ranging so far north as Iceland on the one side and Greenland on the other, the great auk was a migratory species, which in winter wandered so far south as the Bay of Biscay and the shores of Virginia. Both in Greenland and Norway it appears to have been always rare; and its chief or only breeding places were three rocky islands near Iceland, known as the Garefowl Skerries, or Geirfuglasker, and Funk Island off the Newfoundland coast. By the subsidence in the spring of 1830 of one of these islets, which, as being the most inaccessible, was the favourite breeding place, the birds were driven to one nearer the shore, where they were more easily approached; and in the course of the next fourteen years the species became extinct in Europe, the last pair having been killed in the summer of 1844.

The existence of the garefowl on Funk Island was discovered about 1534, when the birds were so numerous as to be reckoned, it is said, by thousands; but incessant persecution for more than two centuries eventually brought about its extermination, which probably took place almost contemporaneously with its disappearance from Europe. It was customary for the crews of several vessels to spend the summer on Funk Island for the sole purpose of killing garefowl for the sake of their feathers.

Although we have only traditions of these expeditions, it is indisputable that stone pens were erected into which the birds were driven like sheep, that they were slain by thousands, and that their bodies were left to rot where they lay, while for some purpose or other frequent and long-continued fires were lighted on the island. The records of this slaughter are still extant in the numbers of garefowl bones to be met with in the soil of Funk Island; such relics, together with a few skins, and a number of egg-shells, being all that remain to us of the finest of the auks.

That the garefowl was generally a gregarious bird, more especially during the breeding season, is evident from the foregoing; but it is stated that solitary pairs were occasionally found nesting with guillemots and razorbills. Although useless for flight, the wings were admirably suited as paddles; and the swimming and diving powers of the bird were probably unrivalled, its migrations being more extensive than those of many of its relatives which possess the power of flight. From the accounts of the natives of Iceland, it appears that the garefowl swam with its head elevated and the neck retracted, and that, when pursued, instead of flapping along the water, it immediately dived.

As in the allied species, the eggs are relatively large in proportion to the size of the bird, often measuring just over 5 inches in length; and they have also the same elongated form, with one end much larger than the other. They have a creamy-white ground colour, marked with black or brown streaks and blotches, with underlying grey patches.

It is of interest to note in this connection that about seventy-two great auk's eggs are known to exist, and of these fifty-one are, it is believed, in Great Britain. A very fine specimen which came into the market some years ago realised 315 guineas.

THE RAZORBILL

The lesser auk, or razorbill (*A. torda*), the only other representative of the genus, differs from the garefowl not only by its greatly inferior size (length about 17 inches), but likewise by its well-developed wings and relatively shorter beak, in which there are only two or three grooves on the lower half, and these indistinctly marked, as well as by the presence of only twelve tail feathers. Lacking the large white spot in front of the eye characterising the great auk, the adult razorbill

GUILLEMOTS

in summer has a narrow white line extending from the beak to the eye. While in summer the chin and throat of the adult are brown, and the head, hind part of the neck, and upper parts black, with the under parts white, in the winter dress the white extends upwards to the throat, chin, and sides of the head, and the plumage of the upper parts is browner.

The razorbill is common to the coasts and islands of both sides of the North Atlantic, ranging so far north as latitude 70° in Greenland, and in winter reaching Gibraltar, whence it wanders a considerable distance up the Mediterranean. Resident throughout the year in the British seas, it breeds on all suitable rocky coasts, from the north of France to the North Cape, generally in large colonies. Concerning its breeding habits, it is stated in the third edition of "Yarrell's British Birds" that "about the middle or latter part of March in the south of England, and early in April in the northern portions of our islands, the razorbills, guillemots, and puffins converge to particular points, where, from the numbers that congregate, and the bustle apparent among them, confusion of interests might be expected. It will, however, be found that, as a rule, the guillemots occupy one station or line of ledges on the rock ; the razorbills another ; the puffins a third ; the kittiwake gulls a fourth ; whilst the most inaccessible crags seem to be left for the use of the herring-gulls. The razorbills generally select the higher and rougher ledges, and they are partial to crevices, their eggs being sometimes deposited so far in, that it is no easy matter

LITTLE AUK IN FLIGHT

to get at them ; at other times they lay their eggs on the broader shelves along with the guillemots, but not so closely together."

GUILLEMOTS

Closely allied, alike as regards structure, the colour and seasonal change of their plumage, and habits, to razorbills, the guillemots differ by the more slender and straighter beak, in which there are no oblique transverse grooves, while the upper half is slightly curved near the point, and has a small notch on the side. The basal nostrils are partially closed by a membrane, which is itself partly feathered. There is still some degree of uncertainty as to the number of species of the typical guillemots, some naturalists recognising but one, while others admit several (page 1354). Whether regarded as species or varieties, all the forms are characterised by the white plumage of the under parts ; this white area in the summer dress stopping short at the base of the throat, but in winter extending upwards, as in the razorbill, to the throat, chin, and sides of the head.

In the typical form of the common guillemot (*Uria troile*), which inhabits both sides of the North Atlantic, the beak is of considerable length, and the head of a uniform smoky brown. It is replaced in the Pacific by a somewhat larger bird, known as the Californian guillemot. Both in the Atlantic and Pacific there are also certain guillemots characterised by the presence of a white streak extending backwards from the eye, and a white ring round the eye itself. Formerly regarded as a distinct species, these ringed guillemots are now considered to be merely sports.

Pallas's guillemot (*U. arra*), of Bering Sea and the North Pacific, is the largest representative of the second modification of the group, in which the beak is much shorter and deeper than in the preceding, while the nape of the neck and back of the head are black like the back. A portion of the base of the cutting edge of the lower half of the beak is light-coloured. Finally, there is the so-called Brünnich's, or Arctic, guillemot (*U. bruennichi*), of the North Atlantic and Arctic Oceans, in which the size is smaller, and the whole of the cutting edge of the upper half of the beak yellowish white. Brünnich's guillemot seems, however, to be so inseparably connected by the Californian form with the common guillemot as to render it doubtful if the two are more than local races of a single species.

Whatever diversity of opinion may obtain as to the distinctness of the above-mentioned forms from the common guillemot, there can be none as to that of the black guillemot (*U. grylle*), which is referred, indeed, by some writers to a separate genus. It is a smaller bird than the common guillemot, from which it is at once distinguished by the whole of the under parts being black in the summer dress ; the beak being relatively short. Typically an inhabitant of the North Atlantic, it is represented in the circumpolar seas by a variety distinguished by the larger size of the conspicuous white patch on the wings.

In the North Pacific the black species is replaced by the pigeon-guillemot (*U. columba*), characterised by the under surface of the wing being grey, instead of smoky white. The typical form of this species has a large white wing-patch ; but there are two varieties (*carbo* and *motzfeldi*), severally distinguished by the presence or absence of white on the head, in which the wing is uniformly black on the outer side.

All the guillemots are very similar in their mode of life, being essentially oceanic birds, which only visit the rocks during the breeding season, and are only found inland when driven by stress of weather, while they are markedly sociable and gregarious. Their food consists of fish, supplemented by various crustaceans ; the common species being especially partial to the fry of herrings and pilchards, which are captured at night in the open sea. Rapid, though heavy and laboured in its flight, the common guillemot is enabled to reach the summits of almost inaccessible cliffs for the purpose of breeding, where, as in the Farne Islands and at Flamborough, it congregates in myriads. On the ledges of the precipitous cliffs near Bempton, Yorkshire—another noted breeding place—the guillemots are sometimes so densely crowded together as to remind the spectator of a swarm of bees.

The breeding season in Britain begins in May and lasts till August ; and while the other species agree with the rest of the family in laying but a single egg, the black guillemot deposits two. The eggs may be laid either on the bare ledges of rock or in fissures ; sometimes several may be found together, but at other times they lie singly. In colouring, guillemots' eggs are remarkable for their extraordinary variability. According to Mr. H. Seebohm, " the ground-colours are cream, white, blue, and yellowish green, dark and clear pea-green, and reddish and purplish brown, with every conceivable intermediate tint. Some are irregularly blotched, others are fantastically streaked with browns, pinks, or greys in endless variety, whilst a few are spotless or nearly so." Some closely resemble those of the razorbill, from which they may always be distinguished by appearing creamy white instead of green when viewed by transmitted light.

Recently a writer in the "Field" newspaper stated that while fishing for pollack off Mullion Cove, Cornwall, with a rod and a phantom-minnow, he felt what seemed to be the pull of a good fish. He had some seventy yards of line out, and the bait was spinning on a single strand of salmon gut. He reeled up, feeling the catch playing well, and, after having to give it line, saw the lead appear, and stood by ready with the gaff to add another fish to the basket, when up came a guillemot, flapping its wings and making more fuss than a fresh-run salmon. The bird must have taken the minnow some four feet under water, judging from the weight of the leads and the pace at which the boat was sailing. After great difficulty the prisoner was freed of the flight of hooks, but its journey through the water had placed it beyond recovery.

CALIFORNIAN EXTINCT AUK

A single bone of the wing, and that imperfect, may nowadays seem but poor material on which to establish a new genus of birds, but Dr. F. A. Lucas appears to be justified in regarding a humerus from the Miocene of Los Angeles, California, as representing a large extinct type of flightless auk, for which the name *Mancalla californiensis* is suggested. It is considered to have equalled the great auk in size, but to have been more nearly allied to the guillemot. The existence of a flightless member of the group at such a comparatively early epoch is remarkable.

SHORT-BEAKED GUILLEMOTS

The North Pacific is inhabited by seven much smaller guillemots, characterised by their very short beaks, of which the tip is not decurved. Although they have been split up into three genera, they may all be included in the typical genus *Brachyrhamphus*. Some of the species, like the marbled guillemot (*B. marmoratus*), have the front of the metatarsus reticulated, but in others, such as the black-throated guillemot (*B. antiquus*), it is covered in front with large plates.

LITTLE AUK

Breeding solely within the limits of the Arctic Circle, the little auk, or rotche (*Alle nigricans*), is an Atlantic species, which only visits the British Isles in winter, and is even then far more common in the

HEAD OF WHISKERED PUFFIN

Orkneys and Shetlands than in the south. It is a very small bird, measuring only about 8½ inches in length, and differing from all the members of the family by the shortness of the front union of the two branches of the lower half of the beak, in which the angle of the chin is much nearer to the tip of the beak than to the nostrils, instead of the reverse. The whole beak is shorter than the head, very thick, and broader than high at the base; the profile being arched, and the tips of both halves notched, while the upper one is faintly grooved. The rounded and lateral nostrils are placed at the base of the beak and partially covered with feathers.

In colour, the little auk very closely resembles the guillemot; the head, chin, and throat, as well as the upper parts, being mostly black, while the remainder of the lower parts, a spot over the eye, the tips of the secondaries, and the margins of the scapulars are white. In the winter plumage, on the other hand, the white area includes the throat, chin, and sides of the head.

The little auk ranges in the Arctic regions from Novaya Zemlya and Spitzbergen to Greenland, migrating southwards in winter as far as New Jersey on the one side of the Atlantic, and to the Canaries on the other. In its breeding places, where it appears in May, it congregates in countless thousands, if not in millions. The single bluish white egg is laid so deep among the loose fragments of rock that it can only be reached with difficulty, and the young leave the breeding places for the open sea before they can fly.

An expert diver and a strong swimmer, the rotche feeds chiefly on crustaceans and marine worms. In spite, however, of its oceanic habits, it appears to be ill-adapted to fight against the storms of winter, during the prevalence of which it is frequently driven far inland.

KNOB-BILLED AUKS

PACIFIC PIGMY AUKS

Related to the rotche are a number of small auk-like birds from the Northern Pacific, all of which differ from that species in having the chin-angle nearer to the nostril than to the base of the beak. Among these are the tufted auk (*Simorhynchus cristatellus*), remarkable for the forwardly-curving tufts of feathers at the root of the beak; the knob-billed auk (*S. pusillus*), taking its name from the presence in summer of a knob at the base of the beak, which disappears in winter; and the parrot-auk (*S. psittaculus*).

RHINOCEROS-AUK

HEAD OF TUFTED AUK

Still more remarkable is the rhinoceros-auk (*Cevorhyncha monocerata*), in which the compressed and curved beak is longer than in the preceding birds, and provided at the base with a single horn-like knob above the nostrils, which is shed in winter. All these birds have much the same habits as the more typical auks, generally frequenting sheltered bays when the weather is rough. The rhinoceros-auk breeds so far south as California and Northern Japan.

PUFFINS

Among the most grotesque of all birds are the puffins, or sea-parrots, whose enormous, compressed, and brilliantly coloured beaks seem out of all proportion to the size of their heads. Represented only by the common Arctic puffin (*Fratercula arctica*) in the Atlantic, the genus attains a greater development in that headquarters of the auk family the Northern Pacific, where lives the horned puffin (*F. corniculata*), characterised by the great development of the horny process arising from the upper-eyelid, and the handsome whiskered puffin (*F. cirrhata*), distinguished by the pendent crest of feathers at the back of the head, and the absence of grooves on the lower half of the beak.

As a group, puffins are distinguished from all the other members of the family by the claw of the second toe being considerably longer and more curved than the other two, as well as by the presence of a rosette-like prominence at the angle of the mouth. They are further character-ised by the circum-stance that the feathers at the base of the beak stop short of the nostrils, and likewise by the peculiarity that the basal portion of the greatly compressed beak is furnished during the breeding season with one or more sheath-like, deciduous pieces of an orange red colour, which are shed in winter. The much com-pressed beak is shorter than the head, and consider-ably deeper than long, with the pro-file of both halves strongly arched, and the ridge of the upper one forming a sharp edge, while there are oblique transverse grooves on one or both halves (page 1354).

The European puffin may be com-pared in size to a teal, the average length in the southern portion of its habitat being about 12 inches, although in the Arctic regions it attains somewhat larger dimensions, and has the beak deeper.

Resembling the guillemot in general colouring, this puffin differs in undergoing no seasonal change of plumage, and in the white area occupying the whole of the sides of the head; the throat being encircled by a dark gorget. The beak has its terminal portion carmine red, behind which are bands of slaty grey and yellow, with a red one on the lower half. With the annual

A PAIR OF PUFFINS

moult both the sheath of the basal half of the beak and the warty red skin at the angle of the mouth are shed.

In Europe the breeding range of this species extends from the North Cape to the mouth of the Tagus, while in winter the birds wander so far south as Gibraltar, and thence pass up the Mediterranean to the Italian coasts. On the opposite side of the Atlantic the winter range reaches so far south as New York.

Essentially oceanic in their habits, puffins are gregarious at all seasons, and fly rapidly, somewhat after the manner of ducks. Swimming easily, and diving with the expertness characteristic of the family, they feed chiefly on the fry of fish. The single egg is laid either in a burrow in the ground or among the deep clefts of rocks. In colour the egg is dull white, faintly spotted with grey and brown; and in the presence of these markings it forms one of many exceptions to the general rule that eggs laid in holes are white.

From this circumstance it has been suggested that these birds have only taken to laying in burrows comparatively recently; the faintness of the markings of the eggs being perhaps indicative that they are in the course of disappearance.

It may be added that the tufted puffin is regarded by many ornitho-logists as typifying a genus by itself, under the name of *Lunda*.

Reference has been already made to a guillemot taking a bait; and a still more extra-ordinary instance of the same nature is recorded in the "Field" news-paper. It is there stated that, on September 25, 1905, while trolling for bass outside the mouth of Water-ville River (Ballin-skelligs Bay, co. Kerry), a gentle-man captured a puffin, which took the bait (a spoon) under water, got hooked, and was safely brought to boat.

On the same day his fisherman, who was with him in the boat, was fly-fishing for bass, when a puffin took the fly and was hauled in. Shortly after, the fisherman took another in the same way. It appears that there was a big shoal of sprats at the spot, on which many puffins were feeding. The fisherman cast his fly, which was an ordinary white one, over the shoal of sprats, hoping to get bass, when instead he caught these puffins. This made three puffins, all captured by rod and line from one boat in a comparatively short time—a strange, if not unparalleled, occurrence.

R. LYDEKKER

ORDER XVII. TUBINARES

GENERAL CHARACTERISTICS OF TUBE-NOSED BIRDS

THE members of this order take their scientific name (Tubinares) from the circumstance that the nostrils are produced into tubes lying on the surface of the beak and directed forwards ; this feature being absolutely peculiar, and serving at once to distinguish them from all other birds. The horny sheathing of their beak is composed of several distinct pieces, separated from one another by more or less marked grooves, and the tip of the beak is sharply hooked. In the skull the palate is of the slit (schizognathous) type ; while its nasal apertures are oval, or holorhinal, and the hind angle of the lower half of the beak is abruptly truncated behind. As in so many sea-birds, the upper aspect of the skull has very deep grooves, which, however, are always separated from one another on the forehead by a wide bar.

The vertebræ of the back are articulated with one another by the usual saddle-shaped surfaces. In the wings, which are generally of great length, the humerus resembles the corresponding bone of the gulls in having a well-marked process on the outer extremity, although the perforations in the basal bone of the second digit of the wing characterising that order are wanting. The tibia, or leg bone, differs from that of all the birds hitherto considered, except the auks, in having a compressed plate-like crest projecting upwards on its front aspect some distance above the level of the head of the bone. The feet are characterised by the small size or even occasional absence of the first toe, and the three front toes are completely webbed.

In the plumage there is a well-defined bare tract on each side of the neck, and the oil-gland is furnished with

THE STORM-PETREL

a tuft of feathers. The young, which are born in a helpless condition, and fed for a considerable period in the nest by the parents, are clothed with down, arranged in a somewhat complex manner.

In habits, all the petrel tribe are marine and carnivorous, subsisting entirely on either carrion, cuttlefish, or crustaceans, together with such refuse as they can pick up. They are all birds of sustained and powerful flight, and, with the exception of the members of one aberrant genus, are swimmers rather than divers. In appearance several of them, more especially the fulmars, present a marked similarity to the gulls ; the plumage in this instance being of the grey-and-white hue distinctive of that group. This resemblance must, however, be regarded as a purely adaptive one, brought about by the needs of a similar mode of existence, there being little structural affinity between the members of the two groups. Generally the members of the petrel tribe have a more or less dusky-hued plumage, and they mostly differ from the chattering and screaming gulls by their comparatively silent habits.

Although found in the seas of all parts of the world, the group is represented by the greatest number of species in the Southern Hemisphere, which may consequently be regarded as its headquarters. Very little is known of the group's geological history, although a species of shearwater has been stated to occur in the lower Miocene strata of France ; the same beds also yielding remains of an extinct genus (*Hydrornis*), tentatively assigned to this order.

CLASSIFICATION OF TUBINARES MENTIONED IN THIS WORK

ORDER
Tube-nosed Birds—Tubinares
FAMILY 1
Albatrosses—Diomedeidæ
GENUS
Albatrosses—Diomedea
SPECIES
Wandering albatross ......Diomedea exulans
Sooty albatross ....D. fuliginosa
Yellow-billed albatross ......D. chlororhyncha

FAMILY 2
Typical Petrels—Procellariidæ
GENUS 1
Ossifraga
SPECIES
Giant petrel ....Ossifraga gigantea
GENUS 2
Fulmarus
SPECIES
Fulmar ....Fulmarus glacialis
GENUS 3
Thalassoica
SPECIES
Silver-grey petrel ....Thalassoica glacialoides
GENUS 4
Pagodroma
SPECIES
Snowy petrel ,............Pagodroma nivea

GENUS 5
Majaqueus
SPECIES
Cape hen ....Majaqueus æquinoctialis
Spectacled petrel.............M. conspicillatus
GENUS 6
Shearwaters—Puffinus
SPECIES
Ashy shearwater ....Puffinus kuhli
Great shearwater ....P. gravis
Manx shearwater ....P. anglorum
Dusky shearwater ....P. obscurus
Sooty shearwater..................P. griseus
Mutton-bird ....P. brevicauda
GENUS 7
Œstrelata
SPECIES
Capped petrel ....Œstrelata hæsitata
GENUS 8
Bulweria
SPECIES
Bulwer's petrel ....Bulweria bulweri
GENUS 9
Daption
SPECIES
Cape petrel ....Daption capensis
GENUS 10
Dove-Petrels—Prion

SPECIES
Dove-petrel ....Prion desolatus
Broad-billed blue petrel ....P. vittatus
GENUS 11
Storm-Petrels—Procellaria
SPECIES
Storm-petrel ....Procellaria pelagica
GENUS 12
Halocyptena
SPECIES
Wedge-tailed petrel .. Halocyptena microsoma
GENUS 13
Fork-tailed Petrels—Oceanodroma
SPECIES
Leach's petrel ......Oceanodroma leucorrhoa
Pacific fork-tailed petrel ....O. furcata
Hornby's petrel ....,,,,,,O. hornbyi
GENUS 14
Oceanites
SPECIES
Wilson's petrel ....Oceanites oceanicus
GENUS 15
Cymodroma
SPECIES
White-bellied petrel......Cymodroma grallaria
GENUS 16
Diving Petrels Pelecanoides
SPECIES
Magellanic diving petrel . Pelecanoides urinatrix

ALBATROSSES

THE albatrosses (family *Diomedeidæ*) are distinguished by their tubular nostrils being placed on the two sides of the beak, and widely separated from one another by the large median portion of its horny sheath. They are further characterised by the extreme length and narrowness of the wing, in which both the humerus and the ulna are greatly elongated ; and also by the large number of quills in the wing, which may vary from thirty-nine to fifty, or more than in any other birds. In the foot the first toe is wanting ; and the skull is characterised by the absence of basipterygoid processes on the rostrum of its inferior surface.

All the albatrosses (which may be included in the single genus *Diomedea*) are of large size, and mainly frequent the southern tropical and subtropical seas, although one species ranges on the Pacific Coast of America so far north as Alaska. The occurrence of remains of a fossil albatross in the Pliocene deposits of the east coast of England is noteworthy.

By far the most familiar representative of the genus is the white, or wandering, albatross (*D. exulans*), which belongs to a group characterised by the absence of a groove in the horny sheath of the sides of the lower half of the beak, and also by the length of the wing being equal to three or four times that of the short and rounded tail. The span of wing varies from 10 to 12 feet, but the average weight of the bird is only some 17 pounds.

The prevailing colour of the plumage is yellowish white, with the quills dusky, and, except in very old birds, the region of the back and the larger wing coverts are irregularly barred with blackish. The beak and feet are whitish. Although the true home of this species is in the South Seas, its wanderings occasionally extend to the north of the Equator (p. 1353).

The smaller sooty albatross (*D. fuliginosa*), of the southern oceans generally and the Pacific, alone represents a second section of the genus, in which the horny sheath of the sides of the lower half of the beak is marked by a longitudinal groove, while the wing is only about twice the length of the graduated tail. In the adult the plumage of the neck, back, and upper parts is dark ashy grey, becoming lighter on the neck and fore part of the back, where the tips of the feathers are nearly white ; the wings and tail are dark slaty ; the beak, with the exception of the grooves, is black ; and the legs and feet are pale reddish. In all the species the young, after passing the white downy stage, are more or less sooty in colour ; so that in coloration the sooty albatross is one of the least specialised forms.

The name albatross, it may be observed, is a corruption of the Spanish word *albatraz*, meaning a gannet ; and was applied to these birds by the old voyagers, in conformity with that propensity to designate newly discovered creatures by familiar titles which is so characteristic of the uneducated. All these birds are strictly oceanic in their habits, rarely visiting the land except for the purpose of breeding, and then selecting remote islands, like Tristan da Cunha, or even isolated rocks.

Much has been written in regard to the flight of the wandering albatross, but these birds are by no means beautiful objects when seen following in the wake of a vessel, as the long and narrow wings seem out of proportion to the body, while five out of every six birds observed are in the brown immature plumage, and look dirty and draggled. After referring to the marvellous powers of flight of the species just mentioned, Mr. J. Gould observes, that although during calm or moderate weather this splendid bird " sometimes rests on the surface of the water, it is almost constantly

ALBATROSSES NESTING

WHITE-WINGED ALBATROS

on the wing—and is equally at ease while passing over the glassy surface during the stillest calm or flying with meteor-like swiftness before the most furious gale ; and the manner in which it just tops the raging billows and sweeps between the gulfy waves has a hundred times called forth my wonder and admiration. Although a vessel running before the wind frequently sails more than two hundred miles in the twenty-four hours, and that for days together, still the albatross has not the slightest difficulty in keeping up with the ship, but also performs circles of many miles in extent, returning again to hunt up the wake of the vessel for any substances thrown overboard."

Professor Moseley states that these birds make the utmost use of the momentum acquired by a few powerful strokes of the wings, taking all possible advantage of the wind, and progressing largely by a gliding movement. Still, however, he adds, they seem to move their wings more frequently than is generally supposed. "They often have the appearance of soaring for long periods after a ship without flapping their wings at all, but if they be closely watched very short but extremely quick motions of the wings may be detected. The appearance is rather as if the body of the bird dropped a very short distance and rose again. The movements cannot be seen at all unless the bird is exactly on a level with the eye."

During the breeding season, when the light-coloured species are in the full beauty of their white plumage, albatrosses resort in large numbers to oceanic islands and rocks. In Tristan da Cunha both the wandering albatross and the smaller yellow-billed albatross (D. chlororhyncha) are found in numbers during the breeding season ; the latter being easily distinguished by its

WANDERING ALBATROSS

yellow gape and the broad yellow stripe on the tip of the otherwise black beak. Commonly known to the sailors as "mollymauks," the yellow-billed albatrosses, according to Professor Moseley, "take up their abode in separate pairs anywhere about in the rookery, or under the trees, where there are no penguins. They make a cylindrical nest of tufts of grass, clay, and sedge, which stands up from the ground. The nest is neat and round, there is a shallow concavity on the top for the bird to sit on, and the edge overhangs somewhat, the old birds undermining it during incubation by pecking away the turf of which it is made."

The nest may be as much as 14 inches in diameter, by 10 in height ; and at the proper season it contains a single white egg, somewhat larger than that of a goose. During incubation the egg is held in a kind of pouch, so that the bird has to be driven quite off the nest before it can be ascertained whether or not an egg is present. In all cases the sitting birds allow themselves to be approached without making the least movement.

The wandering albatross builds a larger and more conical nest than the mollymauk, and its egg is about five inches in length, or about equal in size to that of a swan. At its larger end the egg has some specks of red, but is otherwise white. The males stand or sit near their brooding partners ; and when the latter are approached, they display their displeasure by savagely snapping their beaks at the intruder.

Laysan Island, in the Pacific, is a great breeding place for albatrosses ; and the egg-harvest forms an important industry, the eggs being carried away in carts to a collecting station, where they are broken up for the sake of the oil they yield.

TYPICAL PETRELS

ALL the members of the family *Procellariidæ* differ from the albatrosses by the nasal tubes being laid side by side upon the top of the beak. Generally the wings are long, but the number of quills does not exceed thirty-nine, and is usually about thirty, although occasionally reduced to twenty. The hind toe, although sometimes minute, is generally present ; and there are usually basipterygoid processes on the rostrum of the inferior aspect of the skull.

GIANT PETREL

Next in point of size to the albatrosses is the giant petrel (*Ossifraga gigantea*), the sole member of its genus, distinguished from the other representatives of the family by the length of the beak exceeding that of the metatarsus. The beak is very stout, and has the nasal tubes of great length, and its outer sheath so produced as to reach beyond the proper apertures of the nostrils, thus giving the appearance of a single nostril. The tail is characterised by the presence of sixteen feathers. In general appearance and size this bird is not unlike some of the smaller dark-coloured albatrosses, its total length being about 32 inches, and the span of the wing 66 inches.

Although pale-hued individuals are far from uncommon, the general colour of the plumage is typically dull slaty brown, becoming paler on the face, throat, and under parts, some of the feathers of the upper parts being tinged with chocolate, while those of the back, as well as the wing coverts, have paler greyish margins. The beak is yellowish horn-colour ; and the legs and feet are greyish black.

Commonly known to sailors by the name of nelly, break-bones, or stinker, the giant petrel is widely distributed over the temperate and high southern latitudes, occasionally wandering to a considerable distance north of the Equator, and in power of flight is fully equal to the albatrosses. In habits it differs considerably from the latter, subsisting chiefly on the blubber and flesh of dead seals and whales, as well as the bodies of other birds. Professor Möseley, who compares it in these respects to a vulture, writes that on Kerguelen Island this petrel " soars all day along the coast on the look-out for food. No sooner is an animal killed than numbers appear as if by magic, and the birds are evidently well acquainted with the usual proceedings of the sealers—who kill the sea-elephant, take off the skin and blubber, and leave the carcase. They settled down here all round in groups at a short distance, a dozen or so together, to wait, and began fighting amongst themselves, as if to settle which was to have first bite."

When gorged, they are quite unable to fly ; and, like other members of the family, if disturbed they have an unpleasant habit of disgorging an ill-smelling oily fluid. These birds breed on Kerguelen and Prince Edward's Island, where they lay a single dirty white egg in a natural hollow of the ground. The newly hatched young are covered with long, grey down, and later on the nestlings, when approached, are stated to squirt from their nostrils an oily fluid

to a distance of six or eight feet, the old birds remaining a short distance away.

FULMAR

In the Arctic regions and other parts of the Northern Hemisphere, the place of the giant petrel is taken by the gull-like fulmar (*Fulmarus glacialis*), which is likewise the only well-defined representative of its genus. Of much smaller size than the giant petrel, the fulmar differs by the beak being inferior in length to the metatarsus, and the proportionately shorter and stouter nasal tubes, in which the partition between the two nostrils extends to within a short distance of the

THE GIANT PETREL

orifice ; the tail feathers, moreover, are either twelve or fourteen in number.

The fulmar measures about 19 inches in length, and displays great variation as regards colour. In the typical form, however, the head and neck are white, most of the upper parts, as well as the tail feathers, pearl-grey, the primaries slaty grey, and the breast and under parts white. The eye is dark brown, the beak yellow at the tip, with yellowish white sides, and a greenish tinge at the base above, while the legs and feet are pale grey. A grey phase is also commonly met with, in which the head and neck, as well as the greater portion of both the upper and under parts, are ashy brown, with the back and wings somewhat darker than the rest.

The fulmar breeds in the boreal regions of both hemispheres, but some naturalists consider that in the North Pacific and Bering Sea it is replaced by two distinct species. In autumn and winter the fulmar is a by no means uncommon, although probably involuntary, visitor to the southern shores of Britain, and has been recorded so far south as the Mediterranean.

In habits the fulmar is very like its larger cousin, nesting in hollows in the ground, instead of in deep burrows like the shearwaters, and feeding largely on whale-blubber and refuse. Mr. Scoresby writes that these petrels " are remarkably easy and swift on the wing, flying to windward in the highest storms, and resting on the water in great composure in the most tremendous seas ; but it is observed that in heavy gales they fly extremely low, generally skimming along the surface of the water. They are extremely greedy of the fat of the whale, and though few should be seen when a whale is about to be captured, yet as soon as the flensing process begins they rush in from all quarters, and frequently accumulate to many thousands. They then occupy the greasy track of the ship, and, being audaciously greedy, fearlessly advance within a few yards of the men employed in cutting up the whale."

Highly gregarious during the breeding season, fulmars then collect on the turfy ledges of the St. Kilda cliffs in thousands. The single white egg is laid either in a slight nest of dried grass or on the bare ground ; and although the birds sometimes excavate a hollow of a few inches deep in the turf, they as often nest on its surface.

Nearly allied to the fulmar is the silver-grey petrel (*Thalassoica glacialoides*) of the Pacific and Southern Atlantic, distinguished by its more slender beak, in which the nasal tubes are shorter and more depressed, with their upper border concave. This species extends nearly so far south as the Antarctic pack-ice, where it is replaced by the snowy petrel (*Pagodroma nivea*), a pure white species of the size of a pigeon, with a short and weak bill. The " Cape hen " (*Majaqueus æquinoctialis*) and the spectacled petrel (*M. conspicillatus*) are larger southern species, of the size of the fulmar, with blackish brown plumage.

numerous group of medium-sized dark-coloured petrels known as shearwaters (*Puffinus*) are characterised by the length and slenderness of their beaks, in which the short and depressed nasal tubes open by two separate orifices, generally directed obliquely upwards. The wings are long and pointed, with the first quill the longest ; the graduated tail consists of

FULMAR PETREL

twelve feathers, and the first toe is rudimentary.

Shearwaters may be divided into two groups, according as to whether the under parts are white or are dusky like the back. Among the better-known representatives of the former group may be mentioned the ashy shearwater (*P. kuhli*) of the Mediterranean, Western Europe, and the East Atlantic, characterised by its stout beak, circular nostrils, and brownish grey upper plumage.

The great shearwater (*P. gravis*) of the Atlantic Ocean generally, which measures 18 inches in length, and is an occasional autumn visitor to the British Isles, is a member of the same group, distinguished by its more slender beak, in which the nostrils form longitudinal ovals ; the general colour of the upper parts being sooty greyish brown, with paler tips to the feathers of the back.

The commonest British representative of the group is the smaller Manx shearwater (*P. anglorum*), which measures only 14 inches in length, and has a uniformly blackish upper plumage, without pale tips to any of the feathers ; it frequents the whole of the North Atlantic, although more abundant on the eastern than

MANX SHEARWATER

The beak is longer than in the latter, with shorter nasal tubes, of which the two apertures look directly forwards. While the Cape hen is wholly blackish brown, the spectacled petrel has characteristic white bands across the head and throat.

SHEARWATERS

Nearly cosmopolitan in their distribution, the

on the western side. Another species of this group is the dusky shearwater (*P. obscurus*), which is smaller than the last, with a more slender beak, and a deeper black to the upper plumage. Common to both the Atlantic and Pacific Oceans, this species has been obtained from such widely remote regions as the Bahamas, the Galapagos Islands, and New Zealand.

c

The sooty shearwater (*P. griseus*), which is selected as an example of the second group, may attain a length of 18 inches, and is of a uniform dusky tint above and slightly paler beneath, its range being nearly or quite as extensive as that of the preceding species, and stragglers occasionally reaching the British Isles.

In Australia and New Zealand certain kinds of shearwater (especially *P. brevicauda*), locally known as "mutton-birds," form an important commercial asset. On Barren Island, in Bass Strait, a colony of some 400 half-castes depended during many years on the mutton-birds for their living; and it is estimated that some 1,500,000 of these birds are killed annually on this one island alone. After being boiled and salted, the bodies of the birds are sold for about 10s. per hundred, and the oil fetches from about 10d. to 1s. 3d. per gallon, while a further sum is realised from the sale of the fat and feathers. Those who have eaten mutton-bird speak favourably of it as an article of diet, and the large demand has led to such an increase in the output as to induce the fear that the slaughter is committed on too large a scale.

Recently, owing to the lucrative nature of the trade, white people have come to reside on the island, and are reported to kill two-thirds of the birds taken, much to the dissatisfaction of the half-castes. It has, however, been urged that the latter should be allowed to hold plots of land, and be permitted to prohibit sheep-farming or cattle-ranching on the island. For wherever sheep or cattle go, there the mutton-birds will disappear; and even if the sheep or cattle are confined to certain areas, some would be sure to escape, and soon cause the birds to vacate their haunts. The mutton-bird industry is far more profitable than either sheep-farming or cattle-ranching.

CAPPED PETREL

Allied to the shearwaters is the genus *Œstrelata*, as represented by the capped petrel (*Œ. hæsitata*) and certain other species. It is characterised by the great compression of the rather short beak, in which the terminal curved "nail" is of very large size, and the short and very prominent nasal tubes. The long and pointed wings extend, when folded, considerably beyond the graduated tail, and the hind toe is small and elevated. This species inhabits the warmer parts of the Atlantic, straying occasionally to England and France. While the forehead, the sides of the head, the neck, the upper tail coverts, and the base of the tail are white, the crown of the head has an isolated black cap, and the upper parts are bistre-brown; the whole length being 16 inches.

BULWER'S PETREL

The uniformly blackish brown Bulwer's petrel (*Bulweria bulweri*) is a much smaller bird of some 10½ inches in length, and is one of the two representatives of its genus. It is more slenderly formed and longer-tailed than the last, and frequents the Atlantic in the neighbourhood of the Canaries and Madeira, where it lays in holes or under the shelter of rocks.

CAPE PETREL

From its superficial resemblance to a dark-coloured pigeon, the bird properly known as the Cape petrel (*Daption capensis*) is commonly designated in the colony the Cape pigeon. It represents a genus distinguished from those last mentioned by the presence of fourteen tail feathers; and is further characterised by the beak being broad and depressed, except at its tip, where the nail is small, and occupies less than a third of the total length. The nasal tubes are depressed and concave, and separated by a considerable interval from the terminal nail. In the leg the metatarsus is shorter than the third toe, although much longer than the beak. This bird is of medium size, and easily recognised by the sooty head and neck, the mingled dusky and white plumage of the upper parts, and the immaculate white of that below.

The Cape, or, as it is often called, pintado, petrel is an inhabitant of the South Atlantic and South Pacific Oceans, occasionally straggling northwards of the Equator. In the Antarctic seas these birds are frequently met with in vast numbers; and an observer, who accompanied a whaling expedition in the winter of 1892–93, writes that so eager were they for any scraps thrown over the ship's side that any number of them could have been caught with small hand-nets only large enough to contain one at a time, and many of them were thus captured by the crew. In stormy weather they not infrequently come close to land.

When gracefully hovering in the air, the bird may be seen to make a sudden dart downwards to the water, in order to secure some floating morsel of food it has espied, and on such occasions will dive readily. It is also said to throw up its tail after the manner

CAPE PETRELS SWIMMING

of a duck, and thus to fish up bits of food from slight depths. When caught and placed on deck, it has to run a short distance with outstretched wings before being able to rise; and when first hauled in or handled, invariably ejects from its mouth or nostrils a reddish oily fluid. These petrels breed on Tristan da Cunha and Heard Island, and probably also on some of the Antarctic islands; on Heard Island their nests are made in holes in low basaltic cliffs.

DOVE-PETRELS

The dove-petrels (*Prion*) are much smaller birds, represented by numerous species in the southern seas, and typically characterised by the great breadth of the base of their beaks. One of the best known is the common dove-petrel (*P. desolatus*), which is a small grey species with a broad boat-like beak, furnished with fine horny lamellæ projecting inwards from each side. It flies like a swallow, and may be seen in flocks about a ship, or cruising over the sea, or attendant on a whale to pick up the droppings from its mouth. Hence it is termed by sealers the whale-bird. Its food, like that of all petrels except the carrion ones, seems to consist of the very abundant surface animals of the south seas, especially of small crustaceans. It breeds on Kerguelen, laying its single white egg in a burrow which may be as much as a yard and a half in depth. The broad-billed blue petrel (*P. vitta-tus*) is another representa-tive of this genus.

STORM-PETRELS

The tiny storm-petrel (*Procellaria pelagica*)—the smallest of British web-footed birds—is the first representative of several genera of petrels readily distinguished from all the foregoing by their diminu-tive proportions ; the length of the wing not reaching 7 inches.

The storm-petrels are characterised as a genus by their very small size, by the even or rounded tail, by the length of the metatarsus being approxi-mately equal to that of the middle toe with its claw, and by the presence of a white patch on the rump. The whole length of the true storm-petrel—the "Mother Carey's chicken" of the sailors—is rather less than 6 inches ; the general colour of the plumage being sooty black (page 1354).

This petrel is confined to the more northerly portions of the Atlantic, and, except during the prevalence of severe storms and in the breeding season, is but seldom seen in the neighbourhood of land. Essentially a child of the ocean, it is frequently met with far out at sea, where it will follow vessels for considerable distances.

The breeding places of the storm-petrel include the Atlantic coasts of Europe and portions of the shores of the Mediterranean ; but the bird is not known to nest on any part of America. The single white egg is deposited in a burrow of considerable length ; and in the island of Soa it is stated that the burrows of several pairs often diverge from a common vestibule.

The small wedge-tailed petrel (*Halocyptena microsoma*), of the coast of Lower California, is the sole representative of a genus distinguished from the above by the tail being much rounded, the metatarsus exceeding the length of the third toe, and by the absence of any white on the rump.

Leach's petrel (*Oceanodroma leucorrhoa*) belongs to an allied genus, comprising several somewhat larger species readily characterised by the deeply forked tail, in which the feathers have very broad tips, while there may or may not be a white rump-patch. The species named has a very wide distribution, being common to the Atlantic and Pacific. Other species, such as the Pacific fork-tailed petrel (*O. furcata*) and Hornby's petrel (*O. hornbyi*), of the North Pacific, differ by the feathers of the forked tail being scalloped at the end ; there is no white patch on the rump, and the plumage

STORM-PETRELS

is either uniform bluish ashy or grey, with the forehead, cheeks, a collar on the throat, and the under parts white.

WILSON'S PETREL

The preceding members of the family collectively constitute a subfamily (*Procellariinæ*) characterised by the presence of at least thirteen secondary quills, by the metatarsus being covered with small hexagonal plates, by the sharp and curved claws, and by the leg-bones being shorter than the wing. On the other hand, the small Wilson's petrel, together with some allied species, forms a second subfamily (*Oceanitinæ*), distinguished by the pre-sence of only ten second-aries, by the metatarsus being either booted or covered in front with large oblique plates, by the flat and broad claws, and by the leg-bones exceeding the wings in length. In all the group the aperture of the straight nasal tubes is single and circular.

Wilson's petrel (*Ocean-ites oceanicus*), which is somewhat larger than the storm-petrel, inhabits the Atlantic Ocean and Aus-tralian seas ; but the other members of the group are exclusively southern. The general colour is dusky, with the quills and tail feathers black. As re-gards their muscles, these petrels are highly special-ised, and, in the boot-like plates covering the meta-tarsus, differ from all other water-birds.

The white-bellied petrel (*Cymodroma*, or *Fregetta*, *grallaria*), of the tropical seas, has an even tail, and the metatarsus about twice the length of the third toe, exclusive of the claw. The plumage is parti-coloured.

DIVING PETRELS

Like the albatrosses, all the petrels hitherto mentioned are essentially flying and swimming birds, which dive but little. There is, however, a remarkable aberrant petrel (*Pelecanoides urinatrix*) inhabiting the Straits of Magellan which, together with a second species from the southern Indian Ocean, and a third of western South America, differs from the other members of the order in its short wings and diving habits, and is further distinguished by the nasal tubes being vertical and opening superiorly ; the first toe being absent.

This bird, which many naturalists regard as the representative of a distinct family (*Pelecanoididæ*), is, indeed, in habits and appearance so like an auk that, when seen from a distance, either on the wing or diving and swimming, it would undoubtedly be mistaken for one of these birds. Nevertheless, both in structure and plumage, it is essentially a petrel ; and its auk-like appearance and habits must accordingly be regarded as special modifications for a peculiar mode of life. These birds may be seen in calm weather in Royal Sound floating in immense numbers on the water, the flocks sometimes extending over acres. They dive with extreme rapidity, and, when disturbed, rise and flutter a short distance along the surface, after which they again drop and dive.   R. Lydekker

ORDER XVIII. IMPENNES

PENGUINS

APPROXIMATING to the divers and grebes (to which they also present certain resemblances in the structure of their soft internal parts), in the backward position of their short legs, and their upright posture when on land, the penguins (order Impennes) of the Southern Hemisphere differ from all other members of the class in two important structural features. In the first place, the wings, of which the quills are rudimentary, are transformed into paddles; and, in the second, the short metatarsus is of great width, with its three constituent longitudinal elements incompletely fused together, and separated from one another by small perforations. Consequently, these birds can scarcely be said to have a true cannon-bone. As regards their skulls, penguins agree with the other birds treated in this section in having the palate of the cleft (schizognathous) type, and there are also hollows on the forehead for the reception of glands.

The feathers are provided with after-shafts, the spinal feather tract is not defined on the neck, and the oil-gland is tufted. The young, although born covered with down, are at first helpless, and require to be tended for a long period in the nest. Reference has already been made to the rudimentary condition of the wing quills, and there are no functional tail feathers. It is very noteworthy that the rudimentary scale-like feathers with which the wings are covered are more numerous than the quills and wing coverts of any other birds.

As additional characters of the skeleton, it may be mentioned that the blade-bone, or scapula, is remarkable for its great breadth, while the bones of the wings are flattened; the humerus, which has no process on the outer side of its lower extremity, being very short. Most, if not all, of these peculiarities must, however, be regarded as the result of adaptation to an abnormal mode of life. In habits penguins are marine and carnivorous.

The general appearance of these birds is so well-known that it will be unnecessary to dilate on this point. It may be mentioned, however, that the beak is more or less elongated and straight, with its sides compressed and grooved, and its tip sharply pointed; the slit-like nostrils being situated within the lateral grooves. The sheath of the upper half of the beak is composed of from three to five more or less distinct pieces. The three front toes are of moderate length and completely webbed; while the first toe is very small, and united to the sides of the metatarsus.

It is quite possible that penguins have some affinity to divers and grebes, and, if so, it will be evident that the peculiar structure of the metatarsus is a degraded and not an original feature. Remains of a very large penguin have been obtained in New Zealand from strata of Eocene age, thus showing the antiquity of the group. The members of the order mentioned in this work may be arranged as follows:

FAMILY
Penguins—Spheniscidæ
GENUS 1
Aptenodytes
SPECIES
King penguin .... Aptenodytes pennanti
Emperor penguin .... A. forsteri
GENUS 2
Pygoscelis
SPECIES
Gentu penguin............... Pygoscelis tæniata
Adélie-land penguin .... P. adeliæ
GENUS 3
Crested Penguins—Eudyptes
SPECIES
Rock-hopper .... Eudyptes chrysocoma
Yellow-crested penguin .... E. pachyrhynchus
GENUS 4
Eudyptula
SPECIES
Blue penguin .... Eudyptula minor
GENUS 5
Spheniscus
SPECIES
Black-footed penguin........ Spheniscus demersus
Humboldt's penguin .... S. humboldti
Jackass penguin .... S. magellanicus

As already mentioned, penguins, of which there are about seventeen different species, are confined to the Southern Hemisphere. In the case of most of the more northern species the range does not extend north of the southern tropic, but certain species inhabiting the Galapagos group are practically on the Equator. From the tropic the range extends at least as far

BLACK-FOOTED PENGUINS

as the 80th parallel of south latitude. Penguins are thus found not only on the Antarctic ice, but in South Africa, South America, Australia, and New Zealand, as well as many of the smaller islands of the southern oceans, more especially the Falklands, Kerguelen, and Tristan da Cunha. Although the whole of the penguins are included in the single family *Spheniscidæ*, they are now generally divided into five genera, of which the leading characters may be briefly noticed.

The largest members of the whole group are the king penguin (*Aptenodytes pennanti*), of Marion Island, Kerguelen Land, and other districts in the southern ocean, and the still larger emperor penguin (*A. forsteri*) of the Antarctic pack-ice. In addition to their large size, these species are characterised by the great length and slenderness of the beak, which is slightly arched, and the absence of any crest on the head. In the first of the two the colour of the head, neck, and throat is brownish black, the region behind the ear having a pear-shaped patch of yellow, continued as a streak down the sides of the neck, and meeting on the upper part of the breast; the upper parts are iron-grey, and the under surface is glistening white, faintly tinged with yellow. In the emperor penguin the yellow area is limited to a small patch behind the eye. Specimens of this species stand just under 3½ feet in height, and a large example may weigh as much as 90 pounds.

Closely allied to these is the gentu penguin (*Pygoscelis tæniata*), of Kerguelen Island and the Falklands—a species inferior in size only to the king penguin, and commonly known as the "Johnny." Devoid of a crest, this penguin is distinguished from the preceding by the long and pointed red beak being stouter and more feathered. In colour the plumage of the back is dark blackish and that of the under parts white; the dark of the back being continued on to the head, the summit of which is marked by a conspicuous white patch. A second member of the same genus is the Adélie-land penguin (*P. adeliæ*), of which vast colonies occur in the Antarctic at Cape Adair.

The crested penguins, as represented by the "rock-hopper" (*Eudyptes chrysocoma*) of the Falkland Islands, the yellow-crested penguin (*E. pachyrhynchus*) of New Zealand and the Antarctic, and several others belong to a third genus, characterised by the smaller size of its members, the short, deep, and compressed beak, in which the upper half has a distinctive oval form, and the presence of a pair of yellow crests on the sides of the head, continued forwards as streaks above the eyes to the neighbourhood of the base of the beak. In the rock-hopper these crests are much elongated, attaining a length of from 3 to 5 inches, but in the New Zealand species are shorter, never exceeding a couple of inches in length. The total length attained by the latter species is 27 inches.

Nearly allied is the little blue penguin (*Eudyptula*

GROUP OF PENGUINS

1, rock-hopper; 2, little blue penguin; 3, thick-billed penguin; 4, yellow-crested penguin; 5, young king penguin; 6, adult king penguin

minor), of Southern Australia and New Zealand, which does not measure more than 19 inches in length, and has no yellow streaks or crests on the head; the general colour of the plumage of the upper parts being light blue, with a median black line down each feather, and the under parts dazzling white. Fossil remains of both these genera occur in the superficial deposits of New Zealand.

The remaining penguins are included in the genus *Spheniscus*, of which the black-footed penguin (*S. demersus*), of South Africa, Humboldt's penguin (*S. humboldti*), of western South America, and the jackass penguin (*S. magellanicus*), of the Falkland Islands, are well-known examples. In all these species the beak is straight and moderately short, but very wide and deep, with the tip of the upper half slightly hooked, and that of the lower half truncated. There are no crests, and the metatarsus is relatively long.

In the adult of the black-footed penguin the general colour of the plumage is bluish grey above, and white below; a band the colour of the back extending (as in Humboldt's penguin) from the front of each thigh up the sides to form an arch on the front of the neck,

HUMBOLDT'S PENGUIN

appeared to be a shoal of small porpoises or dolphins. "I could not imagine," he writes, "what the things could be, unless they were indeed some marvellously small cetaceans ; they showed black above and white beneath, and came along in a shoal of fifty or more, from seawards towards the shore at a rapid pace, by a series of successive leaps out of the water, and splashes into it again, describing short curves in the air, taking headers out of the water and headers into it again ; splash, splash, went this marvellous shoal of animals, till they went splash through the surf on to the black stony beach, and there struggled and jumped up amongst the boulders and revealed themselves as wet and dripping penguins, for such they were."

On landing, the penguins always make for certain well-defined tracks leading up to the "rookeries," as their places of assembly are called, and where they not infrequently collect in thousands, these main tracks branching out into a number of diverging paths when they reach the rookery. The nest of the rock-hopper is merely a shallow depression in the black soil, which may or may not be lined with a few stalks of dry grass. In this are deposited two greenish white eggs, about the size of those of a duck, in the incubation of which both male and female birds take their share. The black-footed species deposits, however, but a single white egg, which rests on the bare ground. On the other hand, the jackass-penguin is in the habit of nesting in burrows, which may be as much as twenty feet in depth, and the same is also not infrequently the case with the little blue penguin of New Zealand, although the two eggs of this species are sometimes laid in the crevices of rocks. The breeding time of this species on the islands off the Cape lasts through August, September, and October.

while the white of the throat is continued upwards on the sides of the neck to form a line through the eye. The total length is about 26 inches. In young birds a broad blackish band occupies the whole of the front of the lower part of the throat, and joins directly with the dark area of the back, and there is no white line through the eye. Humboldt's penguin has the white line running just above the eye, and no white area below the same.

Penguins are some of the most strange and bizarre of all birds, alike as regards personal appearance and habits. On land they present a most curious appearance, both when strutting about with their padded feet over the snow, or when gliding on their breasts down a slope toboggan-fashion. When a visitor lands upon the Antarctic ice, the emperor penguins approach fearlessly with their curious duck-like cries, a proceeding which too often leads to their destruction. Their tenacity of life is, however, marvellous, exceeding even that of the proverbial cat ; one observer stating that he has known an emperor penguin to live after its skull had been hopelessly smashed. All the species are gregarious, frequently assembling in tens of thousands, and when on the land during the breeding season are in the habit of ranging themselves in long lines on the ledges of the rocks or ice, thus simulating the appearance of soldiers when seen from a distance.

The food of penguins consists exclusively of fish, which the birds capture beneath the surface by their agility in swimming and diving, when the paddle-like wings are used as the chief instruments of progression. So thoroughly, indeed, are they at home in the water that they are apt to be taken for dolphins rather than birds, as is testified by Professor Moseley, who states that on first approaching the shore of Kerguelen Island he was astonished at seeing what

ROCK-HOPPER PENGUINS

The most extraordinary breeding habits are displayed, however, by the king and the emperor penguins, each of which lays its single egg at the beginning of the southern winter. At this season the king penguins in the Macquarie Islands, near New Zealand, may be seen squatting on stones in a quagmire of mud and water, each with an egg tucked in under the skin and feathers of the abdomen, and held on the feet to save it from the water. In the same way the emperor penguin keeps its egg and chicken from contact with the ice on which it stands. Neither species makes a nest for its eggs and young. It is curious that there is a great difference between the young of these two birds, which in other respects are so remarkably alike. The young are, in fact, wholly different in colour ; the young king penguins, which are still abundant in the rookery when the old birds are sitting on fresh eggs in November, look like young bears in their long brown down, whereas those of the emperor penguin are silvery white, with a head wholly black except for a patch of white on each side, including the check and eye.

Breeding during the darkness and severity of the Antarctic winter, it is only natural to expect that the emperor penguin should lose a number of its eggs and young from the cold, as even momentary contact with the ice is sufficient to cause them to freeze. As a matter of fact, the rigorous climate alone lays claim each year to no less than 77 per cent. of all the chicks hatched, before they are sufficiently old to be independent of their parents. This average mortality among the new-born chicks is no mere guess, but the result of two years' observation, when the dead chickens on the pack-ice were actually counted before the colony broke up at the beginning of the spring. With this tremendous juvenile mortality, it is surprising that the adult birds should be so abundant, for in a rookery it is usual to find only one chick to ten or a dozen adults, each of which is most anxious to nurse that one particular chick.

The penguins inhabiting Tristan da Cunha migrate about April, and return in July or August ; but where they go does not seem to be ascertained, although it is quite certain that they cannot remain at sea for such a protracted period. Although during their aquatic journey they do not travel with anything like the speed of birds on the wing, they have the compensating advantage of a constant supply of food.

Writing of the habits of the little blue penguin, Mr. J. Gould observes that " its powers of progression in the deep are truly astonishing ; its swimming powers are, in fact, so great that it stems the waves of the most turbulent seas with the utmost facility, and during the severest gale descends to the bottom, where, among beautiful beds of coral and forests of seaweed, it paddles about in search of crustaceans, small fish, and

CAPE PENGUIN

marine vegetables, all of which kinds of food were found in the stomachs of those I dissected."

Of the jackass-penguin, Darwin states that " when crawling, it may be said, on four legs, through the tussocks or on the side of a grassy cliff, it moves so very quickly that it might easily be taken for a quadruped. When at sea and fishing, it comes to the surface for the purpose of breathing with such velocity, and dives again so instantaneously, that I defy anyone at first sight to be sure that it is not fish leaping for sport." This species, by the way, derives its popular name from its habit, when on shore, of throwing back its head and giving vent to a cry not unlike a donkey's bray.

On the coast of the Falklands the hard volcanic rocks are in places scored into deep grooves by the sharp claws of millions of penguins passing and repassing for countless centuries.

ORDER XIX. ODONTOLCÆ

THE existence during the Cretaceous epoch in North America of toothed birds (*Ichthyornis*) presenting a considerable resemblance to the modern gulls has been mentioned in an earlier section. In addition to these, there lived at the same epoch a totally distinct type, (order Odontolcæ), in which the teeth, in place of having separate sockets, were implanted in an open groove, while the wings were rudimentary, and the keel of the breast-bone was wanting, although the vertebræ resembled those of existing birds in articulating together by saddle-shaped surfaces.

In general organisation these birds, *Hesperornis*, approximated to the modern divers, with which they agree in the general conformation of the skull and limb bones, as well as of the pelvis. Whereas, however, the

modern divers have the long spike-like knee-cap, or patella, united with the tibia, in the extinct bird these two bones remained distinct. In dimensions *Hesperornis* was a bird of large size, attaining a height of rather more than a yard when in the upright position. That it was thoroughly aquatic in its habits is self-evident, and it may with considerable probability be regarded as a specialised and flightless offshoot from the ancestral stock of the modern divers ; although this would not justify its inclusion in the same order as the latter.

An apparently allied, although very imperfectly known, type of bird (*Enaliornis*) is represented in England, where its remains have been obtained from a thin stratum lying at the base of the Chalk, known as the Cambridge Greensand.

ORDER XX. PYGOPODES
DIVERS AND GREBES

THE divers and grebes (order Pygopodes) differ from all other birds in that the crest of the tibia is prolonged upwards to unite with the knee-cap, or patella, thus forming a spike-like projection at the extremity of that bone, which affords an efficient lever for the muscles in the act of swimming. The group is further characterised by the saddle-like form of the articular surfaces of the vertebræ of the back, by the presence of a small first toe, and the absence of bare tracts on the sides of the neck; while the metatarsus is compressed and knife-like.

The limbs, like those of auks (with which these birds were formerly grouped) are set far behind the middle of the body. The skull has a cleft, or schizognathous, palate; two large grooves, separated from one another by a narrow ridge on the front surface for the reception of glands; and the hind part of the bone of the lower half of the beak abruptly truncated. The humerus, or upper wing-bone, has no transverse projection at its lower end. Dr. R. W. Shufeldt is of opinion that these birds are probably the descendants of toothed divers more or less closely allied to the American Cretaceous *Hesperornis*, the

GREAT NORTHERN DIVER

grebes exhibiting the most marked traces of this relationship. As the flightless *Hesperornithidæ* themselves are doubtless the descendants of flying types, so, in Dr. Shufeldt's opinion, the modern grebes and lories may, if they survive long enough, become in the course of ages incapable of flight.

The following table sets forth the members of the order that are mentioned in this work.

FAMILY 1
Divers—Colymbidæ
GENUS
Divers—Colymbus
SPECIES

Great northern diver	Colymbus glacialis
Black-throated diver	C. arcticus
Red-throated diver	C. septentrionalis

FAMILY 2
Grebes—Podicipedidæ
GENUS 1
Typical Grebes—Podicipes
SPECIES

Western grebe	Podicipes occidentalis
Great crested grebe	P. cristatus
Red-necked grebe	P. griseigena
Sclavonian grebe	P. auritus
Black-necked grebe	P. nigricollis
Little grebe	P. fluviatilis
Least grebe	P. dominicus

GENUS 2
Podilymbus
SPECIES

Thick-billed grebe	Podilymbus podiceps

DIVERS

In the divers (family *Colymbidæ*) the three front toes are fully webbed, and furnished with sharp claw-like nails;

the number of primary quills in the wings is eleven; the tail, although short, is normal; and there are only fourteen or fifteen vertebræ in the neck. The beak is long, sharp, and compressed, and the lores are completely feathered. Although an extinct representative of the family (*Colymboides*) has left its remains in the Miocene deposits of the Continent, the existing divers, of which there are three well-marked species confined to the Arctic and cooler regions of the Northern Hemisphere, are included in the single genus *Colymbus*. Divers, although more slenderly formed, have somewhat the appearance of geese when seen on the water; but on land, owing to the backward situation of their legs, are widely different. In plumage the two sexes are alike; but the winter dress differs considerably from that of summer, as do the young from the adult.

The typical representative of the genus is the great northern diver (*C. glacialis*), a bird attaining a length of about 33 inches, and characterised by its glossy black head and neck, the presence of two gorgets of velvety black and white on the throat, and the belts of white spots of varying size crossing the dark back, the under parts being white. Not uncommon—especially in an immature state—on the British coasts, and thence wandering so far south as the Mediterranean, this diver breeds in Iceland, Greenland, and North-Eastern Canada; but in North-Eastern Asia and Western Arctic America is replaced by a larger variety (*C. g. adamsi*), distinguished by the white or yellow hue of the beak.

Next in point of size is the black-throated diver (*C. arcticus*), which does not exceed 26 inches in length, and is characterised by its light grey head, the purplish black patch surmounted with a black and white striped gorget on the throat, and the presence of two elongated areas on the black back between the shoulders, as well as others on the scapulars, marked by transverse white bands formed by nearly confluent square spots. The breeding area of this species seems to extend from the Hebrides and Scandinavia across Arctic Asia over the greater part of America, although it does not include Greenland, Iceland, or the Orkneys. Some naturalists regard a diver inhabiting the Pacific coast of America as a distinct species.

Finally, the smallest, as well as the commonest, species is the circumpolar red-throated diver (*C. septentrionalis*), so named from the presence of a patch of reddish grey extending down the throat of the adult in breeding plumage. On the upper parts the plumage is blackish brown in colour, with a comparatively small number of spots; the head and sides of the neck being ashy grey, and the nape marked by streaks of black, grey, and white. Young birds, in which the throat-patch is lacking, are much more fully striped. Although it does not breed at the present day in Great Britain to the south of Scotland, remains of this diver discovered in the superficial deposits of the East Coast suggest that it was formerly a resident in this part of England, when the climate was colder.

Feeding almost exclusively on fish, and during the winter being oceanic in their habits, the divers resort to inland lakes for the purpose of nesting. Unlike auks, they are not gregarious, consorting only in pairs, and these generally keeping far apart from one another. Although they are strong on the wing, the backward situation of their legs renders divers extremely ill-adapted for moving on land, where they walk with the greatest difficulty and ungainliness. Accordingly, in order to avoid the necessity of making the attempt, the slight nest is always constructed close to the water's marge, so that the sitting bird can at any moment resort to her native element by merely sliding downwards from her sitting-place.

In contrast to their awkwardness on land is the extreme agility displayed by these birds both on and beneath the surface of the water. They may, indeed, be regarded as almost the diving birds *par excellence*, the great northern diver having been stated to remain below the surface for a period of eight minutes, and all the species readily taking a baited hook while diving. Seldom seen on the wing except during the periods of migration, divers fly in a straight, arrow-like course, somewhat after the manner of ducks. The notes of all the species are harsh and grating.

Arriving at its breeding haunts in the Arctic regions about the end of May or beginning of June, synchronously with the breaking up of the ice, the great northern diver forthwith sets about the work of nesting. For choice, an island is selected, but, failing this, the shelving shore of some lonely lake, or even of a mountain tarn, is taken for a site. The nest, which is constructed of grass and sedge, is placed in an exposed position, where the sitting bird may readily receive warning of approaching danger, upon which it takes at once to the water. This might at first sight seem fraught with danger to the eggs, but it appears that the safety of these is generally sufficiently assured by their protective resemblance to their inanimate surroundings, their colour being dark brown speckled with blackish. The usual number of eggs in a nest is two, and both sexes take their share in the work of incubation.

GREBES

Best known by the little dabchick of the European meres and rivers, the family of the grebes (*Podicipedidæ*) is distinguished from that of the divers by the toes being lobated instead of webbed, and furnished with broad, flat nails, rounded at the tips ; by the presence of twelve primary quills in the wing, by the rudimentary condition of the tail, and by the number of vertebræ

RED-THROATED DIVER

in the neck varying from seventeen to twenty-one. In all of them a bare stripe extends across the lores from the beak to the eye ; the beak, although very variable in form, is always much elongated ; and the nostrils are never protected by an overhanging lobe.

The wings are short and concave, and when closed have the primaries concealed by the secondaries, and in the aborted tail a tuft of downy, soft feathers takes the place of the usual stiff tail feathers. On the lower surface of the body the plumage, which is usually of a pure white hue, is remarkable for its soft, silky texture and brilliant lustre.

Grebes are more addicted to fresh-water than the other members of the order, some of them being inhabitants of lakes and rivers throughout the year, while others are oceanic during a large portion of their existence. They are represented by about sixteen species, ranging over the temperate and subtropical regions of both hemispheres ; five of these being European, and three of the latter breeding in the British Isles, where the others are but winter visitors. The eggs differ from those of the divers in the creamy white colour of their shells, and their green tinge when viewed by transmitted light ; the usual number in a clutch being either three or four.

Using the term *Podicipes* in a wide sense, it will include the greater number of species of the group, or all those in which the length of the slender beak varies from 2½ to 6 times its basal depth. Among the larger members of the genus, the western grebe (*Podicipes occidentalis*) is the sole representative of a group characterised by the great length of the neck and beak, the smooth head, which is devoid of tufts at all seasons of the year, and the absence of any seasonal difference in the general plumage, this species being exclusively North American.

The great crested grebe (*P. cristatus*), which is the largest member of the genus, and attains a length of from 21 to 22 inches, belongs, on the other hand, to a section in which the neck and beak are shorter, and the head of the adult is ornamented, in the breeding season at least, with coloured ruffs, tufts, or patches, while the general plumage in the breeding season differs considerably from that of the adult in winter, and likewise from that of the young. In this particular species the crest, although largest in summer, is borne throughout the year, but in others it disappears in winter completely.

In its summer plumage this bird may be recognised by its chestnut-coloured ear coverts, and the white front of the lower part of the neck and breast, but in winter it has the lores and a stripe over the eye white. Confined to the Eastern Hemisphere, this species is remarkable for the extent of its breeding area, which includes Great Britain and Southern Europe, the whole of Africa, and the greater portion of Southern and Central Asia, as well as Australia and New Zealand,

It is noteworthy that the Australasian birds, though completely isolated, present no differences from the others.

The smaller red-necked grebe (*P. griseigena*), which only measures 16 inches in length, takes its names from the chestnut hue of the front of the lower part of the neck in the summer plumage, in which alone the crests on the head are present. Inhabiting a large portion of Northern Europe and Asia, this species appears to have a circumpolar distribution, although some writers regard the variety occurring in America and North-Eastern Asia as a distinct species, under the name of *P. holboelli*.

A third group of the genus is characterised by the smaller size of its members, in which the neck is short and the beak shorter than the head, while ear-tufts are present in the breeding plumage. Its best-known representative is the circumpolar Sclavonian grebe (*P. auritus*), which visits the British Isles and Gibraltar in winter, and, except in Norway, does not breed north of the Arctic Circle. Measuring a little over 13 inches in length, it is characterised by the compressed beak, and the combination in the breeding plumage of a chestnut fore-neck with black ear-tufts.

On the other hand, the black-necked grebe (*P. nigricollis*) may be recognised by the prevalence of black on both the fore part of the neck and the ear coverts, the black on the neck reaching down to the level of the water when the bird is swimming, which is not the case with the Sclavonian species. This species, which breeds in Great Britain, ranges over the greater part of Europe and Asia (except India and Burma), as well as portions of Africa and the whole of Greenland. It is represented by a variety in North-Western Africa.

Last of all comes the familiar dabchick, or little grebe (*P. fluviatilis*) of the Old World, and the least grebe (*P. dominicus*) of tropical America, together with some southern forms, as the representatives of a fourth group, characterised by the small size of its members, the very short neck and beak, and the absence of tufts or crests in the breeding plumage. By some naturalists these species are regarded as constituting a distinct genus—*Tachybaptes*. The dabchick, which is the commonest of the British grebes, has a wide range in the Old World, its breeding area including the subtropical portions of both the Northern and Southern Hemispheres south of latitude 42°, as well as elevated regions within the tropics, while in Western Europe it extends some 20° further north. Not exceeding 9½ inches in length, the dabchick in breeding plumage is characterised by the chestnut-

red of the cheeks and front of the neck, which in the American species are always ashy grey (page 1354).

With the exception that the dabchick, like its small allies, generally spends the whole year in the neighbourhood of fresh-water, the grebes are very similar in their habits, all of them resorting to rivers and lakes for the purpose of breeding. Their diving powers are such that, when pursued, they seldom take to wing, but nearly always endeavour to escape by disappearing beneath the water, to reappear in the most unexpected places. Indeed, although the larger species fly strongly and well, with the neck stretched out and the wings moving rapidly, the dabchick rarely takes to flight.

The ordinary alarm note of the great crested grebe may be expressed by the syllables "kek-kek," but at the pairing season a guttural sound is uttered. Their food consists of frogs, fish, molluscs, water-insects, and the like, supplemented by the shoots and seeds of aquatic plants. The great crested grebe frequently associates in parties during the breeding season, when, like its congeners, it makes its large nest of decaying water-plants so nearly level with the

THE CRESTED GREBE

surface of the water that it is generally constantly wet.

A colony of these birds, breeding in an immense reed bed on the Zuider Zee, near Danzig, has been described by Mr. H. Seebohm, who writes that "there were dozens of nests, but never very close to each other, and I soon filled my handkerchief with eggs. It was the 5th of June, and only about half the nests contained the full complement of eggs. The birds had evidently seen us long before we approached, and had had ample time to retreat with dignity. In the nests which contained three or four eggs, these were warm and covered with damp moss; but in those containing only one or two, they were uncovered and cold."

Mr. Seebohm was thus led to believe that the eggs are not covered till the female begins to incubate, and the purpose of covering them a protection against chill, and not for concealment, white eggs being quite inconspicuous in the recesses of a dense mass of reeds.

The thick-billed, pied-billed, or Carolina grebe, (*Podilymbus podiceps*), which is an exclusively American bird, clearly represents a distinct genus, characterised by the shortness and stoutness of the beak, in which the length is less than twice the basal depth. The much-arched beak is parti-coloured; the length of the metatarsus is less than that of the third toe without the claw; and the head is not tufted, although the throat is ornamented with a black patch. This grebe inhabits temperate North America and the West Indies, as well as the whole of Central and the greater part of South America.   R. LYDEKKER

ORDER XXI. FULICARIÆ
THE RAIL TRIBE

NATURALISTS are by no means in accord as to the best serial positions to be assigned to the groups classified in this work as rails (order Fulicariæ), pigeons (order Columbæ), and sand-grouse (order Pterocletæ), or, indeed, whether they ought to be allowed to rank as distinct orders. The rails, for instance are classed by some in the same order as the cranes, to which they are undoubtedly related, while the pigeons and sand-grouse are brigaded, together with the auks and gulls, in one group with the plovers, under the name of Charadriiformes. Others, on the contrary, adopt the grouping followed in this work. The members of the group referred to in this work are the following:

ORDER
Rail Tribe—Fulicariæ

FAMILY 1
Rails—Rallidæ

GENUS 1
Typical Rails—Rallus
SPECIES
Water-rail .... Rallus aquaticus
Clapper-rail .... R. longirostris

GENUS 2
Weka Rails—Ocydromus
SPECIES
South Island weka rail .. Ocydromus australis

GENUS 3
Crakes—Crex
SPECIES
Corncrake .... Crex pratensis

GENUS 4
Porzana
SPECIES
Sora-rail .... Porzana carolina
Spotted crake .... P. maruetta

Little crake .... P. parva
Baillon's crake .... P. intermedia

GENUS 5
Pigmy Rails—Corethrura
SPECIES
South African pigmy rail .... Corethrura rufa

GENUS 6
Tribonyx
SPECIES
Mortier's water-hen ...... Tribonyx mortieri

GENUS 7
Moor-hens—Gallinula
SPECIES
Moor-hen .... Gallinula chloropus

GENUS 8
Gallicrex
SPECIES
Water-cock .... Gallicrex cinerea

GENUS 9
Purple Water-hens—Porphyrio

GENUS 10
Porphyriola

GENUS 11
Notornis
SPECIES
Mantell's gallinule ...... Notornis mantelli
White gallinule .... N. albus

GENUS 12
Coots—Fulica
SPECIES
The coot .... Fulica atra

FAMILY 2
Finfeet—Heliornithidæ

GENUS 1
Podica
SPECIES
Senegal finfoot .... Podica senegalensis
Peter's finfoot .... P. petersi
.... P. cameranensis

GENUS 2
Heliornis
SPECIES
American finfoot .... Heliornis fulica

GENUS 3
Heliopais

The rails and their relatives have the cleft, or schizo-gnathous, type of palate common to the cranes and game-birds; and, while evidently allied to the former, are connected with the latter by the so-called bustard-quails. They appear to represent an ancient and generalised type of bird, and are divided into two families, of which by far the greater number are included in the typical *Rallidæ*.

An interesting feature in this group is the large number of species which, from disuse of the wings, have lost the power of flight, several of these having become extinct within the memory of man; while with many others their extinction is but a matter of time.

Among the chief external characteristics of these birds are the long legs and toes, loose and rather hairy plumage, feeble, rounded wings, and short tail. The body is generally narrow and laterally compressed, thereby enabling the birds to thread their way among reeds and grasses with great ease and rapidity; and the neck is long and the head small, with a long or moderate beak.

The young are active from the time they leave the eggs. A large number of genera, including nearly 180 species, constitute the family, but only some of the more important types are here mentioned.

THE WATER-RAIL

TYPICAL RAILS

The members of the genus *Rallus*, as typified by the European water-rail (*R. aquaticus*), are characterised by the beak being longer than the third toe and claw, with the nostrils nearer the feathers at the base than the anterior end of the nasal groove. On the other hand, in all the genera mentioned below the beak is shorter than the middle toe and claw (page 1404).

The clapper-rail (*R. longirostris*) is a North American species, characterised by having the upper parts ashy grey, streaked with blackish brown, the chin and throat white, the fore part of the neck ashy brown, shading into isabelline on the chest and upper portion of the breast, and into whitish on the under parts, the flanks being barred with greyish brown and white. This bird is a resident in many of the south-eastern United States, but only met with in the salt-marshes near the Atlantic, unless driven inland by high tides.

In spring considerable migrations take place at night, and are always conducted in perfect silence. The old naturalist Audubon writes, for instance, that

THE CORNCRAKE, OR LAND-RAIL

" from about the beginning of March to that of April the salt-marshes resound with the cries of the clapper-rail, which resemble the syllables ' cac, cac, cac, cac, cà, cātrā, cātrā.' The beginning of the cry, which is heard quite as frequently during day as by night, is extremely loud and rapid, its termination lower and protracted. At the report of a gun, when thousands of these birds instantaneously burst forth with their cries, you may imagine what an uproar they make. At this period the males are very pugnacious, and combats are rife till each has selected a female for the season. The males stand erect and cry aloud at the least sound they hear, guard their mates, and continue faithfully to protect them until the young make their appearance."

The nest is large and very deep, constructed of marsh-plants and fastened to the stems in the midst of the thickest tufts above high-water mark. This species may be called gregarious, the nests being placed on the most elevated grass-tufts within a few yards of each other. Eight to fifteen eggs, of a pale buff colour thinly spotted with light brown and purple, are laid, of which, on account of the excellent flavour, large numbers are collected for the market. This species can swim fairly well, and traverse partially submerged weeds with great rapidity.

WEKA RAILS

The next group for notice includes the three species of weka rails (*Ocydromus*), only found in New Zealand, and incapable of flight, though provided with ample wings. These birds, of which the South Island *O. australis* is the typical representative, may be recognised by their rather large size, nearly equal to a pheasant, by the beak being shorter than the middle toe and claw, and by the elongate wing coverts, which extend nearly to the extremity

of the quills, as well as by their stout, strong legs. They are semi-nocturnal in habits, and usually remain concealed during the day in thick fern or scrub, taking refuge in hollow logs or other natural cavities. Occasionally they dig a subterranean burrow, the beak only being employed for this purpose, which serves as a retreat as well as a breeding place.

These birds are remarkably bold and fearless, sometimes visiting farmyards and even entering houses. They are also pugnacious and perfectly omnivorous, being well known to plunder and eat the eggs and young of ground-birds, and will even attack full-grown rats. The cry, begun at sunset and continued through the night, is a peculiar and not unpleasant whistle. A pair usually perform together, calling alternately and in quick succession, the male always taking the lead. As already stated, these birds usually breed in burrows, laying two and sometimes three eggs of the usual rail type.

CORNCRAKE

The corncrake, or land-rail (*Crex pratensis*), representing a genus of its own, is found throughout the greater part of Europe and as far east as the Yenisei in Siberia, ranging south in winter to Africa, while it is also an occasional visitor to North America and Greenland (p. 1404).

SORA-RAIL

The American sora-rail (*Porzana carolina*), in which the general colour above is olive brown varied with black middles and white margins on the feathers, represents an allied and widely spread genus with about sixteen species. In this bird the forehead, crown, front of the face, and middle of the throat and neck are black ; the eyebrow stripes, sides of the face and neck, as well as the chest, ashy grey, the breast white, and the flanks barred with black and white. To the same genus belong the spotted crake (*P. maruetta*), the little crake (*P. parva*), and Baillon's crake (*P. intermedia*), all of which visit Great Britain.

CAROLINA SORA-RAILS

PIGMY RAILS

A small group is formed by the pretty pigmy rails of Africa and Madagascar, characterised by the soft tail feathers being almost hidden by the coverts. In the South African species (*Corethrura rufa*) the general colour of the upper parts, sides, and flanks is black, longitudinally streaked with white ; the inner quills, lower back, and tail being also spotted with sable. The head, neck, and chest are rich chestnut, and the breast purplish white streaked and barred with black. Mr. C. J. Andersson, writing from Damaraland, states that " I have only found this species at Omanbondé, where it is not uncommon, and breeds. It frequents stagnant waters, thickly fringed and studded with aquatic herbage, amongst the ever-progressive decay of which it loves to disport itself and to search for food. It is very shy and reserved in its habits, seldom going far from effective cover, and gliding through the mazes of the rank vegetation with astonishing ease and swiftness."

MORTIER'S WATER-HEN

A handsome Australian bird may be taken as the first representative of the group which includes the moor-hens and coots, all characterised by having a bare, fleshy shield on the fore part of the head at the base of the beak. Mortier's water-hen (*Tribonyx mortieri*), the only representative of its genus, may be recognised by its short toes, which do not exceed the metatarsus in length, its large size, and feeble wings, with the primary and secondary quills about equal in length. The general colour is ruddy brown, washed with olive, shading into greenish grey on the wing coverts, which are spotted with white ; the head and neck are dark olive brown, and the rest of the under parts, except a large white patch on each side of the body, greenish grey washed with olive.

In regard to its habits, Mr. J. Gould states that " the localities it affects are marsh-lands and the sides of rivers. It was daily seen by me on the Government demesne at New Norfolk, Tasmania, where it frequently left its sedgy retreat, and walked about the paths and

MORTIER'S WATER-HEN

other parts of the garden, with its tail erect like the common hen. Even here, however, the greatest circumspection and quietude were necessary to obtain a sight of it ; for the slightest noise or movement excited its suspicions, and in an instant it vanished in the most extraordinary manner into some thicket, from which it did not again emerge until all apparent cause

BLACK-BACKED WATER-HEN

for alarm was past. Its habits and general manners are very similar to those of the moor-hen, but it does not dive or swim so much as that bird. It is very easily captured with a common horsehair noose. The nest, which is very similar to that of the moor-hen, is formed of a bundle of rushes placed on the border of the stream ; the eggs are seven in number."

MOOR-HEN

The well-known European moor-hen (*Gallinula chloropus*), easily recognised by its yellow and red beak, green legs, and dove-brown plumage, is the typical representative of a widely spread genus containing half a score of species. In common with all the following genera the toes are long, the third toe and claw exceeding the metatarsus in length. This genus is distinguished by the circumstance that the toes, although not lobed, resemble those of the coots, and have a narrow lateral membrane, while the nostrils are oval and situated in a distinct nasal depression (page 1404).

WATER-COCK

South-Eastern Asia and the Malay Islands are the home of a large species known as the water-cock (*Gallicrex cinerea*), representing a genus by itself, and distinguished by having no lateral membrane on the toes. The male has the plumage black, especially the upper parts, the wing coverts being edged with grey, and the scapulars and the lower portion of the back with brown, while the under tail coverts are buff barred with black. The female is browner and has the wing coverts grey, and the under parts buff with dusky bars, except the throat and middle of the belly, which are white.

PURPLE WATER-HENS

The most striking birds of the group, as regards brilliance of colouring, are the purple water-hens, or gallinules (*Porphyrio*), with their handsome blue and purple plumage, variously shaded with dark green, olive brown, and black. These birds are distributed all over the warmer parts of the Old World, and are represented in America by the members of the allied genus *Porphyriola*, of which one species occur in Africa and Madagascar. Closely allied to these, but much larger, is Mantell's gallinule (*Notornis mantelli*), a native of New Zealand, now nearly, if not quite, extinct ; the white *N. albus*, which formerly inhabited Norfolk and Lord Howe Islands, being another member of the same genus already completely exterminated.

COOTS

A very distinct group is formed by the coots (*Fulica*), all of which are easily recognised by their lobed toes. In habits they resemble ducks and gallinules, being able not only to swim and dive well, but also to thread their way through grass and reeds with ease and swiftness. In rising, they flap along the surface of the water, and fly like rails with their legs dangling ; and their notes resemble those of gallinules, but are more harsh and grating. Thirteen species of the genus are recognised, among which the common *F. atra* ranges from Great Britain to Celebes. The eggs have a creamy or pinkish ground-colour, marked with specks or small spots of purple or brown. Coots have an almost world-wide distribution, and some of the species, especially the European one, associate in flocks of enormous size (page 1404).

FINFEET

Finfeet form a comparatively small group character-

SENEGAL FINFOOT

ised by the grebe-like feet and the absence of after-shafts to the feathers. The family (*Heliornithidæ*) is

COOTS

divisible into three generic groups, of which *Podica* is restricted to Africa, *Heliornis* to Central and South America, and *Heliopais* to South-Eastern Asia.

As a representative of the first genus, mention may be made of the West African finfoot (*Podica senegalensis*), in which the general colour above is dark brown glossed with dark green, the back and wings being ornamented with round ochery spots edged with black, the sides of the face, neck, and throat grey banded with white, and the rest of the under parts white tinged with fulvous and barred with black on the sides.

On the Junk and Du Queah rivers of Liberia these birds are usually met with solitary, more rarely in pairs, slowly swimming about, and very shy and watchful, making for the bank at full speed on the approach of a canoe, and hiding themselves under the thick foliage of the overhanging shrubs. They are very hard to flush unless taken by surprise in the open, when they flutter hastily away, keeping so close to the water that they continually beat the surface with their wings and feet. When swimming, they sit very deep in the water.

Their general habits are much like those of the European coot. Specimens of the closely allied Peter's finfoot (*P. petersi*), of South Africa, are often taken in traps set for otters. These birds have the power of making an extraordinary noise, like the growling of a wild beast, which they effect by drawing the air into their bodies, and forcing it gradually out from their throats ; and they utter this strange noise when taken from the traps, fighting at the same time with all their might. The third species is *P. cameranensis*. Each of the other genera is represented only by a single species. It is noteworthy that the young of the typical American finfoot (*Heliornis fulica*) are stated to be hatched naked and carried about by the old bird, but very little appears to be known at present of the nesting habits of any of these birds.

ORDER XXII. COLUMBÆ
GENERAL CHARACTERISTICS OF THE PIGEON TRIBE

PIGEONS and doves form such a well-marked group (order Columbæ) that there is little difficulty in recognising any of their numerous representatives at a glance. In all of them the moderately large head is set on a gracefully-formed neck, and the body is rather compact and stoutly built. Swollen at the extremity, the beak has its basal portion covered with a soft skin, in which open the nostrils. The legs, which are coated with six-sided scales, are usually more or less thickly feathered on the upper part, although sometimes naked; and the feet have four toes, of which the first is placed on the same level as the others.

With the exception of the ground-doves, in which the wings are short and rounded, most of the existing species have long, powerful quill-feathers, but the extinct dodo and solitaire were incapable of flight. As regards characters of the skeleton, pigeons have a cleft (schizognathous) palate, and the upper bone of the wing, or humerus, is provided with a triangular crest at the upper end (page 1399), to which is attached the great pectoral muscle, thus rendering these birds capable of protracted and powerful flight.

Most pigeons are indeed excellent flyers, capable of traversing enormous distances in an incredibly short time. A distinctive feature of the group is the large size of the crop, which becomes glandular during the breeding season, when it secretes a milky fluid to moisten the half-digested food on which the young are nourished. Pigeons are voracious feeders, and it has been estimated that a single bird is capable of eating a quantity more than equal its own weight. Pigeons drink by thrusting their beaks into the water and retaining them in the fluid till they have quenched their thirst.

All the species pair for life, and both sexes take part in the building of the nest, incubation, and the rearing of the young. When first hatched, the latter are naked and helpless, and thus need care from both parents. The beak in the young is larger and more fleshy than in the adult, and during the operation of feeding, the old bird thrusts its beak inside that of its offspring, and injects the semi-liquid nutriment. The nest is a simple structure composed of twigs, and generally placed in a tree; and the eggs, which are never more than two in number, are invariably pure white.

CLASSIFICATION OF COLUMBÆ MENTIONED IN THIS WORK

GENUS 41
Alopecœnas
GENUS 42
Wonga-wonga Dove—Leucosarcia
SPECIES
Wonga-wonga dove ...... Leucosarcia picata
GENUS 43
Eutrygon
GENUS 44
Otidiphaps
GENUS 45
Blue-bearded Dove—Starnœnas
SPECIES
Blue-bearded dove .. Starnœnas cyanocephala
GENUS 46
Cape Dove—Œna
SPECIES
Cape dove .... Œna capensis
GENUS 47
Tympanistria
SPECIES
White-breasted wood-dove...... Tympanistria
bicolor
GENUS 48
African Ground-dove—Chalcopelia
SPECIES
African ground-dove ....Chalcopelia atra

GENUS 49
Bronze-winged Doves—Chalcophaps
SPECIES
Indian bronze-winged dove . .Chalcophaps indica
GENUS 50
Phaps
SPECIES
Bronze-winged dove ...... Phaps chalcoptera
P. elegans
GENUS 51
Harlequin-Dove—Histriophaps
SPECIES
Harlequin-dove ......Histriophaps histrionica
GENUS 52
Petrophassa
SPECIES
White-quilled rock-dove ....Petrophassa
albipennis
GENUS 53
Geophaps
SPECIES
Pencilled dove .... Geophaps scripta
GENUS 54
Lophophaps
GENUS
Plumed bronze-winged dove ....Lophophaps
plumifera

GENUS 55
Ocyphaps
SPECIES
Crested bronze-winged dove.Ocyphaps lophotes
GENUS 56
Calœnas
SPECIES
Nicobar pigeon .... Calœnas nicobarica
Pelew pigeon.................C. peleuensis
C. maculata
GENUS 57
Crowned Pigeons—Goura
SPECIES
Victorian crowned pigeon......Goura victoriæ
Albertis' crowned pigeon .... G. albertisi
FAMILY 2
Didunculidæ
GENUS 1
Didunculus
SPECIES
Tooth-billed pigeon ..Didunculus strigirostris
FAMILY 3
Dodo and Solitaire—Dididæ (extinct)
GENUS 1
Didus
SPECIES
Dodo......................Didus ineptus
GENUS 2
Pezophaps
SPECIES
Solitaire ....Pezophaps solitaria

PIGEONS IN GENERAL

ALL existing pigeons, with the exception of the tooth-billed Samoan species, may be included in a single family (Columbidæ), which is divisible into several sub-family groups.

WEDGE-TAILED GREEN PIGEONS

The handsome birds known as the wedge-tailed green pigeons, may be taken as the first representatives of the subfamily Treroninæ, which contains a large number of arboreal species inhabiting Africa, South-Eastern Asia, and the islands of the Eastern Archipelago. As a rule, these birds are distinguished by their rather short legs, which are feathered for more than half their length, and are usually shorter than the middle toe and claw ; the soles of the feet being very broad, and the skin of each toe expanded on the sides. The first section of the subfamily contains the green pigeons, characterised by the comparative stoutness of the beak. The plumage, with a few exceptions, in which it is chocolate brown, is mostly green, and most species have a yellow band across the wings. In the first three genera the base of the beak is soft, and the sheath of its upper half does not reach the feathers of the forehead.

The wedge-tailed green pigeons include seven species, with wedge-shaped tails, and the middle feathers more or less pointed, from South-Eastern Asia ; the best known being the Himalayan and Burmese, Sphenocercus apici-cauda and S. sphenurus. The former is distinguished by having the under surface of the tail black, with a broad grey band across the extremity, and the middle pair of tail feathers long and pointed ; while in the latter the under surface of the tail is uniform grey, and the middle pair of feathers is less pointed and much shorter. The males of the various species are not difficult to identify, but females belonging to distinct species often resemble one another very closely.

The above-mentioned S. sphenurus is found in the thick forest-country of Burma, frequenting trees which bear fruit, and going about in flocks. It is a summer visitor to the Himalaya, where it breeds from April to July, constructing a nest of twigs on the outer branches of trees.

In October these pigeons collect in small flocks of six or eight, and quit the mountains. They are rather shy, and their note is a soft, cooing whistle. Their food consists entirely of small fruits, which are swallowed whole.

WALIA PIGEONS

The walia pigeons (Vinago), which take their name from the typical Abyssinian species, are character-ised by the tail feathers forming an almost even line, the outer pair being little shorter than the middle ones, and the feathers on the legs conspicuously yellow, while the feathers on

ABYSSINIAN WALIA PIGEON

forehead in some species is more or less naked.

The Abyssinian walia (V. valia) is met with in the subtropical belt, and but rarely seen on the highlands, being first observed at an elevation of about two thousand feet, and not extending above six thousand. Its call is a liquid whistle, very similar to that of the Indian green pigeon in tone, but with the concluding portion a little harsher and more prolonged. It feeds on fruits, especially figs, in consequence of which it forms, like the other members of the group, an excellent bird for the table.

The allied genus *Crocopus*, of which the members inhabit the Indo-Chinese countries, resembles *Vinago* in having the feathers on the legs yellow, but is distinguished by the first three flight feathers being pointed. The nest is roughly made of sticks, and usually situated rather high up in a mango tree.

MALABAR GREEN PIGEON

A number of green pigeons inhabiting South-Eastern Asia and the Malay Archipelago, and distinguished by having the feathers on the legs of a greenish or whitish colour, constitute the genus *Osmotreron*. Among these is the Malabar green pigeon (*O. malabarica*), which, like its allies, closely resembles the under-mentioned thick-billed species in habits and mode of life. In the male the upper part of the head is grey shading into olive green on the back of the neck, the upper part of the back maroon, and the rest of the upper parts and middle tail feathers olive green; the lateral tail feathers are grey, with a black band across the middle, the quills and wing coverts black, the latter edged with yellow, and the under parts olive yellow. The female has the upper part of the back olive-coloured.

THICK-BILLED GREEN PIGEON

The three remaining genera are distinguished from the foregoing by the circumstance that the sheath of the upper half of the beak reaches the feathers of the forehead. The typical genus *Trevon* includes only two species from South-Eastern Asia, both of which have the third flight feather deeply scooped about the middle of the inner web; the thick-billed green pigeon (*T. nipalensis*) being distinguished by having the grey colour of the cap darker, and not extending over the upper parts of the cheeks. This species, which inhabits the South-Eastern Himalaya, extending to the Malay Peninsula, is very common in Tenasserim, where it occurs in moderately large flocks, feeding on small fruits. These pigeons feed in the morning and evening, and are noisy and quarrelsome. Their flight is rapid, and they frequent dense forest, thin tree-jungle, and even gardens, breeding in February and March, the nests being flimsy little platforms of straw placed about ten feet above the ground.

The five species of the Philippine genus *Phobotreron* are peculiar in having the general colour of the plumage chocolate brown, *P. amethystina* being a handsome bird with the hind portion of the neck a beautiful amethyst colour, and the upper parts with bronze reflections.

PAINTED PIGEONS

The painted pigeons comprise a numerous group of small, brilliantly-coloured species, mostly characterised by having the plumage variegated with patches of different colours, many of them being birds of surpassing beauty. They may be distinguished from the last group of green pigeons by the more slender beak, which is not very distensible at the base; and in this respect, as well as in their smaller size, they differ from the true fruit-pigeons described below.

Of the five genera, by far the largest and most important is *Ptilopus*, which contains at least seventy species, ranging from the Malay Peninsula to Australia and Polynesia.

The numerous species have been arranged under twelve subgeneric groups, the first eight of which are distinguished by having the first flight feather abruptly narrowed at the extremity, except in one species.

Of the subgenus *Leucotreron*, characterised by having no defined cap on the upper part of the head and by the tail being rather long, *P. occipitalis* of the Philippine Islands is a good example, in which the upper part of the head is grey, the cheeks and back of the head purple red, the back of the neck and the rest of the upper parts bronze green, with a grey band at the tip of the tail, and the under surface of the body mostly whitish and grey, with a wide purplish band on the lower part of the breast.

In the second, or typical, group of painted pigeons the tail is moderately long, the feathers on the breast are forked at the extremity, and there is a well-marked cap on the top of the head. Among these, the East Australian *P. swainsoni* has the crown rose lilac, surrounded by a narrow ring of yellow, the upper parts mostly greenish yellow, the inner quills being tipped with deep blue, the breast dull green, each of the forked feathers shading into a silvery grey at the tip, and a lilac band between the breast and the orange abdomen. In the third subgenus, *Lamprotreron*, represented by two species only, there is a broad blue-black band separating the breast from the abdomen. The two species are *P. superbus* and *P. temmincki*, of which the former is very common in New Guinea and Australia.

The other groups of this section of the genus, distinguished by the first primary being narrowed at the extremity, include eleven species, which, unlike those previously mentioned, have the tail rather short: they are all inhabitants of New Guinea or the islands immediately to the east and west. The remaining subgenera differ from those already mentioned in not having the first flight feather narrowed, while none have the head, neck, and breast uniform rose carmine. They include about thirty species, many of which are exceedingly beautiful, but none more so than *P. eugeniæ* of the Solomon Islands. Both sexes of the latter have the head pure white, and the rest of the upper parts bronze green, with a small grey patch on the shoulder and spots on the wing coverts of the same colour, the throat and chest dark purple red, surrounded by a dull purple band, and the breast greyish green, shading into whitish on the abdomen.

SEYCHELLE WART-PIGEON

D

GOLDEN PIGEONS

Another section is represented by the golden pigeons (*Chrysœnas*), distinguished by having the inner webs of the quills yellow or orange yellow, and containing three species from the Fiji Islands, one being the splendid *C. victor*, the male of which has the general plumage bright orange, with the head and throat olive yellow, and the beak and feet green. The female

NUTMEG-PIGEON

has the entire plumage rich green, with the head and throat yellowish green, and the beak and feet black.

WART-PIGEONS

Another genus includes the wart-pigeons (*Alectrœnas*) of the Seychelles, Mascarene, and Comoro Islands, and Madagascar, in which the colour is mostly deep blue, and the feathers of the neck are deeply forked at the extremity. Two of the species have the tail blue, the crown of the head very red in the Seychelle *A. pulcherrima*, and grey in the Comoro *A. sganzini*. In the third and fourth species the tail is red, the former, *A. minor*, of Aldabra, having the head of a light grey colour, while in the latter, *A. madagascariensis*, it is a deep slate-blue. Finally, the one species of *Drepanoptila*, from New Caledonia and the Isle of Pines, is peculiar in having the outer flight feathers divided at the tip, and the legs entirely feathered.

TYPICAL FRUIT-PIGEONS

In a third section of the subfamily are included the true fruit-pigeons, among which, next to the crowned pigeons, are some of the largest members of the family, none being smaller than a rock-dove. The long and basally-expansible beak confers on these birds the capacity to swallow large fruits whole. Their plumage is not much variegated, and in six out of the seven genera there are fourteen tail feathers, the seventh (*Hemiphaga*) having only twelve. The genus *Globicera* contains a few species, differing from the rest by the swollen fleshy knob at the base of the upper half of the beak.

Passing over the peculiar genus *Serresius* from the Marquesas, in which the basal half of the beak is covered with a saddle-shaped production of the skin of the forehead, reference may be made to the members of the typical genus *Carpophaga*, which may be divided

into six subgenera—*i.e.*, groups respectively characterised by the general colour of the plumage and the shape of the flight feathers.

Among these it will suffice to mention a few species in which the general colour is mostly metallic on the upper parts, the tail uniform in colour, and the flight feathers normal in shape. In the nutmeg-pigeon (*C. œnea*), common in the Indo-Burmese countries, Ceylon, and the Andamans, the head, neck, and under parts of the body are grey, the upper parts bronze-green, and the under tail coverts deep chestnut. These handsome birds keep to the larger forest-trees, and live on fruits, especially wild nutmegs, which they swallow whole, although only the mace is digested, the nutmeg being disgorged.

Another closely allied species, the Nicobar nutmeg-pigeon (*C. insularis*), of the islands from which it derives its name, differs in having the under tail coverts mixed with dark green. These pigeons occur singly, in pairs, or in small parties, and their deep, low, cooing notes may be heard all day resounding through the forest. They breed in February and March, and the nest, of which one has been found in a cocoanut-palm about twenty feet from the ground, contains one large white egg.

From all the above the members of the genus *Myristicivora* may be distinguished by their white and black plumage; the white nutmeg-pigeon (*M. bicolor*) being a handsome species found in the Philippine Islands, the Malay Archipelago, and the Andaman and Nicobar Islands. Both sexes have the entire plumage pale creamy white, except the flight feathers, the tip of the tail, and some spots on the under tail coverts, which are black. This bird is not so generally distributed through the Nicobars as the nutmeg-pigeon, and, though occasionally found some distance in the forest, keeps in general to the mangrove-swamps; but on islands such as Treis and Track, where there is little or no mangrove, it occurs everywhere.

Lopholæmus antarcticus, of Eastern Australia, easily recognised by being the only crested species of fruit-pigeon, is the sole representative of its genus. The last genus, *Hemiphaga*, containing a few species from New Zealand and the neighbouring islands, differs from the allied groups by having only twelve tail feathers.

TYPICAL PIGEONS

The green pigeons and their relatives the more typical pigeons, constituting the subfamily *Columbinæ*, are distinguished by having the soles of their feet normal, that is to say, not specially broad, and only the hind toe with the skin prominently expanded on the sides.

The subfamily is split up into three sections, in the first of which the tail is never longer than the wings; this group containing the wood-pigeon, stock-dove, and rock-dove, from the last-named of which domesticated breeds of pigeon are derived. The first genus is represented only by the Albertis' pigeon (*Gymnophaps albertisi*) of New Guinea, distinguished from the allied genera by having the legs feathered for two-thirds of their length, and a naked carmine space in front of and round the eyes.

Next follows the large genus *Columba*, containing about seventy species, whose united range comprises practically the whole habitable portion of the world. In none of them are the legs feathered for more than half their length, while in all the first flight feather of the wing is longer than the sixth.

THE ROCK-DOVE

An important member of the group is the rock-dove or blue rock (*C. livia*), widely spread over Europe, and extending as far as India in the east, where it meets the nearly-allied but grey-rumped *C. intermedia*; while southwards it ranges to the north and west coasts of Africa, where it comes into contact with another closely related bird (*C. gymnocyclus*), differing only in having the plumage dark bluish or blackish slate-colour, and perhaps being nothing more than the descendant of domesticated pigeons.

In the wild blue-rock the general colour is grey, the rump white, the neck and the upper part of the breast are metallic green and purple, and there are two narrow black bars across the wing and a broader one near the end of the tail. This bird frequents districts abounding in caves and deep fissures, and is common along the northern coasts of Scotland and Ireland, wherever suitable resorts occur. The nest is placed on a ledge or in a crevice of the rock in a cavern where little light penetrates.

THE STOCK-DOVE

Another wild British species is the stock-dove (*C. œnas*), often confused with the rock-dove, which it resembles in size and general colour, although distinguished by having the rump grey instead of white (page 1405). This pigeon is especially common in the southern and eastern counties of England, where large flocks may be seen feeding in the fields, frequently in company with wood-pigeons. It breeds on open ledges in the face of a steep cliff (never in caves), and in rabbit-burrows where the soil is light and sandy, or under the shelter of dense furze; and numbers of nests are placed in trees, a favourite situation being among the dense bunches of twigs surrounding the stems of old elms.

THE WOOD-PIGEON

The third and largest British species is the wood-pigeon (*C. palumbus*), distinguished by the broad white patch on each side of the lower part of the neck as well as by its variously-tinted breast and the white band along the edge of the wing (p. 1405). It ranges across Europe into Asia as far east as Northern Persia, and is also found in North-Western Africa, the Azores, and Madeira. The destruction of the larger birds of prey, as well as the extent of land devoted to plantations and green crops, probably accounts for the vast increase in numbers of this pigeon which has taken place in recent years.

That the countless swarms of these voracious birds in parts of the country do an immense amount of damage cannot be denied, but much of their food often consists of the seeds of noxious weeds or useless plants, as may be ascertained by examining the contents of their crops. Mr. Booth remarks that " shortly before harvest wood-pigeons may often be seen flying in small parties to the fields of wheat and barley; after wheeling round for a time, the birds will disappear from view into the standing corn. An examination of the state of the ground on which they were lost sight of would doubtless cause astonishment to those who imagined that the birds were in pursuit of grain; on reaching the spot it would be discovered that for a considerable space the crop was exceedingly scanty, completely choked, in fact, by a mass of weed rank and strong, whose seeds, well-nigh ripe, had proved the sole attraction. Immense flights of these birds arrive on the north-east coast in October and November from the Continent, and about that time of year large flocks, which have recently

BLUE-ROCK AND POUTER PIGEONS

arrived exhausted by their long journey, may be observed fluttering along the coast and about the fir-plantations. In the London parks these pigeons have become some of the commonest birds, and are steadily increasing in numbers, many pairs breeding each year."

BAND-TAILED PIGEON

The band-tailed pigeon (*C. fasciata*), which resembles the wood-pigeon in having a narrow white band or half-collar on the nape, and inhabits the western states of North America as well as Central America, exhibits certain peculiarities in its mode of life In Oregon, for example, it sometimes breeds on the ground, and the normal number of two eggs is laid ; but in Arizona, where it appears to nest in nearly every month of the year, only a single egg is laid.

When frightened, this bird is often in the habit of carrying off its eggs from the nest. Mr. O. C. Poling remarks that "in regard to their carrying the egg about, I have, in addition to the cases noted, shot two other females having the egg embedded in the feathers of the belly, and further held by the legs while flying ; but in such cases they seem simply to alight on the limb of a spruce, and incubate there without any nest. This accounts for the shooting of pigeons having a broken egg smeared over the feathers, as I have done, when no nest was to be seen."

Some of the species of *Columba* found in the islands of the Pacific are more brilliant in colouring than their northern representatives, and among these may be mentioned *C. albigularis* of New Guinea, in which the plumage is blackish slate-colour, with the edges of the feathers metallic purple changing into green, and the cheeks and throat white.

MAURITIUS PIGEON

Among other genera of the group, mention may be made of *Nescœnas*, with one species (*N. meyeri*) from Mauritius, distinguished by having the first flight feather equal to the sixth. The colour is pale pink, darker on the mantle, and shading into brown on the back and wings, while the tail is uniform cinnamon-colour.

SHARPE'S PIGEON

Turturœna contains some half-dozen of the smallest species of this group, none of them exceeding a dove in size, and all having the hind part of the neck brilliantly ornamented with metallic colours. They inhabit Africa, and differ from other *Columbinœ* in that there is often a great dissimilarity in the colours of the plumage in the two sexes. One of the handsomest species is Sharpe's pigeon (*T. sharpei*), from Mount Elgon.

LONG-TAILED PIGEONS

The long-tailed pigeons of South-Eastern Asia and the islands of the Pacific are distinguished by having the tail longer than the wings, in which respect they resemble the passenger-pigeon, the type of the third section of this group, although it differs in having the feathers of the tail broad and round at the tip. Four genera, including some thirty species, are recognised, although little has been recorded of their habits. In *Turacœna* the two species have the beak moderately strong, and the tail somewhat rounded, with the outer feathers much more than half the length of the middle pair. The Celebean *T. menadensis*

has the plumage slate-black, with the face and throat white, while in *T. modesta*, of Timor, it is uniform slate-colour ; both being ornamented with shades of metallic green, lilac, and blue.

The great majority of long-tailed pigeons belong, however, to the genus *Macropygia*, commonly known as cuckoo-pigeons. All these birds have the tail much graduated and wedge-shaped, the outer feathers being less than half the length of the middle pair, and the general colour of the plumage rufous, chestnut, or cinnamon.

In the male of the Indian cuckoo-pigeon (*M. tusalia*) the forehead, chin, and throat are purplish buff, and the top of the head and the rest of the upper parts shining metallic green with purple and bronze reflections, the latter being also irregularly barred with black and purple chestnut ; the under parts are purplish grey shading into buff on the abdomen, and glossed on the chest with golden green and bronze, the quills brown, the middle pair of tail feathers barred with black and vinous chestnut, and the outer pairs mostly grey. The female is less brightly coloured, and has most of the under parts barred with brownish black.

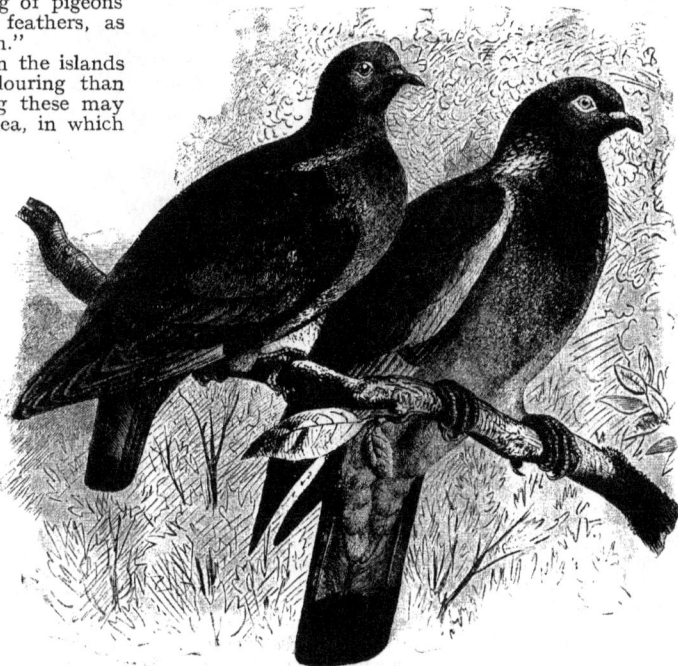

STOCK-DOVE AND WOOD-PIGEON

These birds are of shy disposition, keeping to thick forest and associating in small flocks, which feed chiefly in trees and seldom descend to the ground. In Nepal the two white, or sometimes creamy, eggs are laid in May and June, and the nest, which is the usual loose platform of sticks, is placed on some horizontal branch, at no great height from the ground.

The other genera are characterised by the strong and thick beak and the wedge-shaped tail. The first of these, *Reinwardtœnas*, distinguished by having no crest, contains three species, ranging from Celebes to the Duke of York Island ; Reinwardt's long-tailed pigeon (*R. reinwardti*) having the head, neck, and mantle pale lavender, the back, wings, and middle tail feathers chestnut, and the front of the neck and breast white shading into lavender on the abdomen.

In *Coryphœnas*, containing one species (*C. crassirostris*) from the Solomon group, the plumage is slate-colour, darker on the upper surface, with the head dusky brown and the feathers on the back of the head lengthened into a greyish brown crest.

PASSENGER-PIGEON

Another section of long-tailed pigeons is represented only by the well-known passenger-pigeon *Ectopistes migratorius* of North and Central America, characterised by the length of the wings exceeding that of the tail, and the narrowness of the latter, in which the tips of the feathers are pointed. In the male the head and upper parts of the body are bluish grey, spotted with black on the wings ; the quills brownish black with grey edges, the chin whitish grey, and the breast cinnamon-rufous shading into pale purplish on the rest of the under parts, and white on the under tail coverts. The back and sides of the neck glitter with golden and violet metallic colours. In the female the upper part of the breast is brownish, shading into white on the abdomen and the rest of the under parts.

The vast numbers by which this pigeon was formerly represented have greatly diminished during recent years. Mr. Brewster writes that in Michigan " we found that large flocks of pigeons had passed there late in April, while there were reports of similar flights from almost every county in the southern part of the state. Although most of the birds had passed on before our arrival, the professional pigeon-netters, confident that they would finally breed somewhere in the southern peninsula, were busily engaged getting their nets and other apparatus in order for an extensive campaign against the birds. Our principal informant said that the last nesting of any importance in Michigan was in 1881, a few miles west of the Grand Traverse. It was only of moderate size, perhaps eight miles long.

" Subsequently, in 1886, Mr. Stevens found about fifty dozen pairs nesting in a swamp near Lake City. He does not doubt that similar small colonies occur every year, besides scattered pairs. In fact, he sees a few pigeons about Cadillac every summer, and in the early autumn young birds, barely able to fly, are often met with singly or in small parties in the woods. Such stragglers attract little attention, and no one attempts to net them, although many are shot. The largest nesting he ever visited was in 1876 or 1877. It began near Petosky, and extended north-east past Crooked Lake for twenty-eight miles, averaging three or four miles wide.

" The birds arrived in two separate bodies, one directly from the south by land, the other following the east coast of Wisconsin, and crossing at Manitou Island. He saw the latter body come in from the lake at about three o'clock in the afternoon. It was a compact mass of pigeons, at least five miles long by one mile wide. The birds began building when the snow was twelve inches deep in the woods, although the fields were bare at the time. So rapidly did the colony extend its boundaries, that it soon passed

literally over and around the place where he was netting, although, when he began, this point was several miles from the nearest nest. Nestings usually start in deciduous woods, but during their progress the pigeons do not skip any kind of trees they encounter.

" The Petosky nesting extended eight miles through

THE PASSENGER-PIGEON

hardwood timber, then crossed a river-bottom wooded with arbor-vitæ, and thence stretched through white-pine-woods about twenty miles. For the entire distance of twenty-eight miles every tree of any size had more or less nests, and many trees were filled with them. None were lower than about fifteen feet above the ground.

" Pigeons are very noisy when building. They make a sound resembling the croaking of wood-frogs. Their combined clamour can be heard four or five miles away when the atmospheric conditions are favourable. Two eggs are usually laid, but many nests contain only one. Both birds incubate, the female between 2 o'clock p.m. and 9 or 10 o'clock the next morning ; the males from 9 or 10 o'clock a.m. to 2 o'clock p.m. The males feed twice each day, namely, from daylight to about 8 o'clock a.m., and again late in the afternoon. The females feed only during the forenoon."

MOURNING-DOVES

All the preceding members of the family are more or less arboreal in their habits, and consequently may be collectively termed tree-pigeons. In contradistinction to these is a large group whose members live much on the ground. These ground-pigeons, which form the subfamily *Peristerinœ*, include nearly all the remaining species, and are distinguished from the tree-pigeons by having the legs equal to or longer than the middle toe. Seven sections are recognised, in the first six of which the feathers of the neck are never hackled.

In the first section the so-called mourning-doves, readily distinguished by having a blackish, more or less metallic, spot below the ear coverts, include thirteen rather small American species, placed in four genera, the first of which (*Zenaidura*) resembles the two following ones in having the scapulars and upper wing coverts spotted with black. Its members agree with those of the next genus in having the tail composed of fourteen feathers and the beak nearly straight, but differ in the longer and generally wedge-shaped tail.

All the four species of the former are North and Central American, the best known being the mourning-dove (*Z. carolinensis*) of the United States, Central America, and the West Indies. In the male of this species the crown and upper parts of the body are bluish, mostly washed with light olive brown ; the rest of the head and under parts being cinnamon - buff, tinged with purple on the breast. The female is smaller, and has the under parts brown like the back, but paler. These doves are frequently bred in gardens and shrubberies near dwelling-houses, feeding among domesticated fowls.

Never occurring in large flocks like passenger-pigeons, they are usually found in small parties of from six to a dozen or more, but in autumn, previous to migration, may be met with in flocks of fifty or sixty. They are fond of alighting in roads, where they may often be seen searching for suitable food or

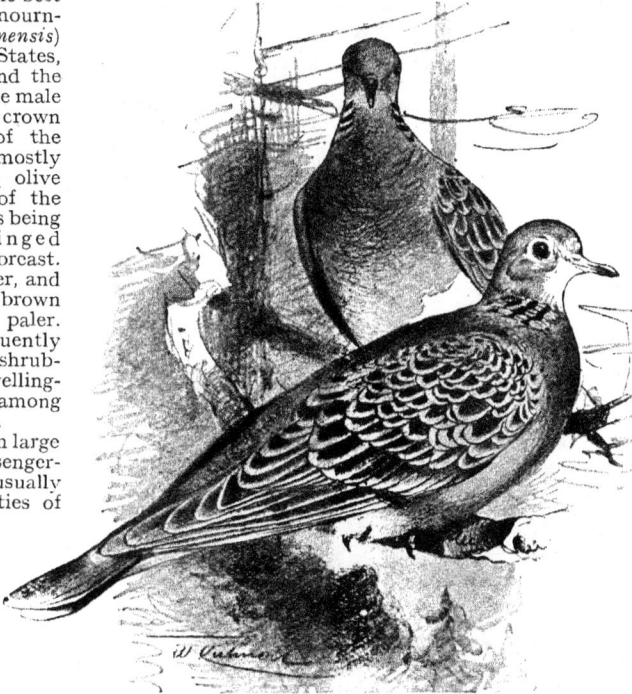

TURTLE-DOVES

gravel, or taking dust-baths, of which they are very fond. In the more arid districts of the west, such as Southern Arizona, where water is scarce, these doves visit regular watering places in the morning and evening, to which they may be seen coming in small parties from all directions.

The nesting sites chosen are variable, and in some localities, such as the Carolinas, these doves nest chiefly on the ground, while they are said occasionally to lay their eggs in the nests of other birds. The old birds are attentive to their young, even long after they have left the nest, and the female has been observed brooding fully-fledged young. As many as four eggs have been found in one nest, but whether these were all laid by one bird is uncertain.

"The mourning-dove," writes Mr. W. H. Dutcher, "is found in all parts of temperate North America from the Atlantic to the Pacific, and north to Ontario ; sparingly in Manitoba and British Columbia. Its southward range extends to Panama and through the West Indies. In the northern portion of its range it is a summer visitor, although an occasional individual may be found quite far north even in the severest winter weather, provided the food supply is suitable and abundant. Between latitudes 30° and 38° it is a permanent resident, but south of 36° it is an abundant resident. It breeds in all parts of the United States, raising two broods and possibly in some instances three in a season. Its nest is usually a frail platform of twigs, if placed in a

bush or tree, but in many instances the eggs are laid on the ground, especially in the treeless west. Unusual nesting sites are frequently selected, such as the abandoned nests of other species of birds, tops of stumps, rocks, sheds, etc."

The species of the genus *Zenaida*, inhabiting the West Indies and Central and South America, differ in having the tail only moderately long and rounded.

The zenaida dove (*Z. amabilis*) of the West Indies has the upper parts brown, with some black spots on the wings, the chin white, the cheeks and throat rufous, two steel-blue spots above and below the ear coverts, and the top of the head, breast, and under parts purplish. The quills are black, edged with white, and there is a conspicuous white band at the end of the secondaries.

More or less solitary in habits, it is never met with in flocks, nor does it breed in communities. Its food consists of small seeds ; the principal part of its time is spent on the ground, and when flushed it flies off in a straight line much like a quail. The nesting habits vary in different localities. In the Bahamas these birds have nested in the fork of a fallen tree about three feet from the ground, while others lay in holes in rocks. Among the islands at Indian Key the nest is placed in a small hole scooped in the sand, and composed of dry leaves and twigs, with a matted inner lining of blades of dry grass, the whole structure being more compact than that of other pigeons.

Writing of another South American species (*Z. auriculata*), distinguished by having no white tips to the secondary flight feathers, Mr. W. H. Hudson observes, that it "is the commonest species of the pigeon tribe in the Argentine country, and is known to every one as the 'torcasa,' probably a corruption of tortola, or turtle-dove. In autumn these doves often congregate in very large flocks, and are sometimes observed migrating, flock succeeding flock, all travelling in a northerly direction, and continuing to pass for several consecutive days. But these autumnal migrations are not witnessed every year, nor have I seen any return migration in spring ; while the usual autumn and winter movements are very irregular, and apparently depend altogether on the supply of food.

"When the giant thistle has covered the plains in summer, incredible numbers of torcases appear later in the season, and usually spend the winter on the plains, congregating every evening in countless myriads wherever there are trees enough to form a suitable roosting place. On bright warm days in August, the sweet and sorrowful sob-like song of this dove, composed of five notes, is heard from every grove—a pleasing, soft, murmuring sound, which causes one to experience by anticipation the languid summer feeling in his veins."

GALAPAGOS PIGEON

The Galapagos pigeon (*Nesopelia galapagoënsis*), which alone represents a genus restricted to the islands from which it takes its name, has the beak long and bent downwards, and the tail rather short, rounded, and composed of twelve feathers.

WHITE-WINGED DOVES

Of the white-winged doves two species are known, the one (*Melopelia leucoptera*) from the south-western United States, Central America, and the West Indies, and the other (*M. meloda*) from Peru and Chile. They are distinguished from the mourning-doves by the absence of black spots on the wings, while there is a white patch on the upper wing coverts. The northern species has a note bearing a close resemblance to the first efforts of a young cock attempting to crow, and this call is frequently uttered and in various keys. In Arizona, towards the end of summer, this bird, which is by no means shy, collects in small parties.

TURTLE-DOVES

The familiar turtle-doves form the second section of the ground-pigeons, and include a large number of Old World species divisible into five subgenera. They differ from the mourning-doves in having no black spot below the ear coverts, and the neck ornamented with a more or less distinct dark collar, or with dark scale-like patches on the sides. In the three subgenera *Turtur, Homopelia*, and *Streptopelia* the feathers of the neck are normal, in the fourth (*Spilopelia*) those of the hind neck, and in the fifth (*Stigmatopelia*) those of the fore neck are forked at the extremity. The first two may also be recognised by having two scale-like patches of dark feathers on the sides of the neck, while *Streptopelia* has a complete black collar on the hind portion of the neck.

In the more typical turtle-doves, *Turtur,* the wing coverts are mottled, with the middle line of the feathers darker than the edges. Of these, the turtle-dove (*T. communis*) is widely distributed over Europe, extending so far east as Yarkand, and ranging southwards in winter to Africa, where it reaches at least so far south as Shoa. The crown and hind part of the neck are bluish grey, with a black patch of white-margined feathers on each side of the neck, the back is pale brown, the inner wing coverts cinnamon-brown with dark middles, and the outer coverts grey washed below with ashy, the chin nearly white, and the throat and breast purplish shading into white on the belly ; the quills and the two middle tail feathers being brown, while the outer pairs are greyish black broadly tipped with white. The male is rather larger than the female, and has the plumage brighter and purer in colour (page 1405).

The turtle-dove, which is a summer visitor to the British Isles, where it is one of the latest migrants, not arriving till the end of April or the beginning of May and departing in September, may be distinguished from other British pigeons by its smaller size. Shortly after its arrival it begins to build its nest, which is loosely constructed of slender twigs and placed in a thick bush, tree, or dense hedge at no great height from the ground. Two small creamy white eggs are laid, and both parents take part in the incubation, which lasts about a fortnight, two broods being sometimes reared in the season.

The turtle-dove is chiefly met with in woods, and is partial to thick coverts and fir-plantations, where its low prolonged coo may be constantly heard, though the bird itself is rarely seen, preferring the seclusion afforded by the thick foliage to the outer branches of the trees. Its flight is always extremely rapid, and when among trees it can turn and twist with extraordinary ease and swiftness. It sometimes flies great distances in search of food and water, and may be often seen in cultivated fields searching for grain and seeds, although seldom in such large flocks as other pigeons. In Britain it is a shy bird, at the slightest sign of danger quickly seeking shelter in the nearest covert.

All the members of the subgenus *Homopelia*, which have the upper wing coverts uniform in colour, inhabit Madagascar and the adjacent islands. The third subgenus, *Streptopelia*, distinguished by having a black collar round the hind neck, includes over a dozen species, but it is uncertain from which of these the domesticated turtle-dove (*T. risorius*) has been derived.

As an example of this group may be mentioned the Tranquebar dove (*T. tranquebaricus*), from India, in which the general colour is red ; the lower part of the back, rump, and flanks grey, the head, under wing coverts, lower portion of the abdomen, and under tail coverts similarly coloured but paler, the outer wing coverts darker, the chin whitish, the purplish red upper parts separated from the grey of the head by a black collar on the hind half of the neck, and the quills blackish with pale edges.

The nearly allied *T. humilis* of the Indo-Chinese countries is distinguished by having the under wing coverts much darker and of a grey colour. In Tenasserim it is found in larger or smaller flocks, very wild and difficult to approach, and keeping to the thin tree and bamboo jungle. Such flocks are often seen in the

DOMESTIC TURTLE-DOVE AND AFRICAN GROUND-DOVE

vicinity of cultivated land, feeding on the ground, and when disturbed rise together and settle on the same tree, sometimes a leafless one, sometimes one with plenty of foliage.

The three species constituting *Spilopelia* are distinguished by having the feathers of the hind part of the neck forked at the extremity, and black ornamented with two white spots at the tip. The Burmo-Malay *T. tigrinus*, recognisable by the dark brown shaftline on the wing coverts, is the common dove of Tenasserim, where it is met with in gardens, fields, and grass-land ; in fact, wherever the country is open, but not in forest or on the higher hills. It is sometimes seen singly or in pairs, at other times in small flocks or in hundreds.

The last group, *Stigmatopelia*, includes three species which may be recognised by having the feathers of the fore neck forked at the extremity, and black with two rufous spots at the end. They have a wide range, the Senegal turtle-dove (*T. senegalensis*) being found all over Africa, while the brown turtle-dove (*T. cambayensis*) ranges from Asia Minor to Central India, and a third species, *T. ermanni*, inhabits Bokhara. The first-named is characterised by the plumage of the upper parts being more or less reddish, and the rump bluish grey.

Yet another section, containing three genera, with a few small species from both the Old and New Worlds, is characterised by the short rounded wings, a rather long tail of fourteen feathers, and no metallic tint on the sides of the neck or elsewhere. The Old World species belong to the genus *Geopelia*, ranging from Burma to Australia, and have the first quill-feather pointed at the extremity. The American genus *Scardafella* includes three doves about the size of sparrows, with the first flight feathers normal, and the tail of twelve feathers ; the Inca dove (*S. inca*) from North and Central America being a well-known example. Only a single species (*Gymnopelia erythrothorax*) is included in the last genus, distinguished by the large naked patch surrounding the eye.

AMERICAN DOVES

The members of six American genera constituting another section are collectively characterised by their small size and the general uniform colouring of the plumage ; the wings being, as a rule, ornamented with metallic spots, and rather short and rounded, while the primaries are not much longer than the secondaries. In five of these genera the tail is rather longer than half the length of the wing ; while the two last may be distinguished by having the first primary pointed at its extremity.

The picui dove (*Columbula picui*), the only representative of its genus, is peculiar in having a steel-blue band across the wing, and by the middle and outer pairs of tail feathers being shorter than the intermediate ones. Restricted to South America, it is the smallest dove of Argentina, where it is a resident, frequenting the neighbourhood of houses ; its song, consisting of a succession of long, rather loud and somewhat monotonous notes, may often be heard in summer or even on warm days in winter. About a dozen small species, ranging from the Southern United States to South America, represent the genus *Chamæpelia*, in

which the tail is shorter than the length of the wing and rounded at the extremity, the middle feathers being longer than the outer ones. The remaining genera may be passed over without notice.

CINNAMON-DOVE

The South African cinnamon-dove (*Haplopelia larvata*) may be taken as a well-known representative of an extremely numerous section of ground-doves represented by somewhat partridge-like birds with no well-defined metallic spots or bands, although a violet patch may be present near the bend of the wing. The metatarsus is stout and longer than the third toe, and in the short and rounded wings the primaries exceed the secondaries in length. Of the nine genera, the one mentioned, which has seven African species, is characterised by the primary feathers being broad and not tapering to a point, while in the remaining eight they are narrow and more or less pointed.

The cinnamon-dove is distinguished from its allies by having the forehead white, the top of the head, hind part of the neck, chest, and upper half of the breast metallic coppery purple, the back and wings olive brown, and the abdomen and under tail coverts pale cinnamon. The zamoen duif, as the Boers call this bird, is common in the dense bush along the coast of Natal, where its brown colouring renders it difficult to detect as it sits motionless among the dense creepers. Never appearing in the open, it is generally seen on the ground beneath trees, silently and busily searching for food.

Of the remaining genera with narrow primaries, two have the outermost pointed at the extremity ; *Leptoptila* being distinguished by the tail equalling more than half the length of the wing. The species range from Texas through Mexico and Central America to Peru and Argentina, and may be divided into a long-tailed and a short-tailed group. Two handsome species form the genus *Osculatea*, in which the length of the tail is less than half that of the wing.

In six allied genera the first primary is not pointed at the tip ; the first five of these agreeing with one another

BLOOD-BREASTED DOVES

BLUE-BEARDED CUBAN DOVES

BLUE-BEARDED DOVE

The striking blue-bearded dove of Cuba and Florida Keys (*Starnœnas cyanocephala*) alone represents the last genus of the present group, in which the fronts of the legs are covered with six-sided scales. The general colour of this bird is olive brown above, and dull rusty beneath ; the top of the head being blue margined with black, and a broad white stripe running below the eye, while the feathers of the throat and breast are black, tipped with blue and narrowly margined with white.

CAPE DOVE

As a well-known representative of another section may be mentioned the Cape dove (*Œna capensis*), the sole member of its genus. The section includes a dozen genera, and about double that number of species, confined to the Old World ; their essential characters being their relatively large size, and the presence of blue or green spots or patches on the wings. Four of these genera are characterised by the presence of two dark bands across the rump. The Cape dove, which may be recognised by its long, wedge-shaped tail, with the middle feathers more than twice as long as the outer pair, ranges all over Tropical and Southern Africa, as well as Madagascar and Arabia. It occurs in pairs, and is chiefly found on the ground, seeking, when disturbed, shelter in low trees and bushes, but rarely in larger trees. The nest is placed in a low bush, and the two white eggs have a rosy tint, caused by the thinness and semi-transparency of the shell.

AFRICAN GROUND-DOVE

Near akin are three genera in which the tail is moderately long and more or less rounded ; *Tympanistria* having the first primary pointed, while in *Chalcopelia* and *Chalcophaps* this feather is normal in shape. The white-breasted wood-dove (*Tympanistria bicolor*), the only representative of the first genus, inhabits the whole of South Africa as well as Madagascar, the Comoro Islands, and Fernando Po. The African ground-dove (*Chalcopelia ara*), one of the two representatives of its genus, in addition to the characters given, may be recognised by the boldly marked patches of metallic steel blue or golden green adorning the wings. Inhabiting the whole of Africa south of about 17 degrees

in having the front of the metatarsus covered with transverse scales, while in the three next mentioned the tail is composed of less than twenty feathers. Of these the Central and South American genus *Geotrygon* is also characterised by its rather short tail of twelve feathers, while *Phlogœnas*, of the Eastern Archipelago and Papua, differs in having fourteen feathers in the tail.

More than twenty species are known, some of which are remarkably handsome birds, but none more so than the blood-breasted dove (*P. luzonica*), of the Philippines, in which the forehead and crown are pale grey, the top of the head, upper parts, and sides of the breast dark grey, with metallic purple and green edges to the feathers, the cheeks, throat, and breast white shading into buff below, and a large patch of blood-red on the middle of the breast. The quills are reddish brown, there are six alternate bands of grey and chestnut across each wing, and a black band near the tip of the outer tail feathers. Two other genera, the South American *Osculatea* and the Sudanese *Alopecœnas*, with less than twenty feathers in the tail, are distinguished by the greater length of that appendage. The first has two, and the second a single species.

WONGA-WONGA DOVE

The handsome Australian wonga-wonga (*Leucosarcia pica'a*) is the only representative of a genus distinguished from the two allied Papuan species classed as *Eutrygon* by the metatarsus being but little longer than the middle toe. This dove, remarkable for its large size, inhabits the brush-country of Eastern Australia, where it spends the greater part of its time on the ground, feeding on seeds and fallen fruits. The noise made by its wings when rising is said to resemble that of a pheasant, and its flight is never long sustained. In the two species of *Eutrygon* the metatarsus is twice as long as the third toe ; while the genus *Otidiphaps*, including three large black species, with chestnut back and wings, from New Guinea and Fergusson Island, is peculiar in having twenty feathers in the tail.

WONGA-WONGA DOVE

CRESTED AND BRONZE-WINGED DOVES

north latitude, it is met with in Abyssinia among bushes and thick underwood, and in the jungles along the banks of water-courses.

BRONZE-WINGED DOVES

The six kinds of bronze-winged doves (*Chalcophaps* and *Phaps*) distinguished by having the upper wing coverts metallic golden green are mostly inhabitants of the Indo-Malay and Australian regions, among which the Indian *C. indica* ranges from India to Western New Guinea. This bird has the top of the head and hind part of the neck dark grey, the white forehead and eyebrow-stripes of the male less distinct in the female, and the middle of the back and shoulder feathers golden green like the wing coverts. This dove is not uncommon in the well-wooded portions of Tenasserim, where it is found singly or in pairs in thick forest or shady gardens.

Omitting mention of several genera, reference may be made to two Australian species in which the beak is feeble, and the tail has sixteen feathers. Of these the common bronze-winged dove (*Phaps chalcoptera*) is generally distributed all over Australia, and differs from its ally, *P. elegans*, in having the throat white instead of chestnut, and the breast purplish in place of grey. It is a plump, heavy bird, weighing fully a pound when in good condition ; and its amazing powers of flight enable it to cross a great expanse of country in an incredibly short space of time. Just before sunset it may often be seen singly, or in pairs, coming swiftly over the plains, or down the gullies to its drinking places. It feeds almost entirely on the ground, picking up various leguminous seeds ; and numbers of old and young are killed in the stubble-fields after the breeding season, which lasts from August to December.

HARLEQUIN-DOVE

A genus characterised by the feeble beak and the tail of fourteen feathers is represented by the peculiarly coloured harlequin-dove (*Histriophaps histrionica*) of North-Western Australia, in which the forehead, a stripe round the ear coverts, and the gorget are snow-white, the remainder of the head, throat, and ear coverts

black, the upper parts of the body and middle tail feathers cinnamon-brown, with a patch of metallic purple on the innermost secondaries, the under parts bluish grey, and the outer tail feathers blackish, shading into grey at the base, and tipped with white.

This species breeds in February, depositing two eggs under any low bush in the middle of the open plains. Towards the beginning of April it collects in large flocks, and feeds on the seeds of rice-grass which the natives collect for food. During the short harvest the flavour of this dove is delicious, but at other times it is indifferent. It flies to water at sunset, when, like the bronze-wing, it only wets its beak ; and it seems astonishing that so small a quantity of water should suffice to quench its thirst in the burning deserts which form the home of this bird.

PENCILLED DOVES

Omitting detailed mention of the white-quilled rock-dove (*Petrophassa albipennis*), which inhabits the sterile districts of North-Western Australia, and has the general plumage almost uniform dark brown, and the bases of the primaries white, reference may be made to two species forming the genus *Geophaps*, characterised, in addition to the characters already mentioned, by the rather stout beak. Of these the pencilled dove (*G. scripta*) has the top of the head, upper parts, and chest light brown, the tips of the upper wing coverts paler, and the inner coverts ornamented with patches of metallic greenish purple,

VICTORIA CROWNED PIGEON

the cheeks, ear coverts, and throat white bordered by a black band in front of the eye, a second black band on the cheek, a third above the eye, the rest of the under parts grey, shading into fawn-colour, the quills brown edged with pale rufous, and the tail reddish brown, with the outer feathers tipped with black.

This bird, which inhabits the plains, is most abundant in the neighbourhood of water, and sometimes observed in pairs, but more frequently in small flocks of from four to six. When approached, instead of taking flight, it runs rapidly in the opposite direction, and crouching down, either on the bare plains or among the scanty herbage, remains till all but trodden on. When it rises, its flight is rapid, and accompanied by a loud, whirring noise of the wings. No nest is made, the two eggs being deposited on the bare ground, and the young are able to run when no larger than quails.

PLUMED BRONZE-WINGED DOVE

The two last genera of bronze-winged doves are distinguished by having the head crested, and they differ from one another in the shape of the tail feathers, which are short and nearly even in *Lophophaps*, while in *Ocyphaps* they are elongated, with the middle pair longer than the outer ones. Of the former, there are three Australian species, all with the general colour cinnamon, the innermost secondaries showing a black band down the middle, beginning on the chin and joining the gorget, which is similarly coloured, and the breast with a grey band. The typical plumed bronze-winged dove (*L. plumifera*), of North-Western Australia, is generally seen perched on a rock, basking in the heat of the

ALBERTIS' CROWNED PIGEON

NICOBAR PIGEONS

sun, and being exceedingly wild, taking flight at the slightest noise.

CRESTED BRONZE-WING

The crested bronze-winged dove (*Ocyphaps lophotes*), the only representative of its genus, is a native of the interior of Northern and Eastern Australia. Its crest is black, the head and under parts are grey, the sides of the neck and breast pinkish salmon-colour, the back and rump olive grey, and the upper tail coverts greyish brown edged with white. The smaller wing coverts are greyish buff, banded with black, and the greater coverts and secondaries edged with white, and mostly ornamented with metallic green and purple. The primaries are greyish black, the middle tail feathers brown, and the outer pairs brownish black, glossed on their outer webs with green, blue, and purple.

In respect of habits, Mr. J. Gould observes that this dove " frequently assembles in very large flocks ; and when it visits the lagoons or river-sides for water during the dry seasons generally selects a single tree, or even a particular branch, on which to congregate before descending simultaneously to drink. Its flight is more rapid than that of any member of the group to which it belongs ; and impetus being acquired by a few quick flaps of the wings, it goes skimming off apparently without any further movement of the pinions. Upon alighting on a branch, it elevates its tail and throws back its head, so as to bring them nearly together, at the same time erecting its crest, and showing itself off to advantage."

NICOBAR PIGEON

An altogether unique and bizarre type is presented by the beautiful Nicobar pigeon (*Calœnas nicobarica*), which has a wide range, inhabiting not only the islands from which it takes its name, but extending through the Mergui and Malay Archipelagoes to the Solomon Islands. Together with a smaller species (*C. pelewensis*), of an indigo-blue colour, from the Pelew Islands, and a third (*C. maculata*), whose home is at present unknown, the Nicobar pigeon represents not only its genus but likewise a special section of the family distinguished from all the preceding groups by the elongated and narrow feathers of the neck, which resemble the hackles of a game-cock.

The general colour of the plumage is metallic green with copper-coloured reflections, but the under parts are less brightly coloured than the beak ; the head, neck, and upper part of the breast are nearly black, the quills much the same colour, and the short tail and its coverts pure white. The beak and the knob at the base are black, and the legs and feet dark purple.

In one of the Nicobar Islands these birds swarm, and in the early morning may be seen flying from the island in flocks out to sea, doubtless to other islands of the group to feed. When well up in the air, the flight is swift and powerful, and somewhat like that of sand-grouse. Having wandered some distance away from the rest of his party, and reached a part of the jungle where the birds had not been disturbed, one explorer found himself surrounded by a flock of at least thirty old and young pigeons. Their gait is pigeon-like, and while digging among the dead leaves in search of food with their bills they are very silent, their only note being a hoarse croak. They are always met with on the ground, and when disturbed invariably perch on the thicker branches along which they often walk. The Nicobar pigeon builds a nest of sticks, in which the single egg is laid.

CROWNED PIGEONS

The magnificent birds included in this genus, which forms a subfamily (*Gaurinæ*) by itself, are the largest existing representatives of the order, and characterised by the erect fan-shaped crest adorning the head, and are further distinguished by having the metatarsus covered all over with small six-sided scales. Exclusively Papuan in their distribution, all the six species inhabiting New Guinea and the Papuan Islands are characterised by the bluish slate general colour, but the Victorian crowned pigeon (*Goura victoriæ*) differs from its allies in having the tips of the crest feathers spatulate or club-shaped. Very little has been recorded concerning the habits of these splendid birds, but they appear to spend much of their time on the ground in search of food, and, being remarkably stupid, may be easily approached within gunshot.

The Albertis' crowned pigeon (*G. albertisi*) is a native of South-Eastern New Guinea, where it was discovered by the explorer from whom it takes its name, while the typical representative of the genus inhabits the western side of Papua and some of the neighbouring islands.

THE TOOTH-BILLED PIGEON

Representing a family (Didunculidæ) by itself, the rare and remarkable tooth-billed pigeon (*Didunculus strigirostris*) is perhaps the most interesting representative of the whole group, on account of its kinship to the dodo. An inhabitant of the Samoan Islands, this pigeon possesses a remarkably heavy and powerful beak and short, strong legs. The head, mantle, and chest are metallic green, the rest of the upper parts chestnut, and the under parts brownish black. The beak is orange and the legs and feet are reddish ochre.

These birds now feed mainly in trees, whereas they formerly procured food on the ground. " I did not attribute much importance to that fact," writes Mr. Whitmee, " because, the bird being wary, I thought its destruction by wild cats to be chiefly in the night when roosting, or when on the nest during the process of incubation, while rats would also destroy the eggs or young in the nest. Hence, I did not see how a change in the place of feeding could alone account for the increase of the bird. I therefore made particular inquiries from natives as to its roosting ; and from the information thus procured I believe the *Didunculi* almost invariably now roost upon the high branches

THE TOOTH-BILLED PIGEON

of trees instead of upon low stumps as formerly."

The nest is so rarely found that few opportunities occur of learning where it is built, but it appears that it is generally situated in the fork of a tree, although formerly the bird nested on the ground. The eggs are white. Verging some years ago on extinction, from the assumption of arboreal habits, the Samoan tooth-billed pigeon has latterly increased rapidly in numbers.

THE DODO AND SOLITAIRE

THE dodo and its near ally the solitaire (family *Dididæ*) are totally extinct members of the order, characterised by their very large size and massive build, accompanied by a complete incapacity for flight. The members of the group were entirely confined to the islands of Mauritius, Réunion, and Rodriguez. A native of Mauritius, and the sole representative of its genus, the dodo (*Didus ineptus*) was somewhat larger than a swan, with rudimentary wings, and a tail composed of short, curly feathers. The beak was very large and hooked, the body remarkably heavy, and the legs and feet short and stout (page 78).

Large, clumsy, and defenceless, the dodo was a bird marked out for early destruction ; and soon after its discovery it fell a prey to sailors, and the animals introduced by them into its island home. A few scattered relics of stuffed specimens, together with bones dug up from the peat of Mauritius, are all that are left of this bird ; but fortunately a good idea of its appearance is given in several contemporary

pictures. It was discovered by Admiral Van Neck in 1598, was still abundant in 1601, and was known to be living eighty years later, although by 1691 it appears to have been exterminated. An allied bird inhabited Réunion, but its precise affinities are unknown.

The gigantic flightless pigeon of Rodriguez known as the solitaire (*Pezophaps solitaria*) survived till a later date than the dodo, having probably lingered on in the more remote parts of the island till 1761. It was much longer in the leg than the dodo, and had a proportionately longer neck, and the males, which were far superior in size to the females, were furnished with a peculiar ball-like excrescence on each wing. The French navigator Leguat, who visited Rodriguez in 1691, and found the solitaire abundant, has given a good account of its habits, and a truthful, if somewhat pre-Raphaelite, portrait ; while of late years numerous fossil bones of the solitaire have been brought to Europe, so that we have now a fair idea of its organisation and affinities.

ORDER XXIII. PTEROCLETÆ
SAND-GROUSE

FORMERLY naturalists were in accord in assigning to the sand-grouse (order Pterocletæ) a position midway between the pigeon tribe and the game-birds, as is done here. This is, however, by no means the case at the present day, many naturalists regarding them as a connecting link between the pigeons and the plover tribe, although admitting that as regards their digestive organs they show distinct signs of affinity with the game-birds.

Among the pigeon-like characters of their skeleton may be specially noted the great triangular deltoid crest of the humerus, or upper bone, of the wing, the peculiar shape of which is so characteristic of the pigeon tribe ; this deltoid crest being the projecting process on the right side of the upper part of the bones figured on this page. To this process is attached the great pectoral muscle which renders these birds capable of sustained and powerful flight. In the game-birds this process (as shown in the same cut) is very differently formed, the edge being rounded and curved inwards instead of nearly flat and triangular. The species may be classified as follows :

FAMILY
Sand-Grouse—Pteroclidæ
GENUS 1
Syrrhaptes
SPECIES

Pallas's sand-grouse .... Syrrhaptes paradoxus
Tibetan sand-grouse .... S. tibetanus
GENUS 2
Typical Sand-Grouse—Pterocles
SPECIES
Black-bellied sand-grouse .... Pterocles arenarius
Painted sand-grouse .... P. fasciatus
Close-barred sand-grouse .... P. lichtensteini
African sand-grouse .... P. quadricinctus
GENUS 3
Pin-tailed Sand-Grouse—Pteroclurus
SPECIES
Pin-tailed sand-grouse .... Pteroclurus alchata
Madagascar sand-grouse .... P. personata
Pyrenean pin-tailed sand-grouse................... P. pyrenaicus
The sand-grouse .... P. exustus

The sand-grouse form a small group, of which all the members may be included in the single family *Pteroclidæ.* Externally they are characterised by the body being rather plump and compact, the neck short, and the head small. The beak is short and shaped like that of the game-birds, although not so strong ; there is never any naked space round the eyes ; the wings are long and pointed, and the legs and toes are so short as to render it impossible for these birds to perch on trees. The general tone of the plumage is suggestive of the sandy arid regions these birds frequent, being a subtle mixture of subdued colours, beautiful in their arrangement and pattern, and well suited to afford protection by their perfect harmony with the surroundings.

The majority of this group inhabit Africa and South-Western and Central Asia, but Pallas's sand-grouse ranges in summer to the north of Lake Baikal, and westwards to Pekin, and the black-bellied sand-grouse extends to the Canary Islands and South-West Europe, while Madagascar contains a species, *Pteroclurus personata*, peculiar to that island, and the Pyrenean

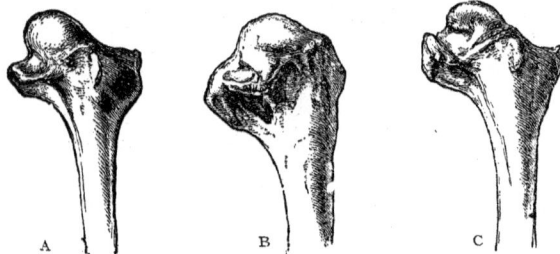

UPPER PORTION OF RIGHT HUMERUS OF (A) SAND-GROUSE, (B) FOWL, (C) PIGEON

pin-tailed sand-grouse (*P. pyrenaicus*), a western form of the Asiatic species *P. alchata*, is met with in South-Western Europe and Northern Africa. The flight of sand-grouse is swift and powerful, and on the wing these birds recall the members of the plover tribe. All are more or less migratory, and some travel immense distances. They are in the habit of repairing in the morning and evening to certain favourite drinking places where large numbers congregate. When drinking, these birds thrust their beaks into the water, where they retain them till their thirst is quenched, after the manner of pigeons.

The young are covered with a close and beautifully patterned down, and are able to run as soon as hatched. The only nest is a hole scratched in the sand ; and the eggs, generally three in number and oval in shape, are double-spotted with brown and pale violet on a ground colour which is generally cream or buff, but more rarely pale red, the faint violet spots being more deeply imbedded in the shell than the others. The eggs are like those of the rails, and unlike those of the game-birds, which are also single-spotted.

PALLAS'S SAND-GROUSE

Turning to the various groups, Pallas's sand-grouse (*Syrrhaptes paradoxus*) represents a genus in which the first toe is absent, while the short front toes are densely covered with feathers, and the two middle tail feathers are produced into long, thread-like points. In Pallas's sand-grouse the first quill of each wing is greatly lengthened and attenuated at the extremity, but in *S. tibetanus*, which is a native of Tibet, these feathers are much less developed. In the former the general colour of the upper parts is pale buff barred with black, the breast being pale grey shading into dirty white, and the under parts black. The male is distinguished by having the throat and a patch of feathers on each side of the neck rust-coloured, the rest of the head and neck uniform greyish, and a band of white feathers barred with black across the chest. In the female, on the other hand, the patches of feathers on the neck and throat are pale yellow, and the throat is bounded below by a narrow black band ; the top of the head, back, and sides of the breast being spotted with black, and the band across the chest absent.

Special interest attaches to this bird on account of the incursions which it has from time to time made into Europe, more often in comparatively small numbers, but in the years 1863 and 1888 in enormous flocks, many of which arrived on the eastern coasts of Great Britain and spread inland to nearly every county, some individuals even reaching the Scilly Islands and Ireland, and others the Shetlands and Faroes. The reason of these migrations is without any satisfactory explanation, especially as the flocks arrive in spring. It seems hardly probable that they are driven back by stress of weather while attempting to reach their breeding haunts, since they are able to endure severe cold, and have arrived in Southern Siberia in the end of March.

BLACK-BELLIED SAND-GROUSE

The black-bellied sand-grouse (*Pterocles arenarius*) represents a genus distinguished by the naked feet, the presence of the first toe, and the normal length of the middle pair of tail feathers. Common during the cold season in the north-western parts of India, this bird does not breed within the limits of the Empire. It prefers the great sandy plains where water is easily accessible, but in places where rivers are too distant it frequents such tanks as are to be found. In the early morning ploughed land is a favourite haunt, where large numbers may be seen squatting close together and basking in the morning rays of the sun.

Like the rest of their kind, these birds are in the habit of taking a midday siesta when the sun is hot, but when the weather is cold and cloudy they are moving about all day. They scratch in loose soil like hens till they have made comfortable depressions that fit them, and there they repose, sunning first one side and then the other, and apparently thoroughly enjoying the scorching heat. During their siesta they are never closely packed, but scattered over the ground singly or in twos and threes.

Mention may also be made of the little painted sand-grouse (*P. fasciatus*) of India, which is one of the most beautiful of all the group, and distinguished by its small size and black and white markings on the head. In habits it differs from the larger sand-grouse, and resembles the nearly-allied close-barred sand-grouse (*P. lichtensteini*), while as regards its plumage it so closely approximates to the small African *P. quadricinctus* that it can only be distinguished by the different arrangement of the black and white bars adorning the smaller feathers of the wings. This species, which is seldom if ever difficult of approach, and can run faster than the other kinds, is usually met with in small packs or in pairs, and frequents the neighbourhood of low, bush-clad or sparingly-wooded hills. These birds are nocturnal, and even in the darkest night they are in the habit of arriving at the edge of the plain at dusk and remaining there to feed and drink for several hours ; parties of six or seven flitting about noiselessly over an opening in the forest may sometimes be noticed long after sunset.

PIN-TAILED SAND-GROUSE

The pin-tailed sand-grouse (*Pteroclurus*, or *Pteroclidurus*, *alchata*) belongs to a genus differing from the last by the elongation of the middle tail feathers, in this respect resembling *Syrrhaptes*. This species ranges from Asia Minor to India, while an allied western form, *P. pyrenaicus*, inhabits Northern Africa and South-Western Europe. Although the two are very similar, the western bird has the wide chestnut band across the breast much darker, and the marginal lines round the smaller feathers of the wing pale yellow instead of white. In both kinds the under parts are pure white, and the males have the throat black and the upper parts dull olive blotched with yellow, but in the females the throat is white and the upper parts are barred with black. The eastern species is a cold-weather visitor to the north-west of India, where some of the sand-grouse habitually associate in countless numbers, flocks of at least ten thousand having been seen in that country as well as in Mesopotamia and on the shores of the Persian Gulf.

A third member of the pin-tailed group is the common sand-grouse (*P. exustus*), which has the general colour of the plumage yellowish buff, shading into dark brown on the under parts in the male, while in the female the breast and upper part of the back are spotted with brownish black, and the rest of the upper surface is barred with the same colour. This bird has a very wide range, inhabiting the whole of India in localities where the rainfall is moderate, the soil fairly dry, and the country open and tolerably level, and extending westwards across Asia and Northern Africa to Senegal.

PALLAS'S SAND-GROUSE

ORDER XXIV. GALLINÆ
GENERAL CHARACTERISTICS OF GAME-BIRDS

THE general appearance of the game-birds (order Gallinæ) is so well marked that the most inexperienced can hardly fail to recognise them. In all of them the body is compact and stout, and the neck rather long, supporting a fairly large, rounded head, with a moderately long, stout beak, the upper half of which is arched and overhangs the lower. Though the legs vary in length, they are always strong and adapted for rapid locomotion, the first toe being present, and the feet, with their powerful, slightly curved claws, specially suited for scratching up the ground. The wings are concave internally, thus fitting close to the body, and the flight, though noisy and somewhat laboured, is often extremely rapid. The tail varies greatly in shape and size, being enormously lengthened and developed in some kinds of pheasants, while in others, as the painted quail, it is rudimentary and hidden by the upper tail coverts.

Among the more striking skeletal characters it may be mentioned that the palate is of the cleft (schizognathous) type, while the breast-bone has two deep notches on each side of the posterior margin, and its superior process perforated to receive the bases of the coracoid bones. The feathers of the body are provided with after-shafts, and the young are in most cases hatched covered with down, although well feathered in the case of the brush-turkeys, and in all cases are able to run soon after they come into the world.

The nesting habits vary, grouse, partridges, and pheasants habitually laying their eggs on the ground with little or no nest, while curassows generally build in trees, and brush-turkeys place their eggs among sand and vegetable remains, where they are hatched by the warmth of the decaying matter and the heat of the sun. In the more typical game-birds the eggs, if spotted at all, are only marked with surface-spots, which are easily scratched off, and never possess the deep, underlying marks characteristic of those of sand-grouse and rails.

THE QUAIL

CLASSIFICATION OF GALLINÆ MENTIONED IN THIS WORK

GENUS 12
Tree-Partridges—Arboricola
SPECIES
Hainan tree-partridge ....Arboricola ardens
Himalayan tree-partridge .....A. torqueolus
GENUS 13
Wood-Partridges—Caloperdix
GENUS 14
Rollulus
SPECIES
Red-crested wood-partridge ..Rollulus roulroul
GENUS 15
Melanoperdix
SPECIES
Black wood-partridge.....Melanoperdix niger
GENUS 16
Quails—Coturnix
SPECIES
Quail ....Coturnix communis
Japanese quail ....C. japonica
South African quail ....C. capensis
GENUS 17
Synæcus
SPECIES
Australian quail...........Synæcus australis
GENUS 18
Painted Quail—Excalfatoria
SPECIES
Painted quail ....Excalfatoria chinensis
GENUS 19
Rufous-breasted Partridge—Ptilopachys
SPECIES
African rufous-breasted partridge..Ptilopachys fuscus
GENUS 20
Bamboo-Partridges—Bambusicola
SPECIES
Chinese bamboo-partridge.......Bambusicola thoracicus
Formosan partridge ....B. sonorivox
Fytch's partridge .... B. fytchi
GENUS 21
Spur-Fowl—Galloperdix
SPECIES
Ceylon spur-fowl .....Galloperdix bicalcarata
GENUS 22
Pheasant-Quail—Ophrysia
SPECIES
Pheasant-quail.........Ophrysia superciliosa
GENUS 23
Blood-Pheasants—Ithagenes
SPECIES
Blood-pheasant........... Ithagenes cruentus
GENUS 24
Tragopans—Tragopan
SPECIES
Crimson tragopan .... Tragopan satyra
GENUS 25
Monals—Lophophorus
SPECIES
Sclater's monal .... Lophophorus sclateri
Himalayan monal .... L. impeyanus
Tibetan monal .... L. l'huysii
GENUS 26
Crestless Fire-backed Pheasants—Acomus
SPECIES
Crestless fire-backed pheasant ...... Acomus erythrophthalmus
A. pyronotus
GENUS 27
Crested Fire-backed Pheasants—Lophura
SPECIES
Vieillot's fire-backed pheasant ..Lophura rufa
GENUS 28
Lobiophasis
SPECIES
Bulwer's pheasant ......Lobiophasis bulweri
GENUS 29
Eared Pheasants—Crossoptilum
SPECIES
Eared pheasant.......Crossoptilum leucurum
Hodgson's eared pheasant ......C. tibetanum
Manchurian eared pheasant .. C. manchuricum
GENUS 30
Kalij Pheasants—Gennæus
SPECIES
Swinhoe's pheasant ...... Gennæus swinhœi
White-crested Kalij pheasant .. G. albocristatus
Black-crested Kalij pheasant .. G. leucomelanus
Dark Kalij pheasant ....G. melanonotus
Silver Kalij pheasant ....G. nycthemerus
GENUS 31
Koklass Pheasants—Pucrasia
SPECIES
Koklass pheasant.........Pucrasia macrolopha
GENUS 32
Chir Pheasants—Catreus
SPECIES
Chir pheasant...............Catreus wallichi

GENUS 33
Typical Pheasants—Phasianus
SPECIES
Pheasant.................Phasianus colchicus
Persian pheasant ....P. persicus
Severtzow's pheasant..........P. chrysomelas
Shaw's pheasant...................P. shawi
Mongolian pheasant..........P. mongolicus
Zungaria pheasant........ P. semitorquatus
Chinese pheasant..........P. satscheunensis
Formosan pheasant...........P. formosanus
Japanese pheasant...........P. versicolor
Elliot's pheasant ....P. ellioti
Hume's pheasant................P. humeæ
Sömmerring's pheasant .....P. soemmerringi
Reeve's pheasant ....P. reevesi
GENUS 34
Golden Pheasants—Chrysolophus
SPECIES
Golden pheasant ....Chrysolophus pictus
Amherst's pheasant ....C. amherstiæ
GENUS 35
Jungle-Fowls—Gallus
SPECIES
Red jungle-fowl ....Gallus banciva
Ceylon jungle-fowl ....G. lafayetti
Grey jungle-fowl ....G. sonnerati
Green jungle-fowl ....G. varius
GENUS 36
Peacock-Pheasants—Polyplectron
SPECIES
Grey peacock-pheasant...Polyplectron chinquis
GENUS 37
Argus-Pheasants—Argusianus
SPECIES
Argus...................Argusianus giganteus
Gray's argus ....A. grayi
GENUS 38
Rheinhard's Argus-Pheasant—Rheinhardius
SPECIES
Rheinhard's argus. .....Rheinhardius ocellatus
GENUS 39
Pea-fowl—Pavo
SPECIES
The pea-fowl ....Pavo cristatus
Burmese pea-fowl ....P. muticus
GENUS 40
Phasidus
SPECIES
Black guinea-fowl ....Phasidus niger
GENUS 41
Turkey Guinea-fowl—Agelastes
SPECIES
Turkey guinea-fowl ....Agelastes meleagrides
GENUS 42
Typical Guinea-fowls—Numida
SPECIES
Guinea-fowl ....Numida meleagris
Abyssinian guinea-fowl.......N. ptilorhyncha
GENUS 43
Crested Guinea-fowl—Guttera
SPECIES
Crested guinea-fowl ....Guttera cristata
Acryllium
SPECIES
Vulture-like guinea-fowl.Acryllium vulturinum
GENUS 45
Turkeys—Meleagris
SPECIES
Mexican turkey ....Meleagris gallopavo
American turkey...............M. americana
Eyed turkey..................M. ocellata
GENUS 46
American Partridges—Dendrortyx
SPECIES
Large American partridge.......Dendrortyx macrurus
GENUS 47
Scaled Partridges—Callipepla
GENUS 48
Mountain Partridge—Oreortyx
SPECIES
Mountain partridge ....Oreortyx pictus
GENUS 49
Lophortyx
SPECIES
Californian quail ......Lophortyx californicus
GENUS 50
Barred Quail—Philortyx
SPECIES
South Mexican barred quail.Philortyx fasciatus
GENUS 51
Crested Quails—Eupsychortyx
GENUS 52
Ortyx
SPECIES
Virginian quail ....Ortyx virginianus

GENUS 53
Full-crested Quails—Cyrtonyx
FAMILY 3
Megapodes and Brush-Turkeys—Megapodiidæ
GENUS 1
Megapodes—Megapodius
SPECIES
Philippine megapode....Megapodius cumingi
Ladrone megapode..............M. laperousii
Nicobar megapode........M. nicobariensis
GENUS 2
Eulipoa
SPECIES
Wallace's megapode ....Eulipoa wallacei
GENUS 3
Lipoa
SPECIES
Mali-fowl ....Lipoa ocellata
GENUS 4
Papuan Brush-Turkeys—Talegallus
SPECIES
Papuan brush-turkey......Talegallus cuvieri
GENUS 5
Australian Brush-Turkey—Catheturus
SPECIES
Australian brush-turkey..Catheturus lathami
GENUS 6
Fleshy-crested Brush-Turkey—Æpypodius
GENUS 7
Megacephalum
SPECIES
Celebean brush-turkey...Megacephalum maleo
FAMILY 4
Curassows and Guans—Cracidæ
GENUS 1
Crested Curassow—Crax
SPECIES
Guiana crested curassow.......Crax alector
Mexican crested curassow.........C. globiera
GENUS 2
Nothocrax
SPECIES
Rufous-coloured urumutu.........Nothocrax urumutum
GENUS 3
Mituas—Mitua
SPECIES
Mitua ....Mitua mitu
GENUS 4
Pauxis
SPECIES
Pauxi Curassow..............Pauxis pauxi
GENUS 5
Oreophasis
SPECIES
Lord Derby's guan ......Oreophasis derbianus
GENUS 6
Guans—Penelope
SPECIES
Brazilian guan..........Penelope superciliaris
Penelopina
SPECIES
Guatemalan guan ....Penelopina nigra
GENUS 8
Ortalis
SPECIES
Chachalaca ....Ortalis vetula
GENUS 9
Aburria
SPECIES
Aburria....................Aburria aburri
GENUS 10
Chamæpetes
SPECIES
Columbian guan ....Chamæpetes goudoti
Costa Rica guan.................C. unicolor

FAMILY 5
Hoatzin—Opisthocomidæ
GENUS
Hoatzin—Opisthocomus
SPECIES
Hoatzin ....Opisthocomus hoatzin
FAMILY 6
Bustard-Quails—Turnicidæ
GENUS 1
Bustard-Quails—Turnix
SPECIES
Andalusian bustard-quail ....Turnix sylvatica
Indian bustard-quail..............T. taigur
GENUS 2
Pedionomus
SPECIES
Australian collared bustard-quail .. Pedionomus torquatus

MONAL

"There are few sights more striking where birds are concerned than that of a grand old cock monal shooting out horizontally from the hillside just below one, glittering and flashing in the golden sunlight, a gigantic rainbow-tinted gem, and then dropping, stone-like, with closed wings, into the abyss below."

THE GROUSE TRIBE

THE members of the grouse tribe (family *Tetraonidæ*) form a group of over thirty species, in which the degree of feathering of the legs and feet varies in the different genera ; the ptarmigan and its allies having the legs and feet entirely covered with feathers ; while in others, such as the blackcock, the toes are naked, and only the legs feathered, while in the hazel-hen and its relatives, not only the foot, but the greater part of the metatarsus is naked. Those groups with naked feet may be distinguished from the members of the pheasant tribe by the horny processes fringing the sides of the toes and producing a comb-like appearance; these being absent in the pheasants, while none of the grouse have spurs on the legs.

PTARMIGAN AND GROUSE

Perhaps no members of this group of birds are more interesting than ptarmigan (genus *Lagopus*) and their allies, on account of the seasonal changes of plumage they undergo in order that they may assimilate themselves to the colour of their surroundings, and be thus protected from their numerous enemies. In the ptarmigan (*L. mutus*) three changes of plumage — summer, autumn, and winter —take place. During winter both sexes of the common species become pure white, with the exception of their outer tail feathers, which are black, the male being distinguished by the presence of a small black patch in front of the eye. It must, however, be remarked that absolutely pure white plumage on the back is not often met with in Scotch examples, except in unusually severe winters, there being generally a few of the greyish autumn feathers left in the plumage of the upper parts, which are neither replaced by white ones nor turn white.

On the other hand, in such examples as inhabit colder climates, like the north of Scandinavia, the male, at least, rarely dons the full summer and autumn plumages, a number of white winter feathers being retained throughout the summer, and in some instances only the head, mantle, and chest change colour, the rest of the plumage remaining white. It would thus seem that, in those countries where the summer is of short duration, sufficient time is not allowed for the full summer and autumn changes to be effected before winter sets in ; and it is probable that this particoloured plumage affords even better protection in such localities than if a complete change to a darker plumage took place. In summer the male ptarmigan has the general colour of the head, upper parts, sides, and flanks dark brown or blackish brown, finely mottled and barred with grey and rusty, the chest and upper part of the breast blackish, sometimes slightly mottled with buff, and the under parts and the middle tail feathers white.

PTARMIGAN IN SUMMER PLUMAGE

The female has the general plumage above, as well as the middle tail feathers, black mixed with rufous buff, most of the feathers being edged with white or buff, and the under parts rufous buff barred with black. This plumage so closely approaches in colour the general surroundings of the nesting places, that the bird when sitting on its eggs is almost invisible. As autumn advances, the darker coloured feathers in both male and female are replaced by a grey plumage finely mottled with black, and sometimes buff, and as the season continues the more or less complete white winter plumage already mentioned is once more assumed.

It is noteworthy that a considerable amount of the changes in colour is due not to moulting, but partly to a rearrangement of the pigment in the feathers themselves. In all the group, except red grouse, the primary feathers—which, like those of the tail, are only cast at the autumn moult—remain white throughout the year.

All ptarmigan are essentially high-ground birds, although the red grouse is an exception, occurring sometimes on low-ground bogs close to the sea. Unlike blackcock, the common ptarmigan and its allies pair with one female, with which they remain throughout the breeding season.

During the nesting season ptarmigan become so tame that they may be approached within a few yards. On the barren hilltops and watersheds, where these birds find a home among the scattered boulders, dwarf Alpine plants, deergrass, and mosses, the wanderer is often startled by the hoarse, croaking cry of the male, as he rises suddenly from the ground, where he was squatting invisible almost at the feet, to settle on some neighbouring rock.

On being again approached, he makes a second short flight to some commanding position, where, with outstretched neck, he watches the movements of the intruder. Soon after this the intruder may nearly walk on the female sitting on her eggs, or in charge of a number of beautiful chicks covered with yellow and brown down. The young scatter in every direction, running with considerable speed, and helping themselves along with their still tiny, undeveloped wings, while the anxious mother, as she runs alongside, goes through a performance intended, by diverting attention, to cover their retreat and convey the idea that at least one of her wings has been broken. In a few seconds, however, she appears to recover, and skulks off among the rocks, and when the young are looked for, they too have vanished.

The ptarmigan inhabits most of the higher mountain ranges of Europe, and possibly extends into Central Asia, where it is represented by the so-called rock-grouse (*L. rupestris*), which differs by the more rufous plumage of the male in summer, though in some

PTARMIGAN IN WINTER PLUMAGE

probability gradually ceased to assume a white winter dress, owing to a milder climate rendering this no longer essential for protection.

Under these circumstances it might be inferred that in the red grouse there would be only two changes of plumage—namely, in summer and autumn ; but this, for some unknown cause, is the case only with the female. In early spring the latter begins to assume the summer dress of black mottled and barred with buff or rufous buff, which harmonises so well with the surroundings of her nest that she is comparatively safe from detection. In the end of June she casts the whole of her plumage, and by the beginning of September the change to the dark buff-spotted autumn dress is complete, though in some examples, probably birds of the year, a few feathers of the back may still be seen in quill so late as December.

The male, on the other hand, makes no spring change, not a single feather being renewed between January and the end of June ; although after the breeding season the entire plumage is replaced by the autumn feathers, which are black, marked, barred, and often edged with buff. Accordingly, while at the end of August the cock bears a considerable resemblance to the female in May, though the buff markings are never so coarse, no sooner is the autumn plumage donned than the dark chestnut winter

localities, such as Newfoundland, a greyer phase, closely approaching the European bird, is met with. The rock-grouse inhabits Northern Asia, extending eastwards to Japan, and through Arctic America to Iceland.

In Spitzbergen occurs a somewhat different species (L. hyperboreus), with more white on the basal part of the tail feathers, the outer web of the outermost pair having only the terminal third black.

The most remarkable member of the whole group is, however, the red grouse (L. scoticus), peculiar to the British Isles, in which the changes of plumage appear unique ; this species differing from all others in having no white winter plumage, and the flight feathers always brownish black (page 1455). The variation in plumage is very great, but the extreme diverseness may be enumerated as the black, red, and white-spotted phases. The first phase has the entire plumage black, and is by far the rarest ; the second, in which the general colour is rufous chestnut, is chiefly met with on the west coast of Scotland, the outer Hebrides, and Ireland ; while the white-spotted phase, in which all the feathers of the breast and under parts, and sometimes also those of the head and back, are widely tipped with white, is apparently dependent on latitude and altitude.

Grouse are remarkable for shedding the sheaths of their toes during the main moult. It is commonly stated that these birds live entirely on heather, but many Yorkshire grouse moors are clothed only with crowberry, and have no heather.

The nearest ally of the red grouse is the circumpolar ripa, or willow-grouse (L. albus), which has three distinct seasonal plumages, those of summer and autumn closely resembling those of the red grouse, although the winter dress is white, in which state the bird can only be distinguished by its large size and thick beak. That the red grouse is only an insular race of the willow-grouse there can be little doubt, and it has in all

SPITZBERGEN PTARMIGAN

feathers, with their mottled black bars, begin to appear, the whole change being completed by December. Some of the autumn feathers are, however, often retained on the back, and may be distinguished from the winter plumage at a glance. The male breeds in this plumage, and very shabby and worn he looks by the end of the nesting season.

It will thus be seen that the male and female have two changes in the season, but while those of the latter are, as might be expected, made in spring and autumn, the male moults in autumn and winter. The Rocky Mountain ptarmigan (*L. leucurus*) is the smallest representative of the genus, distinguished by its pure white tail.

BLACKCOCK AND CAPERCAILLIE

The second group of the grouse tribe is formed by the blackcock, or black grouse, and the capercaillie. These birds have the legs covered with feathers, but the feet naked, and the sides of the toes furnished with horny comb-like appendages. The

WILLOW-GROUSE IN SUMMER PLUMAGE

typical black grouse (*Tetrao tetrix*) of Europe and Asia is replaced to the east by the Caucasian black grouse (*T. mlokosiewiczi*). In the European species, of which the two sexes are respectively known as blackcock and grey hen, the males are distinguished by their general black plumage, and the peculiar shape of the outer tail feathers, which are elongate and curve outwards at the extremity. In the blackcock the under tail coverts are white, although in the species from the Caucasus these parts are black, like the rest of the plumage. The two birds also differ in their changes of plumage; the young male of the European species assuming the black feathers of the adult more or less completely by the first winter, while in the young male of the Caucasian bird a female-like plumage is retained throughout the first winter and spring.

During the autumn moult, when the males are rarely met with, the black plumage of the head and neck is replaced by brownish buff and black feathers, barred like those of the female. Doubtless this temporary change is protective, enabling blackcock to escape observation, when, owing to the heavy moult in their wings and tail, they are rendered almost incapable of flight. Black grouse are polygamous, one male pairing with a number of females, each of which undertakes the entire responsibility of rearing her young. During the pairing season the males are in the habit of resorting to some spot where in the morning and evening they fight for the possession of the females; each challenging the other in turn, and going through a series of skirmishes till the older and stronger birds have driven off the rest and won the females. Black grouse are chiefly found in the neighbourhood of pine and birch forests bordering moorland, where bilberry, cranberry, heath, and bracken flourish, though they may sometimes be seen on the open moor. Although their flight is straight, and their regular wing-beats somewhat laboured, they can fly at a great rate.

Black grouse perch on trees, much of their food consisting of buds and flowers, but in autumn they may be seen on the stubbles in search of grain. During the winter in Scandinavia these birds are reported to make tunnels under the snow, in which they remain for a considerable period; each cock makes its own burrow, but it is probable that in many cases these tunnels unite in a common chamber, where several birds live in company.

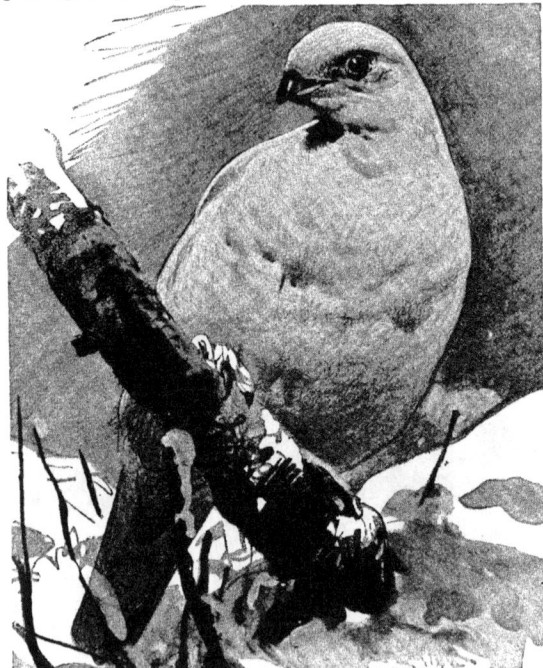

WILLOW-GROUSE IN WINTER PLUMAGE

The blackcock by no means confines his attentions during the pairing season to the hens of his own species, the hybrids produced by a cross between this species and the capercaillie being not uncommon. Sometimes, too, he interbreeds with the red grouse, and more rarely with the willow-grouse, hazel-hen, and pheasant, while crosses with domesticated fowls are known to occur.

The largest members of the genus are the capercaillie and its immediate relatives of the pine forests of Northern and Central Europe and Asia. There are three species, all easily recognisable by their large size, as well as by the rounded tail composed of eighteen feathers. The typical capercaillie (*T. urogallus*) ranges through Northern and Central Europe to Turkestan and the Altai, but in the Urals is represented by a paler form, with the whole of the breast and under parts white in the male. In typical examples the breast and under parts are black, with some of the feathers in the middle of the breast tipped with white, but numerous examples are to be met with in the London market in every intermediate stage of plumage, and are believed to be imported from some of the southern states of Russia, though the exact locality is uncertain (p. 1455).

In North-Eastern Siberia a different species (*T. parvirostris*) occurs, while the third kind (*T. camchaticus*) is confined to Kamchatka. These eastern birds are distinguished from the typical species by the smaller size of the beak, by the scapulars being widely

of Perth, Stirling, and Forfar. The capercaillie is polygamous ; and its nesting habits and eggs are very similar to those of black grouse, the eggs being buff spotted with reddish brown. So many as twelve eggs are sometimes laid, but the capercaillie hen is a bad mother, and seldom succeeds in rearing more than one or two of her somewhat delicate young. The cock is a remarkably wary bird, much more difficult to shoot than the hen, and it is marvellous, considering its large size and weight, how quietly the former can slip out of the far side of a pine tree without being observed. The greater part of his time is spent among the branches of these trees, the needles forming a considerable portion of his food, and giving the flesh a strong flavour of turpentine.

AMERICAN GROUSE

North America is the home of two grouse-like species, namely, the Canadian grouse, or spruce-partridge (*Canachites canadensis*), of Canada and the northern states to the east of the Rocky Mountains, and its near ally, Franklin's grouse (*C. franklini*), inhabiting the coast ranges to the west of the same chain. Both are about the size of a partridge, and have sixteen tail feathers. The cock pairs with only one female, and probably often retains the same mate for more than one season. The male Canadian grouse has the upper parts mostly grey shading into sandy or rufous white on the wings, and barred and mottled with black, the throat, chest, and middle of the breast black, the sides and under parts tipped with white, and the tail black tipped with chestnut. In the female the plumage is barred and mottled with black and rufous yellow. In both the male and female of Franklin's grouse the chestnut band across the end of the tail is absent, and the upper tail coverts are tipped with white instead of grey.

SHARP-WINGED AND DUSKY GROUSE

The sharp-winged grouse (*Falcipennis hartlaubi*) of North-Eastern Siberia and Kamchatka, the sole representative of its kind, may be recognised by having the outer flight feathers narrowed towards the extremity and sickle-shaped.

BLACKCOCK DISPLAYING

tipped with white, and also by the females being much darker on the under parts. From one another they may be distinguished by the white tips of the scapulars in the smaller Kamchatkan species being wide and forming a continuous white band, while in the Siberian bird they constitute an interrupted line of white spots.

Formerly indigenous in Scotland and Ireland, the capercaillie was exterminated towards the end of the eighteenth century, but was reintroduced in 1837 into Scotland, where it is now fairly plentiful in the counties

On the other hand, the dusky grouse (*Dendragapus obscurus*) and its two allies of the pine forests to the east and west of the Rocky Mountains are distinguished by the presence of twenty tail feathers, and an inflatable air-sac on each side of the neck of the cock. The home of the dusky grouse is the southern Rocky Mountains, from New Mexico to Idaho, its place further west being taken by the sooty grouse (*D. fuliginosus*), ranging along the Pacific coast from California to Sitka ; while, on the east side of the Rockies, Richardson's

HEN CAPERCAILLIE AND CHICKS

THE SAGE-GROUSE

The largest American representative of the family is the sage-grouse (*Centrocercus urophasianus*), inhabiting the dry sage-brush plains of the western United States. Distinguished from the allied forms by its long, pheasant-like tail of twenty feathers, with the middle pair elongated and pointed, the male has an inflatable air-sac on each side of the neck, and attains a weight of 8 pounds, the female being smaller. The chief food of this bird, especially during the winter months, is afforded by sage-brush, though during summer it is varied with grasses, berries, insects, and sometimes grain. The stomach of this species is soft, and unlike that of other game-birds, which are provided with muscular gizzards. Major Bendire gives the following account of the sage-cock's courtship:

"Early one morning, I had the opportunity to observe the actions of a single cock, while paying court to several females near him, and I presume he did his very best. His large, pale yellow air-sacs were fully inflated, and not only expanded forward, but apparently upward as well, rising at least an inch above his head, which, consequently, was scarcely noticeable, giving the bird an exceedingly peculiar appearance. He looked decidedly top-heavy, and ready to topple over at the slightest provocation. The few long, spiny feathers along the edges of the air-sacs stood straight out, and the greyish white of the upper parts showed in strong contrast with the black of the breast. His tail was spread out fan-like, at right angles from the body, and was moved from side to side with a slow, quivering movement. The wings were trailing on the ground. While in this position he moved around with short, stately, and hesitating steps, slowly and gingerly, evidently highly satisfied with his performance, uttering, at the same time, low, grunting, guttural sounds, somewhat similar to the purring of a cat, only louder."

SHARP-TAILED GROUSE

Another North American type is formed by the sharp-tailed grouse, of which two kinds are known,

grouse (*D. richardsoni*) is found from Central Montana northwards. Much larger than the Canadian grouse, the males of this species have the upper parts smoky black, mottled with grey, and the under parts grey; while in the females the plumage of the upper parts and breast is barred and mottled with buff. In both the dusky and sooty grouse the tail is somewhat rounded in shape, with a terminal grey band wider (more than an inch wide) in the former. In Richardson's grouse the grey band is absent and the tail square.

PRAIRIE-HENS

The males of the three species of prairie-hen are characterised by an elongate tuft of feathers and an inflatable air-sac on each side of the neck; these tufts being less conspicuous and the air-sacs absent in the females. The typical prairie-hen of the Mississippi Valley (*Tympanuchus americanus*) has the plumage brown above, barred and marked with buff and black, the longer feathers of the neck-tufts black, and the under parts pale brown, barred and fringed with white.

During the pairing season these birds assemble in numbers at early morning on high, dry knolls, when the males go through strange antics to captivate the females. Inflating their orange air-sacs and erecting their long neck-tufts, they utter their strange, booming love-note, which may be heard at a great distance in the still morning air. The females are remarkably prolific, laying eleven to fourteen eggs on an average, while as many as twenty or more are not infrequently found. The females alone undertake the incubation and care of their young.

PRAIRIE-HENS

namely, the large dark northern *Pediœcetes phasianellus*, inhabiting the interior of British North America, and the smaller more rufous and buff southern *P. columbianus*. Both are characterised by the wedge-shaped tail of eighteen feathers, in which the middle pair is larger than the rest; the males not being provided with air-sacs.

RUFFED GROUSE

The last two genera of the family represent a somewhat different group of grouse, the various members of which have the lower parts of the legs as well as the toes devoid of feathers. The ruffed grouse (*Bonasa umbellus*) of North America extends over a wide range, and is subject to local variations regarded by some naturalists as of specific value. This bird may be recognised by the frilled ruffles of black feathers on each side of the neck, though in some specimens the colour of these is dull chestnut. The general colour above is rufous or grey, but every intermediate stage of plumage may be met with.

Major Bendire writes that "the mating season occasionally begins early in February, but usually about the beginning of March, when the familiar drumming of the male may be frequently heard, though the bird is not often seen. This drumming of the ruffed grouse has often been described, and many

different theories have been advanced as to how the sound is produced. It is generally conceded that the sound is produced by the outspread wings of the bird being brought suddenly downward against the air, without striking anything."

HAZEL-HENS

The hazel-hens, forming the last group of the grouse family, differ from the ruffed grouse in having the plumage of the sexes different and lacking the conspicuous ruffles on the sides of the neck, while the tail is composed of sixteen instead of eighteen feathers. Besides the typical species (*Tetrastes bonasia*), ranging from Northern and Central Europe to Kamchatka and from Spain to North China, a second kind (*T. griseiventris*), with the breast and under parts sandy grey narrowly barred with black, comes from Perm in East Russia, while a third (*T. severtzowi*), with the outer tail feathers black barred with white, is found in North-Eastern Tibet. The European species, like its North American ally, has two extreme phases of plumage, a rufous and a grey, between which occurs every intermediate variety of colour. Hazel-hens are essentially forest birds, and when driven from covert invariably perch high up in the branches of trees. They abound in the great forests of Poland.

HAZEL-HEN

THE PHEASANT TRIBE

THE great bulk of the typical game-birds are included in the family *Phasianidæ*, which contains about sixty genera, nearly all of which are restricted to the Old World. In all of the members of the family the legs and feet are naked, the latter never being fringed with horny, comb-like appendages as in the grouse, while in many groups the legs are armed with one or more pairs of spurs. The family may be divided into three subfamilies—the first (*Perdicinæ*) including the partridges, francolins, and quails and their allies; the second (*Phasianinæ*) the pheasants; and the third (*Odontophorinæ*) the American partridges and quails, in which the cutting edge of the upper half of the beak is notched. The division between the first two groups is, however, artificial, for the partridges merge into the pheasants, the bamboo-partridges (*Bambusicola*) and the African and Indian spur-fowls (*Ptilopachys* and *Galloperdix*) being the principal intermediate forms.

The shape of the wing is indeed almost the only character of any importance for distinguishing these groups; all the pheasants, except the typical genus, having the first primary quill shorter than the tenth quill of the same series, whereas in the partridges the

former is equal to or longer than the latter. Unfortunately, the exception among the former is the genus *Phasianus* itself, which has the first primary longer than the tenth; while, on the other hand, in some of the partridges the tenth is somewhat the longer. It is only by using the supplementary character of the length of the tail, coupled with the shape of the wing, that it can be decided to which of these divisions some of the species should be referred. Thus the first group of partridges may be briefly characterised as having the first primary quill longer than or equal to (rarely shorter than) the tenth, and the length of the tail less—usually much less—than that of the wing.

SNOW-PARTRIDGES AND SNOW-COCKS

The snow-partridge (*Lerva nivicola*) of the higher ranges of the Himalaya, and extending eastwards into North and Western China, is somewhat peculiar in having the upper half of the legs feathered, and is further characterised by the plumage of the upper parts in both sexes being black, narrowly barred with white and rufous, while the general colour of the breast and under parts is deep chestnut, and the coral-red legs are armed in the male with stout spurs; the tail having fourteen feathers.

Closely allied to, but distinguished by their larger size, differently shaped wing, and tail of eighteen feathers, are the two rare species of pheasant-grouse (*Tetraophasis*) from the highlands of Central and Eastern Tibet, and from these in but a step we pass to the snow-cocks, the largest members of the group, which are natives of the mountains of Central Asia, where they are found only at high elevations. Among them, the Himalayan snow-cock (*Tetraogallus himalayensis*), one of the largest of the six species, and not much inferior to the capercaillie in size, ranges through the Western Himalaya to the Hindu-Kuh, and northwards through the Altai. In the male the feathers of the upper parts are mostly grey, finely mottled and margined with buff, but there is a large chestnut patch on each side of the nape, and a band of the same colour surrounds the throat, which, together with the chin and eyebrow-stripes, is white. The breast is white barred with black, and the rest of the under parts mostly grey, the sides and flanks being margined with chestnut and buff. The female scarcely differs in plumage, but may be distinguished by her smaller size and the absence of the blunt spurs of the male.

These birds are confined to the snowy ranges above the limits of forest, but are driven by the snows of winter to perform one, or in some places two, annual migrations to the middle regions. In summer they are only seen near the limits of vegetation, but from June till August, however much the sportsman may wander on the highest accessible points in the mountains near the source of the Ganges, only a few are met with, the majority, no doubt, retiring across the snowy range into Chinese Tibet to breed. At the beginning of September they are first seen near the top of the higher grassy ridges, and after the first general severe fall of snow they come down in numbers on some of the bare exposed hills in the forest regions, where they remain till the end of March.

Gregarious, and often congregating in packs to the number of twenty or thirty, snow-cocks never enter forest or jungle, and also avoid spots where the grass is long. When feeding, they walk uphill, picking up tender blades of grass and young shoots of plants, occasionally stopping to scratch up bulbs, of which they are specially fond.

RED-LEGGED PARTRIDGES

Easily recognised by their transversely barred sides and flanks, which contrast strongly with the rest of the plumage of the breast and under parts, red-legged partridges are represented by half a dozen species, characterised by the presence of fourteen tail feathers, and the general external similarity between the two sexes, except that the male is provided with a pair of blunt spurs. In the mountains of Southern Europe, ranging from the Pyrenees to the Balkans is found the so-called Greek partridge (*Caccabis saxatilis*), but in the Grecian Islands and Cyprus its place is taken by the nearly allied chukar (*C. chukar*), distinguished by having the lores, or space in front of the eyes, white instead of black. Its range is extensive, extending across Asia to China, and reaching from the sea-level to an elevation of sixteen thousand feet; while the bird apparently flourishes as well in desert country as in cultivated hills. Their surroundings largely affect chukar both in size and colour; the paler-coloured birds from the Persian Gulf differing widely from the dark type found in Cyprus and the Himalaya.

The common red-legged, or French, partridge (*C. rufa*) is a native of South-Western Europe, and was introduced into England towards the end of the eighteenth century. It is a handsome bird, with the upper parts olive brown shading into chestnut on the top of the head and mantle, the eyebrow-stripe, chin, and throat white, and the throat bounded by a black band. The feathers of the neck and chest are broadly edged with black, while those of the sides and flanks are grey barred alternately with white, black, and chestnut; the legs and feet being bright coral-red (page 1455).

Unlike the common partridge, the red-legged species sometimes perches on trees, walls, and palings, and when pursued always prefers to escape by running. During the pairing season the males are pugnacious, fighting not only among themselves, but with the grey species. The eggs, which vary from ten to eighteen in a clutch, are deposited in a hollow in the ground among rough grass or in growing crops.

SISI PARTRIDGES

The pretty little Bonham's partridge (*Ammoperdix bonhami*), known in India as the sisi, and Hey's partridge (*A. heyi*), inhabiting both shores of the Red Sea, form a small well-marked group nearly allied to the last, but with the colour of the plumage of the two sexes different, the males having the feathers of the sides and flanks margined with black instead of being cross-barred, while the legs are never armed with spurs. The sisi has a wider range, being found from North-West India to the Euphrates Valley, and southwards to Aden.

The male may be recognised by its grey head and neck, with a white black-edged band passing across the forehead and along the sides of the head to the ear coverts, the general colour of the rest of the plumage being purplish buff. In the male of Hey's partridge the head and neck are vinous fawn-colour, and the upper parts much paler. The females of both are almost indistinguishable, having the head and neck isabelline, faintly marked and barred with rufous buff, and the dark marking on the upper parts coarser.

RED-LEGGED, OR FRENCH, PARTRIDGES

RED-LEGGED PARTRIDGE

Bare broken ground is the favourite home of the sisis, which may be met with in suitable localities from sea-level up to about seven thousand feet. Unlike most partridges, they care little for concealment, and when they wish to hide, the colour of their plumage harmonises so perfectly with the sand and stones that it is only necessary for them to keep still to avoid detection.

FRANCOLINS

The next group for consideration is that of the spur-legged partridges known as francolins, which include more than forty species, the great majority natives of Africa. One species, the typical francolin (*Francolinus vulgaris*), is, however, found in Cyprus, and ranges across South-Western Asia to Northern India, while the painted francolin (*F. pictus*) is peculiar to Western and Central India, and the Chinese francolin (*F. chinensis*) inhabits the Indo-Chinese countries. In all except two, the males are provided with at least one, and in certain cases two, pairs of sharp spurs, and in some of the African species the females are almost as well armed. The common and painted francolins are exceptions to the rule, the male of the former having but one pair of blunt, wart-like spurs, while in the latter even these are absent.

The common francolin, or black partridge, once extended over a wider range and inhabited many of the countries bordering the Mediterranean, a few pairs remaining until recently in Sicily. The male has the sides of the head, chin, throat, and under parts deep black, the sides and flanks being more or less spotted with white, according to age ; there is a white band below the eye, and a wide chestnut ring surrounds the neck, while the top of the head and wings are dark brown edged with buff, and the lower part of the back and rump black, narrowly barred with white. The female has the chestnut collar confined to the nape,

the general colour of the upper parts browner, the sides of the head buff shading into white on the throat, and the under parts white mixed with buff and barred with black. This bird, in common with all the francolins, is specially partial to valleys where high grass and jungle are interspersed with cultivated ground.

Although the males and females of this and several other species differ considerably in colour, in the great majority of cases the two sexes are much alike, more especially in the case of a couple of closely allied African species. The male of the Natal francolin (*F. natalensis*) has, for instance, the under parts white, with V-shaped black marks on the feathers, and is similar to the female, whereas in the allied Hildebrand's francolin (*F. hildebrandi*) from Kilimanjaro, and in Johnston's francolin (*F. johnstoni*) from the Shiri Highlands, although the males are very similar to the Natal bird, and only differ in the shape of the black marking on the under parts, the females are different, their general colour below being bright rust-colour.

One of the largest species is Jackson's francolin (*F. jacksoni*) from Masailand, which is only rivalled in size by Erckell's francolin (*F. erckelli*) from Abyssinia.

An allied genus includes the bare-throated francolins (*Pternistes*) of Africa, of which ten species of large size are known, easily recognised by the bare throat covered with red or orange skin. In habits they closely resemble the typical francolins.

TYPICAL PARTRIDGES

The typical partridges are natives of Europe and Asia, where they are represented by several species belonging to two distinct types. Among these the common grey partridge (*Perdix cinerea*) and the bearded partridge (*P. daurica*), with a black horse-shoe mark on the breast, have eighteen tail feathers and the under parts of the body devoid of black cross-bars ; while to the second group belong Hodgson's partridge (*P. hodgsoni*) of Southern Tibet, and the smaller and more northern *P. sifanica* of Kansu, both of which have only sixteen tail feathers and the under parts barred with black.

FRANCOLIN

The grey partridge (page 1406) ranges over Europe and Western and Central Asia, as far, at least, as the Barabinska Steppes and the Altai. For many years the chestnut horse-shoe mark on the breast was considered as distinctive of the male, the female being supposed to have this reduced to a few chestnut spots, or absent. It is true, indeed, that in old birds the differences in this patch are generally characteristic of the male and female, but in the majority of immature females the horse-shoe is well developed, and nearly or quite as large as in the adult male. Immature birds of both sexes may be recognised by having the first flight feather pointed at the extremity, and the legs yellowish brown; but in adults this feather is rounded at the tip and the legs are slate-colour. In the south-eastern counties of England young females rarely show a well-marked horse-shoe, and in some instances all trace of chestnut is absent. On the other hand, in most females from other districts this patch is greatly developed, and occasionally nearly as large as in adult males.

The only trustworthy character for distinguishing the sexes is found in the lesser and median wing coverts; in the male each of these feathers being brownish buff, thickly vermiculated with black, with a chestnut blotch on either web, and a well-defined pale buff shaft-stripe down the middle, while in the female the ground-colour is mostly black with wide-set transverse buff bars, in addition to the buff shaft-stripe common to both sexes.

PARTRIDGES

BUSH-QUAILS

Passing over the Madagascar partridge (*Margaroperdix madagascariensis*), distinguished by having only twelve tail feathers, the next for notice are the small Indian partridges commonly known as bush-quails, none of them being larger than quails, although provided with a much more handsome plumage. The colour of the plumage in the sexes is different, and the tail, which is feebly developed, being less than half the length of the wing, contains twelve feathers.

The jungle bush-quail (*Pedicula asiatica*) and the rock bush-quail (*P. argunda*) represent the typical genus. In the males the breast and under parts are white, barred with black, but while in the former the chin and throat are bright chestnut, in the latter they are dull brick-red. The females of both kinds have the under parts purplish buff, but in the jungle bush-quail the chin and throat are bright chestnut as in the males, while in the other the middle of the chin and throat is whitish. The distribution of these species is complementary to each other—that is to say, though both are found over the greater part of India, the localities they affect are widely different, the jungle-quail being met with on hilly ground covered with moderately thick forest and jungle, while the rock-quail prefers half-barren sandy or rocky plains studded with low scattered bushes. Mr. A. O. Hume describes the former as "little bustling ground-birds, always keeping, according to my experience, in packs or families; never coming out into the open; always feeding in grass, jungle, or stubble long enough to hide their tiny selves."

The painted bush-quails, as typified by the Indian *Microperdix erythrorhyncha*, constitute a genus of three species confined to India and Manipur, and distinguished by the reduction in the number of tail feathers to ten. These birds, which are chiefly found on rocky ground interspersed with bushes, fern, and high grass, are met with in coveys, and prized by the natives on account of their pugnacious habits.

TREE-PARTRIDGES

The members of this group are characterised by the short tail (its length being less than half that of the wing) and their peculiarly long and rather straight nails, that of the first toe being well developed. There are numerous species, ranging from the Himalaya, through the Indo-Chinese countries, to Sumatra, Java, Borneo, Formosa, and Hainan. Unlike the members of the preceding genera, which are all ground-birds, though

TREE-PARTRIDGE

individuals may occasionally perch, these partridges are not infrequently in the habit of perching on trees.

The colouring in most of the species is a mixture of olive brown, black, and rufous, but in *Arboricola ardens* of Hainan the chest is ornamented with a patch of fiery red.

All these birds frequent hill-forest, the Himalayan tree-partridge (*A. torqueolus*) ranging to an elevation of fourteen thousand feet, and being seldom flushed except with dogs ; but when they rise their flight is strong and swift. The eggs differ from those of other partridges in being pure white. The colour of the plumage in all but one species is practically identical in both sexes, but in the tree-partridge of the outer ranges of the Himalaya there is considerable difference in this respect. In both male and female of this species the general colour of the upper parts is olive brown barred with black, and the sides and flanks are grey, widely edged with chestnut and spotted with white ; but the male has the top of the head bright chestnut, the eye-stripe, sides of the face, chin, and throat black, more or less margined with white, and a broad white band across the front of the crop. In the female, on the other hand, the top of the head is brown marked with black, and the sides of the head are rufous spotted with black, the white band being absent.

WOOD-PARTRIDGES

Wood-partridges, of which there are three genera, are peculiar in having the claw of the first toe rudimentary. In the ferruginous wood-partridges (*Caloperdix*) the legs of the males are provided with one or more pairs of spurs. The general colour of the head, neck, and under parts is bright rusty red, the upper portion of the back is black and white the lower parts are black with rusty red markings, and the wings olive brown, spotted with black. The three species respectively inhabit the Malay Peninsula, Sumatra, and Java and Borneo.

The red-crested wood-partridge (*Rollulus roulroul*) of Tenasserim and the Malay Peninsula and Islands is the sole member of a genus characterised by the presence in both sexes of a tuft of long hair-like bristles on the middle of the forehead, and the rudimentary claw of the first toe. The male has also a long, fan-shaped maroon crest of hairy feathers, the rest of the head and neck black with a white band between the eyes, the wings maroon glossed with purplish blue, the remainder of the upper parts rich green with steely blue reflections, the under parts black glossed with blue, and the base of the beak, the naked skin round the eye, and the legs and feet scarlet. In the female the head is blackish grey, the wings chestnut, and the rest of the plumage bright grass-green, shading into

greyish green on the under parts. This green colour is an unusual tint among game-birds, found elsewhere only in the males of the blood-pheasants.

These birds, which ramble about the hillsides at an elevation of about three thousand to four thousand feet, in bevies or parties of six or eight to a dozen, are exceedingly swift of foot, never leaving the jungle, and rarely taking wing. Their note is a soft, mellow whistle, chiefly heard in the morning, or when they have been separated. Their nearest ally is the black wood-partridge (*Melanoperdix niger*), ranging over much the same area, but not found in Java. In this species the male has the whole plumage glossy black, while that of the female is mostly chestnut, the scapulars barred with black, and the upper parts, sides, flanks, and under tail coverts mottled with the same colour.

QUAILS

The supreme development of the partridge type of wing is found in the quails as well as in the snow-partridges, in both of which the first flight feather is but little shorter than the second, and equal to the third, while the tenth is much the shortest. Accordingly, when the wing is expanded the vertical angle of an imaginary triangle is formed by the second quill, instead of by the fifth or sixth.

The common, or migratory, quail (*Coturnix communis*) ranges over an enormous area, being found over Europe, Africa, and Asia north of the Himalaya. In Eastern Asia it is replaced by a distinct species, the Japanese quail (*C. japonica*), although during the breeding season numbers of the common species invade those countries, with the result that the two interbreed. The male of the common quail may be recognised by its white throat with a black anchor-shaped mark down the middle, while in the Japanese species the throat is uniformly dull brick-red (pages 1401, 1455). The females may be distinguished by the former species having the white feathers covering the throat short and rounded, while in the latter they are elongated and pointed, forming a beard.

RED-CRESTED WOOD-PARTRIDGE

In South Africa occurs a race of the quail (*C. capensis*) in typical examples of which the males have the sides of the head, as well as the chin and throat, bright rufous chestnut, and the black anchor-shaped mark characteristic of the migratory species well developed. The ordinary quail interbreeds freely with these birds, and the results are seen in many of the males having the chin and throat partially chestnut.

An Australian quail (*Synœcus australis*), representing an exclusively Malay and Australasian genus, has no anchor-shaped mark on the throat, and the under parts marked with V-shaped black bars; an allied species, with the under parts dull rufous, inhabiting Timor and Flores. The genus includes the smallest and most beautifully marked birds of the group, the tail being very short, hidden by the upper tail coverts, and with only eight feathers.

The painted quail (*Excalfatoria chinensis*), typifying a third genus, inhabits the Indo-Chinese countries, especially the lower hills where the ground is swampy and grass-covered. The male has the upper parts brown, barred and streaked with black, and ornamented with whitish shaft-stripes, the forehead, sides of the head and neck, and wing coverts washed with slaty blue, the chest and flanks slaty blue, and the rest of the under parts rich chestnut. In very old birds the chestnut takes the place of the slate, till very little of the latter remains on the under parts. The genus includes four species, whose united range extends from India to the Malay countries, New Guinea, and tropical Africa.

BAMBOO-PARTRIDGES

The bamboo-partridges represent a section of the family with a wing of the type characteristic of the monal pheasant—that is to say, with the first flight feather shorter than the tenth. As already noted, the typical pheasants are the only exception in this respect, but although the shape of their wings is partridge-like, their long tails at once serve to distinguish them. Little need be said of the African rufous-breasted partridge (*Ptilopachys fuscus*), the sole member of its genus, and ranging from Senegambia and the Gold Coast to Abyssinia, where it inhabits broken ground and stony hillsides. The plumage of both sexes is similar, and the male is never armed with spurs; though these appendages are sometimes developed even in the females of the bamboo-partridges.

The three species of the latter have the plumage alike in both sexes, and in general appearance they recall the male of the common partridge, this being especially the case with the Chinese *Bambusicola thoracicus*, in which the sides of the head, throat, and fore part of the neck are chestnut, while the eyebrow-stripe and crop are grey. The nearly allied Formosan *B. sonorivox* may be recognised by having the sides of the head dark grey, while Fytch's partridge (*B. fytchi*) from India, Burma, and China has the eye-stripe buff. The note of this species is loud, harsh, somewhat fowl-like, and different from the low whistle of the tree-partridges. Found in heavy forest jungle, this partridge generally goes in pairs.

SPUR-FOWL

In general appearance the Indian spur-fowl are more pheasant-like than the last, having a rather long tail, and the colour of the plumage different in the two sexes. The legs in both sexes are armed with spurs, the males having two, and sometimes three, pairs, while the females have one or rarely two, although

occasionally two are present on one leg and one on the other. The three species, which are of relatively large size, are peculiar to India and Ceylon.

The male of the Ceylon spur-fowl (*Galloperdix bicalcarata*) is more striking in appearance than its Indian allies, having the plumage of the head, neck, back, and under parts black and white, the rest of the upper parts dark chestnut, ornamented on the wing coverts with white black-edged spots, and the quills and tail black. The female, on the other hand, has the head blackish, the chin and throat white, and the rest of the plumage chestnut, finely pencilled with black.

Colonel Legge observes that "the shy habits of this bird would prevent its being detected in most places where it is even abundant, were it not for its noisy cries or cackling, so well known to all who have wandered in the Ceylon jungles. It frequents tangled brakes, thickets in damp nalas, forest near rivers, jungle over hillsides, and, in fact, any kind of cover which will afford it entire concealment. "The cocks begin to call about six in the morning, and when one has fairly begun, the curious ascending scale of notes is taken,

COMMON QUAILS

up from one to another until the wood resounds with their cries. They always seem to keep in small parties, which perhaps consist of the young of the year with their parents."

The nest, a mere hollow in the ground, is situated in the forest or in thick jungle, under the shelter of a rock or near the projecting root of a large tree, and it would seem that the full number of the eggs laid is four. but the red spur-fowl lays as many as ten.

PHEASANT-QUAIL

The pheasant-quail (*Ophrysia superciliosa*), which occurs in the north-west of India during the cold season, is probably a native of Tibet, but is so rarely met with that scarcely anything is known of its habits. Never coming into the open, it prefers to skulk in the long grass, from which it can only be flushed with the help of dogs, and when on the wing its flight is slow and heavy. This is the smallest member of the pheasant group, being no larger than a quail, but its affinities are with the blood-pheasants, as is shown by the shape of the wing with its short first primary, the length of the tail, and the long rather loose plumage.

In the male the general colour is grey washed with olive, each feather being edged with black, and the head and throat deep black, the former marked with various white bands. In the female the prevailing colour is warm light brown, paler on the under parts and spotted and marked with black, the chin and throat being whitish.

BLOOD-PHEASANTS

The much larger blood-pheasants are represented by three species from Tibet and North-Western China, of which the males are characterised by the peculiar grass-green colour of the plumage. The typical blood-pheasant (*Ithagenes cruentus*) of the higher regions of Nepal, Sikkim, and Bhutan is a handsome bird, the male having the forehead and a ring round the eye black, the crown buff, and the upper parts grey, washed on the wings with green, and margined on the upper tail coverts and tail with crimson ; the cheeks, throat, and under tail coverts being crimson, and the rest of the under parts mostly green, with some of the feathers more or less margined with crimson. The naked skin round the eye and the legs are bright coral-red, the latter being sometimes armed with no less than four pairs of spurs. The female is mostly brown, lighter on the upper parts, and reddish below, with the back of the head and nape slaty grey.

Living at elevations of from ten to fourteen thousand feet, blood-pheasants are abundant in many of the valleys among forests of pine and juniper. Seldom or never crowing, they utter a weak, cackling noise. When put up, they take a short flight, and then run to shelter. During September flocks of ten to fifteen may often be seen, males and females being in almost equal proportions ; and in December they frequently collect in packs of seventy to one hundred.

TRAGOPANS

The tragopans, or horned pheasants, often miscalled Argus pheasants, include five large, magnificent species, unsurpassed for beauty and the harmony of their tints by any other members of the group. The male is provided with a pair of erectile, fleshy, blue horns, inserted on each side of the crown above the eyes, and during the breeding season the throat is covered with a brightly coloured lappet, hanging down several inches when the birds are excited by passion, but barely visible during the winter. Their habitat includes the higher wooded ranges of Northern India and China.

By far the most brilliantly coloured species is the crimson tragopan (*Tragopan satyra*), ranging in the Himalaya from Kumaon to Bhutan. In the male of this splendid bird the top and sides of the head are black, the neck, mantle, and under parts orange carmine, and the rest of the upper parts olive brown, each feather being ornamented at the tip with a round white spot, partially or entirely margined with black, while the outer wing coverts are edged on each side with dark orange carmine. The throat-wattle is salmon-colour with transverse blue bars, and the legs are pale flesh-colour. The general colour of the female is black above, mottled and spotted with various shades of buff, the chin and throat being whitish, and the under parts sandy, finely marked with black and pale buff or whitish shaft-spots.

In summer these birds, writes Mr. A. O. Hume, " are to be found at elevations of from eight thousand to ten thousand feet, always in thick cover, by preference in patches of the slender reed-like ringal bamboo, in the neighbourhood of water. Although always on hills near to or bordering on the snow, they are never seen amongst it, and seem to shun it as much as the blood-pheasant delights in it."

The nuptial dances of this bird have been described by Mr. A. D. Bartlett, who wrote that the " males can only be seen to advantage in the early morning and in the evening, as they conceal themselves during the day ; the females, however, are less retiring in their habits. When the male is not excited, the horns lie concealed under two triangular patches of red feathers, their points meeting on the occiput ; the large wattle is also concealed or displayed at the will of the bird.

"The male has three distinct modes of showing off. After walking about rather excitedly, he places himself in front of the female, with the body slightly crouching upon the legs, and the tail bent downwards ; the head is then violently jerked downwards, and the horns and wattle become conspicuous. The wings have a flapping motion, and the bright red patch on them is fully displayed. The whole of the neck appears to be larger than usual during this action, so do also the horns, which, moreover, vibrate with every motion. This scene is concluded by the bird suddenly drawing himself up to his full height, with his wings expanded and quivering, the horns erect, and the wattle fully displayed. The second mode consists of simply erecting all his feathers, and elevating one shoulder, thereby exposing a greater surface to view, without, however, showing his head-dress. The third mode is by simply standing boldly erect on an elevated perch,

CRIMSON TRAGOPAN

giving the head one or two sudden shakes, and causing the horns and wattle to appear for a few moments."

MONALS

Quite unsurpassed among the pheasant tribe for splendour of plumage, the half-dozen species of monal are characterised by the males having most of the upper parts glittering with metallic colours, and the head, except in Sclater's monal (*Lophophorus sclateri*), adorned with an elongate crest of racquet-shaped plumes (p. 1403). In place of the crest in the latter species, the crown is covered with beautifully curled feathers. The haunts of these birds are practically the same as those of the tragopans—that is to say, the highest forest regions of the Himalaya and other Asiatic mountains, vegetation and considerable altitude being essentials to their existence.

The male Himalayan monal (*L. impeyanus*) has the crest and head metallic green shot with blue and purple, the back and sides of the neck purple shading into reddish copper and glossed with golden green, the mantle and upper tail coverts shining golden green, the outer wing coverts bluish green, the inner feathers, scapulars, and rump bronzy crimson in some lights, and purple edged with bluish green in others, the lower part of the back snow-white, the tail pale chestnut, and the under parts black slightly glossed with green on the throat.

The female, on the other hand, is dressed in a much more sombre livery, the general colour of the upper parts, chest, and sides of the breast being black, with a buff centre to each feather, the lower part of the back and upper tail coverts irregularly barred with the same colours, the sides of the head reddish buff mottled with black, the chin and throat white, and the rest of the under parts mottled with black and buff, and showing more or less distinct white shaft-stripes. The tail is black, barred with rufous and tipped with white.

This species ranges through the forests of the Himalaya from Afghanistan to Bhutan. Mr. Hume observes that " there are few sights more striking where birds are concerned than that of a grand old cock shooting out horizontally from the hillside just below one, glittering and flashing in the golden sunlight, a gigantic rainbow-tinted gem, and then dropping stone-like, with closed wings, into the abyss below."

Another naturalist, Mr. Wilson, writes that " the monal is found on almost every hill of any elevation from the first great ridge above the plains to the limits of forest, and in the interior it is the most abundant of our game-birds. In summer, when the rank vegetation which springs up in the forest renders

MANCHURIAN EARED PHEASANTS

it impossible to see many yards around, few are to be met with except near the summits of the great ridges jutting from the snow, where morning and evening, when they come out to feed, they may be seen in the open glades of the forest and on the green slopes above. At that time no one would imagine they were half so numerous as they really are; but as the cold season approaches, and the rank grass and herbage die away, they begin to collect together, the woods seem full of them, and in some places hundreds may be put up in a day's walk.

" In autumn they all descend into the forest, frequenting those parts where the ground is thickly covered with decayed leaves, under which they search for grubs; and they descend lower as winter sets in and the ground becomes frozen or covered with snow. . . . Still, in the severest weather, when fall after fall has covered the ground to a great depth in the higher forests, many remain there the whole winter; these are almost all males, and probably old birds.

" It may be questioned whether they pair or not in places where they are at all numerous; if they do, it would appear that the union is dissolved as soon as the female begins to sit, for the male seems to pay no attention whatever to her whilst sitting, or to the young brood when hatched, and is seldom found with them. The call of the monal is a loud, plaintive whistle, which is often heard in the forest at daybreak or towards evening, and occasionally at all hours of the day."

The eggs, which are placed in a depression in the ground, scratched by the female under some sheltering rock or massive root, are usually four or five in number, and dull white speckled with red in colour. Chamba is the home of a second species or race, lacking

SILVER PHEASANTS

and breast purplish blue, the lower back and rump bronze-red, and the middle tail feathers white. All the rest of the under parts are black, the feathers of the sides and flanks show white shaft-stripes, the naked portion of the face-skin and wattles are bright blue, the legs and feet vermilion, and the strong spurs whitish. The female has the crest less developed, the general colour of the plumage chestnut, the throat white, the feathers of the neck and chest margined with white on the sides, the breast and sides of the belly black mottled with chestnut and edged with white, and the rest of the under parts white mottled with black. The soft parts are like those of the male, but paler, and there are no spurs on the legs.

One of the most splendid members of the group is Bulwer's pheasant (*Lobiophasis bulweri*) of North Borneo, the sole representative of a genus in which the male has no less than thirty-two tail feathers, and the female two pairs less. In the adult cock the neck and chest are deep crimson, the rest of the plumage mostly black, each feather being edged with steel-blue at the extremity, and the long, curved tail pure white. The head is almost entirely naked, having only a few feathers down the middle of the crown, and ornamented with three pairs of elongate wattles, all being bright blue; the legs, feet, and spurs are red.

the white on the lower part of the back of the common monal, and having the breast and under parts glossed with green.

In the mountains of North-Eastern Tibet and Western China occurs the equally brilliant and even larger *L. l'huysi*, the male of which may be recognised by the black tail glossed with bluish green and spotted with white. Lastly, the splendid Sclater's monal, already mentioned, inhabits the Mishmi Hills in Assam, and has, in addition to its peculiarly feathered head, the whole of the lower part of the back, rump, and upper tail coverts white, and a white band across the chestnut tail. From the absence of a crest, it is sometimes referred to the distinct genus *Chaleophasis*. The habitats of the other two species of monal are unknown.

FIRE-BACKED PHEASANTS

The crestless fire-backed pheasants of the Malay region are about the size of bantam hens, and also resemble these birds in the shape of their rather short and vaulted tails. In two species, *Acomus erythrophthalmus* and *A. pyronotus*, the males have the lower part of the back and rump fiery bronze-red, while the females are entirely black, glossed with purplish or steel blue, and armed with spurs like the male; but in a third kind (*A. inornatus*), from Western Sumatra, the male has the plumage entirely black, and thus closely resembles the females of the other species.

Closely allied are the crested fire-backed pheasants, inhabiting the same countries, but ranging farther north into Tenasserim, the Shan States, and Cochin China. In this group the cock is adorned with a full, erect crest, composed of bare-shafted feathers, supporting a bunch of plumes at the extremity; the sides of the head are naked and covered with large wattles, and the tail is long and shaped somewhat like that of the jungle-fowl.

The male of Vieillot's fire-back (*Lophura rufa*) is further characterised by having the neck, mantle, and

EARED PHEASANTS

The five species of eared pheasants, natives of Central and Eastern Asia, are birds of large size, their loose, hairy plumage making them look even larger than is really the case. In all except *Crossoptilum leucurum*, in which the male appears to have more white in the tail than the female, the plumage of the sexes is alike, although the females lack the blunt spurs of the males. The top of the head is clothed with soft, curly feathers, the sides of the face are naked and covered with warts, and the ear-coverts produced into long white tufts on each side of the head. The tail is long, full, and rounded, the number of feathers varying from twenty to twenty-four in the different species, and the extremities of the middle pair being much curved, with the webs long and free.

In Hodgson's eared pheasant (*C. tibetanum*), of Tibet and Western China, the whole of the plumage is white, with the exception of the black top of the head, the dark brown quills, and the tail feathers, which are purplish bronze towards the base, shading into dark greenish blue and deep purple towards the extremities. The naked sides of the face are scarlet, and the legs coral-red. The Manchurian eared pheasant (*C. manchuricum*) is a somewhat differently coloured bird, with the back of the head and neck black, shading into brown on the mantle and wings, and dirty white on the lower part of the back, rump, and upper tail coverts; the chin and throat, as well as the ear-tufts, are white, and from the brownish black chest the colour gradually becomes lighter on the rest of the under parts.

The range of this species includes the mountains of Manchuria and Northern China. These birds, which inhabit pine forests at an elevation of from ten to twelve thousand feet, are gregarious in habit, forty or fifty sometimes being met with roosting in company on pine trees. Being remarkably hardy, they do well in confinement, and soon become tame.

KALIJ PHEASANTS

The lower altitudes of the middle ranges of the Himalaya, and thence through the Burmo-Chinese countries, are the home of a group of species of pheasants approaching the crested fire-backed pheasants. Nearly a dozen species belong to this group, which includes the kalij and silver pheasants, as well as the somewhat aberrant Swinhoe's pheasant (*Gennæus swinhœi*). All these birds have a more or less elongate recumbent crest of hairy feathers, the sides of the head naked, and the long tail laterally compressed and vaulted, with the middle pair of feathers at least three times the length of the outer ones. The legs of the male are armed with a pair of stout spurs, but in the females these appendages are wanting.

In the most western member of the genus, the white-crested kalij (*G. albocristatus*), which inhabits the Western Himalaya and Nepal, the cock has the long hairy crest white, the general colour of the upper parts black, glossed with purplish and steel blue, and margined, especially on the rump, with white, and the fore part of the neck dirty white, gradually shading into brown on the under parts.

To the eastwards, in Nepal, occurs *G. leucomelanus*, differing only in having the crest black, glossed with purple; while, still farther east, in Sikkim and Bhutan, the darker *G. melanonotus* has the black crest of the latter, but the white terminal margins on the feathers of the rump and upper parts replaced by deep purplish blue. Bhutan, Assam, and Burma are the home of Horsfield's kalij, which is the darkest of all, the whole plumage being black, glossed with purplish or steel blue, and only the region of the feathers of the lower back and rump edged with white, and we may consider this species as probably representing the ancestral stock from which the others have been derived. Among numerous other species, mention may be made of the lovely silver pheasant (*G. nycthemerus*) of Southern China, noticeable for its white upper plumage, ornamented with dark markings.

KOKLASS PHEASANTS

Another Central Asiatic group is represented by the seven different kinds of koklass pheasants, which range through the Himalaya from Afghanistan to Tibet and Manchuria. These birds may be recognised by the long crest of the cocks, and the elongation of the feathers above the ears to form tufts surpassing the crest in length. The sides of the head are feathered, and there is no extensive naked space surrounding the eye; the tail is elongated and wedge-shaped, and the upper tail coverts are long, extending more than half-way towards the extremity.

In the typical koklass (*Pucrasia macrolopha*), of the Western Himalaya, the male has the crest bright buff, a large snow-white patch on each side of the neck, the rest of the head, including the long feathers above the ears and the throat, black, glossed with dark green, the upper parts, sides, and flanks grey, with black middles to the feathers, the wing coverts similarly marked, but browner, and tinged with rufous, and the middle of the breast and under parts dark chestnut. The middle tail feathers are mostly chestnut, the outer pair black shading into reddish brown towards the base, and tipped with white, the beak is black, and the legs and feet are grey or purplish, and armed with a pair of spurs. The female has a much shorter crest, no ear-tufts, the plumage being black, variously marked with sandy rufous and buff, the throat and under parts white, the latter marked with black, and the under tail coverts chestnut, tipped with white.

Of this species Mr. Wilson remarks that it is " common to the whole of the wooded regions from an elevation of four thousand feet to nearly the extreme limits of forest, but is most abundant in the lower and intermediate ranges. The koklass is of a rather retired and solitary disposition. It is generally found singly or in pairs; and, except the brood of young birds, which keep pretty well collected till near the end of the winter, they seldom congregate much together."

Unlike the great majority of their kind, these birds do not separate after the business of incubation is over, and probably pair for life, since at whatever season one is found, its mate is sure to be met with

COMMON PHEASANTS

somewhere in the neighbourhood. Their flight is extremely rapid, more so than that of any other Himalayan pheasant, and when they dart down the side of the mountains it requires an experienced shot to stop them. The nest is placed at the root of a tree, or under some overhanging tuft of grass, and contains from five to nine eggs, resembling those of the monal in colour.

CHIR-PHEASANT

Before mentioning the typical pheasants, it may be observed that the chir-pheasant (*Catreus wallichi*), of the middle ranges of the Himalaya, alone represents an allied genus. Resembling in general form and the shape of the tail the typical pheasants, this bird lacks the bright metallic plumage of the latter, while the wing is of the monal type, with the first primary shorter than the tenth, and the head is adorned with a full large crest, most developed in the males.

REEVES' PHEASANT

TYPICAL PHEASANTS

Inhabitants of low-lying wooded valleys, and including about a couple of dozen of gorgeously coloured species and varieties, the typical pheasants range from South-Eastern Europe across Central Asia to Japan and Formosa. As already pointed out, the wing in all these birds is partridge-like, and differs from the monal type in that the first flight feather is much longer than the tenth ; but, unlike partridges, the tail is long and wedge-shaped—much longer than the wing. The sides of the head are naked, and there is no crest, but the ear-tufts are considerably lengthened in the cocks and the legs armed with a pair of sharp spurs.

The home of the common pheasant (*Phasianus colchicus*) is South-Eastern Europe and Asia Minor, although the bird has for many centuries been established in Great Britain and various parts of the Continent to the west of its original habitat. Although it

ELLIOT'S PHEASANT

is almost unnecessary to describe such a familiar bird, it may be mentioned that in the cock the top of the head is bronze-green, and the rest of the head and neck dark green, shading into purple on the sides and front of the latter ; the mantle, chest, breast, and flanks are fiery orange red with a purplish green margin to each feather, the middle of the back and scapulars mottled and beautifully marked with buff, black, and orange red, the lower part of the back and tail coverts red glossed with purplish lake, and the wing coverts

sandy brown. The middle of the breast and flank are glossed with dark purplish green, the under parts brown mixed with rufous ; the tail feathers light olive green, the middle pair being barred along the middle with black, the naked areas on the sides of the face scarlet vermilion, and the legs and feet brownish horn-colour. The female is mostly sandy brown, marked and barred with black and buff, shading into chestnut on the mantle and sides of the breast.

The majority of the species allied to the common pheasant may be divided into two groups, namely, those inhabiting that part of Central Asia lying to the west of the meridian of Calcutta, which have the rump and upper tail coverts maroon or rufous, sometimes glossed with green, and those found to the eastward of that line, in which the same areas are greenish or bluish slate-colour. In the most westerly representatives the first group, such as the common and the nearly allied Persian pheasant (*P. persicus*)—the latter of which differs in having the wing coverts white, and inhabits the valleys to the south-east of the Caspian—there is no white ring on the neck, but farther eastwards occur other species, such as Severtzow's pheasant (*P. chrysomelas*), from the Amu-Daria, and Shaw's pheasant (*P. shawi*), of Yarkand and Kashgar, in which the white ring, though absent in the typical examples, is in many individuals distinct or represented by a few white feathers. Farther north, along the valley of the Sir-Daria and ranging east through Turkestan to the valley of the Black Irtish, occurs the Mongolian pheasant (*P. mongolicus*), and still farther eastward, in Zungaria, the allied *P. semitorquatus*, in both of which a wide and nearly complete white collar is present. In the eastern species with the rump slate-coloured a very similar arrangement occurs, the western and more southern species having little or no trace of a white ring ; although in the Chinese pheasant, ranging from the Amur, Manchuria, and Eastern Mongolia through Eastern China, and its ally, *P. satscheunensis*, from the north of the Nan-Shan Mountains, as well as in the Formosan pheasant (*P. formosanus*), the white ring is well developed.

It will thus be seen that the more northern species of both the rufous and grey-rumped groups have a white collar, while in the more southern species of both

CHINESE PHEASANTS IN COVERT

types this is absent, or at best ill-defined. Since it cannot be considered that the individuals with traces of the collar found among the southern species are the results of interbreeding with the northern ringed species, when their ranges are separated by chains of mountains, it may be inferred that the original stock was probably of northern origin, and, like birds now inhabiting higher latitudes, possessed a white ring; that as the species spread gradually southwards this characteristic, from some cause or other, has been lost, but that numerous individuals still show traces of a reversion to the ancestral type.

Of the aberrant species, mention may be made of the strikingly handsome Japanese pheasant (*P. versicolor*), with the under parts uniform metallic green, Elliot's pheasant (*P. ellioti*), from the mountains of South-East China, and Hume's pheasant (*P. humeæ*), from Upper Burma and the Shan Hills. In the two latter the lower part of the back is black barred with white, and there are only sixteen, instead of the normal eighteen, tail feathers. Still more different are Sömmerring's pheasant (*P. soemmerringi*), of Japan, which has the plumage chestnut shot with purplish carmine and fiery gold, and Reeves' pheasant (*P. reevesi*), of Northern China, with its white crown, black collar, tawny plumage, and a tail fully 5 feet in length in old males. All the members of the genus are polygamous, each cock pairing with several hens.

GOLDEN AND AMHERST'S PHEASANTS

By far the most gorgeously adorned members of the whole pheasant family are found in the genus which includes the golden and Amherst's pheasants (*Chrysolophus pictus* and *C. amherstiæ*), of the mountains of Eastern Tibet and Western and Southern China. The characters distinguishing the cocks of this splendid group are the long, full crest of hairy feathers and the cope-like mass of feathers covering the back of the head and neck, as well as the long tail and its greatly lengthened upper coverts (page 1457).

The cock Amherst's pheasant, although possessing fewer brilliant colours than the golden pheasant, has the colours purer and more harmoniously blended. The top of the head, mantle, scapulars, and chest are dark bronze green, the feathers of the long crest blood-red, those forming the cape pure white, margined and barred across the middle with black glossed with steel-blue, the lower part of the back and rump widely tipped with yellowish buff barred with dark green, and the long upper tail coverts white, irregularly barred with black and widely tipped with orange scarlet. The wings and under tail coverts are mostly black with dark purplish green reflections, the long middle tail feathers with arched bars and wavy lines of black, the throat and fore part of the neck brownish black slightly glossed with green, and the rest of the under parts white barred on the flanks with black.

Unlike the golden pheasant, both sexes have a patch of naked blue skin surrounding the eye; but the female has none of the brilliant plumage of the male, the general colour of the upper parts in the former sex being rufous and buff, marked and barred, especially on the wings and middle tail feathers, with dark brown; the outer tail feathers being chestnut mixed with black and barred and tipped with white, and the breast and under parts mostly pale buff, barred on the breast and sides with dark brown. This species is frequently imported from Western China and Eastern Tibet to Europe, where, being of a hardy nature, it thrives in aviaries.

JUNGLE-FOWL

In jungle-fowl which inhabit the Indo-Malay countries and many of the adjacent islands the cocks differ from all the other members of the present group in having a high, fleshy comb extending along the middle of the head from the base of the beak, and the sides of the face, chin, and throat naked and provided with one or two more wattles. The cock of the red jungle-fowl (*Gallus banciva*), with its serrated comb and double-wattled throat, closely resembles its domesticated descendant the game-cock in the colours of the plumage, and is a common denizen of the well-watered jungle country of the lower ranges of the Himalaya, from Kashmir to Assam, and parts of Central India, especially in the vicinity of scattered cultivation. It is also found throughout the Malay Peninsula and eastwards to Cochin China, Sumatra, Java, and the Philippines.

When running or feeding, jungle-fowl droop the tail, but when challenging their rivals or paying their addresses to their mates they carry it erect like domesticated cocks. Of all their kind, these birds, even in a wild state, are the most pugnacious, the cocks often fighting till one or other of the combatants is killed.

Besides the above, several other species are known, such as the Ceylon *G. lafayetti*, the grey jungle-fowl (*G. sonnerati*) of India, which produces the hackles so much in request for making salmon-flies, and the green jungle-fowl (*G. varius*) of Java, Lombok, and Flores, distinguished by having an entire upper margin to the comb, and only a single wattle on the throat.

DOMESTICATED FOWLS
By EDWARD BROWN, F.L.S.

Wide though the distribution is over the whole earth of the domesticated fowl, its original habitat was Southern and South-Eastern Asia, where, from such evidence as is available, it has been domesticated for thousands of years, but where it is still found in its wild state amid the jungles and forests. Thence it has passed to other countries by land and sea, with a value steadily advancing in contribution of food supply. Many modifications have taken place as a result of these forced migrations, for the jungle-fowl is not what is

AMHERST'S PHEASANTS

known as a migratory bird in the same sense as those other members of the poultry yard, the duck and the goose, but the ancestry is assured, the origin fairly certain.

The domesticated fowl belongs to the genus *Gallus*, and all the evidence goes to show that it is descended from the species described above as *G. banciva*, or *ferrugineus*.

For whilst there are variations in colour of plumage and of legs, and, to a lesser extent, in size, the latter species resembles most closely our older type of game fowls, more especially black-reds. Some specimens which the writer possessed recently appeared to be like game in every respect, except that they weighed less than three pounds each, and thus were about midway between the game fowl and the game bantam. In habit, structure of body, and voice this species is closely allied to our common fowls, and, what is of the greatest importance, crosses between the two are uniformly ferile and hardy.

Observations in Asia as well as elsewhere have led naturalists to the conclusion that *G. banciva* is practically alone in its parenthood. To this there is, however, one exception, namely, that the Chinese breeds which came to us about sixty years ago, while they produce, when crossed with wild birds, fertile offspring, yet differ so considerably in structure of wing and in habit that the suggestion has been made that these owe something to a lost species. As to that no proof has been forthcoming, and it is merely a theory with a large possibility. On the other hand, the differences referred to may be the result of a different environment continued over many centuries, for we know that domestic fowls were taken to China at least a thousand years before the Christian era.

The hybrids between the *G. banciva* and the domesticated fowl are so uniformly fertile that the question of origin is scarcely open to question, merely leaving it for reconsideration in case further evidence is forthcoming as to other species.

When the fowl was first domesticated it is impossible with our present knowledge to state. In a Chinese encyclopædia compiled about 1400 B.C., the fowl is referred to, but whether the domesticated fowl, descendant of the wild jungle bird, we cannot tell. The term fowl was probably used for many kinds of birds, as is yet the case.

When, however, we turn to India there is more evidence before us. From the fact that cock-fighting was known as a sport a thousand years before Christ, we have proof that domestication had then taken place. In fact, the sport named has been largely responsible for the popularity of the fowl in many parts of the world and for its dissemination. In this connection it is of interest to note that a reference to the sport of cock-fighting is found in the " Codes of Mann," about 1000 B.C., so that we thus know it has been followed in India for about 3,000 years, and it is still a recognised pastime. In fact, it is only a little more than sixty years since it was prohibited in the United Kingdom, up to which period it held a high place, and it is closely linked with the history of our people and with the national literature for many centuries. Even to-day it is permitted in the north of France.

Whether the earliest domestication of the jungle-fowl was for supply of food or to afford the pleasure of witnessing cocks fight we have no record, and probably never shall be able to determine that point. That the wild males occasionally fight, as do domesticated specimens, is unquestionable, but it is scarcely probable that this fact would lead to taming for that purpose. The most reasonable explanation is that the ancient Asiatics, who had reached at the period named a high standard of civilisation, were accustomed to kill the wild fowls for food, and that learning to know the value of their eggs, nests of young chicks were robbed, and the birds thus obtained brought up in confinement, where, as the numbers increased, the tendency of the cocks to fight, as they could not get away from each other, led them to breed and to rear with the definite end in view of producing birds showing great prowess in this direction.

Were it possible to trace the history of the domesticated fowl right back, we should probably find that from the first breeding in captivity to the time when the sport as such became general, several centuries had elapsed. It was a gradual evolution, growing with the changed conditions of the people as they became centred in villages and towns. This concentration of population we know took place at an early period in the world's history. Our supreme difficulty in finding a solution for the problem is that we know so little of what took place in India, China, and other Asiatic countries. Such meagre records as are available deal more with national than natural history.

Cock-fighting as a sport does not enter into our present study, except that it has considerably modified the fowl and led to the development of meat qualities. But its influence upon distribution was so great that it has had more to do with spreading the fowl far and wide than anything else, at any rate until four or five hundred years ago, when the food or utility value came to be recognised more than had ever been the case before. From such evidence as is available to us, the sport of cock-fighting was very general throughout India and South-Eastern Asia hundreds of years before the Christian era, but so far as yet revealed it was not followed in China, where poultry-keeping has always been mainly for food supply, as is the case to-day.

The migration was westward, though in an unexpected manner. Not by peaceful intercommunication, but by conquest was the fowl started on its westward course. Not by reason of its value as an egg or meat producer, but because of its pugnacity, of its fighting instincts. The natural barriers which hem in the great Dependency to the north and west kept it free for a long period, but its vast treasures tempted the cupidity of the Persian monarchs, who, in the sixth century before Christ, accomplished the invasion and conquest of India. To that event we owe the introduction of the fowl into Persia for its sportive qualities, and ere long it became widely disseminated in that country, as revealed by many old-time records.

Thence it appears to have spread throughout the countries of Western Asia, but how far it is impossible to state. It is essential to bear in mind the wide extent of the Medo-Persian dominion. Probably it was from Persia that the game-fowl reached Phœnicia, the great maritime Power of the Eastern Mediterranean, whence it passed direct to Britain on the ships which came to Cornwall for tin ore, and possibly to Spain and North Africa. Two hundred years after the conquest of India the Greeks invaded Persia, and although it is as yet undecided whether fowls had not been known in Greece before that time, it is evident the enterprise of Alexander the Great made it popular in that kingdom, and again for its fighting quality. There it was termed the Persian bird. Many references are found in ancient literature, from the time of Aristotle onwards, to the domesticated fowl, chiefly dealing with its use for fighting.

GREY JUNGLE-FOWL

Thus we see that in the earlier days the value of the fowl was largely determined by its fighting instincts, though there would be large numbers of both eggs and birds available for food, as only a percentage of those bred would be used in the cock-pit. A further point is that by development of the qualities making for success as fighters the breast muscle was increased. To enable a bird to strike a sharp, hard blow, large wings are all-important, and as the breast meat forms the motor muscles to the wings, increase of one means increase of the other. To that extent, therefore, cock-fighting has been of benefit.

From Greece the fowl had an easy passage to Italy, which was soon to be the mistress of the known world, with free and frequent intercommunication between all the countries owning its sway. But the Roman appears to have been attracted by its value as a contribution to the demand for food luxuries, which led to methods of cultivation and breeding met with nowhere else. It is apparent that in Italy the fowl assumed that

to the connection with Phœnicia already mentioned. From evidence obtained during later years it would appear that whilst all these changes and migrations were taking place in Persia and Greece, in Italy and Western Europe a similar movement was in progression further eastward. Already it has been mentioned that the fowl was known in China many centuries before Christ, but, so far as we are aware, primarily for food supply. Thence through Mongolia and Central Asia migrations evidently took place, but as these lands were not for a long period subject to invasions from far-distant countries, the current was slow, following more the natural course of mankind westward, such as the entry of the Huns into Southern Russia and ultimately to Hungary, and the Mongols into Russia From the class of fowls in Eastern Europe, it may be assumed that those countries received the fowl from Southern Asia rather than Southern Europe.

It will now be of interest briefly to note the result of these various migrations, more especially upon the general characteristics. The *Gallus banciva* is found now in the jungle as three thousand years ago, probably identical in every respect. If, as is generally assumed, all or nearly all our races of domestic fowls have descended from it, whence have so many variations been obtained, not merely in size and shape, ranging from the 12-pound cochin or more to the 12-ounce bantam ; from fowls with 14-feet tail feathers to those with no tails at all ; from the long-legged Malay to the short-legged Courtes Pattes ; from the prolific egg-laying Leghorn to the almost barren Huttegem ; and, finally, whence are derived all the differences in quantity and colours of plumage ? For all these there must be some explanation, which it is our purpose now to give.

That there is a constant tendency to variation in animals and plants even in their original habitat is generally admitted, but that this would have been very restricted is proved by actual facts. Under such conditions a few subraces might have been evolved. These would, however, have differed from the progenitors to a very small extent. We must, therefore, seek for other explanations, which are fully available.

BRAHMA FOWLS

nature which has remained with it throughout all the succeeding centuries.

Thence it passed over Western Europe, even to Britain, for the indications are that we owe what is now called the Dorking fowl, a race which is by no means confined to Britain, as similar breeds are found in France and the Netherlands, to the Romans during the time when their legions were masters of our country When Cæsar, however, invaded Britain he found that the aborigines were keen cock-fighters, due, no doubt,

Assuming that what has already been stated as to the methods of domestication is accepted, it will be evident that fact would tend to production of different types both naturally and by artificial selection. Where natural conditions are equal, uniformity is secured, as is seen in all classes of stock, equally by food obtained and from the fact that such variations as appear are more easily seen by enemies, and cannot, escape with the same facility. Consequently this protective influence is very powerful in maintenance of uniformity. As soon as domestication was secured it ceased to be potent, so that changes would speedily be manifest ; and as they had no real value in the new environment, one being as useful as the other, variation proceeded at an accelerated pace.

Further, the different food supplied, and, in many cases, the nature of the soil on which the fowls were kept, would give an added impetus. In process of time, additional thereto definite selection would take place. The Indian cock-fighters finding that birds with more powerful legs or wings, or having a given colour of plumage, proved to be the best, these were bred from On the other hand, when the rood production became of greater importance, it would gradually be discovered that hens of one kind were the most prolific layers, and of another kind yielded the greatest amount of flesh ; they would consequently be used as breeders. In fact, wherever they are kept in large numbers, compared with the wild state, there must always be involved selection, whether true or false.

Therefore in India itself, whilst the jungle-fowls might retain their original form and type, it will be evident that under domestic conditions variations would be frequently met with, which fact explains partly, at least, the many varieties of the fowl in that and the adjacent countries. When, however, so drastic a change took place as entire removal to other countries where climate and soil were totally different, countries which have produced highly varied types of human beings or animals and plants, it will be apparent that the process of variation would be accelerated to an enormous extent. We may, therefore, accept that the most potent influence in development of new forms, to which the terms races or breeds are applied, has been this forced migration. Therefore we may accept that the chief factor in production of the many breeds and varieties of breeds of the domestic fowls is change from one place to another.

Under ordinary conditions, when the fowl is thus transferred to a fresh environment, the tendency is for variations to take place, some of which may not be met with elsewhere, and in this way new breeds are, in process of time, created. If it were not for what is called artificial selection—that is, arbitrary mating with the object of producing and fixing variations—ultimately one type would be developed, the type that conforms to the special environment found in that district or country, and which would become the barndoor fowl, or *poule commune*. This is arrived at by the elimination of the least fit, which do not survive in the struggle for existence or are killed by enemies. It is here, however, where breeding for special qualities now met with in many countries enters into consideration.

The poultry-keeper, whether his object be sport, or exhibition, or food supply, makes selection of individual birds, and by the mating of male and female seeks to fix whatever is his ideal. We have seen how this would be helpful in producing a better race of fighting-cocks, and it is probable that the Indian cock-fighters were the first to adopt definite and systematic breeding of poultry. Frequently deformities have

JAPANESE
LONG-TAILED BANTAM COCKS

been used in this way, and the quaint forms met with in Burma, China, and Japan were probably due in the first place to abnormalities, neither useful nor ornamental.

Many of the points found in different races of fowls—such as the shape and size of the comb, the colour of the face, the mass of feathers on the head (whether crest or beard), the feathers on the legs of some breeds, and the colour of the plumage—are purely arbitrary, and entirely due to artificial selection, regardless of economic value. These, however, reveal to us the plasticity of the fowl, and how great changes can be attained by mating together specimens showing the same variation or peculiarity. One instance of this will suffice. The dwarf fowl, called the bantam, is an Eastern Asiatic race. From it in this and other countries have been produced, by cross-mating and after selection, a large number of diminutive copies of larger breeds. At first the crosses made are very mixed, but a few show the special characteristics combined with small size of body sought for. By elimination of all those which are farthest from the ideal through several generations it is often possible to attain the end in view, though there are many difficulties and disappointments. Sometimes new races are produced by chance matings, though that is not often the case ; but the majority of modern breeds are due to artificial selection on the basis of natural variations.

Additional to the differences in form, type, and colour of plumage, the combination of which make the respective breeds and varieties, there are changes due to domestication having an economic value. The first of these is, except in bantams, increase in size of body. The jungle-fowl is a small bird weighing $2\frac{1}{2}$ to $3\frac{1}{2}$ pounds. Under domestication, where food is abundant and does not require expenditure of energy to obtain, increase of size would naturally follow, at first by softening of the muscles and fatty deposits, but later by greater weight of skeleton and bone.

Second, is a remarkable advance in the number of eggs produced, more especially in lighter-bodied races, for great fecundity does not accompany large size. The jungle, or wild, hen only lays when she desires to form a nest, and 22 to 24 eggs would be her total for the year. Under domestication, where she has to be fed, such a tale of eggs would not pay the cost of her food. Hence, where profit is desired or household requirements are to be met, selection of the best-laying hens, with a view to increasing the number of eggs per hen, would naturally be adopted as soon as the rudiments of breeding were grasped. Occasionally hens may be found giving 200 and even 250 eggs per annum, but these are abnormal. Averages of 120 to 160 are, however, by no means uncommon.

Third, there has been in many breeds a partial or almost entire suspension of the maternal instinct, in what are called "non-sitters." This is due to disuse and to the fact that the increased egg-production following selection has tended to delay the brooding period and, in some cases, to suspend it.

Finally, it cannot be denied that domestication has tended to reduction of the natural vigour, though more where birds are kept in large numbers or on small areas and where hygienic laws are contravened. That, however, is met with in all animals brought into the service of man. There are other changes as a result of domestication or of altered conditions—namely, colour of shell, varying from deep brown to dead white ; colour of legs, ranging from pure white, through pink and blue and leaden-black, to yellow ; colour of flesh and skin, which is

ANCONAS

white or grey, cream or deep yellow ; and distribution of the edible flesh, in some breeds principally on the breast, in others upon the thighs. But we are unable to more than mention these.

BREEDS

Including bantams, there are known at the present time nearly one hundred distinct races, or breeds, of fowls, in addition to many which have not been accepted as separate breeds. Of these a fair number have several sub-breeds, or varieties, as they are called. For instance, of the English game fowl there are ten varieties, of which five have attained a fair amount of favour. Thus, the total number of breeds and varieties is nearly two hundred, and there are constant additions being made thereto, either by variations in the way already indicated or by definite crossing, or by securing races from other countries hitherto unknown. Examples of the last named are the Leghorn and the Langshan, the former of which was imported from Italy, and the last named from China in the eighth decade of the nineteenth century. The most prominent of the " made " breeds, as they are called, by which is meant that they are a combination of older races, is the Orpington, of which there are four varieties.

In some of these cases, such as certain varieties of the Leghorn, alien blood was introduced with the definite object of securing a different colour of plumage. For some generations the results of such crossing are apparent in that the external characteristics are mixed, derived from the parents on both sides, but by selection for breeding those which in type conform most closely to general appearance of the breed it is desired to maintain, and yet have the coloration of plumage of the other, it is possible in a few years to secure uniformity with the breed of which it is to be a member, and yet be distinct in feather colour and markings. This is, however, a somewhat slow process, and the alien blood frequently reappears. Another and more speedy method has been advocated of late, namely, the Mendelian theory (see Mendel's Law, page 629), which may ultimately alter the entire system of cross-breeding. The practical value of that theory has yet to be proved, though there is abundant evidence to show that it works with considerable certainty in certain directions, chiefly, however, in external characters

CAMPINES

With so large a number of breeds and varieties as already named, it will be seen that there is great diversity of type, not only in respect to size of body and colour of plumage, but also in general structure and productiveness. A considerable number of races have no economic value as regards egg-laying or meat qualities, and are bred for their rarity, peculiarity, or beauty. There can be no question that the system of poultry exhibitions has tended in the direction of popularising breeds which have small value practically, and by exaggerating characteristics which are useless injure the stamina and reduce the productiveness of the breed.

One instance out of many is the Brahma. When first introduced, it was a fair general and good winter layer, and had moderate meat properties. It was in Britain, as it still is in America, a valuable breed for all-round purposes. At that time it had a modest coating of feathers down the outside of the legs and feet. But judges and exhibitors demanded more and more leg and foot feathering, and now show specimens have long, strong feathers, leading to the development of hock feathers and of heavy bone. As a consequence the chickens take much longer to grow, they consume more food, as feather and bone are expensive to produce, when grown they are poor in flesh, and the hens are much inferior as layers compared with the older type. Also in dark Brahmas at one time everything was sacrificed to perfection of pencilling in the hens. Many other instances could be given, but these indicate that the tendency is for exhibition poultry to lose their profitable qualities.

The fancier, as he is called, does not view the question in that light. A hen which only lays twenty eggs in the year, if the majority of chickens bred from these are fit for exhibition, will be more profitable than if she laid 200 eggs capable only of producing table chickens. Thus, we are compelled to relegate a considerable proportion of the known races of fowls to what is called the ornamental class. That breeding for exhibition is a fascinating pursuit cannot be questioned, and that it has done much to extend our knowledge of the workings of Nature is equally true, to which extent it merits commendation, but when carried to the extremes noted it may be destructive of the food qualities, and of the properties for which the great majority of fowls are bred in every country of the world.

It is necessary, therefore, to adopt some method of classifying the races and varieties of poultry, and the method now generally followed is in accordance with the leading characters or qualities. Thus, such as are bred primarily for exhibition, and in which the number of eggs or amount of flesh produced are not sufficient to warrant anyone keeping them for either of these purposes, are included in what is termed the ornamental class, which includes all the varieties of bantams, together with many others, such as the Silkies, Sultans, Malay, Yokohamas, and so forth.

In this connection may be mentioned a point which has never received recognition in Britain, namely, that many of the bantams yield as great, if not greater, return in eggs and flesh *pro rata* to the food consumed than do the larger breeds. In fact, bantam-breeders often declare that they are the most profitable breeds to keep even for food supply, but as the birds and eggs are too small for sale they must be eaten at home. It is their non-marketable size which has prevented recognition of the bantam as a practical fowl. As, however, they can be kept on a small space, many suburban residents would find that they are more than ornamental.

A second large section, or class, are specially valuable as egg-producers, in that they give a higher average than do any others. These are, as a rule, small or small medium in size of body, ranging from $3\frac{1}{2}$ to 7 pounds. This is a very important point. Experience has shown that if, in any breed or family, one hen proves to be more prolific than the others she is less in weight, if not actually smaller in skeleton. And that is true racially as well as individually. The fact is that the food consumed goes to formation of eggs rather than of flesh. Early laying checks growth, and early laying is necessary to prolificacy. Another point is that the heavy-laying breeds are generally non-sitters.

It would appear that nearly all these laying, or non-sitting breeds, as they are called, owe their modern form to Italian influence, which, as already shown, has been great in Western Europe.

In the countries north of the Mediterranean, from Spain to the Transylvanian Alps, the fowls are very largely of one form, variations from which are met with in Germany and the Netherlands. Differences there are, but these are secondary, and are due to changed environment and ideals of breeding. Many varieties have been produced, but the main quality of egg-production is found in all.

Why that should be so is not explained, except that there is always a large and growing demand for eggs, whether in the household or for sale. Hence farmers and others would, consciously or unconsciously, select those which were best in that respect. Light-bodied, active hens are ever the better layers, and with increase of laying the tendency to sitting would be reduced.

HOUDANS

The following is a list of laying races with their varieties, arranged as to their countries of origin :

Breed	Varieties	Country of Origin
Ancona	——	Italy
Andalusian ..	——	Spain
Braekel ..	Gold, silver, black-headed, white, chamois, black, blue	Belgium
Campine ..	Gold, silver	Belgium
Crested Dutch	Black with white crest, blue with white crest	Holland
Du Mans ..	——	France
Hamburgh ..	Black, gold-spangled, silver-spangled, gold-pencilled, silver-pencilled.	Britain
Houdan	——	France
Lakenfelder ..	——	Germany
Leghorn ..	White, brown, black, buff, cuckoo, pile, duckwing, mottled	Italy
Magyar	Black, white, red, speckled	Hungary
Minorca	Black, white	Spain
Polish	Gold, silver, white, chamois, buff	Doubtful
Ramelsloe ..	White, black, speckled	Germany
Redcap	——	Britain
Scotch grey ..	——	Britain
Black Spanish	——	Spain
Transylvanian naked neck	——	Hungary

All the non-sitting breeds lay white-shelled eggs, and the majority are poor in table qualities.

ANCONA—A small-sized fowl, with mottled black and white plumage, yellow flesh and legs, and single comb, the Ancona is a very good layer of average-sized eggs.

ANDALUSIAN—The Andalusian has what is called blue plumage, *i.e.*, slate-coloured; the breast feathers of the cock and body feathers of the hens are edged or laced with black; legs dark; comb single. Prolific layers of large eggs; rather delicate in constitution.

BRAEKEL and CAMPINE—Braekel and Campine are similar in all respects save size of body, which is larger in the former. The comb is single, and the legs dark. Very prolific layers of large eggs.

CRESTED DUTCH—Sometimes called the white-crested Black Polish, the Crested Dutch is kept for its ornamental purposes, is medium in size of body, with dark legs, and has a large globular crest on the head. A good layer, but difficult to keep in good condition.

DU MANS—A large bodied fowl, better in flesh qualities than the majority of this class, the Du Mans has a rose comb, entirely black plumage, and dark legs.

HAMBURGH—One of the most beautiful races of fowls, combining brilliancy of plumage with sprightly carriage, the Hamburgh is small-bodied, has a neat rose comb standing well above the head with a spike behind, and dark legs. The golds have a ground-colour of golden bay, the silvers of silvery white, with black markings (page 1456). A wonderful layer, but eggs small.

HOUDAN—Once very popular in France, the Houdan is large-bodied, with mottled black and white plumage and a good crest, has pinkish white legs and feet, and five toes like the Dorking. Good both in egg and meat qualities.

LAKENFELDER—The Lakenfelder is small in size of body, has black-and-white plumage, and is reputed to be a good layer.

LEGHORNS—Leghorns form the great Italian race; light in body, very active and hardy; have a single comb, yellow legs and flesh; are remarkable egg producers, which are good in size. This breed has proved one of the most useful in America, Denmark, and elsewhere.

MAGYAR—Magyar is a relative of the Leghorn, following it in all respects.

MINORCAS—There are two colours of Minorcas, but blacks are most popular; are large in body, with heavy comb; remarkable layers of very large eggs.

POLISH—Mainly kept for exhibition, Polish are medium in size of body, have dark legs, and carry a large globular crest. Good layers, but difficult to keep in good condition.

RAMELSLOE—The Ramelsloe is not much known, and follows largely the Italian or Leghorn type.

REDCAP—Very like the golden-spangled Hamburgh, except that they are larger in body, Redcaps have bigger combs, and lay larger eggs. Excellent in hilly districts.

SCOTCH GREY—The Scotch grey is a large, somewhat long-legged fowl, with cuckoo—*i.e.*, barred plumage, black on grey; mottled black and white legs; single comb; and a good layer.

BLACK SPANISH—At one time very popular, but no longer so, the black Spanish has an all-white face; a good layer, but delicate.

TRANSYLVANIAN NAKED NECK—A peculiar type in that the neck, from just below the head to the shoulders, is denuded of feathers; the Transylvanian naked neck is hardy, and a good layer.

The third class of fowls are those which are especially good in meat qualities, and, as a rule, these are large, but not very large, in size of body, ranging from 6 to 10 pounds. In some cases the weight is

WHITE LEGHORNS

not much, under ordinary conditions, in excess of the heavier non-sitting breeds, but they have the capacity for laying on a great quantity of flesh naturally in the autumn of the year, or as a result of fattening, while the bones are light and spongy.

Some are fairly good layers, whilst others are poor in that respect. The main point is that the flesh is found more on the sternum, or breast, than on the thighs, and is accompanied by large wings. Consequently, they are prominent in front and deep in body. The better races have white flesh, with either white or dark blue legs, for yellow flesh and legs mean harder meat.

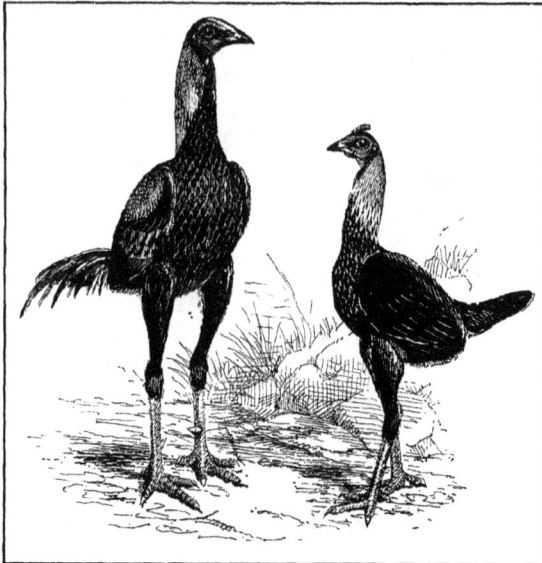

BROWN-RED GAME

They grow quickly in most cases. The following are the leading meat breeds :

Breed	Varieties	Country of Origin
Bresse	Black, white, grey, blue	France
Bruges game..	Red, black, white, blue	Belgium
Courtes Pattes	——	France
Crevecœur ..	——	France
Dorking ..	Dark, silver grey, white, cuckoo red	Britain
Games	Black-red, bright-red, brown-red, blue-red, pile, duckwing, white, black, spangled	Britain
Cornish game		Britain
La Fléche ..	——	France
Sussex	Red, speckled, light	Britain

BRESSE—The great table fowl of France, the Bresse is medium in size of body, which is long, with a single comb and dark legs. Fattens to a great size, and is also a good layer.

BRUGES GAME—A big, heavy-bodied fowl, the Bruges game is slow in growth, but carries a large amount of meat. The colours are as given above. Is used for crossing.

COURTES PATTES—A short-legged fowl, entirely black in plumage, with single comb and dark legs, the Courtes Pattes is deep in body and carries a large amount of fine quality flesh.

CREVECŒUR—The Crevecœur is large in body, heavily feathered, and has a big crest ; dark legs, fine in bone ; is excellent in its meat qualities.

DORKING—Known in Western Europe for two thousand years, from the time of the Roman Empire, the Dorking is square in body, which is long and deep, with large wings ; feet medium in length, pure white in colour, as is the flesh and skin, and with five toes ; single comb, except the white variety, which has rose comb. Requires a dry soil.

GAME FOWLS—Game fowls vary greatly in colours, of which a list is given above. The old fighting types

are medium in size of body, with large wings, and stout, strong legs. These are full in breast and carry a great amount of flesh. The modern exhibition game are long in neck and legs, finer in bone, and have smaller bodies. Of the former especially, some varieties are good layers. All are active in habit, and generally brilliant in colour of plumage.

CORNISH, OR INDIAN, GAME—The Cornish, or Indian, game are very heavy and thick in bone, rather long in the legs, which are deep yellow in colour, as is the flesh and the skin. They are largely brown in plumage, and have a small walnut comb. The breast meat is very abundant, but somewhat hard. For crossing they are very valuable, carrying a large amount of beautifully white flesh. Poor layers of very rich eggs.

LA FLÉCHE—A big fowl, La Fléche is entirely black in plumage, with black legs and a double horned comb. They are chiefly used to produce winter fowls, for which purpose they are excellent, carrying a large amount of beautifully white flesh, and fattening to a great size.

SUSSEX—Sussex is the name applied to a four-toed race found in the South of England, and a relative of the Dorking, which it resembles in table qualities, though not so large and deep. It has pure white legs, feet, and flesh, is fine in bone, and fattens to a great size. The comb is single ; of the colours named above, the red and speckled are the best. The hens are early layers.

The fourth class comprises races which are much more modern, so far as western knowledge is concerned, and owe much to fowls imported from China in the first half of the nineteenth century. Several of the most popular breeds of to-day, notably Malines, Orpington, Plymouth Rock and Wyandotte, were produced by a intermixture of other races with the Asiatics, but they conform more to the latter than to the former. These are among our largest fowls, varying from 8 to 12 pounds, due in large measure to heaviness of bone, as they are long in the leg and massive in build, and as the wings are small in relation to the body, the thighs carry the greater

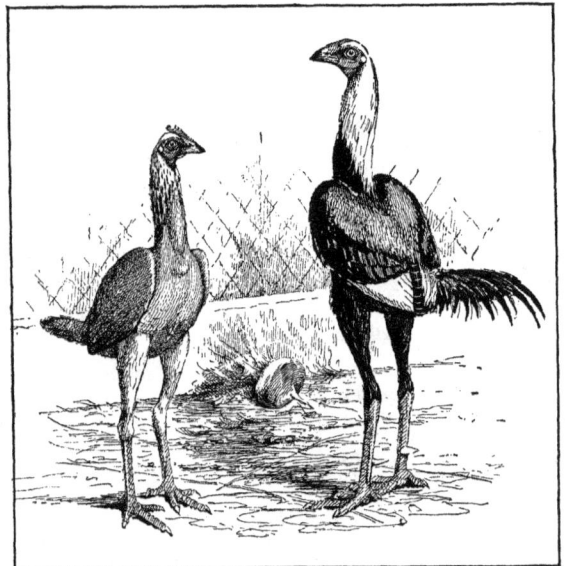

DUCKWING GAME

amount of muscle. In that respect these general purpose breeds do not compare favourably with the class last-named, though there are considerable differences in the various races.

In the greater number the flesh is either yellow or greyish white or creamy white. While they do not give the total number of eggs per annum attained by the non-sitters, in one respect they are superior—namely, that they are better layers in winter, and as their eggs are uniformly tinted in the shells, the value of the produce is considerably enhanced thereby. All the general purpose and nearly all the table breeds are good sitters and mothers, and the fact that they are tame and very hardy explains their popularity. The chief general purpose races known are :

Breed	Varieties	Country of Origin
Bourbourg ..	——	France
Brahma.. ..	Dark, light..	China
Cochin	Buff, black, white	China
Faverolles ..	Salmon, light, black	France
Langshan ..	Black, white, blue	China
Malines	Cuckoo, white, turkey-headed ..	Belgium
Orpington ..	Black, buff, white, spangled	England
Plymouth rock	Barred, white, black, buff	United States
Rhode Island red	——	United States
Wyandotte ..	Silver, gold, white, black, buff, partridge, silver-pencilled, buff-laced	United States

Of the above, the Faverolles, Malines, and varieties of Orpingtons, except the black, have white flesh and skin.

BOURBOURG—A breed found largely in Northern France, said to be related to the Malines, the Bourbourg is large in body, light in colour of plumage, has a single comb, is fair in flesh qualities, and a good winter layer.

BRAHMA—The Brahma was originally brought from Asia, and for many years a most popular breed, as it was very hardy and a good winter layer. It has heavily-feathered legs and feet. Now it is almost entirely an exhibition fowl in Britain, but in America the light Brahma is much used for practical purposes.

COCHINS—Also from the Far East, Cochins are descended from the Shanghaes of sixty years ago. They are large, handsome fowls, rich in colour, with great feather development on body and legs. Poor as layers and for meat properties, they are also chiefly kept for exhibition.

FAVEROLLES—A composite breed of good properties, as it is large, quick in growth, a good layer, and has fair table qualities, the Faverolle in body is upright, rather short, and the favourite colour is salmon, though blacks are found to a considerable extent.

LANGSHAN—As a pure Chinese fowl, the Langshan has evidently been bred in that country for a long time. In the blacks, of which the other varieties are sports, the plumage is brilliant blue black. The legs and feet, which are dark, are slightly feathered. The flesh is creamy white. The hens are excellent winter layers of very rich eggs. There are now two types—the

BLACK LANGSHANS

original Croad, medium in length of leg, and the modern, which is very long in the shanks.

MALINES—This breed is a large, massive bodied fowl, with broad but not heavy bone, tall and deep. The cuckoo and white have single combs. All are white in flesh and legs. They make the famous *poulets de Bruxelles*, as this breed is chiefly kept for meat production. A good winter layer.

ORPINGTON—One of the most popular breeds known, though introduced less than twenty years ago, the Orpington is large in body, which is deep, on medium legs, white in colour, as is the flesh of all save the blacks. The last-named are very dark in legs and feet. The buffs, whites, and spangled are quick-growing fowls, with a fair quantity of flesh, and have the rare combination of white flesh with production of tinted-shelled eggs. Good winter layers.

PLYMOUTH ROCK—A large, tall-bodied fowl, the Plymouth rock has longish yellow legs. The flesh, as in all American races, is also yellow. The barred variety have transverse black bars on steel-grey ground-colour, the others are self coloured. Good winter layers, and fair in flesh qualities, they are very hardy indeed.

RHODE ISLAND RED—A large bird, very rich brown in colour of plumage, with a black tail, the Rhode Island red has the comb single, and in some cases rose. Good layer.

WYANDOTTE—At first there were two varieties of Wyandotte—silvers, in which the ground-colour is silver-white, and golds, with ground-colour of golden bay. In both, the feathers are edged or laced with black, giving a pleasing appearance. Size, medium, having a compact body, on legs of moderate length, which are yellow. The comb is rose, following the line of the skull. Excellent winter layers and fair in meat qualities.

For all the races of fowls which have been taken up by exhibitors standards of excellence are available, giving in detail the respective features and points. These are issued by the Poultry Club, or by the specialist clubs, of which latter there is a large number, formed in the interests of the individual breeds. While the values of external characteristics vary in relation to the difficulty met with in breeding even to a measure of perfection, in the majority of cases they are not antagonistic to the practical qualities except when developed excessively. That there is always a tendency to such excess cannot be questioned, and great harm has resulted wherever found.

The correlation of external characteristics with economic qualities has only been touched very slightly. We know, however, that there are several of the former which are indicative of the latter. For instance, the colour of the leg tells that of the flesh and skin, and to some extent that is true of the beak-colour. Size of comb increases with egg production, and texture of comb tells whether the flesh is fine or coarse.

EDWARD BROWN

POULTRY FARMING

The keeping of poultry as an organised industry has only been recognised for some twenty-five years, but to-day it may be reckoned as the most important of the minor branches of agriculture, and it is believed, moreover, that this industry will continue to be of increasing value to those who make their livelihood on the soil.

As a source of national food supply, and of the greatest good to the largest number, farm poultry keeping must ever take precedence over the more intensified methods as employed on poultry farms. If the greatest benefit is to be derived from the industry it must be followed by all those who are associated with agriculture, for not only will production be greater under such conditions, but the profits accruing will be distributed amongst a larger portion of our rural population.

To regard poultry keeping as a separate industry requires many considerations. Fowls, by their requirements and the effect they have upon the soil, need to be associated with some form of cultivation. If poultry keeping is to continue to hold its present position, and, moreover, to go on increasing its value as a money-making business, it must be regarded as supplemental, and it is along these lines that it will be discussed.

Apart, however, from farm poultry keeping, there are two other sets of conditions under which the work can be followed profitably— namely, by cottagers and suburban residents.

Although the individual production in these cases must of necessity be small, still the accumulation of the produce will in the aggregate be considerable as a means of supplying our home markets, and at the same time, the small wages earned by the classes named will be augmented.

To-day the consumers of England, Scotland and Wales are spending annually upwards of £21,000,000 on poultry produce alone, of which £8,000,000 worth is imported. But, more than this, the demand is increasing yearly, and will continue to do so, for the supplies of foreign eggs and poultry of the better quality are becoming less and less. Never was there a more favourable opportunity for poultry keepers than the present, whether they belong to the farming, cottage, or suburban class.

The meaning of the term farm poultry keeping is one that is frequently misunderstood. It is used simply to describe the system of maintaining fowls on farms and running them in conjunction with the ordinary farm cropping. In one sense this industry has been followed for many generations, but it is only within recent years the belief has become general that poultry kept under such conditions could be made profitable. The birds were formerly kept by the women-folk on the farms, and any profit they were able to make was regarded by them as " pin " money. Although this was considerable in some instances, it must be conceded that, by lack of knowledge and inattention to detail, a full return was not received.

This state of affairs is more common even to-day than is desirable, but all over the country farmers are beginning to realise that fowls, kept under modern hygienic conditions and looked after with the same regularity and care as is bestowed on larger stock, will show as large a proportionate profit as cattle, sheep, or pigs. Before this industry, however, can be raised to a position corresponding to its money-producing capacity, new systems must be adopted and modern methods introduced.

On the production side certain alterations are necessary, the most important of these being with reference to housing, hatching, rearing, and feeding. The common practice of the past has been to house poultry in sheds in the farmyard, or at the best the rick-yard, and in houses frequently much too small for the number of birds kept, which had no pretension as to ventilation or cleanliness. No stock can thrive, much less show profit, when kept under such conditions, and it is a matter of surprise that any returns have been obtained at all.

Fowls are by nature foragers, in that, given the opportunity, they will gather a large proportion of the food necessary for them, but when housed in the

FAVEROLLES

way indicated this was unobtainable. Again, the manurial value of each adult bird is about 1s. per annum, and when the only ground over which they are allowed to wander is the farm or rick yards, this is largely wasted. These two factors together show that the best method is to keep the poultry out on the fields, away from any land that is not being cultivated.

At a low estimate birds at liberty will gather one half of their food, taking the year round, food that is infinitely better for them than any artificial grain or meal that can be supplied. The value of the manure produced being equal to a shilling, housing the fowls on the land reduces the cost by two-thirds as against keeping them in some permanent building within the yard.

By means of portable houses the birds can be moved from one part of the farm to another, as the rotation of crops demands, greatly to their own benefit and the good of the soil. The best type of house that can be used is one measuring 8 feet long by 5 feet wide ; 5 feet high to the eaves, and 7 feet to the gables. The lower half of one of the long sides of the house is boarded, the upper portion being wire-netted, which is covered at will by means of a shutter hinged at the top. The material used in construction is ¾-inch tongued and grooved matching, covered with felt or tarred. A floor of 1-inch wood finishes the house.

The whole is raised on three or four wheels, or on wooden runners, to facilitate moving. Ventilation holes are provided near the gable at each end. The perches are placed lengthwise at the back, and the nest boxes can be fixed under the wire front. Such a house would accommodate twenty-five adult birds or a larger number of growing stock. The initial cost is certainly rather more than when permanent houses are employed, but, considering the increased benefits derived, this is slight. A house, such as that described, can be purchased for from 55s. to 80s., carriage paid, or, if made on the farm, for a less sum.

In the majority of cases when fowls are kept under farm poultry keeping conditions, the work of hatching is carried out in a very haphazard fashion. If fowls are to lay in winter, it is not sufficient that they belong to the

OPEN-FRONTED WINTER LAYING-HOUSE

winter-laying class, for the question as to the time they are hatched is very important. The best period to bring out the chickens must of course depend on local circumstances, but it is usually found that March is the most profitable month.

If the birds are hatched too early they will lay in the summer and in all probability moult the same autumn, not beginning to lay again until after the New Year, and, further, if hatched too late, the stock will be under the laying age when the cold weather sets in, and hence no eggs will be produced until the early spring.

The chief difficulty usually experienced on farms is that there are practically no broody or sitting hens during February, and hence hatching operations are delayed. If hens are not available, then resort must be had to artificial means. Incubators produce such good results and are so reasonable in cost that every poultry-keeper can afford to invest in one or more. The great advantage of a machine is that it is ready for work at all seasons of the year, and, if looked after properly, will not spoil the eggs, as is so often the case with broody

OPEN-FRONTED POULTRY-HOUSE ON WHEELS

hens deserting their nests. It is unnecessary here to give any directions for working incubators, for the makers supply these with each machine.

If sufficient hens can be obtained for hatching during February and the early part of March, they should certainly be used, for it is still a fact that naturally-hatched chickens are better in many ways than those brought out artificially. The appliances necessary are few and inexpensive, and can usually be made at home.

Nest-boxes for hatching are best placed in some quiet shed or outhouse, away from the other stock, and where the hens can be kept warm. It is a mistake to sit hens in the ordinary fowlhouse. A very useful double nest is made 2½ feet long by 15 inches wide and 15 inches high, and divided into two compartments. The walls are made solid, the top is in two halves, and fitted loose to form lids, and the bottom is covered with 1-inch mesh wire netting. Ventilation holes are bored at the top and bottom of the sides to ensure a sufficient circulation of air. To make the nest a few shovelfuls of earth are placed in the bottom, hollowed out saucer-shaped, so as to prevent the eggs rolling away to the corners, and covered with a little broken straw.

All hens should be allowed to sit for at least twenty-four hours on dummy eggs before they are entrusted with good ones, so as to accustom them to their new surroundings. When once settled it is better to place the eggs under each hen in the evening. To feed each day, lift the hen off by the wings and then either tether by the leg to a stake in the ground, or, better still, place in a special feeding-cage, and give a mixture of maize and barley, water, and a dust bath.

Whether the eggs are being hatched naturally or under artificial conditions, on the seventh day it is wise to test them for fertility. This is done by passing the eggs one by one before a candle or lamp and noting their condition. If infertile they will appear translucent, if fertile the centre will be dark, shading off to light at the edges. In those eggs which have white clear shells in all probability the germs and the surrounding network of veins will be discernible. The infertile eggs can be used for feeding the chickens during the first few days of their existence.

The great advantage of testing out on the seventh day is that if a number of the eggs prove infertile, and two or three hens have been set at the same time, the remaining eggs can be placed under one or two hens, as the case may be, and the other hen re-set. This is a great saving of time when hens are scarce. The advantage noticed when the eggs are in a machine is that better results are attained when the infertile eggs are discarded as it leaves that more space in the drawer, thus giving additional air to the live embryos remaining.

When only a limited number of chickens are reared annually—say, one or two hundred in the spring—natural means should be employed. In March there are usually sufficient sitting hens available, and since it is not necessary for these to sit out the entire period before they will take to a brood of chickens, they can be used for rearing alone. Even with incubator-hatched chickens natural rearing can be followed most successfully. After a hen has been sitting on dummy eggs for three or four days, if chickens are placed under her in the evening she will generally take kindly to them.

The most serviceable form of coop for rearing is that known as the double coop. This is 4 feet long, 21 inches wide, 18 inches high in front, and 15 inches high at the back. It is divided into two compartments, one 21 inches long and the other measuring 27 inches. The smaller section is for the hen, the larger for a scratching-room for the chickens. The ends, back, top, and front of the smaller compartment are made solid, the partition and the front of the larger section are fitted with slats 2 inches wide and 2 inches apart. The centre slat in each case is made loose to form a door. A strong movable wooden floor is fitted. The top is best made in two parts, both opening to form lids ; the solid front of the smaller part is also fitted as a door. The advantages of such a coop are many, for the hen can be kept apart from the brood, the latter can be fed alone, and the second compartment forms a good exercise ground for the youngsters during bad weather. The cost of such a coop, if made at home, should not exceed 3s. 6d.

Of the two recognised systems of feeding, that known as the "dry" method is the better. It consists of feeding right from the beginning with a mixture of small seeds and grains, which are scattered in the chaff used as litter on the floor of the scratching-room. This is continued for a month, when one feed of soft food is given first thing in the morning, and the mixture of grains fed for the rest of the day. When six or seven weeks old two feeds of soft food, augmented with two of grain are sufficient. Green food, water, and grit are essential from the beginning. The advantages possessed by this method of feeding over the "wet" system are that, having to scratch for all their food, the chickens obtain plenty of exercise, and it also saves a considerable amount of labour in feeding.

If the birds are kept at liberty the question of feeding the adult stock is rendered more easy, for a small feed of soft food in the morning and a feed of grain at night will prove ample. The most suitable meals to employ for the soft food are barleymeal, toppings (sometimes called middlings, thirds, pollard, and sharps), cooked potatoes, steamed clover-hay chaff, and house refuse.

The evening feed can be made up of barley, oats, buckwheat, and in winter a little maize ; wheat is too expensive a food for poultry unless there is any tail wheat produced on the farm, in which case it makes a very suitable addition. Only a few foods have been mentioned, but practically any crop grown on the farm may be utilised, and attention should be paid to the special feeding value of each food given.

Under ordinary conditions winter egg-production is the best paying branch of farm poultry keeping, but under certain circumstances the preparation of table poultry can be made profitable. In marketing the eggs the following points should be borne in mind :

1. Keep the nests clean, so that the eggs will not require washing, thus losing their bloom.

2. Grade the eggs into two or three sizes after discarding the very large and small ones.

3. Pack them in clean cases, that will travel safely and be easily unpacked ; and

4. Market them when they are fresh. The last-mentioned point is very important, since for the best trade eggs must not be more than three days old when they reach the retailer. All eggs should be marketed twice a week in winter, and three or four times during warmer weather. As stated before, under certain circumstances the fattening of poultry can be made profitable under farm poultry keeping conditions, therefore a few words on this subject may not be out of place. The best demand for chickens is in the spring of the year, the highest prices being realised from February to May. These birds require to be about 3½ pounds in weight, and to be in good condition.

The birds—usually the surplus cockerels and some of the too early hatched pullets — are taken when two and a half to three months old, placed in fattening cages, and fed specially for flesh production. They are given two feeds a day, or three in the case of the youngest, of soft food prepared from ground oats mixed to the consistency of thick cream with soured skim milk. This is continued for some ten days, or until the natural appetite of the birds begins to fail, at which time they can either be killed off in half fat condition, or the process can be continued by artificial means for another period of like duration. If this is done, the same food is used, but in addition a quarter of an ounce of fat per bird per day is given increasing to half an ounce by the end of the time.

Cramming, as it is called, can be carried out in one of three ways—namely, by pellet feeding, funnel feeding, or by means of a cramming machine. If a large number of birds are to be operated on, the last-mentioned method should be employed, but if the number is limited, either of the others will produce as good results. Before killing, the fowls should be starved for twenty-four hours, so as to empty the digestive tract of food. If this is not done the birds will deteriorate rapidly. Killing is best accomplished by dislocation of the neck, and plucking should follow immediately. Dry plucking is preferable to feathering by the wet method, as it leaves the skin in better condition. Packing should never be carried out until the body warmth has been lost, and should then be in well-made wooden cases or shallow baskets. WILLIAM BROWN

PORTABLE POULTRY LAYING-HOUSE

PEACOCK-PHEASANTS

The handsome peacock-pheasants (*Polyplectron*) present a type characterised by the large, full, and rounded tail ornamented with metallic eye-like spots, the sides of the face naked, or nearly so, and the legs of the males armed with two, and sometimes

PEACOCK-PHEASANT

three, pairs of spurs. Having a distribution very similar to that of jungle-fowl, they extend into India only so far east as Darjiling, and inhabit the dense hill-forests ranging from a little above sea-level to an elevation of some six thousand feet.

The grey peacock-pheasant (*Polyplectron chinquis*) of the Indo-Burmese countries is a remarkably handsome bird, the male having the general colour of the upper parts brown dotted all over with dirty white, and each of the feathers of the mantle and wings ornamented with a large, round, dark green eye-spot, showing violet, purple, and blue reflections, and edged with successive rings of black, brown, and dirty white; the upper tail-coverts and tail feathers being similarly ornamented with pairs of oval spots, situated on each side of the shaft at some distance from the extremity, and appearing wholly green in one light and purple in the other. The throat is thinly covered with white feathers, and the rest of the under parts are brown with irregular, mottled, and dotted bars of dirty white; the naked skin on the sides of the face being pale fleshy yellow. The female is darker, and has the eye-spots on the back and wings represented by black spots slightly glossed with purple, while those of the tail are only present on the outer feathers and are much reduced in size.

Mr. Clarke states that this species "is common in the north-east of Cachar, where it is found in dense bamboo-jungle, on the sides of ravines, and on the tops of the lower ranges of hills, wherever there are jamum trees, as well as on the banks of the river Barak, wherever it is well wooded. On the rocky faces of the Barak banks there is a tree which, during the rainy season, is partly submerged, but in

cold weather bears a fruit with seeds like those of a chilli. On these the birds feed greedily in the early morning and towards sunset; insects and worms, with this fruit, form their chief food, but I have on one occasion found small land-shells and pebbles in the stomach of an adult male."

ARGUS-PHEASANTS

Among the most remarkable representatives of the family are the splendid argus-pheasants, distinguished by their large size, enormously developed and eyed secondary quills, which far exceed the outer flight feathers in length, and the extremely long middle tail feathers. The shape of the wing is specially remarkable, and may be regarded as representing the extreme developments of the monal-type; the first flight feather being the shortest, and the tenth the longest, or exactly the reverse of what obtains in quails and snow-partridges.

The typical argus (*Argusianus giganteus*) is a native of the luxuriant forests of Siam, Tenasserim, the Malay Peninsula, and Sumatra, while in Borneo the smaller Gray's argus (*A. grayi*) takes its place. In the cock of the former the naked areas of the sides of the head, throat, and fore part of the neck are dark blue, the feathers on the crown and the short crest black, the upper parts beautifully chequered, mottled, or spotted with black and buff, the chest rufous barred with black, and the rest of the under parts black with wavy bars of chestnut and buff. The primary wing feathers are ornamented on the outer webs with closely approximated rows of black and rufous spots, and the basal part of the inner web shows a rufous band minutely dotted with white and margined by a yellow black-barred line. The outer webs of the enormous secondary quills are adorned with a series of large white, yellow, and rufous eye-like spots, each surrounded by a black ring.

In total length the cock measures 6 feet from the beak to the end of the tail. The female has the general colouring of the male, but lacks the beautiful ornamental marking, as well as the enormously developed secondaries and middle tail feathers. Mr. Davison

THE PEACOCK

writes that these pheasants are quite solitary, every male having " his own 'drawing-room,' of which he is excessively proud, and which he keeps scrupulously clean. They haunt exclusively the depths of the evergreen forests, and each male chooses some open level spot—sometimes down in a dark, gloomy ravine, entirely surrounded and shut in by dense cane-brakes and rank vegetation, sometimes on the top of a hill where the jungle is comparatively open—from which he clears all the dead leaves and weeds for a space of six or eight yards square until nothing but the bare clean earth remains, and thereafter he keeps this place scrupulously clean, removing carefully every dead leaf or twig that may happen to fall on it from the trees above.

" These cleared spaces are undoubtedly used as dancing-grounds, but personally I have never seen a bird dancing in them, but have always found the proprietor either seated quietly in, or moving back-wards and forwards slowly about them, calling at short intervals, except in the morning and evening, when they roam about to feed and drink. The males are always to be found at home, and roost on some tree close by."

A nearly allied pheasant is the still more splendid Rheinhard's argus (*Rheinhardius ocellatus*), from the mountains in the interior of Tonkin, in which the secondary quills are not longer than the primaries, though in the cock the middle pair of tail feathers is enormously lengthened, wide at the base, and tapering to the extremity. The cock of this species measures about 7 feet, from the bill to the end of the tail.

PEA-FOWL

The gorgeously-coloured pea-fowl differ from all the birds already noticed in having the upper tail coverts developed into a long train far exceeding the tail in length. The common species (*Pavo cristatus*) of India, Assam, and Ceylon is too familiar to require description, but in the Indo-Chinese countries, ranging in the north from Chittagong westwards through Siam to Cochin-China, and south through the Malay

ARGUS-PHEASANT DISPLAYING

Peninsula to Java and possibly Sumatra, there occurs the Burmese pea-fowl (*P. muticus*), the cock of which is distinguished by having the crest feathers more elongated and equally webbed on each side of the shafts, while the wing coverts and scapulars are black.

Widely, though locally, distributed over the whole of India, the common species prefers jungly ground in the neighbourhood of water and cultivation, but does not as a rule, range to an elevation of more than four thousand feet, though it has been obtained as high as six thousand. In India the Hindus regard pea-fowl with a superstitious reverence, and object to their being shot; and in native Hindu States, the prohibition being absolute, they are unmolested either by Europeans or natives. In a variety known as the japanned pea-fowl the whole of the wing coverts, scapulars, and second-aries are brownish black, glossed with purple and edged with green, and the thighs black in-stead of buff It closely resembles hybrids between the two species al-ready mentioned, but arises inde-pendently in flocks of the com-mon pea-fowl which have been pure bred for

ARGUS-PHEASANT

years. Possibly it may be a case of reversion to the ancestral type, as it is unknown in a wild state.

GUINEA-FOWLS

A separate subfamily, the *Numidinæ*, is constituted by the guinea-fowls, which are the sole representatives of the pheasant tribe in the African continent. Among their distinctive features is the similarity in the colour of the plumage in both sexes. Before passing to the better-known genera, brief reference may be made to two rare West African forms, of which the black guinea-fowl (*Phasidus niger*), occurring between Cape Lopez and Loango, is smaller than the common guinea-fowl, and has the whole of the plumage blackish brown, obscurely pencilled with brown. With the exception of a band of black feathers from the base of the beak to the occiput, the head and neck are almost entirely naked, the skin being yellow shading into orange on the throat and neck. In the cocks the metatarsus is armed with a pair of stout spurs, thus showing an approach to the pheasants.

The second of the twain is the turkey guinea-fowl (*Agelastes meleagrides*), met with from Liberia to the Gabun, recognisable by the whole head and neck being naked. On the head the bare areas are red, but they become darker on the crown and hind part of the neck, while the lower portion of the neck is milky white; the mantle and chest are white, and the rest of the plumage black finely mottled with white. As in the first-named species the legs of the cocks are each armed with a pair of short stout spurs.

The typical guinea-fowls, represented by about a dozen species, are particularly characterised by the naked head surmounted by a more or less elevated bony helmet, the presence of a pair of wattles at the angles of the gape, and the black and white spotted plumage. These birds are distributed all over Africa, except the more northern parts, as well as Madagascar. The common species (*Numida meleagris*), which is a native of West Africa, where it ranges from Senegambia to the Gabun, may be distinguished by the wide purplish grey collar covering the upper part of the mantle and chest. The bare skin on the sides of the face, neck, and chin, as well as the wattles, is red, and the rest of the neck bluish. Like the rest of their tribe, these birds are highly gregarious, often collecting in large flocks, particularly on grass-covered plains bordering forests. Extremely shy and difficult of approach, they always prefer to escape by running, and in this respect have few equals. The Abyssinian guinea-fowl (*N. ptilorhyncha*), which extends into equatorial Africa, is peculiar in having a bunch of horny bristles at the base of the upper half of the beak. In the right-hand figure of the illustration is represented the crested

guinea-fowl (*Guttera cristata*), one of several species belonging to a group characterised by having a well-developed crest of black feathers, the general colour of the plumage being black spotted with pale blue, and the first four or five secondary quills margined with white, so as to produce a white band along the wing when closed. This handsome species is further characterised by the uniform black collar covering the upper part of the chest, and also by the cobalt-blue colour of the naked skin of the head and neck, that of the chin and throat being red.

This is another West African bird, ranging from Sierra Leone to the Gold Coast; its habits appear to be similar to those of the common guinea-fowl. Allied kinds of both the above-named genera inhabit Southern and Eastern Africa, but require no special mention on the present occasion. A still more striking and remarkable bird is the so-called vulture-like guinea-fowl (*Acryllium vulturinum*), a native of Eastern and possibly Western Africa, characterised by the head and the upper half of the neck being naked, and covered with cobalt-blue skin, with the exception of a horseshoe-shaped band of velvety reddish brown feathers round the nape. The feathers of the neck, chest, and mantle are developed into long black pointed hackles, with white shaft-stripes and cobalt margins, the rest of the upper parts being black, minutely dotted all over with white, and marked with small round black-edged spots :

THE PEACOCK

VULTURE-LIKE GUINEA-FOWLS

the sides and flanks are also similarly marked, but washed with purple, and the breast and under parts are cobalt-blue, and black down the middle. The tail feathers resemble the upper parts in markings, but the middle pair is much elongated and pointed. The male is similar in colour to the female, but larger, and with four or five wart-like knobs on each leg.

That the guinea-fowl has been domesticated for a very long period is proved by the fact that it was known to the ancient Greeks and Romans. It is generally accepted by naturalists that the species *Numida meleagris* of West Africa is the progenitor of our domesticated species, and the name was given after the district in which it is found.

TURKEYS

A second small subfamily (*Meleagrinæ*) is represented by the well-known turkeys, all of which are originally natives of North and Central America, where two or three distinct species are known. The Mexican, or common, turkey (*Meleagris gallopavo*) inhabits the tablelands of Northern Mexico and the neighbouring states, and may be recognised by the broad white tips to the upper tail coverts and tail; while in *M. americana*, or *intermedia*, of the eastern states, which may be only a local race, the same parts are dark chestnut.

The handsomest member of the group is, however, the eyed turkey (*M. ocellata*) of Guatemala, Yucatan, and British Honduras, in which each of the tail feathers is ornamented with a greenish blue eye-spot shot with purple, while the metallic parts of the body feathers are golden or bronze green, and the naked head and neck blue covered with red warts.

Like its allies, the common turkey is polygamous, the female only attending to the duties of incubation, while the male, in addition to neglecting such labours, is even reported to destroy the eggs and young chicks. It seems probable that in the wild state turkeys raise two broods of young in a season, as birds of all sizes may be seen in the masting season (October), when they congregate in large numbers in the cañons to feed on a small, bitter acorn found in the cañons and parks of Southern Arizona and the districts farther southward. Their roosting places are frequently situated in sycamore trees. The name "turkey" appears to have been originally applied to the guinea-fowl, and probably means the bird procured from the Turks or Moors. To the French the turkey is known as *dinde* (the bird of India). Turkeys are heavy, lumbering birds, specially characterised by their tiresome "gobbling" cackle. The head is covered with bare, warty skin, and furnished in the typical species with a large red wattle, generally pendent, but partially erectile during excitement.

DOMESTICATED TURKEYS

Until the discovery of America in 1492 the turkey was unknown in the Eastern Hemisphere, as it is not found in a wild state except on the American continent,

COMMON GUINEA-FOWL AND CRESTED GUINEA-FOWL

where it had been domesticated long prior to the advent of the Spanish discoverers. It would appear that when the early adventurers reached the Western Islands, now known as the West Indies, they found these birds. On returning home, they brought specimens back with them, which they would wish to do as proof that new countries had been reached. It is suggested that the first introduction of this bird was 1499 A.D. Records, however, are somewhat vague as to the actual date. When Cortes invaded Mexico he found the turkey bred in large numbers under domestication, and more common than any other species of poultry. As the continent was further explored, it was found that the wild birds were widely distributed.

The progenitor of nearly all the European and Asiatic races of domesticated turkeys—with which it breeds freely—is the above mentioned *Meleagris gallopavo* or *mexicana*, which is a large-bodied bird on rather short legs. It has a large, well-expanded tail, on which are zones of brown and black, but the ends are white, as are the tail coverts. The body colouring is black, bronze, and white, rather darker towards the flanks, the various colours forming bars across the feathers, which are in many cases tipped with white. In appearance it is very massive, and with large wings carries a great amount of meat or muscle on the breast. The bronze colouring is not so prominent as in the species found further north.

While it cannot be doubted that both this species and *Meleagris americana* are closely allied, such distinctions as are met with being the result of different environment, they are sufficiently distinct to be given separate classification. The latter species are widely distributed in the forests and woods of what are now the United States and Canada, where it is yet found, though not to the same extent. The occupation of the land due to rapidly increasing population and reclamation for cultivation has driven it backward. But there are many States in which it may yet be shot wild in the woods. It is a very large bird, and specimens of 50 pounds have been killed, though such weights are exceptional. The chief differences between it and the *M. mexicana* are that it is longer in the leg, and there is much more bronze and less white in the plumage, giving it a brilliant hue, which in the sunshine glistens very brightly. It is about 100 years ago since its domesticated descendants were first brought over to Britain, and by their size and vigour the Bronze Americans, as they are called, are among the most popular of all our turkeys. *Americana* breeds freely with all the races, thus confirming what has already been stated.

Of the two other members of this family found in Central America, namely, the Honduras (*Meleagris ocellata*) turkey and the crested turkey of Mexico, so far as is known, neither has been used in production of our domesticated races, though both have been kept and bred in a tame state. It is unnecessary to deal with them at greater length.

The peculiar head of the turkey gives it a distinctive character from that of all other poultry, being what is known as carunculated, that is, the head and also the higher portion of the neck are bare, the head blue on top and red below. Where the beak joins on to the head springs a long piece of flesh, capable of elongation, at the end of which are a few feathers. There is not a comb as seen in the fowl, but below the beak are wattles. In the male there is a tuft of feathers springing from the upper part of the breast.

It may be explained that the name given to this species in Anglo-Saxon countries has nothing to do with the origin. It is, however, very misleading, and many writers, including Dr. Samuel Johnson, assumed therefrom that it came from South-Eastern Europe. How such a designation came to be applied was probably that the earliest specimens seen in England were brought by ships trading with the Orient, but, calling at a Spanish port on the way home, these novel birds were purchased. Whether inadvertently or from design the name was given cannot be discovered. In Spain and nearly all other countries the designation means Indian fowl, due to the idea that Columbus had rounded the globe, as he called the lands discovered by him the West Indies.

Change of environment and of conditions have brought about considerable variation, and from the two ancestral types named above we have a fair number of races, which are recorded below. It is necessary to mention that the turkey resembles the fowl, not merely as to structure of body, but also as to habits. It is not a migratory bird in the same sense as the duck and the goose. One main difference is that when desiring to lay it seeks a secluded place, and does not like to sit in an ordinary house, though it will do so if compelled. Probably when it has been domesticated as many centuries as the fowl this may be changed, for it will have learnt that there is not the same danger of discovery by enemies as under the wild conditions.

MEXICAN TURKEY

That there has been some increase in egg production is evident, though small as compared with fowls and ducks, for turkeys would not be profitable as egg-layers. Such increase is due to the custom of taking away the eggs as they are laid. The chief change is reduction in size of body, thus differing from other species of domestic poultry. The reasons for such appear to be varied. As a rule wild turkeys are found where natural food is abundant, congregating in flocks. Thus they had not much to gain in that respect. Second, for ordinary purposes they are too big for a single household, and must either be bred smaller or sold in parts. As the demand increases the need would be more for smaller specimens. Further, this bird does not attain its maturity until it is three years old, but breeding from yearlings has been a common practice, which has tended to reduce the size of body, but also, unfortunately, to lower the vitality. Thus the belief is justified to some extent that the turkey is a delicate bird, especially when young.

VULTURE-LIKE GUINEA-FOWL

CRESTED GUINEA-FOWL

NORTH AMERICAN TURKEY

The best-known races of turkeys at the present time are given in the following table :

Breed	Colour of Plumage	Country
Black Norfolk	Black, with white or brown tips	England
Black	Glossy black	France
Grey	Mixed black and white	Various
White	Pure white	Various
Blue	Slate blue	France
Cambridge bronze	Dull bronze with grey tips	England
Mammoth bronze	Brilliant bronze	America
Narragansett	Black, with grey markings	America
Fawn	Buff	Europe
Ronquieres	Black, grey, and tawny	Belgium

In respect to size, the mammoth bronze are the largest, and they have also proved very hardy. That bird carries a large amount of flesh. The Cambridge bronze stands next, but it is softer and finer in the flesh, and is thought to be the best of all turkeys for meat qualities. The black and the black Norfolk are excellent birds of a medium size, but few of the last named are to be met with in this country. It is the black which produces the finest French turkeys. Of the others the Narragansett and Ronquieres stand high in meat properties, whilst the whites, though beautiful in appearance and largely kept in Southern Europe, have not proved of equal value in Britain, probably because it has been bred for exhibition.

The turkey is essentially a farmer's fowl in that it requires abundance of space, and does not thrive in confinement. Hence there are limitations to its extension, though in few countries have these been reached. An exception, however, is met with in America, where in some of the New England States turkey-breeding forms an important and, for some years, a lucrative branch of live stock. The numbers reared have been in advance of the space available, with the result that a disease known as blackhead, due to attacks of a parasite on the liver, has devastated large areas, probably a result of tainted soil. Upon the farms of East Anglia and Normandy, where the finest European turkeys are raised, it is found that one bird per acre is sufficient, and that the use of fresh ground in rearing every year is essential to prevent disease and consequent loss. Moreover, the soil must be kindly. Heavy clay or damp places are fatal to the young birds. The position should give the maximum of sunshine, and afford shelter against wind and rain. In some cases turkeys are allowed to breed wild in the woods, and under such conditions they thrive, but the number produced in this way is too small to be commercially successful, though useful in yielding strong, hardy breeding stock. Under favourable conditions

and with proper management there is no more profitable branch of poultry-keeping than the rearing of turkeys for market, as these birds realise high prices at Christmas, leaving a good margin between cost and returns. Such result can only be attained when the points laid down are observed.

Where woods are available with trees large enough so that the birds can sleep on the lower branches, houses are not required, save in a severe winter. But in many districts such an arrangement would not afford protection against foxes and other enemies. Under such circumstances the house allotted to the breeding turkeys should be large and lofty, affording abundance of ventilation and plenty of head room. A safe calculation is that for a building at least 7 feet high to the eaves 15 square feet of floor space should be given for each inmate. A thatched roof is best, and if the back and ends are of furze that is desirable, though wood can be employed when the former is not obtainable. The front may be left entirely open if there is no danger of marauders. Otherwise it should be barred, the rods six inches apart, as thus will be secured the circulation of fresh air which is essential. Perches are best made out of fir poles, raised three feet above the ground and three feet apart. Six to eight breeding hens with one male must be kept for every hundred youngsters required, but results vary according to the seasons.

When the time approaches that the hens begin to lay, say in February, arrangements should be made for nests, otherwise they will seek their own. If possible, these should be in the open. In some cases "dug-outs" are made on sloping ground, behind a clump of bushes, so that the entrance cannot easily be seen. Barrels or boxes can be placed in snug retreats hidden out of sight, or hurdles covered with furze to make a quiet corner. It will generally be found that the hens will utilise such opportunities, and when they begin to lay the nests should be visited daily and the eggs removed, leaving a dummy therein. By so doing it is generally found that a greater number of eggs is secured. When it is found that the hens remain on the nest, fifteen eggs may be given to her, and any beyond that number be sat upon by ordinary hens in the usual way. A turkey can brood twenty-five to thirty chicks, but gives better results if she has not more than fifteen or seventeen eggs to sit upon. The period of hatching is twenty-eight days. One very interesting fact is that the turkey cock can be induced to undertake maternal duties, and makes a most faithful sitter and mother.

That young turkeys need special care in rearing is unquestionable, at least until they are nine or ten weeks old, when there takes place what is known as "shooting the red," a sign of sexual development,

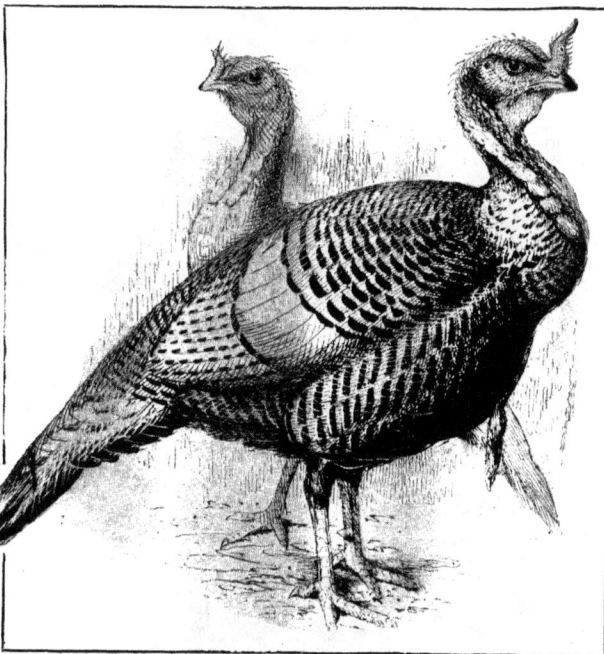

BRONZE TURKEYS

when the head becomes red. Afterwards they are as hardy as ordinary fowls. In the initial state the requirements are—shelter from rain and strong winds, as much sunshine as possible, absolutely fresh ground, abundance of green stuff, and opportunity to find natural food, for which reason arable land is preferable to pasture. The food must be nutritious. Many of the most successful rearers use oatmeal or ground oats very freely, varying with biscuit meal and cooked rice. Perhaps the most important part of the diet is that, unless there is an abundance of grub and insect life available, meat in one form or another must be supplied. During the summer two feeds a day of any of the ordinary meals or grains should be given. If the weather is dry and hot they should be permitted in woods as much as possible, by reason of the fact that there natural food is more abundant. If it is cold and damp feeding will need to be more bountiful. For the Christmas trade, when the nine-months' old bird should weigh 15 to 25 pounds, they require a month's fattening, for which soured skim or butter-milk, mixed with oat or wheat meal are essential to give white-ness and softness to the flesh.

EDWARD BROWN, F.L.S.

AMERICAN PARTRIDGES AND QUAILS

Although the game-birds of North America other than grouse and turkeys are called by Old World names, they are broadly distinguished from their nearest European and Asiatic relatives by the tooth-like processes on the edges of the lower half of the beak. On this account these birds are referred to a separate subfamily, the *Odontophor-inæ*, represented by eleven genera, containing nearly fifty species. The largest species is about the size of an ordinary partridge, while the smallest is inferior in size to a quail. In the majority of the members of the group the beak is stout and grouse-like, and the head of most of the species carries a longer or shorter crest.

The first genus, *Dendrortyx*, includes three so-called partridges inhabiting Central America, from Southern Mexico to Costa Rica. They are rather handsome birds, especially *D. macrurus*, with the tail as long as the wings or nearly so. A second genus, *Callipepla*, is represented by the scaled partridges, easily recognised by their short crests and grey and black-margined plumage, which produces an appearance like scales. They are met with in the south-western United States and Mexico. Specially attractive is the mountain

CALIFORNIAN QUAIL

partridge (*Oreortyx pictus*) of the western states of North America, in which the crest is composed of two very long black feathers; the head, neck, mantle, and breast being grey, the rest of the upper parts olive brown, the throat and fore part of the neck deep chestnut margined with white, and the sides and flanks similarly coloured, but irregularly barred with black and white.

One of the handsomest and most familiar members of the group is the Californian quail (*Lophortyx californicus*), a bird often imported into Europe, and at once distinguished by its conspicuous crest of black club-shaped feathers. In the cock the forehead is buff, with the rest of the head and the throat black edged with a white band; the plumage of the neck, mantle, and chest grey, each feather being margined with black and spotted with white; while the rest of the upper parts are greyish olive brown, and the under parts buff, barred with black and shading into chestnut.

The female has the crest shorter and browner, and the feathers of the head and neck mostly dirty white, with dark middles. This bird, which in-habits the extreme western states, from Washington to California, ranging inland to Nevada, has been introduced into various parts of the world.

A somewhat peculiar type is formed by the South Mexican barred quail (*Philortyx fasciatus*), in which the plumage of the greater portion of both upper and under parts is barred. Central and northern South America are the home of the seven species of crested quails (*Eupsychortyx*), which differ from all the foregoing in the extreme shortness of the tail, as well as by their smaller size.

Closely allied, but lacking the crest, is the genus *Ortyx*, including such well-known birds as the Virginian quail (*O. virginianus*), or "bob-white," as it is called in the States, and several other species, in all of which the greater portion of the under parts is uniform chestnut or brick-red; these birds being natives of south-western states and Mexico. The "bob-white" is now specially protected in the United States on account of its value as an insect and grub destroyer.

Three striking and pecu-

BLACK-THROATED CRESTED QUAIL

liarly marked species of quail constitute the genus *Cyrtonyx*, and inhabit the south-western United States, Mexico, and Central America. The cocks of all three kinds are characterised by the full development of the crest, black-and-white patterned head, and the eyed under parts.

MEGAPODES AND BRUSH-TURKEYS

The second section of the game-birds is represented by two familes, *Megapodiidæ* and *Cracidæ*, in which the first toe is placed on the same level as the others. The members of the first of these two groups are chiefly remarkable on account of their nesting habits, their eggs being deposited in sand or in a mound raised by one

AUSTRALIAN BRUSH-TURKEY ROOSTING

or more pairs of birds, and incubated by the heat caused by the fermentation of the decaying vegetable matter and the warmth of the sun. The young are hatched fully feathered and able to fly almost from birth. The legs and feet of all these birds are remarkably strong and stout, and thus well suited for scratching up the earth and preparing the nesting mounds.

TYPICAL MEGAPODES

The typical megapodes, as the members of the first genus are often called for lack of a proper English name, are represented by some fifteen different species distributed over the islands of the Pacific and Australia ; one kind, *Megapodius cumingi*, inhabiting the Philippines, another, *M. laperousii*, being found in the Ladrone and Pelew Islands, while an isolated western species, occurs in the Nicobars. The plumage is remarkably sombre in colour, generally olive brown or rufous above and grey beneath. The Nicobar megapode (*M. nicobariensis*) during the day frequents dense jungle near the coast, where it associates in pairs or in flocks of thirty or more. It is a difficult bird to flush, usually preferring to escape by running.

The nesting mounds are generally placed near the shore, and average about 5 feet in height and 30 inches circumference. Mr. Davison described one "which must have been at least 8 feet high and quite 60 feet in circumference. It was apparently a very old one, for from near its centre grew a tree about 6 inches in diameter, whose roots penetrated the mound in all directions to within a foot of its summit, some of them being nearly as thick as a man's wrist. I had this

mound dug away almost to the level of the surrounding land, but only got three eggs from it, one quite fresh and two in which the chicks were somewhat developed. Off this mound I shot a megapode which had evidently only just laid an egg. I dissected it, and from a careful examination it would seem that the eggs are laid at long intervals apart, for the largest egg in the ovary was only about the size of a large pea, and the next in size about as big as a small pea.

"These mounds are also used by reptiles, for out of one I dug, besides the megapode's eggs, about a dozen eggs of some large lizard. I made inquiries among the natives about these birds, and from them I learnt that they usually get four or five eggs from a mound, but sometimes they get as many as ten ; they all assert that only one pair of birds are concerned in the making of a mound, and that they only work at night. When newly made, the mounds (so I was informed) are small, but are gradually enlarged by the birds."

WALLACE'S MEGAPODE

An exceptionally marked species, Wallace's megapode (*Eulipoa wallacei*), from Gilolo and some of the islands of the west of New Guinea, is characterised by having the secondary flight feathers much shorter than the primaries, and the feathers of the middle of the back and most of the wing coverts barred with bright chestnut.

MALI-FOWL

Still larger is the mali-fowl (*Lipoa ocellata*) of Southern and Western Australia, distinguished by having the upper tail coverts reaching to the end of the tail, and the plumage of the upper parts mostly grey barred with black.

AUSTRALIAN BRUSH-TURKEY

According to Mr. A. H. E. Mattingley, the mali-fowl does not begin to lay until two years old, and during the first half of the breeding season the eggs are laid regularly every third or fourth day, after which the intervals between the deposition of the eggs increase according to the disposition of the individual birds and the amount of food available. Hot and dry seasons have a noticeable effect on these birds, which under such conditions lay fewer eggs than usual. Laying usually begins early in September, but may be deferred until December is well advanced, and the total number of eggs laid by the individual hens in a season varies from one to a score.

The eggs have unpolished shells of a delicate salmon-pink or pinkish red colour when first laid, but soon fade to earthy brown. They are laid in the mound in tiers, with four in the basement layer; between each tier is a layer of sand 3 or 4 inches thick, and the eggs in the same time are separated from one another by from 6 to 12 inches of the same material, and placed near the solid wall of decaying vegetable matter bounding the egg chamber. The eggs are always placed with the narrow end downwards, so that when hatching the head of the chick, which occupies the larger end, will be uppermost.

Another naturalist, Mr. A. W. Milligan, states that in hot weather the birds remove the top of the breeding heap so as to form a saucer-like depression, which is again filled up when the weather becomes rainy.

BRUSH-TURKEYS

The Papuan brush-turkeys, as typified by *Talegallus cuvieri*, include three or four species of large, dark-coloured birds, with stout beaks, oval nostrils, and the head, throat, and front of the neck thinly covered with small scattered feathers; the genus being confined to New Guinea and some of the adjacent islands.

The Australian brush-turkey (*Catheturus lathami*) represents a second genus distinguished by the presence of a large wattle at the base of the neck, the circular nostrils, and the much longer tail. In both sexes the general colour of the upper parts is dark brownish black, paler on the lower part of the back and rump, the upper parts being dark brownish grey, broadly edged with white, the naked skin of the head and neck pinky red, and the wattle bright yellow. Mr. J. Gould observes that at the beginning of spring these birds scratch together an immense heap of decaying vegetable matter as a depository for the eggs, and trust to the heat engendered by the process of fermentation for the development of the young.

" The heap employed for this purpose is collected by the birds during several weeks previous to the period of laying; it varies in size from two to many cart-loads, and in most instances is of a pyramidal form. The materials composing these mounds are accumulated by the bird grasping a quantity in its foot and throwing it backwards to a common centre, the surface of the ground for a considerable distance being so completely scratched over that scarcely a leaf or a blade of grass is left. The eggs are deposited in a circle at the distance of 9 or 12 inches from each other, and buried more than an arm's depth with the large end upwards."

Another genus (*Æpypodius*), with two species from New Guinea and Waigiou, is characterised by a fleshy crest running from the base of the beak to the crown, a pendulous wattle at the base of the fore neck, and the chestnut upper tail coverts.

THE MALEO

The last genus of the family contains only the maleo (*Megacephalum maleo*), of Northern Celebes and the Sanghir Islands, which is the most remarkable member of the whole group, both as regards structure and habits. In both sexes the head is naked, and the crown surmounted by a large black casque, while the colour of the plumage of the upper parts, chest,

THE MALEO

flanks, thighs, and under tail coverts is dark brown, and that of the under parts beautiful salmon pink.

These birds do not raise mounds in which to lay their eggs, but deposit the latter in holes dug in the sand. Dr. A. R. Wallace describes one of their laying grounds as follows : " The place is situated in the large bay between the islands of Limbé and Banca, and consists of a steep beach more than a mile in length of deep, loose, and coarse black volcanic sand, or rather gravel—very fatiguing to walk over. It is in this loose, hot, black sand that those singular birds, the ' maleos,' deposit their eggs. In the months of August and September, when there is little or no rain, they come down in pairs from the interior to this, or to one or two other favourite spots, and scratch holes three or four feet deep, just above high-water mark, where the female deposits a single large egg, which she covers over with about a foot of sand, and then returns to the forest. At the end of ten or twelve days she comes again to the same spot to lay another egg, and each female bird is supposed to lay six or eight eggs during the season. The male assists the female in making the hole, coming down and returning with her.

" The appearance of the bird, when walking on the beach, is very handsome. They run quickly, but when shot at or suddenly disturbed, take wing with a heavy, noisy flight to some neighbouring tree, where they settle on a low branch, and they probably rest at night in a similar situation. Many birds lay in the same hole, for a dozen eggs are often found together, and these are so large that it is not possible for the body of the bird to contain more than one fully-developed egg at the same time. In all the female birds which I shot, none of the eggs besides the large one exceeded the size of peas, and there were only eight or nine of these."

CURASSOWS AND GUANS

THE second family of game-birds (*Cracidæ*), characterised by the first toe being on the same level as the others, contains a number of large Central and South

GUIANA CRESTED CURASSOWS

American birds, some of which, such as the curassows, are nearly as large as turkeys, while others, like certain guans of the genus *Ortalis*, are considerably smaller than pheasants. All the species have a long and well-developed tail, and in the males the windpipe is long and convoluted, enabling the birds to produce their characteristic loud and harsh cries.

They differ from the megapodes, not only in the structure of the skeleton, but also in having a tuft of feathers on the oil-gland. Their nesting habits are also quite different; the eggs being incubated by the parent in the ordinary manner, though some of the species habitually nest in trees, and lay white eggs. When first hatched, the young are covered with a down showing a pattern like that of the chicks of other game-birds. These birds are arboreal in their habits, the greater part of their time being spent among the highest forest trees.

The various genera may be conveniently grouped into two sections, the first four having the upper half of the beak higher than broad, while in the remaining seven it is broader than high. The typical curassows differ from their relatives in their large size, and also by having the feathers on the top of the head semi-erect and curled at the extremities; in the males the crest being uniformly black, while in the females it is more or less barred with white.

The cocks are all very much alike, the whole plumage being black glossed with purple or dark green, except on the under parts, flanks,

and under tail coverts, which are white; but in two species the tail feathers have also white tips. The colour of the plumage of the females, on the other hand, varies much in the different species, in the crested curassows closely resembling that of the male, while in the remainder the upper parts are variously barred with black, white, rufous, and buff. The distinctive specific characters are accordingly, in most cases, much more strongly marked in females than in males.

CRESTED CURASSOWS

In the Guiana crested curassow (*Crax alector*) the colour of the plumage of both sexes is very similar, but the female has the crest barred with white. This bird, which is a native of the forests of British Guiana and Northern Brazil, but extending into Colombia, is distinguished by the purple gloss on the upper parts, and the absence of a swollen knob at the base of the upper half of the beak, and of a wattle on the base of the lower half of the same in the male. The cere and base of the beak are yellow, with the extremity of the latter horny blue, while the legs and feet are horn-coloured. Being easily tamed, and affording excellent food, these birds are often domesticated. Nearly allied is the Mexican crested curassow (*C. globiera*).

URUMUTU AND MITUAS

Very distinct and also smaller is rufous-coloured urumutu (*Nothocrax urumutum*), of British Guiana and the Upper Amazon, distinguished by its crest of recumbent feathers

MEXICAN CRESTED CURASSOW

and the naked space in front of the eye. Next come the three species of mituas, birds as large as curassows, with the greater part of the plumage black in both sexes. In two of the species the under parts and under tail coverts are chestnut; one of these (*Mitua mitu*) has the tail feathers tipped with white, while in the second (*M. tomentosa*) the tips are chestnut. In the third species (*M. salvini*) both the under parts and the tips of the tail feathers are white. All three may be distinguished from curassows by the elevated and vaulted upper half of the beak and the absence of curling in the crest feathers.

PAUXI CURASSOW

The curious pauxi curassow (*Pauxis pauxi*) of the north-western parts of South America, the only representative of its genus, is remarkable for the large, fig-shaped blue casque on the forehead. The male has the entire plumage black, except the under parts, lower tail coverts, and the tips of the tail feathers, which are white; in the female the back, wings, and breast being chestnut, paler on the flanks, and barred and mottled with black.

LORD DERBY'S GUAN

To the second group, characterised by the width of the beak at the base being greater than the height, belongs Lord Derby's guan (*Oreophasis derbianus*), from the wooded slopes of the Volcan de Fuego in Guatemala. The characteristic features of this bird are the elevated, straight, deep scarlet horn on the top of the head between the eyes, and the densely feathered base of the upper half of the beak. In both sexes the general colour of the head and upper parts is black glossed with dark green, the base of the throat being almost naked, the front of the neck and breast white shading into buff on the sides, with dark shaft-stripes to the feathers, the remainder of the under parts brownish black, and a wide band running across the middle of the tail.

GUANS

A more numerous group is that of the guans, typified by the Brazilian

LORD DERBY'S GUAN

RED-TAILED GUAN

Penelope superciliaris, but including fourteen other species from Central and South America. In all these birds the chin and throat are generally naked and furnished with a wattle, and there is a large naked space surrounding the eye. The allied *Penelopina nigra* represents a genus by itself, characterised by the difference in the colour of the plumage in the two sexes. In the cock of this species, which inhabits the highlands of Guatemala, the plumage is entirely black, glossed with green, but in the female is rufous above barred with black, and beneath sandy mottled with dark brown. In habits all these birds appear to be very similar; during the breeding season they are only found in pairs, while at other times they congregate in large flocks, always frequenting the forest, and passing the greater part of their time in the largest trees, when not engaged in searching for fallen fruits and insects.

In the allied genus *Ortalis*, including seventeen Central and South American species, the throat is naked as in the two last, but there is a thin band of stiff-shafted feathers down the middle. The only member of the family which enters North America is the chachalaca (*O. vetula*), which has a wide range, extending from Southern Texas through Eastern Mexico and Central America to Colombia; the bird varying somewhat in the different parts of this extensive range.

THE ABURRIA

The aburria (*Aburria aburri*), the solitary representative of its kind, is a native of Colombia and Ecuador, and may be recognised by its black plumage glossed with dark green, and the worm-like wattle situated on the naked part of the fore neck. Finally, two species, characterised by the chin, throat, and fore part of the neck being covered with feathers, form the genus *Chamæpetes*; Colombia, Ecuador, and Peru being the home of *C. goudoti*, and Costa Rica and Panama of *C. unicolor*.

PURPLE GUAN

THE HOATZIN

THE very remarkable pheasant-like bird from the northern and western districts of South America known as the hoatzin (*Opisthocomus hoatzin*), appears on the whole to be most nearly allied to the game-birds, and may, therefore, be mentioned in this place. Many naturalists regard it, indeed, as representing a distinct order, but as there is little advantage in unnecessarily multiplying ordinal terms. it is alluded to here merely as forming a family, *Opisthocomidæ*.

The skeleton of this bird, which has many striking peculiarities, notably the form of the breast - bone, with its nearly parallel lateral edges and feebly developed keel, of which the anterior part is cut away, and the posterior portion broad and flattened out. On this flattened surface the greater part of the weight of the body is supported when the bird is at rest. Another striking feature is the shoulder-girdle, the bones of which are completely welded to one another as well as to the breast-bone. The crop is enormous, and occupies the upper portion of the chest, being placed in a deep cavity in the pectoral muscles.

The nest, which is built of sticks and placed in bushes near the water's edge, contains two or three, and sometimes as many as five, oval-shaped eggs of a white colour, doubly spotted with rufous and purple, and remarkably rail-like in character. Unusual interest attaches to the young, which are hatched naked, with the thumb and index finger provided with well-developed claws, enabling them to climb about among the branches soon after they are hatched ; the beak, as well as the legs and wings, being used for holding on to the twigs (page 939). When compelled to do so by falling from the boughs, they swim and dive with facility. Hoatzins spend their existence among the branches, consuming enormous quantities of leaves, and, in spite of their large wings, possess only the most limited powers of flight. They exhale a strong musky odour.

The beak is strong, with the edges of the basal portion of its upper half distinctly serrated ; bristles border the gape, and—a rare feature among birds—the eye-lids are furnished with lashes. In the sharp and rounded wings are ten primary and nine secondary quills, and there are ten feathers to the rather long and nearly even tail. The plumage is olive above with white markings and dull rufous below ; the long and loose crest and the tips of the tail feathers are yellowish, and a naked bluish area surrounds each eye.

THE HOATZIN

BUSTARD-QUAILS

ALTHOUGH the small birds commonly known as bustard-quails (family *Turnicidæ*) are usually included among the game-birds, and have been associated with the quails, they are in reality so markedly distinct from the more typical members of that order that by many naturalists they are regarded as forming an order equal in importance to the Gallinæ. While in some respects they approach both pigeons and game-birds, their affinities with the rails are undoubted, and it is perhaps the wisest course to leave their precise systematic position an open question. Like sand-grouse and rails, they lay double-spotted eggs, quite different from those of any of the typical game-birds ; but the young are covered with down, and able to run soon after they are hatched. A distinctive feature of the group is that the females are always larger and more brightly coloured than the males, and that the latter undertake all the cares of incubating the eggs and tending the young.

In the typical genus *Turnix*, which contains all the species but one, the first toe is entirely absent, but in the Australian collared bustard-quail (*Pedionomus torquatus*) a small first toe is present. The twenty-one members of the typical genus are distributed over Africa, Madagascar, Arabia, and through the Indo-Malay countries to Australia ; while one species, the Andalusian hemipode (*T. sylvatica*), inhabits South Europe. No less than four of these species are found in India, and the Indian bustard-quail (*T. taigur*) may be taken as a typical representative of the group. This bird, besides being found all over India, has a wide range through Burma, the Malay Peninsula, Siam, and South China to Formosa and the Liu-kiu Islands. As might be expected, the plumage of so widely-spread a species shows considerable climatic variation, examples from the dry plains of India having the prevailing colour of the upper parts rufous ; whereas in specimens from the Malay Peninsula, with its heavy rainfall, the general tone is greyish brown. In both sexes the upper parts are barred and marked with black, many of the feathers being margined on the sides with whitish buff ; but the chest and breast are buff barred with black, and the under parts rusty buff. The female has the middle of the throat and chest deep black, whereas in the male these parts are white with narrow black bars.

Mr. A. O. Hume writes that "scrub-jungle, intermixed with patches of moderately high grass on dry ground, is perhaps its natural home ; but it may be met with anywhere in low bush-jungle and on the skirts of forests, and in inhabited districts greatly affects gardens, grass-preserves, and similar enclosures. It strays into stubbles and low crops in the mornings and evenings, even remaining in these at times throughout the day, but more generally retreating during the hotter noontide hours to the cover of some thorny bush or patch of grass upon the margins."

ORDER XXV. CRYPTURI
TINAMUS

SOUTH AMERICA is the exclusive home of a group of birds which, while resembling the game-birds to a great extent in outward appearance and habits, present a peculiarity in the structure of the bony palate of the skull by which they are distinguished from all the groups hitherto described, and thus resemble the ostrich tribe. These birds are the tinamus, order Crypturi, all of which are included in the single family *Tinamidæ*, and are represented by thirty-nine species arranged under nine genera. The members that are mentioned in this work are as follows:

FAMILY
Tinamus—Tinamidæ
GENUS 1
Tinamus
SPECIES

Robust tinamu .... Tinamus robustus
Solitary tinamu .... T. solitarius

GENUS 2
Crypturus
SPECIES

Banded tinamu .... Crypturus noctivagus

GENUS 3
Rhynchotus
SPECIES

Great tinamu .... Rhynchotus rufescens

GENUS 4
Nothura
SPECIES

Spotted tinamu................................. Nothura maculata
Darwin's tinamu .... N. darwini

GENUS 5
Calodromus
SPECIES

Crested tinamu .... Calodromus elegans

GENUS 6
Tinamotis
SPECIES

Pentland's tinamu .... Tinamotis pentlandi

That the tinamus are allied on the one hand to the game-birds, and on the other to the ostrich-like birds, is practically certain; and it is probable that the type of palatal structure they display is very close to the primitive one from which all the others have originated. In this connection it should be noticed that the slit (schizognathous) palate of the game-birds is the one which comes nearest to that of the tinamu and ostrich type.

Partridge or quail-like in general appearance, tinamus have small heads, with short, slender, curved beaks; strong, naked legs and feet, in which the first toe is either small or represented only by its claw; rudimentary tails, which are frequently concealed by the coverts; and the wings relatively short and rounded. They are specially characterised by the circumstance that, while the narrow breast-bone has a well-developed keel, like that of the game-birds, in the palate of the skull, which comes very close to the schizognathous type, the vomer, or median element, is fused with the bones immediately in front and behind it, namely, with the maxillo-palatines in front and with the palatines and pterygoids behind, in which respect the birds come very close to the ostrich tribe. A further resemblance to that group is afforded by the circumstance that the last few vertebræ of the tail do not coalesce to form a ploughshare-shaped bone.

In the skull the apertures of the nostrils resemble those of game-birds in their oval (holorhinal) shape; while on its under surface the rostrum, or rod, on the base of the skull, bears well-developed basipterygoid processes. In the plumage the feathered tracts, both on the neck and elsewhere, are well distinguished from the bare intervening areas, the after-shafts of the feathers are rudimentary, there are ten primary quills in the wings, and the oil-gland is tufted. The young are active almost immediately after hatching, and remarkable for the rapidity with which they acquire their full plumage, being in fact, able to fly more rapidly than the adults. The general colour of the plumage is deep yellowish, marked above with bars of dark brown and black. The eggs of all the species are remarkable for their highly polished surface, which

GREAT TINAMUS OR MARTINETAS

resembles a piece of glazed porcelain, the colour being either wine-red or bluish green.

As regards food, tinamus are vegetable-feeders. Since all the members of the family are very similar in general structure, it will be unnecessary to point out the distinctive characters of the various genera ; although it may be mentioned that the family may

PENTLAND'S TINAMU

be divided into two sections, according to the presence or absence of a distinct first toe.

Of the seven genera with a well-developed first toe, the robust tinamu (*Tinamus robustus*) and the solitary tinamu (*T. solitarius*), of Brazil, are representatives of the typical genus. The banded tinamu (*Crypturus noctivagus*), of Brazil, is a well-known member of the largest genus of the family, which contains some sixteen species ; while the martineta, or great tinamu (*Rhynchotus rufescens*), of Brazil and Argentina, is one of two congeneric species which may be compared in size to a pheasant. Its eggs are of a wine-red colour.

On the other hand, the spotted tinamu (*Nothura maculata*) and the allied Darwin's tinamu (*N. darwini*), both of which are inhabitants of the Argentine pampas, are more nearly the dimensions of a small partridge ; their eggs being either purple red or liver-colour. The two genera in which the first toe is rudimentary are each represented by a single species, of which by far the handsomest is the crested tinamu (*Calodromus elegans*), of Patagonia, in which the plumage of both the upper and lower surface is elegantly mottled, and the head adorned with an upright crest. This species, which is of the size of an English pheasant, lays from ten to a dozen blue-green eggs as large as those of a fowl. Pentland's tinamu (*Tinamotis pentlandi*), the representative of the second three-toed genus, lacks the upright crest.

By ordinary observers tinamus, both as regards general appearance and habits, are considered as game-birds, of which, indeed, they take the place in South America, where they are commonly termed partridges. Endowed with far less powers of flight than ordinary game-birds, the various species of tinamu frequent either the open grassy pampas or seek the shelter of woods. The species inhabiting the Argentine pampas are in the habit of skulking like rails among the luxuriant grass, among which they run from under the very feet of the horses. They can be flushed only with difficulty, although when they do rise their flight is exceeding noisy and violent, and soon leads to the birds exhausting themselves.

Writing of the spotted tinamu in Misiones, Mr. E. W. White states that these birds partake of the colour of the soil, which " is of a ferruginous clay, and differ herein completely from those of Buenos Aires. They are so completely tame and abundant on the fine undulating grass-lands that extend hence southwards, that in the early morning they come right up to the houses, and the boys knock them over with stones."

On account of their confiding disposition, coupled with their general stupidity, and the excellence of their flesh, these birds have been well-nigh exterminated in many districts where they were formerly abundant ; large numbers being taken by riding in a circle and capturing them with a noose.

Generally solitary, many live in close proximity, making their whereabouts known to one another by their soft plaintive cries. Of all the birds of the pampas, the great tinamu, according to Mr. W. H. Hudson, " is perhaps the sweetest-voiced, and sings with great frequency. Its song, or call, is heard oftenest towards the evening, and composed of five modulated notes, flute-like in character, very expressive, and is uttered by many individuals answering each other as they sit far apart concealed in the grass."

The crested species, which is also a regular singer, seems to associate in coveys, and, like game-birds, is in the habit of dusting its plumage in the sandy soil. The young, whose precocity in regard to the development of feathers has been already mentioned, leave their parents at a very early age to shift for themselves. The nests are built in a hollow on the ground, beneath the shelter of a tussock of grass or low bush, and scantily lined with herbage and feathers.   R. LYDEKKER

SOLITARY TINAMU

ORDER XXVI. RATITÆ

GENERAL CHARACTERISTICS OF THE FLIGHTLESS BIRDS

WITH the exception of certain specially modified species like some of the rails, in which the power of flight has been lost, the existing birds hitherto treated are characterised by the circumstance that the breast-bone is provided with a strong vertical median keel, to afford support for the muscles necessary for flight, while both the scapula and coracoid are separate elongated bones, forming an acute or right angle at their junction.

As a rule, flying birds are also characterised by the circumstance that in the pelvis the bones known as the ilium and ischium are united at their outer extremities so as to enclose an aperture, while the head of the quadrate bone, by which the lower half of the beak articulates with the skull, is double, and in the palate the vomer (except in the tinamus) is not fused with the neighbouring bones, or interposed between them and the rostrum of the base of the skull. On the other hand, in the ostrich tribe the breast-bone is devoid of a keel, the coracoid is short and united with the scapula, and the two bones form a very obtuse angle at their junction, while the furcula is incomplete, so that its two branches remain quite separate.

All these birds likewise agree in having their wings greatly reduced, so as to be utterly useless for flight ; and are also characterised by the extremities of the ilium and ischium (except in very old individuals of the rheas and emeus) remaining distinct ; and likewise by the single head of the quadrate bone, as they are by the vomer, which is broad behind, being interposed between the other bones of the palate and the

rostrum of the base of the skull, and frequently fused with some of them.

Other features of the group are the absence of an oil-gland, and of any marked distinction between feathered and unfeathered areas on the skin ; while the hook-like or uncinate processes of the ribs are never more than three in number, and often rudimentary, or even absent. In addition to the agreement in the structure of the skull, they resemble the tinamus in that the terminal vertebræ of the tail are not united to form a plough-share-like bone.

The palate is quite different from that of all the flying birds except tinamus.

On account of the invariable absence of a keel to the breast-bone, the members of the ostrich tribe are collectively designated the Ratitæ—so named from the Latin term for a flat-bottomed boat. This group was formerly reckoned as a subclass of equal rank with one containing all other birds, and termed Carinatæ. It is, however, now evident that most of the structural peculiarities of the ostrich tribe, except those connected with the skull (some of which are common to tinamus), are adaptations due to the loss of flight. Consequently, the Ratitæ are now ranked as an order, albeit of a primitive type. The existing members of the group, which are comparatively few in number and have a remarkably scattered distribution on the surface of the globe, are widely different from one another in structure, and include the largest of all birds, while none of them is very small.

OSTRICHES FEEDING

CLASSIFICATION OF RATITÆ MENTIONED IN THIS WORK

1451

OSTRICHES

THE ostriches (family *Struthionidæ*) are the largest of all existing Ratitæ, and therefore of all living birds, and they are at the same time the most specialised representatives of the order, this specialisation showing itself in the reduction of the number of toes to two, owing to the absence, not only of the first or hind, but likewise of the second toe. In this respect ostriches are perfectly unique among birds. While they agree with the majority of their allies in the short beak, ostriches are further characterised by the short, stunted nails on the toes, the great proportionate length of the humerus of the rudimentary wing, and the absence of after-shafts to the feathers.

In the skeleton the furcula is wanting, the pubic bones of the pelvis unite in a symphysis (as they do in many reptiles but in no other birds), and the lower end of the tibia has no bony bridge over the groove for the extensor tendons ; while there are also certain characteristic features in the base of the skull into the consideration of which it will be unnecessary to enter. In addition to their large size and two toes, ostriches are characterised externally by the small and flattened head, in which the short beak is broad and depressed ; the long, powerful, and practically naked neck ; the full and massive body, provided with short wings ; the muscular and partly bare thighs, and the stout metatarsus and foot. The beak has a very wide gape, reaching back to the line of the eyes ; and the nostrils open near the middle of its length. The third toe is much larger than the fourth, and both are furnished with soft fleshy pads on the under surface.

In the immature state the skin is covered with coarse plumage of a mottled dark brown and yellowish white hue, the neck being completely bare. In the adult female the colour changes to a nearly uniform dusky grey ; but in the male, while the body feathers are black, the tail feathers and quills of the wing are pure glistening white ; the neck in both sexes being clothed with short down. A peculiarity of the feathers of the wings and tail is that their two webs are of equal width.

The young, like the other members of the group, are active as soon as hatched ; and the eggs are polished and pale yellowish white in colour ; those of the southern species being traversed by a number of minute punctures, while those of the northern species are smoother and more ivory-like. Finally, the male is far superior in size to his partner.

Four species of ostriches are now provisionally recognised ; the typical bird being *Struthio camelus* of Northern Africa, Arabia, Syria, and Mesopotamia. In this species the colour of the naked skin of the neck and thighs is flesh-coloured ; whereas in the Somali ostrich (*S. molybdophanes*) it is bluish, and there is a conspicuous red patch on the front of the metatarsus. This Somali ostrich, which lays smooth-surfaced eggs of the northern type but with distinct pits, is replaced in Masailand by *S. masaicus*, characterised by the full red of the neck and legs. Finally the southern *S. australis* of the Cape is characterised by the grey colour of the bare parts and the heavily-pitted eggs. All four kinds are, however, very closely allied, and may well be regarded as local races of a single variable species. In regard to the dimensions attained by ostriches, it may be mentioned that an unusually fine male from the Niger basin measured 4 feet 10 inches in height at the back, and had a total length of 4 feet 3 inches. Ordinary examples of the same sex reach only about 3 feet 8 inches in height.

Although now confined to Africa, Syria, Arabia, and Mesopotamia— and becoming every year scarcer in the three last-mentioned countries—there is a possibility that ostriches formerly existed within the historical period in parts of Central Asia and perhaps in Baluchistan, since there are several allusions to birds which can scarcely be anything else than ostriches in various ancient writings. Quite apart, however, from this, the evidence of its fossilised remains shows that an extinct ostrich, *S. asiaticus*, inhabited North-Western India during the Pliocene period, while a second lived in Central India to a much later date ; and a petrified egg from the province of Cherson, in Russia, points to the former existence of these birds in that part of Europe.

Originally it is probable that ostriches ranged in suitable localities from Senegambia in the west, through Southern Morocco, Algeria, and Egypt to Arabia, Syria, and Mesopotamia in the east ; while they also extended from Algeria through Central and Eastern Africa to the Cape. Being, however, essentially a bird of open, sandy districts, there are many regions in Africa, such as the neighbourhood of Zanzibar and large tracts on the west coast and in the Congo valley, where, owing to the prevalence of forest, the ostrich never existed.

Moreover, the constant persecution with which these birds have been harassed for years on account of their beautiful plumes has led to their almost complete disappearance from Egypt and Nubia, and they are now seldom found to the north of latitude 17°. Ostriches have also disappeared from large tracts in South Africa, although still to be met with in small parties in the great Kalahari Desert, and especially in the part lying to the southward of Lake Ngami. They are likewise still fairly common on the borders of Namaqualand and Damaraland, the great Mabebi flats, and certain parts of Matabililand and Mashonaland, where they are sometimes seen in large flocks.

Always inhabiting more or less desert-like districts, or flats covered with stunted patches of bush, where

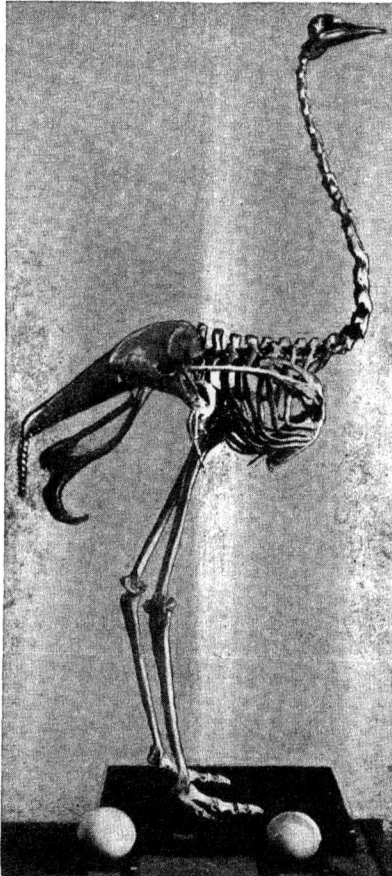

SKELETON AND EGGS OF AN OSTRICH

THE OSTRICH
From the photograph by Gambier Bolton, by permission of the Autotype Fine Art Company, Ltd., 74, New Oxford Street, W.C.

the elevated position of their heads gives them a wide field of vision, ostriches in South Africa generally associate in parties of from ten to twenty individuals, although in the northern parts of the continent the flocks are stated to be smaller. In Southern Africa they frequently associate with herds of gnu and bontequagga ; and their keen sight and wary nature, coupled with their unrivalled speed, render them nearly the most difficult of all animals to capture.

Outstripping the swiftest African antelopes in speed, the ostrich in cool weather could indeed easily escape from any horseman were it not for its foolish habit of running in a circle, and thus allowing shots to be easily obtained. In running at speed, an ostrich spreads it wings, and the distances it can traverse are enormous ; indeed, during the day time it is continually on the move. In the neighbourhood of the sea or lakes ostriches are reported to be in the habit of bathing during the hot season, when parties have been seen standing up to their necks in water ; and salt of some kind seems absolutely essential to their existence. The digestion of an ostrich is proverbial ; and while in their general diet these birds are practically omnivorous, they are likewise in the habit of swallowing stones, sand, bones, or even pieces of metal, to aid in the trituration of their food. In captivity this habit probably becomes abnormally developed ; and there are instances where even the constitution of an ostrich could not resist the effects of some of the substances swallowed. Among the ordinary food of the ostriches are comprised small mammals and birds, snakes, lizards, and insects, as well as grass, leaves, fruits, berries, and seeds. Although they can go for protracted periods without it, and will not wander far out of their way to procure it, yet when water is at hand, ostriches will drink constantly.

Young ostriches are said to be silent, but the old cocks utter a loud cry, which is likened by Livingstone to the roar of the lion, and by Canon Tristram to the lowing of oxen ; this cry being generally uttered in the early morning. The ostrich's chief mode of attack or defence is by kicking with its immensely powerful legs, although in the fights in which the cocks periodically indulge the birds also peck at one another with their beaks.

Much interest attaches to the breeding habits of the ostrich, although, from many of the accounts having been derived from native sources, very erroneous notions are prevalent on this subject. At the pairing season, which takes place early in the spring, each cock, after having gone through various performances to attract their attention, and frequently many contests with his rivals, associates with three or four hens. All these hens lay in a single nest, which consists solely of a large hollow excavated in the sand.

There is still some uncertainty as to the number of eggs laid in a nest, although there is little doubt that

SOMALI OSTRICH

this has been much exaggerated. So many as twenty are, however, frequently incubated ; but in addition to these it appears that a certain number are deposited round the edge of the nest, which are never intended to be hatched, and probably serve as food for the young. It is frequently stated that both sexes take equal shares in the work of incubation, but, as a matter of fact, the cock undertakes the greater part of the task. He sits, for instance, throughout the night, when the nest must be protected from prowling jackals, and in those regions in which the eggs are incubated by day as well as by night is only relieved by the females for short periods during the day in order to procure food. Incubation during the day takes place, however, only in the cooler districts of the ostrich's habitat ; in the hotter regions the eggs being left to themselves, with a covering of sand during the day.

As already mentioned, advantage is taken of the peculiar habits of the ostrich to surround the flocks with parties of mounted men, and by this method many are killed in Africa. There are, however, many other ways of capture. For instance, the Bushmen were formerly in the habit of dressing themselves in the skin of an ostrich, and thus disguised penetrating into the midst of a flock, when the birds were despatched one after another by means of poisoned arrows. The hunter had, however, to take care to keep to the leeward of his victims.

In Somaliland the natives hunt the ostrich on camels ; while in Arabia and the Sahara it is ridden down on horseback. The Bushmen also resorted to the aid of pitfalls, as do the Somalis at the present day ; while the lasso is employed by the Hadendowa Arabs of the Sudan and some other tribes ; and in Senar a curved stick is used in boomerang-fashion for the same purpose. In Namaqualand the birds are either surrounded by a cordon of men on foot, who gradually close in upon the flock, or are driven by mounted hunters past concealed relays of their companions, who in turn take up the pursuit till their victims fall through sheer exhaustion.

Ostrich-farms are now established, not only in South Africa, but in several other parts of the world, especially Arizona and Australia ; the feathers, of which the most valuable are the product of the cock, being cut at regular intervals.

In South Africa attention has of late years been directed to improving the quality of the feathers yielded by these semi-domesticated birds, and especially to eliminate the defect known as " barring," which is stated to cause an annual loss of something like £25,000 to South African ostrich-farmers. Although to the uninstructed eye all South African ostriches appear much alike, an expert will recognise a variation in the length, breadth, density, shape, and lustre of the feathers, and it is with these small but commercially important differences that the breeder has to work.

At present the main object of ostrich-breeders is to secure birds in which all the best feather-characteristics are combined; and such birds will realise very high prices. Pairs of this type are mated in the hope that they will produce equally good progeny; but in some cases all the chicks yield inferior feathers, while in other instances only a percentage of the offspring is as good as the parents. On this account, most farmers who are anxious to maintain their reputation as breeders will not sell their young birds until they are old enough to show the quality of their feathers.

In spite of the uncertainty as to how individual chicks will turn out, it appears that of late years the average standard of the ostrich as a feather-producer has been raised by careful attention to breeding, and a greater number of superior feathers are now produced; but farmers want to be assured that the actual feather-yielding capability of the bird can also be increased.

At the present time attention is being specially devoted to the possibility of ensuring that the progeny shall be as good as their parents, or, in other words, that the birds will breed true, and also to the possibility of producing chicks which shall be even better than

they are valued at from £140 to £200 per pair in America. Exceptionally fine birds sometimes bring as much as £200 each. Good birds will produce from £7 to £10 worth of feathers each year, and exceptional ones from £15 to £18 annually. Probably there are no wild ostriches now killed for plumage, the feathers of the domesticated bird being very much finer and better than those of their wild relations.

Plucking is done by putting the ostrich in a V-shaped corral just large enough to admit its body, with room for the workman. A hood, shaped like a long stocking, is placed over the head of the ostrich, when it becomes perfectly quiet. The operator then raises the wings and clips such feathers as are fully ripe. Great care is exercised at this time, as a premature cutting of the feathers deteriorates the succeeding growth.

No pain is inflicted in plucking an ostrich, not a drop of blood being drawn, nor a nerve touched. The large feathers are cut off, and in two months' time, when it is dried up, the quill is pulled out. An ostrich is first plucked when it is nine months old, and about six feet high. The first crop of feathers is of little value; succeeding crops are taken every nine months, the

AN OSTRICH-FARM AT SOUTH PASADENA, CALIFORNIA

their parents for commercial purposes. With this object in view, selective breeding, cross-breeding, and hybridisation are all being tried. Before the best results can be obtained it is, however, essential that an experimental ostrich-farm should be established in South Africa on a permanent basis.

Ostriches were introduced into the United States for breeding purposes by Dr. Protheroe in 1882, when a number of birds were shipped from Cape Town, some of which reached New York, whence they were transported overland to California. Only twenty-two of them reached their destination, to form the nucleus of the farm started at Anaheim, the corporation being known as the California Ostrich Company. During the next four years three other parties embarked in ostrich-farming, which subsequently proved a great success.

A breeding pair of ostriches will produce from ten to twenty chicks per year, which are worth, when six months old, £20 each; at one year, £30; at two years, £40; at three years, £60 to £70. They begin to breed when four years old, but do not lay satisfactorily until they are six or seven years of age, when, if prolific,

third plucking being the full crop, which will weigh about one pound.

In domestication, at any rate, ostriches mate at four years of age, and remain paired for life. The nest is a hole in the ground scooped out by the breast of the bird, and is about one foot deep by three or four feet in diameter. Eggs are laid every alternate day until twelve or fourteen are deposited, each of which weighs from three to four pounds. The eggs are turned daily in the nest by the birds, and are incubated forty-two days, the male taking the nest at five in the afternoon, where he remains on duty until nine the following morning, when the female takes her turn. The chicks, when hatched, are about the size of an ordinary hen; and their plumage presents a mottled appearance (p. 942). They grow about a foot in height per month until they attain their full stature, which is from about seven to eight feet, when they weigh from 300 to 400 pounds. When fourteen months old the plumage gradually changes, the female assuming a dull grey and the male a glossy black livery, while both grow the long white wing-feathers.

RHEAS

IN South America the place of the ostriches is taken by an allied group of birds known as rheas (family *Rheidæ*), or, as they are often termed, American ostriches, which are distinguished externally by the presence of three toes, furnished with claws instead of nails, the fully-feathered head and neck, and the absence of a tail. The wings also are proportionately longer, and covered with long, slender plumes.

Agreeing with the ostriches in the absence of after-shafts to the feathers, in their pale-coloured eggs, and in the superiority in size of the male over the female, the rheas are further distinguished by certain peculiarities in regard to the bones at the base of the skull, and likewise by the circumstance that the ischia, or hindmost lower bones of the pelvis, unite in the middle line, instead of the pubes doing so. The flattened beak is broad at the base and rounded at the tip, where it has a curved nail-like sheath; and the extremity of the wing has a horny process.

The lores and region round the eye, as well as a ring round the aperture of the ear, are devoid of feathers, and the ear-aperture is clothed with bristles. On the head and neck the feathers are small, thin, and pointed, but those of the body are large, broad, and rounded, although so soft that no distinct vanes are formed. In colour the two sexes are very similar, although the female is generally somewhat paler than her consort.

The best known, and at the same time the most abundant, of the three species by which the single genus is now represented is the typical rhea (*Rhea americana*), inhabiting the pampas of Argentina and Patagonia. This species, although far inferior in size to the ostrich, is the largest of the three. Black on the crown of the head and nape, as well as on portions of the upper part of the neck and the front of the breast, with yellow and bluish grey on the sides and other parts of the neck, the general colour of the plumage on the back, sides of the breast, and wings is brownish ashy grey in the cock, while the remainder of the under parts is dirty white. The coloured part of the eye is pearl-grey, the naked portion of the skin flesh-coloured, the beak horn-brown, and the leg grey. In the female the feathers of the nape and front of the breast are somewhat lighter in hue.

The place of the ordinary species is taken in Eastern Patagonia by the far less common Darwin's rhea (*R. darwini*), distinguished by its smaller size, relatively shorter legs, which are feathered down to the ankle-joint, as well as by the more mottled and less uniformly coloured plumage, and the pale green eggs.

Lastly, we have the long-billed rhea (*R. macrorhyncha*) of Northern Brazil, which is also a small species, characterised by its longer beak, larger and more flattened head-feathers, the longer feathers of the body, and the more slender legs, as well as by the general darker colouring, which is brownish grey mingled with black. Fossil remains of rheas, some of which belong to existing species, are met with in the caverns of Brazil and the superficial deposits of other districts of South America.

HEAD OF THE RHEA

In general habits rheas, although somewhat more gregarious, are very similar to ostriches, and are as thoroughly adapted to a life on the South American pampas as are the latter for existence in the South African veldt and karru. As a rule, each cock rhea associates with from five to seven hens, which he carefully guards from the attentions of other members of his kindred; although after the breeding season such family parties collect together in flocks, which may reach a total of sixty or more birds.

Possessed of a speed but little inferior to the ostrich, the rhea is further protected by the exactness with which the general pale bluish grey hue of its plumage assimilates to the distant haze, thus rendering it invisible even at a moderate distance. Its large form seems, indeed, to melt mysteriously out of sight into the surrounding blue, so that the hunter strains his eyes in vain to distinguish it. A truly noble bird when standing among the tall grasses of its native pampas, the cock rhea summons his scattered consorts by a hollow booming cry, not unlike that of the ostrich, accompanied by

HEAD OF LONG-BILLED RHEA

a kind of sighing or hissing sound.

When running from their pursuers, both sexes have the curious habit of raising one wing above the back in sail-like fashion. In hot weather these birds take readily to water, not only standing in this element with their bodies submerged, but also swimming boldly, though slowly, with their necks bent slightly forward

THE RHEA

From a photograph by Gambier Bolton, by permission of the Autotype Fine Art Co., Ltd., New Oxford Street, W.C.

and scarcely showing any portion of their bodies. In regard to the breeding habits of rheas, Darwin writes that "when we were at Bahia Blanca in the months of September and October, the eggs, in extraordinary numbers, were found all over the country. They lie either scattered and single, in which case they

they are occasionally fierce, and even dangerous, and that they have been known to attack a man on horseback, trying to kick and leap on him."

The truth of the statement that the cock undertakes the whole work of incubation has been demonstrated not only by observations made upon wild birds, but on captive specimens, which in England have bred freely. In the southern rhea the period of incubation lasts from thirty to thirty-one days; and while in the south the usual number of eggs in a nest is from fifteen to twenty, in the north as many as thirty-two have been observed.

The rhea, like the guanaco, is hunted with the bolas, or balls, one method being for a number of mounted men armed with this implement to enclose, with the aid of the female portion of the tribe, a considerable tract of country, and thus slaughter all the game contained within the circle, while the second and more sporting

A GROUP OF RHEAS

plan is for a single horseman to pursue the bird. In the latter case a horse of great endurance and endowed with a fair turn of speed is absolutely essential, while it is requisite that the steed should have learnt to follow all the twistings and doublings of the birds.

The supreme skill and judgment in casting the bolas at the right moment, and with the requisite strength and accuracy of aim necessary to bringing the game to bay, can in general be acquired only by those who have been accustomed to the use of the weapon from their childhood. Rheas have been hunted with the bolas for about two centuries, during which period they have learnt to start off at speed on catching sight of a mounted man; but till some fifty years ago, up to which date they were never shot, they displayed supreme disregard for a person on foot.

are never hatched, and are called by the Spaniards huachos: or they are collected together into a shallow excavation, which forms the nest. Out of the four nests which I saw, three contained twenty-two eggs each, and the fourth twenty-seven. In one day's hunting on horseback sixty-four eggs were found; forty-four of these were in two nests, and the remaining twenty scattered huachos. The Gauchos unanimously affirm, and there is no reason to doubt their statement, that the male bird alone hatches the eggs, and for some time afterwards accompanies the young. The cock when on the nest lies very close; I have myself almost ridden over one. It is asserted that at such times

CASSOWARIES AND EMEUS

Two important features serve at once to distinguish the cassowaries and their near allies the emeus (family *Casuariidæ*) from ostriches and rheas, the first of these being that the feathers have after-shafts of such large size as to make them practically double; while the second peculiarity is to be found in the eggs, which, instead of being light-coloured and smooth, are dark green in colour and granulated in texture. From the observations of Mr. W. Bennett on captive emeus, it also appears that in this group the females are larger than the males. Then, again, the wing is extremely rudimentary, so much so, indeed, that it may be invisible externally, the humerus, or upper bone, being very short.

In addition to certain peculiarities connected with the structure of the bones of the palate, the skeleton is further distinguished by the retention of rudiments of the furcula. In the presence of three toes to each foot, both emeus and cassowaries resemble rheas. By many naturalists the two genera under consideration are regarded as the representatives of as many distinct families, although the amount of difference between them seems sufficiently expressed by referring them to two subfamilies of a single family.

CASSOWARIES

Cassowaries form an extensive genus, containing at least nine well-defined species, confined to Australia, New Guinea, Ceram, and some of the neighbouring islands. They are specially characterised by the bare head being surmounted by a helmet-like prominence, formed by an upward extension of the bones of the skull, and covered with naked skin; by the bare neck, which may or may not be ornamented with pendent wattles, and likewise by the great length of the claw of the second or inner toe. The body is covered with dark-coloured feathers of a peculiarly loose and coarse structure, which are glossy, and appear more like hairs than the plumage of an ordinary bird; and the wing is represented externally merely by some four or five black quills devoid of barbs, presenting the appearance of very coarse bristles.

Although the whole of the numerous species of cassowaries are included in a single genus (*Casuarius*), this has been subdivided into three minor groups, distinguished by the form of the helmet, and the number of the wattles on the neck, or their absence. In the first group, as typically represented by the Ceram cassowary (*C. galeatus*), which appears to be confined to the island from which it takes its name, and was the first of these extraordinary birds made known to science, the helmet is flattened from side to side, or compressed, and the wattles on the neck are either two or double. Other representatives of this group are the Australian cassowary (*C. australis*) of Northern Australia; Beccari's cassowary (*C. beccarii*) of the Aru Islands and New Guinea; and the two-wattled cassowary (*C. bicarunculatus*) of the Aru Islands.

To the second group belongs the single-wattled cassowary (*C. uniappendiculatus*) from the island of Salwatti and the adjacent coast of New Guinea, which, while agreeing with the members of the preceding group in the form of the helmet, differs by having a single undivided wattle. Lastly, there is a third group, characterised by the circumstance that the helmet is flattened from above, or depressed, and wattles are absent; this group, which is exclusively Papuan, includes the Papuan cassowary (*C. papuanus*) of Northern New Guinea, Westermann's cassowary (*C. occipitalis*) from the island of Jobi, the painted cassowary (*C. picticollis*), confined to South-Eastern New Guinea, and Bennett's cassowary (*C. bennetti*), from New Britain.

Fossil remains of an extinct cassowary have been obtained from the superficial deposits of Australia, and in its whole distribution the genus corresponds very closely with the Australasian pouched mammals, none of its representatives occurring to the westward of the island of Celebes.

Other species in addition to those mentioned are

AUSTRALIAN CASSOWARY

known, and a scheme of classification somewhat different to the foregoing has been proposed.

In appearance, owing to the brilliant hues of blue, green, and red on the naked skin of the head and neck, coupled with the glossy sheen of the blue-black plumage, cassowaries are perhaps the handsomest of all the Ratitæ. The largest species of all, and the one in which the horn-coloured helmet attains the greatest development, is the Australian cassowary, which, when erect, stands considerably over five feet in height. Among its distinctive features is the fine cobalt blue tint of the throat and fore neck, and the red terminal flaps of the deeply-divided wattle, the Ceram species having the throat and fore neck dull purple.

Of the species without wattles, Bennett's cassowary —the muruk of the natives—has the neck entirely blue ; while in Westermann's cassowary the fore part of the neck is blue, and the hind portion red, the reverse of this characterising the painted-necked species. Nestlings have the plumage mottled at first, but at a later stage the colour is tawny.

In being forest-birds, cassowaries differ essentially from ostriches and rheas. They appear to be generally shy, and seldom seen in their native haunts; but we are, unfortunately, still in want of good accounts of the habits of these birds in their wild state. Those brought to Europe—where the hens will lay freely—are characterised by their extreme tameness and docility; but this, it is said, is largely due to their being mostly, if not invariably, specimens reared from early chicken-hood in captivity by the natives, among whom these birds are treated almost like domesticated fowls.

The Australian species is reported to frequent rocky wooded districts, where so many as seven or eight may be seen together, keeping almost entirely to the more open portion of the scrub, and seldom venturing out into the plains. From July to September, at least, these birds are known to feed chiefly upon an egg-shaped kind of blue berry, and their entire food is probably of a vegetable nature.

Regarding the muruk, Mr. Bennett was informed that the natives of New Britain capture these birds " when very young, soon after they are hatched, and rear them by hand, but can rarely or never capture the adult, from its being so shy and difficult of approach. They are exceedingly swift of foot, and possessed of great strength in their legs. On the least alarm they elevate the head, and, on seeing danger, dart among the thick bushes, thread localities where no human being can follow them, and disappear with incredible rapidity. The muruk, with its powerful legs and muscular thighs, has an extraordinary power of leaping."

This species utters a kind of chirping cry, susceptible of modulations according to the occasion. Unlike emeus, which kick outwards and backwards, cassowaries invariably kick forwards, at the same time elongating their bodies. In captivity they will not infrequently perform a kind of war-dance around any object that attracts their attention, accompanied by vigorous kickings and many bendings of the neck. In spite of their speed and the rapidity with which they move their limbs, cassowaries do not run after the manner of an ostrich, but may be said rather to trot.

As regards their breeding habits in a wild state very little has been ascertained, although it appears that at this season they associate only in pairs. From native reports concerning the rare Ceram cassowary, Dr. A. R. Wallace wrote that the female laid from three to five eggs, which were brooded by each sex in turn ; but from observations made on menagerie specimens, it appears that all the work of hatching devolves upon the cock, the period of incubation being about seven weeks. Although cassowaries will lay freely, it is but seldom that the eggs are hatched in captivity. In colour, the latter are dark green, with the surface of the shell beautifully granulated, or shagreened. The young cassowaries, in which the position of the helmet is indicated by a flat, horny plate, are carefully tended and fed by the cock bird until able to shift for themselves.

EMEUS

Originally applied indifferently to the members of both the preceding and the present genus, the name emeu (a derivative from the Portuguese word emea, meaning apparently a crane, and then any large bird) is now by common consent restricted to the latter. Agreeing with the cassowaries in the features mentioned above, the emeus—of which all the species are restricted to Australia and some of the adjacent islands —are distinguished by the absence of a helmet, and the complete feathering of the head and neck, and the normal length of the claw of the second toe; the claws of all three toes being much shorter than in the allied genus. They are further characterised by the beak being depressed and broad, instead of narrow, compressed, and keeled ; as they are by the absence of the bare black quills in the still more rudimentary wing.

Standing next in point of size among living birds to the ostrich, the common emeu (*Dromæus novæ-hollandiæ*) of Eastern Australia has the general hue of the plumage light brown, mottled in some parts with grey; the individual feathers being uniform blackish grey, except near the tips, where they are black, with a broad subterminal band of rufous (page 919).

In Western Australia this bird is replaced by the so-called spotted emeu (*D. irroratus*), said to be of more slender build, and having the feathers barred with white and dark grey, and terminating in a black spot with a rufous margin ; doubts have, however, been expressed as to whether it is even entitled to rank as a distinct race.

While the two sexes of the adult are nearly similar, the young of the common emeu have the ground-colour of the plumage greyish white, with two stripes of black down the back, and two others on each side, both divided by a narrow median streak of white, and continued on to the head, where they break up into spots, while there are also others on the fore part of the neck and breast, which terminate on the thighs.

A much smaller species is the black emeu (*D. ater,* or *D. perroni*), which formerly inhabited Kangaroo Island, in Bass Strait, but was exterminated by a squatter during the last century. Emeu bones have also been obtained on King Island, Bass Strait. Colonel Legge has stated that he recalls having seen a pair of Tasmanian emeus in his boyhood, and believes that they were slightly smaller than the mainland species.

SKULL OF AUSTRALIAN CASSOWARY

As it also lays a larger egg, it is regarded as distinct. As in the case of cassowaries, emeus are represented by an extinct species from the superficial deposits of Australia.

At one time abundant in the neighbourhood of Botany Bay and Port Jackson, where it formed as characteristic a feature in the landscape as kangaroos and wallabies, the emeu is now only to be met with in the far interior, where it is yearly becoming scarcer. Unlike cassowaries, emeus are inhabitants of the plains and open forest-country, where, although strictly monogamous during the breeding season, they associate in small parties. Their food consists of fruits, roots, grass, and other herbage; their chief feeding time being the cool of the early morning.

Possessed of great keenness of vision, and swift of foot, emeus rival kangaroos in speed, and afford an exciting chase with dogs. Such hunts do not end till the birds are thoroughly exhausted, when, if seized by the neck, in order to avoid kicks from their powerful legs, they are soon pulled down.

As with most other members of the order, the task of incubation falls to the share of the cock, by whom the eggs, which vary in number from nine to thirteen, are brooded for a period of from fifty-four to sixty-four days. The nest is but a poor affair, consisting merely of a shallow hollow scooped in the sandy soil. In colour the eggs vary from dark bottle-green to light bluish green; their length being just short of 5 inches, and their transverse diameter 3¾ inches.

During the breeding season, at least, hen birds utter a peculiar loud booming sound, produced through the intervention of a pouch communicating with the windpipe, on the front of which it opens by a small aperture; this structure being confined to the female sex. From her larger size, the hen emeu is very liable to be mistaken for the cock.

Writing of the difference in the habits and appearance of the two sexes, Mr. Bennett observed, of a pair in his possession, " one is considerably larger than the other, stouter in limb, and more robust in every feature; it has a slight top-knot, and goes strutting about, especially in damp weather, with its breast feathers fully out, like a pouter-pigeon, or, rather, some huge turkey-cock. It is usually the more courageous and pugilistic. It makes a deep, hollow, guttural boom when under any gentle excitement of pride or pleasure, especially on damp evenings, or in the still hours of the night, sounding like a small gong or distant muffled drum. The other is more agile and graceful in all its movements, corresponding with its slender frame, more docile and inquisitive, fleeter of foot, and with no voice beyond a suppressed hiss when angry, and a sort of grunt when distressed."

The former, although at first regarded as the male, turned out to be the female.

In the wild state emeus take readily to water, and have on more than one occasion been observed swimming wide rivers. Beneath the skin these birds have a thick layer of fat, yielding a pale amber-coloured oil, free from either taste or smell. Very easily tamed, and, in a domesticated state, thriving well in Europe, where it breeds freely, the emeu is noticeable for a curious and somewhat mischievous disposition. It will, for instance, invariably endeavour to inspect every strange object brought into its vicinity, while if a visitor shows any symptoms of fear when brought into a paddock or park containing one or more of these birds, and attempts to escape by flight, he will be certain to be pursued.

On one occasion, at Sydney, a man thus hunted by a tame emeu was much astonished at having his hat removed by the bird. In such chases, emeus appear to be actuated more by a spirit of mischief than anything else; but when they are brought to bay, and take to kicking with their muscular legs, they are formidable adversaries. In kicking, the blow is delivered outwards and backwards.

During the late portion of the Tertiary epoch Australia possessed certain large birds, which have

EMEU AND CHICKS

been named *Dromornis* and *Genyornis*, and apparently indicate an extinct family, *Dromornithidæ*, more or less closely allied to the *Casuariidæ*.

Remains of a very large kind of flightless bird, forming a genus, and perhaps also a family, by itself have been obtained from the lower Tertiary deposits of the Fayum district of Egypt. This bird is the oldest of the whole group.

KIWIS

As the ostriches are the most specialised of the living members of the order, so the kiwis (family *Apterygidæ*) of New Zealand may be regarded as those occupying the most generalised position in the group. The specialisation of the ostriches is shown, among other features, by the gigantic stature of those birds, the reduction in the number of the toes, and the total absence of any trace of a bony bridge at the lower end of the tibia. The kiwis, on the other hand, exhibit their more generalised nature by their comparatively small size—it being obvious that if the Ratitæ were derived from flying birds, the intermediate forms must have been small—by the presence of four complete toes, and by remnants of the bony bridge at the lower end of the tibia.

Whether the long beak of the kiwis is also a generalised feature may be doubtful. If these birds have any close affinity with the tinamus, it cannot be thus regarded ; but if, as some think, they are allied to the rails, then it may be looked upon in this light. Kiwis and roas (as some of the species are called by the Maori) differ from all the other living members of the order by their small bodily size, the presence of four toes to the foot, and the long and slender beak. They are further characterised by the females being much superior in size to the males, the complete lack of after-shafts to the feathers, and the absence of any trace of the furcula. The bones of the wing—especially the humerus—are very small and slender, and externally the whole wing is completely concealed by the plumage of the back.

In general appearance the entire plumage is markedly hair-like, the individual feathers being pointed, and composed of separate filaments towards the end of the shaft, of which the basal half is downy. In build kiwis are very robust, the thighs and legs being remarkably muscular and strong, while the toes are furnished with strong claws. Although in old birds the scales investing the metatarsus have overlapping edges, and form a perfectly smooth surface, in the young they are soft, detached, and reticulated. The general colour of the plumage is mottled grey and brown, the feathers having in some cases light-coloured shafts, and in others dark cross-bars.

In addition to great individual variations of size, kiwis are remarkable for their very large eggs, which are of a creamy white colour, and out of all proportion to the dimensions of the birds by which they are laid. In having the nostrils placed at the tip of the beak, these birds are unique.

At the present day kiwis are represented by several species, of which the first known to science was the South Island *Apteryx australis*. This species is of large size and stout build, with a very long beak ; the general colour of the plumage being lighter, and the individual feathers of a sandier and more greyish brown tinge, than in the next kind.

In the North Island kiwi (*A. mantelli*) the general colour of the plumage of the upper parts is dark rufous streaked with blackish brown, while the under parts are pale greyish brown ; the streaky appearance of the upper surface being caused by each feather having the middle line pale rufous brown, darker towards the tip, and the long hair-like filaments black. The total length of the male, following the curvature of the back, is about 23 inches, and that of the female 27½ inches.

Among the other species, the little grey kiwi (*A. oweni*) of the South Island is characterised by its small size—the length of the male being only 17½ inches—by its moderately long beak, and more slender legs ; the general hue of the plumage of the upper parts being light yellowish brown, mottled and obscurely barred with wavy blackish brown markings, while beneath it is paler, becoming fulvous on the abdomen, where there are faint brown bars.

Some doubt exists as to the right to distinction of the large grey kiwi (*A. haasti*), which has been regarded as merely a hybrid between the South Island and the little grey kiwi. This idea is, however, put out of court by the circumstance that it exists also in the North Island, where the two latter do not occur. It is a large and thick-billed species, of darker colour than the little grey kiwi, the dark bars on the plumage being nearly black, and the fulvous markings tinged with chestnut.

Fossilised remains of the existing species occur with those of the moas, while one is supposed to be extinct, and has been named *Pseudapteryx*. In habits kiwis are purely nocturnal, and when still abundant were commonly found in parties of from six to twelve, their shrill nocturnal cries resounding far and wide throughout the mountainous parts of the country they frequent.

No better account of their general mode of life is extant than one from the pen of Sir W. J. Buller, who,

A KIWI   Photo, W. P. Dando

after mentioning that the kiwi is in some measure compensated for the absence of wings by its swiftness of foot, proceeds to observe that "when running it makes wide strides and carries the body in an oblique position, with the neck stretched to its full extent and inclined forwards. In the twilight it moves about cautiously and as noiselessly as a rat, to which, indeed, at this time it bears some outward resemblance. In a quiescent posture the body generally assumes a perfectly rotund appearance ; and the bird sometimes, but only rarely, supports itself by resting the point of its bill on the ground. It often yawns when disturbed in the daytime, gaping its mandibles in a very grotesque manner. When provoked, it erects the body, and, raising the foot to the breast, strikes downwards with considerable force and rapidity, thus using its sharp and powerful claws as weapons of defence.

"When hunting for its food the bird makes a continual sniffing sound through the nostrils, which are placed at the extremity of the upper mandible. Whether it is guided as much by touch as by smell I cannot safely say; but it appears to me that both senses are called into action. That the sense of touch is highly developed seems quite certain, because the bird, although it may not be audibly sniffing, will always first touch an object with the point of its bill, whether in the act of feeding or of surveying the ground; and when shut up in a cage or confined in a room, it may be heard all through the night tapping softly at the walls.

"The sniffing sound is heard only when the kiwi is in the act of feeding or hunting for food; but I have sometimes observed the bird touching the ground close to or immediately round a worm which it had dropped without being able to find it.

"It is interesting to watch the bird in a state of freedom foraging for worms, which constitute its principal food; it moves about with a slow action of the body, and the long, flexible beak is driven into the soft ground, generally home to the very root, and is either immediately withdrawn with a worm held at the extreme tip of the mandibles, or it is gently moved to and fro, by an action of the head and neck, the body of the bird being perfectly steady. It is amusing to watch the extreme care and deliberation with which the bird draws the worm from its hiding-place, coaxing it out, as it were, by degrees, instead of pulling roughly or breaking it."

On getting the worm fairly out of the ground, the bird throws up its head with a jerk, and swallows its prey whole. The stomachs of specimens that have been dissected contain pebbles, remains of beetles, and the kernels of berries.

In captivity kiwis are dull, listless creatures during the day; lying closely huddled together, and slumbering so soundly that no noise will arouse them. If stirred up with a stick, or suddenly wakened, they make a few drowsy movements, and soon relapse into sleep. From observations made on specimens in captivity, it appears that the female kiwi (unlike the other members of the order) lays but one or two eggs annually, which are deposited in hollows in the ground, and incubated by her partner. When there are two, the eggs, which are placed lengthways side by side, are of such a size as to protrude from the sides of the narrow body of the sitting bird. During the breeding season the kiwi is silent. An egg of the North Island kiwi measured a little over 5 inches in length by 3 in breadth.

MOAS

The fate impending in the case of the kiwis has long since overtaken their gigantic extinct cousins the

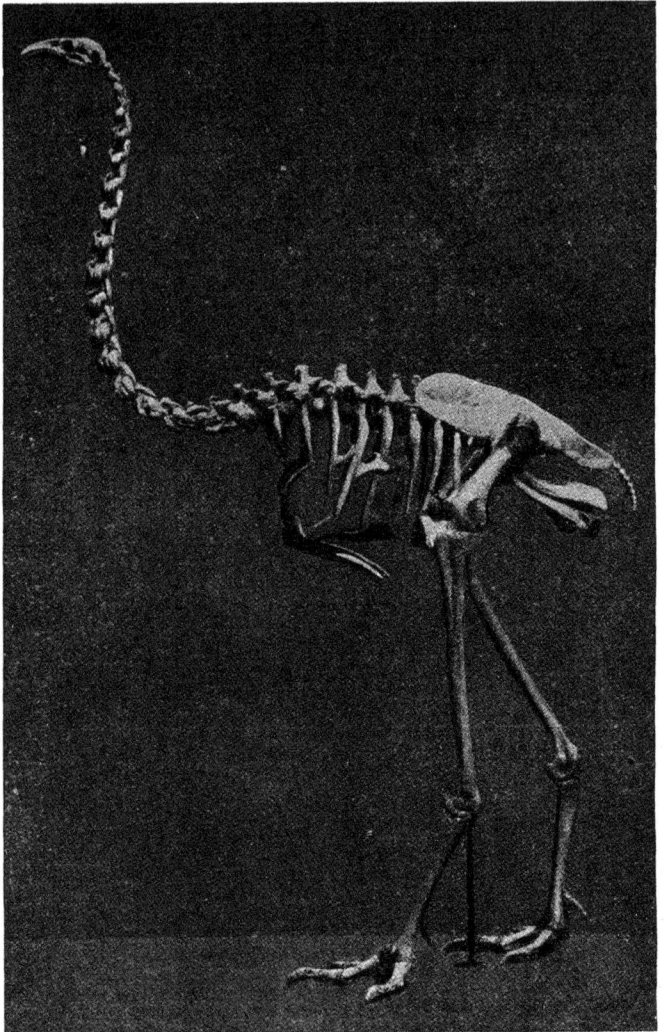

SKELETON OF GIANT MOA

moas (family *Dinornithidæ*), which had already disappeared from New Zealand when those islands were first colonised from Europe, although there is good reason to believe that they lived on till within the last five hundred or four hundred years, if not to a considerably later date. These birds, of which not only the bones, but in some cases the dried skin, feathers, and egg-shells, as well as the pebbles they were in the habit of swallowing, have been preserved in the superficial deposits of New Zealand, attained a wonderful development in those islands, where they were secure from persecution till man appeared on the scene.

Not only did the larger members of the group far exceed the ostrich in size, but they were extraordinarily numerous in species, as they were also in individuals; such a marvellous exuberance of gigantic bird-life being unknown elsewhere on the face of the globe in such a small area. As regards size, the largest moas could have been but little short of 12 feet in height, the tibia being considerably over a yard in length ; while the smallest were not larger than a turkey. In reference to their numbers, it may be mentioned that there are some twenty species, arranged in about six genera ; and the surface of many parts of the country, as well as bogs and swamps, literally swarmed with their bones.

Some of the moas had four toes to the foot, and others three, but all differed from kiwis in having a bony ridge over the groove for the extensor tendons of the tibia. They are, therefore, evidently the least specialised members of the order yet mentioned, seeing that this bridge is present in the majority of flying birds, and has evidently been lost in all the existing Ratitæ (pp. 13 and 914).

While agreeing in some parts of their organisation with kiwis, moas are distinguished by the short beak and the presence of after-shafts to the feathers ; while in the larger forms, at any rate, not only was the wing, but likewise the whole shoulder-girdle, wanting. There is, however, reason to believe that certain pigmy moas—which from their size were evidently the most generalised members of the group—retained some of the bones connected with the wing.

Moas were represented by several very distinct structural modifications ; the largest being the long-legged, or true, moas (*Dinornis*), characterised by the long and comparatively slender leg-bones, and also the large and depressed skulls. In marked contrast to these were the short-legged, or elephant-footed, moas (*Pachyornis*), in which the limb-bones are remarkable for their short and massive form ; the metatarsus being most especially noteworthy in this respect. In these birds the skull is vaulted and the beak narrow and sharp ;

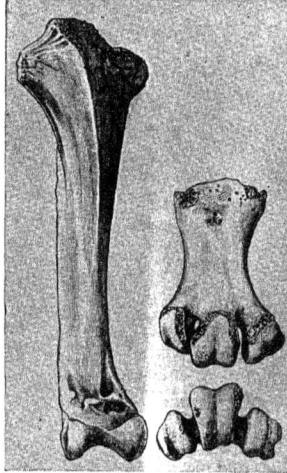

RIGHT TIBIA AND METATARSUS OF SHORT-LEGGED MOA

but in the somewhat smaller and less stoutly-limbed broad-billed moas (*Emeus*) it is broad, blunt, and rounded. The other species, in all of which the beak was sharp and narrow, are of relatively small stature, and include the smallest representatives of the family, some of which were less than a yard in height. The eggs of the moas were of a pale green colour, and probably formed a favourite food of the Maori, by whom these birds were evidently exterminated.

ROCS

For a long period the marshes of Madagascar have yielded the egg-shells of enormous extinct birds, in search of which the natives are accustomed to probe with iron rods ; the largest of these eggs having a longer circumference of upwards of thirty-six inches, and a girth of thirty inches. From these eggs probably arose the legend of the " roc " of the old Arab voyagers ; and it is, at any rate, convenient to adopt that name as the popular designation for the members of the family *Æpyornithidæ*, all of which are included in the genus *Æpyornis*. In the course of time naturalists were rewarded by the discovery of the bones of the birds which laid these gigantic eggs ; some of these remains indicating a bird of larger build than the most gigantic moa, the metatarsus being especially remarkable for its massiveness. Some of these birds appear to have had four toes, and they all differ from moas in the absence of a bony ridge at the lower end of the tibia. The skull was short and moa-like, and the wing seems to have been completely aborted.

ORDER SAURURÆ—LIZARD-TAILED BIRDS

Low down in the geological series in the strata lying below the Chalk and forming the upper part of the Jurassic series, there occur in Bavaria remains of birds departing more widely from existing types than any hitherto mentioned. These birds, of which only a couple of imperfect skeletons, with impressions of the wing and tail feathers, are known, are named *Archæopteryx*, and constitute a group, Archæornithes, with the one order Saururæ, of equivalent rank to one containing all other birds.

In size these lizard-tailed birds were about equal to rooks, with which they agree in being evidently adapted for perching on the boughs of trees. In addition to the possession of a small number of conical teeth in the short jaws, they are characterised by having a long, lizard-like, tapering tail, from each joint of which a pair of feathers takes origin. In this respect they differ from other birds in which the bones of the tail are shortened, so that the tail feathers arise in a fan-shaped manner from its terminal joint (page 915.)

In addition to this, they are further characterised by the first three metacarpal bones of the wing, as well as those representing the corresponding fingers, being perfectly distinct from one another, and each terminal joint of the latter being furnished with a well-developed claw ; all other birds having the metacarpal bones, as well as some of those of the fingers, welded together,

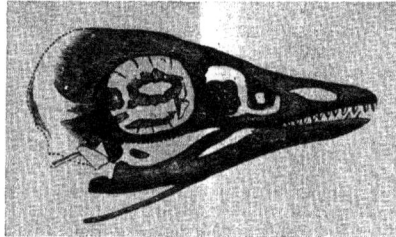

SKULL OF ARCHÆOPTERYX

while there are, at most, but two claws, as in the young of the seriema.

In having cup-shaped articular surfaces to the bodies of the vertebræ, the lizard-tailed birds resemble the above-mentioned *Ichthyornis* of the Chalk period, but they differ from all other members of the class in having the three bones constituting the pelvis perfectly distinct from one another (as in most reptiles), while in the leg the tibia and fibula are likewise separate. As regards the general structure of the wing and leg, these remarkable birds agree, however, with their modern allies ; the foot having a complete cannon-bone, and only four toes, of which the first is directed backwards. In the absence of hook-like (uncinate) processes to the ribs, *Archæopteryx* appears to be more specialised than ordinary birds, seeing that these elements exist in many reptiles.

Finally, it should be noticed that these lizard-tailed birds and their short-tailed but toothed successors of the Chalk period afford a most valuable contribution in favour of the doctrine of evolution, approximating more and more, as we descend in the geological scale, to reptiles, to which birds are evidently nearly related.

Apparently the whole body of *Archæopteryx* was covered with feathers ; and it may be regarded as certain that this strange bird could fly, although its flight may not have been very powerful or prolonged.   R. LYDEKKER

CLASS III. REPTILIA

CHARACTERISTICS AND CLASSIFICATION OF REPTILES

By R. LYDEKKER

In ordinary language the term reptile is applied indifferently to such creatures as crocodiles, tortoises, lizards, snakes, frogs, and salamanders, but by the naturalist it is used in a more restricted sense, and includes only the first four of these groups, together with a host of extinct types; while the frogs and salamanders, with certain other forms, both living and extinct, on account of important structural differences, constitute a class by themselves, known as the Amphibia, and bearing the same rank as the class Reptilia. To an ordinary observer there would seem little in common between a scaled lizard or snake, a cuirassed crocodile, and a carapaced tortoise on the one hand, and a feathered bird on the other. Nevertheless, the connection between reptiles and birds is exceedingly intimate—so close, indeed, that Professor Huxley has termed the latter greatly modified reptiles.

At the present day the two groups are, however, very widely sundered; and it is only by the study of forms long since extinct that it is possible to realise the intimate relationship that really exists between them. That birds are the direct descendants of reptiles may really be taken for granted, although we are still unacquainted with the immediate links connecting the two classes. In another direction reptiles are, however, connected through other extinct forms with amphibians; while from the so-called anomodont, or mammal-like, reptiles mammals themselves have undoubtedly originated.

As will be pointed out later on, amphibians are also intimately connected with the class of fishes, thereby indicating how closely allied are all the classes of the vertebrates, and how difficult is the task of the naturalist to distinguish them satisfactorily one from another when the whole of the extinct forms are taken into consideration. It is, indeed, solely from the still imperfect condition of our knowledge of the past denizens of the globe that it is possible to formulate any definitions at all, for had we the whole chain of organised nature before us, it will be obvious that no breaks would exist, but that every group would pass by imperceptible degrees into the earlier one from which it originated.

Proceeding to the consideration of what constitutes a reptile, as distinct from any other vertebrate, it will be advisable in the first instance to point out some of the features in which reptiles agree with birds, and thereby differ from mammals. In the first place, the skull of a reptile and a bird articulates with the first vertebra by a single knob, or condyle (V of the figure); while each half of the lower jaw is composed of several distinct bones, and the whole lower jaw articulates with the skull by the intervention of a separate quadrate bone. Then, again, reptiles agree with birds in that the appendages developed from the outer layer of the skin

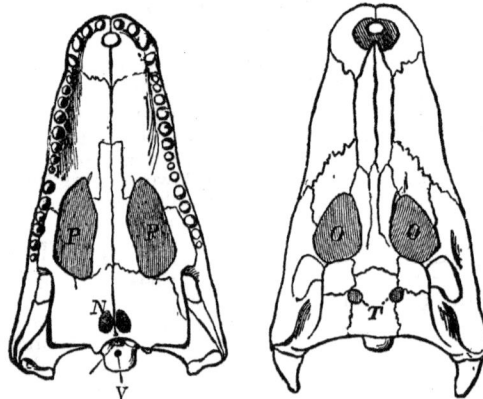

LOWER AND UPPER SURFACES OF SKULL OF A CROCODILE

N, aperture of internal nostrils; O, sockets of eyes; P, vacuities of palate; T, frontal vacuities, or fossæ; V, condyle of occiput

never take the form of hairs, and also by the circumstance that the young are not nourished by means of milk secreted by special glands on the body of the female parent. Neither are gills developed at any period of life, throughout which respiration is effected by means of lungs; this being a feature common to mammals, birds, and reptiles. Reptiles and birds differ, however, from mammals in regard to the position of the ankle-joint, which is situated between the upper and lower rows of small bones entering into the composition of that part of the skeleton, instead of, as in mammals, between the upper row and the lower bones of the leg itself.

In producing their young from eggs (sometimes retained within the body of the parent until hatched), reptiles resemble not only birds, but likewise the lowest mammals; with which they also agree in the nature of the investments surrounding the embryo. As regards the distinction between birds and reptiles, the latter are broadly separated from the former by the absence of feathers; the appendages of the outer layer of the skin taking the form of overlapping horny scales of granules, or of large shields uniting by their opposed edges. Moreover, most reptiles differ from birds in having more than three digits in the fore limb; while in no cases are the collar-bones fused into a furcula, as they are in all flying birds.

A further distinction is to be found in connection with the circulatory system, the blood of all existing reptiles having practically the same temperature as that of the surrounding air, and undergoing oscillations according to changes in the latter; while the aorta, or great propelling blood-vessel of the heart, is double, and crosses both branches (instead of only the left branch) of the windpipe. It will be obvious, however, that these two last characters cannot be verified in the case of extinct reptiles, among which it is quite possible that there may have been some in which the blood had a temperature above that of the surrounding air. A similar remark will apply to the absence among living reptiles of those ramifications of the bronchial tubes throughout the body which form such a characteristic feature in the structure of birds.

As additional features in the skeleton, it may be noticed that reptiles never have the terminal faces of the centra, or bodies, of the vertebræ saddle-shaped; while in those forms in which the number of toes in the hind limb is reduced to three, the metatarsal bones do not unite to form a cannon-bone in conjunction with the lower row of bones belonging to the ankle-joint. Then, again, with the exception of one remarkable extinct group, reptiles, as a rule, are characterised by the three bones of the pelvis remaining distinct from one another through life, whereas in all existing birds they are welded together. There are likewise differences in regard to the form and structure of the breast-bone and sacrum, into the consideration of which it will be unnecessary to enter in this work.

In marked contrast to the general uniformity in outward appearance and internal structure characterising all birds, the various groups of reptiles differ widely from one another both as regards external form and internal structure. Externally, a lizard, a snake, and a tortoise present the most marked differences in general appearance among living members of the order ; while among extinct types there were some which walked on their hind limbs alone, after the manner of birds, and others in which the fore limbs were modified into wings and the digits connected by a leathery membrane like that of bats.

In a typical reptile, such as a lizard or crocodile, both pairs of limbs are well-developed and of approximately equal length, but in snakes all external traces of limbs have disappeared ; while in the extinct flying-dragons, or pterodactyles, the fore limbs much exceed the hind ones in size, and in many of the so-called dinosaurs, or giant extinct land reptiles, the excess in size falls to the share of the hind pair of limbs. In other cases, again, the limbs may be modified into paddles, adapted for progression in the water, as in existing turtles, and the extinct fish-lizards, or ichthyosaurs ; the body in the latter assuming a somewhat fish-like form. In nearly all cases reptiles have long and well-developed tails, although in some of the flying-dragons these become rudimentary.

A large number of reptiles are characterised by the development of bony plates within the deep layer of the skin ; such plates, which are well-displayed in existing crocodiles, being overlain by horny shields, and thus corresponding in every respect with those forming the carapace of the armadillos among mammals. Among certain extinct dinosaurs these bony plates attain a development unparalleled at the present day, and in some they are believed to have occupied the extraordinary position shown in the figure on page 1492.

Still more remarkable differences exist with regard to the form and structure of the teeth ; which, instead of being, as in the two preceding classes, strictly confined to the margins of the jaws, may be spread over the entire palate. In spite, however, of this diversity of form, the teeth of

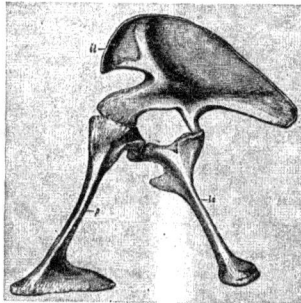

BONES OF LEFT SIDE OF PELVIS OF CARNIVOROUS DINOSAUR
il, ilium ; *p*, pubis ; *is*, ischium

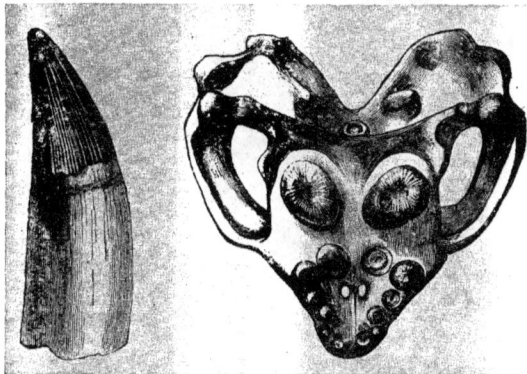

CONICAL TOOTH OF AN EXTINCT PLESIOSAUR AND UNDER SURFACE OF SKULL OF CYAMODUS WITH PAVEMENT-LIKE TEETH ON PALATE

all living and most extinct reptiles differ from many of those of the majority of mammals in that they are never implanted in the jaws by two or more roots ; while in no cases are their crowns complicated by the presence of infoldings of enamel. The simplest type of reptilian tooth is in the form of a cone ; such conical teeth being confined to the margins of the jaws, where, as among crocodiles, they may be implanted in distinct sockets, or, as in the extinct fish-lizards, in an open groove. In other cases, as among lizards, teeth of the same general type may be held by a bony deposit either to the summit or to one side of the margin of the jaw. In place of the one regular replacement characterising the anterior teeth of the majority of mammals, the teeth of most reptiles are replaced irregularly and continuously throughout life ; the successional teeth growing up beneath the bases of those in use, and gradually causing an absorption of their roots.

When teeth are distributed over the whole or a greater portion of the palate, they generally assume a more or less flattened and bean-like shape, so as to form a kind of pavement in the mouth, as shown in the accompanying figure of the under surface of the skull of the extinct *Cyamodus*. Between conical and pavement-like teeth there are various intermediate grades, some of which will be referred to in the sequel. It is, however, by no means all members of the class that are provided with teeth ; tortoises and turtles being living examples of the total loss of these organs, and the consequent conversion of the jaws into horn-clad beaks. Some of the fish-lizards had only a few rudimentary teeth remaining ; and certain representatives of the extinct flying-dragons were likewise devoid of teeth ; and, as in these forms the horn-covered jaws were long and narrow, the resemblance to the beak of a bird becomes most marked. Resemblances of this nature are, however, purely adaptive, and the features on which they are based have been acquired independently in birds and reptiles.

It has been stated that the vertebræ

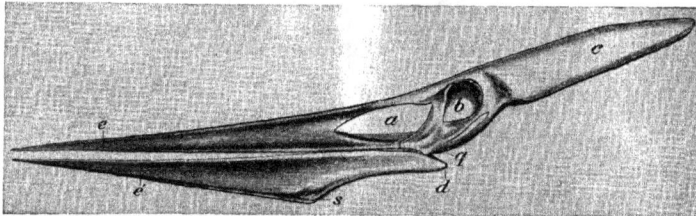

LEFT SIDE OF SKULL OF A BEAKED FLYING-DRAGON
a, vacuity in front of eye ; *b*, socket of eye ; *c*, occipital spine ; *d*, angle of lower jaw ; *e*, extremity of upper, and *é*, of lower jaw ; *q*, articulation of skull proper with lower jaw ; *s*, point where the two branches of lower jaw diverge

of reptiles never articulate by means of those saddle-shaped surfaces so characteristic of birds. They present, however, great diversity of structure in this respect. In some cases, for instance, as in fish-lizards, the bodies, or centra, of the vertebræ are very short from front to back, and have concave surfaces both in front and behind for mutual articulation. In marked contrast to this type are the neck-vertebræ of a dinosaur, where the anterior end of the body of each vertebra forms a convex knob (*b*), received into a cup at the posterior end of the vertebra in advance.

In other instances, as in the existing crocodiles and lizards, an arrangement precisely the reverse of the last is present—that is to say, the ball is at the hind end, and the cup at the front of the body of the vertebra. In a few lizards and in all snakes the vertebræ are further complicated by the development of additional articular facets, taking the form of wedge-like projections from one vertebra received into cavern-like excavations in the adjacent one.

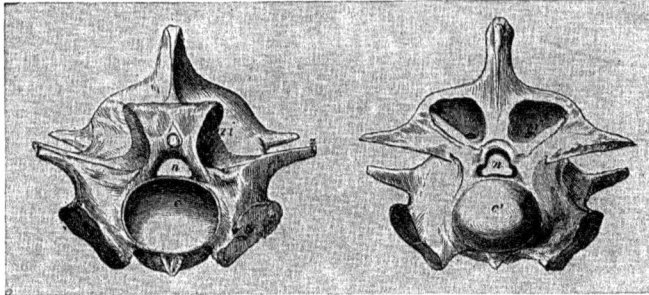

FRONT AND BACK VIEWS OF VERTEBRA OF SNAKE
zi, indicates the additional articular process, which is received into cavity zi '

Omitting mention of certain features connected with their osteology, it may be observed that, while among those existing reptiles furnished with four or five toes to each foot, some, like ordinary tortoises, have the same number of joints in each toe as mammals—that is to say, two in the first toe, and three in each of the others—in the great majority there is a departure from this simple arrangement. In the lizards, for instance, the number of joints in the toes (reckoning from the first to the fifth digit) is 2, 3, 4, 5, 3 in the fore limb, and 2, 3, 4, 5, 4 in the hind limb; while in crocodiles, where there are only four toes in the latter, the numbers are respectively 2, 3, 4, 4, 3, and 2, 3, 4, 4. In this increasing number of joints in the toes from the first to the fourth, such reptiles approximate to birds.

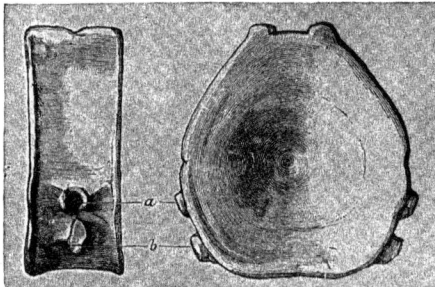

SIDE AND FRONT VIEWS OF BODY OF VERTEBRA OF FISH-LIZARD
a and *b*, attachment of ribs

As regards their soft internal parts, reptiles are characterised by the low development of their brains; which, in conjunction with their cold blood, accounts for the generally sluggish movements of their existing representatives. With the exception of crocodiles, reptiles differ from birds in that the heart has only three, in place of four, complete chambers, thus causing the freshly oxygenated blood returning from the lungs to be mixed with the effete blood which has traversed the body; and even in crocodiles, where the heart has practically four chambers, the fresh and effete blood is partially mingled, owing to a communication between the vessels just outside the heart. The corpuscles of the blood are oval in shape, and furnished with a central nucleus. Like birds, reptiles never have a midriff completely separating the cavity of the chest from that of the abdomen.

In all reptiles the tail passes more or less imperceptibly into the body, of which it seems an integral part;

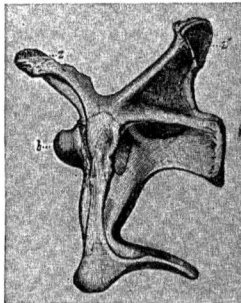

LEFT-SIDE VIEW OF NECK VERTEBRA OF A DINOSAUR
b, anterior ball

and although this feature is perceptible in some of the lower types of mammals, it has completely disappeared in such of the higher forms as retain the tail, in which that appendage is sharply marked off from the body. A fish-like character is displayed among reptiles in the presence of powerful longitudinal muscles in the back and tail; these largely aiding the limbs, of which the feet are short, bent, and flattened. Modern reptiles do not walk or run in the manner characteristic of mammals or even birds. On the contrary, they progress by means of swift rushes of short duration, after each of which they stop to look about them, even if they do not immediately require a more or less lengthened period of repose. Neither the form and structure of the limbs themselves nor the manner in which the weight of the body is superimposed upon them is, indeed, for sustained speed, and the short periods of extreme and violent activity are always succeeded sooner or later by periods of inaction and semi-torpor.

Very characteristic of certain groups of modern reptiles is the tendency to a reduction, or even complete loss, of the limbs; and it is highly noteworthy that extreme reduction in this respect has been attained in groups as distinct from one another as are slow-worms and snakes. And not only so, for even among lizards the loss of the limbs has taken place independently in at least two different families. In a sense, this loss of the limbs is a degradation, although, for the needs of their particular mode of life, snakes are some of the most perfectly organised of all animals.

Some, at least, of the giant land reptiles of the Secondary geological period walked, however, more in the manner characteristic of mammals and birds; while others had wings capable of carrying them in the air; and others, again, had their limbs modified into paddles in whale-fashion (with the exception—in many cases, at any rate—that the hind pair was retained), and vied in speed with the fishes in their own element. Reptiles were, in fact, at that epoch of the earth's history the dominant animals of the world, taking the place now held by the larger mammals on land, the birds in the air (although there were a few contemporary representatives of this order), and the whales and porpoises in the ocean. In the matter of bodily size many of the early giant land reptiles far exceeded any other creatures that have ever walked this earth. But small brains of feeble capacity were unfitted to enable their owners to maintain their status when creatures of a higher grade or organisation appeared in force on the field; and as surely as crossbows and pikes were bound to be driven out of the field and replaced by 12-inch guns and express rifles when once gunpowder made its appearance, the great hulking reptiles of the middle period of the earth's history were doomed to extinction so soon as mammals and birds became forces to be reckoned with.

Reptiles having come into existence at an earlier period than either mammals or birds, and attaining an enormous development during epochs when both those groups were but feebly represented, have, in fact, suffered to a much greater extent by the extinction of types with the lapse of time. This will be made apparent by the statement that while the number of existing orders of reptiles is but four (of which one is represented by a single species) extinct types include at least seven orders.

As regards the serial arrangement of the various orders of reptiles, it has of late years become more and more evident that the remarkable extinct anomodonts of the equivalent of the Triassic epoch in Africa and elsewhere differ very widely from all other reptiles, and approach mammals, of which they were undoubtedly the ancestors. Bearing this in mind, reptiles may be divided into two brigades, the one including the anomodonts and their immediate relatives, and the other all the rest. The former brigade is termed mammal-like reptiles, and the latter (from which birds took their origin) bird-like reptiles. In technical language these two brigades are respectively known as the Theromorpha and the Ornithomorpha.

CLASSIFICATION OF REPTILIA

Extinct Groups are marked with a †

	ORDER	SUBORDER
Bird-like Groups	I. CROCODILIA .. .. Crocodiles	1 Eusuchia 2 †Aëtosauria 3 †Parasuchia
	II. †DINOSAURIA .. .. Dinosaurs	1 Theropoda 2 Sauropoda 3 Ornithopoda
	III. †ORNITHOSAURIA Pterodactyles	
	IV. RHYNCHOCEPHALIA Tuateras	1 †Protorosauria 2 Rhynchocephalia Vera 3 †Acrosauria
	V. †PELYCOSAURIA	
	VI. SQUAMATA .. .. Snakes and Lizards	1 Lacertilia 2 Rhiptoglossa 3 Ophidia 4 †Dolichosauria 5 †Pythonomorpha
	VII. †ICHTHYOPTERYGIA Ichthyosaurs	
	VIII. †SAUROPTERYGIA Plesiosaurs	
	IX. CHELONIA .. .. Tortoises & Turtles	1 Cryptodira 2 Pleurodira 3 †Amphichelydia 4 Trionychoidia
Mammal-like Group	X. †PLACODONTIA Placodus	
	XI. †ANOMODONTIA .. Anomodonts ..	1 Dicynodontia 2 Theriodontia 3 Cotylosauria 4 Pariasauria

Of these groups, by far the most numerously represented at the present day is the one containing the lizards and snakes, all of which are highly specialised, although, in the case of the latter degraded types, occupying a position in the class analogous to that held by the perching birds in the preceding class; the majority being comparatively small or medium-sized forms. Next in point of numbers come the tortoises and turtles, all of which are protected by the presence of a bony shell, or carapace, and some of which attain very large dimensions. The third numerical position

in the reptilian fauna of the present day is held by the crocodiles, of which there are some twenty-four species, all of relatively large size, and all more or less aquatic in their habits. The fourth existing order is now represented only by the lizard-like New Zealand tuatera, although in past times there were a host of allied forms.

Of the extinct orders, the whole, or nearly the whole, of their representatives ceased to exist with the close of the Secondary period—that is to say, soon after the deposition of the Chalk, and previous to that of the overlying London Clay. During that long period, or "world of reptiles," the class attained a development which it never equalled before or since. With regard to the past distribution of the four existing orders, it may be mentioned that the lizards and snakes, with the exception of two extinct suborders, are practically unknown before the beginning of the Tertiary period—that is to say, until after the deposition of the Chalk, so that they may be regarded as essentially the reptiles of the present day, when they attain their maximum development.

The tortoises and turtles, although a much more ancient group, having existed throughout the Secondary period, are, however, still at or about their zenith. But the case is very different with the crocodiles, which were represented during the Secondary period by a host of forms quite unlike those of the present day, and probably more numerous in species than their existing representatives. Many of the extinct crocodiles also exceeded any of the living forms in point of size. Still more markedly is this diminution noticeable in the case of the tuateras, in which, as already stated, a solitary survivor in the Antipodes represents a group once abundant and widely distributed.

As regards the present distribution of the class, it may be observed that while no existing reptiles are denizens of the air, only the turtles and sea-snakes and the Galapagos sea-iguana are habitual inhabitants of the ocean. Of the terrestrial and fresh-water forms, it has been found that the distribution does not coincide very closely with that of mammals and birds, so that the zoological regions into which the globe has been mapped out from the geographical distribution of the latter scarcely hold good for reptiles. This discrepancy may, no doubt, be to a considerable extent explained by the very early period at which certain groups of the class, such as crocodiles and tortoises, spread themselves over the surface of the globe.

The dispersive powers of reptiles in general are but limited. All of them, writes Dr. A. Günther, "are much specialised in their mode of life and propagation, and ill-adapted to accommodate themselves to a change of external conditions. As air-breathing, cold-blooded animals they are unable to withstand prolonged cold; they are therefore entirely absent in the Arctic and Antarctic zones; and such as escape the effects of the winter months in temperate zones by passing them in a torpid condition in well-sheltered places are not peculiarly organised forms, but offshoots from those inhabiting warmer climes. The tropical and subtropical zones are the real home of the reptilian type, which has there reached its greatest development as regards size and variety of forms. In the north, chelonians advance only to 50° latitude in the Western and to 56° in the Eastern Hemisphere; lizards to about 56° in British Columbia, and close to the Arctic Circle in Europe; while snakes disappear some degrees before the lizards. Also in the south, lizards extend into higher latitudes than snakes, namely, to the Straits of Magellan, whilst the latter do not seem to have advanced beyond 40° south latitude, and chelonians to 36°."

Of the various zoological regions into which the globe has been divided, the Oriental or Indian region is characterised by the number of fresh-water soft tortoises and S-necked tortoises, land-tortoises being scarce. Crocodiles, inclusive of the characteristic long-necked gharials, are numerous, as are lizards and snakes—especially pythons. Africa is comparatively poorly off for reptiles, although characterised by its numerous land tortoises, soft tortoises, and side-necked tortoises ; crocodiles being represented only by members of the typical genus, while lizards and snakes are comparatively numerous. Among the lizards, monitors and, among the snakes, pythons are common to the Oriental and African regions, while half of the exclusively Old World group of chamæleons is African. Madagascar is even more remarkable for the number of its chamæleons ; its land and side-necked tortoises are numerous, although soft tortoises, as in South America, are absent ; there is one crocodile ; and among the lizards the South American group of iguanas is represented ; while the snakes, among which none is poisonous, are also of a South American type.

alligator ; while fresh-water S-necked tortoises, as well as soft tortoises, replace the side-necked tortoises of the southern half of the continent. The snapping-tortoises (*Chelydridæ*) are also mainly characteristic of this region, although one genus ranges so far south as Ecuador. As regards its lizards and snakes, this region presents the same relation to the preceding as is held by Northern Europe and Asia to the Oriental and African regions. Lastly, New Zealand stands apart from all other countries in possessing the remarkable tuatera, in addition to which its only reptiles are skinks and geckos.

Hard pressed by the competition of mammals, reptiles have sought to protect themselves in various ways. Crocodiles and alligators, in addition to their large bodily size, powerful teeth, and aquatic habits, have their bodies protected by a panoply of bony plates, covered with horny shields. Tortoises and turtles have a still more solid and inpenetrable cuirass, into which many of the land forms are able to withdraw the head, tail, and limbs, and even to shut them completely in by means of bony flaps to the shell. Snakes in many

ALLIGATOR'S NEST IN FLORIDA, SHOWING YOUNG EMERGING FROM EGG

In the Eastern Holarctic region (that is, say, all Northern and Central Europe and Asia) the reptile fauna is mainly a mixture of Oriental and African types, although there are some peculiar forms. The only non-American alligator inhabits Central China. In the Australian or tropical Pacific region, exclusive of New Zealand, occur side-necked tortoises and a crocodile ; while among the lizards there are skinks, geckos, monitors, and the so-called agamoids ; the latter occurring in all the regions above mentioned, except Madagascar. Venomous snakes here outnumber the harmless ones.

The tropical and South American region is characterised by the presence of land and side-necked tortoises, to the exclusion of soft tortoises. Crocodiles and caimans are numerous (the latter being characteristic) ; while of the abundant lizards the majority are iguanas, the true lizards (*Lacertidæ*) of the Old World being replaced by the tejus (*Teiidæ*) ; snakes are also numerous, among them being rattlesnakes and boas.

In the Western Holarctic, or North American, region there are no caimans, their place being taken by one

cases rely on their threatening attitudes and their hissing voice ; but many of them are armed with the most formidable and fatal of all weapons in the animal kingdom—namely, their poison-fangs and venom.

Other snakes, as well as some lizards, have assumed the form and appearance of large earthworms ; while others, again, are arboreal or aquatic, and some even marine. Pythons and boa-constrictors, on the other hand, are among those which have assumed the offensive, and rely for safety, as well as for food, on the crushing force of their loathsome coils. Lizards conceal themselves to a great extent in holes and crannies, although some are arboreal ; while others—the geckos—are enabled to cling to walls, roofs, and cliffs by the action of suctorial discs attached to their toes.

To the human race reptiles are of comparatively small value, although the chelonian order is the sole source of turtle-soup and tortoiseshell ; while crocodiles and the monitor-lizards afford good and ornamental leather, and turtles' eggs are extensively used as a source of oil.

ORDER CROCODILIA

GENERAL CHARACTERISTICS OF CROCODILES

THE living crocodiles, among which may be included in a general sense not only the reptiles to which that name more properly belongs, but likewise those commonly designated alligators, caimans, and ghàrials, are the only existing representatives of three orders, which comprise among their members not only the most highly organised of all reptiles, and those which approach nearest in their structure to birds, but likewise the largest of all terrestrial reptiles, as, indeed, in the case of the dinosaurs, of any land animals.

Although these three orders possess many characteristics in common, it will be more convenient to describe the leading features of each separately, in the course of which their common attributes will be pointed out.

Sluggish in disposition, hideous in form, and huge in size, crocodiles alone among existing reptiles serve in some measure to recall the giant saurians with which the earth was peopled during earlier periods of its existence. In addition to their large bodily size, crocodiles are characterised by the lizard-like form of their bodies, which are supported on short limbs, and carried close to the ground. The long and powerful tail is much compressed from side to side, so as to be an efficient propeller in swimming ; its superficial extent being increased by a vertical longitudinal crest on its upper surface formed of a double series of horny lobes in the basal half, beyond which it is single. The head terminates in a flattened snout of variable length, and is attached to the body by a short, although muscular, neck, and the bulky body is much depressed. The toes are more or less fully webbed.

Externally the back, tail, and under parts of these reptiles are protected by an armour of quadrangular horny shields of varying size, which are arranged in longitudinal and transverse rows, and are in contact with one another by their edges. In the region of the back, and sometimes also on the under surface of the body, these horny shields are underlain by a corresponding series of pitted or honey-combed bony plates.

In the region of the neck, among existing members of the order, these bony plates are often irregular in form, and vary in number, but on the back they are always quadrangular and broader than long, with a well-marked longitudinal ridge down the middle. Such plates form a considerable number of longitudinal rows ; each plate articulating by its edges with those on either side, while those of each transverse row overlap the ones immediately behind them. When a bony shield is developed on the under surface of the body, the number of longitudinal rows of plates in the existing members of the group is always more than eight ; the transverse rows of plates overlapping, and each plate being composed of two distinct pieces united together by suture.

The limbs are provided with five toes in front and four behind, and the three innermost digits in each foot are furnished with claws. In all crocodiles, whether living or extinct, the conical teeth, which may be of very large size, are confined to the margins of the jaws, where they are implanted in distinct sockets ; while those in use are continually being replaced by fresh ones growing up from beneath the old ones, and gradually absorbing their roots. These reptiles are further characterised by the nostrils opening at the extremity of the snout—which may be either short or long—and by the ears being provided with movable lids.

Such are some of the leading external features of these reptiles, and although these suffice to distinguish them from the other living members of the class, they are insufficient to determine their true affinities. Laying stress upon the above-mentioned characters of their teeth, the naturalist is accordingly compelled to resort to the skeleton and soft internal parts for more distinctive characters. As regards the skull, all crocodiles are characterised by the quadrate-bone (of which the position is indicated in the figure on page 1469) being firmly united with the adjoining bones ; while a further distinctive feature is to be found in the presence of two bony bars on the sides of the skull behind the socket for the eye, the uppermost of these arches being shown immediately below the letter *T* in the figure on page 1486, while the lower and more slender one forms the backward continuation of the inferior margin of the eye-socket. The more anterior ribs (which, as in other reptiles, are present in the neck as well as in the chest) generally articulate with the backbone by means of two distinct heads ; and, while collar-bones are wanting, there is a breast-bone and likewise an inter-clavicle ; the latter being the median bar seen in the lower figure of the illustration on the next page.

A further peculiarity is the presence of seven or eight pairs of abdominal ribs in the wall of the abdomen, which have no connection with the proper ribs, and have their angle of union directed forwards. As regards the soft parts, the heart differs from that of all other living reptiles in having four complete chambers, so that the fresh and the impure blood can only mingle by means of a communication between the great vessels externally to the heart ; and there is also an incomplete midriff, or diaphragm, dividing the cavity of the chest from that of the abdomen ; crocodiles in these respects being more highly organised than any other living reptiles.

TABLE OF LIVING CROCODILIA MENTIONED IN THIS WORK

ORDER

Crocodiles—Crocodilia

FAMILY 1

Crocodilidæ

GENUS 1

Caimans—Caiman

SPECIES

Black, or great, caiman .... Caiman niger
Broad-nosed caiman .... C. latirostris
Spectacled caiman .... C. sclerops
C. trigonatus
C. palpebrosus

GENUS 2

Alligators—Alligator

SPECIES

Chinese alligator .... Alligator sinensis
Mississippi alligator .... A. mississippiensis
A. helois

GENUS 3

Osteolæmus

SPECIES

Short-nosed crocodile .... Osteolæmus tetraspis

GENUS 4

Typical Crocodiles—Crocodilus

SPECIES

Magar .... Crocodilus palustris
C. robustus
Cuban crocodile .... C. rhombifer
Guatemala crocodile .... C. moreleti
Estuarine crocodile .... C. porosus
Timsa .... C. niloticus
Siamese crocodile .... C. siamensis
Sharp-nosed crocodile .... C. americanus
Orinoco crocodile .... C. intermedius
Johnston's crocodile .... C. johnstoni
Long-nosed crocodile .... C. cataphractus

GENUS 5

Rhynchosuchus

SPECIES

Malay gharial ...... Rhynchosuchus schlegeli

GENUS 6

Garialis

SPECIES

Indian gharial .... Garialis gangeticus

CROCODILES AND ALLIGATORS

In addition to the preceding characters, which are common to all members of the order, there are certain others found only in the existing forms and some of their nearest extinct allies. One of the most remarkable of these peculiarities is the extremely backward position of the aperture of the internal nostrils, which in the skull, as shown in the figure on page 1469, is situated close up to the occiput, this being due to the development of special plates by the bones of the palate, which grow beneath the nasal passage, where they unite in the middle line, so as to form a floor to the latter, and thus completely cut it off from the cavity of the mouth. As the summit of the windpipe is continued upwards into this posterior aperture of the nostrils, crocodiles are enabled to breathe while their mouths are wide open and filled with water.

Another distinctive feature of the group, also shown in the figure just referred to, is that the socket for the eye communicates freely behind with the lower temporal pit. Then, again, all existing members of the order are characterised by the bodies of the vertebræ having the articular ball behind and the cup in front; while the ribs of the chest are provided with hook-like, or uncinate, processes resembling those of birds. In the region of the neck the ribs present the peculiarity of having the backwardly projecting and overlapping processes, which effectually prevent these reptiles from turning their heads to one side.

Crocodiles (family *Crocodilidæ*) are denizens of the tropical and subtropical regions of the globe, and are found in such latitudes wherever there are rivers or fresh-water lakes of sufficient size for their mode of life, but one of the Indian species habitually resorts to the sea-coast, where it has been seen floating at a considerable distance from the land. All of them are excellent swimmers, and mainly propelled when in the water by the aid of their powerful tails; the limbs being chiefly used when walking at the bottom of the water or on shore. When in repose, crocodiles lie like logs either in the water or on the banks of the lakes or rivers they inhabit; but when in pursuit of prey in the water they move with great speed, while they are also active on land. The young are, however, decidedly nimbler in their movements than are the adults.

Exclusively carnivorous in their diet, some members of the order feed solely upon fish; while others, in addition to fish, prey upon the flesh of all animals that come in their way. Adult crocodiles, writes Dr. A. Günther, " attack every large animal which accidentally approaches them, and in overpowering it the whole of their powerful organisation is called into requisition. Seizing the victim between their capacious jaws, and fastening their long, pointed, conical teeth into its flesh, they draw it, in one moment, by their weight and with a stroke of the tail, below the water and drown it. Their gullet is, however, much too narrow to allow of the passage of the entire body of the victim; and

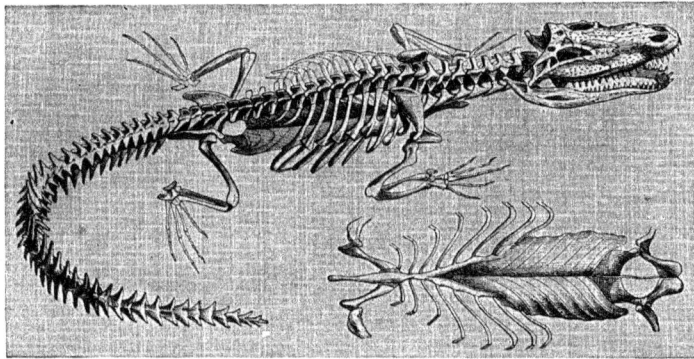

SKELETON AND ABDOMINAL RIBS OF NILE CROCODILE

their teeth being adapted for seizing and holding fast only, and not for biting, they are obliged to mangle the carcase, tearing off single pieces by sudden strong jerks." This rending process is mainly accomplished by lateral movements of the head and front portion of the body.

Too often, human beings who incautiously bathe in crocodile-haunted waters fall victims to these bloodthirsty reptiles; while there are instances of people being seized when merely stooping down to dip water from the river's marge. When seized the only way for an unarmed man to escape is, it is said, to thrust his fingers into the creature's eyes and endeavour to gouge them out. To a considerable extent crocodiles are nocturnal in their habits, and during protracted droughts many of them at least are accustomed to bury themselves in the mud where they become torpid.

As regards their reproduction, crocodiles lay from twenty to sixty eggs, of the approximate size of those of a goose, and invested with a hard, white shell. In some cases these are deposited in hollows in the sand of river-banks, where, after being covered to a greater or less depth, they are left to hatch. In the case of the Malagasy species, at any rate, the parent assists in the incubation of the eggs. In that island the egg-laying season lasts from the end of August to the end of September; the usual number of eggs in a nest varying from twenty to thirty. The nest is excavated to a depth of about two feet in the dry white sand; its lateral walls being undermined so as to allow the eggs to roll into the cavities thus formed from the slightly elevated centre.

Upon the summit of the completed nest, which is not noticeable externally, the parent sleeps; and when the young crocodiles are ready for hatching they utter distinct notes, which are heard by the mother even through a layer of two feet of sand. Digging down to the eggs, the parent crocodile lays them open to the air, upon which the young reptiles make their way out by perforating the shell at one extremity by the aid of a tooth specially developed for this purpose, the whole process occupying as much as a couple of hours. When hatched, the young crocodiles are led to the water by their parent, whose attention they attract by uttering cries, which are, however, of a lower pitch than those emitted while still in the egg. In Brazil the black caiman deposits its eggs beneath a large heap of brushwood, beside which the female parent keeps guard.

CAIMANS, OR SOUTH AMERICAN ALLIGATORS

The existing members of the order are arranged in half a dozen generic groups, of which, in some respects, the most specialised are the caimans (*Caiman*) and alligators, now generally regarded as belonging to distinct genera. Both caimans and alligators are characterised by their relatively short and broad snouts, in which the edges of the jaws are festooned,

L

and the nasal bones extend forwards to the aperture of the nostrils—as is shown in the figure on page 1469, where the nasals are the paired bones on the upper aspect of the skull, of which the narrow points just project into the cavity of the nostrils—while the two halves of the lower jaw are united in front by a very short bony union.

The stout teeth vary considerably in size in different parts of the jaws; the third and ninth in the upper jaw, and the fourth, and frequently the first and eleventh, in the lower jaw, being generally much larger than the others. In these features caimans and alligators resemble many typical crocodiles, from which they are distinguished by the circumstance that, as a rule, both the first and the fourth tooth on each side of the lower jaw are received into pits in the upper jaw, so as to be invisible externally when the mouth is closed, while the upper teeth bite on the outer side of the lower ones. In both groups the number of teeth varies from seventeen to twenty on each side of the upper jaw, and from seventeen to twenty-two in the lower jaw.

Caimans and alligators are characterised by the very small size of the upper temporal pits on the top of the skull, or those marked T in the figure on page 1469, these pits being in some cases completely obliterated. Caimans are specially distinguished by the aperture of the nostrils not being divided by the forward prolongation of the nasal bones, by the presence of a strongly developed bony armour on the inferior surface of the body, and by the bony plates on the upper surface being articulated together. In the black caiman the bony armour also forms sheaths for the limbs.

Caimans, or jacares, as they are called by the natives of Brazil, are restricted to Central and South America, where they are represented by five species. The largest and at the same time the best known of these is the black, or great, caiman (C. niger), from the rivers of tropical South America eastwards of the Andes, which takes its name from the black of the upper surface of the body, the under parts being yellow. This species, which generally attains a length of about 14 feet, is characterised by its partially bony and flat upper eyelids, the presence of upper temporal pits in the skull, and by the number of teeth in each premaxillary, or anterior upper jawbone, being five and the number of lower teeth seventeen or eighteen.

ROUGH-EYED, OR SPECTACLED, CAIMAN
Photo by permission of Dr. J. H. C. Connell

Nearly allied, although of much smaller size, are the broad-nosed caiman (C. latirostris), ranging from the Amazon to the Rio de la Plata, and the spectacled caiman (C. sclerops) of Central and South America, both of which have the upper eyelids rugose, with a small horn-like projection, while in the skull the socket of the eye does not extend so far forwards. Both are uniformly blackish when adult; but in the former the skull is relatively wide, and the number of lower teeth ranges from seventeen to eighteen, while in the latter the skull is narrower, and the lower teeth vary from eighteen to twenty.

The two remaining species, C. trigonatus and C. palpebrosus, are still smaller, and characterised by the colour of the upper parts being yellowish brown, spotted and barred with black; while the upper eyelid is completely bony, the skull has no upper temporal pits, there are only four teeth in each premaxillary bone, and the number of lower teeth is from twenty to twenty-two on each side. In the spectacled caiman the eyes are of a gleaming red.

On the Amazon and Orinoco, as well as other South American rivers, caimans are to be met with in myriads, and appear to be very similar in their habits to the crocodiles of the Old World. Writing of the great caiman, the jacare-uassu of the natives, Mr. Bates observes that "it grows to a length of eighteen or twenty feet, and attains an enormous bulk. Like the turtles, the alligator has its annual migrations, for it retreats to the interior pools and flooded forests in the dry season. During the months of high water, therefore, scarcely a single individual is to be seen in the main river. In the middle part of the Lower Amazon, about Obydos and Villa Nova, where many of the lakes with their channels of communication with the trunk stream dry up in the fine months, the alligator buries itself in the mud and becomes dormant, sleeping till the rainy season returns. On the Upper Amazon, where the dry season is never excessive, it has not this habit. It is scarcely exaggerating to say that the waters of the Solimoens are as well stocked with large alligators as a ditch in England is in summer with tadpoles."

SPECTACLED CAIMANS IN A RIVER

By the natives of these regions the caiman is at once despised and feared, Mr. Bates relating that on one occasion he saw a party boldly enter the water and pull to the shore one of these large reptiles by its tail, while at another time two medium-sized specimens

BLACK CAIMANS AT HOME

that had been captured in a net were coolly returned to the water hard by where a couple of children were playing. Sometimes, however, they have to pay dearly for such temerity. The Macusi Indians of Guiana capture the caiman by means of a baited hook and line, the former being composed of several pieces of wood, which becomes fixed in the creature's jaws. Charles Waterton has given an amusing account of his ride on the back of a caiman caught in this manner, and the feat has been repeated elsewhere. The eggs of the great caiman, which are about the size of those of a turkey, and, as already mentioned, deposited in a heap of dead bushes, are much sought after as food by the natives of Dutch Guiana.

ALLIGATORS

The early European explorers of South America on meeting with a gigantic lizard-like reptile naturally applied to it the name of *una lagarta*, which is the Spanish term for a lizard, and this as naturally became in course of time corrupted into alligator. It would appear, indeed, that this name was first given to the caiman, to which in strict propriety it should therefore belong, but now, by the common consent of naturalists, it is taken as the special designation of the members of the present genus (*Alligator*).

True alligators, as thus restricted, are represented by one species from North America, and by a second from the Yang-tsi-Kiang in China ; although there appears to be a third and imperfectly known species, of which

YOUNG ALLIGATOR'S OPEN JAWS
Photo, S. C. Johnson

the habitat is as yet undetermined. Alligators differ from caimans merely by the forward prolongation of the nasal bones of the skull, so as to divide the aperture of the nostrils into two equal moieties, by the want of articulation between the bony plates of the back, and the absence or extreme thinness of those on the lower surface of the body.

Curiously enough, the Chinese alligator (*A. sinensis*), which is a comparatively small species, is the one coming nearest in structure to the caimans ; this approximation being shown by the great development of bone in the upper eyelids and the presence of thin bony plates on the lower surface of the body. The latter are, however, placed wide apart, without any mutual articulation or overlapping. In this species the front toes are free, the number of plates on the neck is usually six, although they may be reduced to four, while generally there are only six plates in the widest of the transverse rows on the back. The number of teeth in the upper jaw is seventeen or eighteen, against eighteen or nineteen in the lower jaw. In colour the upper parts are greenish black, speckled and streaked with yellow, while the under parts are greyish.

In the much larger Mississippi alligator (*A. mississippiensis*), of which the dimensions exceed those of the great caiman, the front toes are webbed, there are only four plates on the neck, but always eight plates in the widest of the transverse rows of the back.

There are nineteen or twenty teeth on each side of both jaws; and in the adult the colour is dark green or blackish above, and yellowish below. The range of this species embraces the south-eastern United States, from the Rio Grande to North Carolina. The third species (*A. helois*) is a small one, distinguished by the slight compression of the tail, which is scarcely crested.

Information concerning the Chinese alligator, which was first made known to science in 1879, in the living state appears to be mainly or entirely derived from specimens exhibited in the menageries of Europe, and the accounts of the mode of life of the Mississippi species are by no means so full as is desirable. It appears, however, that the latter spends the greater part of its time in the water, where its main diet is formed by fish, although it will seize and drag down such sheep, goats, dogs, deer, or horses as, when drinking, come within reach of its terrible jaws.

During flood-time, when many of the lowlands are under water, the alligators leave the rivers to feed on the fish which abound in the flooded districts, returning to their old quarters with the subsidence of the inundations. To such flooded lowlands, writes the great American naturalist Audubon, " in the early part of autumn, when the heat of a southern sun has evaporated much of the water, the squatter, the hunter, and the planter all go in search of sport. The lakes then are about two feet deep, having a fine sandy bottom. The long, narrow Indian canoe, kept to hunt these lakes, and taken into them during the freshet, is soon launched; and the party seated in the bottom is paddled or poled to look for water-game. Then, on a sudden, hundreds of alligators are seen dispersed all over the lake; their heads and all the upper parts of their body floating like a log, and in many instances so resembling one that it requires to be accustomed to see them to know the distinction. Millions of the large wood-ibis are seen wading through the water, muddying it up, and striking deadly blows with their bills on the fish therein. It is then that you see and hear the alligator at his work. Each lake has a spot deeper than the rest, rendered so by these animals who work at it, and always situated at the lower end of the lake."

By this means a supply of water is ensured; and in these so-called alligators' holes the reptiles may be seen congregating in hundreds. " The fish, that are already dying by thousands through the insufferable heat and stench of the water, and the wounds of the different winged enemies constantly in pursuit of them, resort to the alligators' hole to receive refreshment, with a hope of finding security also, and follow down the little current flowing through the connecting sluices; but no! for, as the water recedes in the lake, they are here confined. The alligators thrash them, and devour them whenever they feel hungry, while the ibis destroys all that make towards the shore.

By looking attentively on this spot, you plainly see the tails of the alligators moving to and fro, splashing, and now and then, when missing a fish, throwing it up in the air. The hunter marks one of the eyes of the largest alligators, and as the hair-trigger is touched the alligator dies. Should the ball strike one inch astray from the eye, the animal flounces, rolls over and over, beating furiously about him with his tail, frightening all his companions, who sink immediately, whilst the fishes, like blades of burnished metal, leap in all directions out of the water, so terrified are they at this uproar."

During the pairing season, which takes place in the spring, the males resort to land, and are but seldom seen; while a short time afterwards the female deposits her hard white eggs, which are said at times to be upwards of one hundred in number. The nest in which the eggs are laid is generally placed among bushes or reeds, at a distance of fifty or sixty yards from the water's edge, the eggs themselves being carefully covered with leaves and other vegetable matter. The heat engendered by the decomposition of the latter aids in the hatching of the eggs, and when the young appear they are conducted to the water by the mother, who has all the time remained on guard near the nest (page 1473).

DOUBLE-TUSKED ALLIGATORS

In the middle and lower Tertiary deposits of both Europe and the United States the present group was represented by certain extinct alligators (*Diplocynodon*) characterised by the presence of a bony armour on the lower surface of the body, coupled with the circumstance that the fourth tooth of the lower jaw was generally received into a notch in the side of the skull, while the third lower tooth was as much enlarged as the fourth. Some of these double-tusked alligators had short snouts, like their existing allies; but in one from the London Clay this part of the skull was much produced as in many crocodiles.

CROCODILE EMERGING FROM AN INDIAN RIVER
Photo, H. C. White Co.

SHORT-NOSED CROCODILE

A small and short-nosed crocodile (*Osteolæmus tetraspis*) from West Africa, in the neighbourhood of Sierra Leone, where it was discovered by the French explorer Du Chaillu, presents much the same relationship to the typical crocodiles as is held by alligators to caimans. The arrangement of the teeth is, for instance, similar to that obtaining in the typical crocodiles, but the nasal bones extend forwards to divide the cavity of the nostrils into two halves. Moreover, the upper eyelids are largely bony, while there are detached bony plates on the lower surface of the body as well as on the throat.

The shield of the neck is distinct from that of the back, and composed of two or three pairs of plates, of which the anterior ones are very large; while that of the back comprises seventeen transverse rows of

TIMSA, OR NILE CROCODILE

YOUNG MISSISSIPPI ALLIGATORS

MISSISSIPPI ALLIGATOR

plates, the broadest row including six of such plates. The ridges on the plates of the neck are strongly marked, but become very obscure in the two middle rows of the back. The fore toes have only rudimentary webs, although those of the hind limbs are webbed for about half their length.

With the exception of parts of the head, tail, and

SHORT-NOSED CROCODILE

back, which are light brown with black markings, the colour of the adult is uniform blackish brown. Young specimens are, however, yellowish brown, spotted with black above, and with bars of the same on the body and tail; while the lower armour is black and yellow. Practically nothing is known as to the habits of this peculiar species, which are, however, probably very similar to those of its allies.

CROCODILES

The typical crocodiles (*Crocodilus*) comprise rather less than a dozen species, ranging over Africa, Southern Asia, Northern Australia, and Tropical America. Having no bony armour on the lower surface of the body, they are distinguished from caimans and alligators by the interlocking of the upper and lower teeth, and by the fourth lower tooth being usually received into a notch on the side of the upper jaw, so as to be partially visible when the mouth is closed, while the number of teeth varies from seventeen to nineteen on each side of the upper jaw, and fifteen in the lower.

From the short-nosed crocodile they are distinguished by the aperture of the nostrils in the skull not being divided by the forward prolongation of the nasal bones. While some of the species resemble alligators in their broad and short snouts, others have elongated, narrow muzzles, approaching those of gharials; but as there is an almost complete gradation from the one type of crocodile to the other, this affords no ground for generic distinction, so that the most that can be done is to arrange them in groups.

MAGAR, OR INDIAN CROCODILE

Commonly known to the natives of India as the magar, and misnamed alligator by Anglo-Indians, the Indian crocodile (*C. palustris*) is the best-known representative of a group of four species which in their broad and short snouts make the nearest approach to caimans and alligators. In all these species the length of the snout does not exceed one and a half times its basal width; while the bony union between the two branches of the lower jaw does not extend behind the level of the fourth or fifth tooth; while on the palate the line of union between the anterior and main jawbones (premaxillæ and maxillæ) extends nearly straight across the skull, as shown in the figure on page 1469.

The Indian crocodile has no bony ridges on the snout, there are usually four longitudinal rows of bony plates on the back, and five teeth in each anterior upper jawbone, or premaxilla. An allied species (*C. robustus*) from the interior of Madagascar differs by having six longitudinal rows of plates on the back; while the Cuban crocodile (*C. rhombifer*), of Central America, and a nearly related species, *C. moreleti*, from Guatemala are distinguished by having a more or less distinct oblique ridge in front of the eye. The habitat of the Indian crocodile includes India, Ceylon, Burma, and the Malay Peninsula and Islands, its most westerly range being Sind and Baluchistan. Inhabiting rivers, lakes, and marshes, it appears to be an exclusively freshwater species, never venturing into estuaries. As to the dimensions attained by this species there is

ESTUARINE CROCODILE

some uncertainty, although it is probable that at the present day specimens seldom grow to the size that was reached before firearms were common. Nowadays from 12 to 14 feet appears to be a large size for this species, but a length of 18 feet has been recorded, while skulls in the Calcutta Museum seem to indicate still larger individuals. Definite information as to the length to which this big crocodile will grow is, however, still lacking.

A nearly allied extinct species has left its remains in the Siwalik Hills of Northern India. Swarming in most of the rivers and marshes of India, except where the current is too swift, the Indian crocodile is stated to be less ferocious than the species next mentioned, generally preying on the smaller animals, and not infrequently dragging down a wounded or dead bird before the eyes of the gunner. When the waters they frequent become dried up, these crocodiles either travel across country by night to another lake or river or bury themselves in the mud.

TIMSA, OR NILE CROCODILE

ESTUARINE CROCODILE

Resembling its compatriot in its pale olive colour, conspicuously spotted with black, the estuarine crocodile (*C. porosus*) of India and other Eastern countries may be at once distinguished by its longer and more slender snout, as well as by the presence of only four teeth in each anterior jawbone, or premaxilla, of the adult. It belongs, indeed, to a group of four species differing from the preceding assemblage in that the length of the snout varies from rather more than one and a half to just over twice its basal width ; and also by the line of union between the anterior and main jawbones running in a V-shape up the palate. The presence of a large ridge running down the skull in front of the eye serves to distinguish this species not only from all the other members of the group, but likewise from the Indian crocodile.

The present species generally, if not invariably, inhabits the tidal portions of rivers, whence it descends to the sea, where it has been observed floating at considerable distances from land. These estuarine and partially marine habits will readily account for the wide geographical distribution of this crocodile, which ranges from India to Australia. Unknown on the western coast of India, the estuarine crocodile is abundant in the lower courses of the rivers of Bengal and other parts of the eastern side of India, as well as in Ceylon and Burma, whence it extends eastwards to Southern China, Northern Australia, and the islands of the Solomon and Fiji groups. In point of size it probably surpasses all other species, one specimen being recorded which reached the enormous length of 33 feet.

In correspondence with its gigantic size, this crocodile appears to be one of the most formidable members of its kind, being exceedingly prone to attack human beings, more especially in the breeding season, which takes place during June and July, when it is stated to attack such small boats as may cross its haunts. Owing to their depredations, these crocodiles are cordially detested as well as feared by the natives of India, and at Dacca, on the north of the Bay of Bengal, crocodile-hunting is pursued as a profession.

The following account of the pursuit of one of these monsters, which had recently carried off a boy, is abridged from an Indian newspaper : " The hunter, having been summoned, moored his canoe hard by the place where the tragedy had taken place, it being well known that a crocodile which has been successful in securing a victim will generally remain for some days about the spot. Soon the crocodile was descried floating on the water, whereupon the hunter and assistant hid themselves in the canoe, while the son of the former entered the water, which he began to beat with his hands. Catching sight of the boy, the crocodile prepared to dive towards him, upon which the boy took refuge in the canoe. In a moment or so the reptile rose to the surface at the expected spot, where it was saluted with a couple of harpoons, one of which secured a firm hold. After a long chase, in which a number of the inhabitants of the village took part in boats, a second harpoon was safely planted in the head of the monster, which was finally dragged to shore. When opened, several gold and silver ornaments—the relics of earlier victims—were found in his stomach."

In Ceylon crocodiles are frequently captured by means of a hook and line, laid overnight in the water and made fast, in the native fashion, by a bunch of fine cords. These cords, becoming fixed between the interstices of the reptile's teeth, are safe from being bitten through, and in the morning the captive is dragged ashore and despatched. When thus captured, crocodiles emit a disagreeable musky smell, due to the secretion of a pair of glands in the lower jaw.

THE TIMSA

Formerly inhabiting the Nile from its mouth to its source, the timsa, or Nile crocodile (*C. niloticus*), owing to the invasion of its haunts by steam vessels and the introduction of rifles, has now well-nigh disappeared from Egypt, even as far back as the year 1870 being but rarely seen below Beni Hassan, and not common till above the second cataract. In the upper reaches of the Nile it still exists in its original numbers, whence its range extends southwards to the Cape, and westwards to Senegal. The species also occurs in Madagascar, while it likewise still lingers in Syria, in the neighbourhood of the Zerka, or Crocodile River, near Cæsarea.

Distinguished from the estuarine crocodile by the absence of the ridge in front of each eye, this species differs from the other two members of the same group by the want of any ridge on the middle of the snout or forehead, so that its whole skull is comparatively smooth. In size it falls little, if at all, short of the estuarine crocodile, although differing from the latter by the uniformly dark olive colour of the adult (page 1559).

As the habits of this crocodile do not differ in any important respects from those of the other members of the genus, they do not require any detailed notice, although a few words may be devoted to its cult by the ancient Egyptians, among whom it was known by the aforesaid name of *timsa*, which was modified by the ancient Greeks into *champsa*. By the former people the crocodile was regarded as the symbol of sunrise—possibly, it has been suggested, on account of the brightness of its eye, or, perhaps, because that is the first part to appear when the reptile emerges from the water. Among the places where the crocodile was specially reverenced were Thebes and the shores of Lake Mœris, as well as Ombi, near Syene.

At Thebes a crocodile was reared from youth in the temple, where it was fed with sacred food, adorned with rings and bangles, and worshipped with divine honours; while after death its mummified body was carefully preserved in the catacombs, where hundreds of embalmed crocodiles are still to be found.

Something analogous to this Egyptian veneration for the crocodile is to be met with in other countries. The Indian crocodile is specially protected by certain religious sects in India, and rendered so tame that it will leave its pond to feed out of its keeper's hand; while at Dix Cove, on the north-western coast of Africa, a pair of tame crocodiles were formerly kept in a pond by priests, dressed in white garments, who fed their charges with snow-white fowls.

On the upper course of the Nile the favourite haunts of the crocodiles are sandbanks, situated in parts of the river where the current is not too strong. There they may be seen at all hours of the day sleeping with widely-opened mouths, in and out of which the black-backed plover walks with the utmost unconcern (page 1340). According to Arab accounts, one and the same crocodile has been known to haunt a single sandbank throughout the term of a man's life, thus leading to the conclusion that these creatures enjoy a long term of existence, during the whole of which they continue to increase in size.

In common with this feature of uninterrupted growth, crocodiles are likewise distinguished by their remarkable tenacity of life, the only shots that prove instantaneously fatal being those that take effect either in the brain itself or in the spinal cord of the neck. It is true, indeed, that a shot through the shoulder will ultimately cause death, but it allows time for the animal to escape into the water, where its body immediately sinks. To reach the brain, the crocodile should be struck immediately behind the aperture of the ear. Although it is commonly supposed that the bony armour of these reptiles is bullet-proof, this is quite erroneous, although if the plates are struck obliquely the bullet will frequently ricochet.

A remarkable instance of boldness and ferocity displayed by a crocodile of this species is narrated by a correspondent of the "Times" during a journey to Mashonaland. On arriving one evening on the banks of the narrow but rocky Tokwi River, a man named Williams rode in with the intention of crossing. During the passage his horse was carried by the stream a few yards below the landing-place, and just as he reached the opposite bank he was seized by the leg by a crocodile, which dragged him from his horse into the stream. There the reptile let go its hold, upon which the man

SHARP-NOSED CROCODILE

managed to crawl on to a small island. Immediately his companion rode in to his assistance, upon which another very large crocodile mounted up between him and his horse's neck, and then slipped back, making a dreadful wound on his side and in the horse's neck with its claws as it did so. The river seemed, indeed, to be absolutely swarming with crocodiles, and it was with the greatest difficulty that the unfortunate man Williams, who ultimately died of his wounds, was brought to the bank.

As an example of the wonderful muscular power of these terrible reptiles, it may be mentioned that a crocodile is stated, on thoroughly trustworthy evidence, on one occasion to have seized by one hind leg a full-grown rhinoceros as it was leaving the river, and, after a prolonged struggle, eventually succeeded in dragging the unfortunate animal stern-foremost into the water, where it was drowned.

SIAMESE CROCODILE

The Siamese crocodile (*C. siamensis*), inhabiting Siam, Cambodia, and Java, may be distinguished from the preceding species by the presence of a longitudinal ridge on the skull between the eyes, although the snout is smooth. It agrees with the latter in having the anterior bony plates of the neck well developed, whereas these are usually absent in the estuarine crocodile.

SHARP-NOSED CROCODILE

The last member of this group is the sharp-nosed crocodile (*C. americanus*) of Central America, which has a longer and sharper muzzle than any of the preceding species, and is further characterised by the presence of a distinct median ridge running down the snout. There are usually four large bony plates on the neck, forming a square, with a smaller pair on the sides of the front ones ; and the plates of the back are arranged in fifteen or sixteen transverse rows, and in either four or six longitudinal bands. In the fore limb the second and third toes are but slightly webbed, while the outer toes of the hind foot are united by larger webs.

In colour the adult is blackish olive above, and yellowish beneath, but the young are pale olive, with black spots. In addition to being widely distributed in Central America and the adjacent regions, such as Ecuador, Colombia, Venezuela, and Florida, this crocodile is also met with in the West Indian Islands.

ORINOCO CROCODILE

Nearly allied to the last, although with a still longer and more slender snout, is the Orinoco crocodile (*C. intermedius*), which is referred to a third group, characterised by the very slender and gharial-like snout, of which the length is equal to at least twice the basal breadth ; and also by the bony union between

the two branches of the lower jaw extending as far back as the sixth, seventh, or eighth tooth, instead of stopping short at the fifth. In this particular species the snout, which has no ridges, varies in length from twice to twice and a half the width at the base ; while the six bony plates on the neck are widely separated from those of the back, and arranged in a square of

LONG-NOSED CROCODILE

four, with a pair on the sides. The colour is olive above and yellowish beneath, and in both this and the preceding species the length is about 13 feet. The Orinoco crocodile appears to be confined to the river from which it takes its name and its affluents.

The best accounts of the Orinoco and sharp-nosed crocodiles are given by the early German traveller Von Humboldt, who states that these reptiles swarm on the Apure, where they may be often seen in parties of eight or ten lying on the open space between the shore of the river and the forest. At the time of his journey the river was, however, still low, and consequently hundreds of crocodiles were lying concealed beneath the mud of the adjacent lowlands. In the stomach of one that was opened were found a half-digested fish and a granite pebble ; such pebbles, and sometimes also balls of hair, being commonly found in the stomachs of the members of the order.

In spite of their comparatively slender jaws, these crocodiles frequently seize the natives while stooping to draw water from the river. A large portion of their prey is, however, afforded by the defenceless carpinchos, which are met in droves of from fifty to sixty head, and fall victims to jaguars on land and to crocodiles in the water.

In their young state—when only from 7 to 8 inches in length—the crocodiles themselves are, however, devoured by vultures, which seize them on the shore or in the shallow water. It was curious, observes Humboldt, to see the address with which the little reptiles defended themselves for a time against their aggressors. Directly they perceived the enemy, they raised themselves on their fore paws, bent their backs, and lifted up their heads, opening their wide jaws. They turned continually, though slowly, towards their assailants, showing their teeth, which, even when the reptiles had but recently issued from the egg, were very long and sharp. Often, while the attention of one of the young crocodiles was wholly engaged by one of the vultures, another seized the favourable opportunity for an unforeseen attack, pouncing on the unfortunate reptile by the neck and bearing it off in the air. Humboldt also relates an anecdote about a native of Calabozo being awakened in the middle of the night by one of these crocodiles suddenly breaking through the mud of the floor of his hut, beneath which it had retired for the dry season.

LONG-NOSED CROCODILE

Omitting notice of the small Johnston's crocodile (*C. johnstoni*) of Northern Australia, the last member of the genus is the curious long-nosed crocodile (*C. cataphractus*) of Western Africa, which forms a kind of connecting link between the other true crocodiles and the gharials. In this species the snout is more elongated and slender than in any of its congeners, its length not infrequently exceeding three times the basal width, while the bony union between the two branches of the lower jaw is likewise of unusual length. In form the snout is convex and devoid of ridges, and the region of the forehead is remarkable for the convexity of its profile.

The great peculiarity about the species is, however, to be found in the arrangement of the bony plates on the neck, which form two longitudinal rows, and are partially if not completely continuous with those of the back ; a somewhat similar arrangement existing in Johnston's crocodile. On the back the number of longitudinal rows of shields is six ; and the lower parts of the legs, as in many other crocodiles, are furnished with jagged horny fringes. In colour the head is olive, spotted with brown ; the back and tail have a brownish yellow ground-colour, with large black spots, and the yellowish white under parts are marked with smaller white spots. In length this species reaches some 18 feet.

The long-snouted crocodile is found in the rivers and marshes of Western Africa, from Senegambia to the Gabun, and also occurs farther to the south in the Congo ; its native name being khinh. Not infrequently found in company with the Nile crocodile, it inhabits the smaller streams and still waters of the interior, generally taking up its position in a deep pool protected by an overhanging bank or rock, and thence sallying forth on its prey, which consists chiefly of fishes, frogs, and aquatic reptiles. The eggs are laid on the bank, where, unlike those of most other members of the family, they are carefully covered with leaves and herbage.

A CROCODILE IN ITS FAVOURITE POOL
Photo, J. A. Dimock

Shy and timid in its disposition, this crocodile is often captured by the natives for the sake of its flesh, which, like that of many of its allies in other regions, is much esteemed as food While very abundant in the fresh waters of the interior, this species likewise haunts the salt-water lagoons of the Guinea coast, and in the delta of the Cameruns may be observed lying on the sandbanks bordering the mangrove swamps, from which, on the approach of a boat, it darts into the water with surprising celerity. When in that element, it often pulls down herons and such other aquatic birds as may be standing or swimming in the water, sailing up to them with the silence of a large fish, to which, when in the water, it presents a considerable resemblance. As in the estuarine and Nile crocodiles, in the adult of this species the second tooth in the fore jawbone, or premaxilla, disappears, leaving only four in place of the normal five teeth on each side.

MALAY GHARIAL

The very long and slender snouted crocodile-like reptile inhabiting Borneo, Sumatra, and the Malay Peninsula, and known as the Malay, or Schlegel's, gharial (*Rhynchosuchus*, or *Tomistoma*, *schlegeli*) represents the first of two genera, each with a single existing species, which differ very remarkably from any of those yet noticed. In both these reptiles the snout is long and slender, with its toothbearing margins nearly straight, instead of being thrown into more or less well-marked festoons ; while the nasal bones never extend forwards to reach the aperture of the nostrils, from which they are separated by a considerable interval. The bony union between the two branches of the lower jaw is also of great length, extending at least as far back as the fifteenth tooth, and including a bone—the splenial—which in other crocodiles remains entirely separate from the symphysis. In neither do the teeth attain the large dimensions characteristic of many other members of the family.

The Malay gharial has the shorter snout of the two, its length not exceeding three and a half times its basal width ; but the species is especially distinguished by the circumstance that the nasal bones extend forwards to articulate with the anterior jawbones, or premaxillæ. The teeth are twenty or twenty-one in number on each side of the upper jaw, and eighteen or nineteen in the lower jaw ; those on the sides of the latter being received in pits between the upper ones, and the first, fourth, and ninth lower teeth being enlarged. The bony plates on the neck and back form a continuous shield, consisting of four longitudinal and twenty-two transverse rows ; and while the fore toes are webbed at the base, the outer ones of the hind feet have larger webs.

In colour the Malay gharial is olive above, with dark spots or bars, and its length may be 12 or 14 feet. In

THE INDIAN GHARIAL

Photo. C. R. Walter

habits this species is probably very similar to the Indian gharial. It is important to notice that several fossil representatives of this genus occur in the Tertiary deposits of Europe, while it is not improbable that the genus is also represented in the underlying Cretaceous rocks. All this is exactly in harmony with what we should naturally have expected to be the case, seeing that the Malay gharial, like the Indian gharial, is evidently a very generalised member of the family.

This gharial is known to the Malays as buaya jinju-long, or long-snouted crocodile, but as it is a very rare reptile, only a few have even heard of its existence. Mr. L. Wray states that a specimen sent to London "was caught at Pulan Tiga, in the Perak River, in June, 1895, and I kept it in a pond till the end of December, when it was killed. For months it would eat nothing but a few small fish, but during the latter portion of the time it would eat freely of any meat or fish given to it. It also became quite tame, and would remain at the surface of the water with its head on the bank while people stood near it.

"The Malay gharial would appear to be essentially a fresh-water animal, and it is said by the natives to often frequent the swamps and marshy lands on the banks of the river. If this is really the case, it differs somewhat in its habits from *Garialis gangeticus*, which is much more aquatic than the crocodile. In the ordinary way, so far as my observations have gone, only the upper part of the end of the nose and the two eyes are above the water. On the approach of anyone, the eyes slowly and quite silently sink beneath the surface, and nothing but a small portion of the nose remains ; on a nearer approach this also quietly disappears. This, doubtless, accounts for the fact that the animal is so very rarely seen."

INDIAN GHARIAL

Apparently owing to a clerical error on the part of its first describer, the slender-snouted crocodile known in India by the vernacular name of gharial is almost always spoken of in Europe as the gavial, while its misspelt name has even been Latinised into *Gavialis*—an error which some writers persist in perpetuating. The Indian gharial (*Garialis gangeticus*) is readily distinguished at a glance from all other crocodiles by the exceeding length and slenderness of its snout ; the length varying from more than five times the basal width in the young to rather more than three in the adult. This narrow snout gives to the reptile a de-cidedly curious appearance ; and it is perhaps note-worthy that both the gharial and the dolphin-like susu, which inhabit the same rivers, and probably feed on the same kind of food, have similarly elongated beak-like snouts, armed with very similar curved and

slender conical teeth ; this resemblance being doubtless due to adaptation to a similar mode of life (page 847).

From the Malay gharial, the Indian species is readily distinguished by the nasal bones being very short, and consequently separated by a long interval from the front upper jawbones, or premaxillæ ; while the teeth—twenty-seven to twenty-nine on each side of the upper, and twenty-five or twenty-six in the lower jaw—are all of nearly uniform size, and those of the lower jaw are not received into distinct pits. The bony union between the two branches of the lower jaw also extends backwards to the twenty-third or twenty-fourth tooth, whereas in the Malay species it stops short at the fourteenth or fifteenth. At its extremity the long and narrow snout becomes much expanded, and in the male this expanded extremity is surmounted by a hollow hump, in the centre of which are placed the nostrils.

The bony plates of the neck form a shield continuous with that of the back, in which the number of longitu-dinal rows is four, while there are twenty-one or twenty-two transverse bands. Externally to the bony shields of the back there occurs on each side a row of soft plates, which are either smooth or but slightly keeled. The toes are well webbed ; and the general colour of the adult is dark olive above, the young being, how-ever, pale olive, with dark brown spots or bars.

The Indian gharial has a somewhat curious geo-graphical distribution, being restricted to the Indus, Ganges, and Bramaputra, with their larger affluents, together with the Mahanadi in Orissa, and the Koladyni River in Arakan. Together with certain tortoises mentioned later on, this reptile is one of the most ancient of living animals, its fossil remains occurring in the rocks of the Siwalik Hills in Northern India in association with those of mammals belonging to extinct species and genera. Attaining a length of fully 20 feet at the present day, and still larger dimensions during the Pliocene period, the gharial subsists chiefly upon fish, for the capture of which its elongated narrow jaws, armed with numerous long, curved teeth, are admirably adapted. There is, however, at least one well authenticated instance of a human being having fallen a victim to these reptiles, but in general they do not attack the larger mammals ; and it is perhaps owing to this generally harmless dis-position that they are held sacred by the Hindus in many parts of India.

In accordance with the nature of its prey, the gharial seems to be more thoroughly aquatic in its habits than most of its allies ; the relatively long hind limbs and the fully-webbed toes being features specially suited to aid in swimming. In the breeding season the female lays about forty eggs in the sand of the river-bank, which are deposited in two layers, and covered to a considerable depth with sand ; the two layers being probably laid on different days. The newly-hatched young, which, from the great proportionate length of their snouts, present a most extraordinary appearance, are very active, and of a greyish brown colour, with five irregular dark oblique bands on the body and nine on the tail.

In addition to those of the existing species, the Siwalik Hills have yielded remains of several extinct gharials, some of which attained gigantic dimensions ; while other species belonging to the living genus have been obtained from the middle Tertiary rocks of England. Possibly, also, certain fossil gharials from the Cretaceous deposits of the United States should find a place in the same generic group. Other Cretaceous species are, however, remarkable for the presence of a vacuity in the skull in front of the eye-socket, in consequence of which they have been separated as a distinct genus, under the name of *Thoracosaurus*.

Mention must also be made of an enormous gharial from the Siwalik Hills, known as *Rhamphosuchus crassidens*, which attained a length of some 50 or 60 feet, and had teeth as large as those of the biggest crocodile ; its upper teeth biting on the outer side of the lower ones, instead of interlocking with them, as in the living species. Another extinct giant of the order is the very imperfectly known *Dinosuchus hatcheri*, from the Cretaceous rocks of Montana, which is estimated to have attained a length of between 35 and 40 feet.

SKULL OF PELAGOSAURUS

O, socket for eye ; *T*, temporal pit or fossa

EXTINCT GROUPS

As already mentioned, all existing crocodiles, together with the species from the Tertiary formations, constitute a single family, characterised by the vertebræ having a ball in front and a cup behind, and by the internal nostrils being situated at the hind end of the skull, as well as by the bony plates of the back being arranged in at least four longitudinal rows. Although a few species found in the topmost beds of the underlying Secondary formations approximate in some respects to the foregoing, the majority of the crocodiles from rocks as old as or older than the Chalk differ very considerably from the existing types.

In the first place, the bodies of their vertebræ articulate with one another by slightly hollowed surfaces at both ends ; while, owing to the want of union between the hindmost bones of the palate beneath the nasal passages, the internal apertures of the nostrils are situated nearly in the middle of the skull. Then, again, when a bony armour is present, the plates on the back are arranged in only two longitudinal rows; while those on the lower surface of the body form two distinct shields. It is remarkable that among these extinct crocodiles some are met with having broad and short snouts like the modern alligators, while others have long and narrow snouts like the gharials.

In the Wealden and Purbeck rocks, underlying the Chalk, some of these crocodiles, such as the short-snouted Swanage crocodile (*Goniopholis*), resembled living types in having the socket of the eye communicating freely with the lower temporal pit, although they were distinguished by the plates of the back articulating together by means of a peg-and-socket arrangement. In still older formations, such as the lower Oolites and Lias, there were, however, many long-snouted crocodiles, such as the steneosaurs (*Steneosaurus*) and pelagosaurs (*Pelagosaurus*), in which the socket of the eye is divided from the lower temporal pit by a bony bar, as shown in the figure. Moreover, in these forms the upper temporal pit, *T*, was larger than the socket of the eye ; whereas in all living kinds the former is much the smaller of the two, and may even be obliterated.

Remains of a remarkable group of extinct crocodiles occur in the Kimeridge and Oxford Clays of Europe, of which *Geosaurus*, *Dacosaurus*, and *Metriorhynchus* are the best-known representatives. These reptiles lacked the bony scutes in the skin of the back characteristic of ordinary crocodiles, and their fore limbs were relatively small and weak, and their eyes furnished with a ring of bony plates similar to those of the ichthyosaurs, or fish-lizards. From the associated remains, it is evident that these crocodiles must have been marine.

The structure of the lower portion of the limbs was, however, long unknown ; but this gap in our knowledge was eventually bridged over by the discovery in the upper Jurassic strata of the Continent of two nearly complete skeletons of members of the group, one belonging to the genus *Dacosaurus* and the other to *Geosaurus*. From the structure of the skeleton, it is quite evident that the fore limbs were short and paddle-like, but the hind pair departed less widely from the ordinary type. The jaws were relatively long and armed with powerful teeth ; and it is considered probable that the long tail terminated in a powerful vertical fin. These reptiles were thus nearly as well adapted for a marine life as whales, and must apparently have been formidable rivals to the ichthyosaurs and plesiosaurs of the Jurassic seas.

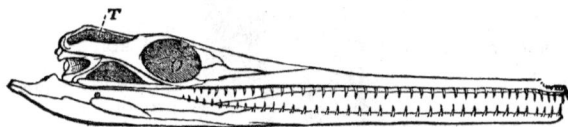

In general, however, the earlier extinct crocodiles were decidedly of a less specialised type than those of the present day ; and as a gradual transition can be traced in these respects from the oldest to the most recent, the group affords a very interesting instance of progressive evolution. In the very oldest of the Secondary rocks—namely, the Trias—there occur, both in Europe and India, certain very remarkable long-snouted reptiles, known as Parasuchia, which appear in some respects intermediate between crocodiles and tuateras. They resemble the former group for instance, in the nature of their teeth, bony armour, ribs, and vertebræ, but approximate to the latter in the structure of the skull, abdominal ribs, and probably of the collar-bones and interclavicle.

The typical representative of the group is the Indian *Parasuchus*, although the European *Phytosaurus*, or *Belodon* is much better known. The most remarkable member of the group is, however, an extraordinarily long-skulled species from the Trias of South Africa, known as *Erythrosuchus*, in which the jaws were armed with sabre-like, serrated teeth, very similar to those of the dinosaurian *Megalosaurus*. As *Erythrosuchus* was of large size it must have been a fell carnivorous monster.

Many naturalists, it should be mentioned, regard the Parasuchia as entitled to rank as a distinct order, intermediate in many respects between crocodiles and tuateras.

Another extinct group of primitive crocodiles, the Aëtosauria, is typified by the genus *Aëtosaurus* of the upper Jurassic strata of Europe, the species of which were quite small in comparison to modern crocodiles.

ORDER DINOSAURIA

DINOSAURS, OR GIANT LAND REPTILES

NEARLY allied to crocodiles are those remarkable extinct reptiles from the rocks of the Secondary period, which include among their number the most gigantic of all land animals, and likewise those members of the reptilian class which make the nearest approximation in their organisation to birds. During that epoch of the earth's history in which the Chalk and underlying Oolitic rocks were deposited, when mammals were represented by a few small forms of lowly type, these strange reptiles were the dominant animals on land; some progressing in the ordinary lizard-like manner, whilst others stalked on their hind limbs like birds. To give some idea of the enormous dimensions attained by some of these creatures, it may be mentioned that the thigh-bone of one species measures 64 inches, while the total length of its skeleton is estimated to have been between 60 and 80 feet. On the other hand, some species were comparatively small, and not more than a couple of feet in length.

Although the whole of these reptiles are markedly distinct from crocodiles, yet they agree with them in the general characters of their skulls, vertebræ, and ribs; but they differ so decidedly from one another that it is not easy to give a definition of the entire order. They are, indeed, divided into four well-marked groups, with so many differences between them that in the opinion of many they are entitled to rank as separate orders; and it will, accordingly, be most convenient to treat these groups seriatim.

INNER AND SIDE VIEWS OF TOOTH OF HOPLOSAURUS

QUADRUPEDAL GROUP

The most stupendous members of the order are included in a group which may be conveniently designated quadrupedal, or sauropodous, dinosaurs, on account of their walking in the ordinary manner, and in having five toes to the feet. The most striking peculiarity of this group is to be found in the circumstance that the vertebræ of the neck and back, as shown in the accompanying figure, had large cavities in their sides, which in the living state may have been filled either with cartilage or with air. These vertebræ resembled those of existing crocodiles, as described on page 1471, in having a ball at one end and a cup at the other; but whereas in crocodiles the ball is at the hind end of the body and the cup in front, in these dinosaurs precisely the reverse of this arrangement obtained.

As regards dentition, these reptiles had their teeth implanted in distinct sockets like crocodiles; but the teeth themselves, as shown in the accompanying figure, were of a peculiar spatulate shape, with the outer side convex and the inner concave. Agreeing in the general structure of their pelvis with crocodiles, these dinosaurs were distinguished therefrom by the circumstance that the bone known as the pubis (*p* in figure on page 1470) enters into the composition of the cavity for the reception of the head of the thigh-bone. The limb-bones are solid throughout. From the nature of their teeth, which are often much worn by use, it may be inferred that these reptiles were

LEFT-SIDE VIEW OF IM-PERFECT VERTEBRA OF HOPLOSAURUS
s, cavity

vegetable-feeders; and it is not improbable that they frequented the margins of lakes and rivers, where their inordinately long necks would enable them to browse with ease on the various aquatic plants. That they must have been very sluggish in their movements and stupid in their ideas, is indicated by the wonderfully small proportionate size of their brains.

These dinosaurs were common both in Europe and the United States, the larger forms having been described under the names of *Pelorosaurus*, *Atlantosaurus*, *Cetiosaurus*, *Brontosaurus*, *Hoplosaurus*, and *Diplodocus*. They also occur in India, Argentina, Africa, and Madagascar.

In Europe one of the best-known of these giant reptiles is *Diplodocus carnegii*, of which the entire skeleton is shown, by means of a cast, in the Natural History branch of the British Museum (page 1488). This monster measures nearly 80 feet in length, and stands some 15 or 16 feet in height at the shoulder. This, however, gives but a very inadequate conception of the general proportions and appearance of the reptile; and it has to be added that while the body is very short, occupying only about one-fifth of the entire length of the skeleton, the rest of the framework is formed by the long neck (with the small head) and the still longer tail (page 1491). What the creature could have wanted with a whip-like tail of 30 feet in length, it is impossible to conjecture; but that the appendage was subject to violent usage seems evident from the circumstance that in a skeleton of an allied species from Peterborough one of the slender terminal vertebræ has been fractured and reunited during life.

In addition to the fact that the aperture of the nostrils is situated at its summit, immediately above the sockets for the eyes, the most remarkable feature about the skull is its relatively small size; it is, in fact, actually smaller than any one of the vertebræ at the base of the neck. If the skull of a crocodile be compared with one of its neck vertebræ, it will be easy to realise what this means.

Equally noteworthy is the fact that the teeth are confined to the front of the jaws, and consist of long, slender pegs, which could have been of little or no use in biting; and even if employed in dragging up plants from the water they would be very liable to be broken. Almost their only use would seem, indeed, to have been raking up soft floating substances from the surface of the water. The form of the vertebræ of the neck serves to indicate that the calibre of the throat must have been very small in comparison with the general size of the reptile.

To carry the enormous bulk of such a monster necessarily requires limbs of great size and strength, and the thigh-bone is accordingly solid throughout, and nearly equal in length to the height of a tall man, with a proportionate width and girth. It is doubtful, however, whether even a limb of the proportions indicated by this enormous bone would not have been crushed by the superincumbent dead weight of the huge carcase had it not been for a special structural modification in the vertebral column.

This picture is from a photograph of a reproduction in plaster of the dinosaurian land reptile (*Diplodocus carnegii*), eighty feet in length, from the upper Jurassic, Wyoming, U.S.A. It stands in the reptile gallery at the Natural History Museum, South Kensington.

Throughout the neck and body, and also in the first half of the tail, the vertebræ are of the type of the one shown on page 1487, having a large aperture on each lateral surface which communicates with a series of honeycomb-like chambers in the interior, and thus reduces the whole structure to a kind of shell. A backbone of this type cannot have weighed more than about one-third as much as that of a whale of equal dimensions, and it is probable that by this means alone was a terrestrial existence possible to these monstrous and unwieldy reptiles.

As regards the probable mode of life of these reptiles, Mr. J. B. Hatcher has written as follows :

" I am inclined toward the opinion that diplodocus was essentially an aquatic animal, but quite capable of locomotion on land. Though living for the most part in the more important rivers and fresh-water lakes, it may not infrequently have left the water and taken temporarily to the land, either in quest of food or in

themselves with a sufficiency of food, unless, indeed, they were eating incessantly from morning to night.

The lightening of the backbone is clearly a feature connected with terrestrial habits (as it would be quite unnecessary in a purely aquatic animal), and it would therefore seem probable that these reptiles passed more of their time on land than Mr. Hatcher considered to be the case.

This is, indeed, the view taken by Mr. E. S. Riggs, who has devoted much attention to the study of these creatures. Arguing both from the peculiarities displayed by the backbone and from the construction of the limbs, he expresses the opinion that in place of being semi-aquatic, or at all events marsh-dwelling, diplodocus and its allies were in the main terrestrial creatures, and this view accords best with what is known of their structure. Still, in the case of the diplodocus it is difficult to imagine how it could have procured food except by entering the water and

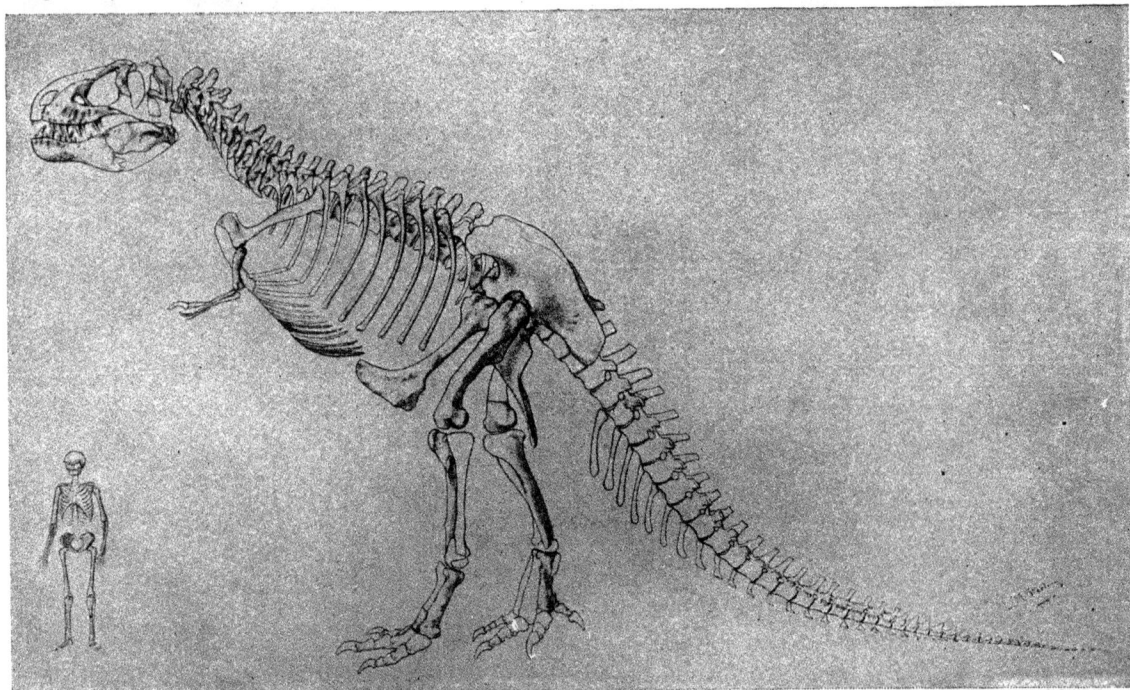

SKELETON OF TYRANNOSAURUS, WITH SKELETON OF MAN TO SHOW RELATIVE SIZES
From the specimen in the Natural History Museum, New York

migration from one to another of adjacent bodies of water. Not only would an aquatic life seem to harmonise best with the anatomical characters of diplodocus as we know them, but such a habitat would also afford these comparatively helpless animals the greatest possible protection from the huge terrestrial carnivorous dinosaurs which were undoubtedly their enemies."

He then goes on to observe that these huge reptiles must have required enormous supplies of succulent and nutritive food ; and further suggests that the country in which they dwelt somewhat resembled Amazonia at the present day, where the flooded and swampy lowlands teem with vegetable products of a nature suitable to the feeble teeth of the diplodocus. Even under such circumstances it is, however, difficult to imagine how those creatures, with their small mouths and narrow gullets, could have managed to provide

devouring soft and luscious vegetation ; and the excessively long and slender neck may have been correlated with such habits.

The skeleton of the diplodocus has been mounted with the body high above the ground and the limbs approximately vertical, but the question has been raised whether this reptile did not crawl on its belly with the limbs splayed out like those of a crocodile. The question is one of considerable importance, as it affects the conception of the general appearance and pose not only of diplodocus, but of the quadrupedal giant dinosaurs generally.

Dr. Hay is of opinion that the reason why a rhinoceros-like carriage has been attributed to diplodocus and its relatives is that the bird-like dinosaurs, such as the iguanodon and the megalosaur, are definitely known to have walked on their hind legs in bird-fashion. After arguing against the probability of diplodocus having a mammal-like carriage from the form of the limb-bones, and especially of the feet, in which the great claws of the outer toes would come much more effectively into use if placed somewhat sideways (instead of vertically) on the ground, he urges that such a bulky creature would inevitably be bogged if it carried itself like a rhinoceros.

"There will be little dissent," he writes, "from the view that these animals inhabited a country in which marshy lands abounded, and that they passed most of their time in the vicinity of bodies of water. As to weight, Marsh estimated that that of the brontosaurus was more than 20 tons. Each footprint was thought to be about a square yard in extent. The pressure was therefore about 1,100 pounds on each square foot of the ground. What progress could such enormous animals have made through morasses and along mud-depositing rivers if they carried themselves as represented in the restorations? Without doubt they would soon have become inextricably mired, and perished miserably."

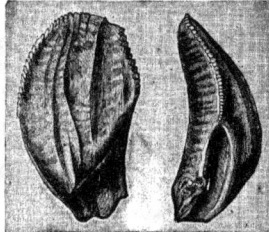

OUTER AND SIDE VIEWS OF TOOTH OF IGUANODON

On the other hand, Dr. Hay's conception of diplodocus and its allies is that they were essentially amphibious creatures, which could swim with considerable ease, and could also creep about on land with their bellies pressed on the ground, although only perhaps in a somewhat laborious manner. The great size of these reptiles seems to be no bar to their being able to creep, as their limb-bones are as large in proportion to the body as in the biggest crocodiles. That they could have reared themselves up on their hind legs seems, however, to be improbable, as it is known to be impossible in crocodiles. When feeding, diplodocus probably swam or crept lazily about in or near the water, gathering in floating plants or those loosely attached to the bottom of lakes. Plants growing at some depth could also be reached by the long neck, after the fashion of a swan, while foliage growing at a height of twenty feet above the surface could likewise have been plucked by the same method. If this view be correct, the skeleton of diplodocus should be reconstructed somewhat upon the

model of a crocodile. The body placed in a crocodilian attitude would be no less imposing than when erect, while the neck might be posed in a graceful curve.

Nearly allied is *Brontosaurus*, the skeleton of which is estimated to have a total length of 62 feet. Contrasted with that of *Diplodocus*, the skeleton of *Brontosaurus* (page 16) is characterised by its relatively shorter body and limbs, and its more massive general structure, the arrangements for lightening its weight being more specialised than in any other member of the group. From the rough terminal surfaces of the limb-bones it is inferred that this reptile was largely aquatic in its habits ; and when sitting down it is supposed that the weight of the body was partly supported by the extremities of the ischia and pubes, which may have been furnished with elastic pads of cartilage or connective tissue.

It may be added that Mr. E. S. Riggs dissents from the view that these gigantic reptiles were semi-aquatic or at least marsh-haunting in their habits. Although the massiveness of the vertebræ recalls cetaceans, yet there is no trace in the latter group of the lightening of this part of the skeleton by means of hollowing and fluting which is so characteristic of these dinosaurs. More important evidence is afforded by the structure of the limbs, which appears to conform strictly to the terrestrial type.

THE CARNIVOROUS DINOSAURS

The carnivorous dinosaurs (family *Megalosauridæ*), of which the English megalosaur (*Megalosaurus*) was the first species discovered, differed from the preceding group in the form of their teeth, which were compressed and sickle-shaped, with sharp-cutting and frequently serrated edges. Their limb-bones also were hollow; while their vertebræ were likewise hollow internally, but had no lateral cavities, and the pelvis (figured on page 1470), although of the same general type as in the quadrupedal group, presented important points of distinction. In place of the short feet of the last-named group, the carnivorous dinosaurs had elongated foot-bones, terminating in sharp claws; the number of functional toes in the hind foot varying from four to three.

RESTORATION OF THE IGUANODON

HORNED DINOSAURS AT HOME

DIPLODOCUS IN LIFE

These life-size restorations stand in Herr Carl Hagenbeck's park, Stellingen, near Hamburg. It must be remembered that the vegetation of the ages in which these dinosaurs lived was different from that existing to-day.

That they habitually walked on the toes of their hind limbs, and not (as was the case with the members of the quadrupedal group) on the whole foot, is evident from the structure of this part of the skeleton, and from the circumstance that the fore limbs were considerably smaller than the hind pair, it may be inferred that progression was at least frequently accomplished by the aid of the latter alone.

The close approximation of the astragalus, or huckle-bone of the ankle, to the lower end of the tibia foreshadows the complete amalgamation which takes place between those bones in birds ; while in one remarkable American form the metatarsal bones of the foot were reduced to three in number, and had nearly the same relationship to one another and to the bones of the ankle as obtains in birds. While the megalosaur attained a height, when erect, of some 15 feet, the little *Compsognathus*, of the upper Jurassic of Bavaria, did not stand more than 2 feet ; and there were other equally diminutive forms, both in England and the United States, in which the whole backbone was so permeated by air-cavities as to be little more than a mere shell of bone.

Megalosaurus was a large reptile, carrying a blunt horn on the muzzle. Other remarkable generic types of carnivorous dinosaurs occur in the Laramie, or topmost Cretaceous, rocks of North America, namely, *Tyrannosaurus rex* and *Dynamosaurus imperiosus*. The former, which appears to have been unprovided with armour, is estimated to have measured 39 feet in length, and walked on the hind limbs only, with the top of the skull raised about 19 feet from the ground (page 1489).

On the other hand, *Dynamosaurus* was an armoured type with about a dozen lower teeth, and a number of curious prominences on the inner margin of the jaw. In this comparatively small number of teeth it seems to differ from the allied *Dinodon*, in which some of the teeth were serrated. A third type, *Albertosaurus sarcophagus*, is based on a skull from Alberta province, Canada. It is apparently more specialised than *Dinodon* in the reduction of the truncated anterior teeth, and more primitive than *Dynamosaurus* in the possession of a larger number of teeth, which are of a less specialised type.

BIRD-LIKE DINOSAURS

All the dinosaurs mentioned above agree with one another in possessing a pelvis approximating to the crocodilian type—that is to say, the pubis, or anterior lower bone of this part of the skeleton, is inclined downwards and forwards, and thus diverges in the form of an inverted *V* from the backwardly and downwardly directed ischium, or posterior lower bone, as shown in the figure on page 1470.

BONES OF RIGHT HIND FOOT OF THE MEGALOSAUR

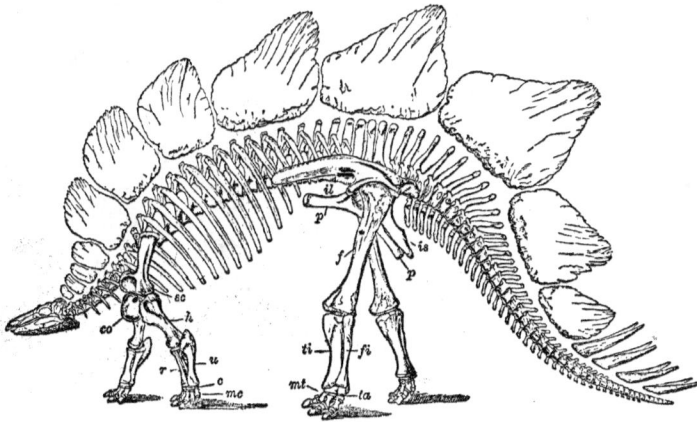

On the other hand, in the bird-like dinosaurs (family *Iguanodontidæ*) the main bar of the pubis is inclined backwards, parallel to the ischium, while it has a secondary plate projecting forwards. In this parallelism of the pubis and ischium these dinosaurs resemble birds, and birds alone ; and from this and other features it has been thought that the latter are derived from reptiles more or less nearly allied to this or the last group of dinosaurs. But it has to be borne in mind that parallelism has probably played a part in producing the resemblance.

All the bird-like dinosaurs are further characterised by the presence of a separate chin-bone (*pd* in the figure on page 1493) at the extremity of the lower jaw ; by the absence of teeth from the front of both jaws ; by the teeth themselves approximating more or less closely to the type of the one represented on page 1490, and by being frequently not implanted in distinct sockets ; and likewise by the vertebræ being completely solid throughout.

The typical representatives of this group are the well-known iguanodons, originally described on the evidence of teeth from the Wealden rocks of England, but now known by entire skeletons from the corresponding deposits of Belgium, which are exhibited in the museum at Brussels. These reptiles, which were represented by allied forms in the United States, habitually walked on their three-toed hind limbs, the largest individuals attaining a length of some 33 feet. They are characterised by the limb-bones being hollow, by the length of the metatarsal bones of the foot, by the first digit of the five-toed fore limb being converted into a large conical spine, and also by the teeth being of the type of the one shown in the above-mentioned figure.

Needless to say, animals with such teeth must have been purely vegetable-feeders, as indeed were all the other members of this group. The hind feet terminated in rather sharp claws, and there was no bony armour on the body. The iguanodons probably stalked about among the palm-forests of the Wealden period, on the leaves and fruit of which they may be presumed to have in great part subsisted. In these reptiles the large flattened and serrated teeth were arranged in each jaw in a single row, but in certain smaller forms known as trachodons (*Trachodon*), which occur in the higher Cretaceous rocks of both Europe and North America, there were several rows of teeth in use at the same time, the edges of these teeth being so flattened and fitted together that a pavement-like structure resulted. These trachodons were all much inferior in size to the gigantic iguanodons. The American claosaur (*Claosaurus*), of which the skeleton

RESTORED SKELETON OF ARMOURED DINOSAUR
sc, shoulder-blade, or scapula : *co*, coracoid ; *h*, upper arm-bone ; *r*, radius ; *u*, ulna ; *c*, wrist, or carpus ; *mc*, metacarpus ; *il*, haunch-bone, or ilium ; *p*, pubis ; *is*, ischium ; *f*, thigh-bone ; *ti*, tibia ; *fi*, fibula ; *ta*, ankle, or tarsus ; *mt*, metatarsus.

is figured on page 1494, differs from the iguanodons in having the fore foot of normal structure.

A theory was long current that the skeletons of the aforesaid iguanodons from the Wealden formation of Bunissart, in Belgium, were deposited in a narrow gorge cut through Carboniferous strata and subsequently filled up with materials of Wealden age. This theory appears, however, to be founded on a misconception. The iguanodons, as a matter of fact, lived on the borders of a lake, where in due course they died, and their bodies became buried in the mud. It is commonly considered that these reptiles always walked in the upright position, but it is more probable that they went frequently on all fours, and commonly assumed this attitude when leaving the lake.

The toothless predentary and premandibular bones at the extremities of the jaws of the iguanodont dinosaurs are considered to have been sheathed in horn, so as to form a beak adapted for nipping off the branches or herbage on which these dinosaurs fed. From time to time footprints of the iguanodon are discovered in the sandstones of the Wealden series at Hastings and in the underlying Wadhurst Clay.

Some years ago an unusually fine series of these impressions, with the track made by the creature's tail as well, was discovered in the formation last mentioned ; and a photograph of the surface of the clay as exposed in the pit enabled Mr. L. Dollo to draw some interesting conclusions with regard to the movements of the reptile by which they were made. The footprints, it may be mentioned, are made by the three-toed hind feet, and are not far short of 12 inches in length ; while the stride of the animal was in some cases from 7 feet to 8 feet.

Three different types of these footprints have been detected, in the first of which the prominences on the sole of the foot form the most conspicuous feature, indicating that the reptile was at rest, an inference confirmed by the fact that footprints of this type—and of this type only—are always accompanied by an impress of the tail, which was resting on the ground at the time the impressions were made. In the second type, on the contrary, the impressions of the toes are alone clearly defined, indicating that the animal was running at the time they were made ; while in the third type the finely-marked impression of the entire foot leaves little doubt that at the time this was impressed on the sandy clay the huge reptile was slowly walking along the sea-shore.

ARMOURED AND HORNED DINOSAURS

Nearly allied to the iguanodons are the remarkable armoured and horned dinosaurs (families *Stegosauridæ* and *Ceratopsidæ*), which constitute a suborder, Ceratopsia, or Stegosauria, characterised by their solid limb-bones, the presence of some kind of bony armour on the body, or horns on the skull, the short foot-bones, frequently terminating in hoof-like toes, and the habitual quadrupedal gait. Beginning in the British Lias, these extraordinary reptiles continued throughout the Secondary period, and seem to have attained their maximum development at the close of the Cretaceous epoch in the United States. Of the armoured forms the huge *Omosaurus* of the English Oxford and Kimeridge Clays, and the allied *Stegosaurus* of the corresponding rocks of the United States, were characterised by the possession of large quadrangular bony plates, arranged as a double series in a vertical position down the middle of the back, while the tail was protected by some formidable spines, as shown in the greatly reduced restoration of the skeleton given on page 1492.

UPPER AND SIDE VIEWS OF SKULL OF A HORNED DINOSAUR

a, nostrils ; *f*, brain ; *h*, horn ; *n*, nasal bones ; *p*, upper chin-bone ; *r*, extremity of upper jaw

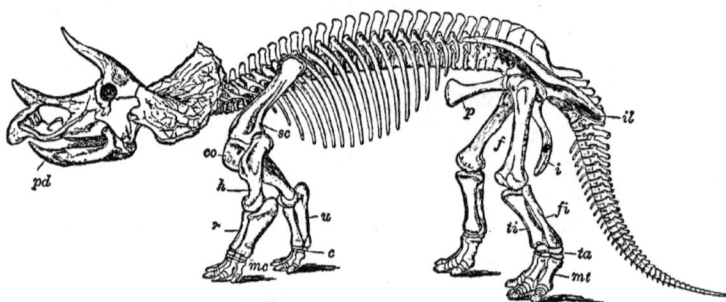

Polacanthus foxi, from the Wealden of the Isle of Wight, appears to have been a long-bodied reptile of about three feet in height at the shoulder, with the hind quarters invested in a solid bony shield, and the upper surface of the rest of the body, the neck, and the tail protected by a double row of large bony plates standing vertically. The creature may, in fact, be regarded as a kind of reptilian armadillo. Still more strange were the somewhat later horned dinosaurs (*Ceratops* and *Triceratops*), of which two views of the skull and a more reduced restoration of the skeleton are here given.

Triceratops is remarkable not on account of its vast bodily size, for in this respect it is not to be compared with the giant *Diplodocus*, neither does it walk on its hind legs with its head raised sixteen feet high in the air like its cousin the iguanodon. On the contrary, *Triceratops*, so far as bodily size and gait are concerned, has no claim to special distinction, its total length falling just short of a score of feet,

RESTORED SKELETON OF HORNED DINOSAUR

b', chin-bone ; other letters as under figure on previous page

while it walked on all fours very much in the fashion of an unusually long-limbed crocodile.

What, however, it lacks in these respects, the three-horned dinosaur, as it may be designated, more than makes up for in the extraordinary conformation and huge size of its skull. It enjoys, in fact, the distinction of having, both absolutely and relatively, the largest

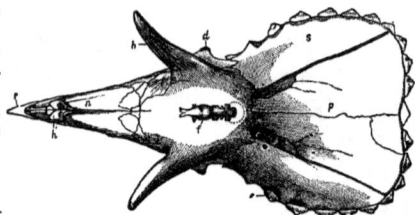

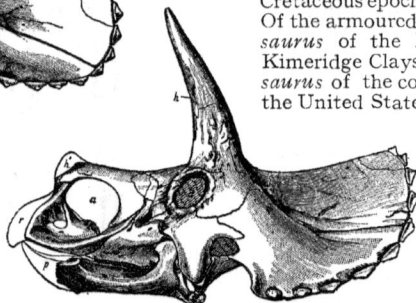

head of any known land animal, living or extinct, and to find a parallel in this respect it is necessary to look to the members of the whale tribe, in some of which the relative size of the head is still greater.

Remains of the horned dinosaurs occur in the upper Cretaceous beds along the eastern flanks of the Rocky Mountains for a distance of about eight hundred miles, but are more abundant than elsewhere in Wyoming, and more especially Converse County. As an indication of the abundance of these remains in the district in question, it may be mentioned that some years ago a single investigator collected bones referable to no less than forty distinct individuals.

The Ceratops-beds, as they are called, are of fresh-water or brackish origin, but rest in some places on marine strata. As they occur some distance below the topmost Cretaceous rocks, their age may be approximately correlated with the highest beds of the English Chalk, so that the horned dinosaurs lived at a much later epoch than the majority of the giant herbivorous dinosaurs, which flourished during the Oolitic and Wealden epochs.

As mounted, the skeleton of *Triceratops* measures 19 feet 8 inches from the front of the curious toothless beak to the tip of the tail ; while at the loins it stands 8 feet 2 inches. In addition to the skull, which is 6 feet long, or nearly one-third the total length, the most noteworthy features of the skeleton are its great relative height at the loins, the extremely short and deep body, shaped more like that of a mammal than that of a crocodile, the tall and massive limbs, and the curious turtle-like flexure of the fore feet.

In addition to its great absolute and relative size, the skull is specially characterised by the presence of three horns, one in front and two behind ; the hind pair being strikingly like the horn-cores of a gigantic ox. Indeed, so ox-like are these horns that a pair was actually described as indicating a Cretaceous bison. Equally note-worthy is the presence of a cutting beak, formed by a separate bone in each jaw. More striking still is the presence of a great bony curtain, or frill, like the flange of a fireman's helmet, over-hanging the neck, and thus rendering the relative length of the skull greater in respect to the rest of the skeleton than is really the case. Both the horns and the curtain were probably covered in life with a thick layer of horn.

Despite the enormous size of the skull, which forms a long triangle in shape, the brain was of the small size of that of dinosaurs in general. Whether this implies slow and sluggish habits is not easy to decide. The powerful armature of the skull is, however, suggestive of activity both in attack and defence, it being difficult to imagine that such structures were not developed for special purposes. That *Triceratops* was herbivorous is evident from the structure of its teeth, which differ, however, from those of the more ordinary dinosaurs in being implanted in the jaws by means of two distinct roots, thereby foreshadowing the mammalian type, although, of course, there is no direct relationship between this group of reptiles and mammals. It has been suggested that the food of *Triceratops* mainly

consisted of soft, succulent vegetation. Although no attempt has been made to reproduce it in the restoration of the skeleton, there is some reason to believe that during life the body was protected to a certain degree by bony armour. This idea is based on the fact that various spines, bosses, and plates of bones have been found in association with the remains of these dinosaurs. To assign these fragments of armour, presuming, of course, that they really belong to *Triceratops*, to their proper position on the body has not at present been found practicable, although it has been suggested that some were probably situated on the back behind the bony curtain of the skull, while smaller ones may have defended the throat. In a restoration of the creature attempted by an American artist the spines are altogether omitted.

That the heavy armature of the head in *Triceratops* and its allies was developed for purposes of attack and defence seems almost a certainty. Presuming this to be the case, it has, however, yet to be definitely demonstrated whether it was for the purpose of aiding these reptiles in encounters with individuals of their own species, or to protect them from the attacks of contemporary carnivorous dinosaurs, such as the *Dryptosaurus* (or *Lælaps*), the Cretaceous representative of the well-known *Megalosaurus* of the Oolites. Professor F. A. Lucas, who has paid much attention to the restoration of extinct animals, is in favour of the former alternative, as is evident from the following passage:

"So long as *Triceratops* faced an adversary, he must have been practically immovable, but, as

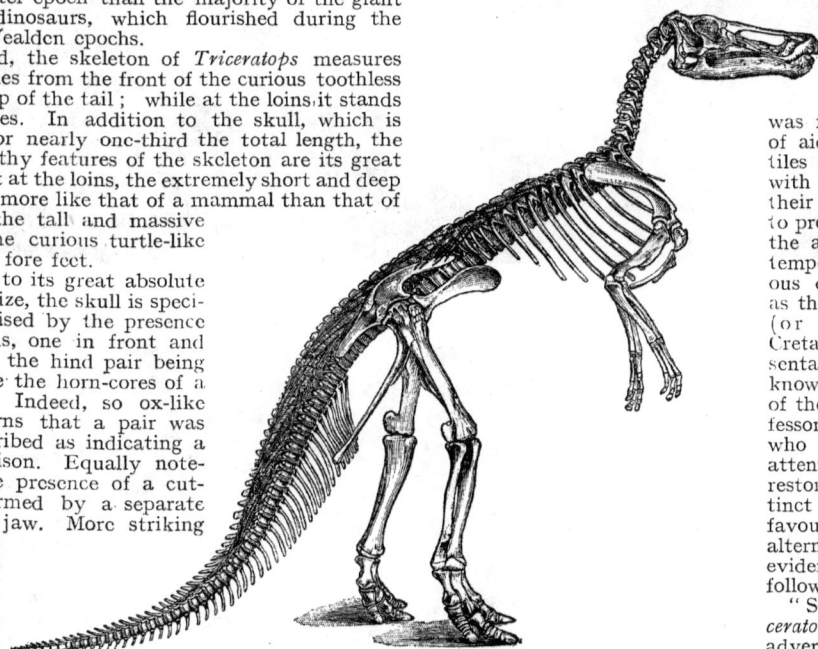

SKELETON OF CLAOSAURUS

he was the largest animal of his time, it is probable that his combats were mainly with those of his own kind, and the subject of dispute some fair female upon whom rival suitors had cast covetous eyes. What a sight it would have been to have seen two of these great brutes in mortal combat as they charged upon each other with all the impetus to be derived from ten tons of infuriate flesh ! We may picture to ourselves horn clashing upon horn, or glancing from each bony shield until some skilful stroke or unlucky slip placed one combatant at the mercy of his adversary.

"A pair of *Triceratops* horns in the National Museum at Washington bears witness to such encounters, for one is broken midway between tip and base ; and that it was broken during life is evident from the fact that the stump is healed and rounded over, while the size of the horns shows that their owner reached a ripe old age."

CARNIVOROUS DINOSAUR AND STEGOSAUR

CARNIVOROUS DINOSAUR FEEDING ON A CARCASE

STEGOSAUR ATTACKED BY A CARNIVOROUS DINOSAUR

These life-size restorations stand in Herr Carl Hagenbeck's park at Stellingen near Hamburg.

Before leaving the subject of dinosaurs attention may be directed for the moment to the wonderful tracks of fossil animals met with in the sandstone of the Triassic strata of the Connecticut Valley, which have been known to the inhabitants for well-nigh a century, and were described years ago by Professor Hitchcock, of Amherst College, who believed most of them to have been made by birds.

Owing to the porous nature of the sandstone, very few of the bones of the ancient creatures which formed these tracks have hitherto been discovered, and for a long time none were known. The few skeletons that have been found indicate, however, that these tracks were made by dinosaurian and kindred reptiles, some of which walked on their hind legs alone, while others went on all fours. One of these bipeds was a large carnivorous species which left three-toed tracks of one type. Another dinosaur, of herbivorous habits, has also left footprints which are in most cases not very dissimilar to those of its carnivorous relative ; but in places we find indications that the giant sat down, resting its tail and small five-toed front feet on the ground ; thus proving that some, at least, of the four-footed tracks were made by animals which were normally biped. One type of four-footed tracks has, however, been proved to have been made by a reptile similar to a crocodile in structure, but with the body carried high above the ground on long mammal-like limbs. For this creature the name of *Stegomus longipes* has been proposed, although it ought to bear the title *Batrachopus gracilis*, bestowed on its tracks.

ORDER ORNITHOSAURIA
PTERODACTYLES, OR FLYING DRAGONS

A T the present day bats and birds are the only vertebrates endowed with the power of true flight, but during the Secondary period, when the former were unknown and the latter but poorly represented, the place of both was taken by the flying-dragons (order Ornithosauria), or, as they are called, from the structure of their wings, pterodactyles. While agreeing with crocodiles in the essential structure of their skulls and in their two-headed ribs, these curious reptiles have the other portions of their skeleton more or less specially modified for the purposes of flight. In the relatively large size of the brain—which is doubtless essential for a flying animal —and general bird-like form of the skull, as well as in the keeled breast-bone and general form of the collar-bones (although these are not welded together into a furcula), the pterodactyles present a curious similarity to birds. Misled by these resemblances, some anatomists have, indeed, been induced to consider that the two groups are nearly related, although a more mistaken notion never existed. Such resemblances as do exist between the two groups are due, indeed, to that parallelism in development to which attention has already been directed, as existing between totally different groups of animals whose mode of life is similar.

The most distinctive feature of pterodactyles is to be found in the modifications of the bones of the fore limbs for the purposes of supporting a wing, which took the form of a membranous expansion of skin analogous to that constituting the wings of bats. This wing was mainly supported by the great elongation of the bones of the outer digit, or finger, of the fore limb, as shown in the accompanying figure of the skeleton,

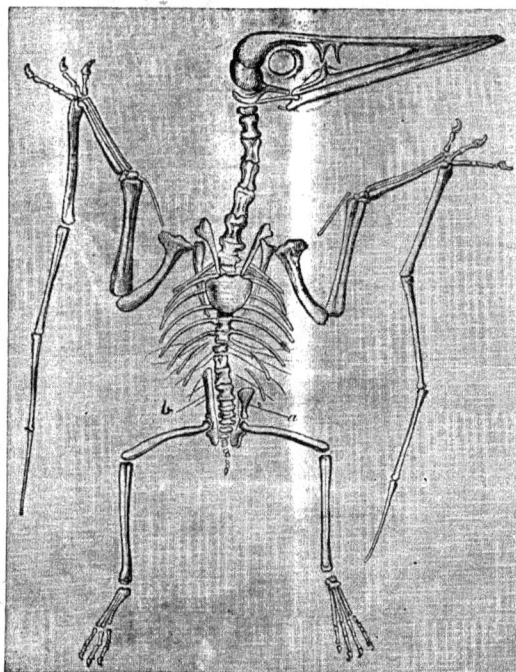

SKELETON OF PTERODACTYLE
The creature is lying on its back, with the head bent to the left. *a* left pelvic bone ; *b*, haunch-bone, or ilium.

and likewise in the restored representation of one of these reptiles. The membrane thus supported seems to have been continued backwards along the sides of the body to include the upper portions of the legs, between which it was extended to embrace the base of the tail in those forms in which the latter appendage was fully developed. Moreover, in the long-tailed species the extremity of the tail itself was provided with a racket-shaped expansion of membrane, which may have served the purpose of a rudder in flight. If it be asked how the presence of such membranes is known, it may be answered that in many of the specimens of these reptiles entombed in the fine-grained lithographic limestones of Bavaria the actual impressions of these membranes have been preserved. The elongated outer finger of the wing had no claw at the extremity, although the other three fingers were thus provided. With regard to the first finger, or the one corresponding to the human thumb, this may have been represented by the small splint-like bone seen depending from the wrist in the figured skeleton.

The hind limbs present no special peculiarities, but, as most of the bones of the skeleton were hollow and permeated by air, like those of birds, it may be inferred that the lungs were probably also constructed after the bird-fashion. The vertebræ of the neck resembled those of living crocodiles in having a ball at the hind end of the body and a cup in front. In general conformation the skull was remarkably bird-like, the snout being produced into a beak, which in some cases was provided with teeth, while in others, as shown in the figure on page 1470, it was toothless, and probably ensheathed during life with horn. Bird-like features

This restoration of the giant pterodactyle, whose remains are found in the upper Cretaceous beds of Kansas, is to be seen in the grounds of Herr Carl Hagenbeck's park at Stellingen.

are likewise shown by the large size of the brain-case, of which the component bones were fused together, and also by the union of the extremities of the two branches of the lower jaw.

Pterodactyles flourished during the greater part of the Secondary period, dating from the epoch of the Lias, and continuing to the close of the one during which the Chalk was deposited. They are represented by several well-marked types, which may be arranged under three family groups.

In the typical pterodactyles (*Pterodactylus*, etc.) the jaws were provided with teeth—which may, however, have been very small in size and few in number—while the skull, as shown in the figure of the skele-

RESTORATION OF A LONG-TAILED PTERODACTYLE

ton, was not produced backwardly, and the tail was reduced to a rudiment. The members of this group, which are common in the Oolitic rocks of the Continent, vary in size from the dimensions of a sparrow to those of an eagle.

The largest member of the group is *Pteranodon longiceps*, of the upper Cretaceous beds of Kansas, the Transatlantic equivalents of the Chalk, which belongs to the group in which the jaws are devoid of teeth, so that during life they were doubtless encased in horn, like the beak of a bird. In size it must have exceeded the largest albatross, the span of wing being no less than 25 feet, while the skull is little short of 40 inches in length. The latter dimension is, however, calculated to give a somewhat exaggerated idea of the size of this monstrous "dragon," for, according to the restoration, the skull is extended nearly as far behind the vertebral column as it is in front. In fact, this part of the skeleton resembles a crutch-stick or a pickaxe, the vertebral column forming the stick itself and the skull the cross-bar.

Another curious feature of the giant pterodactyle is the presence in the socket of the eye of a ring of bones, comparable to those found within the eyes of birds. The existence of this "sclerotic" ring of bones in the eye of the larger pterodactyles is a well-marked instance of a parallel adaptive character, for it is unknown in the smaller members of the group, and since, as already mentioned, pterodactyles have no relationship with birds, it must evidently have been acquired independently in the two groups. Another instance of the same kind is the loss of the teeth in both these groups; the more primitive birds, like the oldest pterodactyles, having the jaws armed with a full series of sharply-pointed teeth.

Lastly, there are the long-tailed pterodactyles (*Rhamphorhynchus*, etc.), which are likewise of Oolitic and Liassic age, and are at once distinguished, as shown in the restoration, from the members of the preceding group by the fully-developed tail. These long-tailed species are evidently the most generalised members of the order; and the retention of the tail in the generalised group and its loss in the more specialised one presents an exact parallelism to the case of ordinary bats and the more highly-developed fruit-bats.

Pterodactyles of all kinds doubtless fed, like gulls and cormorants, on fish, and at first sight it might seem that toothed jaws would be much better adapted to hold such slippery prey than is a smooth, horny beak. It has, however, to be borne in mind that although such toothed jaws would ensure the retention of every fish captured, yet they would prove a decided hindrance to its being swallowed quickly and easily. Possibly, indeed, it would have been necessary for a toothed bird or a toothed ptero-

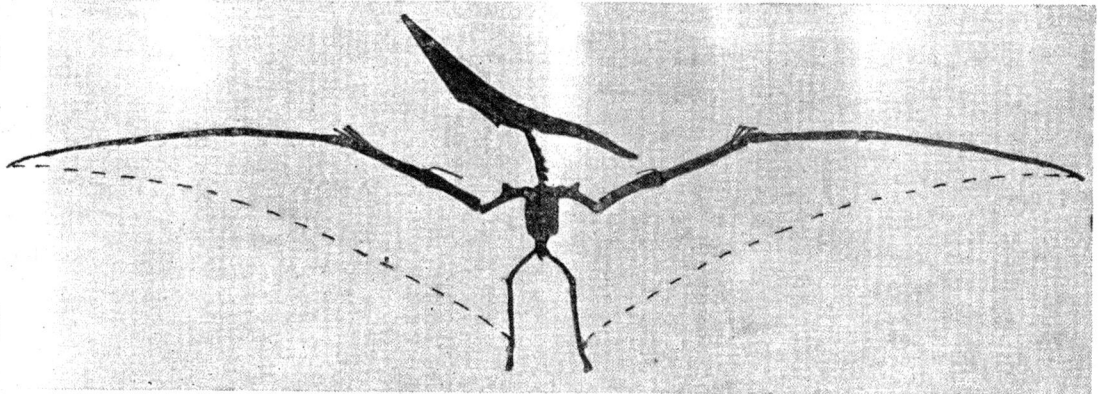

SKELETON OF GIANT PTERODACTYLE

Apparently this backward extension of the occipital region of the skull must have been intended to act as a counterpoise to the huge beak, although if large-billed birds like the adjutant stork can manage to get on without such a structure it is difficult to imagine why it should be necessary in the case of the pterodactyle.

dactyle to resort to the shore before being able to devour its prey; and if this be the case there is a ready explanation of the reason why both birds and pterodactyles discarded teeth in favour of a horny beak, which enabled them to bolt their food while on the wing.

ORDER RHYNCHOCEPHALIA
TUATERAS

THE reptiles included in this and the next order are almost entirely extinct, including, indeed, only a single existing species, and attained their maximum development at a comparatively early stage of the earth's history, being well developed in the Triassic period, which forms the basis of the great Secondary epoch. Although in many cases exhibiting specialised developments, they are essentially a primitive type, exhibiting relationships on the one hand with the parasuchian crocodiles, and on the other with certain groups of early salamanders.

The lizard-like reptile known to the Maori as the tuatera, which is now confined to certain small islands off the north-east of New Zealand, is not only the most remarkable of all existing reptiles to which the term lizard can be applied, but is the sole living representative of a distinct family, as well as of an entire order; and the difference between it and an ordinary lizard immeasurably exceeds that by which the latter is separated from a serpent. As an order tuateras may be characterised as follows.

Externally most of them appear to have been more or less lizard-like; and, as in their living representative, the body was probably covered above with small granular scales intermingled with tubercles. The skull differs essentially from that of lizards in having the quadrate-bone immovably fixed by the upper end to the adjacent bones, and likewise by having both an upper and a lower temporal arch. The hind portion of the palate is formed by the union of the pterygoid bones, which, generally, at least, extend forwards to meet the vomers, and thus divide the palatines; while the anterior upper jaw-bones, or pre-maxillæ, remain separate from each other. The teeth, of which some are on the palate, are not implanted in distinct sockets, and are often welded to the summits of the jaws.

In the trunk the ribs articulate to the vertebræ by single heads, and often have hook-like processes similar to those of birds; while on the lower surface of the body so-called abdominal ribs are always developed, forming a shield composed of a number of segments, and comparable to the plastron of the tortoises. The bodies of the dorsal vertebræ are hollowed at both articular ends. Additional segments, known as intercentra, are developed between the vertebræ, and the tail is provided with V-shaped chevron-bones suspended between adjacent vertebræ.

Each foot carries five toes, and the humerus, or upper arm-bone, is perforated on the inner side of its lower extremity. That tuateras form a very primitive group is clear, not only from their structure, but from their antiquity, representatives of the order occurring in the Permian strata, immediately overlying the Carboniferous, or coal-bearing, rocks.

THE TUATERA

The single existing representative of the order (*Sphenodon punctatus*) forms a family, *Sphenodontidæ*, by itself, and likewise is the representative of a distinct suborder, Rhynchocephalia vera, characterised by each segment of the shield on the lower surface of the body being formed of only three elements, of which the middle one is chevron-shaped, and likewise by the presence of two vertebræ in the sacrum, and the occasional suppression of the intercentra. Typically the fifth metatarsal bone of the hind foot is reduced in length and thickened in the same manner as in lizards.

The group is further characterised by the double nostrils, the union of the two branches of the lower jaw by cartilage, and the deeply-hollowed articular surfaces of the vertebræ. From its extinct allies the family is distinguished by the presence of a perforation on each side of the lower end of the humerus and of hook-like processes to the ribs, as well as of intercentra, or additional segments, between the bodies of the vertebræ; and likewise by the beak-like premaxillary bones carrying a pair of somewhat chisel-like teeth, and the presence of only a single row of teeth on the palate, separated by a groove from the row affixed to the edge of the upper jaw. Into this groove are received the teeth and upper edge of the lower jaw, which in very old individuals becomes as hard and polished as the teeth themselves, the latter being more or less completely worn away in extreme old age. On the upper surface of the skull is a large vacuity, or foramen, in the parietal bones, known as the pineal, or parietal, foramen.

In external appearance the tuatera is lizard-like, the body being

THE TUATERA

slightly and the long tail strongly compressed; while the limbs carry five toes, all furnished with claws, and connected at their bases by webs. There is no external opening to the ear, and the large eye has the pupil vertical. On the upper parts the skin is clothed with small granular scales, intermixed with tubercles; and a crest of spine-like scales runs from the hind part of the head down the middle of the back, continued in a smaller degree of development down the tail; while inferiorly there are large squarish scales arranged in transverse rows.

Attaining a length of about 20 inches, the tuatera is olive or blackish in ground-colour, upon which are small yellowish dots, while the lobes of the crest on the neck and back are likewise of the latter colour. The perforation in the parietal bones of the skull just referred to covers a rudimentary eye, which, although now functionless, was probably a working organ in the ancestors of the vertebrates.

In the adult tuatera this eye is better preserved than in any other animal, and its development has undergone less modification than in other reptiles. The first indication of its appearance is seen at a stage comparable with a two-day-old chick, when a "primary parietal vesicle" buds on the roof of the fore brain slightly to the left of the median line. At a later stage the eye forms a hollow vesicle in front and slightly to the left of its so-called "stalk"—the "parietal stalk," which is a finger-shaped diverticulum of the root of the fore brain, practically in the middle line. The eye is almost or completely separated from the stalk, which contains a prolongation of the cavity of the brain. The "paraphysis" likewise makes its appearance at this stage, as a backwardly-directed outgrowth of the root of the fore brain.

Later still the parietal eye and stalk are conspicuous externally; while immediately before hatching the eye, which is now apparently median, is seen as a white spot with a black border, the latter representing the pigmented margin of the retina and the former the lens. In the adult the eye, though very highly organised, is no longer recognisable externally; but in recently-hatched individuals is still visible as a dark speck through the translucent skin covering the parietal foramen.

The evidence in favour of the originally paired character of the parietal eye is derived principally from the fact that it arises to the left of the median line, while the stalk is practically median, and therefore slightly to the right of the eye. Accordingly the parietal eye in *Sphenodon* is regarded as the left of the original pair, while the right one is represented by the parietal stalk. The origin of the latter appears to be precisely similar to that of the former; while the two have also a very similar structure, although the stalk never acquires the same degree of perfection as the eye.

Both in *Sphenodon* and in lizards the epiphysis, or pineal gland, is a composite structure, in which the paraphysis takes a large share, whereas the parts comparable to the epiphysial outgrowths of fishes form but a small one. In lizards the stalk may represent either the right or the left parietal eye. Beyond that of fellowship, the parietal eye has no real connection with the parietal stalk, being supplied with a special nerve of its own quite distinct from the stalk. Finally, it has been inferred that the ancestors of existing vertebrates were furnished with a pair of parietal eyes, which may have been serially homologous with the existing pair of functional eyes.

Tuateras formerly inhabited the main islands of New Zealand, where they have been destroyed by pigs and other animals. They live in burrows, to which they retire on the approach of danger. They are mainly nocturnal, and are fond of water, in which they can remain for hours without coming to the surface. They feed on living organisms. During the southern summer (November to January) the females lay about ten white, oval, hard-shelled eggs, which are placed in the sand and hatched by the sun's heat. In their movements tuateras are sluggish, and quite unable to run.

EXTINCT TUATERAS

In the Jurassic rocks of Europe there occur remains of reptiles allied to the modern tuatera, but constituting a distinct family (*Homœosauridæ*) typically represented by the genus *Homœosaurus*. These have no tusk-like teeth in the front of the jaws, and the lower end of the humerus has a perforation only on its inner side, and there are no intercentra between the vertebræ of the back, and no hook-like processes to the ribs.

Another family (*Rhynchosauridæ*) is typified by the genus *Rhynchosaurus*, from the Trias, or New Red Sandstone, of England, and is characterised by the beak being toothless and probably sheathed in horn; the palate having two or more longitudinal rows of teeth separated by a groove.

From the preceding families these reptiles differ by having only a single aperture to the nostrils, and by the bony union of the two branches of the lower jaw; while the articular surfaces of the vertebræ are nearly flat. Moreover, there is no vacuity in the middle of the top of the skull for the pineal eye. In the typical genus there is a single row of teeth on the inner side of the groove of the palate, but in the pavement-toothed tuateras (*Hyperodapedon*) there were numerous rows, as is shown in the illustration B. The extremity of the beak in each jaw formed two curved tusk-like processes, which diverged in the lower one. A lower jaw from the Jurassic of Wyoming, described as *Opisthias rarus*, appears to indicate a tuatera allied both to *Sphenodon* and *Homœosaurus*.

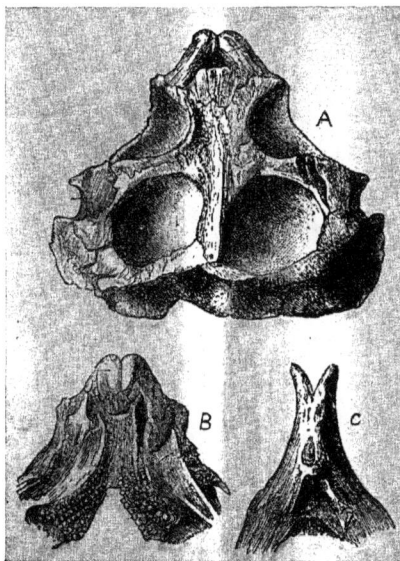

SKULL OF HYPERODAPEDON
A, superior aspect; B, palate; c, under surface of front of lower jaw

PROTOROSAURIA

The Permian rocks of Europe, which underlie the Trias and overlie the Coal-measures, yield remains of genera, such as *Protorosaurus* and *Palœohatteria*, differing markedly from the foregoing, and constituting a second suborder (Protorosauria) of the Rhynchocephalia, characterised by the complex nature of the bones forming the shield on the lower surface of the body, by the fifth metatarsal bone of the hind foot being of an ordinary type, and likewise by the lower bones of the pelvis being expanded into large flattened plates, instead of being comparatively narrow. In the genus first named, the vertebræ of the neck have cup-shaped articular surfaces behind and balls in front, but in the other the articular surfaces of the vertebræ are slightly cupped at each end throughout the series. In both groups intercentra are present.

This group apparently passes imperceptibly into the Microsauria among the primeval salamanders, in which the body was armoured, the vertebræ were completely ossified, and the ribs two-headed.

The Acrosauria, as represented by the Jurassic *Acrosaurus*, may form a third subordinal group of the Rhynchocephalia.

ORDER PELYCOSAURIA
PELYCOSAURS

About 1878 the distinguished naturalist Professor E. D. Cope brought to the notice of the scientific world remains of certain remarkable carnivorous reptiles from the Permian strata of Texas, for which he proposed the group-name of Pelycosauria. This group was at the time regarded as a suborder of the Rhynchocephalia, and was provisionally taken to include the carnivorous mammal-like reptiles of South Africa and Russia. Among the more typical representatives of the pelycosaurs are the members of the genera *Dimetrodon* and *Naosaurus*, extraordinary reptiles in which the dorsal spines of the trunk-vertebræ are so enormously elongated (sometimes with the addition of transverse yard-arm-like projections) that they exceed in height the depth of the body below them. These elongated spines appear to have supported during life a vertical sail-like expansion of the skin of the back.

Dimetrodon was carnivorous, while *Naosaurus* was probably omnivorous. The latter has, perhaps, the most wonderful dentition of any known animal, the incisor teeth being sharp and chisel-shaped, such as might be suited for cutting vegetable substances, while behind these are five pairs of sharp triangular cutting teeth, these being followed by simple cones suited to holding a struggling victim. On the palate and the heavy opposing portion of the lower jaw are plates of bone covered by short, stumpy teeth of a type found in mollusc-eating fish. Although probably omnivorous, *Naosaurus*, instead of possessing a dentition of a generalised type, like that of man or a pig, had a special set of teeth for each kind of food.

As to the systematic position of these reptiles and their kindred, considerable diversity of view has obtained. By many naturalists they are classed with the carnivorous mammal-like reptiles, or anomodonts, but this, according to modern ideas, is altogether unjustifiable, the structure of the temporal arches in the two groups being different. It therefore seems best to revert to the original view that pelycosaurs form a primitive section of the rhynchocephalians.

The group is of special interest as illustrating, perhaps better than any other, the rapid evolution from a generalised type to a complex organisation that may have been the potential cause of early extinction, the life of these reptiles being coterminous with the duration of the Permian epoch. Why these specialised structures

SKELETON OF DIMETRODON

RESTORATION OF NAOSAURUS

were evolved within such a comparatively short time is a subject upon which it is only possible to conjecture. Carnivorous or omnivorous in habit, and easily masters of their contemporaries, these reptiles, it has been suggested, may have developed the tall spines of their vertebræ as the result of a mere exuberance of growth from a utilitarian beginning; but that these structures eventually became useless cannot be doubted.

That pelycosaurs existed elsewhere than in North America is proved by the occurrence of remains of the American genus *Naosaurus* in the Permian of Bohemia and of the exclusively European *Stereorhachis* in that of France, while certain reptiles from Central Germany may also belong to the group. On the other hand, they are unknown in South Africa or India, and it is improbable that they are represented in the Russian Permian. If this be so, pelycosaurs are unknown in any country where anomodonts (in the wider sense of that term) occur, so that the two groups may apparently be regarded as belonging to totally distinct faunas, in some respects complementary to one another.

Certain extinct marine reptiles from the Triassic (Upper New Red Sandstone) deposits of California have been regarded as representing a distinct order, the Thalattosauria, typified by the genus *Thalattosaurus*. In many respects these reptiles resembled ichthyosaurs, or fish-lizards (referred to later on), having the eye similarly furnished with a ring of bony plates. They are, however, broadly distinguished by the structure of the temporal region of the skull, which has an upper and a lower bony arch, and likewise by the character of the dentition, which takes the form of flattened crushing teeth, some of which are situated on the palatine and vomerine bones. From these and other features, the thalattosaurs appear to be most nearly allied to the tuateras, to which they seem to present the same kind of relationship as is borne by the extinct sea-serpents (Pythonomorpha) to lizards. If this be so, the Thalattosauria might, perhaps, be best regarded as a suborder of Rhynchocephalia. Be this as it may, the special interest attaching to the group is the evidence it affords of the independent adaptation of this type of reptile to the exigencies of a marine existence, and this, too, at an early period of the earth's history.

ORDER SQUAMATA
CHARACTERISTICS OF SCALED REPTILES

ALTHOUGH in popular language the term lizard is applied to any four-legged reptile, exclusive of turtles and crocodiles, in scientific usage it is more convenient to restrict it to those members of the great group of scaly reptiles which do not come under the designation of either chamæleons or serpents, whether they are provided with legs, or whether they lack those appendages. Formerly, indeed, lizards and chamæleons were regarded as constituting an order by themselves quite apart from snakes, but the two groups are now known to be so intimately connected as to render any such division inadmissible ; and they are accordingly placed in a single order, known as scaly reptiles, or, technically, Squamata. Structurally this ordinal group differs widely indeed from any of those hitherto treated, and as it is essential to gain a correct idea of such structural differences, they may first be taken into consideration.

Taking their name from the coat of overlapping horny scales with which they are generally invested, the scaly reptiles are primarily distinguished from all the foregoing groups by the circumstance that the quadrate-bone is more or less movably articulated to the skull, and has its lower end projecting freely therefrom, instead of being immovably wedged in among the other bones.

To this primary point of distinction it may be added that the lower temporal arch of the skull is wanting, so that there is no bony bar connecting the lower end of the quadrate-bone with the upper jaw, as there is in the crocodiles ; the absence of this bar being shown in the figure of a lizard's skeleton on the next page. Then, again, the palate, instead of being more or less completely roofed over by bone, is largely open, its bones taking the form of long bars. In some lizards, as in the one of which the skeleton is figured, the upper surface of the skull is covered by bone, so that the temporal pits are roofed over by a superficial layer of bones.

LEFT SIDE OF VERTEBRA OF SNAKE
r, facet for rib ; *zl*, zygosphene, fitting into the zygantrum of the vertebra in front ; *æ* prezygapophysis

Another important feature of the order is to be found in the circumstance that the ribs in the region of the back are single-headed, and articulated to the backbone by means of a facet situated on the body of each vertebra. This feature at once distinguishes the order from crocodiles and dinosaurs, in which the ribs are two-headed, and in the back articulate to a long process arising from the arches of the vertebræ. In most of the members of the order the body of each vertebra has a cup in front and a ball behind, by which it articulates with the adjacent segments of the column – an arrangement paralleled among modern crocodiles. In some lizards, and in all snakes, the vertebræ have additional surfaces on their arches for mutual articulation, thus communicating additional flexibility, and at the same time strength to the backbone.

Another important feature in which the order differs from most of the preceding ones is the absence of any system of true abdominal ribs on the inferior surface of the body. As regards the teeth, these, instead of being implanted in separate sockets, are firmly soldered to the bones of the jaw. In some cases they are attached to the very summit of the jawbones, when the dentition is said to be *acrodont ;* but in others they are affixed to one of the side-walls of the free edges of the jaws, and the term *pleurodont* is then employed. Another divergence from crocodiles is to be found in

the circumstance that the vent opens by a transverse instead of a longitudinal aperture. Finally, in those forms in which the bones of the chest attain their fullest development, there is a breast-bone, or sternum, a pair of collar-bones, or clavicles, and a median T-shaped interclavicle.

The above being the leading characteristics of the entire order of scaly reptiles, it remains to consider how lizards, suborder Lacertilia, are distinguished from the other two suborders into which the existing members of the assemblage are divided. Externally by far the greater number of lizards are four-limbed reptiles of a crocodile-like appearance, with the head, neck, body, and tail well distinguished from one another ; and if there were these alone to deal with, there would be no difficulty in distinguishing between a lizard and a snake.

The matter is, however, complicated by the circumstance that certain lizards, like the familiar slow-worm, have lost all external traces of limbs, and assumed an elongated snake-like form, with the head passing imperceptibly into the body without the intervention of a distinct neck, or any external indication of where the body ends and the tail begins. Externally such snake-like lizards are very difficult to distinguish from snakes, but on opening the mouths of the former it will be found that the tongue cannot be withdrawn into a sheath at its base, as is always the case with the latter. Further help in discriminating between the two groups is afforded by the circumstances that whereas snakes have neither eyelids nor external ear-openings, both these are usually, although not invariably, present in limbless lizards.

As additional distinctive features of the present group, by means of which they can be distinguished from snakes on the one hand and from chamæleons on the other, the following points may be noticed. In all lizards the two branches of the lower jaw are united at the chin by means of a bony suture ; while in all the species furnished with limbs collar-bones are present ; and when the limbs are absent, some traces of the bones forming what is known as the shoulder-girdle persist. In form the tongue is flattened. In most of the members of the suborder the upper surface of the body is clothed with the overlapping scales characteristic of the order in general, these scales being in some cases underlain by bony plates ; but in most geckos the upper scales are granular, although sometimes juxtaposed.

Numerically, lizards are by far the most abundant of all reptiles at the present day, the total number of species not falling far, if at all, short of seventeen hundred, which are arranged under twenty distinct families. In this abundance at the present day, coupled with the specialised features of the greater part of their organisation, lizards may be regarded as occupying a very similar position in the reptilian class to that held by the perching-birds in the preceding class.

With the exception of the polar and subpolar zones, lizards are distributed over the whole globe, ranging in some districts from the level of the sea to the limits of eternal snow, and found alike in fruitful and barren districts, in the neighbourhood of water and in the most arid deserts. Whereas, however, in the colder region they are poor in species and small in size, in the tropics and subtropical regions they attain their maximum development as regards numbers, bodily size, richness of colour, and peculiarity of form.

As regards their distribution over the surface of the globe, lizards present a most remarkable difference from that which obtains among amphibians (frogs, newts, etc.), and, to a less degree, among tortoises. For instance, whereas amphibians, and to some extent tortoises, have their distributional areas defined equatorially, such lines of division, in the case of the present group, must be drawn meridionally. In the case of amphibians, for instance, one great distributional province includes Europe, Asia, and Northern America, and the second embraces the regions lying south of the Equator; whereas in the case of lizards one area marked by peculiar forms includes the Old World and Australia, and the other area comprises the whole of America.

As already noticed above, the distribution of tortoises approximates to the former type, all the side-necked group being confined to the Southern Hemisphere. Again, it is noteworthy that whereas Tropical Africa is closely related to Tropical India as regards its amphibians, while Australia and Africa are near akin to South America in regard to their tortoises, in respect of lizards there is no close connection between India and Africa, but an intimate relationship exists between India and Australia, where members of the same genera occur, while the Australian lizards are totally unlike their South American cousins.

As might have been expected from their great numerical preponderance at the present day, lizards appear to be a comparatively modern group, their remains being rare in the lower Tertiary deposits, while in the Secondary period they are only known by a few species from the rocks of the Cretaceous epoch. That the group has originated from the tuateras, which were so abundant in the earlier strata of the Secondary period, may be regarded as most probable.

Turning to their mode of life, it may be noticed that while a few members of the order resemble crocodiles in spending the greater portion of their time in water, visiting the land only for the purposes of feeding, sleeping, or basking in the sun, by far the great majority of lizards are essentially land-animals, avoiding even damp situations. Although some inhabit trees, the greater number dwell either on the ground or among the clefts of rocks; the conformation of the body generally giving some indication of this diversity of habitat.

Among the land forms, for instance, those with depressed bodies are generally to be found in open sandy deserts, where they seek shelter either beneath stones or in holes; whereas such as have the body compressed are more usually dwellers among bushes or in trees. Those, again, in which the body is more or less cylindrical are in the habit of secreting themselves in the clefts of rocks or the chinks of tree-stems; while the snake-like kinds live on the ground, and those with a more worm-like form beneath its surface.

The movements of the greater number of species—whether they live on the ground, among rocks, on trees, or on cliffs or walls—are agile in the extreme; and while the majority run with their bodies close to the ground, many habitually raise themselves up at times by resting on their hind legs and tails, and are able to spring, either on the ground or from branch to branch, to a considerable distance after their prey.

Of the arboreal species, some make use of their tails to aid in maintaining their hold, while others, together with cliff-dwelling and wall-hunting species like geckos, are enabled to run along the under sides of boughs, or to ascend vertical surfaces by the aid of their expanded and disc-like feet.

The peculiar flying lizards are enabled to take long, flying leaps, supported by a parachute-like membrane borne by the expanded ribs; while all the limbless species move somewhat after the manner of snakes, although making less use of the extremities of the ribs. The few aquatic forms swim and dive without the aid of webbed feet, but many other kinds swim well if thrown into water.

In many cases elegant and graceful in form, although in others rendered more curious than beautiful by the presence of spines or warts, lizards are pleasing rather then repulsive animals; and, with the exception of the American heloderms, none is poisonous, although some will bite sharply. Few lizards possess a distinct voice, the majority merely uttering a low hiss; some, however—especially among those whose habits are nocturnal—emit a clear, sharp cry, which has been likened both to the scream of a frog and to the chirp of a cricket. Of their senses, the most acute is doubtless that of sight, next to which probably comes hearing. In regard to diet, a few lizards are strictly herbivorous, but the great majority are more or less completely carnivorous, the larger species feeding on small mammals,

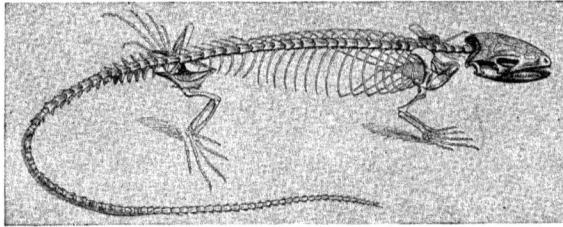

SKELETON OF A LIZARD

birds and their eggs, other reptiles, and, more rarely, frogs and fishes, as well as many kinds of invertebrates. The smaller members of the order, on the other hand, are restricted mainly or entirely to an invertebrate diet, a great portion of which consists of insects, worms, and land molluscs. Nearly all drink by rapidly protruding and withdrawing the tongue; dew affording sufficient moisture to those living on rocks or in trees, while some kinds can exist for long periods, or even entirely, without drinking.

The species inhabiting the warmer regions, save those which are arboreal or aquatic in their habits, pass the hottest and driest season of the year in a state of torpor; but those in colder regions regularly hibernate, such hibernation in the case of some of the species inhabiting the continent of Europe lasting for a period of from six to eight months. As regards their breeding habits, the majority of lizards lay eggs, which may vary from two to thirty in number, and have generally a soft and leathery covering, although sometimes furnished with a hard, calcareous shell.

One peculiarity is the facility with which they are enabled to reproduce lost parts, and more especially the tail. In many lizards, when handled, the tail breaks off without any rough usage, and will readily come in two if pulled when the creature is seeking to escape, this being due to a cartilaginous band across each vertebra of the tail in the case of the common lizard of England. Such missing portion of the tail is speedily reproduced, it may be double; but whereas among the members of the typical family of the order the scaling of the reproduced portion is like the original, in other forms this is not the case; and when the pattern of the scaling of such a new tail differs from the original, it always reverts to that characterising a less specialised and probably ancestral group. In such an assemblage, only a small percentage of species can be mentioned.

CLASSIFICATION OF LACERTILIA MENTIONED IN THIS WORK

GECKOS

Few creatures have given rise to a greater amount of fable and legend than the large group of lizards commonly known as geckos (family *Geckonidæ*),

LOBE-FOOTED GECKO

such legends being probably due to the nocturnal and domestic habits of these reptiles, coupled with the sharp, chirping cry from which they derive their name, and their curiously expanded disc-like toes. Although perfectly innocuous, they have been credited from the earliest times with ejecting venom from their toes, and of poisoning whatever they crawled over; while the teeth of one species have been asserted to be capable of leaving their impression on steel. Indeed, so intense is the dread inspired by these little reptiles that in Egypt the lobe-footed, or fan-footed, species is commonly termed abou-burs, or father of leprosy.

Geckos, of which there are nearly three hundred species, distributed over all the warmer parts of the globe, although more numerous in the Indian and Australian regions then elsewhere, are for the most part small and plumply-built nocturnal lizards, characterised by their depressed form and dull colouring. The rather long and more or less flattened head is broad and triangular in shape; the large eyes are characterised by the absence of movable lids, and by the pupil being, except in a few diurnal forms, vertical, and the aperture of the ear is likewise in the form of an upright slit.

Externally the head is covered with minute granules, or small scales, and the body is devoid of a bony armour, and in most cases covered above with granules, and beneath with small overlapping scales. If to the above features it be added that the tongue is either smooth or covered with tag-like papillæ, and short or moderate in length, and not sheathed at the base, and that the bodies of the vertebræ articulate together by means of cup-shaped surfaces at both their extremities, there is sufficient to distinguish the geckos from all other members of the suborder. As regards their other external characters, the neck is very short and thick, the body, although rounded, markedly depressed, and the tail, which is generally remarkably brittle, usually thick and of moderate length, with its basal portion either cylindrical or laterally compressed, although it may be leaf-like, or even rudimentary. In some cases the tail is known to be prehensile, and it is not improbable that it is frequently endowed with this power.

The limbs are generally remarkable for their shortness, and are always provided with five toes each, the tips or sides of which may be more or less dilated. In those species inhabiting desert regions the toes are of normal form, being often nearly cylindrical, and keeled on their lower surfaces; but in the great majority of the members of the family they are expanded either throughout their length or partially into adhesive discs, of which the under surface is formed by a series of movable symmetrical plates of variable form, by the aid of which these reptiles are enabled to ascend walls and run across the ceilings of rooms. In some cases the claws are retractile, either within the plates of the discs or into sheaths; while in other instances the toes may be united by webs, which are not, however, for the purpose of swimming, all the geckos being land-lizards. The numerous teeth are small, and attached to one side of the summit of the jaw (pleurodont).

It has been stated that diamond-tailed geckos (*Phylurus platurus*) are always found head-downwards on the rocks they frequent. They assume this position,

TURKISH GECKOS

1505

it is asserted, for the purpose of making hawks believe that their heads are their tails; consequently, when seized by one of these birds, which invariably pounce upon what they regard as the head, the brittle tail snaps off, and the gecko wriggles away little or none the worse for the encounter.

With the exception of a certain number of species, geckos, as already mentioned, are nocturnal in their habits; and many are remarkable for uttering shrill cries, probably produced by striking the tongue against the palate, which in some cases are compared to the syllables "yecko," "checko," or "toki," and in others to the monosyllable "tok." A South African sand-gecko is at times stated to occur in such numbers, and to produce such a din by its cry, as to render a sojourn in the neighbourhood well-nigh insupportable.

As regards the situations they frequent, geckos are very variable, some inhabiting arid deserts, where they, in certain instances, burrow in the sand; others are restricted to wooded regions, where they live either among low bushes or on trees, and conceal themselves during the day beneath stones or the bark of the stems; others, again, are found among rocks; while the members of a fourth group have elected to live among human dwellings, where some of them have become as fearless and confiding as domesticated animals. Of the arboreal species, the frilled gecko is peculiar in having a parachute-like expansion of skin, which is used after the manner of that of flying squirrels in aiding its owner to take long leaps from bough to bough. When at rest, the parachute is kept close to the sides of the body by the aid of its intrinsic muscles; and it is stated that this species, like several others, has the power of changing its colour according to the hue of the object in which it is resting.

The species frequenting houses may be divided into those which resort to the interior and those which are content with the outside. Of the former, Sir J. E. Tennent writes that in Ceylon, "as soon as evening arrives, geckos are to be seen in every house in keen and crafty pursuit of their prey; emerging from the chinks and recesses where they conceal themselves during the day, to search for insects that then retire to settle for the night. In a boudoir, where the ladies of my family spent their evenings, one of these familiar and amusing little creatures had its hiding-place behind a gilt picture-frame. Punctually as

the candles were lighted it made its appearance on the wall to be fed with its accustomed crumbs; and if neglected it reiterated its sharp, quick call of "chic, chic, chit" till attended to. It was of a delicate grey colour, tinged with pink; and having by accident fallen on a work-table, it fled, leaving part of its tail behind it, which, however, it reproduced within less than a month.

"In an officer's quarters, in the fort at Colombo, a gecko had been taught to come daily to the dinner-table, and always made its appearance along with the dessert. The family were absent for some months, during which the house underwent extensive repairs, the roof having been raised, the walls stuccoed, and the ceilings whitened. It was naturally surmised that so long a suspension of its accustomed habits would have led to the disappearance of the little lizard, but on the return of its old friends it made its entrance as usual at their first dinner the instant the cloth was removed."

Another Indian observer, Colonel Tytler, writing of these house-geckos, states that although several species "may inhabit the same locality, yet, as a general rule, they keep separate and aloof from each other; for instance, in a house the dark cellars may be the resort of one species, the roof of another, and the crevices in the walls may be exclusively occupied by a third species. However, at night they issue forth in quest of insects, and may be found mixed up together in the same spot; but on the slightest disturbance, or when they have done feeding, they return hurriedly to their particular hiding-places." So far as is known, all the members of the family agree with the house-geckos in being insectivorous. With the exception of two peculiar New Zealand species producing living young, geckos appear to lay eggs, which are enclosed in round and hard shells, and generally two in number.

FRINGED GECKO

LOBE-FOOTED GECKO

Geckos being so numerous in species, which are arranged in about fifty genera, it is of course impossible in a work like the present to do more than notice a few of the better known or more striking. Among these, one of the most familiar is the little lobe-footed or fan-footed gecko (*Ptyodactylus lobatus*), of Northern Africa, Arabia, and Syria, which is one of two species belonging to a genus characterised by the toes being dilated at their summits, where they are furnished inferiorly with two diverging series of plates.

NILE CROCODILE

"On the upper course of the Nile the favourite haunts of the crocodiles are sandbanks, situated in parts of the river where the current is not too strong. There they may be seen at all hours of the day sleeping with widely opened mouths, in and out of which the black-backed plover walks with the utmost unconcern."

P.J. Smit.

P. J. Smit

The toes of the lobe-footed gecko are furnished with claws capable of retraction within notches in the front of the disc. The upper surface is covered with granules, among which are some small keeled tubercles; the colour being greyish or yellowish brown above, with darker and light spots, and below uniform white. The length is a little over 5 inches.

TURKISH GECKO

Equally well-known is the Turkish gecko (*Hemidactylus turcicus*), which is likewise a small species, inhabiting the countries bordering the Mediterranean and Red Seas, but also found in Sind. It belongs to a group of genera with dilated toes and compressed claws, and is specially characterised by the extremities of the toes being free, the plates on the under surface of the discs arranged in double rows, and the presence of some large shields on the under surface of the tail.

Measuring not more than 4 inches in length, this species may be distinguished from the other European geckos by the body being covered with from fourteen to sixteen longitudinal rows of warts, of which some are white and the others blackish, and likewise by the hue of the upper parts being greyish brown spotted with flesh-colour. It is, however, said to be able to change its colour according to circumstances, being of a shiny milky white at night, and dark-coloured during the daytime. The genus to which it belongs comprises over thirty species, ranging from Southern Europe and Asia, Africa, Tropical America, and Oceania.

FRINGED GECKO

A larger and more remarkable species is the fringed gecko (*Ptychozoum homalocephalum*), the sole member of a genus characterised by the presence of an expansion of skin along the sides of the body, continued as lobes on the tail, as well as by the toes being completely webbed, and the inner one devoid of a claw. Attaining a length of nearly 8 inches, this species has a distinctly ringed tail; its colour above being greyish or reddish brown,

marked with undulating dark brown transverse bands, and a dark streak extending from the eye to the first of the bands on the back. This gecko is an inhabitant of Java, Sumatra, Borneo, and the Malay Peninsula

WALL-GECKO

Another species claiming special mention is the wall-gecko (*Tarentola mauritanica*), which is the Mediterranean representative of a small genus ranging from the countries bordering the Mediterranean to West Africa, and including one West Indian species. The genus is readily recognised by all the toes being dilated and only the third and fourth furnished with claws. This particular species varies from rather less than 5 to somewhat more than 6 inches in length, of which one half is formed by the tail. The sides of the neck and body, as well as the upper surface of the limbs, are ornamented with conical tubercles; the back has seven or nine longitudinal rows of larger and strongly-keeled tubercles; and on the anterior half of the tail the ornamentation takes the form of knobs with backwardly-directed spines.

The general colour of the upper parts is greyish brown, with more or less distinct lighter and darker marblings; but a well-marked dark streak passes on each side of the head through the eye. The slender-fingered gecko, shown in the lower figure on this page, represents an allied genus (*Stenodactylus*).

WALL-GECKOS

BARK-GECKO

Very remarkable is the large Malagasy bark-gecko (*Uroplates fimbriatus*), characterised by its crocodile-like head, short trowel-shaped tail, and black skin irregularly marked with white and grey streaks and blotches. This gecko, which measures about 9 or 10 inches in length, dwells on the lichen-clad bark of trees, to which, despite its large, glaring, yellow, bloodshot eyes, it presents so close a resemblance as to be almost impossible of detection at a short distance. It affords, in fact, one of the most

SLENDER-FINGERED GECKOS   Photo, C. R. Walter

remarkable instances of protective resemblance and adaptation to be met with in the whole animal kingdom.

HARDWICKE'S GECKO

A few peculiar geckos, assigned to three genera, and of which the Oriental Hardwicke's gecko (*Eublepharis hardwickei*) is one of the best-known examples,

differ from the ordinary geckos in being furnished with movable eyelids, and also in that their vertebræ are articulated together by means of cup-and-ball joints. Consequently these geckos are referred to a family, the *Eublepharidæ*, by themselves ; some writers also making the bark-gecko the type of a family.

SCALE-FOOTED LIZARDS

To the ordinary observer it might well appear that the whole of the snake-like lizards, or those in which the body has become cylindrical and much elongated, and the limbs either rudimentary or wanting, would pertain to a single family. Such, however, is not the view of modern naturalists, who regard several of these aberrant members of the suborder as having been independently derived from distinct groups of fully-limbed forms, and thus presenting but little relationship among themselves. Of these snake-like groups, one of the most remarkable is that of the scale-footed lizards

thereby resembling those of the monitors. The hind limbs are represented externally by a pair of scaly flaps, which are most developed in the genus to which the common species belongs. The component bones of the limbs may be felt more or less distinctly through the skin, and the skeleton of the common species shows five toe-bones.

The common scale-foot (*Pygopus lepidopus*), which attains a length of about 20 inches, and has a tail twice as long as the head and body, is the typical representative of the few members of this family. The head,

THE SCALE-FOOTED LIZARD

of Australia and New Guinea, which form a family (*Pygopodidæ*) comprising six genera, characterised by the retention of more or less well-marked rudiments of the hind limbs, although the front pair has quite disappeared externally.

Structurally these scale-foots appear to come nearest to the geckos, with which they agree in the essential characters of the skull, as they do in the nature of their tongue, the want of movable eyelids, and the vertical pupil of the eye ; although the last-mentioned character, as being variable in the geckos, cannot be regarded as of much importance. Apart from their external form, they differ from geckos, and thereby resemble the members of the next family, in that the inner extremities of the collar-bones are not expanded into a loop-shaped form, and they are peculiar in that the number of bones entering into the composition of each half of the lower jaw is reduced from six to four.

The small and numerous teeth are closely set, and have generally long, cylindrical shafts, and blunted summits ; although in the genus *Lialis* they are sharply pointed, swollen at the base, and backwardly curved,

which is long, pointed at the snout, and scarcely separated from the body, is covered above with large symmetrical shields, and on the sides with small scales. The ear has an oblique oval aperture, and the rudimentary immovable eyelids are circular and covered with minute scales.

The cylindrical body is slender and of nearly equal thickness throughout, the scales on its upper surface, as on that of the long tail, being keeled. Larger in males than in females, the rudimentary hind limbs have rounded extremities, and are enveloped in overlapping scales.

In general colour this lizard is coppery grey above, sometimes marked with three or five longitudinal rows of blackish dots or elongate spots ; the under parts being marbled grey, with the exception of the throat, which is white. Found both in Australia and Tasmania, and by no means uncommon in the warmer northern parts of Victoria, this lizard, like its kin, is stated to have habits very similar to those of the blindworm, although accurate observations on its mode of life are wanting.

AGAMA LIZARDS

THE southern and eastern portions of the Old World are the home of a very extensive family of lizards (*Agamidæ*) comprising thirty genera and over two hundred species, which may be conveniently termed agamas, from the name of the typical genus. Agreeing with the preceding families in the characters of the tongue and the absence of bony plates beneath the scales, agamas resemble scale-foots in the characters of their collar-bones, but are distinguished from all their allies in having the teeth of the acrodont type— that is to say, situated on the very summit of the edges of the jaws.

While the head is covered with small scales the small eyes have circular pupils and well-developed movable eyelids ; and the scales on the back are of the normal overlapping type. The thick tongue is either completely attached or only slightly free in front, and, at most, has only a very shallow notch in its tip. The teeth may be generally divided into three series, comparable, as regards position, with the incisors, tusks, and molars of mammals ; the last being more or less compressed, and frequently furnished with three cusps, while the tusks, or canines, which may be one or two in number on each side, are of relatively large size in most cases, although occasionally absent.

The fore limbs are always well-developed, and, except in one genus, five-toed. The absence of large symmetrical horny shields, both on the head and under parts, is a noteworthy character of these lizards, many of which develop, either in the males or in both sexes, ornamental appendages, such as crests or pouches. As a rule, the tail is long and not brittle, but in only one genus is it prehensile, although in another it can be curled up at the extremity. The shape of the body is very variable in the different genera, the terrestrial forms being generally depressed, while those that are arboreal in their habits are compressed.

Although the majority of the species are insectivorous, some subsist on leaves and fruits, while others prefer a mixed diet ; but neither the nature of their surroundings nor their food serves to classify agamas, many of the genera of which are very difficult to distinguish. The majority of the species appear to lay eggs, only the members of a single genus being reported to give birth to living young. As regards distribution, agamas are found from the south of Europe to the Cape, and eastwards as far as China, the Malay Islands, Australia, and Oceania, but are unknown in New Zealand and Madagascar. Both as regards genera and species, their headquarters, however, is the Indian region ; Africa possessing only three genera, of which one is confined to the northern part of the continent, while four species alone enter South-Eastern Europe.

FLYING-LIZARDS

Commonly known as flying-dragons, the members of the first genus of the family, *Draco*, are elegant and harmless little reptiles, to which such a title seems singularly inappropriate, and it is therefore preferable to substitute the name of flying-lizards—more especially as the former appellation applies much better to the extinct pterodactyles. These flying-lizards, which are represented by twenty-one species, ranging over the greater part of the Indo-Malay region, are at once distinguished from all their kindred by the depressed body being provided with a large wing-like membranous expansion, supported by the elongated extremities of the six or seven hind ribs, and capable of being folded up like a fan. The throat is furnished with a large membranous expansion, on the sides of which are a smaller pair ; and the tail is long and whip-like.

The best known of the species is the Malay flying-lizard (*D. volans*), which is a rather common form, and belongs to a group characterised by the nostrils being lateral and directed outwards ; this particular species being distinguished by the absence of a spine above the eye, by the aperture of the ear being smaller than the eye, and by the inferior surface of the parachute being ornamented with black spots. In addition to the appendages on the throat, the males have a small crest on the nape of the neck, and in both sexes the back is covered with irregular, large, keeled scales, and its sides have a series of still larger scales, which are also keeled. In length it measures a little over 8 inches.

As regards colour, the upper parts are of a brilliant but variable metallic hue, ornamented with small dark spots and wavy cross-bands ; between the eyes is a black spot, and a similar one occurs on the nape ; the parachute is orange, with marblings or irregular cross-bands of black ; and the throat is mottled with black, its appendage being orange in the male and bluish in the female. This lizard inhabits the Malay Peninsula, Sumatra, Java, and Borneo, and in the living state is stated to be so beautiful as to baffle description.

MALAY FLYING-LIZARD

Essentially arboreal in their habits, flying-lizards generally frequent the crowns of trees, and, as they are comparatively scarce and seldom descend to the ground, are but rarely seen. Describing the habits of the Malay species, Dr. Cantor writes that " as the lizard lies in shade along the trunk of a tree, its colours at a distance appear like a mixture of brown and grey, and render it scarcely distinguishable from the bark. There it remains with no signs of life, except the restless eyes, watching passing insects, which, suddenly expanding its wings, it seizes with a sometimes considerable, unerring leap. The lizard itself appears to possess no power of changing its colours."

When excited, the appendages on the throat are expanded or erected ; and the ordinary movements of these beautiful little reptiles take the form of a series of leaps. After commenting on the fact that both flying-lizards and flying-lemurs inhabit the same countries, and have very similar modes of life, Professor Moseley states that, when springing from branch to branch and from tree to tree, the former pass so rapidly through the air that the expansion of the parachute almost escapes notice. Some examples kept on board ship were in the habit of flying from one leg of a table to another. The females lay three or four oval whitish eggs.

CHANGEABLE LIZARDS

Among a number of genera characterised by their more or less compressed bodies and generally arboreal habits, the numerous tree-lizards constituting the genus *Calotes* are selected for mention. These beautiful lizards belong to a group distinguished from many of

ARMED AGAMA

their allies by the aperture of the ear being open, while they are especially characterised by the absence of any distinct fold of skin across the throat, by the equality in size of the large, keeled scales on the back, and the presence of a large crest on the back and neck ; the tail being very long and whip-like.

One of the best-known species is the changeable lizard (*C. versicolor*), ranging from Baluchistan, India, and Ceylon to the south of China, an exceedingly handsome lizard of some 16 inches in length, with a very large crest, but so variable in colour when alive as almost to defy description. It is one of the commonest of the Eastern Asiatic lizards and derives its name from

its power of changing colour, which is especially marked when the reptile is sitting basking in the sun ; the head and neck being often yellow, flecked with red, the body red, and the limbs and tail black.

When irritated or feeding rapidly, an allied species (*C. ophiomachus*), from India and Ceylon, turns brilliant red over the head and neck, the body at the same time becoming pale yellow ; hence it is popularly known to Europeans as the " blood-sucker."

BLANFORD'S LIZARD

Closely related to *Calotes* is Blanford's lizard (*Charassia blanfordiana*), which in Southern India occupies the place held in other parts of the Old World by the agamas. " In its habits," writes Dr. N. Annandale, " it closely resembles *Agama tuberculata*, being usually seen on rocks, but occasionally entering human dwellings and running about on the walls. The male, in April and May, has the head and fore quarters of a brilliant red colour, and displays his magnificence to the female, which remains concealed, by slowly walking along in some conspicuous position, alternately rising and nodding his head in a very solemn manner. The exact tint of the brilliant parts changes as he does so.

"Both sexes possess considerable powers of temporary colour-change, which seem to be called into play mainly by the amount of reflected light that reaches the integument. The changes do not always assist in concealing the animal. I have seen a temporarily pale individual resting most conspicuously on a red mud wall, and another on a black rock from the surface of which the sun was reflected. On the other hand, other individuals on similar rocks and walls in the shade were much darker and less conspicuous.

" The fact seems to be that the number of different colours that can be brought into play by contraction or expansion of the pigment-cells of this species is a very limited one. Strong reflected light causes certain of the pigment-cells to contract, but does not expose others. The lizard, therefore, can become paler or darker, but cannot change its actual coloration to any great extent. Some of the pigment-cells, however, are probably non-contractile, for the symmetrical markings never disappear, but become more conspicuous as the general tone becomes paler.

" Specimens in spirit from Travancore are much blacker—*i.e.*, less brown—than specimens from Parasnath, on which the rocks are not so black as those of the Western Ghats of Travancore."

CHAMÆLEON-LIZARDS

Nearly allied to the last group, from which they may be distinguished by the presence of a transverse fold across the throat, are the lizards of the genus *Gongocephalus*, ranging from India to Australia. They are mentioned here on account of the remarkable resemblance presented by the typical chamæleon-lizard (*G. chamæleontinus*) to a chamæleon, the resemblance extending to the casque-like form of the head, the ridge of spines down the back, and the long tail, which, however, is not prehensile.

CEYLON HORNED LIZARDS

Three remarkable lizards from Ceylon, constituting the genus *Ceratophora* and belonging to a group in which the aperture of the ear is concealed, derive their name from carrying a more or less elongated horn-like process on the nose, at any rate in the male sex ; the neck and back being devoid of a crest. One of the species, which attains a length of about 10 inches, has a horn measuring half an inch. These lizards appear to be very rare, one of the species being confined to mountain districts.

TYPICAL AGAMAS

For want of a distinct English title, it is necessary to designate the members of the genus *Agama* collectively by anglicising their scientific name. Distinguished from all the groups previously noticed and their allies, with the exception of the flying-lizards, by their more or less depressed bodies, the typical agamas are especially characterised by the exposed aperture of the ear, and the presence of large callous scales in front of the vent in the males. The crest on the back is, at most, small, and may be wanting ; while each side of the throat has a pit, and there is likewise a transverse fold across this part. A sac-like appendage may or may not occur beneath the throat, and the moderately long tail may be either cylindrical or slightly compressed.

Less important characters are to be found in the form of the head, which is short and triangular, very broad behind, and rounded at the muzzle, as well as in the relative length and slenderness of the limbs. The head is covered above with small, smooth scales ; those on the back are overlapping and keeled, while on the tail the scales may be either simply overlapping or arranged in whorls.

The distribution of the genus is somewhat peculiar, impinging on South-Eastern Europe, and embracing the greater part of South-Eastern Asia, as well as the whole of Africa, but excluding India proper, together with Ceylon and Burma, although including the Punjab, Sind, and the Himalaya. As indicated by their depressed bodies, agamas are mainly ground-lizards, generally frequenting barren localities or rocks, although a few species resort to shrubs. The circular pupil of their eyes is equally indicative of diurnal habits ; and a large number of species are fond of basking on rocks in the full glare of the sun. In such situations, as in the valleys around Kashmir, they may be seen in numbers on almost every roadside mass of rock, where their extreme agility renders them very difficult to capture. As regards food, all appear to be insectivorous.

From among rather more than forty representatives of the genus, three are selected for special notice, the first of these being the armed agama (*A. armata*) of South Africa, which attains a total length of some 20 inches, of which rather more than 6 are occupied by the tail. Belonging to the second group of the genus, or that in which the occipital or hindmost median scale on the top of the head is enlarged, this species is characterised by the spiny scales on the back being of unequal size, the aperture of the ear larger than the eye, the fifth toe as long as the first, and the third slightly longer than the fourth, as well as by the scales on the abdomen carrying keels. Both sexes have a low crest on the nape of the neck, whereby the species is distinguished from most of its South African congeners ; and the males have two rows of twelve thickened horny scales in front of the vent.

Although variable, this handsome lizard is strikingly coloured. Generally the upper parts are olive brown, with the enlarged scales lighter, and there is a double series of darker blotches along the back ; the under surface being lighter, and the throat marked with dark longitudinal streaks. Known to the natives of Mozambique by the name of *toque*, this lizard feeds chiefly on beetles, grasshoppers, and ants.

Very different in general appearance to the last species is the spiny agama (*A. colonorum*) of West Africa, which is a rather large species, said to be the

ROUGH-TAILED AGAMA

most common reptile met with on the Gold Coast. It differs from the preceding species by the shields on the back being of uniform size and furnished with spines, as well as by the absence of a crest. The body is not much depressed, and the sides of the head near the ear, as well as of the neck, are ornamented with radiating groups of short spines, which are at least equal to two-thirds the diameter of the ear-opening.

From the allied *A. rueppelli* of North-Eastern Africa, it may be distinguished by the scales on the back being very numerous and considerably larger than those on the tail ; the latter being strongly keeled and arranged in fairly distinct rings. Attaining a length of rather more than 13 inches, this species is noticeable for its brilliant coloration in the living state, although the hues rapidly fade away after death. When alive, the head is flame-red, the throat spotted with yellow, and the body and limbs deep steel-blue, while along the middle of the back generally runs a whitish line. The lower surface of the basal half of the tail is yellowish, and the corresponding upper portion steely blue like the tip, while the remainder is red. In very old specimens, however, both surfaces of the base of the tail are blue, the remainder of the upper surface, except a small blue tip, being red. Females are at all ages much more soberly coloured.

In some spots these agamas are found in swarms, being very fond of climbing up the mud walls and mat roofs of the native huts, at times basking motionless in the sun, and at others running rapidly about in search of insects. When approached by a human being, they raise and depress their heads in a series of nods, which increase in rapidity as the intruder draws near, till, finally, the reptiles lose courage, and disappear with the speed of lightning into some crack or cranny. So brilliant do these gorgeously-coloured lizards appear, when basking in the midday rays of an African sun, that the observer is fain to believe he is gazing on some splendid insect rather than a reptile.

Belonging to a group of the genus distinguished from the one containing the species described above by the absence of enlargement of the occipital scale of the head, the stellion, or rough-tailed agama (*A. stellio*), is interesting as being one of two members of the genus whose range extends into South-Eastern Europe.

Whereas, however, the other members of the group have the tail more or less ringed, the rough-tailed agama, together with the second European species (*A. caucasica*) and a third (*A. microlepis*), is peculiar in that the tail is divided into distinct segments, each composed of a pair of rings of scales. Growing to nearly a foot in length, the species under consideration is distinguished by its stout body and the moderate degree of depression of the head; the cheeks of the male being somewhat swollen. The colour of the upper parts is olive, spotted with black, and generally with a series of large yellow or olive spots down the middle of the back; the throat of the male having fine bluish-grey net-like markings.

In Europe occurring in Turkey and certain islands of the Ægean Sea, the rough-tailed lizard is distributed over the whole of Asia Minor, Syria, Northern Arabia, and Egypt, and is much more common in the latter regions than is the case in Europe. To the Arabs it is known by the name of *hardun*, and it is commonly tamed and kept in captivity by the itinerant snake-charmers of Egypt. As shy and agile in its movements as its congeners, it feeds largely on flies and butterflies, which are captured with remarkable address and agility.

Before taking leave of this extensive genus, it may be mentioned that there is a third group, agreeing with the last in the small size of the occipital scale of the head, but distinguished by the absence of rings on the tail; the agile agama (*A. agilis*) of Persia being a well-known example. The genus *Phrynocephalus* of South-Eastern Europe and Central Asia comprises rather more than a dozen lizards nearly allied to *Agama*, but easily distinguished by the concealed aperture of the ear.

FRILLED LIZARD

Although the swollen, callous scales in front of the vent in the males of the typical agamas have some resemblance to these structures, the whole of the preceding members of the family are characterised by the absence of true pores on this part of the body or on the thighs. In a second group such pores are, however, present in both, or in one or other of these situations; the first example of this group being the remarkable frilled lizard (*Chlamydosaurus kingi*) of Australia, the solitary representative of its genus. This extraordinary-looking reptile, which attains a length of nearly 32 inches, about 11 of which are taken up by the tail, is at once recognised by the curious frill-like membranous expansion surrounding the

throat and extending upwards to the sides of the nape. The frill, which is much more developed in the adult than in the young, has a serrated margin, and is covered with scales of larger size than those on the back; it irresistibly reminds one of the frills with which our ancestors were wont to adorn their throats, and communicates an altogether strange appearance to its owner. In form the body of this lizard is slightly compressed, and although the scales of the back are strongly keeled, there is no distinct crest in this region. The aperture of the ear is exposed, and the tail is either round or slightly compressed, the latter condition occurring in the adult male. The general colour of the upper parts is pale brown, which may be either uniform or mottled with dark brown, or blackish mingled with yellow and shot with blue.

The frilled lizard is an inhabitant of Queensland and Northern and North-Western Australia, as well as some of the islands of Torres Straits; its fossil remains occurring in the superficial deposits of the first-named district. It inhabits sandy districts, where it walks or runs with a swinging gait, on its hind legs alone, with the frill folded up on the sides of the body (page 10). When frightened, it sits on its hind quarters, raises its fore quarters and head as high as possible, expands the frill to its fullest extent, strikes its body with its tail, and shows its teeth at the intruder. Although the reptile is perfectly harmless, this attitude is evidently intended to terrify other animals, and has even been known to frighten people who have seen it for the first time. The frill, which, when fully extended, forms a shield concealing the body, limbs, and tail, is moved by certain special muscles, and is supported by rods of cartilage.

FRILLED LIZARD IN COMBATIVE ATTITUDE
Photo, Cyril Grant Lane

FIN-TAILED LIZARD

Nearly allied to the preceding is the fin-tailed lizard (*Lophurus amboinensis*), which is likewise the sole member of its genus, and takes its name from the presence of a tall fin-like crest on the upper surface of the tail of the adult, supported by a great lengthening of the spines of the vertebræ of that region. The body is markedly compressed, the back carries a low crest, and the throat has both longitudinal puckerings and a transverse fold in the skin, while the aperture of the ear is exposed.

In form the head is short and thick, the compressed tail long and powerful, and the legs and feet are also strong, the toes of the latter being covered inferiorly with small granular scales, and at the sides, especially externally, with a fringe of large united scales, which constitutes one of the distinctive features of the genus. The covering of the upper parts is in the form of small quadrangular scales, which are keeled on the head and back. The dentition comprises six small conical teeth in the front of the jaws, four long tusks, and thirteen cheek-teeth. On the thighs there is a row of pores. Attaining a length of over a yard, the fin-tailed lizard is of a general olive-brown colour, becoming greenish on the head and neck, spotted and marbled with black, while an oblique fold in the skin on the front of the shoulder is deep black,

BACK VIEW

FRONT VIEW

SIDE VIEW WITH FRILL CLOSED

Photos, C. R. Walter

FIN-TAILED LIZARD

Originally brought to Europe from Amboyna, this curious lizard is an inhabitant of the Philippines, Java, Celebes, and the Moluccas ; it is arboreal in its habits, and is generally found in wood or scrub in the neighbourhood of water. Its food consists of seeds, leaves, flowers, and berries, as well as worms, myriapods, and other creatures found in damp situations. If frightened, this lizard immediately dives into the water, and endeavours to conceal itself among the stones at the bottom, where, however, it may be readily captured with a net, or even with the hand, as it makes not the slightest attempt at defence. Its eggs are laid in the sand of the river-banks. By the natives this lizard is hunted for the sake of its flesh, which is white and well flavoured, and consequently much appreciated.

SPINY-TAILED LIZARDS

Quite a different type of tail to that of the last is presented by the spiny-tailed lizards (*Uromastix*), of which there are several species inhabiting arid tracts in Northern Africa and South-Western Asia. From the whole of the foregoing members of the present family, these lizards are sharply distinguished by the circumstance that the front teeth, instead of being small and conical, are large, and in the adult united together into one or two broad cutting teeth, separated from those of the cheek series by a gap ; while externally they are easily recognised by their short tails, covered with well-defined rings of spiny scales. The head is remarkably short and rounded, the body, as in most terrestrial members of the family, is much depressed, and there is no crest along the back. There are no folds or pouches on the neck, but pores are present both in front of the vent and on the thighs, and the aperture of the ear is exposed.

The Arabian spiny-tail, or dabb, as it is termed by the Arabs (*U. spinipes*), is one of the best-known members of the genus, and inhabits Egypt, Crete, and Arabia. It belongs to a group characterised by the rings of spiny scales on the upper surface of the tail being in juxtaposition ; while, in common with two other species, it is specially distinguished by the circumstance that two or more transverse rows of scales on the lower surface of the tail correspond with one on its upper aspect. The Arabian species, which attains a length of about 18 inches, differs from its

two nearest allies in the minute size of the scales covering the body, coupled with the presence of a few scattered somewhat larger tubercular scales on the flanks. Its colour is either sandy grey or greenish above, which may be either uniform or clouded with brown. The ornate spiny-tail (*U. ornatus*), of Egypt and Syria, differs from the other three members of the first group in that the scales of the tail form complete rings, those on the lower surface being as long as those on the upper.

With the exception of one species (*U. microlepis*) inhabiting Persia, the members of the first group are confined to Africa, Arabia, and Syria, whereas the three representatives of the second group are exclusively Asiatic, one (*U. loricatus*) being from Persia, the second (*U. asmussi*) common to Persia and Baluchistan, while the third (*U. hardwickei*) is an inhabitant of Baluchistan and Northern India. In all these three Asiatic species the rings of spiny scales on the upper surface of the tail are separated from one another by rows of smaller smooth scales. In the Indian spiny-tail the spines on the tail are small, with the lateral ones the largest ; there are no enlarged tubercular scales on the back ; and the front surface of the thigh is marked by a large black spot. In size this species is much inferior to its Arabian congener, not exceeding some 11 inches in length. Its colour is either uniform sandy above, or the same spotted or mottled with a darker tint, and whitish beneath, with the aforesaid dark mark on the thigh.

Conforming in their sombre colouring to the desert regions they frequent, the spiny-tailed lizards are entirely vegetable-feeders, and live in burrows resembling those of the smaller rodents, which are excavated by the reptiles themselves. These burrows, which may be as much as four feet in length, sometimes turn almost at right angles to their original course, at a depth of a foot or so from the surface. Generally living solitary or in pairs, these lizards are met with abundantly in parts of Eastern Persia and the Punjab, and when approached at once make for their holes. If they succeed in getting their fore limbs within the aperture of their burrows, it is impossible to pull them out, for, as I know by experience, they will rather suffer their tails to be torn from their bodies than let go their hold.

THE DABB OR ARABIAN SPINY-TAIL

They are generally somewhat heavy and deliberate in their movements, turning their heads from side to side while walking, but are capable of running with tolerable speed. In the cold season, at any rate, they

HARDWICKE'S SPINY-TAIL

never leave their burrows till the sun is well up ; and while in Persia and India they are commonly found on half-desert, gravelly plains scattered over with low bush, the Arabian species is often met with in the clefts of rocks, whence it issues forth to bask on the smooth slabs or boulders. In some places as many as a dozen of these lizards may occasionally be seen on a single slab of rock. All the species appear to be timid and gentle in their disposition, rarely, if ever, attempting to bite when captured.

Their food comprises leaves and flowers, dried fruits, and the seeds of grass, as well as grass itself ; but although in the wild state they seem never to touch animal food, in captivity the Indian species will greedily devour meal-worms. According to Arab reports, the dabb never by any chance drinks, even when water is at hand, and this statement has been confirmed by modern observers. By the Arabs these lizards are frequently tamed and kept in captivity ; and their flesh, which resembles that of young chicken, is much relished by them as an article of food.

Two nearly allied lizards from East Africa—namely, Aporoscelis princeps from Zanzibar and Somaliland, and A. batilliferus from Somaliland—while resembling the members of the preceding genus in general external characters, differ in the absence of true pores either on the under surface of the body or on the thighs, and are consequently referred to a distinct genus. Both appear to be rare, and are of comparatively small size, the first-named measuring only about $7\frac{1}{2}$ inches in length.

THE MOLOCH

Even more strange and uncouth in appearance than the frilled lizard is another Australian species, commonly known as the moloch (Moloch horridus), but termed by the settlers the spiny lizard or thorny devil. This, the last remaining representative of the agamas, differs from all the other members of the family in being covered with large conical spines, as well as by the conformation of its mouth and teeth. In all the groups described above the mouth is large and the teeth of both jaws are erect, but in the moloch the mouth is very small, and the cheek teeth of the upper jaw are placed horizontally, with their summits directed inwardly. About 8 inches in total length, this extraordinary lizard has a small head, with an extremely short snout, on the summit of which are pierced the nostrils ; it has a much-depressed body, a short and rounded tail, and thick, powerful limbs armed with strong claws.

On each side of the head, immediately above the small eye, is a large horn curving outwards and backwards,

while there is a smaller conical spine above the nostril, a second behind the horn over the eye, a third and larger one in front of each ear, as well as one on each side of the occiput. Between these spines the upper surface of the head is protected by small granular tubercles ; while among the spines on the upper surface of the body, limbs, and tail are similar granules intermingled with polygonal scales of which the edges are in apposition. On the back the spines form ten or more longitudinal series, of which the outermost are the largest. The lower surface of the body has a covering of rough and slightly overlapping scales, among which are numerous rounded and keeled tubercles. In general colour this altogether bizarre reptile is yellowish, ornamented with symmetrical chestnut or reddish brown markings defined by darker borders.

Inhabiting Southern and Western Australia, and being not uncommon in several localities in the neighbourhood of Port Augusta, the moloch is found only in districts where the soil is dry and sandy. Occasionally two or three may be observed basking in company on the top of a sandhill ; and it is the frequent habit of this lizard to bury itself in the sand to a small depth below the surface. Its small eyes and general manner indicate pretty clearly that the moloch is diurnal in its habits, although it is possible that it may occasionally move about during the night.

Although generally very slow in its movements, it has been known, when disturbed, to make for a neighbouring hole with considerable speed. In repose it

THE MOLOCH

generally rests with the head so raised as to be on the level of the back. Its chief food appears to be ants, although vegetable substances are sometimes eaten. The female deposits her eggs in the sand. To a certain degree the moloch is endowed with the power of changing its colour to harmonise with its surroundings, such changes taking place very gradually, although not infrequently. The most general change is to a uniform sandy slate or russet colour, when the ornamental markings almost completely disappear. The moloch is a perfectly harmless creature, its formidable-looking armour being never used for attack.

IGUANAS

THE extensive family of lizards (*Iguanidæ*) of which the well-known iguanas of Tropical America and the West Indies are the typical representatives may be regarded as occupying the same position in America as is filled by the agamas in the warmer parts of the Old World. Whereas, however, the agamas are exclusively denizens of the Eastern Hemisphere, the iguana-like group is not absolutely confined to the western half of the globe, two genera occurring in Madagascar, and a third in the Fiji and Friendly Islands. Although, with these exceptions, the family is unknown in the Old World, the same perverseness which causes Anglo-Indians to speak of Indian crocodiles as alligators leads to the monitors of the Old World being commonly termed iguanas, although few lizards are more unlike than the members of these two groups, both as regards external and internal characters.

In their general structural features, iguanas come very close to agamas. Thus in both groups the head is covered with numerous small shields; while the back is clothed with scales of different kinds, often arranged in oblique rows. Similarly, the eyes have round pupils and are furnished with well-developed lids, and the drum of the ear is frequently exposed. Both groups, again, have two pairs of limbs, which may be relatively longer or shorter in the different genera, but are each provided with five toes. The length of the tail is subject to a large amount of variation, although it generally exceeds that of the head and body. Moreover, the two families resemble one another in the form and structure of the tongue, which is thick, short, scarcely notched, and generally fixed to the floor of the mouth throughout its length.

If, however, the teeth of the iguanas are compared with those of agamas, there will be found a striking difference, which at once serves to draw a sharp line of distinction between the two families. As mentioned above (p.1513), in the latter group the teeth are attached to the very summits of the bones of the jaws (acrodont), and are commonly differentiated into front teeth, tusks, and cheek-teeth. In the iguanas, on the other hand, the tall and cylindrical teeth are attached by their sides to the outer wall of the jaws in the so-called pleurodont manner; the whole series being generally more or less uniform in character, and without large projecting tusks.

In the typical iguanas the teeth have somewhat diamond-shaped compressed crowns with serrated edges; and it was from a superficial resemblance to this type of tooth that the teeth of the great dinosaurian reptile from the English Wealden received the name of *Iguanodon*. A few genera, again, have the teeth divided into three lobes, thus resembling a fleur-de-lis. Many species of the family are further characterised by having teeth on the pterygoid bones of the palate, while a single genus is one of the few lizards in which there are teeth on the palatine bones.

The iguanas, which comprise about three hundred species, arranged in fifty genera, may be regarded as especially characteristic of South and Central America, although they extend into the warmer parts of the northern half of the western continent, ranging in the west so far as British Columbia, and in the east to Arkansas and the southern United States, while they are also represented in many of the American islands.

Their occurrence in Madagascar (where, as in America, agamas are wanting) has already been mentioned, and it is possible that this remarkable instance of discontinuous distribution may be explained by the occurrence of fossil remains of species of the family in the upper Eocene rocks of France, where agamas seem to have been wanting.

Very variable in external appearance, iguanas present equal diversity in their modes of life, and it is not a little curious that, with the exception of the flying-lizards, almost every group of the agamas finds a parallel, both as regards structure and habits, in the present family; the two families being thus representative groups. There are, however, certain iguanas, such as the members of the anolis group and the sea-iguanas, which have no representatives in the preceding family. The majority of the iguanas feed on insects, although some, like the true iguanas and the sea-iguana, subsist on a vegetable diet, while one genus is stated to be omnivorous. Only two genera are known to produce living young.

ANOLIS IGUANAS

In the forests, groves, and gardens of all the warmer regions of America live a number of beautiful lizards commonly known by the name of anolis, which is applied in the Antilles to certain members of the group. The distinctive features of these lizards are the pyramidal form of the head, the moderately long neck, the presence of a broad and generally brilliantly-coloured appendage on the throat of the males, the slender body, which may be either compressed, cylindrical, or slightly depressed, the relatively long hind limbs, the large feet, in which the toes are of very unequal length and their middle joints expanded, with smooth transverse plates on the under surface, and the long, curved, and sharp claws raised somewhat above the level of the expanded joints.

The tail is long and hard, although not prehensile; the covering of very minute scales on the back and tail is not infrequently elevated to form a crest; the cheek-teeth are characterised by their distinctly tricusped crowns; and teeth are generally present on the pterygoid bones of the palate. The most distinctive feature of these iguanas is, however, the possession of the power of changing their colour to even a greater extent than is the case with chamæleons.

From among more than one hundred species belonging to the genus, the red-throated anolis (*Anolis carolinensis*) of the south-eastern United States and Cuba is selected as an example. In this species the head, which is long, triangular, and depressed, is nearly smooth in the young, but in the adult has well-marked frontal ridges, and some large rough

RED-THROATED ANOLIS

shields on the crown, and the appendage on the throat of the males is relatively small. The body is not compressed, is flat beneath, and not keeled above; the scales on its upper and lower surfaces being keeled and approaching a hexagonal form, with their edges either in apposition or slightly overlapping. The tail is cylindrical and tapering, with some slightly enlarged scales on its upper surface, and nearly equal to twice the length of the head and body.

In the living animal the colour of the upper surface is brilliant metallic green, and that of the under parts silvery white; the appendage on the throat of the males, which is covered with white scales, is red; there is a large blue eye-like spot above the axilla of the fore limb, and the region of the tail is ornamented with black markings. In some specimens the green colour passes more or less distinctly into brownish or brown; and when excited these reptiles are able to change their general hue from greenish grey, through dark grey and brown of all shades, to the ordinary metallic green. In length this lizard varies from 5½ to nearly 9 inches, according to sex; fully two-thirds of these dimensions being taken up by the tail.

In Louisiana, Carolina, and Cuba the red-throated anolis is one of the most common of lizards, and may be noticed in all suitable spots, such as woods and garden hedges, as well as in the exteriors, and sometimes also the interiors, of dwelling-houses. Like their congeners, they are, however, to be met with most abundantly in the deep woods, where they so closely assimilate to their surroundings that their presence, when at rest on a bough, is generally only revealed by their brilliant eyes. In houses these lizards exhibit but little fear of man, running about with the greatest unconcern in search of flies and other insects; and as, in addition to gnats, flies, butterflies, beetles, and spiders, they kill and eat wasps, scorpions, and other noxious creatures, their visits are encouraged.

In motion throughout the day they display extreme activity and speed, both when hunting among the foliage of trees and on the ground, pouncing upon their insect prey like a cat upon a mouse. In the spring, during the breeding season, the males display a great jealousy of one another, so much so, indeed, that when two meet a combat is certain to ensue, and is often continued till one of the combatants has lost its tail, which appears to be taken as an immediate sign of defeat. During these battles the bag-like appendage on the throat is inflated, and the changes of colour are more rapid than at any other time. With the advent of summer these mutual animosities are, however, forgotten, and these lizards dwell together in perfect amity, sometimes collecting in large companies. Females of some of the species are stated to dig holes for the reception of their few white eggs with

their fore paws, at the foot of a tree or in some moist spot near a wall, afterwards carefully covering them with soil to protect them from the sun's rays. The species represented in the illustration on page 1520 is, however, said to be very careless in regard to the place where its eggs are deposited; these being found either on bare sand or rocks, or even in rooms.

The red-throated anolis, like most of its kindred, can be readily tamed, and makes a most charming pet, which can be, without much difficulty, transported to Europe. Writing of a pair which were at one time in his possession, Professor T. Bell states that he " was in the habit of feeding them with flies and other insects, and having one day placed in the cage with them a very large garden spider, one of the lizards darted at it, but seized it only by the leg. The spider instantly ran round and round the creature's mouth, weaving a very thick web round both jaws, and then gave it a very severe bite in the lip, just as this species of spider usually does with any large insect it has taken. The lizard was greatly distressed, and I removed the spider and rubbed off the web, the confinement of which appeared to give it great annoyance; but in a few days it died, though previously in as perfect health as its companion. The lizard was evidently unused to the wiles of the British spider."

The crested anolis (*A. cuvieri*), belonging to a small group with compressed and crested bodies and tails, is remarkable for the great extent to which the pouch on the throat can be inflated—probably for the purpose of terrifying foes. Another well-known species is the Cuban anolis (*A. equestris*) of Cuba and Jamaica.

Two lizards, respectively from Jamaica and Colombia, differ from all the species of true anolis in having prehensile tails, in consequence of which they are referred to a distinct genus—*Xiphocercus*. In a third genus, *Chamælolis*, the cheek-teeth have smooth and nearly spherical crowns.

THE CUBAN ANOLIS
Photo, R. Ullyett

BASILISKS

The strange form of the members of the genus *Basiliscus* probably suggested to the earlier naturalists the imposition of the name basilisk, a term which appears to have originally denoted a fabulous snake-like reptile before whose deadly glance every living being save the cock perished. Be this as it may, the reptiles now known as basilisks are large, although perfectly harmless, members of the present family, belonging to a group distinguished from the preceding one by the absence of dilatation of the toes, and the more or less marked backward prolongation of the hind portion of the head.

In the presence of a large crest on the upper surface of the tail the basilisks recall the fin-tailed lizard in the agama group, of which, indeed, they may be regarded as the representatives in the present family. As a

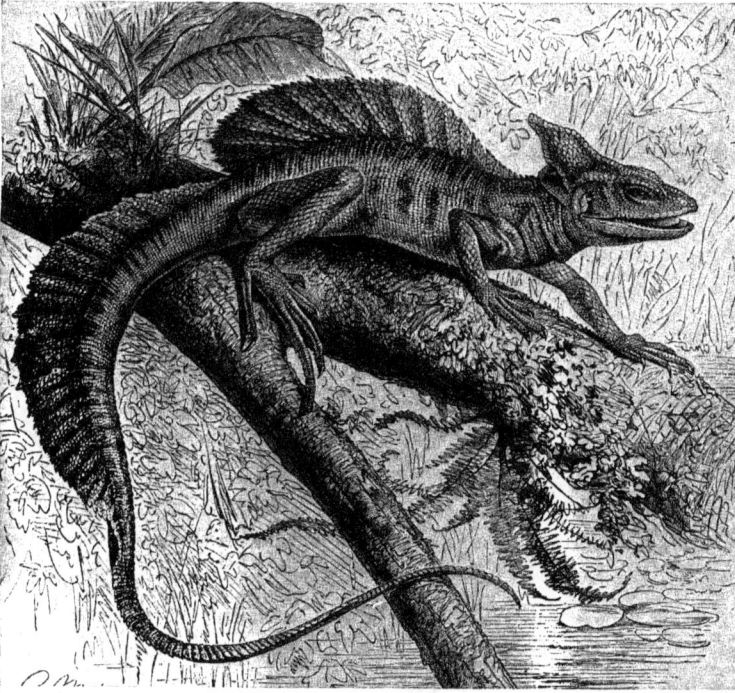

HELMETED BASILISK

a more defined one from the region of the eye to the fore limb.

The banded basilisk (*B. vittatus*), ranging from Mexico to Ecuador, represents a second group of the genus, in which the tail-crest of the males is low, and not supported by bony rays. In this species the scales of the under surface of the body are keeled, whereas in the allied *B. galeatus* they are smooth. In general appearance all the basilisks suggest the idea of lizards upon whose backs has been grafted a fish's fin. As regards habits, all the members of the genus spend their time either on trees or bushes, often basking in the sun on fallen stems, and seldom, if ever, venturing far from the neighbourhood of water. Most numerous in the vicinity of rivers, basilisks are, indeed, so common in Guatemala that the collector has no difficulty in obtaining as many specimens as he may desire, although the rapidity of their movements is so great that some practice is required to effect their capture.

Their food is entirely of a vegetable nature ; and to gather this the basilisks are astir with the first rays of dawn, but during the heat of the day they prefer to rest among the most leafy boughs. At the slightest sound they raise the head, inflate the throat, and elevate the crest ; and as soon as the bright, yellow-irised eye detects the presence of a foe, basilisks throw themselves instantaneously into the water above which they are usually reposing. In swimming, the head and neck are raised, the fore limbs serve the part of propellers, while the crested tail acts as a rudder ; hence the common name of " ferrymen " locally applied to these lizards. At the end of April or beginning of May the female lays from twelve to eighteen eggs in some cranny at the foot of a tree, where they are left for the sun to hatch.

genus, they are characterised by the head in the adult males being produced backwards into a large cartilaginous lobe ; the compressed form of the body and tail, which are covered with small overlapping scales ; and the presence of a crest on the back and tail in the males, such crests being always supported on the back by the prolonged spines of the vertebræ, and frequently also in the tail.

Although there is a transverse fold on the throat, the pouch characterising the anolis iguanas is wanting. The long limbs are covered with keeled scales, and the outer sides of the hind toes have a much-developed lobe of skin. The cheek-teeth have three-cusped crowns, and teeth are borne on the pterygoid bones. Internally basilisks form an exception to the members of this and the two preceding families in that the inner extremities of the collar-bones have a loop-like expansion, as in geckos ; and they differ from the anolis iguanas in the absence of the false abdominal ribs so frequently present in this and the preceding families.

The basilisks are represented by four species from Tropical America, among which the helmeted basilisk (*B. americanus*) is the one most commonly known. The largest representative of the genus, attaining a length of about 31 inches, of which nearly three parts are taken up by the tail, it is one of two species characterised by the great height of the crest of the tail in the males, which is supported by prolongations of the spines of the vertebræ. Inhabiting Panama and Costa Rica, it is specially characterised by the undivided head-crest of the males, and the circumstance that the scales on the under surface of the body are smooth. The natural colour of this reptile is probably green, although specimens preserved in spirit are olive brown above and dirty white beneath. The back is marked with more or less distinct blackish transverse bands, and a lightish streak runs from the temple to the neck, and

RIDGE-HEADED IGUANAS

Nearly allied to the basilisks are the three species of Central American ridge-headed iguanas (*Corythophanes*), characterised by the head being prolonged backwards into a bony, helmet-like projection, while the tail is devoid of a crest, although the neck and back are provided with a low appendage of this nature. On the throat there is both a pouch and a transverse fold. The most interesting of the three species is *C. hernandezi*, in which the head is crowned with a helmet-like prolongation so like that of the chamæleon that the reptile is commonly spoken of under that name by the Mexicans. Like the anolis, these reptiles are in the constant habit of changing their somewhat sombre colours ; and it has been observed in a captive specimen that, whereas the patch on the pouch was white during the day, at night it assumed, like the other light parts of the body, a blackish hue.

STILTED IGUANAS

While agreeing with the basilisks in having the plates on the under surface of the toes distinctly keeled, there are a number of genera in the family distinguished by the absence of any backward prolongation of the

crown of the head. Among these may be mentioned the stilted iguanas (*Uraniscodon*), specially characterised by the large size of the occipital shield of the head, the presence of a vacuity in the breast-bone, the small or moderate-sized scales of the tail, the long and highly-curved toes, and the presence of tusk-like teeth in the jaws.

Of the two representatives of the genus, both of which have a wide distribution in South America, *U. umbra* attains a length of about a foot, two-thirds of which are occupied by the long and cylindrical tail. It has a short and frog-like head, raised into curved ridges over the eyes, with the muzzle very blunt, and the lower jaw longer than the upper. The skin of the neck is curiously puckered inferiorly, the folds forming a pair of pouches on the sides, although there is no pouch on the throat. In form the body is at most but slightly compressed, with a low and slightly serrated crest running from the nape down to the back, and the uniform scales of the back are small and over-lapping, and those on the top of the head enlarged. The long and bent toes are markedly compressed, and furnished with short but strong claws.

In colour this species is one of the handsomest of its tribe. The general ground-colour of the upper parts is reddish or purplish brown, ornamented with more or less distinctly defined blackish transverse bars; a broad black band traverses the fold in front of the shoulder, and may extend across the nape; while

BANDED BASILISK

An inhabitant of the great primeval forests of South America, the stilted iguana has the power of changing colour, and is consequently often designated a chamæleon. It generally associates in pairs, dwelling among trees, and its food appears to be entirely of a vegetable nature. When disturbed, it rushes suddenly up a high branch, where it stands with outstretched head and neck and widely open eyes, gazing steadily at the intruder. Should it be unable to escape otherwise, the reptile raises its neck still higher, inflates the neck-pouches, and, with a sharp cry, springs boldly into the air.

There is a very large number of genera, agreeing with those hitherto noticed in the absence of pores on the thighs, which the limits of space prevent being even mentioned. Attention may accordingly be directed to certain representatives of the second great group of the family, in which such pores are present.

THE SEA-IGUANA

Both as regards their fauna and flora, the Galapagos Islands stand altogether apart from the rest of the world, the greater number of their animals and plants being absolutely peculiar—it may be specifically or it may be generically—while herbivorous reptiles take the place occupied on the continents of the world by vegetable-eating mammals. In no case, however, is this peculiarity more marked than in the occurrence in such a limited area of two distinct genera of the present family, each represented by a single species. Remarkable alike for special features connected with their dentition, as well as for their large bodily size, these two iguanas differ widely from the rest of the family. Whereas, however, the one is a land-reptile, the other is unique among the entire suborder to which

STILTED IGUANA

frequently in front of this band there is a large yellowish orange spot on each side of the neck. On the under surface the colour is brownish or yellowish, either uniform or clouded with brown markings.

it belongs in being marine and subsisting on seaweeds. Agreeing with the great majority of that section of the family characterised by the presence of pores on the thighs, and by the fourth hind toe being longer than the third, the sea-iguana, together with the terrestrial species inhabiting the same islands, differs from all the rest in that the front teeth resemble those of the cheek series in having three-cusped crowns, so that the entire set of teeth is uniform in character.

From its terrestrial ally, the sea-iguana (*Amblyrhynchus cristatus*) is distinguished by its much-compressed and crested tail, as well as by the presence of an incipient web between the toes. This lizard, which is the largest member of the family, and attains a total length of some 53 inches, is characterised by the compressed form of the body and tail, and the extremely short and truncated head. A well-marked crest runs from the nape of the neck to the tip of the tail, and the whole build of the animal is stout and "chubby." The throat is devoid of a pouch, although it has a well-marked transverse fold, and the toes are laterally compressed. In the small and convex head the nostrils are situated near the end of the muzzle, the eye and aperture of the ear are alike small, and the upper surface is surmounted by a number of conical spine-like shields of relatively large size.

The investing scales of the body are small, and, although keeled on the back, smooth below. In the stoutly-made limbs the toes are rather short, the third one in the hind foot being strongly serrated on the inner border of its basal joint. The compressed and crested tail is about equal to one and a half times the length of the head and body, and covered with equal-sized keeled scales. In colour this lizard is black or blackish brown above, with the abdomen and the inner surfaces of the thighs not infrequently dirty white. In the young state, however, the upper parts are brown with paler spots, and more or less distinctly marked with dark cross-bars on the back. In weight full-grown examples reach as much as 20 pounds.

The sea-iguana is extremely common on the rocky coasts of the various islands of the Galapagos group, but is seldom found more than some ten yards from the shore. Of its habits Darwin writes that "this lizard swims with perfect ease and quickness by a serpentine movement of its body and flattened tail—the legs being motionless and closely collapsed on its sides. A seaman on board sank one, with a heavy weight attached to it, thinking thus to kill it directly; but when, an hour afterwards, he drew up the line, it was quite active. Their limbs and strong claws are admirably adapted for crawling over the rugged and fissured masses of lava which everywhere form the coast. In such situations, a group of six or seven of these hideous

GALAPAGOS SEA-IGUANA

reptiles may oftentimes be seen on the black rocks, a few feet above the surf, basking in the sun with outstretched legs."

After mentioning that the stomachs of several examples that were examined contained finely minced seaweed, and also observing that the droves seen swimming out to sea were doubtless in search of food of this nature, the same author proceeds to state that, when frightened, these lizards absolutely refuse to enter the water. "Hence," he continues, "it is easy to drive these lizards down to any little point overhanging the sea, where they will sooner allow a person to catch hold of their tails than enter the water. They do not seem to have any notion of biting, but when much frightened they squirt a drop of fluid from each nostril.

"I threw one several times as far as I could into a deep pool left by the retiring tide, but it invariably returned in a direct line to the spot where I stood. It swam near the bottom, with a very graceful and rapid movement, and occasionally aided itself over the uneven ground with its feet. As soon as it arrived near the edge, but still being under water, it tried to conceal itself in the tufts of seaweed, or it entered some crevice. As soon as it thought the danger was past, it crawled out on the dry rocks, and shuffled away as quickly as it could. I several times caught the same lizard by driving it down to a point, and, though possessed of such wonderful powers of diving and swimming, nothing would induce it to enter the water; and as often as I threw it in, it returned in the manner above described. Perhaps this singular piece of apparent stupidity may be accounted for by the circumstance that this reptile has no enemy whatever on shore, whereas at sea it must often fall a prey to the numerous sharks."

GALAPAGOS LAND-IGUANA

Although originally included in the same genus as its aquatic cousin, there seems no doubt that the land-iguana of the Galapagos (*Conolophus subcristatus*) is entitled to stand as the representative of a distinct generic group; the nearly cylindrical tail and perfectly free toes being distinctive characters which cannot well be overlooked. Not reaching within some 11 inches of the dimensions attained by the last, this iguana is likewise a stoutly-built lizard, with the rather small head slightly longer than broad, the body somewhat depressed, a slight spiny crest on the nape, continued as a low ridge on the back, the scales of the latter small and keeled, but the slightly larger ones on the lower surface smooth.

Although devoid of a pouch, and with a very slight transverse fold, the throat is strongly plicated longitudinally, and covered with minute granules. The stout limbs terminate in very short toes, of which the third in the hind foot is serrated on the inner margin of its basal joint. On the thigh the pores are arranged in a long series, and vary from seventeeen to twenty-one in number. In length the tail scarcely exceeds the head and body, while in form it is slightly compressed, having a low ridge superiorly, and being covered with small keeled scales of uniform size. In general colour this lizard is dark brown, with its head and under parts lighter.

These iguanas are confined to the central islands of the Galapagos group, such as Albemarle and James Islands, where they were formerly found in great numbers in the low, barren districts near the coasts, although also met with in the elevated damp regions of the interior. On James Island Darwin found them so numerous that it was difficult to obtain a spot free from their burrows on which to pitch a tent. Attaining a weight of from 10 to 15 pounds, these lizards are lazy and sluggish in their movements, crawling slowly along with their bellies and tails dragging on the ground, and often stopping for a minute or two to doze with closed eyes, and the hind limbs stretched out on the arid soil.

According to Darwin's account, "they inhabit burrows, which they sometimes make between fragments of lava, but more generally on level patches of the soft sandstone-like tufa. The holes do not appear to be very deep, and they enter the ground at a small angle; so that, when walking over these lizard-warrens, the soil is constantly giving way, much to the annoyance of the tired walker. This animal, when making its burrow, works alternately the opposite sides of its body. One front leg for a short time scratches up the soil, and throws it towards the hind foot, which is well placed so as to heave it beyond the mouth of the hole. That side of the body being tired, the other takes up the task, and so on alternately.

"They feed by day, and do not wander far from their burrows; if frightened, they rush to them with a most awkward gait. Except when running downhill, they cannot move very fast, apparently from the lateral position of their legs. They are not at all timorous; when attentively watching anyone, they curl their tails, and, raising themselves on their front legs, nod their heads vertically, with a quick movement; and try to look very fierce; but in reality they are not so at all; if one just stamps on the ground, down go their tails, and off they shuffle as quickly as they can."

If worried with a stick, these lizards will bite it severely; and when two are held together on the ground they will fight and bite till blood flows. "The individuals, and they are the greater number, which inhabit the lower country can scarcely taste a drop of water throughout the year; but they consume much of the succulent cactus, the branches of which are occasionally broken off by the wind. I several times threw a piece to two or three of them when together; and it was amusing enough to see them trying to seize it and carry it away in their mouths, like so many hungry dogs with a bone."

These iguanas also eat the leaves of several trees, more especially of an acacia, to obtain which they ascend the low, stunted trees, on the boughs of which they may often be observed quietly feeding. The females lay large eggs of an elongated form in their burrows; both these and the flesh of the lizards themselves being eaten by the inhabitants of the Galapagos.

TYPICAL IGUANAS

The typical iguanas, of which there are two closely-allied species from Tropical America and the West Indies, differ from the two preceding genera in that the edges of the crowns of the cheek-teeth are serrated, while the front teeth are simply conical. The distinctive features of these iguanas are to be found in the long and much compressed body, the large four-sided head, covered above with enlarged scales, the short neck, powerful limbs, long-toed feet, and the much-elongated tail, on which the scales are uniform and keeled. The throat is furnished with a large non-dilatable appendage, in front of which is a crest of big, compressed scales; and a continuous crest of long spines runs from the nape along the back, and is continued as a ridge on the tail.

The scales on the back are small, equal, and keeled; the neck has some scattered, large, conical or bluntly-keeled turbercles, and there are also some large tubercular scales on the sides of the throat, more especially one below the aperture of the ear, but on the under parts the scales are either smooth or slightly keeled. The pores on the thigh are numerous, and, in addition to those in the margins of the jaws, there are teeth on the pterygoid bones of the palate.

The common iguana (*Iguana tuberculata*) attains a length of as much as a yard and a half, two-thirds of which are occupied by the tail. The general colour is green or greenish, becoming lighter on the under parts; but the upper surface may be either uniform or variegated with darker brownish bands, the flanks usually having light-edged vertical dark bars, while the tail has more or less distinct dark rings. There is frequently a whitish band in front of the arm, and some of the large tubercular scales on the sides of the throat and neck are often light-coloured.

Both species of iguanas, of which there are several varieties, are essentially arboreal lizards, generally frequenting those regions of the forests where the trees overhang the water. Here they move with great agility, climbing or springing from bough to bough, the

GALAPAGOS LAND-IGUANA

harmony of their colouring to their surroundings rendering them well-nigh invisible. Towards evening they not infrequently descend to the ground to feed, but when frightened immediately rush to the topmost boughs of the trees or plunge headlong into deep water. In the latter element they are, indeed, perfectly at home, and swim strongly and swiftly, with their limbs closely applied to their bodies, and impelled by their powerful tails. They are likewise expert divers, frequently remaining for a considerable time below the surface; their activity in the water being such that they are able to avoid all enemies save crocodiles and caimans.

Their chief food consists of leaves, flowers, and berries, although they will also eat insects; the numbers of small worms sometimes found in their stomachs having probably been swallowed accidentally. Generally seeking to escape at once from human beings, iguanas when unable to flee show fight, erect their heads and assume a fierce aspect, while at close quarters they bite savagely and administer severe blows with their powerful tails. The female deposits from eight to seventeen eggs in a hole dug in sandy soil, but as several individuals not infrequently lay together, as many as ten dozen eggs may be found in a single nest.

In spite of their somewhat repulsive appearance, iguanas are hunted for the sake of their flesh, which is white in colour and delicate in flavour, and is said to resemble the breast of a chicken. In Mexico, where they fetch as much as ten shillings each, live iguanas are offered for sale tied to bamboos by the sinews of their own legs. The eggs also, which consist almost entirely of yolk, are highly esteemed as articles of diet. Iguanas are generally captured by means of nooses, which are thrown over their heads as they repose on the branches.

RHINOCEROS IGUANA

The much smaller rhinoceros iguana (*Metopoceros cornutus*), of San Domingo, constitutes a separate genus, distinguished by the presence of an inflatable pouch on the throat and of a pair of small longitudinally-disposed horns on the nose.

RING-TAILED IGUANA

The West Indian ring-tailed iguana (*Cyclura carinata*) is selected to represent a group of genera distinguished from the foregoing by the crowns of the cheek-teeth being three-cusped or simply conical. While four of these genera—among which is the Fijian iguana (*Brachylophus fasciatus*)—are characterised by the shortness of the row of pores on the thigh, the present species is one of those in which they form a long series; and it is further characterised by the presence of a serrated crest down the back and tail, and also of a pouch and slight transverse fold on the throat. The head is large,

RHINOCEROS IGUANA

swollen below the ears, and furnished with enlarged scales on the snout, while the body and tail are compressed, the body being covered with small scales.

The species derives its name from the rings of keeled scales which form regular segments on the sides of the tail; each segment being composed of from three to five series of small scales, and a single series of larger and somewhat spiny ones. The toes are compressed, and covered below with keeled plates. In total length this iguana reaches about 48 inches, and its general colour is green or dark olive, speckled with darker and lighter tints, and frequently marked with blackish transverse bands.

The ring-tailed iguana is a somewhat local species, occurring most abundantly in Jamaica, on the limestone mountains in the neighbourhood of Kingston Harbour and Goat Island, but also met with on the low grounds lying between the coast ranges and the higher mountains of the interior, where hollow trees occur. Shy and retiring in their habits, these reptiles live in pairs, and display no great partiality for water, although they can swim as well as the true iguanas. They feed mainly on grass, and when disturbed in grazing rush back to the trees with extraordinary speed, sometimes taking flying leaps like a frog, although their movements are generally slow. If unable to escape, they show fight in the same way as the true iguanas. The ring-tailed iguana exhales a disagreeable smell, so objectionable as to cause even ants to forsake a room into which one is brought.

BLACK-TAILED IGUANA

The black-tailed iguana (*Ctenosaura acanthura*), which inhabits Lower California and Central America, is characterised by the shortness of the row of pores on the thigh, the slight compression of the body, and the presence of a crest of spines down the back, and of a pouch on the throat. The females lay their eggs in company, and the flesh, unlike that of the ring-tailed species, is edible (page 1509).

EXTINCT IGUANAS

Here it may be mentioned that the vertebræ of iguanas differ from those of agamas and most other lizards in being furnished with additional articular facets like those of snakes. Vertebræ of this type occur in the upper Eocene rocks of Europe, and have been assigned to *Iguana*, although it is likely that they indicate an extinct genus. Somewhat similar vertebræ from Tertiary strata of the United States have been described as *Iguanavus*.

HORNED IGUANAS

The last, and at the same time the most peculiar, members of the present family are the horned iguanas of southern North America, which may be regarded as the representatives of the moloch lizard among the agamas. From their short, rounded heads, abbreviated bodies, and shortened tails, coupled with a general batrachian appearance, these lizards are commonly termed toads in America, the popular name of the typical species (*Phrynosoma cornutum*) being the Californian, or horned, toad.

Strange and ugly in appearance, these lizards are at once distinguished from all their allies by the presence of several bony spines projecting from the

RING-TAILED IGUANA

back of the shortened head, and of tubercles or spines scattered among the ordinary scales of the body. In form the body is broad and depressed, without any crest down the back, and the tail is very thick at the base, and never longer than the body. The limbs are rather long, with pores on the thighs, and keeled plates on the lower surfaces of the toes.

From most other members of the family these lizards are distinguished by the absence of teeth on the palate. Of the twelve species of the genus, the best known is the common horned toad, which has the tail longer than the head, distinct spines on the back, and the drum of the ear naked. Its general appearance is even more than superficially toad-like, the head being as broad as long, and the body remarkable for its extreme plumpness. Measuring a little over 5 inches in length, this species is rather handsomely coloured. Above, the ground-colour is greyish or brownish, with a more or less well-marked light stripe down the back, and dark brown spots at the bases of the larger spines, while there are likewise markings of the same colour on the nape and head. Beneath, the hue is yellowish, with or without a few small brown spots. In two species of the genus the tail does not exceed the head in length.

The common species is found locally in sandy districts both on the plains and mountains, and is in some places abundant, although from its coloration frequently escaping notice. In spite of its somewhat formidable appearance, it is a harmless creature, not attempting to bite even when captured. Lacking the protrusive tongue of the chamæleon, and being debarred by its clumsy form from running fast, the horned lizard is unable to capture the swifter insects, and consequently

HORNED IGUANAS   Photo, C. R. Walter

preys upon sand-haunting beetles, whose speed is inferior to its own ; such prey being generally captured in the evening, and the creature lying passive on the sand during the day.

Some species of horned lizards are remarkable as being the only members of the family, save one other genus, which produce living young ; the number of young being in some instances as many as twenty-four. Always small feeders, these lizards are capable of undergoing long fasts with impunity ; and as they are habituated to a dry atmosphere, and probably never drink, they may be sent packed in wadding long distances by post.

The most remarkable peculiarity connected with these lizards is their habit of ejecting jets of blood from the eyes, apparently as a means of defence. The following letter from Mr. V. Bailey, written from California in

1891, describes the phenomenon as first observed by him : " I caught a horned toad to-day that very much surprised Dr. Fisher and myself by squirting blood from its eyes. It was on smooth ground, and not in brush or weeds. I caught it with my hand, and just got my fingers on its tail as it ran. On taking it in my hand, a little jet of blood spurted from one eye, a distance of fifteen inches, and spattered on my shoulder. Turning it over to examine the eye, another stream spurted from the other eye. This it did four or five times from both eyes, until my hands, clothes, and gun were sprinkled over with fine drops of bright red blood. I put it in a bag and carried it to camp, where, about four hours later, I showed it to Dr. Fisher, when it spurted three more streams from its eyes." The phenomenon has been subsequently observed in other specimens.

GIRDLE-TAILED LIZARDS

OMITTING mention of a family represented only by one genus (*Xenosaurus*) and species from Mexico, the next group for consideration is that of the girdle-tailed lizards (family *Zonuridæ*), from Tropical and South Africa and Madagascar, of which there are four genera.

such are present. In the South African snake-like genus *Chamæsaura* the fore limbs are wanting, and the hind pair rudimentary, while the tail is of extraordinary length. All the members of the family appear to be carnivorous. As an example of this small family,

GIRDLE-TAILED LIZARD

These lizards, which may be either snake-like in form or provided with four fully-developed limbs, differ from all those hitherto described, with the exception of certain geckos, in having the temporal pits of the skull roofed over with bone, while they are further characterised by a fold covered with small scales running along the sides of the body and marking off the upper from the under parts. The tongue is simple, with its front portion not extensile, and the tip either rounded or but slightly notched, while there are well-developed eyelids, and the drum of the ear is exposed. The back is clothed either with large, shield-like, and mostly keeled, scales, arranged in well-marked transverse zones, or, more rarely, with granules, and the head is protected by large, regular shields. As regards teeth, these lizards conform to the pleurodont type, each tooth having its base widely open.

Resembling in many respects the iguanas, from which they are distinguished by the ossifications in the skull, these lizards also approach the members of the next family, from which they differ by the simple form of the tongue, the hollow bases of the teeth, and the structure of the bony plates underlying the scales, when

special mention may be made of the members of the South African girdle-tailed lizards of the genus *Zonurus*, of which there are several species. These lizards, which differ from the other three genera in having the scales of the back underlain by bony plates of simple structure, resemble in appearance the rough-tailed lizard among the agamas. They have flattened triangular heads, and tails of moderate length. On the upper surface the neck and back are covered with large quadrangular shield-like scales, while beneath there are large, flat shields ; the limbs bearing keeled, overlapping shields, and the tail being protected with whorls of spiny scales. The teeth are small, and the rounded tongue is scarcely notched.

One species (*Z. cordylus*), which attains a length of rather less than 8 inches, generally has the back and tail of a dirty orange-colour, the head and feet of a lighter yellow, and the under parts white, although there are considerable variations from this normal coloration. All the members of the genus inhabit rocky districts, and prefer those where there are ledges, upon which they run in search of food or warmth. They are excellent climbers, and far from easy to catch, often leaving their tails with their would-be captors.

SLOW-WORM TRIBE

NEARLY allied to the preceding family is a small group of lizards (*Anguidæ*) of variable bodily form, typified by the common English slow-worm, or blind-worm. Rigid in their bodies, and having large symmetrical bony shields on the top of their heads, these snake-like lizards resemble girdle-tails in the presence of bony plates beneath the overlapping scales, and also in that the temporal pits of the skull are roofed over with bone. They differ, however, in that the bony plates beneath the scales are permeated by a series of radiating or irregularly-arranged canals, and also in the conformation of the tongue. The latter is composed of two distinct portions, namely, a thick basal half, covered with rough papillæ, and a smaller thin terminal moiety coated with scale-like papillæ, which is extensile and capable of partial withdrawal into a sheath formed by a transverse fold at the front of the basal half. As regards dentition, some of the species have tubercular or conical teeth attached to the sides of the walls of the jaws in the typical pleurodont manner; but in the slow-worms the teeth are long, curved, loosely-attached fangs, very like those of serpents. Instead of hollowing out the bases of the old teeth, as in the preceding family, the new ones grow up beneath them; and there may or may not be teeth on the bones of the palate.

Some of the members of the family agree with the preceding in having a longitudinal fold along the sides of the body, while in others it is absent; and there is a similar variation in external form, some genera having fully-developed five-toed limbs, while in others all external traces of these appendages have disappeared.

In regard to the covering of the head, it should specially be noticed that there is a large occipital shield at its hind extremity. All the species differ from the majority of lizards in changing their skin in a single piece, in the fashion common to most of the snakes.

With the exception of some species of the American genus *Gerrhonotus*, which ascend low bushes, these lizards live on the ground; and all of them are carnivorous, the larger species preying on reptiles and other vertebrates, and the smaller kinds on insects, spiders, slugs, and worms. While slow-worms produce living young, the others lay eggs. Containing seven genera

AMERICAN SCHELTOPUSIK
Photo, Dr. J. H. C. Connell

and about fifty species, this family is most numerously represented in Central America and the West Indies, a few species occurring in North and South America, two in Europe, and one in the Himalaya and Burma; all the forms with functional limbs being American. Attention is here restricted to a couple of the snake-like genera.

SCHELTOPUSIKS OR GLASS-SNAKES

The snake-like reptile termed *Ophisaurus apus* was first discovered by the German naturalist Pallas in the wooded valleys of the steppes bordering the Volga, where it is known, in common with true snakes, by the name of scheltopusik, a term which may be conveniently applied to all the members of the present genus. The species was subsequently discovered in other parts of Russia, as well as in Hungary, Istria, Dalmatia, Greece, Asia Minor, Syria, Persia, Transcaucasia, Transylvania, and Turkestan, and is replaced in Morocco by a more brilliantly-coloured race. Four other species are also known, which extend the range of the genus to North-Eastern India, Burma, and North America, one from the latter continent (*O. ventralis*) being figured on this page.

Agreeing with the American four-limbed genus *Gerrhonotus* in the presence of a fold along the sides of the body, and the more or less conical teeth, scheltopusiks are distinguished by their moderately-elongated snake-like form, and the absence of functional limbs; the European species alone having the hind pair represented by minute rudiments on the sides of the vent. The bodies of these reptiles are covered with squared scales, arranged in straight longitudinal and transverse series; and their mouths are furnished with teeth on the pterygoids, and in certain cases on some of the other bones of the palate.

THE SCHELTOPUSIK
Photo, W. B. Johnson

Underlying the scales is a beautifully-made armour of small bony plates, united to form a continuous sheath. The European species, which, in addition to rudiments of hind limbs, is distinguished by an aperture to the ear, attains a length of rather more than a yard, of which about two-thirds are occupied by the tail. The arrangement of the shields on the head is very much the same as in the slow-worm; and the general colour is brown, becoming lighter on the lower surface. The young are, however, olive grey, with wavy dark brown cross-bands on the back, and bars on the sides of the head.

Dwelling among the dense underwood of thickly-wooded valleys, the scheltopusik harmonises so closely in colour with its surroundings that it can only with difficulty be detected as it glides away among the dead leaves and sticks at the approach of a footstep. Although as free from venom as ordinary lizards, it is frequently mistaken for a snake, and then meets the fate which so often, under similar circumstances, befalls the slow-worm. Preying largely on field-mice, and not even hesitating to attack and kill the deadly viper, the scheltopusik is, however, a fierce and active creature, gliding swiftly and suddenly upon its victims among the moss and leaves of the woods. It also subsists largely on snails, and is reported to eat the eggs and young of birds. Its eggs are laid under thick bushes and leaves. The scheltopusik is believed to be a long-lived reptile, the natives of the countries it inhabits stating that its full period of existence is from forty to sixty years. Fossil scheltopusiks occur in the Miocene deposits of Germany, some of which belong to an extinct genus (*Propseudopus*).

THE SLOW-WORM

The want of a lateral fold along the body distinguishes the slow worm, or blind-worm (*Anguis fragilis*), in common with the remaining members of the family, from the scheltopusiks; the slow-worm being further distinguished from the other genera devoid of this fold by the absence of all external trace of limbs, and the fang-like form of its cheek-teeth. The appearance of the slow-worm, which is the sole representative of its genus, is so well known as not to call for much description. It may be observed, however, that the scales are rounded in form, and arranged on the back in a quincuncial pattern, while on the sides they are disposed in transverse

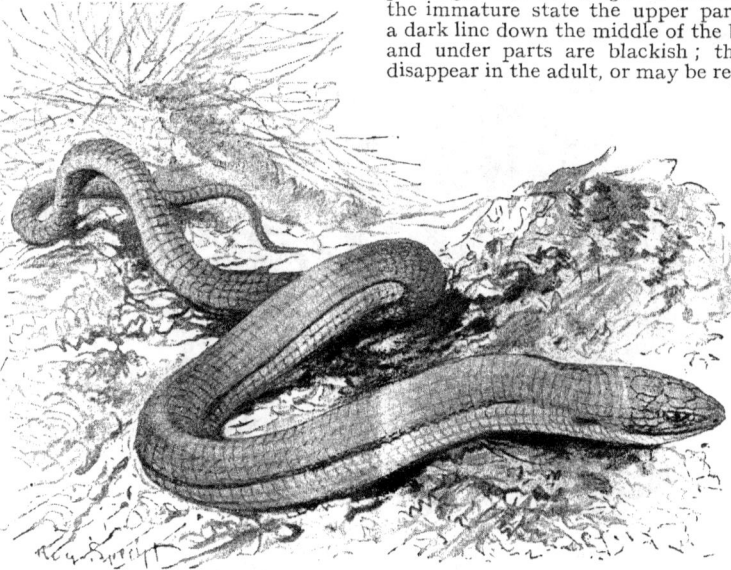

THE SCHELTOPUSIK

rows; the apertures of the ears are usually covered with skin; and the palate is toothless.

Attaining a length of from 10 to 12, or even 14, inches, of which at least half is occupied by the tail, the blind-worm is of almost equal thickness throughout, although tapering slightly at the tail. The head is short and small; the eyes, although minute, are bright and piercing; and the tongue is but slightly notched. In the immature state the upper parts are silvery, with a dark line down the middle of the back, while the sides and under parts are blackish; these markings often disappear in the adult, or may be replaced by dark dots, the upper surface becoming at the same time brown or bronzy. The range of the species includes Europe, Western Asia, and Algeria.

Gentle and inoffensive in its habits, and rarely attempting to bite even when rudely handled, the slow-worm is commonly regarded as one of the most noxious of reptiles, and therefore to be ruthlessly killed at every opportunity. Country people, who ought to know better, are those most convinced of the noxious character of this extremely harmless reptile. When captured, it usually contracts its muscles so forcibly as to become perfectly rigid, in which state it easily breaks if it is bent or struck, thus giving origin to its Latin name.

Generally frequenting woods, heaths, and commons, the blind-worm is one of the hardiest of British reptiles, making its appearance in the spring at an earlier date than any other kind. Early in the autumn it retires beneath masses of decayed wood or leaves, or into soft, dry soil, where it is covered with heath or brushwood, and penetrates to a considerable depth in such situations by means of its smooth, rounded muzzle and polished body. It feeds chiefly upon slugs, supplemented by various insects and worms. In June or July the female produces from seven to twelve or thirteen living young, which are active almost immediately after birth, and soon learn to feed by themselves. Like other viviparous reptiles, the female is much given to bask in the sun during the period of pregnancy, in order that its heat may aid in developing the eggs contained in her body.

THE SLOW-WORM

POISONOUS LIZARDS

Two conspicuously-coloured lizards form the family *Helodermatidæ*. Ranging from the isthmus of Tehuantepec in Central America so far north as New Mexico and Arizona, they stand alone in the suborder in being poisonous, their bite, in certain cases at least, being reputed to be sufficiently severe to produce serious symptoms even in human beings, while smaller animals are soon killed right out. These two species are the Mexican poisonous lizard (*Heloderma horridum*) of Western Mexico, and the Arizona poisonous lizard (*H. suspectum*) from New Mexico and Arizona; the former being known in its native country by the name of silatica.

Nearly allied to the slow-worm, which they resemble in the general structure of the tongue and teeth, although distinguished by certain peculiarities in the conformation of the skull, and by the upper surface being covered with small granular tubercles, externally these lizards are characterised by the depressed head, the plump, rounded body, the moderately long, cylindrical tail, the rather short limbs, in which the third and fourth toes are longer than the others, the exposed drum of the ear, and the transverse arrangement of the rows of tubercles on the upper surface. The curved and fang-like teeth are but loosely attached to the jaws, and have grooves in front and behind for the transmission of the poison, and there are also teeth on the palate. The under surface of the body and tail is covered with squared scales.

SILATICA, OR MEXICAN POISONOUS LIZARD

other animals of the dangerous character of these reptiles.

Inhabiting dry regions from the western side of the Cordillera to the Pacific, and apparently never entering water, the poisonous lizards are nocturnal in their habits, lying during the day hidden among the vegetation in a listless state, and issuing forth at evening. Their movements are at all times deliberate; and as these lizards are most commonly met with in the wet season, and seldom seen during the dry months from November to June, it is probable that they are torpid during part of the latter period. Their food comprises insects, worms, myriapods, and small frogs, as well as the eggs of iguanas.

Regarding the effects of their bite, Sir J. Fayrer writes that he once saw two guinea-pigs bitten by one of these lizards. "The bites were viciously inflicted, and the lizard did not really relinquish its hold. Blood was drawn, the teeth being deeply inserted. Both guinea-pigs were affected; the bitten limb was dragged, and appeared partially paralysed. There were twitchings of the body generally; but these may not have been due to the poison, but to agitation and fear." Both the unfortunate rodents died in the course of the day. Another of these lizards once bit its owner, who was incautiously handling it, with very severe effects, which did not, however, prove fatal. The poison is secreted

ARIZONA POISONOUS LIZARD

In length the Arizona species measures rather less than 20 inches, while the Mexican is somewhat larger. The former has a yellowish or orange ground-colour, marked with a dark network on the head and body, and with blackish rings on the tail. Among the reddish sand, intermixed with dark pebbles, in which these lizards delight to nestle, this colouring, coupled with the granular nature of the skin, appears to be protective; but, on the other hand, it may be designed to warn

in special glands situated near the roots of the teeth. Nevertheless, it has recently been attempted to show that these lizards are not really poisonous. The Arizona species is locally known by the name of Gila monster.

A rare Bornean lizard, *Lanthanosaurus bornëensis*, is nearly allied to the *Helodermatidæ*, although distinguished by the absence of grooves to the teeth and of bony granules in the skin.

MONITORS

No better instance of the essential difference in the distribution of lizards as compared with tortoises is afforded than by those lizards commonly known as monitors (*Varanidæ*). The tortoises of Australia, as mentioned later, belong to a different suborder from those of India, while there are no genera common to Australia and Africa. The monitors, all of which are included in the single genus *Varanus*, are, however, common to the three countries named, while one species actually ranges from India to Australia. That this widespread generic distribution is not a feature of the present epoch is proved by the occurrence of fossil monitors in both the two latter countries, whereas there is no evidence that they ever possessed genera of tortoises in common.

Before proceeding further, it may be well to mention that the Egyptian representative of the group is known to the natives by the name of *ouaran*, which appears to be the Arabic term for lizards in general. Transliterated as *waran*, this word has been confused with the German *warnen*, to warn, whence these reptiles have been termed *warn-eidechsen*, or warning lizards; this, again, having been translated into monitors—a name which, however erroneous in origin, is too well established to be superseded.

Monitors are distinguished from all the lizards hitherto described by the long and deeply-forked tongue, which is capable of being protruded far in front of the lips, and is furnished at the base with a sheath, into which it can be withdrawn, as in snakes. Including the largest members of the suborder, monitors are further characterised by the long body, the broad, uncrested back, the well-developed, five-toed limbs, and the long tail, which is very frequently markedly compressed. The head is covered with small polygonal scales; the eyelids are well developed; the opening of the ear is distinct; and the head is covered with small scales.

As regards the skull, the absence of a bony roof over the temporal pits and of teeth on the palate, as well as the union of the two nasal bones into a single ossification, are notable features. The teeth are large and pointed, with expanded bases fixed to the sides of the jaws. On the back the scales are rounded and bordered by rings of minute granules, so that they do not overlap; while on the under surface the squared scales are arranged in cross-rows. Pores are absent both on the under surface of the thigh and in front of the vent. A peculiarity of the group is the presence of an imperfect midriff, found elsewhere among reptiles only in crocodiles.

Monitors inhabit Africa, Southern Asia, Oceania, Papua, and Australia, and are represented by nearly thirty living species, the largest of which attains a length a little short of 7 feet. A fossil species from Northern India was, however, probably 12 feet long, while one from Australia could not have fallen much, if at all, short of 30 feet. The group is an isolated one, without near relationship to any other family.

The genus may be divided into four distinct sections, the first of which is represented solely by the desert monitor (*V. griseus*) of North-Western Africa and South-Western Asia, where it extends from Arabia and the Caspian to North-Western India. This species differs from all the rest in that the nostrils are in the form of oblique slits, while the tail, except sometimes near its tip, is cylindrical. Attaining a length of 4 feet 2 inches, this species takes its name from its greyish yellow colour, which may be relieved by brown cross-bars on the back and tail, and streaks of the same hue along the sides of the neck, the young always having yellow spots and dark bars. In accordance with its sombre colouring, this species is an inhabitant of sandy deserts.

A far handsomer lizard than the last is the Cape monitor (*V. albigularis*) of Southern and South-Eastern Africa, where it is commonly known to the Boers as the "adder." It is the first representative of the second group of the genus, in which, while the nostrils are in the form of oblique slits, the tail is compressed and keeled. Belonging to a subgroup characterised by the smooth scales of the abdomen, it is further distinguished by the absence of large (supra-ocular) scales above the eyes, by the nostril being three times as far from the snout as from the eye, and by the small size of the scales. It is slightly inferior in size to the last, and has the upper parts greyish brown, banded and spotted with yellow, and the under parts yellowish. It generally frequents

CAPE MONITOR

cliffs or low, rocky hills, in the interstices of which it delights to hide, coming out to bask on the flat surfaces. Gray's monitor (*V. grayi*) is an example of a second subgroup in which the abdominal scales are keeled.

In the third great group, of which the great water-monitor (*V. salvator*) is the largest member—as it is, indeed, of the whole genus—round or oval nostrils are accompanied by a compressed tail. In the species mentioned there is a series of transversely elongated scales above the eyes, the oval nostril is situated as far from the eye as from the tip of the snout, there are more than eighty transverse rows of scales between the fold on the throat and the groin, and the scales on the nape are not larger than those of the back.

This fine species, which ranges from India through the Malay region and China to Australia, attains a length of nearly 7 feet. In colour it is dark brown or blackish above, with yellow rings; the snout being generally lighter, with transverse black bars, and a dark band, bordered by a yellow one, running backwards from the eye, while the under surface is uniformly

THREE SPECIES OF MONITORS

DESERT MONITOR

WATER-MONITOR

GOULD'S MONITOR

yellow. The water-monitor frequents marshy localities, being often found on trees overhanging rivers, and taking readily to water, either fresh or salt (page 1510).

Another member of this group is the well-known Nile monitor (*V. niloticus*), whose range extends all over Africa except a portion of the north-western regions. Belonging to the same group as the last, it represents a second subgroup characterised by the equality in the size of the scales above the eyes, and is distinguished from its allies by the nostril being rather nearer the tip of the eye than the snout. In size it is somewhat larger than the desert-monitor. The colour of the adult is brownish or greenish grey, with darker reticulate markings, and more or less distinct yellowish eye-like spots on the back and limbs; while beneath it is yellowish, crossed by some dark bands. This species is likewise found in the neighbourhood of water, generally building itself a nest among the bushes on the banks, especially of those streams that dry up in the hot season (page 1562).

The Papuan monitor (*V. prasinus*) of New Guinea and the islands of Torres Straits is an example of the fourth group, in which, while the nostrils are round, the tail is nearly cylindrical. In colour it is bright green, with black markings.

As will be gathered from the foregoing remarks, monitors present considerable diversity of habitat, although the majority prefer the neighbourhood of water. The Papuan species is, however, believed to be arboreal, as, in fact, is made almost certain by its colour. All are carnivorous in their diet, feeding on frogs, snakes, the smaller mammals, and birds, as well as the eggs of both birds and reptiles, especially crocodiles.

Their movements are extremely rapid, both on land and in water; and many a sportsman in his first day's snipe-shooting in the rice-fields around Calcutta has been startled by the sudden rush of the common Indian *V. bengalensis* as it darts among the herbage close to his feet.

Those species in which the tail is most compressed are the best swimmers, this appendage serving as a powerful propeller in the water, and being also used as a weapon of offence on land. In order to enable these lizards to remain under water for some time, the nostrils are expanded into large cavities within the snout, and when the apertures are closed these pouches serve as reservoirs of air.

Writing of the great water-monitor Dr. Cantor states that it is "very numerous in hilly and marshy localities of the Malay Peninsula. It is commonly, during the day, observed in the branches of trees overhanging rivers, preying upon birds and their eggs and smaller lizards, and when disturbed throws itself from a considerable height into the water. It will courageously defend itself with teeth and claws and by strokes of the tail. The lowest castes of Hindus capture these lizards commonly by digging them out of their burrows on the banks of rivers for the sake of their flesh."

Sir V. Ball gives the following account of a meeting with a lizard of the same species in the Nicobars: "As I did not care to shoot him, though I wanted to capture him, I threw stones at him, whereupon he hissed and lashed his tail in a manner that might prove alarming to anyone not knowing the harmless nature of the beast. As I was pressing him into a corner, he made a rush into the waves, but returned, apparently not liking the surf. Just as I thought he could not escape, he made a sudden dart into the water, dived through the surf, and disappeared."

From observations made on specimens in captivity, it appears that these lizards eat eggs by taking them in their mouths, raising their heads, and then breaking the shells, so that the contents are allowed to flow down their throats.

Although but little is ascertained regarding their breeding habits, monitors are known to lay white, soft-shelled eggs, which are sometimes deposited in the nests of white ants. As many as twenty-four eggs, of a couple of inches in length, have been taken from the body of a single female. By the Burmese these eggs are much relished as articles of food, and command a higher price in the market than hens' eggs.

Gould's monitor (*V. gouldi*) is a well-known Australian species. In colour it is brown above, with yellow spots on the back and limbs and yellow rings on the tail; the under parts being yellowish, occasionally with black spots.

THE WATER-MONITOR

TEJU LIZARDS

In America the place of the typical lizards of the Old World is taken by a nearly allied group which may be termed the teju lizards (*Teiidæ*), some of which rival medium-sized monitors in dimensions. In common with the remaining members of the suborder, these

TEJU OR TEGUEXIN

lizards are distinguished from all the foregoing by their tongues, which are slit at the tip and frequently shaped like an arrow-head, being either covered with overlapping scale-like papillæ or marked by oblique folds. In all of them the head is covered with large symmetrical shields, very different from the small scales of the monitors. They further differ by the collar-bones being dilated, and often loop-shaped, at their inner extremities.

The teju lizards are specially characterised by the absence of a bony roof to the temporal pits of the skull, and by the shields of the head being completely free from the underlying bones ; while there are no bony plates in the skin of the body. On both the body and tail the scales are arranged in transverse rows. The teeth, although very variable, differ from those of the typical lizards of the Old World in not being hollow at the base ; the replacing teeth being developed in small sockets at the roots of those in use. In some cases these teeth, which may be either pointed or of a flattened crushing type, are placed near the summits of the jaws, and in others somewhat on the side, so that the dentition is intermediate between the typical acrodont and pleurodont

modifications ; the front teeth are always conical. On the palate teeth are seldom present, and, if developed at all, are small. The long tongue, which is frequently retractile within a sheath, is generally covered with overlapping scales, the drum of the ear is exposed, and the eyes are usually furnished with lids. The majority of the species resemble the typical lizards in general appearance, although in some the number of toes is reduced to four. In others, however, the limbs take the form of mere stumps, and the hind pair may be wanting, in which case there is a near approach to the amphisbænas.

The present family comprises over a hundred species, arranged in thirty-five genera, and distributed over the warmer parts of America, although most numerous in the equatorial regions. Varying in their habitat, some frequent dry, sandy plains, others dwell among the herbage of meadows, while others prefer woods, and a few are partially or wholly subterranean ; these latter either taking possession of some empty hole, or digging one for themselves. In their general mode of life they resemble the monitors and true lizards, although some are more like the amphisbænas. They are generally swift and active in their movements ; and the larger kinds are thoroughly carnivorous, subsisting not only on insects, worms, slugs, and snails, but likewise hunting such of the smaller vertebrates as they are able to overcome. Most species deposit their eggs in the hollow stems or among the

ANOTHER SPECIMEN OF THE TEGUEXIN   Photo, C. R. Walter

roots of trees. A few of the larger species are hunted for the sake of their flesh, which is stated to be tender and well-flavoured

TEGUEXINS

The typical and at the same time the largest representative of the family is the lizard termed the teju, teguexin, or jacuaru (*Tupinambis teguexin*), which ranges over a large portion of South America and the West Indies, and belongs to a genus comprising three species. These lizards may be recognised by the tail being cylindrical at the root and slightly compressed near the middle, the double fold of skin on the neck, the uniform scales of the back, the rather small, squared shields of the under surface of the body—arranged in more than twenty rows—the want of teeth on the palate, the compressed tricusped cheek-teeth of the young, and the long tongue, which is of nearly equal width throughout, and sheathed at the base. In old individuals the crowns of the cheek-teeth become obtuse.

The teju, which attains a length of about a yard, is a bulky and strikingly-coloured lizard. On the upper parts the ground-colour is olive, upon which are markings and bands of black, and more or less distinct rows of lighter spots, while the under surface is yellowish, with interrupted black bars, the lines of division between the shields of the head being black.

Ranging from Guiana to Uruguay, the teju is very common in the forests of the Amazon, where it may be observed in numbers during the midday stillness scampering, apparently in sport, over the dead leaves; but in certain other districts it haunts sugar-plantations.

Although frequently found in the neighbourhood of water, it apparently never enters that element, and generally dwells in wide-mouthed holes situated beneath the roots of trees. Shy and retiring to a degree in inhabited districts, when driven into a corner it shows fight, hissing at and striking with its muscular tail the dogs employed in its pursuit. When sitting, the head

BLACK-POINTED TEGUEXINS CLIMBING
Photo, C. R. Walter

is generally raised, while the forked tongue is in constant motion. Its diet comprises such living creatures as it can capture, together with eggs. The female lays from fifty to sixty hard-shelled eggs about the size of those of a pigeon, generally placed in the hillocks of white ants.

THE DRACÆNA

The dracæna (*Dracæna guianensis*), of the Guianas and Amazonia, is a somewhat smaller lizard, distinguished by its compressed and doubly-keeled tail, the intermixture of keeled tubercles among the scales of the back, and the extremely broad crowns of the cheek-teeth.

AMEIVAS

The Surinam ameiva (*Ameiva surinamensis*) is a well-known representative of a genus of nearly twenty species distributed over Central and South America, where they take the place occupied by the typical lizards in the Old World. Ameivas are distinguished by their round, keelless tails, the presence of less than twenty rows of large, smooth scales on the under surface of the body, and the compressed bicusped or tricusped cheek-teeth. The tongue can be withdrawn into a sheath. The Surinam species, which is found over South America as far as Nicaragua, attains a length of from 15 to 20 inches, and is very variable in colour.

The young are olive brown, with darker markings or white dots, and a black, white-edged band running along the side of the body and extending on the tail; but these bands generally disappear with age, although sometimes retained in the females. In the adult the upper surface is usually greenish, with some black and a few white spots; but the under parts are greenish white, spotted with black on the sides. Ameivas are generally found in dry districts—more especially near the coasts—and in general habits are not very different

SURINAM AMEIVA

from the teju, usually living in holes, among old wood, or the herbage of gardens. In fact there is nothing worthy of special notice in regard to their general mode of life.

AMPHISBÆNAS

Among the most remarkable of all lizards are those whose typical representatives have the power of moving equally well either backwards or forwards, from which peculiarity they derive the name (*Amphisbænidæ*) by which the whole group is commonly designated. Very nearly related to the preceding family, through the members of the latter with aborted limbs, amphisbænas are distinguished by the simple and degraded characters of the skull, in which all the arches have been lost, and the two premaxillary bones are fused into one.

All are adapted to a purely subterranean existence, and have long, worm-like bodies, devoid, except in one species, of any external trace of limbs; while even the bones of the shoulder-girdle and pelvis are more or less rudimentary. The eyes are concealed beneath the skin, the mouth is small and frequently inferior in position, and the ear is completely wanting. Although the head is covered with large symmetrical shields, the skin of the body is divided into squared segments forming regular rings, like those of worms; from which character the members of the group are sometimes spoken of as ringed lizards. In all the species the tail is short; and the large teeth are few in number, and fixed either in the inner or upper edges of the jaws.

MEXICAN AMPHISBÆNA

The amphisbænas, which are arranged in eleven genera, including between sixty and seventy species, are most numerously represented in America south of the Tropic of Cancer, although also occurring in the West Indies, while Africa possesses over twenty species, and four are found in the Mediterranean area. All the members of this family are burrowers, and may live in ants' nests. They bore narrow galleries in the earth, in which they are able to progress backwards as well as forwards. On the ground their progress in a straight line by slight vertical undulations, not by lateral movements, as in other limbless reptiles; and the tail of many species appears to be more or less prehensile. The food of these lizards consists of small insects and worms. As regards their breeding habits, it is known that one species lays eggs, which are deposited in ants' nests.

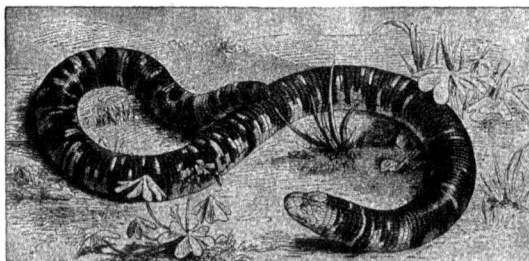

SPOTTED AMPHISBÆNA

The marked resemblance of these lizards to earthworms is a most curious instance of the similarity produced in the external form of different groups of animals by adaptation to similar modes of life; the remarkable feature in this case being the occurrence of this resemblance in creatures so widely sundered from one another as are worms and amphisbænas. Fossil members of the family have been discovered in the Tertiary rocks of North America.

MEXICAN AMPHISBÆNA

The one member of the family which exhibits evidence of its relationship to less specialised lizards in the retention of rudimentary fore limbs is the Mexican amphisbæna (*Chirotes canaliculatus*) of Mexico and California; this being one of the two species found on the continent of America to the north of the Tropic of Cancer. This reptile, which attains a length of about 7 inches, and is of a brownish flesh-colour, is distinguished by the presence of a pair of small depressed fore limbs, placed close to the head, to which they are about equal in length; each of these being provided with four well-developed and clawed toes, of which the outermost is the shortest.

TYPICAL AMPHISBÆNAS

The typical members of the family constitute a genus common to Tropical America and Africa, and represented by nearly thirty species. Belonging, like the last genus, to the group in which the teeth are attached to the inner edges of the jaws, these limbless amphisbænas are specially characterised by the anterior body-rings not being enlarged, by the laterally-placed nostrils being pierced in a special nasal shield, the rounded or slightly compressed snout, the obtuse, cylindrical tail, and the presence of pores in front of the vent.

The spotted *Amphisbæna fuliginosa*, which is from Tropical America and the West Indies, derives its name from its pied skin, and attains a length of about 18 inches. Writing of the habits of a member of the genus, Mr. Bates observes that their "peculiar form, added to their habit of wriggling backwards as well as forwards, has given rise to the fable that they have two heads, one at each extremity. They are extremely sluggish in their motions, and live habitually in the subterranean chambers of the saüba ant; only coming out of their abodes occasionally in the night-time. The natives call the amphisbæna the mai das saübas, or mother of the saübas, and believe it to be poisonous, although it is perfectly harmless. It is one of the many curious animals which have become the subject of mythical stories with the natives. They say the ants treat it with great affection, and that if the snake be taken away from a nest the saübas will forsake the spot.

"I once took one quite whole out of the body of a young jararaca, whose body was so distended with its contents that the skin was stretched out to a film over the contained amphisbæna. I was, unfortunately, not able to ascertain the exact relation which subsists between these curious reptiles and the saüba ants. I believe, however, that they feed upon the saübas, for I once found the remains of ants in the stomach of one of them."

TYPICAL LIZARDS

THE typical lizards form a large family (*Lacertidæ*), with seventeen genera, distributed over Europe, Asia, and Africa (exclusive of Madagascar), but most abundant in Africa, and comparatively rare in the Indo-Malay countries. Taking the place in the Old World occupied in the New by the teju lizards, these reptiles are readily distinguished from the latter by the temporal pits of the skull being roofed over with bone (as shown in the figure of the skeleton on page 1503), and likewise by the shields of the head being firmly attached to the underlying bones, as well as by the union of the two premaxillary bones, the latter feature being common to this family and the amphisbænas.

All the members of the family have well-developed limbs, each furnished with five toes, the body plump, and separated from the head by a well-marked neck, the tail long and brittle—the drum of the ear exposed, and the eyelids distinct and generally freely mobile. The skin contains no bony plates, and the scales of the back are either overlapping or in apposition, but those of the under surface generally longer, and arranged in longitudinal and transverse rows. The teeth are always attached to the sides of the edges of the jaws (pleurodont) and differ from those of the teju lizards in their hollow bases; those of the check series having two or three cusped crowns. The flat and scaly tongue is of considerable length, and cleft both in front and behind, so as to assume the form of an arrow-head. As a rule, pores are present on the hind surface of the thigh.

Out of about one hundred species of typical lizards, two are found in the British Isles, where, with the exception of the slow-worm, they are the only representatives of the suborder; but many others inhabit Southern Europe. Lizards of this family are veritably creatures of the sun, delighting to bask in its rays on some warm sandy bank, wall, or rock, and retiring to their holes and crannies in cloudy or rainy weather. The more powerful and bright is the sun the more active, indeed, do these reptiles become, since most of them are dull and listless in the mornings and evenings, and only wake to full activity in the midday glare.

Over the greater part of Europe these lizards spend a large portion of their time in their holes, and with the beginning of October retire for their winter sleep, from which they do not awake till spring is well advanced. Comparatively rare in Northern Europe, in the south of the Continent lizards are common enough to form an attractive feature in the landscape, their burnished metallic green and bronzy scales flashing in the sunlight on every wall, and in every road and path. The darting movements of these pretty reptiles, as they are in pursuit of the flies and other small insects which constitute their chief prey, are familiar to all. While the majority lay eggs, the viviparous lizard produces living young.

EYED LIZARD

The handsome eyed, or pearly, lizard (*Lacerta ocellata*) of Southern Europe, also represented by a variety in Algeria, may be taken as the first example of the typical genus *Lacerta*, of which there are over twenty species, inhabiting Europe, North and West Asia, Africa north of the Sahara, and the Atlantic islands. The members of this group are distinguished by the following features. The body is cylindrical or slightly depressed, the head pyramidal, with upright sides, the neck not very well defined, and the tail cylindrical, tapering, and long. The throat is furnished with a well-marked collar of enlarged scales; the scales on the back are smaller than those on the tail, and at most but slightly overlapping; while the shields of the under surface are squared and slightly overlapping. The rounded or compressed toes have either smooth, tuberculated, or indistinctly-keeled pads on the lower surface, and the thighs are provided with pores. In common with several other genera, the nostrils are placed close to the so-called labial scales, from which they are separated at most by a narrow rim; and if there be a transparent disc in the lower eyelid, it is smaller than the eye itself.

Among the most beautifully coloured members of the suborder, the eyed lizard, which attains a length of from 16 to 23 inches, claims a foremost place. Belonging to a large group of the genus, in which the edge of the throat-collar is strongly serrated, this species agrees with certain other members of the genus in its smooth tail, and in the scales on the sides of the body not being smaller than those on the back. As marked characters of the species, it may be noted that the scales are smaller than in the allied forms, and that there are not less than seventy scales round the middle of the body, eight or ten of which belong to the under surface. The head is very large in the male, and characterised by the great width of its hindmost, or occipital, median shield. In colour the upper parts are either green with black dots or network, or blackish olive with yellowish netting, the sides are marked with a row of about a dozen eye-like blue spots, and the under surface is uniform greenish yellow. The olive-coloured young are, however, dotted all over with white or pearly blue black-edged spots.

Common in Spain, and also occurring in the south of France and North-Western Italy, or wherever the olive-tree grows, the eyed lizard is generally to be met with in the neighbourhood of hollow trees,

EYED LIZARD

GREEN LIZARDS   Photo, C. R. Walter

frequently ascending some distance up their trunks, or even climbing among the branches. The males are somewhat quarrelsome, and the females lay from six to ten eggs, generally deposited in a hollow olive-tree.

GREEN LIZARD

Another well-known European species is the green lizard (*L. viridis*), which attains a length of about 12 inches in Germany, but in the more southern portions of its habitat measures as much as 17 inches; fully two-thirds of this length being occupied by the long tail. Having not more than sixty-six scales in a row at the middle of the body, this lizard is distinguished by the general presence of two small superimposed scales behind each nostril, the small size and triangular form of the occipital shield, and the arrangement of the abdominal scales in six longitudinal rows; the collar being serrated. Usually the nostrils are in contact with the front, or rostral, shield of the head, and in the female and young the foot is longer than the head.

As regards colour, the males, which may be distinguished from females by the larger and taller head, the thickened root of the tail, stouter hind limbs, and generally superior size, are some shade of green olive, passing below into yellow. Black dots, expanding into large spots, generally adorn the upper surface, whereas the under parts, except for a blue patch on the chin and throat, are uniform. The females, in which the blue on the throat is less constantly present, have a more brownish tinge, with the sides ornamented with black-bordered yellowish spots. The young are generally leather-brown in colour, with one or two yellow side-stripes. Both sexes vary, however, considerably according to age, and southern specimens are more brilliantly coloured than those from the north.

The green lizard is an inhabitant of the countries lying to the east and north of the Mediterranean, and extending thence eastwards to Persia. Very common in Portugal and Spain, where it is represented by a variety, it extends in France so far north as Paris, although it is unknown in Sardinia. In place of

resorting, like the eyed lizard, to trees, this species is usually found on the ground, more especially in districts where the subsoil is rocky, ranging from the sea-level to a height of some three thousand feet, and being equally at home on the plains or among the mountains, in stony or sandy districts, on bare rocks, or among thick bush.

As rapid as lightning in its movements, it feeds chiefly upon large insects and their larvæ, together with slugs and worms; living in grassy districts almost entirely upon grasshoppers, but at times attacking smaller species of its own tribe. In Switzerland and Germany the female usually deposits her eight to eleven white eggs during June, these being hatched in the course of a month or so, and it is generally during the breeding season that the blue on the throat is assumed by this sex. The East European race is known as *L. viridis major*.

SAND-LIZARD

The third European representative of the genus is the much smaller sand-lizard, or hedge-lizard (*L. agilis*), a more northern form, ranging into the British Isles and Scandinavia. Usually not more than 8 inches in length, although occasionally measuring nearly 10, this lizard may be recognised by its short, thick, and blunt-snouted head, and by the tail being considerably less than twice the length of the head and body. Never having more than fifty-eight scales in a row at the middle of the body, it is further distinguished by the rostral shield of the head being separated by a small interval from the nostrils, the trapezoidal shape of the small occipital shield, the absence of the row of small granules which occur between the shields of the eyelids (supraoculars) and eyebrows (supraciliaries) in the green and wall lizards, and by the foot being not longer than the head.

Although there is great variation in this respect, the general colour of the male is greenish, and that of the female grey or brown; the crown of the head, a streak down the back, and the tail being mostly brown, while the chin and under parts are greenish or yellowish. The streak down the back, and in the females the sides also, are ornamented by rows of white spots, which are sometimes large and eye-like, and the under surface is marked with black. Some individuals, especially males, closely approach the green lizard in colouring.

VIVIPAROUS LIZARD

The range of the sand-lizard embraces North, Central, and Eastern Europe, and extends eastwards to Western Siberia and Asiatic Russia. In England it is generally found on sandy heaths, where it may often be seen running across the open paths with a speed less rapid than that of the more common viviparous species. It is more timid and less easily tamed than the green lizard, generally pining and refusing to feed in captivity. The female lays her eggs, to the number of twelve or fourteen, in hollows in the sand, which is excavated for the purpose, and having covered them carefully with sand, she leaves them to be hatched by the solar heat.

VIVIPAROUS LIZARD

A still smaller, and at the same time a more slightly built species is the common English viviparous lizard (*L. vivipara*), which varies from 6 to just over 7 inches in length. The scales, which are larger than those of the last, are not more than forty-five in number in each row round the middle of the body, and the foot generally exceeds the head in length; granules being absent above the eyes. The absence of teeth on the palate is another feature in which this species differs from the sand-lizard. The colour of the adult is brown, yellowish, or reddish, ornamented with small dark and light spots, and often with a dark streak down the back, and another, edged with yellowish, on each side. In the male the under surface is orange or vermilion, spotted with black; and in the female pale orange or yellow, sparsely spotted with black, or uniform. The young are nearly black, and this hue occasionally persists.

Unknown to the south of the Alps, the viviparous, or, as it is sometimes called, the mountain, lizard is spread over the greater part of North and Central Europe, and the whole of Northern Asia, as far as Amurland, ranging in the Alps to a height of nearly ten thousand feet. At this elevation it is, however, dormant for fully three-quarters of the year, being active for only two or three months. In Britain it extends to Scotland, and it is one of the few reptiles found in Ireland. Generally similar in its habits to its allies, it is more fond of water, and is a good swimmer, usually frequenting heaths and banks. When active, its movements are beautifully graceful as well as rapid.

This lizard usually comes out of its hiding-place during the warm parts of the day from the early spring till autumn has far advanced. When out, it is fond of basking in the sun; when an insect comes within its view, it turns its head with a rapid movement, and darting like lightning upon its prey, seizes it with

SAND-LIZARD   Photo, W. B. Johnson

its little sharp teeth and speedily swallows its victim. Unlike its kin, this species produces living young, varying from three to six in number, which are active as soon as born, and remain in the company of their parent for some time.

WALL-LIZARD

The last representative of the typical genus noticed here is the beautiful wall-lizard (*L. muralis*), which inhabits the countries bordering both sides of the Mediterranean, and extends eastwards into Persia.

WALL-LIZARD   Photo, W. B. Johnson

It belongs to a group in which the edge of the collar on the neck is even or but slightly serrated, and the scales of the back are granular. Attaining in Germany a length of from 7 to $7\frac{1}{2}$ inches, but reaching from 8 to $9\frac{1}{2}$ inches in Italy, this species has a series of granules between the shields above the eyes, while the scales of the abdomen are arranged in six (rarely eight) rows, and those on the upper surface of the leg are larger than those on the back; and there is only a single (postnasal) scale behind each nostril.

In colour the wall-lizard presents such an astonishing variation that it is almost impossible to give any general description. In German examples the ground-colour of the back is, however, often brown or grey, with bronze-green reflections in sunlight, upon which are blackish streaks, marblings, and spots; while the flanks have a row of blue spots, and the under parts vary from milk-white to copper-red, frequently variegated by spots or marblings (page 1508).

In Southern Europe these lizards may be seen basking on almost every wall, old building, or face of rock, where they delight all beholders with their activity and tameness. "Scarcely two," writes Dr. Leith-Adams, "are marked alike; the brightness and variety of their hues are most beautiful and attractive, and, like the chameleon, they change colour with the coruscations of sunshine, but, of course, not to the same extent. During an excursion to the islet of Filfla, on the southern coast of Malta, in the month of June, I was surprised to find that all the lizards on the

rock were of a beautiful bronze-black, and much tamer than their agile brethren on the mainland. Many individuals were so tame that they scrambled about our feet, and fed on the refuse of our luncheon."

ALGERIAN KEELED LIZARDS

Whereas in the Southern Tyrol these lizards remain active till December, and reappear by the middle of February, in Germany the winter sleep is considerably longer. Like its congeners, this species has an exceedingly brittle tail; and it was observed some years ago that on a certain road in Madeira all the lizards belonging to a nearly allied species (*L. dugesi*) were without tails. The circumstance was explained by the spot being the favourite resort of the midshipmen landing from the ships visiting the island, who amused themselves by knocking off the lizards' tails.

KEELED LIZARDS

The members of the genus *Lacerta* are collectively characterised by the presence of a well-marked collar on the neck, by the scales of the back being smaller than those on the tail, or by the toes being without fringes on their sides, or keels on their soles. The allied genus *Algiroides*—represented by three species from the eastern coast of the Adriatic, Greece, Sardinia, and Corsica—differs by the strongly overlapping scales of the back being nearly as large as those of the tail.

On the other hand, four species inhabiting South-Western Europe and the opposite coast of Africa constitute a third genus, *Psammodromus*, in which the collar is indistinct or wanting, the toes are not fringed, though generally more or less distinctly keeled inferiorly, while the overlapping scales of the back bear strong keels. Among these the Spanish keeled lizard, or sand-runner (*P. hispanicus*), retains a trace of a collar, and has strongly-keeled soles;

whereas in the Algerian keeled lizard (*P. algirus*) the collar is wanting, and the soles are at most but feebly keeled. The latter species, which inhabits not only North-Western Africa, but likewise Portugal, Spain, and the south of France, reaches nearly $10\frac{1}{2}$ inches in length, and has a tail almost twice as long as the head and body. It is specially distinguished by the scales of the abdomen being of nearly equal width and arranged in six rows, as well as by the presence of from thirty to thirty-six scales round the middle of the body.

In colour this lizard is bronzy green above, with one or two golden, dark-edged streaks along the side; the male being ornamented with a pale blue eye-like spot above the shoulder, sometimes followed by one or two behind, while the under parts are whitish.

Abundant in Algeria and the neighbourhood of Montpellier, this lizard is found in the former region both in hedges and on limestone rocks, whereas in France it frequents hedges only. Preferring dry, open, and warm districts, and thriving well in captivity, it presents nothing noteworthy as regards habits.

FRINGE-TOED LIZARDS

The fringe-toed lizards (*Acanthodactylus*), of which there are about half a score of species, ranging from Southern Spain and Portugal and Northern Africa through South-Western Asia to the Punjab, differ from the preceding group by the toes being both fringed on the sides and keeled below, a more or less distinct collar occurring on the throat. On the head the occipital shield is wanting, and the nostrils are pierced between one labial and two nasal shields. Pores are present on the thigh, and the tail is nearly cylindrical.

The common fringe-toed lizard (*A. vulgaris*) is a species of from $4\frac{1}{4}$ to $4\frac{3}{4}$ inches in length, agreeing with most of its kindred in having the hind scales of the back little enlarged, but specially characterised by the strong keeling of the scales on the upper surface of the tail, and the slight fringing of the toes. It is represented by two races, one occurring in Spain and Portugal, and rarely in the south of France, characterised by the smooth or slightly keeled scales of the back; and an African form, in which these scales are very strongly keeled, and the colour is brighter. The colour of the adult is greyish or brownish, with faint longitudinal series of light and dark spots and lines, and sometimes eye-like blue spots on the flanks; but the young are longitudinally

FRINGE-TOED LIZARDS

streaked with black and white, and show white spots on the limbs. All these lizards inhabit dry, sandy districts, and are remarkably shy in their habits, seldom venturing forth from their retreats except when the sun is shining brightly.

SEPS

STUMP-TAILED SKINK

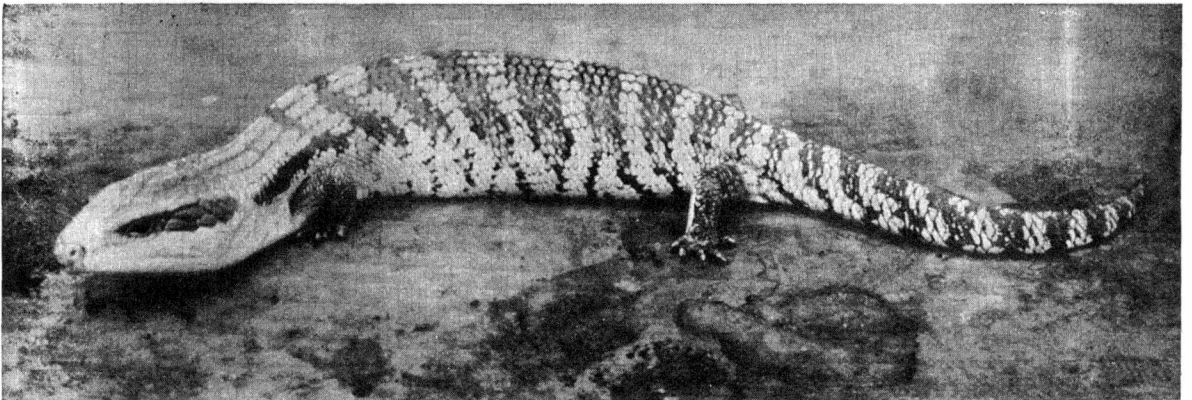

BLUE-TONGUED SKINK

SKINK TRIBE

THE preceding family is connected with the one now to be considered by a small group of five African genera constituting the family *Gerrhosauridæ*, the members of which, while resembling typical lizards in having a single premaxillary bone and the presence of pores on the thigh, agree with the skinks in possessing bony plates of peculiar structure beneath the scales.

The skink tribe (*Scincidæ*), taking their title from the lizard commonly known by that name, form a very numerous family, comprising upwards of twenty-five genera and nearly four hundred species, and presenting great variety of bodily form, some kinds being four-limbed, while others are more or less snake-like.

Agreeing with the typical lizards in the characters of the tongue and teeth, as well as in the roofing-over of the temporal pits by bone, skinks differ in having two distinct premaxillary bones in the skull, in the presence of bony plates traversed by symmetrical tubules beneath the scales, and in the invariable absence of the pores so generally present in the thighs of the members of the *Lacertidæ*.

The limbs, when present, are relatively short, and in some cases are reduced to a single pair, and in others wholly absent; the number of toes is very variable, even among the members of a single genus; the short and scaly tongue is free, and but slightly notched in front; and the drum of the ear is generally covered with scales. The eyes have round pupils, and well-developed and generally mobile lids, the lower one of which is furnished with a relatively large transparent window. The teeth, which are attached to the sides of the jaws, may have either conical, bicuspid, or broad and spheroidal crowns. The head is covered by large symmetrical shields, among which an unpaired occipital is generally wanting; and the overlapping scales of the body are generally subhexagonal in form, and arranged in a quincuncial manner.

World-wide in distribution, the members of the skink tribe are most numerously represented in Australia, Oceania, the Indo-Malay region, and Africa, while very few occur in South America, and there are not many in North America and Europe. Although their habits are not fully known, it appears that, with the exception of two genera, skinks bring forth living young, varying from two to ten in number. The majority are terrestrial, a few only being able to climb, while none are aquatic. They sedulously avoid the neighbourhood of water, frequenting dry situations, and more especially those where the soil is sandy with an admixture of pebbles or fragments of rock. Moreover, they generally possess the faculty—rare among lizards—of burrowing in the ground with the dexterity, if not with the power, of moles. From this habit the members of the group are sometimes spoken of as the burrowing lizards; and it may be remarked that their spindle-shaped bodies, covered with highly-polished scales, their short legs, and frequently abbreviated tail, as well as the transparent window in the lower eyelid, are all features specially adapted for such a mode of life. From among the numerous genera, limits of our space render it necessary to restrict attention to examples of a few very divergent types.

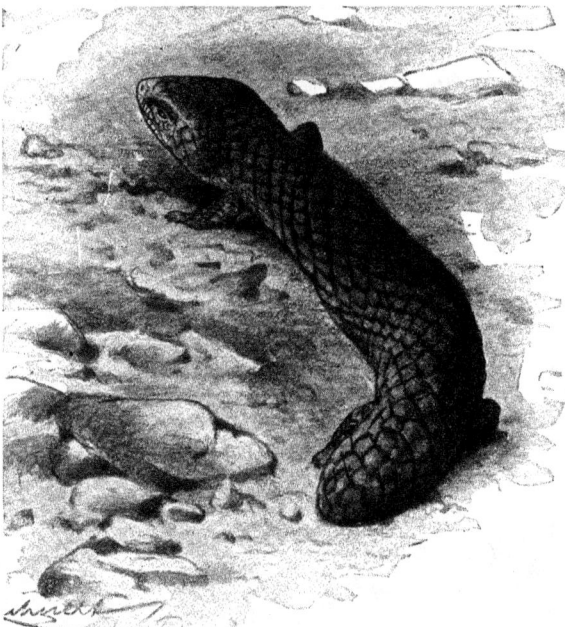

STUMP-TAILED SKINK

STUMP-TAILED SKINK

First described so long ago as the year 1699, the stump-tailed skink (*Trachysaurus rugosus*) of Australia is the sole representative of one of the most remarkable genera in the entire suborder. With a short, pyramidal, depressed head of great width, a short but distinct neck, a long, thick, and flattened body, and a very wide and stumpy tail, this strange reptile is clothed with an armour of rough, thick, brown, horny scales, which give it very much the appearance of a living pine-cone. On the lower surface the scales are smooth and much smaller.

The small and stout limbs are widely separated, and terminate in five short toes, each provided with strong, curved claws. In length this skink measures about 14 inches, and its colour above is brown, with spots or irregular bands of yellow; while beneath it is yellowish, with brown spots, marblings, or longitudinal and transverse streaks. The cheek-teeth have subconical crowns. Its habits are still imperfectly known; but it is slow and lethargic in its movement, and its chief food consists of worms and insects, although fruit and vegetables are occasionally eaten. Like so many other members of its tribe, it is a burrower.

BROAD-TOOTHED SKINKS

The Austro-Malay skinks of the genus *Tiliqua* are nearly allied to the last, but have smooth, overlapping scales on the body; and the short tail is smooth and shiny; while the cheek-teeth have broad, flat crowns. The giant skink (*T. gigas*), which is ochre-yellow, with irregular chocolate-coloured bars, grows to 2 feet in length, and the blue-tongued *T. Scincoides* is another well-known Australian species. *Mabuia* is an allied genus, with a large number of species, inhabiting Africa, South Asia, and Tropical America. The Australian Cunningham's skink (*Egernia cunninghami*) and the New Zealand lined-eye skink (*Lygosoma lineoocellatum*) represent other genera.

SNAKE-EYED SKINKS

Another lizard (*Ablepharus pannonicus*) belongs to a genus containing a number of small species distributed over Australia, South-Western Asia, South-Eastern Europe, and Tropical and South America, one of which (*A. boutoni*) ranges irregularly over the hotter parts of both the Eastern and Western Hemispheres. These lizards differ from all their kin in having no movable eyelids, their place being taken by a transparent disc of skin stretched over the eye after the manner of snakes. In these skinks the aperture of the ear may be either open or concealed by scales; and while some of the species have well-developed limbs, in others they are more or less aborted, the number of toes being also highly variable.

The typical species, which ranges in Europe from Hungary to Greece, and is also spread over Asia Minor, Syria, and Northern Arabia, measures only 4 inches in length, of which fully half is occupied by the tail. Its general colour is bronzy olive, becoming darker on the sides, with a blackish, light-edged streak passing through the eye along each side of the body, while the under parts are greenish. The European species is found alike on slopes covered with short grass or in sandy spots, and does not appear to be a burrower. Feeding on small insects and worms, it does not generally venture forth from its lurking-places till four or five o'clock in the afternoon, and retires before night. In common with the other members of its genus, it differs from the majority of its family in laying eggs.

THE SKINK

TYPICAL SKINKS

While the genera mentioned above belong to a group characterised by the palatine bones meeting in the middle of the palate, the typical skinks (*Scincus*) represent a second and smaller group in which those bones are separated from one another. Skinks are neatly-made, somewhat short-tailed lizards, with short limbs provided with five toes, which are serrated on their sides. The tail is conical, the head and snout are wedge-shaped, the ears more or less concealed, and the nostrils pierced between an upper and a lower nasal shield.

Of the nine species of the genus, which range from Northern Africa through Arabia and Persia to Sind,

CUNNINGHAM'S SKINK
Photo, C. R. Walter

the most familiar is the common skink (*Scincus officinalis*) of the Sahara and Red Sea littoral. This species, which attains a length of 8¼ inches, has smooth, shining, rounded scales of great breadth, and is of a yellowish or brownish colour above, with each scale marked by small brown and whitish spots and streaks, and the sides of the body often ornamented with dark, transverse bands, the under parts being uniformly whitish.

Not uncommon in Egypt, and abundant in the Algerian and Tunisian Sahara, the common skink derives its specific name from having been at one time extensively employed in medicine as an infallible remedy for almost every disease under the sun; its reputation as a healing agent still surviving among the Arabs, by whom the flesh of the reptile is used both as a drug and as an article of food.

The exclusive haunts of the skink are sandy districts, where it generally moves in a slow and deliberate manner, and, when frightened, buries itself in the soil instead of attempting to seek safety in flight. Indeed, the celerity with which the reptile sinks into the sand is described as being little short of marvellous, suggesting the idea of its escaping into some hole already existing rather than of excavating a fresh burrow for itself, such a burrow not infrequently extending to the depth of several feet.

During the daytime the skink, if quietly approached, may be observed reposing in the sun by the side of one of the small hillocks or ridges raised in the sand at the base of trees by the wind; and from such a state of idleness it is only roused by the approach of a beetle or a fly, upon which it darts with unerring aim. In spite of their strong teeth or claws, when captured, skinks never make any attempts to defend themselves, beyond struggling vigorously. Of their breeding habits little or nothing definite appears to be known. The flesh of a few well-boiled skinks is said to form a dish not to be despised even by a European palate.

SEPS GROUP

Under the title of chalkis, the ancient Greeks designated a remarkable snake-like lizard inhabiting Italy, Sardinia, and Sicily, as well as Algeria and Tunis, which was known to the Romans by the name of seps; the latter being in allusion to the poisonous properties with which this perfectly harmless reptile was supposed to be endowed. The true seps (*Chalcides tridactylus*) is the typical representative of a genus of some twelve species belonging to the present family, which exhibit a most interesting example of the gradual degradation of limbs, some species having five toes to each foot, while in others, as the figured example, the number of digits is reduced to three; and in one kind the limbs are represented merely by undivided rudiments.

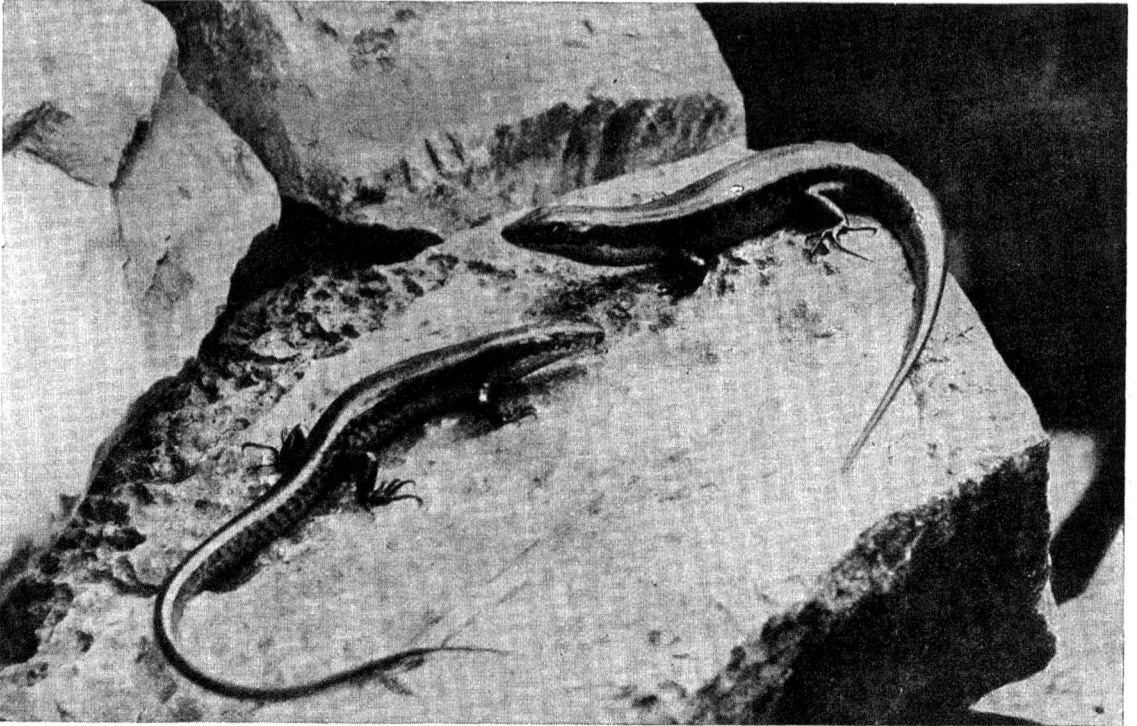

LINE-EYED SKINKS

Photo, C. R. Walter

All the species of seps, as the members of the genus may be collectively termed, belong to an assemblage of genera differing from all those already noticed in that the nostrils are pierced either in or close to the terminal rostral shield of the head, instead of being more or less widely separated therefrom. In the case of the present genus the nostrils are situated in notches cut in the hind border of the shield in question; while the body is greatly elongated, and the limbs are either short or rudimentary. The species figured below is one of two species with three-toed limbs, and attains a length of 13½ inches, of which about half is occupied by the tail.

In colour it is olive or bronzy above, and may be either uniform or marked with an even number of darker and lighter longitudinal streaks. In the south of France, Spain, and Portugal it is replaced by the smaller striped seps (C. lineatus), in which the body is marked with nine or eleven longitudinal stripes. The range of the whole genus embraces Southern Europe, Northern Africa, and South-Western Asia, from Syria and Arabia to Sind.

The three-toed seps much resembles a slow-worm in general appearance and habits, frequenting damp places, where abundance of its favourite worms, snails, slugs, insects, and spiders are to be met with. Here it moves with a wriggling, serpentine motion similar to that of the slow-worm, which it likewise resembles in producing living young and in retiring into a burrow for its winter sleep. When not feeding, this skink, like most of its kind, delights to bask on sandy spots in the full glare of the sun. The seps was believed to inflict death on cattle by biting them during the night, its bite filling their veins with corruption (hence the word septic), and in consequence of this belief the unfortunate creature is still persecuted with the same hatred as is the slow-worm in most parts of England.

THREE-TOED SEPS

OTHER FAMILIES

The two remaining families, Anelytropidæ and Dibamidæ, of lizards are represented by worm-like burrowing species allied to the skinks (of which they may be regarded as degraded types), but with no bony plates beneath the scales, no external ear-openings, and eyes concealed beneath the skin. The former family is represented by three genera, of which two are African, and the third is from Mexico; while of the latter there is only a single genus, with one species from Papua, the Moluccas, and Celebes, and a second from the Nicobars.

SQUAMATA—SUBORDER RHIPTOGLOSSA
CHAMÆLEONS

THE skinks and their allies are the last reptiles which, in the zoological sense, are included under the title of lizards, and attention has now to be directed to the second subordinal group, represented by those strange creatures known as chamæleons (Rhiptoglossa). From the lizards proper these reptiles are at once distinguished by their worm-like extensile tongues, which are club-shaped and viscous at the extremity, and capable of being protruded with the rapidity of lightning to a distance of from four to six inches in front of the mouth. Hence the name of worm-tongued lizards has been suggested for the group. Internally chamæleons differ from all lizards provided with well-developed limbs in having no collar-bones (clavicles); while there are likewise certain distinctive features in connection with the skull into the consideration of which it will be unnecessary to enter.

Another important feature by which these reptiles differ from lizards is the structure of the feet, in which the toes are divided into two opposing branches, thus forming grasping organs of great power. In the fore limb the inner branch of the foot includes three, and the outer two, toes, but in the hind foot precisely the reverse arrangement obtains; and from this peculiar hand-like structure of the foot—which, by the way, recalls the feet of the parrots and many picarian birds —chamæleons have been termed four-handed lizards. Yet another peculiarity in the structure of these reptiles is presented by the eye, which is in the form of a very large and prominent globe covered by a thick granular lid, in the centre of which is a minute perforation for the pupil. The deliberate way in which a chamæleon rolls round one of these extraordinary eyes until it has focussed it on the fly about to be caught by the tongue is very striking and weird.

The foregoing are the essential features by which chamæleons are distinguished from the lizards proper; those remaining for mention not being such as would be regarded by naturalists as of subordinal importance. Among these may be noticed the triangular helmet-like form generally assumed by the hind part of the head, which often has three longitudinal ridges, connected together posteriorly by a cross-ridge, and each ornamented with tubercles. The teeth, which are small, triangular, and compressed, are placed on the summits of the jaws in the acrodont fashion, none being present on the palate.

The body is much compressed, and the neck short; the slender limbs are so much elongated as to raise the body high above the ground in a manner different from ordinary lizards; the tail is long and prehensile, thus acting as a fifth hand; and, in place of scales, the head and body are covered with small tubercles or shagreen-like granules. The larger species attain a length of some 15 inches, but the dwarf chamæleon of Madagascar (*Brookesia nasus*) is less than 2½ inches in length.

The members of the suborder mentioned in this work are as follows:

FAMILY
Chamæleons—Chamæleontidæ
GENUS 1
Typical Chamæleons—Chamæleon
SPECIES

The chamæleon	Chamæleon vulgaris
Arabian chamæleon	C. calyptratus
Pigmy chamæleon	C. pumilus

GENUS 2
Brookesia
SPECIES

Dwarf chamæleon	Brookesia nasus

GENUS 3
Rhampholeon

Chamæleons are represented by about fifty species, all of which are comprised in the single family *Chamæleontidæ*, and by far the greater majority in the typical genus *Chamæleon*. Indeed, of the two aberrant genera, *Brookesia* is represented by three species from Madagascar, while *Rhampholeon* comprises two tropical African kinds. The true home of the group is Africa and Madagascar, together with the neighbouring islands, each of these areas comprising nearly half of the known species. The common chamæleon (*C. vulgaris*) is, however, found on the African and Asiatic coasts of the Mediterranean, entering Europe in Andalusia; and a second species inhabits the Isle of Socotra, a third Southern Arabia, and a fourth India and Ceylon.

While most of the species are more or less uniformly coloured, the Arabian *C. calyptratus* is a most brilliantly coloured reptile, the head being mainly bright chrome-yellow, while the back is marked with alternate vertical bands of blue and yellow, and the tail ringed with the

THE CHAMÆLEON

same two colours. On the flanks the yellow bands of the back become bright green, while the blue ones are replaced by mottled maroon and blue, the under parts and limbs being olive flecked with yellow.

What may be the purpose of such a remarkable departure from the ordinary chamæleon type of colouring is not easy to conjecture without seeing the reptile in its native haunts. It may be suggested, however, that this chamæleon frequents trees with much less dense foliage than those to which the ordinary green species resort, and that its colouring is of the " breaking-up " type analogous to that of zebras, and thus adapted to render the reptile more or less nearly invisible at a short distance.

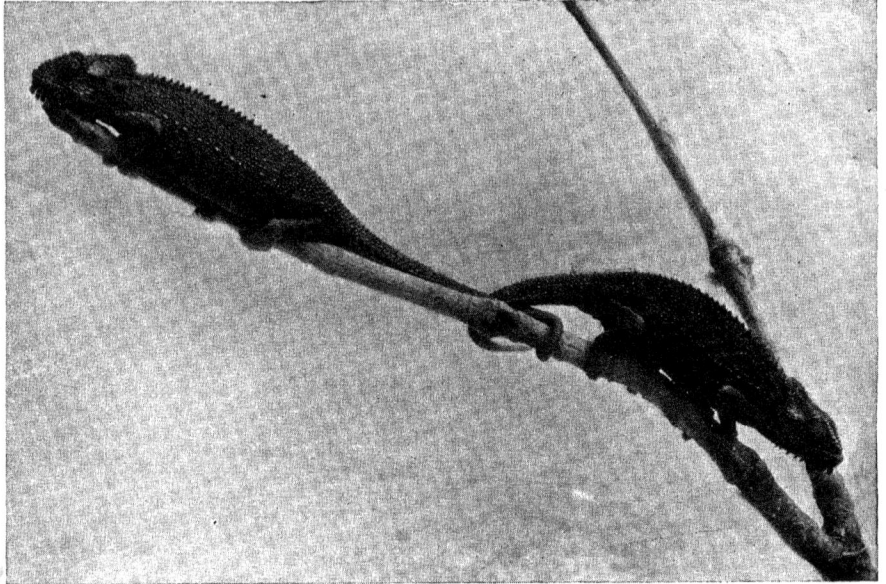

PIGMY CHAMÆLEONS

Evidently extremely specialised reptiles, chamæleons stand altogether apart from lizards, not only as regards their anatomical structure, but likewise in their power of moving one eye independently of the other, in the enormous extensibility and protrusive power of their tongues, and in their slow and deliberate movements. According, however, to those who have had the opportunity of observing them in their native haunts, chamæleons do not move quite so slowly as in confinement, where they take half a minute in determining which limb to move, or on which bough to replace it.

Passing the whole of their lives in trees, like most of their Malagasy compatriots the lemurs, chamæleons are chiefly found in regions where foliage is abundant, and where the fall of rain or dew is sufficient to supply them with the amount of moisture they need. Consequently, they are most numerously represented in coast districts and islands. A few, however, frequent such parts of desert regions as come under the influence of the sea moisture, and support a more or less scanty vegetation. Needless to say, all the species live on insects, and more especially flies of various kinds, which are caught by the viscid secretion of the tip of the protrusile tongue.

Being utterly defenceless creatures, and having a large number of enemies, chamæleons depend entirely upon their resemblance to their environment for protection ; and for this end they have the power of changing colour, although not, apparently, to such an extent as is the case with some lizards of the genus *Calotes*. At night the European species appears generally to be of a whitish yellow hue, but with the first dawn of day it assumes the dark green colour characteristic of most of the species, which exactly assimilates to the surrounding leaves, and continues to grow brighter and brighter with advancing day. When resting on a bough, or when captured in the hand, the colour changes, however, to brown ; this change in the latter case taking place with exceeding rapidity, and the skin sometimes becoming nearly black, with the disappearance of all the bright marking.

This change appears to be due to anger, the creature at the same time emitting a sound something between a hiss and the chirp or squeak of a very young bird, and trying to bite its captor. " Meanwhile," writes Miss C. C. Hopley, " it is all impatience to ascend, no matter where, so that it climbs upwards. Up, up, always up ; it may be your dress, or whatever is near. It seems to think it can be safe only at the top of something. And yet they are not found invariably on the upper branches of their bush, though generally rather high. Released from the hand, its anger soon subsides, so does the dark hue, and the creature assumes the tint of the surface on which it is placed, greyish, reddish, darker or lighter, green or yellow, as may be."

Several individuals are not infrequently met with on the same bush, where they cling tightly to the stems among the crowded leaves, being alike difficult to detect and to detach, and always exhibiting their displeasure at being disturbed by the aforesaid hissing sound. "Absolutely still they remain," continues the writer just quoted, " hour after hour, the only evidence of life about them being that revolving little globe of an eye, with its pupil turning as an axis, now up, now down, forwards or backwards, while its owner clings motionless as death. In repose, the long tongue is folded up within the dilatable skin of the chin, where it has a special sheath for its reception ; but it can be darted out with such speed as to take a fly at a distance of fully six inches."

Although the majority of the species lay eggs, the pigmy chamæleon (*C. pumilus*) of the Cape, together with five nearly allied African species, produces living young, which may be as many as eleven in number In confinement chamæleons quickly become tame. and, if allowed to rest in peace, after a few days cease to bite and hiss when handled, and soon venture to take a fly from their owner's hand.

When chamæleons see a foe approaching, they always move to the opposite side of the bough, where, by compressing the body to more than its usual extent, they are more or less completely concealed from view.

SQUAMATA—SUBORDER OPHIDIA
GENERAL CHARACTERISTICS OF SNAKES

FORMERLY regarded as representing a distinct ordinal group of the reptilian class, snakes are now generally considered to form merely a suborder of the great assemblage of reptiles which includes both lizards and chamæleons ; and, from their close structural resemblance to the limbless lizards, there can be do doubt that the more modern view is the true one. As a matter of fact, it is by no means easy to draw a satisfactory distinction between lizards and snakes ; and such characters as naturalists rely on for the differentiation of the two groups are mainly such as are due to adaptation to the special needs of the latter group.

Agreeing with lizards in their external covering of scales, snakes are characterised by their exceedingly elongated and slender bodily conformation ; the head, which is generally more or less flattened, being often not defined from the body by a distinctly marked neck, while external limbs are wanting, and the body passes posteriorly by imperceptible degrees into the tail. Occasionally, however, external vestiges of the hind limbs may persist, in the form of a pair of small spur-like processes near the vent (p. 1551) ; and internally there may be traces not only of the pelvis, but likewise of the thigh-bone, or femur. None of these characters will, however, serve to distinguish snakes from the limbless lizards, and it is therefore necessary to point out how the two groups may be separated.

The most characteristic peculiarity of snakes, as distinct from lizards, is the absence of a firm union at the chin between the two branches of the lower jaw, which are connected together merely by an elastic ligament ; this arrangement permitting of the separation of the two halves of the jaw, and thus allowing the mouth to be dilated so as to be capable of swallowing prey of much larger dimensions than the normal width of its aperture. In addition to this arrangement, in the majority of snakes the bones of the upper jaw and palate are likewise movable, thus further increasing the capacity of the gape.

From the great majority of lizards snakes are, of course, widely distinguished by the absence of functional limbs ; while from the comparatively few limbless representatives of the former group they differ in having the tongue completely retractile within a basal sheath, as well as by the presence of additional articulations to the vertebræ, which are described below. Moreover, none of the limbless lizards have the large shields on the inferior surface of the body characterising the majority of snakes, while most of them possess eyelids and an external aperture to the ear.

No snake has movable eyelids ; that portion of the skin representing the lids extending as a convex transparent disc across the eye, and covering it as a watch-glass covers the face of a watch. When a snake changes its skin, which it does several times during the year, the discs over the eyes peel off with the rest, and appear as lenses in the dry slough. Equally characteristic is the absence of any external aperture of the ear.

Resembling that of lizards in its flattened form, the tongue of snakes is narrow and smooth, and terminates in a fork formed by two long thread-like points, while at its base it is inserted into a sheath from which it can be protruded at will. The head, although not very large, is generally, and especially in the viper tribes, wider than the body, from which, as already stated, it is but seldom separated by a recognisable neck, and is usually oval or triangular in shape, with a more or less well-marked depression on the forehead. Near the sides of its extremity, and sometimes at the very tip, are situated the nostrils ; while posteriorly the gape in some cases extends almost to the back of the head. Superiorly, as well as on its under surface, the head is generally covered with a number of large symmetrical shields, having their edges in apposition, and varying in relative size in the different groups.

Although the blind snakes have a uniform cuirass of polished scales all round the body, while some of the sea-snakes also have the scales of the under part similar to those of the back, in the great majority of the members of the suborder the under surface of the body is protected by large transverse shields, extending completely across it from side to side. These broad shields often extend so far backwards as the termination of the body proper, but at the beginning of the tail, and thence backwards to the extremity, they are replaced by a double row of smaller shields. These large inferior shields take an important part in the progression of snakes on land, so that the reason for their absence in the burrowing and marine forms is self-apparent.

In all snakes the number of joints in the backbone is very great ; and each of these, with the exception of a few near the extremity of the tail, is provided with a pair of rather long, slender, and curved ribs, the extremities of which correspond to the large inferior shields of the body in the species where these are present. Superiorly the ribs, as shown in the figure on page 1502, articulate by a single head with a facet on each side of each vertebra, in the same manner as in lizards. Only certain groups of lizards have the vertebræ with the additional articular facets on the front and back surfaces known as zygantra and zygosphenes, but in snakes these are invariably present ; and it is owing to this complicated system of articulation that a snake is able to make the wonderful foldings and contortions characteristic of its kind without fear of dislocating its spine. It may

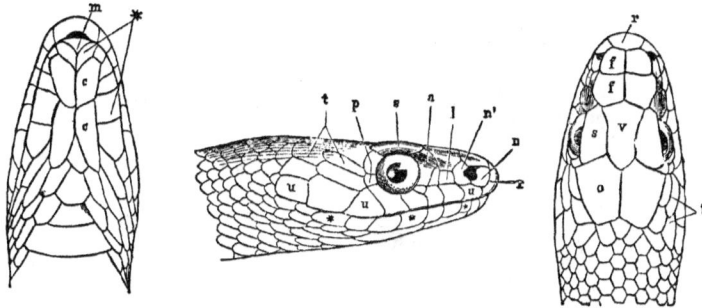

INFERIOR, LATERAL, AND SUPERIOR VIEWS OF THE HEAD OF A SNAKE

r, rostral shield ; *ff*, anterior and posterior frontal shields ; *v*, interparietal shield ; *s*, supraocular shield ; *o*, parietal shield ; *nn'*, nasal shield ; *l*, loreal shield ; *a*, preocular shield ; *p*, postocular shield ; *uu*, upper labial shields ; *tt'*, temporal shields ; *m*, mental shield ; **, lower labial shields ; *cc*, chin-shields

be added that no snake has any trace of a breast-bone, nor any vestige of a pectoral arch, there being no rudiments of either blade-bone, coracoid, or collar-bone.

When progressing on a firm surface, an ordinary snake, in common with the limbless lizards, walks entirely by the aid of its ribs, which are but loosely articulated to the vertebræ, and thus readily admit of a large amount of motion. In describing their mode of progression, Dr. A. Günther remarks that "although the motions of snakes are in general very quick, and may be adapted to every variation of ground over which they move, yet all the varieties of their locomotion are founded on the following simple process. When a part of their body has found some projection of the ground which affords it a point of support, the ribs, alternately of one and the other side, are drawn more closely together, thereby producing alternate bends of the body on the corresponding side. The hinder portion of the body being drawn after, some part of it finds another support on the rough ground or a projection, and the anterior bends being stretched in a straight line, the front part of the body is propelled in consequence. During this peculiar kind of locomotion, the numerous broad shields of the belly are of great advantage, as, by means of the free edges of these shields, they are enabled to catch the smallest projections on the ground, which may be used as points of support. Snakes are not able to move over a perfectly smooth surface."

It may be added that a snake is only able to move by lateral undulations of its body in a horizontal plane, and that those pictures in which these reptiles are depicted as advancing with the folds of the body placed in a vertical plane are altogether erroneous. In conformity with their elongated bodies the internal organs of snakes are long and narrow; and it is remarkable that, as a rule, only one of the lungs is functionally developed, the other being reduced to a useless rudiment.

Resembling the other members of the order to which they belong in that the teeth are never implanted in distinct sockets or grooves, snakes exhibit some considerable degree of variation with regard to the number and structure of their teeth. In the ordinary harmless kinds, such as the English water-snake, there are generally two rows of short, slender, and sharply-pointed teeth in the upper jaw, the innermost of which are attached to the bones of the palate, while the lower jaw carries only a single row of such weapons. One or two of the outer row of upper teeth, either at the front or back of the series, may, however, be enlarged beyond the rest, and grooved or tubular; and it is probable that all snakes with such a dental armature are more or less venomous.

Some of the most deadly poisonous serpents have, on the other hand, a type of dentition of their own, and there is no doubt that all snakes with teeth of this nature are extremely venomous. In such snakes the fore part of the very short maxillary bone of each side of the upper jaw is armed with an elongated tubular tooth, which ordinarily lies nearly flat on the surface of the palate, but can be erected, by means of a peculiar

mechanism of the bones, when the jaws are opened, so as to assume a nearly vertical position. Although in this group the poison-fangs are always tubular, in some of the other venomous serpents they are merely grooved for the conveyance of the venom from the secreting gland; but there is a more or less complete transition between the two types, as the closed tube is formed merely by the edges of the groove being elevated until they unite in the middle line.

On each side of the upper jaw, below and behind the eye, is situated, in poisonous snakes, the poison-gland, which is merely an ultra-development of an ordinary salivary gland; these glands in some cases being so large as to extend far back along the sides of the body. The gland is overlain by a layer of muscles for the purpose of forcing the secretion into the tooth (the base of which is always open) when required; this action automatically taking place when the snake opens its mouth to bite. The poison then flows along the channel or tube to the tooth, and is discharged at its extremity into the wound. Considerable force is used in the emission of the venom, and when a snake is irritated the fluid may be seen to spirt for some distance from its point of discharge.

In some of the less specialised poisonous snakes the venom-tooth, which has an open channel, is not greatly longer than the others, and is placed nearly vertically when the mouth is closed. Although the poison-teeth are commonly regarded as purely defensive weapons, their chief use is for the destruction of the prey of their owners, which is always killed before being swallowed. The venom-tooth of the more specialised poisonous snakes is exceedingly likely to be broken off during use, but to take its place there are always several others lying on the gum behind

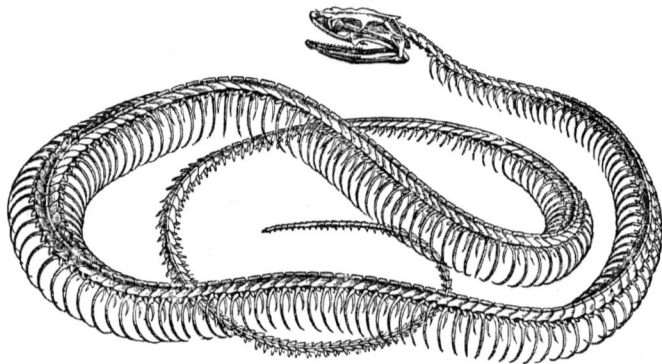

SKELETON OF SNAKE

it in different stages of development.

One African snake has the power of spitting out its venom to a considerable distance; and so powerful is the secretion that, if it happens to strike the eye, severe and painful inflammation ensues.

Before the doctrine of parallelism in development received the attention it merits, snakes were generally divided into harmless and poisonous groups, but since naturalists have become better acquainted with that important factor in evolution, it has been recognised that such a distinction is a purely artificial one, and has nothing to do with real affinity. Certain groups of snakes, such as the members of the viper tribe, may, however, be wholly poisonous, while in other groups, such as the typical snakes, some species may be venomous and others innocuous.

Many attempts have been made to draw up a list of characters by means of which the harmless members of the suborder can be distinguished at a glance from those which are hurtful, but this, as might have been expected, has proved impossible. As Mr. G. A. Boulenger observes, "There is no sure method of distinguishing the two kinds by external characters; except, of course, by a knowledge of the various forms. And even then a cursory examination is not always sufficient, since there is, in some cases, a striking resemblance between snakes of totally different affinities, by which even specialists may at first be deceived." Moreover, it has now been ascertained that the saliva of certain reputedly non-venomous species, like the rat-snake, has poisonous properties.

The mortality due to snake-bite in India and other tropical countries where the natives are in the habit of going about with unprotected ankles is simply appalling, but in the case of Europeans, whose legs are well-protected, death from snake-bite is comparatively rare ; and an English sportsman in India thinks nothing of walking over ground which may be swarming with snakes. In regard to the comparative rarity of death from snake-bite among Europeans in Australia, Messrs. Lucas and Le Soeuff observe that it is not owing to lack of venomous properties in the snakes themselves. On the contrary, in many Australian snakes, " the virulence of the poison itself is actually much greater than that of cobra-venom. But there are several mitigating circumstances. The snakes are, as a rule, more anxious to avoid man than he is to avoid snakes. You may traverse miles of virgin bush in what seems to be good snake-country without seeing one of them. With the exception of the death-adder, they quickly withdraw when they hear your step on the ground or the crackling of the dead sticks as you pass along. It is only in January or February, when they are mating, that the snakes lie about, and are too pre-occupied to make off at your approach.

"In the more populous parts of the continent the snakes are only in evidence during the summer months ; in the winter they are hibernating in retreat. The quantity of venomous fluid ejected by our snakes is nothing like the quantity ejected by the cobra or rattlesnake. The proportion of the toxic principles present in the fluid, too, has been found to vary very greatly, doubtless depending on the condition of the snake, its age and vigour. The poison-glands may be exhausted by previous bites. In cold weather the reptiles are nearly dormant ; while changing the skin they are sluggish. The fangs of our largest venomous snakes are not nearly so long as are those of the rattlesnake and cobra, and consequently do not penetrate so deeply, and have less chance of perforating a vein. The points are solid for a quarter of the length of the fang, and consequently the poison is not conducted to the depth of the full length of the fang."

Till the discovery that an anti-toxin might be made from the venom itself—in regard to which more is stated under the heading of sea-snakes—bites from the more venomous kinds of serpents nearly always produced fatal effects.

Snakes very generally assume threatening attitudes when encountered, and endeavour to frighten enemies by hissing loudly. An idea has been long current that snakes have the power of " fascinating " their intended victims, or reducing them to a condition of helplessness through abject terror, but recent experiments tend to show that the whole story of " fascination " is a popular notion, the great majority of animals taking no notice of the presence of snakes. It is, however, very significant that monkeys and apes form a remarkable exception in this respect, gibbering and trembling with terror whenever snakes are brought near them, and it has accordingly been suggested that the antipathy to snakes which is so deeply implanted in human nature is an inheritance from ape-like ancestors.

Snakes have adapted themselves to almost all kinds

of situations. Many inhabit grass-jungles, swamps, and forests ; some are arboreal ; others are inhabitants of rocky situations or sandy deserts, while certain kinds spend much of the time in fresh-water, and the members of one group are wholly marine. Others, again, burrow like worms. One Malay species, *Coluber tæniurus*, lives in caves, where it preys on bats. The cave-haunting examples of this species are much paler in colour than the typical form, which ranges from China to the Malay Archipelago, and attains a length of $7\frac{1}{2}$ feet. In the cave specimens the top of the head is bluish grey, and a black line about an inch long runs through the eye towards the neck. The neck and back are pale ochery, each scale being tipped with tawny, becoming paler towards the tail ; the middle of the back is yellowish, and the under side pale yellowish white. The tail has a white bar along the upper surface, and the under part is pure white ; along the sides runs a purplish grey bar, becoming darker or black towards the tail. This colouring is suited to the habits of the snake, which frequents the darkest portions of the caves, often living at a considerable distance from the mouth, although it may sometimes be met with at the mouth.

This snake has a habit of resting on the ledges of rocks in the neighbourhood of the mouth of the caves, with the head hanging over the edge, so as to capture the bats as they fly in and out. The walls of the caves, though of white crystalline limestone, are not pure white, but pale ochery yellow, with black veins running down the sides. The colouring of the snake is so exactly that of the walls, the black line on the tail representing the shadow of a crack or projecting vein, that the reptile when at rest on the walls is difficult to see, and readily escapes observation. The snakes are usually found in the darkest parts of the caves, but though they seem dazed and purblind when suddenly brought out into a bright light, they soon regain normal vision. A specimen of a second species, *Coluber moellendorffi*, has been captured at Caïkim, Tonkin, in a limestone cave, were the most complete darkness prevails. Compared with *C. tæniurus*, this species is quite differently coloured, its upper parts being grey with a median series of large dark, black-edged spots, and a lateral series of smaller spots, and the tail encircled by coral-red, black-edged bands ; the lower parts being red, with black spots.

Certain snakes are also stated to take short journeys in the air, the natives of Borneo attributing the power of flight to three species, namely, *Chrosopelea ornata*, *C. chrysochlora*, and *Dendrophis pictus*, and from a note by Mr. R. Shelford it appears that these reptiles really can descend from a height in a manner analogous to that practised by flying-squirrels. In the ventral scales of these snakes there is a suture on each side, and by muscular contraction these scales can be drawn inwards, so that the whole lower surface becomes concave, and the body may be compared to a bamboo split longitudinally. Experiments showed that the snake, when made to fall from a height, descended with the body rigid, and that the line of fall was at an angle from the point of departure to the ground. Mr. Shelford considers it probable that the concave lower surface buoys the reptile up in its fall, since it is well-known that a longitudinally bisected rod of bamboo falls through the air perceptibly slower than an undivided rod of equal weight.

Among snakes even partial albinism is rare, while

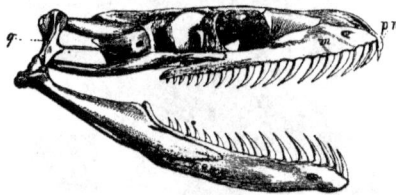

SKULL OF A PYTHON

m, maxillary ; *pm*, premaxillary ; *q*, quadrate-bone

wholly white individuals appear to be unknown. A semi-albino of the ringed snake has, however, been described, which was obtained in the neighbourhood of Berlin, and is preserved in the museum of that city. The ground-colour of the head and back is yellowish white, passing into bluish white on some parts of the region of the forehead. On the back are a number of brownish red spots, some of which are united, and where they are thickly approximated tend to form longitudinal and oblique lines. There is also a pale yellowish neck-band, with a blackish brown band behind it, while the lip-shields show a yellowish ground-colour, with a blackish grey border. In the eye the iris presents a number of yellowish brown pigment-spots, and the pupil is mainly dark red. Very noteworthy is the fact that the tongue is white, instead of the normal black. It may be added that the Berlin Museum also possesses a ringed snake from North-Western Germany in which the general ground-colour of the upper surface is light grey, while the network-pattern seen in embryos is indicated by blackish lines.

Geologically speaking, snakes are a comparatively modern group, being scarcely known below the lowest portion of the Eocene division of the Tertiary period, although one or two kinds have been described from the underlying Cretaceous rocks, and one species has been recorded from the Gault of Portugal—a formation underlying the Chalk. A very large python (*Gigantophis garstini*) occurs in the middle Eocene deposits of Egypt, and remains of big sea-snakes (*Palæophis*) are found in the lower Eocene London Clay.

It is noteworthy that one of the North American lower Eocene snakes has the additional articular facets of the vertebræ very imperfectly developed; and there can be little doubt that the whole group is an offshoot from the lizards. From the beginning of the Tertiary period, the group seems to have gone on steadily increasing in numbers, and it is now represented by some fifteen hundred species, ranging all over the world except New Zealand.

Snakes are, however, much more abundant in the moist tropical regions of the globe than in colder regions, and it is there only that they attain their maximum development in point of size. India and the Malay countries, where there are representatives of the whole of the nine families into which the suborder is divided, are the home of a greater number of both genera and species of snakes than any other part of the world. Tree-snakes are very common in this region, while the gigantic pythons are shared by it in common with Africa.

The proportion of poisonous to innocuous species is likewise very high in the Indo-Malay region, and has been estimated at about one in ten. Africa has scarcely half the number of snakes found in the Indian region; and it is noteworthy that the species inhabiting Madagascar have little in common with those of the mainland; the so-called lycodonts, which are so common in Africa, being unknown in Madagascar, while some of the species from that island are closely allied to South American types. Whereas pit-vipers are absent, an especial feature of Africa is the number of typical vipers which inhabit that country; and Australia, which differs so remarkably from India in tortoises, possesses snakes (and likewise lizards) closely allied to Indian forms.

Next to the Indian region, Tropical America is richest in ophidians, although the number of generic types is not so great. The proportion of poisonous species is, however, high, and has been estimated at as much as one in eight. In Southern Argentina and Patagonia snakes become scarce. Unlike the chelonians, the snakes of North America present a resemblance to those of Central America. Indeed, a feature of the whole of America is the absence of typical vipers, and the abundance of pit-vipers, although several genera of the latter are common to Asia. Europe and Northern Asia are comparatively poor in snakes, but (next to Africa) are characterised by the number of typical vipers and colubrine water-snakes.

Although a few members of the suborder subsist on eggs, snakes, as a rule, capture and devour living animals, which are in all cases swallowed whole, as these reptiles have no apparatus for rending or masticating their food. And it is in order that they may be able to swallow larger animals than would otherwise be possible that they have the power of dilating their jaws in the manner already indicated. Not only can the jaws be thus enlarged, but the throat and stomach are likewise capable of dilatation, owing to the circumstance that the lower ends of the ribs, from the absence of a breast-bone, are quite free; and in swallowing a snake seems gradually to draw itself over the object to be devoured.

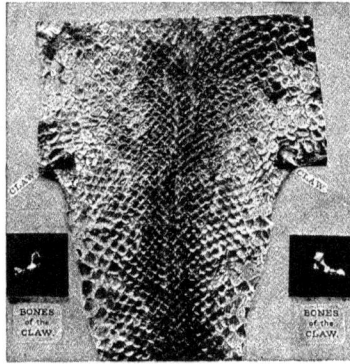

PART OF FLATTENED SKIN OF AN AFRICAN PYTHON
Showing claws representing hind limbs, together with their supporting bones

The majority of snakes devour their prey alive, and a frog may be seen struggling in the stomach of a common English ring-snake long after it has been swallowed. Other snakes, however, kill their prey either by striking it with their poison-teeth, after the manner of vipers, or by encircling and smothering it in the folds of the body, like boas. Although the process of digestion is very rapid, snakes feed but seldom; and it has been asserted that two or three frogs are sufficient to supply the needs of the European ring-snake for a whole year. All snakes drink much, water being absolutely essential to their existence.

As might have been expected from their numbers, snakes exhibit great diversity in their modes of life; and while those of the tropical regions remain active throughout the year (unless they lie by during periods of drought), the species inhabiting colder regions hibernate during the winter. The most remarkable diversity from the ordinary mode of serpent life is displayed by the blind-snakes, which lead a completely subterranean existence, very seldom making their appearance above the surface.

The great majority of serpents are terrestrial in their habits, seldom entering the water or climbing trees; and these ground-snakes, as they may be called, are characterised by their cylindrical form and the width of the shields on the inferior surface of the body. Tree-snakes, on the other hand, which are mostly remarkable for their brilliant colouring, lead an almost completely arboreal life. Frequently they have the body very slender, or the shields on its under surface may be keeled in order to afford a firmer hold in climbing, while in other instances the tail is prehensile. It is among this group that the egg-eating species are found.

The fresh-water snakes swim and dive with facility in the waters of rivers and lakes, where they spend a large portion of their time, feeding on such aquatic creatures as they can capture therein. As a rule, these snakes are distinguished by having the nostrils placed at the top of the muzzle, and likewise by the tapering form of the tail. Finally, the sea-snakes, which have the nostrils situated as in the last group, are distinguished by the lateral compression of their tails. In all cases extremely poisonous, these snakes are almost entirely pelagic in their mode of life, seldom approaching land, and generally dying if washed ashore, although in one genus the shields on the under surface of the body are sufficiently developed to admit of terrestrial progression.

By far the great majority of the members of the suborder lay eggs, of an oblong form enclosed in soft leathery shells, which are hatched by the natural heat of the places where they are deposited. Pythons, however, incubate their eggs, which are spherical, although it does not appear that at such periods a temperature perceptibly above the normal is developed. On the other hand, both in the fresh-water and sea snakes, the eggs are retained within the body of the mother until hatched.

TABLE OF LIVING OPHIDIA MENTIONED IN THIS WORK

BLIND-SNAKES

THE blind-snakes, which are arranged under two families (*Typhlopidæ* and *Glauconiidæ*), are small, worm-like reptiles, with cylindrical bodies and short heads and tails, entirely adapted for a subterranean burrowing life. Lacking the large inferior transverse shields characterising ordinary snakes, blind-snakes have the body and tail covered on all sides with round, overlapping scales of equal size on both the upper and lower surfaces ; and there are large shields on the fore part of the head, one of which on each side covers the rudimentary eye. The cleft of the mouth, which is very small, is placed on the lower surface of the head, and the jaws admit of scarcely any dilatation.

An important point of difference from all other members of the suborder is that teeth are absent in either the upper or the lower jaw, and in all cases larger or smaller vestiges of the pelvis remain. The most important distinction is, however, to be found in the palate of the skull, which differs from that of all other snakes in lacking the so-called transverse or transpalatine bone, which connects the pterygoid, or hindmost bone of the palate, with the posterior extremity of the jawbone, or maxilla. In the first, or typical, family of blind-snakes, the upper jaw, which is but loosely attached to the rest of the skull, is furnished with teeth, while the lower jaw is toothless ; the pelvis being represented merely by a single bone on each side.

On the other hand, in the second family (*Glauconiidæ*), while the upper jaw is devoid of teeth, there are a few teeth in the lower one, the pelvis being represented by a pair of bones on each side, of which the two anterior ones meet in the middle line. As regards their origin, it seems probable that blind-snakes have little or no near relationship with any other members of the suborder to which they belong.

The typical blind-snakes, or those belonging to the family *Typhlopidæ*, are inhabitants of all the warmer regions of the globe, and represented by nearly a hundred species arranged under three genera. By far the greater number of these species belong to the typical genus *Typhlops*, which has a distribution coextensive with that of the family ; the other two genera—namely, *Helminthophis*, with five species, and *Typhlophis*, with one—being confined to Central and South America. The second family contains only the single genus *Glauconia*, of which there are about thirty species, inhabiting America, Africa, and South-Western Asia.

BLIND-SNAKE

Very little has been recorded in regard to the habits of these curious snakes, although it is ascertained that they lay eggs, which are few in number, large in size, and elongate in form. Although they generally remain in their subterranean burrows, in showery weather these snakes not infrequently come to the surface for a short time. The remains taken from their stomachs show that they feed largely upon millipedes and ants, and they probably also consume the larvæ of many insects. Captive specimens have been observed to drink freely. The European blind-snake (*Typhlops vermicularis*) is an inhabitant of Greece and several of the adjacent islands, Asia Minor, Syria, Arabia, Petræa, and the Caucasus as far as Transcaspia.

PYTHONS AND BOAS

INCLUDING the largest of living and extinct snakes, the family *Boidæ* is regarded as the most generalised of the entire suborder (exclusive of the blind-snakes), all the others presenting such characters as would admit of their having taken origin from ancestral types belonging to the one under consideration. In common with the remaining families, pythons and boas differ from the blind-snakes in that both jaws are fully toothed, and likewise in the presence of a transverse bone to the palate.

The characters specially distinguishing the present from the other families of the suborder are largely derived from the structure of the skull, and therefore require some degree of anatomical knowledge for their proper appreciation, and cannot be described without the use of a considerable number of technical terms. It may be mentioned, however, that the lower jaw has on the inner side of each branch a thin bone known as the coronoid ; while on the top of the skull the pre-frontal bones, which lie on the outer side of the fore part of the frontals, articulate with the nasal bones, or those roofing the front of the cavity of the nose. In the hind part of each side of the skull lies a large bone termed the supra-temporal, from which is suspended the quadrate bone for the articulation of the lower jaw ;

while a further important characteristic is to be found in the presence of vestiges of the pelvis and hind limbs, the latter usually taking the form of a claw-like spur situated on either side of the vent (p. 1550-1).

The family, which contains a very large number of genera and species, has an extensive geographical distribution, being represented in South-Eastern Europe, Central and Southern Asia, Africa, Australia, the West Indies, western North America, and Central and South America ; it is thus essentially characteristic of the warmer regions of the globe. Pythons belonging to extinct genera lived on the Continent and in England during the earlier part of the Tertiary period.

TYPICAL PYTHONS

The large snakes to which the term python properly belongs are the typical representatives of the first of the two subfamilies into which the *Boidæ* are divided ; the essential feature of this subfamily (*Pythoninæ*) being the presence on the upper aspect of the skull of a supraorbital bone lying on each side of the frontal bones, and forming the upper border of the socket of the eye.

Agreeing with three other less important genera in the presence of teeth in the premaxillæ, or anterior upper jawbones, and also in generally having two rows of shields on the under surface of the tail, pythons are specially characterised by the distinctly prehensile tail, and likewise by the presence of deep pits in the rostral and anterior upper labial shields of the head.

As minor characteristics, it may be mentioned that the teeth, none of which is grooved, gradually decrease in size from the front to the back of the jaws, and that the eye is of moderate size, with a vertical pupil. The head, which is distinct from the neck, has the extremity of the snout covered with large shields, and its hind portion may be overlain either with symmetrical shields or with small scales. Each nostril is placed in a half-divided nasal shield, separated from its fellow on the opposite side by a pair of internasal shields. The body in these snakes is more or less compressed, the scales on the upper surface and sides are small and smooth, and the prehensile tail is of moderate length, or short, with the whole or greater part of the inferior shields arranged in two rows.

Pythons, or, as they are frequently termed, rock-snakes, are represented by about half a score of species, and range over Tropical and Southern Africa, South-Eastern Asia, and Australasia. With the exception of the American anaconda, some of the pythons are the largest of all snakes, and it is now ascertained that the Malay python (*Python reticulatus*), although generally much smaller, occasionally attains a length of 30 feet, while the South African python (*P. sebæ*) is stated to reach 23 feet. It is, however, seldom that pythons of more than from 15 to 20 feet in length are met with ; but even these are sufficiently formidable creatures, since they have a circumference as large as a

man's thigh, and can easily kill such animals as small deer, full-grown sheep, and dogs of considerable size. They appear, however, to be unable to devour animals of larger dimensions than a half-grown sheep. Very much larger dimensions have recently been assigned to some members of the group.

A python destroys its victim in much the same manner as do many of the smaller snakes, gradually smothering it by throwing over it coil after coil of its body. In swallowing, writes Dr. A. Günther, pythons " always begin with the head, and as they live entirely on mammals and birds, the hairs and feathers offer a considerable impediment to the passage down the throat. The process of deglutition is, therefore, slow, but it would be much slower except for the great quantity of saliva discharged over the body of the victim.

"During the time of digestion, especially when the prey has been a somewhat large animal, the snake becomes very lazy ; it moves itself slowly when disturbed, or defends itself with little vigour when attacked. At any other time the rock-snakes will fiercely defend themselves when they perceive that no retreat is left to them. Although individuals kept in captivity become tamer, the apparent tameness of specimens brought to Europe is much more a state of torpidity caused by the climate than an actual alteration of their naturally fierce temper."

In their general habits snakes of this genus are nocturnal, and they generally live on or among trees in the neighbourhood of water, frequently swimming in the water. The handsomely marked Malay python (*P. reticulatus*), of Burma and the Malay Archipelago, not infrequently takes up its abode in buildings, whence it issues forth at night to capture such prey as it can find.

It had long been reported by travellers in India that pythons incubated their large spherical eggs, which

THE DIAMOND-SNAKE

AFRICAN PYTHON SWALLOWING A BIRD

or these subgroups is the Australian diamond-snake, or carpet-snake (*P. spilotes*), characterised by the crown of the head being covered with scales or small irregular shields, and the presence of pits on two or three of the upper labial shields of the snout. This snake, formerly referred to a genus apart (*Morelia*), is an inhabitant of New Guinea and Australia, and is of comparatively small size, attaining a total length of only about 6½ feet; its colour being extremely variable. The variety in which the skin is most spotted was long regarded as a distinct species, under the name of the carpet-snake.

The other two members of this group are the amethystine python (*P. amethystinus*) and the Timor python (*P. timorensis*), both distinguished by the presence of large symmetrical shields on the crown of the head, and by four upper labial shields being pitted. The former, which grows to a length of about 11 feet, ranges from the Moluccas and Timor to New Guinea, New Ireland, New Britain, and the north of Queensland, while the latter is restricted to the islands of Timor and Flores.

The second subgroup, comprising the species with from sixty-one to ninety-three scales round the body, includes the Malay python (*P. reticulatus*), with from sixty-nine to seventy-nine scales in a row, and four upper labials with pits. This species, which ranges from Burma and the Nicobar Islands to the Malay region and Siam, is one of the largest of the genus, occasionally reaching 30 feet in length. In colour it is light yellowish or brown above, ornamented with large circular rhomboidal or X-shaped dark markings, while the head has a median blac kline, and the under parts are yellowish, with small brown spots on the sides. It is, however, subject to considerable variation, a specimen from Siam, formerly in the London Zoological Gardens, showing bright yellow lines on the sides. Young specimens show three longitudinal rows of light spots with black edges along the back.

Somewhat smaller is the South African python (*P. sebæ*), of Tropical and Southern Africa, which attains a length of 23 feet, and has from eighty-one to ninety-three scales in a row on the thickest part of the body, and only two of the labial shields pitted. This species occurs typically in Western Africa, and it was long considered that the South African python, or Natal rock-snake, was distinct. Its colour is pale brown above, with dark brown, black-edged, and more or less wavy cross-bars, usually connected by an interrupted or continuous dark stripe running along each side of the back; while the sides are marked with large black spots and small dots. On the top of the head is a large triangular dark brown blotch, bordered on each side by a light stripe, beginning above the nostril at the end of the muzzle and passing above the eye; and there is a dark stripe on each side of the head, and a somewhat triangular blotch beneath each eye. The upper surface of the tail has a longitudinal

may be compared in size to lawn-tennis balls, and although such reports were received with incredulity, their truth was established in 1841, when a female African python in the Jardin des Plantes, Paris, laid fifteen eggs on May 6, which she subsequently proceeded to incubate. When first laid, the eggs, which were completely separate, were soft, oval, and ashy grey, but they soon assumed a rounder form and a clear white tint, at the same time hardening. The parent collected them into a cone-shaped pile, around which she rolled herself in such a manner as to conceal the whole number, with her head forming the summit of the cone.

For upwards of six-and-fifty days this position was maintained without movement, except when persons attempted to touch the eggs. On July 2 the shell of one of the eggs split, revealing a fully-formed baby python within, and on the next day the little reptile came forth into the world. During the four succeeding days eight more snakes made their appearance, but the rest of the eggs were spoilt. In from ten days to a fortnight the young pythons changed their skins, after which they caught and devoured some live sparrows, seizing and smothering them in the manner in which full-grown individuals destroy prey of larger size. A second instance of a python brooding its eggs has been subsequently observed.

Pythons may be divided into two groups, according as to whether the number of pairs of shields on the lower surface of the tail exceeds or falls short of fifty; the members of the first group being further divisible into two sections, according as to whether the number of scales in a row round the thickest part of the body varies from thirty-nine to sixty, or from sixty-one to ninety-three. The first representative of the former

light stripe, bordered on each side by a dark one; and the under parts are spotted and dotted with dark brown.

In India, Ceylon, the south of China, the Malay Peninsula, and Java the last-named species is replaced by the Indian python (*P. molurus*), which agrees with *P. sebæ* in having only two of the labial shields pitted, but differs in that there are from sixty-one to seventy-five scales in a row, and likewise in that the rostral shield is broader than long, instead of with these two diameters equal. In colour this python is greyish brown or yellowish above, with a series of large, elongated, squared, reddish brown, black-edged spots down the middle of the back, flanked by a series of smaller ones.

The head and nape of the neck show a characteristic spear-shaped brown mark, a brown band runs on each side of the head

INDIAN PYTHON CRUSHING ITS PREY

through the eye, and there is a vertical one of this colour beneath the latter. The under parts are yellowish, with the sides spotted with brown. Known in India by the name of adjiga, this python ranges through peninsular India, Rajputana, and Bengal to the foot of the Himalaya, and is not uncommon, although in Ceylon, the Malay Peninsula, and Java it is rare. It does not commonly exceed about 12 feet in length.

Three other species of the genus form the second main group, in which there are less than fifty pairs of shields on the lower surface of the tail; the number of shields in a row at the thickest part of the body varying from fifty-three to sixty-three, and none of the species being of very large size. The best known of the three is the royal python (*P. regius*), of Senegambia and Sierra Leone; the other two being the rare Anchieta's python (*P. anchietæ*), of Benguela, and the Sumatran python (*P. curtus*), the latter easily recognisable by the large amount of red in its colouring.

OTHER PYTHONS

The *Pythoninæ* are represented by six other genera, which demand merely a brief reference; the first three of these agreeing with the typical genus in the presence of teeth in the premaxillary bones, while in the remainder that portion of the upper jaw is toothless. From the typical pythons the first three genera may be distinguished by the tail being but very slightly, if at all, prehensile, and by the rostral shield of the head being either devoid of pits, or with only very shallow ones.

The first genus, *Loxocemus*, as represented by a single comparatively small Mexican species, *L. bicolor*, has no pits in the labial shields, no loreal shield, and the nostril situated in a single nasal shield. *Nardoa boa*, of New Ireland, alone represents the second genus, and may be distinguished by the presence of pits in the lower labial shields, and by the laterally-placed

nostril being situated between two nasal shields. On the other hand, the third genus, *Liasis*, is represented by several species, ranging from Flores and Timor to Papua and the north of Australia, and may be distinguished from the second by the nostril being placed more superiorly in a half-divided nasal shield. Finally, three genera in which the anterior jawbones, or premaxillæ, are toothless are *Chondropython*, with one Papuan species, *Aspidites*, represented by two species from the north of Australia, and *Calabaria*, with a single West African representative.

The interest attaching to these snakes is the connection which they form between the pythons and the boas. Thus while the two first differ from typical pythons and resemble boas in the presence of teeth on the palate, the second and third likewise agree with the latter in having the shields on the lower surface of the tail mostly or entirely single; the tail itself being but slightly, if at all, prehensile.

TREE-BOAS

The tree-boas of Tropical America may be taken as the first examples of the second subfamily, *Boinæ*, of the assemblage of snakes under consideration. The members of this subfamily are distinguished from the preceding group solely by the absence of a supraorbital bone on the upper surface of the skull above the socket of each eye. They further differ from all the pythons, with the exception of two of the three genera last mentioned, in having teeth on the palate; and, with the exception of the three connecting genera, in the absence of teeth in the anterior upper jawbones, or premaxillæ. The boas and their allies further differ from the typical pythons in having the shields on the lower surface of the tail for the most part single, thereby agreeing with the genera *Aspidites* and *Calabaria*; and thus showing that the small group to which the two latter belong forms such a close connection between pythons and boas as to preclude their reference to separate families.

In common with the majority of the thirteen genera into which the subfamily is divided, tree-boas are characterised by having the head distinctly defined from the neck, and the tail more or less prehensile. They are specially distinguished by the anterior teeth being much larger than the hind ones, by the smooth scales of the body, the presence of shields on the head, and by the labial shields being either devoid of pits or with shallow ones only. In form the body is more or less compressed, and the tail either moderate or long; the eye is of medium size with a vertical pupil, and the shields on the head may be either small and irregular or large and symmetrical.

These snakes are represented by about nine species, the largest of which is the pale-headed tree-boa (*Epicrates angulifer*) of Cuba, attaining a length of about 7 feet; another well-known species being the streaked tree-boa (*E. striatus*), from San Domingo and the Bahamas. A third kind is the thick-necked tree-boa (*E. cenchris*), ranging from Costa Rica to the northern districts of Peru and Brazil. The streaked species, which attains a length of about 5 feet, is either pale brown above with dark olive brown spots separated by narrow intervals from one another, or brown with wavy or zigzag yellowish cross-bands, not infrequently margined with blackish brown. Each side of the head usually has a more or less distinct streak behind the eye; while the under parts are pale olive or yellowish, more or less spotted with brown or black.

Closely allied to the last, the five species of the genus

DOG-HEADED TREE-BOA

Corallus are distinguished by having deep pits in the labial shields of both the upper and lower lips. The body is compressed, with small smooth scales, and the prehensile tail is either short or more or less elongated. The members of this genus have a somewhat remarkable geographical distribution, four of them being inhabitants of Tropical America, while the fifth (*C. madagascariensis*), which is distinguished from the rest by the shortness of its tail, is restricted to Madagascar. The dog-headed tree-boa (*C. caninus*), a native of the Guianas and Brazil, which usually attains a length of some 5 feet, although it may be considerably larger, belongs to a group of two American species distinguished from the other kinds inhabiting the same countries by the relatively shorter tail, with only from sixty-four to eighty-two shields on its inferior surface; whereas in the true tree-boa (*C. hortulanus*) and another species there are at least a hundred of these shields.

The dog-headed species is specially characterised by the scales being arranged in sixty-one or seventy-one rows, and by the number of shields on the under surface of the body ranging from one hundred and eighty-two to two hundred and nineteen,

STREAKED TREE-BOA

while those on the tail vary from sixty-four to seventy-nine. In colour this snake is decidedly handsome, the upper parts of the adult being bright green, ornamented with irregular spots and cross-bars of white, and the under parts bright yellow. In the young the ground-colour is yellowish, and the white markings are edged with dark green or purplish black.

Most abundant in the neighbourhood of the Amazons, this species becomes more rare in Guiana, while southwards it likewise diminishes in numbers in lower Brazil. Feeding principally upon birds, the dog-headed boa is an excellent swimmer, and has been observed swimming both in the Rio Negro and in the salt water of the harbour of Rio de Janeiro. Although it frequently visits the huts of the Brazilian negroes in search of prey, it does not appear that this snake ever voluntarily attacks human beings. If, however, it be driven to bay and unable to escape, it is capable of inflicting very severe bites with its long front teeth, such wounds being difficult to heal.

A third genus of tree-boas (*Enygrus*) is distinguished from both the preceding by the scales having distinct keels; the labial shields of the head being devoid of pits, and the tail short and prehensile, with a single row of shields on its inferior surface. This genus is represented by four species, inhabiting the Moluccas, the Papuan region, and Polynesia.

THE ANACONDA

The gigantic anaconda (*Eunectes murinus*), the sole representative of its genus, is the typical member of a group of several genera distinguished from the tree-boas by the teeth gradually decreasing in size from the front to the back of the jaws without any marked enlargement of those in the fore part. Merely mentioning the allied Tropical American genera *Trachyboa, Ungalia,* and *Ungaliophis,* the first and last of which are each represented only by a single species, it may be observed that the anaconda is specially distinguished as a genus by the large size of the rostral shield of the head, behind which one pair of the nasals come in contact with one another in the middle line, and also by the very small size of the smooth scales of the body. The head is markedly distinct from the neck; the nostrils are directed upwards and placed between three pairs of nasal shields, of which the hindmost are those which meet in the middle line; the small eye has the pupil vertical, the body is cylindrical, and the tail is short and slightly prehensile, with a single row of shields inferiorly.

In colour the anaconda is greyish brown or olive above, with either one or two series of large blackish transverse spots, and a single or double row of lateral eye-like spots having whitish centres and blackish rims. The upper part of the head is dark, and divided by a black streak terminating in a point on the muzzle from the lighter cheeks; another oblique black streak runs on each side behind the eye, and the under parts are whitish with blackish spots.

The anaconda, which is a native of the Guianas, Brazil, and North-Eastern Peru, is essentially an inhabitant of the tropical forest region. That it is the largest of all living snakes there can be little doubt, but the precise limits of size to which it may occasionally attain cannot be ascertained. A stuffed example in the British Museum has a total length of 29 feet, and the species is commonly stated to reach 33 feet; while, if recent reports are to be trusted, individuals of much larger size are occasionally met with. Although many naturalists are indisposed to credit the existence of monsters of 40 feet or more, it is difficult to share their incredulity, as it is improbable that the largest specimens have come under European observation.

From all accounts it appears that the anaconda generally spends much more of its time in water

THE ANACONDA AT HOME

than on land, frequently floating down rivers with the current, and at other times lurking in quiet pools with only its head raised above the surface. In such situations, or resting on rocks, stranded tree-trunks, or sandbanks, it lies in wait for its prey. It, however, frequently leaves the water to pass a longer or shorter period on shore, when it may be found either in trees, among rocks, or even on hot sand; and it appears that when in a tree this snake will often dart down its head from a considerable height to seize a passing peccary or other animal. Mr. Bates mentions that the anaconda will occasionally seize human beings, and this statement is fully confirmed by other observers.

AGAMA, HORNED VIPER, AND SACRED BEETLES

" When about to attack, the horned viper moves rapidly forward with a sideways motion, unlike that of any other serpent; and as it will attack when quite unprovoked, the horned viper is more dreaded than any other North African snake, men frequently dying from its bite within half an hour."

P. J. Smit.

P. J. Smit

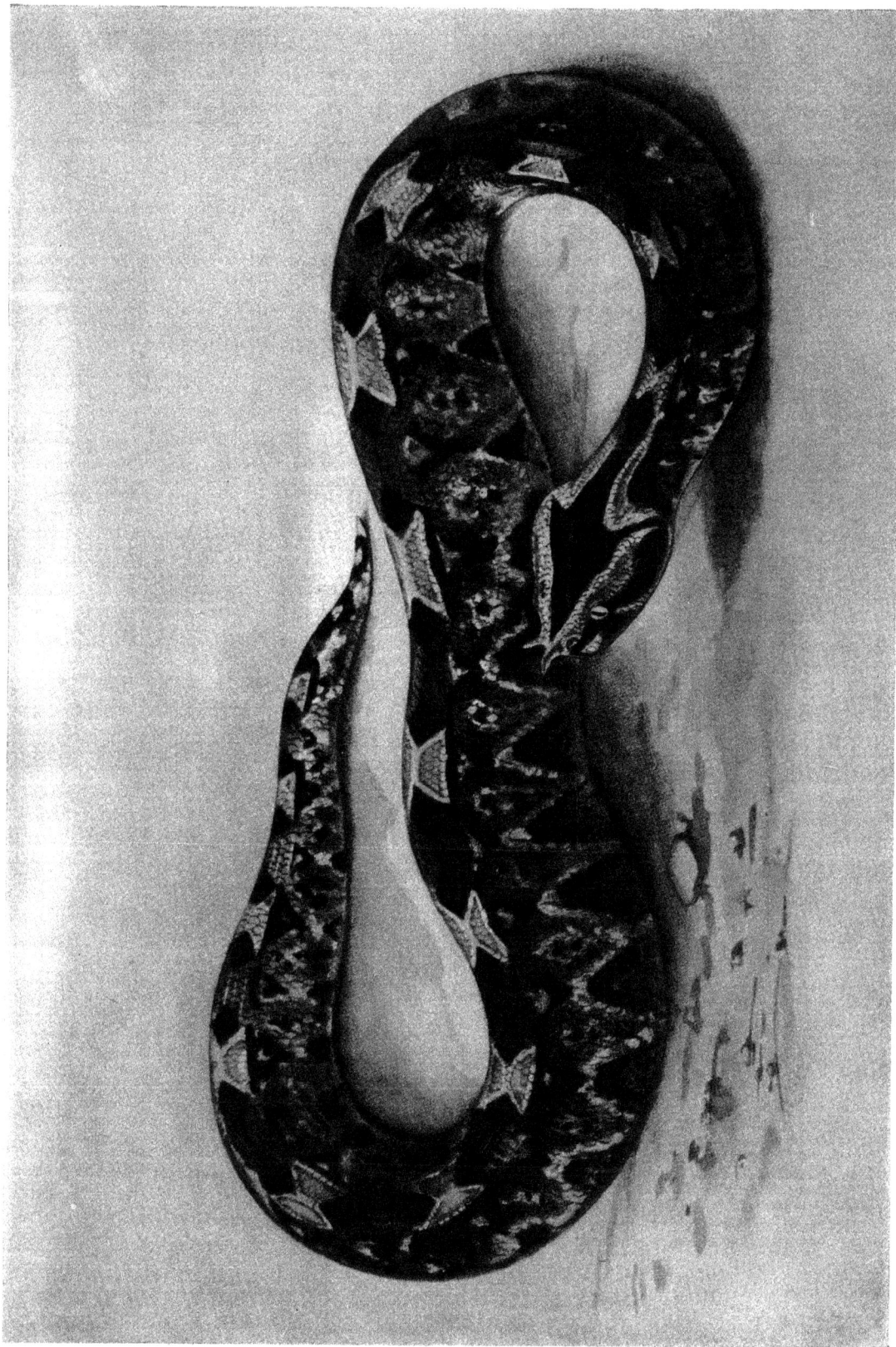

NOSE-HORNED VIPER

"The young of the Gabun viper are perfect living gems in the matter of colouring. Even more gorgeous in this respect is the West African nose-horned viper, especially just after changing its skin."

In Brazil, where water is abundant throughout the year, the anaconda is active at all seasons, although it is stated to display the most activity during the hot months of December, January, and February.

In other districts, however, during the dry season, it is in the habit of burying itself deep in the mud of the dried-up rivers, where it is sometimes disinterred by the natives in a torpid condition. Very little is known with regard to the breeding habits of the anaconda, but as females have several times been killed containing eggs with embryos far advanced inside them, it would seem that the young are born alive. When they first make their appearance in the world, the young are reported to take to the water, although they soon leave it to pass a large portion of their time in trees (page 18).

BOAS

Long supposed to be exclusively a Tropical and South American group, the boas (genus *Boa*) are now known to be common to the hotter regions of America and Madagascar. From the anaconda, typical boas may be distinguished by all the pairs of nasal shields being separated from one another in the middle line by small scales. The body may be either cylindrical or slightly compressed, and the short and more or less prehensile tail may have either the whole or a portion of the shields on its lower surface arranged in a single series. In America the genus is represented by five species, two of which range so far south as the inland districts of Upper Argentina.

All the species are characterised by having the loreal region of the head covered either with a single small shield or with small scales, and by the number of rows of shields on the under surface of the tail ranging from forty-five to sixty-nine. On the other hand, in the Malagasy boas (*B. madagascariensis* and *B. dumerili*) there are several shields on this particular region of the head, and the number of rows of shields beneath the tail is only from twenty to forty-one.

The best-known representative of the genus is the typical boa, or boa-constrictor (*B. constrictor*), which ranges in South America from Venezuela to Upper Argentina. At times reaching as much as 12 feet in length, it has the muzzle slightly prominent in the adult, although obliquely truncated in the immature state. In general colour it is pale brown on the upper parts, with from fifteen to twenty dark brown cross-bars, which expand inferiorly, sometimes to such an extent as to become connected on the sides of the body, and thus to surround oval or elliptical spots of the light ground-colour; the expanded portion of each bar having a light longitudinal line.

On the sides is a series of large light-centred dark brown spots, most of which alternate with the cross-

THE BOA

bars; and on the tail all the markings become relatively larger, of a brick-red colour, margined with black, and separated by yellowish intervals. From the muzzle to the nape runs a dark brown median streak, widening posteriorly, where it may be looped; another bar of the same colour passes on each side of the head through the eye, there is a third bar below the latter, and the lips are marked by short bars; the rostral shield of the snout being also ornamented with a crescentic blackish mark. The under parts are yellowish, with spots and dots, or merely dots, of black. The whole tone of colouring is dull, sombre, and adapted to harmonise with the shades of brown, black, and yellow on the bark of tropical forest trees.

Could we see the boa during the night in the depths of its native forests—at which time alone it is thoroughly active—we should doubtless obtain a very different idea of the creature than that gathered from an inspection in the daytime of the lethargic specimens in menageries. Lying coiled on the branch of some large tree, with its head projecting ready to be darted on its prey with the rapidity of lightning, the boa is generally unobserved by the passing traveller, unless it happens to make a dart at an unfortunate dog belonging to his party. Feeding generally on such mammals as agutis, pacas, rats, and mice, which are destroyed in the manner which has given rise to its trivial name, the boa, when it attains unusually large dimensions, is also capable of killing deer and large dogs; while it is always ready for such birds as it can capture, and does not disdain, when in captivity, a meal of eggs. The stories of its killing adult human beings and horses are, however, mere fabrications.

Nothing seems to be known of the breeding habits of this snake and its kindred in a wild state; but from observations made on specimens in captivity, it appears that the eggs are generally hatched within the body of the parent, although one instance is on record where young and eggs were produced simultaneously. To European palates, snakes would probably be highly unacceptable as food, however temptingly they might be dressed; but in eastern South America the flesh of the boa-constrictor is regarded as a most dainty dish, while its fat is reputed to be highly efficacious in the healing of various diseases. The skin is used to ornament saddles and bridles, and for other decorative purposes. None of the other members of the genus attains dimensions equal to those of the boa-constrictor, the Malagasy species being the smallest of all.

KEEL-SCALED BOA

The last representative of the section of the subfamily in which the head is well defined from the neck, and the tail more or less prehensile, is the keel-scaled boa (*Casarea dussumieri*), of Round Island, near Mauritius, distinguished as a genus by the keeling of the scales, and the long tail ; its other general characters being similar to those of the typical boas, except that the nasal shields of the head are separated by a pair of internasals. This snake, which attains a length of about 4 feet, and has a prominent and obliquely truncated muzzle, is either uniform pale brown above, or brown with two dark stripes and a lateral series of small spots down the body, a dark streak on each side of the head through the eye, and the under parts either plain yellow or yellow spotted with black, the under side of the tail always having such spots.

SAND-SNAKES

The snakes of the genus *Eryx*, together with those of three allied genera, which constitute the remaining members of the family, may be distinguished at a glance from the boas and their allies by the gradual passage of the head into the body without any constriction at the neck, and are further characterised by the tail being, at most, only slightly prehensile. From their allies sand-snakes are distinguished by the small scales being either smooth or singly keeled, and by the head being covered with small shields, of which the rostral is enlarged. The eye is small, and sometimes minute, with a vertical pupil, the body cylindrical, and the very short tail, which is frequently without any power of prehension, has a single row of shields on its lower surface. These snakes are represented by seven species, with a geographical distribution including Northern and Eastern Africa and Southern and Central Asia, as well as a part of the extreme south-west of Europe.

The best-known species is the Egyptian sand-snake (*Eryx jaculus*), which has a length of about 2 feet, and is an inhabitant of the Ionian Islands, Greece, South-Western and Central Asia, and the north of Africa. In colour it is very variable, the upper parts being in some examples pale greyish, reddish, or yellowish brown, ornamented either with dark brown or blackish transverse blotches or alternating spots, although in other cases the general colour is brown with pale spots. A dark streak runs from each eye to the angle of the mouth, the under parts are either uniform white, or white with blackish dots, and there is a more or less distinct dark streak along each side of the tail.

This species is exceeded in size by the Indian sand-snake (*E. johni*), which attains a length of over a yard, and inhabits the plains of North-Western, Central, and Southern India. This snake is generally banded, but the young may be of a uniform pale coral-red colour. Although resembling boas in being nocturnal, sand-snakes are quite different in their mode of life, inhabiting open sandy plains, and feeding on small

mammals, lizards, and worms. In search of their prey, they frequently enter holes and crevices among rocks, and they will also burrow in the sand. They are perfectly harmless, and generally make no attempt to bite, but are somewhat unsatisfactory reptiles in captivity, owing to their habit of lying concealed among the gravel of their cages, where, however, they illustrate the perfect harmony existing between their colouring and their desert surroundings.

The Indian species is frequently carried about by snake-charmers, who are in the habit of mutilating the short tail so as to make it look like a head, from which arises the legend of two-headed snakes. A second Indian species, *E. conicus*, was formerly referred to a separate genus (*Gongylophis*), on account of having a series of keeled scales between the eyes.

Of the remaining members of the family, *Lichanura*, with one Californian species, differs from the sand-snakes by the smaller size of the rostral shield of the head, which is longer than wide ; while *Charina*, which is likewise Californian, has the head covered with large shields. On the other hand, *Bolieria*, as represented by a single species from Round Island, near Mauritius, differs from all the other members of the group in having three or four keels on the scales, the muzzle being covered with large shields.

EXTINCT SNAKES

One of the largest of all snakes is an extinct python (*Gigantophis garstini*), from the middle Eocene formation of the Fayum district of Egypt, which appears to have attained a length of about 30 or 40 feet. In the Lower Eocene London Clay of Europe occur remains of smaller but still gigantic snakes, which have been described under

EGYPTIAN SAND-SNAKE

the name of *Palæophis*, and are represented by closely allied, if not generically identical forms in the corresponding strata of North America. In these snakes the vertebræ differ from those of pythons by the much greater height of the upper or neural spine, which has not the backwardly-directed process at its summit characterising modern pythons. From the shape of these vertebræ it is evident that these snakes had compressed bodies like modern sea-snakes, while from the nature of the deposits in which their remains occur, there can be little doubt that they were marine. Whether they were allied to the pythons and boas may be doubtful, but in any case it is probable that they indicate a separate family.

A still larger marine snake (*Pterosphenus schweinforthi*) has left its remains in the lower Eocene of Egypt, and species of the same genus also occur in the North American Eocene. In both cases these snakes are associated with remains of the whale-like zeuglodonts, which at that period abounded in the ocean, where they represented the toothed whales of modern times.

CYLINDER-SNAKES

AGREEING with pythons and boas in the retention of vestiges of the hind limbs, the small group of cylinder-snakes (*Ilysiidæ*) appears to form a connecting-link between the two former and the under-mentioned family of shield-tailed snakes; their essential point of distinction from the *Boidæ* being that the supra-temporal bone of the skull is of small size, and included in the walls of the brain-case, instead of standing out as a support for the quadrate bone, which is much shorter than in boas and pythons. Teeth are present on the palate as well as in the jaws; and the vestiges of the hind limbs usually take the form of a spur on each side of the vent. In general appearance and in the arrangement of the scaling these snakes approximate to the boas; while as regards the structure of the skull they are intermediate between the latter and the next family. The distribution of the group is remarkable, being restricted to Ceylon and South-Eastern Asia in the Eastern, and to Tropical America in the Western Hemisphere. Three genera, of which two have one species, while the third has three, represent the family.

The single representative of the typical genus of the family is the beautiful coral-snake (*Ilysia scytale*), inhabiting the Guianas and Upper Amazonia, and attaining a length of something over 2½ feet. The distinctive features of the genus are the presence of two teeth in the anterior upper jawbones, or premaxillæ, and the situation of the eye in the middle of an ocular shield. The colour is a splendid coral-red, ornamented with black rings, or incomplete ring-like black bands. From the little that is known concerning its habits it appears that this snake is sluggish in its

CORAL SNAKE

movements, and never wanders far from its retreat, which is situated under the roots of a tree or in a hole or cleft in the ground. It feeds on insects and blind-snakes, and produces living young.

The true cylinder-snakes, as typically represented by the red snake (*Cylindrophis rufus*), differ from the coral-snake by the absence of teeth in the anterior upper jawbones, and likewise by the eye not being included in any of the head-shields. This genus, which has three representatives, is distributed over Ceylon and South-Eastern Asia to the eastwards of the Bay of Bengal; the common red snake ranging from Burma and Cochin-China to the Malay region. This snake, which attains a length of about 2½ feet, is either brown or black above, with or without light alternating cross-bars, the under parts being either white with black transverse bars or spots, or black with white bands, and the under surface of the tail brilliant vermilion.

All the snakes of this genus are burrowing reptiles, seldom showing themselves above the surface of the ground, and feeding on insects, worms, and the smaller mammals.

In common with their allies, they have the body covered with polished, rounded scales, which (in conformity with their burrowing habits) are scarcely larger on the lower than on the upper aspect, although becoming wider on the inferior surface of the tail.

The third genus of the group *Anomalochilus*, represented by a single species from Sumatra, differs from *Cylindrophis* by the absence of a groove on the chin.

SHIELD-TAILS

THE snakes of the family *Uropeltidæ* agree with boas and pythons in the structure of the lower jaw, but are sharply distinguished by the loss of all traces of the limbs, and likewise by the complete disappearance of the supratemporal bone in the skull. They have been regarded as specially modified direct descendants of the *Ilysiidæ*. The skull is remarkable for the firm union of its constituent bones; and although both jaws are toothed, the teeth are small and feeble, and very rarely present on the palate.

Externally these snakes are characterised by their cylindrical bodies, short, narrow heads, passing imperceptibly into the neck, and the extremely short, truncated, or slightly tapering tail, which generally ends in a rough, naked disc, although in one genus it is covered with keeled scales. On the body the scales are small and polished, those on the lower surface being always somewhat larger than those above; the eye is minute, and the cleft of the mouth comparatively small, and incapable of much dilatation. These snakes are represented by seven genera, some of which comprise

SHIELD-TAILED SNAKE

a large number of species, and are restricted to Ceylon and the mountains of Peninsular India.

All shield-tails are purely burrowing reptiles, generally living in soft earth at a depth of several feet below the surface, and consequently seldom seen unless specially searched for. They are, however, frequently dug up in the cultivation of tea and coffee plantations, and may be found beneath logs and stones. On the mountains these earth-snakes, as they are frequently called, may be met with in the open grasslands, and during the rainy season not infrequently leave their burrows to travel some distance on the surface.

Of relatively small size, many of them are beautifully coloured with red and yellow, while those that are black display an iridescence like that of some of the smooth-scaled skinks among the lizards. The food of these reptiles appears to consist solely of earth-worms, and the eggs are hatched before quitting the body of the parent. Curiously enough, these snakes burrow by means of their abruptly truncated tails, instead of with their muzzles.

COLUBRINE SNAKES

THE skulls of all the remaining groups of snakes are markedly distinguished from those of the preceding families by the total absence in the lower jaw of the bone known as the coronoid ; while in all cases a supratemporal bone is present on the upper surface of the skull. The present family, *Colubridæ*, which includes by far the great majority of the species of the suborder, and comprises both harmless and noxious kinds, is specially distinguished from those to be mentioned later by the circumstance that in the skull the upper jawbone, or maxilla, is fixed in a horizontal position, and also that the pterygoids reach either to the quadrate-bone or the lower jaw.

Before discussing the colubrine family, it should, howeeer, be mentioned that there is one remarkable snake, *Xenopeltis unicolor*, from South-Eastern Asia, which retains in the structure of its skull traces of affinities with the boas and pythons. This affinity is displayed by the fact that the prefrontal bone, which lies immediately behind the nasal aperture of each side, is of large size, and extends forwards and inwards to articulate with the nasal bone in the same manner as in the boas. Accordingly, this snake is regarded as the representative of a distinct family, *Xenopeltidæ*, which is considered to have originated from the *Boidæ* independently of the colubrines.

From *Xenopeltis* the colubrines are distinguished by the small size of the prefrontal bone of each side, which articulates merely to the outer front angle of the frontal bone without having any contact with the nasal bone.

In such a large group it is highly important to have some means of division into subgroups of higher value than genera ; and, according to the modern classification, three such serial divisions may be indicated by the characters of the teeth. The first and most primitive of these series, which may be termed the solid-toothed colubrines (Aglypha), is characterised by

poisonous, while many of those of the second are noxious in a minor degree. All three sections contain species adapted to particular modes of life, so that there may be two or three snakes which, although externally very similar, are only distantly allied to one another.

THE WART-SNAKE

The wart-snake (*Acrochordus javanicus*) may be taken as a well-known representative of the first subfamily, *Acrochordinæ*, of the solid-toothed colubrines, which includes five genera, distributed over South-Eastern Asia and Central America. Unfortunately, the characters distinguishing this subfamily from the next are connected with the bones of the skull, and cannot therefore be verified without dissection, but in the study of snakes, according to the modern system, the student must accustom himself to such difficulties. The essential feature of the skull in the present group is the production of the postfrontal bone above the cavity of the eye ; but, as a secondary feature, the scales of the body overlap one another only slightly, if at all.

The wart-snake, the sole representative of its genus, is characterised by the absence of lower shields, by the head being covered with uniform granules, and by the very slight compression of the body. The head is rather short and broad, with the muzzle wider than long, and the small eyes directed forwards, while the nostrils are placed close together on the tip of the muzzle. The nearly cylindrical tail is short and prehensile. The colour is brown above and yellowish on the sides ; the young having large irregular dark brown spots, which coalesce into bands on the back, and gradually tend to disappear in the adult. In size this snake may measure upwards of 8 feet. It is distributed over the Malay Peninsula, Java, and New Guinea ; and, although it has been stated to be terrestrial, modern observations indicate that it is essentially aquatic, seldom even leaving the water, where it feeds on fishes and frogs. A female in the possession of Dr. Cantor gave birth to twenty-seven young ones in less than half an hour, which were active and bit fiercely as soon as they came into the world.

An allied genus, represented by the single species, *Chersydrus granulatus*, ranging from Southern India to New Guinea, differs by the marked compression of the body and tail, and thus closely resembles the sea-snakes of the front-fanged series of the family, and likewise resembles them in habits, frequenting the mouths of rivers and the coast from Southern India to New Guinea, and being often found far out at sea. It produces living young, and subsists on fish. A third Oriental genus, likewise known by one species, *Xenodermus javanicus*, has large shields on the under surface. In the other two genera—

the whole of the teeth being solid, without any trace of grooves : all its representatives are harmless. On the other hand, in the second series, or hind-fanged colubrines (Opisthoglypha), one or more of the hinder teeth of the upper jaw is grooved ; while in the third series, or front-fanged colubrines (Proteroglypha), the front teeth of the upper jaw are grooved or tubular. Of the last series the whole of the members are

Stoliczcaia from India, and *Nothopsis* from Central America—not only are large lower shields present, but the granules on the head are replaced by large shields.

WATER-SNAKES

The large group of water-snakes (*Tropidonotus*) forms the first representatives of the second and by far the largest subfamily of the solid-toothed colubrines, known as the *Colubrinæ*, and distinguished from the preceding group by the supratemporal bone not being produced over the region above the socket of the eye ; as well as by the scales usually overlapping one another and the teeth being present throughout the entire length of the upper and lower jaws. The water-snakes belong to a large assemblage of genera of the subfamily

GARTER-SNAKE   Photo, C. R. Walter

characterised by the circumstance that in the skeleton inferior projections or spines are present throughout the length of the backbone, the vertebræ in the hind region of the body having these spines represented by more or less well-developed crests or tubercles.

From their allies, water-snakes are distinguished by having the hinder upper teeth larger than those in front, the equality in the size of the lower teeth, the rather large size of the eye, in which the pupil is round, the presence of a pair of internasal shields between the nostrils, the regular longitudinal series formed by the scales throughout the body, and by the teeth in each hind upper jawbone varying in number from eighteen to forty and forming a continuous series.

Represented by over forty species, the water-snakes have an almost cosmopolitan distribution, although they are unknown in South America, and in Africa south of the Sahara are less abundant than in other regions, and in Australia they occur only in the northern districts. Dr. A. Günther writes that the typical water-snakes " are easily recognised by their stoutish cylindrical body, keeled scales, flat head covered with regular shields, wide cleft of the mouth, and numerous teeth, the strongest of which are at the hinder end of the maxillary bone. They frequent the neighbourhood of fresh water, and feed on aquatic animals—frogs, toads, and fishes. They do not overpower or kill their prey by throwing a coil of the body round it but, having seized it, they at once begin to swallow it. They are excellent swimmers, but more frequently live near water than in it, in agreement with which habit the position of their nostrils is not on the upper surface of the head, as in the true fresh-water snakes, but on the side."

RINGED SNAKE

The best known and at the same time the typical representative of the group is the common ringed snake (*Tropidonotus natrix*), inhabiting Europe, Algeria, and West and Central Asia, and attaining a maximum

length of 6½ feet. Belonging to a group of the genus in which the number of teeth in the hind upper jawbone does not exceed thirty, this snake has a single anterior temporal shield on the head, usually seven upper labial shields, of which the third and fourth enter the aperture of the eye, and from one hundred and fifty-seven to one hundred and ninety shields on the lower surface of the body. The eye is of moderate size, and most of the scales are strongly keeled.

The colour is usually grey, olive, or brown above, with spots or narrow transverse bands ; the labial shields being white or yellowish, with their dividing lines black, while the under parts are mottled black and white, or grey. There are, however, several variations as regards the colouring of the neck. In the ordinary race, for instance, there is a white, yellow, or orange collar, usually divided in the middle, behind which is a broad black collar ; the latter being sometimes alone present. In another race, mostly from the south of Europe, the collar is altogether wanting, or reduced to a small black patch on each side of the nape ; while in the south-eastern race the collar, although well marked, is divided in the middle, and there is a yellowish streak along each side of the back.

In England the ringed snake is one of the most common reptiles, inhabiting woods, heaths, and hedges, especially where water is abundant. Although its chief food consists of frogs, it also preys upon field-mice, young birds, and fishes, and is stated occasionally to consume eggs. When a frog is pursued by one of these snakes, it seems paralysed with fear, and, instead of making any effort to escape, sits still and gives vent to a piercing scream, sometimes also heard when one of these amphibians is trodden upon. Generally the frog is seized by the hind leg, and gradually swallowed by the snake without its position being changed.

On this point Professor T. Bell observes that " when a frog is in the process of being swallowed in this manner, as soon as the snake's jaws have reached the body, the other hind leg becomes turned forwards, and as the body gradually disappears, the three legs and head are seen standing forwards out of the snake's mouth in a very singular manner. Should the snake, however, have taken the frog by the middle of the body, it invariably turns it by several movements of the jaws, until the head is directed towards the throat of the snake, and it is then swallowed head-foremost."

As a rule, the frog remains alive during the swallowing process, and it may sometimes be heard to croak when buried in the stomach of its captor, while instances are on record where a frog has returned after being thus entombed. When swimming, the ringed snake carries its head and neck raised above the surface of the water. The skin, as in the case of other serpents, is shed several times during the year, and is drawn off turned inside-out, so that the lenses covering the eyes appear concave instead of convex. Previous to changing its coat the reptile becomes almost if not completely blind, and evidently ill at ease, and the change is accomplished by the old skin bursting at the neck, and being gradually pulled off by the owner wriggling its body between brushwood or dense herbage.

From about sixteen to twenty eggs are annually deposited by the female of the ringed snake, these being attached together by a viscid substance. Although they are sometimes hatched solely by the heat of the sun, at other times the process of development is hastened by their being placed in a heap of decaying vegetable matter or manure. When the cold of autumn makes itself felt, these snakes retire for the winter, passing their time in a state of torpor ensconced in holes in hedge-banks, under the roots of trees, or some such place, where they remain till awakened by the returning warmth of spring. Not infrequently several snakes occupy the same hole for the winter, and occasionally a considerable number have been found coiled up together in a mass.

TESSELATED AND VIPERINE SNAKES

The ringed snake, as already mentioned, belongs to the typical section of the genus, in which the teeth of the maxilla, or hind upper jawbone, do not exceed thirty in number, and are gradually enlarged towards the hind end of the series, while the eyes and nostrils are lateral, and the internasal shields broadly truncated in front. Of the second section, in which the number and characters of the teeth are similar, but the small eyes and nostrils are directed upwards and outwards, and the internasal shields usually much narrowed in front, the tesselated snake (*T. tesselatus*) and the nearly allied viperine snake (*T. viperinus*) may be taken as examples; both of these being found in Europe, although the former is a more southerly type than the latter, and extends eastwards into South-Western and Central Asia.

The tesselated snake, which never grows quite so large as the ringed species, is olive or olive grey above, and may be either uniformly coloured, or marked with dark spots, usually arranged quincuncially, on the back. The nape of the neck is ornamented with a dark chevron; the upper labial shields are yellowish, with dark lines of division between them, and the under parts are either yellow or red mottled and marbled with black, or almost wholly black.

The viperine snake is rather smaller, having the upper surface grey, brown, or reddish, with a zigzag black band down the back, and a row of yellow-centred black spots down each side. There is a more or less distinctly marked oblique dark band on each side of the top of the head, and another on the nape of the neck, but the labials and under parts are coloured like those of the tesselated snake. In general habits both these species are very similar to the ringed snake, but in spring they are more generally found concealed in pairs beneath stones, and only take to the water in summer.

As other well-known North American representatives of the genus, reference may be made to the garter-snake

(*T. ordinatus*) and moccasin-snake (*T. fasciatus*); the former belonging to the first, and the latter to the second section. As an example of the third section, in which the last two or three upper teeth are suddenly enlarged, the Indian two-banded snake (*T. stolatus*) may be mentioned.

OBLIQUE-EYED SNAKES

Among a large number of other generic types belonging to the present series, it must suffice to mention the genus containing the numerous species of oblique-eyed snakes (*Helicops*). Generally having smaller eyes than the water-snakes, the members of this genus are distinguished by possessing only a single internasal shield; while they are further characterised by the nostril being placed in a half-divided nasal shield, by the nearly equal size of the teeth of the lower jaw, and by the scales lacking the pits found in those of an allied genus. There are from eighteen to twenty-five teeth in the maxilla, or hind upper jawbone; the head is, at most, but slightly distinct from the neck, the body cylindrical, the tail, which has two rows of shields beneath, of moderate length, and the scales are usually striated and keeled. The genus is represented by about eleven species, some of which are found in the New World, while others inhabit South-Eastern Asia, and others Tropical Africa.

The keel-tailed snake (*H. carinicauda*), which inhabits Brazil, and attains a length of between 3 and 4 feet, is characterised by having the scales on the back of the head smooth, and those on the body keeled and arranged in nineteen rows, the frontal shields being nearly or quite as long as the parietals, while there are from one hundred and twenty-six to one hundred and fifty-five shields on the lower surface of the body (page 1570).

The general colour is dark olive brown above, with four more or less distinctly defined blackish stripes, and a yellow stripe along the two lower rows of scales; on the under parts the ground-colour is yellow or red, with black spots or stripes on the body, and a black stripe on the tail. In the neighbourhood of the Rio Grande do Sul this species is one of the commonest of snakes; and while its general habits appear to be very similar to those of the water-snakes, like all the other members of its genus, it produces living young.

PIGMY SNAKES

The pigmy snakes, although belonging to the typical subfamily (*Colubrinæ*) of the solid-toothed series, differ from the preceding group in that inferior spines are developed only in the vertebræ of the anterior half of the backbone, and are further characterised by the nasal bones being fully as large as the prefrontals. The members of the preceding group are more or less aquatic in their habits, but those of the present assemblage are terrestrial or arboreal. The pigmy snakes have the hind borders of the shields on the lower surface of the body entire, the front lower teeth larger than the hinder ones, the eyes relatively small, and no internasal or temporal shields on the head.

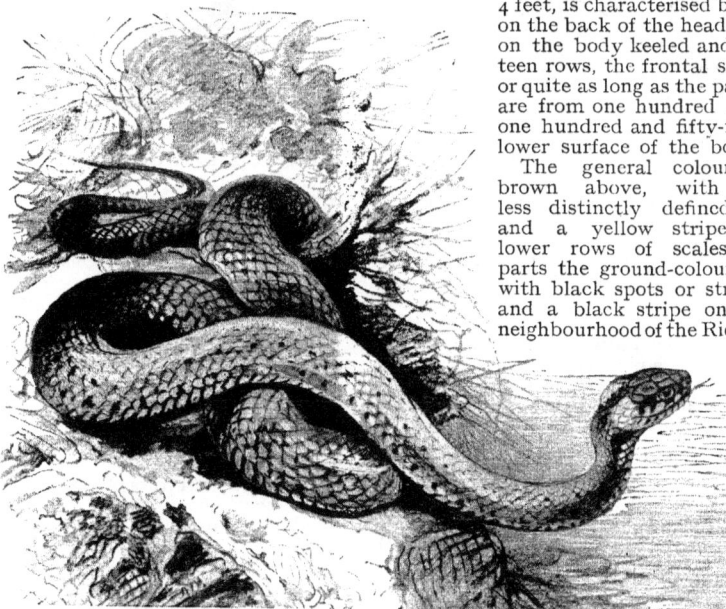

RINGED SNAKE

The head is not distinct from the neck, each nostril is pierced in a very small nasal shield, the body is cylindrical, with the smooth scales arranged in thirteen rows, and there are two rows of shields on the lower aspect of the tail.

These snakes, which are represented by about thirty species, have their headquarters in the islands of Java, Sumatra, and Borneo, and are all of small size, frequently not exceeding a foot in length. Most, or all of them, are in the habit of hiding themselves among stones, beneath fallen tree-trunks, or in grass. Their small dimensions, together with the relatively narrow cleft of the mouth, and a want of dilatibility in the throat and body, indicate that they do not prey upon other reptiles. Gentle and harmless themselves, these snakes are often attacked and killed by kraits and other venomous members of their own tribe. *Calamaria linnæi* is the scientific name of the Javan species represented in the picture on page 1570.

VIPERINE AND TESSELATED SNAKES

SMOOTH SNAKE

The well-known European smooth snake (*Coronella austriaca*, or *C. lœvis*) represents a genus with about twenty known species ranging over Europe, Western Asia, Africa, and America, while one (*C. brachyura*) occurs in India. They belong to a group of genera in which the whole of the lower teeth are nearly equal in length, and are specially distinguished by the presence of from twelve to twenty teeth in the hind upper jawbone, which increase in size towards the back of the series. The head is short, and scarcely distinct from the neck ; the eye being rather small, with a round pupil, and the head-shields normal. The body is cylindrical, and covered with smooth scales arranged in from fifteen to twenty-five rows, and furnished with pits at their tips ; the tail is of moderate length, and the shields on the inferior aspect of the body are rounded, with those beneath the tail arranged in a double series.

The smooth snake, which attains a length of about 25 inches, is very variable in colouring, but the ground-colour of the upper parts is generally brown. The most distinctive features are a large dark spot on the neck, often extending into a stripe, and two rows of dark brown spots arranged in pairs, and running down the body ; there is also a dark stripe passing through the eye and the side of the neck, while the under parts are either steely blue, or reddish yellow and white, in some cases spotted with black.

This snake is found over the greater part of Europe, and is occasionally met with in some of the southern counties of England, more especially in the neighbourhood of Ringwood, Hampshire, where, however, it is less common than formerly. Although now and then found in damp or swampy localities, it frequents dry stony places where there is plenty of sunshine, resorting sometimes to old stone bridges and heaps of building material. Like its congeners, this snake is chiefly terrestrial in its habits ; in disposition it is fierce, and

its prey consists of other snakes and lizards. In the end of August or beginning of September, the smooth-snake lays from three to thirteen eggs, which are so far developed that the included young almost immediately break the shells and escape.

As a third representative of the genus, the king-snake (*C. getulus*) may be mentioned.

RAT-SNAKE GROUP

Nearly allied to the last is the genus *Zamenis* of which the Indian rat-snake is a well-known example ; this group having several South-European representatives, collectively known in Germany as *Zornschlangen*. From the members of the preceding genus these snakes may be distinguished by the more slender form of the body, and the presence on the head of one or more suboculars below the preocular shield ; while the arrangement of the longitudinal rows of scales in odd numbers differentiates them from an allied genus. The number of teeth in the maxilla, or hind upper jawbone, varies from twelve to twenty, and these increase in size towards the hind end of the series ; the head is long and distinct from the neck, with the eye of moderate size or large, and its pupil round. The body is elongated and cylindrical, with the smooth or slightly keeled and pitted scales arranged in from fifteen to thirty-one rows. On the lower surface of the body the shields are rounded, or obtusely keeled on the sides ; and the long tail has two inferior rows of shields (page 1561).

These snakes are represented by about twenty species, ranging over Europe, Asia, and Northern Africa ; several of them occurring on the Continent, although none is met with in the British Isles. Their headquarters may be considered to be the countries surrounding the Mediterranean basin. Deriving their German name from the fierce and bold demeanour of the majority of their representatives, these snakes are terrestrial or partially arboreal in their habits, and feed chiefly on small mammals and birds. Of the European forms, a well-known example is the dark green snake (*Z. gemonensis*), inhabiting Hungary and the Mediterranean countries, and extending so far north as the

KEEL-TAILED SNAKE

south of Switzerland; in the east it is represented by a race known as the Balkan snake, which attains a larger size than the typical form.

This snake is distinguished from its allies by the regular arrangement of the shields on the head, the presence of two preorbital shields, of which the lower one is small and placed in the line of the labials, and by the relative shortness of the tail, which scarcely reaches a fourth of the total length. The smooth scales are arranged in from seventeen to nineteen rows. The typical race may attain a length of about 4 feet, but is generally smaller. The ground-colour of the head and nape is greyish yellow, of the back and tail greenish, and that of the under parts yellow. Upon this are black markings, irregular above, but forming regular oblique bars inferiorly, and on the hind part of the body, arranged in longitudinal stripes which continue to the end of the tail. In some specimens, however, the ground-colour of the upper parts is a beautiful yellowish green, and on the lower surface canary yellow; in a third variety the whole upper surface is uniform olive brown, and in some cases it is completely black, the under surface of the body being grey, with a steely blue lustre on the sides and the whole of the under parts.

This snake is very abundant in Italy, and may be met with in most gardens in the neighbourhood of Rome. Its habits vary to a certain extent according to locality; and while in the Russian steppes it frequents the hottest and driest spots, in Dalmatia and the Tyrol it is found in sunny, although by no means dry, situations, either in woods or among old buildings.

The other European species is the horseshoe-snake (*Z. hippocrepis*), common both to Southern Europe and Northern Africa, and represented in the lower figure of the illustration on page 1572. From its allies it is distinguished by the presence of a series of small suborbital shields beneath the eye, which completely separate the latter from the upper labials, by the divided anal shields, the presence of from twenty-five to twenty-nine

longitudinal rows of scales on the body, and likewise by the constancy of the colouring. Measuring nearly 6 feet in length, this handsome snake has the ground-colour of the upper parts varying from greenish or greyish yellow through orange to reddish brown.

As a rule, the head is marked by a dark oblique band between the eyes, behind which is a second band, convex in front, and reaching to the neck, and a third marked with light spots, so that a horseshoe pattern is formed between the spots and bands. On the back runs a row of yellow-edged dark oval patches, which tend to unite towards the hind extremity; and on each side of this is a series of smaller spots, beneath which, again, are more upright dark marks, extending downwards to the lower surface. As the upper dark patches are very large, the ground-colour is generally reduced to a series of rings, forming a very regular and pretty pattern. The under parts are yellow or orange red, spotted with black.

The aforesaid Indian rat-snake (*Z. mucosus*), which is a large species, attaining a length of 6 feet or more, is brown above, frequently with more or less distinctly defined black cross-bands on the hind part of the body and tail, the under surface being yellowish, often with black edges to the shields of the hind portion of the body and tail.

The range of this well-known species extends from Transcaspia to Java. Common everywhere in India, and feeding on mammals, birds, and frogs, the rat-snake derives its name from its habit of entering houses in search of rats and mice. Like all its allies, it is fierce and always ready to bite, and old specimens brought to Europe never become tame. When irritated, it utters a peculiar sound, which has been compared to that produced by gently striking a tuning-fork.

JAVAN PIGMY SNAKE

Although formerly supposed to be quite innocuous, the saliva of this species is known to possess slightly poisonous properties. A smaller allied Indian species (*Z. corrus*) differs by having the scales arranged in fifteen, instead of seventeen, rows.

The black snake (*Z. constrictor*) is a large and well-known North American representative of this group.

THE PANTHER-SNAKE

The handsome panther-snake (*Drymobius bifossatus*) is the typical representative of a small Tropical American genus to a great extent intermediate between *Zamenis* on the one hand and *Coluber* on the other, being just separable from the former by the larger number of teeth (twenty-two to twenty-eight), and connected with the latter by two species in which the teeth are nearly equal in size. In addition to the more numerous teeth, most of these snakes differ from *Zamenis* by the larger size of the eyes. They are all large and powerful reptiles, with cylindrical bodies, clearly defined heads, large eyes, regularly tapering tails, which are at least equal to a fourth of the total length, the scales smooth or keeled, with apical pits, and arranged in from fifteen to seventeen rows, no subocular head-shields, unkeeled inferior shields, and the twenty-two to thirty-eight maxillary teeth increasing either in height or thickness at the hind end of the series.

The panther-snake, which is an inhabitant of Paraguay and Brazil, and is especially common in the neighbourhood of Rio de Janeiro, is characterised by having fifteen rows of scales on the body, the lack of the small lower preorbital shield, and its general form and colouring; its length being as much as 7 feet. The ground-colour is yellowish grey on the upper parts; on the front of the head are three dark cross-bars; two broad longitudinal stripes run along the hind part of the head and neck; the ornamentation of the back takes the form of a row of large greyish brown black-edged spots, which are lozenge-shaped on the neck, but further back become irregular, and confluent with two lateral rows of spots. The yellowish white shields of the edges of the jaws have black lines of division, and behind each eye a blackish brown streak runs to the angle of the mouth. This snake frequents swampy situations

THE KING-SNAKE

are distributed over the greater part of Europe, Asia, and North and Tropical America. These snakes differ from *Zamenis* in having the teeth in the maxilla of nearly equal size throughout, although those in the front of the lower jaw are larger than the rest. In number the maxillary teeth range from twelve to twenty-two. These snakes are also distinguished by having the scales of the body, which may be either smooth or pitted, arranged in from fifteen to thirty-five longitudinal rows, and furnished with pits at their extremities, those in the middle line of the back not being larger than the others. The long head is well defined from the neck, with a moderate-sized eye, of which the pupil is circular, and the shields normally arranged; the elongated body is slightly compressed, with its scales either smooth or keeled; and while the shields on the lower surface of the body may be either rounded or furnished with a more or less well-marked keel on the side, those of the tail are arranged in a double row.

All these snakes are fierce in their disposition, and while all can climb well, some are almost entirely arboreal; others, again, frequent the neighbourhood of water, and are good swimmers. The food of all consists of small mammals and birds.

Among the European representatives of the genus, the yellow, or Æsculapian, snake (*Coluber longissimus*) may be recognised by the small head, imperfectly distinguished from the neck and rounded at the muzzle, as well as by the stout body, rounded tail, and the nature of the scaling. On the head there is no small preorbital shield, and of the eight upper labials the fourth and fifth enter the circle of the eye; the body has from twenty-one to twenty-three rows of smooth scales, and the anal shield is divided.

Generally the upper surface is brownish yellow with a tinge of grey, and the lower aspect whitish, the hind part of the head having a yellow spot, while the back and sides are marked with small whitish dots, which in some places are very distinctly defined, and assume the form of the letter X. There is, however, great individual variation in colour, and a dark and a light variety

DARK GREEN SNAKE   Photo, C. R. Walter

well covered with trees and bushes, and is remarkably swift and active in its movements. In its general habits it appears to resemble the ringed snake, feeding almost entirely on frogs and fishes.

THE ÆSCULAPIAN SNAKE

The typical representatives of the family *Colubridæ* are the numerous snakes of the genus *Coluber*, which

may be recognised. In the south of Europe, where it attains a length of about 4 feet, this snake prefers rocky, or at least, stony districts abundantly covered with bushes, but in Schlangenbad, the only German locality where it is found in any numbers, old walls are its favourite resorts. As it feeds chiefly on field-mice, it is a decided benefactor to the agriculturist and gardener. It also consumes, however, a certain number of lizards, as well as such birds as it can contrive to capture, and occasionally plunders a nest and sucks the eggs. It is very fond of climbing bushes and low boughs or stumps of trees, as represented in the illustration; and in thick forests will go from bough to bough, and then from tree to tree, without descending to the ground. Indeed, it is such an adept in climbing that it frequently captures swift-running lizards on the stems of trees.

THE LEOPARD-SNAKE

Another southern species is the leopard-snake (*C. leopardinus*), a handsomely-marked reptile belonging to a group characterised by the absence of a subocular shield below the preocular on the head, the smooth or faintly-keeled scales arranged in twenty-three or more rows, and the circumstance that the third labial shield does not enter the circle of the eye. As special characters of the leopard-snake may be mentioned the presence of from twenty-five to twenty-seven rows of perfectly smooth scales, and of from two hundred and twenty-two to two hundred and sixty ventral and sixty-eight to eighty-nine subcaudal shields.

There are two phases or varieties of this snake, in the first and rarest of which the ground-colour is brownish grey, usually with four black longitudinal stripes, here and there interrupted, although these may sometimes be replaced by two dark or blood-red lines. On the sides are small blackish spots; the under surface of the head and fore part of the body are either yellowish white or bright yellow, but each under shield is marked with four or five irregular blackish spots, which become so large posteriorly that the whole surface appears steel-blue, the yellow only showing on the edges of the shields. In the second, or typical, phase the ground-colour is mahogany-red, mottled on the upper surface with blood-red black-edged spots, which may either be arranged in two rows, or coalesce

into transverse bands, while on each side there is a row of smaller, blackish, crescentic spots alternating with those of the back. The range of the species is bounded to the west by the mountains of Southern Italy and Sicily, and to the east by Asia, both phases occurring together in most districts between these limits, although in Greece and Dalmatia only the typical leopard-snake is known.

FOUR-LINED SNAKE

Among the largest of European serpents is the four-lined snake (*C. quatuorlineatus*), which attains a length of between 6 and 7 feet, and is olive brown or flesh-coloured above, often marked with a pair of longitudinal blackish brown stripes, a black line running from the eye to the mouth, and the under parts being straw-yellow. There are, however, many variations from this typical coloration; some specimens being entirely black, while the young generally have black cross-bands on the head, three rows of large brown spots on the back, the sides likewise spotted, and the under parts with a blackish steel-grey tinge. The distinctive specific characters are the presence of a small preorbital shield on the head, the arrangement of the scales of the middle of the body in from twenty-three to twenty-five longitudinal rows — these scales being smooth in the young but distinctly keeled in the adult—the divided anal shield, and the presence of from one hundred and ninety-five to two hundred and thirty-five ventral and sixty-three to ninety subcaudal shields.

The distributional area of this snake includes the whole of Southern and South-Eastern Europe, from Lower Italy and Dalmatia to Turkey, as well as Greece and the adjacent islands, and extends to the interior of Asia Minor, as well as to the Transcaucasian region. All observers are in accord that the four-lined snake is not only harmless, but useful, since it destroys rats, field-mice, and smaller snakes. It also preys upon moles, lizards, and small birds.

Yet another European species of the same large genus is the black-marked snake (*C. scalaris*), which belongs to a separate group characterised by the following features. The rostral shield of the head is of a large size, convex, and pointed in front, while it extends backwards between the prefrontal shields, where it terminates in a point. The tail is relatively shorter than in the

THE ÆSCULAPIAN SNAKE

BLACK-MARKED AND HORSESHOE SNAKES

typical group. The black-marked snake, formerly separated as *Rhinechis*, and represented in the upper figure of the illustration on page 1572, has the cylindrical body relatively thick, the tail short and blunted, and the flattened head broad behind and sharp in front. The body-scales, which usually form twenty-seven, although occasionally from twenty-five to twenty-nine, rows, are long, four-sided, and smooth; the shields on the under surface of the body, from two hundred

taking its Latin name from its marbled black and yellow colouring; and the South American corais snake (*C. corais*), which is a very large species, varying in colour from brownish orange to nearly black.

SIPOS, OR WOOD-SNAKES

Unlike all the above-mentioned members of the colubrine family, which only climb trees in search of food, the American wood-snakes (*Herpetodryas*) are

Photo, W. B. Johnson
THE LEOPARD-SNAKE

Photo, C. R. Walter
THE BULL-SNAKE

and one to two hundred and twenty in number, are bent at the edges, those beneath the tail form a double series of from forty-eight to sixty-eight, and the anal shield is usually divided.

The ground-colour varies from bright grey or greenish grey, through reddish or yellowish brown, to olive or reddish yellow; and in the young the markings of the head take the form of a perpendicular black streak through the eye, and another from the eye to the mouth, forming with its fellow a V-shaped mark; the neck having a dark cross-band, and a row of H-shaped black markings running down the back, below which is a series of small spots, followed inferiorly by a third and fourth row. With age these markings tend gradually to disappear, till finally there remain only two dark brown or blackish rows running from the neck to the tip of the tail. In length, this snake measures rather more than 4 feet. Everywhere rare, the black-marked snake is confined to Spain, Portugal, and the south of France. In addition to other food, it preys upon grasshoppers, and will follow mice into their burrows.

Photo, C. R. Walter
CORAIS SNAKE

A good climber, it is stated to be more rapid in its movements than any other of the European snakes, and its keenness of vision is remarkable.

As well-known New World representatives of the same large genus, mention may be made of the North American corn-snake (*C. guttatus*), with reddish blotches on a yellow ground, the chicken-snake (*C. obsoletus*), and the bull-snake (*C. melanoleucus*), the latter

purely arboreal reptiles, especially adapted by their colouring to such a mode of life. Compared with the species of *Coluber*, the hind maxillary teeth tend to be stouter, if not taller, than those in front; the total number of these teeth varying from twenty-eight to thirty-two. They are further distinguished by their larger eyes, which may be of very great size, the distinctly compressed and more slender body, and the small number of the longitudinal rows of scales, which does not exceed from ten to twelve. The known species, about five in number, are inhabitants of the West Indies and the forest-districts of Central and South America, all being characterised by their more or less uniform olive green coloration.

In the forests of Brazil, the Guianas, and Venezuela, as well as in the Lesser Antilles, lives the sipo, or Brazilian wood-snake (*H. carinatus*), which may be selected as a well-known example of the genus. Frequently attaining a length of about 7 feet, and remarkably beautiful in colouring, this snake generally has the upper parts of a bright verditer or olive green, shot with a tinge of brown on the back, while the under parts are greenish or bright yellow; the greenish hue prevailing in the middle of the body, and the yellow elsewhere. Throughout there is a shimmering play of colours of all shades of green passing into metallic brown, and the middle line of the back displays a brighter longitudinal streak, frequently bordered on each side by a darker band.

In the West Indies this species undergoes a remarkable change of hue, becoming blackish brown or black above, with the under parts steel-grey; the upper lip and edges of the jaws alone preserving the original yellowish green. The scales are arranged in twelve rows, and are

FOUR-RAYED SNAKE

mostly smooth, although the two middle rows on the back are keeled, and the eye is of very large size.

Next to the small-scaled snake, the sipo is the most abundant of Brazilian serpents, and may be met with on sandy jungle-clad ground close to the shore both at Rio de Janeiro and Cape Frio, where specimens of upwards of 10 feet in length have been observed. In addition to sandy localities, it also frequents swampy spots near the sea. In its movements it is so rapid that, when startled, it seems to disappear like a flash of lightning. It feeds largely on frogs, as well as on lizards and young birds, and lays only five eggs, remarkable for their cylindrical and slender form.

TREE-SNAKES

In the Old World and Australasia the wood-snakes are replaced by the typical tree-snakes (*Dendrophis* and *Dendrelaphis*), distinguished from all the preceding types by the hind border of each of the shields on the lower surface of the body having a notch on each side, corresponding to a suture-like lateral keel; the scales of the body being arranged in from thirteen to fifteen rows. While in the first-named of the two genera all the teeth in the maxilla, or hind upper jawbone, are approximately equal in length, and the scales of the row in the middle line of the back larger than the others, in the second genus the foremost teeth in the hind upper jawbone are enlarged, but the scales of the middle row of the back are similar to the rest.

All these snakes have large eyes, and elongated and often compressed bodies, and their general colour is some shade of green or olive, often with a bronzy tinge. In habits they are mostly arboreal. Of *Dendrophis* about nine species are known, ranging from India to Australia and New Guinea; while *Dendrelaphis* is represented by about five species, with a range extending from India and the Malay region to the Philippines.

EGG-EATING SNAKE

The last representative of the solid-toothed series of the colubrines that will be mentioned here is the remarkable little egg-eating snake (*Dasypeltis scabra*), of South Africa, which represents a subfamily (*Dasypeltinæ*) by itself. The essential character of the subfamily is the rudimentary condition of the dentition, the front of both the lower and upper jaws being devoid of teeth. To compensate for this lack of ordinary teeth, the egg-eating snake is, however, provided with a series of about thirty of what may be termed throat-teeth; these being formed by the lower spines of the vertebræ projecting into the œsophagus, and being tipped with enamel. The scales are strongly keeled.

This little snake is about a couple of feet in length, and has a body not much thicker than a man's finger Although it lives in trees, and feeds on the eggs of small birds, it will, when pressed by hunger, descend to the ground and rob hens' nests. That such a tiny reptile should be able to swallow a hen's egg seems incredible, but nevertheless a specimen has been taken with the egg actually within its jaws, and the whole head so swollen as to render the mouth incapable of being closed; and an example in the London Zoological Gardens swallowed pigeons' eggs without any apparent difficulty. When swallowed,

SIPO, OR BRAZILIAN WOOD-SNAKE

the egg is split longitudinally by the row of teeth in the throat, and the whole of the contents secured. After being thus broken, the two halves of the shell, generally fitted into one another, are rejected. Before the shell is broken the throat of the reptile becomes distended to an extraordinary degree, and is quite tense, with the scales standing nearly upright. Directly the shell is broken, the swelling begins to subside

MOON-SNAKES

The moon-snakes (*Oxyrhopus*), as they are called in Brazil, may be taken as the first representatives of the

CROWNED MOON-SNAKE

second of the three great parallel series into which the colubrine family is divided. This back-fanged series, or Opisthoglypha, is characterised by having one or more pairs of the hind upper teeth longitudinally grooved, and thus capable of acting as poison-fangs. Many of these snakes are indeed extremely venomous, their bite being capable of producing death in a few minutes. They are divided into two subfamilies, of which the first, or *Dipsadomorphinœ*, is characterised by the lateral position of the nostrils; these snakes being either terrestrial or arboreal in their habits, with a world-wide distribution.

Belonging to the first of the two subfamilies, the moon-snakes are characterised by the slender and somewhat compressed form of the body, the flattened head, imperfectly differentiated from the neck, and broad behind and narrow in front, although somewhat pointed at the muzzle, and the upper jaw projecting considerably over the lower. The scales, which are arranged in from seventeen to nineteen rows, are smooth, with apical pits, the anal shield and the shields on the lower surface of the tail may be either single or double, and the eye, as in most of the other members of the sub-family, has the pupil vertical. The species of moon-snakes, about seventeen in number, range from Mexico, through Central America, to Tropical South America; the species figured (*O. coronatus*) being an inhabitant of Brazil and the Guianas.

In size this snake is comparatively small, measuring only about 2 feet in length. It is one of three species characterised by the subcaudal shields being undivided, and differs from the other two in having the subcaudal shields arranged in seventeen in place of nineteen rows. In colour the upper parts vary from reddish or pale brown to blackish, and pass into pale brown or yellowish on the sides; the upper surface of the head

and nape is blackish, with or without a yellowish band across the temple and occiput; and the under parts are yellowish white. Very common in the neighbourhood of Bahia, this snake, like the other members of the subfamily, is almost exclusively nocturnal; and its food consists solely of lizards. Although their fangs are large, these reptiles never attack human beings.

THE CAT-SNAKE

One of the few European representatives of the group the cat-snake (*Tarbophis fallax*) belongs to an Old World genus characterised by the solid maxillary teeth gradually decreasing in length from back to front, the absence of enlargement in the scales of the back and the subocular shields, the oblique arrangement of the scales, the spindle-shaped body, the clear distinction between the flattened head and the neck, the relatively short tail, and the small size of the eyes. In place of a lower preocular shield, the elongated loreal extends backwards to the eye, so as to come in contact with the upper preocular; this arrangement being unknown in any other European snake. In the lower jaw the front teeth are much longer and more bent than those which follow; while the fangs in the hind part of the upper jaw are also elongated and much curved.

Sometimes reaching a little over a yard in length, this snake is of a dirty brownish yellow or grey ground-colour, with small black dots and a chestnut-brown spot on the shields of the head, while the neck has a large blackish or reddish brown patch, and rows of smaller spots of the same colour ornament the back. There is also a dark band from the eye to the corner of the mouth; each side of the body has a row of small spots; and the under parts are whitish with a

THE CAT-SNAKE

grey or brown marbling. The cat-snake ranges from the shores of the Adriatic to the neighbourhood of the Black and Caspian Seas, Asia Minor, and Northern Syria. It inhabits rocky and sunny spots, and feeds mainly, if not exclusively, on lizards. Although slower than in water-snakes, its movements are more rapid than those of vipers. The virulence of its poison is such that a lizard bitten died in a minute and a half.

NOCTURNAL TREE-SNAKES

The tropical regions of the Old World are the home of the typical genus, *Dipsadomorphus*, of the subfamily, the members of which are characterised by the long and compressed body and tail, the sharp distinction of the head from the neck, the moderate or large size of the eye, with its vertical pupil, and the normal arrangement of the shields on the head, in which the hind nasal is more or less markedly hollowed. The number of teeth in the hind upper jawbone varies from ten to twelve, the two or three hinder pairs being elongated and grooved, while in the lower jaw the front teeth are the largest. The scales on the body are arranged in from seventeen to twenty-seven longitudinal rows, those of the middle row of the back being larger than the rest, and the medium-sized or long tail has its inferior shields in two rows.

These snakes are represented by about twenty-one species, inhabiting Southern Asia, New Guinea, Northern Asia, and Africa. The majority are inhabitants of forests or scrub-jungle, and are almost entirely arboreal, but a few are terrestrial, and frequent open country. Many of these snakes attain a length of 6 or 7 feet, and their prevalent ground-colours are brown and black. The Indian species are purely nocturnal, and their food consists of small mammals, birds, and, more rarely, lizards, and occasionally birds' eggs. It is noteworthy that some species prey entirely on mammals, while others confine their attention to birds. Eight species of the genus are recorded from India, Ceylon, and Burma, while a well-known Malay form is the ularburong (*D. dendrophilus*).

GREEN SNAKES AND WHIP-SNAKES

Among tree-snakes with fangs in the back of the jaws may be mentioned the species of *Philodryas* characteristic of South America, and the whip-snakes (*Dryophis*) of India and the Malay countries. In the American genus the hind fangs are not very large, being not double the height of the solid teeth in front of them. The body and tail are elongated and more or less compressed, the eyes large, and the smooth or keeled scales arranged in from seventeen to twenty-one rows, while the prevailing colour is green. The genus is represented by some thirteen species, among which the green snake (*P. viridissimus*) attains a length of nearly three feet, and has from two hundred to two hundred and twenty-eight shields on the lower surface of the body. It is a native of the Guianas, Eastern Peru, and the intermediate countries.

In the Indian whip-snakes the teeth in the posterior upper jawbone vary in number from twelve to fifteen, one or two near the middle being much enlarged and fang-like. After these comes an interval devoid of teeth, and at the hind end of the jaw the two last teeth are grooved. In the lower jaw the third or fourth tooth is enlarged and fang-like ; those in the

hind part of the series being small and uniform. The head is long, and markedly distinct from the neck ; and the eye rather small, with a horizontal pupil. The scales investing the elongated and compressed body are smooth and without pits, and arranged in fifteen oblique rows, those down the middle of the back being slightly enlarged. The shields on the under surface of the body are rounded, and those beneath the tail form two rows. The Indian whip-snake (*D. mycterizans*) ranges from India, Ceylon, and Burma to Siam.

Deriving their name of whip-snakes from the extreme elongation and slenderness of the body and tail, these serpents move awkwardly enough on a flat surface, although when coiling and climbing among the branches of trees their rapid movements are graceful in the extreme. While retaining their hold by means of a few coils of the tail thrown round a branch, the length of their body enables them with ease to reach another at a considerable distance, or to dart forth their head in order to seize any hapless bird or lizard that may be within striking distance.

SHARP-NOSED SNAKES

Nearly allied to the preceding are the sharp-nosed snakes (*Oxybelis*), of which the four species inhabit Central and South America ; the Tropical African *Thelotornis kirtlandi* being a nearly-related type. These snakes have small heads, with the snout narrow and elongated, and the rostral shield projecting considerably beyond the lower jaw. The neck is thin and slender, the body greatly elongated and laterally compressed, and the long and thin tail tapering to a fine point. The upper jaw carries seventeen solid teeth of nearly equal size, and four large fangs. In appearance and habits these snakes closely resemble whip-snakes.

ORIENTAL FRESH-WATER SNAKES

Brief reference may be made here to a group of ten genera of aquatic snakes from India, Burma, China, New Guinea, North Australia, and the adjacent countries, which constitute a second subfamily (*Homalopsinæ*) in the hind-fanged series. From the members of the preceding subfamily these may be readily distinguished by the position of the nostrils on the upper surface of the muzzle ; while they are further differentiated by their thoroughly aquatic habits.

As none of the species is of general interest, it will be unnecessary to particularise the various genera, but it may be mentioned that the typical *Homalopsis* belongs to a group in which the two nasal shields of the head are in contact, while in a second group, as represented by *Cantoria*, they are separated by an internasal shield.

Most of these snakes are of small size, few of them exceeding a yard in length, while many are considerably smaller. Although mainly fresh-water snakes, seldom coming to shore, a few members of the group enter the sea. Many of them are furnished with

SMALL-SCALED SNAKE

prehensile tails, by means of which they attach themselves to convenient objects, and the majority feed exclusively on fish, though a few prefer crustaceans. Their young are produced alive in the water. The single representative of the type genus is *H. buccata.*

ELAPINE SNAKES

The beautiful but venomous smallscaled snake (*Elaps corallinus*) is the best-known representative of a genus belonging to the third and last series of the colubrine family. All the members of this front-fanged series (Proteroglypha) are characterised by having the front teeth of the hind upper jawbone, or maxilla, grooved, and the posterior ones simple and solid. These snakes, which are all poisonous, are divided into two subfamilies, according to their habits and the conformation of the tail. In the first, or elapine subfamily (*Elapinæ*) the tail is cylindrical; the snakes themselves being either terrestrial or arboreal in their mode of life.

These elapine snakes are distributed in larger or smaller numbers over Asia, Africa, and America, and are especially abundant in Aus-

INDIAN WHIP-SNAKE

tralia, where they form by far the greater part of the serpent fauna. All of them—doubtless on account of the immunity from attack conferred by their poisonous character—are remarkable for the beauty of their coloration.

The small-scaled snake and its allies constitute a genus restricted to the warmer regions of America. They are small, although rather long and plump serpents, with the body cylindrical, the head flattened and scarcely differentiated from the neck, and the tail short. The small eye has a circular pupil, the mouth is narrow, and the jaws admit of but slight dilatation. Superiorly the body is clothed with equal-sized, smooth, small scales, arranged in fifteen rows; while inferiorly the body-shields are rounded, the anal one being undivided, and the shields beneath the tail arranged in a double series. Behind the fangs the teeth are all small.

One of the handsomest members of a beautiful group is the small-scaled snake which inhabits a large part of South America, and also occurs in the West Indies. Attaining a length of from 2 feet to 2¼ feet, this snake has its ground-colour a brilliant cinnabarred, with a special lustre on the under parts. On the body this red colour is divided into sections of equal length by broad black rings, bordered by more or less distinct greenish white margins, all the red and greenish portions showing black spots on the tips of the scales.

The front of the head, as far back as the hind end of the frontal shields, is bluish black; at the back of the parietal shields there begins a greenish white cross-band, running behind the eye, and occupying the whole of the lower jaw; and after this comes a black neck-ring, followed by one of the red spaces of the body. As a rule, instead of being red, the tail has alternations of black and whitish rings, with its tip whitish. The smallscaled snake is generally found in forests: the neighbourhood of human dwellings it strictly avoids. Somewhat slow in its movements, it is unable to climb trees; and its food consists of other snakes, lizards, insects, and centipedes. In South Africa these snakes are represented by two species of the closely-allied genus *Homorelaps.*

MASKED ADDERS

In Asia the place of the small-scaled snake and its allies is taken by a group of species which includes the well-known masked adder *Callophis macclellandi.* From the genus *Elaps* these are distinguished by the presence of a distinct groove along the whole of the front surface of the upper fangs, and also by the scales being arranged in thirteen rows. None of the teeth behind the fangs is solid, and the shields on the head (among which the loreal is wanting) are of large size. A further difference from the American genus is to be found in the presence of postfrontal bones in the skull. These snakes, which are mostly less than 3 feet in length, are represented by species ranging from India to Southern China and Japan.

The masked adder, which attains a length of 26 inches, and ranges from Nepal to the south of China, is generally reddish brown above, with regular black light-edged transverse rings placed at equal distances from one another; the under parts being yellowish with black cross-bands or squarish spots. These snakes resemble the small-scaled snake, which is referred to above on this page, in the slowness of their movements, and their inability to ascend trees, their favourite resorts being hilly districts.

Curiously enough, the masked adders closely resemble the harmless snakes of the genus *Calamaria*, upon the different species of which they chiefly feed; these latter, as described on page 1569 of this work, being of small size, frequently not exceeding a foot in length.

LONG-GLANDED SNAKES

Nearly allied to the preceding are four snakes from Burma, the Malay region, and the Philippines, which merely differ in that the poison-glands, instead of being confined to the back part of the head, extend along each side of the body for about a third of its total length,

LONG-GLANDED SNAKE AND MASKED ADDER

gradually thickening till they end in front of the heart in club-shaped expansions. The heart being thrown further back in the body than ordinarily, these snakes may be recognised externally by the thickening of that region.

The species (*Doliophis intestinalis*), which is shown in the illustration, is an extremely elongated and slender snake, inhabiting Burma and the Malay Islands, and attaining a length of 2 feet. It is generally brown above with a yellowish black-edged line running down the middle of the back, and a nearly similar one on each side of the body; the under parts being banded with yellow and black.

KRAITS

Although the native name krait applies properly only to a single member of the genus *Bungarus*, it may be conveniently extended to include the whole of the six species, which range from India to the south of China, five occurring in India and Ceylon.

Closely connected with the masked adder and its relatives by the genus *Hemibungarus*, in which a solid tooth is present behind the fangs, the kraits have from one to three small solid teeth behind these; and the smooth scales are arranged in thirteen or fifteen rows, with the middle row of the back larger than the others. The head resembles that of the last genus in being imperfectly distinguished from the neck, as well as in the size and number of its shields; and the small eye has a similar round pupil. The tail is of moderate length, or short, with the

shields on its lower surface arranged in either a double or single series.

BANDED KRAIT, OR RAJ-SAMP

The banded krait, or raj-samp, meaning king-snake (*B. fasciatus*), belongs to a group in which the shields on the lower surface of the body are very large, and broader than long; those of the tail being arranged in a single series. The species is distinguished by the presence of a distinct ridge along the back, the obtuse extremity of the tail, and by the front temporal shield of the head being scarcely longer than deep; these three features distinguish it from the typical krait (*B. candidus*) and the nearly allied Ceylon krait (*B. ceylonicus*).

The banded krait, which ranges from Bengal to Java and Celebes, commonly measures about 4 feet in length, although it grows to 6 feet. In colour it is bright yellow, with black rings equal to or exceeding in length the light interspaces; on the head a black band begins between the eyes and widens towards the nape of the neck, and the tip of the muzzle is brown. The Burmese and Indian race is distinguished as *B. c. cœruleus*.

The typical krait (*B. candidus*) is of a dark, almost steel-blue black, or chocolate-brown colour, with narrow cross-bars, streaks, or rings of white; the under surface being of a dark livid hue, or whitish or yellowish. It inhabits the whole of India, but is not so large as the raj-samp.

The krait frequently insinuates itself into houses, where it conceals itself in bath-rooms, verandas, cupboards, or between the bars of shutters; but an instance is on record where one was discovered coiled up beneath the pillow of a palki in which a lady had made a night's journey. Next to the cobra, the krait is credited with killing more human beings in India than any other snake. Regarding the character of the banded krait, Major F. Wall writes as follows: " The

BANDED ADDER

statement that the banded krait is one of the most deadly of all snakes is inaccurate ; in fact, it reckons among the least dangerous of poisonous snakes, and 1 cannot find one single authentic case of a human being succumbing to the bite of this snake, though it is common enough. The Burmans even go so far as to assert that it is not poisonous. Major Leonard Rogers, as a result of direct experiment, estimated that the virulence of this poison is one-fourteenth that of the cobra and one-twentyeighth that of the common krait. Messrs. Martin and Lamb estimate that the venom of *bungarus fasciatus* is one-seventh as potent as that of the cobra and hamadryad, and one-thirtieth that of the common krait."

COBRAS

The name "cobra da capello," or hooded snake, was applied by the Portuguese in

INDIAN COBRA

Ceylon to the common Indian representative of a genus of deadly serpents distinguished from the kraits by their power of inflating the neck, and likewise by the scales in the middle of the back not being larger than the rest. By Europeans these snakes are now generally known by the name of cobras. Agreeing with the kraits in having the fangs furnished with a complete groove on the front surface, and likewise by the presence of from one to three solid teeth behind them, the cobras have the head distinct from the neck and covered with large shields, among which the loreal is wanting, the eye being rather small, with a round pupil. The body is cylindrical, with the smooth scales disposed in fifteen or more oblique rows, and the tail is of moderate length, with its inferior shields in either a single or a double series. The dilatation of the neck, which always takes place when they are excited and about to strike, at once serves to distinguish the cobras from all other snakes (page 1561).

Cobras are confined to Africa and Southern Asia, and are represented by ten species, two of which are found in India and a third in the Philippines, while the others are African. Of the Indian representatives of the group by far the most abundant is the typical, or true, cobra (*Naia tripudians*), known to the natives of India as the kala-nag or kala-samp (black snake). Distinguished by having no large shields on the head behind the parietals, and by the whole of the shields on the under surface of the tail being arranged in a double series, this snake is a very variable species as regards colour-pattern, some examples having a dark spectacle-like mark on the back of the hood, while others show a

single eye-like spot, and others, again, have no mark at all in this region.

In regard to colouring, it has been remarked that the hue of the upper parts may be greyish brown or black, with or without a spectacle—or loop-shaped black light-edged marking on the neck—or with light spots or cross-bands on the body ; while beneath it varies from whitish, through brownish, to blackish, sometimes with cross-bars on the fore part of the body. Occasionally attaining a length of a few inches over 6 feet, while one instance is on record where a specimen measured upwards of 7 feet 3 inches, this cobra is distributed over the whole of India and Ceylon, ranging westwards through Afghanistan to the Caspian, and to the east of the Malay region and the south of China.

The second Indian species is the king-cobra, or giant cobra, sometimes known as the hamadryad (*N. bungarus*), a larger snake, distinguished by the presence of a pair of large shields on the head behind the parietals, while the shields beneath the tail usually form only a single series. When adult, its colour is yellowish or brown, with more or less distinctly marked dark cross-bands ; but young specimens are usually black, with yellow rings on the body and bars on the head, and in some instances there are light spots on the upper surface, and the inferior shields are whitish with black margins. In size the king-cobra is known to measure as much as 14 feet 7 inches, and possibly grows larger. Fiercer than the common species, this cobra is fortunately far less abundant ; its range extending from India through Burma and Siam to the Malay region and the Philippines. This cobra feeds to a great extent on other snakes.

On this subject Major F. Wall writes as follows : " It has been known for many years that this redoubtable snake does not discriminate between harmless and poisonous species, but devours anything that chance has to offer. Major G. H. Evans mentions a specimen in Burma 11 feet 4 inches long, which he encountered carrying a snake in its jaws as a dog would a stick. The latter proved to be another hamadryad 8 feet 6 inches in length. The same authority on another occasion found one in the act of swallowing a cobra (*N. tripudians*). Mr. Craddock has recorded one at Pahang, Federated Malay States, 9 feet 8 inches long, which had swallowed a 4-feet cobra, and Mr. Primrose found one 7 feet 5½ inches in length eating a banded krait (*Bungarus fasciatus*) 3 feet 1¼ inches. A specimen killed by Mr. Branson contained a Russell's viper, which was swallowed tail-first."

A third species is the African cobra (*N. haie*), the asp (so called from the hood, which is compared to a shield, *aspis*) of the Bible, while the spy-slange, or spitting-snake, of the Boers represents a fourth kind (*N. nigricollis*). The range of the former includes nearly the whole of Africa, and it presents great local variations in colour. Somewhat exceeding in size the true cobra, the asp is distinguished by the sixth upper labial shield of the head much exceeding the others in length, and uniting with the temporal, so as to form a large plate, which anteriorly comes in contact with the postocular shield. In most Egyptian examples the colour of the upper parts is uniformly straw-yellow, while the under parts are light yellow ; but there may be dark cross-bands on the under surface of the region of the neck, which sometimes unite into a patch.

YOUNG KING-COBRA

The straw-colour may, however, shade into blackish brown and occasionally the hues may be brighter. As a rule, the spectacle-like marks on the hood are absent.

Although frequently seen in motion during the day, cobras are most active during the night, and in India feed chiefly on the small mammals, birds' eggs, frogs, fish, and even insects. The king-cobra, as already mentioned, subsists, however, almost entirely on other snakes, and the other species will occasionally rob hens' nests, swallowing the eggs whole. In captivity, cobras will live weeks and even months without tasting food of any kind or touching water.

Essentially terrestrial in habits, cobras readily enter water, in which they swim well, and they occasionally climb trees in search of food, and are often found, more especially during the rainy season, in old buildings and walls, or in wood-stacks and heaps of rubbish. It is when collected in such situations that they are most commonly trod upon by the natives—and more frequently at night than at other times—with the well-known fatal results. These snakes lay from eighteen to twenty-five oval eggs about the size of those of a pigeon.

Ascending to a height of some eight thousand feet in the Himalaya, the common cobra, according to Sir J. Fayrer, " is equally dreaded and fatal wherever met with ; fortunately it is not naturally aggressive, unless provoked, at which times its aspect is most alarming. Raising the anterior third or more of its body, and expanding its hood, with a loud hissing, it draws back its head prepared to strike, and, when it does so, darts its head forwards, and either scratches, or seizes and imbeds its fangs in the object of attack. If the grasp be complete and the fangs imbedded in the flesh, dangerous and often fatal effects result ; but if the fangs only inflict a scratch, or if the snake be weak or exhausted, the same great danger is not incurred. If the poison enter a large vein and be quickly carried into the circulation, death is very rapid ; men having been known to perish from cobra-bite within half an hour.

"The largest and strongest as well as the smallest and weakest creatures succumb ; but, fortunately, all who are bitten do not die. In the first place, some human beings, as well as lower animals, have greater tolerance than others of this or of other poisons—a result, doubtless, of idiosyncrasy or varying degrees of nervous energy, which enables one to resist that to which another would yield ; or a wound may have been inflicted and yet but little of the poison inoculated ; or, in the third place, the snake may be weak or sickly, or it may have been exhausted by recent biting, and thus have become temporarily deprived of the power of inflicting a deadly wound. But when a cobra in the full possession of its powers bites, and injects the poison into man or beast, it is almost surely fatal, and all the remedies vaunted as infallible antidotes are futile."

THE RINGHALS

Nearly allied to the cobras is the African ringhals—that is, banded neck (*Sepedon hæmachetes*)—which has a similar dilatable hood, but differs by the absence of small teeth on the maxilla behind the fangs, and also by the strongly-keeled scales, which are arranged in nineteen rows. The general colour on the upper surface is black, more or less distinctly banded with yellow ; the bands being most clearly defined on the neck. The under parts are almost wholly black, although generally showing one or two whitish bands across the lower part of the neck.

PURPLE DEATH-ADDER OR BLACK SNAKE

MALE AND FEMALE OF SHORT DEATH-ADDER
Photos, Cyril Grant Lane

The black-necked cobra (*N. nigricollis*) often does not move until it is approached pretty closely, when it will creep into a hole if not molested, but if frightened stands up and with neck much dilated, and if, in trying to hit it, you miss, it comes straight at you rather rapidly, and will spit with remarkable accuracy for your face. One taken alive by Mr. Fitsymonds measured five feet. The fluid seems acrid, and may blister a tender skin ; if it goes into the eye, it gives rise to a good deal of smarting for perhaps a day. These snakes are said to be very poisonous ; but Mr. Fitzsymonds states that he knows of no authenticated case of anyone having been bitten. And, according to his experience, they very rarely bite, but always spit until the saliva is exhausted.

DEATH-ADDERS

Among the deadliest of Australian snakes is the purple death-adder (*Pseudechis porphyriaca*), typically representing a genus characterised by the great elongation and slenderness of the cylindrical body, the sharply-pointed tail, the small head, imperfectly differentiated from the neck and clothed with large shields, the smooth scales, arranged in from seventeen to twenty-three rows, the divided anal shield, and the arrangement of the shields on the under surface of the tail at first in a single, and posteriorly in a double series. Behind the fangs are one or two solid teeth in the upper jaw, the pupil of the eye is round, and the neck cannot be dilated. This snake, which grows to a length of about seven feet, is very variable in colouring. Generally, however, the colour of the back varies from shining purplish black to dark olive brown, the under parts being red and the sides carmine, but the latter colours not occupying the centres of the scales, which are black, as are the hinder borders of the shields of the under surface. Generally known to the settlers by the name of the black snake, this reptile is dreaded alike by natives and Europeans, although, fortunately, it nearly always endeavours to escape when discovered.

The short death-adder (*Brachyaspis curta*) is selected as a well-known example of a group of Australian genera, which include a large number of species. Closely resembling the harmless snakes in general appearance, these death-adders are distinguished from the other members of this section by the presence in the upper jaw of a row of small, curved, solid teeth behind the fangs. The head is unsymmetrically four-sided, flattened, and rounded at the muzzle, the body massive, and the tail either moderate or short. The smooth and equal-sized scales are arranged in from fifteen to twenty-one rows, those on the middle of the back not being larger than the rest, and there is but a single row of shields on the under surface of the tail.

All these species are peculiar in the group for producing living young. The species shown on page 1581 varies from 3 to 4 feet in length, has a short tail, and nineteen rows of scales. Although very variable as regards colouring, the head is generally uniform black, the body olive-colour, with broad brown or black crossbands, the hind part of the body and the upper surface of the tail uniformly blackish, and the whole of the under parts light yellow. Some specimens have, however, no dark bands on the back.

The spine-tailed death-adder (*Acanthophis antarcticus*) represents a genus easily recognised by the horny appendage with which the tail terminates ; the middle row of scales in the fore part of the body being more or less distinctly keeled. In addition to Australia and New Guinea, this snake also inhabits the Eastern Moluccas, as well as Ceram and Amboyna. It feeds chiefly upon frogs and young birds, and is regarded by Europeans as most deadly, although the natives believe that no one ever dies from a death-adder's bite.

SPINE-TAILED DEATH-ADDER

SEA-SNAKES

Although often referred to a separate family, the sea-snakes are now considered to represent merely a subfamily (*Hydrophiinæ*) of the front-fanged colubrines. From the preceding subfamily (*Elapinæ*) they are distinguished, not only by their marine habits, but likewise by their strongly-compressed and oar-shaped tails, in the skeleton of which both the superior and inferior spines of the vertebræ are very strongly developed. With the exception of the broad-tailed sea-snakes, which form a kind of transition between the present and preceding subfamilies, these snakes never leave the water ; and the inferior surface of the body and tail is either covered with scales similar to those on the upper parts, or, if shields be present, they are of small size. All are very poisonous, and produce living young.

Their headquarters are the coasts of the Indian Ocean and the tropical districts of the Western Pacific, their range extending from the Persian Gulf to New Guinea and Northern Australia. The parti-coloured sea-snake has, however, a more extensive distribution, ranging from the western coast of Africa to the western shores of Tropical America, and extending so far north as Japan and Manchuria, and so far south as New Zealand. All of these snakes have relatively small heads, jaws, and fangs ; and while in some cases the body is short and thick, in others it is very thick only in the region of the tail, and elsewhere disproportionately elongated and attenuated.

Always varied, the colouring is often brilliant and beautiful ; and the oar-like form of the tail and hind part of the body is obviously an adaptation to an aquatic life, as indeed is the colouring, which is often of the mackerel type. Living in the sea, or in tidal waters, their movements in the clear blue water are agile and elegant, but when thrown ashore, as frequently happens, the majority are helpless. Their food consists of fishes and such other creatures as they can capture in the sea. In parts of the Bay of Bengal sea-snakes are sometimes seen congregating in large shoals. The group is divided into nine genera, no less than six of which are represented in Indian waters.

The broad-tailed sea-snakes, of which there are four species, constituting the genus *Platurus*, in general appearance closely resemble some of the kraits, especially as regards the shape of the skull and the scaling of the head and body, but are distinguished by the compression and depth of the tail. In the upper jaw, which is very short, there is in the maxilla of each side a pair of large grooved fangs, followed by a single very small solid tooth. The arrangement of the shields of the head is normal, each nostril being pierced in a laterally-placed nasal; the scales on the body are smooth and overlapping, and the inferior surface is covered with large shields.

Of the four species, the banded sea-snake (*P. laticaudatus*) is distinguished by the absence of a keel on the lower surface of the hinder part of the body, and also of an unpaired shield on the muzzle; the scales being arranged in nineteen rows. In colour it is olive above and yellowish beneath, with black rings fully equal in width to the light interspaces. Attaining a length of a little over a yard, this species ranges from the Bay of Bengal and the China Sea to Polynesia. An allied but larger species (*P. colubrinus*), with the same distribution, is distinguished by the presence of an unpaired shield on the head, and the arrangement of the scales in from twenty-one to twenty-five rows; while the third species (*P. schistorhynchus*), from the China Sea and Western Pacific, differs in having a keel along the hinder half of the lower surface of the body.

That the broad-tailed sea-snakes are the direct descendants of terrestrial forms allied to the kraits is proved by their retention of large inferior shields, and by their habits. Not only are these snakes frequently found at some distance from water, but in Sumatra a specimen was captured nearly a day's march inland.

In common with all the other members of the subfamily, the parti-coloured sea-snake (*Hydrus platyurus*) has the nostrils placed on the upper surface of the muzzle; and the under surface of the body and tail in this species is scaled like the rest, although in some of the genera traces of enlarged shields still persist. In the skull the maxilla is considerably longer than the transverse bone, and carries a pair of short fangs, followed, after an interval, by seven or eight solid teeth; the muzzle is elongated; the head-shields are large, the nasals being in contact with one another, and the scales on the relatively short body hexagonal in form and with their edges in apposition.

This snake attains a length of a yard; and in colour is either yellowish with symmetrical black transverse bands or spots, or uniformly black above and yellow with or without black spots below, the yellow tail being ornamented with either black spots or bars. It is the sole representative of its genus, and has a wider distribution than any other member of the group, ranging over the whole of the Indian Ocean and the tropical and subtropical portions of the Pacific (p. 1584).

The typical sea-snakes, forming the large genus *Hydrophis*, differ in having from seven to eighteen solid teeth in the maxilla, by the longer body, on the anterior part of which the scales are imbricating, and by the presence of more or less distinct small shields on the lower surface. The black-banded sea-snake (*Distira cyanocincta*) may be taken as an example of another large genus differing from the preceding in that the fangs are followed in the maxilla by from four to ten solid teeth with their front surface grooved. In these snakes the body is more or less elongated, and generally has the scales of its front portion slightly overlapping, while the under surface carries small shields. The typical species, which grows to a length of 6 feet, is of a greenish olive above, with black transverse bars or rings, which are sometimes connected by a longitudinal stripe on the under surface. This snake ranges from the Persian Gulf to the Malay Archipelago and Japan, and is one of the most abundant in the Indian seas.

There are several points in which the sea-snakes differ from their land cousins as regards habits in addition to those already noticed. In the first place, the skin is changed piecemeal, instead of entire; the casting taking place at very frequent intervals. Moreover, the tongue is very short, and only the extreme tips of its two extremities are exserted through small notches on either side of the rostral shield of the head, which is prolonged downwards so as to close the mouth. When, however, these snakes are cast ashore and almost blinded by the unaccustomed light, the tongue is used in the ordinary manner as a feeler.

As already mentioned, it has long been a matter of common knowledge that the sea-snakes which are met with in such numbers in Indian waters are highly poisonous, and several instances are on record where human beings have succumbed to the effects of bites from these venomous reptiles. For instance, Sir Joseph Fayrer mentions a case in which a ship's captain was bitten by a sea-snake with fatal results while bathing in the Bay of Bengal. The native fishermen on the Orissa coast, where these snakes are very abundant, are well aware of their poisonous properties; and take such good care to avoid being bitten that deaths from this source appear to be rare among them.

Although the fact of the venomous nature of sea-snakes has been so long known, it has only recently been recognised that the poison of these reptiles is far more virulent than that of land-snakes. And the reason of this is not far to seek. The majority of terrestrial poisonous snakes of large size, such as the

BANDED SEA-SNAKE

cobra, are in the habit of preying on warm-blooded mammals and birds, to which a comparatively small amount of poison proves fatal. Sea-snakes, on the other hand, prey mainly, if not entirely, on fish, some of which are as much as a foot in length; and fish require a much larger injection of poison in order to produce fatal effects than is the case with warm-blooded animals; the amount of venom necessary to cause death varying, of course, with the size of the fish.

A large number of specimens have been experimented upon by Dr. L. Rogers at Puri, on the Orissa coast, where these snakes are taken by the fishermen during the cold season in their nets with a frequency which is in proportion to the number of fish captured. As soon as taken, they were transferred to a salt-water tank, in which the majority lived only for a few days, although a few individuals survived for several weeks. The venom was obtained by inducing the snakes to bite on a watch-glass with a thin layer of gutta-percha tissue stretched tightly across; the poison being by this means ejected as clear drops into the glass without any mixture of saliva.

The symptoms produced in animals by the injection of sea-snake venom were found to be, with one notable exception, practically the same as those caused by cobra-poison. The exception was that no appreciable effect is produced on the blood of a land animal bitten by a sea-snake. When fish were bitten, the symptoms were very much the same as in the case of warm-blooded animals, although they were somewhat more difficult to observe.

Previous to the beginning of the experiments it was, as already mentioned, well known

PARTI-COLOURED SEA-SNAKE

that a much larger dose of cobra-venom is necessary to kill a fish than is required to destroy a mammal or bird of the same weight, and it was likewise well known that the *Enhydrina* feeds on fish. As the result of experiments made on the common mud-fish, it was found that fifty times as much cobra-venom is required to kill a fish as is necessary to prove fatal to a warm-blooded animal of the same size.

When, however, the poison of the *Enhydrina* was employed, the fatal dose for a fish was only ten times as great as that required for a mammal. In other words, the poison of the *Enhydrina* is much more deadly for fish than is cobra-venom, even when full allowance is made for the greater potency of the former in the case of warm-blooded animals. Sea-snake poison is, in fact, specially adapted for the needs of the reptiles by which it is secreted, being about fifty times as effective for the destruction of fish as is cobra-venom. The advantage to the *Enhydrina* of being thus enabled to kill its slippery prey with great suddenness is sufficiently obvious. Even more deadly is the venom of *Hydrus platyurus*. It may be added that it is just these two species—the *Enhydrina* and the *Hydrus*—which are regarded by Puri fishermen with the greatest dread, probably from the fact that human beings have in times past succumbed speedily to their bites.

It has been found out that sea-snake venom has no appreciable effect in causing coagulation of the blood of the animals into which it is injected, being in this respect markedly different from cobra-poison. Nevertheless, the former kills by producing symptoms almost identical with those caused by the latter. Now, there is very strong reason for believing that these symptoms, in the case of sea-snake bite, are due to injury to the nervous system of the victim. And this affords a very strong presumption that Sir Joseph Fayrer was right in his view that death by cobra-bite was due to lesion of the nervous system, and not to the effects of the poison on the blood. Traces of action on the blood are, however,

exhibited by sea-snake poison, and it would thus seem probable that the more marked effects produced by cobra-venom are likewise inherited from ancestors which did actually kill by blood-poisoning instead of by nerve paralysis.

"In this connection it is interesting to observe," writes Dr. Rogers, "that all through the poisonous snakes we find evidence of an action on the blood and on the nervous system in different degrees. Thus, beginning with the viperine snakes, we first have the duboia, *Vipera russelli*, which appears to be the purest blood-poison of the known venomous snakes, killing by producing intravascular clotting in large doses, and the opposite effect of total loss of coagulability in repeated sub-minimal lethal ones. Then we come to the class of pit-vipers, of which the rattlesnakes of America have been most closely investigated by Messrs. Mitchell and Reichert. They also found a very marked effect on the blood, apparently similar to that produced by the duboia; but, combined with this, we have a marked paralytic effect on the nervous system, and especially on the respiratory centre, for the authors mentioned conclude that although death may occur through the effect on the blood, yet they add : ' There can be no question, however, that the respiratory centres are the parts of the nervous systems most vulnerable to the poison, and that death is commonly due to their paralysis.'

"Leaving the viperine snakes and passing on to the poisonous colubrines, we first come to the Australian *Pseudechis*, and we find again a combination of the two effects to such a marked degree that, when the venom is administered intravenously, death results from intravascular clotting, as in the viperine snakes, while if minimal lethal doses are given subcutaneously, death results through paralysis of the respiratory centres.

"Next we come to the cobra, and here we find the nerve-symptoms quite predominate, although some considerable effect on the blood, in the form of reduction of coagulability and dissolution of the red corpuscles, still survives, although it now takes quite a secondary position to the effect on the nervous system.

"Lastly, we have the *Hydrophiinæ*, which, morphologically considered, are but colubrines modified for an aquatic existence, and here we find a practically pure nervous poison, although there still persists a a trace of action on the blood if strong solutions of the venom are employed, although it can have no actively poisonous effect.'"

It has been already stated that sea-snake poison is much more virulent than cobra-venom, and it may be added that it appears to be intermediate between the latter and the poison of tetanus. Moreover, the action of the poison of tetanus and of the venom of colubrine snakes is so similar that the suggestion naturally arises of an intimate relationship between the two. And if this be so, both these snake-venoms probably act by being taken up in the circulation and fixed in the nerve-centres until such time as a quantity sufficient to paralyse the action of the latter has been accumulated.

Omitting mention of the small and unimportant family of harmless snakes known as blunt-heads (*Amblycephalidæ*), represented by two Asiatic and two Tropical American genera, attention may be directed to the viper tribe (*Viperidæ*), which includes the whole of the remaining members of the suborder. The distinction between a colubrine and viperine snake is that in the latter the maxillæ, or hinder upper jaw-bones, are capable of being erected in a vertical plane at right angles to the transverse bones, while in form they are short and thick, and they always carry a single pair of large tubular fangs.

All vipers are poisonous, and nearly all of them produce living young, while they are more or less nocturnal and terrestrial in their habits, although a few ascend trees. The thick body, the flat, large, and often triangular head, sharply defined from the neck, the short and stumpy tail, the reduction of the maxillary teeth to a single pair of fangs, and the vertical pupil of the eye, are all features distinguishing vipers as a whole from the poisonous colubrines; but, as already mentioned, it is frequently necessary to examine the structure of the skull itself before any particular snake can be assigned to its proper serial position. That the vipers form a highly specialised group is self-evident; and it seems very probable that they are descended from the hind-fanged colubrines. The family is divided into two groups, namely, the true vipers of the Old World, which attain their maximum development in Africa, and the American and Asiatic pit-vipers.

TYPICAL VIPERS

The first representatives of the Old World vipers, forming the sub-family *Viperinæ*, are the typical vipers, which constitute a genus (*Vipera*) with about ten species, ranging over Africa (exclusive of Madagascar), Europe, and a large portion of Asia, one of them reaching India. In common with the other members of the subfamily, they have no pit in the loreal shield of the head; while they are specially distinguished by the upper surface of the head being covered either with scales or small shields, and by the keeled scales of the body running in straight longitudinal rows, which vary in number from twenty-one to thirty-eight, and likewise by the double row of shields beneath the tail.

The common viper (*V. berus*), which is happily the only British poisonous snake, is one of the smallest representatives of the genus, and distinguished by the mixture of scales and shields on the head (three of the latter being larger than the rest), and the general presence of only a single row of scales between the eye and the upper labial shields beneath. In colour and markings the common viper is extremely variable; but as a rule a dark zigzag stripe runs down the whole length of the middle of the back.

With regard to colour, in some specimens the ground is nearly olive, in others deep rich brown, and in others dirty brownish yellow; while a mark between the eyes, a spot on each side of the hind part of the head, the above-mentioned zigzag line formed of confluent quadrangular spots on the back, and a row of small irregular triangular spots on each side of the body are of a darker hue than the ground-colour, and frequently nearly black. In some examples the under parts are lead-colour, with lighter or darker spots, but in others they are almost wholly black.

A specimen has been recorded in which the ground-colour was nearly white and the markings black; and in one variety the ground-colour is brick-red, with ferruginous markings, while in a second the under parts acquire a more or less marked blue tinge, and in a third the whole skin, with the exception of that beneath the jaw and throat, is black, the usual markings being visible in certain lights. The average length of the common viper is about 10 inches. Its geographical distribution is greater than that of any other European snake, extending from Portugal eastwards to the island of Saghalin, while northwards it reaches to the Arctic Circle, and southwards to Central Spain.

Some years ago Mr. G. A. Boulenger described a viper from Pembrokeshire, remarkable for the absence of the characteristic zigzag marking down the back. This specimen was dark brown in colour, with a rather ill-defined, straight-edged band of darker brown down the middle of the back, and an oblique dark bar on each side of the occiput.

A second specimen of very similar character was described in 1842, and a third taken more recently near Ullswater. Apparently the nearest approach to this peculiar colour-phase is found in Spanish specimens, in which the zigzag dorsal band is replaced by a dark brown vertical stripe, three to five scales wide, bordered on each side by a series of subtriangular or crescentic black spots placed opposite to each other. This band may be relieved on each side by a pale brown or yellowish streak.

SOUTHERN VIPER

In South-Western Europe the true viper is replaced or accompanied by a closely-allied species which may be called the southern viper (*V. aspis*). As it was doubtless to this snake that the Latin term *Vivipara* was applied, German writers restrict the name viper to the southern form, and use the term *Kreuzotter* for the common viper. In the latter the front of the upper surface of the head is covered with three distinct small

THE VIPER

LONG-NOSED, OR SAND, VIPER

the group, attaining a length in some rare instances of just over a yard.

The sand-viper ranges from Italy to Armenia. In Carinthia it is the commonest of snakes, in the Tyrol it is local, but it is abundant in the south of Hungary and Dalmatia. Mainly nocturnal it is much more commonly found in hilly than in level districts, ascending in the mountains to a height of between three thousand and four thousand feet. Except during the pairing season, when it is found in couples, it is a solitary reptile subsisting on other snakes, field-mice, birds, and lizards.

RUSSELL'S VIPER

As being one of the deadliest of Indian snakes, special mention may be made of the beautiful Russell's viper, or duboia (*V. russelli*), of India, Ceylon, Burma, and Siam. From the other viper inhabiting Kashmir (*V. lebetina*) this species may be distinguished by having the rostral shield of the head as long as broad, and the scales on the body arranged in from twenty-seven to thirty-one rows. Sometimes known as the chain-viper, this snake attains a length of 4 feet. Its ground-colour is pale brown, with three longitudinal series of black light-edged rings, sometimes replaced by faint dark spots; the lower parts being yellowish white, either with or without small crescentic black spots. In young specimens, the black rings on the upper parts enclose dark reddish brown spots, which in the middle series are in contact with one another; the species in this condition being an exceedingly handsome snake.

Sir J. Fayrer regards this viper as being, next to the cobra, the most dangerous in India, stating that fowls bitten by it sometimes expire in less than a minute. "It is nocturnal in its habits, is sluggish, and does not readily strike unless irritated, when it bites with great fury; it hisses fiercely and strikes with vigour. Its

shields, but in the southern form it is clothed only with smooth or slightly-ridged scales, among which seldom more than a single polygonal roundish one can be regarded as representing a frontal shield; while instead of the single row of small scales generally separating the eye of the common viper from the upper labial shields, the southern form always has two such rows. There is likewise a difference in the shape of the muzzle in the two species.

The southern viper may be considered characteristic of the Mediterranean countries, occurring in North Africa as well as in Europe. It is noteworthy that in the borderland of the distributional areas of the two forms, such as Northern Spain and Italy, it is often difficult to say to which of these any specimen may belong.

More numerous in Scotland than the ringed snake, but, like the latter, unknown in Ireland, the viper generally frequents heaths, dry woods, and sandy banks. Although its bite produces severe effects, it is seldom, unless the sufferer be very young or in ill-health, that death ensues. During the winter months vipers generally hibernate in small parties for the sake of mutual warmth, several being often found twined together in a torpid condition.

SAND-VIPER

Another well-known poisonous European snake is the long-nosed, or sand-viper (*V. ammodytes*), easily recognised by the presence of a soft horny appendage at the end of the nose, covered with scales, and not unlike a conical wart in appearance. It is also distinguished from the common viper by the absence of any large shield except the supraoculars on the top of the head, although in colouring the two species are very similar. In size it is the largest European representative of

RUSSELL'S VIPER

long, movable fangs are very prominent objects, and with them it is capable of inflicting deep as well as poisoned wounds. When disturbed, its loud hissing is calculated to warn those who approach it; and it does not appear to cause many human deaths, although it may be that its misdeeds are sometimes ascribed to the cobra. This viper is said to frequently kill cattle while grazing, by biting them about the nose or mouth. In proof of its sluggish nature, there is a well-authenticated tale of a young person having picked one up, and, mistaking it for an innocent snake, carried it home ; its true character being only discovered when it bit a dog."

PUFF-ADDERS

In Tropical and Southern Africa the typical vipers are replaced by the large and brilliantly-coloured serpents known as puff-adders (*Bitis*), which differ from the typical vipers by the presence of a series of scales separating the nasal from the rostral scales, as well as by the large size of the post-frontal bone of the skull. The typical puff-adder (*B. arietans*), which inhabits the greater part of Africa and reappears in Southern Arabia, grows to about six feet in length, and is specially characterised by the upward direction of the nostrils, the presence of only one or two rows of scales between the nasal and rostral scales, and the arrangement of the body-scales in from twenty-nine to forty-one rows (p. 1588).

In appearance most hideous and repulsive, this snake has the large and flattened head triangular in shape, very broad and blunt at the muzzle, and sharply defined from the body, the latter being thick and almost triangular in section. The scales of both head and body are keeled and overlapping, and differ from one another only in size. The colouring and marking vary to a certain extent individually, and there is a great change in the brightness of the tints immediately after the changing of the skin. As a rule, however, the ground-colour of the upper surface varies from yellowish to orange brown with a series of regular chevron-shaped dark bars or other markings; the whole type of colouring being admirably adapted to render these reptiles inconspicuous when reposing amid rocks and stones. The under parts are uniformly yellowish white. The enormous size of the hideous head is largely due to the great dimensions of the poison-glands.

The puff-adder, which is everywhere dreaded on account of its deadly nature, and inhabits dry and sandy places, derives its name from its habit, when angry or alarmed, of drawing in a full breath and causing the body to swell visibly. Then the air is allowed to escape gradually, producing as it does so a prolonged sighing or blowing sound, which continues till the lungs are emptied, this process being repeated so long as the provocation lasts. Usually this reptile lies half hidden in the sand, with its head fully exposed, and when approached merely rises without attempting

A FAMILY OF VIPERS

to escape, and so virulent is its bite that even horses have been known to die within a few hours after being struck. The poison was formerly used by the Bushmen of the Cape for their arrows, to the tips of which it is made to adhere by being mingled with the viscid juice of the amaryllis.

The other members of the genus, such as the Gabun puff-adder, or viper (*B. gabonica*), of Tropical Africa generally, and the West African nose-horned viper (*B. nasicornis*), differ by the nostrils being directed outwards as well as upwards, the two species named being specially characterised by the presence of four or five series of small scales between the nasal and rostral scales, and the large size and more or less raised or horn-like character of the scales between the supranasals ; the body-scales being arranged in from thirty-three to forty-one rows (p. 1562).

The Gabun viper is marked on the back with oblong yellow patches arranged in a median row, as well as by a pattern of other colours. The young, of which there may be as many as a dozen or fifteen at a birth, each about ten inches in length, are perfect living gems in the matter of colouring. Even more gorgeous in this respect is the nose-horned viper, especially just after changing its skin. The head is ornamented with a large chocolate arrow-head mark bordered with yellow, while down the middle line of the back runs, at considerable intervals, a series of whitish, yellow-bordered, somewhat butterfly-like patches. Chevrons of chocolate-brown with white edges form a series all along the sides of the body, and the general ground-colour is deep carmine ; minor markings occurring between the lateral chevrons and the central patches.

HORNED VIPERS

Next to the southern viper, or asp, no serpent was more feared by the ancients than the Egyptian cerastes, or horned viper (*Cerastes cornutus*). As a genus, the two species of *Cerastes* are characterised by the small crescentic nostrils situated on the sides of the muzzle, the presence in the male, and sometimes in the female, of a pair of scale-covered, horn-like processes above the eyes, the arrangement of the scales of the body in oblique rows, and the short keels on the scales, which stop short of their tips.

The common horned viper (p. 1559) may be immediately recognised as an inhabitant of desert places from the general sombre and mottled tone of its colouring, which is admirably adapted to such surroundings. Usually attaining a length of about 2 feet, this serpent is of a light-brownish ground-colour, more or less tinged with yellow, upon which are six longitudinal rows of circular or quadrangular dark markings, increasing in size from the middle of the back towards the sides. Beneath

arranged in from twenty-nine to thirty-three rows, the anal shield is single, but the shields beneath the tail form a double series.

The range of this snake includes Northern Africa east of Morocco, as well as Kordofan, Southern Palestine, and Arabia ; the second species, *C. vipera*, inhabiting the northern fringe of the Sahara from Algeria to Egypt. Canon Tristram writes that the usual habit of the horned viper is "to coil itself on the sand, where it basks in the impress of a camel's footmark, and thence suddenly to dart out on any passing animal. So great is the terror which its sight inspires in horses that I have known mine, when I was riding in the Sahara, suddenly start and rear, trembling and perspiring in every limb, and no persuasion would induce him to proceed. I was quite unable to account for his terror, until I noticed a cerastes coiled up in a depression two or three paces in front, with its basilisk eyes steadily fixed on us, and no doubt preparing for a spring as the horse passed."

When about to attack, this snake moves rapidly forward with a sideways motion, unlike that of any other serpent ; and as it will attack when quite unprovoked, the horned viper is more dreaded than any other North African snake, men frequently dying from its bite within half an hour. Its food consists of deserthaunting rodents, together with lizards, and perhaps birds. The second species is commonly believed to have been Cleopatra's asp

EJA, OR SAW-VIPER

While agreeing with the horned vipers in having the lateral body-scales arranged in oblique rows, the two species of the nearly-allied genus *Echis* are distinguished by the presence of only a single series of shields beneath the tail, as well as by the absence of horns. The upper surface of the head is covered with scales, and the keeled scales of the body form from twenty-five to thirty-five rows.

The common saw-viper, or, as it is called in Egypt, eja (*Echis carinata*), attains a length of about 2 feet, and has the keels on the lateral scales of the body strongly serrated. In colour it varies from pale buff to greyish, reddish, or pale brown on the upper parts, with three series of whitish spots edged with dark brown, in addition to which there may be a dark brown zigzag band along each side, while the head is ornamented with a cross or arrowhead mark, and the under parts are whitish, either with or without brown dots. This species inhabits the desert regions of Western and Northern Africa, South-Western Asia, including Seistan, and India and Arabia ; but the second species, *E. colorata*, is restricted to Arabia, Palestine, and the island of Socotra.

THE PUFF-ADDER

the eyes runs a dark-brown band, while the middle of the head is marked by a light brownish yellow streak, dividing posteriorly, and uniting on the sides of the neck with another stripe coming from the chin. The scales surrounding the mouth are bright sandy yellow, the shields on the under surface being also either bright yellow or whitish. The scales of the body are

The most remarkable peculiarity of this viper (which, however, it may possess in common with the horned vipers, since the scales of the latter have a similar structure) is its power of making a curious, prolonged, almost hissing sound, produced by rubbing the folds of the sides of the body one against another, when the

HORNED VIPERS IN THE SAND

serrated lateral scales grate together. This species is a very fierce and vicious viper, always throwing itself into an attitude of defence and offence, coiled up like a spring, and rustling its carinated scales as it moves one fold of the body against another. It is aggressive, and does not wait to be attacked before darting its head and body at its enemy, the mouth wide open, and the long fangs vibrating, so as to present a most menacing appearance. It is very poisonous, and there can be little doubt that it destroys many human lives, as Indian natives are much more exposed to contact with this species than with Russell's viper.

RATTLE-SNAKES

The dreaded rattle-snakes of the New World form the first representatives of the subfamily of pit-vipers (*Crotalinæ*), which are common to Asia and America, and characterised by the presence between the nostril and the eye of a deep pit in each loreal shield, the physiological significance of which is still unknown. All have triangular, broad heads and short, thick bodies. The Asiatic representatives of the group are less deadly serpents than their American relatives, and the only vestige of the rattle of the latter to be found in the former is a small horny spine at the end of the tail of one species. Many of the Indian species are arboreal in their habits, and in correlation with this their colouring assimilates to that of the foliage and boughs among which they dwell.

As regards their geographical distribution, pit-vipers present a curious similarity to bears and deer, and since they are most abundant in the Indo-Malay region, and are also more numerous in North America than in South America, it has been suggested that the group originated in the Indo-Chinese countries, and thence spread northeastwards to North America, and so onward to the southern half of the New World, which area, having been the last to receive the group, has not had time, in spite of its extreme fitness for the same, to allow reptilian life to attain its full development. Rattle-snakes are sufficiently distinguished from their allies by the pointed,

horny appendage at the end of the tail from which they derive their name. In the young rattle-snake the tail terminates in a somewhat nail-like " button," which in a perfect rattle remains at the tip, the various rings, reaching in some instances to twenty or more in number, being gradually interpolated between this and the scaly portion of the tail.

More or less symmetrical in form, the rattle is composed of hollow, horny rings, somewhat like quills in substance, which are interlocked with one another, and are yet so elastic as to allow a considerable amount of motion between them. The various rings do not appear to be formed with any regularity, sometimes several being added in a single year, while at other seasons one only is developed; neither does there seem to be any relation between the growth of the rattle and the changing of the skin. That very large rattles must belong to old snakes is, however, obvious; and that this is really the case is shown by the circumstance that at the present day rattles with twenty rings are very seldom met with, since with the advance of cultivation it is only rarely that these noxious reptiles are suffered to attain their full age. The body is thick, and, for poisonous snakes, somewhat long, and the poison-glands attain very large dimensions.

Since the different species of rattle-snakes are extremely variable in colouring, reliance has to a great extent to be placed on the arrangement of the shields covering the fore part of the head for the purpose of discrimination. In the common rattle-snake of North America (*Crotalus horridus*) the distinctive character is the presence of one pair of internasal shields, and of a large shield between the latter and the supraocular, the paired shields being separated by a series of small ones in the middle line. Between the supraocular

EJA, OR SAW-VIPER

shields begin the long keeled scales covering the body, which are arranged in from twenty-five to twenty-seven longitudinal rows. The ground-colour of the upper surface is a dull greyish brown, upon which are two rows of large, irregular spots, which may unite into zigzag cross-bands, and are gradually lost on the dark tail; the under parts being yellowish white, marked with small black dots. Generally about $4\frac{1}{2}$ feet in length, this species may grow to 6 feet.

In the south-eastern United States the commonest member of the genus is the diamond rattle-snake (*C. durissus*), which is not only the most beautiful, but likewise the largest species, adult females—which in this group are always larger than the males—not infrequently measuring 8 feet in length. From the common rattle-snake this species may be distinguished by the large and narrow head, on which there are two pairs of internasal shields, by the rostral shield being deeper than broad, by the scales of the body being always arranged in twenty-seven rows, and also by the colouring. After shedding, the new skin is of a beautiful greenish, or occasionally golden brown, ground-colour, upon which is a triple lozenge-shaped chain-pattern on each side of the back, the golden yellow lines of which stand out in marked contrast to the dark diamonds of the ground-colour. A blackish brown band runs from the muzzle through each eye to the corner of the mouth; and the top of the head is either uniformly coloured, or ornamented with irregular markings.

In the south-western United States the group is represented by *C. confluentus*, distinguished by the broader, transversely striated, supraocular shields. Of the eleven species of the genus, ten are confined to North America, and only one ranges to the southward of the Isthmus of Panama. The latter species (*C. terrificus*), shown in the lower figure of the illustration on page 1591, approaches the common species as regards the arrangement of the shields on the head, but in colouring is like the diamond rattle-snake. From the former it may be distinguished by the circumstance that the two pairs of shields between the rostral and the supraocular usually have no small shields between them, so that they come into contact with one another in the middle line; while from the latter the larger size of the lozenges on the body, and the presence in each of a light-coloured centre, will serve as a sufficient distinction, in addition to the different arrangement of the head-shields. In the latter there is a pair of prefrontal shields behind the internasals, which are wanting in the other species. The range of this species extends from Arizona to Argentina.

As already mentioned, rattle-snakes chiefly frequent dry and sandy localities, more especially when they are covered with bushes; but they also frequently take up their abode in the burrows of the prairie-dog, or prairie-marmot. Formerly it was believed that the snakes and the marmots lived together in complete harmony, but it is now ascertained that the former prey on the young of the latter. The general food of rattle-snakes consists of small mammals, birds, lizards, and frogs, the latter being especial favourites; but mammals as large as minks have occasionally been taken from their interiors.

The most extraordinary peculiarity connected with the common species is its habit in the colder regions of North America of collecting in enormous numbers for the winter sleep. In some districts the snakes used to assemble in hundreds, or even thousands, from all sides to sleep in the ancestral den, some of them, it is said, travelling distances of twenty or even thirty miles. Huddled together in masses for the sake of warmth, the serpents passed the winter in a state of more or less complete torpor, until the returning warmth of spring once more started them to spread over the country. When rattle-snakes were abundant, annual or biennial hunts used to take place at these dens, the fat of the slaughtered reptiles being used as a valuable supply of oil.

Catlin relates how, when a boy, he once assisted at one of these hunts at a place known as Rattlesnake Den, whence the snakes used to come forth on to a certain ledge of rock in swarms. At one time, he writes, there was a knot of them "like a huge mat wound and twisted and interlocked together, with all their heads like scores of hydras standing up from the mass," into which he fired with a shot-gun. Between five hundred and six hundred were killed with clubs and other weapons, but hundreds more escaped to the den. Fortunately, one large one was taken alive, and was made the means of destroying the rest, a powder-horn with a slow fuse being applied to its tail, and the reptile allowed to crawl back to the cave, where a loud explosion soon told the tale of the destruction that had taken place.

The most interesting point in connection with rattle-snakes is the use to which the appendage from which they derive their name is put—for use it must surely have. The old view was that it was intended to warn creatures preyed on by these reptiles of the approach of their enemy; but, in regard to this supposition, Darwin well observes that "I would almost as soon believe that the cat curls the end of its tail when preparing to spring in order to warn the doomed mouse. It is a much more probable view that the rattle-snake uses its rattle, the cobra expands its frill, and the puff-adder swells while hissing so loudly and harshly, in order to alarm the many birds and beasts which are known to attack even the most venomous species. Snakes act on the same principle which makes a hen ruffle her feathers and expand her wings when a dog approaches her chickens."

In this passage the writer commits himself to the view that the rattle is an instrument of intimidation. It may, however, be observed that the sound would be quite as likely to attract enemies as to repel them. Moreover, it is now a well-ascertained fact that rattle-snakes do not possess the power of hissing; and as that faculty seems more closely connected with fear than any other emotion, it would be quite reasonable to suppose that the rattle stands in place of the hiss.

Another feature in the controversy is the circumstance that the sound of the rattle of one snake causes all

THE RATTLE-SNAKE

its kindred within hearing to sound their own, and the organ therefore probably serves as a means of communication. What is known as the "dinner-bell" theory—that is, that a rattle-snake attracts insects like grasshoppers and cicadas within striking distance by the resemblance of the sound of its rattle to their own stridulating utterances, has been pretty clearly disproved; while if it required a further quietus, the circumstance that these reptiles do not appear to prey habitually upon insects would be sufficient. On the whole, while admitting that fear has probably some share in the matter, it seems better to suspend judgment before definitely committing ourselves to any particular view.

That rattlesnakes are some of the most deadly of all venomous serpents may be freely admitted; and it seems that we must almost concede that they possess the mysterious power of "fascinating" their victims before striking. Moreover, the assertions as to the power possessed by vipers of swallowing their young are equally numerous in the case of the serpents under consideration. Pigs are deadly foes to rattle-snakes, killing and eating them greedily whenever opportunity occurs.

It should be added that certain rattle-snakes, such as *Sistrurus miliaris*, from North America east of the Rockies, are separated as a distinct genus, owing to the circumstance that the head is covered with large shields instead of a number of small ones.

THE BUSHMASTER

The formidable South American snake known to the Dutch settlers of Guiana as the bushmaster, but by the Brazilians termed the surukuku (*Lachesis muta*),

DIAMOND AND SOUTH AMERICAN RATTLE-SNAKES

differs from the rattle-snakes by the presence of a distinct keel-like ridge down the back, and, in place of a rattle, in having the under surface of the tip of the tail covered with from ten to twelve transverse rows of small, spiny, sharp scales, while the extremity terminates in a spine. This snake, which attains a length of from 9 to 12 feet, has the ground-colour of the upper parts reddish yellow, upon which is a longitudinal row of large blackish brown lozenges, each having two light spots on either side of the middle line, while the under parts are yellowish white, with a porcellaneous glaze. The large size and enormous poison-fangs of the bushmaster render it one of the most formidable of the pit-vipers; its bite being apparently fatal to human beings in a few hours. Fortunately it is far from common, and inhabits only the secluded portions of the primeval forest, where it lies coiled up on the ground. Unlike most snakes, when disturbed it makes no attempt to flee, but strikes with the rapidity of lightning at the disturber of its slumbers. It lays eggs.

Although formerly referred to a separate genus (*Trimeresaurus*), a number of pit-vipers from Tropical America and Asia are now included in *Lachesis*. Like the bushmaster, these vipers

THE BUSHMASTER

are long-bodied snakes, characterised by the whole of the upper surface of the triangular head being covered with scales instead of shields, the tail, which is frequently prehensile, ending in a sharp point, and having either one or two rows of shields on its lower surface. In all the Asiatic species there are two rows of these subcaudal shields, and it is only in a few of the New World members of the group that they are reduced to a single series. The number of longitudinal rows of scales on the body is very variable in the different species, ranging from as few as thirteen to as many as thirty-one. In Asia these snakes range from India to the South of China and the Liu-Kiu Islands; and while some species are terrestrial and normally-coloured, others are arboreal, and in the greenish tints assimilate to the colour of their surroundings.

GREEN TREE-VIPER

Among the latter, the green tree-viper (*L. graminea*) belongs to a group of four allied Indian and Burmese species characterised by their prehensile tails and the arrangement of the scales on the body in from thirteen to twenty-three rows; this particular species usually having twenty-one rows of scales, while there are from seven to thirteen scales in a transverse series on the head between the supraoculars. The temporal scales are smooth, and the shields on the lower surface of the tail vary in number from fifty-three to seventy-five.

Attaining a length of 2½ feet, this snake usually has the upper parts bright green, although in some specimens they may be yellowish, greyish, or purplish brown, and may or may not be marked with black, brown, or reddish spots. Generally there is a light-

CLIMBING PIT-VIPER

coloured or reddish streak along the outer row of scales, and the end of the tail is frequently red or yellow, the under parts being green, yellow, or whitish. Ranging from Bengal to China and Timor, this species is thoroughly arboreal in its habits.

These snakes are very common in the limestone-hills near Moulmein, Burma, where they are exactly of the same green colour as the foliage amongst which they hide themselves. Dr. Stoliczka states that he saw small specimens very often on low umbelliferous plants growing about a couple of feet high. One of the snakes had its tail wound below round the stem of the flower on the top of which it was basking. All were very sluggish, and did not make the slightest attempt to escape when approached, and even allowed themselves to be removed from the top of the plant. Neither did they offer to bite, unless when pressed to the ground with a stick, but when thoroughly aroused, they turned round and bit furiously.

THE FER-DE-LANCE

Much dreaded by the negroes of the West Indies is the terrible rat-tailed viper, or fer-de-lance (*L. lanceolata*), one of several American species with non-prehensile pointed tails, whose habits are terrestrial. Reaching a length of nearly seven feet, with a body as thick as a man's arm, this snake is very variable in colouring, although the ground-colour of the upper parts is generally reddish yellow brown. The distinctive markings take the form of a black stripe, which is but seldom absent, running from the eye to

HALYS VIPER

the neck, and of two rows of irregular dark cross-bands on the body. In some specimens the sides of the body are, however, bright red. The form and arrangement of the scales on the head, the presence of seven upper labial shields, and the arrangement of the body-scales in not more than twenty-nine rows, together with the uniformly-coloured under surface of the body, serve to distinguish this species from its congeners. This snake is an inhabitant of the Antilles and Tropical America. During the daytime it lies curled up in repose, with the head within the middle of the coils of the body, ready to dart out with the rapidity of lightning on the approach of an enemy.

THE LABARIA

The South American snake locally known as the jararaca is now considered to be inseparable from the fer-de-lance, but the labaria (*L. atrox*) is regarded as a distinct species. In common with the typical form of the fer-de-lance, the first of these two closely allied snakes, which ranges from Amazonia southwards to San Paulo and westwards to Ecuador and Peru, has eight or nine upper labial shields on the snout, and from twenty-five to twenty-seven rows of scales on the body, the general colour of the upper parts being grey or greyish brown, with small dark brown cross-bands, bordered by darker edges ; while the under parts are grey, with two or four irregular longitudinal rows of whitish or yellowish spots.

The labaria differs in having only seven upper labials, as well as in certain details of colouring, the back showing dark lozenges alternating with X-shaped markings, while the under parts are darker, with sometimes two rows of white spots, and from the eye to the corner of the mouth runs a broader dark brown stripe. Inhabiting Eastern Brazil, this species extends so far north as Guiana, but its southward range is less than that of the jararaca.

Writing of the latter, Mr. T. Bates states that in Brazil it is far more dreaded than the jaguar or the alligator. " The individual seen by Lino lay coiled up at the foot of a tree, and was scarcely distinguishable, on account of the colours of its body being assimilated to those of the fallen leaves. Its hideous, flat, triangular head, connected with the body by a thin neck, was reared and turned towards us ; Frazao killed it with a charge of shot, shattering it completely, and destroying its value as a specimen. In conversing on the subject of the jararaca as we walked onwards, every one of the party was ready to swear that this snake attacks man without provocation, leaping towards him from a considerable distance when he approaches. I met, in the course of my daily rambles through the woods, many jararacas, and once or twice very narrowly escaped treading on them, but never saw them attempt to spring. On some subjects the testimony of the

natives of a wild country is worthless. The bite of the jararaca is generally fatal."

HALYS VIPERS

The so-called halys vipers (*Ancistrodon*) are specially characterised by having the upper surface of most or the whole of the front of the head covered with large shields, by the body being rather long and clothed with from seventeen to twenty-seven rows of keeled scales, and by the very short tail having its lower shields arranged in either a double or a single series. Some of the species are furnished with a small spine at the extremity of the tail, which is regarded as a rudimentary rattle. The genus is common to Asia and North and Central America, and is represented by about half a score of species, two of which are found in India, while one ranges so far east as the Urals, where it just enters the confines of Europe. In habits they are all terrestrial.

Of the Indian species, in both of which at least the majority of the shields on the lower surface of the tail are arranged in two rows, the Himalayan halys (*A. himalayanus*) is distinguished by having two pairs of large shields on the muzzle, the extremity of which is but little turned upwards. In colour it is brown, with black spots or transverse bands, but sometimes a light festooned stripe runs down the back ; from the eye to the angle of the mouth is a black streak edged with white, and the under parts are either dark brown, or variegated with black and white.

This snake, which grows to nearly a yard in length, is abundant in the North-Western Himalaya, at elevations of between five thousand and eight thousand feet, although it sometimes ascends considerably higher. The carawila (*A. hypnale*), of Ceylon and Western India, is a much smaller species, not exceeding 20 inches in length, and characterised by the extremity of the upturned muzzle being covered with small scales.

RAT-TAILED VIPER, OR FER-DE-LANCE

The typical member of the group is the Siberian nalys (*A. halys*), which is somewhat superior in size to the common viper, and may be recognised by the small portion of the head that is covered with shields, and also by the circumstance that each shield, or pair of shields, overlaps with its hind edge the shield immediately behind it, thus producing a more or less marked imbrication of the whole of the head-shields. Another characteristic is to be found in the small size of the anterior frontal shields, which have collectively a crescentic shape and somewhat saddle-shaped upper surface. The head is very distinctly defined from the compressed neck, and the body is rather long, of a rounded triangular form in the middle, and covered with twenty-three rows of triangular scales ; the very short tail, which is much thinner than the hinder part of the body, is conical, and armed with a forked horny appendage, representing a rattle.

The ground-colour of the middle of the back is a dark brownish yellow grey, while that of the under parts is a yellowish white, with more or less well-defined black spots on the hinder shields. The yellow ground of the labial shields of the head has chestnut-brown markings ; and the crown of the head bears a large quadrangular blotch, forming an interrupted transverse band on the frontal shields, and a temporal band running from the hinder border of the eye to the angle of the mouth and the side of the neck. Somewhat similar markings ornament the back, and are more or less clearly margined with yellow. Along the back and the ridge of the tail are a number of yellowish or yellowish white black-edged irregular blotches or cross-bands ; and on the sides are two rows of blackish brown spots with white edges, which frequently run one into another, the first dark spot on the neck differing from the rest by its horseshoe form. The distributional area of this snake extends eastwards from the Volga to the Yenesei.

In North America, one of the best-known and most widely distributed members of the genus is the strikingly-coloured and deadly copper-head, or moccasin snake (*A. contortrix*), which seldom much exceeds a yard in length. The body is strong and thick, the short tail provided with one row of shields inferiorly and with a horny appendage at the end, while the elongated triangular head is markedly distinct from the neck, with the pits on the snout rather shallow, the gape of the mouth very wide, and no small smooth shields behind the large parietals.

A beautiful coppery brown, becoming lighter on the sides, forms the ground-colour of the upper parts, and on this are about sixteen reddish brown dark-edged bands, becoming wider on the flanks. On the under parts the shields are copper-red, marked on the sides with large polygonal or rounded alternating dusky spots. The flattened head is generally lighter-coloured than the body, and marked by a broad stripe running from the snout along the side to the angle of the mouth. The distribution of the copper-head extends from the 45th parallel of north altitude to the extreme south of the eastern United States. Its favourite haunts are damp situations, more especially shady meadows covered with tall grass, and its food consists of mice, birds, and probably frogs.

THE COPPER-HEAD

Another well-known North American representative of the same genus claiming brief mention is the water-viper (*A. piscivorus*), which inhabits marshes, rivers, and lakes, and attains a length of nearly 5 feet. From the copper-head this species may be distinguished by the presence of two small smooth supplemental shields behind the parietals, and of numerous small scales between the hinder frontal and temporal shields. The colour is very variable, but in the majority of specimens on a shining greenish grey ground are a larger or smaller number of dark bands somewhat similar to those of the copper-head.

Always found in the neighbourhood of water, this snake extends southwards from North Carolina over the whole of North America and westwards so far as the Rocky Mountains. Feeding chiefly on fishes and frogs, it will also devour all animals that may happen to fall into the water and are not too large for its maw, while in the rice-fields it is the dread of the negroes. Not only is the water-viper feared by man, but it is shunned by all animals dwelling in or near water.

THE WATER-VIPER

DOLICHOSAURIA AND MOSASAURIA

EXTINCT SEA-SERPENTS

DURING the Chalk, or Cretaceous, period there existed two groups of marine reptiles which, although clearly referable to the order Squamata, come under the designation neither of lizards or snakes, and are accordingly regarded as representing suborders by themselves. The first of these groups is represented by a small snake-like reptile from the English Chalk, described under the name of *Dolichosaurus*, and typifying the suborder Dolichosauria.

and the quadrate very loosely attached to the temporal region. Teeth were present on some of the bones of the palate, as well as on the margin of the jaws ; those of the latter series being large, sharply pointed, and attached by expanded bases. The bones of the shoulder-girdle and pelvis were more or less imperfectly developed, and the limbs, although conforming in general structure to those of lizards, were modified into paddles or flippers, with the toes encased in a common skin, and devoid

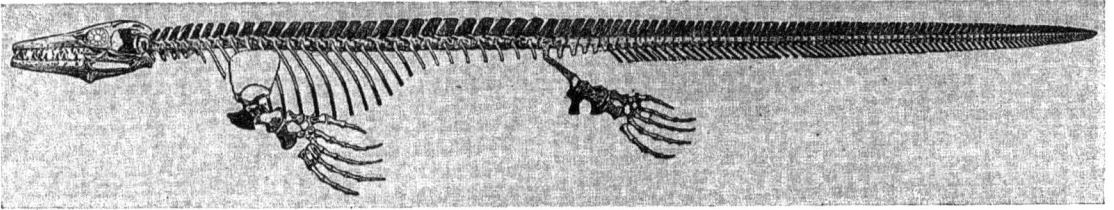

SKELETON OF A MOSASAUR

Unlike ordinary lizards, which have not more than nine vertebræ in the neck, this strange reptile has from fifteen to seventeen of these segments, while its hind limbs are characterised by having all the five metatarsal bones of the foot well-developed. The feet retain a generally lizard-like structure, the head is relatively small, and the whole structure of the skeleton reveals a very generalised type of organisation. The vertebræ have additional articulations like those of snakes. Remains of a nearly allied genus have been discovered in the Dalmatian representative of the Chalk, which takes the form of a hard, compact limestone.

Whether such a creature as a "sea-serpent," in the proper sense of the word, exists at the present day is a question it would be unprofitable to discuss in this place; but that the Upper Cretaceous seas were inhabited by gigantic snake-like reptiles, for which no better designation could possibly be suggested, is an absolute certainty. They derive their name of mosasaurs from the circumstance that the remains of the typical *Mosasaurus camperi* were first discovered at Maastricht, in Holland, on the bank of the Meuse.

of claws. There were either nine or ten vertebræ in the neck ; and while in some cases the vertebræ resemble those of snakes, in other instances they lack the additional articulations distinguishing the latter. The jaws were loosely articulated together in front, to admit of these voracious reptiles swallowing prey of large size ; and the white of the eye was furnished with a ring of bones comparable to that of ichthyosaurs. In addition to the typical *Mosasaurus*, there are numerous other generic types, such as *Platecarpus, Liodon, Tylosaurus,* and *Clidastes*. Mosasaurs seem to have inhabited the Cretaceous seas of all parts of the world, having been obtained from regions so far apart as England, New Zealand, and Argentina ; and while some attained a length of between 30 and 50 feet, others were not more than 8 or 10. Then, again, while in some cases the jaws were armed with powerful teeth to their extremities, other kinds had long, toothless beaks.

These marvellous sea-serpents, which must have been the tyrants of the later Cretaceous seas, came to the fore just about the time that the fish-lizards, or ichthyosaurs, were on the wane, and flourished

SHARP-NOSED TYLOSAUR
From a painting in the American Museum of Natural History

All these reptiles of the suborder Mosasauria are characterised by the elongated form of the body, the great relative size of the skull, which approximates in structure to that of the monitors among existing lizards, the nasal and premaxillary bones being welded together,

for a short period after the latter group had disappeared. Their reign was, however, of short duration, as they were soon destined to give place to the ancestors of the modern Cetacea, among which the killer can alone vie in voracity with these tyrants of the Cretaceous seas.

ORDER ICHTHYOPTERYGIA
FISH-LIZARDS

Prior to the reign of the sea-serpents of the Upper Cretaceous epoch the oceans of the world formed the hunting-grounds of two ordinal groups of marine reptiles, neither of which presents near kinship to any of the existing reptilian orders, while the two are in point of structure widely sundered from one another. These two groups are the fish-lizards, or ichthyosaurs, and the plesiosaurs, both of which made their first appearance in the Trias, or lowest strata of the Secondary epoch, and lived on till the close of the Chalk period, so that, like ammonites, they are

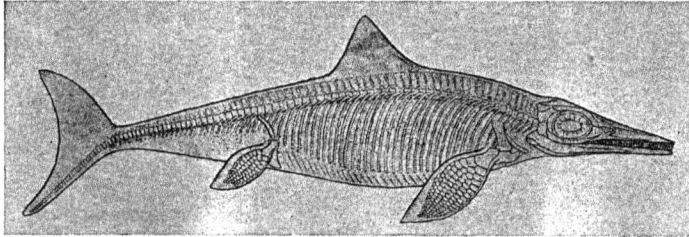

DIAGRAM OF FISH-LIZARD, SHOWING SKELETON AND FORM OF BODY AND FINS

absolutely characteristic of the Secondary epoch. Starting with insignificant beginnings — for their Triassic representatives were quite small reptiles—by the time of the succeeding Liassic period many of the species had attained huge dimensions. Both groups were carnivorous, and both (like other marine reptiles) produced their young alive.

That ichthyosaurs and plesiosaurs are alike descended from land-reptiles is demonstrated not only by the circumstance that the earlier forms of each group are less specially modified for a marine life than is the case with their successors, but likewise by the absence in both groups of gills, which would certainly have been retained had their ancestors been aquatic. Owing to this absence of gills the members of both groups had to come periodically to the surface to breathe. Although in the English Lias remains of ichthyosaurs and plesiosaurs are found together, it is probable that while some of the former reptiles swam in the open sea like whales, the plesiosaurs lurked in estuaries and sheltered bays near the coast. As to origin, it is possible that ichthyosaurs were more or less distantly related to the ancestral tuateras, but nothing can be affirmed with regard to the relationships of the plesiosaurs.

More or less familiar to all from the beautifully preserved skeletons obtained from the Lias of England and the Continent, specimens of which are exhibited in almost every museum, fish-lizards, as typified by the members of the genus *Ichthyosaurus*, were large marine reptiles, with the naked body thick and whale-like, the neck extremely short, and the limbs modified into paddles differing from those of other

members of the class in the structure of their skeleton. The skull is produced into a long snout, generally furnished with a full series of sharp teeth, and mainly formed in the upper jaw by the premaxillary, or front jawbones ; and the nostrils are consequently placed close to the eyes, which, like those of birds, were provided with a ring of movable bony plates. Superiorly the skull has a hole, or foramen, in the parietal bones for a median parietal eye, while posteriorly the upper and lower temporal arches are connected behind the socket of the eye by a bone known as the supratemporal, so that this portion of the skull is completely roofed over, as in the labyrinthodont amphibians. Then, again, the quadrate bone, with which the lower jaw articulates, is firmly united to the adjacent elements of the skull ; while in the general relations of this bone

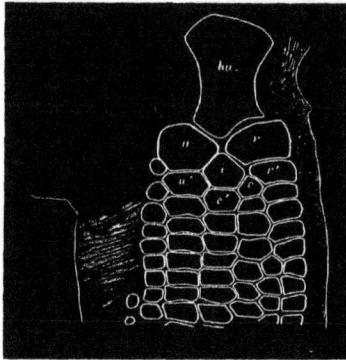

PART OF FORE PADDLE OF FISH-LIZARD

and the bones of the palate there is a marked agreement with the tuateras.

The teeth are confined to the edges of the jaws, where they are implanted in distinct sockets, and generally have conical and fluted crowns, although more rarely they are compressed and smooth, with sharp cutting edges at the front and back. The backbone presents a nearly similar structure, the vertebræ, as shown in the figure on page 1471, consisting of short discs, which may be either deeply cupped or nearly flat at the two ends. In the body and neck each of these vertebræ carries a pair of tubercles on each side for the articulation of the forked ends of the ribs, but in the tail there is but one such tubercle, the ribs being single-headed. The vertebræ are further remarkable for the absence of any bony union between the body or centrum (the part represented in the figure), and the arch enclosing the spinal marrow, so that these two portions are always found detached. The bones of the shoulder-girdle much resemble those of lizards, the collar-bones being well-developed, and the T-shaped interclavicle resting on the lower surface of these and the coracoid bones.

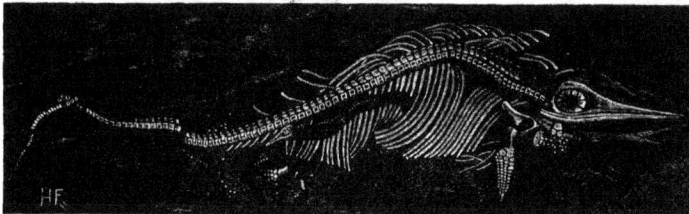

SKELETON OF FISH-LIZARD, SHOWING YOUNG WITHIN CAVITY OF RIBS

The limbs are quite unlike those of any other reptiles, the upper bone (humerus in the fore limb) being very short and thick, while below this the whole of the bones, as shown in the accompanying figure, are polygonal, and so articulated with one another that the skeleton of the paddles assumes a kind of pavement-like or mosaic structure. In most of the species the front paddles were much larger than the hind pair, and while, in some cases, two longitudinal series of bones originate from the bone marked *i* in the figure, thus producing a very broad type of paddle, in other forms (as shown in the skeleton in the figure) only a single series articulates with that bone, and the whole paddle is consequently much narrower.

Specimens like the one figured here show that while the soft parts of the paddle extended but a short distance in advance of the front edge of the bones, on the hind side they terminated in a wide fringe, thus forming a structure admirably adapted for swimming. Other examples indicate that the back of these reptiles was furnished with an upright triangular fin somewhat like that of a porpoise, while the extremity of the tail

shore as a regular dwelling-place, it still resorted thereto on occasion, and probably swam in shallow water in place of frequenting the open sea. In contrast to this are the highly specialised representatives of the Middle Oolitic *Ophthalmosaurus* which were evidently adapted to play the part in the Jurassic oceans of the whales of the present day. Even these, however, display great simplicity of structure in all parts of their organisation except those specialised for swimming, and it is thus evident that ichthyosaurs trace their ancestry to an extremely generalised type of reptile, while it is equally clear that the group is one of the oldest in its class.

Our knowledge of the mode of life of the fish-lizards has been summed up by the great American naturalist, Professor H. F. Osborn, who, after tracing the ichthyosaurian paddle into a limb of the type of that of

SKELETON OF A FISH-LIZARD FROM THE LOWER LIAS OF STREET, SOMERSETSHIRE
From the specimen in the British (Natural History) Museum

was expanded into a vertical fin, of which the lower lobe contained the terminal portion of the vertebral column.

Many of these reptiles attained a length of from 30 to 40 feet ; and the group flourished throughout the whole of the Secondary epoch, that is to say, from the period of the Trias, or New Red Sandstone, to that of the Chalk, most or all of the forms from the first-named deposits being of a more generalised type than those of later date.

The Triassic representatives of the Ichthyopterygia differ from their successors of the Jurassic and Cretaceous epochs for the most part by the less special adaptation to the exigencies of a purely aquatic mode of life, thereby bringing them into closer connection with less specialised land-reptiles. What their terrestrial ancestor may have been is, however, still unknown, but it probably existed at least so early as the Lower Trias.

By the middle portion of that period we find an undoubted aquatic form (*Cymbospondylus*), which retains, however, sufficient indications of affinity with a land-reptile to give a clue to the origin of the group. Although this reptile had probably abandoned the

the existing terrestrial tuatera (*Sphenodon*) of New Zealand, which is regarded as nearly related to the ancestral stock of the group, points out that these reptiles had a dorsal and a caudal fin, a naked scaleless skin, and a spiral valve to the intestine, similar to that of sharks ; while, from the inclusion of skeletons of fœtuses within the ribs of full-grown individuals, it is evident that they produced living young. This viviparous condition is, of course, an adaptive modification, similar to that which occurs in the sea-snakes of to-day, rendered necessary by the pelagic habits of these reptiles.

The similarity in bodily form existing between sharks, dolphins, and fish-lizards is another instance of such an adaptive modification. One of the small American Triassic representatives of the group was originally called *Shastosaurus*, but the name has been changed, on account of pre-occupation, to *Merriamia*.

The restored specimen shown on page 1596 is the typical broad-paddled *Ichthyosaurus communis*. On the other hand, the skeleton shown on this page is that of one of the species, *I. tenui-rostris*, with paddles of a longer and more slender type.

ORDER SAUROPTERYGIA
PLESIOSAURS, OR LONG-NECKED MARINE LIZARDS

THE extinct marine reptiles known as plesiosaurs, or long-necked marine lizards, whose range in time embraced the whole of the great Secondary period, during which were deposited the vast series of strata extending from the Chalk downwards through the Oolites to the Lias and Trias, resemble tortoises in that all or nearly all of the ribs of the back are articulated to the vertebræ by single heads, and in the absence of hook-like (uncinate) processes to the ribs, as well as in the want of a breast-bone, or sternum. In the skull the quadrate bone is immovably fixed, and the palate more or less completely closed; while the lower bones of the pelvis are expanded into large flat plates.

The lower surface of the body was protected by a series of abdominal ribs, each composed of three pieces, forming a forwardly-directed angle, and the whole corresponding to the plastron of tortoises and turtles. The skull has only a lower temporal arch, and the jaws are furnished with a number of pointed and grooved teeth, implanted in distinct sockets; one of such teeth being figured on page 1470. The neck was generally much elongated, and its vertebræ are peculiar in

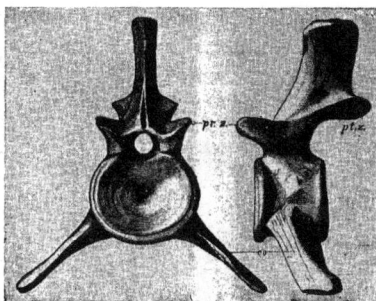

FRONT AND SIDE VIEWS OF NECK-VERTEBRA OF A PLESIOSAURIAN
r. z, anterior, and *pt. z*, posterior articular surfaces o. the arch; *co*, rib

that the ribs which may have either single or double heads, are articulated only to the body of each vertebra (as shown in the accompanying figure). In this respect they differ widely from those of crocodiles, which always have two heads, the lower articulated to the body, and the upper to the arch of the vertebra.

Throughout the backbone of the plesiosaurs the bodies of the vertebræ have either nearly flat or slightly cupped articular surfaces; and in the region of the back each pair of ribs is articulated to a process arising from the arch of each vertebra. Although there are

reptiles. In the typical plesiosaurs (*Plesiosaurus*) of the Lias the ribs of the neck articulate to the vertebræ by two heads; whereas in the genera *Cryptoceidus* and *Murænosaurus* of the Oolites and *Cimoliosaurus* of the Chalk, the ribs, as shown in the figure of a neck-vertebra, are single-headed.

Some of these reptiles were of huge size, attaining a length of between 30 and 40 feet; and in certain of the species from the Cretaceous strata the neck much exceeded the body and tail in length, containing as many as forty vertebræ. Marine and carnivorous in their habits, these formidable reptiles probably lurked in shoal-water, whence they darted their long necks to seize passing fishes in their jaws. In all the groups mentioned above the head was comparatively small, but in the huge pliosaurs (*Pliosaurus*) of the Upper Oolitic strata the skull was of enormous size, attaining in some instances a length of 6 feet, and the neck proportionately short and thick. The teeth in this genus had more or less triangular crowns, and in some cases, inclusive of the root, measured quite a foot in length. In the Chalk the pliosaurs were represented by *Polyptychodon*, characterised by the heavily-fluted conical teeth. There were other and smaller representatives of the pliosaurs in the Upper Oolites, such as *Peloneustes* and *Simolestes*, in which the neck was longer.

As is the case with all the higher aquatic vertebrates, there is evidence to show that plesiosaurs were originally derived from land-animals; the representatives of the group found in the earlier (Triassic) Secondary rocks having limbs departing much less widely from the ordinary type than those of the Liassic forms, and bearing claws at the extremities of their digits. In the

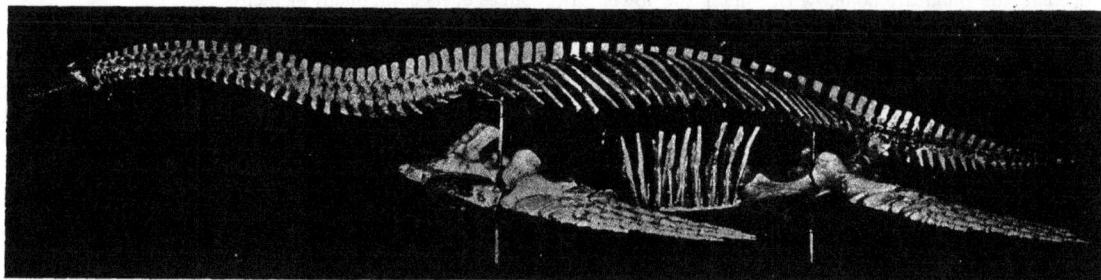

SKELETON OF A PLESIOSAUR
From the Oxford Clay of the Jurassic period, near Peterborough

other interesting features in these reptiles, those mentioned distinguish them from crocodiles and dinosaurs on the one hand, and tortoises and turtles on the other.

With regard to the various groups into which the order is divided, it may be mentioned that in the typical forms, constituting the family *Plesiosauridæ*, the limbs, as shown in the figure above, are converted into flattened paddles, with a shortening of the bones of the upper segments, and an increase in the number of bones corresponding with those of the toes of ordinary

small *Lariosaurus*, which measured about a yard in length, the limbs appear to have been somewhat intermediate in structure between the clawless paddles of the true plesiosaurs and those of more ordinary reptiles; and these early plesiosaurs were probably amphibious in their habits, spending part of their time on land, and part in the water. The Triassic *Neusticosaurus* had limbs of very similar type, but in the contemporaneous *Nothosaurus* and *Simosaurus* the limbs were better adapted for walking, from which it may be inferred that their owners were still more terrestrial in habits.

ORDER CHELONIA

GENERAL CHARACTERISTICS OF TORTOISES AND TURTLES

AMONG all the existing reptiles the most easily defined are those commonly known as tortoises and turtles, and technically as Chelonia, since the presence of a more or less fully developed bony shell investing the body, and containing within it the upper portions of the limbs, at once separates them from all other members of the class. Indeed, so utterly strange is the conformation of these extraordinary reptiles, that if they were met with only in the fossil state they would inevitably be regarded as among the most marvellous of all creatures. Here, however, as elsewhere, the time-honoured proverb holds good, and familiarity from childhood with the common European land-tortoise undoubtedly tends to render us inappreciative of the marvellous bodily conformation of this group of reptiles.

Although the presence of a bony shell is of itself sufficient to distinguish the members of the group from other living reptiles, it is necessary to add somewhat to this in order to give a comprehensive definition of the order. The skull resembles that of the crocodiles in that the quadrate-bone, with which the lower jaw articulates, is firmly wedged in among the adjacent bones, to which its relations are, however, somewhat different. Unlike those of all crocodiles, the jaws are entirely devoid of teeth and encased with horn, so as to form a cutting beak, which is invariably short. A further peculiarity in the skull of a tortoise is to be found in the presence of a greatly developed median spine, marked *sup* in the figure, projecting backwards from the hind region, externally to which is a pair of shorter processes.

In other respects the skull is extremely variable, the sockets of the eyes being sometimes surrounded by bone, while in other cases they are open behind. Sometimes, moreover, the bony roof behind the eye-socket may be prolonged backwards so as to cover the whole of the region marked *par*. There is an equal degree of variation in regard to the position of the posterior or inner nostrils, which sometimes open on the palate close behind the beak, while in other cases they may be situated, as in living crocodiles, close to the hind extremity of the skull.

A most important feature in the structure of these reptiles is to be found in the circumstance that the ribs have but a single head apiece, and that the anterior ones articulate at the junction between two of the vertebræ, so that one portion of the head is applied to one vertebra and the other portion to the adjacent vertebra. The bony shell consists of an upper portion or carapace, and of an inferior portion, covering the lower aspect of the body, which is termed the plastron.

When the shell attains its fullest development, the upper and lower moieties are completely connected together, as shown in the figure of the skeleton of a land-tortoise on page 1600; but in certain groups the two remain more or less separate, and in some cases the lower shell is but very slightly developed. Moreover, while the carapace is generally immovably welded to the vertebræ of the back and the ribs, in the leathery turtle it is completely separate from both. In its fullest developed form, the shell consists of a series of bones articulating with one another at their edges by finely interlocking sutures, and thus forming a continuous whole, capable of increasing in size by growth at the edges of its component elements.

UPPER SHELL OF CHAIBASSA TERRAPIN, AND FOSSIL SPECIMEN OF SAME IN WHICH THE HORNY SHIELDS ARE WANTING

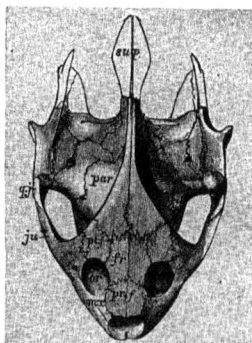

UPPER VIEW OF SKULL OF A SOFT TORTOISE

In the carapace the bones forming the middle of the back are formed by expansions growing from the spines of the vertebræ, while the large lateral plates grow upon the ribs, from which they are inseparable. Within the cavity thus formed are placed the bones of the shoulder and pelvis, to which are respectively articulated the arm-bone and thigh-bone, so that, as shown in the figure of the skeleton, these bones actually come within the ribs, instead of being external to them, as in all other living vertebrate animals. At the fore and hind extremities of the shell are left large apertures, through which are protruded the head and neck, the fore and hind limbs, and the tail A large number of tortoises are able to retract the head, neck, limbs, and tail within the margins of the shell, the apertures of which are then completely blocked; such portions of the head and limbs as are exposed being protected by horny shields.

With the exception of the marine leathery turtles, the fresh-water soft tortoises, and one other species, in which it is invested merely with a continuous leathery skin, the shell of chelonians is covered with a number of horny plates, which, in the adult state at least, are in contact with one another by their edges. As these horny shields are very important in determining the different species of tortoises, it is essential to enter in some detail into their mode of arrangement, and the names by which they are known. In the carapace of an ordinary tortoise, such as the one represented in the left-hand illustration on this page, it will be found that the middle line of the back, exclusive of the margins, is occupied by a single row of large polygonal shields, symmetrical in themselves ; these shields, which are marked *v* in the accompanying diagram, being known as the *vertebrals*. On each side of this median series of vertebral shields is another row of shields (*c*), which are not symmetrical in themselves, and are termed *costals*. The extreme margins of the carapace are formed by a large series of much smaller shields, of which the anterior unpaired one (*nu*) is termed the *nuchal*, and the posterior one (*ca*), which may be either single or double, the *caudal*. Between the nuchal and the caudal runs a series, generally eleven in number on each side, designated the *marginals* (*m*). These same marginal shields, being angulated, pass over the edges of the middle portion of the shell, and thus cover the sides of the middle of the plastron, or lower shell, as shown in the right-hand figure of the accompanying diagram. The shields of the plastron proper are generally arranged in pairs, which may be termed, beginning at the front, *gulars* (*gu*), *humerals* (*hu*), *pectorals* (*pc*), *abdominals* (*ab*), *femorals* (*fe*), and *anals* (*an*). In some cases, as will be illustrated in the sequel, the two gulars may, however, be separated by a single *intergular ;* while, as in the accompanying diagram, there is frequently an inguinal shield immediately in advance of each notch for the hind limbs. Sometimes there is also a series of *inframarginals* separating the marginals from the plastron.

This disposes of the superficial horny shields, which constitute commercial tortoiseshell, but a few words are necessary with regard to the bony elements underlying the shields and forming the shell of a tortoise. On stripping off the horny shields from the carapace of a tortoise, the underlying solid shell, as shown in the right-hand illustration on page 1599, will be seen to be marked by a series of channels corresponding to the borders of these same shields. If the shell be not that of a very aged animal, there will be seen, in addition, a number of finely jagged sutures, marking the divisions between the component bones ; and it will be noticed that in their general plan of arrangement, although not

in number, size, or shape, these underlying bones correspond very closely with the overlying horny shields.

In the middle line of the carapace, for instance, is a series of polygonal plates, symmetrical in themselves, and attached to the summits of the vertebræ, which are known as *neurals*, these being clearly indicated in the figure referred to. In front the series is completed by a large *nuchal* plate, having no connection with the backbone, while, behind, it terminates in one or two *pygals*, which are likewise perfectly distinct from the underlying vertebræ. Externally to the neurals are placed on each side the eight *costal* plates, so named from being attached to the ribs ; the inner halves of these plates being alone visible in the shell figured on page 1599, which belonged to a rather aged animal. Finally, the edges of the carapace are formed by the *marginal* plates, which, like the horny shields similarly named, are angulated, and form the lateral borders of the middle portion of the plastron.

In the plastron itself the anterior portion is formed by a pair of plates, known as the *epiplastrals*, corresponding to the collar-bones, or clavicles, of other vertebrates ; while between or behind there is a single unpaired *entoplastral* element, which may be either dagger-shaped or rhomboidal, and represents the interclavicle of less specially modified reptiles. The remainder of the plastron is formed by three pairs of plates, respectively known as the *hyoplastrals, hypoplastrals,* and *xiphiplastrals*, of which the latter or hindmost and generally more or less deeply notched or forked. These three elements appear to correspond to the so-called abdominal ribs of crocodiles, and it will thus be evident that chelonians have no representative of the breast-bone, or sternum. In certain groups there may be an additional pair of bones—the *mesoplastrals*—intercalated between the *hyoplastrals* and *hypoplastrals*.

As regards their limbs, the members of this order present a great amount of variation, some of them, like the land-tortoises, having the feet adapted for walking, while in the turtles the entire limbs are modified into paddles for swimming. In some cases each of the five toes may be furnished with strong, curved claws, but in other groups, like the soft-tortoises, only three are thus armed. As a general rule the number of joints in the toes of the fore limb, counting from within outwards, is 2, 3, 3, 3, 3, while in the hind limb they are more generally 2, 3, 3, 3, 2, although in a few species the number is the same as in the fore limb. In both limbs

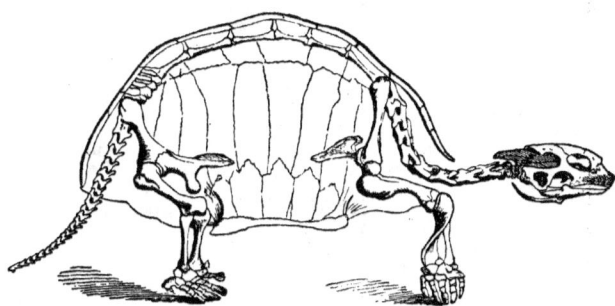

SKELETON OF A TORTOISE

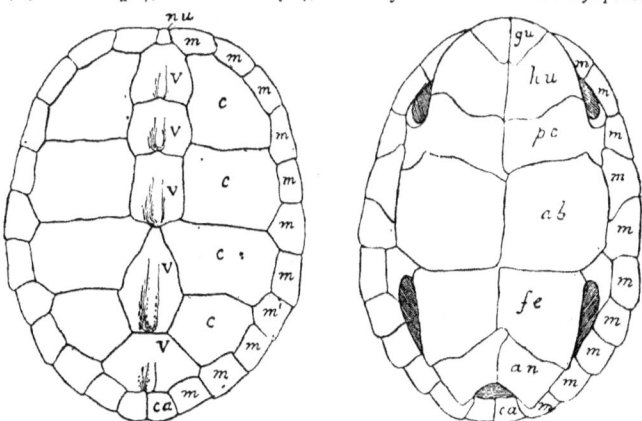

DIAGRAM OF HORNY PLATES ON SHELL OF FRESH-WATER TORTOISE
v, vertebrals ; *c*, costals ; *nu*, nuchal ; *ca*, caudal ; *m*, marginals ; *gu*, gulars ; *hu*, humerals ; *pc*, pectorals ; *ab*, abdominals ; *fe*, femorals ; *an*, anals.

the number of these joints may, however, be reduced, but, except among the soft-tortoises, they are never augmented. Very generally, the surfaces of the limbs, especially the anterior ones of the front pair, are protected by horny plates of variable size, which, among the land-tortoises, may be underlain by nodules of bone.

In habits the members of the order display as much diversity as in structure; some being carnivorous and other herbivorous, while some are marine, others fresh-water, and others, again, more or less exclusively inhabitants of dry land. All, however, are fond of water, and even the most strictly terrestrial species can swim. In nearly all cases the eggs are hard-shelled; and these are invariably deposited on land, the turtles resorting to the shore at certain seasons for this purpose. Land-tortoises and turtles lay round eggs, but those of most of the fresh-water terrapins are cylindrical.

As regards distribution, tortoises are especially characteristic of the warmer parts of the globe, only six species inhabiting Europe, and these being confined to the more southern and eastern countries. The various groups and families are, however, by no means equally distributed over the different regions of the globe. The side-necked tortoises, for instance, are now exclusively confined to the Southern Hemisphere, and in Australia are the only representatives of the order;

LEFT HALF OF PLASTRON OF CHAIBASSA TERRAPIN

whereas the S-necked group attains its greatest development in the opposite half of the world, although represented in many countries lying to the south of the Equator. The soft river-tortoises, again, are confined to the waters of Asia, Africa, and North America, being totally unknown both in South America and Australasia.

Giant land-tortoises within comparatively recent times have been mainly confined to what are known as oceanic islands, although they formerly occurred on most of the large continents; while the smaller members of the same genus (*Testudo*) are far more numerous in South Africa than they are in Asia. Geologically the order is a very ancient one, being represented throughout the whole of the Secondary period, and thus beginning at a date when true crocodiles are not known to have come into existence. As to the origin of the group, nothing definite is at present known.

Chelonians may be divided into three main groups, or suborders, severally designated S-necked tortoises (including the turtles), side-necked tortoises, and soft-tortoises.

The only members of the order of any real commercial value are two species of marine tortoises, one of which takes an important place in culinary affairs, while the other is the source of tortoiseshell. A small amount of tortoiseshell furnished by a few other species is, however, used.

TABLE OF LIVING CHELONIANS MENTIONED IN THIS WORK

SPECIES
Maw's terrapin ....Dermatemys mawi
GENUS 2
Staurotypus
GENUS 3
Claudius

FAMILY 5
Snappers and Alligator-Terrapins—Chelydridæ
GENUS 1
Alligator-Terrapins—Chelydra
SPECIES
Alligator-terrapin ...... Chelydra serpentina
Mexican alligator-terrapin...... C. rossignonii
GENUS 2
Snapper—Macroclemmys
SPECIES
Temminck's snapper .... Macroclemmys temmincki

FAMILY 6
Turtles—Chelonidæ
GENUS 1
Chelone
SPECIES
Green turtle .... Chelone mydas
Hawksbill turtle .... C. imbricata
GENUS 2
Thalassochelys
SPECIES
Loggerhead .... Thalassochelys caretta
Mexican loggerhead ....T. kempi

FAMILY 7
Leathery Turtles—Dermochelyidæ
GENUS
Dermochelys

SPECIES
Leathery turtle ....Dermochelys coriacea

FAMILY 8
Chelydæ
GENUS 1
Chelys
SPECIES
Matamata tortoise ....Chelys fimbriata
GENUS 2
Snake-necked Tortoises—Hydromedusa
SPECIES
Snake-necked tortoise .... Hydromedusa maximiliani
Argentine snake-necked tortoise .. H. tectifera
GENUS 3
Australian long-necked Tortoises—Chelodina
GENUS 4
Rhinemys
GENUS 5
Hydraspis
GENUS 6
Platemys
GENUS 7
Emydura
GENUS 8
Elseya

FAMILY 9
Pelomedusidæ
GENUS 1
Podocnemis
SPECIES
Arau tortoise .... Podocnemis expansa
P. sextuberculata

GENUS 2
Pelomedusa
GENUS 3
Sternothærus

FAMILY 10
Carettochelydæ
GENUS
Carettochelys
SPECIES
Fly River turtle .... Carettochelys insculpta

FAMILY 11
Soft Tortoises—Trionychidæ
GENUS 1
Trionyx
SPECIES
Gangetic soft tortoise .... Trionyx gangeticus
Nile soft tortoise .... T. triunguis
Phayré's soft tortoise .... T. phayrei
American soft tortoise .... T. ferox
GENUS 2
Pelochelys
SPECIES
Cantor's soft tortoise .... Pelochelys cantori
GENUS 3
Chitra
SPECIES
The chitra .... Chitra indica
GENUS 4
Emyda
GENUS 5
Cycloderma
GENUS 6
Cyclanorbis

LAND-TORTOISES AND TERRAPINS

THE land-tortoises (*Testudinidæ*), together with the greater number of the fresh-water tortoises, or terrapins, of the Northern Hemisphere, as well as their southern allies, collectively constitute one of several families belonging to the first great group of the order. From the circumstance that all its members are so constructed as to be able to withdraw their heads within the margins of the shell by bending and retracting the neck in an S-like manner in a vertical plane, the group are designated S-necked tortoises ; their scientific designation being Cryptodira. Since, however, the soft-tortoises likewise retract their heads and necks in a similar manner, it is obvious that this character alone will not suffice to define the group, and it must accordingly be supplemented by others.

Although the degree of ossification of the shell is very variable in the group, the carapace and plastron being in some cases welded into a complete box, and in other instances separate, yet with the exception of the leathery turtle and some of its extinct relatives, there is invariably a complete series of marginal bones, connected with the ribs ; the presence of the full series of marginals, together with the S-like retraction of the neck, being sufficient to distinguish the group. A peculiarity in which the members of the group differ from those of the next one, is to be found in the circumstance that the bones of the pelvis remain throughout life unconnected with the plastron ; while in the greater number of cases the latter comprises only six pairs of horny shields, there being no intergular shield between the first pair, or gulars.

The skull is characterised by the tympanic ring (*t* in the accompanying figure) having a notch in its hind border, and also by the condyle of the quadrate bone fitting into a hollow at the hind end of the lower jaw. This S-necked group includes the marine turtles, and all the tortoises of the Northern Hemisphere, with the exception of the soft river-tortoises, and thus comprises by far the greater number of the living representatives of the entire order. Although well represented in Africa and South America, the group is quite unknown in Australia.

The land-tortoises and terrapins, constituting the family *Testudinidæ*, are characterised by having the shell well-developed and of a more or less ovoid shape ; by the plastron being connected with the carapace either by a straight articulation or by means of sutures, and by the absence of an intergular shield in the front of the latter. The limbs are adapted more or less completely for walking, and are never modified into paddles ; and the head is capable of complete retraction within the margins of the shell. A very important structural feature in the shell is that the nuchal, or unpaired median bone in the front of the carapace, does not send back processes underlying the marginal bones of the same ; while in the tail each vertebra has a cup in front of its body or lower portion, and a ball behind.

SIDE VIEW OF SKULL OF LAND-TORTOISE WITH LOWER JAW REMOVED
t, tympanic ring.

None of the members of the family is marine, but while some are inhabitants of the land, others are more or less exclusively dwellers in fresh-water. There are, moreover, equally important differences in regard to their food, all the land forms being herbivorous, while those frequenting the water some subsist on vegetable and others on animal substances.

LAND-TORTOISES

By far the most numerously represented genus of the whole family is the one including the typical, or land-tortoises (*Testudo*), of which there are rather more than forty existing species (counting a few that have been exterminated within the historic period). These tortoises, of which a few are more or less aquatic in their habits, have the upper and lower portions of the shell completely welded together, the former being

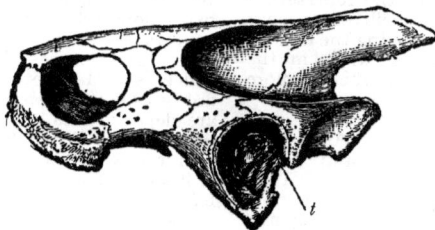

frequently very convex and much vaulted ; while the top of the head is covered with large horny shields. The limbs, which are entirely adapted for walking, are of a club-like form, and covered with large horny scales or tubercles ; their toes being unwebbed and furnished with strong, claw-like nails. The tail is always short, and its proportionate length is not greater in the young than in the adult.

More important characters are, however, furnished by the bony shell and skull, to observe the former of which it is of course necessary that the horny shields should be stripped off. In a shell thus treated it will be seen that the unpaired median neural bones of the carapace are relatively short and wide, and so arranged that a four-sided one is interposed between two that are octagonal, although in some cases the majority of these bones are hexagonal ; the costal, or lateral, plates are alternately narrow above and broad below. As a rule, the line dividing the costal horny shields from the marginals coincides with the suture between the corresponding bones of the carapace, whereas in the other members of the family one is above the other ; while a further peculiarity of most species of the genus is that there is but a single caudal horny plate at the hind end of the carapace.

In the skull the palate is provided with one or two ridges on each side, and the hind aperture of the nostrils is situated on the line of the eyes. It may also be mentioned that, as in the majority of the representatives of the order, the form of the shell differs considerably in the two sexes ; the male having the central region of the plastron deeply concave, while in the female it is flat or slightly convex. The shell is more or less vaulted, and the eggs are spherical.

The typical land-tortoises are distributed over Southern Europe and Asia, the whole of Africa, the southern portions of North America, and South America (inclusive of the Galapagos Islands). They are strictly herbivorous in their diet, and certain species now confined to oceanic islands attain gigantic dimensions, and are by far the largest representatives of the family. The species inhabiting colder regions hibernate during the inclement season by burrowing in the ground, whereas those found in more genial climates are active throughout the year. All the species appear to be diurnal in their habits, and although they are all fond of water, the common European species always withdraws into its shell at the slightest shower. These reptiles will live to an enormous age, which, in some instances at least, may be reckoned by centuries. The species of this extensive genus may be arranged under seven groups, of each of which representative species are noticed below.

GOPHER TORTOISE

The land-tortoises of North America include three species, of which one of the best known is the gopher tortoise (*T. polyphemus*) of the south-eastern United

States. All these species may be easily recognised by the anterior extremity of the palate of the skull having a median longitudinal ridge, instead of the deep pit characterising all other members of the genus.

In the gopher tortoise, as well as in the allied Agassiz's tortoise (*T. agassizi*), the length of the shell is more than twice its height, while the beak is not hooked, and the fore limb is broadest at its extremity.

On the other hand, in Berlandier's tortoise (*T. berlandieri*), from Mexico and Texas, the shell is proportionately shorter, the beak is hooked, and the fore limb widest at the elbow. These species are all of small size, not exceeding 10 inches in length.

BRAZILIAN TORTOISE

The Brazilian tortoise (*T. tabulata*) represents a group by itself, characterised by the carapace being much elongated and somewhat depressed, with its margins not everted, and the general colour dark brown or black, with a yellowish centre to each of the shields on the back. The nuchal shield of the carapace is wanting, and in the plastron the gular shields, although well-developed, are prolonged anteriorly into horn-like processes. The head and limbs are marked with orange or red spots on a dark ground.

This handsome tortoise, which attains a length of nearly 22 inches, is an inhabitant of tropical South America to the east of the Andes, and also of the Windward Islands, ascending to an elevation of about two thousand feet. In many wooded districts it appears to be very abundant, feeding not only on leaves and grasses, but likewise on the fallen fruit which is to be met with in great quantities. In the hot season it constructs a nest of dry leaves, wherein are deposited its eggs, which may be a dozen or two in number. When first hatched, the young are of a uniform yellowish brown colour, with their shells still soft. The young, and to a less degree the adults, have numerous enemies.

Against the puma and jaguar the stout shell of even the adult seems to be no defence, since, according to native reports, those animals, on finding one of these tortoises, will set it up on end and scoop out the flesh with their paws ; while from the occurrence of broken shells in the forest it seems that in some cases they are actually able to tear the plastron away from the carapace. As the flesh is devoid of smell, it is likewise eagerly sought after by both Indians and Portuguese, who are in the habit of keeping these tortoises—known in Brazil by the name of schabuti—in stews, where they are fattened for the table. They are also allowed to run about the houses, where they are fed chiefly on plantains.

BROWN AND LEOPARD-TORTOISES

The five species belonging to the third group, of which the Burmese brown tortoise (*T. emys*) is a well-known example, are characterised by the presence

BRAZILIAN TORTOISE   Photo. W. B. Johnson

of some very large conical, bony, spur-like tubercles on the lower portion of the hind leg, and the circumstance that the length of the union in the middle line of the anal shields of the plastron is considerably less than that of the abdominal shields, the colour of the carapace in the adult being either uniform brownish or yellowish brown closely spotted with black. The Burmese brown tortoise, which attains a length of 18 inches, while agreeing with the species above noticed in the possession of a nuchal shield on the front of the carapace, differs in that the caudal shield at the hind extremity of the same is divided, as in the terrapins.

The shell of this species is much depressed, with the anterior and posterior borders of the carapace serrated, the adult being dark brown or blackish in colour, while in the young the carapace is yellowish brown, with dark brown markings. In addition to the spur-like tubercles on the back of the heel, the whole of the front of the fore limb is covered with overlapping bony tubercles, arranged in four or five longitudinal rows, and there are some conical ones on the back of the thigh, as well as others on the lower surface of the hind foot. This species is an inhabitant of Assam, Burma, Siam, the Malay Peninsula, B o r n e o, and Sumatra, where it frequents moist wooded districts, and is aquatic in habits. The association of a divided caudal shield, with habits resembling those of terrapins, is distinctly noteworthy. The allied *T. pseudemys* is from the Malay Peninsula.

In the other four members of this group the nuchal shield is wanting, and the caudal

BURMESE BROWN TORTOISE

single. Of these, the Argentine tortoise (*T. argentina*), of South America, and the spurred tortoise (*T. calcarata*), of Africa, are characterised by their flattened and uniformly brownish-coloured carapaces.

A specimen of the spurred tortoise from the Sudan preserved in the British Museum has a shell measuring 35 inches along the curve and 29 inches in a straight line, and still larger examples are believed to exist. These tortoises almost completely bury themselves in the sand of the desert during the day, and wander about at night, so that they are seldom captured.

On the other hand, the handsome leopard-tortoise (*T. pardalis*), of Southern Africa, has the carapace highly vaulted, and closely spotted with black upon a yellowish brown ground, its anterior margin being very deeply notched. This species also grows to a large size.

Here it may be mentioned that the remains of a giant land-tortoise (*T. ammon*) have been obtained from the Eocene of the Fayum district of Egypt. The especial interest of this form is its antiquity, which far exceeds that of all other known members of the group. It is probably the ancestral form of the giant tortoises met with in several European Tertiary

horizons, and it is possible that the African *T. pardalis* may be a small survivor of the group, to which the extinct Indian *T. atlas* and *T. cautleyi* and the existing *T. sumeirei* (the well-known giant tortoise of Port Louis referred to on page 1607) may also pertain, all these species agreeing in the horn-like prolongation of the epiplastrals.

STARRED TORTOISES

The fourth group comprises about ten very beautifully coloured small or medium-sized tortoises, the majority of which are confined to South Africa, although the species *T. elegans* is an inhabitant of India and Ceylon. All these tortoises are easily recognised by the carapace being extremely convex, and either black in colour, with yellow lines radiating from the centre of each of the shields of the back, or yellow or brownish marked with radiating black lines. Frequently, moreover, the shields of the back are swollen, so as to form more or less prominent bosses.

The Indian species, together with the allied *T. platynota* of Burma, is distinguished from the other members of the group by the absence of the nuchal shield at the front of the carapace. Of the other species nearly all are South African, although the radiated tortoise (*T. radiata*) is from Madagascar. One of the best-known members of the group is the common geometric tortoise (*T. geometrica*) of the Cape, which attains a length of some 5½ inches.

In the eyed tortoise (*T. oculifera*) the pectoral shields of the plastron may not meet in the middle line, as is the case in some individuals of the Burmese brown tortoise. While the Indian starred and the African geometric tortoise have the carapace black, with narrow yellow rays, in the eyed tortoise the markings take the form of brownish yellow and dark brown rays of nearly equal width.

The Indian starred tortoise is comparatively common in dry, hilly districts, where it inhabits the high grass jungles at the foot of the hills. Nevertheless, these reptiles are by no means easy to find, owing to their colour and appearance harmonising so closely with the rocky ground, and from their habit of remaining in concealment beneath shrubs or tufts of grass during the heat of the day. They are tracked by the Bhils of Meiwar to their hiding-places by following the trail of their footsteps in the dry sand, the same method being employed by some of the wild tribes of South Africa in the case of the allied species inhabiting that continent.

In the rainy season the starred tortoise is, however, extremely active, and wanders about in search of food at all hours of the day. At the approach of the cold weather these reptiles select a sheltered spot, where they conceal themselves by thrusting their shells into

LEOPARD AND RADIATED TORTOISES

LEOPARD TORTOISE

RADIATED TORTOISE
Photographs by C. R. Walter

1605

thick tufts of bushes or shrubs, in order to be better protected from the cold. There they remain in a kind of lethargic, although not truly torpid, state till the hot season, when they issue out to feed only after sunset and in the early morning. Specimens kept in captivity have been observed to be very fond of plunging into water during the hot season, where they often remained for half an hour at a time. They also drank large quantities of water at this period of the year, which they took by thrusting in their heads and swallowing in gulps. About November the female lays her eggs in a shallow pit excavated by herself.

One of the aforesaid captive specimens in the course of about two hours succeeded, according to Captain T. Hutton, " in making a hole six inches in depth and four inches in diameter ; in this she immediately deposited her eggs, four in number, filling up the hole again with the mud she had previously scraped out, and then treading it well in, and stamping upon it with her hind feet alternately until it was filled to the surface, when she bent it down with the whole weight of her body, raising herself behind as high as she could stretch her legs, and suddenly withdrawing them, allowing herself to drop heavily on the earth, by which means it was speedily beaten flat ; and so smooth and natural did it appear that, had I not detected her in the performance of her task, I should certainly never have noticed the spot where she had deposited her eggs. She did not immediately leave the place after finishing her work, but remained inactive, as if recovering from her fatigue."

In disposition these tortoises are decidedly pugnacious, this being especially the case with the males. These combats seem to be chiefly trials of strength, " one male confronting the other, with the hind and fore legs drawn into the shell, and the hind feet planted firmly on the ground, and in this manner striving against each other until one or both becomes fatigued. This was done chiefly when they wanted to pass each other in any narrow space ; and sometimes if the one could succeed in placing his shell a little beneath the other, he tilted him over on his back, from which he had great difficulty in recovering himself ; and I have frequently found them sprawling thus, making desperate efforts with head and feet to throw themselves back to their natural position, which they were unable to effect unless the ground chanced to be very uneven, so as to assist them."

GIANT TORTOISES

During the Pliocene, or later division of the Tertiary period, gigantic land-tortoises were, as attested by their petrified remains, widely distributed over the continents of the world ; species having been obtained from India, France, Egypt, and North and South America. The largest of these was the well-known atlas tortoise (*T. atlas*) from the Siwalik Hills of Northern India, in which the length of the shell was about 6 feet ; the species, together with the Eocene Egyptian *T. ammon*, being, as already mentioned, probably related to the African spurred tortoise and the Burmese brown tortoise. Probably more or less abundant during the epoch in question, with the advent of the ensuing Pleistocene epoch giant tortoises seem to have disappeared entirely from the continental areas, to survive on certain oceanic islands where they were free from the competition of large animals of higher organisation. Some of these insular species, like those of Madagascar and Malta, did not apparently survive the Pleistocene

epoch ; while in other regions they flourished and multiplied till the action of man led to their partial or total extermination.

At the present day the few survivors of these monstrous reptiles are being rapidly reduced in numbers, and, unless special means be speedily taken for their preservation, they will ere long entirely cease to exist. During the historic period the islands where giant tortoises are known to have existed constitute four distinct groups. Three of these are situated in the Indian Ocean, and comprise Aldabra, to the northwest of Madagascar, the Mascarene group—including Réunion, Mauritius, and Rodriguez—lying to the east of the same, and the Seychelles ; while the fourth, or Galapagos group, taking its name from the Spanish word for tortoise, is situated in the far distant South Pacific, off the western coast of South America.

During the sixteenth and seventeenth centuries the tortoises are stated to have existed in enormous numbers in the above-named islands ; but as they afforded a most valuable supply of food, and could be kept alive on board ship, their numbers were rapidly reduced in those of the Indian Ocean, and South Aldabra is now the only island in that area where they still exist in a wild state. Many of these tortoises have, however, been exported to the Seychelles, where the North Aldabra species is preserved ; and it is believed that five specimens carried thence to the Mauritius were the only survivors of the species that formerly inhabited the Seychelles. Regarding the former abundance of these tortoises in Rodriguez, the French traveller and naturalist François Leguat, writing in 1691, observes that " there are such plenty of land-turtles in this isle that sometimes you see three thousand of them in a flock, so that you may go above a hundred paces on their backs."

In Mauritius they were still abundant in 1740, but about 1761 they had probably become scarcer, as thousands were then imported from Rodriguez as food

INDIAN STARRED TORTOISE

for the patients in the hospitals of Mauritius. The continued exportation—some ships taking as many as four hundred tortoises at a time—coupled with the destruction of their eggs and young, finally led to their extermination in both Mauritius and Rodriguez ; this extirpation having probably taken place early in the nineteenth century. The Réunion tortoise, of which little is known, seems to have disappeared earlier.

The total number of species of giant tortoises known to have existed within the historic period is at least twenty, all of which are characterised by their large size, their long necks, and the uniformly dark brown or black colour of their shells. They may be divided

NORTH ALDABRA GIANT
TORTOISES

into four groups, according to their geographical distribution, each characterised by certain structural peculiarities. The first group comprises the Aldabra tortoises, characterised by the presence of a nuchal shield on the front of the carapace, and the distinctness of the gulars on the front of the plastron. On the other hand, in the four best-known Mascarene species, constituting the second group, the nuchal shield is wanting, while the two gulars have coalesced into one; the plastron being characterised by its extreme shortness. The Seychelle tortoise agrees with the latter in its long neck and the absence of the gular shield. Lastly, the fourth, or Galapagos, group presents a condition intermediate between those existing in the two others, the nuchal shield of the carapace being absent, while the gulars of the plastron remain double.

Of the two species formerly inhabiting the islands of the Aldabra group, the North Aldabra tortoise (*T. gigantea*) now survives only in the Seychelles, where a certain number of individuals are preserved in a semi-domesticated condition, and in Ceylon, where there is one specimen. In some specimens of this species the caudal shield is divided, as in the Burmese brown tortoise. This tortoise is one of the largest of the whole group, measuring nearly four feet in total length. The South Aldabra *T. vicina* still inhabits its native island, and is even superior in size to the last.

Of the Mascarene tortoises, the three species from Mauritius (*T. indica, trisserrata,* and *inepta*), all of which are extinct, are characterised by the thinness of the carapace, of which the margins are thickened.

The Rodriguez tortoise (*T. vosmœri*) has a still thinner carapace, which in the male does not shelve down in front in the usual manner. Allusion has already been made to the numbers in which these tortoises existed in Leguat's time. In the Artillery Barracks of Port Louis in the Mauritius lives a very ancient tortoise, which is one of five imported into that island by the navigator Captain Marion du Fresne from the Seychelles in 1766; one of these having been subsequently presented to the London Zoological Gardens in 1832 by Sir C. Colville. The latter weighed 289 pounds, and its shell measured 4 feet 4½ inches in length along the curve, and 4 feet 9 inches in width; while in the Port Louis specimen the circumference of the shell is 9 feet 3 inches, and its height 2½ feet.

Marion's tortoise, as the Port Louis example is called, is thus definitely known to have lived for a hundred and twenty-seven years, and as it was doubtless of large size when brought from the Seychelles, and since all these tortoises take an immense time to reach large dimensions, it is probable that it is the solitary survivor of the enormous herds of these reptiles which existed in Leguat's day. It is believed to represent the original Seychelle species, and is named *T. sumeirei*.

From the length of their necks, and also on account of a peculiarity in the structure of the hinder vertebræ of the neck, it appears that the tortoises of this and the Mascarene species have the power of raising their necks to a nearly vertical position, which would give them a wide range of vision. This elevated range of vision would accord well with the account given by Leguat, who writes concerning these tortoises as follows: "There's one thing very odd among them; they always place sentinels at some distance from their troop, at the four corners of the camp, to which the sentinels turn their backs, and look with the eyes, as if they were on the watch."

The different islands of the Galapagos group, such as Abingdon, Duncan, Albemarle, Chatham, Hood, James, and Charles, are the respective homes of one or more species of giant tortoise. Of the various species inhabiting these islands, the South Albemarle *T. vicina* has the horny shields of the carapace concentrically striated in the adult, and the plastron notched instead of truncated behind. *Testudo nigrita*, of James Island, is an allied species. In other species the shields of the back are smooth, while the plastron has its hind end truncated. In the Central Albemarle *T. microphyes* the width of the bridge connecting the upper and lower shells is of considerable length, and the shell itself stout.

On the other hand, in the Duncan Island tortoise (*T. ephippium*) and the Abingdon tortoise (*T. abingdoni*) the same bridge is relatively short, and the shell remarkable for its thinness, the carapace being much narrowed anteriorly, where it is so pinched in at the sides as to have a sharp ridge on the back. In the former of these saddle-backed species the shell still retains the normal bony framework, but in the second it is soft and leathery. Both have very long necks, which are carried nearly vertically; and in the Abingdon

1607

species the notches in the front end of the shell are so large that in a front view the animal appears merely to have a kind of mantle thrown over the body. It is hard to see what can be the object of this softening and atrophy of the shell; but it is quite clear that it renders the animals very liable to injury, and thus probably accounts for the fact that none of them has been brought alive to Europe. The carapace of this species attains a length of 38½ inches, and the weight of one individual was just over 200 pounds

One of the best accounts of the habits of the Galapagos tortoises is given by Darwin regarding the one inhabiting South Albemarle. These tortoises frequent in preference the high, damp parts, although they likewise live in the lower and arid districts. Very numerous in individuals, some grow to such a size that it requires six or eight men to lift them, while they will yield as much as 200 pounds of meat. "The old males are the largest, the females rarely growing to so large a size; the male can be readily distinguished from the female by the greater length of its tail. The tortoises which live on those islands where there is no water, or in the lower and arid parts of the others, feed chiefly on the succulent cactus. Those which frequent the higher and damp regions eat the leaves of various trees, a kind of berry which is acid and austere, and likewise a pale green filamentous lichen, that hangs in tresses from the boughs of the trees.

"The tortoise is very fond of water, drinking large quantities, and wallowing in the mud. The larger islands alone possess springs, and these are always situated towards the central parts, and at a considerable height. The tortoises, therefore, which frequent the lower districts, when thirsty, are obliged to travel from a long distance. Hence, broad and well-beaten paths branch off in every direction from the wells down to the sea-coasts; and the Spaniards, by following them up, first discovered the watering-places. When I landed at Chatham Island, I could not imagine what animal travelled so methodically along well-chosen tracks. Near the springs it was a curious spectacle to behold many of these huge creatures, one set eagerly travelling onwards with outstretched necks, and another set returning after having drunk their fill. When the tortoise arrives at the spring, quite regardless of any spectator, he buries his head in the water above his eyes, and greedily swallows great mouthfuls, at the rate of about ten in a minute. The inhabitants say that each animal stays three or four days in the neighbourhood of the water, and then returns to the lower country; but they differed respecting the frequency of these visits."

After mentioning that some tortoises live on islands where the only water they obtain is that which falls as rain, and also that the inhabitants of the Galapagos Islands, when overcome with thirst, are in the habit of killing a tortoise and drinking the water contained in its interior, the writer proceeds as follows: "The tortoises, when purposely moving towards any point, travel by night and day, and arrive at their journey's end much sooner than would be expected. The inhabitants, from observing marked individuals, consider that they travel a distance of about eight miles in two or three days. One large tortoise, which I watched, walked at the rate of sixty yards in ten minutes, that is three hundred and sixty yards in the hour, or four miles a day—allowing a little time for it to eat on the road.

"During the breeding season, when the male and female are together, the male utters a hoarse roar or bellowing, which, it is said, can be heard at a distance of more than a hundred yards. The female never uses her voice, and the male only at these times; so that when the people hear this noise they know that the two are together. They were at this time (October) laying their eggs. The female, when the soil is sandy, deposits them together, and covers them up with sand; but where the ground is rocky she drops them indiscriminately in any hole. Mr. Bynoe found seven in a fissure. The egg is white and spherical; one which I measured was 7⅜ inches in circumference, and therefore larger than a hen's egg. The young tortoises, as soon as they are hatched, fall a prey in great numbers to the carrion-feeding buzzard (*Polyborus*). The old ones seem generally to die from accidents, as from falling down precipices; at least, several of the inhabitants told me that they never found one dead without some evident cause.

"The inhabitants believe that these animals are absolutely deaf; certainly they do not hear a person walking close behind them. I was always amused

GALAPAGOS GIANT TORTOISES

GIANT TORTOISES

MARION'S TORTOISE

AN ALDABRA TORTOISE

when overtaking one of these great monsters, as it was quietly pacing along, to see how suddenly, the instant I passed, it would draw in its head and legs, and, uttering a deep hiss, fall to the ground with a heavy sound, as if struck dead. I frequently got on their backs, and then, giving a few raps on the hinder part of their shells, they would rise and walk away—but I found it difficult to keep my balance.''

Like their Mascarene allies, the Galapagos tortoises are much esteemed as food ; and, in order to see whether they are sufficiently fat to be killed, the inhabitants are accustomed to make a slit beneath the tail, through which the interior of the body can be seen. With the usual hardihood of reptiles, the rejected individuals appear to recover completely from this operation.

From several of the islands the giant tortoises have already disappeared, and it is to be feared that they will soon cease to exist throughout the Galapagos group.

EUROPEAN TORTOISES

The common *Testudo græca* typifies the sixth main group of the genus, which comprises seven Old World species of small or medium size, characterised by the carapace being brown or olive, either uniform or spotted with black, or black and yellow ; by the gular shields of the plastron being distinct ; and by the slight prominence and shortness of the ridge on the palate of the skull. The common tortoise belongs to a section of the group in which the anal, or hindermost, shields of the plastron meet in the middle line by a suture of considerable length ; and it is further characterised by the presence of five claws on the fore foot. From its nearest allies it may be distinguished by the fifth vertebral shield of the carapace being much broader than the third ; by the caudal shield being usually double ; and by the absence of a large tubercle on the inner side of the thigh.

The shell of this species is moderately vaulted, not much expanded behind, and its margins are not serrated. The nuchal shield is very long and narrow ; in the male the divided caudals are much incurved, and the shields of the back show a strongly-marked concentric striation. In ground-colour the shell is bright yellow, with the shields of the carapace spotted and bordered with black, and a broad band of black running along each side of the plastron. The length of the shell is about 5½ inches. Mainly a South European species, the common tortoise inhabits the Balearic Islands, Corsica, Sardinia, Sicily, Italy, Dalmatia, the Balkan Peninsula, and the Greek Archipelago, while it also occurs in Syria.

The larger but allied Algerian tortoise (*T. ibera*), in which the shell attains a length of about 9 inches, may be distinguished by the fifth vertebral shield being not broader than the third, by the single caudal shield, and the presence of a large subconical tubercle on the inner surface of the thigh. In colour this species differs from the last in having the plastron more or less spotted with black, while in some examples the carapace is

uniformly brown. Its range includes Southern Spain, North-Western Africa, Syria, Asia Minor, Transcaucasia, and Persia.

A third species often represented among the shiploads of these reptiles imported into England is the Greek tortoise (*T. marginata*), which attains a length of 11 inches, and appears to be confined to Greece. The absence of an enlarged tubercle on the thigh serves to distinguish it from the preceding species, from which it also differs by the longer and more depressed shell, in which the hind margin is much expanded, and more or less serrated. Usually the carapace of the adult is black, with a small yellow or greenish spot on each shield ; while the ground-colour of the plastron is yellowish, each of its shields being marked by a black patch, which generally takes a triangular form. This species appears to be confined to Greece, but in Lower Egypt and Syria it is replaced by the smaller Leith's tortoise (*T. leithi*), in which the carapace is relatively shorter and more deeply notched in front, while the form and arrangement of the tubercles on the fore limb is different. Gilbert White's tortoise, which lived for nearly forty years in a Sussex garden before it came to Selborne, was a remarkably large specimen of *Testudo marginata*, as is proved by its shell, which is exhibited in the Natural History branch of the British Museum at South Kensington.

All these tortoises appear identical in their habits, frequenting dry and sandy places, and being extremely fond of sunshine, in which they will bask by the hour together. In some of the Greek islands and the south of Italy the common tortoise is found in great numbers, and in the markets of Sicily and Italy it is regularly exposed for sale as an article of food. At the approach of winter it buries itself deep in the earth, where it remains during the cold months, usually reappearing in April, but in Sicily as early as February. Although its main food consists of plants and fruits, it will likewise consume such snails, worms, and insects as it may meet with during its wanderings.

GREEK TORTOISE

In captivity, where they have been known to live for a great number of years, these tortoises display great partiality for milky plants, such as lettuce, and they are always fond of a bath. At the approach of rain they always hide themselves, but in fine weather remain abroad throughout the day. In excavating a burrow for the winter's sleep, the earth is dug up by the strong fore limbs, and thrown out from the hole by the hind pair. The pairing season begins immediately after the awakening from the winter sleep ; and in May or June the female lays from eight to fifteen hard-shelled, spherical white eggs, of about the size of a hazel-nut. These are deposited in a hole in the earth in some sunny spot, and, after being carefully covered up, are left to hatch. By September the young tortoises are about the size of half a walnut-shell, and present an exceedingly comic appearance. The common tortoise occasionally produces eggs in England.

SOFT TORTOISES

"Fiercer and more spiteful than any other members of the chelonian order, these tortoises, owing to a peculiarity in the structure and mode of articulation of some of the vertebræ of the neck, have the power of darting out the head with inconceivable rapidity."

R.J. Smit

GREY AND STREAKED GURNARDS

"Of a decidedly ugly appearance, the gurnards are easily recognised by their enormous square and elevated heads, and likewise by the finger-like first three rays of the pectoral fins, which serve not only for walking on the sea-bottom, but also as organs of touch."

FISHES IN THEIR NATURAL COLOURS

A. Beaked Chætodon
B. Scarlet Holacanthus
C. Banded Chilodactyle
D. Eyed Blenny

E. Trigger-fish
F. Cobbler-fish
G. Red-finned Apistus
H. Japanese Dragonet

I. Two-spotted Chilinus
J. Fire-fish
K. Banded Paracirrhites
L. Electric Eel

M. Many-banded Muller
N. Muræna
O. Zebra Sole
P. Semicircular Chætodon

BURMESE AND HORSFIELD'S TORTOISES

There are certain other species belonging to the same group as the common tortoise which demand a brief notice. Among these is the handsome Burmese tortoise (*T. elongata*), from Bengal, Burma, Cambodia, and Cochin China, taking its name from the great length of the depressed shell of the males; the females being much smaller, with a relatively shorter and wider shell.

These tortoises differ from the European species by the anal shields of the plastron having a very short line of union in the middle, even if they meet at all. The ground-colour of the shell is greenish yellow, with an irregular black patch on each shield, which may occupy nearly the whole area of such shields, leaving merely a narrow yellow margin, or may be much broken up and indistinct. The male attains a length of between 10 and 11 inches. Forsten's tortoise (*T. forsteni*), from Celebes and Gilolo, may be distinguished by the want of a nuchal shield in the front of the carapace.

Horsfield's tortoise (*T. horsfieldi*), which, although allied to the European species, differs in having only four claws on the fore as well as on the hind feet, inhabits the deserts, oases, and even mountains of Central Asia, where it ranges from the Aralo-Caspian region and the Kirghiz Steppes to Afghanistan. The shell, which is considerably depressed and not much longer than broad, has a brown or olive ground-colour above, which may be either uniform or blotched with black; while beneath it has large patches of black, which sometimes almost cover the whole surface.

The Burmese tortoise is active in its habits, and the male in captivity is very confiding, eating readily from the hand, although the female when touched at once withdraws within the shell. Captive specimens were observed to be very restless at night; they fed freely on plantains, but a female on one occasion ate prawns and fish procured to feed some soft tortoises. Horsfield's tortoise, although equally fond of immersing its lower shell in water, is said to be more brisk in hot weather than are the European species; it is purely diurnal in its habits, not venturing forth till after sunrise, and retiring to rest before sunset. Its food in the wild state is stated to be entirely of a vegetable nature; snails and worms being never eaten.

ANGULATED TORTOISE

The angulated tortoise (*T. angulata*), of South Africa, and the allied *T. yniphora*, from an island near the Comoros, constitute the last and seventh group of the genus, and are distinguished from all the others by the great prolongation of the anterior extremity of the plastron, which is covered by a single gular shield only. The former species attains a length of about 7½ inches, and has an elongated and very convex carapace, of which the hind margin is at most but slightly serrated. In colour the shell is yellow above, each shield being bordered with black, and usually ornamented with a black spot in the centre,

ANGULATED TORTOISE

while the plastron is black in the middle, or has some large black blotches.

AREOLATED TORTOISE

Nearly related to the typical tortoises, with which it agrees in the general structure of its shell, the areolated tortoise (*Homopus areolatus*) of South Africa, together with three allied species from the same continent, differs by the absence of the median ridge on the front of the palate characterising all the former, and is on this account referred to a distinct genus. If the horny shields be stripped from the carapace, it will be found that the underlying neural bones, instead of being alternately octagonal and quadrangular, are irregularly hexagonal, with the shorter of the two lateral surfaces placed posteriorly; since, however, the same feature occurs in some of the typical tortoises, it is not absolutely characteristic of the African genus.

The areolated tortoise is a small species, with a shell of about 4 inches in length; and characterised by having only four claws on the front feet, and its depressed carapace, which is of equal width throughout, with even margins. On the back the shields are more or less inflated, and separated from one another by deep channels; the centre of each shield having a depressed areola, or nucleus, surrounded by concentric grooves. In colour the carapace is olive, with a reddish brown centre to each shield, but the plastron is brown in the middle and yellow at the edges.

A second species, *H. femoralis*, differs by having the hind margin of the shell serrated, and a conical tubercle on the posterior surface of the thigh; while in a third, *H. signatus*, there are five toes on each fore foot. Lastly, *H. nogueyi* differs from all the others in its vaulted carapace, which is gibbose behind; this species being a native of Senegal, while the other three are South African. In general habits it is probable that the members of this genus closely resemble typical tortoises.

HINGED TORTOISES

Three remarkable tortoises inhabiting Tropical Africa constitute a genus distinguishable at a glance from the other members of this section of the family by the circumstance that the hind portion of the carapace is articulated to the anterior moiety by a ligamentous hinge, upon which it is freely movable, so that when the reptile is withdrawn into its shelter the posterior extremity of the shell can be completely closed. This hinge runs between the fourth and fifth costal bones and the seventh and eighth marginals of the shell. The skull agrees with that of the preceding genus in the absence of a median ridge on the front of the palate, the neural bones of the carapace are hexagonal and short-sided behind and the caudal shield is undivided. The costal bones of the carapace differ, however, from those of the tortoises described above in being of nearly equal width throughout, instead of alternately narrow at one end and broad at the other.

Of the three species of the genus, the common hinged tortoise (*Cinixys erosa*), from Guinea and the

Gabun, is characterised by the front and hind margins of the carapace being everted and strongly dentated, by the absence of a nuchal shield, the projection of the extremity of the plastron in front of the carapace, and the sloping contour of the hind extremity of the latter. In length the shell measures 9 inches ; its general colour above being dark brown, with lighter

AREOLATED TORTOISE

centres to the shields, and the lower sides of the costal shields yellowish, while on the plastron the shields have dark brown centres and yellowish margins.

In the nearly allied Home's hinged tortoise (*C. homeana*), from the same region, there is a nuchal shield, the extremity of the plastron does not project in advance of the carapace, and the hind extremity of the latter descends vertically. On the other hand, in Bell's hinged tortoise (*C. belliana*), which ranges right across Tropical Africa, the margins of the carapace are neither everted nor serrated, and a nuchal shield is present on the front of the carapace. In length the last-named species does not exceed 7½ inches.

In habits the hinged tortoises show a complete transition from the land-tortoises to the terrapins, and thus fully justify the conclusion arrived at from structural considerations that both groups should be included in a single family. Bell's hinged tortoise is, in fact, essentially a land-reptile, inhabiting regions formed of gneiss rocks or other dry localities, where it is active during the hot, rainy season, but in the cooler portion of the year, from May to October, according to native reports, lies deeply buried in the earth. Both the other species, on the contrary, seem to be mainly aquatic in their habits ; the dentated hinged tortoise, which is fairly common in Guinea, being stated to spend a large portion of its time in water, where one specimen remained for upwards of a month.

This species is found in rivers, even close to the sea, whence it emerges to lay its eggs on the banks. In spite of its club-like feet, it dives and swims with facility ; captive examples having been observed to descend to the bottom of a deep vessel in which they were kept. On land its motions are, however, slow and deliberate in the extreme, and have been compared to those of the minute hand of a clock. Its food is of a vegetable nature, one captive specimen displaying great partiality for cherries. By the inhabitants of Guinea these tortoises are eagerly sought after as food, and are thus difficult to obtain by Europeans.

SPIDER-TORTOISE

The last member of this section of the family is the spider-tortoise (*Pyxis arachnoides*) of Madagascar, the sole representative of a genus characterised by the presence of a transverse hinge across the front of the plastron, by which means the anterior lobe of the latter can be bent upwards so as to close the front of the shell. In having the neural bones of the carapace alternately octagonal and tetragonal, this species approaches the typical tortoises nearer than is the case with the hinged tortoises. In length the shell is only just over 4 inches ; its coloration is yellow, with radiating black bands from the centres of the shields of the back.

LAND-TERRAPINS

The whole of the tortoises hitherto described are collectively characterised by the absence of all trace of webbing in the toes, by the presence of not more than two joints, or phalanges, in each toe, by the metacarpal bones of the fore foot being but slightly, if at all, longer than wide, and also by the majority of the bony neural plates of the carapace being hexagonal, with their shorter lateral surfaces posteriorly placed, or alternately octagonal and tetragonal. On the other hand, in the remaining members of the family the digits are usually furnished with webs, or at least a rudiment thereof ; while the middle toe of each foot has three joints, and the metacarpal bones are elongated.

Their first representatives form a small group, mainly confined to the Indo-Malay region, which both in structure and habits tends to connect this section of the family with the preceding one. These species, as shown in the right-hand figure of the illustration on page 1599, agree with the hinged tortoises in that most of the hexagonal neural plates of the carapace have the shorter of the two lateral surfaces placed posteriorly and the longer anteriorly. Moreover, if the horny shields from the plastron be removed, it will be found that the entoplastral, or median unpaired bone of that part of the skeleton, is crossed by the groove marking the boundary between the humeral and pectoral shields. Terrapins lay elongated eggs.

The spiny land-terrapin (*Geoëmyda spinosa*) may be taken as a well-known example of the first genus, characterised by the absence of a hinge in the plastron, and of a bony temporal arch on the sides of the skull. The three species of this genus are large-sized tortoises,

BELL'S HINGED TERRAPIN

confined to Burma and the Malay region ; the spiny land-terrapin having a shell of 8 inches in length, while that of the great land-terrapin (*G. grandis*) of Burma and Siam measures upwards of 16 inches. In the former of these two species both the front and hind margins of the shell are deeply serrated, but in the latter, as well as in the third representative of the genus, only the hind border is thus ornamented. The colour of the carapace in these terrapins is brown or blackish, frequently with darker markings. Together with the other members of the group, they differ from the majority of terrapins in having the head covered with a continuous skin, instead of with small shields. The small size of the webs of these terrapins indicates that in habits they are probably in part aquatic and in part terrestrial.

CHAIBASSA TERRAPIN

The Chaibassa terrapin (*Nicoria tricarinata*), taking its name from a district in Bengal, is selected to represent a genus common to the Indo-Malay region in the east, and Central and South America in the west, distinguished from the preceding by the presence of a bony temporal arch to the skull. Of the various species of this genus, the smallest has a shell of only 5 inches in length, but in a larger one it may measure as much as 16 inches. While in the figured Chaibassa terrapin both fore and hind margins of the shell are entire, in other species either one or both of these may be deeply serrated.

The Chaibassa species, which ranges from Bengal to Assam, has the carapace dark brown or black in colour, with the three longitudinal ridges from which it takes its Latin name yellow; the plastron being uniformly yellow, and the neck and limbs blackish. From the larger three-keeled terrapin (*N. trijuga*) of India and Burma this species is further distinguished by its more convex shell, which descends very abruptly behind, as well as by the rudimentary condition of the webs between the toes ; on both of which grounds it may be regarded as more exclusively terrestrial in its habits. A fossil shell of the Chaibassa terrapin, represented in the right figure on page 1599, has been obtained from the Pliocene rocks of the Siwalik Hills of Northern India, thus indicating the antiquity of the species. In some individuals the hind half of the plastron is connected with the upper shell merely by ligament, while in others there is the usual bony union.

HINGED TERRAPINS

The third genus of this group (*Cyclemys*), which is confined to India, Malaya, and the south of China, is represented by about half a dozen species, which, while agreeing with the members of the foregoing genus in the presence of a bony temporal arch to the skull, differ by having a well-marked transverse ligamentous

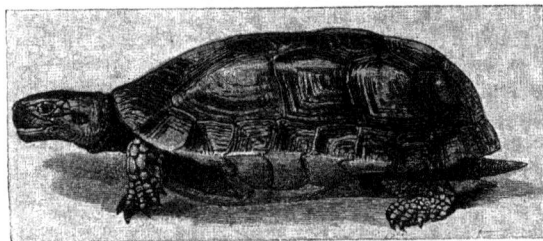

CHAIBASSA TERRAPIN

hinge across the middle of the plastron, whereby its hind lobe is rendered movable, and capable of more or less completely closing the posterior aperture of the shell. None of the species has a shell of more than 8 inches in length.

The genus may be subdivided into two groups, each containing three species. In the former, as represented by *C. dhor* of Northern India and the Malay region, the plastron, which is notched behind, cannot completely close the shell, and the hind margin of the

carapace is serrated. In the second group, on the other hand, of which the Amboyna hinged terrapin (*C. amboinensis*) is a typical example, the plastron is capable of completely closing the hind aperture of the shell, and the posterior margin of the carapace is not

DENTATED HINGED TERRAPINS

serrated. These species also have the shell keeled on the back in the young state. In the Amboyna species, as also in *C. flavomarginata*, the hind end of the plastron is entire, although in a third, *C. trifasciata*, it is notched.

BOX-TORTOISES

Agreeing with the hinged terrapins in the presence of a transverse ligamentous hinge across the middle of the plastron, by the aid of which the openings of the shell can be closed, the two North American species of box-tortoises, together with all the remaining members of the family, differ from the former in that the hexagonal neural bony plates of the carapace have the shorter of their two lateral surfaces placed anteriorly, instead of posteriorly, this arrangement being shown when the shell is stripped. The presence of the hinge in the plastron serves to distinguish the box-tortoises from all the members of the second group, with the exception of the pond-tortoises, while from the latter they are separated by the beak being hooked, and the absence of a bony temporal arch to the skull. In the box-tortoises the head is covered with smooth skin above, the toes having only a rudimental web, and the tail being short.

The Carolina box-tortoise (*Cistudo carolina*) is a somewhat variable species as regards size, the length of the shell ranging from a little over 4 to somewhat more than 5 inches. The highly convex carapace is almost hemispherical in shape, and attached to the plastron solely by ligament, so that the whole shell can be completely closed. As a general rule, the upper shell is dark brown or blackish with yellow spots, or brownish yellow with dark brown spots or rays, but there may be an interrupted yellow streak down the middle of the back. The plastron may be either a uniform dark brown or blackish, or may have irregular yellowish blotches on a ground of the same, although in some instances it is yellowish with dark blotches of variable size.

The range of this species embraces the southern and south-eastern United States and Mexico. In the ornate box-tortoise (*C. ornata*) of Nebraska and some of the neighbouring states the shell is more depressed, and the plastron and carapace are connected together by a very short bony bridge, so that the shell cannot be completely closed. The toes, moreover, have no distinct webs.

The vaulted carapace of the box-tortoises, with the abruptly descending hind profile and the rudimentary condition of the webs of the toes, at once proclaims the partially terrestrial habits of these reptiles, which form, indeed, one of the connecting links between the typical tortoises and the fresh-water terrapins. Although mainly, if not entirely, carnivorous—as is indicated by the absence of a median ridge in the front of the palate—the box-tortoises appear to resemble the typical tortoises very closely in their general mode of life. According to some observers, they are more frequently to be met with in dry and even hilly districts than in swamps, but they are partial to spots where colonies of night-herons are in the habit of nesting, owing to the quantity of insects, snails, worms, and fragments of fish to be met with in such localities ; and they are frequently found in woods where the ground is either moist or swampy. At times they enter the water of their own will, and they have been seen half buried under loose earth or moss in search of worms and insects.

Unlike most members of the family, box-tortoises shun the light and are most active during the evening and night, shutting themselves closely up in their shells when the sun is shining brightly. The closure of the shell is also effected at the approach of any large animal, and when thus securely boxed up there are few creatures these tortoises need fear. Like most other terrestrial tortoises, the females lay their eggs in holes dug in the ground by themselves ; the number laid being usually only five or six, whether the parents be half grown or adult. Each individual egg is carefully covered with earth, and the time taken before the young are hatched is said to vary from eighty-eight to a hundred days.

When first hatched, the young are well developed and of great relative size and strength, although their shells are still soft and cartilaginous, and the remnant of the yolk-sac depends from the plastron. In Pennsylvania both young and old bury themselves deep in the ground about the middle of October, where they remain till the latter part of April ; the spot selected having a dry soil, and being protected from the cutting blasts of the north. Many individuals which have not buried themselves sufficiently deeply are, however, frozen to death during the winter slumber. On account of the strong and disagreeable flavour of their flesh, doubtless engendered by the nature of their food, box-tortoises are not eaten.

POND-TORTOISES

In marked contrast to the vaulted and abruptly descending carapace of the box-tortoises is the depressed and shelving shell of the pond-tortoises, this difference indicating a distinction in the habits of the two genera. Box-tortoises, as already mentioned, are mainly land reptiles, whereas pond-tortoises are as decidedly aquatic in their mode of life. In addition to the difference in the form of the shell, the members of the present genus are readily distinguished from those of the last by the beak not being hooked, and by the presence of a bony temporal arch in the skull.

In the shell the carapace is united to the plastron solely by ligament, and the plastron itself is more or less distinctly divided by a ligamentous transverse hinge upon which its two lobes are freely movable. Agreeing with the box-tortoises in having the top of the head covered with undivided skin, pond-tortoises differ by having the toes fully webbed, and also by the more elongated tail, which, while very long in the young, is of moderate length in the adult.

Although the genus *Emys* was formerly made to include many of the fresh-water terrapins, it is now restricted to the European pond-tortoise (*E. orbicularis*) and the nearly allied North American *E. blandingi*. The former, which is familiar to most visitors to Southern Europe, is characterised by the short oval form of its carapace, which is widest posteriorly, and in the young state has a more or less distinct median keel. In colour the upper shell of the adult is dark brown or black, ornamented with a variable number of light, usually yellow, flecks or radiating streaks ; the plastron being either yellow, brown and yellow, or almost wholly blackish brown. In the young, on the other hand, the upper shell is dark brown, and the lower black ; all the shields of the latter, as well as the marginal ones of the former, having a large yellow spot.

CAROLINA BOX-TORTOISE

The skin of the head, neck, body, and limbs is marked with yellow and blackish in varying proportions ; the head of the male having brownish dots on a darker ground, while in the female the dots are yellow. When fully grown, the shell attains a length of 7½ inches, but in most of the specimens imported into England it is not much more than half that size. At the present day the pond-tortoise is found, in suitable localities, in South and East-Central Europe, and South-Western Asia as far as Persia, and in Algeria.

During the Pleistocene period, when the climate of Northern Europe must at certain times have been much more genial, the pond-tortoise had a much more extensive distribution, its fossilised remains having been found in the superficial deposits of Belgium, Denmark, Germany, Lombardy, Norfolk, Sweden, and Switzerland. The North American species, which inhabits the north-eastern United States and Canada, has the carapace rather more elongate, and the tail shorter ; the former being black with pale yellow or brownish circular spots, and the plastron yellow with a large black patch on each shield.

The European species, which is specially abundant in the forest districts of Poland, inhabits both stagnant and running waters, and may be found alike in slow or swift-flowing streams, or in open lakes. During the daytime it leaves the water to bask in the sun on sequestered spots of the banks, where it remains without moving by the hour together, but towards sunset it begins to move, and remains active throughout the night. At the beginning of winter it constructs an underground chamber, in which it remains buried in slumber till spring, usually reappearing, if the weather be favourable, about the middle of April, at which time it reveals its whereabouts by a peculiar whistling cry characteristic of the breeding season. An excellent swimmer and diver, the pond-tortoise disappears beneath the water at the slightest sound, and when on land its motions are far more active than those of the typical tortoises.

Agreeing with other carnivorous terrapins in the absence of a median ridge on the fore part of the palate, this tortoise feeds chiefly upon worms, water-insects, crustaceans, frogs, newts, tadpoles, and fish. In devouring fish, pond-tortoises reject the air-bladder, which floats on the surface of the water ; and from the number of such floating air-bladders some idea may be formed as to whether a pond is numerously tenanted by these tortoises. In captivity, where they will live for years, pond-tortoises, in addition to their natural food, will sometimes eat raw meat, and in this state they frequently become so tame as to take food from the hands of their masters. The cylindrical eggs, varying from nine to fifteen in number, are laid at night during

SCULPTURED TERRAPIN

May in hollows dug by the female in dry soil, at a considerable elevation above the bank, where they are carefully covered up and left to develop. These tortoises are eaten by the inhabitants of all the countries in which they occur.

TYPICAL TERRAPINS

The remaining members of the family *Testudinidæ*, which may be collectively known as terrapins, have the plastron without any transverse hinge and firmly connected by bone with the carapace, so that the whole shell is solid and immovable. They comprise a large number of species, arranged under eleven genera, and all that can be attempted in a work of the present nature is to select for special notice one or two species of such genera. Although many of these terrapins are exceedingly unlike one another externally, yet they are all so closely connected that the genera can only be distinguished by the characters of the skull and the bony plates of the shell, so that their description must of necessity be somewhat technical.

The sculptured terrapin (*Clemmys insculpta*), of eastern North America, is selected as a fairly well-known representative of a genus with about eight species. This genus, it must be premised, forms one of a group of four agreeing with the two last noticed in the absence of a longitudinal ridge on the fore

part of the palate, and in the carnivorous habits of its various members. From the three allied genera, *Clemmys* may be distinguished by the aperture of the inner nostrils in the skull being situated between the eyes, by the unpaired entoplastral bone of the lower shell being traversed by the groove formed by the junction between the humeral and pectoral shields, and by the upper part of the head being covered with a continuous smooth skin.

This particular terrapin belongs to a group of five species collectively characterised by the median union of the anal, or hindmost, shields of the plastron being longer than that between the femoral shields ; and while four species of this group are confined to North America, Beale's terrapin (*C. bealei*) inhabits China, thus showing a distribution analogous to that of alligators. On the other hand, the Caspian terrapin (*C. caspica*), ranging from the Caspian Sea to the Persian Gulf, the Iberian terrapin (*C. leprosa*), of Spain and North-Western Africa, and the Japanese terrapin (*C. japonica*) resemble one another in having the median union of the anal shields shorter than that of the femorals.

EUROPEAN POND-TORTOISE

The sculptured terrapin, which attains a length of about 7 inches, is specially characterised by the toes being webbed only at their bases, by the upper jaw having a notch in the middle, on the sides of which are a pair of tooth-like projections, and by the serration of the hind border of the carapace. The shell is much depressed, with a raised keel down the middle of the back, and the shields of the carapace ornamented with the radiating and concentric striæ from which the species takes its name. While the ground-colour of the carapace is blackish, the radiating lines are yellow ; the plastron being yellow, with a large black blotch on each of its shields. The soft parts are dark brown or olive, and the sides of the head speckled with red.

This species is exceedingly abundant on the Atlantic side of the United States, from Maine to Pennsylvania and New Jersey. Frequenting both marshes and rivers, it leaves the water for much longer periods than its European congeners, and is sometimes found for months at a time in perfectly dry places. In wandering from one stream to another, it makes regular tracks through the woods, and is hence frequently termed in America the wood-terrapin. In its feeding and general mode of life this terrapin presents no features distinguishing it from other carnivorous kinds.

The Iberian terrapin (*C. leprosa*) is remarkable for the fact that an alga grows in the shell, frequently eating through the horny shields and penetrating the underlying bones, so as to cause them to have a spongy and rotten texture.

THICK-NECKED TERRAPIN

Nearly allied to the sculptured terrapin and its relatives is the thick-necked terrapin (*Bellia crassicollis*) from Tenasserim, Siam, the Malay Peninsula, and Sumatra, which, with a second species from Borneo (*B. borneensis*), constitutes a genus distinguished by the greater development of the bony buttresses connecting the upper with the lower shell, and by the hind part of the head being covered with small horny shields. The feet are fully webbed, and the anterior vertebral shields of the carapace more or less distinctly balloon-shaped.

The typical species, which measures rather more than 6½ inches in length, is of a general dark brown or black colour, usually with some yellow markings on the plastron, and some large spots of the same colour on the head. Several extinct representatives of this genus are met with in a fossil state in the Pliocene deposits of North-Western India.

HAMILTON'S TERRAPIN

The very handsomely coloured Hamilton's terrapin (*Damonia hamiltoni*) of India, conspicuous for its black and yellow, highly-vaulted, and three-keeled carapace, is the best-known representative of a third genus, distinguished from the foregoing by the hind aperture of the nostrils opening behind the line of the eyes, and the great breadth of the palate. As in the two preceding genera, the entoplastral bone of the plastron is traversed by the groove formed by the union between the humeral and the pectoral shields, and the hind part of the head is covered with small shields.

Hamilton's terrapin has the elevated carapace marked with three interrupted longitudinal keels, or rows of nodose prominences, which are most pronounced in the young, the colour of the shell being dark brown or blackish, with spots and streaks of yellow, and the soft parts having a similar coloration. While in young individuals the hind border of the carapace is strongly serrated, in the adult it becomes nearly smooth. This species attains a length of nearly 9 inches at the present day, but fossil examples found in the Pliocene rocks of Northern India were still larger. These fossil specimens lived with numbers of mammals belonging entirely to extinct species. There are four other species of the genus, ranging over Malaya, Southern China, and Japan.

SALT-WATER TERRAPIN

The last representative of the group with a smooth palate and carnivorous habits is the North American genus *Malacoclemmys*, distinguished from *Damonia* by the head being covered with continuous skin, and by the groove formed on the plastron by the junction between the humeral and pectoral shields being situated in advance of the entoplastral bone. While two of the species inhabit the valley of the Mississippi, the salt-water terrapin (*M. terrapin*) is a frequenter of the salt-marshes of the Atlantic coast. The last-named species has an oval and much-depressed carapace, which attains a length of nearly 7 inches, and is characterised by the great width of the first and second vertebral shields; its general colour being either olive, with black concentric lines, or uniform blackish. The plastron is yellowish or reddish, with variable black markings.

It is this species that generally forms the celebrated New York dish known as terrapin; but it would seem that other species are also used, as several accounts refer to terrapins taken high up the rivers. The best terrapins go by the name of "diamond-backs," and do not generally exceed some 7 inches in length, although they may rarely measure as much as 10 inches, but all terrapin of larger dimensions belong to the inferior kinds, ordinarily designated "sliders."

Formerly caught in shoals, the diamond-back has now become very scarce, and is, indeed, in some danger of extermination. The terrapin furnished in hotels is almost invariably "sliders," diamond-backs being sold to private houses only.

PAINTED TERRAPIN

The remaining genera of the family *Testudinidæ* constitute a distinct group, distinguished from the one including the six genera last mentioned by the circumstance that the broad front portion of the palate of the skull is marked by one or two longitudinal ridges, and likewise by all the species being mainly or exclusively herbivorous in their diet. Among these, the large and exclusively American genus *Chrysemys*, with a dozen

PAINTED TERRAPIN

species, of which the painted terrapin (*C. picta*) is one of the best known, belongs to a subgroup of three genera characterised by the bony buttresses connecting the upper with the lower shell being short or of moderate size. From its allies *Chrysemys* is distinguished by the opening of the posterior nostrils being placed between the eyes, and by the entoplastral bone being situated in advance of the groove on the plastron formed by the junction of the humeral with the pectoral shields.

The painted terrapin of eastern North America, which attains a length of 6 inches, and has a much-depressed shell, takes its name from its brilliant colouring, the carapace being olive or blackish, with yellow lines bordering the shields, and its marginal shields red, with black concentric or crescentic markings; while the plastron is yellow, sometimes with small streaks of black on the middle line, and the bridge red, with black markings. The soft parts have a brown or blackish ground-colour, with lighter bands, which are yellow on the head and red elsewhere.

EYED AND CHINESE TERRAPINS

The eyed terrapin (*Morenia ocellata*) of Burma, together with an allied species from Bengal, constitutes a genus distinguished from the preceding by the aperture of the posterior nostrils opening behind the line of the eyes. The typical species, in which the shell measures nearly 9 inches in length, takes its name from the eye-like black spots ringed with yellow adorning each shield of the back portion of the carapace.

On the other hand, the Chinese terrapin (*Ocadia sinensis*), which is the sole existing representative of its genus, differs from *Chrysemys* in having the entoplastron intersected by the groove formed by the junction between the pectoral and humeral shields. The genus is of special interest on account of its being represented by several extinct species in strata belonging to the upper part of the Eocene division of the Tertiary period in the south of England and the Continent.

BATAGURS

The remaining members of the family, which are arranged under six genera, and may be collectively known as batagurs (from the native name of one or more of the Indian species), are exclusively confined to India, Burma, and the Malay region. They comprise the largest fresh-water representatives of the family, and are readily characterised by the great development of the vertical bony buttresses connecting the carapace with the plastron, which project as walls within the shell, so as partially to divide it into compartments.

Of the six genera, *Cachuga*, which is represented by about seven species from India and Burma, is readily recognised by the great elongation of the fourth vertebral horny shield of the carapace, which extends over four or five of the underlying neural bones. The smaller members, such as Smith's batagur (*C. smithi*), and the black-and-yellow batagur (*C. tectum*), of the Ganges and Indus, are characterised by the fourth vertebral shield terminating in front in a narrow point. Whereas the former of these has a depressed and feebly keeled shell, the latter, especially when young, has the carapace much vaulted, and the third vertebral shield produced behind into a conical elevation forming the highest part of the shell. The name of black-and-yellow batagur is derived from the irregular black patches on the bright yellow plastron; the carapace being brown. Like the undermentioned dhongoka, it occurs fossil in the Pliocene deposits of Northern India.

The larger species of the genus, such as the Indian dhongoka (*C. dhongoka*), which grows to over 14 inches, has the fourth vertebral shield broad in front, instead of being narrowed to a point. Of the remaining genera, *Callagur*, *Batagur*, and *Hardella* differ from the preceding in that the fourth vertebral shield of the carapace is not longer than the third, but it will be quite unnecessary to refer in detail to the features distinguishing these genera from one another. The largest of all is the true batagur (*Batagur basca*), in which the carapace measures no less than 20 inches in length. The two genera *Liemys* and *Brookia*, each with a single large species, are confined to Borneo.

All the batagurs are exclusively vegetable-feeders, and the larger species are thoroughly aquatic in their habits, spending by far the greater portion of their time in the water. They abound in the larger rivers of India and Burma, where their huge shells form conspicuous objects as they rise to the surface to breathe. Describing the habits of a captive specimen, Dr. John Anderson states that when it rose to breathe " its nostrils were simply protruded above the surface of the water, and retained in that position for about half a minute, during which it made a long expiration, followed by a deep inspiration, the creature then slowly subsiding, tail-backwards, to the bottom. The animals, unless they were much irritated, never attempted to bite, but when so treated they sluggishly seized any object put in their way, holding it between their jaws with considerable tenacity, at the same time withdrawing the head into the shell. They moved about on the ground with considerable agility, supporting their heavy bodies erect on their legs, like a land-tortoise."

Another species will occasionally snap, when, owing to the friction of its serrated jaws against each other, a peculiar kind of barking sound is produced. Batagurs are eaten in Lower Bengal by some of the inferior castes of Hindus, and are kept for this purpose in tanks. They are believed to be highly dangerous to bathers. Fossil remains of batagurs occur in the Pliocene Tertiary deposits of India and Burma, but are unknown beyond the limits of the Indo-Malay region.

BIG-HEADED TORTOISE

This extraordinary reptile (*Platysternum megacephalum*), which is an inhabitant of the south of China, Siam, and Burma, is the sole representative, not only of a very remarkable genus, but likewise of a distinct family (*Platysternidæ*), which appears to be to a great extent intermediate between that of the tortoises (*Testudinidæ*) and that of the snappers (*Chelydridæ*). The most peculiar feature about this tortoise is the disproportionately large size of its head, in which the beak is much hooked; and in the skeleton the temporal pits of the skull differ from those of all the members of the preceding family in being roofed over with bone, as in the following family of the snappers. The tail also resembles that appendage in the latter in its great length, and also in the circumstance that the articular surfaces of most of its vertebræ have the cup behind and the ball in front, whereas in the tortoise family just the reverse occurs. On the other hand, the carapace resembles that of the *Testudinidæ* and differs from that of the snappers in the absence of a rib-like process from each of the posterior angles of the nuchal bone passing backwards beneath the marginal bones.

The carapace is characterised by its extreme depression and oval form, and the plastron is of moderate size and connected with the carapace solely by ligament, so that bony buttresses are totally lacking. The enormous head is covered with a continuous horny shield, and the hooked jaws are of great power. The toes are of moderate length, and but slightly webbed; all except the fifth in the hind foot being furnished with claws. The long and cylindrical tail becomes compressed at the end, and is covered with rings of quad-

BIG-HEADED TORTOISE

rangular shields. In size this tortoise is small, the length of the carapace being only about six inches, and that of the tail some three-quarters of an inch more. In the adult the colour is olive brown above and yellowish brown beneath, but the young is more brilliantly coloured. Of the habits and mode of life of this tortoise little appears to have been ascertained.

MUD-TERRAPINS

THE mud-terrapins (*Cinosternidæ*) form the first of two nearly related families confined to the New World, all the members of which differ from those previously noticed by the circumstance that the nuchal bone of the carapace gives off from each of its hind angles an elongated rib-like process which underlies the marginal bones. From the second family the mud-terrapins, of which there are about eleven species, all inhabiting America north of the Equator, are broadly distinguished (as, indeed, they are from all other members of the order) by the fact that there are only eight bones in the plastron, owing to the absence of the unpaired entoplastral element.

As regards their other characters, mud-terrapins resemble the *Testudinidæ* in the conformation of the vertebræ of the tail and in the absence of a roof to the temporal pits of the skull, as well as in the extreme shortness of the tail. The carapace is more or less depressed, and articulated by a bony suture with the plastron; the latter having the gular shields fused into one, or wanting, and its fore and hind lobes more or less movable. The toes are fully webbed, and, with the exception of the fifth in the hind foot, strongly clawed.

The best-known representative of the one genus is the Pennsylvanian mud-terrapin (*Cinosternum pennsylvanicum*), which attains a length of about 4½ inches, and inhabits eastern North America from New York to the Gulf of Mexico. In colour the shell is brown or brownish above, and either yellow or brown beneath, the lines of junction between all the shields being dark brown or blackish, while the head and neck are brown with yellowish spots. From other species of the genus it is distinguished by the large size of the plastron, in which the anterior lobe is narrower than the mouth of the shell.

PENNSYLVANIAN MUD-TERRAPIN

In general habits the mud-terrapins seem to be very similar to the fresh-water members of the tortoise family, although they prefer swamps and marshes to running waters. Carnivorous in their diet, they subsist chiefly on small fishes, insects, and worms, but they have been observed to capture newts. They will readily take a baited hook, and when thus caught sink rapidly and heavily to the bottom, thereby causing the angler to believe that he has hooked a weighty fish. At the beginning of winter they bury themselves in moss, where they remain dormant till the following May. An extinct genus nearly allied to the mud-terrapins occurs in the Tertiary rocks of Baden.

MAW'S TERRAPIN AND ITS ALLIES

MAW'S terrapin (*Dermatemys mawi*) may be taken as a good representative of the family *Dermatemydidæ*, all the three genera of which are restricted to Central America. This family connects the preceding one with the snappers, agreeing with the latter in the presence of an entoplastral bone, and with the former in the characters of the vertebræ of the short tail, which have the cup in front, and the absence of a roof to the temporal pits of the skull. Maw's terrapin and its allies further agree with the mud-terrapins in the incompleteness of the series of neural bones of the carapace, the hinder ones being wanting, and thus allowing the costal plates to meet their fellows in the middle line of the shell.

Externally, the members of the present family may be distinguished from the *Testudinidæ* by the presence of an additional series of infra-marginal shields between the marginals and those of the plastron—a feature which they possess in common with the big-headed tortoise and the snappers. Maw's terrapin, which attains a length of about fifteen inches, and is the sole representative of its genus, has the plastron large and connected with the carapace by an elongated bridge, the gular shield being single, and the usual five other pairs of shields present on the plastron. Unlike most other tortoises, there are twelve pairs of marginal shields, in place of the usual eleven.

In the other two genera of the family—*Staurotypus* and *Claudius*—the plastron is reduced to a cross-like shape, and has merely a short connection with the carapace; the number of paired shields on the former is only four or three, and the chin is provided with a pair of wattle-like appendages of which there is no trace in Maw's terrapin. While in the two species of *Staurotypus* the plastron is connected with the carapace by a bony bridge, in the single representative of *Claudius* the junction is entirely ligamentous.

This family is represented by several extinct genera in the Tertiary and Cretaceous strata of North America, one of which (*Baptemys*) had the full series of neural bones; and there appear to have been allied tortoises in the European Tertiaries.

SNAPPERS AND ALLIGATOR-TERRAPINS

RESEMBLING the big-headed tortoise in the great relative size of their hook-beaked heads and their elongated scaly tails, the snappers and alligator-terrapins of North and Central America constitute a well-marked family (*Chelydridæ*) by themselves. In the first place, they differ from the species last named in that the nuchal bone at the front of the carapace gives off a pair of rib-like processes underlying the marginals, while the temporal region of the skull is only partially covered with a bony roof. The majority of the tail-vertebræ have the articular cup in front and the ball behind. The members of the American family are further characterised by the relatively small size of the carapace, of which the hind border is strongly serrated, while the cruciform plastron is likewise small and loosely articulated with the upper shell by a very narrow bridge. Both the upper and lower shells are not completely ossified till very late in life, vacuities remaining for a long time between the costal and marginal bones in the former, and in the middle line of the latter.

Then, again, the plastron is peculiar in that the abdominal shields, which are separated from the marginals by an inframarginal series, do not meet one another in the middle line, although they may be connected by some small, irregular, unpaired additional shields. The enormous head cannot be completely retracted within the carapace, of which the anterior margin is deeply excavated in order to afford it room, and the chin is provided with one or more pairs of pendant wattles. With the exception of the fifth in the hind limb, the toes are furnished with claws, and the long tail is crested above.

ALLIGATOR-TERRAPINS

The alligator-terrapin, or snapping turtle (*Chelydra serpentina*), is a giant among river-tortoises, and takes its name from a fancied resemblance to an alligator surmounted by a chelonian shell. It is one of two species belonging to a genus characterised by the eyes being directed upwards and outwards, so that their sockets are visible in a top view of the skull, and by the tail being furnished with large horny shields on its lower surface, as well as by the absence of the supramarginal shields found on the carapace of Temminck's snapper. The carapace, which may attain a length of at least 20 inches, is characterised by its rugose surface, bearing three well-marked tuberculated keels, which tend to become smoother with advancing age, its vertebral shields being remarkable for their great width. The snout is short and pointed, with a very narrow space between the eyes; the skin is warty, and on the chin developed into a pair of wattles or barbels. In the young the tail is as long or even longer than the shell, but it becomes relatively shorter in the adult; its upper surface has a crest of large, compressed tubercles, and the shields on the lower surface have been already mentioned. As in the other members of the family, the colour is uniform olive brown. The alligator-terrapin inhabits the rivers of North America to the eastward of the Rocky Mountains, from Canada to Mexico, and Central America.

A second living species (*C. rossignonii*), distinguished, among other features, by the presence of four wattles on the chin, is met with in the mountains of Ecuador. Nearly allied to this is a third and extinct species (*C. murchisoni*) from the Miocene rocks of Baden; and since, as already mentioned, the mud-terrapins, and probably also Maw's terrapin, were represented in the Tertiary strata of Europe, it is not improbable that the Eastern Hemisphere may have been the original home of the present group of families.

THE ALLIGATOR-TERRAPIN

TEMMINCK'S SNAPPER

Attaining considerably larger dimensions than the alligator-terrapin, the species commonly known as Temminck's snapper (*Macroclemmys temmincki*) is distinguished by the lateral position of the eyes, the sockets of which are invisible in a front view of the skull, as well as by the presence of three or four additional, or supramarginal, shields on the sides of the carapace, and by the under surface of the tail being covered with small scales. The triangular head is proportionately even larger than in the alligator-terrapin, and the carapace carries three very strongly marked longitudinal ridges. In length the shell may measure at least a couple of feet, the tail being somewhat shorter. This species inhabits North America from Western Texas to Florida, extending northwards to Missouri.

Since the alligator-terrapin and Temminck's snapper appear to be very similar in their mode of life, their habits may be treated of collectively. Both these tortoises frequent alike the rivers and larger swamps of the United States, occurring in certain localities in enormous numbers, and most commonly in waters that have a muddy bottom, not even disdaining the most malodorous pools. As a rule, they lie in deep water near the middle of the river or swamp, although at times they show themselves on the surface, where, with outstretched necks, they float with the current. In populated districts the least sound is, however, sufficient to send them at once to the bottom, although in more remote regions they are less shy. At times they may be observed at considerable distances from the water, probably in search of food or of suitable spots to deposit their eggs.

Temminck's snapper well deserves its name, since, from the moment of its escape from the egg, it begins to snap and bite at everything within its reach; and an adult has been known to make a clean perforation with its powerful beak through the blade of an oar half an inch in thickness; and these chelonians have been known to inflict terrible wounds on persons who have incautiously entered waters where they abound. In the water the movements of these reptiles are more rapid than those of most of their kin, and when in pursuit of prey they swim with surprising speed.

Their food consists largely of fishes, frogs, and other water-animals; but they will also frequently seize and drag down large aquatic birds, more especially ducks and geese. Tame specimens kept in a pond in the United States proved terrible foes to the stock of fish contained in the same. The eggs, which vary from twenty to thirty in number, and are about the size of those of a pigeon, are deposited on the ground near water, and carefully covered over with leaves.

In captivity these tortoises thrive well in Europe, if the water be kept at a sufficiently high temperature; and a specimen of Temminck's snapper, which lived for more than thirteen years in the Brighton Aquarium, grew to a length of between 4 and 5 feet from beak to tail, whereas, on its arrival, it measured less than a foot. Although the flesh of the adult of this species has such a strong musky flavour as to be uneatable, that of the young is stated to be tender and palatable. The eggs are also sought after as articles of food; and when two or three females have laid together, as many as from sixty to seventy may be taken from a single nest.

TURTLES

THE members of all the families of chelonians hitherto mentioned have their feet more or less fully adapted for walking on land, and the majority of the toes furnished with well-developed claws or nails, while the carapace is generally of a somewhat oval form. The true turtles (*Chelonidæ*), on the other hand, while agreeing with the foregoing groups in having their shells covered with horny plates, are at once distinguished by the limbs being converted into flattened paddles, in which, at the most, only two of the toes are furnished with claws. They are further characterised by the heart-like form of the carapace, within which the head can be only partially withdrawn; as well as by the circumstance that the plastron is never united by bone to the carapace, and vacuities remain in the latter between the costal and marginal bones either throughout life or for a very long period. The skull has its temporal pits completely roofed over by bone, and the vertebræ of the very short tail have the articular cup in front and the ball behind.

Entirely marine in their habits, and resorting to the shore only for the purpose of breeding, turtles lay spherical eggs with thin, parchment-like shells, on sandy shores. In their entire conformation these chelonians are admirably adapted for an aquatic life, the body being depressed to facilitate rapid progress through the water, both the skull and shell being of unusually light and porous structure; while the limbs form most perfect paddles, capable of propelling their owners with great speed through the water. The head is placed upon the neck in such a manner as to allow of the nostrils being rapidly raised above the surface of the water for the purpose of breathing, and the nostrils themselves can be hermetically closed by means of fleshy valves. The three best-known species of turtles, which are assigned to two genera, are inhabitants of all tropical and sub-tropical seas; one species—the loggerhead—occurring in the Mediterranean, and occasionally wandering northwards.

Widely celebrated as the source of turtle-soup, the green turtle (*Chelone mydas*) is one of two species belonging to a genus characterised by the presence of four pairs of costal shields on the carapace, and by the persistence of the vacuities between the costal and marginal bones of the latter throughout life. The plastron is, moreover, distinguished, by the presence of an intergular shield between the two gulars; while, as in the second genus, there is a row of inframarginal shields between the marginals and the proper shields of the plastron. The skull is of moderate size in comparison to the shell, with the sockets of the eyes placed nearly vertically, and separated by a broad bar of bone. Such are the characters common to the two species of the typical genus of the family.

The green turtle is specially distinguished by its short beak, which is devoid of a hook at the tip, and by the shields of the carapace being in contact by their edges all through life. In the young the carapace shows a faint median keel; and its hind margin is at most only feebly serrated at all ages. Generally there is only a single claw on each paddle, although, in some instances, young specimens also have a claw on the second digit. In colour the shell of the adult is olive or brown, with yellowish spots or marblings; but in the young it is uniform dark brown or olive above, and yellow beneath, the limbs being bordered with yellow on the upper surface, and inferiorly yellow with a brown spot near the extremity. The food of this species consists of seaweeds, especially the sea-wrack, upon which the turtles graze at the bottom of the water, rising occasionally to the surface to breathe.

Generally rejected as food, the hawksbill turtle (*C. imbricata*) enjoys thereby no respite from persecution, since it is eagerly hunted for the beautifully mottled horny shields of its shell, which are the chief source of the tortoiseshell of commerce. In its young state the hawksbill may be readily distinguished from the preceding species by the circumstance that the horny shields on the back of the three-ridged shell overlap one another like the tiles on a roof. With advancing age the shields gradually, however, become smooth, and in very old specimens they meet at their edges, as in other members of the order. At all ages the hind margin of the carapace is more or less strongly serrated; and the compressed and sharply-hooked beak will always serve to distinguish at a glance a hawksbill from a green turtle. Moreover, the limbs are always furnished with two claws each.

In the adult the shields of the carapace are beautifully marbled and mottled with yellow and dark reddish brown, while the plastron is yellow, and the shields on the head and paddles are brown with yellow margins. In size this species is somewhat inferior to the green turtle, the carapace attaining a length of about 32 inches, against 42 inches in the latter. In habits the hawksbill differs markedly from the green turtle, being exclusively carnivorous.

The third, and probably the largest, species of turtle is the loggerhead (*Thalassochelys caretta*), easily recognised by its enormous head and the presence of at least five costal shields on each side of the carapace, which differs from that of the two preceding species by becoming completely ossified in the adult state.

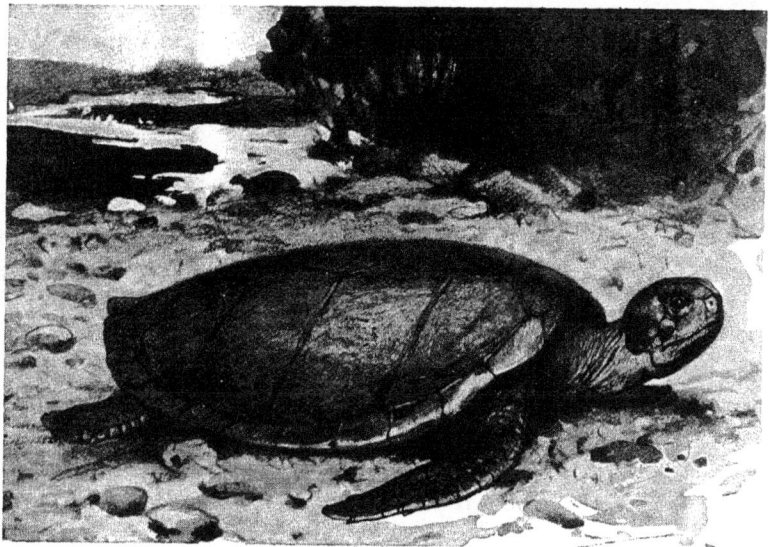

THE GREEN TURTLE

The beak is strongly hooked; and while in the young there are usually two claws to each paddle, one of these frequently disappears in the adult. In colour the adult is brown above and yellowish beneath; but the young are uniformly dark brown or blackish.

The Mexican loggerhead (*T. kempi*), from the Gulf of Mexico, differs in having a median ridge on the bone of each jaw, whereas in the ordinary species such ridges are confined to the investing horny sheath. Loggerheads appear to feed on cuttles and other molluscs, their powerful beaks enabling them to crush strong conchshells as easily as a man can crack a nut.

Apart from the difference in their food, all turtles appear to be similar in their general mode of life, never leaving the sea except for the purpose of laying their eggs, and then shuffling along in an awkward, ungainly manner. During the laying season they resort in vast numbers to low, sandy coasts, especially unfrequented tropical islands, and if once turned on their backs, while on shore, are unable to right themselves again. This habit of resorting to the land to lay their eggs clearly proves, it may be observed, the descent of turtles from freshwater members of the order.

Writing of the green turtles at Aldabra, near the Seychelles group of islands, Mr. Spurs remarks that the males permanently frequent the bays of that island, the females when they attain full maturity (twenty or twenty-five years) disappearing altogether. When the latter come to the shore for the purpose of laying, their shells are covered with barnacles of two or three weeks' growth. Commercially the females are more valuable than the males, and, as they are more easily captured, the proportion found on the island is one female to every ten males, although for one of the latter about ten of the former sex are hatched.

Turtles usually come ashore on fine moonlight nights, displaying great caution in landing, and then generally uttering a loud hissing noise which serves to disperse many of their enemies. Once landed, the female turtle, writes Dr. Audubon, " proceeds to form a hole in the sand, which she effects by removing it from under her body with her hind flippers, scooping it out with so much dexterity that the sides seldom, if ever, fall in. The sand is raised alternately with each flipper, as with a large ladle, until it has accumulated behind her, when, supporting herself with her head and fore part on the ground fronting her body, she, with a spring from each flipper, sends the sand around, scattering it to the distance of several feet. In this manner the hole is dug to the depth of eighteen inches, or sometimes more than two feet. This labour I have seen performed in the short space of nine minutes.

" The eggs are then dropped one by one, and disposed in regular layers, to the number of one hundred and fifty, or sometimes nearly two hundred. The whole time spent in this part of the operation may be about twenty minutes. She now scrapes the loose sand back over the eggs, and so levels and smooths the surface that few persons on seeing the spot could imagine that anything had been done to it. This accomplished to her mind, she retreats to the water

HAWKSBILL TURTLES SWIMMING

with all possible despatch, leaving the hatching of the eggs to the heat of the sand."

During a season each female will lay three clutches of eggs, at intervals of from a fortnight to three weeks, usually from one hundred and twenty-five to one hundred and fifty in number. No sooner are the young turtles hatched than hosts fall victims to landcrabs, frigate-birds, and other sea-birds, while, when they reach the sea, they are attacked by swarms of predaceous fishes. To escape the latter, the young reptiles allow themselves to be carried out by currents into deep water, where they are less readily seized. During the breeding season the males fight desperately with one another, to the great joy of the sharks, by whom the disabled ones are seized.

When first laid, the round eggs of turtles are never quite full, but before hatching become completely distended. In describing the breeding habits of the turtles kept in a pond near the dockyard in Ascension Island, Professor Moseley states that in the breeding season the females dig great holes as large as themselves in a bank of sand in which to deposit their eggs. The sand in which the eggs are laid does not feel warm to the hand, but during the daytime is rather cool, and at all times is moist. Its temperature appears to undergo no material variation, owing to the depth at which the eggs are deposited, such medium amount of heat being sufficient for the hatching.

Although a large number of green turtles are captured by being turned on their backs while on shore, in the Seychelles and Bahamas they are harpooned. In Keeling Island the method of capture is described by Darwin as follows : " The water is so clear and shallow that, although at first a turtle dives quickly out of sight, yet, in a canoe or boat under sail, the pursuers, after no long chase, come up to it. A man, standing nearly in the bows at this moment, dashes through the water upon the turtle's back, then, clinging with both hands to the shell of the neck, he is carried away, till the animal becomes exhausted, and is secured."

In China and Mozambique turtles are captured by

means of sucking-fishes, which are taken to a spot where the reptiles are basking upon the surface of the water. Each fish has a ring round its body, to which a line is attached, and as soon as it securely fastens itself by its sucking-disc to the back of a turtle, both captor and captured are drawn ashore. Although those of the loggerhead have a somewhat musky taste, the eggs of the other species of turtle are much esteemed as articles of food, while all yield a valuable oil.

As already mentioned, tortoiseshell is mainly a product of the hawksbill turtle. As the raw tortoise-shell is very unlike the finished article, with which all are familiar, Professor T. Bell's brief account of the process of manufacture may be quoted. As the horny shields, when removed from the turtle, are highly curved, " the uneven curvature has first of all to be removed, and the plate rendered perfectly flat. This is effected by immersing it in hot water, and then allowing it to cool under heavy pressure between smooth blocks of wood or metallic plates. The surface is then rendered smooth, and the thickness equal, by scraping and filing away the rough and prominent parts. In this way each plate receives an even and smooth surface. But it is in many cases desirable to employ larger pieces than can be obtained from single plates, and two pieces are then united together in the following manner. The edges are bevelled off to the space of two or three lines, and the margins, when placed together, overlap each other to that extent. They are then pressed together by a metallic press, and the whole is submitted to the action of boiling water ; and by this means the two pieces are so admirably soldered together as to leave no indication of the line of union. By the application of heat, also, the tortoiseshell may be made to receive any impression by being pressed between metallic moulds." Necklaces, and so forth, are made by pressing the fragments and dust in moulds.

Turtles more or less closely allied to the existing kinds abound in marine strata of the Tertiary and Cretaceous epochs, some belonging to extinct and others to the living genera. Among the latter, the gigantic Hoffmann's turtle (*Chelone hoffmanni*) from the Chalk of Holland appears to have been allied to the hawksbill, but had a shell of some 5 feet in length. Extinct loggerheads occur in the London Clay ; and an allied extinct genus (*Lytoloma*), common to the same formation and the upper Cretaceous deposits, is

YOUNG LOGGERHEAD TURTLE

remarkable for the great length of the bony union between the two branches of the lower jaw, and also for the circumstance that the aperture of the internal nostrils is placed quite at the hind extremity of the palate, as in modern crocodiles. In strata older than the Chalk, such as the Purbeck and other Oolitic rocks, are found remains of turtles having heart-shaped shells, but clawed limbs, and a vacuity in the centre of the plastron, these forming an extinct family (*Acichelyidæ*), from which the modern turtles have probably originated.

A large extinct turtle from the Cretaceous rocks of Kansas approximates in the structure of its shell to the above-mentioned *Lytoloma*, but the skull comes nearer to that of the *Chelydridæ*. That the genus should be classed with the true turtles Dr. G. R. Wieland, who has described several specimens, is fully convinced, although he believes the limbs to have been independently modified for swimming. The most interesting feature in the organisation of this turtle is the presence of certain bony elements overlying the junctions between the neural bones of the carapace, which may have been more extended in other types, and may represent the mosaic-like shell of the leathery turtles.

LEATHERY TURTLE

The remarkable leathery turtle, or luth (*Dermochelys coriacea*), which is the solitary survivor of a series of extinct forms, is one of those animals whose serial position is a matter of dispute among naturalists ; some of whom regard it as so different from all other chelonians that it ought to represent a suborder by itself, while others believe it to be merely a highly specialised form allied to the true turtles and give it family rank as *Dermochelyidæ*. From the evidence afforded by extinct species, the latter view appears more likely to be true. The essential peculiarity of the leathery turtle is to be found in the nature of its carapace, which is a mosaic-like structure composed of a number of irregular discs of bone closely joined together and entirely free from the back bone and ribs.

In certain extinct species the carapace, on the other hand, is represented merely by a row of marginal bones, while in others there appears, as mentioned above, to be evidence of a superimposed carapace corresponding to that of the modern luth, and of a second subjacent one representing the true carapace of ordinary turtles. From this it has been inferred that these reptiles were derived from true turtles by a gradual disintegration and breaking up of the carapace.

In the living genus the carapace is completely bony, and marked by seven prominent longitudinal keels ; but the plastron is much less fully ossified, and carries five similar keels, the unpaired entoplastral bone being wanting. The head, which is covered with small shields, is remarkable for its relatively large size and globose form ; the beak bearing two triangular cusps situated between three deep notches. The jaws differ from those of the true turtles in being sharp-edged from end to end, without any expanded bony palate ; and there is also an important difference in the structure of the skull itself, which may, however, be apparently the result of specialisation. As in the true turtles, the limbs are converted into flattened paddles, which are, however, completely destitute of claws ; the front pair being much elongated, narrow, and pointed, while the hind ones are short and truncated. The humerus, or bone of the upper arm, has the same general form as in the true turtles, and is thus very unlike the corresponding bone of other members of the order. The process marked *b* in the figure on page 1627 is more developed than in the typical turtles ; and the foramen (*e*), at the lower end, is unique in the order. Largest of living turtles, the leathery turtle exceeds

6 feet in length; and while in the young the front flippers are equal in length to the shell, in the adult they become relatively much shorter. The general colour is dark brown, which may be either uniform or relieved with yellow spots; the longitudinal tuberculated keels on the shell, as well as the margins of the limbs, being invariably yellow in the young.

This turtle is generally distributed throughout the tropical portions of the Atlantic, Indian, and Pacific Oceans, whence it occasionally wanders to the coasts of cooler regions, including those of the British Isles. Yearly becoming scarcer, it seems to be one of those species which stand a fair chance of extermination at no very distant date. Although little is known as to the mode of life of this turtle, it appears that its food is chiefly of an animal nature, comprising fish, crustaceans, and molluscs. In the breeding season it appears in numbers on the Tortugas Islands, off the coast of Florida, and sometimes in still greater abundance on the sandy shores of Brazil.

Arriving somewhat later than the true turtles, it deposits its eggs in a similar manner, laying as many as three hundred and fifty, in two batches; while at times, when three or more females have a nest in common, upwards of a thousand eggs may be found in a single spot. When hatched, the young turtles immediately seek the water, where, however, they have almost as many foes as on land; so that it is probable that only a very small percentage arrive at maturity. The strength and weight of a full-grown individual are very great; one captured some years ago on the coast of Tenasserim requiring the combined efforts of ten to twelve men to drag it on to the beach. The flesh has an unpleasant flavour, and is not, therefore, generally eaten. The name luth refers to a fanciful resemblance between the carapace of this species and an ancient type.

EXTINCT TURTLES

Gigantic as is the existing leathery turtle, it was considerably exceeded by some of its extinct allies. Among these the huge *Eosphargis*, from the London Clay, with a skull of nearly a foot in length. apparently had a carapace consisting only of one median row of broad-keeled body plates, and a border of marginal bones; while in *Psephophorus*, from the higher Eocene and Miocene strata of the Continent, both upper and lower shells were formed of mosaic-like bones, which, it is thought, were overlain by horny shields. In the earlier *Protostega* and *Protosphargis*, from the Cretaceous rocks of North America and Europe, the upper shell appears to have been represented merely by a row of marginal bones, while the lower one was very stoutly ossified; some of these early forms probably attained a length of from 10 to 12 feet.

LEATHERY TURTLE

Another generic type of marine turtle (*Eochelone brabantica*) has been described from the middle Eocene of Belgium. In the forward position of the aperture of the inner nostrils, as well as in the shortness of the union between the two branches of the lower jaw, this species differs from typical turtles, and approximates to the luth. And as both the upper and lower shells are much reduced, and show many vacuities, this species seems to confirm the view that the luth is a specialised form derived from the ancestors of the true turtles. Nevertheless, on account of its peculiar unattached "mosaic" carapace, its describer, Mr. L. Dollo, considers that, together with *Psephophorus*, it should represent a family by itself. On the other hand, all the other turtles, including the luth-like *Eosphargis* of the London Clay, are placed by this naturalist in the family *Chelonidæ*, on account of their possessing, albeit in some, cases in a rudimentary condition, a carapace attached to the ribs.

Further evidence of kinship between the luth and typical turtles is afforded by the American Cretaceous turtles classed in a separate family, the *Protostegidæ*. These extinct generic types of turtles, as represented by *Protostega* and *Archelon*, attained gigantic dimensions, and, in accordance with the needs of a marine existence, lightened the carapace by a great reduction in the size of the costal plates, which are more aborted in the first-named genus than in the modern genus *Chelone*, thus leaving very large intercostal vacuities. This reduction of the costals is carried to a still greater degree in *Archelon*, the absorption process being also extended to the neural bones, many of which appear to be reduced to thin films. Upon the neurals in this genus are, however, superimposed a series of irregularly shaped epineural dermal bones, which undoubtedly correspond to the neural keel of *Dermochelys*, and discharge the function of the aborted neurals.

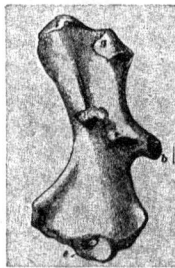

LEFT HUMERUS OF AN EXTINCT LEATHERY TURTLE

In life *Archelon* apparently possessed a leathery hide, with a system of keels similar to those of the leathery turtle. In regard to the relationship between these and other turtles, Dr. G. R. Wieland observes "that of the two camps which have attacked the difficult and highly attractive problem of the origin of *Dermochelys*, those favouring the view of a close relationship to ordinary turtles and a comparatively recent origin have rather the best of the argument." It seems, in fact, that *Dermochelys* and its allies, having become less pelagic in habits than ordinary turtles, found the reduction of the bony framework of their carapace a disadvantage, and they accordingly developed a secondary structure of overlying dermal bones to take the place of the proper carapace, which then underwent a still further reduction, and finally vanished.

SIDE-NECKED TORTOISES

In place of withdrawing the head into the shell by means of an S-like flexure of the neck in a vertical plane, as in all the groups hitherto described, the remainder of the living tortoises with complete shells bend the neck sideways in a horizontal plane (as shown in the illustration on page 1630), and thus bring the head within the margins of the shell. Accordingly, the group is collectively spoken of as the side-necked tortoises, or Pleurodira. This character is alone amply sufficient to separate the group from the foregoing assemblage of S-necked, or cryptodiran, tortoises, but since there are also certain features by which the skulls and shells of the two groups can be identified, it is important that these should be noticed.

As regards the skull, this is distinguished in the first place by the tympanic ring surrounding the aperture of the ear being complete, as may be seen by comparing the accompanying figure with the one on page 1602, and also by the circumstance that the lower jaw articulates by means of a knob-like condyle with a corresponding cavity in the quadrate bone, whereas in the preceding group the positions of the condyle and cup are reversed. The shell, which is always fully developed, and forms a solid box, presents the peculiarity that both the carapace and the hind part of the plastron are immovably welded to the bones of the pelvis; its upper and lower moieties thus having a bond of union which is totally lacking among the S-necked tortoises. Further, the vertebræ of the neck are furnished with projecting lateral or transverse processes, which are absent from the latter group.

In addition to these absolutely characteristic features, there are certain other points connected with the anatomy of the side-necked tortoises which demand a brief notice. With the exception of one species, which lacks horny shields on the shell, all these tortoises are characterised by the presence of an intergular (*i.gu*) shield between the two gulars (*gu*) on the front of the plastron; such intergular shield being, as already mentioned, rarely present in the S-necked group. Very generally among the present assemblage one or more of the pairs of costal bones of the carapace may meet in the middle line, owing to the absence of some of the median unpaired series of neural bones; in certain cases the whole of the costals thus meeting, owing to the absence of all the neural bones. It may be added that while in the family *Chelyidæ* the plastron contains the same nine bones as in the side-necked tortoises, in a second family, *Pelomedusidæ*, there are eleven bony elements in this part of the shell, owing to the presence of an additional (mesoplastral) pair between the normal hyoplastral and hypoplastral bones.

The side-necked tortoises, of which the great majority may be included in the two families *Chelyidæ* and *Pelomedusidæ*, are all of fresh-water habits, and at the present day are exclusively restricted to the Southern Hemisphere, while they are the only members of the order found in Australia and New Guinea. During the earlier portion of the Tertiary period they extended, however, into the Northern Hemisphere, and in the preceding Secondary period were abundantly represented in Europe. These facts show that the group is a very ancient one, and by the presence of the additional mesoplastral elements in the lower half of the shell of some of its representatives it is allied to a third and totally extinct group, the Amphichelydia, which disappeared before the close of the Secondary period.

THE MATAMATA

The extraordinary reptile known as the matamata (*Chelys fimbriata*) is the typical representative of the first of the existing families of the group; the various genera included in this family (*Chelyidæ*) being collectively characterised by the presence of the normal nine bones in the plastron, by the neck being incapable of complete retraction within the margins of the shell, and the absence of a bony temporal arch to the skull. Eight living genera are included in the family, the range of which is restricted to South America, Australia, and New Guinea.

The matamata, which is an American species inhabiting Guiana and Northern Brazil, and is the sole living representative of its genus, is easily recognised by its broad and elongated neck, of which the sides are fringed with peculiar fimbriated projections, and the depressed and triangular head, terminating in a proboscis-like nose, and furnished with very small eyes. Not less characteristic is the equally depressed and much-corrugated shell, in which the carapace bears three longitudinal ridges, subdivided into knotty protuberances by cross-valleys; the horny shields of the same being extremely rugose, and marked with deep, radiating lines. The vertebral shields are broader than long, the hind marginals are more or less strongly serrated, and there is a distinct nuchal shield on the front edge of the carapace. On the removal of the horny shields from the carapace, it is seen that only the last pair of costal bones meet in the middle line, owing to the presence of only seven neural bones.

The plastron is narrow and deeply notched behind, the tail is very short, and the toes are fully webbed. In addition to the rows of fimbriated appendages on each side of the neck, there is a similar outgrowth of skin on the chin, and a larger pair of appendages above the ears. In colour the adult is uniform brown, but the young are prettily marked with bands of brown and yellow on the chin and neck, and the shell is ornamented with black and yellow spots. The species is of comparatively large size, the shell attaining a length of 15 inches.

Unfortunately, little is known as to the mode of life of this strange tortoise. When in its native element,

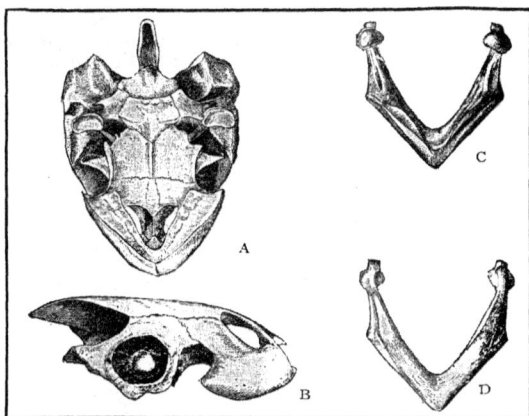

(A) LOWER AND (B) SIDE VIEWS OF SKULL, AND (C) UPPER AND (D) LOWER ASPECTS OF LOWER JAW OF ARRU TORTOISE

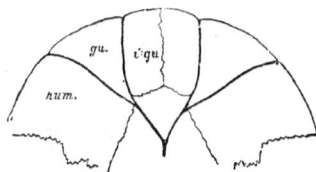

FRONT PORTION OF PLASTRON OF A SIDE-NECKED TORTOISE, WITH HORNY SHIELDS REMOVED
The thicker lines indicate the boundaries of the shields

the warty appendages on the neck float in the water like some vegetable growth, while the rugged and bossed shell strongly resembles a stone, and it is thus probable that the whole appearance of this reptile is advantageous either in deluding its enemies or in attracting to it the animals on which it feeds—the latter being the more likely hypothesis. Although it appears that the matamata will occasionally eat vegetable substances, its chief food consists of fishes, frogs, and tadpoles, some of which are very probably attracted within reach by mistaking the appendages on the neck for plants or animals on which they feed. The matamata is, however, stated to capture some of its prey by swimming swiftly among water-plants, diving immediately that a fish or frog is seized in its beak.

SNAKE-NECKED TORTOISES

The snake-necked tortoises, of which there are two South American species (*Hydromedusa maximiliani* and *H. tectifera*), agree with the matamata in their long necks and weak jaws, but differ in the smooth shell, the absence of a proboscis to the nose, and the presence of only four claws on each foot—the matamata having five claws on the fore feet and four on the hind pair. The flattened shell in the young state has an interrupted median ridge, and presents the unique peculiarity that the broad nuchal shield of the carapace is placed behind the first pair of marginals (which consequently meet in the middle line), and thus simulates a sixth vertebral shield.

H. tectifera, the species represented in the illustration on page 1630, ranges from Southern Brazil to Buenos Aires, and has a shell measuring about 8 inches in length and its feet largely webbed. In colour the carapace is dark brown and the plastron yellowish, with brown spots in the young; the head and neck being olive, with a curved white streak on each side of the throat, and a broader white band, edged with black, running along the sides of the head and neck.

Nocturnal and carnivorous in their habits, the snake-necked tortoises appear to agree in their general mode of life with the majority of fresh-water species. During the daytime they are generally to be found lying asleep on some dry spot near the water with the neck bent on one side, and the head, like the limbs and tail, retracted within the margins of the carapace. When disturbed, the head and neck are, however, shot out with marvellous rapidity, reminding the observer of the sudden dart of a snake.

AUSTRALIAN LONG-NECKED TORTOISES

In Australia and New Guinea the place of the preceding group is taken by another genus of long-necked tortoises known as *Chelodina*, the members of which may be recognised by the presence of a normally-placed nuchal shield on the carapace, coupled with the circumstance that the intergular shield of the plastron, instead of being placed between the gulars, as in the figure on page 1628, is situated behind the latter, which consequently meet in the middle line. The vertebral horny shields are longer than broad, and the whole of the shields remarkable for their extreme thinness. On removing the latter from the carapace, it will be found that, owing to the absence of the neural bones, all the pairs of costal bones meet in the middle line—a peculiarity shared with one American and two other Australian genera of the family. There are four species of these long-necked tortoises, three of which are found in Australia, while the fourth is Papuan.

OTHER GENERA

In addition to the foregoing, there are four other genera belonging to the family *Chelyidæ*, all of which are distinguished by their shorter necks, the length of which is inferior to that of the back. Of these the American *Rhinemys*, *Hydraspis*, and *Platemys* are characterised by the narrow anterior extremity of the lower jaw, and by the first vertebral shield of the carapace being wider than either of the others. The second of these genera, of which a member is represented in the figure on page 1631, is by far the most numerous in species, and is noteworthy on account of being represented by a fossil species in the Eocene deposits of India. The third genus differs from the other two in the absence of neural bones to the carapace. On the other hand, the two Australian genera—*Emydura*

MATAMATA TORTOISE

and *Elseya*—both of which present the feature last mentioned, are distinguished by the broad anterior extremity of the lower jaw, and by the first vertebral shield of the carapace not exceeding the others in size.

ARRU TORTOISES

Certain tortoises which may be collectively designated by the native name of the large Brazilian species take their scientific title, *Podocnemis*, from the presence

SNAKE-NECKED TORTOISES

side. In colour the upper shell is brown or olive, with darker markings, but the plastron is yellowish, spotted with brown; the young being olive above and yellow beneath, with some yellow spots on the head. All the other members of the genus are of greatly inferior dimensions, a second Amazonian species (*P. sextuberculata*) having a shell of scarcely more than a foot in length, and being distinguished from its larger relative by the presence of only a single wattle on the chin.

One of the best accounts of the habits of these tortoises is given by the old German traveller Baron von Humboldt, who states that on the Orinoco the period of egg-laying coincides with that of the lowest level of the waters of the river, or from the end of January till the latter part of March. During January the tortoises collect in troops, which soon leave the water to bask on the warm banks of sand exposed by the lowering of the river. Throughout February they may be found on such banks during the greater part of the day; but early in March the several troops collect in larger bodies, and then make their way to the comparatively few islands where the eggs are habitually deposited. At this time, shortly before the egg-laying begins, thousands of the tortoises may be seen arranged in long strings around the shores of these islands, stretching out their necks, and holding their necks above water, in order to see whether there is anything to prevent their landing in safety.

As the reptiles are exceedingly timid, and especially averse to the presence of human beings or boats, the Indians, to whom the harvest of tortoise-eggs is of the utmost importance, take every precaution to prevent them being disturbed, posting sentinels at intervals along the banks, and warning all passing boats to keep in the middle of the river. When the tortoises have landed, the laying of the eggs takes place at night, and begins soon after sunset; the females digging holes about three feet in diameter and two feet in depth, with the aid of their powerful hind limbs. So great is the contention for space that one tortoise will frequently make use of a pit dug by a neighbour, and in which one set of eggs has already been deposited, although not yet covered over with sand; two layers of eggs thus occupying one area. The crowding and jostling of the reptiles necessarily leads to an immense number of eggs being broken; the number being estimated at a fifth of the whole, and the contents of the fractured shells in many places cementing the loose sand into a coherent mass. The number of tortoises on the shore during the night being so large, many of them are unable to complete the work of egg-laying before dawn; and these belated individuals become quite insensible to danger, continuing there even in the presence of the Indians, who repair to the spot at an early hour.

The great assemblage of these chelonians takes place on one particular island in the Orinoco, hence known as the Boca de la Tortuga; and, according to native accounts, no other spot is to be met with on the river, from its mouth to its junction with the Apure, where eggs can be found in such abundance. On this particular island the number of eggs deposited is enormous; a large stretch of smooth sandy beach being underlain with an almost continuous layer. To determine the position and extent of the deposit, a long pole is thrust down by the natives at intervals into the sand; the sudden want of resistance to its descent proclaiming

of a pair of large shields on the outer side of the hind foot of the typical species. These tortoises belong to the family *Pelomedusidæ*, which contains three genera, and is now confined to Africa, Madagascar, and South America. This family is broadly distinguished from the last by having eleven elements in the plastron, owing to the presence of a pair of mesoplastral bones, and the neck is completely retractile within the margins of the shell. The skull differs from that of the preceding family in having a bony temporal arch, as shown in the figure on page 1628, and lacks the distinct nasal bones generally found in the former.

AMAZONIAN ARRU TORTOISES

The largest and best-known representative of the whole family is the great Amazonian arru tortoise (*P. expansa*), which considerably exceeds in size all other living members of the entire group, having a shell which may measure as much as 30 inches in length. It belongs to a genus including seven existing species, of which six are South American, while the seventh is an inhabitant of Madagascar. This extremely anomalous distribution is to some extent explained by the occurrence of fossil representatives of the genus in the Eocene rocks of India and Egypt, which indicates that these tortoises were at one time widely spread.

As a genus, these tortoises are characterised by the skull having a roof over its temporal region, coupled with the presence of five claws on the fore feet and four on the hind pair, and likewise by the circumstance that the mesoplastral bones are small and confined to the edges of the plastron, so that they are widely separated from one another in the middle line. The toes are broadly webbed, and the tail is remarkable for its extreme shortness.

The typical species, which inhabits Tropical South America to the eastwards of the Andes, and is extremely abundant in the upper part of the Amazonian system, has the shell expanded posteriorly, and much depressed in the adult, although at an earlier stage it presents a roof-like form. The chin is furnished with two small wart-like appendages, and the hind foot characterised by the presence of two very large shields on its outer

when the loose layer containing the eggs has been reached.

According to measurements taken by Humboldt, the stratum extended to a distance of one hundred and twenty feet from the water, and averaged three feet in depth. The whole area is regularly parcelled out among the Indians, who proceed to work the layer with the regularity of miners. The sand having been removed, the eggs are carried in small baskets to the neighbouring encampment, where they are thrown into long wooden troughs of water. Here they are broken and stirred up with shovels, and the mass then left in the sun till all the oily matter has collected at the surface, from which it is continually ladled off, and taken away to be boiled over a quick fire. The result of this process is a limpid, inodorous, and slightly yellow substance known as " turtle-butter," which can be used for much the same purpose as olive-oil. In spite of the enormous quantity of eggs thus taken, numbers are hatched, and Humboldt saw the whole bank of the Orinoco swarming with small tortoises of an inch in diameter, that escaped only with difficulty from the pursuit of the Indian children. All these tortoises are vegetable-feeders ; and the females greatly exceed the males in size.

On the Upper Amazon the large species, according to Mr. Bates, is captured either by means of nets or by shooting with arrows. On such occasions, after the net is set in a semicircular form at one extremity of a pool, the party spread themselves around the swamp at the opposite end, and begin to beat with poles in order to drive the tortoises towards the middle. This process, on the occasion referred to, " was continued for an hour or more, the beaters gradually drawing nearer to each other, and driving the hosts of animals before them ; the number of little snouts constantly popping above the surface of the water showing that all was going on well. When they neared the net, the men moved more quickly, shouting and beating with great vigour. The ends of the net were then seized by several strong hands and dragged suddenly forwards, bringing them at the same time together, so as to enclose all the booty in a circle. Every man now leapt into the enclosure, the boats

GREAT ARRU TORTOISE

were brought up, and the turtles easily captured by the hand and tossed into them." Altogether, about eighty individuals were captured in the course of twenty minutes or so.

In shooting tortoises, the arrow employed has a strong lancet-shaped steel point, fitted to a peg which enters the tip of the shaft. To the latter the peg is secured by a hank of twine some thirty or forty yards in length, and neatly wound round the body of the arrow. When a tortoise is struck, the peg drops out from the shaft, and is carried down by the diving animal, leaving the latter floating on the surface. Thereupon the sportsman paddles up to the arrow, and proceeds to " play " his victim until it can be drawn near to the surface, when it is struck with a second arrow, after which, by the aid of the two cords, it can be safely drawn ashore. In many villages on the Amazon every house has a pond, in which a number of these tortoises are kept for food.

OTHER GENERA

The other two living genera of the family *Pelomedusidæ*, namely, the typical *Pelomedusa* and *Sternothærus*, differ from the first by the absence of a bony roof to the temporal region of the skull, and likewise by the presence of five claws in both the front and hind feet. The former has the mesoplastral elements of the plastron small and similar to those of the arru tortoises, but in the latter these are as well developed as the other elements of the plastron, meeting in the middle line. *Pelomedusa* is represented by a single species common to Africa and Madagascar, but of the six species of *Sternothærus* five are exclusively African, while the sixth inhabits Eastern Africa and Madagascar. *S. derbianus* is one of the species.

Reference may be made here to *Stereogenys cromeri*, a large species from the Eocene of Egypt characterised by the backward position of the opening of the inner nostrils, due to the development of a secondary bony floor below the palate. A similar feature occurs in the contemporary turtles of the genus *Lytoloma*.

THE FLY RIVER TURTLE

A remarkable chelonian (*Carettochelys insculpta*) from the Fly River, New Guinea, differs from all other members of the side-necked group in the absence of

A SOUTH AMERICAN SIDE-NECKED TORTOISE

horny shields on the shell and the conversion of the limbs into paddles, each of which carries only two claws. The neck is not retractile. In the carapace there are six very small neural bones, which are not in contact with one another, thus allowing each pair of costals to meet in the middle line, and the plastron has only the usual nine bones. A wavy sculpture ornaments the whole of the external surface of the shell, which attains a length of about 18 inches. The head is large, and the tail relatively short. The species, which represents a separate family, *Carettochelyidæ*, is still very imperfectly known; and it has been suggested that it does not belong to the side-necked group at all. It is not improbable that a chelonian, *Hemichelys*, from the Eocene rocks of India indicates a second member of the same family, as its shell was similarly devoid of horny shields.

STERNOTHÆRUS DERBIANUS   Photo, W. B. Johnson

EXTINCT FAMILIES

The most aberrant members of the entire chelonian order are certain extinct gigantic tortoises (*Miolania*) from the later Tertiary deposits of Australia and Patagonia, characterised by the presence of several pairs of horn-like protuberances on the skull, and also by the investment of the tail in a bony sheath, recalling that

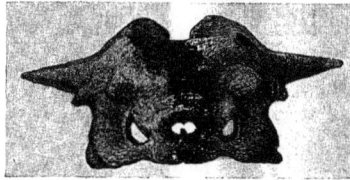

SKULL OF A HORNED TORTOISE

of an armadillo. Unfortunately, the shell of these strange reptiles is known only by fragments; but, from the conformation of the bones of the feet, it appears that they were terrestrial, while the structure of the palate indicates that they were herbivorous. They clearly constitute a fourth family, *Miolaniidæ*, of side-necked tortoises. The remarkable geographical distribution of this group is probably due to a former land connection between South America and Australia.

The Secondary rocks of Europe contain the remains of a number of extinct tortoises which may be referred to a fifth family, *Plesiochelyidæ*, of the side-necked group. While agreeing with the existing *Chelyidæ* in having

only nine bones in the plastron, these extinct forms differ by the much greater thickness of their shells, and also by the circumstance that only one of the lower bones of the pelvis is welded to the upper surface of the plastron, whereas in the existing families both are thus united. Abundant in both the Oolitic and Wealden rocks, the majority of these tortoises are referred to the genus *Plesiochelys*, although some are separated as *Hylæochelys*, these being distinguished by the enormous width of the vertebral shields, in which the breadth may be three times the length. Nothing approaching this conformation is to be met with among living representatives of the order.

Certain extinct tortoises, such as *Pleurosternum* from the Purbeck Oolite of Swanage, and *Baëna* of the Eocene rocks of the United States, indicate the existence of an extremely generalised group of the order known as the Amphichelydia, the members of which present many characters common to the existing S-necked and side-necked groups, so that they may have been the ancestral stock of both the latter. All of them have eleven bones in the plastron, owing to the presence of mesoplastrals, and an intergular shield, but the pelvis may or may not be connected with the plastron.

In the first of the genera named, the mesoplastral bones extend right across the shell to meet in the middle line, and one of the bones of the pelvis articulates to a smooth oval facet on the plastron. On the other hand, in the second genus the mesoplastral bones are incomplete, as in the existing arrau tortoises, and there is no union between the pelvis and the plastron. Since it is probable that the plastron of the chelonians has originated from a system of abdominal ribs similar to those of the tuatera, it is interesting to notice that these generalised tortoises had a larger number of plastral elements than are to be found in the majority of the existing representatives of the order.

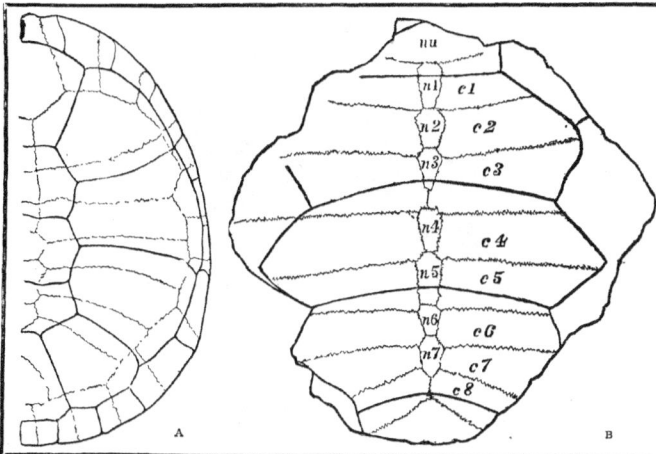

(A) RIGHT HALF OF CARAPACE OF BLACK STENOTHERE WITH HORNY SHIELDS REMOVED. (B) IMPERFECT CARAPACE OF WIDE-SHIELDED WEALDEN TORTOISE

SOFT TORTOISES

THE last group of the chelonian order comprises the soft river-tortoises, which are now confined to the warmer regions of Asia, Africa, and North America, but during the middle portion of the Tertiary period appear to have been extremely abundant in the rivers of England and other parts of Europe. The whole of these tortoises are included in a single family, *Trionychidæ*, which forms a group, Trionychoidea, of equivalent value to the S-necked and side-necked sections ; and it is not a little remarkable that while in the greater part of their organisation they approximate to the former group, in certain features connected with the skull they come nearer to the latter.

The most striking peculiarity of the soft tortoises is to be found in the nature of their shells, which are covered with a thin continuous leathery skin overlying the sculptured bones of the shell, and are entirely devoid of horny shields. The lower shell, or plastron, is always very imperfectly ossified, and completely separate from the carapace ; and the carapace itself never has a complete series of marginal bones, and passes at its borders into a soft expansion of skin, from which the name of the group is derived. If marginal bones occur at all, they are confined to the hind border of the shell, and are unconnected with the ribs, having, in fact, nothing in common with the bones so named in other tortoises, and being doubtless of independent origin. In being unconnected with the plastron, the pelvis resembles that of the S-necked group, and the head is retracted by a similar S-like flexure of the neck in a vertical plane.

In regard to the mode of articulation of the lower jaw with the skull, and likewise in the presence of a notch in the hind border of its tympanic ring, the soft tortoises again resemble the group last mentioned ; although in the general form of the skull and the conformation of the palate they come nearer to the side-necked group. A distinctive peculiarity of the skeleton is to be found in the presence of at least four joints in the fourth toe of each foot. Externally soft tortoises are characterised by their long, snake-like necks, which, together with the head, can be completely withdrawn into the shell, and also by the proboscis-like snout, and the thick, fleshy lips concealing the jaws. The ear is completely concealed ; and each foot, as indicated by the scientific name of the group, carries only three claws, which are borne by the three inner toes. As a rule, the colour of the skin is greenish olive, with small yellow or orange spots, passing into streaks on the under surface of the head, but some species have a few large

YOUNG AMERICAN SOFT TORTOISES

eye-like spots on the back of the shell. Although all the soft tortoises are included in one family, they are arranged in six genera, three of which are nearly allied to one another, as are likewise the remaining three among themselves.

The first and largest genus, the typical *Trionyx*, contains fifteen living species, with a distribution coextensive with that of the family. The members of this genus are collectively characterised by the absence of a fold of skin on the hind part of the under shell, beneath which the leg may be concealed, by the sculpture on the shell being generally in the form of wavy raised lines, and by the hyoplastral and hypoplastral bones of the lower shell remaining distinct from one another. In the skull, as shown in the figure on page 1628, the sockets of the eyes are placed relatively far back, and widely separated from the aperture of the nose.

Among the better-known species, mention may be made of the Gangetic soft tortoise (*T. gangeticus*), now confined to the river system from which it takes its name, but formerly found, as shown by fossil specimens, in the Narbada valley ; the length of the shell and fleshy disc reaching as much as 2 feet. Like all the Old World representatives of the genus, this species has eight pairs of costal bones in the carapace ; but it belongs to a subgroup characterised by having two neural bones between the first pair of costals, and by the absence of a pronounced ridge in the middle of the upper surface of the extremity of the lower jaw.

The soft tortoise of the Nile (*T. triunguis*), ranging over Africa and Syria, and attaining still larger dimensions, belongs to a second subgroup, distinguished by having only a single neural bone between the first pair of costals ; while Phayre's soft tortoise (*T. phayrei*) of Burma may be taken to represent a third section, differing from the last by the presence of a median ridge in the front of the lower jaw. On the other hand, all the American soft tortoises, of which *T. ferox* is a well-known example, differ from all the preceding in having only seven pairs of costal bones. Numerous representatives of the genus occur in the Miocene and Eocene strata of Europe, as well as in the Tertiary rocks of India and the United States.

Two other members of the first subfamily, confined to Asia, represent as many genera. Of these Cantor's soft tortoise (*Pelochelys cantori*), from India, Burma, and Malaya, is characterised by the more forward position of the sockets of the eyes, as compared with the type genus. This forward position of the eye-sockets is still more marked in the much-elongated skull of the great Indian chitra (*Chitra indica*), where they are placed close up to the nose.

The three remaining genera of the family are characterised by the sculpture of the shell generally taking the form of small pustules, and thus resembling shagreen ; while the hyoplastral and hypoplastral bones of the lower shell are united, a n d there is a flap of skin on each side of the under surface, beneath which the hind limbs can be concealed. All the members of this group are confined to the Old World, and while one of the three genera is Indian, the other two are African. The Indian genus *Emyda* is readily characterised by the p r e s e n c e of a complete series of neural bones in the c a r a p a c e, coupled with a semicircle of marginal bones at its hind extremity.

UPPER AND LOWER VIEWS OF INDIAN SOFT TORTOISE

Photos, Dr. J. H. C. Connell

In neither of the three living species of this genus does the length of the shell and its soft disc exceed 10 inches, but much larger fossil forms are found in the Pliocene rocks of India. Both the African genera lack marginal bones, but whereas in one, *Cycloderma*, there is a full series of neural bones to the carapace, in the other, *Cyclanorbis*, these form an incomplete and interrupted series.

All the soft tortoises are thoroughly aquatic, most of them rarely leaving the water except for the purpose of laying their eggs, and, in consequence of these habits, comparatively little is known as to their mode of life. Although confined as a rule to rivers, a few of the species frequent estuaries, and Cantor's soft tortoise has been found some distance out at sea. Occasionally, again, specimens of the Indian granulated soft tortoises of the genus *Emyda* have been met with wandering on land far from the neighbourhood of water.

Fiercer and more spiteful than any other members of the chelonian order, these tortoises, owing to a peculiarity in the structure and mode of articulation of some of the vertebræ of the neck, have the power of darting out the head with inconceivable rapidity, the great Indian chitra being by far the greatest adept in this practice. Owing to this habit, the larger species are dangerous creatures to approach incautiously, as their bite is very severe, and the natives are not infrequently bitten by them in India and Burma while bathing.

All the members of the typical genus, together with Cantor's soft tortoise and the chitra, are known to be carnivorous, and it is commonly believed that the same is the case with the other members of the group. According,

CANTOR'S SOFT TORTOISE

however, to Dr. J. Anderson, this is incorrect with regard to the granulated soft tortoises of India, which he states to be exclusively vegetable and grain feeders. The larger species probably feed both on fish and other aquatic animals, as well as on the flesh of such carcases as may be floating in the rivers they inhabit.

In correlation with their asserted herbivorous habits, the small granulated species do not snap and bite after the manner of t h e i r larger cousins. On shore, when left to themselves, t h e s e species will slowly and cautiously extend their necks, a n d when approached, instead of attempting to escape, withdraw rapidly into their shells, of w h i c h the upper a n d lower front margins then meet, owing to the soft flaps of skin with w h i c h they are covered. All the species are chiefly nocturnal, remaining d u r i n g the day time partially or completely buried in the mud at the bottom of the water, and not beginning to swim till sundown. Such species as inhabit marshes or swamps liable to be dried up during the hot season bury themselves in the mud, at no great depth below the surface, during the period of drought.

As these tortoises are known to remain frequently for a period of from two to ten hours, and occasionally as much as fifteen hours, beneath the water without coming to the surface to breathe, it is obvious that they must have some special means of oxygenating their blood. It is probable, indeed, that certain filamentous appendages of the mucous membrane of the throat found in these tortoises subserve the office of gills, and thus enable the blood to be renovated by means of the atmospheric air dissolved in the water they inhabit. With regard to their breeding habits, it appears that the females of the granular - shelled species scrape a shallow hole in the mud, in which the small, round eggs are laid and then covered up.

ORDER PLACODONTIA

Brief mention may be made of the pavement-toothed, or placodont, reptiles, which may possibly have some relationship to the tortoises, and are characterised by the presence of broad, flattened teeth on the palate and jaws, as shown in the figure on page 1470 ; the skull being very short and more or less triangular, with the double nostrils situated near the extremity of the muzzle, some distance in advance of the sockets of the eyes, which occupy a nearly central position in the skull. These reptiles are confined to the period of the Trias, being represented by the genera *Placodus* and *Cyamodus*.

ORDER ANOMODONTIA
MAMMAL-LIKE REPTILES

THE last order of reptiles, which is entirely extinct and confined to the Triassic and Permian epochs, is of especial interest to the evolutionist as being the one which apparently includes the ancestral stock from which mammals have originated, and also as being closely related to certain extinct amphibians noticed in the sequel, which doubtless gave rise to the anomodonts themselves. It should be observed that anomodonts show the nearest affinity with the egg-laying mammals, and until we know the true relationship of the latter to the other members of the same class it is impossible to attempt to define the genealogy more exactly. Anomodonts, which, as stated in the introductory section, form a group of equivalent value with the one containing all the other reptilian orders, are the only reptiles which agree with the egg-laying mammals in having three distinct bones on each side of the true shoulder-girdle—that is to say, a blade-bone, or scapula, above, and a precoracoid and coracoid below. The pelvis is also very mammal-like, not only in that its three elements are united, but likewise in the small size of the vacuity, or foramen (of), between the pubis and ischium. It will be seen from the figures how close is the resemblance between the pelvic and shoulder girdles of these reptiles, each having one bone above and two below.

Even still more marked is the similarity between the upper arm-bone, or humerus, of the anomodonts and that of the egg-laying mammals; each having a perforation on the inner border of the lower end, whereas in those existing reptiles which possess such a perforation (with the exception of the tuatera, where there is one on each side) it is situated on the outer border. As a rule, anomodonts further resemble mammals in the absence of abdominal ribs; and there are important similarities in the structure of the skull, as well as, in many cases, in the teeth.

Special mammalian resemblances are also exhibited in the structure of the wrist (carpus) and ankle (tarsus) joints. On this point Dr. R. Broom points out that in the two anomodont groups respectively known as dicynodonts and theriodonts the mammalian approximation is most marked. "In these latter we find more or less approximation to the mammalian type, but if we take into consideration the extreme mammalian specialisation—

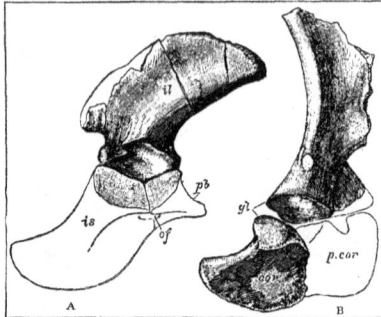

(A) PELVIS AND (B) SHOULDER-GIRDLE OF AN ANOMODONT

il, haunch-bone, or ilium; *is*, ischium; *pb*, pubis; *of*, foramen between ischium and pubis; *sc*, blade-bone, or scapula; *p.cor*, precoracoid; *cor*, coracoid; *gl*, cavity for head of upper arm-bone, or humerus

UPPER SURFACE OF SKULL OF A PARIASAURIAN

the presence of a large tibiale and fibulare with a centrale which is not in the centre, but comes between the tibiale and the first tarsale—then we are driven to the conclusion that the mammalian ancestor must have been a dicynodont, a theriodont, or a form belonging to a closely-allied order. From the examination of the skull we have good reason to believe that the ancestor was a theriodont, and the evidence of the tarsus fully confirms that drawn from the skull and other parts of the skeleton; and the carpus, while it does not add any very strong evidence, certainly does not afford any evidence that is not in harmony with this conclusion."

Anomodonts, which are met with most abundantly in the Triassic and Permian strata of South Africa, India, and Russia, are represented by several well-marked subordinal types. In the first group, known as mammal-toothed, or theriodont, reptiles, the teeth are differentiated into incisors, tusks, and cheek-teeth; the latter frequently having three cusps arranged in a longitudinal series.

Another modification is presented by the dicynodonts of Africa and India, in which the jaws formed a horny beak, either destitute of teeth, as in tortoises, or provided with a huge pair of tusks in the upper jaw. Some of these reptiles were of gigantic size, and certain of the theriodonts were likewise large; but it is obvious that the groups from which mammals sprang must have been represented by small species. In both theriodonts and dicynodonts, as well as in certain other groups, the temporal pits of the skull are open on the upper surface; but in the pariasaurian anomodonts the hind part of the skull is roofed over by bone, in the manner characterising the labyrinthodont amphibians, to which these reptiles are allied; a peculiar sculpturing of the surface of the skull being another point of resemblance. In the species of which the skull is figured a number of spines surmount the head; but these are wanting in the African *Pariasaurus* and *Propappus*, gigantic creatures, with a somewhat frog-like head, an apology for a tail, and powerful, short limbs, of which the toes were armed with long claws. Both wore defensive

SKELETON OF PARIASAURUS

armour; but it appears that while in *Pariasaurus* this is restricted to three longitudinal rows of ossicles in the neighbourhood of the spines of the vertebræ, *Propappus* was provided with a complete shield of bony plates.

STAGES IN THE DEVELOPMENT OF A FROG

This series of pictures shows the mass of eggs which swells until a tadpole emerges from each, and, gradually losing its fish-like characteristics, develops limbs, and finally passes into a fully-formed frog

Photographs by W. B. Johnson

CLASS IV. AMPHIBIA
CHARACTERISTICS AND CLASSIFICATION
By R. LYDEKKER

IN popular estimation frogs and toads, together with their near relatives the newts and salamanders, are regarded as reptiles, but they are really very different creatures, and constitute a class by themselves, being in many respects intermediate between reptiles and fishes. From the mode of life of its members, the very appropriate name of amphibians has been proposed for the class, and is the one which should be adopted, although the term batrachians, which more properly applies to frogs and toads alone, is not infrequently used in the same sense. Agreeing with the higher vertebrates in the structure of their limbs, which are divided into the same number of segments as in mammals and reptiles, and supported by corresponding bones, existing amphibians are distinguished from reptiles by the absence of any ossification in the basioccipital region of the lower surface of the hind part of the skull, in consequence of which the latter is articulated to the first vertebra by means of two distinct knobs, or condyles, formed exclusively by the exoccipital bones. A further important point of distinction is afforded by the absence in the embryo of those membranous structures known as the amnion and allantois, which are present in the higher groups of vertebrates.

Moreover, the great majority of amphibians pass through a metamorphosis, or, rather, a series of metamorphoses, beginning their existence immediately after leaving the egg in a larval condition, during which they breathe the air contained in water by means of gills, whereas in the adult state they breathe atmospheric air direct by means of lungs. Varying much in external form, these animals nearly always have the body covered with a soft, naked skin; but in a few instances among existing forms scales are embedded in the skin, and most of the extinct types had a well-developed armour of scales and bony scutes. In some kinds a longitudinal fin is developed down the middle of the back and tail, but this is always soft, and lacks the supporting spine-like bones characterising that appendage in fishes.

In passing through a metamorphosis, amphibians are more like the inferior groups of animals than the higher vertebrates; and while in the earlier stages of their existence, during which they breathe by gills, they may be regarded as very closely allied to fishes, in the adult state they come much nearer to reptiles. The extinct labyrinthodonts, which are themselves not very widely removed from fishes, and have the basioccipital bone ossified, serve to connect other members of the class with the anomodont and beaked reptiles.

As already mentioned, the skin of most existing amphibians is soft and naked; and it is invested with a colourless epidermis, which is periodically shed entire, while the deeper layer is often coloured with blotches or streaks of yellow, red, brown, or black. Other colours, however, such as green and blue, are produced by pigment-cells, which generally make their appearance under special conditions of warmth and moisture. As a rule, the colour of amphibians varies to a great extent with the nature of their surroundings, as is well exemplified in the case of the frog, which changes its hue according to the nature of its habitat, while tree-frogs harmonise with the foliage among which they dwell. It is, however, very remarkable that in Costa Rica a certain toad simulates to an extraordinary degree the colouring of the snakes—both poisonous and harmless—of the same country, while in North Sumatra amphibians of various groups are spotted with carmine-red. One West African frog (*Trichobatrachus robustus*) is peculiar in having hairs on the skin.

In all amphibians the skin is furnished with glands secreting a more or less milk-like fluid; these glands being generally distributed all over the body, although sometimes they are confined to the sides of the neck behind the eyes. In many toads and land-salamanders some of the larger glands appear as prominent warts, pierced with large pores. The viscid, milky fluid secreted by these glands is exuded during excitement, and is endued with more or less poisonous properties, being intended to serve as a means of defence.

Although some irritation of the skin may be produced by handling many of the species in which these poisonous properties are most developed, the stories of toads or salamanders spitting venom are, it is almost needless to observe, pure fabrications. When introduced into the circulation, batrachian venom acts, however, as a powerful poision, influencing the heart and central nervous system; and the secretion of one South American species is employed by the Indians to poison the spears and arrows used in killing monkeys.

In the economy of amphibians the naked skin and its glands play a most important part, since none of them drink, in the proper sense of the word, but imbibe moisture through the pores of their integument. Moisture is, indeed, essential to their existence, and if they be confined in a dry atmosphere they soon perish. It is true that frogs may be seen basking in the sun's rays, and apparently enjoying the warmth as much as lizards, but they only do this in the neighbourhood of water, to which they retire when necessary. Such members of the class as inhabit dry localities are mostly nocturnal, avoiding sunshine, and wandering abroad when they can obtain moisture from dew.

The skeleton of amphibians presents many peculiarities, and in some cases shows numerous fish-like characters. For instance, in certain of the species furnished with permanent gills the vertebræ are scarcely distinguishable from those of fishes, being disc-like, with a cup at each end, and thus quite unlike those of the typical newts, each of which has a rounded knob at the front of the body and a cup at the hind extremity, and is closely articulated with its fellows. In the long-tailed groups the number of vertebræ is considerable, but in frogs and toads those of the back are reduced to seven or eight, the hind end of the backbone terminating in a long style, extending between the greatly produced extremities of the haunch-bones, or ilia, which articulate with the lateral processes of the sacral vertebræ. The transverse processes of all the vertebræ are well developed, and in some cases very long; and these take the place of ribs, which, at the most, are represented by some small rudiments. In consequence of this absence of ribs, amphibians are unable to breathe in the ordinary way by alternate expansion and contraction of the cavity of the chest; and they, so to speak, swallow air, taking in a large gulp, and then closing the mouth.

Some years ago it was discovered that certain salamanders, both in Europe and America, are devoid in the adult condition of both lungs and gills, and since that date speculation has been rife as to the manner in which these creatures breathe. These lungless salamanders, it may be observed, are not individual abnormalities, but are the normal representatives of the species to which they severally belong. The question at issue has been whether these salamanders absorb oxygen and give off carbonic acid by means of the skin alone, by the mouth and gullet, by the combined action of the two, or by the aid of some other part of the body. Experiments on both dead and living salamanders has, however, shown that the skin plays an important part in the respiration of all species, and that its action is relatively greater in those without lungs. On the other hand, there seems to be no doubt that lungless species also breathe very largely, and perhaps mainly, by means of the mouth and gullet, the mucous membrane of which is better adapted for permitting the air to pass into and out of the blood than in the skin.

In addition to the peculiarities connected with its condyles and the basioccipital region, the skull is distinguished by its flattened, broad, and more or less semicircular form ; the sockets for the eyes being generally large and ill-defined. In front of the condyles the under surface of the middle of the skull is overlain by a large parasphenoid bone, which is frequently dagger-shaped ; this bone being generally but slightly, if at all, developed in the higher vertebrates, although very large in fishes. The lower jaw, which articulates with the skull by the intervention of a quadrate bone, is composed of at least two pieces on each side, and may contain more elements. The palatine bones and vomer, and more rarely the parasphenoid, may be armed with teeth, like the upper jaw ; but in the frogs and toads the lower jaw is very generally toothless. In all cases the teeth are small,

inwards nearly to the middle line of the back ; and in the frogs each coracoid has an inward cartilaginous expansion, which may either meet or overlap its fellow, and is of much importance in classification.

In advance of the coracoid bones is another pair of transverse bars commonly known as the precoracoids ; while in front of these is a single median rod termed the omosternum ; the proper sternum, or breast-bone, occupying a similar position behind the coracoid bones. In the fore limb the radius and ulna may be united, and the wrist cartilaginous ; and the number of toes among living forms never exceeds four, and may in some instances be reduced to three.

More variation exist in the hind foot, the number of toes in the long-tailed group ranging from two to four, whereas in the frogs and toads it is always five. Only in a few frogs and newts are the toes furnished with claw-like nails ; in the greater number of forms these being naked, although often connected by webs, and sometimes carrying adhesive discs on their lower surfaces.

Throughout the class the brain is of a very low type, its component portions lying in a line one behind the other, without overlapping. All the members of the group possess the three chief organs of sense, although in some instances the eyes may be very minute and covered with an opaque skin. In frogs and toads the eye is large and very highly developed, generally possessing two lids, of which the lower one is larger and thinner than the upper, and more or less transparent. Greater variation exists in the structure of the ear, which is simplest in the long-tailed forms.

The nose opens externally in a pair of nostrils situated near the muzzle, and by another pair of apertures into the mouth ; the latter character distinguishing

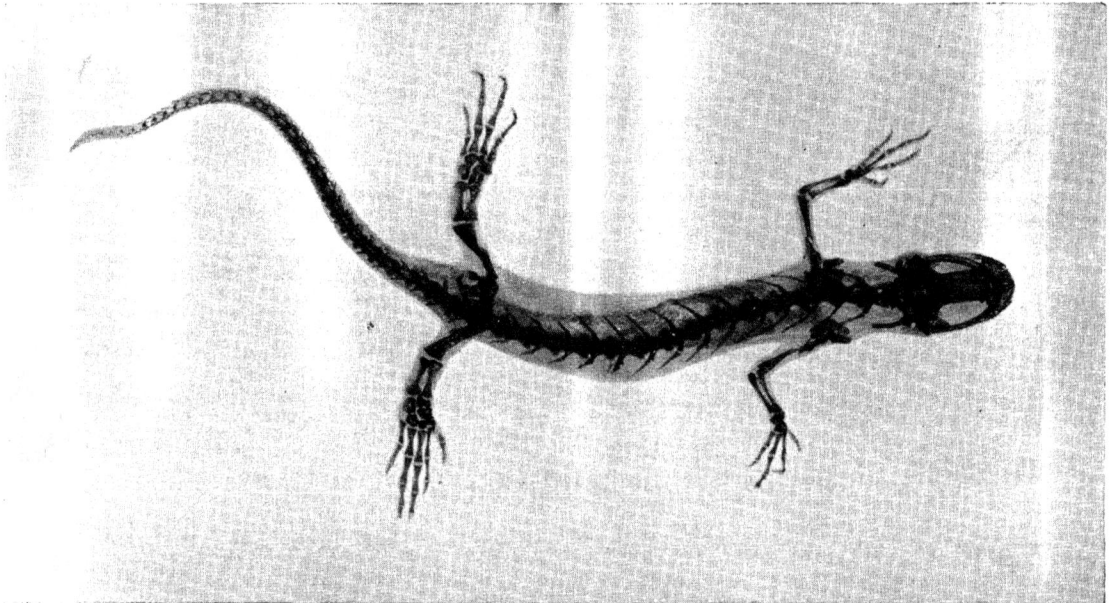

AN X-RAY PHOTOGRAPH OF THE SKELETON OF A NEWT Photo, J. Leadbeater

simple, and pointed, being adapted for holding, and not for masticating.

The shoulder-girdle, which is largely cartilaginous, is placed very close to the head, and comprises the usual elements. Each scapula, or shoulder-blade, has, however, an upper cartilaginous portion, extending

amphibians from the majority of fishes. The tongue, which acts only in the very slightest degree as an organ of taste, and is wanting in one group of frogs, is generally well developed and thick, filling the whole space between the jaws, and being capable of a large amount of motion ; but it differs essentially from that of the higher vertebrates in that it is affixed to the inner side of the front of the lower jaw, with its tip pointing down the throat.

All amphibians lay eggs, which are generally, although by no means invariably, deposited in fresh water, and fertilised as they are extruded from the female. As a rule, these eggs, which much resemble those of fishes, are small, very numerous, and connected together by muci-lage, forming either a string or jelly-like mass, in which the dark yolks are very conspicuous. Some of the tree-frogs, however, lay un-usually large eggs, within which the larvæ undergo the whole of such trans-formation as takes place ; and in one genus, instead of the usual gills, a temporary breath-ing organ is de-veloped on the tail.

A land-frog in the Solomon Islands also lays large eggs, like small marbles, which are deposited in the crevices of rocks, and from which emerge fully-developed frogs ; and other instances of such abnormal development are noticed below.

With these and certain other exceptions, the eggs, after being deposited in water, are hatched by the heat of the sun ; and it appears that the dark colour of the yolk is for the purpose of absorbing as much solar heat as possible. Such eggs as are laid during the late spring and summer are less darkly coloured and have thinner coats than those deposited in the early part of the spring, and while the former are placed on the ground at the bottom of the water, the latter float on the surface ; the reason of this difference being that in the early part of the year the lower strata of water are too cold to admit of the development of the ova.

In ordinary cases, when the larva has reached a certain stage, it bursts the investing membranes of the egg, and comes into the world adapted for an aquatic life, and always possessing a long, compressed tail composed of zigzag-shaped masses of muscles, similar to those of fishes. The first process is the sprouting forth of branching external gills from the sides of the neck, which in the larvæ of frogs and toads are subsequently replaced by internal gills, but in the long-tailed group persist for a longer period, or even throughout life. After the disappearance of the external gills the water is expelled from the gill-chamber by one or two tubes, generally discharging by a single orifice, which may be situated either on the lower surface of the body or on the left side.

So soon as the external gills have made their appearance, development is concentrated on the tail and the absorption of the remainder of the yolk. The vertical, fin-like expansions of the tail rapidly increase, and the body becomes relatively smaller and more slender ; while about the same time the limbs begin to make their appearance as buds, although the

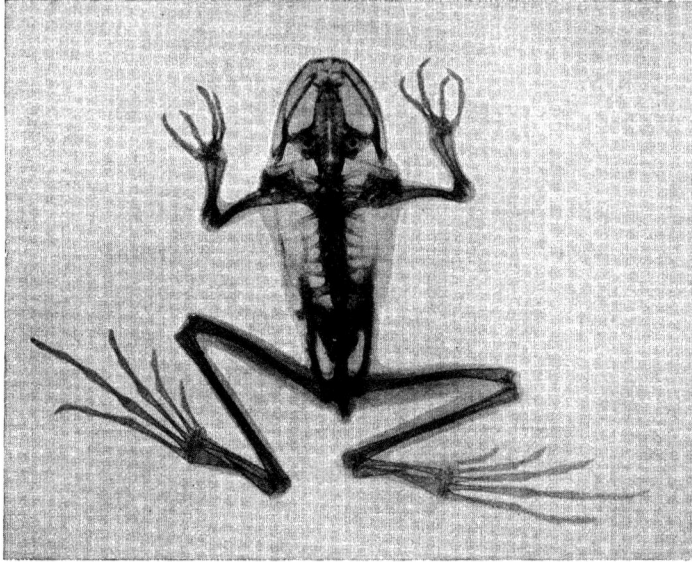

AN X-RAY PHOTOGRAPH OF A FROG   Photo, J. Leadbeater

date of development of the front and hind pair varies in different groups. In the newts the front pair of limbs is the first to appear, in the frogs the reverse is the case. In the latter the hind limbs appear some considerable time before the front pair, the fish-like tail persisting till the sprouting of these, when the change from a herbivorous fish-like animal to one of a carnivorous and reptile-like type begins.

The jaws are at first invested with horny teeth, and subsequently with horny sheaths, which eventually disappear ; while the tail gradually diminishes in size, and finally is lost. It may be observed that no vertebræ are developed in the frog's tail, and that the long spine in which the backbone of the adult ter-minates is an out-growth from the hindmost vertebra. Not less remarkable is the shortening of the intestinal canal of a frog or a toad, as the creature changes its herbi-vorous for carnivor-ous habits.

To trace in detail the development of the soft parts would occupy a very large space, and it must suffice to mention that in one group of tailed amphibians the external gills of some individuals may be retained permanently, while in others of the same species they are cast at an early period. Then, again, the number of these gills is by no means constant, for in the Cingalese cæcilian and one salamander there are three pairs of these organs, in the tadpoles of some frogs there are two, and in others, as well as in one genus of cæcilians, there is only a single pair.

In regard to the abbreviation of the early develop-mental stages in certain species, Mr. J. S. Budgett states that in a West African tree-frog of the genus *Phyllomedusa* the male and female hold together the edges of a leaf (which afterwards become united by the jelly of the egg-mass) during oviposition so as to form a funnel for the reception of the eggs. In the short period of six days the embryo leaves the egg as a pellucid tadpole of a bright green colour, whose only conspicuous part is formed by its eyes. The tadpole, which may have to travel several inches in order to reach the water, is hatched without a trace of yolk, and with the loss of external gills, breathing taking place by means of a median spiracle, and the lungs being distinctly visible through the body-wall. Pigment is locally developed next day ; and at the end of about five weeks the hind limbs appear. When both pairs of limbs are developed, the young frog lands, and sits quietly among the grass till its tail is completely absorbed, when it is practically adult.

Two other species of frogs, one from West Africa and the other from Brazil, deposit their spawn in nests formed of leaves stuck together, the tadpoles moving in a mass of froth, recalling that of the cuckoo-spit insect. In both these instances the spawn is deposited in the neighbourhood of water, into which the tadpoles ultimately fall; but in a tree-frog from Rio de Janeiro, in which the eggs are likewise hatched in a frothy mass among leaves, the larvæ actually die if they are put into water. In another Brazilian tree-frog the tadpoles frequent cracks in rocks, and adhere to the surfaces of the latter by means of an abdominal sucker.

Full reference to the mode of development in the Surinam toad, and also to that of the marsupial frogs, in which the young are developed in a dorsal pouch, is made later on, but attention may be specially directed to one of the most extraordinary " nursery " arrangements existing in the entire group—namely, those of the Chilian *Rhinoderma darwini*—in which the tadpoles undergo their development in an enormous pouch on the throat of the male, which appears to be a modification of the vocal sacs found in many species of frogs.

A specimen of the large West African tree-frog known as *Hylambates rufus* produced in confinement a number of eggs remarkable for their large size, the mother being about the same size as a full-grown common frog. Some of these eggs were placed in water, but failed to develop. Judging from the large size of the yolk-mass (for a large yolk means an abbreviated development), it may be inferred that the young of these frogs undergo a considerable part of the metamorphosis within the egg. But nothing is yet known with regard to the mode of development of the frogs of the genus *Hylambates*. The mouth of a female *Hylambates breviceps* from the Camerunscontained several large yellow eggs, very similar, except for their size (*H. breviceps* being a smaller species), to those of *H. rufus*. This mouth-nursing by the female thus affords an addition to the list of extraordinary breeding habits in batrachians, the nearest known case to that of *Hylambates* being that of the above-mentioned Chilian *Rhinoderma*.

Geologically, amphibians are a very ancient group, remains of their oldest representatives occurring in the Carboniferous and Permian rocks of Europe and North America. All these ancient representatives of the class belong, however, to the group of labyrinthodonts, which survived till the period of the Trias, and are structurally very different from the modern forms, approximating in certain respects to fishes. Indeed, since no amphibians have hitherto been discovered between the Trias and the Wealden, or lower Cretaceous, rocks of Belgium, it is difficult to assert that the modern representatives of the class are the direct descendants of the labyrinthodonts, although there is little doubt that such is really the case. Beginning in the Belgian Wealden, newts and salamanders occur throughout the greater part of the Tertiary rocks, but frogs and toads are first known in North America from Eocene beds, while in Europe they are not met with before the Oligocene.

At the present time amphibians are distributed over all parts of the world except the polar regions, although they are more dependent upon the presence of water and warmth than any of the preceding classes of vertebrates. They are, accordingly, most abundant in the tropical and subtropical regions; and as none of them are marine in their habits, even a narrow arm of the sea is generally sufficient to limit their range. When they occur on islands, it is probable either that their eggs have been carried by birds, or that there has been a comparatively recent separation from the mainland.

In absolutely desert districts amphibians are unknown; while in countries where there is a long dry season followed by a period of rains they are in the habit of becoming torpid during the former, the length of the sleep in one Javan species being upwards of five months.

In cold climates all the members of the class become torpid during winter. As regards their general distribution, amphibians closely resemble fresh-water fishes, and differ widely from lizards. Indeed, from an amphibian point of view, the globe may be divided into two great regions—namely, a northern one, characterised by the abundance of newts and salamanders and the absence of cæcilians; and a southern one, distinguished by the scarcity of the former and the presence of the latter group.

In their mode of life it is probable that very few amphibians are diurnal, most of the terrestrial forms making their appearance abroad with the first shades of evening, and retiring to their hiding-places at dawn. In wet or cloudy weather frogs and toads—especially in South America — frequently appear in great numbers during the day; and both these groups are in the habit of making night hideous with their croakings. Although in all cases the adults are carnivorous, the larvæ subsist more or less exclusively on vegetable substances, some confining themselves to that kind of diet, while others also consume animalcules and other minute creatures.

It is estimated that out of many thousands of eggs deposited annually by each female of the common American *Bufo lentiginosa* only two develop into adult toads.

Amphibians may be arranged in the four following orders, viz.:

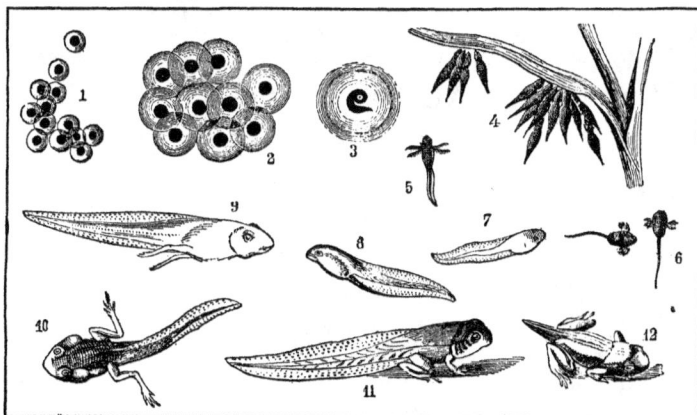

DIAGRAMS SHOWING DEVELOPMENT OF FROG
1, Eggs when first laid; 2, eggs at a later stage; 3, egg containing embryo; 4, newly-hatched tadpoles; 5, 6, tadpoles with external gills; 7-12, later stages in the development of tadpoles

ORDER	SUBORDER
I. ANURA, OR ECAUDATA Frogs and Toads	1 Firmisternia—Frogs, etc. 2 Arcifera—Toads, etc. 3 Aglossa—Surinam Toad, etc.
II. CAUDATA, OR URODELA Salamanders and Newts	
III. CÆCILIA, OR APODA Worm-like Amphibians	
IV. LABYRINTHODONTIA, OR STEGOCEPHALIA Extinct Primeval Salamanders	

ORDER I. ECAUDATA
CHARACTERISTICS OF FROGS AND TOADS

Frogs and toads are distinguished from their near relatives salamanders and newts by the presence of four limbs and the absence of a tail in the adult state, the latter feature giving origin to the name Ecaudata, by which the order to which they belong is scientifically designated. They all have short and frequently thick bodies, in which the backbone comprises, at most, only eight vertebræ in advance of the sacrum, those behind the latter being fused into a long rod-like bone, as shown in the figure of the skeleton on this page

In the limb, as shown in the same figure, the bones of the forearm (radius and ulna) are completely fused together; and the same is the case with regard to the tibia and fibula in the hind leg. Moreover, the hind limb contains a kind of additional segment, owing to the elongation of the calcaneum and astragalus in the ankle-joint, which form a pair of long bones lying parallel to one another.

As a rule, frogs and toads undergo a lengthened larval period; the "tadpoles," as shown in the figure

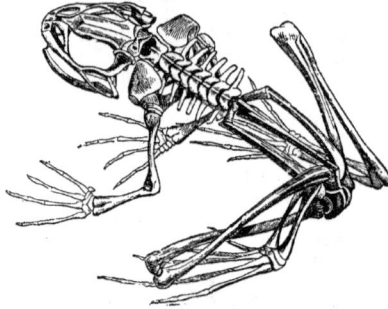

SKELETON OF A FROG

on page 1640, having a globular head and body, a fish-like tail, external or internal gills, and no limbs in the first stages of their existence. The hind limbs are the first to appear, and after the front pair are developed the tail is gradually absorbed, upon which the young for the first time leave the water.

Represented by over a thousand species, frogs and toads have a world-wide distribution, although they are more abundant in tropical and subtropical than in temperate regions, and are especially numerous in India and South America. It is not a little remarkable that some of the largest species are inhabitants of islands, although others are found in Western Africa, and another is a native of Brazil. From the nocturnal habits of the adults it is frequently difficult to find out whether in any locality they are abundant or the reverse; but in the spring this may generally be ascertained by observing the tadpoles in the rivers and ponds, since all of these show specific differences, to the full as well marked as those in the adult.

CLASSIFICATION OF ECAUDATA MENTIONED IN THIS WORK

ORDER
Frogs and Toads—Ecaudata
SUBORDER 1
Firmisternia
FAMILY 1
Frogs—Ranidæ
GENUS 1
Typical Frogs—Rana
SPECIES

Indian frog	Rana breviceps
Edible frog	R. esculenta
English frog	R. temporaria
Moor-frog	R. arvalis
Agile frog	R. agilis
Iberian frog	R. iberica
Lataste's frog	R. latastei
Bull-frog	R. catesbyana
Montezuma's frog	R. montezumæ
Camerun frog	R. goliath
Guppy's frog	R. guppyi

GENUS 2
Oxyglossus
GENUS 3
Flying Frogs—Rhacophorus
SPECIES

Malay flying frog	Rhacophorus nigro-palmatus
Javan flying frog	R. reinwardti

FAMILY 2
Dendrobatidæ
GENUS 1
Madagascar Tree-Frog—Mantella
GENUS 2
Dendrobates
SPECIES
Variable tree-frog .... Dendrobates tinctorius
FAMILY 3
Narrow-mouthed Frogs—Engystomatidæ
GENUS 1
Breviceps
SPECIES
African short-headed frog .... Breviceps mossambicus
GENUS 2
Rhinoderma
SPECIES
Darwin's frog........... Rhinoderma darwini
FAMILY 4
Ceratobatrachidæ
GENUS
Ceratobatrachus
SPECIES
Günther's frog...... Ceratobatrachus guentheri
SUBORDER 2
Arcifera

FAMILY 1
Leptodactylidæ
GENUS 1
Horned Frogs—Ceratophrys
SPECIES

Brazilian horned frog	Ceratophrys boiei
Argentine horned frog	C. ornata

GENUS 2
Leaf-Frogs—Hylodes
SPECIES
Antillian frog .... Hylodes martinicensis
GENUS 3
Leptodactylus
FAMILY 2
Dendrophryniscidæ
FAMILY 3
Toads—Bufonidæ
GENUS 1
Typical Toads—Bufo
SPECIES

The toad	Bufo vulgaris
American toad	B. lentiginosa
Green toad	B. viridis
Natterjack toad	B. calamita
Moorish toad	B. mauretanica

GENUS 2
Nectes
GENUS 3
Nectophryne
GENUS 4
Rhinophrynus
SPECIES
Sharp-nosed toad .... Rhinophrynus dorsalis
FAMILY 4
Tree-Frogs—Hylidæ
GENUS 1
Acris
SPECIES
Grasshopper-frog .... Acris gryllus
GENUS 2
Typical Tree-Frogs—Hyla
SPECIES

European tree-frog	Hyla arborea
Brazilian tree-frog	H. faber
Goeldi's tree-frog	H. goëldii
	H. nebulosa
Australian tree-frog	H. cœrulea

GENUS 3
Marsupial Frogs—Nototrema
FAMILY 5
Toad-Frogs—Pelobatidæ
GENUS 1
Toad-Frogs—Pelobates

SPECIES

Brown toad-frog	Pelobates fuscus
Black-spurred toad-frog	P. cultripes

GENUS 2
American Toad-frogs—Scaphiopus
GENUS 3
Pelodytes
SPECIES
Punctured toad-frog...... Pelodytes punctatus
GENUS 4
Batrachopsis
GENUS 5
Leptobrachium
GENUS 6
Megalophrys
SPECIES
Burmese frog ....Megalophrys robusta
FAMILY 6
Disc-tongued Frogs—Discoglossidæ
GENUS 1
Discoglossus
SPECIES
Painted frog ....Discoglossus pictus
GENUS 2
Bombinator
SPECIES
Fire-bellied frog ......Bombinator igneus
GENUS 3
Alytes
SPECIES

Midwife-frog	Alytes obstetricans
Spanish midwife-frog	A. cisternasii

FAMILY 7
Amphignathodontidæ
FAMILY 8
Hemiphractidæ
GENUS
Ceratohyla
SPECIES
Ceratohyla bubalus
SUBORDER 3
Aglossa
FAMILY 1
Spur-toed Frogs—Xenopodidæ
GENUS
Spur-toed Frogs—Xenopus
SPECIES
Smooth spur-toed frog .... Xenopus lævis
FAMILY 2
Pipidæ
GENUS
Pipa
SPECIES
Surinam toad .... Pipa americana

1641

SUBORDER I. FIRMISTERNIA

FROGS

THE true frogs (*Ranidæ*), together with four other families, constitute a suborder (Firmisternia) characterised by the presence of a tongue, and by the firm union of the two large coracoid bones of the chest by means of a single cartilage uniting their free edges. From the other members of the group the typical frogs are distinguished as a family by the presence of teeth in the upper jaw, and by the transverse processes of the sacral vertebra being either cylindrical, or only very slightly dilated at their extremities. These characters are sufficient to distinguish the typical frogs from the other families of the suborder, but it may be added that the vertebræ are cupped in front and hollowed behind, while there are no ribs, and the terminal style of the backbone is articulated to the sacrum by two condyles.

The terminal joints of the toes may be either simple or pointed, T-shaped, Y-shaped, or even claw-like, the species in which these joints are thus expanded having the soft parts similarly expanded and flattened. For a long time it was considered that the shape of the tips of the toes was connected with the mode of life of their owners, and although this is so to a great extent, it is now ascertained that several of the species in which the toes are somewhat expanded are as aquatic as those in which they are pointed, and species presenting both modifications are included within one and the same genus. The typical frogs are divided into over twenty genera, only two of which are noticed in this work.

EDIBLE FROG

TYPICAL FROGS

Under the general title of typical frogs may be included all the members—about 150 in number — of the genus *Rana*, to which belongs the common English frog. The distinctive characters of these frogs are to be found in the horizontal pupil of the eye ; the more or less deeply notched and free tongue ; the presence of teeth on the vomerine bones of the palate ; the absence of webs in the toes of the fore feet, and their presence in those of the hind limb ; and the separation of the outer metatarsal bones of the hind foot by a web, the extremities of the fingers being simple or expanded.

With the exception of the southern part of South America—where the whole family is unrepresented—Papua, and New Zealand, these frogs have a world-wide distribution. Although the great majority of the species are probably aquatic during the breeding season, at other times much diversity of habit is displayed by the different representatives of the genus, some being aquatic, others terrestrial, and others, again, burrowing, or even more or less arboreal.

The existence of burrowing habits is indicated by the great development of a tubercle on the inner side of the metatarsus, which in one Indian species (*R. breviceps*) has a sharp edge, and is used in a shovel-like manner to excavate the burrow. Such burrowing species are further characterised by the shortness of the hind limbs, and thus assume a more or less toad-like appearance. Large discs at the end of the toes, on the other hand, are usually indicative of arboreal habits, although, as already mentioned, smaller discs are met with in certain purely aquatic species.

Selecting some of the European representatives of the genus for special mention, reference may be made in the first place to the edible frog (*R. esculenta*), characterised by the pointed tips of the toes, the smooth under surface of the body, the presence of a broad glandular fold along the sides, and the marbling of the thighs. Exceedingly variable in colour, this frog generally has the upper parts olive or bronzy brown, more or less spotted or marbled with dark brown or black ; there are usually three light stripes along the back, and the sides of the head and ground-colour of the flanks are sometimes green, the marbling on the thighs occupying their hind surfaces, and being black in colour. The males are specially characterised by the presence of a globular sac, connected with the production of the croaking, on each side of the head, opening by a slit behind the angle of the mouth.

Inhabiting Europe, Asia as far east as Japan, and North-Western Africa, the edible frog is common in England, the dark race occurring in the fens of Cambridgeshire, and the green variety in Norfolk. The use of the flesh as food probably led to the introduction of this species into Cambridgeshire by the monks ; but the Norfolk colony was imported between 1837 and 1842.

MOOR-FROGS

From this species the common English frog (*R. temporaria*) is readily distinguished by the incomplete webbing of the hind feet, and the presence of a dark temporal spot extending from the eye to the shoulder, as well as by the absence of external vocal sacs in the males. Moreover, if the skulls of these two species are compared, it will be found that while in the edible frog the teeth on the vomers do not extend behind the line of the apertures of the posterior nostrils, they do so to a small extent in the present species. In colour the upper parts of the common frog are greyish or yellowish brown, more or less spotted with dark brown or black; the temporal spot being always dark, and a light line running from below the eye to its extremity, while the sides of the body are profusely spotted, the limbs transversely barred, and a larger or smaller number of spots is present on the under parts. The vocal sacs of the males are internal. This species is spread over Europe and Northern and Temperate Asia; its range including the west of Ireland as well as England.

Closely allied is the moor-frog (*R. arvalis*), of Eastern Europe and Western Asia, which may be distinguished by the tubercle on the inner metatarsal being compressed instead of blunt, and by the pointed, in place of obtuse, muzzle. The colouring is very similar to that of the common species, but there is sometimes—as in the right-hand figure of the illustration—a light stripe bordered by two black ones down the middle of the back, while the under parts are uniform.

A third European species is the agile frog (*R. agilis*), which belongs to a group distinguished by the greater length of the hind limbs; the whole form being slender, and the muzzle pointed. Its general colour is greyish brown, with dark spots; the temporal spot being dark and distinct, with a light line running from its extremity to the snout, while the hind limbs are regularly barred, and the under parts unspotted. Two other European species, the one (*R. iberica*) from Spain and Portugal, and the other (*R. latastei*) from the neighbourhood of Milan, differ by the spotted lower surface of the body. Even the tadpoles of these more or less nearly allied species present differences by which they can be readily distinguished from one another.

The common frog, whose habits may be taken as typical of the allied members of the genus, is found in most parts of Europe where there is a sufficiency of moisture and shelter for its existence; the presence of water being essential during the breeding season. All are probably familiar with the manner in which a frog swallows air, but it is perhaps less generally known that if the mouth of one of these creatures be kept forcibly open, death must inevitably ensue, owing to the impossibility of breathing while in this state. The croaking of the frog, which is common to both sexes, is principally uttered during the breeding season;

MALE AND FEMALE ENGLISH FROGS

BULL-FROG

and when large numbers of these amphibians are collected in a pond together, the volume of sound produced is considerable, and can be heard from long distances, although it is nothing compared to that of the bull-frog and many tropical species.

Frogs subsist entirely on slugs, snails, insects, and the like, swallowing large beetles whole, and devouring several at a meal. The frog captures its prey by suddenly throwing forward the tip of its tongue, which is invested with a viscid secretion, upon the insect or slug, and then as quickly withdrawing it to its normal inverted position. So rapid is the whole movement, that it requires a sharp eye to detect it, the insect seeming to disappear as if by magic. "Frogs retire," writes Professor T. Bell, "on the approach of winter to their hibernating retreats, where they pass the dreary season in a state of absolute torpidity. This is generally in the mud at the bottom of the water, where they are not only preserved, though at low degree, but also secured from external injury. Here they congregate in multitudes, embracing each other so closely as to appear almost as one continuous mass. On the return of spring they separate from each other, emerge from their places of retirement, and begin again their active life by exercising the important function of reproducing their species."

During the breeding season a warty protuberance is developed on the thumb of the male frog to assist in holding the female, and in some foreign species the whole forearm becomes enlarged at this time. The spawn is deposited at the bottom of the water, but soon rises to the surface in the well-known glairy masses; and in due season the tadpoles make their appearance. During the tadpole-stage they are devoured in large numbers by newts and the smaller fishes, as well as by the larvæ of dragon-flies and water-beetles; while in the adult condition numbers fall a prey to the weasel and polecat, the heron and other wading birds, and the common snake, whose food is almost entirely composed of them.

Although the common frog is to a large extent aquatic, it is much less so than the edible species, which inhabits indiscriminately running or still waters, the borders of rivers, rivulets, or streams, lakes, or ponds, salt or fresh-water marshes, or even ditches and pools of water. Owing to the presence of the external sacs, the croaking of the male of this species is louder than in the common frog. Both species, like all the more typical representatives of the genus, progress on land by means of leaps, and in water swim by means of the hind limbs alone.

Compared to the bull-frog (*R. catesbyana*) of eastern North America, European frogs are mere dwarfs. The bull-frog is one of those species in which the tips of the toes are pointed, and is specially characterised by the web extending to the tip of the fourth toe of the hind foot, the large size of the aperture of the ear, and the relative length of the hind leg; the two latter characters distinguishing it from Montezuma's frog (*R. montezumæ*) of Mexico. The body has no lateral glandular fold, and the vocal sacs of the males are internal. In colour the bull-frog is brown or olive above, with darker marblings; the under parts being either uniformly coloured, or marbled with brown. In length it measures from 7 to 7½ inches, exclusive of the legs.

More abundant in the southern than in the northern portion of its habitat, the bull-frog is generally met with in rivers and streams well shaded with trees or bushes, where it may be seen in numbers basking in the midday sun. Its croaking is said to be louder than that of any other species, and can be heard for a distance of several miles. In the Southern States of America, although most intense during the spring and summer, the croaking is continued throughout the year; but in the north it is confined to the spring and summer, being especially loud during the breeding season. It is a remarkable fact that in Canada, at any rate, the bull-frog passes its first winter in the larval condition, and takes two years to attain its full growth. A well-known African species is *R. aspersa*.

The largest known species of frog is *R. goliath*, from the Cameruns, which measures no less than 10 inches from the tip of the muzzle to the hind end of the body. Next in point of size comes Guppy's frog (*R. guppyi*), from the Solomon Islands, measuring 8¼ inches from snout to vent, while the third place is taken by the North American bull-frog, of which the largest specimen in the United States National Museum is stated to measure barely 8 inches in length. Among the toads are two giants—namely, the Malay *Bufo asper* and the South American *Bufo marinus*, of both of which the largest specimen in the British Museum measures 8½ inches from the snout to the end of the body.

Without referring to the numerous other generic representatives of the group, it may be of interest to mention that the small Indian frogs forming the genus *Oxyglossus* differ from the members of the typical genus by the absence of any notch in the tongue, and the want of vomerine teeth. They are specially interesting on account of being represented by fossil species in the Eocene rocks of Bombay. Fossil frogs belonging to the typical genus *Rana* occur in the lower Miocene rocks of Europe.

FLYING FROGS

Certain frogs inhabiting the Malay countries and Madagascar, and constituting the genus *Rhacophorus*, are mostly characterised, among other features, by the complete webbing of both fore and hind feet, as well as by the presence of enlarged adhesive discs to the tips of the toes. Owing to the large size of the webs, the Chinese coolies in Burma state that these frogs are in the habit of flying down from the trees in which they habitually reside, and such

accounts, after being at first accepted and then discredited, appear in the main to be founded on fact. These frogs are usually found on the broad-spreading leaves of the banana, but sometimes on those of other trees; and it is in such situations that their spawn is habitually deposited and undergoes its development. When in the open, the colour of these frogs is generally green, but when in deep shade, or on faded leaves, the tint changes to dusky or chocolate, in harmony with the tone of their surroundings.

In a well known Malay and Bornean species (*R. nigropalmatus*), which measures about 4 inches in length, the ground-colour of the back and legs is deep shining green, but the under parts and inner toes are yellow, and the webs of the toes black rayed with yellow.

In the allied *Rhacophorus reinwardti* of Java the sexes differ in size, the female being one-third larger and twice as broad as her partner, while she is further distinguished by the smaller development of her voice organs and somewhat less brilliant colouring.

Writing of the aforesaid *R. reinwardti*, Professor Siedlecki states that it is about 4 inches long, of a deep, shining green above and yellow beneath, and, like the European tree-frog, is arboreal in its habits, even breeding amid the foliage. It is, moreover, remarkable for its power of changing the colour of its skin. A specimen captured in the daytime and examined in strong sunlight will, for instance, be found of a brilliant greenish blue; towards evening it will, however, change to green, and, finally, to nearly black, the transformation taking place more rapidly in males than in females. Night is the time when these frogs are really active, and it is then that they awaken from their diurnal torpor and begin to search for grasshoppers and other orthopterous insects. Like other tree-frogs, they hold on to leaves and boughs by means of adhesive discs to their toes, but these differ in structure from those of true tree-frogs. In regard to the leaps from which they take their name, flying frogs will leap to a height of about a foot in an arc of a circle and alight two or three yards distant in a characteristic attitude, with their bodies inflated to the greatest possible degree and their toe-membranes fully extended. During these flying leaps, which are for the purpose of escaping foes, the webs perform the part of a parachute. Each leap is of extreme rapidity, lasting only a fraction of a second.

AGILE FROGS

VARIABLE TREE-FROGS

Out of over forty species of the genus more than thirty occur in South and East Asia, and the remainder in Madagascar. While allied in most respects to the typical frogs, all the members of this genus differ by the presence of a small additional bone between the terminal and penultimate joints of the toes, and likewise by the penultimate joints of the latter being distinctly marked externally as a kind of ridge, while, as already mentioned, they are for the most part further characterised by the webbing of the toes of the fore feet, although the degree to which this is carried is variable. The tips of the toes are always expanded into round discs, and very generally their terminal joints are forked. The males are provided with one or two internal vocal sacs. The tadpoles of these arboreal frogs are remarkable for the possession of an adhesive disc behind the mouth on the under surface; and they also have the muzzle prolonged into a proboscis, and the single breathing-pore situated on the right side of the body, nearer to the tail than to the muzzle.

Writing of the habits of one of the Cingalese members of the genus (formerly separated as *Polypedates*), in which the front toes are only half webbed, Sir J. Emerson Tennent observes that it " possesses, in a high degree, the faculty of changing its hues ; one as green as a leaf to-day will be found grey and spotted to-morrow. One of these beautiful little creatures, which had seated itself on the gilt pillar of a lamp on my dinner-table, became in a few minutes scarcely distinguishable in colour from the ormolu ornament to which it clung."

FOREST-FROGS

As is shown to be the case with the snakes, among frogs the members of two totally distinct families have taken to an arboreal life, and have thus become so like one another that it is necessary to depend on anatomical differences for their distinction. In the family *Dendrobatidæ*; while the structure of the bones of the chest is of the same solid structure as obtains in the typical frogs, and the extremities of the transverse processes of the sacral vertebra are not expanded, an important difference presents itself in the absence of teeth in the upper jaw and on the palate. The toes of both feet are quite free from webs, and have their tips expanded into rounded discs.

These forest-frogs, as they may be called in order to distinguish them from the typical tree-frogs of the family *Hylidæ*, are represented by two genera, one of which, *Mantella*, is confined to Madagascar, and is distinguished by the tip of the tongue being notched, while in *Dendrobates* of Tropical America the tongue is entire.

The American genus is represented by seven species, among which the variable tree-frog (*Dendrobates tinctorius*), which measures barely an inch and a half in length, is widely distributed in Tropical America, and is remarkable for its variability in colour, some examples being uniformly black, others grey above and black on the sides and beneath, and others grey with large black blotches. This, however, is by no means the limit of variation, since certain individuals are black above, with two or three longitudinal white or pink stripes, and grey with black spots on the under surface of the body; in other cases the ground-colour is black with white spots and streaks above, and spots or marblings of the same beneath. From the small size of the discs on its feet, which do not admit of its clinging to upright stems, this frog seems to be less arboreal than some of its allies; and, as a matter of fact, it is generally found among fallen leaves on the ground in forests. Like its kindred, it displays remarkable care and attention to its young. The secretion from its skin is employed by the Indians as an arrow-poison.

NARROW-MOUTHED FROGS

An important family (*Engystomatidæ*) of the suborder is that of the narrow-mouthed frogs, a group represented by more than a score of genera, distributed over Africa, Madagascar, India and the adjacent countries, Southern China, Papua, and America. While agreeing with the members of the preceding family in the absence of teeth in the upper jaw, these frogs are distinguished by the broad expansion of the extremities of the transverse processes of the sacral vertebra.

EAST AFRICAN SHORT-HEADED FROGS

The other vertebræ are similar in conformation to those of the typical frogs, and there is the same absence of ribs. There is, however, considerable variation in regard to the bones of the chest, several of the genera lacking the transverse bars known as precoracoids ; and the terminal joints of the toes may be either simple or T-shaped. Although there are no arboreal forms, the family comprises terrestrial, aquatic, and burrowing representatives ; the last having either the front or the hind limbs specially strengthened and furnished with horny sheaths.

In some genera the mouth is extremely narrow ; but although it is convenient to take this character as the basis of the name of the family, it must not be considered that it is applicable to all its members. These narrow-mouthed forms feed exclusively or mainly on ants and termites, and thus exhibit a modification of structure approximating to that characteristic of ant-eating mammals. More than half the members of the family are nocturnal, and may be recognised by the vertical pupil of the eye.

SHORT-HEADED FROGS

The exceeding plumpness of the body serves not only to distinguish the short-headed frogs of the genus *Breviceps* from all their allies, but also makes them some of the most peculiar representatives of their class. Indeed, when the body is puffed out to its fullest extent, they more resemble indiarubber balls than frogs. The genus belongs to a group in which a pair of precoracoid bones are present and the coracoid bones much dilated, and is specially distinguished by the horizontal position of the pupil and the absence of teeth on the palate. These frogs are restricted to Tropical and Sub-tropical Africa, where they are represented by a small number of species, of which the Mozambique *B. mossambicus* inhabits the eastern districts. Generally having a perfectly smooth skin, this frog is of a brown or blackish hue on the upper parts, with a dark oblique streak below the eye. The narrow mouth and long tongue indicate that its food consists of white ants.

DARWIN'S FROG

Darwin's frog (*Rhinoderma darwini*), which likewise belongs to the present family, and inhabits Chile, alone represents a genus remarkable for the throat-sac of the males being enlarged and modified so as to form an extensive chamber on the under surface of the body, in which the eggs and tadpoles undergo their develop-ment. This chamber is entered by two apertures situated on the floor of the mouth on each side of the tongue ; and when the eggs, generally from eleven to fifteen in number, are laid by the female, they are taken and swallowed by her consort, who passes them into his pouch, where they remain until hatched. When the tadpoles are sufficiently developed, they enter the

world by escaping through the parental mouth. It appears that at no stage of their existence do the tadpoles possess external gills.

GÜNTHER'S FROG

Omitting detailed mention of the small and un-important family of *Discophidæ*, characterised by the

SOLOMON ISLAND SHARP-NOSED FROG

presence of teeth in the upper jaw and the expansion of the extremities of the transverse processes of the sacral vertebra, the last representative of the first suborder is Günther's frog (*Ceratobatrachus guentheri*), of the Solomon Islands, which constitutes the family *Ceratobatrachidæ*. The essential characteristics of this family are the presence of teeth in both the upper and lower jaws (a feature found elsewhere only in two families of the next suborder), coupled with the absence of expansion of the extremities of the sacral vertebra.

This frog has a very large triangular head, ornamented with prominent ridges, and terminating in front in a pointed flap of skin ; similar flaps occupying the eyelids, and the mouth having an enormous capacity. In the eye the pupil is horizontal, and teeth are present on the vomers. The hind limbs are rather short, and all the toes have simple terminations, and are devoid of webs. In colour this frog is variable. It appears to be abundant in the Solomon Islands, and is remarkable for laying very large eggs, from which the young emerge nearly fully developed.

SUBORDER 2. ARCIFERA

Iᴺ most parts of South America and nearly the whole of Australia the typical frogs are replaced by a family which, for want of a better name, may be called the southern frogs (*Leptodactylidæ*). These, together with seven other families, differ essentially from the forms hitherto considered in regard to the conformation of the bones of the chest, and thus collectively constitute a second suborder, known as the Arcifera. It will be remembered that in the preceding suborder the pair of coracoid bones are connected together in the middle

of the chest by a single cartilage joining their free edges ; in the present group, on the other hand, each coracoid bone terminates in a large cartilage in such a manner that one cartilage overlaps its fellow of the opposite side.

SOUTHERN FROGS

The southern frogs resemble typical frogs in having the upper jaw alone toothed, and in the transverse processes of the sacral vertebra being cylindrical or but slightly expanded; and as they also agree in the characters of the vertebræ and the absence of ribs, it is evident that the two groups form parallel or representative series. It must, however, always be remembered that it is only an assumption that the conformation of the bones of the chest is the character of primary import; and that it is quite possible that there may have been parallelism in this case also, in which event the present family would have to be placed next the typical frogs. The members of the group are confined to the countries mentioned above, where they are represented by about twenty-five genera and nearly a couple of hundred species. While most of the American species have the pupil of the eye horizontal, this condition occurs but rarely in those from Australia.

HORNED FROGS

Among the best-known representatives of the family are the horned frogs, or horned toads, *Ceratophrys*, all of which are remarkable alike for their large size and brilliant colouring, as well as for the enormous dimensions of their mouths and their fierce and carnivorous habits. Represented by about half a score of species from Tropical and South America, they belong to a group characterised by the more or less marked union of the outer metatarsal bones of the hind foot, the absence of a bony style to the breast-bone, and the webbing of the hind toes; while, as a genus, they are distinguished by the horizontal position of the pupil of the eye and the notching of the tongue. The webbing of the toes varies in extent in the different species, but there is never any expansion of the extremities. The outer metatarsals are completely united, and the skull is remarkable for the extent to which ossification is carried out. In some species, such as the Brazilian horned frog (*C. boiei*), the upper eyelid is produced into a horn-like appendage; but in others, like the Argentine horned frog (*C. ornata*), this is so little developed as to be scarcely noticeable.

The largest representative of the genus is the above-mentioned Brazilian horned frog, which measures as much as 8 inches in length, and is one of the most handsomely ornamented of the genus. The smaller Argentine species represented in the illustration differs by the upper eyelid being only slightly pointed and triangular, as well as by the presence of a bony shield on the back. The skin is covered with tubercles above and granules below; the general colour of the upper parts being yellowish or greenish, with large olive spots surrounded by light-coloured or golden margins, while there are sometimes wine-red lines between the spots.

These frogs, or escuerzos, as they are locally called, are abundant in many parts of Argentina, and in damp weather may be met with crawling about among the grass in numbers, after the manner of European toads. They are exceedingly bold and ferocious, flying fiercely at anyone who attacks them, and maintaining their hold with the tenacity of a bulldog, at the same time uttering a kind of barking cry. On other occasions they give vent to a peculiarly deep bell-like note. When in repose, escuerzos are in the habit of burying themselves in the soil with only the top of the back exposed. In this position they lie in wait for their prey, which includes other frogs, birds, and small mammals.

LEAF-FROGS

Another American genus, containing a very large number of species, is that of the leaf-frogs, *Hylodes*, which deserves mention on account of the peculiar mode of reproduction of one of its representatives, the so-called Antillian frog (*H. martinicensis*). These frogs differ from the group to which the last genus belongs by the absence of a bony style to the breast-bone and the unwebbed hind toes, and are further characterised by the expansion of the tips of the toes into smooth discs, the horizontal pupil of the eye, and the presence of teeth on the vomer.

The Antillian frog, or, as it is locally termed, the coqui, is an inhabitant of several of the West Indian Islands, and may be recognised by its warty under surface; the general colour of the upper parts being grey or brownish, with indistinct darker markings on the head and back, and cross-bars on the hind legs, while there is a large dark mark on the temporal region, and another near the muzzle. In common with several frogs, this species lays large eggs, within which take place such transformations as are undergone by the tadpoles, which after emergence undergo no alteration, except the absorption of the remnant of the tail.

PIPING FROGS

As the typical representatives of the family, brief mention must be made of the piping frogs (*Leptodactylus*) of Central and South America, which differ from the preceding genera in having a dagger-like bony style to the breast-bone. They have the pupil of the eye horizontal, and the teeth on the vomers placed behind the apertures of the inner nostrils. Externally these frogs closely resemble the typical European frogs, with the exception that the hind toes are not webbed. In the males the humerus is expanded into a large flange-like plate, and in the breeding season the whole fore limb in this sex becomes much swollen.

These frogs derive their name from their loud pipe-like croaking, which varies in tone and intensity

ARGENTINE HORNED FROGS

according to the species. Some are noteworthy from their habit of digging a hole in the ground near water, and lining it with a layer of scum, upon which the eggs are deposited, and left to hatch. The nests seem, however, always to be so placed that at a certain season they will be flooded by the rise of the neighbouring water.

I A

TOADS

Passing over the unimportant family of the *Dendro-phryniscidæ*, which includes two small South American genera, the next representatives of the arciferine suborder are the true toads, which constitute a family (*Bufonidæ*) distinguished by the absence of teeth in both jaws, and the expansion of the extremities of the transverse processes of the sacral vertebra. The other vertebræ resemble those of the typical frogs, and there is the same absence of ribs as in the latter. The terminal joints of the toes are either blunt or T-shaped, and in only two out of the eight genera is the pupil of the eye vertical. Two of the genera approximate in character to the preceding family.

Toads have an almost cosmopolitan distribution, and while the more typical species are characterised by their terrestrial habits, rough skin, and creeping gait, so unlike that of the frogs, others are burrowing, and others again (*Nectes*) thoroughly aquatic. On the other hand, the disc-footed toads (*Nectophryne*) of Western Africa and the Indo-Malay region, in which the toes terminate in disc-like pads, appear to be arboreal, while the one Mexican representative of the allied genus *Rhinophrynus* is distinguished by its ant-eating habits.

TYPICAL TOADS

The common European toad (*Bufo vulgaris*) is the typical representative of a large genus, with between eighty and ninety species, ranging over the whole world, with the exception of Madagascar, Australia, New Guinea, and the islands of the Pacific. As a genus these typical toads are distinguished by the entire tongue, the horizontal pupil of the eye, the freedom of the toes of the fore foot, and the partial webbing of those of the hind limbs, as well as by the breast-bone being either cartilaginous or with only a partially ossified style. The degree of webbing of the hind toes varies, and while the tips of the toes are generally simple, they are sometimes expanded into small discs. The head may or may not have bony ridges.

The common toad belongs to a group characterised by the absence of these head-ridges, and by the hind toe being at least half-webbed, but it is specially distinguished by the absence of a fold on the ankle, and by the tubercles beneath the joints of the hind toes being mostly double. On the upper parts grow a number of more or less prominent warts, which, although frequently spiny, are not distinctly porous ; and the glands behind the eyes are remarkably prominent, and of an elongated elliptical form. In colour the upper parts are brownish, with darker spots or marblings, but the lower surface is whitish, more or less thickly spotted with black. A black line runs on the outer side of the gland behind the eye, and in specimens from China and Japan this line extends along the upper sides of the flanks. The distributional area of the toad includes Europe, Asia (exclusive of India and adjacent regions), and North-Western Africa ; but the species is unknown in Ireland, as also in the Balearic Islands.

Few animals have suffered more from popular superstition than the toad, which, although practically harmless, has been almost universally shunned and detested. It is, however, true that the secretion from its skin is acrid and irritating, as may be seen by the foaming lips of dogs which attempt to meddle with these amphibians. Sluggish and terrestrial in its habits, the toad needs not the long and fully-webbed hind limbs of its active cousin the frog ; its usual pace being a kind of crawl, although, when disturbed, it can execute an imperfect hop or leap. When alarmed or threatened with danger, a toad immediately stops and puffs out its body to its utmost capacity, at the same time causing the acrid secretion to exude from the pores of its skin, and likewise discharging a pure limpid fluid from a special reservoir.

Of its general habits, Professor T. Bell writes that the toad " becomes torpid during the winter, and chooses for its retreat some retired and sheltered hole, a hollow tree, or a space among large stones, or some such place, and there remains until the return of spring calls it again into a state of life and activity. Its food consists of insects and worms of almost every kind. It refuses food which is not living, and, indeed, will only take it at the moment when it is in motion. When about to feed, the toad remains motionless, with its eyes turned directly upon the object, and the head a little inclined towards it, and in this attitude it remains until the insect moves ; when, by a stroke like lightning, the tongue is thrown forward upon the victim, which is instantly drawn into the mouth. When the prey is taken, it is slightly pressed by the margins of the jaws ; but as this seldom kills it, unless it be a soft, tender larva, it is generally swallowed alive. Toads will also take earth-worms of considerable size, and it is a curious sight to watch the manner in which the powerful and writhing worm is secured. If the toad happen to take it by the middle, the extremities of the worm are twined with great force and activity around the muzzle of its captor in every direction in its attempts to escape ; but the toad pushes one portion after another into its mouth, by means of the fore feet, until it disappears, when it is swallowed whole."

EUROPEAN TOAD

The eggs of the toad differ from those of the frog in that, instead of forming an irregular mass with their enclosing jelly, they are arranged in a regular, double, and alternating series in the form of a string, which may be a yard or more in length. These strings are generally deposited in the water about a fortnight later than the spawn of the frog, and it is not till autumn that the young toads complete their metamorphosis, and forsake the water. From that of the frog, the tadpole of the toad is distinguished by its smaller size and blacker colour.

NORTH AMERICAN TOAD

In North America the common toad is *Bufo lentiginosa*, a species which, like its European relative, is more abundant near towns and villages than in the open fields, owing to the destruction to which it is subject in the latter from agricultural implements. Writing of this species, Mr. N. Miller asks the question why, in spite of the great fertility of the female, the numbers of the species remain practically stationary. Taking the

FROGS AND TOADS

GREEN TOAD

COMMON TOAD

NATTERJACK TOAD

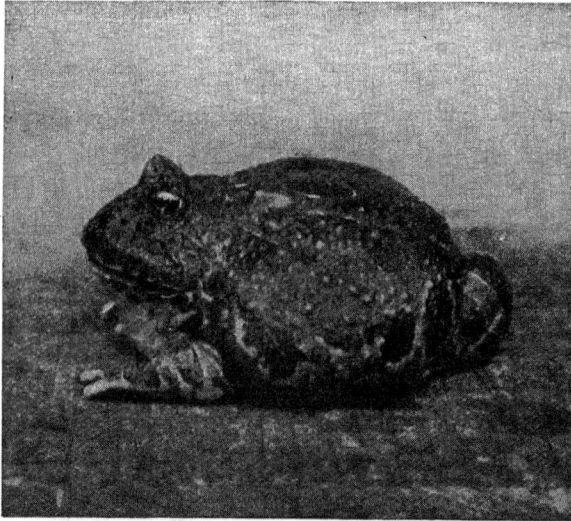

ARGENTINE HORNED FROG

SOUTH AFRICAN FROG

MOORISH TOAD

SOUTH AMERICAN GIANT TOAD

Photographs by W. B. Johnson and others

low figure of eight thousand eggs as the number in one spawn, it appears, that with the exception of two, all these, as well as the whole of the eggs in the other spawns of the same female, must perish if the species keep, as appears to be the case, at the same numerical level. Various water-animals, such as dragon-flies

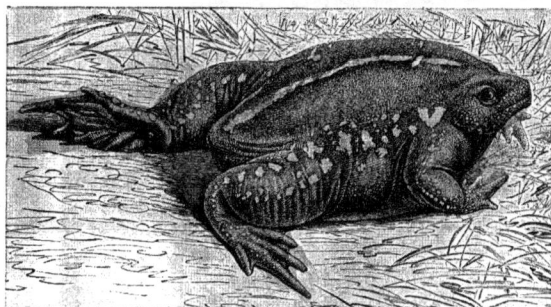

MEXICAN SHARP-NOSED TOAD

and water-beetles and water-bugs, together with their larvæ, newts, and crayfish, appear to be the chief agents in carrying on the work of destruction.

GREEN TOAD

The green toad (*B. viridis*) of Europe, Asia, and Northern Africa is a far handsomer species, distinguished by the presence of a fold on the ankle, and also by the simple structure of the tubercles on the lower surface of the toes of the hind foot. There is likewise a vocal sac beneath the throat of the male, which is wanting in the common toad. The upper parts carry a number of irregular, flattened, and porous warts; and the glands behind the eyes, although sometimes enormously developed, are generally of moderate size and more or less kidney-shaped. The colour is olive or greenish above, generally spotted or marbled with a darker shade, although occasionally nearly uniform, while the under parts are either uniformly whitish or whitish with dark spots.

NATTERJACK TOAD

A third European species, which, unlike the last, is locally represented in England, is the natterjack toad (*B. calamita*), easily recognised by the yellow or whitish line running down the middle of the back. From both the preceding European species the natterjack is distinguished by the much smaller extent of the webbing of the hind toes; while there is a fold on the ankle, and the tubercles on the joints of the lower surface of the hind toes are to a large extent double. The hind limb is usually short, the flattened warts on the back are distinctly porous, the glands behind the eyes are small, depressed, and either oval or triangular, and there is an additional gland on the leg.

The general colour of the upper parts is light olive, with darker marblings or spots, the above-mentioned light line being generally present, but the light under parts are more or less spotted with black. In its movements the natterjack is less sluggish than the common toad, its pace being often quickened to a kind of run, during which the body is raised considerably above the ground. It is likewise less intolerant of drought, being frequently found in hot, sunny situations, and only resorting to the neighbourhood of water during the breeding season. The Moorish toad (*B. mauretanica*) is another interesting species of the genus.

MEXICAN SHARP-NOSED TOAD

The Mexican sharp-nosed toad (*Rhinophrynus dorsalis*), already referred to as subsisting on white ants,

is the only other member of the family that can be mentioned here, and is generically distinguished by the long and narrow tongue being free in front, by the vertical pupil of the eye, and by the rudimentary breastbone. The front toes are free, and those of the hind limb webbed, with simple tips; the general form of the body is extremely stout, the head is small, with a long, truncated muzzle and narrow mouth, and the eyes are small and the limbs remarkably short. In colour this toad is olive brown or bluish grey above, frequently with yellowish spots on the flanks and middle of the back, those on the back sometimes uniting to form a line.

TREE-FROGS

The numerous, and for the most part arboreal, species which, in contradistinction to the flying-frogs and the forest-frogs, may be designated tree-frogs, constitute the family *Hylidæ*, comprising some ten genera, very abundant in Australia and America, and more sparingly represented in Europe, Asia north of the Himalaya, and Northern Africa, with one species ranging into North-Eastern India and Burma. While resembling toads in the expansion of the processes of the sacral vertebra, tree-frogs differ by the presence of teeth in the upper jaw, and they are also peculiar in the claw-like form of the terminal joints of the toes. The bodies of the vertebræ are cupped in front and spherical behind, and there are no ribs.

GRASSHOPPER-FROG

The pretty little grasshopper-frog (*Acris gryllus*) of North America is the sole representative of a genus characterised by the horizontal pupil of the eye, the webbing of the hind toes, of which the tips are but little expanded, and the slight expansion of the process of the sacral vertebra. In form this little frog is slender, with a narrow head and rather sharp muzzle; the skin of the upper parts is either smooth or slightly tuberculated, and that of the under parts granulated. The mottled and striped colouring is very variable, the ground-tint ranging from reddish brown to green; but there is generally a large, triangular dark brown spot between the eyes, and sometimes a light stripe down the back. Locally very abundant in Eastern

GRASSHOPPER-FROGS

and Central North America, the grasshopper-frog derives its name from its piercing, strident cry, which resembles the noise of its insect namesake. It frequents stagnant waters, and is fond of resting on the leaves of aquatic plants. Unlike most of its allies, it lurks among plants, and seldom, if ever, ascends bushes or trees.

TYPICAL TREE-FROGS

Closely allied to the last are the numerous species of typical tree-frogs (*Hyla*), which are by far the most beautiful representatives of the entire order, and are best known by the common European species. In this genus the pupil of the eye is horizontal, the toes of both limbs are dilated into adhesive discs, and those of the hind foot more or less extensively webbed; the tongue is either adherent or partially free behind, and the expansion of the transverse processes of the sacral vertebra is more or less strongly marked. As in the last genus, there are teeth on the vomers.

Represented by about one hundred and fifty species, this genus has a distribution coëxtensive with that of the family, the sole Indian member of the latter being included. The under surface of the body of these frogs is very different to that of the terrestrial species; for the skin, instead of being smooth, is covered with granular glands, pierced by numerous pores, through which the dew or rain spread on the surface of the leaves is rapidly absorbed into the system, and reserved to supply the moisture necessary for cutaneous respiration. Except during the breeding season, when the greater number of them seek the water, or when they retire before the cold of winter or drought of summer under mud, beneath stones, the bark of trees, or in other safe spots, these frogs spend their lives among the leaves of trees, where they find alike their dwelling-places and their hunting-grounds.

As in the case of the Indo-Malay flying-frogs, which may be considered to represent this group in those countries, their colour harmonises exactly with their natural surroundings, and changes even more rapidly than that of the chamæleons. So exactly, indeed, do they resemble the foliage among which they hide that it is often difficult to tell frogs from leaves; and it has been noticed that where there is the greatest variety and brilliancy of colour among the forest trees, the tree-frogs attain their most brilliant and varied tints.

The European tree-frog (*H. arborea*), which is one of three species inhabiting the Old World proper, has a wide geographical distribution, inhabiting the greater part of Europe, Asia north of the Himalaya as far east as Japan, and North Africa. With the exception of the

EUROPEAN TREE-FROG

higher mountain ranges and the extreme north, as well as Norway and the British Isles, it is spread over the whole of Europe, although varying locally to a considerable degree in colouring and habits. The males are furnished with a large external vocal sac on the throat, and the skin is smooth above and granulated

beneath. The general colour may be described as greenish above and uniform whitish beneath, but there are many variations in regard to the markings on the upper parts; the typical form having a greyish or

AUSTRALIAN TREE-FROG

black light-edged streak extending from the nostril through the eye and ear along each side of the body, and sending a branch upwards and forwards on the loin, while a whitish line descends from the upper lip to the shoulder, and then runs upwards to the eye, thus enclosing an elongated green area.

In habits this frog is most active; and while in swimming it is nearly equal to the common frog, in leaping it is its superior, in addition to which it is a most expert climber. When croaking, the sac on the throat of the males becomes so inflated as to make this appendage nearly as large as the body. Like toads, tree-frogs do not appear to touch the insects on which they prey until these begin to move. Flies, spiders, beetles, butterflies, and smooth caterpillars appear to form their favourite food, although, they have been known to attack and kill humble-bees. The forest districts of Central, Southern, and Eastern Europe constitute the chief stronghold of these frogs; and the noise made by these tiny batrachians comes as a surprise to those who hear it for the first time. The European species is of very small size, but some of the American and Australian species attain comparatively large dimensions, one of the largest members of the genus being *H. faber*, of Brazil, which measures as much as $3\frac{1}{2}$ inches in length.

An interesting account of the breeding habits of the frog last mentioned, which in Brazil is known as the ferreiro, or smith, is given by Dr. E. Goeldi, whose observations were made in the Organ Mountains adjoining the bay of Rio de Janeiro. This frog constructs regular pools of a circular form in the shallow borders of ponds and swamps, such pools being surrounded by a narrow mud wall. In 1894 one swamp contained nine of these pools, which serve as nests for the tadpoles. "On the night of February 18," writes the describer, "between nine and eleven o'clock, we approached the pond, occupied, as we could hear from a distance, by at least a dozen of the large tree-frogs. The moon was shining brightly, and much favoured our undertaking, but even under these circumstances we had to accustom our sight to discern the details in the marginal vegetation, and the portion somewhat hidden in the shadow. By-and-by we discovered the ferreiros, some at work, others drumming together on the walls of some pool, or in the middle of the pond, sitting upon some floating object, such as water-plants. The vocalists, of which we could distinguish the moderately inflated gular sacs, were all males."

After stating that he was posted on a side of the pond where five nests were already situated, Dr. Goeldi observes that he and his companion were fortunate enough to see the rising of a new nest. In a certain spot he writes that " we first saw some slight movement in the water, produced by something stirring below the surface. We then soon saw a mass of mud rising to the surface, carried by a tree-frog, of which no more than the two hands emerged. Diving again, after a moment's time, the frog brought up a second mass of mud, near the first. This was repeated many times, the result being the gradual erection of a circular wall. From time to time the head and front part of the body of the builder appeared suddenly with a load of mud at some point ; but what astonished us in the highest degree was the manner in which the frog used its hands for smoothing the mud wall, as would a mason with his trowel. And by examining the hands of this tree-frog, it will readily be understood how they are most serviceable trowels, their terminal joints bearing large expansions. This careful process of smoothing could be better observed as the wall gradually heightened, until it reached about four inches, when the frog was compelled to come out of the water.

" The parapet of the wall receives the most careful smoothening, the outside being neglected, and the levelling of the bottom attained by the action of the lower surface of the creature's body, aided by the hands. The aspect of the pool may be compared to the crater of a volcano, or a vessel of a foot in diameter filled with water. Although the female undertakes the entire task of building, she is incommoded the whole time by the male sitting on her back. Should he be frightened from his post, he will soon emerge from the water at a distance of a few feet, when, if signs of danger be wanting, he will climb the walls of the nest and regain his original seat."

Another Brazilian tree-frog of the same genus (*H. goëldii*) breeds in the water contained in the central cup of certain trees belonging to the *Bromeliaceæ*. Of this species, Dr. Goeldi states that the first specimen found was a female, carrying on her back a large globular mass of whitish eggs. When put in a vivarium, " for a few days the egg-mass remained attached to the mother's back. But suddenly it fell away, and simultaneously I saw in the glass some small, nearly black-coloured frogs, all provided with the anterior and posterior legs, together with a larval tail of medium or rather small size."

Yet another tree-frog from Brazil (*H. nebulosa*) has acquired the remarkable habit of depositing its eggs in the sheaths of old and decaying leaves of bananas. Dr. Goeldi states that this frog " glues its lumps of eggs on the edges and on the inside of banana-leaves,

where, even during the hot hours of the day, sufficient coolness and moisture are preserved. These lumps are enclosed in a frothy, whitish substance, comparable to the scum formed by certain *Cicadidæ*. Sometimes the tailed larvæ are seen struggling in this frothy mass. If put into fresh-water, all will die in a few hours." Australia possesses an interesting species in *H. cœrulea*.

POUCHED FROGS

On account of the peculiarity of their reproduction mention must be made of the curious marsupial or pouched frogs (*Nototrema*), all of which are distinguished from the typical genus *Hyla* by the presence of a backwardly opening pouch at the hinder end of the back in the females. These frogs are represented by some half-dozen species, mainly confined to Central and Western Tropical America, although one of their number is found on the eastern side of that continent at Pernambuco. The pouch of the female is extended beneath the skin of the back and sides to form a very large chamber, in which the eggs and tadpoles undergo the whole of their transformations. The eggs, generally about fifteen or sixteen in number, appear to be placed in the pouch by the male, who employs his hind feet for the purpose; and they are remarkable for the large relative size of the yolk. The tadpoles, when first hatched, are peculiar in having a bell-shaped structure for the protection of their two pairs of external gills.

TOAD-FROGS

The fifth family (*Pelobatidæ*) of the order belonging to the section with overlapping cartilages to the coracoid bones comprises eight genera, which may be collectively termed toad-frogs, since they come under the designation neither of toads nor frogs. Agreeing with the tree-frogs in the presence of teeth in the upper jaw, the members of this group may be distinguished by the much greater expansion of the processes of the sacral vertebra ; ribs being absent, and the terminal joints of the toes simple. In all the species the pupil of the eye is vertical, and while the majority of the genera agree with the preceding groups in having the articular cup at the front end of the bodies of the vertebræ, and the ball behind, in a few this arrangement is reversed. The family is distributed over Europe, the Indo-Malay region, North America, and New Guinea ; the various genera having a more or less restricted geographical range.

The brown toad-frog (*Pelobates fuscus*) is the typical representative of a genus containing two European species, neither of which is found in Great Britain.

BROWN TOAD-FROG

PAINTED FROG

BURMESE FROG

They are characterised by the rod at the end of the backbone being welded to the sacral vertebra, as well as by the extensive webbing of the hind toes : the presence of a bony style to the breast-bone, coupled with the want of an externally visible ear-membrane, serving to distinguish them from the allied North American genus *Scaphiopus*.

The brown toad-frog is a rather large species, usually measuring from 2½ to 3 inches in length, and having a smooth brown skin, marbled on the upper surface with darker markings ; a spur which is present on the metatarsus being yellowish brown. The males have no vocal sacs, but are furnished with a large elliptical gland on the upper surface of the fore limb. This species is decidedly local, and in some districts replaced by the allied *P. cultripes*, easily distinguished by the black spur on the metatarsus. Spending only a few days during the breeding season in the water, it is essentially a land frog, generally frequenting spots with a sandy soil. Here, with the aid of its metatarsal spur, it rapidly excavates hollows in the ground, throwing out the earth backwards, and soon partially concealing itself. An aperture is, however, always left to the excavation, and should the rays of the morning sun reach its occupant, the burrow is quickly deepened.

In its movements the toad-frog is more active than typical toads, approaching in this respect frogs, as it takes considerable leaps, swims strongly, and burrows with rapidity. The breeding season takes place in April, during which time the males utter a loud croaking, accompanied in a lower tone by the females. The eggs, which are laid in strings of about a couple of feet in length, are taken from time to time by the male and carefully deposited round reeds, grass, or other plants growing near the edge of the water. In from five to six days the small black tadpoles are hatched out; and in the course of four months these have completed their development and leave the water. When an adult toad-frog is suddenly seized or pinched, it utters a cry like the mewing of a kitten, at the same time emitting a pungent vapour with a strong odour of garlic, both these being apparently intended as a means of defence.

Of the remaining genera, *Pelodytes*, as represented by the punctured toad-frog (*P. punctatus*) of Western Europe, and the Papuan *Batrachopsis* differ from the preceding in that the sacral vertebra has two condyles for articulation with the rod forming the termination of the backbone, the hind toes being slightly webbed. In the Indo-Malay genus *Leptobrachium* there is, however, only a single condyle for the articulation of the rod-like bone. The Burmese frog (*Megalophrys robusta*, is a representative of another genus of this family.

In the Miocene rocks of Europe occur remains of numerous frogs which are assigned to an extinct genus, *Palæobatrachus*, regarded as representing a family, *Palæobatrachidæ*, connecting the present one with the undermentioned *Xenopodidæ*. In these extinct frogs the upper jaw is toothed, the transverse processes of the sacral vertebra have expanded extremities, the sacral vertebra itself articulates with the terminal rod of the backbone by means of two condyles, the bodies of the other vertebræ have the articular cup in front, and there are no ribs.

DISC-TONGUED FROGS

The disc-tongued frogs, as the members of this group may be called, form a small family (*Discoglossidæ*) represented by four genera and about seven species, inhabiting the northern half of the Old World and New Zealand. As a family, these frogs are characterised by the presence of teeth in the upper jaw, the expansion of the processes of the sacral vertebra, the presence of short rudimentary ribs, and the circumstance that in the bodies of the vertebræ the articular cup is placed at the hind end, and the ball in front. In both the latter respects these frogs resemble salamanders and newts, and they may accordingly be regarded as some of the least specialised representatives of the order. Their fossil remains occur abundantly in the middle Tertiary rocks of Europe. The family derives its name from the disc-like form of the tongue, which may be either free or adherent. From those of all the groups hitherto described, the tadpoles, after shedding the external gills, differ in having the breathing pore situated in the middle of the under surface of the body, instead of on the left side.

PAINTED AND FIRE-BELLIED FROGS

From the painted frog (*Discoglossus pictus*) of Southern Europe and North-Western Africa, which alone represents the typical genus of the family, the fire-bellied frog (*Bombinator igneus*) is distinguished by the absence of an external tympanic membrane to the ear, and is further characterised by the adherent tongue, the triangular form of the pupil of the eye, and the great expansion of the extremities of the transverse processes of the sacral vertebra. This frog, which inhabits Europe and Asia, although unknown in the British Isles, has the skin very warty on the upper parts, while beneath it is quite smooth. In colour, it is olive above, with or without black marblings, but beneath is orange or yellow, marbled with black. The males, which are unprovided with vocal sacs, during the breeding season develop black rugosities on the inner side of the forearm, as well as on the inner tubercle of the metacarpus and on the two innermost front toes.

There are two varieties of this frog (reckoned by some naturalists as distinct species), of which the one with orange-coloured under parts is to be found in streams or marshes in the lowlands, while the yellow-bellied phase lives at considerable elevations in the mountains. They are essentially aquatic frogs, only leaving the water for a short time in spring, when they may be seen hopping on the land on their long hind legs. In water they generally take up their position at some distance from the bank, sitting with their heads slightly raised above the surface, and disappearing with lightning-like speed at the slightest noise, to seek safety in the mud at the bottom. The tadpoles grow to an unusually large size, and are especially characterised by the great development of the tail-fin.

MIDWIFE-FROG

The third European representative of this family is the so-called midwife-frog (*Alytes obstetricans*), of which the typical race inhabits France, Switzerland, Belgium, and Western Germany, while a variety occurs in Spain and Portugal; Spain being also the home of the second member of the genus, *A. cisternasii*. From the fire-bellied frog these two species are distinguished by the presence of a distinct external tympanic membrane to the ear, the elliptical and vertical pupil of the eye, and the moderate expansion of the transverse processes of the sacral vertebra. The common species has the

FIRE-BELLIED FROGS

skin of the upper parts warty, and that of the under surface granular; the glands near the head are small or indistinct, but there are large ones on the limbs; and the males have no vocal sacs. The colour of the upper parts is olive-grey, with darker dots and irregular spots.

Essentially an aquatic species, this frog derives its name from the circumstance that the male takes charge of the eggs during their development. The breeding season lasts for upwards of six months—namely, from

March to August—although the eggs are laid only from March till June. These are deposited by the female in the form of long chains, which may be upwards of a yard and a half in length. These chains are taken by the male, and wound round his legs and thighs; and when thus loaded he retires to some burrow or convenient hollow near the bank, where, at least during

MIDWIFE-FROG

the daytime, he remains in concealment until the tadpoles are ready for hatching. He then enters the water, and the tadpoles soon come forth, and swim away to take care of themselves; the hatching of the tadpoles taking place from June to September. After the cares of the nursing period are over, the male loses his voice, which is not resumed till the following February, when it is continued till August. The males are more numerous than the females, and during the breeding season their loud croaking is almost continuous. From September till the beginning of March the habits of this sex are similar to those of other frogs.

The lower Miocene strata of the Continent have yielded remains of an extinct frog belonging to the same genus; while in the rocks of the upper part of the same division of the Tertiary period occurs a gigantic frog belonging to the same family, which has been referred to an extinct genus, under the name of *Latonia*.

ALLIED FAMILIES

The other two families—*Amphignathodontidæ* and *Hemiphractidæ*—belonging to the suborder Arcifera are not of much importance, and are represented only by a small number of genera and species from Central and South America. They are, however, of some interest from the circumstance that both the upper and lower jaws are furnished with teeth, in which respect they agree with the sharp-nosed frog among the members of the first suborder.

A rare frog (*Ceratohyla bubalus*) belonging to the second of these families, and a native of the Peruvian Andes, is remarkable for its breeding habits. In a female specimen there were nine eggs tightly adhering to the skin of the back, in which the spines of the vertebræ were so prominent as to leave indentations in the egg-membranes. Fully-formed young frogs were contained in the eggs.

SUBORDER 3. AGLOSSA

ALL the members of the order Ecaudata hitherto considered are furnished with a well-developed tongue, but in the suborder Aglossa, which comprises the two families *Xenopodidæ* and *Pipidæ*, this organ is totally wanting. The bodies of the vertebræ resemble those of the disc-tongued frogs in having their articular cups at the hind ends, but ribs are wanting. The coracoid

SURINAM TOAD

bones correspond in structure to those of the suborder Arcifera, although the cartilages at their edges do not overlap. The tadpoles of these remarkable frogs exhibit the peculiarity of having a pair of breathing-pores, after the loss of the external gills, situated symmetrically on each side of the body. Each family is represented by a single genus respectively confined to Tropical Africa and Tropical South America.

SPUR-TOED FROGS

The spur-toed frogs (*Xenopus*), of which there are three species from Tropical Africa, are characterised as a family by the presence of teeth in the upper jaw ; but are further distinguished by the circular pupil of the eye, the absence of an external tympanic membrane to the ear, the free front toes, and the webbed hind foot, in which each of the three inner toes is furnished with a sharp, spur-like nail.

The smooth spur-toed frog (*X. lævis*), which is the species here represented, and has a wide geographical distribution, ranging from Abyssinia to the Cape, is characterised by its smooth skin, marked round the body with more or less distinctly defined tube-like lines. In colour it is dark brown above, and whitish beneath ; some individuals being uniform, but others spotted with brown on the under surface. The spur-toed frogs are exclusively aquatic, pursuing even their prey beneath the surface of the water, and capturing it with their fore feet. The pairing season takes place in August, and the large eggs are laid singly. The tadpoles, which at birth have already lost their external gills, on the third day after leaving the eggs develop a pair of barbels hanging down from the corners of the mouth.

SURINAM TOAD

Among all the strange nursing habits met with among frogs and toads, none is more remarkable than those characteristic of the large aquatic species commonly

known as the Surinam toad. This species represents a family *Pipidæ* distinguished by the absence of teeth in both jaws. Agreeing with the spur-toed frogs in its circular pupil, smooth palate, and absence of a tympanic membrane to the ear, the Surinam toad (*Pipa americana*) has the extremities of the free front toes dilated into radiating appendages, and the fully-webbed hind toes devoid of nails. In form the head is triangular and much depressed, with the eye minute, one or two short tentacles on the lip in front of the eye, a large flap at each corner of the mouth, and sometimes a third at the tip of the muzzle. The skin, which is covered with small tubercles, is olive brown or blackish on the upper parts, but beneath it is lighter, being sometimes ornamented with white spots, and at others with a black stripe down the middle line.

The Surinam toad is an inhabitant of the damp forests of the Guianas and Brazil, and the females deposit their eggs after the usual manner in the water. At this period the skin of the back of the female becomes extremely soft and much thickened, and the eggs, as soon as laid, are taken by the males and embedded one by one in this softened skin, which soon closes over, so as to enclose each in a separate cell. In these cells the eggs undergo the full course of development, the juvenile toads issuing forth from their confinement in a perfect condition, although their dimensions are, of course, small, and no gills are developed at any stage. Although there may be so many as one hundred and twenty cells in the back of a single individual, the more usual number is from sixty to seventy. The period from the deposition of the eggs to the appearance of the young toads is eighty-two days, and the young, when first bursting through the covering of their cells, generally protrude the head or one limb. Soon after the birth of her offspring the female changes the superficial layer of her skin by rubbing it off against

SMOOTH SPUR-TOED FROG AND ITS TADPOLE

stones or plants, the place occupied by each cell being then indicated by a small pit. Except during the breeding season, the pipa appears to be completely aquatic.

The occurrence of a somewhat similar type of nursing habits in a member of the family *Hemiphractidæ* is referred to above.

ORDER II.   CAUDATA

CHARACTERISTICS OF SALAMANDERS AND NEWTS

THE salamanders and newts are readily distinguished from frogs and toads by the retention of the tail throughout life—a feature from which they are collectively designated tailed amphibians. Although they have generally two pairs of limbs, in a few instances the hind pair is wanting ; and in all cases the bones of the limbs are of a normal type, the radius and ulna in the front pair, and the tibia and fibula in the hind pair, remaining distinct from one another. In the skull the frontal bones are not united with the parietals, and the palatine bones are distinct from the hinder upper jawbones, or maxillæ. Generally more or less lizard-like in form, the tailed amphibians undergo a less marked metamorphosis than the tailless group, some even retaining gills throughout life.

As regards their geographical distribution, the salamanders and newts, of which there are rather more than one hundred and twenty existing species, are mainly characteristic in the Northern Hemisphere, being represented only by a few scattered types in the Southern Hemisphere, and quite unknown in Africa south of the Sahara and Australasia. The northern part of the Old World is the home of the true newts, of which four species extend into Northern Africa, and this area likewise contains one of the fish-like salamanders and the olm.

True newts are very abundant in the western portion of this region, but further eastwards they become less numerous, and an approximation to American types of the order may be noticed, although only two genera are common to the Old and New Worlds. North America is especially rich in tailed batrachians, containing more than half the representatives of the entire order, and being the sole habitat of the two-legged salamanders (*Sirenidæ*). Axolotls are here especially abundant, and there are also peculiar genera belonging to the same families of the fish-like and gilled salamanders.

The Indo-Malay region possesses very few species—namely, a peculiar genus (*Tylotriton*) of newts in Yunan, the Eastern Himalaya, and the Liu-Kiu Islands, an allied type (*Pachytriton*) from Southern China, and an axolotl in Siam and Burma. Tropical America, on the other hand, has numerous species, among which may be specially noted the newts of the genus *Spelerpes*, which are also represented by one species from Central America and the West Indies, and others from the mountains of Colombia, Ecuador, and northern Peru, the allied *Plethodon* extending southwards into Argentina. Geologically the group is by no means old, its earliest representative (*Hylæobatrachus*) occurring in the Wealden of Belgium ; and these animals do not become abundant until the Tertiary epoch.

SMOOTH NEWT ON LAND DURING SUMMER
Photo, W. B. Johnson

Although the black Alpine salamander forms an exception, nearly all newts and salamanders appear to be inhabitants of water during at least some period of their existence, some frequenting muddy swamps, and others deep lakes or subterranean waters, while a few are found in mountain tarns at elevations of several thousand feet above the sea. Without exception nocturnal in their habits, spending the day in slumber, either concealed in hiding-places on land or at the bottom of the water in their aquatic haunts, and venturing abroad only at evening or after heavy rain, they are all difficult of observation, and consequently much still remains to be learnt with regard to their mode of life.

The terrestrial species generally frequent soft, shady, damp spots, but occasionally narrow valleys or forests, where they conceal themselves under stones or fallen trunks of trees, or in holes in the earth. During their permanent or temporary sojourn in the water, the adults of those species unprovided with external gills are obliged to come periodically to the surface in order to breathe ; and while in that element all are less completely nocturnal than when on land. Such species as are inhabitants of cold regions undergo a period of torpor during the winter months, while in the tropical region others become quiescent when their haunts are dried up. They exhibit a wonderful tenacity of life, and when dried up in mud, or frozen in ice, will awaken at the first shower of rain, or when their icy bonds are dissolved by the sun's rays. They have also the capacity of reproducing lost limbs, apparently any number of times.

Although on land the majority of species are slow and sluggish in their movements, some salamanders from the south and west of Europe, belonging to the genera *Salamandrina* and *Chioglossa*, run with the celerity of lizards ; while others, again, climb sloping or perpendicular faces of rock like geckos. In the water all swim quickly, mainly by means of serpentine movements of the tail, although the water-newts are perhaps the most expert swimmers.

Through long familiarity we have become so accustomed to the idea of frogs dwelling in trees that the strangeness of the assumption of such arboreal habits by batrachians does not strike us. When, however, we hear for the first time of a species of salamander habitually taking up its abode and breeding in holes in tree-stems, such a peculiar departure from the normal mode of life of animals of this group at once forces itself upon our attention. For many years very little was known with regard to the mode of life of a salamander from the Pacific side of North America familiar to naturalists under the name of *Autodax* (or *Anaides*) *lugubris*.

It is true that the eggs were detected so long ago as 1899, but since subsequent search in the same locality failed to reveal others, it became evident that the normal breeding station had not been discovered. The eggs found in 1899 were placed in the ground, but later on it was accidentally discovered that normally the egg-chains, at least in California, are deposited in chinks and holes in the stems of a species of oak, and also that the adult salamanders dwell in the same situations. The discovery was made by workmen employed to clean and

MALE   SMOOTH NEWTS IN WATER   FEMALE   Photos by W. B. Johnson

dress the stems of a grove of these oaks in the grounds of the California University in order to preserve them from decay. Nearly one hundred adult salamanders were thus taken, together with a dozen bunches of eggs. Some specimens were captured at a height of about thirty feet from the ground, and in some of the larger holes as many as a dozen examples were found, although in most cases there were only two, and occasionally one, in each cavity. When a large number were found, they appeared to be all members of a single family.

They are carnivorous in their diet, feeding chiefly upon molluscs, worms, spiders, and insects. Their breeding habits are peculiar in that there is usually no union between the two sexes, the females seizing the packets of spermatozoa deposited by the males and conveying them to their own reproductive chambers. While some species lay eggs, in other cases the eggs are hatched within the bodies of the female parent, and the tadpoles born alive, sometimes in a highly advanced stage of development.

In the case of the common salamander, during the breeding season the male enters the water first, and is followed shortly afterwards by the female, who gives birth to her tadpoles, but in the Alpine salamander the young are born on land. The water-newts, on the other hand, lay eggs attached to the stems and leaves of aquatic plants. The majority of the terrestrial forms pass the earlier stages of their existence in the water, not leaving this element till their lungs have become fully developed. In the tadpole-stage all the members of the order are remarkably alike ; and this resemblance forbids any wide separation of species like the olm, in which the external gills are retained, from the typical newts and salamanders, in which these appendages are lost at an early period.

Although some of the larger kinds prey upon small fishes, none of the newts and salamanders can be said to be harmful to man. The terrestrial species are defended against all foes, except fishes, frogs, and snakes, by the poisonous secretion exuded by the glands of their skins, but water-newts are devoured by aquatic birds and mammals. The reputed noxious characters of the common salamander, and its alleged immunity to the effects of fire, are, of course, purely fabulous. The existing members of the order are divided into four families.

CLASSIFICATION OF CAUDATA MENTIONED IN THIS WORK

ORDER
Salamanders and Newts—Caudata
FAMILY 1
Salamanders—Salamandridæ
GENUS 1
Typical Salamanders—Salamandra
SPECIES
Spotted salamander .... Salamandra maculosa
Alpine salamander .... S. atra
Caucasian salamander .... S. caucasica
GENUS 2
Chioglossa
SPECIES
Spanish salamander ...... Chioglossa lusitanica
GENUS 3
Newts—Molge
SPECIES
Crested newt .... Molge cristata
Blasius's newt .... M. blasii
Marbled newt .... M. marmorata
Alpine newt .... M. alpestris
The newt .... M. vulgaris
Webbed newt .... M. palmata
Banded newt .... M. vittata
Waltli's newt .... M. waltlii
GENUS 4
Salamandrina
SPECIES
Spectacled salamander .... Salamandrina per-
GENUS 5   spicillata
Tylotriton
GENUS 6
Pachytriton

GENUS 7
Axolotls—Amblystoma
SPECIES
Mexican axolotl ......Amblystoma tigrinum
GENUS 8
Hynobius
GENUS 9
Salamandrella
GENUS 10
Onychodactylus
GENUS 11
Ranidens
GENUS 12
Batrachyperus
GENUS 13
Dicamptodon
GENUS 14
Cave-Salamanders—Plethodon
SPECIES
Argentine cave-salamander .... Plethodon
       platensis
Carolina cave-salamander .... P. shermani
GENUS 15
Spelerpes
SPECIES
Spelerpes fuscus
S. maculicaudus
GENUS 16
Desmognathus
GENUS 17
Thorius
GENUS 18
Typhlotriton

FAMILY 2
Fish-like Salamanders—Amphiumidæ
GENUS 1
Cryptobranchus
SPECIES
Giant salamander .... Cryptobranchus
       maximus
Mississippi salamander .... C. lateralis
GENUS 2
Amphiuma
SPECIES
Three-toed salamander .... Amphiuma means
FAMILY 3
Proteidæ
GENUS 1
Proteus
SPECIES
Olm .... Proteus anguineus
GENUS 2
Necturus
SPECIES
Furrowed salamander .... Necturus maculatus
       N. punctatus
FAMILY 4
Two-legged Salamanders—Sirenidæ
GENUS 1
Siren
SPECIES
Mud-eel..................... Siren lacertina
GENUS 2
Pseudobranchus
SPECIES
Georgian salamander .... Pseudobranchus
       striatus

1657

SALAMANDER TRIBE

COMPRISING the typical members of the order, the family *Salamandridæ* is specially characterised by the absence of gills in the adult condition, the presence of upper jawbones, or maxillæ, as well as of teeth in both the upper and lower jaws, and likewise by the development of distinct eyelids. The family, which includes by far the great majority of the species of the order, is divided into four subfamilies, the first of which is characterised by having the teeth on the

SPOTTED SALAMANDER

palate of the skull arranged in two longitudinal series, diverging posteriorly, and inserted on the inner margin of two backwardly prolonged processes of the palatine bones. The median parasphenoid bone on the base of the skull is devoid of teeth, and the bodies of the vertebræ are convex in front and concave behind.

TYPICAL SALAMANDERS

The typical genus of the first subfamily (*Salamandrina*) is represented by three species, ranging from Central and Southern Europe to the Caucasus, Syria, and Algeria. As a genus, these salamanders are characterised by the large and suboval tongue being free on the sides, and to a small degree also behind, by the palatine teeth forming two curved series, by the presence of four front and five hind toes, and likewise by the nearly cylindrical section of the tail.

SPOTTED SALAMANDER

The spotted species (*Salamandra maculosa*), which varies in length from seven to nine inches, may be recognised by the length of the tail being slightly less than that of the head and body, and still more readily by its brilliant black and yellow colour. The head is depressed and nearly as broad as long, the stout body is likewise somewhat depressed, without any crest along the middle of the back, and the short toes are devoid of connecting webs. The smooth and shining skin is covered on the upper parts with pores, from which exudes a viscid and acrid secretion endowed with decidedly poisonous properties. The yellow markings on the head, back, and tail are arranged in two longitudinal series, broken up into more or less irregularly-shaped patches. The species is an inhabitant of Central and Southern Europe, Algeria, and Syria, and is the one which from time immemorial has been dreaded, not only on account of its undoubtedly poisonous properties, but likewise owing to the extraordinary superstition that if thrown on a fire it would not be consumed. Frequenting moist and shady spots, either in the mountains among rocks, or in valleys and forests, this salamander passes the daytime in a kind of torpid condition, only issuing forth from its hiding-places among stones or roots of trees either during rainy weather or after nightfall, its skin being quickly dried up if exposed to the direct rays of the sun. Its movements on land are slow and sluggish, its gait being a crawl with a marked lateral movement; but in water the creature swims strongly, mainly by the aid of its tail.

Although frequently found in the neighbourhood of its fellows, this salamander can scarcely be termed a sociable creature, and it is only during the breeding season that the two sexes live in company. From the sluggishness of its own movements, it is only slow-moving creatures such as snails, worms, and beetles that the salamander can capture for its food, although it is stated to occasionally kill small vertebrates. Generally a large quantity of food is consumed, after which there is a long fast, sometimes lasting for as much as a month.

During the pairing season, which takes place in April or May, both sexes resort to the water, when the females collect the masses of spermatozoa deposited by the males. Although the young are usually born alive, it occasionally happens that eggs are laid by the female, from which the young almost immediately make their escape. The number of tadpoles produced at a birth is very large, as many as fifty eggs being frequently found within the body of the female, while an instance is on record where upwards of forty-eight young were born within four-and-twenty hours. More generally, however, from eight to sixteen, and less commonly from twenty-four to thirty, tadpoles make their appearance into the world during a period of from two to five days. Generally all these are in an equally advanced stage of development, but sometimes in captivity both eggs and tadpoles are produced simultaneously, the former being translucent and showing the young tadpoles curled up inside.

The tadpoles, which are generally produced in clear, running water, are blackish grey in colour, with a more or less well-marked greenish tinge, but there

ALPINE SALAMANDER

are small golden spots on the back, which slowly increase in size with advancing age. The skin also gradually becomes less shining and smooth, and at the same time the gills shrink, till about August or September the young salamanders quit the water for a terrestrial life. A few may, however, remain till so late as October. It is remarkable that the young salamander is rather inferior in size to the tadpole in the latest stage of development, and it is not yet

MALE AND FEMALE OF MARBLED NEWT

known for how long a period it continues to grow after leaving the water. In aquaria these salamanders develop more quickly, and specimens have been known to leave the water within three weeks. The winter sleep generally takes place in moss-lined crannies, well protected from the frost, and may endure till the beginning of April.

ALPINE SALAMANDER

The black Alpine salamander (*S. atra*), inhabiting the Alps at elevations of from three to ten thousand feet, is a smaller species than the last, and ranges from the Alps into Styria, Carinthia, and some of the mountains of Würtemberg and Bavaria. It inhabits moist woods or the banks of mountain streams, where it is generally found in small family parties, the members of which conceal themselves after the manner of their kind beneath stones and moss, or at the roots of the Alpine rose.

Although resembling the spotted salamander in producing living young, this species differs in that never more than two are born at a time. The most remarkable circumstance connected with the reproduction of the species is, however, that from thirty to forty eggs are found in the oviducts of the females, out of which only one develops in each oviduct, at the cost of the remainder, which form a glutinous mass surrounding the developing egg, and in which the liberated tadpole can afterwards freely move. There are also some fifteen unimpregnated eggs in each oviduct, which serve as the food of the newly-hatched tadpole.

The tadpole, which does not attain its full size till after birth, lies in the oviduct of the female with its tail curled up, but is capable of moving and even turning round. At first the gills are of unusual length, being nearly half as long as the whole body, but before birth they shrivel up and are represented by mere knobs, so that the whole of the tadpole stage is passed through within the maternal body. Tadpoles that have been taken from the oviduct before completing their

development will, however, live in water like those of the other species, thus proving that the species originally went through a temporary aquatic existence. Although the two young salamanders are generally born at the same time, occasionally one develops more rapidly than the other, so that there may be an interval of several days between the births of the two. At the pairing seasons these salamanders enter the water for a few hours, but are otherwise purely terrestrial.

CAUCASIAN SALAMANDER

The third representative of the genus is the Caucasian salamander (*S. caucasica*), distinguished from both the others by the tail being longer than the head and body. In colour this species is black, with irregular rows of round yellow spots down the back.

SPANISH SALAMANDER

The Spanish salamander (*Chioglossa lusitanica*) is the sole representative of a genus distinguished from the last by the tongue being supported on a central protrusile pedicle, and consequently free everywhere except on the front half of the middle line. Considerably smaller than the spotted salamander, this species is dark brown in colour, rather lighter above than below, with two broad reddish golden bands along the body, separated from one another by a dark line along the middle of the back. It inhabits the north-western districts of Spain and the whole of Portugal.

NEWTS, OR EFTS

The newts, all of which are thoroughly aquatic in their habits, form an extensive group, spread over Europe, Northern Asia, and North America, and are the only members of the order found within the limits of the British Isles. Having the same number of

MALE CRESTED NEWTS

toes as the salamanders, they are distinguished by the highly compressed and rudder-like tail, as well as by the frequent presence of a fin-like crest down the middle of the back, which often attains a special development in the males during the breeding season. With the exception of the crested newt, the skull differs from that of the salamanders by the presence of a ligamentous or bony arch connecting the frontal with the squamosal bone, and the palatine teeth are arranged in two nearly straight or slightly curved series. The tongue is free along the sides, but may be either attached or more or less free behind.

The genus is divided into two main groups, according to the presence or absence of a crest down the middle of the back of the males ; and each of these may be further subdivided and have the characters of the so-called fronto-squamosal arch of the skull.

Belonging to the group in which the males are provided with a dorsal crest, the crested species (*Molge cristata*) differs from all the others in the absence of a fronto-squamosal arch to the skull, and is further characterised by the serration of the crest, and the orange and black-spotted colouring of the under parts. The total length varies from 5 to 5¾ inches, and the toes of both limbs are free. The colour of the upper parts is brown, blackish, or olive, with more or less distinct black spots, the sides are white-spotted, and the under parts orange with black spots or marblings.

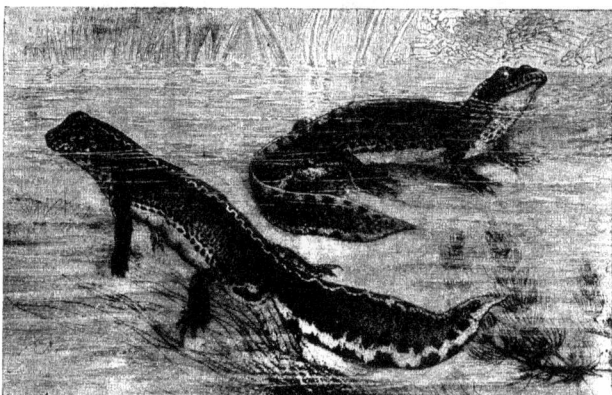

ALPINE NEWTS

During the breeding season the head of the male is marbled with black and white, and there is a silvery band along the sides of the tail, and in the female the under surface of the tail is uniformly orange. The toes are yellow with black rings.

An inhabitant of Great Britain, but in common with other newts unknown in Ireland, this species is spread over the greater part of Europe, extending as far north as Sweden, but unknown in Italy, and ranging eastwards to Greece, Turkey, and Russia. Not improbably the so-called Blasius's newt (*M. blasii*), from North-Western France, is a hybrid between the present and the next species, having the form and colouring of the former, but the fronto-squamosal arch of the latter.

Of the other European species one of the handsomest is the marbled newt (*M. marmorata*) from France. Spain, and Portugal, which has a ligamentous fronto-squamosal arch to the skull, and is specially distinguished by the smooth dorsal crest of the male and by the under parts being generally dark with white dots. The total length is about 5¼ inches. In general colour the upper parts are green with black marblings ; the crest of the male being ornamented with black and white vertical bars, while the female displays an orange streak running down the middle of the back. The sides of the tail show a silvery white band, most distinctly marked in the male during the breeding season ; the under parts are brown or greyish, with more or less distinct darker spots, and dotted with white ; and the green toes are marked with black rings. Rare in France, this species is common in Spain and Portugal, where it lives in ponds and streams only in the early spring, spending the remainder of the year on dry land.

WEBBED NEWT   Photo, W. B. Johnson

The next species for notice is the Alpine newt (*M. alpestris*), which differs from the last by the much lower dorsal crest of the males, and likewise by the uniformly orange colour of the under parts. In size it is a comparatively small species, varying from 3½ to 4 inches in length.

In colour the upper parts, which may be either uniform or with darker marblings, vary from brown or greyish to purplish ; the sides have a series of small black spots on a whitish ground, beneath which in the male during the breeding season runs a sky-blue band ; the crest on the back and tail is white with round black spots ; the throat is frequently dotted with black ; the under parts are uniform orange or red, and the lower edge of the tail of the female is orange spotted with black. The Alpine newt inhabits France, Belgium, Holland, Germany, Switzerland, Austria, and the north of Italy.

A fourth European representative of the genus is the small common newt (*M. vulgaris*), which belongs to the same group as the preceding, and is specially distinguished by the festooning of the dorsal crest, the lobate hind toes of the male, and the black-spotted under parts. Abundant in almost every English pond and ditch where the water is sufficiently clear, this species ranges all over Europe, with the exception of the south of France, Spain, and Portugal, and is likewise widely distributed in temperate Asia. It measures about 3¼ inches in length, and has a nearly smooth skin (pages 1656-7).

The upper parts are brown or olive in colour, with darker spots, larger and more rounded in the male than in the female ; the head is marked with five longitudinal dark streaks ; the under parts are yellowish, with a median orange or reddish zone, and marked with black spots in the male, and dots of the same in the female. In the latter the lower edge of the tail is uniformly orange, whereas in the male it is red, bordered with blue, and interrupted by vertical black bars.

The last of the European species that can be noticed at length is the webbed newt (*M. palmata*), distinguished from all the preceding members of the genus by the bony fronto-squamosal arch to the skull, and likewise by the webbed hind toes of the male. This is the smallest species yet noticed, its length not exceeding three inches. The colour of the upper parts is brown or olive, with small dark spots on the body and longitudinal streaks on the head. In the male there are also minute brown speckles on the head ; and the dorsal crest, as well as the upper part of the caudal crest, together with the hind feet, are blackish. Except for a median orange zone, the under surface is uncoloured, although there may be a few small blackish dots ; there is a series of spots along the upper and lower borders of the tail, and the crest on its lower surface is orange in the female and bluish grey in the male. The webbed newt has been recorded from Great Britain, France, Belgium, Holland, Switzerland, Western Germany, and the north of Spain.

With the exception of the banded newt (*M. vittata*) of Asia Minor and Syria, distinguished by the presence of a black band along each side of the body, all the other members of the genus are devoid of a crest along the back in the male. One of the most remarkable of these is Waltli's newt (*M. waltlii*), from Spain, Portugal, and Tangier, distinguished by the elongation of the ribs, which in some instances actually perforate the skin, so as to form a row of sharp points on each side of the body. In a fossil state the genus has been recorded from the lower Miocene paper-coal deposits near Bonn.

Since the general habits of all newts are very similar, one account will serve for the entire group; but it must be remembered that whereas the whole of them are aquatic during the breeding season, at the close of that period some species leave the water and live for the rest of the summer on land, and nearly all seem to pass some portion of the year out of the water. Newts generally prefer clear and running water, with plenty of aquatic plants on which to deposit their eggs.

On land they are somewhat awkward and slow, but in water they swim with great rapidity by the aid of their oar-like tails, their hind legs being pressed close to the sides of the body, so that their mode of progression is exactly the opposite to that of a frog. They often stand upright in the water when coming to the surface to breathe, after which they sink to the bottom with a snake-like movement in search of prey. When on land they seek shelter beneath stones and roots, or in holes in the ground, and in such situations often undergo their winter sleep, although such as live in deep water pass the cold season of the year in a kind of torpor at the bottom. All newts are carnivorous or insectivorous, and the crested newt feeds largely on the tadpoles of the common frog, while the larger species will prey on the smaller members of their own genus.

Although there is considerable difference in the spawning time of the various species, the eggs are generally deposited during May or June, the female laying each egg singly on the edge of the leaf of some water-plant, which is folded together by her hind feet, and thus held by the viscosity of the egg. In the course of a few days after its deposition, the white embryo assumes an elongated form within the egg, and soon after is seen to be folded upon itself, with the gills well developed, and in advance of them a pair of lobes by which the liberated tadpole affixes itself to aquatic plants. When about a quarter of an inch

in length, and while the gills are still simple, the tadpole bursts its envelope; the front limbs at birth being represented merely by a pair of small knobs behind the gills. When hatched, it swims about in an aimless kind of way till it strikes against some object to which it can easily attach itself, and after a short time starts on another voyage.

Development now proceeds apace, and in the course of two or three weeks the tadpole will have attained a length of about half an inch, the gills will have become elegantly branched, and the fore limbs well-developed. At this period the eyes assume their permanent character, and the mouth has become terminal, while the lobes for attachment to plants are well-nigh absorbed. Still later the front feet, which had previously been only digitated, acquire four distinct toes, the hind limbs make their appearance and gradually assume their full proportions, and the gills have become still more complex. From this date the latter appendages gradually diminish in size, and shrivel, while the lungs are at the same time developed, until finally, about the latter part of the autumn, the newt has completed its metamorphosis, and passed from the condition of a fish to that of a reptile. Although in most cases newts shed their skin piecemeal, in the crested newt it has been observed to be cast entire.

SPECTACLED SALAMANDERS

SPECTACLED SALAMANDER

The presence of only four toes to each foot, and of a bony fronto-squamosal arch to the skull, are the most distinctive features of the little spectacled salamander (*Salamandrina perspicillata*) of Italy, the sole representative of the genus to which it belongs. It is, however, further distinguished by its slender form, and also by its somewhat compressed and rapidly tapering tail, furnished both above and below with a longitudinal keel, as well as by the palatine teeth being arranged in two parallel series diverging posteriorly. The tongue is very similar to that of the genus *Chioglossa*.

Reaching from rather more than 3 to nearly 4 inches in length, this pretty little salamander has a warty skin, and is generally black on the upper parts, although there is a triangular or chevron-shaped yellow mark on the top of the head. On the under side the chin is white, the throat black, and the rest of this aspect white, usually marked with black spots; the lower surface of the tail and adjacent part of the body is, however, bright carmine. The tarantolina, as this salamander is termed in Italy and Sardinia, inhabits cool, shady spots on the flanks of the mountains, where it feeds chiefly upon ants and spiders, and is active at all seasons of the year, having been seen abroad even in January.

WALTLI'S NEWT

Although it appears that the pairing takes place on land, the females resort to the water in March to deposit their eggs, those that are the first to arrive taking the best places, such as sheltered corners of rock, where the spawn will be less likely to be washed away by floods. The young are hatched in about three weeks, and generally leave the water in June. In its movements on land this salamander is as active as a lizard.

OTHER GENERA

There are two other existing genera of the subfamily under consideration, both differing from the preceding groups in the circumstance that the maxilla, or upper jawbone, is more or less fully in contact with the pterygoid bone. Both have a fronto-squamosal arch, but whereas in *Tylotriton* this is bony throughout, it is ligamentous posteriorly in *Pachytriton*, which has also the tail cylindrical at the base, instead of compressed throughout. The former genus, in which the skin is extremely warty, is represented by one species from Siam and the Eastern Himalaya and a second from the Liu-Kiu Islands, while the latter is known only by a single Chinese form.

AXOLOTLS

Although, properly speaking, the term axolotl applies only to the permanent larval form of the Mexican representative of the genus *Amblystoma*, it is convenient in practice to make it include all the members of that group, whether mature or immature. Together with certain other genera, *Amblystoma* constitutes a second subfamily, *Amblystomatinæ*, distinguished from the *Salamandrinæ* by the teeth on the palate forming a transverse or posteriorly converging series, and being inserted on the hind portion of the bones known as the vomers, as well as by the bodies of the vertebræ being cupped at each end.

The typical genus is specially characterised by the palatal teeth forming a nearly straight or angulated series, not separated by a space in the middle line ; and likewise by the radiating folds of skin on the tongue, which are oval or nearly circular in form, with the sides completely and the front partially free. There are five hind toes, and the tail is more or less compressed. Represented by a number of North American species,

one of which ranges as far south as Mexico, the genus has also one Asiatic member, inhabiting the mountains of Siam, probably at a great elevation, and also Burma.

The majority of axolotls pass from the tadpole to the salamander stage in the ordinary way, but this is not the case with the typical race of the Mexican axolotl (*A. tigrinum*), a species which likewise extends over a large area in the United States. In the adult form, which is shown in the upper illustration, the head is large and depressed, and has a broad and blunt muzzle, the limbs being stout, with short toes, and the rather long tail distinctly compressed, and keeled above and below near the extremity. The shining skin is finely granulated, and the general colour brown or blackish, with more or less numerous yellow spots, which may be arranged in transverse bands.

ADULT OF MEXICAN AXOLOTL

In the United States the transformation from the larva to the adult apparently goes on in the ordinary manner ; but the case is very different in Mexico, where the city bearing that name is surrounded by an extensive lake, while the country itself is characterised by its extreme dryness. In this lake dwell the creatures represented in the lower illustration, which are known to the natives by the name of axolotl. It will be seen from this picture that they resemble the tadpole stage of ordinary salamanders and newts in having large branching gills and a deep rudder-like tail ; and the natural conclusion would be that they are larval forms. However, in the Mexican lakes, the axolotls remain permanently in the water, retaining their gills throughout life, and laying eggs, as if they were adult ; and it was consequently long considered that they belonged to a type with persistent gills.

It was not until the year 1865 that light was thrown on the history of these creatures by six examples which had been living for more than a year at Paris. These comprised five males and one female, and in February the latter began to lay eggs, which in a month

LARVAL STAGE OF MEXICAN AXOLOTL

hatched into tadpoles like their parents. In September the gills and crest of the tail of one of these began to shrivel, while the head increased in size, and yellow spots appeared upon the dark skin. Towards the early part of October similar changes took place in the others, and soon afterwards the four assumed the appearance of the terrestrial salamander described as *Amblystoma tigrinum*.

P J Smit

P.J. Smit

Subsequently experiments were made with other young axolotls by placing them in a glass vessel filled with water, but with rocks at one end, so that the creatures could creep out and expose themselves to the air as much as they pleased. After a day's interval the amount of water in the vessel was diminished; and almost immediately the gills of the axolotls began to shrink, and in the course of time, during which they dwelt chiefly in damp moss, they gradually developed into air-breathing salamanders. It has been inferred from these remarkable experiments that the Mexican axolotl, like the other members of the genus to which it belongs, originally went through the normal series of transformations; but that, owing to the dry nature of the country it inhabits, it has acquired the habit of retaining the larval condition permanently. From its being able to breed in this state, it may further be inferred that the tadpole stage was originally the permanent condition of all members of the order, and that the salamander stage is a later development.

The belief that the Mexican axolotl remains usually in the larval state throughout life because it has no opportunity for a terrestrial existence and breathing air, and that, conversely, when axolotls develop at an unusually early period of their existence into adult tiger-salamanders it is owing to their being deprived of an aquatic habit, has been challenged by Mr. J. H. Powers, who believes the acceleration or retardation of development to be due rather to differences of nutrition. After mentioning that, in his opinion, differences of temperature are not the all-important factor in the change, he proceeds to observe that all the numerous instances of accelerated development in axolotls that have come under his notice have been the result of partial starvation. For instance, among a number of axolotls kept in a tank of cold spring-water, in which there was no visible food-supply, several developed into salamanders, three under the size at which the metamorphosis usually takes place.

Prolonged starvation is, however, by no means necessary to stimulate the change, and when full-fed axolotls of a certain age are kept without food for a few days, some of them are almost certain to undergo the metamorphosis. Liability to change, under favourable circumstances, seems also to depend to a certain extent on the temperament of the individual, restless and excitable axolotls being prone to undergo metamorphosis, while their more phlegmatic brethren remain *in statu quo*. If these observations and interpretations be correct, it follows that the converse should be true, that is to say, a constant food-supply should tend to check the metamorphosis.

In axolotls from Colorado and Dakota two kinds of metamorphosis occur during the passage from an aquatic to a terrestrial existence. First of all is the development of the limbs and lungs, the alteration of the circulatory system, and the maturation of the reproductive organs. But there are also secondary changes, which may occur either early or late in life. In some districts axolotls pass into the adult amblystoma state when quite small; but in Mexico, as stated above, these secondary changes never take place at all.

ALLIED GENERA

There are six other genera belonging to the subfamily *Amblystomatinæ*, of which *Hynobius* is represented by several Japanese species. *Salamandrella*, distinguished by having only four hind toes, is a Siberian type, with two species; *Onychodactylus*, which may be recognised by its black claws, is known by one species from Japan; while *Ranidens*, from Eastern Siberia and North-Eastern China, *Batrachyperus* from North-Western China, and the Californian *Dicamptodon*, all of which have the palatal teeth arranged in two arches, with their convexity forwards, and separated by a wide space in the middle, are likewise respectively represented by a single species.

CAVE-SALAMANDERS

Five genera represent a third subfamily, the *Plethodontinæ*, mainly American in distribution, and characterised by the series of palatal teeth being transverse and situated on the hind part of the vomers, and also by the presence of teeth on the parasphenoid bone, as well as by the bodies of the vertebræ being cupped at both ends. Of the five genera, *Plethodon*, with the tongue attached along the middle line to the anterior margin, and five hind toes, is mainly confined to North America, where it is represented by several species, but has one outlying member (*P. platensis*) in Argentina. *P. shermani* of Northern Carolina is characterised by its brick-red legs and lead-coloured body.

On the other hand, the large genus *Spelerpes*, which has the tongue attached only by a central pedicle, and all its edges free, ranges into Central America and the West Indies, with an outlying species (*S. fuscus*) on the mountains round the Gulf of Genoa and Sardinia. The fact that some, if not all, salamanders of this genus are devoid, when adult, of lungs and gills has been mentioned in the introduction to the class.

The typical American cave-salamander (*Spelerpes maculicaudus*) appears to be confined to the Mississippi valley, where, although commonly found in caves, it may occasionally be met with in woods. When in caves, it is generally to be found at no great distance from the entrance, usually but little beyond the twilight-zone. For breeding purposes, however, these salamanders penetrate deeper into the recesses of the caves, where the larvæ are produced, such fullgrown larvæ as are met with in the open country having probably been washed out by freshets. Within caves the adult salamanders are usually to be found in crevices or upon rock-shelves.

From an artesian well 188 feet deep, bored at San Marcos, in Texas, were thrown up, in the year 1894, about a dozen specimens of a strange salamander previously unknown to science. The largest specimen was only 4 inches in length, and all were elongated, semi-transparent creatures, with external gills, long, slender limbs, and functionless eyes embedded in the skin. Two years later the species was described as the representative of a new genus under the name of *Typhlomolge rathbuni*, and from a superficial resemblance to the olm of Carniola, it was long considered as a member of the family *Proteidæ*. Later investigation of the anatomy of the Texas species has, however, convinced Miss E. Emerson that such a reference is incorrect, and that the creature is really the persistent larval form of a salamander allied to *Spelerpes*, and that its superficial resemblance to the olm is merely the result of the similarity of its habitat. Whether in other surroundings it would, like the axolotl, undergo a metamorphosis into a gill-less adult form can be only a matter of conjecture. Dr. Kingsbury has suggested that the olm and *Necturus* are likewise persistent larval forms, and related to *Spelerpes*. Without denying the possibility of this, Miss Emerson points out that they can scarcely be regarded as allied to the latter genus, since they possess lungs, which are wanting in the group containing *Spelerpes*.

A fourth subfamily, the *Desmognathinæ*, which differs from the last by the bodies of the vertebræ being cupped behind and convex in front, is represented by *Desmognathus* from North America generally, *Thorius* with one Mexican species, and *Typhlotriton* with another from a cave in Missouri which is stone-blind.

FISH-LIKE SALAMANDERS

THE members of the family *Amphiumidæ*, which, for want of a better name, may be collectively designated by a translation of their German title, *fisch-molche*, differ from the *Salamandridæ* in the absence of eyelids. The bodies of their vertebræ are always cupped

THE HELL-BENDER OR MISSISSIPPI SALAMANDER

at both ends. They are all characterised by the weakness of the limbs in comparison to the body, and the wide separation of the front from the hind pair. They live chiefly or entirely in the water, and breathe by means both of lungs and internal gills in the adult state. Only two genera are here recognised, although many naturalists distinguish three.

GIANT SALAMANDERS

The earliest known record of this family is afforded by a skeleton from the upper Miocene of Oëningen, in Baden, described by Dr. Scheuchzer in the year 1726, under the name of *Homo diluvii-testis ;* the learned doctor believing that he had to do with a human skeleton, which, like all fossils at that time, was considered to have been buried by the Noachian deluge. This fossil species, which was fully as large as the existing giant salamander, together with a smaller extinct species from lower Miocene strata near Bonn, probably belongs to the same genus as the living species.

The living giant salamander (*Cryptobranchus maximus*) was first discovered in 1820 by Dr. Siebold in the rivers of Japan. It is characterised by having four front and five hind toes, the absence of a gill-opening, and the presence of two internal gill-arches. The tongue covers the whole of the floor of the mouth, to which it is completely adherent ; while the palate has a curved series of teeth on the vomers, parallel to those on the margin of the upper jaw. In form the giant salamander is very stoutly built ; the head being relatively large, wide, and flattened, with the muzzle regularly rounded, the small nostrils situated near the extremity, and the eyes minute. The body is likewise broad and depressed ; the legs and toes are short, the outer ones,

as well as the outer side of the hind leg, having a membranous fringe, and the short tail is strongly compressed, with a fin above and below, and its tip rounded. The skin, which forms a thick fold along each side of the body, is very warty, especially on the head, and the general colour is brown with black spots, becoming lighter on the upper parts. The ordinary length of this salamander is about 35 inches, but it is stated at times to grow to as much as 44 inches.

Although the original specimens were purchased by Dr. Siebold in the market of Nippon, the giant salamander is now ascertained to inhabit not only the mountain streams of that island, but likewise those of several parts of the Japanese mainland, as well as of Western Central China. Nowhere very abundant, this monstrous eft generally frequents the upper courses of small mountain-streams at elevations of from seven hundred to five thousand feet above the sea-level ; some of these streams being not more than a foot in width, and completely covered over with grasses and other herbage. The water is clear, and while the full-grown salamanders usually curl themselves round masses of rock in the bed of the stream, the younger ones live in holes.

Except in search of food, which consists of worms, crustaceans, fishes, and frogs, these amphibians do not leave their hiding-places, and then only at night, and they never venture on land. In confinement they are extremely slow and sluggish in their movements, only exhibiting any marked activity when they rise to snap at a worm or other tempting morsel. In spite of her large size, the female lays very minute eggs, which are generally deposited in August and September. The first two living examples brought to Europe were obtained in 1829 by Dr. Siebold, and were fed on fresh-water fish brought from Japan, but when these began to fail, the male devoured his unfortunate partner. When suitable food was procured, the male, however, flourished and increased rapidly in size, surviving till the year 1881, when it died in Amsterdam.

THREE-TOED SALAMANDER

Although many other examples have been subsequently kept in zoological gardens and aquariums in Europe, some for a great length of time, till 1902 none had been observed to breed, and the little that was known concerning this function was due to the investigations made in Japan. In that year, however, the director of the Amsterdam Zoological Gardens, Dr.

Kerbert, was so fortunate as to witness the oviposition and nursing habits in specimens which had been kept there since 1893. No pairing was observed to take place. The eggs were deposited on September 18 and 19, 1902, these forming two rosary-like strings, much as in the midwife-toad, and numbering upwards of 500.

The male witnessed the lengthy operation of spawning in great excitement, fiercely chasing away such intruders as the small fish sharing the aquarium. He afterwards took charge of the eggs, and for the ten weeks which elapsed until the release of the last larvæ he kept close to them, at times crawling among the coiled mass of strings or lifting them up, evidently for the purpose of aëration. The larva on leaving the egg is about an inch long, and is provided with three pairs of large fringed external gills, like those of newt-larvæ or axolotls. The fore limbs are two-toed at their extremities in the newly-hatched tadpole. When the young salamander has attained the length of 6 inches it has acquired all the characteristics of the adult.

THE HELL-BENDER

The ill-sounding name of hell-bender is applied in its native country to the Mississippi salamander (*Cryptobranchus lateralis*), which differs from its Asiatic cousin by the presence of a gill-opening, at least on the left side of the neck, and likewise by the presence of four pairs of gill-arches, and by the anterior border of the tongue being free. In general form this salamander closely resembles its larger relative, the skin being porous and rather smooth, and the head covered with scattered wart-like tubercles. The colour is brown or greyish, with darker blotches; but the tips of the toes are yellowish. In length this species reaches about 16 or 17 inches, and it inhabits all the tributaries of the Mississippi, and ranges into North Carolina. In these streams it crawls or swims in a sluggish manner, seldom leaving the water, although it can exist on land for twenty-four hours or so at a stretch, feeding on crustaceans, worms, and fishes, and being not infrequently taken on the angler's hook. From the circumstance that the tadpoles have not been observed, it would seem that the larval stage must be of very short duration, and the only thing known about the development of the species is that the eggs are of relatively large size.

Although perfectly innocuous, the hell-bender is regarded by American fishermen as a most noxious and poisonous reptile. It was first brought alive to Europe in 1869, since which date it has been frequently exhibited; and if fed on meat or the heads of fishes will rapidly increase in size, although it appears to voluntarily undergo long fasts. While in the water it has been observed to make the air from its lungs pass over the gills, with the apparent object of more fully oxygenating the blood in the latter.

THREE-TOED SALAMANDER

The eel-like, or three-toed, salamander (*Amphiuma means*) alone represents a North American genus ranging from the Mississippi to South Carolina, and distinguished by its extremely elongated and eel-like form, and the small size of the limbs, each of which terminates in three or two minute toes. The tongue is indistinctly defined, covering the whole of the floor of the mouth, to which it is everywhere adherent; there is a gill-aperture on each side of the neck, and four internal gill-arches are present. The head is relatively small,

GIANT SALAMANDER

with a rather long and narrowing muzzle, at the extremity of which are the small and widely-separated nostrils; the eyes are likewise minute, the lips are unusually thick and fleshy, and the short compressed tail is keeled superiorly. The smooth and slimy skin is of a uniform blackish brown colour, although rather lighter below than above. In total length full-grown examples measure about 31 inches. From the difference in the number of the toes it has been thought that there are two species, but since the two-toed and three-toed forms are in other respects similar, it seems preferable to regard them as varieties, or possibly local races, of a single species.

These salamanders are inhabitants of muddy waters, frequently burying themselves in the mud at the bottom, in one instance to the depth of a yard or more in thick, clayey mud of the consistence of putty, in which they burrowed like worms. They also frequent the irrigation-channels in rice-fields, and occasionally venture on land. Their food comprises fresh-water mussels, fishes, beetles and other insects, and crustaceans. Beyond the fact that the female lays eggs, in which the tadpole lies coiled up until it attains several times the length of its chamber, little appears to be known with regard to the breeding habits of this species.

OLM TRIBE

REPRESENTED only by the curious olm of the subterranean waters of Carniola and other parts of Europe, and two allied but generically distinct species in North America, the members of the family *Proteidæ* are characterised by the permanent retention of external gills, on which account they may be regarded as some of the lowest representatives of the order. In addition to this primary feature, they are characterised by the absence of the upper jawbone, or maxilla, although the premaxilla is present, and, like the lower jaw, furnished. with teeth. There are no eyelids, and the bodies of the vertebræ are cupped at both extremities. There are differences in the external form of the two genera; the olm being a long, snake-like creature with small limbs, whereas the two American species resemble salamanders in general form.

THE OLM

Known for more than a couple of centuries, the remarkable creature to which Dr. Oken gave the name of olm is the sole representative of its genus, and is technically known as *Proteus anguineus*. From its American relatives it is distinguished by its snake-like body and small and widely separated limbs, of which the front ones are provided with three toes, while the hind pair carries only two toes. The eyes are concealed beneath the skin, the small tongue is free in front, and the palatal teeth are small and arranged in a double series. In the typical race from Carniola the head is elongate, with a long and narrow muzzle, truncated at the tip, and the mouth is small, with large lips. The short and much-compressed tail is provided with a fin, and rounded or bluntly-pointed at the tip.

The smooth skin is marked by twenty-six or twenty-seven grooves, corresponding to the ribs, and is uniformly flesh-coloured, with coral-red gills. In a variety from Dalmatia the snout is longer and narrower, and the number of costal grooves only twenty-four; while in a third variety, inhabiting Carinthia, the whole form is stouter, the head shorter, with a rounded muzzle, and the number of costal grooves twenty-five. There is also a certain variation as regards colour, apparently largely depending upon the amount of light to which the creatures have been exposed, some examples being reddish brown, and others darker with bluish black spots. The usual length is about 10 inches.

Totally blind, the olm is found solely in the subterranean waters of the caverns of the Alps of Carniola, Dalmatia, and Carinthia, and has long been an object of the greatest interest to naturalists. It has been thought that the waters in which the olm lives were all connected together underground, and that the salamanders only came up during flood-time, but the great distance from one another of the various localities where they are found is somewhat against this view.

It is, however, only when the subterranean waters are at their greatest height that the olms are captured by the peasants, by whom they are placed in glass jars half-filled with water, and sold to tourists. In confinement, where they have been known to survive from six to eight years, they lie sluggishly all day at the bottom of their tank, only moving if a ray of light impels them to seek a darker corner. When in small vessels where the water is not often renewed they frequently come to the surface to breathe, opening their mouths and letting air pass through their gill-openings; but in deeper or in frequently-changed water they breathe entirely by means of their gills.

Many experiments have been made, with the view of ascertaining whether the olm will, under any circumstances, lose its gills, but hitherto without result. In captivity the food of these amphibians consists of molluscs, worms, and the minute creatures to be found among the leaves of water-plants.

In spite of having been kept for many years in captivity, it was not ascertained till 1875 that the olm lays eggs, and it was thirteen years later before any tadpoles were hatched in captivity. In April, 1888, upwards of seventy-six eggs were laid by a single female, which, after a period of three months, developed into tadpoles. These were very similar to the adult, but the tail-fin extended three-quarters down the back, the eye was larger, and apparently more susceptible to light, and the hind limbs were in the form of small knobs.

So long ago as 1831 the olm was reported to produce its young alive, and although the statement was long overlooked, it appears that under certain conditions, at any rate, this is actually the case. In 1907 Dr. Nusbaum recorded the fact that the olm is sometimes viviparous, although he believed this to be abnormal, owing to the specimen in his possession having been kept for thirteen months in a comparatively warm place.

Later on the viviparous nature of the reproduction was verified in several examples kept in Vienna by Dr. Kammerer, who, differing from the former writer, regards the viviparous reproduction as normal in caves with a constant temperature of 11–12° C.; but in captivity, where so low a temperature cannot be maintained throughout the year, the olm has a tendency to get rid of its eggs as soon as fertilised, instead of retaining them in the uterus until the young have reached a more or less advanced state of development. The case observed by Dr. Nusbaum in contradiction with this theory might be explained by the fact that the captive female which gave birth to young was too old to adapt herself to new conditions.

The subterranean Texas salamander (*Typlomolge rathbuni*), which has been regarded as a permanently gilled type allied to the olm, is mentioned above under the heading of the family *Salamandridæ*.

FURROWED SALAMANDERS

A very different looking animal to the olm is the furrowed salamander (*Necturus maculatus*), of Eastern

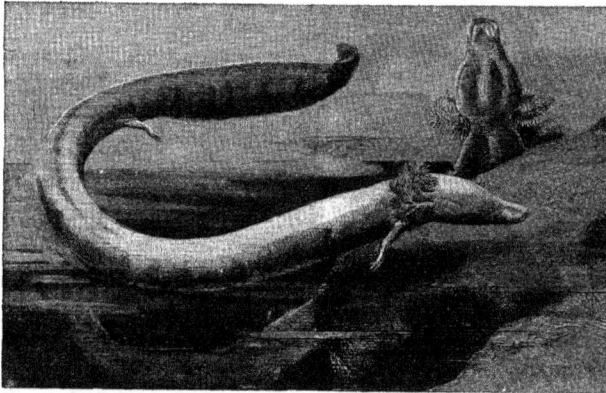

THE OLM

North America and Canada, which takes its name from the strongly-marked fold of skin on the throat. In addition to its shorter and more lizard-like form, and relatively longer limbs, it differs from the olm by having well-developed eyes, and four toes to each foot. The tongue is large, with the front border free; and the palatal teeth are large and form a single series. In colour the smooth skin is brown, with more or less well-defined circular blackish spots, and lighter on the under parts than on the back. The total length is about a foot.

An allied species (*N. punctatus*) inhabits the rice-fields of the Southern States. The food of these salamanders is similar to that of their relatives, and in winter they seek protection from frost by burrowing deep in the mud. They come at times to the surface to breathe, and will even venture on land; but they chiefly respire by means of their gills, and if the latter become entangled, they are carefully rearranged by means of the fore foot. These salamanders, it is important to notice, possess on the sides of the tail a slime-canal, corresponding to the so-called lateral line of fishes.

TWO-LEGGED SALAMANDERS

THE sole representatives of the last family (*Sirenidæ*) of the tailed amphibians are the two-legged salamanders of North America, of which there are two species, arranged under the genera *Siren* and *Pseudobranchus*. While agreeing with the members of the preceding family in the permanent retention of external gills, they are distinguished by the total loss of the hind limbs, and likewise by the absence of teeth in the margins of the jaws.

THE MUD-EEL

The so-called mud-eel (*Siren lacertina*), which inhabits the South-Eastern United States, and may be compared to a snake furnished with a pair of short fore legs and external gills, is specially distinguished by the presence of three pairs of gill-openings on the sides of the neck and the four-toed feet. The smooth skin is either uniformly blackish, or marked with small white dots, and the total length reaches to as much as 28 inches.

GEORGIAN TWO-LEGGED SALAMANDER

The Georgian two-legged salamander (*Pseudobranchus striatus*), on the other hand, has only a single pair of gill-openings on the neck, and but three toes to the feet. These salamanders are stated to frequent swampy localities, especially pools of water beneath the roots of old trees, up the stems of which they will sometimes climb.

A living example of this salamander was received in England in 1825, where it lived till 1831. This specimen was fond of coming out of the water to rest on sand or among moss, and in summer ate worms, tadpoles, and various other small creatures, but became torpid from the middle of October till the end of April. That these salamanders can breathe entirely by means of their lungs is proved by a specimen in an aquarium whose gills had been eaten off by a fish.

THE MUD-EEL
From a specimen in the Natural History Museum, New York

ORDER APODA
CŒCILIANS OR WORM-LIKE AMPHIBIANS

THE remarkable worm-like and blind amphibians forming this group are generally regarded as the representatives of a distinct order, although they are considered by some writers to be merely a degraded branch of the tailed amphibians, to which they are allied through the fish-like salamanders. Be this as it may, the group is readily distinguished by the total absence of limbs, and the general worm-like appearance of the head and body, the tail being either rudimentary or wanting. In the skull the frontal bones are distinct from the parietals, but the palatines are fused with the maxillæ.

As regards their reproduction, these amphibians differ from the newts and salamanders in that the two sexes come together in the ordinary manner. Some of them are peculiar in having overlapping scales embedded in the skin, like fishes ; and in all of them the eyes are either wanting, or are so deeply buried beneath the skin as to be entirely useless. The whole of the members of this curious group are burrowing in their habits, and in the adult state are completely terrestrial, laying eggs from which are developed gilled scales are developed, at least in some portion of the body, being the Asiatic *Ichthyophis* and the South American *Cœcilia*. One of the species of the latter genus is represented in the illustration.

The common Sinhalese species, *Ichthyophis glutinosus*, which ranges from Ceylon and the Eastern Himalaya to Sumatra and Java, inhabits damp situations, and usually burrows in soft mud. In some hollow near the water the female—which measures about 15 inches in length—lays a cluster of very large eggs, arranged in pyramidal form, round which she coils her body and broods them after the manner of a python. When the young are hatched out, they remain in the egg-mass until they have lost their external gills, after which they take to the water, to lead for a time an aquatic life. During this stage of their existence the head is fish-like, with large lips, and the eyes better developed than in the adult. There is a gill-opening on each side of the neck, and the tail is distinctly defined, much compressed, and furnished both above and below with fins. Of the group without scales, the genus *Gegenophis*

A CŒCILIAN OR WORM-LIKE AMPHIBIAN

tadpoles that do not take to the water till some time after birth. The fourteen genera into which the group has been divided may all be included in the single family *Cœciliidæ*. Geographically these amphibians are spread over the Indo-Malay region, Africa south of the Sahara, and Central and South America ; but it is not a little remarkable that they are unknown in Madagascar, although two species occur in the Seychelles. The members of the order mentioned in this work are as follows :

FAMILY
Cœcilians—Cœciliidæ
GENUS 1
Ichthyophis
SPECIES

Sinhalese cœcilian.........................Ichthyophis glutinosus
GENUS 2
Cœcilia
GENUS 3
Gegenophis
GENUS 4
Siphonops
GENUS 5
Typhlonectes
GENUS 6
Chthonerpetum
GENUS 7
Herpele

Cœcilians may be divided into two main groups, according to the presence or absence of scales in the skin ; two of the best-known representatives of the group in which is from Southern India, *Siphonops* from Tropical America, and *Typhlonectes* and *Chthonerpetum* from South America.

Special interest attaches to the geographical distribution of the genus *Herpele*, which is represented in India, Panama, and West Africa ; such a distribution, in the case of a worm-like burrowing group, appearing altogether inexplicable on the theory that continents and ocean-basins are permanent, or, indeed, anything like permanent. On the other hand, the distribution of *Herpele*, together with that of certain sublittoral hermit-crabs, which is curiously similar, affords strong support to the now generally accepted view that India and Africa were connected by land at a comparatively recent epoch of the earth's history.

These instances also add one more link to the chain of zoological evidence which points to a former land-connection between Africa and South America across the Atlantic. The Indo-African connection would explain the presence of cœcilians in the Seychelles as well as the absence of the above-mentioned littoral hermit-crabs from the east coast of Africa. The only alternative view to the Transatlantic connection between West Africa and America (apart from one by way of the Pacific) would be that these salamanders travelled from a common northern home down the Eastern and Western Hemispheres.

ORDER LABYRINTHODONTIA
PRIMEVAL SALAMANDERS

THE remaining amphibians are extinct, and form an order characteristic of the upper Palæozoic and Triassic periods. They derive their name of labyrinthodonts from the complex structure of the teeth of the higher forms, these displaying a peculiar pattern, caused by infoldings of the outer layer, which penetrate nearly to the centre of the crown in festooned lines. Most of these amphibians have the general form of a salamander, with the front limbs shorter than the hind pair, the latter having always five toes, although in the former the number may be reduced to two.

The most characteristic feature of the group is, however, to be found in the structure of the skull, in which the bones are generally covered with a pitted or radiated sculpture, somewhat similar to that of crocodiles. From the figure of the skull of the mastodonsaur, on page 1672, it will be seen that the whole of the upper surface behind the sockets of the eyes is covered by a complete bony roof, extending continuously from the bone marked *P*, which immediately covers the brain-cavity, to the sides of the hinder part of the jaws (*QJ*), whereas in all the modern salamanders this region is more or less open.

This roofed skull of the primeval salamanders, from which they take their alternative scientific title of Stegocephalia, presents an approximation to the earlier

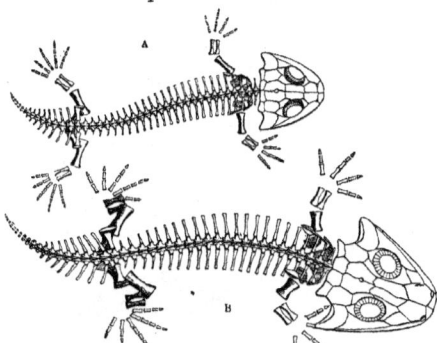

SKELETONS OF PRIMEVAL SALAMANDERS
A, Protriton; B, Pelosaurus

fishes; and a resemblance to that group is also shown by the paired supraoccipital bones (*SOc*), which in all the higher vertebrates are fused together. Nearly all these salamanders are further distinguished by having the chest protected by three sculptured bony plates, one of which is central, while the other two are lateral, the position of these plates being shown in the figures of the skeleton, where they are seen on the lower surface of the body, immediately behind the head, underlying the backbone and ribs. Besides this armour, some species had the whole of the under surface of the body protected by a series of bony scales, arranged in a chevron-pattern, and in a few instances similar scales also invested the upper surface of the body.

The majority of the members of the order had the vertebræ of the backbone in the form of simple doubly-cupped discs, similar to those of fishes; but in some of the more primitive types each vertebra (page 1672) consists of four distinct pieces—namely, a single basal piece (*i*), a pair of lateral pieces (*pl*), and a single arch and spine (*s*). Among certain reptiles the basal piece remains between two adjacent vertebræ as the intercentrum, but in the higher forms the other elements coalesce. Since a similar type of vertebra occurs in certain extinct fishes, this structure presents another bond of union between the latter and the primeval salamanders.

Brief reference must also be made to the small aperture in the roof of the skull of the primeval salamanders in the bone marked *P*, since this corresponds to one in the skull of the tuatera of New Zealand. In that reptile, as stated on page 1500, the aperture overlies the rudiment of an eye sunk deep down in the brain and now totally useless, but probably functional

in the tuatera's ancestors. The large size of the aperture in the primeval salamanders suggests that the central eye may still have been capable of receiving impressions of light, although it may be necessary to go back to earlier forms before it was of any functional importance as an organ of vision. As in many existing amphibians, teeth frequently occur on the bones of the palate as well as in the margins of the jaws.

Another feature of the skulls of many members of the order is the presence of what are called mucous canals in the bones of the upper surface, as is shown in the figure of the skull of *Mastodonsaurus* on page 1672, these canals also occurring in certain fishes. In connection with these head-canals, it may be mentioned that of the five groups into which the labyrinthodont amphibians are divided, a lateral-line system, corresponding to that of fishes, is found in all except the Aistopoda. As a rule, the system presents itself in the form of the mucous channels of grooves constituting the "lyra" on the skulls of the typical labyrinthodonts, the smoothness of the bottom of these canals, which is most developed in the Stereospondyli, being apparently a feature distinctive either of age in the individual or of specialisation in the group.

While these canals differ in some degree from the slime-canals of certain fishes, such as the bow-fin (*Amia*), yet a certain degree of homology between the two types of structure can be traced. For these canals on the labyrinthodont skull definite names have been proposed. In the branchiosaurian group the head-canals are lacking, and their place taken by a true "lateral line" on each side of the tail, similar to that of the modern salamander *Necturus*. An important result of recent investigation is the determination that the bone originally termed the squamosal in the labyrinthodont skull is really that element, and not, as it has been attempted to prove, the supratemporal. So far as can be ascertained, both external and internal gills generally disappeared in the adults.

Labyrinthodonts have been classified as follows:

(1) Branchiosauria, or Lepospondyli, represented by the tiny Permian *Branchiosaurus*, or *Protriton*, *Melanotriton*, and *Micrerpetum*, in which the vertebræ form sheaths to the notochord, the ribs are short, thick, and straight, and there are four front toes, in which the number of joints is 3, 3, 4, 3. In these and most other respects the Branchiosauria resemble newts and salamanders, to which they are perhaps ancestral.

(2) Microsauria, represented by the Carboniferous *Hylonomus* and the Permian *Hyloplesion*, in which the ribs are long, slender, and curved, and there are five front toes. This group probably includes the ancestors of the Rhynchocephalia, and thus of other orders of reptiles.

(3) Aistopoda, characterised by the snake-like form, and represented by several Permian genera.

(4) Stereospondyli, or Labyrinthodontia Vera, in which the vertebræ are fully ossified and the dentine of the teeth is usually thrown into complex foldings, radiating from and including prolongations of the pulp-cavity. Most of the genera, like *Mastodonsaurus* (of which the

type species is as large as a crocodile), *Trematosaurus*, and *Rhytidosteus*, are Triassic, but *Loxomma* is Carboniferous. This group probably gave rise to the anomodont reptiles.

(5) Temnospondyli, or Archegosauria, including *Archegosaurus* of Europe, *Gondwanosaurus* of India, and *Eryops* of North America, in which each vertebra is of the above-mentioned rhachitomous type, consisting of four pieces—two lateral, one superior, and one inferior.

In the opinion of Dr. R. L. Moodie, who has devoted special attention to the study of this exceedingly interesting order, labyrinthodonts should be split up into two distinct groups—namely, the Branchiosauria of the Carboniferous and Permian, on the one hand, and a second group, embracing the Microsauria, Aistopoda, and the typical Labyrinthodontia and Archegosauria on the other. The first group is regarded as representing the ancestral stock of the modern tailed amphibians, whereas the second is closely related to reptiles, and should perhaps be included in that class.

From the Microsauria, in which the ribs are long and curved, the Branchiosauria differ by their short ribs, which articulate with the transverse processes of the vertebræ. They agree, in fact, with the modern Amphibia Caudata in their short, straight ribs, the stout transverse processes arising from the bodies of the vertebræ, practically in the number of the presacral vertebræ, as well as in the structure of the skull and pectoral and pelvic girdles, in the number of the toes (four in front and five behind) and of their component segments, as well as in the structure of the long bones, the shape of the body, and the existence of a lateral-line system. In skull-characters, as well as in the shoulder-girdle, the modern tailed amphibians exhibit marked signs of degeneration, and they may accordingly be regarded as degenerate derivatives of the Branchiosauria. Similarly, the Aistopoda are provisionally regarded by Dr. Moodie as a degenerate branch of the Microsauria.

Labyrinthodonts appear to have been spread over nearly the whole globe. The true labyrinthodonts, typically represented by the gigantic *Mastodonsaurus* and the somewhat smaller *Metoposaurus* of the Trias, were crocodile-like animals, generally with disc-like vertebræ in the adult, the teeth more or less plicated, and the surface of the skull marked with sculpture and mucous canals. In the Permian *Archegosaurus* the vertebræ, as already mentioned, were of the complex primitive type. The gilled labyrinthodonts, or Microsauria, are a group of much smaller forms, characterised by their barrel-shaped vertebræ, pierced by a remnant of the canal of the primitive notochord, short and

straight ribs, articulating by a single head, simple teeth, and the absence of ossification in the occipital region of the skull, as well as in the wrist and ankle joints, a further point of distinction being the development of internal gills in the young.

The Permian and Carboniferous snake-like labyrinthodonts, Aistopoda, are characterised by the snake-like form of the body and the apparent absence of limbs. The vertebræ were elongated and without spines, while the ribs were slender and barbed like those of fishes, and the teeth smooth and simple. Probably external gills persisted throughout life. In Britain the group is represented by the small *Dolichosoma*, but *Palæosiren* of Bohemia is estimated to have been over forty feet long.

The Microsauria include small lizard-like forms, such as *Ceraterpetum* and *Hylonomus* from the Carboniferous of Europe and Nova Scotia, which appear more highly organised than the preceding, and thus connect the amphibians with the tuatera. Their vertebræ are long and constricted, with traces of the notochord; the ribs are generally long, curved, and two-headed; the teeth have large central pulp-cavities, but no plications; the occiput is ossified, but the wrist and ankle are either ossified or cartilaginous, and in some cases the back is covered with bony scales. In several forms the bony scales on the under surface are so slender as to assume the appearance of abdominal ribs like those of the tuatera.

As regards the relationship of the amphibians to the lower vertebrates, Mr. W. E. Kellicott, at the conclusion of a memoir on the development of the vascular and respiratory systems of the Australian lung-fish (*Ceratodus fosteri*), remarks that " it is impossible to believe that the amphibian resemblances seen in *Ceratodus* in the development of the vascular, respiratory, and urinogenital systems, as well as throughout the earlier processes of development, are in the nature of parallelisms. In the light of their embryology it is impossible to believe that the Dipnoi and the Amphibia are not closely related, and that they have not travelled for a time along the same path at some period during their history." If this view is to be accepted, the early lung-fishes must apparently be regarded as the direct ancestors of the primeval salamanders. A further inference would seem to be that the gills of modern salamanders are directly inherited from fishes, and not, as has been suggested by some naturalists, a new and independent development.

It is very remarkable that in certain branchiosaurians (*Eumicrerpetum*) from the Carboniferous of Illinois the greater part of the intestine has been preserved in a petrified state.

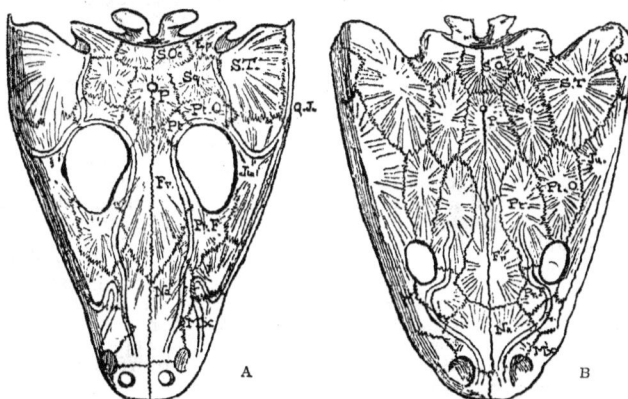

SKULLS OF (A) MASTODONSAURUS AND (B) METOPOSAURUS

SOc, supraoccipital; *Ep*, epiotic; *P*, parietal; *Sq*, squamosal; *ST*, supratemporal; *QJ*, quadratojugal; *Ju*, jugal; *Pt*, postfrontal; *PtO*, postorbital; *Fr*, frontal; *PrF*, prefrontal; *L*, lachrymal; *Na*, nasal; *Mx*, maxilla. The premaxilla has no letter

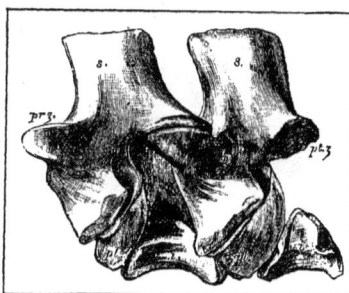

TWO VERTEBRÆ OF A PRIMEVAL SALAMANDER

prz, anterior end; *ptz*, posterior end

HOW LIFE STEPPED ASHORE

BY SIR H. H. JOHNSTON

THAT which we call life, which starts with animated protoplasm and proceeds in time from complex to more complex forms of conjoined and differentiated cells, was most probably born in the water, and of course, with its present potentialities, is absolutely dependent on water for its existence. But whether it came into being in the uttermost depths of the ocean—if, indeed, at that day, there were deep oceans on the earth's surface—or whether it was generated and evolved in shallow water, is not yet determined by our finite knowledge. The generally-accepted theory is that living matter originated in the ooze or in shallow water, thence sending a few specialised forms to inhabit the ocean depths, and thence ever and again crawling up on to the shore, endeavouring to have done with the water, to inhale the life-stimulating atmospheric air directly, and to conquer the land for its own. Indeed, a fanciful person might almost see a progressive movement from the water to the land, to the air, and at last, eventually, the ether of the universe.

Already there are types of men and other mammals, and of birds, which cannot live well or healthily at low altitudes, but flourish most when breathing the more rarefied atmosphere of high mountains, or, at any rate, of elevated tablelands. In the course of our human history the races of the hills have again and again conquered peoples of the plains, and themselves been conquered in turn by fresh vigorous peoples generated on the heights.

Living beings—on this planet—require oxygen and other atmospheric gases to maintain their vital activities. The earliest forms of life, and many of their descendants at the present day, obtain the oxygen they need from the water, but far back in the Primary epoch, many a founder of orders and classes was possessed by an impulse to get out of the water, at any rate for a short time, and to oxygenate its blood directly from the air. In this way sea-worms became land-worms, even earth-worms which drown in water. Sea-gasteropods became land-snails; wormlike, aquatic arthropods developed into air-breathing, flying insects; from water-dwelling, isopod crustaceans were evolved terrestrial wood-lice; and from the marine king-crabs emerged the land-dwelling scorpions, and, in subsequent descent, the spiders and ticks.

In the same way sea-weeds and still more primitive vegetable organisms left the water to grow on the land and to become fungi, mosses, ferns, mare's-tails,

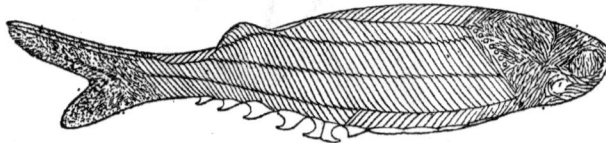

RESTORATION OF A SMALL OSTRACODERM OF THE UPPER SILURIAN AGE

palms, lilies, conifers, and the vast range of the dicotyledonous flowering trees and plants.

And among the early forests of—how far back shall we say?—eighteen, twenty million years ago?—before flowers were invented by Nature, there crawled centipedes and millipedes, worms and leeches, and there flew cockroaches and earwigs, termites and primitive beetles, stick-insects, dragon-flies—some of them two feet long—moths, and bugs.

That was the heyday of the insect class, the Carboniferous ages of the Primary epoch.

So many references are made in this article to the Primary, or, as some people call it, the Palæozoic,

epoch that some concise information as to its distance in time and its (arbitrary) divisions might be helpful to the reader. It was preceded by the Archæan epoch, or youth of the earth, in which, so far as we know, no life—or, at any rate, no complex forms of living matter—existed; unless it were in the vast ages of the Laurentain period, which some separate from the Archæan ages of unstratified rocks, and reckon as the beginning of the vast primary division of life-history.

It has been roughly estimated—by calculating the depth of strata and guessing at the time it would take to form them from sedimentary deposits—that the Primary epoch, including the Laurentian period, may have lasted for some forty-two millions of years; whereas to the Secondary epoch, which succeeded it, are assigned only about nine millions, to the Tertiaries less than three millions, and to the Quaternary, or present age—since the beginning of the Pleistocene, or Cold, period—about four hundred thousand years.

After the vaguely vast Laurentian era (at a guess a duration of 18,000,000 years), the different phases or periods of the Primary epoch are in order of remoteness: (1) the Cambrian, (2) the Silurian, (3) the Devonian, (4) the Carboniferous, and (5) the Permian. The old Red Sandstone is Devonian, the New Red Sandstone Permian and Triassic. As early as the Cambrian there were crustacea, and possibly the lowest types of starfish, worms, and molluscs. In the Silurian age the first vertebrates came into existence, those recorded out of a million vanished types being seemingly allied to lampreys, or of an unclassified heavily-armoured type —with, in some cases, a single pair of front limbs— intermediate between the lamprey and the fish.

In the Devonian period true flying insects seem to have made their appearance, and this age is likewise noteworthy for the definite establishment of true fish (ganoids) and of a land vegetation of ferns and calamites. The insects became prominent as land-animals—and the first creatures to fly—in the Carboniferous age, and the spiders were also in existence then, having been preceded by the scorpions as far back as the Silurian times. It was also in the Carboniferous age that land-vertebrates — amphibians — are first recorded in the rocks. This, most of all, was the time when the four-limbed fishes crept on shore and took on the semblance of a new class; though Huxley thought this great event might have occurred even in the preceding Devonian period.

The earliest reptiles emerged from the amphibian type in the beginning of the last term of the Primary epoch—the Permian age; and before that age was finished the first mammalian forms, and possibly the first birds, had come into being from reptilian ancestors.

But already in the sea, towards the close of the Cambrian period, fresh developments had been taking place. Some other branch of the "worm idea" had developed a continuous chord of nerve-matter extending throughout the length of the ringed body—those annelid rings or segments of which there are still traces even in the human organism, perhaps in the separate vertebræ, the segments of bone which grow up to protect the spinal cord.

The worm type became an ascidian, and from this group emerged creatures like the headless, eyeless, limbless amphioxus, with a notochord—forerunner of the vertebræ extending the whole length of the animal from snout to tail-tip. These represent the class of the Cephalochordata, and from these most primitive vertebrates arose the fishes, represented imperfectly at first by the lampreys. This class or subclass (Cyclostomata) has a head and a cartilaginous skull, a pair of well-developed eyes, but no biting jaws, merely a round, buccal funnel.

From something like the lampreys evolved the first true fish, quite small animals, so far as we may judge by their fossil remains, only one to seven inches long, and obliged, like their crustacean contemporaries, to devise as soon as possible an armour which would shield their toothsome bodies from being devoured by more powerful, better organised invertebrates.

But Nature had once again hit on an excellent idea in this development of backboned animals. The fishes of the Silurian period throve and increased vastly in numbers, size, and elaboration of development. This stage is represented very imperfectly by fossil forms classified as the Ostracodermi, Antiarchi, and Arthrodira, remarkable for the heavy armour of the fore part of the body, the typical fish-like tail, but not clearly provided with *two* pairs of lateral fins or limbs, though in one or more the median anal fin is present. But in one example (*Pterichthys* of the Arthrodira) there were long front flippers very like those of a turtle.

In the late Silurian and early Devonian periods the ganoid fishes are well established, with their scale armour, their two pairs of lateral fins, and accessory fin-fringe along the back and tail. From this well-developed yet generalised group arose by divergence all the other types of true fish, from the Carboniferous period onwards to the present day ; but so far as the progress towards a land life was concerned, carrying with it the germ of Man, the classification of ancient and modern fishes might be thus stated. From this ganoid basal order arose the sharks and rays (Elasmobranchii), with a cartilaginous skeleton, but with several features in their internal anatomy and limb-structure which suggest the foreshadowing of the amphibian. Then came the Dipnoi (or Dipneusti), or " double-breathing " fishes, which developed both lungs as well as gills for oxygenating their blood, and lastly the great group of Teleostomi, or " perfect-mouthed " fishes. These, again, in later ages specialised into two main divisions—those with fringed fins (Crossopterygii),

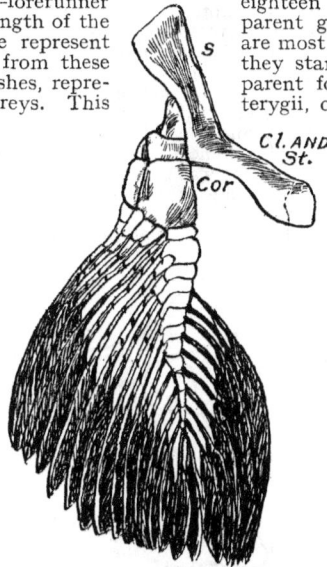

FORE LIMB, OR PECTORAL FIN, OF PLEURACANTHUS

A generalised fish of the lower Permian formations. This fin is on the biserial plan : *s*, rudimentary scapula ; *cor*, beginning of coracoid process ; *cl* and *st*, rudimentary clavicle and sternum

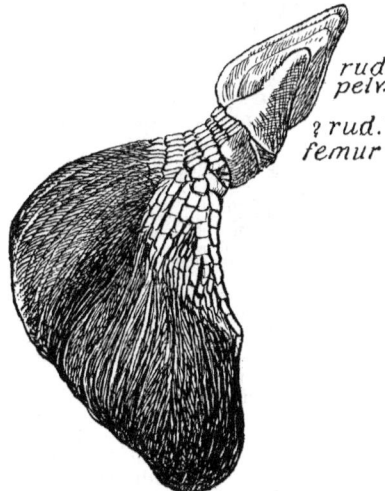

HIND LIMB, OR PELVIC FIN, OF PLEURACANTHUS

This fin is on the uniserial plan, showing rudimentary pelvis, and possible rudimentary femur

and those with fins arranged like fans, Actinopterygii ; the fins, as to the structure of which reference is made, being the two pairs of lateral fins.

The Crossopterygii have remained through some eighteen millions of years very like the original parent ganoid group of the Primary epoch, and are most important creatures to the biologist, as they stand nearest of all living fishes to the parent forms of the amphibians. The Actinopterygii, or fan-finned fishes, are again subdivided into the Chondrostei (sturgeons and their allies), the Holostei (bow-fins, bony pikes and their allies), and finally the order of the Teleostei. This last-named single order includes the great majority of living fishes, which would be described in French as " all that there is of most fish." It is from them that the deep-sea fishes are derived ; also the most typical fishes, and fishes of the most amazing beauty and of the highest intelligence and completest specialisation for marine life. But their interest to the biologist who specially devotes himself to the study of the Tree of Life is merely secondary, for the Teleostei have nothing to do whatever with the evolution of the land-vertebrates.

Palæontological research has as yet not put its finger on the actual fish ancestor of the amphibian, of the first vertebral type which deliberately left the water to take up a land life. The nearest in structure to such a form is perhaps the Pleuracanthus fish of the lower Carboniferous. This might very well have represented the parent genus from whose subsequent variations were derived the parent forms of the sharks, the dipnoi, and the crossopterygians, from which, in turn, would be derived the other orders of the subclass Teleostomi.

Indeed, it is interesting to note that, so far as research has been able to carry us in our survey of the past, the further we go back into fish-forms up to the limit of the late Devonian or early Carboniferous, the nearer we get to suggestions of affinity with the amphibians ; and this line of research is assisted by the continued existence down to the present day of dipnoan and crossopterygian fishes, the internal anatomy of which is probably as little changed from Carboniferous times as is the skeleton.

From the evidence of the rocks, we may imagine that many ganoid fishes of the middle Primary epoch frequented water that was shallow, and in time took to pursuing their insect or crustacean prey out of the water on to the sand or mud. Like many modern fishes, they must have become so dependent on access to the atmosphere that at last they could not live in deep water. Indeed, their direct descendants —the Polypteri—at the present day are fresh-water fishes in tropical rivers.

So soon as these fishes, in pursuit of their insect or crustacean food, began to rely more on progressing over the ground in shallow water or above water, than by movements through the water itself, they would naturally lay more and more stress on the lateral fins on either side, which through this stress would develop into strong limbs with terminations adapted to gripping the ground, and upper bony parts so connected with the internal axis of the fish's body as to exercise a sufficient leverage to drag or lift it over obstacles.

At this stage in our inquiry as to how life stepped ashore, it may be as well to consider the origin and the types of fishes' fins. It has been deduced from evidence collected that the original fish of the lamprey type was limbless, but soon began to develop a continuous fringe of membranous fin along the median ridge of the back, round the extremity of the tail, and along the under side of the tail as far as the vicinity of the anal orifice. When this failed to suffice for the creature's movement through the water, it developed lateral flaps of skin and muscle—gradually strengthened by cartilaginous or bony rays—which acted as an additional swimming, undulating surface, much as we may see the same movements, differently derived, in the modern flat-fish.

By degrees, in many forms, the median fin membrane along the back and round the tail broke up into separate fins, of which the representatives in ancient and modern fishes are the dorsal fins, the characteristic fin at the end of the tail, and the anal fins, one or two in number. These *median* fins must be carefully distinguished from the paired *lateral* fins, which, in their turn, are specialised remnants of the continuous folds growing out on each side of the lower part of the body. The median fins are more ancient than the lateral paired limbs. They are the first to make their appearance in ancient fishes and in the embryos of modern fishes. They have disappeared completely from all land-vertebrates, but, without a bony skeleton, they have reappeared by adaptation in the great sea-reptiles (*Ichthyosaurus*) and the mammalian whales; perhaps also in the armature of certain extinct reptiles. They are of little interest to us in our study of the development of the fish into an amphibian, as they have no homology with the limbs of vertebrates. But it is interesting to note that in this early type of perfect fish (*Pleuracanthus*) the skeleton of the two median anal fins bore an extraordinary resemblance to the bones of amphibian limbs and feet, so that when this type of fish began to perfect its paired fins to crawl over the land, it already possessed the idea of the formation of what is called an appendicular skeleton.

It is to the tracing of the evolution of this limb skeleton from the truncated, muddled arrangement of the typical fish to the well-devised arms and legs of some salamander or reptile that the greatest interest and, at the same time, the greatest difficulties are attached in the process of reconstructing by imagination the change of a fish into a land-vertebrate. It

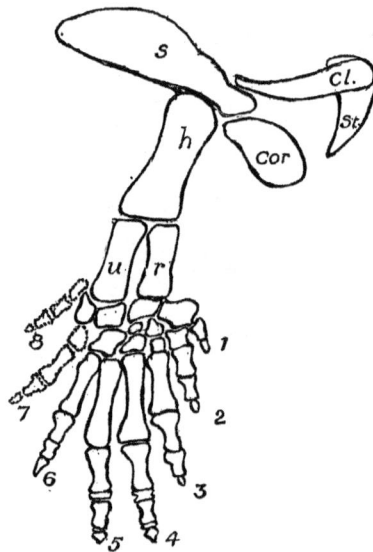

THEORETICAL SKELETON OF FORE LIMB OF SOME EARLY TYPE OF AMPHIBIAN-REPTILE

This creature would not be far removed from the fish-ancestors of the land-quadrupeds. *s*, scapula; *cl*, clavicle; *cor*, coracoid; *h*, humerus; *u*, ulna; *r*, radius. 1, Pre-pollex, or finger before thumb; 2, thumb; 3, fore-finger, or second digit; 4, third digit; 5, fourth digit; 6, fifth digit, or little finger; 7, reappears in six-fingered people, which, with 8, may have existed in the original fish-amphibian

is easy to understand the transition of the air-bladder, beginning originally as a separate small intestine, into air-breathing lungs. We see that process accomplished in the African mudfish (*Protopterus*), and less perfectly in the other existing dipnoi. The tongue, which plays such an important part in the economy of amphibians, reptiles, birds, and mammals, is already sketched out in the sharks, the dipnoi, and the modern ganoids. In existing and in ancient fishes there is sufficient evidence to show how the nostrils, from being mere smelling-holes, gradually communicated by pervious channels with the palate, and later the throat. The formation of the land-vertebrate's skull, biting jaws, and teeth, the rise in his blood temperature, his methods of breeding, even mammalian viviparity and milk-producing, his vocal organs (grunting, growling, clacking, whistling), are all adumbrated in living or in extinct fishes. But it is not at first sight easy to realise how the limbs of the land-vertebrate grew out of the paired fins of the fish.

The plan of the fore limbs in any generalised type of land-vertebrate—such as man himself, in this respect—consists firstly of a pectoral, or " shoulder," girdle, made up of a shoulder-blade, collarbones, or clavicle, and a coracoid process on either side; which, together with the two clavicles, unites the shoulder-blades with some median aggregation of bone, the sternum, or breast-bone. The parallel to the shoulder-girdle in regard to the hinder limbs is the pelvic girdle, which is much more closely attached to the backbone. The pelvic girdle consists in its essence of the twin bones of the ilium—at the lower ends of which each thigh-bone is inserted—and the two ischial bones, which are a prolongation of each ilium and unite in front in the ventral median line to form the pubes.

The shoulder-girdle is connected with the external front limbs by means of an arm-bone (humerus), which articulates with two parallel bones (radius and ulna), and these again, by means of the carpal bones with the metacarpals and the finger-bones. In the case of the legs the pelvic basin—as it becomes by the union of the iliac bones—articulates with the thigh-bone (femur), and this at its lower end with the twin tibia and fibula, and these, again, with the tarsal bones of the foot, which, with the metatarsals and toe-bones, complete the skeleton of the hinder limbs.

Now, in the case of the most perfect fish known to us from the point of view of appendicular skeleton—and this is the *Pleuracanthus* of eighteen to fifteen millions of years ago we find there is a curved shoulder-girdle, which apparently represents the scapula, clavicle, coracoid, and omosternum, fused more or less into a single bone. When allusion is made to "bones" in these early fishes and amphibians, it must be more frequently understood as cartilage more or less calcified. Then follow two or more broad, short sections of bone, which may be the origin of the humerus, or arm-bone. These are succeeded downwards by twin bones, broad and narrow, which may well be the homologues of the radius and ulna.

Then comes a closely compacted mass of small nodules, which may stand for the carpal bones, and these are prolonged into a single long finger, or phalange, consisting of thirteen or fourteen small bones tapering to a point. But from all the supposed carpal bones, much more on the tailward and lower edge, spring the skeletons of a great many fingers, seventeen on this hinder edge of the central phalange, and about eight on the headward and upper edge. The greater number of these fingers, which are tipped with long, flattened membranes, arises more from the bones of the long,

extend the jointed fin-rays. The big phalange is not, as in the fore limb, more or less in the middle of the radiating fin, but situated at its front edge, as though it represented the first in a series of fingers, analogous, in fact, to the disproportionately developed big toe in the human foot. As the very numerous other fin-rays, or fingers, in this pelvic paddle arise, all of them, from the same hind and uppermost edge of the tarsal bones, this type of fin is called uniserial ; whereas the front fins of the same fish (*Pleuracanthus*) are called biserial, because they arise from both sides of the carpal bones.

As the predilection for a shore life increased in some primitive ganoid or dipnoan fish, the appendicular

CLARIAS PLATYCEPHALUS, A CAT-FISH WHICH REQUIRES CONSTANT ACCESS TO THE AIR

central phalange than from what might be called the carpal bones at the lower end of the radius and ulna. Comparing it to man, it would be as though he had eight or nine fingers growing from the hinder edge of the lower arm and hand in continuation backwards from the little finger.

As regards the hind limb of the *Pleuracanthus*, though this has a more elaborate skeleton than in any other known fish, it is not so near as the fore limb is to the plan of the land-vertebrate. The pelvic girdle is represented by a large pear-shaped bone, which, however, lies well *below* the backbone and without any contact with it. Indeed, this is a marked feature

skeleton of the limbs must have finally resembled the rudiments of the amphibian-reptilian land-vertebrate's arms and legs ; and the very numerous terminal rows of radiating fin-bones were gradually reduced to the number of eight or seven, then six, and, finally, five at the extremity of each of the paired limbs.

The paired fins of fishes, especially of ancient types, are, as already stated, classified as uniserial or biserial. Uniserial, if the bones of the fin-rays radiate from the hinder edge of the limb—somewhat like the quills in a bird's wing—with the line of stress carried downwards along the forward-facing edge of the fin ; biserial, if these metacarpal or metatarsal rays grow out from all

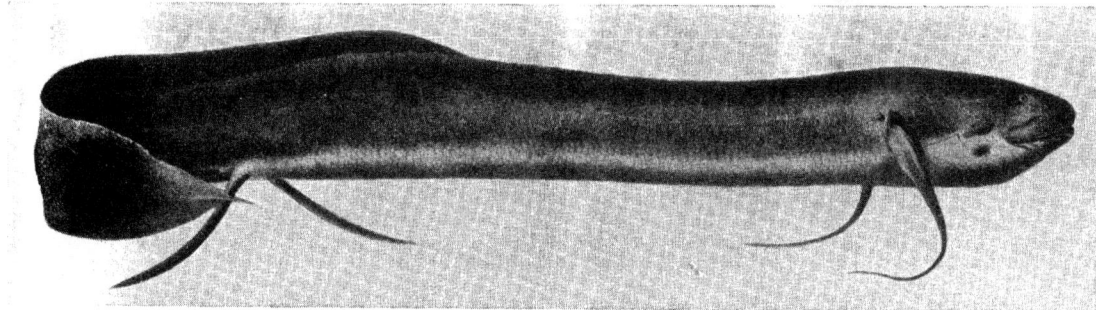

PROTOPTERUS DOLLOI, THE LUNG-FISH OF CENTRAL AFRICA

in all known fishes, and in direct contradiction to the construction of land-vertebrates. The hinder of the two paired fins is rarely, if ever, as prominently developed as the front limbs, and its pelvic girdle, if it exists, never comes in contact with the spine.

On the other hand, in the land-vertebrates, from the earliest onwards, the hinder limb is usually longer than the front limb, and is closely attached to the backbone. In *Pleuracanthus* there is a broad, three-cornered bone which one might take to be the femur, and below this are possibly two square bones which might represent the tibia and fibula. Then comes an interval of a single row of " metatarsal " nodules, from which

round the base of the external limb, with a long, central, jointed finger in the middle. This last " biserial " condition is characteristic of the living lung-fish (*Neoceratodus*) of Australia.

It is probable that the fish ancestors of the amphibians and reptiles conformed to the biserial type, with the stress carried down the middle of the limb's extremity. At first the bones of the other fingers or toes may have radiated from this big central digit, but as the arm and leg bones lengthened, and the finny hands and feet shortened, so there was a tendency for the fingers and toes on either side of the central one to attach themselves rather to the carpal and tarsal bones than

POLYPTERUS ORNATIPINNIS, A MODERN GANOID FISH OF AFRICAN RIVERS

to the sides of their big brother finger. And if this surmise is correct, that is why men—in common with the many other mammals that are equally primitive in their limb-bones—have the middle finger the longest. Had the fish ancestors of the amphibians leaned more to the uniserial type, it is probable that the stress would have lain along the line of the thumbs or big toes, the first digit of the series ; and the proof that they did not lies in the fact that in the majority of land-vertebrates, with a few marked and specialised exceptions, the first and the fifth digits are the shortest and the soonest lost in evolutionary development.

But there are some marked exceptions which might be cited. Occasionally in reptiles and mammals all the five fingers and toes are of equal length. In the flying pterodactyls the fifth, or "little," finger became elongated to an extraordinary degree, until it sustained the whole length of the great flying membrane. In the

human beings. They are six-fingered (or toed) because, although they possess the seventh digit, they have not simultaneously developed a pre-pollex or pre-hallux. Cases—less common—are known in man and other mammals, and still more so in reptiles, where the six-fingered condition is produced by the growth of the real first digit, the one which precedes the thumb or the big toe. In one, at least, of the beast-like reptiles— *Theriodesmus*, of the later Permian primaries—this " pre-pollex " seems to have been functional.

Yet not only were the early land-frequenting amphibians restricted to five fingers and five toes, but the new class of which they became the type was strongly disposed to content itself with only four fingers on the hands, though it developed persistently five-toed feet. In only the earliest types of quadrupedal Amphibia are the five fingers present. Most of their contemporaries are not only reduced to four, but have not even that

POLYPTERUS WEEKSII

feet of the human species the big (first) toe is becoming the longest and strongest of the digits. In the sea-lions, and some of the seals, the first and fifth digits are becoming longer and more important than the others.

I have alluded to the early land-frequenting fishes having modified their fins into *seven*-fingered hands and feet. There are many good reasons for supposing that fishes may have stepped on shore with as large an assortment as this. These additions to the penta-dactyle system took the form of a pre-pollex or a pre-hallux—that is to say, a short finger coming before the thumb or before the big toe—and another digit beyond the fifth of our series.

This outermost seventh finger or toe was very soon dropped in any known form of land-vertebrate, but although it may be eighteen million years since it was in being, the tendency for it to come back is not wholly lost, and it reappears in the hands and feet of six-fingered

vestige of the thumb which is left in the frog order of the existing amphibians, among a few genera of which the remaining one or two nodules of bone, representing the lost first finger, are becoming so useful as to constitute an abortive, resuscitated thumb.

Unless further discoveries are made, revealing the existence of numerous forms of five-fingered amphibians in the lower Carboniferous period of the Primary epoch, we are almost driven to conclude that Nature must have hurried through the amphibian stage in her preparations for the creation of man ; and not only have rapidly evolved at the very beginning of the amphibian experiment a real five-fingered reptile, no longer having any connection with the water in the larval stage of its young, but also that she had scarcely finished turning out a reptile than she began to think of transforming it into a mammal. Once vertebrate life had stepped ashore and adapted itself to progression

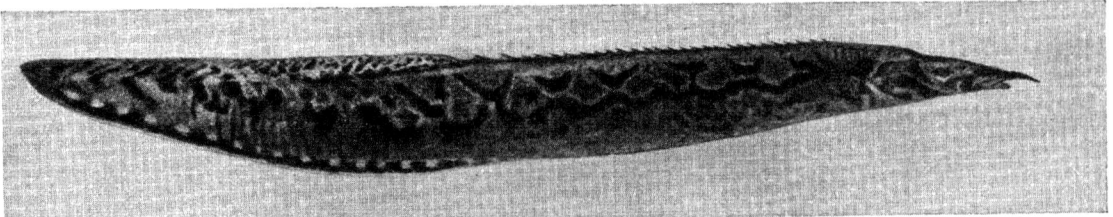

MASTACEMBELUS CONGICUS, A TELEOSTEAN RIVER-FISH WHICH WOULD DROWN IF DEPRIVED OF ACCESS TO THE AIR

on four limbs, evolution upwards towards more complex creatures clad in bony armour, in hair, and in feathers seems to have been rapid.

The normal type of amphibian has not definitely parted with an aquatic existence ; it feels obliged to return to the water to deposit its eggs ; and its young, the tadpoles, lead a fish-like existence in fish-like form, and breathe by means of gills until they have reached a certain age, have developed lungs, and are ready to adopt a land life. But in some modern amphibians, such as tree-frogs, certain toads, and *Cæciliidæ*—burrowing, limbless amphibians—the larval and aquatic stage in the development of the young is considerably abrogated, and the creature is ready to lead the terrestrial existence of its parents in a form scarcely differing from the adult.

So we may imagine that not many thousands of years after the successful establishment of a fish-out-of-water as a predaceous animal [like a huge salamander or newt, which pursued and devoured insects, and crustaceans, and no doubt had terrific struggles with the two-feet-long dragon-flies], the amphibian - reptile began to lay its eggs in moist places, and to expect its tadpoles to be born very like itself when they issued from the egg and made ready to forage for their own living on land.

But there have been other amphibians to whom in course of time the meaning of the Greek term, " double-lived," no longer applies, in that their far-back ancestors have recoiled from the more strenuous life of the land, and have returned altogether to the water. Of such are the giant salamander, the axolotl, proteus, siren, and newts ; and the spur-toed, tongueless frogs (*Xenopus*).

The call back to the waters, owing to competition on land and the attraction of fish and crustacean diet, has also affected the reptile class very largely. There have been the great fish-lizards *Ichthyosaurus*, *Plesiosaurus*, *Mososaurus*, and extinct crocodilians, so wholly aquatic in habit that they were probably no more able to exist on land than is a whale or a porpoise. Then, again, from out of the creodont Mammalia have evolved these very whales and porpoises, and the primitive ungulate order of the Sirenia : neither of which can ever quit the water completely, but must pass an absolutely fish-like existence from birth to death, except that they are dependent on atmospheric air like the land vertebrates.

On the other hand, the modern fishes of the Teleostomi subclass are not necessarily contented with the water, or devoid of ambitions to leave the water for the land, or unprovided with the means of locomotion over the ground.

Many fresh-water and even some marine fishes of the present day could be drowned by confining them below the surface of the water and preventing their direct access to atmospheric air. Of such are the genera *Mastacembelus, Ophiocephalus, Clarias* (cat-fish), *Anabas* (climbing " perch "), *Misgurnus* (the roach), *Poypterus* (of the Nile and Congo), and, of course, the order of the *Dipnoi*, or lung-fishes.

As regards the development of limbs for pedestrian locomotion, several modern types of teleostean fish have independently modified their pectoral or even pelvic fins to serve this purpose. There are the angler-fishes (*Lophius*), which can walk along the sea-bottom on the tips of their pectoral fins, the gurnards (*Trigla*), which have developed accessory " fingers " in front of the anterior fins, making most effective legs for tripping over the submarine sands, and at the same time serving as feelers in search of prey. The eels glide like serpents at night-time over the cool grass from one pond or stream to another.

A genus of the goby family, the jumping-fish (*Periophthalmus*), lives chiefly on the wet mud of tropical rivers, estuaries, or coasts, and leaves the mud frequently to ascend the trunks or branches of trees in its search after insects. The pectoral fins of this fish have a superficial resemblance to the arms of land-vertebrates, with an actual elbow. The pelvic fins, however, join externally to form a kind of sucker-pad immediately below the " arms." On this the *Periophthalmus* rests, and from this it jumps. Its superficial resemblance to a frog about the

PERIOPHTHALMUS KOELREUTERI, A TELEOSTEAN FISH WHICH LIVES OUT OF WATER

head is very noticeable. Like causes produce like effects.

The *Periophthalmus* only returns to the water for breeding purposes, or for temporary escape from an enemy. It may be observed to retain the lower part of the tail sometimes in the water, and it is thought that through the vascular surface of the tail-fins the downward-prolonged air-bladder receives oxygen through the water. But anyone who has seen the jumping-fish, as I have, on the lower parts of the trees in the mangrove swamps of West Africa—looking very much like malformed salamanders—realises well what must have been the outward semblance of the adventurous ganoid fishes of the Primary epoch when they stepped or hopped out of the water to try a life on shore.

The climbing perch (*Anabas scandens*) of Ceylon, and several other fresh-water fishes of the Indian and Malay regions, such as the snakehead or *Ophiocephalus*, crawl about over the sand, mud, or herbage, and can exist for days or weeks without submerging themselves in water. The snakehead, like the unrelated lung-fish (*Protopterus*) of Africa, can live for months in mud that is merely damp. The Cuchia " eel " of India (*Amphipnous*) has small lung-like air-bladders, and lives almost as much out of the water, on the grass, as in the stream or pond. The allied *Symbranchus*, an eel-like fish of India and South America, has no air-bladder, but, nevertheless, lives nearly all its adult life out of water amid damp marsh vegetation.

CLASS V. PISCES
CHARACTERISTICS AND CLASSIFICATION OF FISHES

Although lampreys and hags are popularly included among fishes, while until comparatively recently the lancelet was commonly placed in the same class, it now seems more correct to make these the representatives of distinct classes. The true fishes, as thus restricted, may be described as cold-blooded vertebrate animals, adapted to a purely aquatic life, and breathing almost invariably by means of gills alone. The heart generally consists of only two chief chambers (three in the lung-fishes); the limbs, if present, are fins; there are also unpaired median fins, supported by fin-rays; and, as in all higher classes, the mouth possesses distinct jaws. The skin may be naked, or provided with scales or bony plates. As a rule, fishes lay eggs, and the young do not undergo so revolutionary a metamorphosis as those of amphibians.

Although the bony fishes of the present day form a specialised side-branch, which has lost many of the characters common to the two classes, it will be evident that fishes and amphibians are very closely allied groups, the latter of which has been directly derived from the former. Geologically, fishes are older than any of the classes hitherto described, their fossil remains occurring in strata belonging to the upper part of the Silurian division of the Palæozoic epoch.

The structure of the skeleton, both external and internal, being of the utmost importance in the classification of fishes, it is essential that the attention of the reader should be more fully directed to this point than has been done in the case of the higher vertebrates. It should be first mentioned that fishes are divided into four subclasses—namely, the sharks and rays (Elasmobranchii), the chimæroids (Holocephali), the end-mouthed fishes (Teleostomi), and the lung-fishes (Dipnoi). These may be further subdivided into orders as follows :

Subclass 1
Sharks and Rays—Elasmobranchii
Order 1
Primitive Sharks—Proselachii, or Pleuropterygii (extinct)
Order 2
Fringe-finned Sharks—Pleuracanthodii, or Ichthyotomi (extinct)
Order 3
Spiny Sharks—Acanthodii (extinct)
Order 4
True Sharks and Rays—Selachii
Subclass 2
Chimæroids—Holocephali
Subclass 3
End-mouthed Fishes—Teleostomi
Order 1
Lobe-finned Ganoids—Stylopterygii, or Crossopterygii
Order 2
Fan-finned Fishes— { Flat-finned Ganoids—Astylopterygii
Actinopterygii { Order 3
{ Ordinary Bony Fishes—Teleostei
Subclass 4
Lung-fishes—Dipnoi

The shape of a typical fish represents a more or less perfect adaptation to rapid movement in water, as may be seen by examining a mackerel, herring, or perch. Here the form may be described as a rounded wedge,

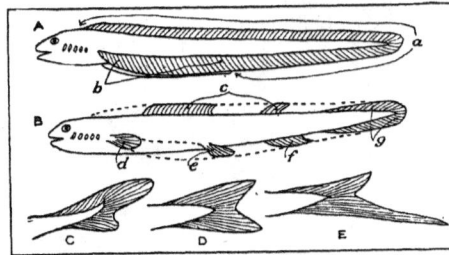

EVOLUTION OF FINS

A, ancestral type; B, primitive type; c, heterocercal; D, homocercal; and E, flying fish's caudal fins. a, continuous unpaired; b, left lateral; c, dorsal; d, left pectoral; e, left pelvic; f, anal; and g, caudal fins

corresponding very closely to that adopted for submarines as the result of mathematical calculation. It is the shape that reduces friction with the surrounding water to a minimum, this practical end being also furthered by the slimy nature of the skin, and the absence of projections calculated to hinder progress.

There is often a departure from the typical shape in fishes which do not spend most of their time in actively swimming about. Eels, for example, are cylindrical in form, their long, narrow bodies enabling them to secure food that may be present in crannies, or to make their way through tangled masses of aquatic vegetation. There are also many bottom fishes that live on the sea-floor, and here the body is commonly flattened, either from above downwards, as in skates and rays, or from side to side, as in the ordinary flat-fishes, such as sole, turbot, or flounder.

Globe-fishes, which inflate themselves with air, and drift with the currents, have rounded contours, and the little sea-horses, that live attached to seaweeds by their curly tails, deviate greatly from the typical shape. The long axis of the body is kept vertical, and the head, instead of being in the same line with it, as usual in fishes, has been bent down into a horizontal position, an obvious convenience under the circumstances. Other instances of fishes which deviate from the normal type will be found in the sequel.

The fins are flat expansions of the body-wall which have to do with balancing, steering, and swimming. They are of two kinds, (a) unpaired, in the median plane, and (b) paired, the former having been evolved before the latter. The unpaired fins were at first in the form of an expansion on the upper side of the body, which continued round the tail and then extended some distance forwards on the under surface. Such a continuous fin is suitable enough for a fish moving slowly along by wave-like lateral movements of the body, and is exemplified at the present day by eels, which have reverted to the old-fashioned method of progression.

But with increased rapidity it has proved advantageous for parts of the originally continuous fin to disappear, the remnants enlarging and acquiring a firmer skeleton. These remnants have become the dorsal, caudal, and anal fins of such a fish as the perch. The first and last of these are chiefly useful as balancers, but the caudal, or tail, fin has become a powerful propeller. It originally formed a continuous and symmetrical fringe (diphycercal tail, see page 1685), but in sharks and sturgeons became greatly enlarged, the backbone being bent up into the bigger upper lobe (heterocercal tail, see page 1700). Such a tail-fin gives a downward bias when swimming is unsteered, and its possessors seek most of their food at or near the sea-floor.

With increased swimming powers, the tail ultimately became outwardly symmetrical in shape (homocercal), the continuation of the backbone into the upper lobe shrivelling up, and the lower lobe becoming of larger size. Such a propeller—the most efficient so far evolved—enables unsteered swimming to be straight ahead. In flying-fishes we find a curious reversal of the old state of things, for here the lower lobe is the larger, on which account unsteered swimming is in an upward direction, an obvious advantage when we remember that these fishes often escape from their enemies by gliding into the air.

There are two kinds of paired fins, the pectoral in front and the pelvic behind, corresponding to the fore and hind limbs of land-vertebrates. The first stage in the evolution of these would appear to have consisted in the development of a continuous lateral fin on either side, serving to prevent the body from falling over to right or left. By suppression of the middle parts of these fins, and the enlargement of their front and back ends, pectoral and pelvic fins came into existence, serving at first simply as balancers. Later on they acquired the power of independent movement, and thus were able to steer as well as balance the body. To begin with, they were lobate, possessing a thick central part supported by firm gristly pieces, and a thinner marginal fringe strengthened by horny epidermal fin-rays. The final stage in their evolution is seen in ordinary bony fishes, where the thick central part has sunk, as it were, into the body, and the fringe has become a sort of readily movable fan (page 1682). Thus, in the brief outline of classification given above, "fringe-finned" applies to the earlier and more primitive condition; "fan-finned" to the later and more specialised one.

The large majority of fishes possess an external skeleton made up of scales, or plates, and here, too, belong the fin-rays, so far as these are bony or horny —e.g., those of the fish represented in the figure on the next page. The most primitive scales are those characteristic of sharks and rays, to which the term "placoid" is often applied. The rough surface of shagreen, or shark-leather, used for polishing and also for ornamental purposes, owes its peculiar character to the presence of innumerable minute scales of this kind.

It is interesting to note that a placoid scale resembles a tooth in structure, consisting of true bone covered by ivory or dentine at the base, while the projecting part is invested by a hard layer of enamel. The first two kinds of substance are formed in the dermis, or deep part of the skin, while the enamel is of epidermal origin. Scales of this kind may be either regularly or irregularly arranged, and may either terminate in a sharp spine or a rounded knob.

In some of the older types of fishes, a very complete armour of rhomboidal bony plates, covered by a harder substance (ganoin) is developed in the dermis. They constitute a fairly flexible, but rather cumbrous armour, the firmness of which is increased by an arrangement of pegs and sockets. Plates of this kind are often known as "ganoid scales."

In typical bony fishes (*Teleostei*), such as cod, perch, salmon, and so forth, the strong armour of more primitive forms has been discarded in favour of thin horny scales, the hinder parts of which overlap the scales behind them. These, again, are purely dermal structures, and exhibit lines of growth. A distinction is here drawn between "cycloid" and "ctenoid" scales, the latter being distinguished by their toothed

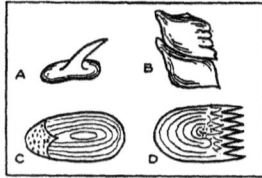

SCALES OF FISH
A, placoid; B, ganoid; C, cycloid; D, ctenoid

posterior edges. Thin horny scales of the kind described afford a certain amount of protection, while at the same time they do not stand in the way of rapid locomotion. It is a matter of importance to be able to determine the age of food-fishes, especially in their earlier years, and recent work renders it probable that an approximate estimate can be formed, at least in some species, by examination of the thin scales. Besides the fine concentric lines that can be seen upon these, there are also rather more distinct "rings," each of which would appear to correspond to a year's growth.

The scales of ordinary bony fishes may be degenerate, or even absent, while in some instances they are replaced by a firm armour, as in the coffer-fish (*Ostracion*).

The fin-rays, to which allusion has been made above, are chiefly useful as supports to the fins, but some of them may be transformed into protective spines, which in certain cases are poisonous. The earliest fish-remains so far known largely consist of skin defences of the kind which, together with teeth, are of extreme durability, and lend themselves to preservation in the fossil state.

Before proceeding to consider the internal skeleton, it is necessary to make a preliminary statement about the breathing organs of fishes, as without some knowledge of these the structure of part of the internal skeleton will be unintelligible.

Throughout life, fishes breathe in the way which was evolved in the earliest vertebrates—i.e., by taking water in at the mouth and causing it to pass to the exterior again through slits or clefts on either side of the region immediately behind the head. As a general rule, there are five pairs of these gill-clefts, but two existing sharks possess six and seven pairs respectively. The partitions between such clefts are known as gill-arches, and these are supported by parts of the internal skeleton. In sharks and rays we find a small hole, the spiracle, in front of the first gill-cleft, and there is every reason to believe that it was once a fully-developed cleft of the same kind.

The arch between the spiracle and the first gill-cleft is called the hyoid arch, and the hard parts supporting it make up the hyoid part of the internal skeleton. In front of the spiracle again is the opening of the mouth,

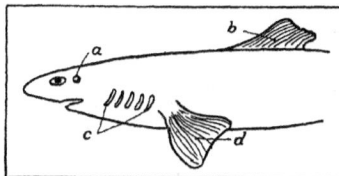

GILL-CLEFTS AND ARCHES OF FISH
a, spiracle; b, dorsal fin; c, gill-clefts; d, pectoral fin

and it has been suggested that this is really a pair of gill-slits that have fused together. However that may be, the arch between it and the spiracle is known as the mandibular arch, and the hard parts by which it is supported ultimately become the framework of the jaws. It is usual to speak collectively of the mandibular, hyoid, and gill-arches as the visceral arches, and the slits between them as the visceral clefts, thus getting over the obvious difficulty of using the terms gill-arch and gill-cleft for parts which may now have no gills connected with them, whatever may have once been the case. The backbone is divisible only into trunk and caudal sections. In the fringe-finned ganoid fishes the primitive notochord persists, although it may be partly surrounded by rudimentary arches; while in sharks and higher bony fishes the column is divided into segments, forming vertebræ with doubly-cupped bodies. In sharks and rays the arches and bodies of the vertebræ remain separate, but in the other groups are fused together; in the tail there is also an inferior arch and spine to each vertebra.

Although the characters of the skull have been largely employed in classifying fishes, it is in many cases of so extremely complicated a nature that no useful purpose would here be served by attempting a detailed description, and only such general features will be dealt with as can be easily explained. The simplest case with which to make a start is that of a shark or dog-fish, where the entire skeleton is composed of gristle and fibrous tissue. We can distinguish here between the skull proper and the firm supports of the visceral arches (page 1687), on this account termed the visceral skeleton.

The skull proper consists of a gristly brain-case of capsules in which the organs of smell are contained, and of denser capsules in which are lodged the essential organs of hearing. Between the olfactory and auditory capsule on either side is a sort of excavation, the *orbit*, in which the eyeball is situated. As a certain amount of movement is necessary for this it has not fused with the brain-case like the other sense-capsules.

In order that the part of the digestive tube (pharynx) perforated by the visceral clefts may be prevented from

that support the gill-arches, become an elaborate jointed framework supporting the pharynx, keeping this well expanded, and affording attachment to muscles effecting the movements of breathing.

On coming to bony fishes we find that the primitive gristly skull and visceral skeleton are partly replaced by distinct bones (cartilage bones), and partly covered by other bones (membrane bones) developed in the investing membranes, and apparently evolved from plates that originally belonged to the external skeleton, but which have sunk inwards and been grafted on the skull.

In the earlier stages the unpaired and paired fins appear to have evolved on similar lines, but the former, as being the older, may conveniently be taken first. In one of the earliest extinct sharks such a fin was supported by gristly rays (radials) attached to short rods (basals) sunk within the body.

An advance upon this may be seen in existing sharks, where only the base of such a fin is supported by the radials, while the freely projecting part is a fold of skin strengthened by flexible horny dermal rays. Turning

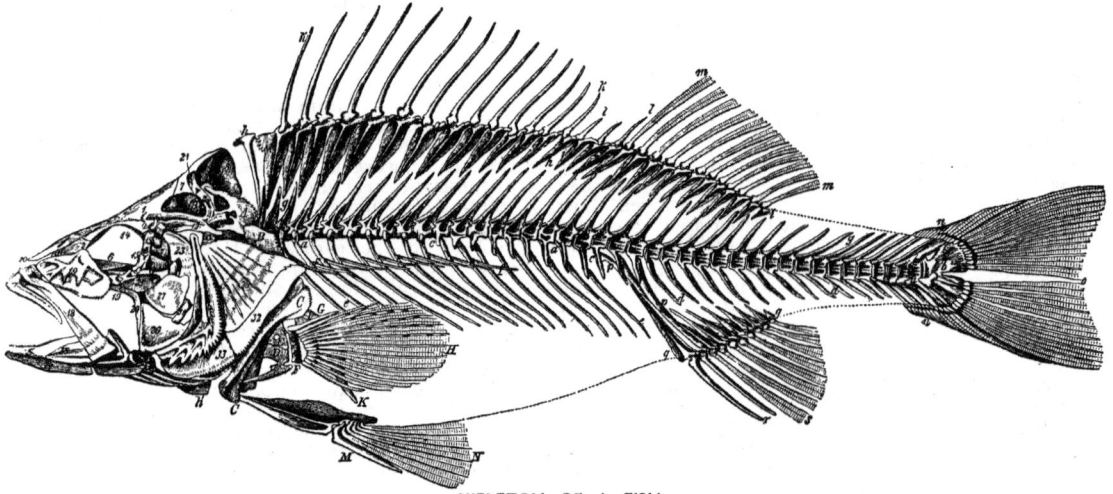

SKELETON OF A FISH

1, frontal ; 2, prefrontal ; 3, ethmoid ; 4, postfrontal ; 5, basioccipital (concealed) ; 6, parasphenoid ; 7, parietal ; 8, supraoccipital ; 9, paroccipital ; 10, exoccipital ; 11, alisphenoid (concealed) ; 12, mastoid ; 13, squamosal ; 14, orbitosphenoid ; 15, presphenoid (concealed) ; 16, vomer (hidden by 19) ; 17, premaxilla ; 18, maxilla, or upper jawbone ; 19, 19', infraorbital ring ; 20, nasal ; 21, supratemporal ; 22, palatine (concealed) ; 23, hyomandibular ; 24, ectopterygoid ; 25, entopterygoid ; 26, quadrate ; 27, metapterygoid ; 28, opercular ; 29, stylohyal (concealed) ; 30, preopercular ; 31, symplectic ; 32, subopercular ; 33, interopercular ; 34, dentary ; 35, articular ; 36, angular ; 37, hyals (concealed) ; *a*, vertebræ ; *b*, confluent tail-vertebræ ; *c*, transverse processes ; *d*, inferior arches and spines of tail-vertebræ ; *e*, ribs ; *f*, barbs of ribs ; *g*, superior arches and spines of vertebræ ; *h*, anterior interspinals ; *i*, posterior interspinals ; *k*, fin-rays of anterior dorsal fin ; *l*, hard, and *m*, soft rays of hinder dorsal fin ; *n*, *o*, hard and soft rays of caudal fin ; *p*, interhæmal spines ; *q*, interspinals of anal fin ; *r*, *s*, hard and soft rays of anal fin ; *A*, post-temporal ; *B*, supraclavicular ; *C*, clavicular ; *D*, coracoid ; *E*, scapula ; *F*, basals ; *G*, *H*, rays of pectoral fin ; *I*, *K*, postclavicular ; *L*, pelvis, *M*, *N*, hard and soft rays of pelvic fin

collapsing, firm pieces of gristle have been developed in the partitions, or visceral arches, between the clefts. On looking at an embryo dog-fish (as a transparency) from the side the successive arches and their internal supporting rods will be clearly seen (page 1682), and by comparing this with the visceral skeleton of the adult (page 1687) we shall understand the subsequent changes that take place.

Supporting the first visceral (mandibular) arch, between the mouth and the spiracle, is a rod shaped like an inverted L, the upper end of which runs forward above the mouth, and will later on be separated, as the upper jaw, from the remainder of the rod, which will then be the lower jaw. A somewhat similar rod supports the second (hyoid) visceral arch between the spiracle and the first gill-cleft, and the upper part of this becomes separated off as a hyomandibular cartilage, by which the jaws are attached to the skull. When the jaws are suspended in this way the skull is said to be "hyostylic." The lower part of the hyoid rod, and the branchial rods

to the paired fins, we find the front, or pectoral, of the extinct shark already mentioned was a stiff triangular projection supported by radials, though at its hinder side a fringe of skin containing horny rays was beginning to develop. Such a fin was merely a balancer, and the evolution from it of a movable paddle, as in a recent shark, involved some marked specialisations. In order to provide a narrow attachment to the body, without which free movement would be impossible, the basals, instead of being imbedded in the body-wall, projected to support the narrow attached region, then came a part supported by radials, and finally a fringe of skin containing horny fin-rays.

In ordinary bony fishes like the perch, which move with great rapidity, and require delicate steering fins rather than clumsy paddles, the basals and radials have sunk into the body, undergoing at the same time more or less reduction and fusion, and the free fin is now nothing but a fold of skin supported by dermal rays, which, however, are now of bony nature. The hind or pelvic fin has passed through similar stages to those outlined for the pectoral, but it never or rarely attains the same importance.

Mr. E. S. Goodrich has recently made a careful study of the dermal fin-rays (dermotrichia) that support the thin part of most fins as already described. He distinguishes between three kinds. (1) Horny unjointed ceratotrichia, typical of sharks and chimæras ; these contain no bone-cells and are not connected with the placoid scales scattered through the skin. (2) Those found in Teleostomi (bony fishes and ganoids), and consisting of small unjointed horny actinotrichia at the edges of the fins, probably to be regarded as the remains of cerato-trichia, and of branching bony lepidotrichia, resembling the scales of the body in primitive forms, and probably derived from these. (3) In lung-fishes (Dipnoi) there are jointed bony camptotrichia, containing bone-cells, and comparable to the lepido-trichia of the second type.

The teeth of fishes may be regarded as modified scales, and in a shark or dogfish gradations between the two are found in the neighbourhood of the mouth. They present a large amount of variation, in adaptation to food of different kinds. Cutting and crushing teeth are the two chief types. While in some cases they may be totally wanting, in others they may be developed on all the bones of the mouth, and even on the hyoid bones and gill-arches ; or they may be attached only to the membrane lining the mouth. Frequently they are welded to the underlying bone or cartilage by a broad basis ; or, as in the saw-fishes, they may be implanted in distinct sockets. The coating of enamel is usually very thin ; and the ivory, or dentine, is more vascular than in other vertebrates. In rare instances the ivory may be penetrated by branching prolongations from the central pulp-cavity, as well as by similar infoldings from the exterior, thus producing a structure similar to that obtaining in those of the primeval salamanders. As a general rule, the teeth are being constantly renewed throughout life, but in a few instances a single set persists.

For beauty, variety, and changeability, the colours of fishes cannot be exceeded by those of any of the other vertebrate classes ; metallic tints and almost all the colours of the rainbow being very commonly displayed ; while the beauty of the coloration is often enhanced by rapid changes (page 1614). In many cases the colours are protective. An example is afforded by the colouring of the upper surface of many flat-fishes, which exactly harmonises with the tints of the sea-bottom ; another instance is that of pelagic fishes, like mackerel and flying-fish, which live near the surface, and have the under parts silvery white, while the back is mottled with dark green and black. Viewed from below against the sky such a fish is practically invisible, and is equally inconspicuous when seen from above among the dark waters. The vivid hues of some spiny fishes such as trigger-fishes (*Balistes minacanthus*) and globe-fishes, are probably examples of mating coloration, while courtship colours are seen in the males of some species, as in the sticklebacks (*Gastrosteus*), where the male becomes bright scarlet during the breeding season

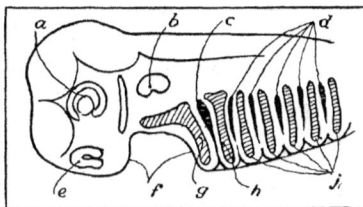

HEAD OF EMBRYO DOG-FISH

a, eye ; *b*, ear ; *c*, spiracle ; *d*, gill-clefts ; *e*, nose ; *f*, mouth ; *g*, mandibular arch ; *h*, hyoid arch ; *j*, branchial arches

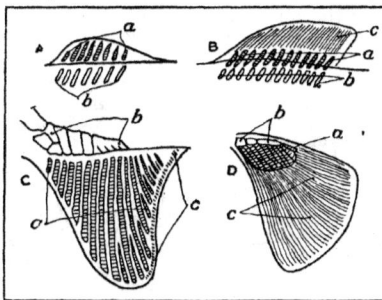

SKELETONS OF FINS

A, extinct, and B, existing shark's dorsal fins ; C, extinct shark's, and D, existing perch's pectoral fins
a, radials ; *b*, basals ; *c*, horny fin-rays

The body-wall of fishes is made up of a great lateral muscle on each side, divided into a number of segments, and also separated into a dorsal and ventral moiety by a median longitudinal groove. On its surface the lateral muscle is marked by a number of white zigzag stripes, generally forming three angles, of which the middle one is directed forwards ; these stripes being formed by the edges of the tendinous divisions between the segments. Generally the muscles are glistening white in colour, but in some instances they are " salmon-coloured," this tint being due to the colouring matter of the crustaceans on which such fish subsist, which is turned red by the action of the digestive fluids in the same manner as by boiling. The electric organs possessed by certain fishes are nearly always formed from specially developed muscles.

With regard to the brain, it will suffice to say that it is of a low type, but in sharks and rays is specialised as compared with bony fishes, and correlated with considerable intelligence. In lung-fishes the brain presents many points of resemblance to that of Amphibia.

The skin of fishes is provided with small groups of sense-cells, many of which are arranged in a characteristic way. Probably everyone has noticed a streak running along each side of a fish's body, and contrasting in colour with the adjacent skin. This is the *lateral line*, which can be traced forward to the head, where it divides into a number of branches. In a few primitive forms, it is a groove in which successive groups of sense-cells are sheltered, but in ordinary cases it has become converted into a canal, opening by pores to the surface. Specialised scales are found in its neighbourhood.

The use of this arrangement is not definitely known, but it is undoubtedly related to the conditions of aquatic life, and is affected by certain vibrations in the surrounding water. Very possibly it has to do with the perception of position in space and the balance of the body. There are other even more problematic sense-organs in ganoids and shark-like forms.

Fishes undoubtedly possess the sense of touch, and special tactile organs are not infrequently to be found near the mouth in the form of tentacle-like barbels, as in cat-fishes, cod, and many deep-sea forms. Some of the groups of sense-cells in the skin closely resemble the taste-buds of higher vertebrates, but how far they minister to the sense of taste is not known. It is only in lung-fishes that they are limited to the mouth cavity.

Well-developed organs of smell are present in the form of a pair of nasal sacs developed as in-pushings of the skin. In ordinary bony fishes, the nasal aperture is divided into two, apparently that there may be a regular circulation of water through the sac. Many of the predaceous fishes hunt their prey entirely or largely by smell. Lung-fishes differ from all other members of their class in the possession of internal nasal apertures, situated just within the margin of the upper lip, a character in which they agree with Amphibia.

In higher vertebrates the essential part of the organ of hearing is a complicated membranous bag, the labyrinth, sheltered within an auditory capsule. A similar arrangement is found in fishes, but there are no specialised sound-conducting structures, although the spiracle, when present, would appear to serve as a short cut for the passage of vibrations from the surrounding water. It is definitely known that the labyrinth, in all vertebrates, has to do with space-perception and the maintenance of equilibrium, and in this connection it is especially interesting to find that the fish-labyrinth develops from a part of the lateral line, which is pushed in from the surface. Whether fishes are able to hear in the ordinary sense of the word is not absolutely certain. We know, however, that many fishes make characteristic sounds, produced in various ways, and possibly these help to keep shoals together, especially when spawning. This suggests that such fishes do really hear.

With regard to sight, it need only be said that the eyes of fishes are constructed in much the same way as those of other vertebrates, but the crystalline lens is spherical as an adaptation to the dense nature of the surrounding medium. Probably most persons in the pursuit of breakfast-table anatomy have dislodged the lens, looking like a sugar-coated pill, from the eye of a whiting or mackerel. In the uncooked condition it is, of course, transparent. The eyes of deep-sea fishes are commonly either of enormous size, or, on the other hand, reduced to useless vestiges, not infrequently being absent altogether. In many such fishes—as, too, in deep-sea invertebrates—phosphorescent organs are present, and but for the light these produce, the abysses of the ocean would be profoundly dark.

In most of the more primitive fishes, the mouth is placed on the under side of the head, but in the higher types it is terminal, a more convenient situation under ordinary circumstances. The tongue is ill-developed and not protrusible. The large mouth cavity passes back into a capacious pharynx, perforated by the spiracular cleft and the gill-slits. The inner edges of the latter are commonly fringed with more or less well-developed filaments, which prevent food from passing out to the exterior. These are especially necessary in fishes which feed on the minute forms of life making up the plankton or floating population of the sea and lakes. The pharynx is continued into a gullet, this into a stomach, which is commonly U-shaped, and this again into a rather short intestine. In sharks and rays, chimæroids, ganoids, and lung-fishes, the lining of the intestine projects inwards as a spiral fold (spiral valve) that largely increases the surface for the absorption of digested food. The intestine terminates either by opening into a cloaca, a chamber also receiving the ducts of the kidneys and reproductive organ (sharks and rays, lung-fishes), or else communicates directly with the exterior (chimæroids, ganoids, teleosts).

In all fishes except sharks and rays, an outgrowth from the pharynx becomes a sac known as the swim-bladder, usually situated just below the backbone, and

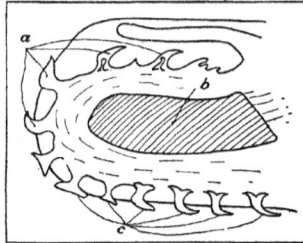

TEETH OF FISHES
a, teeth ; *b*, jaw cartilage ; *c*, scales

filled with a mixture of gases. Its primary use is to serve as a hydrostatic organ, helping to maintain the balance of the body in the water, but in some instances it also helps in breathing. The origin of the swim-bladder is obscure, but some fishes of the shark kind possess a small, probably glandular, pouch, opening into the upper side of the pharynx, and this may perhaps be regarded as the first step in the evolution of the organ in question.

In all members of the class except lung-fishes, the heart contains only impure blood, which is poured into a thin-walled auricle, thence passing into a muscular ventricle, by which it is pumped to the gills for purification (page 1684). After being oxygenated, this blood supplies the body, ultimately reaching the heart again, by this time having parted with the greater part of its oxygen and taken up a large amount of the waste product, carbonic acid gas. There are differences in detail which will be alluded to in the sequel. In lung-fishes the auricle is divided into right and left halves, of which the former receives impure blood from the body, while the latter serves as a receptacle for the blood which has been purified by the swim-bladder. The arrangement is not unlike what has elsewhere been described as characteristic of the Amphibia.

The gills of sharks, rays, and chimæroids are contained in pouches, usually five in number on each side, each pouch opening externally by a slit, and into the pharynx internally. The gills are delicate vascular ridges on the walls of the pouches. In the embryos filamentous gills protrude externally from the slits. The spiracles found on the side of the head in fishes belonging to the groups mentioned are the external openings of a canal leading on each side into the pharynx, and represent what is known as the first visceral cleft in the embryo. In the bony fishes the comb-shaped gills, generally four in number, are attached to bar-like gill-arches, and project into a cavity on each side of the head, which is covered over by the gill-cover, behind which is a slit. During ordinary respiration water enters the mouth, is driven over the gills, and then expelled by the gill-opening. Allusion must be made in this place to certain organs known as false gills, or *pseudo-branchiæ*. These are the vestiges of gills situated in front of the persistent ones, which were once functional. Although most fishes breathe as indicated, many forms can utilise atmospheric air for the purpose.

The most important accessory breathing organ is the swim-bladder, which in the lung-fishes, as their name indicates, acts as a lung. This is also true, to a lesser extent, for certain other forms.

Although in a few fishes which produce living young, as well as in the sharks and rays, an actual connection takes place between the two sexes, in the great majority of cases the ova are deposited by the female, after which they are fertilised by the male. The bony fishes lay numerous eggs of relatively small size, or even extremely minute, those of the eel being almost microscopic ; but

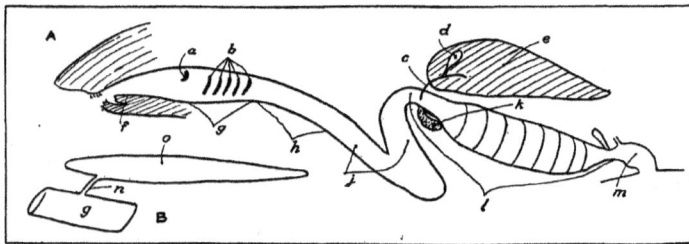

A, DIGESTIVE ORGANS, AND B, SWIM BLADDER OF FISHES
a, spiracular cleft ; *b*, gill-clefts ; *c*, bile-duct ; *d*, gall-bladder ; *e*, liver ; *f*, tongue ; *g*, pharynx ; *h*, gullet ; *j*, stomach : *k*, pancreas ; *l*, intestine ; *m*, cloaca ; *n*, duct of swim-bladder ; *o*, swim-bladder

there is a considerable degree of variation in this respect. Dr. Wemyss Fulton has calculated the number of eggs produced by the British food-fishes in individuals of different size, and the following are a few of his figures : Ling (54 pounds weight), 28,361,000 ; turbot (17 pounds 5 ounces), 9,161,000 ; cod (21½ pounds), 6,652,000 ; herring (11½ inches long), 47,000.

The female does not usually take any care of her eggs after spawning, but there are exceptions to this rule ; for instance, in catfishes (*Aspredo*) they are pressed into the skin of the under surface of the body, while in a kind of pipe-fish (*Solenostoma*) they are carried in a pouch formed by the fusion of the broad pelvic fins with the skin of the body. Among bony fishes there are several instances where the young are tended by the father ; some, like the sticklebacks, building a nest, while others, like certain pipe-fishes, have an abdominal pouch in which the eggs are hatched.

The eggs of sharks, rays, and chimæroids differ remarkably from those of bony fishes, being large in size, few in number, and laid singly instead of in masses. They are invested in a hard, horny envelope, which is generally oblong in form, with the four corners elongated, or even drawn out into tendrils by means of which the egg is moored to some foreign substance. The males of these fishes are armed with organs known as claspers, processes of the pelvic fins, evidently connected with reproduction.

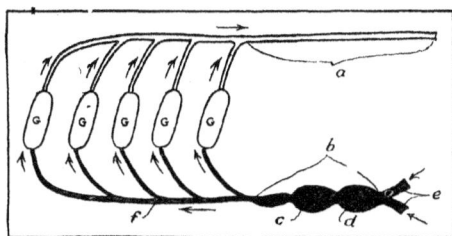

CIRCULATORY ORGANS OF FISHES
a, dorsal aorta ; *b*, heart ; *c*, ventricle ; *d*, auricle ; *e*, great veins ; *f*, ventral aorta ; *g*, gills

The young of many fishes differ markedly from the adult, this being especially the case with those of the eels, which have long ribbon-like bodies and small heads. Formerly these so-called " Leptocephali " were regarded as the young of shore-fishes which had been carried out to sea and undergone abnormal development.

The changes which take place in the flat-fishes during development may be more conveniently noticed under that group. Although male and female rays differ remarkably from one another in the structure of their teeth, while both in this group and in the sharks and chimæroids the males are distinguished by the possession of claspers, there is generally but little sexual differences among fishes. In the bony fishes, however, the females are larger than the males ; among the cyprinodonts the difference between the two being occasionally as much as six times.

Fishes differ markedly in regard to their power of bearing alterations in their normal environment. On this subject Dr. Günther writes that " some will bear suspension of respiration—caused by removal from water, or by exposure to cold or heat—for a long time, while others succumb at once. Nearly all marine fishes are very sensitive to changes in the temperature of the water, and will not bear transportation from one climate to another. This seems to be much less the case with some fresh-water fishes of the temperate zone, since carp may survive after being frozen in a solid block of ice, and will thrive in the southern parts of the temperate zone. On the other hand, some fresh-water fishes are so sensitive to a change in the water that they perish when transferred from their native river into another apparently offering the same physical conditions.

"Some marine fishes may be abruptly transferred from salt into fresh water, like sticklebacks; others survive the change when gradually effected, as many migratory fishes; whilst others, again, cannot bear the least alteration in the composition of the salt water (all pelagic fishes). On the whole, instances of marine fishes voluntarily entering brackish or fresh water are very numerous, whilst fresh-water fishes proper but rarely descend into salt water."

The foregoing remarks lead naturally to the subject of the distribution of fishes. In the first place, we find that many marine fishes have a much narrower range than might be expected ; while, on the other hand, we find families and genera, and even species of fresh-water fishes inhabiting widely-separated areas of the earth's surface. The primary division into fresh-water and marine fishes is not sharply defined, a transition being formed by brackish-water types, species or even individuals of which can accustom themselves to live in either salt or fresh water. Then, again, we have certain essentially fresh-water fishes, like salmon and some cat-fishes, which pass a certain part of their life in the sea ; while, on the other hand, some marine forms, such as sturgeons, periodically ascend rivers for the purpose of spawning. To a certain extent such habits will tend to explain the occurrence of peculiar families of fresh-water fishes, such as the cichlids of Africa, South America, and India, in widely separated areas, although this must probably be supplemented by dispersal from a common northern centre.

After the separation of the fresh-water and brackish-water types, the marine fishes are divided by Dr. Günther into littoral, pelagic, and deep-sea groups, although no hard and fast lines can be drawn. The littoral or shore-fishes are those found close to land or sunken shoals, the majority living near the surface, and very few descending as deep as three hundred fathoms. Their distribution depends not only on the temperature of the surface water, but also on the nature of the neighbouring land and its organisms ; some such fishes being suited to flat coasts with muddy or sandy bottoms, while others frequent rock-bound shores where the water is deep, and others, again, affect coral-reefs. Cod, rays, and flat-fish are well-known examples of this group.

Pelagic fishes, such as tunnies, flying-fish, sword-fishes, and sunfishes, inhabit the surface layers of the open sea, approaching the shores only by accident, in search of food, or to spawn.

In marked contrast to the last are the deep-sea fishes, inhabiting the abyssal depths of the ocean, where they are undisturbed by tides or currents, and live for the most part in total darkness, their organisation, owing to the great surrounding pressure, preventing them coming to the surface in a healthy condition. From the similarity in the physical conditions of the ocean depths in all parts of the world, there seems no reason why a single species of deep-sea fish should not range from the Equator to the Poles.

Not the least remarkable feature about the carnivorous deep-sea fishes is the enormous size of their stomachs, which enable them to swallow creatures nearly their own size. Although when brought to the surface deep-sea fishes are soft, flabby creatures, their scales standing out at right angles, and their eyes starting from their sockets, yet at their own proper level, under enormous pressure, their bodies are doubtless as firm and compact as those of ordinary fish. Deep-sea fish certainly live at a depth of two thousand seven hundred and fifty fathoms. There are fully nine thousand known species of living fishes, while considerably more than one thousand fossil forms have been already described.

SUBCLASS I. ELASMOBRANCHII
SHARKS AND RAYS

THE first subclass of fishes includes existing sharks and rays, together with more or less closely allied extinct forms; some of the latter being the most primitive members of the class yet known. Indeed, taking these primitive types into consideration, and remembering that sharks and their allies are the oldest fishes with which we are acquainted—dating from the upper Ludlow beds of the Silurian epoch—it seems probable that the present subclass may have been the stock whence all other fishes were derived. Agreeing with the bony fishes and ganoids in having the suspending apparatus of the lower jaw movably articulated to the skull, sharks and rays have the skeleton entirely gristly, though the gristle is often calcified; membrane-bones—except in one extinct group—being entirely wanting. The gills open by separate external clefts, and have no cover. When bony elements are developed in the skin, these agree in structure with teeth. In all living members of the subclass the optic nerves cross without giving off any mutually interlacing fibres, the arterial bulb of the heart is furnished with three valves, the intestine has a spiral valve, the eggs are large and detached, and a swim-bladder is wanting.

The subclass is divided into the following four orders, the members of the first three being extinct:

Order 1.—Primitive Sharks (Proselachii, or Pleuropterygii).
Order 2.—Fringe-finned Sharks (Pleuracanthodii, or Ichthyotomi).
Order 3.—Spiny Sharks (Acanthodii).
Order 4.—True Sharks and Rays (Selachii).

ORDER I. PROSELACHII—PRIMITIVE SHARKS

THE only member of the order Proselachii so far known is the primitive form, *Cladoselache*, from the Cleveland Shale of Ohio, which is either upper Devonian or lower Carboniferous in age, and also represented in the lower Carboniferous rocks of Scotland. There are several species, none exceeding about 6 feet in length, and some considerably smaller.

Cladoselache is of considerable interest because it is probably the most primitive creature of the shark kind known, and some of its characters throw light on the evolution of fishes in general. The rounded body is covered with minute lozenge-shaped scales, and the mouth is terminal—that is, situated at the front of the head, and not on its under surface, as in almost all modern members of the subclass. There are two low dorsal fins, and the upwardly-bent tail is surrounded by a caudal fin, which is fairly symmetrical externally, suggesting that most of the swimming was in a horizontal plane.

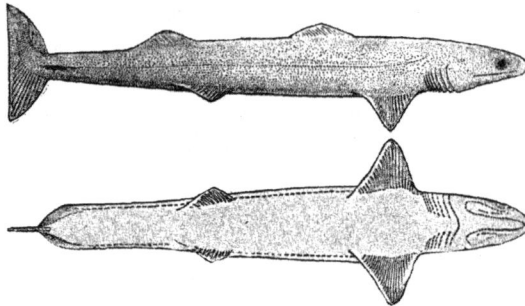

RESTORATION OF CLADOSELACHE

The paired fins, both pectoral and pelvic, resemble the dorsals in shape and structure, and lend support to the theory that paired fins are the remains of continuous lateral fins that existed in the stock from which all fishes have been derived (see page 1679). They possessed little, if any, power of movement, their use no doubt being to balance the body in the water. A small horizontal keel-like projection on either side of the tail probably had the same function.

There were certainly five pairs of gill-slits and probably seven, while spiracles are to be seen in front of the first pair of these slits. The eyes were protected by rings of flattened scales, and the nostrils were near the tip of the snout. The lateral lines were in the form of open grooves.

So far as can be ascertained, the internal skeleton was very primitive in character, and the notochord would seem to have persisted throughout life.

ORDER II. PLEURACANTHODII—FRINGE-FINNED SHARKS

RANGING from lower Carboniferous to lower Permian, the extinct sharks of the order Pleuracanthodii, or Ichthyotomi, are little less primitive than those of the previous one.

The essential characteristic of this group, as shown in the figure, is the fringed structure of the pectoral fins, which consist internally of a long, tapering segmented axis, giving off a double series of gristly rays. The internal skeleton of these sharks shows granular calcifications; but the notochord is never or but seldom constricted into distinct vertebræ, the calcification, except in the tail, stopping short at an

SKELETON OF EXTINCT FRINGE-FINNED SHARK SHOWING DIPHYCERCAL TAIL

incomplete stage, when the body of each segment of the backbone consists of three separate pieces. The upper and lower arches and spines of the backbone are tall and slender, the upper spines having no intercalary cartilages between them.

As represented by the genus *Pleuracanthus*, common to the Carboniferous, Permian, and Triassic rocks of both sides of the Atlantic, these sharks are further characterised by the slender and slightly depressed form of the body, the terminal position of the mouth, and the diphycercal tail. The long and low dorsal fin is continued along the whole of the back from a short distance behind the head, and its cartilages are more numerous than the subjacent spines of the vertebræ; immediately behind the head is a long barbed spine, and the body was probably devoid of shagreen. The teeth are peculiar, consisting of two divergent and generally unequal cones, on an expanded base.

ORDER III. ACANTHODII—SPINY SHARKS

WHILE the two preceding orders contain the most primitive and generalised representatives of the subclass, the one now to be considered, Acanthodii, also confined to the Palæozoic epoch, comprises sharks of a more specialised type than any existing forms. Indeed, these fringed sharks bear much the same relationship to the spiny-finned group as is presented by the bony fishes to the fringe-finned ganoids. An essential feature of the group is the development of membrane-bones covering the original cartilaginous skull; the socket of the eye is also often surrounded by a ring of bones of similar origin. In the internal skeleton the notochord is persistent, and the cartilages are superficially calcified.

Teeth, when present, are firmly fixed upon membrane-bones overlying the cartilages corresponding to the jaws of other sharks. The gill-arches bear a series of

RESTORATION OF A SPINY SHARK

appendages which during life were probably furnished with membranous expansions similar to those of the existing frill-gilled shark. In the fins the cartilages of the internal skeleton are greatly reduced, and the membranous portions are almost destitute of cartilaginous rays; while each of the paired and most of the median fins are provided with a large front spine. The tail is heterocercal, and the males lack the claspers characterising existing sharks and rays. Externally the body is covered with small and closely-arranged quadrangular granules, between two series of which runs the lateral line. Three families constitute the order; the first of these, as represented by the genus *Acanthodes*, having but a single dorsal fin, while in the other two—respectively typified by *Ischnacanthus* and *Diplacanthus*—there are two of these fins.

TABLE OF EXISTING SHARKS & RAYS MENTIONED IN THIS WORK

ORDER IV. SELACHII—EXISTING SHARKS AND RAYS

THE whole of the existing representatives of the sub-class Elasmobranchii form the order Selachii, which is characterised by the gristly internal skeleton being usually only superficially calcified; while, except in some early extinct types, the notochord is constricted at the centre of each vertebra. The upper and lower arches of the vertebræ are short and stout, and special pieces (intercalary cartilages) are very generally developed between them. The pectoral fin does not possess a jointed central axis, but its gristly rays form a fan-shaped structure radiating from an abbreviated base; and the axis of each pelvic fin is drawn out in the males into a "clasper," connected with reproduction.

In the skull, the hyomandibular usually intervenes between the upper jaw and the cranium proper; but in the genus *Notidanus* the hyomandibular takes no share in supporting the jaws, the upper jaw articulating directly with the cranium by means of a facet behind the eye; this being probably the original arrangement. The gill-pouches usually open to the exterior by five vertical slits on each side of the neck, but sometimes the number of these clefts is increased to six or seven.

The mouth is commonly situated on the under side of the head; and the teeth may be either sharply pointed and separate, or blunt and crowded together into a crushing surface. In the former case there is a constant succession of new teeth to replace the old ones as they are worn away and shed. The tail-fin is usually heterocercal, with the upper lobe greatly elongated; the pelvic fins are always abdominal; and the dorsal fins of many extinct and a few living types bear large spines on their front edges.

Spiracles are frequently developed; and ovoid or diamond-shaped intercalary cartilages are situated between the upper arches of the vertebræ. The eggs are generally contained in horny rhomboidal capsules, often furnished at the four corners with long, tendril-like filaments, by which they attach themselves to the stems of seaweeds and other bodies. In some species, however, the eggs are hatched internally; and in all cases the embryos are furnished with external gills, which are shed before birth.

All the members of the order are carnivorous, but while the typical sharks are highly predaceous, devouring everything they come across, some of the largest species only possess small teeth, and feed on molluscs and other invertebrates. The rays, too, are largely shell-fish eaters, and mostly differ from many sharks in living on or near the bottom, instead of swimming about actively at or just below the surface.

The species are typically marine, but many ascend tidal rivers, and in the Viti Levu Lake in Fiji, as well as in Lake Nicaragua, there are sharks living permanently in fresh water. The species inhabiting the former lake, which is cut off from the sea by a cataract, is *Carcharias gangeticus*, common in the Ganges and in the Tigris, and ascending in the latter three hundred and fifty miles from the sea. A species of saw-fish is found in a fresh-water lake in the Philippines.

It is commonly stated that sharks scent their prey from a distance, since they rapidly congregate whenever animal refuse is cast overboard; but it may be suggested that such assemblages, as in the case of vultures, are rather due to one shark following the movements of another. The order was formerly divided into two subordinal groups, based upon the conformation of the body; the one group including all the sharks and dog-fishes, and the other the rays and their immediate allies. Although this difference in bodily form is of considerable importance in classification, yet it does not constitute the essential line of distinction, which is based upon a difference in the internal structure of the bodies of the vertebræ.

SKULL AND PARTS OF SKELETON OF DOG-FISH

A, side view; B, vertebræ; *a*, nose capsule; *b*, upper jaw; *c*, lower jaw; *d*, hyoid arch; *e*, ear capsule; *f*, spiracle; *g*, hyomandibular; *h*, branchial arches; *j*, intercalary cartilage; *k*, vertebræ; *l*, dorsal pieces

COMB-TOOTHED SHARKS

A very remarkable family, (*Notidanidæ*) is now represented by four species of comb-toothed sharks (*Notidanus*) and the frill-gilled shark (*Chlamydoselachus anguineus*). In all other existing sharks the gill-clefts are five in number, but in the present family they are six or seven; while there is a further peculiarity in regard to the structure of the skull. It has been already stated above that in the more typical sharks the lower jaw is suspended from the cranium by a hyomandibular element; but in the present family the upper jaw articulates directly with the cranium, and the hyomandibular takes no share in its attachment. In addition to their more numerous gill-slits, the comb-toothed sharks are distinguished externally from all those hitherto considered by having only a single dorsal fin, which is situated far back on the body and has no spine. The eye is devoid of a nictitating membrane, the spiracles are small, and the teeth, of which several series are in use at the same time, have sharply-pointed cusps.

In the typical genus the body is moderately elongated, the mouth inferior in position, and the six or seven gill-clefts devoid of flaps.

LOWER TEETH OF EXTINCT COMB-TOOTHED SHARKS

The principal teeth consist of a series of cusps placed upon a long base, all inclining in one direction, and decreasing in size from the front to the back. With the occasional exception of some portions of the tail, the notochord persists throughout life.

At the present day the range of the existing members of the genus includes most temperate and tropical seas, some of the species reaching as much as 15 feet in length. While in the grey comb-toothed shark (*Notidanus griseus*), of the Atlantic and Mediterranean, the number of gill-clefts is six, in each of the other three species it is seven. Fossil species undoubtedly range from the Trias to the Pliocene; many of these, like the one of which two teeth are shown in the illustration, being of much larger

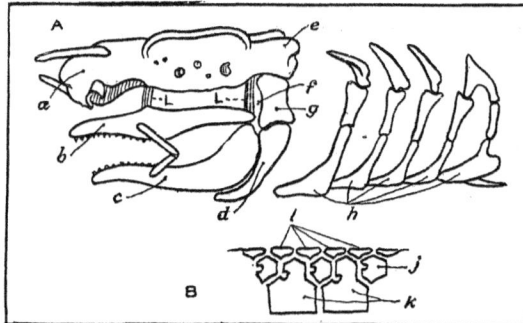

dimensions than any of the existing forms. There are also some Carboniferous forms which are probably referable to the family. As to the habits of these sharks there appears to be practically no information.

FRILL-GILLED SHARK

The Japanese frill-gilled shark (*Chlamydoselachus anguineus*) differs from the typical genus by its greatly elongated and slender body, and by each of the six

FRILL-GILLED SHARK

gill-clefts being protected by a frill-like flap of skin, much as in *Acanthodes* (page 1686). The teeth are of somewhat simpler structure, being similar in both jaws, and each consisting of three slender, curved, and subconical cusps, separated by a pair of rudimentary ones; while there is an unpaired median series at the extremity of the lower jaw only, instead of in both the upper and the lower. Although mainly persistent, the notochord is in part replaced by ill-developed vertebræ of the type characteristic of the suborder. Fossil teeth from the Italian Pliocene have been assigned to this genus.

PAVEMENT-TOOTHED SHARKS

The well-known Port Jackson shark (*Cestracion philippi*) and three allied species are the sole existing representatives of a family (*Cestraciontidæ*) which was exceedingly abundant during the Secondary epoch. They possess a strong spine on the front edge of each of the two dorsal fins. The first dorsal fin is situated above the space between the pectoral and pelvic pairs; and the teeth, of which several series are in use at the same time, are more or less blunt and broad, more especially in the hinder part of the jaws, although those in each oblique row are never fused together into continuous plates. In the existing genus there is no nictitating membrane to the eye; the body is moderately elongated, with the second dorsal fin in advance of the line of the anal; and the mouth is almost or quite terminal.

The front teeth are small, numerous, and sharp, while the hinder ones are broad and flattened. The spines of the dorsal fins are smooth, covered on the sides with a thick layer of ganoin; the shagreen is fine; and the head is devoid of spines. As in the last family, the lower jaw is directly attached to the skull, though a small hyomandibular helps to suspend it. In the existing species the egg-capsules assume a remarkable screw-like form, quite unlike that of any other member of the family.

The living members of the genus, none of which exceed 5 feet in length, have been recorded from the seas of Japan, Amboyna, Australia, the Galapagos Islands, and California; while remains of extinct forms occur in Europe from the upper Jurassic period onward. Very little appears to be known as to their habits, but their food is stated to consist principally of molluscs and crustaceans, the hard shells of which are crushed by the pavement-like hinder teeth.

EXTINCT TYPES

One of the earliest of the numerous fossil genera of the family is the Carboniferous *Orodus*, with teeth very like those of the later *Hybodus*, ranging from the Trias to the lower Cretaceous. In the last-named genus the notochord is persistent, the bluntly conical or cusped teeth have a central and two or more lateral cusps, the fin-spines are ridged, and there are two hook-like spines below each eye. *Acrodus*, with a nearly similar range, has, on the other hand, blunt teeth; while the Jurassic and Wealden *Asteracanthus* differs from *Hybodus* by its rhomboidal, roughened, and flattened teeth, and the star-like ornamentation of the spines of the dorsal fins.

In *Synechodus*, of the Chalk, all the teeth are cusped, the anterior ones having a tall central cusp, flanked with from three to five small lateral pairs. An allied extinct family (*Cochlidontidæ*), confined to the Carboniferous rocks, differs by the component teeth of at least one of the oblique rows being fused into a continuous curved plate, which may be either smooth or ridged. Many of the extinct representatives of these families were larger than the Port Jackson shark.

DOG-FISHES AND THEIR ALLIES

In the dog-fishes and their allies (family *Scylliidæ*) there are no spines to the dorsal fins, the first of these being placed rather far back. The eye does not possess a nictitating membrane, and the teeth are small, with several series generally in use at the same time. In all there are distinct spiracles.

TRUE DOG-FISHES

Represented in British waters by the lesser (*Scyllium canicula*) and larger spotted dog-fish (*S. catulus*), this

PORT JACKSON SHARK

genus is characterised by the first dorsal fin being above or behind the line of the pelvic pair; by the origin of the anal being in advance of the line of that of the second dorsal; the absence of serration of the upper edge of the caudal fin; and the small and delicate teeth, which are arranged in numerous series, and generally have a long central cusp, flanked by one or two small ones on each side. About half a score of species have been described, ranging over the coast regions of most temperate and tropical seas, and all comparatively small. The majority have prettily spotted skins. The food consists mainly of crustaceans and molluscs; and their flesh is eaten not infrequently by fishermen, while in the Orkneys, where the British species are more abundant than elsewhere, it is regularly dried for winter consumption. The shagreen of their skins is also employed in wood-polishing.

The lesser spotted dog-fish is often extremely abundant on our coasts, and may be a serious nuisance, as it steals the bait from deep-sea lines. These sharks lay eggs enclosed in horny cases drawn out into tendrils. Fossil dog-fishes date from the period of the Chalk; and they are represented in the Kimeridge Clay by the extinct *Palæoscyllium*, in which the origin of the second dorsal fin is placed in advance of that of the small anal.

ZEBRA-SHARK

Among several allied genera is the zebra-shark (*Stegostoma tigrinum*) of the Indian Ocean, 10 to 15 feet long, and marked with black or brown transverse bars or round spots on a brownish yellow ground.

In this fish the first dorsal fin is above the line of the pelvic pair, while the second is in advance of the line of the anal, which is approximated to the caudal, the latter being greatly elongated and equal to half the total length. Young specimens of this shark are generally met with near the coast, but the adults are more or less pelagic. Their tail-fins are relatively larger, and their colours brighter than in adults.

OTHER GENERA

Smaller dog-fishes from the Indian Ocean constitute the genus *Chiloscyllium*, in which the first dorsal fin is either above or behind the line of the pelvics; while the anal is far behind that of the second dorsal and close to the caudal; the teeth being small and triangular, with or without lateral cusps. Existing species are very handsomely ornamented with dark bands and spots. In a fossil state the genus has been recorded from the Miocene.

Three bottom-haunting sharks from the Japanese and Australian seas have been described under the name of *Crossorhinus*, and are remarkable for the presence of leaf-like expansions of the skin on the

ELFIN SHARK

sides of the head. As in the case of other fish similarly adorned, these structures are probably for the purpose of attracting prey; and in order that they may be well concealed, these sharks have a coloration closely assimilating to that of a rock covered with seaweed or corallines.

PORBEAGLE GROUP

In the porbeagle group (family *Lamnidæ*) there is considerable difference in size between the two spineless dorsal fins. The first is very large, and placed above the interval between pectoral and pelvics. The second is small, and situated well behind the pelvics and above the anal. There is no nictitating membrane, while the fully-formed teeth are pointed, solid in structure, and in most genera relatively large. The gill-slits are generally wide, and the spiracles either minute or absent. This family dates from the Chalk period, where there occur remains of species some of which are referable to genera still existing, such as the porbeagles, while others indicate extinct generic types. The fox-sharks and the gigantic *Carcharodon* are, however, unknown before the Tertiary period.

PORBEAGLES

The porbeagle (*Lamna cornubica*)—a name supposed to be derived from its porpoise-like appearance and active predatory habits—is the type of a genus containing three extinct species, and characterised by the very small size of the second dorsal and anal fin, by the presence of a pit at the root of the caudal fin—of which the lower lobe is much developed—and also of a keel along the sides of the tail. The teeth are narrow and slender, with one or two pairs of small accessory cones at their bases, the edges of the main cone being smooth. The common porbeagle wanders all over the North Atlantic, and has also been taken

PORBEAGLE SHARK

in Japan. It does not commonly exceed 10 feet in length, and its colour is dull grey above and whitish beneath. Its food chiefly consists of fishes, which are apparently swallowed whole, the lancet-like teeth being apparently more adapted for seizing and holding than for tearing prey. The porbeagle is said to be a viviparous species.

GREAT WHITE SHARK

The most formidable of all existing members of the group is the gigantic white, or man-eater, shark (*Carcharodon rondeletii*), distinguished from porbeagles by the huge broadly triangular teeth, which have strongly serrated edges, and may possess basal cusps.

BASKING-SHARK

The existing species, a purely pelagic creature ranging over all the warmer seas, is known to attain a length of 40 feet, and one of the teeth of a specimen 36 feet long measured 2 inches along the edge of the crown and $1\frac{3}{4}$ inches across the base.

Similar teeth are found in the Crag deposits of Suffolk, and are referred to the existing species ; but from these same beds, and also from the bottom of the Pacific, between Polynesia and Australia, there are obtained other teeth of much larger dimensions, some of them measuring upwards of 5 inches along the edge and 4 inches in basal depth. These teeth evidently indicate sharks beside which the existing form is a comparative dwarf ; and it is not a little remarkable that the specimens dredged from the bed of the Pacific indicate that these giants must in all probability have survived to a com-

long flexible horizontal blade projecting from the back of the skull in front of the muzzle so as to leave both jaws quite free. It is long and slender in shape, and grey in colour.

FOX-SHARK

The fox-shark, or thresher (*Alopecias vulpes*), not uncommonly met with in British waters, is the only species of its genus, and is easily recognised by the great length of the upper lobe of its tail-fin, from which it derives its name. Growing to a length of 15 feet, of which more than half is taken up by the tail, this shark has very small second dorsal and anal fins ; the caudal fin extremely elongated, and without a pit at its root ; no keel on the sides of the tail ; and small smooth-edged teeth, compressed and triangular in shape.

Like most sharks, the thresher has a wide range, being abundant throughout the Atlantic and Mediterranean, and also found off New Zealand and California. The comparatively small size of the teeth indicates that it is not adapted for killing large prey, and it chiefly feeds upon the various species of the herring tribe and mackerel, inflicting terrible destruction.

It derives its name of thresher from its habit of beating the water with its long tail in order to drive the members of the shoals on which it preys into a compact mass, when they can be the more readily seized ; and its voracity may be inferred

FOX-SHARK

paratively recent date. Observations are still required as to the mode of life and breeding habits of this great shark.

BLADE-SNOUTED SHARK

A very remarkable form from deep waters in Japan, described under the name of the elfin shark (*Mitsikurina owstoni*), is characterised by the presence of a

from the fact of no less than nineteen mackerel and two herring having been taken from the stomach of a single individual. It is commonly reported by sailors that threshers, in company with killers and sword-fish, make attacks on whales by leaping high in the air and belabouring the unfortunate cetaceans with powerful blows of their tail as they descend.

BASKING-SHARK

The largest North Atlantic member of the suborder is the basking-shark (*Cetorhinus maximus*), which now alone represents a genus with very small second dorsal and anal fins, a pit at the root of the caudal fin, a keel on each side of the tail, very large and wide gill-clefts, and very small, numerous, and conical teeth without basal cusps, and seldom serrated at the edges. This shark, which grows to a length of over 30 feet, is regularly hunted on the west coast of Ireland for the sake of the oil from its liver, of which a single fish may yield considerably more than a ton.

This shark derives its name from its habit of lying motionless during calm, warm weather on the surface of the water, with the tall first dorsal fin and a considerable portion of its back exposed, several individuals consorting together. The gill-arches are provided with very long rakers bearing granular tooth-like structures, and in the young the snout is relatively longer and more pointed than in the adult. Unless attacked, when it can stave in the sides of a boat with its tail, this shark is perfectly harmless, its food consisting entirely of small gregarious fishes and various invertebrates. Remains of an extinct species occur in the Pliocene deposits of Belgium, while others from older Tertiary beds have been tentatively assigned to the genus.

WHALE-SHARK

Although resembling the true basking-shark in the large size of its gill-clefts and the structure of its gill-rakers, the gigantic whale-shark (*Rhinodon typicus*) differs in having the mouth and nostrils situated near the end of the snout, as well as in the

WHALE-SHARK

caudal fin is well developed. In its varied coloration this fish differs markedly from the majority of sharks, being ornamented with buff spots and stripes upon a dark ground. It represents a separate family, *Rhinodontidæ*.

Although probably widely distributed within the tropics, this monster has hitherto been met with but locally. For many years the sole evidence of its existence rested upon a specimen, 15 feet long, brought ashore in Table Bay in April, 1828, described and figured by the late Sir Andrew Smith, and now in the Paris Museum. Forty years later Dr. Percival Wright met with this shark in the Seychelles, and obtained the first authentic information about it. It does not seem to be rare in that archipelago, but is very seldom obtained on account of its large size and the difficulties attending its capture.

It is reported to attain a length of 60 feet, being thus, among living animals, only exceeded in size by the largest whales ; but, in spite of its colossal dimensions, it is quite harmless to human beings. Dr. Wright saw specimens which exceeded 50 feet in length, and one that was actually measured proved to be more than 45 feet long. Nothing more was heard of the species until 1878, when the capture of another specimen was reported from the Peruvian coast near Callao; finally, in the "nineties" it was discovered on the west coast of Ceylon, where two or three specimens were obtained. One of these is now exhibited in the Fish Gallery of the British Museum, where it forms one of the most striking objects, although only a young example, measuring 17 feet from the end of the snout to the extremity of the tail. It has been stated that this fish feeds on seaweeds, but probably its food is similar to that of the basking-shark.

BLUE SHARK

backward position of the small first dorsal fin, which does not reach to the level of the highest point of the back, instead of standing immediately above it. Instead of being subcylindrical, the whole body of this shark is markedly depressed, and the huge mouth forms a nearly oblong aperture, armed with bands of very small and numerous teeth. The sides of the tail bear a well-defined keel, and the lower lobe of the

BLUE SHARK AND ITS ALLIES

The well-known blue shark (*Carcharias glaucus*) may be taken as the typical representative of its family *Carchariidæ*. From other sharks the members of the present family may be distinguished by the absence of spines in both the dorsal fins, of which the first is situated above the interval between the pectoral and pelvic pairs ; by the presence of a nictitating membrane to the eye ; and by the teeth, when fully formed, being hollow, and usually pointed. The bony elements in the skin take the form of minute granules, thus constituting the well-known " shagreen," as the dried skin is termed.

In all the members of the typical genus (*Carcharias*) the snout is produced forwards. The crescentic mouth is situated on its under side, and armed with large, flat, triangular, single-coned teeth, which differ in shape in the upper and lower jaws. Spiracles are absent, and there is a pit at the root of the caudal fin, which has a distinct lower lobe. At the present day these sharks are represented by between thirty and forty species, of which the blue shark is one of the commonest and most widely distributed ; while in a fossil state the genus is known from the Tertiary formations. The blue shark is often from 12 to 15 feet long, but some of the other species are stated to grow to as much as 25 feet.

In common with the other large members of the suborder, all these sharks are more abundant in tropical than in temperate seas ; but the blue shark is by no means an uncommon visitor to British waters, more especially on the southern and western coasts of Ireland. It is a somewhat remarkable fact that in places like Aden, where sharks of various kinds abound, the natives will swim and dive fearlessly in the open sea, where a European would be almost instantly devoured by these monsters. The blue shark has the whole of the upper parts slaty blue, and the under surface white.

TOPE

Our next representative of the family is the tope (*Galeus canis*), a small shark belonging to a

HAMMER-HEADED SHARK

genus including only two species and characterised as follows.

The snout is short and the mouth crescentic ; very small spiracles are present ; there is no pit at the root of the caudal fin, which has only a single notch ; and the teeth have serrated edges, and a notch on the hinder border. The common tope, which is usually about 6 feet in length, although it may grow to 7 feet, ranges over all temperate and tropical seas, visiting the shores of such widely separated localities as California, the British Islands, and Australia. In colour it is dark grey above and dirty white beneath. The second living species inhabits the Japanese seas ; and teeth from

the European Tertiary have been referred to the genus.

In habits the tope is a bottom-haunting species— especially during the winter months—and devours other fish, crustaceans, and star-fishes. It is not infrequently taken by the line, and is thus a great source of annoyance to fishermen, especially on the Norfolk coast, where considerable numbers are sometimes hooked. The young are produced alive, and it is stated that there have been instances of as many as fifty individuals in a single brood.

HAMMERHEADS

Having teeth very similar to the true sharks, the five species known as hammerheads, or hammer-headed sharks, form a genus (*Sphyrna*) unique among fishes in the extraordinary conformation of the head, the front part of which is broad, flattened, and expanded on each side into a process bearing the eye. This is quite sufficient to distinguish the genus ; but it may be added that the caudal fin has a single notch and a pit at its root, there are no spiracles, the nostrils are situated on the front edge of the head, and the mouth is crescentic. The teeth differ from those of the true sharks in being similar in both jaws ; their margins being either smooth or serrated.

Hammerheads range over all the warmer seas, the common species being sometimes taken on the British coast ; and an extinct form occurs in Miocene strata. Growing to a length of some 14 or 15 feet, the common hammerhead is one of the most formidable and voracious of its tribe, and is much feared in the Indian seas.

HOUNDS

This name is given to two small British sharks, externally not unlike the tope, but with a blunter snout. As a genus they are characterised by a crescentic mouth, a rather short snout, minute spiracles, the absence of a pit at the base of the caudal fin, which has scarcely any lower lobe, and the slight difference in the size of the two dorsal fins. The teeth are small and numerous, being either blunt or with indistinct cusps, and forming a kind of pavement-like structure ; those in the upper jaw being similar to those in the lower.

The smooth hound (*Mustelus lævis*) is generally about 4 feet long, although it may reach to 6 feet. The sides of the back are marked by a series of whitish spots, more distinct in the young than in the adult. Feeding on molluscs and crustaceans, this species (which ranges over most warm seas) produces about a dozen young at a birth, these being attached by a placental structure to the walls of the uterus. Curiously enough, such connection is totally wanting in the other British form (*M. vulgaris*). In habits the hounds are bottom-haunting species, as might be inferred from the nature of their food. On the English coast the smooth hound generally makes its appearance during the summer in pursuit of shoals of pilchard and herring.

SPINY DOG-FISHES AND THEIR ALLIES

Although the members of the family *Spinacidæ* approximate in their external conformation more to the typical sharks than to the rays, yet in the structure of their vertebræ they agree with the latter. They

SPINY DOG-FISH AND SMOOTH HOUND

include somewhat generalised forms, in which the body is cylindrical or triangular, and but very slightly depressed ; the mouth being gently arched, and the snout blunt. The pectoral fins have no forward prolongation, and are not notched at their point of origin ; and the small and lateral gill-clefts may be either in the line of the pectorals or half below. The spiracles are large ; there is no nictitating membrane, and the two dorsal fins may or may not be provided with spines.

SPINY DOG-FISH

The common spiny, or piked, dog-fish (*Acanthias vulgaris*) is the most familiar representative of a very small genus characterised by the presence of spines to the dorsal fins, and by the peculiar form of the teeth, which are similar in the two jaws, and small, triangular, and compressed, with the points much turned aside, and the cutting edge formed by the inner margin.

The common species measures from 3 to 4 feet in length, and is slaty blue above and yellowish white beneath. It is very abundant on the British coasts, and, like the small spotted dog-fish, is a great pest on account of its habit of stealing bait. In common with an allied species (*A. blainvillei*), this dog-fish presents the peculiarity of inhabiting the two temperate zones but being unknown in the intervening tropical seas. The eggs are hatched within the body of the female, and a considerable number of young are produced at a birth. Somewhat dangerous wounds may be inflicted by the spines.

GREENLAND SHARK

The Greenland shark (*Læmargus borealis*) of the Arctic seas, which occasionally strays so far south as Britain, represents a genus

characterised by the small size of all the fins and the want of spines to the dorsals, the first of which is situated considerably in advance of the pelvic pair. The skin is uniformly covered with small tubercles. The upper teeth are small, narrow, and conical ; but those of the lower jaw, which are numerous and form several series, have their points so much bent to one side that their inner margins form the cutting edge, which is not serrated. Growing to a length of 15 feet, the Greenland shark is a determined enemy to the right-whale, and when feeding on the carcase of one of these is easily harpooned without attempting to escape. Four living young are said to be produced at a birth.

SPINY SHARK

Finally, we have the spiny shark (*Echinorhinus spinosus*) of the Mediterranean and Atlantic, which, while agreeing with the last in the small size of the fins and the absence of spines to the dorsals, differs by the teeth being alike in both jaws, and by the presence of large rounded tubercles scattered over the skin. The body is very bulky, and the tail short. This shark lives at considerable depths, rarely coming to the surface.

ANGEL-FISH

The sole existing representative of its family (*Rhinidæ*), the angel-fish, or monk-fish (*Rhina squatina*), constitutes, so far as external form is concerned, a connecting link between the sharks and the rays. The body is as much depressed as in some of the latter, but the mouth is nearly terminal, and although the basal portion of the pectoral fins is much produced forwards, it does not extend so far as to join the head. The wide gill-clefts are lateral, and partly covered by the base of the pectoral fins ; the spiracles are wide ; and the teeth are conical and pointed. The spineless dorsal fins are situated on the tail, and the skin is studded with tubercles.

Not infrequently growing to a length of at least 5 feet, the angel-fish is almost cosmopolitan, and by no means uncommon on the British coasts, more especially off Scotland. In colour it is mottled chocolate-brown above and whitish beneath, and, except that it produces living young, which may number as many as twenty at a birth, its general habits resemble those of the rays.

SIDE-GILLED SAW-FISHES

In the members of the family *Pristiophoridæ* the upper jaw is produced into a long, flattened beak, furnished on either edge with a series of large, sharp, and pointed teeth, set in distinct sockets at a considerable distance from one another. These fishes approximate to the sharks in the lateral position of the gill-clefts. There is but a single existing genus (*Pristiophorus*), containing four species, one of which (*P. japonicus*)

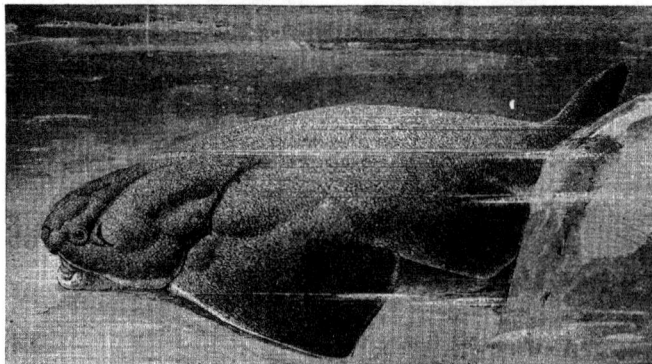

ANGEL-FISH

is shown in the illustration. They are comparatively small fishes confined to the Japanese and Australian seas. Having the body scarcely depressed, and the pectoral fins of moderate dimensions, and not extending forwards to the head, these forms are distinguished from true saw-fishes by the lateral position of the gill-clefts, but also by the full development of the so-called prepalatine cartilage, and the presence of a pair of long barbels on the lower aspect of the jaw.

TRUE SAW-FISHES

The true saw-fishes (family *Pristidæ*) possess a similar weapon to that found in the members of the last family, but the gill-clefts are situated on the under surface of the body, as in all rays, of which they must be regarded as the least-specialised examples. The elongated and slightly depressed body merges insensibly into the powerful tail, and the pectoral fins are only of moderate size. The first of the two spineless dorsals are about opposite the bases of the pelvics. There are no barbels.

The teeth of the saw are lodged in sockets of calcified cartilage, while those in the jaws are minute and blunt. The saw consists internally of three, or sometimes five, hollow calcified cartilages, in the form of long, tapering tubes, placed side by side, and held together by integument, which is likewise more or less hardened by the deposition of calcareous matter.

There is but one genus (*Pristis*), including several existing species which are most abundant in the tropical seas. The most familiar species (*P. antiquorum*) inhabits the Mediterranean and Atlantic. These fishes not uncommonly grow up to 20 feet in length, but Day records one of 24 feet; in such monsters the saw may be fully 6 feet in length, with a basal width of 1 foot. Some of the Indian species ascend rivers to a considerable distance beyond the influence of the tides. Saw-fishes use their weapon of offence by striking sideways through the water, and thus inflict terrific injuries; and it is stated that in the Indian estuaries large ones have been known to cut bathers completely in two. After tearing off pieces of flesh, or ripping up the body of their victim with the saw, these fishes seize and swallow the smaller fragments thus detached. In the Malay region the flesh of one species is highly esteemed as food; and its fins, like those of sharks, are, after due preparation, exported to China.

Fossil remains of extinct species of the genus occur throughout a large portion of the Tertiary; and in an allied Eocene genus, *Propristis*, the teeth of the saw are not implanted in calcified sockets. A very remarkable type (*Sclerorhynchus*) has left its remains in the upper Cretaceous rocks of Mount Lebanon. Not only does this fish differ from living forms by the distinctly depressed form of the relatively short and broad body, and the backward extension of the pectoral fins, which almost reach the pelvic pair, but the teeth, instead of being implanted in sockets, are merely attached to the skin by an expanded and crimped base. The central of the three rods in the interior of the saw extends to the saw's extremity, instead of stopping short; and from the smaller teeth at the base of the saw a complete gradation can be traced to the tubercles dotting the skin.

Assuming, as is most probably the case, that saw-fishes are highly-specialised sharks, it is remarkable to find that the earliest known member of the family has a somewhat skate-like form of body, and a type of dentition which could not apparently be very readily modified into that of the existing forms.

BEAKED RAYS

With the family of the beaked rays (*Rhinobatidæ*) we come to the first of rays and skates proper, in all of which the pectoral fins are so extended forwards as to form, with the head, the body, the so-called "disc." The dorsal fins are always situated on the tail, while the mouth is generally, and the gill-clefts always, on the under side of the body. The tail is long and powerful, with two well-developed dorsal fins, and a longitudinal fold on each side; the disc is not excessively dilated, the rayed portion of the pectoral fins stopping short of the beak; and there is no electric organ. Skates and rays in general are among the most hideous and repulsive of all fish, and many—especially in the warmer seas—attain enormous dimensions; while some inflict dangerous wounds with their tail-spines. The tooth-like tubercles on the skin frequently attain a great development, and are aggregated into prominent bosses or longitudinal ridges.

Dr. Günther writes that the mode of life of these fishes is quite in accordance with the form of their body, the true rays leading a sedentary life, moving slowly on the bottom of the sea, and rarely ascending to the surface. Their tail has almost entirely lost the function of an organ of locomotion, acting in some merely as a rudder. They swim slowly by means of the pectoral fins, the broad and thin edges of which are moved in an undulating manner, like the dorsal and anal fins of ordinary flat-fishes (*Pleuronectidæ*).

They are exclusively carnivorous, but being unable to pursue and catch rapidly-moving animals, feed chiefly on molluscs and crustaceans. The colour of their skins harmonises so closely with their surroundings that other fishes approach near enough to be captured by them. The mouth of the rays being on the lower surface of the head, the prey is not directly seized by the jaws; but the fish darts over the victim so as to cover and hold it down, when it is conveyed by some rapid motions to the mouth. Rays do not descend to the same depth as sharks; with one

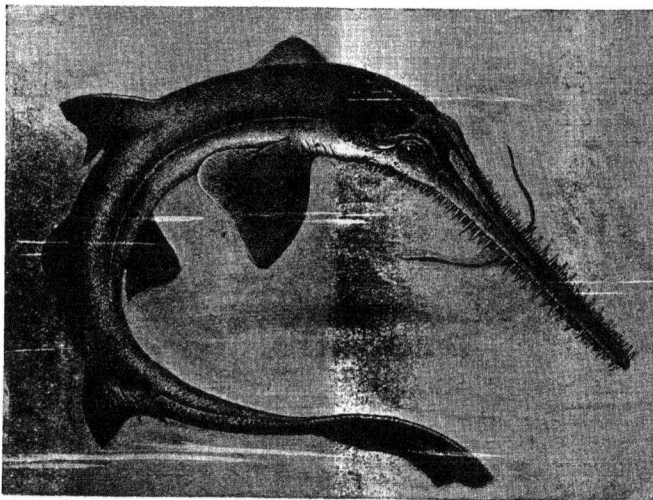

JAPANESE SAW-FISH

exception, none have been caught in more than one hundred fathoms. Most are coast fishes, and have a comparatively limited range, none extending from the northern into the southern temperate zone. Some of the eagle-rays are, however, more or less pelagic, though when met with swimming in the open sea it is probable that shoal-water is fairly near.

As may be observed in many of the western lochs of Scotland, skates and rays are more or less gregarious. They frequently arrive suddenly on oyster-beds, where they appear to remain so long as any of the molluscs are obtainable. Writing of the species armed with caudal spines, Day observes that they "lie concealed in the sand, and are reputed to be able to suddenly encircle fish or other prey swimming above them with their long, whip-like tails, and then wound them with their serrated tail-spines." Many rays ascend rivers to considerable distances, and some, especially in Tropical America, are exclusively inhabitants of fresh water. Nearly all lay eggs.

HALAVI RAY

To illustrate the typical genus of the family, which is represented by about a dozen species from the warmer seas, we take the halavi ray (*Rhinobatis halavi*), ranging from the Mediterranean and West African coasts to China. The depressed body passes imperceptibly into the tail; the snout is produced into a long beak, the space between which and the pectoral fin is occupied by a membrane; and the wide nostrils are oblique, with their front valves separate. The blunt teeth are marked by an indistinct transverse ridge; the spineless dorsal fins are situated far behind the pelvics, and the caudal has no lower lobe. Fossil species are found from the upper Jurassic onwards.

TRUE RAYS, OR SKATES

Represented by ten British species, all of which belong to the type genus (*Raia*), the true rays (family *Raiidæ*) are characterised by the broad rhombic form of the disc, the skin of which is generally marked with tooth-like rugosities. The tail has a longitudinal fold on each side, the degree of development of the median fins is variable, and the rayed portion of the pectoral fins extends to the snout. Except in the tail, electric organs are wanting. Of the type genus we take the common British thornback (*R. clavata*), so named

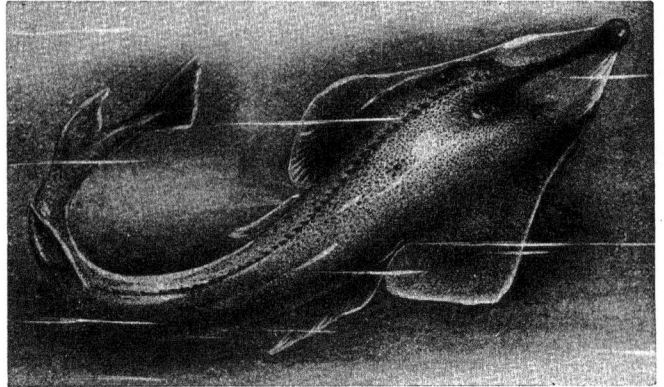

HALAVI RAY

because the whole of the upper surface is studded with claw-like spines; the tail being also armed with longer spines, of which a row runs along the middle of the back. The prevailing colour of the upper surface is brown, with numerous lighter spots, while beneath it is pure white.

Fossil skates of this genus are first known from the upper Cretaceous. At the present day the family is represented by three genera, each with but few species, from the warmer seas; and there are certain generic types. In this genus the tail is very sharply defined from the disc, which is generally covered with rugosities; the pectoral fins stop short of the end of the snout, the pelvics are deeply notched, with a stout front cartilaginous ray; the tail carries two dorsal fins, and the caudal is rudimentary or wanting.

Most of these skates present sexual differences, which in the thornback and several other species display themselves in the teeth. These, in the male, are sharp and pointed, but in the female blunt and flattened. While the males of all species are armed with patches of claw-like spines lying in grooves on the upper surface of the pectoral fins, and frequently also on the sides of the head, the females of some species have a kind of buckler of asperities on the disc, which is wanting in the other sex. In other cases the variation takes the form of a difference in colour. The numerous members of this genus are in the main characteristic of the cooler seas, and while they are more abundant in the Northern than in the Southern Hemisphere, some of them approach nearer to the Arctic and Antarctic Circles than is the case with any other rays. The flesh of all is edible, and many species are commonly sold as food.

The common skate (*Raia batis*), which is ordinarily of from 2 to 4 feet in length, is greyish white in colour, with black specks, the whole upper surface being more or

SAW-FISH

less granulated. Buckland records an unusually large specimen, which weighed 90 pounds.

EAGLE-RAYS

Known as devil-fishes, the eagle-rays (family *Myliobatidæ*) include the largest representatives of their tribe, and are characterised by the extreme width of the disc, owing to the great development of the pectoral fins, which are, however, interrupted at the sides of the head, to reappear as one or two small cephalic fins on the snout. The tail is slender and whip-like, the mouth straight, and the teeth, when present, take the form of a solid pavement, adapted for crushing the shells of molluscs and other hard substances.

The eagle-rays are inhabitants of tropical and temperate seas ; and the members of some of the genera are remarkable for the development of the so-called cephalic fins into a pair of horn-like appendages, which are said to be employed in capturing prey and helping to convey it to the mouth. Five genera are included in the family, all the members of which appear to be viviparous.

The type genus (*Myliobatis*) is represented by a smaller number of existing species, two of which are European ; one of these (*M. aquila*) occasionally visiting the British coasts, where it is often termed the whip-ray. In this group the head is free from the disc, and the fin on the snout single. The large, flat, hexagonal teeth form a tesselated pavement, highly convex in the upper but flat in the lower jaw ; the individual teeth are arranged in seven longitudinal rows. The whip-like tail, in addition to a dorsal fin near the root, is generally armed with a large barbed spine about the middle of its length. This almost cosmopolitan species may attain a length of more than 15 feet, with a weight of about 800 pounds. When captured, these rays lash out with their tails, inflicting severe wounds with the spine.

Fossil species of this genus occur in most of the Tertiary strata ; and among these one from the Eocene of Egypt is remarkable for its enormous size, the teeth of the middle row being rather more than 5 inches in width. Although it is difficult to form an estimate of the exact size of the fish to which these teeth belonged, it is thought that the width of the disc must have been about 15 feet.

In an allied genus (*Aëtobatis*), now represented by a single widely-spread tropical species, but common in the Tertiary formations, the snout carries two fins, and the dentition comprises only a single series of transversely elongated teeth, corresponding to the central row of the typical genus. In the third genus (*Rhinoptera*), of which there are seven living and several Tertiary species, the so-called fins on the snout are also double ; while the tesselated teeth form five or more series.

The largest existing members of the family belong to the mainly tropical genera *Dicerobatis* and *Cephaloptera*,

THORNBACK SKATES

to which the name of devil-fish might well be restricted. In the former the pectoral fins do not extend on to the sides of the head, which is truncated in front, and furnished with a pair of forwardly-directed appendages containing fin-rays, the nostrils being widely separated. Both jaws contain numerous rows of flat or tuberculated teeth ; and the whip-like tail has a single dorsal fin above and between the pelvic pair, and may be armed with a spine. In the second genus the mouth is terminal, and teeth are present only in the lower jaw.

One of the Indian representatives of the first genus is known to measure fully 18 feet across the disc, and a weight of over 1,200 pounds has been recorded. Sir W. Elliot states that the horn-like appendages " are used by the animal to draw its prey into its mouth, which opens like a huge cavern between them. The fishermen [in India] say they see these creatures swimming slowly along with their mouths open, and flapping these great sails inwards, drawing in the smaller crustaceans on which they feed." The capture of such hideous monsters is a work of no little difficulty and danger, as they are quite capable of overturning a boat ; and the danger is said to be the greatest in the case of a female accompanied by her single offspring.

Curiously-ridged quadrangular teeth from the Chalk, described under the name of *Ptychodus*, appear to have belonged to an extinct type of eagle-ray. In these teeth the highly-polished crown is ornamented with large transverse or radiating ridges, surrounded by a more finely-marked marginal area of variable width. They are arranged in longitudinal rows.

ELECTRIC RAYS

Like the electric eel, the members of the family *Torpedinidæ* are characterised by their power of giving galvanic shocks ; the organs from which this power is derived taking the form of a series of vertical hexagonal prisms, situated on each side of the front of the disc between the head and the pectoral fins. In addition to this distinctive feature, these rays are characterised by the broad and smooth disc, in which the rays of the pectoral fins do not extend in advance of the base of the snout; while the median fins are well developed. The family is represented by several genera, ranging over the Mediterranean, and Atlantic, and Indian Oceans.

MARBLED ELECTRIC RAY

A well-known example of the typical genus is the marbled electric ray (*Torpedo marmorata*). The hexagonal prisms forming the electric organs are sub-divided into a series of cells by a number of delicate transverse partitions, the cells at the two ends of the prisms being in contact with the skin, and the whole structure liberally supplied with nerves. Internally, each cell is lined by a nucleated structure, within which is a mass of jelly-like substance.

"The fish," writes Dr. Günther, "gives the electric shock voluntarily, when it is excited to do so in self-defence, or intends to stun or kill its prey; but to receive the shock the object must complete the galvanic circuit by communicating with the fish at two distinct points, either directly or through the medium of some conducting body. It is said that a painful sensation may be produced by a discharge conveyed through the medium of a stream of water." Specimens measuring from 2 to 3 feet across the disc are stated to be able to disable a man by the discharge of the battery.

STING-RAYS

Apparently the most specialised members of the entire group are the sting-rays (*Trygonidæ*), in which the pectoral fins are continued round the end of the snout.

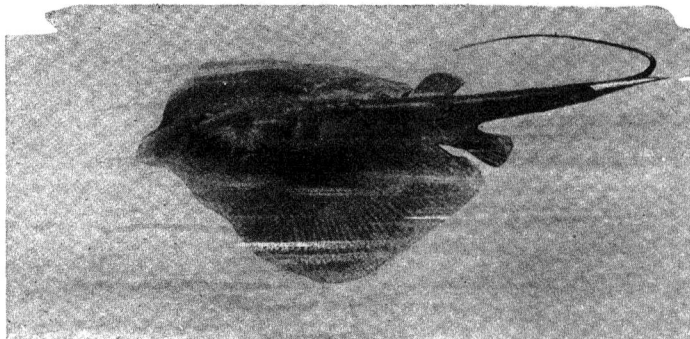

COMMON SKATE AND MARBLED RAY

so that the whole of the edge of the very wide disc is formed by these fins, in the centre of which are the more elevated head and body. The long and slender tail, frequently armed with a serrated spine, is sharply defined; and the median fins, if present, are either imperfectly developed or modified into serrated spines. The forms with armed tails, the sting-rays proper, inflict very severe wounds, dangerous not merely from the actual lesion, but apparently also from the presence of poison. In the larger kinds these spines may be as much as 7 or 8 inches in length, and are from time to time shed and replaced by new ones growing from behind. Very numerous in species, and arranged under several genera, the sting-rays are more abundant in tropical seas, although some range in temperate waters.

The typical genus includes some twenty-five species one of which (*Trygon pastinaca*) ranges from the south of England westwards to America, and eastwards to Japan. In this group the greatly elongated and tapering tail is armed with a barbed, arrow-shaped spine, while the skin is either smooth or dotted over with tubercles; the nasal valves unite into a quadrangular flap; and the teeth are flattened. Mainly characteristic of tropical latitudes, these rays are most abundant in the Indian and Atlantic Oceans, although some are inhabitants of fresh-water lakes in Eastern Tropical America.

THE STING-RAY

SUBCLASS II. HOLOCEPHALI
CHIMÆROIDS

REPRESENTED by three existing marine genera and a number of extinct types, the quaint-looking chimæroids probably represent an early offshoot from the ancestral sharks. The skeleton is cartilaginous, with the notochord either persistent or constricted and surrounded by cartilaginous rings, which are sometimes partly calcified. The upper jaw is fused with the skull, much as in lung-fishes. In the adult the skin is frequently quite naked, although in the young it may bear on the back a series of structures similar in structure to teeth, some extinct forms having plates of the same nature. In existing members of the group the optic nerves simply cross one another, and the intestine has a spiral valve; while further resemblances to sharks are shown by the presence of claspers in the males, and by the large size and small number of the single eggs.

The four gill-clefts on each side are close together, and protected by a fold of skin containing a cartilaginous operculum. There are no spiracles in the adult. The mouth is situated at the end of the snout, and the teeth on the palate and lower jaw are molar-like, while there is also a small pair of cutting vomerine teeth in the front of the upper jaw; the whole dentition superficially resembling that of the lung-fishes, although there are two pairs of upper palatal teeth, which present certain hardened areas known as *tritors*. The intestine possesses a spiral valve, as in sharks, but does not terminate in a cloaca, as in those fishes. The heart is shark-like in structure. The first dorsal fin may have a movable spine articulated to the spinous processes of the vertebræ, and the tail tapers into a thread. The existing members mentioned in this work are as follows:

FAMILY
Chimæridæ
GENUS 1
Chimæra
SPECIES

Chimæra .... Chimæra monstrosa

GENUS 2
Callorhynchus
SPECIES

Southern chimæra...................... Callorhynchus antarcticus

GENUS 3
Harriotta

The form variously known as the chimæra, king of the herrings, and rabbit-fish (*Chimæra monstrosa*), with two other existing species, typically represents the

HARRIOTTA

family *Chimæridæ*, which alone has survived to the present day. The family is characterised by the presence of a spine to the first dorsal fin and a prehensile spine-like structure on the head of the male; there are no superficial plates on the skull and only a single pair of lower teeth. The family, which contains a number of extinct genera, dates from the Lias; the typical genus being, however, unknown before the latter part of the Tertiary period. The living chimæras do not probably exceed 5 feet in length, and have the soft snout devoid

THE CHIMÆRA

of an appendage. The dorsal fins occupy most of the back, and the axis of the long filamentous tail is nearly continuous with that of the back, its end being provided above and below with a long, low diphycercal fin. The common species (page 1664) is found in the Mediterranean and East Atlantic; while a second occurs in the North Pacific, a third off Portugal and in the West Atlantic, and a fourth off Japan.

The common species takes its name of " king of the herrings " from being so frequently in pursuit of shoals of the herring, upon which species and other kinds of small pelagic fishes it chiefly subsists. It is said, however, also to feed on jelly-fish and crustaceans.

The southern chimæra, or elephant-fish (*Callorhynchus antarcticus*), from the southern temperate seas, differs from the preceding genus by the presence of a cartilaginous prominence, ending in a flap of skin, on the snout, and likewise by the upward direction of the extremity of the tail, which has no fin on its upper surface. The third genus, *Harriotta*, distinguished by the extreme elongation of the snout, is represented by one species from the Atlantic, and a second from Japanese seas.

As well-known extinct types of the family, we may refer to the Cretaceous and Tertiary genera *Edaphodon* and *Elasmodus*, the former including fishes of gigantic dimensions. The members of the extinct family *Myriacanthidæ*, of the Jurassic rocks, differ by having a few bony plates on the head, and three lower teeth; while the *Squaloraiidæ*, as represented by *Squaloraia* of the Lias, were somewhat ray-like forms, with a depressed trunk and elongated muzzle, and no spines to the dorsal fins.

SUBCLASS III. TELEOSTOMI
END-MOUTHED FISHES

ALTHOUGH there is still some degree of uncertainty as to the best mode of arranging certain groups of the bony fishes, the following scheme may be adopted :

Fan-finned Fishes—Astinopterygii

The typical bony fishes of the present day were formerly regarded as indicating a primary group (Teleostei) of equal rank with a second one known as the Ganoidei ; the latter containing the American bony pike and the African bichir, together with a host of extinct genera possessing a similar armour of hard ganoid scales. A fuller study of these and other allied fossil forms has,

however, shown the existence of such a complete transition from these so-called ganoids to the typical bony fishes that it has become necessary to include the whole of them in a single subclass, under the title heading this chapter.

The subclass agrees with sharks and rays in the structure of the skull, which is formed on what may be termed the hinged type (hyostylic)—that is to say, the upper jaw remains separated from the cranium proper, to the hinder part of which it is movably articulated by the hyomandibular. The internal skeleton is more or less ossified, with the development of membrane-bones on the jaws ; the gills are comb-like ; the gill-clefts are but slightly separated from one another, and are fully protected by a gill-cover, or operculum ; the membrane-bones of the shoulder-girdle (that is to say, the scapula, claviculars, etc.) are connected with the hinder part of the skull ; and the external skeleton takes the form either of plates of bone or of calcified overlapping scales. In existing forms the eggs are small, numerous, and generally massed together ; the two optic nerves may either simply cross one another or may give off mutually interlacing fibres ; a swim-bladder—with or without a duct—is very generally present ; and the intestine is sometimes furnished with a spiral valve.

ORDER I. STYLOPTERYGII—LOBE-FINNED GANOIDS

THE first two suborders, Actinistia and Rhipidistia, of the order Stylopterygii, or Crossopterygii, have no existing representatives ; but of the third, Cladistia, the following members are dealt with in this work :

The suborder Actinistia is represented by the hollow-spined ganoids (*Cœlacanthidæ*), which range from the Carboniferous to the Jurassic, and are best known by the genera *Cœlacanthus* and *Undina*. In these fishes the body is deeply and irregularly spindle-shaped, with the scales overlapping, rounded, and more or less coated with ganoin. There is a gill-cover and a single pair of jugular plates ; the paired fins are obtusely lobate ; the tail is diphycercal, frequently with a small supplemental fin at the extremity ; and the swim-bladder is ossified. The suborder Rhipidistia includes Devonian forms, differs in the structure of the unpaired fins, and is represented by three well-defined families. In the first, which is typified by the genus *Holoptychius*, the lobes of the pectoral fins are long and acute, while the teeth have complex infoldings of the outer layer, somewhat after the manner of those of the primeval salamanders, and the scales are thin and cycloidal.

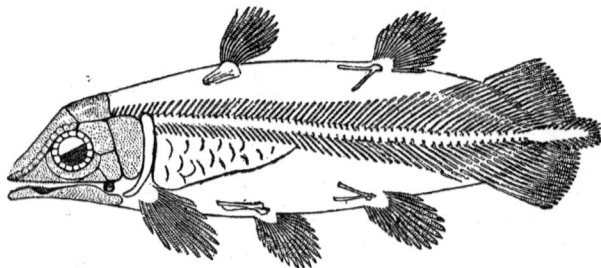

SKELETON OF A HOLLOW-SPINED GANOID

The second family, of which *Rhizodus* is the typical genus, differs by the lobes of the pectoral fins being shorter and blunter, and also by the less complicated infoldings of the teeth. To this family belongs *Gyroptychius*, from the Devonian or Old Red Sandstone of Scotland. While agreeing with the last in the obtusely lobate pectoral fins, the third family, as represented typically by *Osteolepis* of the Old Red Sandstone, is characterised by the walls of the teeth being slightly infolded only at their bases, and by the scales being of the true quadrangular ganoid type. Remains of these fishes occur in extraordinary abundance in the Old Red Sandstone of Scotland ; and as this deposit is of fresh-water origin, it is evident that they were either fluviatile or lacustrine forms.

BICHIRS AND THEIR ALLIES

The survivors of the suborder Cladistia are the various species of bichir of the rivers of Tropical Africa, of which *Polypterus bichir* of the Nile is the typical representative, and the reed-fish (*Calamoichthys calabaricus*) from Old Calabar ; these constituting the family *Polypteridæ*.

Here the notochord is more or less constricted and replaced by ossified vertebræ ; the baseosts, or superior supporting elements, are rudimentary or wanting in the median fins ; whereas the axonosts, or inferior supports, form a regular series equal in number to the dermal fin-rays with which they articulate. The scales are ganoid, and the fins without fulcra. The dorsal fin is divided into a number of finlets, each formed by a spine in front and a series of rays behind the anal fin being situated close to the diphycercal caudal, and the vent near the end of the tail, while the whole caudal region is very short. In the bichir the body is moderately elongated ; the teeth are rasp-like, and arranged in broad bands in the jaws and on the vomers and palatines, the jaws also bearing an outer series of larger pointed teeth ; and the pelvic fins are well developed, but do not show the obtusely lobate structure characterising the front pair. The large air-bladder is double.

The bichirs inhabit the Upper Nile and the rivers on the west of Tropical Africa, examples being occasionally carried down into the Lower Nile. The number of finlets varies from fifteen to eighteen in the typical Nile species, which grows to three or four feet in length, but in all the species from the Congo the number is reduced, being only six or seven in *P. retropinnis*.

Bichirs live close to the mud of the river-bottoms, upon which they appear to walk with their pectoral fins. From time to time, when the water is foul, they rise swiftly to the surface to take in a mouthful of air ; so that we may assume the air-bladder acts in part as a breathing organ. They feed on fishes, frogs, and crustaceans, and lay green eggs of the size of millet-seed. The reed-fish is a smaller form, characterised by the elongation of the body and the absence of pelvic fins.

ORDER II. ASTYLOPTERYGII—FLAT-FINNED GANOIDS

In this order the paired fins do not possess a thick central muscular lobe, but are flat, and supported by diverging rays. Owing to the fan-like arrangement thus constituted, it is usual to designate the members of the present order, together with those belonging to the next order (Teleostei), fan-finned fishes (Actinopterygii). The existing members of the order mentioned in this work are as follows :

SUBORDER 1
Sturgeon Tribe—Chondrostei
FAMILY 1
Toothed Sturgeons—Polyodontidæ
GENUS 1
Polyodon
SPECIES
Spoon-bill sturgeon.........Polyodon folium
GENUS 2
Psephurus
SPECIES
Sword-bill sturgeon........Psephurus gladius
FAMILY 2
Acipenseridæ
GENUS 1
Toothless Sturgeons—Acipenser

SPECIES
Common sturgeon .... Acipenser sturio
Sterlet....................... A. ruthenus
Giant sturgeon .... A. huso
GENUS 2
Shovel-beaked Sturgeons—Scaphirhynchus
SPECIES
Mississippi shovel-beaked sturgeon .. Scaphirhynchus platirhynchus
GENUS 3
Parascaphirhynchus
SPECIES
White sturgeon .... Parascaphirhynchus albus
SUBORDER 2
Ætheospondyli

FAMILY
Lepidosteidæ
GENUS
Lepidosteus
SPECIES
Bony pike .... Lepidosteus osteus
SUBORDER 3
Protospondyli
FAMILY
Bow-fin—Amiidæ
GENUS
Amia
SPECIES
Bow-fin .... Amia calva

The dermal rays of the dorsal and anal fins are more numerous than their supports in the suborder Chondrostei, and the baseosts, or superior row of supporting ossicles of the pelvic fins, are well developed. In living representatives of the sturgeon tribe the fibres of the optic nerves do not interlace at their crossing, and there is a spiral valve to the intestine. In both the living and extinct types the tail is of either the diphycercal or heterocercal type.

The sturgeon tribe is further characterised by the more or less completely persistent notochord, by the inferior and superior supporting ossicles (axonosts and baseosts) of the dorsal and anal fins forming a simple and regular series, and also by the presence of a pair of infraclavicular plates in the pectoral girdle. In all the known forms there is a single dorsal and an anal fin, both of which are well separated from the caudal ; while in the existing members the swim-bladder is furnished with a duct. Although represented at the present solely by the sturgeons and their allies, the group was very abundant during the Secondary epoch ; and while the sturgeons, together with certain extinct families, form what may be termed a degenerate specialised series characterised by the absence of

ganoid scales, in a second and normal series the body was covered with such scales.

TOOTHED STURGEONS

The toothed sturgeons, of which there are two existing representatives, each forming a genus by itself, constitute the family *Polyodontidæ*. While agreeing with the other members of the group in having the cartilaginous skull invested with a series of superficial bony plates, these fishes are specially distinguished by possessing a median unpaired series of bones in this shield, by the absence of branchiostegal rays, the presence of minute teeth in the adult,

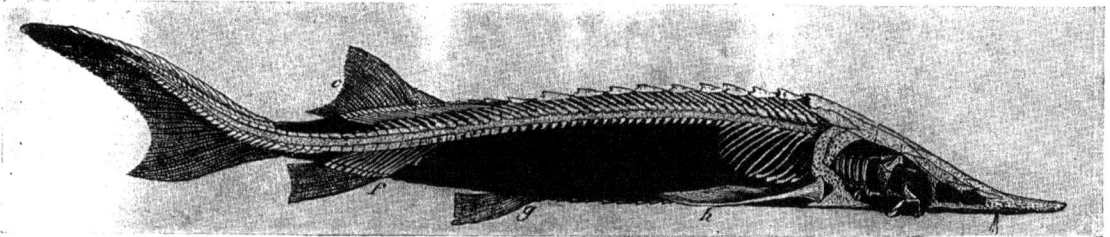

SKELETON OF STURGEON, SHOWING HETEROCERCAL TAIL
c, dorsal fin ; f, anal fin ; g, pelvic fin ; h, pectoral fin

the heterocercal tail, and by the skin being either naked or with some scales on the upper lobe of the tail.

SPOON-BILL STURGEON

The first of the two existing genera is represented by the spoon-bill sturgeon, or paddle-fish (*Polydon folium*) of the Mississippi, which grows to a length of 6 feet, and is characterised by the production of the upper jaw into a very long spoon-like beak, with thin, flexible margins. The gill-cover ends in a long, tapering flap ; the upper lobe of the tail bears a numerous series of narrow fulcra ; and the swim-bladder is cellular.

These great fishes feed on the muddy bottoms of rivers and lakes, using their spatula-shaped snouts to stir up the surface layer and secure small organisms as food. These chiefly consist of small crustaceans, which are strained out by means of the gill-rakers. As in the case of whales, the throat is extremely small, and this is no doubt correlated in both instances with the minute character of the prey.

A considerable fishery of spoon-bill sturgeons has been established in the lakes of the Lower Mississippi, chiefly for the sake of the roe, which is used as a substitute for caviare. It is strained, salted, and packed in kegs of 150 pounds each, and by mixing with it a proportion of the roe of the true sturgeons its quality s greatly improved. The flesh of these fishes is sent to the northern cities of the United States, and after being dried or smoked is sold as sturgeon. Only about one fish in every twenty-five yields caviare.

SPOON-BILL STURGEON

SWORD-BILL STURGEON

The sword-bill sturgeon (*Psephurus gladius*), from the Yang-tse-kiang and Hoang-ho rivers of China, differs in the more conical form of the bill, and in the large size and small number of the caudal fulcra. Growing to an enormous length—it is said as much as 20 feet—this fish agrees with the preceding in the very small size of its eyes, from which it may be inferred that neither hunts by sight. Indeed, in the muddy waters of the rivers they inhabit, eyes can be of little use, and no doubt this form makes use of its bill in the same way as the last species. Its flesh is used as an article of food.

TOOTHLESS STURGEONS

The typical sturgeons (*Acipenseridæ*) may be distinguished from members of the preceding family by the absence of teeth in the adult, and the presence of five longitudinal rows of bony plates on the naked subcylindrical body, as well as by the presence of four barbels in a transverse line on the under surface of the snout. The latter is somewhat elongated, and either subspatulate or conical in form, with the small, transverse mouth on its lower surface. All the vertical fins are armed with a single series of fulcra on their front edges ; the dorsal and anal are situated at a moderate distance from the caudal ; and the large swim-bladder is simple. Confined to north temperate regions, sturgeons are either exclusively or partially fresh-water fish, some of them only ascending rivers for the purpose of spawning, after which they return to the sea. With the sword-bill sturgeon, they include the largest fresh-water fishes of this region, several of the species commonly growing to 10 feet, while some are much larger. The females deposit enormous numbers of extremely minute eggs, the product of a single individual having been estimated at upwards of three millions during a season. This wonderful fecundity easily accounts for the vast numbers in which sturgeon, in spite of constant persecution, still crowd the northern rivers during the spawning season.

In addition to the excellence of their flesh, sturgeon are valued for their roe, from which is manufactured caviare, and for their swim-bladder, the inner coat of which forms the basis of isinglass. In a fossil state sturgeons are unknown before the upper part of the Eocene period. All the members of the genus are exceedingly voracious, and the majority are mainly carnivorous. During the winter many or all of them crowd together, either in inlets of the sea, estuaries, or the deep pools of rivers, where they undergo a kind of hibernation ; and it is stated that in some localities they bury their noses in the mud, with their bodies and tails standing vertically upwards like a series of posts. They increase very rapidly in size ; and the eggs are hatched in five days.

Although still abundant in the northern rivers, in those of Central Europe sturgeon have greatly decreased in numbers, and few really big fish are now taken. In the beginning of the year, when they are still torpid, sturgeon are captured by breaking the ice, and stirring up the mud at the bottom by means of very long poles armed with barbed prongs. As they seek to escape, some are stabbed with spears ; and it is said that ten large fish may be thus taken by a single fisherman. In summer regular fishing-stations are established on the Russian rivers, where the approach of a shoal is heralded by a watchman. Upwards of fifteen thousand sturgeon have been taken in a day at one of these stations ; and when the fishing is suspended for a short time, a river of nearly four hundred feet in width and five-and-twenty in depth has been known to be completely blocked by a solid mass of fish.

TRUE STURGEONS

The common sturgeon (*Acipenser sturio*) is the typical representative of the first genus, in which the rows of bony plates remain distinct from one another on the tail, spiracles are present, the upper tail-lobe is completely surrounded by fin-rays, and the snout is either short or developed into a narrow beak of moderate length.

There is some doubt as to the exact number of species of sturgeons, as these fish vary considerably according to their age, but it is probable that there are nearly twenty different kinds. Among the better-known forms one of the most esteemed is the sterlet (*A. ruthvenus*), which, although rarely exceeding a yard in length, yields better-flavoured flesh and finer caviare than any of the others. It is characterised by its narrow, pointed snout, and the great number of bony plates (sixty to seventy) on the sides of the body. Common in the Black Sea and Caspian, and their rivers, the sterlet is also found in the Siberian streams, while it ascends the Danube as far as Vienna.

In contrast to the sterlet may be noticed the giant sturgeon, or hausen (*A. huso*). Possessing from forty to forty-five lateral bony plates, this species may be readily distinguished by the absence of shields on the short, pointed snout. It is found in the Black Sea, Caspian, Sea of Azov, and their tributaries, and occasionally enters the Mediterranean. At one time this sturgeon was to be met with in the Danube by thousands, specimens over 24 feet in length being by no means uncommon ; but relentless slaughter has greatly reduced not only their numbers, but likewise their size, although even now fish of from 1,200 to 1,500 pounds weight are occasionally taken. These, however, are mere pigmies to certain Russian examples, one of which is stated to have weighed 2,760, and a second 3,200 pounds.

Migratory in its habits, this sturgeon crowds into the Russian rivers as the ice is breaking up. It appears

THE STERLET

that only full-grown fish ascend some rivers, as no small ones are found in the Danube ; but in the Volga these sturgeon are stated to remain during the winter in a semi-torpid condition. Although extremely powerful, the hausen is an inactive and timid fish, fleeing even from the diminutive sterlet, and passing much of its time on the mud at the river-bottom, though rising occasionally to swim near the surface. It is omnivorous, feeding on vegetable substances, other fish, especially various kind of carp, and even water-fowl. Its isinglass is inferior to that of the common sturgeon.

Rarely visiting the British coasts, where it is a " royal " fish, the latter species has only from twenty-six to thirty-one lateral plates, and from eleven to thirteen down the middle of the back ; the snout being pointed, and about equal to one-half the length of the head. It is a widely distributed form, frequenting both Atlantic coasts, but absent from the Caspian, although found in the Black Sea. In Italy it ascends the rivers from March to May ; and while in that country it does not commonly exceed 5 or 6 feet in length, specimens of upwards of 18 feet are on record.

SHOVEL-BEAKED STURGEONS

The four species of the genus *Scaphirhynchus* (which must not be confused with the toothless sturgeons) differ from those of the preceding genus by the production of the snout into a spatulate beak, by the narrow and depressed hinder portion of the tail being completely covered by the bony plates, by the absence of spiracles, and by the fin-rays not surrounding the extremity of the upper lobe of the tail, which terminates in a long filament. Of the four species, one (*S. platy-rhynchus*) is restricted to the Mississippi river-system, while the others inhabit the rivers of Central Asia, all being exclusively fluviatile in their habits.

The white sturgeon of Illinois (*Parascaphirhynchus albus*) is a new species described comparatively recently, and taken as the type of a new genus. It is distinguished from the ordinary shovel-beak by its uniformly light colour, long small eye, long and narrow snout, bare under parts, small and numerous plates, and a larger number of fins.

BONY PIKES

The bony pikes of the fresh waters of North America constitute a family (*Lepidosteidæ*) which forms the sole

existing representative of a distinct suborder (Ætheospondyli). The tail is of the abbreviated heterocercal type—that is to say, while its fin is more or less nearly symmetrical, the vertebral column, which retains its primitive tapering extremity, runs into the upper half. The scales are either typically ganoid or rounded and distinctly overlapping. The swim-bladder is connected with the œsophagus by a duct, the optic nerves simply cross one another, without any interlacing of their fibres, and there is a spiral valve to the intestine. The suborder including the bony pike may be distinguished from the next by the full ossification of the internal skeleton ; the scales being always of the typical quadrangular ganoid type, and the branchiostegal rays having no gular plate in advance of them.

As a family, the bony pikes, as represented by the common species (*Lepidosteus osteus*) are distinguished from all other fish by having the bodies of the vertebræ convex in front and concave behind, instead of being biconcave. The fins are furnished with fulcra, the dorsal and anal consisting of soft rays only, and being placed far back, near the caudal, which is of the abbreviated heterocercal type ; while the trunk is much longer than the abdominal portion of the vertebral column, and the branchiostegal rays are comparatively few. The body is elongate and subcylindrical ; the long snout spatulate or beak-shaped ; the cleft of the mouth wide ; and the palate and jaws armed with bands of rasp-like teeth, and also with larger conical ones. There are four gills and three branchiostegal rays on each side, and the air-bladder is cellular.

Bony pike, of which there are three existing species, are now confined to North and Central America and Cuba ; but they are represented in the European Eocene, and by allied extinct genera in the Eocene and Miocene strata of the United States, one of these also occurring in the French Eocene. The existing forms grow to a length of 6 feet, and are carnivorous, feeding upon smaller fishes. They are often known by the name of gar-pike, although that title is best restricted to a totally different group.

These fishes are very inimical to the spoon-bill

THE BONY PIKE

sturgeon fishery, getting enclosed in the nets, and breaking their way out with much damage. When it is remembered that they often attain a length of 8 feet and a weight of 200 pounds or more, the extent of the mischief can readily be imagined. As many as 30,000 are said to have been taken at one time in Lake Washington, their carcases being left on the banks. It has been suggested that their mail-clad skins might with advantage be used for the manufacture of ornamental leather, as in the case of the crocodile.

BOW-FINS

The bow-fin (*Amia calva*) of the fresh waters of the United States is the sole existing representative of a second and larger subordinal group (Protospondyi), differing from the last by the imperfect ossification of the skeleton, the notochord being either persistent throughout life, or, if partly replaced by vertebræ, those in front of the caudal region have their bodies composed of three distinct elements (pleurocentra and intercentrum), which remain separate and alternating even when fully developed. The lower jaw is complex, and composed of several pieces; in the pectoral arch the infraclavicular plate is absent; and the pectoral fin has more than three basal elements belonging to the true internal skeleton; while the tail is always abbreviated heterocercal.

GIANT AND COMMON STURGEONS

Together with three extinct genera, the bow-fin constitutes a family (*Amiidæ*), characterised as follows: The lower jaw has its suspending arrangement directed backwards, and the cleft of the mouth is wide; the degree of ossification of the vertebræ is variable, although these often form complete discs; the body is elongate or fusiform; the margins of the jaws are armed with an outer series of large and conical teeth, internally to which are smaller ones; fulcra to the fins are either wanting or of minute size; and the dorsal fin is of variable, although usually of considerable length. Having the scales thin, somewhat rounded, and overlapping, the bow-fin represents a genus in which there are no fulcra, and the long dorsal fin occupies three-fourths the length of the body, while the anal fin is short, the caudal rounded, and the throat furnished with a single gular plate, followed by a number of branchiostegal rays.

The single existing species of the genus (*Amia calva*), which attains a length of 2 feet, is confined to the fresh waters of the United States, where it is exceedingly abundant in some of the northern lakes; but

THE BOW-FIN

remains of extinct species have been obtained, not only from the Eocene rocks of the same country, but also from the upper Eocene and Miocene strata of Europe. Carnivorous in its diet, preying upon other fish, aquatic crustacean, sand insects, the bow-fin is capable of living for fully an hour out of water, and when in its native haunts, especially where the water is foul, comes frequently to the surface to breathe. There can be no doubt that the large swim-bladder serves as a lung. When near the surface, it often utters a bell-like note, probably due to the passage of air from the swim-bladder.

The breeding season, during which the colours of the fish are more brilliant, begins early in the spring, when the bow-fins leave the deep waters of the lakes and resort to sunny shallows, where rushes, flags, and other water-plants are abundant. Before spawning, the shoals divide into small parties, each consisting of one female and several males, a curious reversal of the usual order of things. The nest is made some time before spawning, but the way in which it is constructed is not definitely known. It is stated by fishermen to be formed by the fish swimming in circles among the weeds and rootlets, pressing these together into a mass comparable to a roughly-made bird's nest.

When the nest is ready, the female visits it several times in company with one or more of her partners, until spawning is completed. The fertilised eggs are sticky and cling firmly to the materials of the nest. The eggs being safely fixed, a male takes charge of the nursery, and by constantly swimming round it sets up currents which no doubt keep up a constant supply of oxygenated water, thus promoting hatching, which takes place within twenty-four hours after the eggs are laid. The fry leave the nest in a compact swarm, in charge of the male, round whom they constantly swim. He gradually leads his numerous offspring further and further from the nest, looking after them with exemplary care, and guarding them as far as possible from the attacks of the numerous foes that prey on young fishes. They are ultimately left to their fate, but the parental solicitude bestowed upon them in their tenderest hours must promote a larger survival than would otherwise be the case.

Megalurus, from the upper Jurassic, is an allied extinct genus with a short dorsal fin and fulcra; while the Jurassic *Eurycormus* and *Liodesmus* likewise belong to the same family. The family is represented by *Amiopsis* in the Cretaceous rocks, while the type genus *Amia* is first found in the Eocene of both Europe and North America, persisting in the latter until the present time.

ORDER III. TELEOSTEI—BONY FISHES

GENERAL CHARACTERISTICS

I N the order Teleostei are included the vast majority of existing fishes, including most of the familiar species used as food or valued for purposes of sport. The general shape of the body in typical cases—such as salmon, mackerel, and cod—is eminently adapted to rapid progression, and the externally symmetrical (homocercal) tail facilitates swimming at the same level without steering. A study of the development and internal anatomy, together with an examination of extinct types, shows that the tail is internally unsymmetrical, and that the stock has been derived from fishes with unequally lobed (heterocercal) tails like those of sharks and sturgeons.

The external part of the paired fins, which are used as delicate steering organs, is flat, and almost entirely supported by bony fin-rays developed in the skin, while the basal supports (pterygia) are much reduced. Except in the latter character these fins are similar to those of the flat-finned ganoids (Astylopterygii), on which account the two orders are sometimes grouped together as fan-finned fishes (Actinopterygii, see page 1679). The unpaired fins resemble the paired ones in structure.

In most cases the scales are thin, flexible, and over-lapping, lending themselves to rapid sinuous move-ments and diminishing friction, which is further reduced by the slimy nature of the skin. The internal skeleton is well ossified, and the vertebræ distinct, their bodies being usually biconcave.

The intestine (except in *Chirocentrus*) has no spiral valve, and it opens directly to the exterior, not into a cloaca; there is no spiracle, and the deeply-cleft comb-like gills are protected by a gill-cover supported by a very complete set of bones; there is no arterial cone to the heart (except in *Albula*); and the optic nerves simply cross without interchange of fibres. These fishes hatch out as larvæ, which are often very unlike the adult.

CLASSIFICATION OF TELEOSTEI MENTIONED IN THIS WORK

CLASSIFICATION OF BONY FISHES

SUBORDER 2
Carp, Cat-fishes, etc.—Cyprini-siluriformes
FAMILY 1
Electric Eels—Gymnotidæ
GENUS
Gymnotus
SPECIES
Electric eel ....Gymnotus electricus
FAMILY 2
Cat-fishes—Siluridæ
GENUS 1
Sac-gilled Cat-fishes—Saccobranchus
GENUS 2
Silurus
SPECIES
Wels ....Silurus glanis
GENUS 3
Bagarius
SPECIES
Yarrell's cat-fish ....Bagarius yarrelli
GENUS 4
Arius
SPECIES
Queensland cat-fish ....Arius australis
GENUS 5
Pimelodus
SPECIES
Leopard cat-fish ....Pimelodus pati
GENUS 6
Bagrus
SPECIES
Bayad ....Bagrus bayad
GENUS 7
Electric Cat-fishes—Malapterurus
GENUS 8
Mailed Cat-fishes—Callichthys
GENUS 9
Armoured Cat-fishes—Loricaria
GENUS 10
Doras
FAMILY 3
Carp Tribe—Cyprinidæ
GENUS 1
SPECIES
Blacksucker ....Catostomus elongatus
GENUS 2
True Carp—Cyprinus
SPECIES
Common carp ....Cyprinus carpio
GENUS 3
Carassius
SPECIES
Crucian carp ....Carassius vulgaris
Golden carp ....C. auratus
GENUS 4
Barbels—Barbus
SPECIES
European barbel ....Barbus vulgaris
Mahseer ....B. mosal
GENUS 5
Gudgeons—Gobio
SPECIES
British gudgeon ....Gobio fluviatilis
GENUS 6
Leuciscus
SPECIES
Roach ....Leuciscus rutilus
Chub ....L. cephalus
Dace ....L. vulgaris
Ide ....L. idus
Rudd ....L. erythophthalmus
Minnow.......................L. phoxinus
GENUS 7
Tench—Tinca
SPECIES
Tench ....Tinca vulgaris
GENUS 8
Beaked Carp—Chondrostoma
SPECIES
Beaked carp ....Chondrostoma nasus
GENUS 9
Rhodeus
SPECIES
Bitterling ....Rhodeus amarus
GENUS 10
Abramis
SPECIES
Common bream ....Abramis brama
White bream...................A. blicca
GENUS 11
Aspius
SPECIES
Rapfen ....Aspius rapax
GENUS 12
Alburnus
SPECIES
Bleak ....Alburnus lucidus
GENUS 13
Pelecus
SPECIES
Sichel ....Pelecus cultratus

GENUS 14
Misgurnus
SPECIES
Giant loach...............Misgurnus fossilis
GENUS 15
Nemachilus
SPECIES
Common loach ....Nemachilus barbatulus
GENUS 16
Cobitis
SPECIES
Spiny loach ....Cobitis tænia
GENUS 17
Gastromyzon
SPECIES
Bornean loach .....Gastromyzon borneënsis
FAMILY 4
Characinoid Fishes—Erythrinidæ
GENUS
Serrasalmo
SPECIES
American piraya ....Serrasalmo piraya

SUBORDER 3
Eels and their Allies—Anguilliformes
FAMILY 1
Murænidæ
GENUS 1
Muræna
SPECIES
Muræna .... Muræna helena
GENUS 2
True Eels—Anguilla
SPECIES
Common eel..............Anguilla vulgaris
GENUS 3
Congers—Conger
SPECIES
Common conger.............Conger vulgaris
GENUS 4
Deep-sea Congers—Synaphobranchus
GENUS 5
Saccopharynx
FAMILY 2
Symbranchidæ
GENUS 1
Amphipnous
SPECIES
Amphibious eel.........Amphipnous cuchia
GENUS 2
Monopterus
GENUS 3
Symbranchus
SPECIES
Bengal short-tailed eel .... Symbranchus
bengalensis

SUBORDER 4
Pike-like Fishes—Esociformes
FAMILY 1
Galaxiidæ
GENUS 1
Galaxias
SPECIES
Southern pikelet.......Galaxias attenuatus
GENUS 2
Neochanna
FAMILY 2
Southern Salmon—Haplochitonidæ
GENUS
Haplochiton
SPECIES
Zebra-salmon .... Haplochiton zebra
FAMILY 3
Killie-fishes, or Cyprinodonts—Cyprinodontidæ
GENUS 1
Cyprinodon
GENUS 2
Anableps
SPECIES
Double-eye.........Anableps tetrophthalmus
FAMILY 4
Pike—Esocidæ
GENUS
Esox
SPECIES
Pike ....Esox lucius
FAMILY 5
Scopeloids—Scopelidæ
GENUS 1
Scopelus
SPECIES
Phosphorescent sardine....Scopelus engraulis
GENUS 2
Ipnops
GENUS 3
Plagyodus

GENUS 4
Harpodon
SPECIES
Bummalow .... Harpodon nehereus
FAMILY 6
Blind-fish—Amblyopsidæ
GENUS 1
Amblyopsis
SPECIES
Kentucky blind-fish......Amblyopsis spelœa
GENUS 2
Chologaster
GROUP 2
Closed-bladdered Fishes—Physoclisti
SUBORDER 1
Spine-finned Fishes—Acanthopterygii
FAMILY 1
Perch Tribe—Percidæ
GENUS 1
True Perches—Perca
SPECIES
Common perch.............Perca fluviatilis
American perch .... P. flavescens
Turkestan perch .... P. schrenki
GENUS 2
Pike-Perches—Lucioperca
GENUS 3
Ruffes—Acerina
SPECIES
British ruffe .... Acerina cernua
GENUS 4
Percarina
FAMILY 2
Centrarchidæ
GENUS
Centrarchus
FAMILY 3
Bass and Sea-Perches—Serranidæ
GENUS 1
Centrogenys
GENUS 2
South American Perch—Percichthys
GENUS 3
Bass—Morone
SPECIES
Common bass .... Morone labrax
GENUS 4
Centropristes
GENUS 5
Anthias
GENUS 6
Typical Sea-Perches—Serranus
SPECIES
Sea-perch .... Serranus scriba
GENUS 7
Polyprion
SPECIES
Stone-bass .... Polyprion cernuum
GENUS 8
Lates
SPECIES
Nile perch .... Lates niloticus
Indian perch .... L. calcarifer
GENUS 9
Psammoperca
FAMILY 4
Scaly-finned Fishes—Chætodontidæ
GENUS 1
Chætodon
SPECIES
Bristly chætodon ....Chætodon setifer
Three-banded chætodon........ C. trifasciatus
Single-banded chætodon .... C. fasciatus
GENUS 2
Heniochus
SPECIES
Heniochus ....Heniochus macrolepidotus
Horned heniochus .... H. varius
GENUS 3
Holacanthus
SPECIES
Emperor-fish ....Holacanthus imperator
Indo-Malayan zebra-fish...... H. diacanthus
FAMILY 5
Archer-Fishes—Toxotidæ
GENUS
Toxotes
SPECIES
Archer-fish .... Toxotes jaculator
FAMILY 6
Red Mullets—Mullidæ
GENUS 1
Mullus
SPECIES
Common red mullet ....Mullus barbatus
Striped red mullet .... M. surmuletus

FAMILY 29
Barracuda-Pikes—Sphyrænidæ
GENUS
Sphyræna
SPECIES
Barracuda-pike............Sphyræna vulgaris

FAMILY 30
Sand-Smelts—Atherinidæ
GENUS
Atherina
SPECIES
Common sand-smelt ......Atherina presbyter

FAMILY 31
Grey Mullets—Mugilidæ
GENUS
Mugil
SPECIES
Common grey mullet ....Mugil capito

FAMILY 32
Gar-Fish and Flying-Fish—Scombresocidæ
GENUS 1
Belone
SPECIES
The gar-fish................Belone vulgaris
GENUS 2
Scombresox
SPECIES
Saury, or skipper.........Scombresox saurus
GENUS 3
Half-Beaks—Hemirhamphus
GENUS 4
Flying-Fish—Exocœtus
SPECIES
The flying-fish ....Exocœtus evolans

FAMILY 33
Sticklebacks—Gastrosteidæ
GENUS 1
Spinachia
SPECIES
Sea stickleback...........Spinachia vulgaris
GENUS 2
Gastrosteus
SPECIES
Three-spined stickleback ....Gastrosteus
aculeatus
Four-spined stickleback.........G. spinulosus
Nine-spined stickleback.........G. pungitius
United States stickleback ..G. novæboracensis

FAMILY 34
Sucker-Fishes—Gobioescidæ
GENUS
Lepadogaster
SPECIES
Two-spotted sucker-fish ....Lepadogaster
bimaculatus

FAMILY 35
Serpent-Heads—Ophiocephalidæ
GENUS 1
Ophiocephalus
SPECIES
The serpent-head ......Ophiocephalus striatus
GENUS 2
Channa

FAMILY 36
Labyrinth-Gilled Fishes—Anabantidæ
GENUS 1
Anabas
SPECIES
Climbing perch ....Anabas scandens
GENUS 2
Polyacanthus
SPECIES
Paradise-fish ......Polyacanthus opercularis
GENUS 3
Osphromenus
SPECIES
Gurami..............Osphromenus olfax
Indian gurami ....O. nobilis
GENUS 4
Betta
SPECIES
Fighting-fish ....Betta pugnax

FAMILY 37
Pomacentridæ
GENUS
Pomacentrus
SPECIES
Silver-dotted pomacentrus ......Pomacentrus
scolopsis

FAMILY 38
Wrasses—Labridæ
GENUS 1
Labrus
SPECIES
Striped wrasse..............Labrus mixtus
Ballan wrasse................L. maculatus

GENUS 2
Crenilabrus
SPECIES
Gold sinny ....Crenilabrus melops
GENUS 3
Tantoga
SPECIES
Black-fish...................Tantoga onitis
GENUS 4
Scarus
SPECIES
Mediterranean parrot-wrasse ..Scarus cretensis

FAMILY 39
Viviparous Wrasses—Embiotocidæ
GENUS 1
Ditrema
SPECIES
Silvery viviparous wrasse..Ditrema argenteum
GENUS 2
Heterocarpus

FAMILY 40
Cichlids—Cichlidæ
GENUS
Chromis
SPECIES
Buti ....Chromis niloticus
Tristram's cichlid............C. tristrami

SUBORDER 2
Tuft-Gilled Fishes—Lophobranchii

FAMILY 1
Mailed Tube-mouths—Solenostomatidæ
GENUS
Solenostoma
SPECIES
Blue-finned tube-mouth ....Solenostoma
cyanopterum

FAMILY 2
Pipe-Fishes and Sea-Horses—Syngnathidæ
GENUS 1
Syngnathus
SPECIES
Great pipe-fish............Syngnathus acus
GENUS 2
Siphonostoma
SPECIES
Deep-nosed pipe-fish ....Siphonostoma typhle
Australian pipe-fish..........S. intestinalis
GENUS 3
Doryichthys
GENUS 4
Nerophis
SPECIES
Straight-nosed pipe-fish....Nerophis ophidium
GENUS 5
Gastrotokeus
GENUS 6
Typical Sea-Horses—Hippocampus
SPECIES
Short-snouted sea-horse.......Hippocampus
brevirostris
GENUS 7
Australian Sea-Horses—Phyllopteryx
SPECIES
Seaweed-like sea-horse ....Phyllopteryx eques

SUBORDER 3
Comb-Gilled Fishes—Plectognathi

FAMILY 1
File-Fishes and Coffer-Fishes—Balistidæ
GENUS 1
Triacanthus
SPECIES
Indian file-fish.......Triacanthus brevirostris
GENUS 2
Trigger-Fishes—Balistes
SPECIES
Drumming trigger-fish.....Balistes aculeatus
GENUS 3
Coffer-Fishes—Ostracion
SPECIES
Four-horned coffer-fish..Ostracion quadricornis

FAMILY 2
Globe-Fishes and Sun-Fishes—Diodontidæ
GENUS 1
Triodon
SPECIES
Sac-fish ....Triodon bursarius
GENUS 2
Small-spined Globe-Fishes—Tetrodon
GENUS 3
Large-Spined Globe-Fishes—Diodon
SPECIES
Porcupine globe-fish ....Diodon hystrix
Spotted globe-fish.............D. maculatus

GENUS 4
Sun-Fishes—Orthagoriscus
SPECIES
Rough sun-fish ....Orthagoriscus mola
GENUS 5
Ranzania
SPECIES
Oblong sun-fish ....Ranzania truncata

SUBORDER 4
Soft-Finned Fishes—Anacanthini

FAMILY 1
Lycodidæ

FAMILY 2
Gadidæ
GENUS 1
Gadus
SPECIES
Cod ....Gadus morrhua
Haddock·.....................G. œglefinus
Whiting ....G. merlangus
Whiting-pout ....G. luscus
Power-cod ....G. minutus
Whiting-pollack..............G. pollachius
Coal-fish ....G. virens
GENUS 2
Merluccius
SPECIES
Hake...............Merluccius vulgaris
Southern hake ....M. gayi
GENUS 3
Lota
SPECIES
Burbot or eel-pout ....Lota vulgaris
GENUS 4
Molva
SPECIES
Ling ....Molva vulgaris
GENUS 5
Rocklings—Motella
SPECIES
Five-bearded rockling ....Motella mustela
Four-bearded rockling ....M. cimbria
Three-bearded rockling ....M. tricirrhata

FAMILY 3
Sand-Eels and their Allies—Ophidiidæ
GENUS 1
Cave-Fish—Lucifuga
GENUS 2
Brotula
GENUS 3
Typhlonus
GENUS 4
Aphyonus
GENUS 5
Snake-Fishes—Ophidium
GENUS 6
Genypterus
GENUS 7
Parasitic-Fish—Fierasfer
GENUS 8
Encheliophis
GENUS 9
Sand-Eels—Ammodytes
Lesser sand-eel ....Ammodytes tobianus
Greater sand-eel ....A. lanceolatus

FAMILY 4
Flat-Fishes—Pleuronectidæ
GENUS 1
Psettodes
SPECIES
Adalah....................Psettodes erumei
Hippoglossus
SPECIES
Halibut ....Hippoglossus vulgaris
GENUS 3
Rhombus
SPECIES
Turbot....................Rhombus maximus
Brill ....R. lævis
Mary-sole ....R. aquosus
GENUS 4
Pleuronectes
SPECIES
Plaice....................Pleuronectes platessa
Flounder.........................P. flesus
GENUS 5
Soles—Solea
SPECIES
Sole.......................Solea vulgaris
Lemon sole...................S. aurantiaca
Banded sole ....S. variegata
Dwarf sole....................S. minuta

TELEOSTEI—GROUP I. PHYSOSTOMI
TUBE-BLADDERED FISHES

Iℕ the group Physostomi the swim-bladder, when present, has a duct communicating with the stomach or œsophagus, while the pelvic fins are always abdominal, and the parietal bones are usually in contact with each other. All the fin-rays are soft and jointed, with the exception of the first in the

TARPON LEAPING OUT OF THE SEA

dorsals and pectorals, which may be ossified into spines. The members of the first suborder (Salmoni-clupeiformes) are the most primitive existing Teleostei, approaching such ganoids as the bow-fin (*Amia*) in some respects. The maxillary bone generally helps to form the margin of the upper jaw. None of the fins possesses spiny rays, and the pelvics are abdominal—placed far back.

Three extinct Mesozoic families (*Leptolepidæ, Pholidophoridæ,* and *Oligopleuridæ*) embrace a number of small primitive fishes, which help to fill the gap between Teleostei and simpler types.

TARPONS AND THEIR ALLIES

The family *Elopidæ* includes the most generalised members of the suborder. The scales are not easily shed, as in herrings ; the small dorsal fin is placed in the middle of the back, and the pectoral fins are set on close to the under side of the body. There are 10 to 16 rays in the pelvic fins. The mouth is large, and the very small teeth are all similar. The remains of fishes belonging to the family abound in the Cretaceous rocks (*Osmeroides* and *Elopopsis*), while the two existing genera (*Elops* and *Megalops*) first make their appearance in lower Eocene strata.

The ten-pounder (*Elops saurus*) is a handsome silvery fish widely distributed through tropical and subtropical seas, and particularly abundant off the coast of the southern United States. The young are extremely unlike the adults, being fragile ribbon-shaped creatures. An allied species (*E. lacerta*) is found on the west coast of Africa, and ascends rivers, having been found far up the Congo.

The second existing genus (*Megalops*) has larger

scales than the preceding, and the last ray of the dorsal fin is drawn out into a long filament. The gigantic tarpon (*M. atlanticus*), which attains a length of six feet and a weight of 110 pounds, ranges from the south-east coast of North America to Brazil. It is caught with rod and line from boats, and affords notable sport, leaping out of water like the grey mullet. The scales are over two inches broad, and are effectively used in Florida for fancywork. The allied but smaller ox-eye (*M. cyprinoides*) of India is much prized as food, and is sometimes kept in tanks for culinary purposes.

LADY-FISHES

The fishes included in the family *Albulidæ* are blunt-snouted, with small mouths and thick lips. Small, sharp teeth are borne on the edges of the jaws, and there are large, blunt crushing teeth within the cavity of the mouth. The pectoral and pelvic fins closely resemble those of the last family. The family is undoubtedly represented in Eocene strata, and possibly dates from the Cretaceous period. There are only two existing genera (*Albula* and *Bathythrissa*).

The lady-fish (*Albula conorhynchus*) is widely distributed on the shores of tropical and subtropical seas, and its food consists chiefly of bivalve molluscs, which are crushed by its powerful teeth. It is almost unique among Teleostei in the fact that the heart possesses a small arterial cone, a chamber of the heart

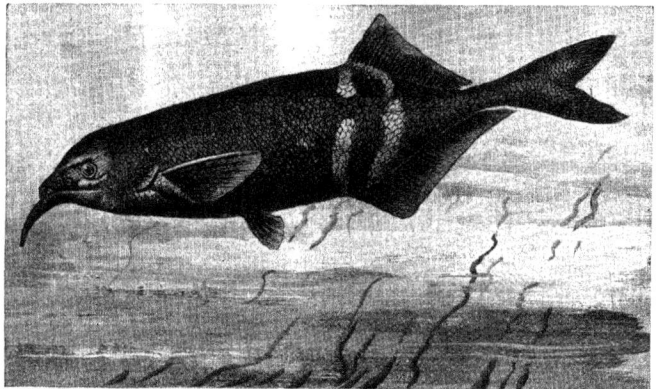

PETER'S BEAKED FISH

which succeeds the ventricle and is characteristic of the lower groups of fishes. In this case the conus, a waning structure, is short and provided with two rows of valves. The lady-fish hatches out as a transparent band-shaped larva, something like that of the eel. Adults are from 20 to 30 inches in length, and weigh from 3 to 10 pounds. They are fished for with living bivalves, affording good sport.

The second existing genus of the family (*Bathythrissa*) includes a form, the long-finned herring (*B. dorsalis*), from the deeper parts of the Japanese seas.

AFRICAN BEAKED FISH

The very remarkable fish *Mormyrus petersi* is a representative of a large group of fresh-water fishes (family

BORNEAN FEATHER-BACK

Mormyridæ) confined to Tropical Africa. These fishes are especially distinguished by a large air-filled space on either side of the brain-case in the auditory region. The anterior vertebræ are simple and unmodified ; and a subopercular bone is present in the gill-cover. Externally both the body and tail are covered with scales, but the head is naked, and the snout has no barbels. In the upper jaw the middle portion is formed by the united premaxillæ, and the sides by the maxillæ ; the gill-opening is reduced to a small slit ; there are no false gills ; and the swim-bladder is simple. A fatty fin is wanting ; and whereas in the typical and several other genera all the other fins are well developed, in *Gymnarchus* the caudal, anal, and pelvic fins are wanting, and the tail tapers to a point, instead of terminating in a deeply-forked fin. The eye are ill-developed, and sometimes covered by thick translucent skin. As a set-off against this, the sense of touch is unusually keen, while a relatively, large brain is correlated with considerable intelligence. The flesh of these fishes is much esteemed as food.

The beaked fishes are divided into numerous genera, distinguished from one another by the characters of the dentition, the length of the dorsal fin, and the form of the snout, and so forth. The figured species belongs to a group in which the dorsal fin is relatively short, scarcely exceeding the anal in length, while the snout is long and bent down. From its nearest allies the species in question is distinguished by the production of the extremity of the lower jaw into an elongated, conical, dependent, fleshy appendage, nearly equal in length to half the head. In colour the skin is dark brown, relieved by two lighter cross-bands between the dorsal and anal fins.

In the genus *Genyomyrus*, from the Congo, both jaws are prolonged into a tube-like beak ; and in some species of the allied *Gnathonemus* this beak, which may be bent downwards, exceeds half the length of the body. On the other hand, in *Mormyrops*, *Marcusenius*, and *Myomyrus*, of the Congo, there is no distinct beak. Some of these fishes grow to a length of between 3 and 4 feet, but others are comparatively small. In form the above-mentioned West African *Gymnarchus* is eel-like ; its jaws being armed with a series of incisor-like teeth, and its length

reaching to upwards of 6 feet. The eggs are deposited in a floating nest, guarded by the male. All the genera are furnished with a pair of organs on the sides of the tail, which are electric in function. Each consists of an oblong capsule, divided by vertical partitions into a number of chambers filled with a gelatinous substance. One of the species (*Mormyrus oxyrhynchus*) from the Nile is frequently depicted in the frescoes of the ancient Egyptians.

The forms with large mouths, like *Mormyrus* and *Gymnarchus*, feed on fish and crustaceans, the rest on smaller prey.

SILVERY MOON-EYE

Externally, the silvery moon-eye (*Hyodon tergisus*), the sole representative of the family *Hyodontidæ*, has the body covered with cycloid scales, the head naked, and no barbels. The margin of the upper jaw is formed by the premaxillæ in front and the maxillæ at the sides, the latter bones being articulated to the former at the point of junction ; and all the elements of the gill-cover are present. There is no fatty fin, and the short dorsal is placed in the caudal region, above the fore part of the longer anal ; the caudal fin being forked. The gill-openings are wide ; the stomach is horseshoe-shaped, the intestine short, and the swim-bladder simple ; false gills being absent. The body is oblong and compressed, with a part of the lower surface forming a sharp edge ; and the cleft of the mouth is somewhat oblique ; all the bones of the palate bearing small teeth, and the edges of the tongue carrying a larger series. Before their extrusion, the eggs fall into the cavity of the abdomen. This fish grows to a length of from 1 foot to 18 inches, and is confined to the fresh waters of North

CHISEL-JAW AND MOON-EYE

America, being abundant in the lakes and rivers east of the Rockies.

FEATHER-BACKS

From the peculiar form of the dorsal fin certain fresh and brackish water fishes from West Africa and the Oriental region, one of which (*Notopterus borneënsis*) is shown in the illustration, have received the not inappropriate name of feather-backs. They constitute a family (*Notopteridæ*) differing from all the others in this section by the tail being tapering and fringed inferiorly by a continuation of the anal fin. Both the body and the head are covered with small scales; barbels are wanting; the margin of the upper jaw is formed in front by the premaxillæ and at the sides by the maxillæ; and the opercular bones are incomplete. There is no fatty fin, and the dorsal, when present, is very short, and situated in the caudal region; the pelvic pair being rudimentary or wanting. The swim-bladder is divided internally into several compartments, and terminates at each end in a pair of narrow prolongations, of which the anterior ones are in communication with the organ of hearing. A further peculiarity is that the spawn falls into the cavity of the abdomen previous to its extrusion

ARAPAIMAS

The large tropical fresh-water fishes, *Osteoglossidæ*, collectively known as arapaimas (although this name properly belongs to the Brazilian species) present a curious general similarity in their geographical distribution to existing lung-fishes, although in the present instance one of the genera has a much wider range than either of the lung-fishes. As suggestive of the northern origin of the present group, it is noteworthy that an arapaima exists in Sumatra, and also that an extinct genus (*Dapedoglossus*) occurs in the Eocene strata of the United States.

Externally these fishes have the body covered with large, hard scales, of a mosaic-like structure; the lateral line being formed by wide openings of the mucus canal, and the scaleless head nearly covered with roughened ossifications of the skin. The margin of the upper jaw is formed both by the premaxillæ and maxillæ, the gill-openings are wide, and false gills are wanting. The long dorsal closely resembles the anal fin, over which it is placed in the caudal region of the body; both coming very close to the tail-fin, with which they may unite as an abnormality. In structure the swim-bladder may be either simple or divided into cells.

The true arapaima (*Arapaima gigas*) of Brazil and the Guianas, the sole representative of its genus, is the largest fresh-water bony fish, its length not infrequently exceeding 15 feet, while its weight may be more than 400 pounds. As a genus it is distinguished by the broad cleft of the mouth, in which the lower jaw is very prominent, and the absence of barbels, as well as by the rounded lower surface of the body and

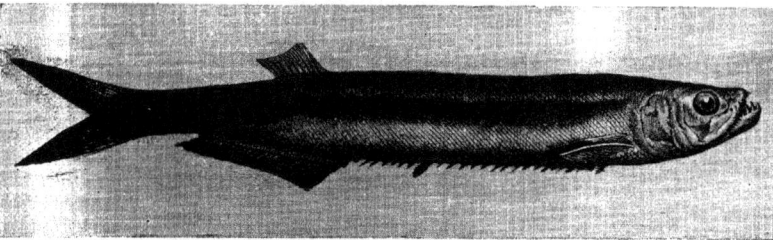

THE DORAB

the moderate length of the pectoral fins. Besides small conical teeth on the margins of the jaws, there are rows of rasp-like teeth not only on all the bones of the palate, but on the tongue and hyoid bones.

In spite of its enormous dimensions, the arapaima is captured by the natives of Brazil with a hook and line; its flesh being highly esteemed as food, and in a salted condition largely exported. It is also taken by being struck with an arrow, to which a line is attached. A party go out in a boat, and row about until a fish is

sighted, when the bow and arrow are brought into requisition.

The four representatives of the typical genus *Osteoglossum* may be distinguished from the last by the presence of a pair of barbels to the lower jaw, the obliquity of the cleft of the mouth, the sharp lower surface of the body, and the greater length of the pectoral fins. Of the four species one is American, and has the same distribution as the true arapaima, a second occurs in Sumatra and Borneo, and the other two are Australian. The two latter have, however, but a very local distribution, one (*O. leichardti*) being confined to the rivers of Queensland, where it is known to the natives as the barramundi, or Dawson River salmon; while the second (*O. jardinei*), distinguished by the absence of a spine to the anal fin, inhabits the rivers discharging into the Gulf of Carpentaria. The flesh of both these species is highly esteemed as an article of food.

The third genus of the family, including only a single species (*Heterotis nilotica*), differs from both the foregoing in the comparatively small size of the cleft of the mouth, and also in the approximate equality of the length of the jaws, on which there is only a single series of teeth. These are also present on the pterygoid and hyoid bones, but wanting on the vomer and palatines. A further peculiarity is to be found in the presence of a peculiar spiral organ on the fourth gill-arch; and the swim-bladder differs from that of the other members of the family in its cellular structure, while the stomach comprises a membranous and a muscular portion. This fish is found alike in the upper Nile and in the rivers of Western Africa, and grows to about 2 feet in length.

CHISEL-JAWS

The chisel-jaw (*Pantodon buchholzi*), of the family *Pantodontidæ*, is native to the rivers on the west coast of Africa, and takes its name from the strong dental armature of the jaws. Closely related to the arapaimas, this fish has the body covered with relatively large scales, and the sides of the head with bony plates; the margin of the upper jaw being formed in front by the united premaxillæ, and at the sides by the maxillæ. The short dorsal fin is situated still further back than in the last genus, its front margin being considerably behind that of the rather longer anal; both the pectoral and pelvic fins are very tall, the rays of the latter forming isolated filaments, and the caudal is long and pointed, with some of its rays projecting. The snout is blunt, and the cleft of the mouth directed upwards. In the gill-cover there is only an opercular and a preopercular bone, the gill-openings are wide, and the branchiostegal rays are numerous. False gills are absent; the swim-bladder is simple; and the ovaries of the female, and the corresponding organs of the opposite sex, are furnished with a duct. This fish, which measures only about 3 inches in length, is in the habit of "flying," or, rather, parachuting, after the manner of the flying-fishes, and is the only fresh-water species endowed with this power.

HERRING TRIBE

Of great commercial importance, the herring tribe, family *Clupeidæ*, are remarkable for the enormous number of individuals by which several of the species are represented rather than for the multiplicity of the species themselves; this being one of the chief reasons for their value as a food supply. Although the existing representatives of the family may be readily distinguished from salmonoids by the absence of a fatty fin, yet extinct forms indicate such an intimate connection between the two groups as to induce some naturalists to include both in a single family.

In addition to the absence of the fatty fin, most herrings are characterised by small ventral bony plates. The whole body is scaled, with the lateral line mostly wanting, while the head is generally naked, and the snout always without barbels. The under surface is more or less compressed, and generally so much so as to form a sharp edge, usually serrated. In the gill-cover the four elements are present, and the gill-openings are generally very wide. Both premaxillæ and maxillæ enter into the formation of the margin of the upper jaw, but each of the latter bones is peculiar in being composed of three separate pieces. The single short dorsal fin has a small or moderate number of weak rays, and the anal may be many-rayed. The stomach is furnished with a blind sac; the swim-bladder is more or less simple; and well-developed false gills are usually present.

Distributed through all temperate and tropical seas, herrings are mainly shore-fishes, none of them inhabiting deep water, and none being truly pelagic. Although most are marine, many will enter fresh water, and some live permanently therein, while it is probable that all can be acclimatised to such conditions. As might have been expected from their generalised structure, herrings are an ancient group, the typical genus dating from the period of the Chalk, while anchovies and other existing generic types are known from the Eocene. A number of allied Cretaceous genera connect the family with the higher ganoids. Unlike that of other food-fishes, the spawn of herrings sinks.

THE DORAB

The dorab (*Chirocentrus dorab*), which ranges from the Red Sea to the Malay Archipelago, is the first of what we may term the herring and salmon group. In common with the herrings, this fish has but a single true tail-vertebra. Externally the body is covered with thin deciduous scales; barbels and a fatty fin are alike lacking; but the elements of the gill-covers are fully developed. The margin of the upper jaw is formed partly by the premaxillæ and partly by the maxillæ, which are firmly welded at their junction; the short dorsal fin is situated in the caudal region of the vertebral column above the much longer anal, the tail is deeply forked, the pelvic fins are minute, the lower surface of the body is sharp, the gill-opening wide, and false gills wanting. The upward direction of the cleft of the mouth, which is armed with formidable teeth, coupled with the elongation of the lower jaw, gives a rather peculiar expression to the head, and the eyes are remarkable for being covered with skin. The stomach is furnished with a blind appendage,

BRAZILIAN ARAPAIMA

the intestine is short, and the air-bladder cellular. As this fish attains a length of fully a dozen feet, it is a sufficiently formidable monster, and when captured is said to bite viciously at every object within its reach. Its flesh is of poor quality.

TYPICAL HERRINGS

The common herring (*Clupea harengus*) belongs to a group of genera characterised by the equality in the length of the two jaws, the presence of free fatty lids to the eyes, and the serration of the lower border of the hinder part of the body; typical herrings being distinguished from allied genera by the anal fin being of moderate length, with less than thirty rays, and the serration of the under surface beginning from the chest or point of origin of the pectoral fins. Usually the scales are of moderate or large size, although they may be small; the cleft of the mouth is of medium width; and if teeth are present at all, they are rudimentary and deciduous. In position the dorsal fin is opposite the anal, and the caudal is deeply forked.

Represented by some sixty species, the genus has a distribution coextensive with that of the family; but while the flesh of the majority of its representatives is of excellent quality, that of some tropical forms may be poisonous. In the case of such a well-known fish as the common herring, it will be superfluous to give any description, but it may be mentioned that this species may be distinguished from its allies by the presence of a patch of small ovate teeth on the vomer. It has also the dorsal fin situated exactly midway between the extremity of the snout and the longest ray of the caudal fin, and the pelvic fins are directly under the dorsal. Whitebait are mostly the young of this species and the sprat.

Common to both sides of the cooler regions of the Northern Atlantic, the herring ranges eastwards to the seas on the north of Asia. Associating in shoals numbering millions of individuals, it feeds upon crustaceans, worms, insects, and the young and eggs of other fishes, as well as those of its own kind. "During the day," writes Mr. J. M. Mitchell, "the shoals are sometimes observable near the surface, and may be seen playing on the water, as the fishermen call it, making a ripple—a dark roughness similar to what we may see at the beginning of a slight breeze, this being somewhat observable without the appearance of either whales or birds. The passing near or over them of a boat or ship makes them instantly dart off in every direction, leaving the appearance of long trails of light, if at night. We have been informed by fishermen of Newhaven that the herrings take considerable flights out of the sea; off Stonehaven, in the month of September, one of these men having seen a shoal, after the spawning season, rise up out of the water in a vast mass of many yards in extent, sparkling and flashing and flying several feet above the surface.

"On some of the coasts, as on those of Norway, the herring-shoals are frequently accompanied or pursued by numbers of whales and aquatic birds, which are all occupied in preying on them. The large dark masses of the whales rising and blowing and throwing up great quantities of the herring into the air, sparkling and glittering in the clear winter day, the constant movements of the birds, with their shrill notes, actively engaged in seizing their easily-obtained food, vying with man in their attacks on the countless myriads of herrings, form a most wonderful sight. When the herrings swim near the surface, if it is calm weather, the sound of their motion is distinctly heard at a small distance, and at night their motion, if rapid, causes a beautiful bright line from the phosphorescent quality of the skin.

The much smaller sprat (*C. sprattus*), so abundant on the Atlantic coasts of Europe, differs by the absence of vomerine teeth; while the shad (*C. finta*), shown in the lower figure of the illustration on this page, may be distinguished by having one or more black blotches on the sides. In this species, which not only frequents the European coasts, but ascends rivers, and is abundant in the Nile, the bony gill-rakers, of which there are from twenty-one to twenty-seven on the horizontal portion of the outer gill-arch, are short and stout. On the other hand, in the similarly spotted allice-shad (*C. alosa*) the gill-rakers are very long and fine, and number from sixty to eighty on the part mentioned. Both the shads are considerably larger than the herring. While in both herring and sprat the opercular bone is smooth, in the pilchard or sardine (*C. pilchardus*) this part is marked by ridges radiating towards the sub-opercular. This species is abundant in the English Channel, the seas of Spain and Portugal, and the Mediterranean; Vigo Bay being noted for its sardine fishery.

HERRING, SPRATS, AND SHAD

FRESH-WATER HERRINGS

Especial interest attaches to the Australian fresh-water herrings (*Diplomystus*), which differ from the typical genus in having a series of bony plates similar to those on the lower surface between the back of the head and the dorsal fin, since a similar type of fish has been long known in a fossil state, having been obtained from the Cretaceous rocks of Brazil and Syria, and the lower Tertiary of the United States and Britain. The persistence at the present day of this ancient type of herring in the fresh waters of Australia is an instance of the survival of primitive forms of life in that region.

ANCHOVIES

The common anchovy of the Mediterranean (*Engraulis encrasicholus*) is the typical representative of a second widely-spread genus, with over forty species, differing from the last by the more or less nearly conical snout projecting beyond the lower jaw, and also by the eyes being covered with skin; while the cleft of the mouth is deep and the tail-fin forked. In most cases each side of the body is ornamented with a broad longitudinal stripe. The common anchovy is met with off the south-western coasts of England, but wanders still further to the north, and serves to supply the markets of the world. Some species have the rays of the pectoral fins produced, and thus lead on to the allied Oriental genus *Coilia*, in which the foremost rays of these fins are filamentous, and the exceedingly long anal fin extends backwards to join the caudal.

SALMON TRIBE

In the salmon tribe (family *Salmonidæ*), which includes the finest and "gamest" of all fresh-water fish, the margin of the upper jaw is formed by the premaxillæ in front and by the maxillæ at the sides. As a rule, the body is scaled, while the head is invariably naked; the under surface of the body being rounded. Inhabiting both salt and fresh waters, those species which spend a part or the whole of their existence in rivers or lakes are in the main confined to the temperate and Arctic zones of the Northern Hemisphere, although one outlying genus (*Retropinna*) occurs in New Zealand.

The majority of the purely marine forms are deep-sea fishes, and two genera are entirely pelagic in their habits. A considerable number of the species inhabiting fresh waters descend periodically or occasionally to the sea; and in some cases it is perhaps rather difficult to say whether these fishes should be regarded as marine of fresh-water.

All the salmonoids are remarkable for the excellent quality of their flesh, which in many forms is of a more or less strongly marked pinkish hue, due to the crustaceans on which these carnivorous fishes so largely feed. Fossil marine salmonoids are possibly represented by some Cretaceous fossils, but the oldest undoubted representatives of the family occur in the Miocene strata.

The more typical members of the family have the parietal bones of the skull separated from one another by the supraoccipital, but in *Coregonus* and *Thymallus* they unite together in front of it. There is, however, a genus (*Stenodus*) in which both conditions exist, so that there is no justification for making the union of the parietals a reason for referring *Coregonus* to a separate family. In all cases the supraoccipital extends forwards to join the frontals, and is thus quite different from the condition obtaining in the carps and characinoids.

COMMON CARP

MIRROR-CARP

LEATHER-CARP

VARIEGATED GOLD-FISH

BARBEL

ROACH

DACE

TYPICAL SALMON

Having the dorsal nearly or quite opposite the pelvic fins the members of the typical genus *Salmo* are characterised by the small size of the scales of the body, the strong and fully-developed teeth, and the presence of not more than fourteen rays in the anal fin, and of numerous blind appendages to the intestine. The cleft of the mouth is always deep, the maxilla extending up to or beyond the line of the eye. Conical teeth are present not only on the margins of the jaws, but likewise on the vomer and palatine bones, as well as on the tongue, although there are none on the pterygoids. The eggs are remarkable for their relatively large size, and the young, like those of most or all the other genera, are marked with dark cross-bars. In the male the lower jaw is more developed than in the females, and at certain seasons may be developed into an upturned hook. The genus is confined to the colder portions of the Northern Hemisphere, its southern limits in the Old World being the rivers of the Hindu Kush and the Atlas range, and in America the rivers flowing into the head of the Gulf of California.

The food value of salmonoids has led to various attempts to introduce them into the Southern Hemisphere, the fertilised eggs being kept during transport in a state of arrested development by the use of ice. British and American firms have in this way been successfully established in Tasmania and New Zealand. In the case of the latter country, Mr. A. J. Rutherford concludes that greater success is likely to attend the introduction of the North Pacific salmonoids than that of *Salmo salar*, which is a more delicate fish, unlikely to find what it requires in an ocean so far removed from its native habitat. In regard to trout, the author is of opinion that " whatever variety we liberate of the ordinary species of trout, it will develop into a *Salmo novæ-zealandiæ*, suited to the water in which it is liberated, and corresponding with trout in similar localities in the Northern Hemisphere more closely than with the varieties found in the more northern latitudes of our mother country."

Few zoological subjects have given rise to a greater amount of discussion than the life-history of the members of this genus, and the number of species by which it is represented. As regards the latter point, great difference of opinion still prevails among experts. Thus, for instance, Day considered that all the indigenous British salmonoids might be arranged under three specific types—namely, the salmon, the trout, and the charr ; while other authorities admit an almost endless amount of species. We shall confine our attention to the salmon, the typical sea and river trout, and the charr.

As regards the variations of these fishes we may, however, quote a passage from Dr. Günther, who writes that " these are dependent on age, sex, and sexual development, food, and the properties of the water. Some of the species interbreed, and the hybrids mix again with one of the parent species, thus producing an offspring more or less similar to the pure breed. The coloration is, first of all, subject to variation ; and consequently this character but rarely assists in distinguishing a species, there being not one which would show in all stages of development the same kind of coloration. The young of all the species are barred, and this is so constantly the case that it may be used

as a generic, or even as a family, character, not being peculiar to *Salmo* alone, but also common to *Thymallus*, and probably to *Coregonus*. The number of bars is not quite constant, but the migatory trout have two (and even three) more than the river-trout. In some waters river-trout remain small, and frequently retain the parr-marks all their lifetime ; at certain seasons a new coat of scales overlays the parr-marks, rendering them invisible for a time. When the salmonoids have passed this ' parr ' state, the coloration becomes much more diversified. The males, especially during and immediately after the spawning time, are more intensely coloured and variegated than the females, specimens which have not attained to maturity retaining a brighter silvery colour, and being more similar to the female fish.

" Food appears to have much less influence on the coloration of the outer parts than on that of the flesh, the more variegated specimens being frequently out of condition, whilst well-fed individuals with pinkish flesh are of a more uniform, though bright, coloration. . . . The water has a marked influence on the colours. Trout with intense ocellated spots are generally found in clear, rapid rivers, and in small open Alpine pools ; in the large lakes with pebbly bottom the fish are bright silvery, and the ocellated spots are mixed with or replaced by X-shaped black spots ; in pools or parts of lakes with muddy or peaty bottom the trout are of a darker colour generally ; and when enclosed in caves or holes they may assume an almost uniform blackish coloration."

A change of colour also takes place in the migratory species with the renovation of the scales, which occurs during their residence in the sea, the newly-grown portion of the silvery scales concealing the spots ; and this change of coloration varies greatly according to the habitat of the individuals of some of the species. Variations of size are also common, these being for the most part dependent upon the abundance or otherwise of the food, and the extent of the area in which

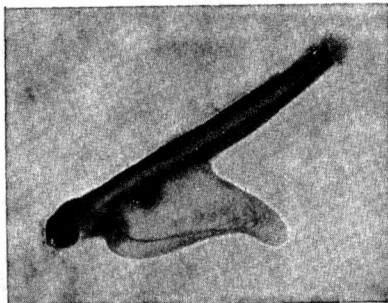

YOUNG TROUT WITH YOLK-SAC
Photo S. C. Johnson

the fish dwell ; but differences in this respect also occur among the fish hatched from the same batch of spawn, and living under the same conditions. The variations in the form and proportions of the body, and more especially in the head and jaws, according to age, sex, and season, are likewise very important, but cannot be noticed fully.

THE SALMON

The true salmon (*S. salar*), together with the kindred species and trout, belongs to a group of the genus characterised by the presence of teeth on the whole length of the vomer during at least some period of life. Dr. Günther gives certain characters by which this fish may always be identified, and among these the following may be noticed : The scales on the tail are of relatively large size, and each transverse series running from behind the fatty fin towards the lateral line contains only eleven, or occasionally twelve, whereas in trout there are from thirteen to fifteen. Secondly, the main part or body of the vomer carries a single series of small teeth, which, with advancing age, gradually disappear from behind forwards, so that half-grown and adult individuals have but a few remaining.

Having a circumpolar distribution, the salmon ranges southwards in America to 41° north latitude, and in the Old World to 43°, being unknown in any of the rivers flowing into the Mediterranean. Salmon will grow to a length of between four and five feet, and commonly reach as much as 40 lb. Much heavier fish are, however, occasionally captured. Among these may be mentioned a salmon of 60 lb. from the Severn in 1889; one from the Tay of 62 lb. in 1891; a third of 63 lb. from the Esk in 1890; another of 68 lb. from the Tay in 1893; and a fifth from the same river taken in 1870, which weighed a fraction under 70 lb. There is an earlier record of a British salmon of 83 lb. weight, while a Russian fish is stated to have scaled upwards of 93 lb.

For the following brief sketch of the life-history of the salmon we are indebted to a paper by Mr. G. Rooper, from which the following extracts, with some verbal alterations, are taken. After mentioning the well-known periodical migration of salmon, the writer observes that the eggs are deposited by the female "some time during the winter months, in beds of gravel, over which a rapid stream flows, principally in the upper reaches of the river, where the water is more aerated and free from pollutions of any sort—since clay,

RIVER-TROUT

earth, or any extraneous substance would choke and destroy the embryo fish. Indeed, from the time of entering the river, the object of the fish seems to be to arrive at its source. Until they have spawned they never descend, but, resting at times in favourite pools, continually struggle upwards. Only the late fish spawn in the lower waters.

"To such as have only seen the salmon in prime condition, the appearance of the fish when on the eve of spawning would come as a surprise. The female is then dark in colour, almost black, and her shape sadly altered for the worse from that which she presented when in condition. As for the male, he is about as hideous as can well be imagined, his general colour being a dirty red, blotched with orange and dark spots. His jaws are elongated, and the lower one furnished with a huge beak, as thick and nearly as long as a man's middle finger; while his teeth are

SEA-TROUT

sharp and numerous, and his head, from the shrinking of the shoulders, appears disproportionately large. His skin also is slimy and disagreeable to handle, and, in fact, scarcely a more repulsive creature in appearance exists.

"Arrived on the spawning ground, the female, then called a baggit, alone proceeds to form the nest, or 'redd,' as it is termed. This she effects by a sort of wriggling motion of the lower part of her body working on the loose gravel. Many authors state that this is effected by the action of the tail, but I think the convex

formation of the body at that period would prevent the tail touching the gravel, unless the fish stood at an angle of 45°, in which case the stream would carry her down. The redd, a deep trench, being formed, the female proceeds, attended by the male fish—frequently by two kippers, as they are then called—to deposit her eggs. This she does, not all at once, but in small quantities at intervals, frequently returning to the redd for the purpose. The eggs are at once fecundated by the milt of the kipper, this process going on for two or three days, the fish sinking down occasionally into the pool below to rest and recover their strength.

"The effect of the fertilisation of the ova is to add greatly to their specific gravity; the eggs sink, and are at once covered with gravel by a similar motion on the part of the baggit to that used in the formation of the redd. Here, the process being completed, the eggs remain during a period of from one hundred and twenty to one hundred and forty days, according to the temperature of the water. At the expiration of that time the little fish come into existence, and after a few days wriggle out of their gravelly bed, and seek refuge under an adjacent rock or stone, where they remain in safety for some twelve or fourteen days longer. The appearance of the young fish at that time gives little promise of the beautiful form to which they subsequently attain. These alevins are, indeed, shapeless little monsters, more like tadpoles than fish, each furnished with a little bag of nutriment forming a portion of the abdomen. On this, for two or three weeks, they subsist, until it is absorbed, when they take the form of fishes. They are then about one inch in length, and are known as salmon-fry, or samlets. A portion of the eggs are washed down the stream during the process of spawning, and become the prey of trout and other fish which attend the redds for the purpose of feeding on them. In this they do no harm whatever, for these eggs, being uncovered and unfecundated, could never arrive at maturity.

"The kippers, when not actually engaged in the spawning process, swim rapidly about the redd, fighting fiercely with one another. The use of the beak appears then to come into operation. Many authors erroneously describe this beak either as a weapon of offence, or as a sort of pickaxe used in digging out the redd; but it seems to me that Nature has provided this singular excrescence as a protection and safeguard against the savage attacks made on each other. So large is its size, and so closely does it fit into the hole or socket formed in the upper jaw, that it would appear almost impossible for the fish even to open his mouth; but he does so, to some extent, at least, and with his cat-like teeth inflicts deep, and sometimes dangerous, wounds on his antagonist. As to its alleged use as a digging implement, the substance of the beak is cartilaginous, not horny, and by no means hard; it would be worn down in the process of digging in ten minutes, and, moreover, the female alone prepares the redd.

" After leaving the stone or rock under which it has sought protection, the young fish grows very rapidly, as is natural in one destined to attain such huge dimensions as the salmon. In the course of a month or six weeks the fry have attained to the length of four inches, and are then called ' parr '; when they bear conspicuously on their bodies transverse marks or bars, which are common to the young of every member of the salmon family. The time of their remaining in the parr stage is also a subject of dispute; and while some say two, three, or sometimes four years, my opinion is that they remain one year only.

" In the second April of their existence a change in the appearance of the parr occurs, which assumes the silvery scales of the adult fish, wearing his new apparel over his old barred coat. He is now called a smolt, and perhaps, with a wish to exhibit himself in his new and beautiful apparel, evinces a daily increasing restlessness and desire to quit his home. With the first floods in May myriads of these lovely little fishes start on their downward journey toward the sea. It is a beautiful sight to watch their movements when descending, and for many days the river teems with them, not a square foot of water being without one when the stream is at all rapid.

" As fry the smolts were exposed to many dangers, but they were nothing to those which beset them as parrs on their journey towards the sea. Their enemies are legion. Trout and pike devour them; gulls swoop down and swallow them wholesale; herons, standing mid-leg deep in the water, pick them out as they pass; and even their own kindred devour them without scruple. Unluckily, too, for them, a certain number of great, hungry kelts (as the fish are called after spawning), having recovered to a great extent their condition, accompany them on their seaward journey, and prey upon their young companions as they travel; and I believe that a hungry kelt will devour upwards of forty or fifty smolts in a day.

" Arrived at the sea, the little fish are met by a fresh array of enemies. The army of gulls is always with them, and these are reinforced by cormorants, divers, and other sea-birds, besides which shoals of ravenous

MAY-TROUT AND HUCHO

fish await their arrival, and assist in thinning their ranks. It is wonderful that any should escape, and but for the extraordinary fecundity of the salmon they would speedily be annihilated; but such is their prolific nature that a remnant always survives to return to the spawning beds and keep up the supply. Buckland calculated that the number of eggs laid by a salmon was about one thousand to the pound weight, a fish of 15 pounds therefore producing fifteen thousand eggs. The food of the smolt during his sojourn in the sea is abundant, consisting chiefly of sand-eels, molluscs, and marine insects."

Many of the smolts return to the river as grilse, or salmon-peel, during the spring or summer of the following year, their weight being then from 2 to 10 pounds. Some of them, however, remain in the sea for more than a year before resuming life in fresh water. The name salmon is not correctly applied until they are at least four years old.

Valuable information about the movements of fishes have been obtained by marking them in various ways, and this has been done at various times with salmon from a very early period. A series of very interesting observations were initiated by Messrs. Calderwood and Malloch under the auspices of the Tay Salmon Fisheries Company in 1905. No less than 6,500 smolts were successfully marked in that year by pushing a bit of fine silver wire into the front of the dorsal fin and twisting it up into a loop.

A number of the smolts so marked were recaptured the following year as grilse (the first on June 1), and the average weight of sixteen of these was 5¼ pounds, equivalent to a gain of 6 ounces per month during life in the sea. It was further demonstrated that some of the marked smolts did not return to fresh water until the lapse of a year and a half, in the condition of " small spring fish " that had spent two winters in the sea.

The older stages in the life-history of the salmon have also been marked, the most successful method being by means of a silver stud, on the principle of a paper fastener, bearing a definite number. In this way the fact has been established that the duration of stay in the sea is shorter or longer in individual cases throughout the life of the salmon.

With reference to the statement in the account that salmon always return to the river of their birth, it may be observed that although this is generally the case, the circumstance that salmon occasionally make their appearance at the mouth of the Thames and other rivers which they have ceased to inhabit shows that there are exceptions to the rule. The obstacles that salmon will surmount in their ascent of rivers during the return from the sea are too well known to require notice; but it is probable that the height to which they can leap has been exaggerated.

The period of spawning varies with the country, taking place in the south of Sweden and North Germany at the latter part of October or early in November, while in Denmark it may be deferred till February or the beginning of March, November and December being the usual spawning months in Scotland.

The salmon may be considered as a sea fish which ascends rivers for spawning purposes, a protective measure enhancing the safety of the young. In the sea the food consists very largely of herrings and sandeels, but a much debated question is whether salmon feed in fresh water. Hundreds of post-mortem examinations have been made in Britain, Holland, Germany, and America, the result being that only in a very small percentage of cases is any food to be found in the stomach or intestines, the conclusion being that the fish do not feed regularly in fresh water. Admittedly, however, a salmon will sometimes take worms, or even swallow trout or roach, but that does not affect the general truth of the statement. Why these fishes rise so readily to the fly or other bait is certainly difficult to explain, but this is not quite the same thing as the continuous and deliberate seeking of food, and it has been plausibly suggested that during the breeding season salmon are in what may be termed a "highly strung" condition, so that their behaviour with reference to the skilfully displayed lure of the angler is of abnormal character.

TROUT

In spite of their diversity of habitat, and likewise of coloration and structure, Day is of opinion that the migratory sea-trout, or salmon-trout (*S. trutta*), and the stationary river-trout (*S. fario*), as well as the various forms from the British lakes, are nothing more than varieties of a single variable race; and it must be confessed that no one has hitherto been able to define all the nominal British species with anything like definiteness. Some of the characters distinguishing the salmon from the trout have been already indicated on page 1717; and it will suffice to note very shortly some of the reasons given by Day for regarding all the British trout as referable to a single species.

It is well known that sea-trout—as represented not only by the typical form, but likewise by the so-called sewen (*S. cambricus*) of the Welsh rivers—are silvery in colour, with black spots during their sojourn in the sea; when, however, they enter the rivers for the purpose of spawning, an orange margin appears on the upper and lower edges of the caudal fin, and also on the fatty one, while spots of the same colour show themselves on the body. On the other hand, the non-migratory forms may be arranged under two types of coloration, some loch-trout (which may have been originally migratory, but are now land-locked) being mainly silvery during the smolt stage, and subsequently golden and spotted; while the estuarine, lake, and river trout are all golden, with purplish reflections, and more or less fully marked with black and vermilion spots. It appears, indeed, that a long residence in fresh water generally leads to the disappearance of the silvery sheen characteristic of the salmonoids while in the sea (and which is probably their primitive type of coloration), and to the promotion of colour.

As a partially transitional type between sea-trout and river-trout may be taken the Lochleven trout, which is somewhat silvery during the smolt stage, with the spots generally black, and no orange border to the fatty fin, but at a later stage assumes the general coloration of the river-trout, although lacking the white black-based front margin to the dorsal, anal, and pelvic fins characteristic of the latter. Silvery trout do, however, occasionally occur in fresh waters, where there is no possibility of their having migrated from the sea. In concluding his observations concerning the coloration of trout, Day writes that "reasons have been shown for admitting that sea-trout might breed in fresh waters without descending to the sea. That they can be traced step by step, and link by link, into the brook-trout, and vice versa; that the Lochleven

THE CHARR

trout, which normally possesses a smolt or grilse stage, passes into the brook-trout; and also that breeding any of these two forms together sets up no unusual phenomena."

Later on Day observes that some of the chief distinctions between the sea and fresh water forms of trout consist in the comparatively more complete system of dentition in the fresh-water races, their generally longer head, blunter snout, and stronger upper jaw, irrespective of the smaller number of blind appendages to the intestine. The dentition is, however, excessively variable; and specimens with the coloration and form of the river-trout taken in estuaries, or even in the sea, usually have the small number of vomerine teeth characteristic of the migratory forms; while, on the other hand, fresh-water examples with the coloration of the migratory type may have a dentition of the non-migratory type.

The ordinary sea-trout, which is essentially a North European fish, much more common in Scotland than in England, grows to a length of 3 feet; while, as an example of a spotted form, we take a variety of the Continental lake-trout (*S. lacustris*). Known on the Continent as the Maiforelle (May-trout), this fish has the sides of the body marked with irregular angular or X-shaped black spots, between which are red spots, these spots becoming less numerous beneath the lateral line, while the under surface may be tinged with red. On the gill-cover the spots are larger and more rounded. In the typical variety of this trout, from the Lake of Constance, the spots do not extend below the lateral line; this form being known as the Schwebforelle. The migrations of the sea-trout are very similar to those of the salmon, in Sutherland the great run of these fish to the sea taking place in June, while they reascend the rivers in autumn to spawn.

CHARR

Much the same difference of opinion as obtains with regard to the number of species of trout exists in the case of charr, Dr. Günther recognising five British lacustrine species, which he regards as distinct from *S. umbla* of the Swiss lakes; while Day includes the whole of these under the latter, which is also taken to embrace the sæbling (*S. salvelinus*) of the mountain lakes of Bavaria and Austria, as well as the migratory northern charr (*S. alpinus*), ranging from Lapland and Scandinavia to Iceland and the northern parts of Scotland.

Charr are typically an Arctic group of fishes with the general habits of the salmon—that is to say, they spend the summer in the ocean, but at the beginning of winter ascend rivers for the purpose of spawning. Even in the high north there are, however, some charr which remain throughout the year in their native lakes, although they could reach the sea if so disposed. Others live in land-locked lakes, from which they cannot possibly escape, and this condition obtains with all the charr of the British Isles, which are, in the main, restricted to lakes where the water is cold and deep

By fishermen charr are often termed " fresh-water herrings " or " whiting," from their generally brilliant silvery appearance. Two of the Irish varieties—namely, one formerly inhabiting the lakes at the source of the River Lee, in County Cork, and another Lough Neagh, have been long extinct, the former since 1837, and the latter since 1839. Some others still survive, namely, the Lough Owel charr, of Westmeath, the charr of Lough Finn, Donegal, those of Lough Esk, and certain other lakes in Donegal, Mayo, and Galway, and the races respectively inhabiting Lough Melvin, Fermanagh, Lough Cumasaharn, Kerry, and Lough Luggala, Wicklow. These varieties differ from one another in the shape of the muzzle, the size of the teeth, and the number of rows of scales.

All charr differ from salmon and trout in having the teeth at all ages confined to the head of the vomer, instead of being distributed over its whole length ; and all the forms mentioned above, which have a very uniform type of coloration, agree in having median teeth on the hyoid bone.

In the spawning season the upper parts of this fish are brownish green, and the sides lighter ; the under surface passing through all shades of orange to vermilion, from the throat to the pelvic fins, where the colour attains its greatest intensity. The sides are ornamented with rounded spots varying from white to red in colour ; the dorsal fin has dark markings, and the pectoral and pelvic fins are brilliant red. This form commonly grows to a length of 8 or 9 inches, but the northern charr attains much larger dimensions.

The North American charr (*S. fontinalis*), which has been successfully introduced into British waters,

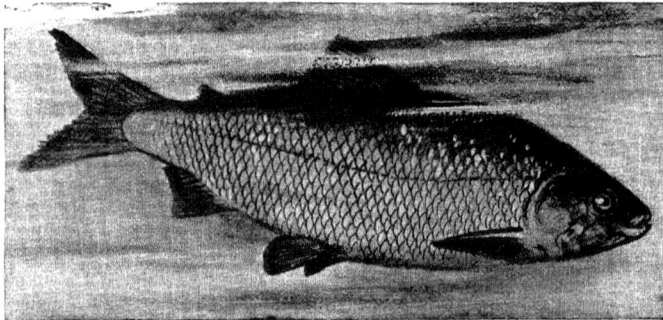

WARTMANN'S WHITE-FISH

together with the hucho (*S. hucho*) of the Danube, differ from the foregoing in the absence of median teeth on the hyoid bone. The general colour of the American charr is greenish—lighter above than beneath—beautifully shot with purple and gold, ornamented with numerous dark spots above, and fewer below the lateral line, many of which in front of the dorsal fin

coalesce into streaks, and also with red spots above the aforesaid line. Most of the fins have dark markings ; and in the breeding season the male assumes a black tint along the under surface. These fish usually range in size from 2 to 3 pounds, although they may be larger. The hucho, on the other hand, which is readily characterised by its elongated, slender, and almost cylindrical form, attains dimensions equal to those of the salmon.

QUINNAT SALMON

The North Pacific, or Quinnat salmon (*Onchorhynchus*) inhabiting the North American and Asiatic

SMELT

rivers flowing into the Pacific differ from the typical genus in having more than fourteen rays in the anal fin ; while their kelts are remarkable for the degree to which the jaws are hooked, and the humping of the back. An average fish weighs about 22 pounds, but greater weights, up to 100 pounds, are recorded. Quinnat salmon from British Columbia and Alaska are tinned and exported to a large extent.

SMELTS

The beautiful and delicately flavoured little smelts are represented by three species, one of which (*Osmerus eperlanus*) is an inhabitant of the seas and many fresh waters of Northern and Central Europe, while the second (*O. viridescens*), perhaps only a variety, is confined to the opposite side of the Atlantic, and the third (*O. thaleichthys*) is found on the coasts of California. These fish form a kind of connecting link between the salmon and its allies and the under-mentioned *Coregonus*, but internally differ from both, the appendages to the intestine being short and few in number, and the eggs small, while the teeth are strongly developed. The scales are of moderate size ; the cleft of the mouth is wide, with the maxillary bone extending nearly or quite to the hinder margin of the eye ; the teeth of the upper jaw are much smaller than those of the lower ; the vomer is armed with a transverse series of teeth, several of which are tusk-like ; the palatines and pterygoids bear conical teeth ; while there are also tusk-like teeth on the front of the tongue, and several longitudinal series of small ones on its back part. In length the pectoral fins are medium. Growing to a length of 7 or 8 inches in the sea, the common smelt is also found in rivers and land-locked lakes, where its size is always considerably less. The allied candle-fish (*O. thaleichthys*), of the Pacific coasts of North America, distinguished by its rudimentary teeth, has flesh of such an oily nature that it can be burnt as a candle, although it is likewise used as a food.

COREGONOIDS

Among an extensive group of mostly fresh-water salmonoids (" white-fish ") remarkable for their herring-like form, the gwyniad (*Coregonus pennantii*) of Bala Lake ; the closely allied or identical powan (*C. clupeoides*), of the Lake District ; the vendace (*C. vandesius*), of Lochmaben ; and the pollan (*C. pollan*) of the Irish lakes, are well-known British forms. They are the only British fresh-water fish that cannot be taken with hook and line. In these fish the scales are not strikingly large ; the cleft of the mouth is of moderate size, with a broad maxilla, either short or of medium length, and not extending beyond the front margin of the socket of the eye ; the teeth, if present, are minute and deciduous, in the adult usually persisting only on the tongue. The dorsal fin is not over long, and the caudal is deeply forked. While in the small size of their eggs these fish resemble smelts, they differ in having about one hundred and fifty blind appendages of nearly uniform length and attached to the intestine.

Represented by over forty species, ranging over Northern Temperate Europe, Asia, and North America, coregonoids are for the most part entirely fresh-water fishes, although a few make periodical migrations to the sea, while the European schnæpel (*C. oxyrhynchus*) is as much a marine as a fresh-water form.

Local in their distribution in Europe, although as many as three different species may inhabit the same lake, coregonoids (" white-fish ") are extremely abundant in all the fresh waters of North America. Though the British forms are small, some of the Continental species may attain a length of fully 2 feet. The figure represents a species (*C. wartmanni*) from the deeper lakes of the Northern Alps.

The genus may be divided into groups, according to the conformation of the snout and jaws. Of these, the first is represented solely by the schnæpel, or houting (*C. oxyrhynchus*), which frequents the coasts and rivers of Belgium, Holland, Germany, and Sweden, and occasionally wanders into British waters. It is

PIGMY MARANE AND MARANE

easily distinguished by the production of the extremity of the upper jaw into a conical fleshy snout projecting beyond the lower, while its scales are more or less nearly circular. In length this fish grows to a foot and a half. As an example of the group in which the snout is obliquely truncated, with the nose projecting, we may take the marane (*C. lavaretus*), which is widely distributed in the lakes of the Continent, where its flesh is highly esteemed. Living at great depths, this fish feeds on worms, insects, and water-snails.

While the gwyniad belongs to another group characterised by the vertical truncation of the snout, the pollan and vendace are assigned to yet another division in which the lower jaw is longer than the upper. As a representative of this latter group we take the pigmy marane (*C. albula*) of Northern Europe. Pollan, which grow to a length of about 6 inches, are largely sold in Belfast during the spawning season, at which time

THE GRAYLING

they come up from the deep waters of Lough Neagh to the shallows.

GRAYLING

The last salmonoids to be noticed are the grayling. Nearly allied to the coregonoids, they are readily distinguished by the greater height and length of the dorsal fin, which includes from thirteen to twenty-three rays. The cleft of the mouth is also smaller, and the maxilla of small size. Small teeth are present on the jawbones, as well as on the palatines and the head of the vomer, but they are wanting on the tongue. The blind appendages of the intestine are less numerous than in either the salmon or the coregonoids, and the swim-bladder is usually large. The range of the genus includes a large portion of Europe, Northern Asia, and the colder regions of North America. Its name, *Thymallus*, has reference to the thyme-like odour that is exhaled.

The common species (*T. vulgaris*) is found locally over a great part of Europe, ranging from Lapland to Venice, and from England to Russia. It is, however, unknown in Ireland, and has only been introduced of late years into Scotland ; while in England it is most abundant in the rivers flowing from the limestone Pennine chain in the north, and the Red Sandstone districts of the central counties, and likewise in the chalk streams of the south. In the latter area grayling occasionally run to nearly 4 pounds in weight, but in Northern Scandinavia they may reach 1 pound more. In Switzerland they are found in Lake Constance and other large pieces of water. An elegantly-shaped fish, the grayling varies considerably in colour according to the season, the back being generally greenish brown, passing into grey on the sides, while the under parts are silvery. The sides of the head are yellow, with black spots, which also occur on the fore part of the body ; and brownish grey longitudinal stripes run in the direction of the rows of scales. The pelvic and anal fins are violet, frequently marked with brown cross-bars ; the pectorals are yellow, turning to red in the breeding season ; while the black-bordered dorsal and caudal are generally red, although sometimes blue ; the former,

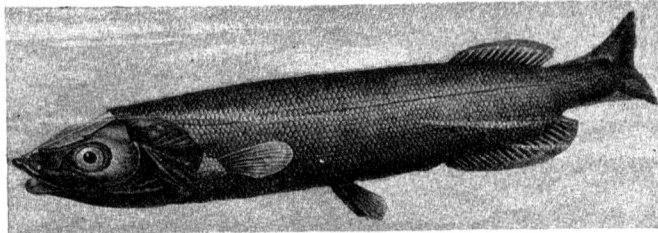

BLACK SMOOTH-HEAD

and sometimes also the latter, being ornamented with longitudinal dark bands or rows of spots.

THE CAPELIN

The capelin (*Mallotus villosus*) is native to the Arctic coasts of the North Pacific, and is from 6 to 9 inches long. The scales are small, and in the male four longitudinal bands of them project in such a way as to give a shaggy or villous appearance. The eggs are laid in vast numbers among the sand of the shore.

SMOOTH-HEADS

The family called smooth-heads (*Alepocephalidæ*) includes about thirty-five species of remarkable deep-sea fishes related to salmon and herring, and of wide distribution. The dorsal is placed near the caudal, and there is no fatty fin. The skeleton is fragile. A typical form is the black smooth-head (*Alepocephalus niger*).

WIDE-MOUTHS

In the wide-mouths (family *Stomiatidæ*) are included some extraordinary deep-sea forms, in which scales are extremely thin or absent. A series of phosphorescent organs is

SILVERY-LIGHT FISH AND BARBED HEDGEHOG-MOUTH

usually present, and there is often a barbel on the hyoid region. The eyes are large, and the gill-openings wide, and a fatty fin is sometimes present.

The silvery-light fish (*Photichthys argenteus*) possesses phosphorescent organs, and its other characters are as follows : The body may be either covered with thin deciduous scales, or entirely naked ; barbels are wanting ; and the fatty fin is either rudimentary or very minute. Both the premaxillæ and maxillæ take a share in the formation of the margin of the upper jaw, and bear pointed teeth of variable length. The bones of the gill-cover are not fully developed ; the gill-opening is of great width ; false gills may or may not be developed ; and, when present, the swim-bladder is simple. While in the figured species the teeth are small, in the allied genus *Chauliodus* they are greatly elongated, indicating highly predaceous habits.

HEDGEHOG-MOUTHS

The barbed hedgehog-mouth (*Echiostoma barbatum*) may be readily distinguished from the preceding by the presence of a long barbel to the hyoid, the skin being either naked or covered with exceedingly delicate scales, and the fatty fin frequently wanting.

When a fatty fin is present, as in the genus *Astronesthes*, the rayed dorsal is of considerable length, and placed in advance of the anal ; but in the other genera both the anal and dorsal are short, and placed opposite to one another a short distance in advance of the forked caudal. In the genus represented by the hedgehog-mouth the body is naked, and the pectoral fins are filamentous ; but in the allied *Stomias* there are exceedingly small scales which scarcely overlap one another. Occasionally met with floating in a helpless condition, these fishes have been dredged from depths of 1,800 fathoms ; and although dwelling in total darkness, they, like most of their allies, have well-developed eyes.

The genus *Malacosteus* possesses peculiar plates below the eye, and a specimen of this fish captured before death had ensued was observed to emit a yellowish light from the uppermost plate beneath the eye, while that from the lower plate had a greenish tinge. In the genus *Stomias*, says Filhol, " the sides of the body present a double longitudinal series of phosphorescent plates, which emit light in such a manner as to cause the whole fish to be bathed in a brilliant luminous halo. This fish must, indeed, be a formidable creature to the other inhabitants of the ocean abysses, being in every way constructed and armed for strife, and its powerful teeth admirably fitted to seize and tear the flesh of the other fishes upon which it preys "

THE BEAKED SALMON

The so-called beaked salmon (*Gonorhynchus greyi*) from the seas of the Cape, Japan, and Australia constitutes a family (*Gonorhynchidæ*) by itself. Both the head and body are completely covered with scales, of which the free edges are spinose ; and the margin of the upper jaw is formed entirely by the short premaxillæ, which are continued downwards over the maxillæ. The short dorsal fin is situated far back on the body, above the pelvic pair, the still shorter anal having a more posterior position ; and the tail-fin is slightly forked. There is no fatty fin. Barbels are present, the gill-openings are narrow, the swim-bladder is wanting, and the stomach simple. Measuring from 12 to 18 inches in length, this fish seems to be partly pelagic and partly littoral in its habits, being found in New Zealand, where it is known as the sand-eel, in bays with a sandy bottom, while elsewhere it has been taken in the open sea.

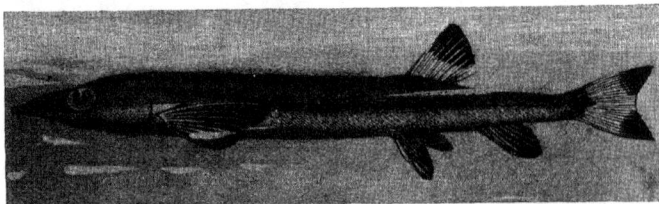

THE BEAKED SALMON

ELECTRIC EELS

The fishes of the suborder Cyprini-siluriformes are distinguished by the fact that the front vertebræ, usually to the number of four, fuse together, and parts of them are detached on either side to make up a chain of little bones (*Weberian ossicles*) running from the swim-bladder to the internal ear. The fins are usually spineless, though dorsals and pectorals may have one spine each. By far the greater number of fresh-water fishes are here included.

Together with four other genera from the fresh waters of Tropical America, the well-known electric eel (*Gymnotus electricus*) constitutes the family *Gymnotidæ*, of which the leading characteristics are as follows : The jaws are formed in the same manner as in the true eels, and the head is scaleless and without barbels. The dorsal fin is either totally wanting or reduced to a fatty rudiment ; the anal is extremely elongated ; pelvics are wanting ; and the caudal is likewise generally absent, the tail terminating in a point, which, when broken off, can be renewed in the same manner as in the blind-worms. The vent is situated in or near the throat ; the gill-openings are rather narrow ; a swim-bladder is present ; the stomach has a blind appendage ; and the ovaries are provided with ducts. In the skeleton the pectoral girdle is attached to the skull.

As a genus, the electric eel is characterised by the absence of the caudal and dorsal fins, by the anal extending to the extremity of the tail, the absence of scales, the single series of conical teeth, and the minute eyes. Abundant in the rivers and lagoons of certain parts of Brazil and the Guianas, the electric eel grows to a length of fully 6 feet, and is capable of giving a more powerful shock than any of the other fishes endowed with electric power. The electric organs form two pairs of longitudinal structures lying between the skin and the muscles, one pair being situated on the back of the tail, and the other along the sides of the base of the anal fin. They are made up of modified muscular tissue. That these organs are capable of giving shocks sufficient to kill other fish and small mammals is undoubted, while they cause annoyance to pack-animals.

CAT-FISHES OR SHEATH-FISHES

Although represented only by a single European species confined to the rivers east of the Rhine, the family of cat-fishes (*Siluridæ*) is one of great importance in tropical and subtropical countries, its members being extremely abundant in the fresh waters and estuaries of the Oriental region and South America, and including somewhere about 1,000 species. An essential characteristic is the invariable absence of scales, the skin being either smooth or covered with bony tubercles or plates ; and this character, together with the presence of the barbels from which they derive their popular title, will always serve to distinguish the cat-fishes from the other great fresh-water family of the carps. In the skull an essential feature is the absence of a subopercular element to the gill-cover ; while

ELECTRIC EEL

the margin of the upper jaw is formed mainly by the premaxillæ, the maxillæ being more or less rudimentary. A rayed dorsal fin may be absent, but the fatty dorsal is generally present ; and when a swim-bladder is developed it may be either free in the abdominal cavity or enclosed in bone, but always communicates with the ear by the intervention of the Weberian ossicles, which are somewhat lenticular in form.

The skull is characterised by the full ossification of its lateral region, the septum between the eyes being also bony ; and in many instances the skull is prolonged backwards by the development of a kind of bony helmet over the nape of the neck, formed by dermal ossifications overlying some of the bones of the pectoral girdle. Frequently this shield, as well as the hinder bones of the skull, are ornamented with a tuberculated sculpture. Many of these fishes have also a powerful spine at the front of the dorsal fin, which can be locked into a fixed, erect position by a rudimentary spine acting as a kind of bolt at its base, and is itself articulated to the vertebræ, and also joined by a ring to a second spine, in a manner similar to that obtaining in the angler-fish. To support this spine certain special modifications exist in the structure of the pectoral girdle.

Some of the genera, such as the one represented by the eel-like cat-fish, have additional breathing organs, in this particular instance taking the form of a branched structure attached to the gills. On the other hand, in the sac-gilled cat-fishes (*Saccobranchus*) there is a long sac running down the muscles of the back behind the proper gill-chamber. Through this breathing-sac blood is carried from and returned directly to the heart ; and in consequence of this arrangement these fishes can remain alive for hours or even days apart from water, so that they are able to traverse spaces where aquatic respiration is impracticable.

Most cat-fishes inhabit the fresh waters and estuaries of the tropical and subtropical regions of the globe ; but, as we have seen, one species is found in those of Eastern Europe, while a considerable number enter the sea, although generally keeping near the coasts. They are found not only in rivers, but also in lagoons and marshes.

The streams of many parts of North America, from New York and New Jersey westward to Montana and Wyoming, and southwards to Georgia, Alabama, and Texas, are inhabited by certain small cat-fishes locally known as "stone-cats" and "mad Toms." From other cat-fishes they may be distinguished by the presence of a keel-like fatty fin joined to the back, and more or less continuous with the tail-fin. That the mad Toms can inflict a painful wound with the spines of their pectoral fins has long been known ; but some difference of opinion has prevailed as to whether they possess a special poison-gland with the secretion of which these spines are anointed. The effects of their sting have been likened to those of a bee's sting, although in one species, at any rate, the pain is less intense and usually confined to the immediate neighbourhood of the wound. In the case of a very severe sting on the tip of a finger, pain may, however, be experienced all over the hand and wrist, while it may sometimes extend as high up as the elbow. The pain lasts from one to several hours, but is seldom attended by swelling. Nevertheless, the slight character of the puncture is strongly suggestive of poison ; and, as a matter of fact, all these cat-fishes are furnished with a pore situated just above the root of the pectoral fin, while some also possess glands on the pectoral and dorsal spines.

Careful examination of several species by Mr. H. D. Reed has rendered it certain that the pectoral pore is the opening of a gland, and that the secretion of this gland is highly poisonous. The form of the pore is more or less slit-like, and so situated that when the fin is depressed its spine lies either parallel with or straight across the pore. It seems, however, doubtful whether the spine is anointed with the poison when in this position, and it is more probable that the secretion flows on to the spine when the fin is extended, at which time the mouth of the pore is more or less widely open.

It is worthy of note that spine-glands are only developed in species in which the spines are smooth or but slightly serrated. A serrated dorsal spine without a gland cannot, apparently, inflict a stinging wound ; but the non-glandular pectoral spines are supplied with their poison from the skin-gland at the root of the fin. The presence of serrations makes practicable the infliction of a large number of wounds, and, consequently, the introduction of a large amount of poison at one and the same time. This would render species so armed more formidable, although the poison is secreted in smaller quantities, and is not more virulent than in others.

THE WELS

The wels (*Silurus glanis*) is the typical representative of a subfamily in which the rayed dorsal fin is but little developed, and, if present at all, occupies only the hinder region of the trunk ; the fatty portion being small or wanting. The anal fin is not much shorter than the caudal region of the backbone, and the pelvic fins are behind or below the dorsal. In the wels and its congeners the short dorsal has no pungent spine ; the fatty fin is wanting ; there are two upper and two or four lower barbels ; the head and body are naked ; and the tail-fin is rounded.

The wels itself, which is confined to the European rivers east of the Rhine, has six barbels, of which the upper pair are considerably longer than the head, and commonly attains a length of from 6 to 9 feet, although it occasionally grows to 13 feet. In colour the head, back, and edges of the fins are bluish black, the sides greenish black spotted with olive-green, and the under parts reddish or yellowish white with blackish marblings. Frequenting rivers and lakes with muddy bottoms, the wels feeds on fishes, frogs, and crustaceans, but it will also seize and pull down ducks, geese, or other birds swimming on the surface. The spawning time is in the middle of summer, when these fish resort to the shallows in order to deposit their eggs on the stems and leaves of water-plants.

In a very interesting article on parental care among fresh-water fishes published by the Smithsonian Institute of Washington, Mr. Gill shows, by means of

figures, that the glanis differs from the wels by having only four (in place of six) barbels on its great ugly snout. In the parental care displayed for its eggs it resembles some of the American cat-fishes ; and it is quite probable that the same habit may characterise some of its Asiatic relatives.

YARRELL'S CAT-FISH

Another gigantic species is Yarrell's cat-fish (*Bagarius*

THE WELS

yarrelli), from the large rivers and estuaries of India and Java, which attains a length of fully 6 feet, and from its huge head and mouth is one of the ugliest fishes in existence. The only member of its genus, it belongs to a subfamily in which the rayed dorsal fin is short, and situated in the hinder part of the body in advance of the pelvics ; and there is always a fatty fin, which may, however, be short ; and the anal is shorter than the caudal region of the backbone. When nasal barbels are developed, they belong to the hinder nostrils. In the group of genera to which Yarrell's cat-fish belongs the two divisions of the nostrils are placed near together, with a barbel between them ; and in this particular form there are eight barbels, and the upper surface of the head is naked.

The well-known genus *Arius*, from all the tropical regions of the world, belongs to another group of the same subfamily, in which the divisions of the nostrils are close together, but have no barbel, although the hinder pair are provided with a valve. The species of this genus are noted for care of eggs and young. Semon has described a Queensland fish (*A. australis*) that makes a bowl-shaped excavation some 20 inches in diameter for the reception of the eggs, which are covered by layers of stones. Marine and estuarine species of this and some other genera adopt a much more remarkable method of protection, for in them the eggs are carried in the mouth and pharynx of the male, or more rarely of the female.

LEOPARD CAT-FISH

The Tropical American genus *Pimelodus* is the typical representative of a third group of the same subfamily, in which the two pairs of nostrils are equally devoid of barbels, but are placed at a considerable distance apart. The largest species is the leopard cat-fish (*P. pati*) from the rivers of Argentina and Uruguay, growing to a length of 6 or 7 feet, and having the yellowish skin marked with a number of black spots, like a hunting-leopard. Somewhat curiously, this genus is represented by two outlying species from West Africa.

THE BAYAD

The best-known representative of the fourth and last group of genera in this subfamily is the bayad (*Bagrus bayad*) of the Nile ; the group being easily recognised by the circumstance that while the two divisions of the nostrils are remote from one another, the hinder have barbels. Both species are confined to the Nile.

ELECTRIC CAT-FISHES

Brief mention must be made of the electric cat-fishes (*Malapterurus*) of Tropical Africa, belonging to a subfamily in which the rayed dorsal fin, when present, is short and confined to the hinder region of the body, while the pelvic fins are inserted behind. The electric organ is of unique character, being formed by a modification of the skin. Very powerful shocks can be given ; hence the name "raad" (*i.e.*, thunder) given by the Arabs to these forms. From their allies they are distinguished by the total absence of the rayed dorsal, so that they have only a fatty dorsal immediately in front of the tail (which is rounded), and opposite the anal. The head and body are smooth, the pectoral fins spineless, and there are six barbels. The Nile species grows to about 4 feet in length.

MAILED CAT-FISHES

The mailed cat-fishes (*Callichthys*, *Loricaria*, etc.), constitute a subfamily mainly confined to Tropical and South America, although represented by a few Oriental forms. In all these fishes there is always a rather short rayed dorsal fin, beneath or in front of which the pelvics are generally inserted. The gill-membranes are confluent with the skin of the isthmus, and the gill-openings constricted to small slits. The pectoral and pelvic fins are placed horizontally ; and the vent is in front of, or only slightly behind, the middle of the length of the body.

Among these fishes the species of the genus *Callichthys*, confined to the rivers on the Atlantic side of South America, belong to a group characterised by the nearness of the divisions of the nostrils, between which there is generally a short flap, and by the expansion and reversion of the lower lip to form a broad flap more or less deeply notched in the middle. It appears that they are in the habit of anchoring themselves to stones in the river-bed by means of the sucker-like mouth ; respiration being at such times effected by taking in water through the gill-openings and expelling it again by the same aperture in the opposite direction. In the genus mentioned, the head is covered with bony plates, and the body encased in two rows of transversely elongated overlapping shields on each side ; all the species being of small size. Like certain other South American forms belonging to another subfamily, of which the members of the genus *Doras* are perhaps the best known, these mailed cat-fishes are in the habit of making nocturnal journeys during the hot season, when the pond they inhabit is about to dry up, to another of greater capacity, and they likewise construct nests for their eggs.

The nests, which are made at the beginning of the rainy season, are formed of leaves, beneath which the eggs are deposited and watched over by both parents ; the whole structure being sometimes placed in a hole on the margin of the river or pond. In the armoured cat-fish, forming the genus *Loricaria*, the body is remarkable for its elongated and slender form ; while the head is depressed, with a more or less produced and spatulate snout, on the under surface of which the mouth is situated at a considerable distance from the extremity, its margins being surrounded by large folds, and each corner having a barbel. Both the dorsal and anal fins are short and elevated, and the entire head and body enveloped in a bony cuirass.

CARP TRIBE

As in the last family, the margin of the upper jaw of the *Cyprinidæ* is formed by the premaxillæ, and the whole mouth is toothless, teeth being developed on the pharyngeal bones alone. While the head is invariably naked, the body is generally covered with scales, and although it may be scaleless it is never invested with bony plates. False gills may be developed, and, if so, are glandular. When a swim-bladder is present, it is always of large size ; and it may be divided into two lateral moieties enclosed in an ossified capsule, or constricted into an anterior and posterior portion which are not thus protected. The numerous members of this family are fresh-water fish, confined to the Old World and North America, being quite unknown in the southern half of the New World, and also in Australia. Showing much less diversity of form and habits than the cat-fishes, the carp tribe are mostly omnivorous, although a few are purely vegetarian. Some of them prefer muddy situations, where their barbels are probably of assistance, but they mostly differ from cat-fishes in selecting clear waters. The Indian forms seem to be more carnivorous than their European relatives, many of the larger kinds preying upon their smaller brethren.

On account of their more cleanly feeding habits the flesh of the carps is superior to that of the cat-fishes. The family is represented by over a hundred existing genera, arranged under two subfamilies.

TRUE CARPS

The common carp (*Cyprinus carpio*) claims our attention as the typical representative of the subfamily *Cyprininæ*, characterised by the swim-bladder not being enclosed in bone, and divided into an anterior and posterior moiety.

Belonging to a group in which the anal fin is short and usually furnished with five or six branched rays, true carp have the lateral line running along the middle of the tail, the dorsal fin placed opposite the pelvics, and containing a more or less strongly saw-edged bony ray, and more than nine branched rays, while the pharyngeal teeth are arranged in three series, with those of the outermost one molar-like. The snout is

THE GUDGEON

rounded and blunt, with four barbels, and the rather narrow mouth at its extremity. The true carps form a small genus confined to the temperate parts of Europe and Asia, the common species being a native of the latter continent, and abundant in a wild state in China, where it has also long been domesticated. Thence it was introduced into Germany and Sweden, and subsequently into Britain—it is said early in the seventeenth century. Besides the ordinary form, there are many domesticated varieties, differing either in the form of the body or the size and arrangement of the scales.

Among the latter, one of the most remarkable is the so-called mirror-carp. In this variety, found only in ponds, the scales are three or four times the normal size, and, instead of covering the whole body, are arranged in from one to three longitudinal rows, with bare skin between them. The leather-carp is another form in which the scales are either few and enlarged or absent altogether.

In Western Europe the carp has taken kindly to its new habitat, not infrequently attaining as much as a yard in length, with a weight of 25 pounds, while very much larger specimens are on record. Preferring still waters, with a soft, muddy bottom, in which it grubs with its snout for food, the carp feeds on various vegetable substances, as well as on insects and other small aquatic invertebrates. When the surface of their haunt is locked in ice, carp lie deeply buried in holes in the mud, frequently consorting in numbers, and undergoing a partial hibernation, which is not broken till the returning warmth of spring. Their growth is extremely rapid, and their fecundity extraordinary, nearly three-quarters of a million eggs having been counted in the roe of a medium-sized specimen.

GOLD-FISH AND TELESCOPE-FISH

CRUCIAN AND GOLDEN CARP

Easily distinguished by the absence of barbels, the Crucian carp (*Carassius vulgaris*), and the golden carp or gold-fish (*C. auratus*) are the best-known representatives of another closely allied genus ; the former being a native of Central and Northern Europe, but also found in Italy and Siberia, while the home of the second is China and the warmer parts of Japan. Both are comparatively small species, and have long been domesticated ; but while the Crucian carp always retain the original brownish colour, the domesticated variety of the golden carp has assumed the well-known golden tinge from which it takes its name ; an albino form being also known. Among the numerous varieties of this fish the most curious is the so-called telescope-fish, taking its name from the prominence of the highly movable eyes, and also characterised by the great development of the caudal fin.

BARBELS

Represented by some two hundred tropical and temperate Old World species, the barbels are best known by the common European form (*Barbus vulgaris*) and the gigantic mahseer (*B. mosal*) of India and Ceylon. Agreeing with carp in the structure of the anal fin, and the position of the lateral line and dorsal fin, they belong to a subgroup of genera in which there are generally not more than nine rays in the dorsal fin, the pharyngeal teeth being arranged in three rows, the greater part of the cheek not covered with bone, the anal scales not enlarged, and the eye unprovided with a fatty lid ; while they are specially characterised by the arched mouth—devoid of internal folds—and by false gills. The anal fin is frequently tall, the lips are devoid of any horny covering, and the barbels, if present, may be either two or four in number. The scales are either small or very large, and the body is frequently much more elongated than in true carp. While some species are not more than 2 inches in length, the mahseer, and some other kinds, may grow to at least 6 feet. The common

form, which has four barbels, not uncommonly grows to a length of 2 feet, with a weight of from 8 to 10 pounds, but may attain much larger dimensions.

GUDGEONS

We have next to mention the gudgeons (*Gobio*), which may be distinguished from the foregoing by the pharyngeal teeth being arranged in a double or single series, the body being entirely covered with scales, and the snout having two small barbels, with the mouth inferior in position, and the premaxillary bones protractile. The scales are of moderate size, the short dorsal fin has no spine, and the intestine is remarkable for its shortness. These small fishes are represented only by two species, of which *G. fluviatilis* is British ; and, like the barbels, they are purely carnivorous.

WHITE-FISH

From the whole of the forms just described, the so-called " white-fish " differ in the circumstance that the anal fin is short or of medium length, with from eight to eleven branched rays, and does not extend forwards beneath the line of the dorsal ; the lateral line, when complete, running nearly or quite in the middle of the tail. From certain allied forms they are distinguished by the short dorsal fin having no bony ray ; and the pharyngeal teeth form a single or double series, the margin of the lower jaw is not cutting, and there are no barbels. As distinctive peculiarities of the white-fish may be mentioned the protractile premaxillary bones, the overlapping scales, and the smooth outer surface of the pharyngeal teeth.

The roach (*Leuciscus rutilus*) agrees with several other species in having a single series of pharyngeal teeth, at least ten rays in the anal fin, and the dorsal nearly opposite the pelvic fins, its deep body being silvery, and the lower fins of the adult generally tinged with red. Its range is confined to Europe north of the Alps. On the other hand, the chub (*L. cephalus*) is an example of a second group in which there are two series of pharyngeal teeth. This fish has a somewhat wider distribution than the last, extending south into Italy and east into Asia ; it is uniformly coloured, with grey-edged scales,

To the same group of the genus belongs the dace (*L. vulgaris*), with the same distribution as the roach, to which it represents a considerable external resemblance, although smaller and longer in form ; its sides being silvery, but the fins not tinged with red. Roach and dace are commonly found in company, and have identical habits. The ide (*L. idus*) is confined to the central and northern countries of the Continent, and is a uniformly-coloured species nearly allied to the last. The orfe is a golden-coloured domesticated variety bred in Germany. Another member of the same group is the rudd, or red-eye, (*L. erythophthalmus*) which ranges all over Europe and Asia, and may be distinguished by its scarlet lower fins, the general hue of the scales being coppery.

The familiar and diminutive minnow (*L. phoxinus*) differs from all the foregoing members of this group by the incomplete lateral line, its range being limited to Europe, although it is represented by an allied species in North America.

MINNOWS

TENCH

Representing a genus by itself, the European tench (*Tinca vulgaris*) differs from the white-fish by the presence of a small pair of barbels to the mouth ; the pharyngeal teeth forming a single series. The small scales are deeply imbedded in the thick skin ; there is a complete lateral line ; both the dorsal and anal fin are short ; and the caudal is but slightly indented. The terminally-situated mouth has its lips moderately developed.

While white-fish prefer clear running steams, the tench frequents ponds, lakes, and other more or less stagnant water ; its colour, which is sometimes bronzy golden, and in other cases olive-green, with a more or less blackish tinge, is stated to vary with the purity or otherwise of the water in which it lives. Tench always keep near or in the mud, beneath which they entirely bury themselves during the colder months. A good tench will weigh 4 pounds, but examples of 5 pounds, and even over, are not very uncommon. It is probably owing to the abundant supply of mucus secreted by the skin that this fish was considered to be endowed with healing powers. Tench are exceedingly prolific, and, as they bear transport easily, are admirably adapted for stocking ponds.

BEAKED CARP

By the name of beaked carp may be distinguished a small genus, containing seven species, from Continental Europe and Western Asia, and differing from the two foregoing by the margin of the lower jaw forming a cutting edge, overlaid by a brown horny layer ; one of the species (*Chondrostoma nasus*) being represented on page 1729. These fishes are further characterised by the medium or small size of the scales, the termination of the lateral line in the middle of the

THE TENCH

deepest part of the tail, by the dorsal fin having not more than nine branched rays, and being situated opposite the root of the pelvics, and also by the rather elongate anal bearing ten or more rays. The mouth is inferior in position, and transverse ; and there are no barbels. Commonly known in France as *le nez*, the figured species does not usually exceed 18 inches in length, with a weight of about 3 pounds. It is generally found in deep water, where it feeds on various vegetable substances, especially on the green confervoid growth covering submerged stones, which is neatly mown off by a scythe-like action of the horny margin of the transverse lower lip.

BITTERLING

The bitterling (*Rhodeus amarus*), is a small roach-like European fish representing one of four genera of small carp mainly native to Eastern Asia, and having the following distinctive features. The anal fin is of moderate length, and extends forwards to below the line of the dorsal ; the lateral line, when fully developed, runs on or near the middle of the tail ; and there is but a single series of pharyngeal teeth.

The bitterling belongs to a genus characterised by the incomplete lateral line and the small size of the scales, and is locally distributed in Central Europe, not infrequently in hot springs. It is one of the smallest of European fishes, the females being generally about 1½ inches long, while the males do not exceed twice this size. The name is derived from the bitter taste of the flesh ; and only perch and eels will take it as a bait. In common with its allies, the bitterling is remarkable for the fact that in the spawning season the oviduct of the female is produced into a long projecting tube, which is introduced within the shells of fresh-water mussels, where the eggs will be protected from the attacks of enemies.

BREAM

The common European bream (*Abramis brama*) is the type of a large group of genera, characterised by a long anal fin, and by part or all of the abdomen being compressed into a sharp edge. In the type genus the body is deep or oblong in form, with the scales of moderate size, and the lateral line running below the middle of the tail ; the short spineless dorsal fin being situated opposite the interval between the pelvic and anal fins. The lips are simple, the upper being protractile, and generally longer than the lower. The pharyngeal teeth are in either a single or double series ; and the scales do not extend across the sharp edge of the lower surface of the hinder part of the body. Distributed over Europe north of the Alps, portions of Western Asia, and North America, breams are represented by about fifteen species, of which the common bream and the white bream (*A. blicca*) are found in Britain. The white bream has the general colour of the sides bluish white, without any trace of the golden yellow lustre distinguishing the common species, often termed the carp-bream. They may also be distinguished by the iris of the eye in the latter being yellow, and in the former silvery white, tinged with pink. In some of the Irish lakes bream run to as much as 12 or 14 pounds in weight ; and, as they are greedy, great numbers can be taken by the aid of ground-baiting.

RAPFEN

Rapfen is the Austrian name of the typical representative (*Aspius rapax*) of a small genus of carps containing four species from Eastern Europe and China, and somewhat intermediate in structure between breams and bleaks. Agreeing with the former in the shortness of the gill-rakers, these fishes always have the lower jaw projecting considerably beyond the upper, which is but slightly protractile ; the anal fin never has less than thirteen rays ; and the sharp lower edge of the abdomen behind the pelvic fins is crossed by the scales. Common in Eastern and Northern Europe, but unknown in Britain, the rapfen is generally found in lakes or rivers flowing through level country, as it requires clear but tranquil waters. In colour it is bluish black above, with the sides bluish white and the under surface white ; the dorsal and anal fins being blue, and the others tinged with red. In weight this fish does not exceed a dozen pounds, and in length never measures more than a yard.

BLEAK

Especial interest attaches to the beautiful little fish known as the bleak (*Alburnus lucidus*), on account of the use of the pearly matter from its scales in the manufacture of artificial pearls.

The Chinese are supposed to have employed the scales of various fishes for ornamental purposes from time immemorial, and to have invented the artificial pearl process, which was introduced into Germany and France about the middle of the seventeenth century. The scrapings of the scales are placed in an aqueous solution of ammonia, constituting what is known as " essence d'orient." The pearls consist of hollow spheres of glass, which are first lined with the essence and then filled with a kind of wax.

There are fifteen species, ranging over Europe and Western Asia ; the common British species being found only to the north of the Alps, although represented by an allied form in Italy. From both the

BITTERLINGS

(1) SICHEL, (2) RAPFEN, AND (3) BEAKED CARP

preceding genera these fish are distinguished by the slender and lanceolate form of the closely-set gill-rakers. The body is more or less elongate, with the scales of moderate size, and the lateral line running below the middle of the tail. The fins are generally similar to those of the last genus ; and the lower jaw projects more or less beyond the upper, which is protractile. In the hinder part of the abdomen the scales do not extend across the sharp lower edge.

Generally about four or five inches in length, and never exceeding seven, the common bleak is steel-blue in colour above, with silvery white sides and under surface, and the dorsal and caudal fins grey, the others being colourless. It is found in rivers, lakes, and ponds, preferring clear water ; and in calm, warm weather swimming rapidly about near the surface in search of flies and other insects. During the spawning season (May and June) bleak collect in large shoals, which are preyed upon not only by perch, but also by gulls and terns.

SICHEL

The sichel (*Pelecus cultratus*) is a curious-looking fish, forming the sole representative of its genus. It is characterised by the whole of the abdominal surface of the oblong and compressed body forming a sharp cutting edge ; the scales being small, and the lateral line making a sudden descent behind the pectoral fin towards the lower surface. The cleft of the mouth is always peculiar in having a nearly perpendicular direction. The pectoral fins are unusually tall, and the dorsal is placed far back, and above the anal, which resembles that of the bream in its numerous rays. On the pharyngeal bones the teeth are arranged in a double series, and are strongly hooked. In profile this fish, which generally ranges from six inches to a foot in length, is remarkable for the straightness of the line of the back and the convexity of its lower border. It is widely distributed in Eastern Europe, being common in the Black and Caspian seas, as well as in their affluent rivers.

LOACHES

With the small fishes known as loaches, of which there are three European genera, we come to the second subfamily (*Cobitinæ*) of the carp tribe, which is characterised by the swim-bladder being either partially or entirely enclosed in a bony capsule ; false gills being always absent. The intestine may act as an accessory breathing organ. The swim-bladder sometimes loses its original balancing function, and becomes transformed into what may be called a " meteorological " sense organ, being connected with the skin by a special passage and reacting to changes in atmospheric temperature and pressure ; on which account the Germans give the name of " weather fishes " to these forms.

In these fishes the body may be elongate, oblong,

(1) COMMON LOACH, (2) GIANT LOACH, AND (3) SPINY LOACH

compressed, or cylindrical, but is never depressed ; the snout and lips are fleshy ; and the small, inferiorly-placed mouth is furnished with from six to twelve barbels. The median fins are spineless, the dorsal having a variable number of rays, but the short anal possessing but few, while the pelvic pair may be wanting ; scales small, rudimentary, or absent, and, when present, cycloid, and usually immersed in mucus ; in one Oriental genus, developed upon the back and sides of the head. The loaches of this subfamily are confined to Europe and Asia ; and while some European species are partial to swift clear streams with a stony bottom, the Indian forms delight in muddy tanks, where they bury themselves in the mud. All are carnivorous, and, in spite of their small size, the European species are esteemed as food.

The giant loach (*Misgurnus fossilis*) is the largest European member of the group, and belongs to a genus of four species, common to Europe and Asia north of the Himalaya. The genus is characterised by the elongate and compressed form of the body, the absence of an erectile spine near the eye, and the presence of from ten to twelve barbels, four of which belong to the lower jaw ; the dorsal fin being placed above the pelvic pair, and the caudal rounded. The European species, which grows to a length of 10 inches, is found in stagnant waters in Southern and Eastern Germany, and North-Western Asia ; being replaced by an allied form in China and Japan.

The true loaches (*Nemachilus*), on the other hand, have six upper barbels, and none on the lower jaw. They are represented by some fifty species from Europe and Temperate Asia ; the common British loach (*N. barbatulus*) being found in clear streams all over Europe with the exception of Denmark and Scandinavia. The spiny loach (*Cobitis tænia*) is the typical representative of a third genus, distinguished from the last by the presence of a

small, bifid, erectile spine below each eye. It is recorded from Cambridgeshire, Gloucestershire, and Surrey. The figured species is locally and sparingly distributed in Britain, but more common on the Continent. An extraordinary little loach, between 3 and 4 inches long, from North Borneo (*Gastromyzon borneënsis*) is specially adapted to life in mountain streams and torrents, for the paired fins with the under side of the body are converted into a sucker, by which the fish can hold on to stones and escape the danger of being washed away.

THE PIRAYA

As an example of a very extensive family (*Erythrinidæ*) of fresh-water fishes confined to Tropical America and Africa south of the Sahara, we select the American piraya (*Serrasalmo piraya*). As in the carps, the body is scaled and the head naked ; but barbels are invariably wanting, and the jaws may be either toothless or furnished with a powerful dentition. In most cases there is a small fatty fin behind the dorsal ; the swim-bladder is always transversely divided into halves, and there are no false gills

The peculiar geographical distribution of the family is very similar to that of the cichlids, and there can be little doubt that the ancestral types originally inhabited the great land-mass of the Northern Hemisphere, from whence they migrated southwards to their present isolated distributional areas. In their migration to Africa they have been accompanied by members of the carp tribe ; in Tropical America they entirely take the place of that family. The numerous genera, none of which are common to the two hemispheres, are ranged under eleven groups or subfamilies, the majority of which are confined to either the one or the other half of the distributional area, although a few have representatives of both. As regards their habits, some of these fishes are strictly carnivorous, while others are as exclusively vegetable-feeders.

The piraya belongs to the last subfamily, which includes four exclusively American genera, represented by some forty species, and characterised by the some-

THE PIRAYA

what elongated dorsal fin, behind which is a small fatty fin ; by the gill-membranes being free from the isthmus, and also by the distinct serration of the middle line of the under surface of the body. The piraya and its allies attack human beings, and inflict formidable bites with their razor-edged teeth. The smell of blood is said to attract them in very large numbers.

EELS AND THEIR ALLIES

We now come to the consideration of the suborder Anguilliformes, including several families.

The members of the two families *Murænidæ* and *Symbranchidæ* are characterised by the elongated "eel-like" form of the body, but this is not a safe guide to affinities. The first family, which includes the true eels, murænas, and congers, is characterised by the normal structure of the upper jaw, which is formed in front by the premaxillæ (more or less fused with the vomer and ethmoid) and laterally by the toothed maxillæ. The median fins, when present, are either confluent or separated by the projecting tail; the pectorals may or may not be developed, but the pelvic pair is invariably wanting. There are no accessory breathing organs; the stomach has a blind appendage; the vent is generally situated far back, but may be near the pectoral fins; and the ovaries have no ducts. The skin may be either completely naked or may contain rudimentary scales.

Eels are found in the fresh waters and seas of the greater part of the temperate and tropical regions, some living at abyssal depths in the ocean. The young of all are pelagic for a portion of their existence, and many or all of these larvæ, which are known as *Leptocephali*, or glass-fishes, are very much larger than the fully-developed young eels.

MURÆNAS

The murænas are large marine eels, remarkable for their bright spotted or mottled coloration, and taking their name from the species here figured (*Muræna helena*), which was so called by the ancient Romans. Belonging to a small section of the family characterised by the gill-openings into the pharynx being in the form of narrow slits, they are specially distinguished by the median fins being well developed, and the total absence of pectorals. The skin is scaleless, the mouth is well

THE EEL

furnished with teeth; and there are two nostrils on each side of the snout, the front pair being tubular, while the hinder ones may be either tube-like or mere flat openings.

The murænas, of which there are more than eighty species, are distributed over all tropical and temperate seas, and a few ascend tidal rivers. The majority of them are armed with formidable teeth—which

frequently alter considerably with age—adapted for seizing the fish on which they feed. They are often very savage. The figured species, which ranges from the Mediterranean to the Indian Ocean and Australia, has the ground-colour a rich brown, upon which are large yellowish spots, each dotted with smaller spots of brown.

TRUE EELS

The typical eels, familiar to all in the form of the common European species (*Anguilla vulgaris*), agree with the great majority of the family in having the

MEDITERRANEAN MURÆNA

gill-openings into the pharynx as wide slits. The skin contains small scales imbedded in its substance; the upper jaw does not project beyond the lower; the small teeth are arranged in bands; the narrow external gill-openings are situated at the base of the well-developed pectoral fins, and the dorsal fin begins at a considerable distance behind the back of the head.

Eels, of which there are numerous species, appear to be distributed throughout the fresh waters of the habitable portions of the globe, being reputed to be absent only from those of the Arctic regions, and probably also from cold, elevated districts like Turkestan and Tibet. The common European eel is spread over the greater part of Europe and the Mediterranean area—although unknown in the Danube—and reappears in the United States. About a yard is a good size for an eel, although much longer specimens are on record. Few subjects have given rise to more discussion than the mode of propagation of eels, and it is only comparatively recently that the puzzle has been solved.

The matter has excited interest and engaged attention from very early times; allusions to the subject will be found in some of the early publications of the Royal Society; and Gilbert White, in one of his letters (1768) embodied in "The Natural History of Selborne," speaks of the "dubiousness and obscurity attending the propagation" of eels, among other animals. Male eels were only definitely recognised in 1873 by Syrski, at Trieste, while ripe eggs were not known until 1888, when Raffaele discovered some in the Bay of Naples. It was, of course, known that full-grown eels migrate to the sea, and that in the spring and early summer innumerable little "elvers" ascend rivers, which naturally suggested that spawning takes place in salt water.

We now know that the curious little glass-fishes (*Leptocephali*), once supposed to be distinct species, are the larvæ of various fishes of the eel kind. J. T. Cunningham describes them as follows: "The

Leptocephali (which means, literally, small-heads) have some of the characters of the larvæ of other fishes, but they have also many striking peculiarities of their own. Like other fish-larvæ, they are transparent, and have no scales or silvery layer in the skin, and the bony skeleton is not developed. Their chief peculiarities are the great depth of the body from the back to the belly, its remarkable thinness from side to side, the small size of the head, and the large size of the whole creature. The largest specimen described was 10 inches long, but the more usual length is 5 or 6 inches, and some are smaller than this. The vent is usually near to the hinder end of the body.

"It is remarkable that the great vertical depth belongs to the body itself, and is not due to the presence of a broad fin-membrane like that of other fish-larvæ. On the contrary, the fin-membrane is very narrow or absent, and in cases where a longitudinal fin with rudimentary rays is present, this also is narrow. In the thin specimens there is no red blood, but specimens have been described in which the body was more rounded and red blood was present; in these the whole development was obviously more advanced."

The first recorded specimen of these *Leptocephali* was captured near Holyhead in 1763, and received the name of morris-fish (*Leptocephalus morrisii*). Mr. T. N. Gill, in 1864, concluded that the morris-fish is really a larval conger, and this was actually proved by Yves Delage, at Roscoff, in 1886. The details were worked out in South Italy during 1887-1892 by Professor Grassi and Dr. Calandruccio, and a brief summary of their results follows.

The common eel matures in the deep sea, where it acquires larger eyes than are ever observed in individuals which have not yet migrated to deep water, with the exception of individuals found in the disused Roman cloacæ. The abysses of the sea are the places of spawning, and the eggs float in the sea-water. In developing from the egg, the eel undergoes a metamorphosis—that is to say, passes through a larval stage (*Leptocephalus brevirostris*). How long this development takes is very difficult to establish, but the following data are available : (1) The eel migrates to the sea from October to January. (2) The currents, such as those of the Straits of Messina, throw up from the abysses of the sea specimens which, from the beginning of November to the end of July, are more advanced in development than at other times, though not completely mature. (3) Eggs which most probably belong to the common eel are found in the sea from August to January inclusive. (4) *Leptocephalus brevirostris* abounds from February to December. There is some uncertainty about the other months of the year, because, during these, the only natural fisherman, the sun-fish (*Othagoriscus mola*) appears very rarely. (5) The elvers ascending rivers are probably already one year old, and aquarium specimens of *Leptocephalus brevirostris* can transform themselves into elvers in the course of a month.

As to the actual facts from which these conclusions are drawn, reference may first be briefly made to the marine eels (*Muræna*), which abound in the Mediterranean. They ordinarily live at great depths —at least 500 metres—but are at times brought to the surface by currents. To such currents are due the capture of individuals containing ripe eggs, the eggs themselves, and the larval forms. When first hatched, the larva is not a typical glass-fish (*Leptocephalus*), but soon becomes one, sometimes diminishing in size to an extent of an inch or an inch and a half.

The larvæ of the common eel, when captured, vary in length from 2½ to 3 inches. In the course of development the tail-fin gradually assumes the adult character, but the eyes are larger than in the youngest elvers yet observed. Although a complete transition from *Leptocephalus* to elver, or young eel, was not observed in the original research, yet the transition from one to the other actually took place in Grassi's aquarium. But, apart from this, the identity as to the number of muscle-segments in the larva and adult, together with certain peculiarities of coloration, seemed sufficient to prove the identity of the two forms, and the identification has been confirmed by later observations.

The stage into which the first larvæ developed need not be described, except to say that, as in other members of the family in a similar grade of development, the teeth were extremely minute and few in number. In this stage the eel can be found in the sea during winter.

The next developmental form can be captured when migrating from the sea into fresh water. When kept in an aquarium they assume the characters of elvers, taking no food, and diminishing more or less in size.

As the change into the elver proceeds, the temporary teeth in the upper jaw are shed, the pointed snout becomes rounded, and the vent becomes gradually shifted forwards. The back and belly fins are also pushed farther and farther forwards, the former at a greater rate than the latter. The depth of the body likewise becomes reduced until it becomes nearly cylindrical, while in the later stages the length becomes reduced. Pigment is developed in the skin, the eyes also becoming slightly reduced in size during the change. After breeding in the deep sea, eels apparently die ; at all events, they are not known to re-enter fresh water.

For some time it has been known that eels put on a special nuptial dress, the skin acquiring a silvery hue, devoid of the usual yellow tinge, the pectoral fins becoming blackish in colour, and the eyes increasing in size. That these changes have reference to spawning was inferred from the fact that they only took place in eels with enlarged reproductive organs, which had ceased to take food, and were migrating to the sea.

Individuals brought up by currents in the Straits of Messina exhibited these features to an exaggerated degree, the pectoral fins being deep black, and the eyes still larger, and round instead of elliptical. The reproductive organs were fully developed in the males.

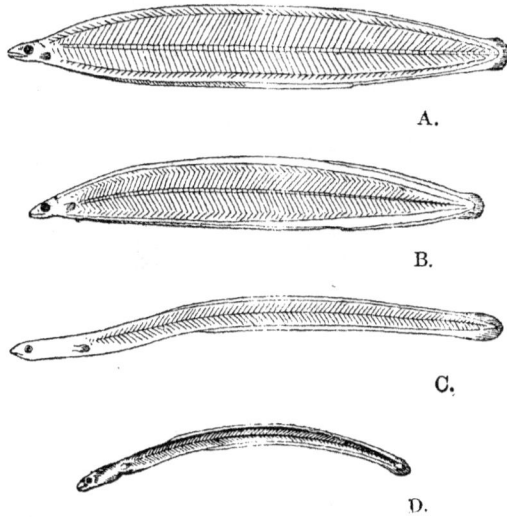

METAMORPHOSIS OF AN EEL

A, glass-fish ; B, beginning of metamorphosis ; C, glass-elver ; D, elver from fresh water

The migrations of these fishes may be said to be two annually, adults descending seawards to breed, as they do in the Severn, about the month of September, although this migration in Norfolk is asserted to begin as early as July. There is likewise an up-stream migration of young eels, or elvers, in the earlier months of the year up to May or June, or even later, during this period the banks of the rivers being in places black with these migrating little fishes. These young eels have been observed to ascend floodgates of locks, to creep up water-pipes or drains ; in short, mechanical difficulties scarcely obstruct them, and they will even make a circuit over a wet piece of ground in order to attain a desirable spot.

In order to give some idea of the vast numbers of young eels that take part in these migrations, or, as they are popularly called, "eel-fares," it may be mentioned that upwards of three tons of elvers were despatched in a single day from the Gloucester district in the spring of 1886, and that it has been calculated that over fourteen thousand of these fish go to make a pound weight. In the previous year the annual consumption of eels was estimated at a minimum of 1,650 tons, with a total value of £130,000. The annual value of the yield of the Danish eel-fisheries is estimated at £100,000. The largest species of eel occur in the islands of the South Pacific and New Zealand, where they inhabit lakes. Specimens from these regions have been recorded to measure from 8 to 10 feet in length.

CONGERS

Resembling the true eels in the presence of pectoral fins, in the tail being surrounded by the median fin, and the free tongue, the gigantic marine forms known as congers differ in being scaleless, in the deep cleft of the mouth, in the presence of a set of teeth on the outer sides of the jaw, placed so close to one another as to form a cutting edge, and by the dorsal fin beginning at a point just behind the base of the pectorals.

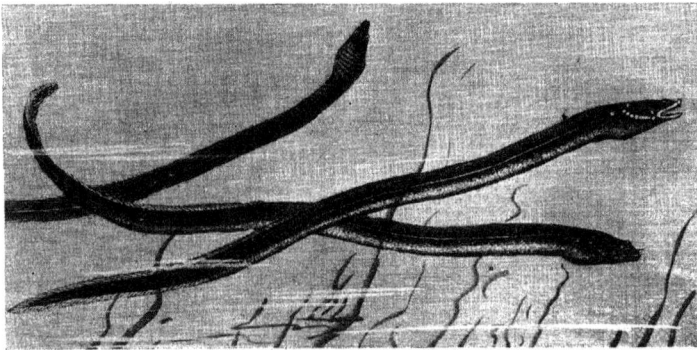

CONGER EEL

BENGAL SHORT-TAILED EELS

The common conger (*Conger vulgaris*), the female of which may grow to a length of 8 feet, appears to be almost cosmopolitan in distribution, being as abundant in the seas of Tasmania and Japan as it is in British waters. Congers to a weight of about 70,000 cwt., and the value of almost £50,000, are caught annually in British waters. These fishes feed chiefly by night, and prey upon crustaceans, cuttles, and various kinds of fish, such as pilchards and herrings. Their favourite resorts are either hollows or crevices in the rocks, or sandy bottoms, in which they can bury themselves, and in such situations they are sometimes left by the ebbing tide. The flesh of these eels is of a highly gelatinous nature, and is said to be largely employed in soups.

DEEP-SEA EELS

Only a few words can be devoted to the deep-sea members of the family, which are represented by several genera. Among these are certain congers (*Synaphobranchus*) occurring in the Atlantic, Pacific, and Indian Oceans at depths of from 200 to 2,000 fathoms, and characterised by the gill-openings being united into a single longitudinal slit on the under surface of the body between the pectoral fins, the gape being very wide, the teeth small, and the body scaled. In these forms the muscular system is well developed ; but in another genus (*Saccopharynx*) it is extremely feeble, except on the head, and the bones are soft and spongy. The head and gape are of immense size ; the snout is short and flexible ; the weak jaws are armed with long, slender, curved teeth, placed at intervals ; and the gill-openings are wide and situated on the lower part of the sides at some distance from the head, the narrow gills being free and exposed.

The long and band-like tail ends in a long, tapering filament, and the dorsal and anal fins are rudimentary. As in the last genus, the stomach is capable of great distention, and specimens which had swallowed fish of many times their own weight have been found floating in the Atlantic with this organ dilated to its utmost.

SINGLE-SLIT EELS

It has been already noticed that in one of the deep-sea eels the gill-openings are confluent into a longitudinal slit on the under surface of the body, and a very similar condition characterises the second family of eels (*Symbranchidæ*), only in this case the slit is transverse. A better distinction is, however, afforded by the structure of the upper jaw, the margin of which in the present family is formed entirely by the premaxillæ, on the inner side of which lie the maxillæ. The paired fins are rudimentary, and the vertical ones wanting ; while the scales, if present, are minute ; and accessory breathing organs may be developed. A swim-bladder is wanting, the stomach has no blind appendage, and the ovaries are furnished with ducts, the vent being situated far behind the head. Although the majority of these eels inhabit

SLENDER PIKELET

fresh and brackish waters in Tropical Asia and America, they are also represented in Australia, where one genus is marine.

Of the fresh-water forms, the most remarkable is the amphibious eel (*Amphipnous cuchia*) of Bengal, in which there is an accessory breathing apparatus, and the body is scaled. There are only three gill-arches with rudimentary laminæ, separated from one another by narrow slits ; and the additional breathing organ takes the form of a lung-like sac on each side of the neck communicating with the gill-chamber. Day states that " this amphibious fish, when kept in an aquarium, may be observed constantly rising to the surface for the purpose of respiring atmospheric air direct. It usually remains with the snout close to the surface, and in like manner lies in the grassy sides of ponds and stagnant pieces of water, so that without trouble it may obtain access to air." Indeed, the chief respiration of this fish is carried on by means of the two sacs on the sides of the neck, which can be inflated and emptied at will.

In the other two fresh-water genera, one of which (*Monopterus*) is confined to the Oriental region, while the other (*Symbranchus*) has a distribution coextensive with that of the family, there is no additional breathing organ, the body is naked, and the pectoral girdle is attached to the skull. Whereas in the former of these genera the gills are rudimentary, in the latter they are well developed ; and, in the absence of an accessory apparatus, it seems strange how the one species with rudimentary gills manages to breathe at all. The Bengal short-tailed eel (*Symbranchus bengalensis*) has been selected to illustrate the external form of the members of this very remarkable family.

SOUTHERN PIKELETS

In the group *Esociformes* the fins are generally spineless, and there are no barbels. The pelvic fins are abdominal. With the exception of certain deep-sea fishes, most of the species inhabit fresh water.

We may designate by the name of southern pikelets (family *Galaxiidæ*) certain fishes, mostly fresh-water forms, from the Southern Hemisphere, one of which (*Galaxias attenuatus*) is represented in the figure. These fishes are distinguished by having the base of the cranium simple, and a round or forked tail. The body is naked, and there are no barbels ; the fatty fin is absent, and the medium-sized dorsal opposite the anal. Internally the swim-bladder is large and simple ; and the eggs, as in the last family, fall into the abdominal cavity. Represented by about two dozen species, the largest of which seldom exceeds 8 inches

in length, these fish are of especial interest on account of their geographical distribution.

They inhabit the fresh waters of the southern extremities of the great land-masses, and associated islands, being recorded from Cape Colony, the southern parts of Australia, Tasmania, New Zealand, Chili, Patagonia, and the Falkland Islands. One species (*Galaxias attenuatus*) is believed to occur in all those areas, except the first. This typical example of " discontinuous distribution " is probably explained by the fact that the species in question is not purely fresh-water, but lives on the coasts of the Falklands (there known by the name of " smelt "), while a marine species has been found off the Chatham Islands. We have, in fact, a family formerly marine, but now chiefly of fresh-water habit.

From their spotted bodies, the New Zealand representatives of the genus were formerly known as "trout" by the colonists, while the young of *Galaxias attenuatus* were eaten as " whitebait." An allied New Zealand genus (*Neochanna*), represented by a single species, differs in the absence of pelvic fins, all the known specimens of this singular form having been found buried in burrows of clay or hard mud at a considerable distance from the water.

SOUTHERN SALMON

By the name of southern salmon may be designated two genera of fresh-water fish, constituting a family (*Haplochitonidæ*) which represents the salmonoids in the Southern Hemisphere, the zebra-salmon (*Haplochiton zebra*) being figured as an example of the typical genus. Devoid of barbels, like the salmon and herrings, these fish agree with the former in the presence of a fatty fin, but differ in having the margin of the upper jaw formed solely by the premaxillary bones. The body may be either naked or covered with scales ; the gill-opening is wide ; false gills are present ; and the swim-bladder is simple. The ovaries are in the form of plates, and, in the absence of a duct, the eggs fall into the abdominal cavity. The species of the typical genus, which, although devoid of scales, are externally very similar in appearance to trout, are confined to the lakes and rivers of Chili and the extreme south of Patagonia and the Falkland Islands.

KILLIE FISHES OR CYPRINODONTS

The cyprinodonts (family *Cyprinodontidæ*) are specially distinguished by the margin of the upper

ZEBRA-SALMON

jaw being constituted solely by the premaxillæ, and the enlargement of the third upper pharyngeal bone. Externally they may be readily distinguished from the carps by the head being scaled as well as the body, and they have no barbels. Both jaws are toothed, and the pharyngeals are also furnished with teeth, which are heart-shaped. There is no fatty fin, and the dorsal is placed far back. The swim-bladder is simple, and false gills are wanting.

Inhabiting fresh, brackish, or salt water, these fish are distributed over the south of Europe, Africa, Asia, and America ; some being purely carnivorous, while others feed on the organic substances in mud. Most are viviparous ; and the much smaller males frequently have the anal fin specially modified to aid in the reproductive process. The fins are usually relatively larger in the males, and there is also some difference in the coloration of the sexes.

Cyprinodonts are represented by about a score of genera, which may be divided into two subfamilies, according to the nature of the food. In the first of these, which includes the typical genus *Cyprinodon*, and has a distribution coextensive with that of the family, all the forms are carnivorous or insectivorous, and are characterised by the firm union of the two branches of the lower jaw in front, and likewise by the shortness or slight convolution of the intestines. On the other hand, in the second subfamily, which is exclusively restricted to Tropical America, the species seek their food in mud, and have the two branches of the lower jaw but loosely joined together, while the intestine is highly convoluted. It is in this group that the sexual differences are most strongly marked.

As an example of the family we take a remarkable genus belonging to the first subfamily, represented by three species from Tropical America, one of which (*Anableps tetrophthalmus*) is shown in the accompanying illustration. Having a broad and depressed head, with the region over the eyes much raised, the elongate body compressed in front and depressed behind, a protractile snout, and the cleft of the mouth horizontal and of moderate size, these fish are specially characterised by the structure of the eye, which is unique. In each eye the integuments are divided into an upper and a lower moiety by a dark-coloured transverse band in the outer layer, the pupil being also bisected in the same plane by means of a lobe projecting from each side of the iris. The scales are of small or moderate size ; the dorsal and anal fins are short, the latter being placed in advance of the line

DOUBLE-EYES

of the larger rivers, many of them being often left stranded by the retiring tide, where they progress on the slime by a series of leaps. After birth the young are carried about by the female in a thin-skinned sac divided by a partition, until they are able to take care of themselves. These fishes frequently swim at the surface with the eye half in and half out of the water ; and it is in accordance with this habit that the eyes are divided, the upper portion being able to see in the air and the lower in water. This is proved by the structure of the lens. In land-animals the lens is biconvex, but in ordinary fishes it is spherical. In the double-eye the upper and lower halves of the lens are respectively of these two shapes. In Brazil the flesh of these fish forms an article of consumption.

PIKE

The pike (*Esox lucius*) represents, with other members of the same genus, a family (*Esocidæ*) by itself. The head is scaleless, but the body is covered with cycloid scales ; there are neither barbels nor a fatty fin ; and the dorsal is situated in the caudal region, in the position of the fatty fin of the salmon tribe. The stomach has no blind appendage, the false gills are glandular and concealed, and the gill-opening is unusually wide. In the upper jaw sickle-shaped teeth are borne by the premaxillæ, palatines, and vomer, the maxillæ being toothless, while the lower teeth are of variable shape. The long, narrow body terminates in a forked caudal fin, and the long, broad, and depressed snout has a projecting lower jaw.

Confined to the fresh waters of the temperate regions of the three northern continents, pike may be considered a western rather than an eastern type, seeing that whereas the common species has a range equivalent to that of the family, the other six species are confined to the United States. In Europe the pike inhabits all the Russian rivers, except those of the Crimea and Transcaucasia, and is also found in Siberia. In Lapland it extends even beyond the limits of the birch, while to the south it is common in the Venetian lagoons. Growing very rapidly, the pike not uncommonly attains a length of

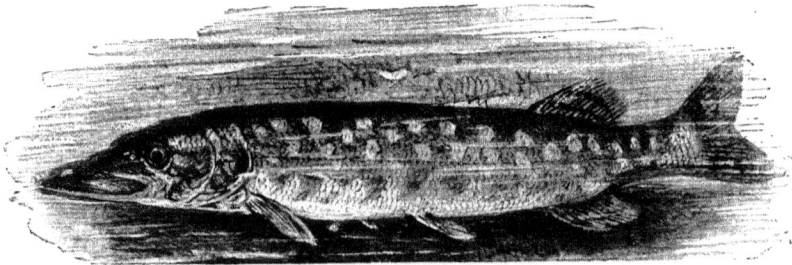

THE PIKE

of the former, and in the male (here larger than the female) modified into a long, thick, scaly organ, with an aperture at the end.

These fishes are the largest existing members of the whole family, growing nearly to a foot in length. They are abundant in North Brazil and the Guianas, frequenting mud-banks on the coast and in the estuaries

45 or 46 inches, with a weight of 35 or 36 pounds; while 45 pounds is recorded; when under 4 pounds they are known as jack.

It is pretty well ascertained that fish of 45 inches are not commonly more than about fifteen years old, and the stories of examples living for a century or more appear to be legendary.

Pike are among the most predaceous and greedy of all fresh-water fish, nothing coming amiss to their

PHOSPHORESCENT SARDINE

voracious appetites, since not only will they devour worms, leeches, frogs, trout, carp, and other fishes, but they pull under the young, and often even the adults, of all kinds of water-birds, and have no objection to an occasional water-vole. Their habit of lying like a log in the water, as well as the sudden rush they make after their prey, are well known to all; and the damage these fish do to trout-streams is almost incredible.

Pike are also great devourers of the smaller members of their own kind. Frequenting alike ponds, lakes, and rivers, pike in Ireland spawn as early as February, but in England a month or two later, while in some parts of the Continent the season lasts till May. Males, which are inferior in size to their consorts, are said to be more numerous than the latter, and it is not uncommon for a female in spawning time to be attended by three or four of the opposite sex, who crowd round as she deposits her eggs.

SCOPELOIDS

The so-called phosphorescent sardine (*Scopelus engraulis*) may be taken as an example of an important family (*Scopelidæ*) of mostly pelagic or deep-sea fishes. The members of the family are devoid of barbels, but possess a small fatty fin some distance behind the dorsal. The scales, when present, are spineless. The margin of the upper jaw is always constituted solely by the premaxillæ; the gill-cover may be incompletely developed; the gill-opening is wide; false gills are present; but a swim-bladder is wanting. The intestine is extremely short; and the eggs are enclosed in the sacs of the ovaries, whence they are extruded by means of ducts.

In the typical genus (*Scopelus*) the body is oblong, large-scaled, and more or less compressed. Along the sides run series of phosphorescent spots, while similar glandular structures may in some species occur on the front of the body and on the back of the tail. The cleft of the mouth is unusually wide; the premaxillary bones being long, slender, and tapering, and the maxillæ well developed. The teeth are villiform, and the eye is relatively large. The pelvic fins are inserted just in front of or immediately below the line of the foremost rays of the dorsal (which is situated nearly in the middle of the length of the body), and are composed of eight rays; the fatty fin is very small; the anal is generally long; and the caudal forked. There are from eight to ten rays in the branchiostegal membrane.

Among several other remarkable forms of the family may be noticed a very curious fish (*Ipnops*) obtained at great depths during the voyage of the Challenger. Possessing an extremely elongated and cylindrical body, covered with large, thin, deciduous scales, this fish has a depressed head and an elongate, broad, spatulate snout, of which the whole upper surface is occupied by a luminous or visual organ, divided longitudinally into two halves, and representing the highly modified eyes. The whole length of this strange fish does not exceed between 5 and 6 inches.

Another deep-sea fish (*Plagyodus*) is noteworthy on account of its large dimensions and formidable teeth: the scaleless body being long and compressed, the snout much produced, and the teeth of the jaws and palate of very unequal size, some forming long and sharply pointed tusks.

The Indian luxury called " Bombay duck " consists of the salted and dried bodies of the bummalow (*Harpodon nehereus*), a scopeloid with uniformly phosphorescent skin that occurs in large numbers in the surface of the Indian Ocean.

BLIND-FISH

The celebrated blind-fish (*Amblyopsis spelœa*) from the Mammoth Cave in Kentucky, the Wyandotte Cave in Indiana, and the subterranean streams which appear to connect the waters of the two, is generally regarded as the typical representative of a family (*Amblyopsidæ*) closely allied to the last. This fish, which does not exceed 5 inches in length, and breeds viviparously, closely resembles the genus *Cyprinodon* in that certain specimens lack the pelvic fins. All traces of external eyes are wanting, and the skin is colourless.

In order to enable the creature to find its way about in dark subterranean waters, its head is provided with a large supply of organs of touch arranged in a series of transverse ridges on each side, and its sense of hearing is also stated to be very highly developed. Professor Cope writes that if these fish " be not alarmed, they come to the surface to feed, and swim in full sight, like white aquatic ghosts. They are then easily taken by the hand or net, if perfect silence is preserved, for they are unconscious of the presence of an enemy except through the medium of hearing. . . . They must take much of their food near the surface, as the life of the depths is apparently very sparse. This habit is rendered easy by the structure of the fish, for the mouth is directed partly upwards, and the head is very flat above, thus allowing the mouth to be at the surface."

Nearly allied to that variety of the blind-fish in which pelvic fins are absent is a small fish known as *Chologaster*, in which small external eyes are retained and the body

KENTUCKY BLIND-FISH

is coloured, the front of the head being provided with a pair of horn-like appendages. These small fish were first known from three examples taken in the ditches of the South Carolina rice-fields, but a fourth specimen was captured in a well in Tennessee in 1854. The retention of the eyes and their dark colour indicates that these fishes have taken to a partially subterranean life more recently than the blind-fish.

TELEOSTEI—GROUP 2. PHYSOCLISTI
CLOSED-BLADDERED FISHES

THE fishes of the group Physoclisti are distinguished from tube-bladdered fishes (Physostomi) by the general absence of a duct to the swim-bladder (when present), by the parietal bones of the roof of the skull being always separated from one another by the intervention of the supraoccipital, and by the pelvic fins being in most cases either thoracic or jugular in position. They are said to be thoracic when in the same vertical line as the pectoral fins, and jugular when in advance of them.

The enormous suborder Acanthopterygii includes most marine fishes. The front rays of the dorsal and anal fins are generally spiny, and the toothless maxillary bones do not help to form the edge of the upper jaw. The pelvic fins are situated far forward, and the large gill-opening is placed in front of the pectoral fin.

PERCH TRIBE

The first representatives of the spine-finned fishes are the perches (*Percidæ*), which, with several allied families, belong to a sectional group (*Perciformes*) of the suborder, are characterised by the lower pharyngeal bones being generally separate, and the scales usually of the ctenoid type. The preopercular bone of the gill-cover has no bony stay connecting it with the eye ; the spinous portion of the dorsal fin is well developed ; none of the additional rib-like bones known as epipleura are attached to the bodies of the vertebræ ; the pelvic fins are thoracic in position, and have usually five (rarely four) branched rays ; and the supporting bones (pterygials) of the pectoral fins are longer than broad, and of a more or less distinctly hour-glass form.

The perch family is distinguished by the following characters. In the skeleton the anterior vertebræ have no transverse processes, but in the dorsal part of the series all or most of the ribs are attached to such processes. There are two nostrils on each side ; the gill-membranes are free from the isthmus, or space between the two branches of the lower jaw and gill-openings ; there are four pairs of gills, with a slit behind the fourth ; the gill-rays, or branchiostegals, vary from six to eight on each side ; more or less fully developed false gills are generally present ; the soft portion of the dorsal fin is not very much more developed than the anal, and the latter has either one or two spines.

In common with the two following families, the perches are further characterised by the general presence of a lateral line continuous from the head to the tail, the usual absence of scales from the median fins, the simple conical teeth, and the absence of barbels round the mouth. The body is more or less elongate, compressed, and cylindrical, rarely it may be slightly compressed. The family includes a dozen genera inhabiting the fresh waters of North America, Europe, and Western Asia ; but the members of the genera *Lucioperca* and *Percarina* enter salt water. All are carnivorous.

TRUE PERCHES

The common perch (*Perca fluviatilis*), which is a fish of wide distribution, is the type of a small genus,

agreeing with eight others in the following characteristics. In the head the mucus or slime canals are but moderately or slightly developed on the top and at the sides ; and the spinous and soft portions of the dorsal fin are separate. In common with six other genera, the body is more or less compressed ; the perches and pike-perches being specially distinguished by having usually seven (rarely eight) gill-rays ; by the premaxillæ, or anterior upper jawbones, being capable of protrusion, and by the serration of the preopercular bone of the gill-cover.

As a genus, the true perches are distinguished from the pike-perches by the small and uniform size of the marginal teeth, and the close approximation of the pelvic fins. There are teeth on the palatine and vomerine bones, but none on the tongue, and there are thirteen or fourteen spines in the first dorsal fin and two in the anal. The scales are small, the upper surface of the head naked, and the preorbital as well as the preopercular bone is serrated. There are seven branchiostegal rays, and more than twenty-four vertebræ.

THE PERCH

As in most of the members of the family, the mouth is capable of a certain degree of protrusion.

The common perch, which seldom exceeds 5 pounds in weight, is distributed over the rivers of Europe (except Spain) and Northern Asia so far east as Lake Baikal ; two others being known—namely, *P. flavescens* from the eastern United States and *P. schrenki* from Turkestan. Generally preferring still waters, and occasionally descending into estuaries, the perch is one of the most voracious of fishes, feeding indiscriminately upon worms, insects, and small fishes. The spawning season in England is at the end of April or May, when the female deposits her eggs in net-shaped or elongated bands on the leaves of aquatic plants. The eggs are very numerous, upwards of two hundred and eighty thousand having been taken from a fish of half a pound in weight.

PIKE-PERCHES

The pike-perches (*Lucioperca*) are inhabitants of many of the lakes and rivers of Europe, Western Asia, and eastern North America, and take their name from their somewhat elongated and pike-like form. From the true perches they differ by the presence of more or less enlarged tusks in the marginal series of teeth, and by the wider interval between the pelvic fins. The two dorsal fins are rather low, the first having from twelve to fourteen spines, and the scales are small. The common species is confined to Eastern Europe, where it is much esteemed as a food-fish, grows to a length of 3 or 4 feet, and attains a weight of from 25 to 30 pounds. Its extreme voracity and destructiveness to other fish render it an undesirable inhabitant of preserved waters.

RUFFES

Under this name, which belongs properly only to the British form (*Acerina cernua*), may be included a few small perches. From the other members of the family this and the allied genus *Percarina* differ by the large size of the slime-cavities on the sides and top of the head ; the ruffes being specially distinguished by the dorsal fin being undivided, and also by the maxilla being covered by the preorbital bone. The fishes of this genus have the body somewhat low, and the scales somewhat small, the continuous single dorsal fin carrying from thirteen to twenty-nine spines ; and there are two spines in the anal fin. There are no tusks among the small teeth of the jaws, and the tongue and palatine bones are devoid of teeth, although these are present on the vomer.

The genus is confined to the cooler portions of the Northern Hemisphere, the common species ranging from Britain through Central Europe to Siberia. The "pope," as the ruffe is frequently called in England, is common in most of the rivers and canals of that country, generally preferring slow, shaded streams, with a gravelly bottom, and closely resembling the perch in its mode of life.

Here may be mentioned a small family (*Centrarchidæ*), with ten genera of perch-like fishes, distinguished from the *Percidæ* and the following family by the mode of attachment of the ribs, which, with the exception of the last, or last two or four, are inserted on the bodies of the vertebræ behind the transverse process, instead of on the process itself ; and all the vertebræ in front of the tail, save the first two or three, having such processes. Externally these fishes differ from the perches in the presence of at least three spines in the anal fin. The family is typified by the North American genus *Centrarchus*, of which there is but a single species.

THE PIKE-PERCH

All are carnivorous fresh-water fishes, sometimes entering estuaries, and many are in the habit of building nests for the protection of their young.

BASS AND SEA-PERCHES

Agreeing with the *Percidæ* in the structure and relations of the vertebræ and ribs, the sea-perches and their allies (family *Serranidæ*), differ in the circumstance that the second suborbital bone develops an internal plate for the support of the eye. The number of spines in the anal fin is variable ; and in one genus (*Centrogenys*) the lower pharyngeal bones are united. The family is a very extensive one, and may be divided into several subfamily groups.

SOUTH AMERICAN PERCH

Together with the sea-perches, the bass represents a subfamily (*Serraninæ*) presenting the following character-

istics. The upper jawbone, or maxilla, is exposed, its upper border not being entirely concealed by the overlapping preorbital ; the scales are not shed ; there is no scaly process at the base of the pelvic fins ; the anal fin has three spines ; the gill-membrane is free behind ; and the false gills are well developed. In distribution the subfamily is cosmopolitan, and while most of the forms are marine, a few inhabit fresh water. Among the latter may be mentioned the South American perch (*Percichthys*), of Chili, Western Argentina, and Patagonia, which, in common with five other genera, has a divided dorsal fin. From an ordinary perch these fish may be distinguished by the scaly upper surface of the head, and the presence of nine or ten spines in the first dorsal and three in the anal fin.

BASS

To this section of the subfamily also belong the bass (*Morone*), which are partly marine and partly fresh-water fishes, easily distinguished from the true perch by having only nine spines in the dorsal fin, while there are usually three in the anal. There are also teeth on the tongue ; and while the preopercular bone is serrated, with denticulations on its lower border, the front border of the preorbital bone is entire. The scales are rather small, and extend all over the head.

Of the three European and Atlantic species which are almost entirely marine, the best known is the common bass (*M. labrax*), characterised by its extreme voracity and fierceness. Elsewhere, the genus is represented by fresh-water species from the rivers of the United States and Canada. Generally not exceeding a foot or eighteen inches in length, the common species may grow to three feet ; but its flesh is then much less delicate than that of ordinary specimens. Bass frequent the coast in shoals, spawning in summer, generally near the mouths of rivers, up which they not infrequently ascend for considerable distances.

SEA-PERCHES

In the other genera of the subfamily the dorsal fin is undivided, although it

THE RUFFE

may be deeply notched, the number of its spines being generally nine or eleven. Under the common title of sea-perches may be included the members of several allied genera, such as *Centropristes* and *Anthias*, although the name is often restricted to those of the typical genus *Serranus*, one of which (*S. scriba*) is represented in the middle of the illustration on the next page.

In the sea-perches the body is oblong or compressed, and covered with small ctenoid or cycloid scales ; there are large tusks among the villiform teeth of the jaws ;

and teeth are also present on the palatines and vomers, although absent from the tongue. The preopercular bone is serrated behind and at the angle, but not below, and the tail-fin may be either rounded, squared, or emarginate.

The sea-perches of the genus *Serranus*, of which there are an enormous number of species, range through the seas of all the tropical and temperate regions, occasionally ascending tidal rivers for short distances in pursuit of prey, but being otherwise strictly marine. Many of the species vary considerably, both in colour and in the form of their fins, with age, so that specific distinctions are difficult to establish.

STONE-BASS

The stone-bass (*Polyprion cernuum*) is one of two species constituting a genus distinguished from the last by the absence of large tusks in the jaws, and the presence of teeth on the tongue; the single dorsal fin having eleven or twelve spines, and the anal three. The preopercular bone is denticulated, and there is a strongly marked, rough, longitudinal ridge on the opercular. The common species is abundant on the European coasts, while the second is from the seas of Juan Fernandez. Both attain a very large size, ranging in weight to 80 pounds or more, their flesh being of excellent quality. The European stone-bass frequents the neighbourhood of floating wood, probably for the purpose of feeding on the creatures to be met with around such objects.

ORIENTAL AND AFRICAN PERCHES

The two species of the genus *Lates*, one of which (*L. niloticus*) inhabits the mouth of the Nile, while the second (*L. calcarifer*) ranges from the shores of Baluchistan through the Indo-Malayan seas to China and Australia, may be taken as representatives of another subfamily (*Centropominæ*), with three genera, this subfamily differing from the last by the extension of the lateral line on to the tail-fin, the presence of a scaly process at the bases of the pelvic fins, and the small size or absence of the false gills. Having no teeth on the tongue, and a divided dorsal fin, these fish may be distinguished externally from the true perches by the presence of seven or eight dorsal and three anal spines. Both the preopercular and preorbital bones are serrated, and the latter denticulated at the angle, the finely pectinated scales being of moderate size.

The Indian perch, which may grow to a length of 5 feet, is the only Oriental member of the family which commonly ascends rivers to any distance. When taken in the larger rivers its flesh is excellent for the table, great quantities being sold in the Calcutta market, where it is commonly known by the name of cock-up. The allied genus *Psammoperca* is represented by two species, one ranging from Australia to China, while the other is exclusively Australian. There are many other generic representatives of this extensive family, which are far too numerous to mention, no less than twenty-seven occurring within the limits of British India.

SCALY-FINNED FISHES

Nearly allied to the perches, the beautiful tropical species designated scaly-finned fishes are so named on

(1) COMMON BASS. (2) SEA-PERCH, AND (3) STONE-BASS

account of the median fins being more or less thickly covered with small scales. These fishes (family *Chætodontidæ*) are also characterised by the deep and compressed form of the body, on which the scales are either ctenoid or entire, and the continuous lateral line, which stops short of the tail-fin. The usually small mouth is placed at the extremity of the snout, and has a distinct lateral cleft. The small teeth are arranged in bands, and there are neither tusks nor incisors. The soft portion of the single dorsal fin is rather longer than the spinous; the anal has three or four spines; the lower rays of the pectorals are branched; the pelvic pair is thoracic in position, with one spine and five soft rays; and the scaling of the median fins causes them to pass imperceptibly into the body. Most of these curious and beautiful fishes are inhabitants of tropical seas, and are common in the neighbourhood of coral-reefs; and it may be remarked in passing that many reef-haunting animals exhibit brilliant hues, harmonising with those of the coral-polypes themselves.

Some forms ascend estuaries and tidal rivers for a short distance. All are carnivorous, and of relatively small size, while they are but seldom used for food. The three genera of which examples are represented on page 1741 are those in which the zebra-like coloration attains its most marked and striking development; and for the beauty and singularity of their adornment these fishes are almost unequalled. Out of a large number of existing genera it is to these that our attention will be chiefly directed; and it may be remarked that the whole of them are met with in the Indian seas.

CHÆTODON

The typical genus *Chætodon* belongs to a group of genera in which there are no teeth on the vomers or palatine bones, while the spines of the dorsal fin are not separated from the soft rays by a hollow or notch, and there is no spine to the preopercular bone, the genus in question being particularly distinguished by the short or moderately long snout, and the approximately uniform length of the spines of the dorsal fin. These fishes are common in the tropical Atlantic and Indo-Pacific, where they are represented by some seventy species. Nearly all are ornamented with bands or spots, a dark or two-coloured band passing through the eye, and then inclining backwards, being very characteristic.

Of the species represented, *C. setifer*, (1, p. 1741) ranging from the Red Sea to Polynesia, is readily recognised by the elongation of the fifth ray of the dorsal fin, behind the base of which is a large dark spot with a light rim; *C. trifasciatus* (3), which also has a similar range, but reaches the coasts of India, is marked by numerous fine longitudinal stripes on the body, and several dark bands across the head. On the other hand, in *C. fasciatus* (2), of the Indian and Malayan seas, the body-stripes are oblique, and there is a single dark band across the head.

HENIOCHUS AND HOLACANTHUS

The fish (*Heniochus macrolepidotus*) numbered 4 in the illustration on page 1741, is a common Indo-Pacific member of a genus differing from *Chætodon* by the more or less marked elongation of the fourth spine of the dorsal fin, which in the figured species assumes the form of a whip-lash. Broad, dark bands across the body are very characteristic of the genus; and in the young the head is armed with numerous horn-like processes, which are permanently retained in a species named *H. varius*.

The two large fishes shown swimming towards the left in the illustration belong to a genus (*Holacanthus*) distinguished from all the foregoing by the presence of a large spine on the hinder edge of the preopercular bone; the dorsal fin having from twelve to fifteen spines. The genus includes some forty species, with the same range as the typical representative of the family. The splendidly-coloured emperor-fish (*H. imperator*), shown on the right side of the illustration (6), ranges from the east coast of Africa to the Indian and Malayan seas, and has the ground-colour of the body a deep blue, upon which are some thirty longitudinal golden-yellow stripes. The eye-stripe and a patch above the pectoral fin are black edged with yellow, and the tail-fin is yellow. This species, 15 inches long, is eaten in India.

Beautiful as the emperor-fish is, it is exceeded by the Indo-Malayan zebra-fish (*H. diacanthus*). In this species (5) the general colour is yellowish, with from eight to twelve vertical brown-edged blue bands; the caudal fin is yellow, and the anal marked with bluish lines running parallel to its margin.

ARCHER-FISHES

In the members of the small family of archer-fishes (*Toxotidæ*) there are teeth on the palatines and vomers, and the body is oblong and much less deep than in the typical forms, with the undivided and five-spined dorsal fin situated in its hinder half. It is represented by five species from the seas and estuaries of the East Indies, North Australia, Polynesia, and New Zealand. The family includes only one genus, *Toxotes*. The archer-fish (*T. jaculator*) is so called because it ejects drops of water at insects flying near the surface, and thus secures them as food.

ARCHER-FISH

RED MULLETS

Two long, erectile barbels on the lower jaw serve at once to distinguish the red mullets (family *Mullidæ*) from all the preceding families, with which they agree in the characters already mentioned. The body is rather low and somewhat compressed, with large thin scales, of which the edges may be very finely serrated. The lateral line is continuous, and the moderate-sized eyes are situated on the sides of the head. The terminal mouth has a rather short lateral cleft, and the teeth are very feeble. There are two short dorsal fins, placed at a considerable distance from one another; the spines of the first being weak, and the second being placed above the anal, which it resembles in form. The ventrals have one spine and five rays, and the pectorals are short. In place of the seven branchiostegal rays of the perches, the red mullets have but four.

Represented by something like forty species, the red mullets, which range over the seas of Europe and the tropics, are typically represented by the genus *Mullus*, of which there appear to be but two European species.

The ordinary European red mullet (*Mullus barbatus*), which does not usually exceed 6 inches in length, is coloured carmine-red on the upper parts, the under parts being silvery white. On the other hand, the striped mullet (*M. surmuletus*) has three or four yellow longitudinal stripes on the sides; and is also stated to differ slightly in the number of the fin-rays. This kind is common on the Cornish coast, whereas the plain-coloured form is but seldom met with in the British seas, although abundant in the Mediterranean. Mullets live chiefly on small crustaceans, frequenting coasts where the bottom is more or less muddy. Occasionally they visit the British coasts in

STRIPED RED MULLET

GROUP OF SCALY-FINNED FISHES

1, Chætodon setifer ; 2, C. fasciatus ; 3, C. trifasciatus ; 4, Heniochus macrolepidotus ; 5, Holacanthus diacanthus ; 6, H. imperator

vast shoals. The barbels are used to feel in the sand for food. They are folded back into grooves during swimming.

SEA-BREAMS AND SNAPPERS

The sixth family (*Sparidæ*) of the present section is especially characterised by the peculiarity of the dentition, the palate being generally devoid of teeth, while either cutting or conical incisor-like teeth are developed in the front of the jaws, or crushing molars on their sides ; in some cases both these types being coexistent. The oblong body of the sea-breams is markedly compressed ; and the scales are either but very slightly serrated or smooth. The terminal mouth has a distinct lateral cleft ; and the medium-sized eyes are also lateral. The single dorsal fin is composed in about equal moieties of a spinous and a soft portion ; the anal is three-spined ; as a rule the lower rays of the pectorals are branched ; and the pelvics, which are ventral in position, are furnished with one spine and five rays. The number of branchiostegal rays varies from five to seven. They are not related to the true breams, which belong to the carp family.

Sea-breams include a large number of genera, and are of sombre colouration and medium size ; the flesh of the majority being used for food.

The black sea-bream, or old wife (*Cantharus lineatus*), not uncommon on the south coast of Britain, may be cited as a well-known example of the typical genus of the first subfamily, in which the extremities of the jaws are furnished with broad, cutting, and occasionally lobate incisor-like teeth ; while there are no vomerine or molariform teeth, and the lower rays of the pectoral fins are branched.

The second subfamily is represented by *Haplodactylus*, from the temperate South Pacific, in which both jaws are furnished with flat and generally tricuspid teeth ; vomerine teeth being present, but molars wanting ; while the lower pectoral rays are simple. These fish are vegetable-feeders.

Better known than the last is the third subfamily, containing only the single genus *Sargus*, with some twenty species from the Mediterranean, Atlantic, and Indian seas, among which the common sargo (*S. annularis*), is a familiar fish on the Continent. The essential features of the group are the single series of cutting teeth in the front of the jaws, the presence of several rows of molars on the sides of the same, the toothless palate, and the simple lower pectoral rays. The figured species is a uniformly-coloured fish ; but in the larger "sheep's-head" (*S. ovis*), from the Atlantic coasts of the United States, which attains a weight of 15 pounds, and is highly esteemed for the table, the body and tail are marked by a number of broad vertical bands. The strong molars of these fish indicate that their food consists of hard-shelled molluscs, crustaceans, or sea-urchins.

GILT-HEADS

The first example to be mentioned of the fourth subfamily, which contains several genera, is the gilt-head (*Pagrus auratus*), so called on account of the golden spots between the eyes. It is common in the Mediterranean, where it is used as food, and occasionally wanders to the British coasts. The group is characterised by the presence of conical teeth in the front of the jaws, and of molars on their sides, the palate being toothless ; while the genus under consideration is distinguished by having scales on the cheeks, and at least three rows of upper molars.

The gilt-head is a handsome fish, with a short and elevated head, the body deepest at the beginning

SARGO AND GILT-HEADS

the smaller ones. In the single dorsal fin the spinous and soft portions are of nearly equal extent; the anal fin, which is generally smaller than the soft dorsal, carries three spines; and the pelvics, which, although thoracic in position, are situated at a considerable distance from the root of the pectorals, have one spine and five rays.

These fishes are inhabitants of all tropical seas and the temperate zone of the South Pacific. They may be divided into two groups, according to the presence or absence of teeth on the vomers, the first group including the small prettily-coloured fishes known as *Cirrhites*, and *Chorinemus*, of which the former is characteristic of the Indian and Pacific Oceans, while the latter is confined to the Australasian seas.

FIRM-FINS

Of the group with teeth on the vomers the spotted firm-fin (*Cirrhitichthys marmoratus*) belongs to a genus differing from the typical one by the presence of teeth on the palatine bones, and by the spiny opercular bone, the preopercular being serrated in both. These fish have six branchiostegal rays, tusks in the jaws, and ten spines in the dorsal fin. Five to seven of the lower pectoral rays are unbranched, the scales are of moderate size, and there is no swim-bladder. The spotted firm-fin, which ranges from the Red Sea, through the Indian and Malayan seas, to the Sandwich Islands, is one of those in which there is no elongation of a ray of the pectoral fins; while it is specially characterised by the spotted coloration.

LONG-FINS

In the group with toothless vomers, one of the most notable genera is that of the long-fins (*Chilodactylus*), so named on account of the elongation of one of the rays of the pectoral fins. Most of the species are inhabitants of the temperate region of the South Pacific, although some are found round the coasts of Japan and China. The species figured on the next page (*C. macropterus*) is an Australian one, and both in Australia and at the Cape these fishes form a valuable food supply, since they attain a weight of from 5 to 25 pounds,

of the dorsal fin, the iris yellow, a semilunar golden spot between the eyes, and a violet patch on the gill-cover. The back is silvery grey, with a tinge of blue, and the under surface steely, with longitudinal golden bands on the sides. In length it seldom exceeds a foot. Fully adult examples show a perfect pavement of teeth on the jaws; and with these the fish crunches up mussels and other shell-fish with such vigour that the noise thus made sometimes reveals its presence to fishermen. In order to obtain food, it is stated to stir up the sand of the sea-bottom with its tail.

Another species (*P. vulgaris*) is common in the Mediterranean; the snapper (*P. unicolor*) is esteemed as a food-fish in Australia and New Zealand, as is the scup (*P. argyrops*) in the United States; while the Japanese red tai (*P. major*) is of a beautiful scarlet colour. The common sea-bream (*Pagellus centrodontus*) is well known on the southern British coasts, and ranges from Norway to the Mediterranean.

THICK-RAYED PERCHES

The small family *Cirrhitidæ*, which, for want of a better English name, we designate as above, comprises several genera from the Indo-Pacific and Australasian seas, some members of which are of the first importance as food-fishes in the British colonies. Closely allied to the next family, they differ therefrom—and thereby resemble the preceding groups—in the absence of a bony connection between the preopercular bone and the infraorbital ring of the skull; while they are specially distinguished by the thickened and undivided lower rays of the pectoral fin, which in some cases are elongated so as to aid in the movements, while in others they may perhaps serve as additional organs of touch.

The body is oblong and compressed, and covered with cycloid scales; the mouth is terminal, with a lateral cleft; and the eyes are lateral. As a rule, the branchiostegal rays are six in number, although they may be reduced to five or three. The teeth are villiform or pointed, and in some cases there are tusks among

SPOTTED FIRM-FIN

and are easily captured. An allied genus (*Latris*), distinguished by the absence of any elongation of the pectoral rays and the deeply notched dorsal, is well known in Tasmania and New Zealand in the form called trumpeter-fish (*L. necatia*), which claims the first place among the fish products of those colonies, ranging in weight from 30 to 60 pounds, and being at the same time most excellently flavoured.

SCORPÆNAS

With the family *Scorpænidæ* we come to a group easily distinguished from all the preceding representatives of this section by the articulation of the preopercular bone with the orbit by means of a projecting process from the infraorbital ring. Some of the other bones of the head are also armed, and the dentition is but feebly developed. These fishes, which are represented by a large number of genera, are found in most seas, and are all carnivorous. Dr. Günther writes that " some resemble the sea-perches in their form and habits, as *Sebastes, Scorpæna,* etc. ; whilst others live at the bottom of the sea, and possess in various degrees of development those skinny appendages resembling the fronds of seaweeds, by which they either attract other fishes, or by which they are enabled more effectually to hide themselves. Species provided with these appendages have generally a coloration resembling that of their surroundings, and varying with the change of locality. Some of the genera live at a considerable depth, but apparently not beyond three hundred fathoms. Nearly all are distinguished by a powerful armature, either of the head, or fin-spines, or both ; and in some the spines have been developed into poison-organs." The group is scarcely known in a fossil state, although remains of a species of the typical genus occur in the Eocene of Algeria.

The family is divided into two sections, according as to whether there are distinct scales on the body, or whether these are rudimentary or wanting. In the former are included *Sebastes* and the typical *Scorpæna.* Most curious of all are the members of the Tropical Indo-Pacific genus *Pteröis,* in which the spines of the dorsal and the rays of the pectoral fins are more or less produced—so much so, indeed, that in the case of one species, at least, *P. volitans,* it was formerly thought that they indicated the possession of flying powers in their owner. The fins and body of this extraordinary-looking fish are most beautifully marked with alternating light and dark transverse bands.

TEUTHIS

The single generic representative of the *Teuthididæ,* of which a species (*Teuthis striolata*) is shown in the illustration on page 1744 is characterised by the toothless palate, and the presence of a series of narrow

AUSTRALIAN LONG-FIN

serrated incisor teeth in the front of each jaw. The scales on the oblong and compressed body are very small, and there is a continuous lateral line. In the single dorsal fin the spinous considerably exceeds the soft portion in length, the anal has seven spines, and the thoracically-placed pelvic fins have an outer and inner spine, between which are three rays. These fishes have a large swim-bladder, forked at both extremities, and they also display several peculiarities

in the structure of the skeleton, the abdomen being surrounded by a complete ring of bones, owing to the backward prolongation of certain elements of the pectoral arch, and the unusual development of the pelvis.

A considerable number of species have been described from the Indo-Pacific, where their eastward range stops about the longitude of the Sandwich Islands. The largest of them is not more than 15

THE HOG-SCORPÆNA

inches in length, and all are vegetable-feeders. In the figured species, which is from the New Hebrides, the general colour is brownish red, marked with narrow meandering blue lines, the spines of the fins also bearing white spots.

SLIME-HEADS

With the slime-heads we come to fishes (*Berychidæ* and related families) distinguished from the whole of the preceding, and forming a group by itself characterised by the presence on the head of large mucus-bearing cavities, covered with a thin skin, and by the thoracically-situated pelvic fins having one spine and five rays (save in *Monocentris,* where the latter are reduced to two). The compressed body may be either oblong or deep in form, but is always short ; and the scales (rarely wanting) are ctenoid. Lateral in position, the eyes are almost always large in size ; the lateral cleft of the mouth slopes obliquely upwards ; the teeth in the jaws are villiform ; teeth are in most cases developed on the palatines ; the bones of the gill-cover are more or less fully armed ; and there are nearly always eight branchiostegal rays, although these are sometimes reduced to four. There are no scales on the head, and false gills are present.

The almost cosmopolitan slime-heads, which comprise a considerable number of living and extinct genera, are nearly all marine, many of them living at great depths. They are also a geologically ancient group, represented by a large number of generic types, both existing and extinct in the Chalk and other Cretaceous deposits. The fact that some of the berycoids have persisted without much change from the Chalk period to the present day marks them as a primitive group, and this is confirmed by their structure. In some of them (*Beryx* and *Holocentrum*) there is a duct to the swim-bladder, as in the tube-bladdered fishes, and they may be regarded as forming a link between such fishes and the closed-bladdered forms (Physoclisti).

The only existing forms that frequent the higher strata of the ocean belong to the genera *Holocentrum* and *Myripristis ;* but even some species of the latter may descend as deep as one hundred and fifty fathoms. The typical genus *Beryx* has been taken from between three and four hundred fathoms ; and from the small size of the eye the forms known as *Melamphæes* must, in Dr. Günther's opinion, inhabit still lower levels. Another sign of their deep-water habits is afforded by the high development of the slime-secreting apparatus of these fishes. An exception to the marine habit typical of the group is afforded by a small species known as the pirate perch (*Aphredoderus sayanus*), native to fresh water in North America.

" The fishes of the genus *Myripristis* are very pugnacious, and always ready to quarrel with their own kind. In Hawaii the natives take advantage of this trait to catch the un (*M. murdjan*). Having obtained one alive by a net or other means, they attach a string to it and put it back into the water in front of the crevices in the rock in which these fishes lurk. The other fishes soon come out to fight it, and the crowd is brought to the surface of the water by slowly drawing in the string ; a net is passed cautiously beneath and the whole crowd captured."

Of other genera the most peculiar is *Monocentris*, represented by a single small and rare species from the seas of Japan and Mauritius, and distinguished by the absence of armatures on the gill-cover, the large scales, which are articulated together so as to form a solid armour, and the reduction of the ventral fins to a single long spine, and a few rudimentary rays.

The fish figured in the illustration (*Trachichthys trailli*) represents a genus, with a few species from New Zealand and Madeira, characterised by the short and blunt snout, the prominent chin, the strong spine at the angle of the preopercular, the rather small scales, and the serration of the lower border of the body. The palatines and vomers carry villiform teeth, the single dorsal fin has from three to six spines, the anal six rays, and the tail is strongly forked. The allied *Anoplogaster*, of the Tropical Atlantic, is devoid of scales. In both genera the eye is very large.

The typical genus *Beryx*, which has likewise but a single dorsal, may be distinguished by the smooth abdomen, and the lack of a spine on the preopercular. At the present day this genus is known from the Tropical Atlantic, Madeira, and the seas of Australia and Japan, while in a fossil state it is abundant in the Chalk. Two barbels at the throat serve to distinguish *Polymixia* ; while in several of the other genera, such as *Holocentrum*, the dorsal fin is double. Spread over all tropical seas, the latter genus is likewise one of those dating from the Cretaceous epoch.

Each of the two unimportant families *Curtidæ* and *Polynemidæ* represents a section of equal rank with the perch-like division of the suborder, the first being characterised by having the single dorsal fin much shorter than the long and many-rayed anal. The compressed body, as in *Pempheris mangula*, is oblong in form, deep in front, and sharply narrowing towards the tail. If developed at all, the spines of the short dorsal are few in number, the scales are small or medium in size, and both the jaws, palatines, and vomers bear villiform teeth. While the typical genus *Curtus* is confined to the Indian seas, *Pempheris*

SPINE-FINNED FISHES
1, Teuthis striolata ; 2, Pempheris mangula ; 3, Polynemus plebejus

ranges over the Indian Ocean, the Malayan seas, and the tropical parts of the Pacific.

The presence of a number of filaments, which may be enormously long, is the most distinctive feature of the second family, as shown in the type genus (*Polynemus plebejus*) ; while a second characteristic is to be found in the two rather short dorsal fins, situated at a considerable distance from one another, and a third in the well-marked mucus-bearing canals on the head. The body is oblong and somewhat compressed, with smooth or slightly ciliated scales, and a continuous lateral line. The snout projects somewhat beyond the mouth, which is inferior in position, with a lateral cleft, and the large eyes are lateral. There are villiform teeth in the jaws and on the palate, and the pelvic fins are thoracic, with one spine and five rays.

These fishes include three genera, containing a number of tropical shore forms, which sometimes enter brackish or fresh water. Their filaments, which sometimes exceed twice the entire length of the head and body and can be moved independently of the pectoral fins, serve as feelers ; and as they live in muddy water, and generally have their large eyes obscured by a film, the use of such accessory organs of touch is easily understood.

SCIÆNOIDS

Of more general interest than the last is the family of Sciænoids (*Sciænidæ*), among which the umbrine of the Mediterranean and the widely distributed meagre are well-known examples. In this group the spinous dorsal is abbreviated at the expense of the more or less elongated soft dorsal, which also exceeds the anal fin in length ; and, although mucus canals are well developed on the head, there are no filaments near the pectoral fin. The somewhat elongated and compressed body is coated with ctenoid scales, and the uninterrupted lateral line sometimes continued on to the tail-fin. The long mouth is terminal, the medium-sized eyes are lateral, and, in addition to bands of villiform teeth, the jaws may carry tusks, although they are never provided with incisors or molars, and the palate is toothless. The preopercular bone is smooth, and without any bony connection with the orbit, and the thoracic pelvic fins carry one spine and five rays. Frequently the swim-bladder is provided with a number of appendages.

These fishes have a rather curious geographical distribution, being unknown in the Pacific and the Red Sea, but widely spread in the Atlantic and Indian Oceans, and especially common round the shores of India, where many species enter estuaries and rivers. Some have taken completely to fresh water. Nearly all are eaten as food, and the swim-bladders of many of the Indian forms are extensively used as a source of isinglass.

THE DRUM

The North American fish known by the name of "drum" (*Pogonias chromis*) represents a genus characterised by the upper jaw of the convex snout overhanging the lower, the presence of numerous small barbels on the chin, and the absence of tusks. Ten stout spines form the first dorsal fin, and there are two spines in the anal, the hindmost of which is very strong. The scales are of moderate size ; and there are a number of large flattened molar-like teeth on the pharyngeal bones. In length the drum often exceeds 4 feet ; while it may scale upwards of 1 cwt. In what manner the extraordinary drumming sounds uttered by this fish, in common with other members of the family, are produced does not appear to be ascertained, although it has been suggested that they may be due to the clapping together of the upper and lower pharyngeal teeth.

UMBRINES

The umbrine of the Mediterranean (*Umbrina cirrhosa*), ranging southwards to the Cape, was well known to the ancients, and belongs to a genus containing about twenty species distributed through the Mediterranean, Atlantic and Indian Oceans. Having an overlapping upper jaw, it differs from the last genus in the presence of but a single short barbel on the

TRAILL'S TRACHICHTHYS

chin ; while the first dorsal fin has ten flexible spines, and the anal either one or two. In size the typical species reaches about 3 feet.

MEAGRES

The third genus that we notice (*Sciæna*) differs from both the preceding in the absence of any barbels; the cleft of the mouth being oblique and deep, and the

eyes situated rather wide apart. The genus includes a very large number of species, with a geographical range equal to that of the family ; one of the best-

THE MEAGRE

known being the typical meagre (*S. aquila*), ranging from the British coasts to those of the Cape and Australia. Although most of the species are smaller, this fish may attain a length of upwards of 6 feet. Yarrell states that the flesh of the meagre " appears always to have been in great request with epicures ; and, as on account of its large size it was always sold in pieces, the fishermen of Rome were in the habit of presenting the head, which was considered the finest part, as a sort of tribute to the three local magistrates who acted for the time as the conservators of the city." It is certain members of this genus that have taken to a fresh-water existence.

SWORD-FISHES

With the small and well-defined family of sword-fishes (*Xiphiidæ*), all the members of which attain very large dimensions, we come to our first representatives of purely pelagic fishes. Sufficiently distinguished from all their allies by the production of the upper jaw into a long, sword-like weapon, they are further characterised by the elongate and compressed body, the laterally-placed eyes, and the deep cleft of the mouth. Teeth are either absent or rudimentary ; and scales are either absent or represented rudimentarily. The dorsal fin is either single or divided, but has no distinct spinous portion ; and the pelvics, if present, take the form of long, rod-like, thoracic appendages. There are seven branchiostegal rays, and a swim-bladder is present. In the adult the sword is formed by the coalescence of the premaxillæ, vomer, and ethmoid, and is rough on the under surface from the presence of rudimentary teeth.

The sword-fishes are divided into the genera *Xiphias* and *Histiophorus*, according to the absence or presence of pelvic fins, these appendages in the latter being in the form of from one to three rays. There is considerable variation in the height of the dorsal fin, which is frequently so lofty as to project some distance above the water when the fish is swimming near the surface, and even, it is said, to answer the purpose of a sail. In the young this fin is much higher in proportion to the length of the body than it is in the adult. In very young examples of the typical genus the beak is comparatively long, there are conical prominences on the edge of the supraorbital, the occiput is devoid of a spine, and there are two short, tooth-like processes at the angle of the preopercular.

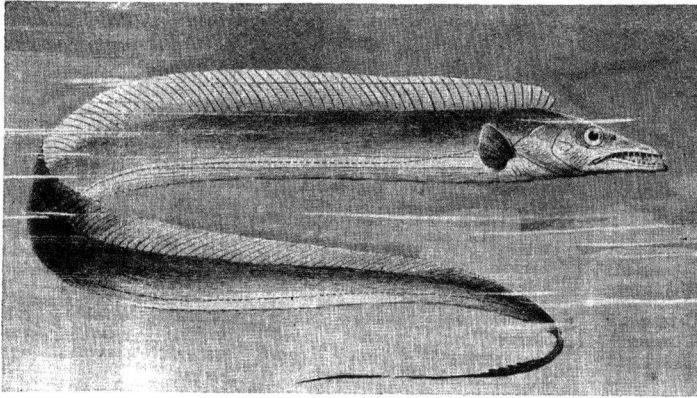

THE SCABBARD-FISH

a bony spine on each side of the occiput and at the angle of the preopercular. Although they are frequently not more than 4 to 6 feet in length, sword-fishes may measure as much as from 12 to 15 feet, and the sword itself may exceed a yard in length.

The common European sword-fish (*Xiphias gladius*), occasionally taken on the British coasts, ranges from the European seas to the opposite side of the Atlantic ; while to the southward it occurs off the northern and western coasts of Africa. *Histiophorus*, on the other hand, seems to be confined to the Pacific and Indian Oceans, ranging eastwards to Japan. Of the three Indian species, the spotted Indian sword-fish (*H. gladius*) is distinguished by the dorsal fin being much higher than the body, and marked with dark blue spots on a lighter ground of the same colour, the body being bluish grey above and lighter beneath. On the other hand, in the black-finned sword-fish (*H. immaculatus*) the general colour of the body is dull grey, and the dorsal and anal fins are blackish. The third species (*H. brevirostris*) has the dorsal fin lower than the depth of the body, the general colour being grey, but the dorsal and pectoral fins tipped with black.

Mainly pelagic, sword-fishes are extremely predaceous and savage, transfixing their ordinary prey, which includes cod and tunny, with their formidable sword, and, also attacking whales. In such conflicts the sword-fish, after making repeated stabs, is generally victorious. Occasionally, however, one of these fishes appears to mistake a ship's bottom for a whale, and promptly charges it, sending the sword crashing through several inches of solid timber. In such cases it may happen that the sword-fish cannot withdraw its weapon, which is then broken off short in the struggles of its owner to escape. Instances are on record of these fish attacking and transfixing bathers. When it feels the hook or spear, a sword-fish takes tremendous leaps in the air, and may jump into the boat of the fishermen. In the South Sea Islands young sword-fish are caught in strong nets.

One of the most recent instances of a sword-fish attacking a ship occurred in the year 1874, on the voyage between Bombay and Calcutta. On this subject Frank Buckland writes that there is in the museum of the College of Surgeons a section of the bow of a South Sea whaler, in which " is seen the end of the sword of a sword-fish, measuring 1 foot in length and 5 inches in circumference. At one single blow the fish had lunged his sword through, and completely transfixed, thirteen and a half inches of solid timber. The sword had, of course, broken off in the hole, and thus prevented a dangerous leak in the ship.

In the British Museum is a second specimen of a ship's side with the sword of a sword-fish fixed in it, and which has penetrated no less than twenty-two inches into the timber."

SCABBARD-FISHES AND HAIR-TAILS

Another group of equal rank with the perch-like section is formed by the family *Trichiuridæ*, characterised by the elongate and compressed or even band-like form of the body ; the mouth having a wide cleft, and several large conical teeth either in the jaws or on the palate. The dorsal and anal rays are long and many-rayed, with the spinous nearly equal in length to the soft portion, finlets sometimes occurring behind the latter ; the pelvic fins, if present, are thoracic in position and the caudal is sometimes wanting but, when developed, forked. In all cases the scales are either rudimentary or wanting, but the swim-bladder is constant. These fishes are distributed over all tropical and subtropical seas, but while some are surface forms, never found at any great distance from the coasts, others descend to considerable depths in the open sea ; all are carnivorous, and many very powerful.

SCABBARD-FISH

Among the better-known forms, the scabbard-fish (*Lepidopus caudatus*) represents a genus characterised by the absence or rudimentary condition of the pelvic fins, the long single dorsal, and the distinct but small tail-fin. Although it may attain a length of 5 or 6 feet, the attenuation of the body is so great that the whole weight does not exceed as many pounds. The fish has a very wide geographical distribution, ranging from the Mediterranean and warmer regions of the Atlantic to the Cape, and thence to New Zealand and Tasmania, while it occasionally wanders to the British coasts. This wide range may possibly be taken as an indication that the scabbard-fish is a comparatively deep-sea form.

In New Zealand, where it is known as the frost-fish, the scabbard-fish is highly esteemed for its flesh, which is white, rich, firm, and tender, with an excellent flavour. On calm and frosty nights, during the autumn and winter months, numbers of frost-fish, as they are called, come ashore alive through the surf on the beaches of the Pacific, and there wriggle up on to the firmer sands above, to be devoured by the watchful sea-birds, or picked up by the fortunate fishermen.

HAIR-TAILS

The scaleless fish, the hair-tails (*Trichiurus*), take their name from the absence of a caudal fin, the body tapering posteriorly into a fine point. The single dorsal extends the whole length of the ribbon-like body ; the pelvic fins are represented merely by a pair of scales, or are completely wanting, and the anal is rudimentary, its spinous portion being reduced to a number of very small spines scarcely projecting above the skin. The jaws are provided with long tusks, and there are teeth on the palatine bones, although none on the vomer.

Essentially tropical fishes, generally found in the vicinity of land, they appear to be sometimes carried by currents out to sea. These fishes attain a length of from 3 to 4 feet, and one of the Indian species is described as extremely voracious, preying on crustaceans and various fishes, among which members of its own kind are included.

BARRACUDAS

The local name, barracuda, given to a New Zealand representative (*Thyrsites atun*) of another genus may be taken as the popular title of all its members. These fishes, in which the rather elongate body is covered with minute scales, are characterised by having from two to six finlets behind the dorsal and anal, and the presence of teeth on the palatines. Barracudas, which grow to as much as 5 feet, form important food supplies in the Cape, South Australia, New Zealand, and Chili.

SURGEONS

With the family *Acronuridæ*, we come to an almost entirely marine group of fishes, including some thirteen others, which present the following characteristics in common. The dorsal fins are either placed together or continuous, the spinous portion being, when fully developed, shorter than the soft part, while it may be modified into tentacles, detached spines, or an adhesive disc, and the anal is similar in characters to the soft dorsal, and in some instances both these fins are modified posteriorly into finlets. The pelvic fins, if developed, are always thoracic or jugular in position, and are never modified into a sucker.

The first family is typified by a genus (*Acronurus*), the representatives of which are called " surgeons," owing to the presence of a sharp lancet-like spine on each side of the tail in the adult. In addition to the presence of one or more such spines or bony plates, the family is further characterised by a single dorsal fin, with a very small number of spines. The body is compressed, and oblong or deep in form, with a covering of minute scales ; the moderate-sized eyes are lateral in position, the small mouth is furnished in front with a single series of more or less compressed upper and lower incisors, which may be either pointed or serrated ; but the palate is toothless. The pelvic fins are thoracic, and the hinder extremity of the swim-bladder is forked. These fishes are inhabitants of all tropical seas, and are most common near coral reefs and islands, where some feed on the coral-polyps, and others on various vegetable substances.

In the true surgeons (*Acronurus*) there is an erectile spine situated in a groove on each side of the tail, and the pelvic fins are generally furnished with a single spine and five rays. These fishes are represented by a

EUROPEAN SWORD-FISH

large number of species, the largest of which does not exceed 18 inches in length, and they are distributed over all tropical seas with the exception of the Eastern Pacific. The true surgeons use their spines as formidable weapons of attack, by erecting them and striking sideways with their tails.

HORSE-MACKERELS

Although the name horse-mackerel properly applies only to a single British fish, *Caranx trachurus*, otherwise known as the scad, it may serve for the whole of the members of the family *Carangidæ*. Having the body more or less compressed, these fishes are specially distinguished by the teeth, when present, being villiform or conical. The spinous portion of the dorsal fin is sometimes rudimentary ; the hinder rays of both the dorsal and anal may be broken up into separate finlets, and, when present, the pelvic fins are thoracic in position. In the skeleton, there are ten trunk and usually fourteen tail vertebræ. The gill-openings are wide, the eyes lateral, and there is no bony stay connecting the preopercular with the infraorbital ring. The scales, which are usually small, may be altogether wanting, and in many cases the lateral line is wholly or partially armed with shield-like overlapping plates. There is always a swim-bladder. Carnivorous in their diet, the horse-mackerels are distributed at the present day over all temperate and tropical seas.

In the typical genus *Caranx* the body is generally more or less compressed, although sometimes almost cylindrical ; the hard dorsal fin, which may be rudimentary, is continuous, with about eight weak spines ; while in a few species the soft portion of both this and the anal is broken up into finlets. The scales are very small, and while in the British horse-mackerel (*C. trachurus*) the whole of the lateral line is protected by bony plates, in many other species these plates are restricted to its hinder moiety. The genus is represented by nearly a hundred species, some of which have teeth on the palate, while in others these are wanting.

Ranging over almost all temperate and tropical seas, many of them swim out to considerable distances from the shore, and thus acquire a very large distributional area. The larger forms may measure fully a yard in length, and the flesh of all is edible.

SPOTTED INDIAN SWORD-FISH

Horse-mackerel sometimes appear in enormous shoals on the British coasts.

PILOT-FISH

Another genus is represented by the pelagic pilot-fish (*Naucrates ductor*), so called from a supposed habit of guiding and protecting the sharks and ships which it accompanies. Having no plates on the lateral line, this fish is further characterised by the rounded under surface of the body, by the first dorsal fin being composed in the adult of detached spines, by the absence of

THE HORSE-MACKEREL

finlets, and the presence of a keel on each side of the tail. When adult, the pilot-fish measures about a foot in length. In colour it is bluish, with five or six dark vertical bands; the tail-fin sometimes having the ends of its two lobes dark, as also a band across the middle third.

Ranging over all temperate and tropical seas, pilot-fish were regarded as sacred by the ancients, by whom they were known as *pompili*, the common belief being that when the ship neared land, the fish suddenly disappeared, and thus gave warning to the sailors of impending danger. Many legends have grown in later times as to how pilot-fish will prevent sharks from taking a bait by swimming round them and enticing them away, but all these appear to be pure fictions, and perhaps the best account of the habits of the fish is one by Dr. Meyen, from which the following summary is taken. It appears that the pilot-fish constantly swims in front of the shark, sometimes coming close to its snout or front fins as it approaches a ship, and sometimes darting sideways or forwards for a short distance, and then returning to the side of the larger fish. In one instance, where a baited hook was thrown over the ship's side, the pilot-fish rushed up, and, after swimming close to the bait, returned to the shark, and, by swimming and splashing round it, appeared to be attracting its attention. Soon after the shark began to move, with the pilot-fish in front, and was almost immediately hooked.

Instead of the pilot-fish taking care of the shark, it would rather seem to frequent the company of the latter for the sake of the fragments of food and other substances to be found in its neighbourhood; and it is doubtless for the same reason that these fishes follow ships.

DORIES

The deep form of the compressed body, the division of the dorsal fin into two distinct moieties, and the circumstance that the number of trunk-vertebræ exceeds ten and that of the tail fourteen, form the leading features by which the small family of the dories (*Cyttidæ*) is distinguished from the other members of the group under consideration. The body may be invested either with small scales or bony plates, or may be naked. The eyes are lateral, and the teeth conical and small. There is no connection between the preopercular and the orbit; the gill-opening is wide, and the pectoral fins are thoracic in position.

The John dory (*Zeus faber*), which gives the name to the family, and is said to derive its own title from a

corruption of an old Gascon name (jan dorée) meaning "gilded cock," represents a genus with few species, characterised by a series of bony plates at the base of the dorsal and anal fins, and another on the under surface; the anal having four spines. The eight or nine spines of the first dorsal fin, which is not much shorter than the second, are produced into long, slender filaments; and there are but few or no scales. The genus ranges over the Mediterranean, the eastern coasts of the temperate zones of the Atlantic, and the Australian and Japanese seas. An exceedingly ugly creature, with a huge protruding mouth, the common dory is olive brown tinged with yellow in colour, showing blue and metallic reflections. The sides bear a large black spot, surrounded by a white ring; a similar mark occurring in some of the other species. A somewhat migratory fish on the British coasts, the dory has been long esteemed by epicures.

The dory feeds on sprats or small fishes such as sprats, and swims in a slightly lopsided way. It is related to the flat-fishes, and gives an idea of what the ancestors of these forms were like. The use of the flattened form in the dory is suggested by J. T. Cunningham, who says: "In the aquarium at Plymouth I have noticed that the dory has a peculiar and interesting method of securing its prey. It does not overtake it by superior speed like the mackerel, or lie in wait for it like the angler, but stalks it and approaches it by stealth. It is able to do this in consequence of the extreme thinness of its body, and the peculiar movement of its hinder dorsal and ventral fins. The dory places itself end on towards the fish it desires to devour, and in this position it is evident that it excites no alarm on the part of its prey. The appearance of the dory seen in this way is a mere line in the water, to which no particular significance can be attached."

STROMATEIDS AND CORYPHÆNAS

The two families *Stromateidæ* and *Coryphænidæ* are collectively distinguished from the preceding by the absence of any distinct spinous portion to the dorsal fin; the compressed body being either oblong or very deep, and there being more than ten vertebræ in the trunk and more than fourteen in the tail. In the first of the two the dentition is feeble, the palate being devoid of teeth, but there are horny barbed processes projecting into the œsophagus which take the place of

THE PILOT-FISH

oral teeth. The scales are very small, the eyes lateral, and the dorsal fin long. The typical genus *Stromateus*, which includes about half a score species from most tropical and temperate seas, is characterised by the absence of pelvic fins in the adult; the dorsal and anal fins being long, with their points curving backwards in several of the species, and the caudal deeply forked. These fishes are partly pelagic.

The second of the two families is represented typically by the well-known pelagic coryphænas (*Coryphæna*), miscalled dolphins. As a family the *Coryphænidæ* are readily distinguished from the *Stromateidæ* by the absence of tooth-like processes in the œsophagus. In the typical genus the body is somewhat elongated and compressed, the adults having an elevated crest on the top of the head ; and the cleft of the mouth is wide. The single dorsal fin extends in a nearly straight line from the back of the head almost to the deeply-forked caudal ; the anal resembles the dorsal in having no distinct spinous portion ; and the well-developed pelvic fins are thoracic in position, and can be received into a groove in the abdomen. Teeth are present in the jaws, as well as on the vomer, palatines, and tongue ; the cycloid scales are small, and there is no swim-bladder.

The coryphænas, of which there are some half-dozen species, are purely pelagic fishes, ranging over all temperate and tropical seas, and remarkable for the beauty of their fleeting colours. Dr. Günther observes that so " far as the colours are capable of description, those of the common species (*C. hippurus*), which is often seen in the Mediterranean, are silvery blue above, with markings of a deep azure, and reflections of pure gold, the lower parts being lemon-yellow, marked with pale blue. The pectoral fins are partly lead-colour, partly yellow ; the anal is yellow, the iris of the eye golden. These iridescent colours change rapidly whilst the fish is dying, as in the mackerel. The form of the body, and especially of the head, changes considerably with age.

This species ranges over all tropical seas, and attains a length of from 5 to 6 feet ; although its flesh is unpalatable to Europeans, it is eaten by the natives of Madras. Powerful swimmers, and associating in large shoals, coryphænas are determined enemies to flying-fish, pursuing them as they skim from wave to wave, and capturing them as they again fall into the water.

MACKERELS

This family (*Scombridæ*) is typically represented by the true mackerels (*Scomber*), and is characterised by the oblong or slightly elongated form of the body—which is but very slightly compressed, and either covered with very minute scales, or naked—and the structure of the dorsal fins. The first of these may be either modified into free spines or an adhesive disc, or the posterior dorsal, together with the anal, is split up into finlets. There may or may not be a swim-bladder.

Characterised by their beautiful protective coloration, which is some shade of bluish green, mottled or barred with black above, and iridescent silver beneath, the members of this family are all pelagic and carnivorous fish, associating in shoals, which may be of immense size, and frequenting all tropical and temperate seas. To enable them to keep up their constant rapid movements, their muscles, which are consequently red in colour, receive a much more abundant supply of blood than is the case with other members of the class. Although spawning in the open sea, at certain times of the year they make periodical migrations towards the shore in pursuit of the shoals of herrings and their fry, on which they so largely subsist. When this kind of prey is not available, they are able to live on the minute crustaceans which make up a large part of the floating population of the sea (plankton).

The true mackerels are characterised by a continuous first dorsal fin, with feeble spines ; five or six finlets behind the dorsal and anal ; very small evenly distributed scales ; small teeth, and two short ridges on each side of the caudal fin. Although there are but very few species of mackerel, these have a very wide range, and the genus is represented throughout the temperate and

JOHN DORY

tropical seas, with the exception of the Atlantic seaboard of temperate South America. Of the three European forms the common mackerel (*S. scombrus*) has no swim-bladder, while the southern mackerel (*S. pneumatophorus*) possesses that organ, as does the Spanish mackerel (*S. colias*). Of the common mackerel, Yarrell writes that " the ordinary length varies from 14 to 16 inches, and their weight is about 2 pounds each ; but they are said to attain the length of 20 inches, with a proportionate increase in weight. The largest fish are not, however, considered the best for table."

TUNNIES

Under this general title may be included not only the fish to which the name tunny (*Thunnus thynnus*) properly pertains, but also those called bonitos and albicores. The genus, comprising some of the largest of all pelagic fishes, differs from the true mackerels by the greater number (six to nine) of finlets, by the scales forming a kind of corselet on the anterior part of the body only, and the presence of only a single longitudinal fringe on each side of the tail. The tunnies have a geographical distribution coextensive with that of the family.

The common species, which attains a length of over 10 feet and a weight of half a ton, is an occasional visitor to the British coasts, and is abundant in the Mediterranean, where it has been regularly fished for since very early times. At the present day specimens of a hundredweight each may often be seen in the Lisbon market ; their flesh, which is as red as beef, being cut up and sold by weight. The bonito (*T. pelamys*) is a smaller and more slender fish, rarely exceeding a yard in length, and frequenting all temperate and tropical seas ; while the name of albicore is applied to species like *T. albicora* of the Atlantic, characterised by the great length of their pectoral fins, some of these fish attaining a length of 6 feet. Albicore and bonito will follow in the wake of sailing-ships for weeks together. They prey largely on flying-fish. Passing over several allied genera, we proceed to a more interesting group of the family.

SUCKING-FISHES

The remarkable adhesive disc on the upper surface of the head at once serves to distinguish the sucking-fishes, not only from their immediate relatives, but likewise from all other members of the class. The genus *Echeneis*, to which all the half-score species of sucking-fish pertain, differs from all those noticed above in the absence of finlets ; the sucking-disc being formed by a modification of the dorsal spines, and being composed of a number of transverse plates, varying from twelve to twenty-seven, according to the species.

The body of the sucking-fish is elongate and pear-shaped ; the eyes are lateral, or directed downwards and outwards ; and the cleft of the mouth is deep. Villiform teeth are present, not only on the jaws and the bones of the palate, but generally also on the tongue ; the scales are minute ; and there is no swim-bladder. The second dorsal and anal fins are elongated, and the pelvics thoracic. Both in this genus and *Elacate* the shape of the caudal fin is subject to considerable change with age ; the middle portion in the young being produced into a long filament, which gradually shortens until a rounded margin is produced. At the time of the full development of the fish the corners of the tail have, however, grown out, so as to convert the rounded fin into an emarginate or forked one. One of the most common members of the genus, *Echeneis remora*, is comparatively small, growing only to a length of about 8 inches ; whereas the slender *E. naucrates* may reach a yard in length.

Sucking-fishes are commonly found attached to sharks, although they may affix themselves either to turtles or ships ; and as they are carried by their involuntary hosts through a much greater extent of water than their limited powers of swimming would admit of their traversing by themselves, they naturally obtain more food than would otherwise be possible. The erection of the plates constituting the sucker produces a series of vacua, by means of which adherence is affected ; so strong is this that it is very difficult to remove one of these fishes except by sliding it along the surface to which it is attached. Moseley remarks that in shark-

reversed. No doubt the object of this arrangement is to render the fish less conspicuous on the brown back of the shark. Were its belly light-coloured, as usual, the adherent fish would be visible for a great distance against the dark background. . . . When one of these fish, a foot in length, has its wet sucker applied to a table, and is allowed time to lay hold, it adheres so tightly that it is impossible to pull it off by a fair vertical strain." When they have lost their shark these fish often attach themselves to a ship, which they probably mistake for a large individual of that tribe. Certain races are in the habit of employing sucking-fishes for the capture of turtles. This curious mode of fishing is practised by the natives of Zanzibar, Cuba, and Torres Straits, the sucking-fish being prevented from escaping by the attachment of a thin cord to its tail.

STAR-GAZERS AND WEEVERS

The family *Trachinidæ* is here taken to include not only the typical weevers, but likewise the star-gazers and several other more or less nearly allied types, these being split up into five subfamilies, though some, if not all of these subfamilies are sometimes regarded as distinct families. In this wider sense the family is characterised by the more or less elongated and narrow form of the body, which may be either naked, or have scales. A spinous dorsal, or a spinous portion of the dorsal, is generally distinct, in which the spines are connected by membrane ; there are no finlets ; the caudal (except in the tile-fish) is not forked ; the pelvic fins include a single spine and five rays ; and the gill-openings are more or less wide. The number of vertebræ in the trunk is generally ten or more, and there are always more than fourteen in the tail.

Carnivorous in their habits, the majority of these fishes are of small size, with but feeble swimming powers, and living on the bottom of shallow seas. The tile-fish and its allies are, however, large deep-water forms ; and the genus *Bathydraco* has been taken from depths of over 1,200 fathoms. They inhabit all seas except the Arctic, where they are almost unknown.

STAR-GAZERS

The star-gazers, as typically represented by the species *Uranoscopus scaber*, form the first subfamily, and take their name from the upward direction of their small eyes. They are further characterised by the continuous lateral line, and by the spinous portion of the single or double dorsal fin being less developed than the soft part, which is similar to the anal. The members of the typical genus are distinguished by the large, broad, and massive head being partly covered with bony plates ; the vertical cleft of the mouth, and the minute size of the scales. The first of the two dorsal fins has from three to five spines, and the rays of the pectorals

COMMON MACKEREL

fishing the suckers sometimes drop off as the shark is hauled on board, and sometimes remain attached ; and that when a shark is hooked and struggling in the water, they may often be seen to shift their position.

He adds that as it is the back of the sucking-fish that is applied to the body by which it is transported, this " being always less exposed to light is light-coloured, whereas the belly, which is constantly outermost and exposed, is of a dark chocolate colour. The familiar distribution of colour existing in most other fish is thus

are branched. Villiform teeth are present in the jaws and on the bones of the palate, but there are no tusks. The gill-cover is armed, and there is generally a long filament below or in front of the tongue, but there is no swim-bladder. While the figured species is from the Mediterranean, the others range from the Indo-Pacific to the Atlantic.

Rarely measuring a foot in length, these fish can raise or depress their small eyes at will, and are generally found lying sluggishly on the sea-bottom in wait for their prey, frequently concealed among stones. The filament in front of the mouth, which is moved by the stream of water continually passing through the latter, doubtless acts as a lure to entice prey.

WEEVERS

The greater English weever, or sting-bull (*Trachinus draco*), is the best known representative of the typical genus of the second subfamily, in which the eyes are more or less lateral, the lateral line continuous, and the hinder part of the premaxillary bones devoid of an enlarged tooth; the dorsal fins being one or two in number. In this particular genus the cleft of the mouth is very oblique; the eyes have an upward inclination; the cycloid scales are very small; and there are villiform teeth both on the jaws and on the bones of the palate. Of the two dorsal fins, the first is very short and furnished with six or seven spines, and the lower rays of the pectorals are simple. In the head both the preorbital and preopercular bones are armed.

The weevers have a somewhat peculiar geographical distribution, being found in the European seas, but unknown on the Atlantic coasts of America, although reappearing in Chilian waters. In the British seas they are represented by the greater weever (*T. draco*), frequently measuring about a foot in length, and the lesser weever (*T. vipera*), which seldom exceeds 6 inches. The poisonous secretion is lodged in a deep double groove in the spines of the dorsal fin and gill-cover. Weevers are probably an example of " warning " coloration, which advertises unpleasant qualities and keeps off enemies—in this case, other fishes—for when buried in the sand, a common habit, the projecting dorsal fin, being black in colour, is very conspicuous.

TILE-FISH

The third subfamily—regarded by many writers as a distinct family—has long been known by the genera *Latilus* and *Pinguipes* from various tropical and subtropical seas, and is characterised by the body being covered with small scales, the lateral position of the eyes, the continuous lateral line, and the presence of a large tooth on the hinder part of the premaxillary bones. Especial interest attaches to the group, on account of the discovery of a new member living in warm Gulf Stream water off No Man's Land, Massachusetts, in 1879, which received the name of tile-fish (*Lopholatilus chamæleonticeps*). The tile-fish is one of the most brilliantly-coloured fishes out of the tropics, and remarkable for the presence of a soft dorsal fin, resembling that of the salmon, which is placed on the neck in advance of the regular dorsal fin instead of behind it, as in the salmon family.

ANGLER-FISHES AND THEIR ALLIES

The angler-fishes and their allies (family *Lophiidæ*) are remarkable for their extreme ugliness and strange forms. Possessing the group characters already noticed, they are specially distinguished by having the spinous dorsal fin placed far forwards on the head, and generally modified more or less completely into tentacles, although it may be represented by isolated spines. The head and fore part of the body are of enormous relative size,

and the teeth in the capacious mouth are either villiform or rasp-like. When present, the pelvic fins consist of four or five soft rays; and the pectorals are supported by a prolongation of some of the superior bones. The gill-opening is reduced to a small aperture situated near the pectoral fin; and the gills themselves are either two and a half or three and a half in number, false gills being generally absent.

These fishes are distributed over all seas. Dr. Günther

THE SUCKING-FISH

writes that " the habits of all are equally sluggish and inactive; they are very bad swimmers; those found near the coasts lie at the bottom of the sea, holding on with their arm-like pectoral fins to seaweeds or stones, between which they are hidden; those of pelagic habits attach themselves to floating seaweed or other objects, and are at the mercy of wind or current." A large proportion have, therefore, found their way to the abysses of the ocean, retaining all the characteristics of their surface ancestors, but assuming deep-sea modifications.

The small number of species constituting the typical genus (*Lophius*) include the weirdest forms, among these being the British angler-fish (*L. piscatorius*), known also as the fishing-frog, frog-fish, or sea-devil (p. 1753). We may note the enormous size of the broad, depressed, and rounded head, near the middle of the upper surface of which are situated the small eyes; and the great width of the cleft of the mouth, which looks like a yawning chasm. Both the jaws and palate are armed with unequal rasp-like teeth, capable of being raised and depressed. The body is naked; the first three spines of the dorsal fin form long tentacles on the head, and the next three are connected; the soft dorsal and anal fins being of small length.

The whole of the few known forms are coast-hunting fishes, the common species ranging from the European and South African seas to those of the western side of North America; while a second is found in the Mediterranean, a third in Chinese and Japanese waters, and a fourth in those of the Admiralty Islands. In the British species the general colour of the upper surface is uniform brown, becoming darker on the fin-membrane; while the under parts, as well as the pectoral and pelvic fins, are white; the tail being dark blackish brown. The colour varies according to that of the surroundings. Although commonly not more than a yard in length, specimens of this ugly monster have been known to measure more than 5 feet

The angler affords us an example of a creature admirably adapted to its particular mode of life. Living on the mud or sand of a shallow sea-bottom, it is protected not only by its power of adapting its colour to local surroundings, but also by the seaweed-like appearance of the fringed head-appendages.

BULL-HEAD, OR MILLER'S THUMB

The structure of the paired fins enables the fish to walk on the sea-bottom ; and with these limbs it can also stir up sand and mud to attract prey, and at the same time to aid in concealment. Fish and other prey are also attracted by the constant movement of the first head-tentacle, which terminates in an expanded lappet ; and no sooner is the victim well within reach than it is engulfed with one snap of the capacious mouth, the erectile and backwardly directed teeth preventing any chance of escape. The angler-fishes that live in the deep sea have similar habits, but in some of them the " lure " is phosphorescent, an adaptation to the gloom of the ocean abysses.

BULL-HEADS AND GURNARDS

The next family of the present section (Cottidæ) differs from all the foregoing, with the exception of the genus Pseudochromis and its allies, in the presence of a bony process arising from the infraorbital ring of the skull to connect it with the spine at the angle of the preopercular bone. The shape is more or less elongate and subcylindrical ; the cleft of the mouth is transverse, and the weak teeth are generally arranged in villiform bands. As a rule, there are two dorsal fins, of which the spinous is less developed than the soft, both the latter and the anal being elongated ; the pectorals may be provided with filamentous appendages, and the pelvic pair have not more than five rays. The body may be either naked, scaled, or protected by a single row of plate-like scales. The members of this family are distributed over almost all seas, while a few inhabit fresh waters.

Of comparatively small or medium size, they swim feebly, and spend their time swimming or crawling at the bottom of the sea in shallow water, not far from land. A Japanese bull-head is stated, however, to have been dredged in five hundred fathoms of water.

BULL-HEAD, OR MILLER'S THUMB

The familiar bull-head, or miller's thumb (Cottus gobio), of the streams of Britain and many other parts of Europe, belongs to a genus containing some forty species, mostly distributed over the fresh waters and coasts of the temperate zone of the Northern Hemisphere. All are of small size, and characterised by the broad, depressed, and rounded head ; the subcylindrical body, somewhat compressed posteriorly ; the absence of scales ; the distinct lateral line, and the rounded pectoral fins, in which some or all of the rays are simple. Villiform teeth are present on the jaws and vomer, although there are none on the palatine bones. In the majority of the fresh-water species the spine on the preopercular bone is simple, but becomes branched in many of the marine forms.

The common fresh-water species, which ranges over Central and Northern Europe to Northern Asia, seldom exceeds 4 or 5 inches in length, and is more generally found in small streams than in large rivers. It has a well-known habit of concealing its broad and flat head beneath loose stones on the river-bottom, and in this position will lie motionless for hours, but when disturbed swims swiftly away. Its food consists of the larvæ of water-insects and crustaceans, as well as the eggs and fry of other small fish. In some part of Germany it is used for making soup.

The other British representatives of the genus are marine, and include the sea-scorpion (C. scorpius) and father-lasher (C. bubalis), both of which are also found on the opposite side of the Atlantic, as well as two other less common species. The males of the common marine species are stated to build a nest of stones and seaweed for the reception of the spawn, and to guard and defend the young fry when hatched. The four-horned cottus (Cottus quadricornis) possesses four roughened projections on the head. It is an Arctic and Baltic species, which is occasionally taken on the British coast.

FOUR-HORNED COTTUS

GURNARDS

Of a decidedly ugly appearance, the gurnards (Trigla) are easily recognised by their enormous, square, and elevated heads, in which the upper surface and sides are entirely bony, and likewise by the finger-like first three rays of the pectoral fins, which serve not only for walking on the sea-bottom, but likewise as organs of touch, and are employed in the search for the small crustaceans, etc., that constitute their food.

SAPPHIRINE GURNARD

There are two dorsal fins, of which the spinous is tall, and the soft one long, low, and similar to the anal ; the tail-fin being slightly rounded. The teeth are villiform ; and the swim-bladder may be divided into two longitudinal halves. Their colours are frequently brilliant, and the fins highly decorated. The genus is represented by some forty species, distributed over all temperate and tropical seas, out of which no less than seven are found in British waters. Their flesh, which is firm and flaky, and of a pale orange-pink tinge, is extensively used as food.

One of the best known of the British species is the red gurnard (*T. pini*), which seldom exceeds 12 or 14 inches in length, and, when freshly caught, is of a bright red colour, with the sides and under parts silvery white, and the fins reddish-white. Its food consists of crustaceans, which give the pinkish tinge to its flesh, and the spawning season is May or June. The sapphirine gurnard, or tub-fish (*T. hirundo*), another British species, takes its Latin name from the length of the pectoral fins, and its English title from the beautiful azure tint of their inner surfaces. More abundant than the other species, this gurnard may reach a couple of feet in length, its general colour being brownish red.

A third British form is commonly known as the piper (*T. lyra*), and may be recognised by the unusually large head, the more projecting snout, and the greater length of the spines of the gill-cover. The general colour is brilliant red, with the under parts white. It attains a length of a couple of feet, and is supposed to take its name from the grunting sound which, in common with other species, it emits when first handled. The European forms are rarely found on the other side of the Atlantic, where their place is taken by other species.

Two British species are figured in the coloured plate on page 1613—namely, the grey gurnard (*T. gurnardus*), above, and the streaked gurnard (*T. lineata*), below.

THE ANGLER-FISH

FLYING-GURNARDS AND THEIR ALLIES

The so-called flying gurnards (*Dactylopteridæ*) are easily recognised by their armour of bony keeled plates or scales. In form the body is elongate and subcyclindrical; the teeth are weak; and there is a bony stay connecting the preopercular with the infraorbital ring.

These fishes are all marine, some being pelagic, and are found in the warmer parts of the Atlantic and Indian Oceans. As in ordinary flying-fishes the pectoral fins are enormously enlarged, but in this case the "flights" are shorter. It appears, however, that flying gurnards are able rapidly to vibrate their pectoral fins, which, therefore, have a better claim to be called "wings" than the corresponding fins of common flying-fishes, which are parachutes pure and simple. The four known species all belong to a single genus (*Dactylopterus*), the most familiar being a form (*D. volitans*) native to the Mediterranean and Atlantic.

DRAGON-FISHES

Here may be noticed the curious little dragon-fishes (*Pegasus*), from the Indian, Chinese, and Australian seas. In these strange little fishes the broad and depressed body is covered with bony plates, which are movable, although those investing the tail are firmly welded together. The narrow gill-opening is situated in front of the pectoral fin; the gill-cover is formed of a single plate, and the gills themselves are four in number. The single short dorsal fin is placed opposite an anal of similar size; the pectorals are long, horizontal and composed of simple rays, some of which are spinous; and the pelvic comprises one or two rays, the outer one being elongated. Both teeth and swim-bladder are wanting. The species (*P. natans*) figured on page 1755 is an Australian one, and is less well-known than the Indian *P. draco* and the Chinese *P. volitans*.

LUMP-SUCKERS

With the lump-suckers we come to a small section characterised by the spinous dorsal fin being short, and either composed of flexible spines, or much less developed than the soft dorsal, or soft portion of the same; the soft dorsal being equal in extent to the anal. If present, the pelvic fins are either thoracic or jugular in position, with one spine, and generally five (rarely four) soft rays. In no case is there a bony stay to the preopercular from the infraorbital ring.

As a family, the lump-suckers (*Cyclopteridæ*) are characterised by the thick or oblong body, either naked or tuberculated; small teeth; narrow gill-openings; and the presence of a circular adhesive disc on the lower surface of the chest, surrounded by a fringe of skin, and supported by the rudimentary pelvic fins, the gill-opening being narrow. All the members of the family are carnivorous coast-dwellers, restricted to the colder seas of the Northern Hemisphere, and ranging into the Arctic Ocean. They derive their name from their habit of attaching themselves to rocks by means of the adhesive disc.

The members of the typical genus *Cyclopterus* are thick, short fishes, with a viscous tuberculated skin; the large head has a very short, blunt snout; and there are rows of villiform teeth on the jaws, but none on the palate. The skeleton is remarkable for its softness. In the British species (*C. lumpus*), the skin is so thick as almost to conceal the first dorsal fin; and in the adult the large rough tubercles are arranged in four longitudinal series on each side of the body.

Although these fishes may reach a length of a couple of feet, they do not usually measure more than 12 or 14 inches. These voracious forms feed chiefly on the fry of other species. The lump-sucker is able to change its colour with fair rapidity to harmonise with its surroundings. The smaller male is distinguished from the female by bright red and yellow sides and belly, and he is particularly brilliant in hue during the breeding season. He makes a hole in which the eggs are laid, and guards them jealously, moving his pectoral fins up and down so as to set up aerating currents. When hatched, the young attach themselves for a time to his body by means of their suckers.

GOBIES AND MUD-SKIPPERS

The gobies and their allies (family *Gobiidæ*) differ from lump-suckers in possessing distinct rays to the pelvic fins ; although in some cases the two fins may be joined in the middle line. The elongated body may be either scaled or naked ; and the teeth are generally small, but may have enlarged tusks among them. The spinous portion of the dorsal (whether separate or continuous with the soft dorsal) is always composed of flexible spines, and shorter than the soft dorsal. The gill-opening is more or less narrowed, and there is usually

LUMP-SUCKER

no swim-bladder. This very extensive family comprises small carnivorous shore-fishes, a few of which have accustomed themselves to a fresh-water life. It contains a large number of genera, some of which are extremely numerous in species, as are the latter in individuals ; and their range includes the coast-regions of all the temperate and tropical seas.

GOBIES

The gobies form a very large genus (*Gobius*), with a geographical distribution as extensive as that of the family, but especially well represented in tropical and subtropical seas, no less than forty different kinds being recorded from those of India alone. These fishes have the body generally scaled ; two dorsal fins, of which the first is usually furnished with six flexible spines ; the pelvic fins united to form a sucker, which, however, is at most only partially adherent to the abdomen ; the teeth in more than a single row ; and the vertical gill-opening of moderate width. The form of the body is subject

FRESH-WATER GOBY

to considerable specific variation ; and in some species the head, and in others a part or even the whole of the body, is devoid of scales. In some cases there may be barbels or warts on the head, and in others a crest on the occiput. There are likewise considerable differences in the dentition, some species having large tusks among the ordinary teeth.

The gobies, of which there are several British marine species, are specially partial to rocky coasts, where they protect themselves against waves and storms by adhering to rocks by means of the pelvic sucker. Some prefer brackish estuaries or lagoons, while others again, for example, a Russian species (*G. fluviatilis*), live in rivers. In many of them the male constructs a nest, in which the spawn is hatched. In the case of the spotted goby,

or polewig (*G. minutus*)—a species found for some distance up the Thames—the male, when in tidal pools, generally chooses one of the shells of a cockle or some other bivalve for its nest ; the shell being placed on the sand with its concave surface downwards, beneath which the sand is hollowed out and cemented by a special mucilaginous secretion from the skin of the fish ; a cylindrical tunnel giving access to the nest, and the whole structure being covered over with loose sand. The female having deposited her eggs, which are fixed to the shell, in this nest, the male mounts guard over them, maintaining his watch during the whole period of incubation, which lasts from six to nine days.

A European goby (*Aphia pellucida*), belonging to a distinct genus, and characterised by its translucent body, is almost peculiar among vertebrates in that it lives only a year. In June and July the spawn is deposited, the eggs are hatched in August, while in the late autumn or winter the fishes become fully mature ; these, however, die off in the following July or August, so that in September only the fry survive. The giant goby (*G. capito*) of the Mediterranean and English Channel grows to a length of 9 or 10 inches.

MUD-SKIPPERS

The mud-skippers (*Periophthalmus*) are remarkable for their prominent eyes and strange habits. They frequent the coasts and estuaries bordering the Indo-Pacific Ocean, and the shores of West Africa. Their name is taken from the prominent eyes, which are set close together somewhat below the line of the profile, and are not only capable of protrusion and retraction, but are furnished with a well-developed outer eyelid. The elongate body is covered with cycloid or slightly pectinated scales, extending on to the bases of the pectoral fins ; the cleft of the mouth is nearly horizontal, with the upper jaw projecting somewhat beyond the lower, and the conical teeth are vertical. The first dorsal fin includes a variable number of flexible spines ; the base of the pectorals are muscular ; the pelvic fins are united for a portion of their length ; and the caudal fin has its lower border obliquely truncated.

The species figured (*P. koelreuteri*) has a wide range, being found in the Red Sea, on the coasts of India, where it ascends tidal rivers and estuaries, as well as in the Andamans, the Malay Archipelago, and the islands of the Pacific. Concerning their habits, Day writes that " these fishes, from the muscular development at the base of the pectoral fins, are able to use them for progression on mud or for climbing. It is a most curious sight to see *P. schlosseri* along the side of the Burmese rivers ; at a distance the fishes appear like large tadpoles, stationary, contemplating all passing objects, or else snapping at flies or other insects ; suddenly, startled by something, away they go with a hop, skip, and a jump, either inland among the trees or on to the water like a flat stone or a piece of slate sent skimming by a schoolboy. They climb on to trees and large pieces of grass, leaves, and sticks, holding on by their pectoral fins exactly as if these were arms. Now and then they plant these firmly as organs of support, the same as one places one's elbows on a table, then they raise their heads and take a deliberate survey of surrounding objects."

The external opening of the gill-cavity is very narrow, so that the gills are prevented from drying up, and the lining of this cavity very likely acts as a sort of lung. Mr. S. J. Hickson writes concerning mud-skippers : " Their position is usually one of clinging to the edge of the rocks or mangrove roots by their fins, with their tails only in the water. . . . The fact that they live the greater part of their lives with their head and gills out of water suggested to me an investigation of their respiratory organs, as I thought it possible that they might possess some interesting modifications of the swimbladder to enable them to breathe the air. It was not, however . . . until 1887 that an explanation of the mystery of their respiration occurred to me—namely, that the respiration is mainly performed by the tail. Since then Professor Haddon has been carrying on some experiments in Torres Straits, and has shown that this explanation is correct."

DRAGONETS

Concerning the precise systematic position of the beautifully coloured dragonets (*Callionymus*), there is some difference of opinion ; though conveniently placed with the gobies, they are often referred to a separate family. They are characterised by the head and anterior portion of the naked body being depressed, while the remainder is cylindrical. The pointed mouth has a narrow horizontal cleft and a very protractile upper jaw; the large eyes have a more or less upward direction; small teeth are present on the jaws, but none on the palate; and there is a strong spine at the angle of the preopercular.

Of the two dorsal fins, the foremost has from three to four flexible spines ; the widely separated pelvics are five-rayed ; and the gill-opening is very narrow, and generally reduced to a perforation on the upper border of the gill-cover. There is a large amount of sexual difference, the adult males having the fin-rays produced into filaments and the intervening membranes brightly coloured, while the females wear a much more sombre livery ; and it is due to this variation that there were long supposed to be two British representatives of the genus—namely, the jewelled and the sordid dragonet ; the former being the male and the latter the female.

In the adult male of the common dragonet, or sculpin (*C. lyra*), the first dorsal spine is greatly elongated ; the general colour of the smooth skin being yellowish, beautifully banded and spotted with lilac ; the first dorsal fin bears several lilac spots, and the second carries lilac bands. In length the male measures about 10 inches. The yellow sculpin, as the male is called in some parts of Britain, is generally found in comparatively deep water, whereas the female often approaches the margin of the tide. Both sexes feed on molluscs and other hard-shelled creatures, as well as on worms. Out of some forty species, the majority are inhabitants of the coast-regions of the north temperate zone of the Old World, although a few are found in the Tropical Pacific.

AUSTRALIAN DRAGON-FISH

BLENNIES

The pelvic fins in the family *Blenniidæ*, when present at all, are jugular in position and composed of a very few soft rays. In the anal fin the spines are few or wanting. The low and elongate body is more or less cylindrical, and either naked or covered with scales, which are generally of small size. The dorsal fin, which may be either single, double, or triple, occupies nearly the whole length of the back ; and when it has a distinct spinous portion, this is at least as much developed as the soft part, while in some instances the whole fin may be spiny ; the anal being elongate. In most cases false gills are present.

All the marine members of the family are littoral forms, and the majority are of small size, while some are among the smallest of all fishes. They are abundant throughout all tropical and temperate seas ; and whereas some forms inhabit brackish water, others are exclusively fresh-water.

The blennies of the typical genus *Blennius*, of which there are some forty species, are found in the northern seas, the Tropical Atlantic, the coasts of Tasmania, and the Red Sea. They are characterised by the moderate elongation of the naked body, the short snout, the single continuous dorsal fin, and the presence of one spine and two rays in the pelvics. The cleft of the mouth is narrow; the jaws contain a single series of fixed teeth, behind which there is generally one larger curved tooth, at least in the lower jaw. Above each eye is the longer or shorter tentacle, and the gill-opening is relatively wide.

Among British species we have the eyed, or butterfly, blenny (*B. ocellatus*), distinguished by the dark spot on the elevated spinous portion of the dorsal, the smooth blenny or shanny (*B. pholis*), and the large *B. gattorugine*, which may grow to a foot in length. Male blennies guard the eggs and young. As an example of a species living in inland lakes we may cite the freshwater blenny (*B. vulgaris*) of Southern Europe. Most can be readily accustomed to a fresh-water life, and many of the marine species attach themselves to floating objects, while some are found far out at sea among patches of drifting seaweed.

MUD-SKIPPER

VIVIPAROUS BLENNIES

The fish *Zoarces viviparus* is one of two species of a genus remarkable for producing living young. With an elongate body, rudimentary scales, and conical teeth in the jaws, these fish have an extremely elongated dorsal fin, separated from the caudal merely by a depression formed by a series of spines much shorter than the rays; these spines being the only ones throughout the fins. The pelvic fins are composed of

VIVIPAROUS BLENNIES

three or four rays; and the long anal fin is continuous posteriorly with the caudal. The gill-openings are wide. While the figured European species, not uncommon on the British coasts, does not exceed a foot in length, its Transatlantic cousin (*Z. anguillaris*) may measure two or three times as much. The general colour of the adult fish is pale brown, with the dorsal fin and upper parts mottled and barred with darker brown.

The butter-fish, or gunnel (*Pholis* [*Centronotus*] *gunnellus*), though generally associated with the blennies, is the type of a distinct family (*Pholididæ*). It is an elongated form, sometimes attaining a length of as much as a foot, which is common on the British coasts. The eggs are rolled into a ball by the united efforts of both parents, and they are deposited in a hole in the rock, where the mother—and sometimes the father also—keeps guard over them.

WOLF-FISHES

Easily recognised by the powerful tuberculated and molar-like teeth, the wolf-fishes (*Anarrhichas*) may be regarded as gigantic and somewhat specialised blennies. In this genus, represented by a small number of species from the northern seas of both the Eastern and Western Hemispheres, the elongate body is covered with rudimentary scales; the snout is rather short and the cleft of the mouth wide; and the jaws are armed with strong conical teeth, those of the lateral series carrying several pointed cusps, while a double row of large molar-like teeth runs down the middle of the palate. The long dorsal fin has flexible spines, and there is a distinct caudal, but the pelvic pair is quite wanting. The gill-opening is wide. The common wolf-fish (*A. lupus*), often known as the sea-wolf or sea-cat, like two allied species, ranges so far north as Norway and Greenland; in both of which countries its flesh forms a staple article of food. It may attain a length of 5 or 6 feet. It is impossible to imagine a more voracious-looking animal than the sea cat-fish, with the massive head and long, sinuous, muscular body, its strongly-rayed fins, its vice-like jaws, armed with great pavements of teeth, those in front long, strong, pointed, like those of a tiger. It has been known to attack furiously persons wading at low tide among the rock-pools.

BARRACUDA-PIKES

Four families form a sectional group differing from those we have been considering by the position of the pelvic fins, which are abdominal and have one spine and five soft rays. The two dorsal fins are situated more or less remote from one another, the first being either short, like the second, or composed of weak spines.

The large and ferocious fishes known as barracuda pikes (*Sphyræna*), of which a species (*S. vulgaris*) is figured, are the sole existing representatives of the family *Sphyrænidæ*, distinguished by the elongated and subcylindrical form of the body, the large cutting teeth, the continuous lateral line, and the presence of only twenty-four vertebræ in the backbone. The scales are small and cycloidal, the cleft of the mouth wide and the medium-sized eyes lateral.

Represented by less than a score of species, barracuda pikes are distributed over all temperate and tropical seas, but generally prefer the neighbourhood of the coast to the open ocean. They are all carnivorous and fierce in their disposition, and since they frequently grow to 6 or 8 feet in length, they are as much or even more dreaded by bathers in seas where they are as common as sharks but not so easily driven away.

SAND-SMELTS

The second family (*Atherinidæ*) of the present section is typically represented by the sand-smelts; the common British species (*Atherina presbyter*) being familiar. Its body is clear green, with a white lateral stripe. As a family, the *Atherinidæ* are distinguished from the barracudas by the indistinct lateral line, the feeble or moderately developed dentition, and by the number of vertebræ being usually in excess of twenty-four. The body is more or less elongate, with but slight compression. The scales are smooth and cycloid, and the teeth minute; the first dorsal fin is short and completely separate from the second; and the snout is blunt, with the cleft of the mouth straight, oblique, and extending at least as far back as the line of the border of the eye.

These fish derive their popular title from their resemblance to the true smelts, from which they may be distinguished at a glance by the small spinous first dorsal fin. While most are coast forms, associating in large shoals, others are fresh-water, although these also retain the same habit. The genus has a wide

BARRACUDA PIKE

distribution in temperate and tropical seas, some of the species ranging from Eastern Africa to India. Atherines are very abundant in the Mediterranean, where the fry cling together in enormous masses for some time after hatching. These fish are caught in great abundance on the south coast of Devonshire, in creeks and estuaries. The British species seldom exceed 6 inches in length, and are marked by a broad silvery stripe along each side of the body.

GREY MULLETS

The grey mullets (*Mugilidæ*) are distinguished by the total absence of a lateral line, the presence of only four stiff spines in the first dorsal fin, and the limitation of the number of vertebræ in the skeleton to twenty-four. The more or less elongate and somewhat compressed body is covered with cycloid or slightly ctenoid scales of moderate size ; the cleft of the mouth is

GREY MULLET

small or medium ; the teeth are feeble or wanting ; the lateral eye is of moderate size, and the gill-opening wide. In some species there may be a fatty lid to the eye. The grey mullets (*Mugil*) include a very large number of species, and are distributed over all temperate and tropical coast-regions, frequenting brackish-water estuaries, sometimes ascending rivers for considerable distances. Feeding chiefly upon the animals and organic matter found in sand and mud, they possess a special straining apparatus in the pharynx. Sand or mud are triturated between the pharyngeal bones, in order to extract nutriment, the mineral part being rejected. Another peculiarity is to be found in the structure of the gullet and stomach, the former being lined with long thread-like papillæ, while the second part of the latter is muscular like the gizzard of a bird.

The figure represents the common grey mullet (*M. capito*), one of several species frequenting the British coasts. Although this mullet only grows to the weight of about 4 pounds, some of the foreign species may scale three times as much. This mullet has been kept in a fresh-water pond, where it seemed to thrive better than in the sea. The flesh of all the grey mullets is of good quality, but bears no comparison to that of their red namesakes. The natives of Honolulu construct mullet-ponds by building walls across the mouths of narrow inlets of the sea, leaving numerous small openings, through which the young fish enter, soon, however, growing too large to make their escape.

GAR-FISHES AND FLYING-FISHES

In this place may be noticed a family (*Scombresocidæ*), in regard to the serial position of which there is some difference of opinion, some writers placing it among the tube-bladdered fishes, while others consider that its true position is here. The inclusion of the group among the tube-bladdered fishes utterly spoils the definition of that suborder, since in those members of the present family provided with a swim-bladder that organ lacks a duct.

Agreeing with the preceding section in the abdominal position of the pelvic fins, these fishes differ from those yet described, with the exception of certain perches, in the union of the lower pharyngeal bones ; while they are further characterised by the absence of a spinal dorsal fin, and the deeply forked caudal. The single dorsal is situated opposite to the anal fin in the caudal region, the swim-bladder is generally present, the false gills are hidden and glandular, and the simple stomach merely forms a dilatation of the intestinal tract. Although the majority of the members of this family are marine, some being pelagic, a few have taken to a fresh-water existence ; and while many of the latter are viviparous, the whole of the others deposit eggs in the usual manner. Distributed over all the temperate and tropical seas, these fish are strictly carnivorous in their habits.

Gar-fishes are represented by nearly fifty species from temperate and tropical seas, among which the figured one (*Belone vulgaris*) is common on the British coasts, also ranging over the whole of the seas of Northern Europe. As a genus, these fishes are easily recognised by the production of the jaws into a long, slender beak, formed in the upper one exclusively by the premaxillary bones ; while they are further characterised by the whole of the rays of the dorsal and anal fins being connected by membrane. Both jaws are beset with a number of rugosities, and also with a series of long, conical teeth placed at considerable intervals. A peculiarity of these fish is to be found in the green colour of their bones.

Although British species do not exceed a couple of feet in length, some of the foreign representatives of the genus may grow to as much as 5 feet.

The saury, or skipper (*Scombresox saurus*), is the British representative of a much smaller genus, differing from the gar-pikes by the minute size of the teeth, and likewise by the presence of a number of small finlets behind the dorsal and anal fins. On the other hand, the half-beaks (*Hemirhamphus*), some of which inhabit fresh water, have the lower jaw larger than the upper throughout life. They mostly feed on green seaweed.

Represented by over forty species from tropical

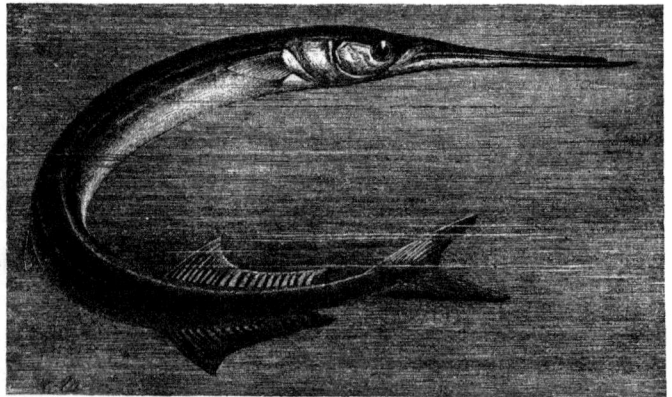

THE GAR-PIKE

and subtropical seas, the flying-fishes, of which the common *Exocœtus evolans* is figured, form a genus recognisable by the great length of the pectoral fins. They are further characterised by the blunt and short-jawed head, and the moderately long, oblong body invested in a coat of rather large-sized scales ; the teeth, when present at all, being minute or rudimentary. The ordinary length of a flying-fish is from 10 inches to a foot, but specimens are occasionally met with half as

FLYING-FISH

long again; and although the common form ranges round the world, the distribution of some other species is extremely restricted, one being recorded only from the seas on the Pacific side of the Isthmus of Panama. The species differ considerably in the length of the pectoral fins; those in which they reach to the tail-fin being capable of taking the longest flights, whereas in some others they do not extend beyond the anal. Associating in shoals, which are sometimes of immense size, all these fish are pelagic in their habits, and all are capable of taking the skimming flight from which they derive their name. That these fish take their flights primarily to escape from their enemies may be regarded as certain; and it is equally well ascertained that the continuance of the flight is due to the original impetus of the leap from the water, and is not prolonged by any flapping of the fins.

The following account is abridged by Dr. Günther from one published by Dr. Möbius, and runs as follows: " Flying-fish are more frequently observed in rough weather and in a disturbed sea than during calms; they dart out of the water when pursued by their enemies or frightened by an approaching vessel, but frequently also without any apparent cause, as is also observed in many other fishes; and they rise without any regard to the direction of the wind or waves. The fins are kept quietly distended, without any motion, except an occasional vibration caused by the air whenever the surface of the wing is parallel with the course of the wind. Their flight is rapid, but gradually decreasing in velocity, greatly exceeding that of a ship going ten miles an hour, and extending to a distance of five hundred feet. Generally, it is longer when the fish fly against than with or at angle to the wind. Any vertical or horizontal deviation from a straight line is not caused by the will of the fish, but by currents of the air; the fish retaining a horizontally straight course when flying with or against the wind, but being carried to one side whenever the direction of the latter is at an angle to that of their flight. It may, however, happen that in the course of its flight a fish may dip its tail in the crest of a wave, thus changing its direction to the left or right.

" In calm weather the line of flight is always also vertically straight, or, rather, parabolic, like the course of a projectile, but in a rough sea, when the fish are flying against the course of the waves, it may become undulating. In such instances the flying-fish frequently overtop each wave, being carried over by the pressure of the disturbed air. Flying-fish often fall on board

vessels, but this never happens during a calm, or from the lee side, always taking place in a breeze, and from the weather side."

STICKLEBACKS

Taking their name from the presence of a variable number of isolated spines in advance of the soft dorsal fin, sticklebacks (*Gastrosteidæ*) have the body more or less elongate and compressed, the cleft of the mouth oblique, and the teeth villiform. The gill-cover is unarmed, and the cheek covered by the infraorbital bone; and, in place of scales, there are generally large plates along the sides of the body. The pelvic fins, although abdominal in position, are connected with the pectoral girdle by means of the pelvic bones, and consist of but one spine and a single ray; and there are but three branchiostegal rays. Confined to the temperate and Arctic zones of the Northern Hemisphere, where they are represented by some half score species of small size, sticklebacks are mainly fresh-water fishes, although the sea-stickleback (*Spinachia vulgaris*) is a marine or brackish-water form, and all the rest can live as well in salt as in fresh water.

The British fresh-water forms are distinguished by the number of dorsal spines, and are known as the three-spined (*Gastrosteus aculeatus*), four-spined (*G. spinulosus*), and nine-spined sticklebacks (*G. pungitius*); while in the United States *G. novæboracensis* is the most familiar kind. The three-spined stickleback was formerly supposed to exist in three different varieties, but these have now been shown to be seasonal changes in the same individuals. Very different in appearance from the others is the fifteen-spined, or sea, stickleback, in which the body is very long and thin, this species ranging so far north as Norway and the Baltic. It has recently been ascertained that all the individuals of this stickleback die within a year of their birth; so that we have here a second example of an annual vertebrate, the first being the one mentioned on page 1754.

Sticklebacks are extremely pugnacious, and at the same time highly voracious, fishes, the males engaging in fierce conflicts with one another; while both sexes consume a vast quantity of the fry of other fishes, and are therefore most objectionable denizens of preserved waters. In fighting, the males use their formidable dorsal spines, with which they have been seen to rip open the body of an antagonist. The most interesting trait in the economy of sticklebacks is, however, the nest-building habit of many species. In the sea-stickleback the nest is composed of a mass of pendent sea-weeds, bound together by a silk-like thread into a pear-shaped form, in the centre of which are deposited the eggs. Such a nest has been known to be guarded for a period of upwards of three weeks by the male, and when it sustained any damage by which the eggs were exposed to view, the watchful guardian set about repairing the mischief with the greatest despatch and energy, thrusting his nose deep in the structure, and pushing and pulling the materials till all was once more sound.

The following account of the nesting of the three-spined stickleback in an aquarium was forwarded by a correspondent to Frank Buckland. On this occasion the male " selected a spot nearly in the centre of the trough, and busily set to work to make a collection of delicate fibrous materials, placed on the ground, and matted into an irregularly circular mass, somewhat depressed, and upwards of an inch in diameter, the top being covered with similar materials, and having in the centre a rather large hole. His work was begun at noonday, and was completed, and the eggs deposited, by half-past six in the afternoon. Nothing could exceed the attention from this time evinced by the male fish. He kept constant watch over the nest, every now and then shaking up the materials and dragging out the eggs, and then pushing them into their receptacle again, and tucking them up with his snout, arranging the whole to his mind, and again and again adjusting it till he was satisfied ; after which he hung or hovered over the surface of the nest, his head close to the orifice, the body inclined upwards at an angle of about 45°, fanning it with the pectoral fins, aided by a side motion of the tail. This curious manœuvre was apparently for the purpose of ventilating the spawn ; at least, by this means a current of water was made to set in towards the nest, as was evident by the agitation of particles of matter attached to it.

" This fanning or ventilation was frequently repeated every day till the young were hatched ; and sometimes the fish would dive head foremost into his nursery and bring out a mouthful of sand, which he would carry for some distance and discharge with a puff. At the end of a month the young ones were first perceived. The nest was built on April 23, and the young appeared first on May 21. Unremitting as had been the attention of this exemplary parent up to the time of the hatching of the eggs, he now redoubled his assiduity. He never left the spot either by day or night, and during the daytime he guarded it most pertinaciously, allowing nothing to approach."

SUCKER-FISHES

The small fish *Lepadogaster bimaculatus*, of which three examples are shown in the illustration below, is one of three British representatives of a genus belonging to a small family (*Gobioescidæ*) which constitutes a sectional group by itself. Long confounded with the lump-suckers, which they resemble in having an ovate adhesive disc on the under surface of the body, the sucker-fish differ from that group not only in the structure of that disc, but likewise in several other respects. They have no spinous dorsal fin ; the soft dorsal and anal are short or of medium length, and situated far back, at the root of the tail ; the pelvic fins

TWO-SPOTTED SUCKER-FISHES

are almost jugular in position, and have the adhesive disc placed between them, while the body is covered with a naked skin. While in lump-suckers the pelvic fins are close together, and actually constitute the base of the sucking disc, in the present family they are widely separated from each other, and only help to form part of the edge of the sucker, which is completed by a cartilaginous expansion of the bones of the girdle. In size the shoulder-disc is relatively large, its length being sometimes as much as one-third that of the whole fish, and it is divided into an anterior and a posterior moiety,

SEA-STICKLEBACK, NINE-SPINED STICKLEBACK, AND TWO THREE-SPINED STICKLEBACKS

of which the second may or may not have a free front margin. All these fishes are shore forms of small size, ranging over both temperate zones, where they are more numerous than in the tropics. The figured species is generally carmine-red above and pale flesh-colour below, with a light patch between the eyes and two more or less distinct spots on the sides. It has been obtained adhering to stones and shells in deep water off Torquay.

SERPENT-HEADS

Mainly characteristic of the Oriental region, although also represented in Africa, the fresh-water fishes known as serpent-heads (family *Ophiocephalidæ*) are interesting not only on account of their structure, but likewise from their peculiar habits. They form a single family, constituting a sectional group by itself, and represented by two genera, in one of which (*Ophiocephalus*) pelvic fins are present, while in the second (*Channa*) they are wanting. As a family, the serpent-heads are characterised as follows. The body is elongated and covered with medium-sized scales ; all the fins are spineless, the anal and single dorsal being long and low ; and there is an additional cavity above the proper gill-chamber, although this is not furnished with supplementary gills. The depressed head is covered with somewhat plate-like scales, and has the eyes lateral and the gill-openings wide ; each gill-chamber containing four gills, while teeth are present on the jaws, palatines, and vomer. If present, the pelvic fins are thoracic in position, and composed of six rays. The lateral line is sharply curved or almost interrupted, and a swim-bladder is present.

Of the typical genus there are some thirty existing species, having a distribution coextensive with that of the family, and in Asia ranging over Baluchistan, Afghanistan, India, Ceylon, Burma, China, Siam, and the Malay Archipelago ; the figured species (*O. striatus*) being common to such distant localities as India and the Philippines, and at times reaching as much as a yard in length. The second genus, *Channa*, is represented only by a single species from Ceylon and China.

In India the serpent-heads are found in rivers, ponds, tanks, and swamps, many of them seeming to prefer stagnant to running waters. Day writes that these fishes, "having hollow cavities in their heads, and an amphibious mode of respiration, are able to exist for lengthened periods out of their native element, and can travel some distance over the ground, especially when it is moist. They are able to progress in a serpentine manner, chiefly by means of their pectoral and caudal fins, first one of the former being advanced and then the other. These fishes appear to be monogamous, some breeding in grassy swamps or the edges of tanks, some in wells or stone-margined receptacles for water, and others again in holes in river-banks."

SERPENT-HEAD

When living in muddy water they rise to the surface from time to time to take in air, and captive examples prevented from doing this have been known to die.

LABYRINTH-GILLED FISHES

In the members of the two families of *Anabantidæ* and *Luciocephalidæ*—rather small estuarine and fresh-water fishes, which constitute a sectional group by themselves —the apparatus for enabling them to exist for a considerable time out of water is carried to a greater degree of complexity than in the last, and takes the form of a laminated accessory gill-like organ, situated in a chamber on each side of the head above the one containing the true gills. The body is compressed, oblong, and elevated, with medium-sized ctenoid scales. The eyes are lateral, the gills four in number, the gill-opening rather narrow, and false gills either rudimentary or wanting. The single dorsal fin, as well as the anal, has a variable number of spines ; and the pelvic fins are thoracic in position. While in some cases the lateral line is interrupted, in others it is altogether wanting ; and the swim-bladder may be either present or absent, but when developed it is generally very large, sometimes even extending into the tail.

These fishes are confined to Southern Asia and South Africa, and are all capable of existing for some time out of their native element, when they oxygenate their blood directly from atmospheric air by means of the accessory gill-like organ. Some are carnivorous, and others vegetarian. All are capable of domestication, in which state they are subject to considerable variation, and several have been acclimatised in countries other than their own. The flesh of all is said to be eatable, and that of some is of excellent quality. On account of their brilliant coloration and the curious habits of some of them, these fish have always attracted more than ordinary interest.

CLIMBING-PERCH

The somewhat inappropriate name of climbing-perch (*Anabas scandens*) has long been applied by Europeans in India to the sole representative of a genus characterised by the presence of teeth on the palate, and the serration of the free margins of the opercular and preorbital-bones. The body is compressed and oblong ; the lateral line interrupted ; the single dorsal fin has its spinous portion much longer than the soft part ; and in the anal fin the spines are less

numerous than those on the back. The caudal fin is rounded, and the scales are rather large. In length the climbing-perch may reach at least $8\frac{1}{2}$ inches, and in the adult state its general colour is dark green, usually marked with dusky bands, which disappear soon after death. It frequents estuaries, rivers, and tanks, and is distributed over India, Ceylon, Burma, the Malay Archipelago, and the Philippines. That this fish can travel long distances on land, where it drags itself along by hitching its pectoral fins round the stems of grass and other herbage, is perfectly well ascertained, but the statement that it climbs trees lacks confirmation. When breathing on land by means of the accessory respiratory organ, air is taken into the mouth, passing out again in the same way.

PARADISE-FISH

The Oriental region is the home of another allied genus of fishes (*Polyacanthus*), represented by several species, and differing from the climbing-perch by the absence of teeth on the palate, and the smooth margins of the preorbital and opercular bones ; the mouth being small and slightly protractile. The spinous part of the single dorsal fin is much longer than the soft portion, the anal being similar ; the pelvic fins have one spine and five soft rays, some of which are usually elongated ; and the caudal is rounded or pointed. The lateral line, which is never complete, may be wanting. These fishes inhabit fresh waters and estuaries along the coast of South-Eastern Asia, but are seldom found any great distance inland.

The pretty and brightly coloured paradise-fish (*P. opercularis*) is an inhabitant of China and Cochin-China, and was long regarded as the representative of a distinct genus. It is, however, now known to be merely a domesticated variety of a species of *Polyacanthus*, although we are not aware that the normal form has hitherto been discovered. Throughout China this fish is kept in confinement ; and is even more suited to captivity than the gold-fish, as it will breed

CLIMBING-PERCH

in very small vessels. It is even stated to live in water strongly impregnated with acid, and its tenacity of life is very great. When kept in dark or muddy waters the colour is generally a dull uniform brown ; and it is only when living in clear water, exposed to the sunlight, that the golden hue and red transverse bands make their appearance, these showing at an earlier period in the males than in the females. The male constructs a fairy-like floating nest by cementing air-bubbles together with a sticky material secreted by the lining of the mouth. He guards eggs and young with zealous care.

THE GURAMI

On account of the excellent quality and taste of its flesh, mention must be made here of the gurami (*Osphromenus olfax*) as a well-known representative of a third genus belonging to this family. Agreeing with the members of the preceding genus in the absence of teeth on the palate, the smooth border to the preorbital and opercular, and the structure of the pelvic fins, these fishes differ by the smaller number of spines in the dorsal or anal fins, which are either fewer than the soft rays, or but very slightly exceed them. The body is moderately elevated and compressed; the small and oblique mouth is capable of a considerable degree of protrusion; and the first ray of the pelvic fins is elongated into a slender filament, the remainder being generally rudimentary. When present, the lateral line is continuous, and there is always a swim-bladder.

Distributed over the rivers of South-Eastern Asia, these fishes are represented in India only by a small species (*O. nobilis*), of some 4 inches in length, inhabiting North-Eastern Bengal and Assam. The gurami, which is a native of the rivers of China and the Malayan Archipelago, has, however, been introduced into several parts of India, and has also been naturalised in Mauritius, Cayenne, and Australia. It is easily recognised by its large size, great convexity of the profile of the under surface, and greenish brown colour, marked in the immature condition by four or five dark vertical bands. It attains a weight of fully 20 pounds, and, when kept in clean water, is stated to be the best flavoured water-fish in South-Eastern Asia. As it is extremely tenacious of life, and almost carnivorous, it is in every way admirably adapted for transportation and acclimatisation.

THE FIGHTING-FISH

A fourth genus (*Betta*), distinguished by the short dorsal fin occupying the middle of the back, and without any pungent spine, the long anal, and the production of the outer ray of the five-rayed pelvic

THE GURAMI

fins, must also be mentioned on account of its containing the so-called fighting-fish (*B. pugnax*), which is bred by the Siamese for the sake of the sport afforded by its pugnacity.

POMACENTRUS

In almost all the families of spine-finned fishes hitherto described the lower pharyngeal bones are

distinct, but in the four remaining families these are united. It has been considered that this difference was of sufficient importance to justify the reference of the families with united pharyngeals to a subordinal group of equal rank with one containing those in which

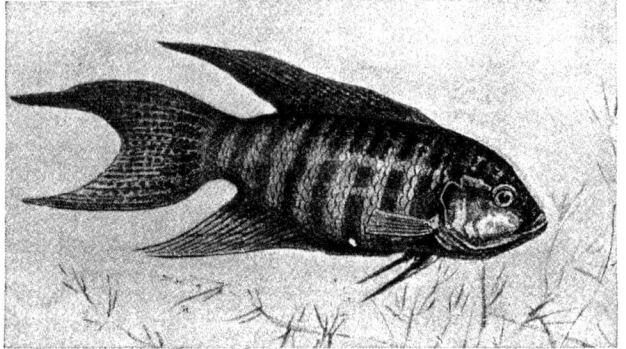

PARADISE-FISH

these bones remain distinct; but we prefer to follow Day in regarding the group now to be considered merely as a section of the suborder which includes all the other spine-finned fishes. That this is the correct view is proved by the circumstance that in one aberrant genus of perches (*Gerris*) some of the species have the lower pharyngeal bones separate, while in others they are united. In the three families (*Pomacentridæ, Labridæ, Cichlidæ*) constituting the present group there is a single dorsal fin, in which the number of spines and soft rays is nearly equal; while the anal is usually similar in character to the soft dorsal; and the pelvic fins are thoracic, and include one spine and five soft rays.

The first of the families of the present section takes its name from the genus *Pomacentrus*, which, together with the allied genera, includes tropical fishes mainly frequenting the neighbourhood of coral reefs and islands, and thus closely resembling the scaly-finned fishes; a few species of the family range, however, into temperate seas. As an example of the typical genus, we figure *P. scolopsis*, from the Malayan seas and Polynesia. As a family, these fishes are specially characterised by the presence of false gills and ctenoid scales. The body is more or less short and compressed; there are weak teeth on the jaws but none on the palate, and there is a swim-bladder.

The family is represented by eight genera and considerably over a hundred species; and the genera may be divided into groups, according as to whether all or some of the opercular bones are serrated at the edges or are all simple; *Pomacentrus* belonging to the intermediate group, in which the preopercular is serrated, while the edges of the other bones of the gill-cover are entire. *Pomacentrus* is the largest genus of the family, its representatives ranging over the tropical seas of both hemispheres.

Curiously enough, not only do these fishes resemble the scaly-finned fishes in their mode of life, but they are very similarly coloured, so much so, indeed, that in some instances actually the same pattern of coloration is common to members of the two families. This, as remarked by Dr. Günther, is one of many instances showing that the coloration of animals depends to a great extent on their mode of life and natural surroundings. All these fishes are

SILVER-DOTTED POMACENTRUS

carnivorous, subsisting on various small marine animals, those furnished with compressed teeth probably browsing on the coral-polyps.

WRASSES

Distinguished from the preceding family by their cycloid scales, the wrasses form an extensive group (*Labridæ*), many of the members of which may be easily recognised by their greatly thickened lips, sometimes provided with an internal fold. False gills are present, and the true gills three and a half in number on each side. The body is oblong or elongate, and while teeth are present in the jaws, they are absent on the palate. In the single dorsal fin the number of spines is usually equal to that of the rays; the anal is similar to the soft dorsal, and a swim-bladder is present.

Littoral in their habits, the great majority of the wrasses are found in tropical and temperate seas, none being polar. Rocks and coral-reefs are their favourite haunts, most of them feeding chiefly on molluscs and crustaceans, for crushing the shells of which their teeth are specially adapted. In many kinds there is an additional pointed curved tooth at each angle of the upper jaw, used for holding a shell against the front and side teeth, by which it is crushed.

The majority of the wrasses are beautifully coloured, decorated not only with transient iridescent hues on the scales, but also with permanent colours formed by pigment. Some species grow to a large size (upwards of 50 pounds); and it is these larger species which are most esteemed as food-fishes the flesh of the smaller kinds being inferior.

TRUE WRASSES

The numerous genera of wrasses are arranged in groups according to the structure of the front teeth. In the genus *Labrus*, of which the striped, or red, wrasse (*L. mixtus*) may be taken as a well-known British example, the body is compressed and oblong in form, with the moderate-sized scales arranged in over forty transverse rows; the snout is more or less sharply pointed; the cheeks and opercular bones are covered with imbricating scales, which are, however, wanting or but few in number on the interopercular; and the conical teeth are arranged in a single row. The spines of the dorsal fin are numerous (13 to 21), and all of approximately equal height; there are three spines in the anal fin; and the lateral line is continuous. In the young, the edge of the preopercular bone is serrated.

These wrasses are chiefly characteristic of

the Mediterranean area, gradually diminishing in the more northern seas of Europe, and being quite unknown in those of India. The striped wrasse exhibits a remarkable sexual variation of colour; the males usually having the body marked with blue streaks or a blackish band, while in the females the back of the tail shows two or three blackish blotches. The other British species is the Ballan wrasse (*L. maculatus*), in which the general colour is bluish green, the scales being margined with reddish orange, and the fin-rays also of the latter tint.

The gold sinny (*Crenilabrus melops*) is a British species, distinguished by the serrated edge of the preopercular. Another well-known form is the blackfish (*Tautoga onitis*), of the Atlantic coast of North America, so named on account of its blackish brown colour, and the sole representative of a genus characterised by the naked opercular, the rudimentary scales on the cheek, and the double row of teeth in the jaws. The true wrasses make seaweed nests in rock crevices, both parents taking part in the building operations. The eggs are not deposited in a special cavity, but are scattered among the pieces of weed.

PARROT-WRASSES

The parrot-wrasses are inhabitants of the tropical Atlantic, except a Mediterranean species (*Scarus cretensis*), which was held in high estimation among the ancients. These fishes are easily recognised by their sharp beak, formed by fusion of the teeth; and also by the projecting lower jaw. The splendidly-coloured Mediterranean species feeds on seaweed; and the mastication required to reduce this to a pulp probably gave rise to the old idea that it was a ruminant.

VIVIPAROUS WRASSES

The members of the small family *Embiotocidæ* may be termed viviparous wrasses, on account of their peculiar reproductive arrangements. Agreeing with the wrasses in the presence of false gills and the cycloid scales, they differ in having four gills, and the anal fin furnished with three spines and numerous soft rays. The compressed body is elevated or oblong, and the lateral line continuous. The single dorsal fin has a spinous portion in front, and a scaly sheath along the base, separated by a groove from the body-scales. Small teeth are present in the jaws, but the palate is toothless. Generally not exceeding a pound in weight, these fishes are confined to the temperate North Pacific, much more numerous on the American than on the Asiatic side.

They are commonly known as "surf-fishes," living as they do in the rough water close to the shore. While the majority belong to the genus *Ditrema* (*D. argenteum*, is figured), one species constitutes the genus *Heterocarpus*, distinguished by the number of dorsal spines being

STRIPED WRASSE

from sixteen to eighteen, instead of only from seven to eleven. All these fish produce living young.

CICHLIDS

Although some members of the preceding family sometimes enter rivers, the cichlids (*Cichlidæ*) differ from all the other fish with united lower pharyngeals in being exclusively fresh and brackish forms. Their distribution is somewhat peculiar, and very similar to that of the lung-fishes (exclusive of the Australian form). Thus they are found in Tropical America and Africa, together with Madagascar, Syria, and Palestine, one outlying genus occurring in India. They are particularly abundant in Lake Tanganyika.

All the genera from the New World are distinct from those of the Old World. Mostly of comparatively small size, although one species of the type genus from the Nile grows to a length of about twenty inches, the cichlids may be distinguished from all the other three families of the present group by the absence of false gills. The body is somewhat variable in form, and generally covered with ctenoid scales, although in some cases these may be cycloid ; and the lateral line is more or less interrupted. In the single dorsal fin the spinous portion usually exceeds the soft in extent ; the anal fin having three or more spines, and its rayed portion being similar to the soft dorsal. The jaws are provided with small teeth, but the palate is smooth ; and the number of gills is four. In some species the teeth are lobate and the intestines complicated by many foldings, these types being vegetable-feeders while all the remainder are carnivorous.

The "bultis" are remarkable for the care which they bestow on their eggs and young. Most fishes lay eggs, which are shed at random, either floating on the surface of the water or sinking to the bottom, a system which of course entails a terrific waste, which Nature has provided for by endowing the fish which behave thus with a marvellous fecundity. By making proper provision for their progeny the "bultis" have been able to greatly reduce the number of eggs which are usually laid. The male fish constructs a simple basin-shaped nest in the sand among reeds or tamarisk-bushes near the shore. On the bottom of this excavation the female deposits her eggs, which she afterwards takes into her mouth, retaining them there until the young fish have completely developed, and can take care of themselves.

Among the best-known representatives of the typical genus *Chromis* is the so-called bulti of the Nile (*C.*

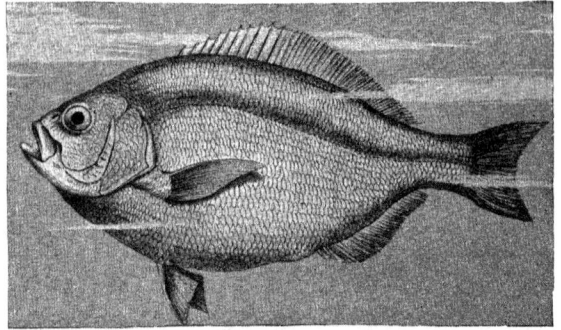

SILVERY VIVIPAROUS WRASSE

niloticus), which is one of the largest members of the family ; while Tristram's cichlid (*C. tristrami*), here figured, is from salt and other lakes in the Sahara and Ashanti. As a genus, *Chromis* is distinguished by its lobate teeth, a three-spined anal fin, and a scaly gill-cover, and it therefore belongs to the vegetable-feeding group.

MAILED TUBE-MOUTHS

The two small subordinal groups (Lophobranchii and Plectognathi) now to be noticed are often placed after the soft-finned fishes, but from their anatomy it appears more probable that they are highly specialised spine-finned types.

A few small fishes from the Indian Ocean (*Solenostoma*) are the sole representatives of the first family, *Solenostomatidæ*, of the suborder Lophobranchii, the distinctive features of that subordinal group being as follows.

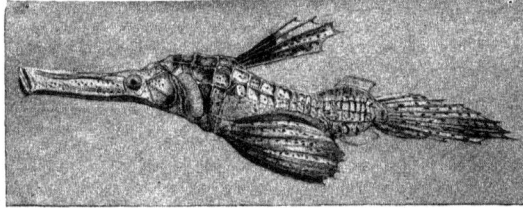

BLUE-FINNED TUBE-MOUTH

The body is invested in a segmented bony dermal skeleton, and the bones of the gill-cover are reduced to a single plate. The gill-openings are small, and the gills themselves consist of small, rounded tufts springing from the gill-arches, while the muscular system is feeble. The simple swim-bladder, when present, has no duct, and the prolonged snout terminates in a small toothless mouth.

In the mailed tube-mouth the gill-openings are wide, the rays of the first of the two dorsal fins are not articulated, and the whole of the other fins are well developed. The name has reference to the great elongation of the tube-like snout, the compressed body having a very short tail, and, like the head, being covered with a thin skin, beneath which are the large bony plates, marked with a radiate pattern. The soft dorsal and anal fins arise from boss-like elevations of the hinder part of the body, the pelvic fins, which are placed close together in the same vertical line as the tall first dorsal, and have seven rays, are separate from one another in the males, but in the opposite sex have their inner edges joined to the skin of the chest so as to form a brood-pouch. The swim-bladder is wanting.

A female of the blue-finned species (*S. cyanopterun*) is figured, the range of this form extending from the coast of Zanzibar to China and Ceram. The female takes the whole charge, not only of the exceedingly minute eggs, but of the newly-hatched fry. Like the members of the next family, these fishes generally swim in a more or less nearly vertical position, the dorsal fin being the chief propeller.

TRISTRAM'S CHROMID

PIPE-FISHES AND SEA-HORSES

From the members of the preceding family the pipe-nshes (*Syngnathidæ*) may be distinguished by the reduction of the gill-opening to a very small aperture at the upper angle of the gill-cover, as well as by the single soft dorsal fin, and the absence of the pelvic fins ; some of the other fins being likewise wanting in certain genera. Mainly marine, although frequently entering brackish, and more rarely fresh waters, these strange fishes are to be found on the coasts of tropical and temperate seas in such situations as, from the abundance of seaweed, offer them sufficient shelter. They are poor swimmers, and if carried away from protective covert may be borne helplessly out to the open ocean by the action of currents. Unlike the tube-mouths, the males take charge of the eggs and young, being often provided with a pouch formed by a fold of skin arising from each side of the body and tail, and joined together in the middle line.

Including several genera, the pipe-fishes are characterised by the absence of prehensile power in the tail, which generally terminates in a fin. In the typical genus *Syngnathus*, as represented by the great pipe-fish (*S. acus*), the body is marked with more or less distinct longitudinal ridges, among which the one down the back is not continuous with that on the tail. The pectorals are well developed, the caudal present, and the dorsal fin placed nearly or exactly above the vent. The great pipe-fish is a common Atlantic species, and grows to a foot and a half.

As an example of a second genus, we may mention the deep-nosed pipe-fish (*Siphonostoma typhle*) of the British seas, distinguished by the upper ridge on the tail being continuous with the lateral line, but not with the dorsal ridge. An Australian species (*S. intestinalis*) lives inside sea-cucumbers (holothurians), as do species of *Doryichthys*. The slender straight-nosed pipe-fish (*Nerophis ophidium*), which may not infrequently be seen served up among a dish of whitebait, is a British example of a fourth genus, in which the body is rounded and nearly smooth, and the caudal fin either rudimentary or wanting. All the pipe-fishes are carnivorous in their diet ; they swim about slowly in a very peculiar manner, more generally vertically or in an inclined position than horizontally, contorting their bodies into every conceivable kind of posture, and poking their long snouts inquisitively into bunches of seaweed in their search for food.

It has long been known that the males of the strange-looking pipe-fishes take charge of the eggs as soon as they leave the bodies of the female parents, and nurse them in a special pouch on the under side of their own bodies ; but it appears to have been reserved for an American naturalist to observe the actual manner in which the transfer of the eggs takes place. From his account, it seems that the male and female fishes entwine their bodies in the form of a double letter S, and that in this position the eggs are passed from the mother to the pouch of the male. As might have been expected, all the eggs are not transferred at once.

SEAWEED-LIKE SEA-HORSE

After the first transference all the eggs of this batch are in the upper part of the pouch, where no more can be received until these are shaken down into the lower end. In some genera (*Nerophis*, *Doryichthys*, and *Gastrotoceus*) the male is devoid of a pouch, and the eggs are pushed into the soft skin lining a broad groove in the under side of his abdomen.

The prehensile structure of the tail is the chief difference between sea-horses and pipe-fish, although in all the existing representatives of the former group there is no caudal fin. The sea-horses are divided into several genera, of which the typical one is best known by the short-snouted sea-horse (*Hippocampus brevirostris*), ranging from the Atlantic and Mediterranean to Australia, and occasionally found in the British seas. The body is more or less compressed and deep, with its investing bony shields raised into tubercles or spines of variable length, while the back of the head is compressed into a crest, terminating in a well-marked knob. Small pectoral fins are present, and the males have a brood-pouch beneath the tail, with its aperture near the vent.

Sea-horses are represented by about a score of species. A remarkable instance of protective resemblance to their natural surroundings is afforded by the three representatives of an Australian genus, as in *Phyllopteryx eques*, shown in the accompanying illustration. The body may be either compressed or as broad as deep ; some or all of its smooth bony plates being furnished with long spine-like processes projecting from its edges, and many of these terminating in irregular leaf-like appendages. There are a pair of spines on the snout, and others above the eye ; pectoral fins are present, and the tail is about equal in length to the body. In the absence of a pouch, the eggs are imbedded in soft membranous skin lining a groove on the under surface of the tail. These sea-horses closely resemble the colour of the seaweeds to which they attach themselves, while the filamentous appendages of their spines appear as if they were actually a part of the vegetable growth. They are of relatively large size, attaining a length of as much as a foot.

FILE-FISHES AND COFFER-FISHES

With file-fishes and their allies (*Balistidæ*), we come to the first of the two families constituting the suborder Plectognathi, of which the following are the distinctive characters. The bones are completely ossified only in the head, the number of vertebræ being few. The small gill-openings are situated in front of the pectoral fins, and the gills themselves are pectinate ; the mouth being narrow, with some of the bones of the upper jaw united, and in certain cases both jaws prolonged to form a beak. There is generally a single soft-rayed dorsal fin, placed far back immediately above the anal, and there may be remnants of a spinous dorsal, while the pelvic fins, when retained at all, take the form of simple spines. The skin may be either entirely naked, covered with rough scales, invested in plates, or studded with bony spines. The swim-bladder has no duct.

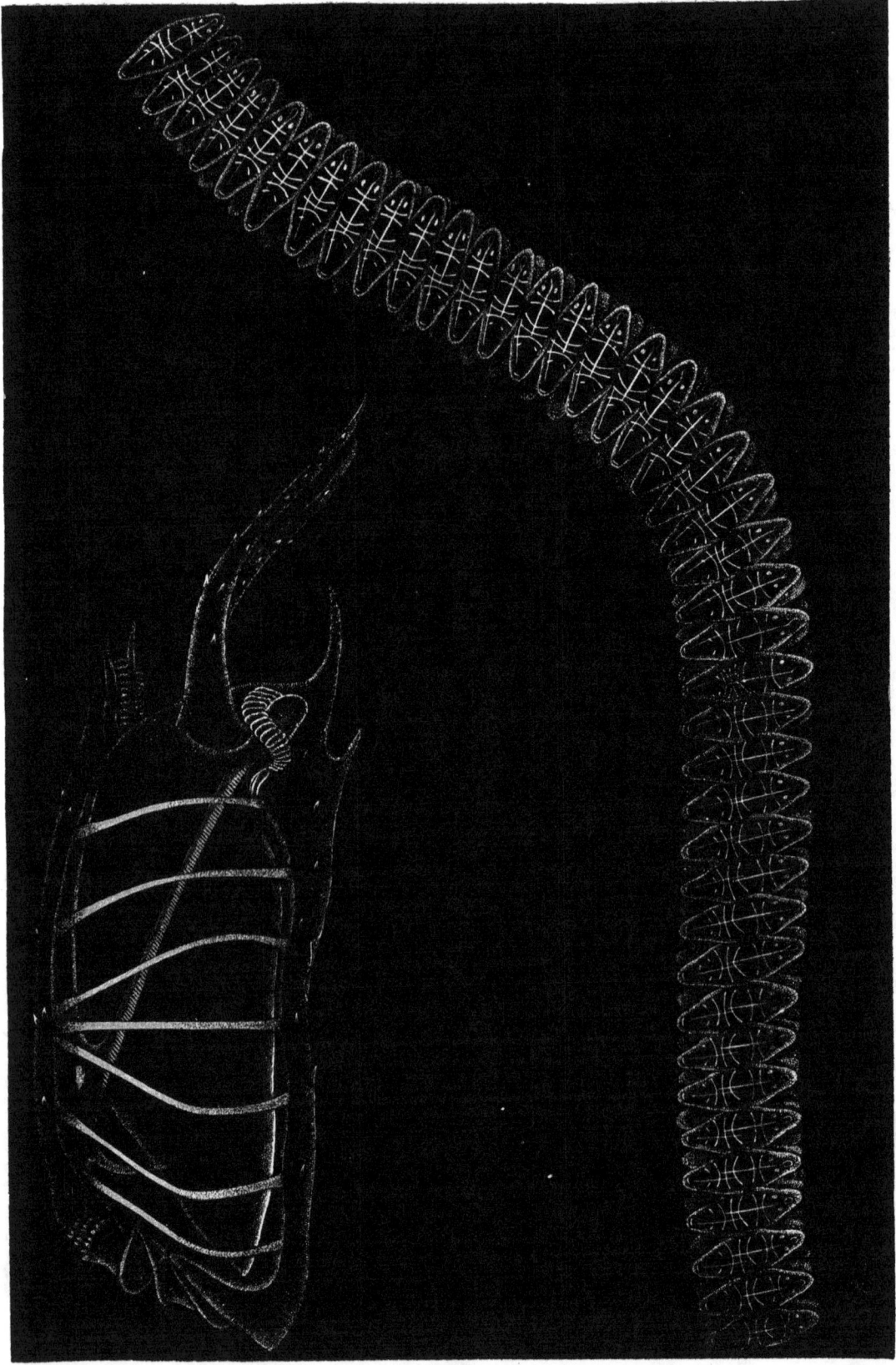

SOLITARY SALPA AND CHAIN-SALPA

"Pelagic in habit and transparent in structure, salpæ have been not inaptly compared to a barrel with both ends knocked out, and really consist of a huge pharynx swimming through the water and taking in large mouthfuls of the same at each contraction of its muscles."

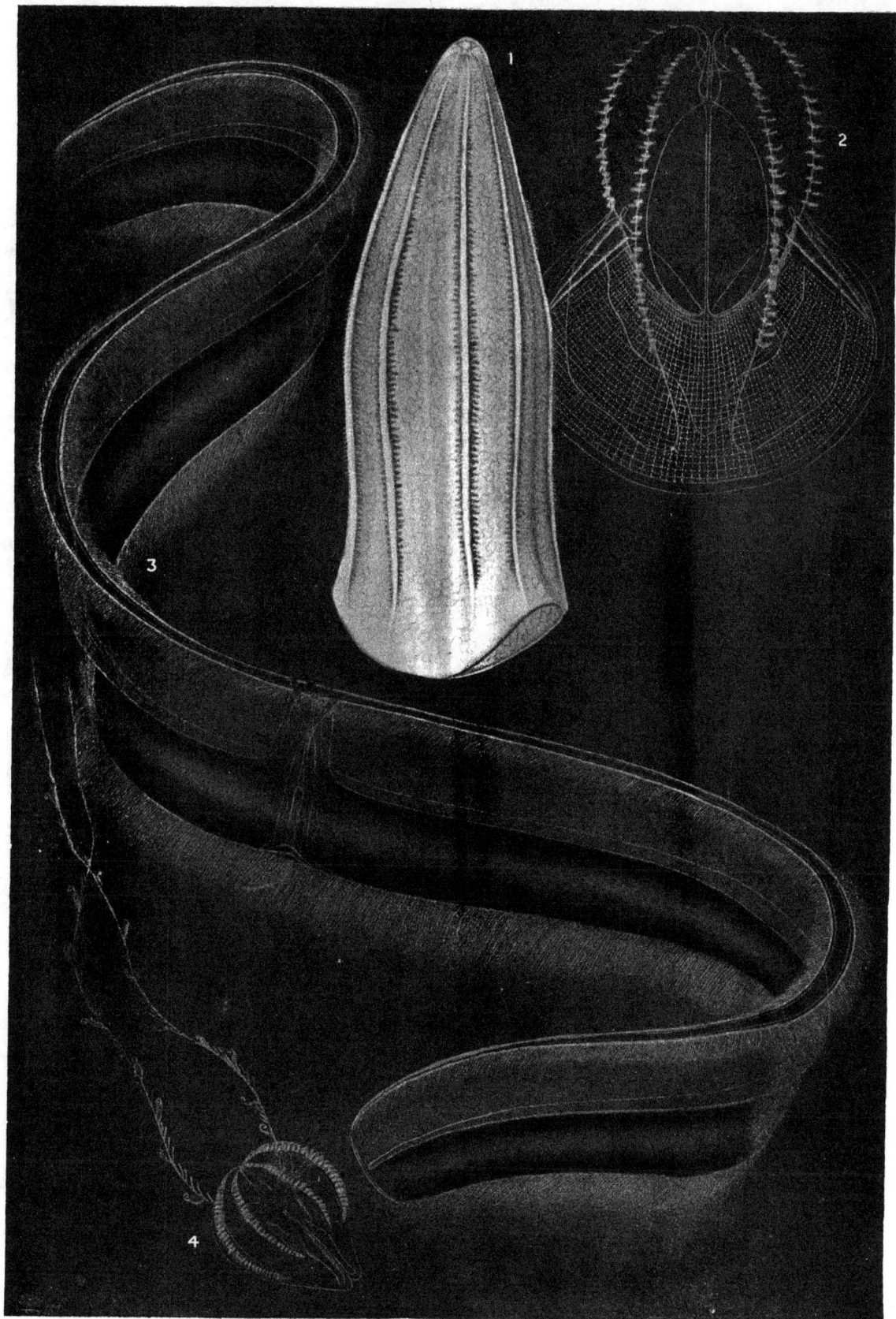

CTENOPHORES

I, BEROË FORSKÅLII ; 2, DEIOPEA KALOKTENOTA ; 3, CESTUS VENERIS ; 4, HORMIPHORA PLUMOSA

"Ctenophores feed upon all kinds of small pelagic animals, especially crustaceans, while they themselves fall a prey to the disc-shaped jelly-fishes and sea-anemones. The largest specimens are, as a rule, found in waters sheltered from the wind."

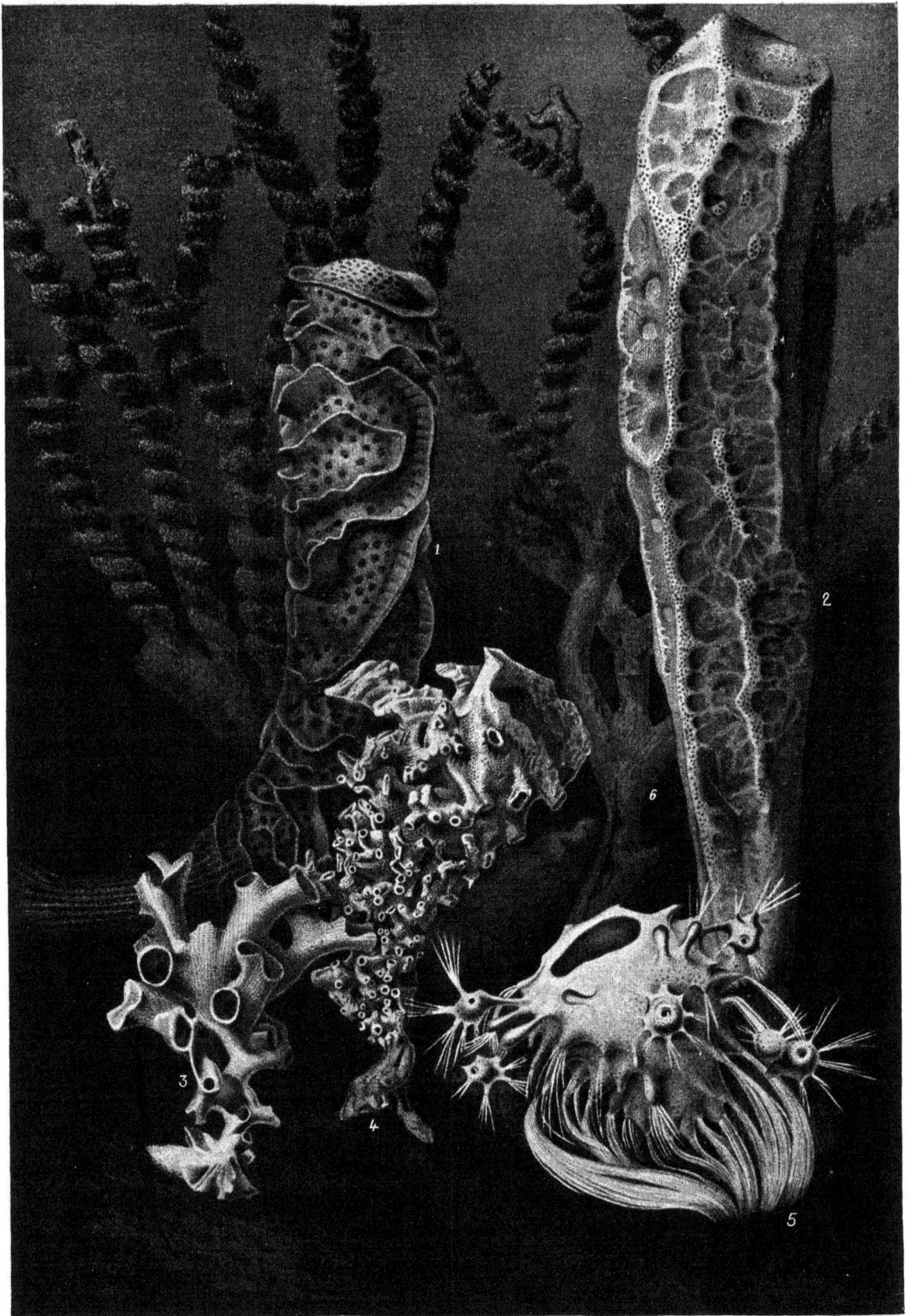

GLASS-SPONGES

1, VENUS FLOWER-BASKET, EUPLECTELLA ; 2, SEMPERELLA ; 3, FARREA ; 4, PERIPHRAGELLA ;
5, POLYLOPHUS ; 6, SCLEROTHAMNUS

"Glass-sponges, with one or two exceptions, have been obtained in deep water from ninety to two thousand nine hundred fathoms. Previous to the deep-sea dredging expeditions specimens had been found in only a few localities."

As a family, the file-fishes and their allies are specially distinguished by the presence of a small number of distinct teeth in the jaws. Their bodies are either compressed or angulated, with a somewhat produced snout; more or less distinct vestiges of spinous dorsal and pelvic fins generally occur, and the skin may be either rough or spiny, or the whole body invested in a bony cuirass. The colours are often brilliant, and may possibly be of a "warning" kind, keeping off foes by advertising the possession of unpleasant characters, in this case bony spines. A peculiar nervous disease called "ciguatira" is produced by eating most species. These fishes, which are of medium size, range over all tropical and temperate seas, being more numerous in the former, and may be divided into three subfamilies.

The first subfamily is typically represented by *Triacanthus brevirostris*, from the Indian Ocean, the other genera being Australian. The special characters of the group are to be found in the compressed form of the body, and its covering of rough, scale-like plates, as well as in the presence of a pair of strong spines representing the pelvic fins; the type genus being distinguished by having from four to six spines in the spinous dorsal fin. The typical file-fishes, or trigger-fishes (*Balistes*), belong to a group of three genera in which the body is compressed and covered either with a rough skin or movable scale-like plates, and the pelvic fins are either wanting or represented merely by a single median swelling on the abdomen. These fishes are distributed over all tropical and subtropical seas, the first two genera including a very large number of species.

Many species are beautifully ornamented with symmetrical markings, and while most are of small size, some attain as much as a couple of feet in length. Of the members of the typical genus (*Balistes*) Dr. Günther writes that "both jaws are armed with eight strong incisor-like and obliquely truncated teeth, by means of which these fishes are enabled to break off pieces of the corals on which they feed, or to chisel a hole into the hard shells of molluscs, in order to extract the soft parts. They destroy an immense number of molluscs, thus becoming most injurious to the pearl fisheries. The first of their three dorsal spines is very strong, roughened in front like a file, and hollowed out behind to receive the second much smaller spine, which, besides, has a projection in front at its base, fitting into a notch of the first. Thus these two spines can only be raised or depressed simultaneously, and the first cannot be forced down unless the second has been previously depressed. The latter has been compared to a trigger, hence a second name—trigger-fish—has been given to these fishes."

This quotation also serves to show how an opinion may require to be modified by the acquisition of further knowledge. It is true that the trigger-fishes destroy pearl oysters, and it was at one time proposed to wage a war of extermination in the Gulf of Manaars (Ceylon). Professors Herdman and Hornell, however, have shown that Oriental pearls are due to the irritation of the embryos of dead tapeworms, of which one stage is harboured by the trigger-fish. The extermination of the tape-worm would follow that of the trigger-fish, and no more Oriental pearls would be formed.

The box-like coffer-fishes (*Ostracion*), of which there are rather more than a score of species from the tropical and subtropical seas, alone represent the third and last subfamily, and are easily recognised by the enclosure of the angulated body in a complete cuirass formed of six-sided bony plates with their edges in juxtaposition, thus forming a mosaic-like pattern. Both the spinous dorsal and the pelvic fins are wanting, although their position may be indicated by prominences. In the whole backbone there are but fourteen vertebræ, of which the last five are very short, while those in the front of the series are much elongated, and the ribs are entirely wanting. In some of the species the cuirass is marked by three, and, in others, by four or even five, ridges, but in other cases it is armed with long spines, which vary in length according to the age of their owner. One species (*O. quadricornis*) is represented in the figure and in the coloured plate on page 1713.

GLOBE-FISHES AND SUN-FISHES

Unlike as they are in external appearance, the spine-clad globe-fishes and the huge flattened sun-fishes are referred to a single family (*Diodontidæ*), distinguished from the last by the bones of the jaws being confluent and modified into a cutting beak, which may or may not have a median suture, the dentition taking the form of plates composed of thin parallel layers. The body is more or less shortened; spinous dorsal, anal, caudal and pectoral fins are developed, but the pelvics are wanting. The external covering may take the form either of a number of small or large spines, or of plates, and the swim-bladder may be either present or absent. Inhabitants of tropical and subtropical seas, with the exception of a few found in the fresh waters of the same regions, the members of this family are mostly small or medium-sized forms, although this is by no means the case with the sun-fishes.

COFFER-FISH

THE SAC-FISH

Like the preceding, the present family may be divided into three groups or subfamilies, the first of which is represented only by the sac-fish (*Triodon bursarius*) of the Indian seas, which takes its name from the sac formed by the dilatable skin of the abdomen; this sac being supported by the pelvic bone, and filled with air at the will of the fish, although its lower portion consists merely of a flap of skin into which no air can enter. The dental plate of the upper jaw is divided by a median suture, while that of the lower jaw is continuous. The elongate tail terminates in a forked fin, and the body is invested with spiny, bony plates, which do not overlap one another. The single species, which may attain a length of 20 inches, ranges over the Indian and Malayan seas, and is of a general brown colour, with a spot of variable colour on the sac and yellow fins.

GLOBE-FISHES

The essential characteristics of the globe-fishes, which form the second subfamily, dating from the Upper Eocene, are that the tail and its fin are distinct and well developed, and that a portion of the œsophagus is highly distensible and capable of being inflated with air All the globe-fishes are easily recognised by the short and cylindrical or rounded form of the body,

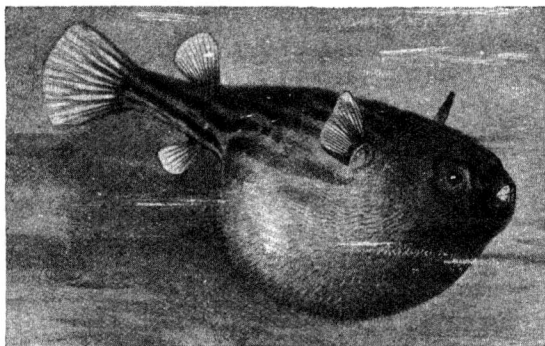

THE GLOBE-FISH

which is generally covered with a scaleless skin bearing a number of spines of variable size. When these spines are large, they are spread uniformly over the whole body, but when small they are partial in their distribution.

These fishes are divided into two groups, according to the nature of the dental plates. In the first, or small-spined group, as typified by the genus *Tetrodon*—of which a species is represented in the lower figure of the coloured plate—the dental plate of each jaw is divided by a median suture, and the spines are frequently very small, and may be even altogether absent; many of the species being very brilliantly coloured. One member of the genus inhabits the rivers of Brazil, and a second those of West Africa and the Nile, while a small form is found in the brackish-water estuaries of India.

PORCUPINE GLOBE-FISH

In the second group, of which the porcupine globe-fish (*Diodon hystrix*) is shown in the upper figure of the coloured illustration on page 1713, the dental plates in the jaws are undivided, and the spines are large and frequently erectile. In addition to the undivided dental plates on the edge of the jaws, in the members of this group there is another crushing plate in the middle of the palate, opposed by a similar one in a corresponding position in the lower jaw; these plates being divided by a median suture, and from their laminated structure forming most admirable triturating instruments.

The porcupine globe-fish, which may measure fully two feet in length, is distributed over both the Atlantic and Indo-Pacific Oceans, where it is accompanied by the smaller spotted globe-fish (*D. maculatus*). In their normal state the globe-fishes have rather elongated cylindrical bodies, but they are able to assume a globular form by swallowing air, which passes into the œsophagus and blows out the whole animal like a balloon, with the spines standing out at right angles from the tense skin. In this condition the fish naturally floats back downwards, and it is then driven to and fro on the ocean-surface by waves and currents in a perfectly helpless condition, although the bristling spines render it perfectly safe from all attack.

SUN-FISHES

The gigantic sun-fishes (*Orthagoriscus* and *Rangania*) —pelagic forms distributed throughout the whole of the temperate and tropical seas—alone represent the third subfamily, and are distinguished by the extremely short and truncated tail, the confluence of all the median fins, and the short and highly compressed body, the dental plates of the jaws being undivided. The skin is either rough or smoothly tesselated, and incapable of distension with air; there are no pelvic

fins; the swim-bladder is wanting; and there is an accessory opercular gill. As in the globe-fishes, there are no pelvic bones in the skeleton, and the vertebral column is remarkable for its extreme shortness, there being only seventeen segments in the whole series, of which seven belong to the tail. In all the members of the suborder the spinal cord is noticeable for its shortness; but in the sun-fishes this abbreviation is extreme. According to Dr. T. W. Bridge, " in a sun-fish (*Orthagoriscus*) $2\frac{1}{2}$ metres long, and weighing about a ton and a half, the cord was only 15 mm. in length ($\frac{3}{5}$ inch), or shorter than the brain."

The rough sun-fish (*O. mola*), which has a rough, finely granulated skin, attains very large dimensions, an example caught off the coast of Dorsetshire in 1846 measuring $7\frac{1}{2}$ feet in length. The large dorsal and anal fins can be moved from side to side, and are the chief organs of propulsion, but as the former often projects above the surface of the water the anal appears to be able to act either alone, or in conjunction with the tail-fin. The food largely consists of larval fishes, those of eels being particularly noticeable.

Far rarer is the oblong sun-fish (*Ranzania truncata*), one of the scarcest objects in museums. It is readily distinguished by its smooth, tesselated skin, and the more elongated form of the body, the entire length being nearly three times the breadth. An example of this fish, weighing 500 pounds, was taken in Plymouth Sound in the year 1734.

THE COD TRIBE

The suborder of soft-finned fishes (Anacanthini), which includes the important families of the flat-fish and cods, is characterised by the median and pelvic fins being entirely composed of soft-jointed rays, the pelvic fins, if present, being either jugular or thoracic in position. It should, however, be mentioned that a fresh-water Australian fish (*Gadopsis*) forms an exception as regards the structure of its fins, having spines in the anterior portion of both the anal and

THE SUN-FISH

dorsal. The suborder is divided into two sections, according to whether the head and body are symmetrical or distorted.

Passing over the unimportant family *Lycodidæ*, the affinities of which are doubtful, and which includes small shore-fishes much resembling blennies in general appearance, and mostly characteristic of high latitudes, although a few occur within the Tropics, we come to the cod tribe (family *Gadidæ*).

Equalled only in this respect by the mackerels, flat-fishes, salmon, and herrings, the cod tribe forms a family of the utmost importance from a commercial point of view, and therefore demands a somewhat detailed notice. They are specially characterised by the pelvic fins being generally composed of several rays ; and by the caudal being either free, or, if united with the median fins, by the first dorsal being divided into two moieties. More or less elongate and subcylindrical in form, the body is covered with small cycloid scales ; there are either one, two, or three dorsal fins, occupying nearly the entire length of the back, the rays of the hindmost being well developed ; the anal is either single or divided ; and the jugular pelvic fins are usually formed of several rays, but if reduced to filaments there is always a double dorsal. The gill-opening is wide and the gill-membrane in most instances not attached to the isthmus ; while if false gills are present at all they are either glandular or rudimentary. As a rule, there is a swim-bladder.

Mostly marine, the members of the cod family are comparatively shallow-water fishes mainly characteristic of the Arctic and Temperate seas. There are, however, a certain number of deep-water types, and these have a much more extensive distribution, some occurring in the tropical Indian seas. The fresh-water forms are limited to two or three. Although the flesh of the cod tribe is by no means remarkable for its delicacy or flavour, it affords a most wholesome and substantial food, and as it takes salt readily it is more valuable as a food-supply than would otherwise be the case. The liver of the cod is the source of a well-known oil, greatly increasing the value of the fishery of this species. The family is divided into numerous genera, but in this work our attention will be chiefly concentrated on those containing species of commercial importance.

THE COD

The common cod (*Gadus morrhua*) is the typical representative of a genus primarily characterised by the presence of three dorsal and two anal fins, and of teeth on the vomer, the palatine bones being toothless. The degree of elongation of the body is moderate, and the narrow pelvic fins include six or more rays. In most of the eighteen species recognised by naturalists there is a single chin barbel, but in some forms this is absent. The species are distributed over the Arctic and North Temperate regions. The common cod belongs to a group of several species characterised by the upper jaw being the longer, and the outer series

I, HADDOCK ; 2, WHITING ; 3, YOUNG COD ; 4, ADULT COD

of upper teeth stouter than the inner ones ; its barbel is relatively long. Cod from the British seas and German Ocean are usually greenish or brownish olive in colour, with a number of yellowish or brown spots. As a rule, cod vary in length from 2 to 4 feet, and may weigh as much as 100 pounds.

The range of the cod includes the coasts of Northern Europe, Iceland, and Greenland, whence it descends on the American coast as far as the latitude of New York ; the depth at which the fish is found extending as low as one hundred and twenty fathoms. In Britain the spawning-time is in January, when these fish resort to the shores in great numbers, although at other times of the year they are only found in the neighbourhood of land singly. In America cod do not deposit their spawn till May. Cod feed on various crustaceans, worms, molluscs, and small fish ; and since they always frequent comparatively deep water, they are caught by means of lines.

THE HADDOCK

Belonging to the same group of the genus as the common cod, the haddock (*G. æglefinus*) may be always recognised by the blackish patch on each side of the body above the pectoral fin and the black lateral line. Generally haddock vary in weight from $\frac{1}{2}$ to 4 pounds, but in northern seas they attain a larger size than further south, and measure as much as a yard in length. In England the largest haddock are taken in winter, when they come near the coast to spawn. They generally associate in large shoals, and in stormy weather seek shelter in deep water among seaweeds, when it is useless to attempt fishing for them. In addition to crustaceans and other invertebrates, their food comprises small fishes of various kinds. They range across the Atlantic.

THE WHITING

By far the most delicately flavoured British representative of the genus is the whiting (*G. merlangus*), which differs from all the preceding species in the absence of a barbel on the chin, and is specially distinguished by a black spot near the root of each pectoral fin. The usual weight is about 1½ pounds; 4 pounds being nearly the maximum attained. The distributional area of the whiting is restricted to the seas of Northern Europe, where it is found in vast shoals. Very shy in its habits, the whiting is a voracious fish, Yarrell stating that several sprats have been taken from the stomach of one, while in another of 4 pounds weight were found four full-grown pilchards. The same writer states that it appears to prefer sandy banks, but frequently shifts its ground in pursuit of the fry of various other fishes on which it chiefly feeds.

THE POUT

Another species with a black spot near the pectoral fin is the pout, or whiting-pout (*G. luscus*), which may be at once distinguished from the whiting by the barbel on the chin, and the greater depth of the body, which during life is marked with dark crossbands. Seldom exceeding 5 pounds in weight, this fish ranges from Scandinavia to the Mediterranean, but does not cross the Atlantic. The name of pout is derived from the power possessed by this fish of inflating the membranes covering the eyes and adjacent regions into a bladder-like form.

THE POWER-COD

Nearly allied is the much smaller power-cod (*G. minutus*), which seldom exceeds half a dozen inches in length, and may be further distinguished by the smaller proportionate depth of the body. Found in vast shoals in the Baltic, the power-cod, although of little or no value, is always welcomed as the harbinger of the advent of its larger cousins.

THE POLLACK

The pollack, or whiting-pollack (*G. pollachius*), is a British representative of the group in which the lower jaw is the longer, and all the upper teeth are of equal size; it has a dark spot near the pectoral fin, but no barbel. This fish is an inhabitant of European seas as far as the western portions of the Mediterranean. Haunting rocky ground, pollack are famous for their power of withstanding strong tides and currents; they are very common in many parts of the south coast of England, as Devonshire, but become scarcer to the north. Being free biters, they afford good sport with the line.

THE COAL-FISH

The coal-fish (*G. virens*) is a closely allied but somewhat larger form, more northern in its distribution, and taking its name from the black colour it frequently assumes. This fish is very common in the Baltic and other northern seas, numbers being captured in the Orkneys. The largest specimen on record measured just over 3½ feet in length, and scaled 25 pounds.

HAKES

The hake (*Merluccius vulgaris*) is the British representative of a genus belonging to a large group, characterised by having two dorsals, a single anal, and a separate caudal fin. As a genus, the hakes are distinguished by the strong development of the pelvic fins, which are broad at the base, as well as by the presence of strong teeth on the jaws and vomer, and the absence of a barbel. The common hake is found on both sides of the North Atlantic and other European seas; and is represented in the colder seas of South America, as well as in those of New Zealand, by the allied *M. gayi*. In size the common hake is a rather large fish, reaching 2 or even 3 feet in length. On the Cornish coasts, which they frequent in numbers in pursuit of the shoals of pilchards, hakes have been taken in vast quantities. Although the flesh is coarse and of inferior flavour, large numbers of hakes are dried and salted.

THE BURBOT

As an example of a fresh-water representative of the cod family, we may refer to the well-known burbot, or eel-pout (*Lota vulgaris*), the sole member of its genus. It is common in the rivers of Central and Northern Europe and North America. Belonging to the group with two dorsals, one anal, and a distinct caudal, the genus *Lota* has the first dorsal fin well developed, with from ten to thirteen rays, the pelvics with several rays, the head flattened, the body much elongated, and villiform teeth in the jaws and on the vomer. The chin is furnished with a barbel. In length the burbot exceeds a yard, and its flesh ranks high among fresh-water fish.

In Britain found only in the east of England, where it is not uncommon in the Cam and Ouse, the burbot is widely distributed on the Continent, frequenting alike large rivers, small streams, lakes, and pools. It prefers, however, deep to shallow water, being

THE BURBOT OR EEL-POUT

found in large lakes at a depth of from thirty to forty fathoms; its colour being then paler than is the case with specimens from shallower water. Its food consists of the fry of other fishes, or the adults of the smaller kinds; and it is stated to be particularly destructive to the perch. In the spawning-season, which varies considerably according to localities, burbot are in the habit of congregating in large numbers; and in some of the German rivers masses of these fishes, including as many as a hundred individuals, may be found knotted together after the fashion of eels. While some burbot spawn in November and December, in others the function is delayed till March; and it is during the spawning-season that the fish is in the best condition for the table. The burbot is a dangerous article of food, unless well-cooked, for its flesh is often infected with the bladder-worm stage of a tapeworm that attains to full maturity in the human intestine.

LING

Distinguished from the burbot by the presence of several enlarged teeth in the lower jaw and on the vomer, the ling (*Molva vulgaris*) may be regarded merely as a marine representative of that genus. The common ling, which generally measures from 2 to 3 feet in length, is a northern form, ranging from the coasts of Greenland and Iceland to those of Britain and other parts of Northern Europe. In this fish the upper jaw is the longer, but the reverse condition obtains in a second Scandinavian species, and also in a third from the Mediterranean, which are the only other representatives of the genus. The ling-fishery is an important industry, large quantities of these fish being cured and dried.

ROCKLINGS

Belonging to the same group of the family as the ling, the rocklings (*Motella*) are readily distinguished by the reduction of the first of the two dorsal fins to a narrow-rayed fringe, with the first ray elongated, more or less completely received in a longitudinal groove. There is a band of teeth in the jaws, and another on the vomer, and all the species have barbels, not only on the chin, but likewise on the snout, the number of these appendages affording the readiest means of specific discrimination. They are all of small size, and while ranging over the same seas as the ling, likewise extend to those of Japan, the Cape, and New Zealand.

The British representatives of the genus include the five-bearded rockling (*M. mustela*), with four upper barbels, the four-bearded rockling (*M. cimbria*), and the common three-bearded rockling (*M. tricirrhata*); the little fish commonly known as the mackerel-midge, and formerly regarded as the representative of a distinct genus, being only the young of the rocklings.

SAND-EELS AND THEIR ALLIES

In this rather small family (*Ophidiidæ*), almost all the members of which are marine, the pelvic fins, if developed at all, are rudimentary; there is no separate anterior dorsal or anterior anal, and the caudal is generally confluent with the median fins. In form the body is more or less elongate, but it may be either naked or scaled. The dorsal fin occupies the greater portion of the back; the rudimentary pelvics are jugular in position; the gill-openings are wide; and the gill-membranes are not attached to the isthmus. While some species are deep-sea forms, others are littoral. There are five subfamilies.

CAVE-FISH

The most remarkable representatives of the first subfamily (in which pelvic fins, attached to the pectoral girdle, are always present) are two small fishes from the subterranean fresh waters of certain caves in Cuba constituting the genus *Lucifuga*. They are totally blind, with the eyes rudimentary and covered with skin, or wanting, and always live in perpetual darkness. These cave-fish are closely allied to certain small fishes from the Tropical Atlantic and Indian Oceans forming the genus *Brotula*, and characterised by the elongate body being covered with minute scales, the moderate-sized eyes, the reduction of each pelvic fin to a single filament, the villiform teeth, and the presence of barbels on the snout, these barbels being reduced in the cave-fish to small tubercles. With the exception of these cave-fish, all the members of this family are marine forms; and it is very curious that among the latter there are two very rare species, respectively constituting the genera *Typhlonus* and *Aphyonus*, found at great depths in the southern oceans, which

are also completely blind, and apparently unprovided with any phosphorescent organs.

SNAKE-FISHES

The typical genus *Ophidium*, constituting, with an allied form, the second subfamily, has the pelvic fins replaced by a pair of barbel-like filaments; the elongated and compressed body being covered with very minute scales, while the eyes are medium, and the teeth small. The few species of this genus range over the Atlantic and Pacific. In the South American, South African, and Australasian seas there occur three much larger but nearly allied fishes, which have been referred to a second genus (*Genypterus*), on account of the outer row of teeth in the jaws, as well as those of the single palatine series, containing some enlarged tusks. These fish are of considerable commercial importance, and are known at the Cape as klipvisch, and in New Zealand as Cloudy Bay cod, or ling.

PARASITIC FISH

Some half-score species of very small eel-like fishes, scientifically known as *Fierasfer* and *Encheliophis*, and inhabiting the Mediterranean, Atlantic, and Indo-Pacific, have an especial interest on account of their curious mode of life. They constitute a subfamily, readily characterised by the total absence of pelvic fins and by the vent being situated at the throat; and are commensal in other marine animals, frequenting the hollows in the bodies of jelly-fish, the breathing chambers of star-fishes and sea-cucumbers, and sometimes insinuating themselves between the layers of the mantle of pearl-mussels or other bivalve molluscs. Occasionally they may become imbedded in the substance of the shell of the pearl-mussel by the deposition of pearly matter over their bodies; an instance of this peculiar mode of preservation being shown in the accompanying illustration.

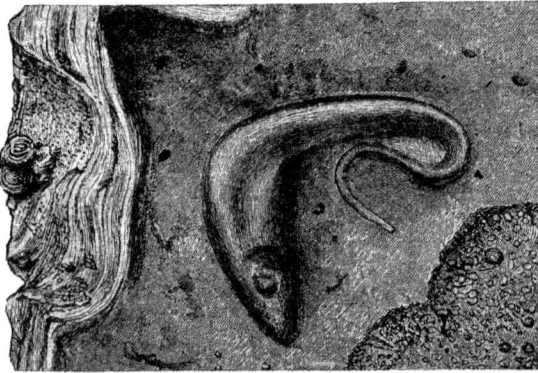

PARASITIC FISH IMBEDDED IN A PEARL-MUSSEL

SAND-EELS

The third subfamily is represented by the well-known sand-eels, or launces—of which a British species, the lesser sand-eel, or launce (*Ammodytes tobianus*), is figured on the next page—so abundant on sandy shores in Europe and North America, as well as by an allied genus from Madras. While agreeing with the preceding group in the want of pelvic fins, they differ in having the vent situated far back in the body; and are further characterised by the great width of the gill-openings, the gill-membranes of opposite sides not being united. The lower jaw exceeds the upper in length, the dorsal fin occupies nearly the whole length of the back, and the anal is likewise elongated. The figured species, which is by far the commoner on the British coasts, generally measures from 5 to 7 inches in length, whereas the greater sand-eel, or launce (*A. lanceolatus*), may grow to a foot and a half.

Sand-eels feed on marine worms and very small fish; and when buried in the sand are captured in some parts of England by raking the sand with a long-pronged rake; their chief use being for bait. They are, however, by no means restricted to this kind of life, frequently swimming near the surface in large shoals, when they will at times suddenly descend to the bottom, where they bury themselves with surprising rapidity by the aid of the elongated horn-like extremity of the elongated lower jaw. During ebb-tide, numbers remain buried at the depth of five or six inches in the sand till the next flood ; and it is then that they are dug out with rakes or other implements. When swimming, they are followed by shoals of mackerel and porpoises

FLAT-FISHES

Distinguished by the unsymmetrical conformation of the head and anterior region of the body in the adult, in consequence of which both eyes are brought on to one side of the body (in some cases the right, and in others the left), the flat-fishes (family *Pleuronectidæ*) differ from all other members of their class. The body is strongly compressed and flattened, with the eye-bearing side, which is turned upwards, coloured dark, while the opposite, or eyeless side is, as a rule, colourless. The bones of the head are unequally developed and unsymmetrical ; and the dorsal and anal fins are of great length, and undivided, the former often extending forwards so as to separate the blind from the eyed side of the head. In the most specialised forms the teeth and jaws are more developed on the lower or blind side than on the other, and there is no swim-bladder.

These fishes are ground-feeders, and their peculiarities all have relation to their mode of life. At first sight they have a superficial resemblance to the skates and rays, which also are bottom-fish, but these are flattened from above downwards. If, as now supposed, flat-fishes are descended from forms resembling the dories, it is particularly interesting to notice that at least one of these swims in a slightly tilted position. Why this should be so is difficult to understand, but if we suppose the tilt to have increased, and a ground-feeding habit gradually assumed, the further stages are comprehensible. In the course of its life-history an animal commonly repeats, in an imperfect and generalised way, what may be supposed to have been the stages in the evolution of its group. Flat-fishes begin life as perfectly symmetrical larvæ, which at first swim about like ordinary fishes. They then gradually tilt over, take to ground-feeding, and become unsymmetrical, the eye of the shaded side migrating to the surface exposed to the light. That this gives some idea of the stages passed through by the ancestral forms is supported by geological evidence. In the upper Eocene rocks are found the fossil remains of a small fish (*Amphistium paradoxum*), the only member of a special family, and intermediate in structure between dories and flat-fishes, being, however, symmetrical like the former. In these same rocks we also find the remains of the oldest known undoubted flat-fish (*Psetta*). This was but slightly unsymmetrical, and closely resembles the least modified of living species, the adalah (*Psettodes erumei*), which is described below. Markedly specialised flat-fishes do not make their appearance until a good deal later in the geological record.

When lying on the sandy bottom of the sea—and they prefer sandy to muddy situations—flat-fishes are

LESSER SAND-EEL

almost indistinguishable from their inanimate surroundings, the spots with which the bodies of many of them are marked harmonising exactly with the bright-coloured pebbles strewing the sand. This resemblance is, perhaps, carried to the fullest extent in the flounder, as anyone who visits a large aquarium may ascertain for himself. Occasionally rising to the surface, they swim with a graceful undulating lateral movement of the body ; and they are found in shallow water, or at moderate depths. They are inhabitants of all seas, except those of the polar regions, and where the coast is precipitous and rocky ; and although more numerous in the tropics, they attain their greatest size in the temperate regions. Many species, such as flounders, ascend rivers to a considerable distance ; and a few have become accustomed to a fresh-water existence. The whole of the species are exclusively carnivorous. As a food-supply the flat-fishes are of especial value, not only on account of the large size, and abundance of their numerous representatives, but likewise from the excellent quality and flavour of the flesh of the majority of these. It is on the coasts of the temperate regions of the Northern Hemisphere that the pursuit of these fishes is carried out with the greatest energy and success.

PSETTODES

The least specialised member of the family, *Psettodes erumei*, which ranges from the Red Sea through the Indian Ocean to China, and also occurs on the West Coast of Africa, belongs to a group in which the teeth and jaws are nearly equally developed on both sides, and is specially distinguished by the dorsal fin beginning on the nape of the neck, whereas in all the other forms it starts from above or in front of the eyes. In the Indian fish, which attains a length of about 16 inches, the eyes are as frequently on the right as on the left side, and the transposed one is situated nearly in the line of the dorsal fin. This species is a connecting link between the other members of the family and ordinary fishes, and is reported to swim at times in a vertical position.

HALIBUT

Having the jaws nearly equally developed on both sides, and the dorsal fin beginning above the eyes, the halibut (*Hippoglossus vulgaris*) is one of two species forming a genus characterised by the eyes being on the right side, and the teeth of the upper jaw arranged in a double series, those in the front of the upper and on the sides of the lower jaw being enlarged. The mouth is relatively wide. In colour the halibut is dusky brown, frequently inclining to olive, on the dark side ; the opposite side being white and smooth. It is the largest member of the family, ranging usually from 3 feet to 6 feet in length ; one specimen with the latter length having had a breadth of 30 inches, and a weight of 161 pounds. A length of 7½ feet and a weight of 320 pounds have been recorded on good evidence. It feeds on fishes and crustaceans.

Halibut are found near all the northern coasts of Europe, as well as those of Kamschatka and California, generally frequenting banks at some distance from the shore, in water of from fifty to one hundred fathoms, in depth, where they often associate in considerable numbers. The flesh is coarse, and of inferior flavour. This fish is shown in the upper figure of the coloured picture on page 1716.

THE TURBOT

In the genus typically represented by the turbot (*Rhombus maximus*) the dorsal fin begins on the snout in advance of the eyes; the eyes are on the left side; the mouth is wide; and the jaws are furnished with a single series of equal-sized villiform teeth, while there are also teeth on the vomer. Scales are either very small or wanting. It feeds almost exclusively on other fishes, and is in the habit of covering itself with sand, being thus made inconspicuous to its prey. The genus includes seven species, ranging over the North Atlantic and Mediterranean, but those on the two sides of the Atlantic are different. The turbot, which attains a yard in length, and is by far the best food-fish of its tribe, is exclusively European, and has the pelvic fins distinct from the anal, and no scales; the general colour being greyish or brownish, sometimes spotted with a darker tint.

THE BRILL

On the other hand, the brill (*R. lævis*), which is likewise European, is a smaller fish, of more oval shape, with the body and all the head, except the snout, covered with minute scales; its colour being greyish brown, with reddish brown spots. Turbot commonly weigh from 5 to 10 pounds, and occasionally reach 20 or even 30, while considerably greater weights have been recorded. Another British representative of the genus is the Mary-sole (*R. aquosus*), which may be distinguished by its ciliated scales. A turbot is shown in the central figure on page 1716.

PLAICE AND FLOUNDER

The plaice (*Pleuronectes platessa*) and flounder (*P. flesus*), which are shown in the right lower corner of the coloured picture, are examples of a genus pertaining to a group characterised by the narrowness of the cleft of the mouth, and by the jaws and teeth being much more developed on the light than on the dark side. Unlike the turbot and its allies, where the upper is somewhat behind the lower, the two eyes are in the same transverse line, and generally situated on the right side. The dorsal fin begins above the eyes; the scales are minute or wanting; and there are no teeth on the palate, while those in the jaws are of medium size, and may be arranged in either a double or a single row. The genus, which is common to the temperate and Arctic seas of the East and West Hemispheres, contains over a score of species.

The plaice, which ranges from the French coasts to Iceland, and is represented by an allied form on the opposite side of the Atlantic, belongs to a group with compressed, lanceolate, or truncate teeth, and no fewer than ninety dorsal rays; the brownish upper surface is marked with bright orange spots. The ordinary teeth are broad, flat, and straight-edged, and there are crushing teeth in the throat. These peculiarities have reference to the food, which consists mainly, though by no means exclusively, of bivalve molluscs. The plaice is exclusively marine, but the flounder is almost as much a fresh-water as a sea fish, ascending rivers to a considerable distance. Distinguished from the plaice by the dark mottlings on the brownish or brownish yellow skin of the upper surface, it belongs to a group in which the teeth are conical; the lateral line being very slightly curved in front, and the scales minute. Its distribution is practically the same as that of the plaice, and it is represented by an allied species in the Mediterranean.

SOLES

In the plaice and its allies the pectoral fins are always well developed, but in the group to which the common sole (*Solea vulgaris*) belongs these may be wanting, while the upper eye is always somewhat in advance of the lower one, both being on the right side. As a genus the numerous species of soles (somewhere about forty) are characterised by the median fins being separated from one another, and the ctenoid scales; the dorsal fin beginning on the snout, and the lateral line being straight. By means of the ctenoid scales with their rough comb-like edges, easily seen with a pocket lens, the sole can easily be distinguished from the "lemon-sole," which has smooth-edged cycloid scales. The latter is far less desirable for food, and is often sold to the uninstructed as a true sole. The mouth is very narrow, and twisted round to the left, or blind, side; and it is on this side only that the villiform teeth are developed in the jaws, the palate being toothless. The sole feeds chiefly on worms, which it hunts by smell and touch, aided by delicate tactile threads in the side of the head. In accordance with this it is able to find its food in the dark, and it is not surprising to find that its eyes are not particularly well developed. This fish seeks protection by burying itself in the sand.

With the exception of the lower south temperate zone, soles are distributed over all temperate and tropical coasts in localities suited to their habits; many of the species entering, or even dwelling permanently, in fresh waters. The common sole, which is found on the coasts of the greater part of Europe, has both pectoral fins well developed, and the nostrils of the blind side very narrow; the general colour being dark brown, with the tips of the pectoral fins blackish. Large specimens may weigh as much as 5 or 6 pounds, and a fish of 9 pounds in weight is on record. Soles are taken by trawling; the best ground in England being along the south coast from Dover to Devonshire.

THE TURBOT

The lemon-sole (*S. aurantiaca*), which is a more southern form, ranging from the south of England to Portugal, and, living in deep water, is one of a group characterised by one of the nostrils of the blind side being dilated and surrounded with a fringe of papillæ. It is smaller and wider than the common species, and orange or light brown in colour, dotted over with numerous small brown spots. Other British species are the banded sole (*S. variegata*) and the dwarf sole (*S. minuta*), both belonging to a group characterised by the small size of the pectoral fins. The common species is shown in the left lower corner of the coloured picture.

SUBCLASS IV. DIPNOI
LUNG-FISHES

THE lung-fishes constitute a group of which the few existing members are so specialised that by some authorities they are placed in a class of their own.

The paired fins are lobate—that is, they possess a thick central part supported by a firm axis, and the tail fin is either unsymmetrical (heterocercal) or symmetrical both externally and internally (protocercal). In the latter case we probably have to do with a character which is not primitive, but due to secondary modifications. The scales are cycloid. The upper jaw is firmly united to the cranium proper, so that there is no separate structure for the suspension of the lower jaw. To this type the name of solid-skulled (technically, autostylic) fishes may be applied. In the lung-fishes the skeleton is partially ossified, with well-developed membrane-bones; the gill-clefts are but slightly separated, and open into a single cavity, protected by an external cover; and the external skeleton consists of true bony tissue. There are large crushing dental plates on the palates and lower jaw.

In the existing members of the group the optic nerves (or those proceeding from the brain to the eyes) simply cross one another, without any interlacing of the constituent fibres; the intestine has a spiral valve; the swim-bladder is elongated, and performs the functions of a lung; and the nostrils open behind by two apertures into the cavity of the mouth, after the manner of the higher vertebrates.

The fishes of this class formed a cosmopolitan marine group during an early part of geological time (newer Palæozoic), but since then have rapidly declined, and are now represented by a very small number of fresh-water species in Africa, South America, and Australia.

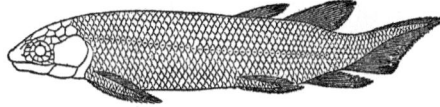

RESTORATION OF DIPTERUS

This is one of the best instances known of "discontinuous distribution," where related forms exist at widely separated points. In this case the explanation is not difficult. The original marine Dipnoi were hard pressed by competition with more specialised fishes and other sea-animals, and only those which took refuge in the fresh waters of the land have survived to the present day. It is these which, having become adapted to breathing atmospheric air by means of a sort of lung as well as air dissolved in water by means of gills, in ordinary fish fashion, that have earned the name of "Dipnoi"—double-breathers—for the sub-class. The extinct marine forms, no doubt, breathed by means of gills alone, and the title given to the subclass is therefore somewhat misleading. It is divided as follows:

ORDER 1
Sirenoidei
FAMILY 1
Comb-toothed Dipnoi—Ctenodontidæ (extinct)
FAMILY 2
Thread-tailed Dipnoi Uronemidæ (extinct)
FAMILY 3
Antler-toothed Dipnoi—Ceratodontidæ
GENUS
Neoceratodus
SPECIES
Australian lung-fish . . . . Neoceratodus forsteri
FAMILY 4
Eel-like Dipnoi—Lepidosirenidæ
GENUS 1
Protopterus
SPECIES
African mud-fish . . . . Protopterus annectans
GENUS 2
Lepidosiren
SPECIES
South American mud-fish . . . . Lepidosiren paradoxa
ORDER 2
Arthrodira (extinct)

These groups will now be successively considered.

ORDER I. SIRENOIDEI

THE first order (Sirenoidei) includes all fishes that unquestionably belong to the Dipnoi.

The family Ctenodontidæ includes the large majority of the extinct Palæozoic forms, in which the dental plates present a radiating set of ridges studded with conical projections, these being the tips of numerous teeth which have fused together. A fanciful resemblance to the teeth of a comb gives the name to the family. The oldest and most familiar form is Dipterus, from the Old Red Sandstone of Scotland. This was covered with rounded, overlapping scales, the paired fins were acutely lobate, and the tail heterocercal.

The family Uronemidæ includes a few Carboniferous and Permian fishes, with tapering protocercal tail. Uronemus, from the Lower Carboniferous of Scotland, gives the name to the group.

ANTLER-TOOTHED DIPNOI

In the family Ceratodontidæ the body is elongated and somewhat flattened from side to side. It is covered by large, overlapping cycloid scales. The paired fins are large paddles. The swim-bladder is undivided and serves as a lung.

For a great number of years there were known from the Triassic strata of various parts of Europe dental plates of the remarkable type of the specimen repre-

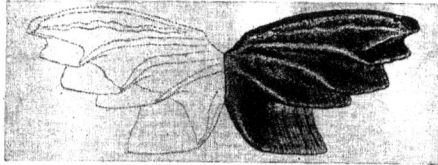

UPPER PALATAL TEETH OF CERATODUS

sented in the accompanying figure; and from the fancied resemblance to a deer's antler, presented by these teeth, the name of Ceratodus was suggested for the otherwise unknown fishes to which they pertained. Similar teeth were subsequently obtained from Secondary rocks in India and also in South Africa, but it was not until the year 1870 that a fish was discovered in Queensland having teeth of a similar type.

AUSTRALIAN LUNG-FISH

Known to the natives, in common with other large fresh-water species, by the name of barramundi, the Australian lung-fish (Neoceratodus forsteri) agrees very closely with the extinct forms. Its mouth is furnished in front with a pair of chisel-like teeth situated on the vomers, behind which come a pair of palatal teeth carrying six complete ridges, and an incomplete seventh; while there are a pair of similar teeth in the lower jaw, carrying only six ridges each. In the living species the teeth of opposite sides are separated by an interval; but in the fossil forms they were in contact, and had fewer ridges.

The existing Australian lung-fishes, of which two species have been described, are said to attain a weight of 20 lb., and a length of upwards of 6 feet. The body is elongated and much compressed, with very large scales; the paddle-shaped limbs have very broad fringes;

AUSTRALIAN LUNG-FISH

and the flesh is salmon-coloured. From the occurrence of masses of leaves in its stomach it is evident that the Australian lung-fish crops the vegetation with its great teeth ; but it is believed that the most important part of its food consists of the small creatures living on and between the leaves of the various water-plants. The female lays her rather large eggs loosely and singly among the vegetation, and in the embryo the fore limbs make their appearance in about a fortnight, but the hinder pair not before two and a half months. In the course of its development this fish presents marked resemblances to the amphibians, and also to the lampreys ; but it is noteworthy that there is no trace of a sucking mouth, or of external gills. As might have been inferred from the study of allied extinct forms, the large palatal teeth are formed by the fusion of a number of separate small denticles.

According to Dr. Semon, the Australian lung-fish is confined to the middle portion of the Burnett and Mary Rivers of Queensland. Living among the mud and leaves at the bottom, it rises at intervals to the surface to obtain more complete oxygenation of its blood by the inhalation of atmospheric air into its lung-like swim-bladder, although its general breathing is carried on by the gills. A grunting noise sometimes uttered by this fish is probably produced by the expulsion of the air from the lungs when it rises to the surface.

There can be little doubt that the swim-bladder of fishes has been evolved as a hydrostatic organ, helping in the maintenance of equilibrium in the water. But as it is richly provided with blood-vessels it is by no means surprising to find that in various cases it has assumed a breathing function. In the Queensland lung-fish we find a special reason why this should be the case, for during the dry season its native rivers are reduced to a series of water-holes, with attenuated bits of stream between. In these water-holes the decay of vegetation produces a foul condition that is fatal to most of the fishes present, which have only gills to breathe with, but the lung-fishes, being able to breathe atmospheric air, are by no means incommoded. It is quite likely that under some such conditions the earliest backboned animals of the land (the ancestors of amphibians) took origin. It would, however, be extremely rash to suppose that the lung-fishes are in the direct line of descent of these terrestrial forms, for a similar environment is known to lead to similar adaptations in groups which are not closely related, a fact which is expressed by the term " convergence of characters."

The breeding season is at its height in September and October, but lasts from April till the beginning of November ; and the eggs, which are enveloped in a gelatinous coat and are heavier than water, take some ten days to hatch.

EEL-LIKE DIPNOI

In the family *Lepidosirenidæ* are included the rest of the existing lung-fishes, in which the body is elongated and rounded (an adaptation to life in mud), and there are little round scales embedded in the skin. The paired fins are reduced to narrow, awl-shaped appendages, and the swim-bladder is modified into a pair of lungs. The young form is a larva with suckers and external gills, something like those of tadpoles.

The family includes the African (*Protopterus*) and South American (*Lepidosiren*) mud-fishes, the latter being the most specialised members of the subclass.

AFRICAN MUD-FISH

In the African mud-fish (*Protopterus annectans*), widely spread over the tropical regions of Africa, the filamentous paired fins retain a small fringe containing rays. There are six gill-arches, with five intervening-clefts, and three small, tentacle-like appendages above the small gill-opening on each side. A length of as much as 6 feet may be attained.

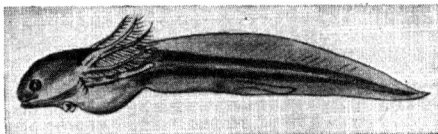

LARVA OF PROTOPTERUS

In the Gambia River, where they are very abundant, these fishes are in the habit of burying themselves during the dry season, making a kind of nest, in which they pass a period of torpidity. Here they may remain for a considerable period, only resuming their normal aquatic life with the return of the wet seasons. During the dormant period, which usually lasts from August to September, they are nourished by the absorption of the fat stored up in various parts of the body. They are most active at night, and appear to keep chiefly to shallow water, where they move deliberately about the bottom, alternately using the peculiar limbs of either side, though their movements do not seem to be guided by any strict regularity. The powerful tail forms a most efficient organ for rapid swimming.

It is well known that this fish comes to the surface to breathe at short intervals, and thus it is evident that the double swim-bladder performs an important, if

AFRICAN MUD-FISH

not the chief, part in respiration during active life. The air passes out again through the opercular aperture, and the movements of the operculum itself indicate the fact that branchial as well as pulmonary respiration takes place.

The investigations of Budgett have revealed some very interesting facts regarding the life-history. Close to the edge of a swamp a hole of about a foot deep is excavated to serve as a nest, and within this the eggs are deposited on the mud. The male sedulously guards these until they hatch, on about the eighth day, keeping them aerated meanwhile by constantly stirring up the water with his tail. The larvæ are very much like tadpoles, as may be seen from inspection of the illustration on page 1777.

SOUTH AMERICAN MUD-FISH

The mud-fish of the Amazons (*Lepidosiren paradoxa*) is the sole representative of a genus in which the limbs are reduced to mere tapering filaments, owing to the disappearance of the marginal fringe. The vomerine teeth are conical and pointed, and the palatal teeth have strongly marked cusps supported by vertical ridges. There are five gill-arches, with four intervening clefts, but there are no external appendages above the gill-opening. In adult males the upper surface of the hind limb is beset with crimson tufts of tentacle-like papillæ during the breeding season. This mud-fish grows to a length of about 4 feet, and

SOUTH AMERICAN MUD-FISH

occurs not only in the Amazon and its tributaries, but likewise in the swamps of the Chako country, forming the tributaries of the upper Paraguay River. These fish feed chiefly on the large snails known as ampullariæ, which are found in great masses in the Chako swamps, their shells being easily crushed by the powerful teeth of their devourers.

Our knowledge of the habits and life-history of the South American mud-fish are very largely due to the observations of Prof. Graham Kerr in the Chako swamps. During the dry season of the year these dry up, and the result has been to bring about a series of adaptations much like those already described for the African form. While there is still plenty of water the fish possesses an abnormal appetite, the result being that a large amount of fat is stored up in the body as a reserve of nutriment for use when no other food is available. The dormant stage is passed through in a deep burrow, the mouth of which is closed by a plug of clay. The plug is perforated by several holes to secure the access of air.

The eggs are laid in a long burrow at the bottom of a swamp, and are guarded by the male. It is suggested that the red outgrowths on his pelvic fins at this time serve as special gills which enable him to remain in the burrow without danger of suffocation.

The larvæ are very much like those of the African mud-fish.

ORDER II. ARTHRODIRA—BERRY-BONE FISHES

THE extraordinary Palæozoic group, of which the affinities are very doubtful, typically represented by the berry-bone fish (*Coccosteus*) of the Scottish Old Red Sandstone, differs from the true lung-fishes in that in place of scales the fore part of the body is protected by large bony plates, of which one pair is articulated by a hinge to the hinder part of the skull, which is likewise invested with bones bearing a berry-like sculpture. The fore limbs were either rudimentary or wanting ; but a pair of pelvic fins were developed. Many of the forms may be included in the single

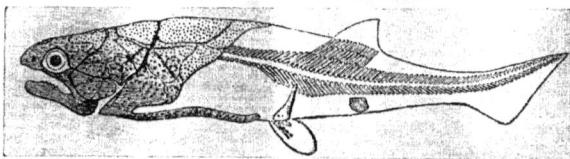

PARTIAL RESTORATION OF THE BERRY-BONE FISH

family *Coccosteidæ* ; and among these the typical genus is distinguished by the absence of any pectoral fin, while in the allied *Brachydirus* this appendage is represented by a hollow spine. In both these the sockets of the eyes form notches on the sides of the skull ; and the same is the case with the gigantic *Dinichthys* of the North American Devonian, a fish that is believed to have been at least 10 feet long, while its head-shield undoubtedly attained a breadth of over a yard. These dimensions were greatly exceeded by the related form *Titanichthys*.

PALÆOSPONDYLUS

THE Old Red Sandstone of Caithness has yielded the skeletons of a small, limbless, fish-like creature (*Palæospondylus*), which was at first regarded as one of the forerunners of the modern lampreys, and has also been compared to the larvæ of lung fishes. Measuring only about a couple of inches in length, these skeletons show a well-calcified skull, while the notochord is surrounded by a series of calcified rings, and the tail has a large fin, of which the supports on the upper side are forked

PALÆOSPONDYLUS

like those of lampreys. The front of the head has a circular opening surrounded with a ring of tentacles, and the researches of Prof. William Sollas have led to the discovery of what appear to be gill-arches and a lower jaw. Living as it did at a time when several groups of fishes were being evolved, this problematic little form may stand at the " parting of the ways," and resemble the ancestors of more than one subclass.

CLASS VI. CYCLOSTOMATA

THE LAMPREY GROUP

Till within recent years both the lampreys and the strange little creature known as the lancelet were generally included among the class of fishes, which was also taken to comprise a number of armoured extinct forms, of which a brief notice is given below. On the other hand, the marine animals commonly termed sea-squirts, but technically known as ascidians or tunicates, together with certain worm-like creatures, were classed with the great assemblage of so-called invertebrates.

Anatomical and palæontological investigations have, however, revolutionised our ideas concerning the creatures in question, with the result that while the lampreys are now separated from the fishes to form a class by themselves in the vertebrate subkingdom, the lancelet and sea-squirts, together with the above-mentioned worm-like creatures, are now regarded as forming a division by themselves, known as the primitive vertebrates or Protochordata, which agree with the vertebrates in the possession of a hollow dorsal central nervous system, of a notochord, and of gill-slits.

ORDER I. PETROMYZONTES—LAMPREYS

As a class, the lampreys and their near allies the hag-fishes are distinguished not only from the fishes, but likewise from all the vertebrates hitherto described, by the absence of a hinged lower jaw, by the single nostril, and by the rasping tongue; there being no limbs or ribs, and a persistent notochord. In the existing members of the group the skeleton is cartilaginous; the skull, as in the chimæroid fishes and some of the sharks, is immovably joined to the vertebral column; and the gills are in the form of fixed pouches, without gill-arches, and with their small external apertures usually opening on the sides of the neck. Anterior in position, and adapted for sucking, the mouth is surrounded by a circular or subcircular lip supported by cartilages. The naked body is provided with median fins, having cartilaginous rays like those of many fishes. Internally, the heart is devoid of an arterial bulb; the intestinal canal is straight and simple; and the reproductive organs discharge into the cavity of the body. The place of teeth is taken in some forms by horny structures, while in others the mouth is completely unarmed.

TRUE LAMPREYS

The true lampreys, of which the sea-lamprey (*Petromyzon marinus*), river-lampern (*P. fluviatilis*), and the small, or Planer's, lampern (*P. planeri*) occur in Britain, are the typical representatives of a family (*Petromyzontidæ*) characterised by the nasal passage terminating in a closed sac behind, without perforating the palate. As in all the other members of the group, the naked body is eel-like in form; but the family is peculiar in that its members undergo a metamorphosis, the young being devoid of teeth, and furnished with a single median fin, while in the adult the sucking-mouth is furnished with horny teeth resting on a soft cushion, and the median fin is divided. In the adult the tongue is furnished with rasping teeth, while above and below the aperture of the mouth there are a series of upper and lower teeth, and the sucker is likewise provided with smaller isolated teeth. Eyes are present in the

adult; and the nostril is situated in the middle of the head. The seven gill-pouches open externally by as many apertures on each side of the neck, but communicate internally with a respiratory tube that opens in front into the throat. The intestine is furnished with a spiral valve, and the eggs are minute.

The true lampreys are characterised in the adult condition by having two dorsal fins, the hinder one continuous with the caudal; and also by the upper series of oral teeth consisting either of a doubly-cusped transverse ridge or of two closely-placed separate teeth; while the teeth on the tongue are serrated. When open, the mouth of lampreys is nearly circular in shape, but when closed forms a narrow slit. They are represented in Britain by three species, which are confined to the coasts and fresh waters of the Northern Hemisphere, ranging as far south as West Africa. The largest of these is the sea-lamprey (*Petromyzon marinus*), which may grow to as much as a yard in length, and is common to Europe, North America, and West Africa. On the other hand, the river-lampern (*P. fluviatilis*), which at certain seasons ascends the rivers of Europe, North America, and Japan in innumerable hosts, is somewhat less than two feet in length, and differs from the last species in being uniformly coloured instead of marbled with black.

(1) SEA-LAMPREY, (2) RIVER-LAMPERN, AND (3) PLANER'S LAMPERN

Still smaller is the small lamprey (*P. planeri*), also known as the pride or sand-piper, which is also common to Europe and Western North America, and scarcely reaches 1 foot in length; its coloration being uniform. The young of this form was long regarded as a distinct genus, under the name of *Ammocœtes*, but its true nature was discovered by watching the transformation into the adult. "The larva," writes Dr. Günther, "requires three or four years for its full development. At first the head is very small, and the cavity of the mouth surrounded by a semicircular upper lip, the separate lower lip being very small. There are no teeth, but several fringed barbels surround the mouth. The extremely small eyes are hidden in a shallow grove; but there is a median single nasal opening, and seven gill-openings, as in the adult. The vertical fins form a continuous fringe, in which the later divisions are more or less distinctly indicated.''

Much has still to be learned regarding the habits of lampreys, but it appears that all ascend rivers for the purpose of spawning, and that some of them pass the whole of their larval conditions in fresh waters. They are all carnivorous, and in the adult state attach themselves by their mouths to the bodies of fishes, from which they rasp off the flesh with their horny teeth; fish being not infrequently met with bearing the scars of wounds thus inflicted, and a salmon has been taken high up in the Rhone with a sea-lamprey tightly adhering to its side. Bathers have also been known to be attacked by the same species. Commonly keeping to the bottom, the sea-lamprey may at times be seen swimming near the surface with a serpentine movement of the body. In the Severn the capture of this species lasts from February to May, while in the Thames the season is May and June; but in the Scottish rivers the lampreys do not ascend till the end of June, remaining till the beginning of August. During the spawning season these fishes excavate furrows in the river-bottoms for the reception of their eggs, and are said to remove impeding stones by lifting them up with their sucking-mouths. Being much exhausted by the function of spawning, at its conclusion they make their way with all speed to the sea.

The river-lamprey was at one time thought to be a permanent inhabitant of fresh waters, but it has been taken in the sea, and it has even been considered that it may undergo its metamorphosis in salt water. Always restricted to low-lying countries, this lamprey may be found alike in rivers, streams, lakes, and marshes, although it only spawns where the water is clear and flows swiftly over a stony bed. During the spawning season, which takes place in March and April, the lampreys acquire a brilliant metallic lustre; while at the conclusion of the function they generally perish. Their chief use is as bait for cod and other fish; for which they are specially adapted on account of the ease with which they can be kept alive.

SOUTHERN LAMPREYS

In the Southern Hemisphere the family is represented by four genera, in one of which there is a single species (*Mordacia mordax*) common to the coasts of Chili and Tasmania; while in a second (*Geotria*) there is one Chilian and another South Australian species. The first of these two genera agrees with the typical representatives of the family in the continuity between the second dorsal and caudal fins, but differs in having two groups of three-cusped teeth above the aperture of the mouth; while in the second genus the two fins above mentioned are separate, and there is a four-lobed plate above the mouth. Some of these lampreys grow to a length of a couple of feet; and in the adults of some or all of them the skin of the throat is so much expanded as to form a kind of pouch.

The third genus (*Exomegas*) appears to be known only by two examples from the Atlantic side of South America, one of which was picked up in the streets of Buenos Aires in 1867, while the second was obtained from the Bay of Monte Video in 1890. With the exception that the dentition is of a peculiar type, very little is known as to the structure of this rare form.

The fourth genus is represented by *Macrophthalmia chilensis*, from the coast of Chili. It will not fail to be noticed that the remarkable geographical distribution of these southern lampreys is paralleled by that of certain fresh-water fishes already described, with the exception that there is no instance among the latter where a species is common to Australia and South America.

ORDER II. MYXINOIDES—HAG-FISHES

THE hag-fishes are distinguished from lampreys by the nasal sac having a posterior passage which perforates the palate; the single external nasal aperture being situated above the mouth at the extremity of the head, which is furnished with four pairs of barbels. The mouth is devoid of lips, the palate is provided with a single median tooth, and there are two comb-like series of rasping teeth on the tongue. The gill-apertures, or aperture, are situated at a considerable distance from the head; and each gill-pouch has a separate passage opening into the œsophagus. The sides of the abdomen carry a row of mucous sacs, and there is no spiral valve to the intestine. The large eggs are invested in a horny envelope, furnished with threads for adhesion. In the true hag-fishes (*Myxinidæ*), of which the common species (*Myxine glutinosa*) is found on the coasts of Europe and North America, there is but a single gill-opening on each side of the abdomen, leading by means of six passages to as many gill-pouches. Another species has been recorded from the extremity of South America; and the range of the genus also includes Japan.

The Pacific hag-fishes (*Bdellostomatidæ*) include a single genus (*Bdellostoma*), of which there are two species from the coasts of the South Pacific; there are six or more gill-openings on each side, each communi-cating by a separate passage with a gill-chamber. All these creatures are marine, and are frequently found deeply buried in the bodies of fishes, more especially members of the cod family, into which they bore for the purpose of feeding on the flesh. They are totally blind, and secrete vast quantities of slime, which seriously interferes with fishing in localities where these creatures abound. Met with in the fjords of Norway at a depth of about 70 fathoms, hag-fishes have been dredged from depths of nearly 350 fathoms.

The affinities of lampreys and hags have been the subject of much debate, and as they are not adapted for preservation in the fossil state, the geological record throws no light on their evolution, though they are doubtless of vast antiquity. In some respects they are undoubtedly primitive, in others (such as the rasping tongue) highly specialised, and they may be degenerate. It has even been maintained that they have descended from ancestors that possessed biting jaws, but this view has not been proved. They are certainly not fishes, and their peculiarities are but expressed by assigning them a special class of their own. Some think that the mouth of ancestral vertebrates was of a different nature from that now possessed, and that the nasal aperture in hags corresponds to the old mouth.

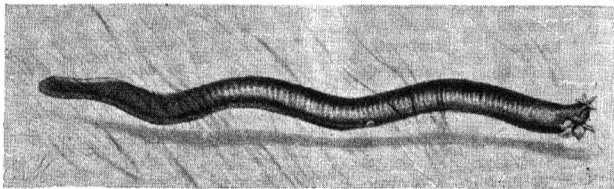

THE HAG-FISH

CLASS VII. OSTRACODERMI
PRIMITIVE ARMOURED VERTEBRATES

A GROUP of remarkable armoured creatures from the Palæozoic rocks formerly placed among the fishes is now regarded as probably constituting a distinct class. Strange in form, and utterly unlike any living animals, these primeval armoured vertebrates are characterised by the great development of the external skeleton, the head and fore part of the body being protected by large bone-like plates. There are no hard structures to the mouth, and there are, at most, but rudimentary indications of arches for the support of limbs, while the notochord is persistent. It is very generally held that these problematic creatures come near the jawless ancestral stock from which higher vertebrates have doubtless taken origin.

PTERASPIS

The simplest of these armoured forms is typified by the genus *Pteraspis* of the Devonian rocks, a partial restoration of which is given in the annexed figure. In these creatures the head and fore part of the body are protected both above and beneath by shields; while the tail, in some cases at least, is scaled. The structure of the shield is curious, each plate consisting of an outer and inner hard layer, between which is a thick stratum of polygonal chambers, perforated by delicate tunnels of the sensory canal system of the skin; all the layers lacking the elements of true bone, and the outer surface being marked with fine concentric striæ.

The eyes are lateral and widely separated, and towards the hinder end of the back shield (which is provided with a spine) there is an aperture on each side for the escape of water from a gill-cavity. Although nothing is known as to the form and structure of the nose and mouth, the nostrils must evidently have been placed near the mouth on the under surface of the head. A pit between the eyes probably marks the site of a rudimentary median eye; and the structure of the under surface of the shield indicates the presence of separated gill-pouches, which were probably supported by arches.

CEPHALASPIS

The next family of the group is typified by the genus *Cephalaspis*, in which the front shield appears to be

PARTIAL RESTORATION OF PTERASPIS

confined to the head and gill-region, and consists of a single piece, rounded or pointed in front, abruptly truncated behind, and with the rounded margin bent inwards below to form an ornamental flattened rim. Of the triple-layered shield, the inner layer is bony, the thick middle one solid, although traversed by a network of blood-vessels, while the upper one is tuberculated and tooth-like in structure.

The eyes are placed close together in the middle of the shield, the nostrils must have had much the same position as in *Pteraspis*, and at the back of the shield there occurs on each side a small flap, which must be regarded as a gill-cover. Immediately behind the shield begins the ordinary scaling of the body, without any signs of arches for the support of limbs. Paired fins appear, indeed, to be totally absent, although a dorsal and a caudal fin, stiffened by little elongated scales in place of rays, are present. The large, deep, quadrangular scales covering the body form a series of interlocking rings, doubtless corresponding in the living state to the underlying muscle-plates of the body.

PTERICHTHYS

The third modification of the group, as represented by the Devonian *Pterichthys*, agrees in the general structure of the shield with certain members of the last section, in which there is no dividing line between the head-shield and the united scales of the body. The head is, however, sharply defined from the body; and the armour, instead of being simple, consists of a number of overlapping plates arranged symmetrically.

An important point of distinction from all the preceding forms is to be found in the presence of a pair of hollow limb-like pectoral appendages, jointed near the middle. A small movable plate between the eyes seems to have lodged a median eye; another movable plate on the cheek appears to represent the gill-cover; and a pair of loose jaw-plates on the lower surface of the front of the head, in some forms at least, are finely toothed on the hinder border; but nothing definite is known with regard to the nature of the nose, mouth, and jaws. The arrangement of the median fins is generally similar to that obtaining in the second family.

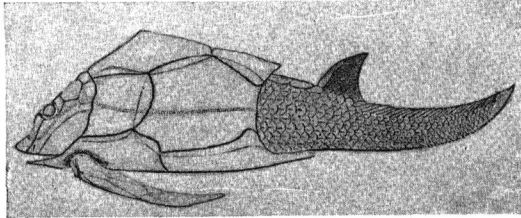

RESTORATION OF PTERICHTHYS

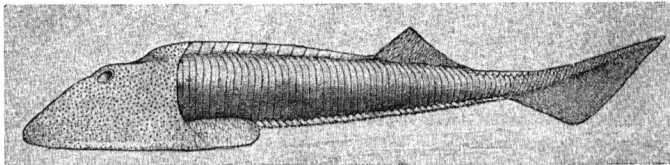

RESTORATION OF CEPHALASPIS

DIVISION PROTOCHORDATA
CLASS I. CEPHALOCHORDA—LANCELETS

WITH the semitransparent little creatures known as lancelets, forming the only family (*Branchiostomatidæ*), with two genera, of the class to which they belong, we leave the vertebrates proper and come to the lower group of protochordates ; all of which retain the three essential vertebrate features mentioned on page 1779. First described by the Russian naturalist Pallas in 1778, from a specimen captured on the Cornish coast, the common lancelet (*Branchiostoma lanceolatum*) was referred to the Mollusca, where it remained till 1834, when it was rediscovered by Costa, on the Neapolitan coast, who gave it the name of *Branchiostoma*, and placed it among the fishes, in the neighbourhood of the lampreys and hags. It was again discovered by Yarrell in 1836, who assigned to it the title of *Amphioxus*, and was the first to recognise the existence of a gelatinous notochord.

The upper figure of the illustration shows the pointed extremities of the body, and also a number of chevron-shaped lines, with their angles directed forwards, these being the partitions dividing the lateral muscles into a series of segments. It

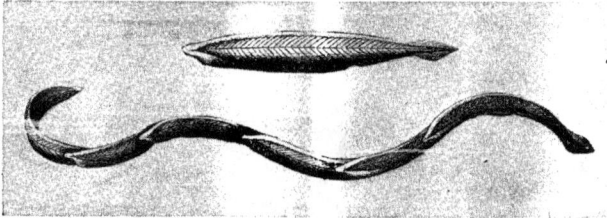

COMMON LANCELET, SINGLY, AND IN A CHAIN

is due to this segmented structure that the lancelet is enabled to swim so speedily as it does, its progress being effected by serpentine movements of the body. Paired fins are wanting ; but the back is provided with a continuous dorsal fin, expanded posteriorly into a caudal fin, which passes below into a ventral fin. In front of this a lateral fold on either side is continued forwards to join the ring of feelers, or tentacles, growing from the margin of the hood-like expansion of skin which surrounds the mouth.

The notochord extends to the anterior and posterior extremities of the body, reaching beyond the muscle-plates, and in advance of the front extremity of the overlying nerve-cord ; the latter feature being peculiar to the lancelet (hence the name "Cephalochorda"). An aperture distant about two-thirds of the whole length from the head, and opening in the middle line of the lower surface of the body, is the outlet of a large cavity, or atrial chamber, surrounding most of the internal organs, and especially the large pharynx ; and the vent is situated high up on the left side, near the hinder end of the body. The reproductive organs, which form oval structures lying below the muscle-plates, differ from those of the vertebrates in that they consist of a large number of perfectly distinct chambers.

In connection with the fins, it should be observed that, except at its two extremities, the dorsal fin is supported by a series of gelatinous rays, each lying in a chamber of its own ; while the ventral portion of the caudal fin has a paired series of similar supports. In young and transparent examples, the pharynx, or that portion of the alimentary tract immediately behind the mouth, is distinctly visible through the walls of the body, and can be seen to be perforated on each side by a very large number of vertical gill-slits, opening into the atrial chamber. In the living creature an almost

continuous current of water is drawn, for the purpose of breathing and feeding, through the mouth into the pharynx, whence it escapes by means of the gill-slits into the atrial chamber, from which it is discharged through the pores.

Unlike even the lowest vertebrates, lancelets have no cartilaginous skull ; the only solid structure in the head taking the form of a ring of cartilage in the hood surrounding the mouth, which gives off a series of processes for the support of the feelers. Although paired eyes, as well as organs of hearing, are totally wanting in these strange little creatures, a pigment-spot at the front end of the nerve-tube represents a median eye ; behind which is a small nasal pit on the left side, communicating in the larva by means of a small pore with the front of the nerve-tube. With regard to the other soft parts, it will suffice to mention that the anterior extremity of the nerve-tube is not expanded to form a true brain ; and that the heart is represented merely by a series of pulsating dilatations of the great blood-vessel, the blood itself being devoid of colour.

Lancelets are represented by a considerable number of species, which may be arranged in two genera, mainly distinguished, as mentioned below, by the arrangement of the reproductive organs. Essentially littoral forms, inhabiting shallow water, especially where the bottom is sandy, these creatures have an almost universal distribution on the temperate and tropical coasts, although they are often curiously local. The European form has been recorded from Scandinavia, Heligoland, the English Channel, France, the Mediterranean, and Chesapeake Bay. Others species occur on the Atlantic and Pacific shores of North and South America, as well as on the coasts of Australia, Japan, Ceylon, and the Fiji Islands.

Dr. A. Willey remarks that the lancelet "possesses an extraordinary capacity for burrowing in the sand of the seashore or sea-bottom. If an individual be dropped from the hand on to a mound of wet sand, which has just been dredged out of the water, it will burrow its way to the lowest depths of the sand-hillock in the twinkling of an eye. Its usual *modus vivendi* is to bury the whole of its body in the sand, leaving only the mouth with the expanded buccal cirri [tentacles] protruding. When obtained in this position in a glass jar, a constant inflowing current of water, in which food-particles are involved, can be observed in the neighbourhood of the upstanding mouths. The food consists almost entirely of microscopic plants (diatoms, desmids, etc.) and vegetable débris. Occasionally after swimming about for some time it will sink to the bottom, and there recline for a longer or shorter period upon its side on the surface of the sand. When resting on the sand, it is unable to maintain its equilibrium in the same position as an ordinary fish would do, but invariably topples over on its side—indifferently, the right or left."

CLASS II.　UROCHORDA—SEA-SQUIRTS

EXTERNALLY, scarcely any creatures are more unlike the lancelet than those fixed marine animals commonly known as sea-squirts, and technically as tunicates, or ascidians. Nevertheless, the relationship is probably closer than that existing between the former animal and the larva of a lamprey, in spite of the much greater external resemblance between the two latter. It is, however, when we dissect a sea-squirt that we meet with structures recalling certain features in the anatomy of the lancelet ; while to find evidence of the chordate affinities of the former we have to go back to its larval condition.

As there are both simple and compound fixed ascidians, so there are two similar types among the pelagic forms ; but some of the latter are complicated by an alternation of generations, the one generation being a simple form, whereas in the other generation the units are aggregated into chains, as shown on page 1765 of the creatures known as salpæ. Among the compound fixed types the colonies, as they are termed, consist of a number of individuals produced by budding from a single parent stock ; such colonies frequently attaining very large dimensions, and being remarkable for their brilliant coloration, although in other cases they merely form thin incrustations on the surface of various marine objects. Other forms, on the contrary, are merely connected at their bases by a common creeping root-like base, from which new buds are from time to time given off, the individuals being otherwise free.

Externally, a simple sea-squirt resembles a leather bottle with two spouts ; these spouts forming funnel-shaped projections, one of which—generally situated at a higher level than the other—takes in water, which is discharged from the second. The whole organism is invested in an external tunic, varying much in structure, but being frequently warty, and generally opaque, although in the salpæ it is transparent. A remarkable feature connected with this outer tunic is that it contains a substance—cellulose—identical in composition with that forming the cell-walls of plant-tissues. On cutting through the outer tunic, we come to an underlying muscular tunic, forming the true body-wall, and consisting externally of an epidermis underlain by interlacing muscular fibres.

In the illustration, a indicates the inhalent and b the exhalent orifice of this inner tunic. On cutting into the inner tunic, we find a large so-called atrial cavity, enclosing to a great extent the viscera, and communicating with the exterior by means of the exhalent orifice. The inhalent orifice, or mouth, communicates, on the other hand, directly with the exceedingly large pharynx or branchial chamber (resembling that of the lancelet in many details), which extends nearly to the hinder end of the body, and is perforated by a vast number of gill-openings, through which the water taken in at the mouth passes into the atrial chamber. Instead of passing directly into the latter chamber with the water, the food-particles are caught up in a mass of slime, and carried round the base of the mouth-tube until they reach the entrance to the œsophagus, which lies near the hinder end of the

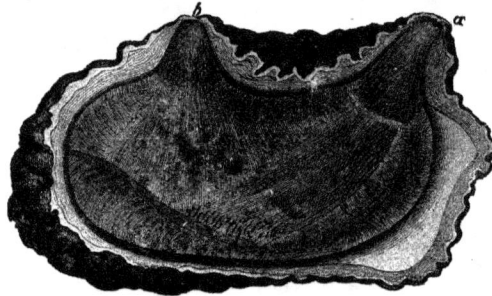

LEATHERY SEA-SQUIRT WITH ONE SIDE OF THE OUTER TUNIC REMOVED

a, Inhalent, and b, exhalent orifices

dorsal surface of the branchial chamber. Lancelet and larval lampreys feed in much the same way. Hence the food passes into the stomach, and along the intestine, which forms a U-shaped curve turned away from the dorsal aspect ; the vent opening on the same aspect into the atrial cavity below the exhalent orifice.

With regard to the nervous and circulatory system, it will suffice to say that there is a large nerve-ganglion imbedded in the tissue of the inner tunic, and lying on the dorsal surface of the body between the inhalent and exhalent orifices ; and true blood-vessels are wanting, the blood merely flowing through a series of spaces in the muscles and other tissues of the body and between the viscera, and the heart forming a dilated tube. Unlike the higher chordates, all the ascidians are hermaphrodite ; the reproductive organs frequently lying within the loop of the intestine, and discharging into the atrial cavity alongside of the vent. A remarkable physiological feature of the group is to be found in the periodical reversal of the action of the heart ; the blood being driven for a certain time in one direction, after which the heart makes a short pause and then propels it in an opposite course.

All ascidians go through a free-swimming larval stage, during a part of which they develop a tail containing a notochord (hence the name "Urochorda") and nerve-tube. Generally the larval condition lasts but a short time ; and this may be the reason for the development of the tail, as a powerful swimming organ would seem to be essential in order to enable the creature to reach a spot suitable for its permanent existence. During its development a groove makes its appearance on one surface of the ascidian embryo, the large cells on the side of which grow inwards so as to enclose a tube, corresponding to the nerve-tube of vertebrates, beneath which is the notochord. When of an oval shape, and while still contained in its investing membrane, the embryo assumes a ventral curvature, and at the same time produces a long tapering tail, which eventually becomes coiled round it.

In addition to certain other structures, this outgrowing tail includes the nerve-tube and the notochord, and in some forms contains the only muscles developed at all. Subsequently a rudimentary brain, corresponding to a simple structure in the lancelet, makes its appearance, and also an unpaired eye. After certain other changes, among which the development of a stomach and intestine are included, the larva is ready to burst from its membranes, which it does by spasmodic jerkings of the tail ; and it thereupon starts on a free-swimming existence. Before long the cellular structure of the notochord in the tail begins to disappear by the formation of vacuities ; and eventually the whole structure becomes filled with gelatinous matter. After a brief free existence it fixes itself by its front end to some submarine object, with the tail stretched out and generally motionless. In a short time this appendage begins to shorten, and finally disappears. Ascidians are probably the degenerate descendants of permanently free-swimming forms provided with a complete notochord and nerve-tube.

ORDER ASCIDIACEA—TYPICAL ASCIDIANS

ACCORDING to the classification adopted by Professor Herdman, the tunicates may be divided into three orders, the first of which is known as the Ascidiacea. This group includes both fixed and pelagic, simple and compound types, none of which are provided in the adult state with a tail and retain no trace of a notochord; the free-swimming forms constituting colonies, and the simple types being generally fixed. The outer tunic is permanent and well developed, generally increasing with the age of the individual; and the muscular structure of the inner tunic takes the form of an irregular network, and never of hoop-like bands.

The walls of the large branchial chamber are perforated by numerous slits, opening into a single atrial cavity, which in turn communicates with the exterior by means of the exhalent aperture of the tunic, the vent opening into the atrial cavity. Many of the forms reproduce their kind by budding, and in most the sexually-produced embryo develops into a tailed larva.

The order is divided into three subordinal groups, of which the first—Ascidiæ Simplices—includes fixed (rarely unattached, but never free-swimming) and generally solitary forms, which very rarely reproduce by budding. When colonies occur, each of their individual members has a distinct outer tunic of its own, the whole society never being buried in a common investing mass. Four families are contained in this suborder, each represented by a large number of genera.

Omitting mention of the first family, we may take as an example of the second (*Cynthiidæ*) the genus *Microcosmus*, of which specimens are shown in the figures. As a family, these sea-squirts are characterised by being usually attached, and sometimes stalked, although rarely free. The outer tunic is generally membranous or leathery, but occasionally cartilaginous or covered with sand; while the inhalent aperture is usually, and the exhalent aperture invariably, provided with lobes, meeting together at the centre. The branchial chamber is longitudinally folded, with its gill-slits straight, and the tentacles may be either single or compound. In the figured genus the body is attached and sessile, and the tunic is thin, leathery, and tough, both its apertures having four lobes and the tentacles being compound.

As an example of the family *Ascidiidæ* we may take the well-known *Phallusia mammillata*, from the seas of North-Western Europe and the Mediterranean, which is the sole representative of its genus. In the family to which it belongs, the body is attached and usually fixed, although rarely stalked; the inhalent aperture generally has eight and the exhalent six lobes; and, as a rule, the outer tunic is either gelatinous or cartilaginous, although it may be horny. The branchial chamber is devoid of folds, with the gill-slits either straight or curved; and the tentacles are simple and thread-like. In the figured genus the body is erect and attached, and the outer tunic of a cartilaginous nature, its surface being

mammillated in a very characteristic manner. It may be mentioned here that all the simple sea-squirts of this group, when touched, emit a jet of water, and that some of them, like the one here represented, are used as articles of food. To the same subfamily as *Phallusia* also belongs the extensive genus *Ascidia*, in which the outer tunic is soft and flexible instead of being cartilaginous.

A totally distinct subfamily is, however, indicated by the remarkable deep-sea genus *Hypobythius*, of which the two known species were obtained at depths varying from six hundred to two thousand nine hundred fathoms, during the voyage of the Challenger. Here we find the cup-shaped or pear-like body attached, by a longer or shorter stem; while the apertures are circular and not closed by lobes. The outer tunic is cartilaginous, but soft and thin, although thickened in places to form plates. The internal longitudinal bars usually found in the branchial chamber are wanting in this genus; the gill-slits are small and irregularly placed; and the viscera form a compact irregular mass on the dorsal side of this chamber. In the species here figured (*H. calycodes*), which is from the North Pacific, the stem is of great length, and the outer tunic thickened so as to form a number of nodules or plates; but in the South Atlantic form (*H. moseleyi*) the stem is much shorter, and there is only a single plate, situated on the dorsal side. Of very large dimensions, these deep-sea ascidians are decidedly the most beautiful members of the class, and present some resemblance to the glass-sponges.

A totally different type of structure is presented by the last family (*Clavelinidæ*) of the suborder, in which the body of each individual is attached by its posterior end, and usually by means of a stalk, to a creeping basal stolon, or common mass, from which young individuals are produced by budding. The outer tunic, which is usually thin and transparent, is in most cases gelatinous, although occasionally cartilaginous; and its circular apertures are but seldom distinctly lobed. Folds are wanting in the branchial chamber, but longitudinal bars may be present, although these lack the papillæ found in the preceding family; and the gill-slits are straight. The tentacles resemble those of the last family in their simple, thread-like form; but the digestive tract is usually extended behind the branchial chamber to form an abdomen. In addition to the ordinary sexual reproduction, colonies may be formed by budding from the common stolon. Ten genera are included in the family, from among which the typical *Clavelina* is selected for illustration. Here the body is elongated and club-shaped, and is attached to a delicate, branched, creeping stolon, from which arise the buds. The thin outer tunic is gelatinous or cartilaginous, with its circular apertures devoid of lobes. The inner tunic is likewise thin, with its muscles mainly longitudinal; and the intestinal tract is extended to form a well-

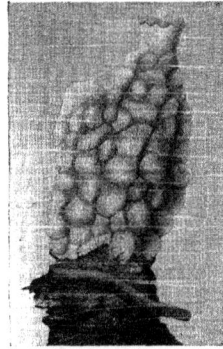

PHALLUSIA

MICROCOSMUS

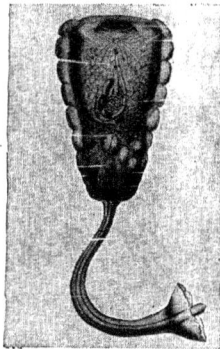

HYPOBYTHIUS

marked abdomen. In its restricted sense, the genus includes only half a dozen small species from North-Western Europe and the Mediterranean, the one here figured (*C. lepadiformis*) being characterised by the yellow or brown lines on the region known as the thorax.

The second suborder of the typical sea-squirts—A s c i d i æ Compositæ—i n c l u d e s fixed forms which reproduce by buds so as to constitute colonies in which the individuals are buried in a common investing mass, and thus possess no separate tunics. The group includes seven families ; and Professor Herdman remarks that as many of these have originated independently from simple forms, the whole assemblage is to a certain extent an artificial one.

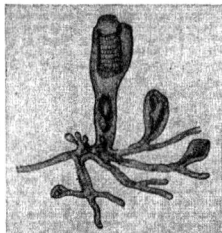

CLAVELINA
A creeping ascidian

In the first family (*Botryllidæ*) the colonies usually form thin incrustations on seaweeds or stones, although they occasionally take the shape of thick fleshy masses, the individuals being arranged so as to form either circles or ellipses, or in branching lines. The common apertures of discharge are distinct, and usually furnished with lobes ; the individual units are short, and show no division of the body into regions ; and the outer tunic, which is usually soft, is traversed by numerous vessels with large terminal knobs. Internal longitudinal bars are present in the large and well-developed branchial chamber, in which the gill-slits are numerous ; and the simple tentacles do not exceed sixteen in number.

Budding may take place either from the sides of the units constituting the colony or from the vessels in the tunic. Among the five genera constituting the family, the typical and well-known *Botryllus* has the colony thin and incrusting, with the individuals arranged in a circular manner, whereas in *Botrylloides* they form ellipses or branching lines. In the figured species of the latter (*B. albicans*), from North-Western Europe and the Mediterranean, the colony is pure white in colour, but in some it is purple with yellow or green markings, and in others yellowish green. As an example of forms in which the colony is thick and massive, we may cite the genus *Polycyclus*.

AMAROUCIUM
A compound ascidian in winter condition

Passing over the second family of the suborder, we come to the third (*Polyclinidæ*), in which the colony is usually massive, being sometimes incrusting, but in other cases lobed, or even stalked. The arrangement of the individuals is highly variable, and the common apertures of discharge are usually inconspicuous. Although of an elongated form, the individuals usually differ from those of the family last noticed by being divided into three regions, the inhalent aperture having six or eight lobes, while the exhalent is frequently provided with a tongue-like process. The gelatinous or cartilaginous outer tunic is frequently stiffened by imbedded grains of sand ; and the branchial chamber is usually small and poorly developed, with minute gill-slits and no internal longitudinal bars. The tentacles are small and not numerous ; the digestive tract is extended posteriorly to a considerable distance beyond the extremity of the branchial chamber ; and budding takes place from the end of the postabdominal region.

The family is represented by well nigh a score of genera, among which *Amaroucium* may be selected as an example, on account of its numerous species. Here the colony is massive, being sometimes lobed or stalked ; the mode of arrangement is usually compound and irregular, and the individuals are elongated, with six lobes to the inhalent orifice, and the postabdominal region elongated. The species here figured (*A. densum*) is from North-Western Europe, and is characterised by its greyish yellow colour, and the abundance of sand in the tunic ; but other kinds may be black, orange, rosy red, or white. In our figure, *a* shows fully active individuals ; while those in the outer ring, indicated by *b*, assume a kind of torpid condition during the winter, but give rise to fresh buds in the spring.

Omitting mention of the remaining families of the group just considered, we come to the third and last suborder of the typical ascidians—namely, the phosphorescent ascidians—*A. luciæ*. These are represented solely by the genus *Pyrosoma*, which is thus the only member of the family *Pyrosomatidæ*. These ascidians

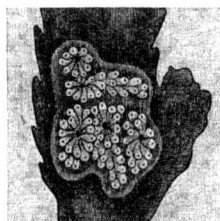

BOTRYLLOIDES
An incrusting ascidian on seaweed

are free-swimming, pelagic forms, reproducing by buds in such a manner as to form colonies in the shape of a sac, such colonies sometimes attaining huge dimensions.

In the tube thus formed the constituent individuals are imbedded in such a manner that all their inhalent apertures open on its outer surface, while their exhalent orifices are situated within the cylinder, the mouth of the sac forming the common discharging aperture. The apertures of the units are not lobed ; and the outer tunic is gelatinous and transparent, containing no hard spicules, but provided with numerous minute cells. The branchial chamber is well-developed, and the tentacles are simple. The first four individuals of the colony grow in the form of buds from a rudimentary, sexually developed larva, the subsequent increase taking place by budding from a ventral posterior stolon.

The genus is represented by only four species, in one of which (*P. elegans*) the individuals form regular oblique rows in the walls of the sac, while in the other three they are arranged irregularly. The largest of all is *P. spinosum*, from the Atlantic, in which the total length of the colony may be upwards of four feet, this species being distinguished by the surface of the sac being provided only with short, sharp spines, instead of with large processes of the tunic. It is to these ascidians that the most beautiful phosphorescence of

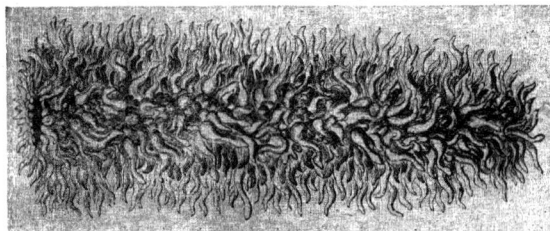

A PYROSOMA COLONY

tropical seas is due, each colony, when stimulated by a touch or shake of the water, giving forth a brilliant ball of bluish light, which lasts for several seconds, as the organism floats along beneath the surface, and then suddenly disappears.

ORDER THALIACEA—NON-LUMINOUS PELAGIC ASCIDIANS

WITH the exception of the family just mentioned, and also of a second one which constitutes another order, the present ordinal group—termed the Thaliacea—includes the whole of the free-swimming pelagic representatives of the class. Either simple or compound in structure, these ascidians lack both a tail and a notochord in the adult, but have a persistent outer tunic, which may be either feebly or fully developed. In the inner tunic the muscles are arranged in the form of more or less nearly complete circular bands, the contraction of which forms the motive agency of the creatures. The branchial chamber has either two large openings, or a number of smaller gill-slits, leading to a single atrial cavity, the latter communicating with the exterior by the exhalent aperture, and the vent opening within it. In all the members of the group an alternation of generations takes place, and this may be further complicated by the individuals of a single generation being unlike one another. During one period of existence temporary colonies may be formed, but these never increase by the budding of the constituent units, which eventually separate from one another and disperse.

The well-known salpæ form a suborder—Hemimyaria—characterised by the formation of temporary colonies in the sexual generation, and represent a family (*Salpidæ*) distinguished by the muscular bands of the inner tunic being incomplete on the lower surface of the body. Pelagic in habit, and transparent in structure, salpæ have been not inaptly compared to a barrel with both ends knocked out; and

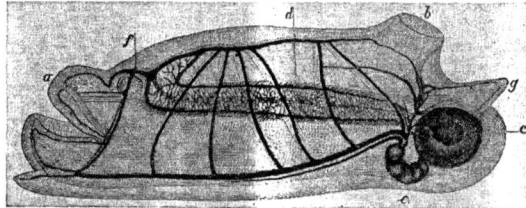

AN INDIVIDUAL OF A CHAIN-SALPA
a, inhalent, and *b*, exhalent, orifice ; *d*, gill ; *c*, *e*, viscera ; *f*, eye (?) ; *g*, pedicle of union

really consist of little more than a huge pharynx, swimming through the water, and taking in large mouthfuls of the same at each contraction of its muscles. Through the hollow, to below the hinder aperture, runs obliquely a rod-like gill (*d*) from above the mouth, although this is too narrow to interfere with the free flow of the water ; while the lower surface of the interior of the creature |is furnished with a ciliated slime-secreting band, corresponding to the structure known in other ascidians and the lancelet as the endostyle. It may here be well to mention that in the lancelet the structure in question is an elongated gland situated at the base of the pharynx, and against which the ends of the gill-bars abut. The only part of the salpa that is not transparent is the thick mass of viscera (*e*, *c*) at the hinder end of the body ; while the muscular bands, by the contraction of which the water is driven through the barrel, may be compared to the hoops of the latter. Externally, the whole animal is invested with a thick, tough, transparent tunic ; and in some forms there are two tail-like appendages to the hinder part of the body.

Such is the structure of a salpa ; but there are two generations in the life of these creatures—namely, the simple form, and the chain-like or aggregate form, the first being shown in the upper, and the second in the lower figure of the picture on page 1765. It will be observed that in the chain the individuals are attached to one another by their upper and lower surfaces, and thus have these two apertures free ; and when taken from the water the whole chain, which is several feet in length, can be easily resolved into its component units. The specimen represented in the figure below is one of these detached units from a chain, the projection marked *g* being for the purpose of attachment to the neighbouring individual.

The solitary salpa is born from an egg carried within the body of one unit of the aggregate form, the embryo being nourished by means of a placenta from the blood of the parent. On the other hand, the chain-salpæ are produced assexually by budding from a stolon within the body of the solitary form. In the chain-salpa the eggs arise, however, at an exceedingly early period of its development, with the curious result that three generations are present at one time in a single individual.

Thus a solitary salpa has within it the buds of a chain-salpa, the units of which may each contain eggs which will ultimately develop into the next solitary form. And, as a matter of fact, in a solitary salpa the germ cells of the embryo of the next solitary form are actually visible before the development of the stolon which is to give rise to the chain-form. As the stolon forms in the body of the latter, it includes within it the mass of germinal cells ; and while the former elongates to form the chain of units, the mass of germ-cells likewise lengthens, with the result that a single egg-cell is shut off in each unit of the chain. Simple salpæ vary in size from a quarter of an inch to upwards of eight inches ; and in some parts of the ocean-surface are met with in incalculable swarms. The food of these creatures consists of minute marine organisms, both animal and vegetable. In swimming, chain-salpæ progress by an undulating, snake-like movement.

The second suborder—Cyclomyaria—of the free-swimming, non-luminous ascidians takes its name from the muscular bands of the inner tunic forming perfect rings, and is typically represented by the genus *Doliolum* of the family *Doliolidæ*. The life-history is complicated by polymorphism, the tailed larva developing into a sexless form, the buds from which give rise to nutritive units, fostering units, and reproductive units. In the typical genus all the muscles form encircling hoops, and the three forms of the sexual generation occur together on one stolon, or outgrowth ; but in *Anchinia* there are only two complete muscular rings, and the three forms of the sexual generation are produced successively.

ORDER LARVACEA—TAILED ASCIDIANS

THE free-swimming form known as *Appendicularia* is the type of the third and last order—Larvacea—of the class, all the members of which are characterised by the possession in the adult state of large tail-like appendages, furnished with a skeletal axis. These creatures, which are of minute size, have not undergone the degeneration so noticeable in the adult of the other tunicates, and thus correspond much more closely to the larval stage of the latter. A curious feature is the rapid production of a temporary outer tunic, which may be shed at any time, and replaced by a second one. There is no separate atrial cavity ; and the branchial chamber is simply an elongated pharynx, with two openings on the lower surface, which correspond to the gill-slits, and are well furnished with cilia. The nervous system consists of a large ganglion placed in the anterior part of the dorsal surface, followed by a long cord, provided with smaller ganglia, and extending backwards over the intestine to reach the tail, where it runs along the left side of the skeletal axis. The intestine itself is situated behind the branchial chamber, and the vent opens on the inferior, or ventral, aspect of the body in advance of the gill-slits. Neither budding, metamorphosis, nor alternation of generations takes place ; and the reproductive organs are situated at the hinder end of the body. The group comprises only the single family *Appendicularidæ*, which contains five genera.

These curious little free-swimming creatures very probably resemble the original ascidian stock more than any other forms, for they are undoubtedly less degenerate. It has also been suggested that they have descended from fixed forms in which the adult stage has been eliminated, and are " permanent larvæ " that have acquired reproductive organs, like the axolotl among amphibians.

CLASS HEMICHORDA

THE last, and at the same time the lowest, group of the protochordates is typically represented by the marine *Balanoglossus*. Living buried in the sand or mud of the seashore, these worm-like creatures exhale a peculiar odour resembling that of iodoform, and secrete a copious supply of slime, to which adhere particles of sand, thus forming a protective tube for their bodies. "At the front extremity of the creature," writes Dr. Willey, " there is a long and extremely sensitive proboscis, which is capable of great contraction and extension, and is in the living animal of a brilliant yellow or orange colour. Behind the proboscis follows a well-marked collar-region, consisting externally of a collar-like expansion of the integument, with free anterior and posterior margins overlapping the base of the proboscis in front and the anterior portion of the gill-slits behind. In the ventral middle line, at the base of the proboscis, and concealed by the collar, is situated the mouth. Following behind the collar is the region of the trunk or body proper, which, in the adult of some species, reaches a relatively enormous length, even extending to two or three feet. The ectodermal covering of the body consists in general of ciliated cells, among which are scattered unicellular mucous glands ; the cilia, however, appear to be more prominent on the proboscis than elsewhere.

"In the region of the trunk, which immediately follows upon the collar-region, there are a great number of paired openings on the dorsal side of the body placing the anterior portion of the digestive tract in communication with the outer world. These are the gill-slits, and they are arranged strictly in consecutive pairs to the number of upwards of fifty in the adult. In their structure, and more especially in the possession of tongue-bars, they bear a remarkable resemblance to the gill-slits of the lancelet. This is particularly striking in young individuals."

On dissection, a rod-like structure, which arises as an outgrowth of the alimentary canal above the mouth, is seen projecting into the interior of the proboscis ; and this rod has been identified with the vertebrate notochord. Above this rod, and extending farther back, is a dorsal nerve-cord, corresponding to the vertebrate nerve-tube, and having, as in the latter, a central canal, at least during the earlier stages of growth. Some distance behind the notochord this nerve-tube gives off a descending branch, connecting it with a similar cord lying on the ventral aspect of the animal.

We thus have evidence of the existence in this strange worm-like creature of three essentially vertebrate characteristics—namely, gill-slits, a notochord, and a nerve-tube ; and it is not a little remarkable that while in the sea-squirts the notochord is found in the transitory tail, in *Balanoglossus* it is situated in the anterior extremity, where it extends some distance in advance of the mouth. Careful examination has also shown that the tornaria-larva of one species of *Balanoglossus*

also possesses an endostyle (see page 1786), comparable to that of the lancelet and sea-squirts. That the creature under consideration is closely allied to the other protochordates, and thus to the vertebrates, may be considered fairly certain ; but there are also indications of affinity with non-chordates.

In the first place, while certain species of *Balanoglossus* pass through the earlier stages of their existence without undergoing a metamorphosis, in other kinds such a transformation takes place ; the young making its first appearance in the world in the form of what is known as a tornaria-larva, or one closely resembling that of a star-fish. And it is held by competent naturalists that this resemblance must be indicative of some kind of genetic relationship between *Balanoglossus* on the one hand, and star-fish and sea-urchins on the other. In the second place, there are not wanting indications of affinity with the so-called nemertine worms, described in the sequel ; these resemblances presenting themselves in the structure of the outer layer of the skin, the presence of a proboscis (kept retracted in the nemertines), as well as in regard to the nervous system, the reproductive organs, and the alimentary canal.

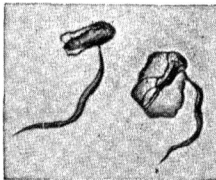

APPENDICULARIA
One is forming its house

OTHER FORMS

Among the other forms included in the Protochordates two are respectively known as *Cephalodiscus* and *Rhabdopleura*, and bear the same relation to the last as is presented by the compound ascidians to the lancelet. Both these curious creatures are fixed forms, living in societies, reproducing their kind by means of buds, and having a U-shaped, instead of a straight, intestine. Both are likewise deep-water creatures, the former having been dredged in the Straits of Magellan at a depth of two hundred and forty-five fathoms, while the latter has been taken off the Shetlands in ninety, and off the Lofoten Islands in two hundred fathoms.

Extremely minute in size, *Cephalodiscus* lives in colonies, the individuals wandering about the tubes of a common house, the walls of which are composed of a gelatinous material, covered with spiny projections and perforated by numerous apertures for the free circulation of water. The mouth is overhung by a large shield-like plate, surmounted by the row of plume-like tentacles ; while on the side of the body is a pedicle from which grow the buds ; locomotion being probably effected by means of this pedicle and the mouth-plate. The latter contains a short notochordal rod ; and there is a single pair of gill-slits opening from the pharynx, water being passed into this from the mouth by the action of the tentacles.

In the allied genus *Rhabdopleura* the individuals which go to form a colony are connected with one another by means of a common stem, representing the remnants of their original contractile stalks. Each polype has but a single plume-like tentacle ; and the buds arising from the soft part of the common stem never become detached.

INVERTEBRATES

CHARACTERS & CLASSIFICATION OF BACKBONELESS ANIMALS

THERE are three different ways of studying in succession the various groups of the animal kingdom. One is to begin with the lowest forms and work upwards, which is the logical method, because it follows the course of evolution. It is, however, attended with the disadvantage that the simplest forms of life are mostly unfamiliar to all but professional naturalists. The second method is to begin at the top and work downwards, the only advantage of which is that it begins with mammals, a group in which all are more or less interested. The third method is a compromise, and is adopted in this work. So far, the plan of working from above downward has been adopted, and all the various groups of backboned animals and their allies have been passed in review. Fishes, however, as a class, were treated according to the second method, following to some extent the probable course of evolution, for in no other way would it have been possible to discuss intelligently the adaptations to aquatic life exemplified by members of the class.

In dealing with backboneless, or invertebrate, animals, we propose to reverse the plan pursued at starting, and, instead of beginning with the highest and most specialised groups, we shall begin with the lowest, which have an interest of their own, and this will be a convenient point at which to do something by way of review, while at the same time the opportunity will be afforded of considering some general principles.

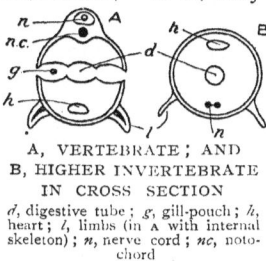

A, VERTEBRATE; AND B, HIGHER INVERTEBRATE IN CROSS SECTION

d, digestive tube; *g*, gill-pouch; *h*, heart; *l*, limbs (in A with internal skeleton); *n*, nerve cord; *nc*, notochord

The backboned, or chordate, animals are taken as the highest subkingdom, not merely on account of their high degree of specialisation—for in this respect they are rivalled by insects—but also because they include human beings and other forms distinguished by a relatively high degree of mental attainment.

All members of the subkingdom are distinguished by the possession of three structural characters: (1) A hollow dorsal central nervous system; (2) an elastic rod, the notochord, which underlies part or all of this nervous system, either permanently or during some stage in the development; (3) gill-slits on the side of the throat for some part of the life-history, if not for the whole of life.

The hollow character of the central nervous system is due to its method of development. It arises as a thickened part of the outer layer of the skin (epidermis), which is folded off from the exterior, and therefore assumes a tubular character.

The notochord apparently came into existence as an adaptation to swimming, serving to stiffen the body and give attachment to lateral muscles, producing undulations from side to side. It develops as a thickened strip of the dorsal wall of the digestive tube. In the large majority of the chordate forms (the term "chordate" has reference to the possession of a notochord) this structure is more or less completely replaced by a backbone, and it is to such animals that the name vertebrate is properly applied. It may be well to remember that lampreys and their allies, together with the shark-like fishes, possess a backbone which is mainly or entirely composed of gristle or cartilage and fibrous tissue.

The gill-slits, which persist throughout life in aquatic chordates, are an adaptation to breathing the oxygen dissolved in water. This is taken in at the mouth and passed out through the gill-slits, typically passing over delicate outgrowths known as gills contained in gill-pouches. A gill is richly provided with blood-vessels, separated only by a thin membrane from the water that bathes it, so that dissolved oxygen can diffuse into the blood, and the waste product, carbon dioxide, diffuse out. The origin of gill-pouches and gill-slits is not altogether easy of comprehension. It may be said, however, that they undoubtedly came into existence in ancestral jawless forms which fed on minute organisms and organic particles contained in the surrounding water, these being brought into the mouth by ciliary action, a common means of setting up currents which will be elsewhere described.

This method of feeding is still exemplified by the most primitive chordates, such as the lancelet and sea-squirts. It involves the taking in of a relatively large amount of water, and the gill-pouches and gill-slits probably arose as a means of getting rid of the superflous liquid. If so, they were not in the first instance evolved as breathing organs. But in all cases, among aquatic forms, where the covering or lining of the body is thin, some breathing takes place, provided there are blood-vessels or blood-spaces near the surface, and a case of this has already been described (page 1754), in the little shore-haunting fishes called mud-skippers, which use their tails as accessory organs for this purpose. We may therefore suppose that the gill-pouches and gill-slits, primarily evolved in the interests of feeding, became secondarily adapted to the function of breathing. Such *secondary* adaptations are met with in many groups of the animal kingdom, and the principle of "change of use or function" involved enables us to understand many things which would otherwise be permanent enigmas. An excellent example was described in dealing with lung-fishes (page 1776), forms in which the swim-bladder, originally evolved in connection with balance of the body in the water, has been converted into a lung, namely, a pouch suited for breathing ordinary atmospheric air.

Adult land-vertebrates, such as reptiles, birds, and mammals, do not possess gill-pouches or gill-slits, but the last are present in their embryos, though they soon close up and do not seem to be any use. They are undoubtedly to be regarded as indicating that the land forms in question have descended from aquatic fish-like ancestors. In the frog and other amphibians, we find a striking confirmation of this view, for such animals begin life as tadpoles, which possess gill-pouches and gill-slits, in this respect closely resembling fishes. Later on these tadpoles become perfectly adapted to land-life, developing lungs and land-limbs, and often losing their gill-pouches and gill-slits altogether, at least as breathing organs.

It is worth while noticing that the great bulk of land-vertebrates have retained the first gill-pouch for the purpose of conducting sound-waves to the internal organs of hearing, which are imbedded in the wall of the skull. In a human being, for example, a passage known as the eustachian tube leads from the pharynx, or throat cavity, to the "drum" of the ear, closed externally by a membrane (tympanic membrane) and traversed by a chain of minute bones (auditory ossicles) that conduct vibrations of the air from the membrane to the internal ear. The eustachian tube and drum together represent the first gill-pouch, of which the external opening, or gill-slit, has become closed. This is another admirable instance of the principle of change of function ; indeed, we may say two instances, for the little auditory ossicles just mentioned, or some of them, probably correspond to structures originally used in hanging up the lower jaw to the skull.

From what has been said it will be realised that corresponding structures in different kinds of animals—structures that agree in relative position and in mode of origin—do not necessarily perform the same functions. It is a great convenience to have words expressing this kind of correspondence, and those in use are homology, homologous, and homologue (Greek *homologos*—conformable). We may say, therefore, that the eustachian tube with the drum of the ear are homologous to the first gill-pouch of a fish, or that the two structures are homologues, or again that they display homology. But it is also convenient to have a set of words having reference to similarity of use or function, and those selected for this purpose are analogy, analogous, and analogue (Greek *analogos*—similar). The wings of bat, bird, and insect, being all of them flying organs, are analogous, display analogy, and may be termed analogues. It must be remembered that homology and analogy often go together. In the example given, the wings of bat and bird are not only analogous, but also homologous, for both are fore limbs. But they are not homologous with the wings of insects, which have a totally different origin. Were it not for the changes of function that have often taken place, all homologous organs would be analogous.

The vast assemblage of creatures to which the term invertebrates is conveniently applied do not constitute a single subkingdom, comparable to that of the vertebrates, for the word invertebrates simply means non-vertebrate. But this, of course, involves negative characters—*e.g.*, invertebrates possess a central nervous system that is solid, *not* hollow, and the greater part of it is ventral, *not* dorsal ; notochord and gill-slits are *not* present. These and certain other points of difference between a vertebrate and a higher invertebrate are represented on page 1788. The other points are as follows :

VERTEBRATE	HIGHER INVERTEBRATE
1. Body a double tube, the upper (dorsal) one containing the central nervous system.	1. Body a single tube.
2. Heart ventral.	2. Heart dorsal.
3. Not more than two pairs of limbs, taking origin from same side of body as heart.	3. Often more than two pairs of limbs, taking origin from same side of body as central nervous system.

By carefully looking at the diagram it will be seen that the one for the vertebrate is not unlike that for the invertebrate upside down, hence the well-known epigram of Professor von Dohrn : " A vertebrate is a worm lying on its back." But such a theory leaves a great deal to be explained.

Invertebrates are divided into a number of sub-kingdoms, or phyla. The most important of these are as follows, beginning with the lowest : (1) PROTOZOA, or animalcules, including a vast number of simple animals, mostly minute or microscopic. (2) PORIFERA, or sponges. (3) CŒLENTERA, or zoophytes, of which sea-anemones, corals, and jelly-fishes are familiar examples. (4) PLATYHELMIA, or flat-worms, including flukes, tapeworms, etc. (5) NEMATHELMIA, or thread-worms, embracing many notorious parasites. (6) ECHINODERMA,

or hedgehog-skinned animals, such as star-fishes and sea-urchins. (7) ANNELIDA, or segmented worms, such as earthworms and leeches. (8) GEPHYREA, or siphon-worms. (9) POLYZOA, or moss-polypes. (10) BRACHIOPODA, or lamp-shells. (11) ROTIFERA, or wheel-animalcules. (12) NEMERTEA, or nemertine worms, unfamiliar marine forms with un-segmented bodies. (13) MOLLUSCA, or shellfish, such as cockles, snails, and cuttlefishes. (14) ARTHROPODA, or jointed-limbed animals, embracing lobsters, centipedes, insects, and spiders.

If we take the words "lower" and "higher" to mean simpler and more complex with reference to bodily structure, the three first phyla are decidedly "lower" as compared with the rest. We should also be justified in considering Mollusca and Arthropoda as "higher" phyla. But to arrange the fourteen phyla mentioned in a "linear series," or straight line, as it were, according to lowness or highness is quite impossible. Every rational attempt at arranging the phyla according to their resemblances and differences takes the form of a tree, which expresses our knowledge as to actual affinities or blood relationships, being, in fact, an imperfect genealogical tree. This approaches more and more towards correctness as our knowledge increases. The earliest forms of life with which we are acquainted belong to a great variety of invertebrate phyla, upon the origin and mutual relationships of which we can only speculate more or less plausibly.

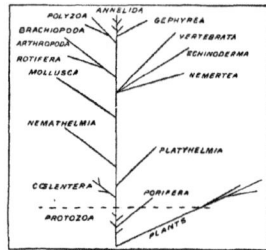

TREE OF ANIMAL LIFE

It has been aptly remarked that the construction of a genealogical tree of animals is as difficult as it would be to make a correct drawing of a tree almost entirely submerged in water by the aid of the twigs projecting above the surface. Subjoined is an exceedingly rough diagram, giving some idea of the possible relationship between the chief animal sub-kingdoms, including the vertebrates.

To distinguish a higher animal from a higher plant is extremely easy, for the former is of compact form, and requires food of complex character and partly solid nature, which is taken into an internal digestive cavity. The higher animal also possesses marked powers of locomotion, for it has to move about in search of food. The higher plant, on the other hand, is of a branching form, and does not move about. Its food consists of water with mineral matters in solution absorbed from the soil by the branching roots, and of carbon dioxide absorbed from the air by the leaves borne on the branching stem. As none of this food is solid no internal digestive cavity is necessary, and powers of locomotion are absent because the food is everywhere present. That the higher plant is able to subsist on such an apparently innutritious diet is due to the fact that the green pigment (chlorophyll) present in the leaves and young stems enables the living substance of the body to use the energy of sunlight for building up the simple food into more complex substances that repair waste, and supply the material for growth.

When we come to the lowest and microscopic forms of life these differences are not in all cases discernible, and it is sometimes impossible to say whether a given organism is a plant or an animal. In fact, some of these simplest organisms are described both in works on botany and those of zoology.

It is therefore not an unreasonable conclusion that animals and plants have sprung from common ancestors, which most likely were able to live on simple food, and in this respect resembled existing plants more than the vast majority of existing animals.

SUBKINGDOM PROTOZOA
THE LOWEST ANIMALS, OR ANIMALCULES

THE animalcules mostly belong to a world invisible to the naked eye, a world unknown two hundred years ago. In 1755 Rösel von Rosenhof saw sticking on the side of a glass vessel of water and weed a tiny particle of jelly, the movements of which attracted his attention. "It fastened itself," he writes, "on the side of the glass; and since, like animals, it moves, although very slowly, from place to place, and thereby continually alters its form, and as I frequently examined the water with a magnifying-glass, the creature was necessarily discovered; as soon as I touched it, it contracted itself into a sphere and fell to the bottom." Rösel removed the specimen to a watch-glass, and observed it continually changing its shape. In consequence of this peculiarity, he named the animal "the small Proteus," after the monster of fable.

Later the animal was named *Amœba*, as the name *Proteus* had been bestowed on another animal. This minute creature affords excellent material for the study of comparatively pure living matter or protoplasm. It is a minute speck of semifluid substance found living on the surface of mud or aquatic plants in ponds, and also in the sea. The body consists of a single cell, or unit of structure, in the interior of which may be seen the nucleus, a particle of specialised protoplasm, which appears to regulate the life of the animal and, as we shall see later, has much to do with propagation. It crawls actively about by constantly pushing out blunt protrusions of the body (pseudopods), the absence of a bounding membrane rendering this possible.

If an amœba be kept under observation it will be found to feed constantly, flowing round small plants and the like, partly digesting them, and flowing away from the undigested remnants. The spaces containing these matters are called food-vacuoles. Food is necessary, as in all organisms, because the unstable living substance of the body is always breaking down into simpler chemical compounds, some of which as waste products are excreted, or passed out to the exterior, in the form of carbonic acid gas, water, and nitrogenous matters of simple kind. In fact, the body of an organism is continually passing through a cycle of chemical change (metabolism), and it is this which is the essential character of living as distinguished from non-living matter. Huxley has aptly compared an organism to an eddy in a stream, which, though of fairly constant shape, never consists of precisely the same particles of water for two moments together. A mass of mineral substance, for instance a diamond, may remain chemically unchanged for an indefinite period.

The complex and unstable nature of living matter

adapts it eminently to undergo the constant chemical change to which allusion has been made, and such change, in so far as it consists of the transformation of complex compounds into simpler ones, renders possible movement and other manifestations of life. In order to render this plain it will be convenient to represent the cycle of chemical change in the form of a diagram usually called the "metabolic staircase."

The left-hand side of the diagram represents the gradual up-building of food into more and more complicated compounds until protoplasm is ultimately reached, while the right-hand side indicates successive stages in down-breaking, ending in the formation of comparatively simple waste products, which, being of no use, are excreted, or cast out of the body. Without this cycle of change movement and other life-manifestations would be impossible, for they involve the employment of actual or kinetic energy, and this is produced from stored or potential energy, as represented by complex unstable chemical compounds. The up-building compensates for this waste, and renders growth possible. If the two sides of the staircase were of equal length there would be no advantage, for as much kinetic energy would be transformed into potential during up-building as would be generated from potential in the process of down-breaking. But animals live on comparatively complex food, and hence the left-hand side of the staircase is shorter than the other, so that more kinetic energy is produced than used up.

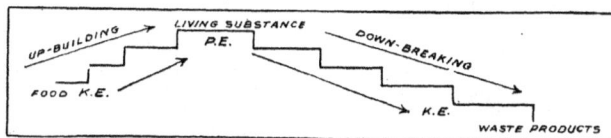

THE METABOLIC STAIRCASE
K E, kinetic energy; P E, potential energy

We must, however, inquire as to the source of the kinetic energy by which the complex substances of animal food were constructed. Directly or indirectly animals depend on green plants for their food, and in such plants the two sides of the metabolic staircase are equal; for the food consists of water, carbonic acid gas, and some simple mineral compounds. But in a way that is as yet only imperfectly understood green plants are able to use the kinetic energy of the sun's rays in the up-building process, this process depending upon the green colouring matter (chlorophyll) by which they are characterised. Both plants and animals therefore depend upon the sun, by the help of which green plants are able to bridge the gulf between non-living and living matter, using exceedingly simple chemical compounds of the former kind as building material.

An amœba not only feeds, but breathes, a process which consists in taking in oxygen, and at the same time getting rid of, or excreting, the waste product, carbonic-acid gas. The down-breaking, or "local death," that continually goes on in the body is essential to life, because it furnishes the requisite kinetic energy

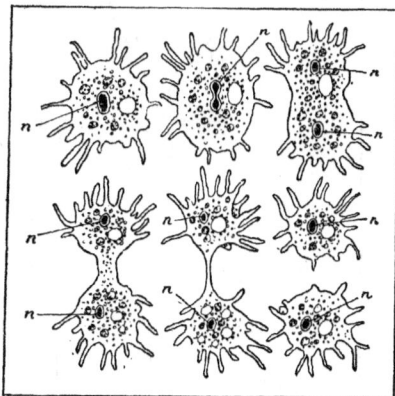

AMŒBA UNDERGOING FISSION
n, nucleus

for effecting movement, etc., and the oxygen taken in during breathing promotes this process, which is essentially one of oxidation.

Close examination of an amœba will reveal the presence of a clear space, or vacuole (pulsating vacuole), in the protoplasm, which alternately appears and disappears, probably assisting breathing, and perhaps helping to get rid of, or excrete, waste products in general. It is easy to prove that this minute creature is sensitive to external influences or stimuli. Contact with external bodies, alterations of temperature, the presence of chemical irritants, weak electric currents, and so forth, modify the shape of the body for the time being, and either accelerate or retard locomotion. We must further note that the characteristic flowing or creeping movements associated with the protrusion of pseudopods are due to alteration in shape of the body. This property of changing shape, while the volume remains the same, is technically known as contractility, and in one form or another is characteristic of living matter or protoplasm.

Under unfavourable conditions an amœba becomes encysted, contracting into a sphere and surrounding itself by a firm membrane. After reaching a certain size, an amœba propagates in a peculiar way, splitting into two—undergoing fission. Here the nucleus takes the lead, elongating and becoming dumb-bell shaped, and finally dividing into two, as shown on page 1790.

To sum up: An amœba, like any other living organism, feeds, breathes, and excretes, exhibits contractility and is sensitive, and reproduces its kind. Living matter as contrasted with non-living matter is characterised by the cycle of activities indicated, to which, in all but the lowest forms, death and decay may be added.

In the earlier days of scientific work, when exact observation was the exception, and vivid imagination the rule, it was very generally believed that even such highly organised animals as snakes, insects, and so on, might be generated "spontaneously" from decaying matter. Though this was gradually disproved for such creatures as those named, it was believed until comparatively recently that animalcules and bacteria could be formed directly from non-living matter. The exact experiments, however, of Tyndall and Pasteur proved conclusively that this is not the case. So far as we know, in this stage of the world's history, organisms are derived from pre-existing organisms, but if the evolution theory be carried to its logical conclusion, the first origin of life on the globe must have been due to the direct building-up of non-living matter into the earliest organisms. It would be out of place here to enter into the many ingenious speculations that have been advanced on this subject, but the first forms of life must have been of extremely simple structure, and one can scarcely avoid concluding that they possessed, like green plants at the present day, the power of utilising the sun's rays in their feeding processes.

For the purposes of this work it may suffice to divide the Protozoa into three great groups, roughly distinguished by their means of locomotion—thus, (1) Infusoria, in which the body is drawn out into a small number of

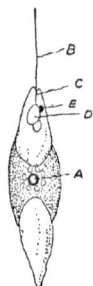

EUGLENA
a, nucleus
b, flagellum
c, mouth
d, pulsating vacuole
e, eye-spot

lash-like threads of protoplasm (flagella), or is more or less covered with a larger number of shorter threads (cilia) of similar nature ; (2) Rhizopoda, which possess pseudopods, either blunt like those of amœba or slender in form ; (3) Sporozoa, the adult forms of which do not possess any of the organs of locomotion mentioned, but typically wriggle about in a somewhat worm-like fashion. They are essentially parasitic in nature.

INFUSORIA

The members of the first group, Infusoria, were first discovered in putrefying infusions of hay and other organic substances, a circumstance to which they owe their name. They broadly fall into three sub-divisions:

(1) Flagellated Infusoria (Flagellata).
(2) Ciliated Infusoria (Ciliata).
(3) Tentacle-bearing Infusoria (Tentaculata).

FLAGELLATED INFUSORIA

Stagnant pools of rain-water are often found to contain immense numbers of microscopic, spindle-shaped animalcules, which, when examined under the microscope, are seen to move about like minute worms, changing their shape considerably meanwhile. They do not do this, however, in the same irregular fashion as amœba, for the body is covered by a thin elastic membrane or cuticle which prevents the formation of pseudopods. These tiny organisms are green in colour, and have received the name of *Euglena viridis*. The pigment they contain is identical with the leaf-green (chlorophyll) of green plants, and possesses the power in the presence of sunlight of bringing about a reaction between carbonic acid gas and water, with formation of organic substance and liberation of free oxygen. In water swarming with euglena bubbles of oxygen are given out to a marked extent, and it is interesting to note that by collecting and testing them Dr. Priestley was led to the discovery of oxygen. That euglena is a single cell is attested by the presence of a rounded nucleus, and it also possesses a pulsating vacuole comparable to that of amœba. Close to the latter is a red spot.

From one end of the body springs a long whip-like thread of protoplasm (*flagellum*), which, by its complicated lashing movements, is able to propel its possessor through the water. From the manner of its feeding, as so far described, euglena might be regarded as a plant, but close examination shows that the flagellum springs from within a funnel-like depression that internally opens abruptly into the soft protoplasmic interior of the body, and serves for the inward passage of minute particles of solid food. The external opening of the funnel may therefore be regarded as a mouth, and we are probably justified in considering euglena as an animal which to some extent feeds as a plant.

Euglena propagates (like amœba) by a process of splitting, or fission, and at times it contracts into a rounded form and becomes encysted, again like amœba. Within this cyst the body may divide into two, four, or more parts, each of which becomes a new animal when the wall of the cyst ruptures.

The members of the great army of flagellates differ as

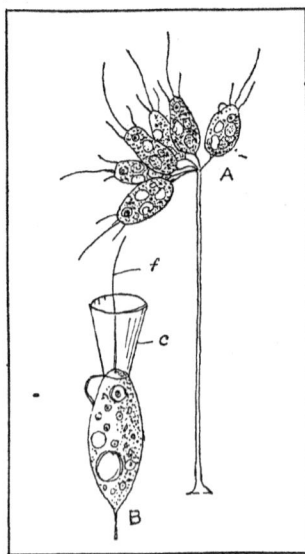

COLLAR ANIMALCULES
A. Group   B. Individual
f, flagellum ; c, collar

to the number of flagella present. Among those with one only, as in euglena, are the deadly parasites known as trypanosomes, which spend part of their lives in insects (or ticks) and part in human beings or other backboned animals, causing such diseases as sleeping sickness and nagana, or fly disease (fatal to ox and horse), which are among the scourges of Tropical Africa.

Another very interesting group is that of the collar animalcules, fixed forms in which the base of the flagellum is surrounded by a collar-like projection. The movements of the flagellum set up currents of water by which food-particles are brought to the animalcule. These stick to the outside of the collar and are carried down to a soft spot that serves as a mouth. Some of these forms are colonial, as a result of imperfect fission. In one particularly interesting case (*Proterospongia*) the members of the colony are imbedded in a kind of jelly, and some of them draw in their collars, assume the form of amœbæ, and break up into a number of actively moving zoospores, from which fresh colonies arise. We shall find later on that sponges possess collar-cells very much like the animalcules just described, and they have very likely taken origin from creatures of the kind.

Many of the smallest flagellates found in putrefying infusions are vaguely termed "monads," some of which are of particular interest. An example is the springing monad (*Bodo saltans*), a pear-shaped creature only about one three-thousandth of an inch long, and possessing two flagella, one of which is attached to the pointed end, and executes swimming movements, while the other one, which takes origin a little behind the first, is trailed behind. This second flagellum is also often used to anchor the animalcule to some firm surface, under which circumstances it alternately coils up and straightens out in a jerky fashion; hence the name of the monad.

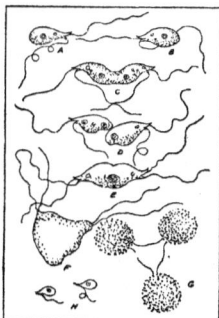

The springing monad, like many of its kind, has no mouth, and therefore cannot subsist on solid particles, nor, in the absence of chlorophyll, can it feed like a green plant. Its food, in fact, consists of liquid substances absorbed from the infusion in which it lives. Propagation is by longitudinal or transverse fission, and also by means of spore formation. Two individuals fuse together or conjugate, and become an angular mass surrounded by a firm coat, the contents of which break up into an enormous number of minute particles or spores, which escape by rupture of the firm wall, and grow into new individuals.

Some of the members of this group bearing two flagella are green in colour, and feed like plants, being,

indeed, regarded as such by some authorities. These forms may be aggregated into colonies, one of the most beautiful of which is the fresh-water *Volvox*, which is a hollow sphere about the size of a small pin's head, and slowly revolves as it swims by the lashing action of its flagella. New colonies are formed by division of some of the cells, and these pass into the cavity of the parent sphere, which ultimately bursts to liberate them, sinks, and dies. There is also another method of reproduction, in which egg-cells are formed (as in higher animals), and these, after fertilisation, grow into new colonies.

In certain species three or four flagella are present, but it is characteristic of these whip-like threads that they should not occur in large numbers.

One of the common causes of the phosphorescence of the sea is a peach-shaped flagellate (*Noctiluca*) attaining the comparatively gigantic size of about one-fiftieth of an inch in diameter. It possesses one very large flagellum, by which swimming is effected, and which springs from within a depression like that described for euglena and serving the same purpose of permitting the entry of solid food-particles into the body. A second and much smaller flagellum takes origin near the first. Noctiluca multiplies by fission and also by the formation of spores preceded by fusion or conjugation of two individuals, somewhat as in the springing monad.

Noctiluca was first described by Baker in 1753, and his quaint account runs as follows: "A curious inquirer into Nature, dwelling at Wells, upon the coast of Norfolk, affirms from his own observations that the sparkling of the sea-water is occasioned by insects. His answer to a letter wrote to him on that subject runs thus: ' In the glass of sea-water I send with this are some of the animalcules which cause the sparkling light in sea-water; they may be seen by holding the phial up against the light, resembling very small bladders or air-bubbles, and are in all places of it, from top to bottom, but mostly towards the top, where they assemble when the water has stood still some time, unless they have been killed by keeping them too long in the phial. Placing one of these animalcules before a good microscope, an exceedingly minute worm may be discovered [the large flagellum] hanging with its tail fixed to an opaque spot in a kind of bladder, which it has certainly a power of contracting or distending, and thereby of being suspended at the surface, or at any depth it may please in the including water.'

" The above-mentioned phial of sea-water came safe, and some of the animalcules were discovered in it, but

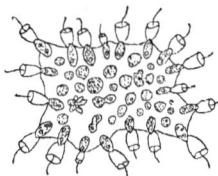

PROTEROSPONGIA   SPRINGING MONAD   NOCTILUCA

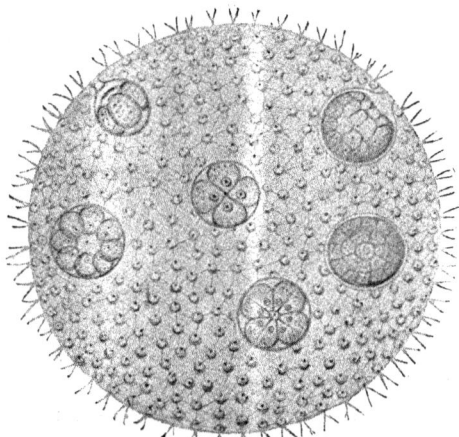

VOLVOX GLOBATOR

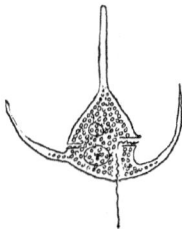

CERATIUM TRIPOS

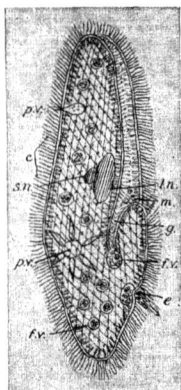

SLIPPER ANIMAL-
CULE

pv, pulsating vacuole;
c, cilia; *fv*, food vacuoles; *ln*, large nucleus;
sn, small nucleus; *m*,
mouth ; *g*, gullet; *e*, anus

they did not emit any light, as my friend says they do upon the least motion of the phial when the water is newly taken up. He like-wise adds that at certain times, if a stone be thrown into the sea, near the shore, the water will become luminous as far as the motion reacheth; this chiefly happens when the sea hath been greatly agitated, or after a storm."

Another phosphorescent flagellate, not much inferior in size to noctiluca, is *Ceratium tripos*, a triangular animalcule with its corners drawn out into spines. There are two flagella, one of which wraps round a transverse groove. Ceratium and some of its immediate allies are abundantly present among the minute or small forms that make up the floating population (plankton) of the sea and lakes.

CILIATED INFUSORIA

The highly organised animalcules, Ciliata, like the members of the last group, may be either free-swimming or fixed, and the term Infusoria is often restricted to them and the tentacle-bearing forms to be considered later. A very typical form is the slipper animalcule (*Paramæcium*), which can just be seen by the naked eye as an actively moving white spot in decomposing decoctions of hay and similar substances.

The movements are due to the rowing action of innumerable short threads of protoplasm (cilia) which project beyond the elastic membrane or cuticle covering the body. Each such thread or cilium alternately bends and straightens, and is not thrown into the lash-like undulations that are characteristic of flagella.

Unlike these last, cilia are associated in large numbers on the same cell, their movements being correlated to bring about some common end, in this case locomotion. In one way or another cilia play an important part in almost all the animal sub-kingdoms, arthropods (insects and so forth) excepted, and are present in even the highest backboned forms.

The living substance making up the body of the slipper animalcule is divided into a firmer outer layer (ectosarc), from

which the cilia spring, and a semi-fluid internal layer (endosarc), which is in a state of constant flowing movement. There are two pulsating vacuoles, one near either end, and instead of a single nucleus, as in amœba and the flagellates, there are two, one large (meganucleus) and the other small (micronucleus). There is a groove-like depression on the body (suggesting the opening of a "slipper") which leads to a mouth that opens into a short tube (gullet) which ends abruptly in the soft interior of the body.

All these parts are lined by cilia, which set up food-bearing currents. As in amœba, there are food-vacuoles, and the undigested remnants of food are cast out at a soft spot near the pointed hinder end of the body. The ectosarc is packed with elongated capsules (trichocyts), from which long threads

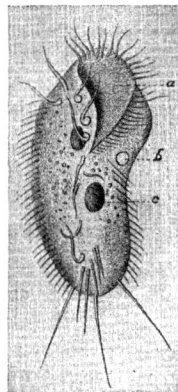

UNDER SURFACE
OF MUSSEL ANI-
MALCULE

a, mouth ; *b*, pulsating
vacuole ; *c*. nucleus

are shot out if the animal is subjected to the action of chemical stimuli.

Like so many other animalcules, *Paramœcium* propagates by fission. There is also a remarkable process of conjugation. Two individuals become applied together, the large nuclei disappear, and the small ones undergo several divisions. Each of the conjugating animalcules receives a fragment from the small nucleus of the other one, and ultimately large and small nuclei are reconstituted. After exchange of nuclear material the two paramœcia separate and undergo active fission, the process of conjugation seeming to have a revivifying action.

MUSSEL AND TRUMPET
ANIMALCULES

The mussel animalcule (*Stylonychia mytilus*) is more specialised than the last described form, and effective cilia are limited to the flat under surface of the body, on which it is able to creep by means of some curious bristles that have been formed by the fusion of groups of cilia. The trumpet animalcule (*Stentor*) is an interesting fresh-water form, of comparatively large size, being about one twenty-fifth of an inch long when fully extended. It is commonly found attached by its narrow end to duckweed or other aquatic plants. A depression at the broad end leads to a mouth and gullet, as in the slipper animalcule. The large nucleus is much elongated and looks like a bead necklace, and there is also a small nucleus or rounded form (not shown in the figure).

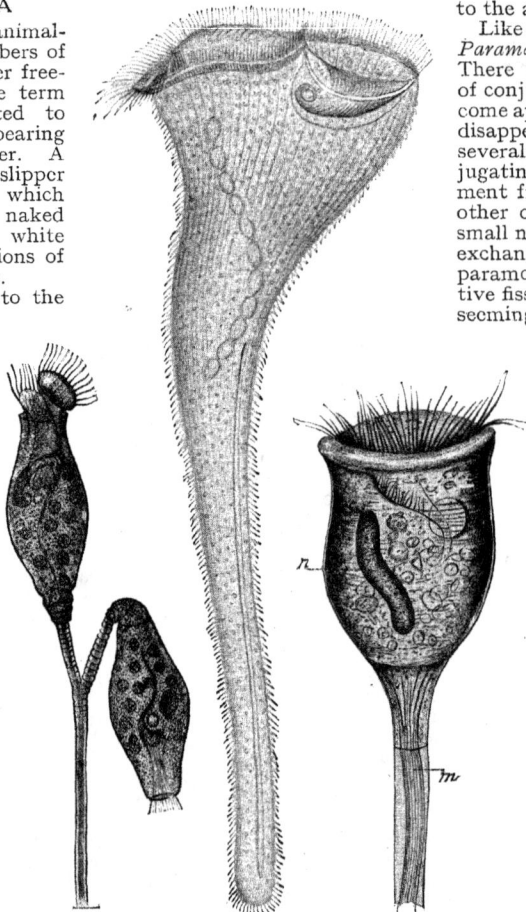

EPISTYLIS   STENTOR   VORTICELLA

n, nucleus ; *m*, muscle band

In some species of trumpet animalcules a short gelatinous tube surrounds the attached end of the body, and serves as a refuge into which the little creature can withdraw itself. In other ciliates there may be definite shells. In *Pyxicola*, for example, there is an investment in the form of a goblet, and when the animalcule retracts into this, a little lid, or operculum, on the other side of its body closes the shell. There is a somewhat similar arrangement in *Thuricola*, but here the goblet has no stem, and the operculum is hinged to its side, being drawn down when required by a contractile filament.

BELL ANIMALCULES

Among the most familiar types of this group are the bell animalcules (*Vorticella*), the bodies of which are attached by a long stalk containing a wavy thread, which on the approach of danger is able to contract, coiling up the stalk like a corkscrew. When fully expanded the bell-shaped body is seen to be provided at its broad end with a short spiral of cilia, which set up food-bearing currents that flow into a groove, at the bottom of which the mouth is situated. This opens into a gullet which, as in the slipper animalcule, ends abruptly within the soft endosarc. The large nucleus is a curved band, while the small nucleus is rounded. A pulsating vacuole is present in the neighbourhood of the mouth.

The bell animalcule propagates by longitudinal fission, one of the two individuals produced keeping the stalk, while the other swims away, becomes attached, and grows a new stalk. Sometimes a fixed vorticella rapidly divides several times to produce a number of free-swimming bells, which have to do with a remarkable process of conjugation. One such little bell permanently fuses with a large fixed one, and remarkable nuclear changes then take place, comparable to those described for the slipper animalcule. Although in this case conjugation actually reduces the number of individuals, since two fuse into one, this is made up for by the increased vigour resulting, which shows itself in active fission.

In some of the close allies of the ordinary bell animalcule, colonies are produced by imperfect fission. *Carchesium*, for instance, is a tree-like colony of the kind, each individual of which resembles a vorticella. Another colonial form is the nodding bell animalcule (*Epistylis*), so called because the bells droop down when irritated.

TENTACLE-BEARING INFUSORIA

The group Tentaculata appears to be an offshoot from the Ciliata. They are solitary or colonial, in either case being distinguished by the possession of curious tentacles, by which various kinds of animal prey are pierced and sucked. A simple form (*Acineta*) and a bud-producing type (*Hemiophrya*) are figured.

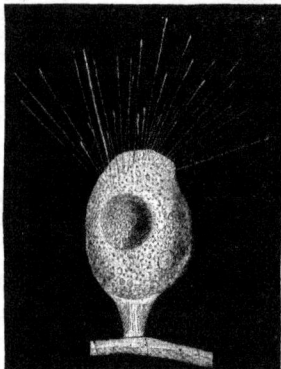

ACINETA

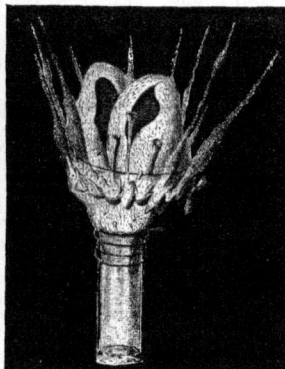

HEMIOPHRYA

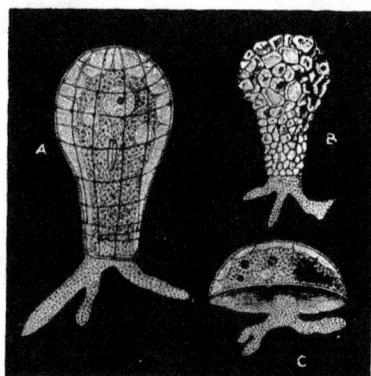

A, QUADRULA; B, DIFFLUGIA; C, ARCELLA

RHIZOPODA

Possession of pseudopods, blunt or thread-like, distinguishes the large group of animalcules termed Rhizopoda, and amœba has already been described as a typical example. A pseudopod differs from a cilium or flagellum in being a temporary instead of a permanent structure, and is simply an overflowing of the soft protoplasmic body.

The following orders may be distinguished: (1) Amœbas (Lobosa). (2) Foraminifera. (3) Sun Animalcules (Heliozoa). (4) Ray Animalcules (Radiolaria). (5) Fungus Animalcules (Mycetozoa).

LOBOSA

Although different kinds of proteus animalcules (*Amœba*, etc.) are found almost everywhere, the members of the order Lobosa are predominatingly fresh water, or inhabitants of damp places on land. They are also found as parasites in the intestine, giving rise to a kind of dysentery.

Some forms are characterised by the possession of a shell. In the simplest case (*Difflugia*) the body secretes a sticky covering to which grains of sand adhere, while in other forms there is a shell of horny nature. In *Quadrula* this resembles a short flask in form and is made up of square plates, while the shell of *Arcella* is hemispherical, with a rounded aperture below. The latter animalcule is able to secrete bubbles of gas (gas vacuoles) in its protoplasm, and these enable it to float in water.

FORAMINIFERA

The order Foraminifera includes a great variety of shell-bearing animalcules, most of which are marine. In the majority of cases the shell is composed of carbonate of lime, and many of the limestones which help to build up the solid framework of the globe are largely composed of the shells of these minute creatures. The pseudopods are slender and branching; they often fuse into networks.

It is convenient to divide the order into two groups, Imperforata and Perforata. In the former group the shell, when calcareous, looks like white porcelain, and the pseudopods are protruded through one or a few comparatively large apertures. But in the perforate forms the calcareous shell is of glassy appearance and is pierced by innumerable little pores. It is this circumstance which has given the name to the order (Lat. *foramen*, an aperture ; *fero*, I bear).

IMPERFORATA

A simple type (*Gromia*), common to salt, fresh, and brackish water, is represented in the figure. Here there is an ovoid membranous shell, from the single aperture of which numerous slender pseudopods project, while others arise from a thin layer of protoplasm covering the shell.

As in amœba, the pseudopods are used to secure food, and also to effect creeping movements.

Some of the imperforate Foraminifera construct variously shaped shells of sand-grains cemented together. Two kinds with irregular shells of this kind are represented below. The calcareous shells of imperforate species possess varying number of chambers, arranged in different ways. The form represented (*Miliola*) is one of the several closely allied types which make up the important "Miliolite limestone," largely employed as a building stone in Paris and its neighbourhood. A spiral species (*Peneroplis*) also is shown on page 1796, the front view (on the right) exhibiting the narrow aperture through which the pseudopods are protruded.

Some of the imperforate types are very complicated. The coin-shaped *Orbitolites* figured below may attain a breadth of more than half an inch, and is made up of a very large number of compartments which have been successively built up round an original spiral chamber (see figure). The most complex of these compartments are on the outside, and here, too,

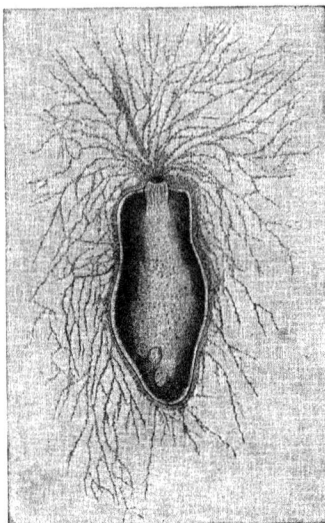

EGG-SHAPED GROMIA

is often applied to them. The shells of the same form are also an important constituent of chalk, which is simply a consolidated foraminiferal ooze, deposited on the floor of an ocean that stretched from the southern part of what is now North America westward through Europe into Asia. *Polystomella*, illustrated on pages 1796 and 1797, is a spiral form of complex structure. An idea of its internal form may be gathered from the illustrations on page 1796, which represents the soft portoplasm after the calcareous investment has been dissolved away by the action of acid. A closely allied coin-shaped form (*Nummulina*), some species of which attained the size of half-a-crown, was exceedingly abundant during the early part of the Tertiary epoch, its shells building up the "nummulitic limestones," of which Sir Charles Lyell says: "The nummulitic limestone, with its characteristic fossils, plays a far more conspicuous part than any other Tertiary group in the solid framework of the earth's crust, whether in Europe, Asia, or Africa. It often attains a thickness of many thousand feet, and extends from the Alps to the Car-

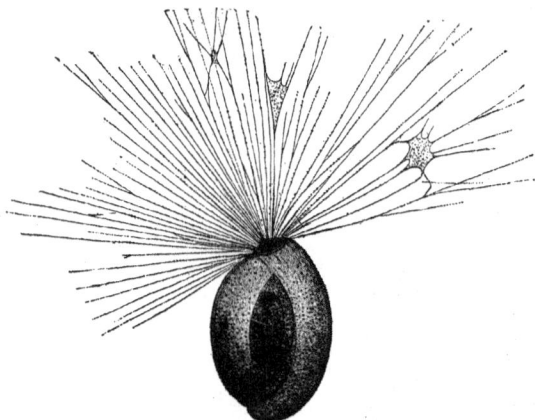

MILIOLA TENERA
Showing network of pseudopods

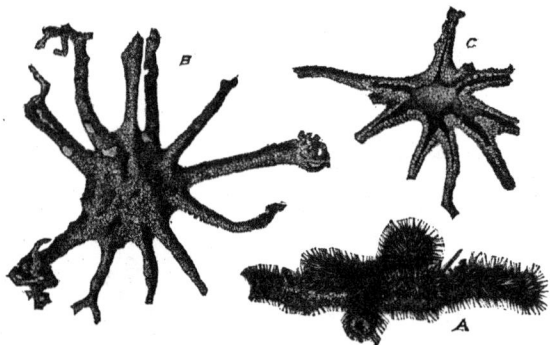

A, HYPERAMMINA RAMOSA; B & C, ASTRORHIZA LIMICOLA
B, entire; C, cut open

are the openings through which the pseudopods project to the exterior.

PERFORATA

Here, again, we have many different kinds of shell, varying in the number and arrangement of chambers. One of the most interesting forms is perhaps *Globigerina*, which is particularly abundant in the calcareous deposits which cover a large part of the ocean-floor at the present day, so much so that the name of "Globigerina ooze"

pathians, and is in full force in the north of Africa, as in Algeria or Morocco. It has also been traced from Egypt, where it was largely quarried of old for the building of the Pyramids, into Asia Minor, and across Persia, by Bagdad, to the mouths of the Indus. It occurs not only in Cutch, but in the mountain ranges which separate Scinde from Persia, and which form the passes leading to Cabul; and it has been followed still further eastwards into India, as far as Eastern Bengal and

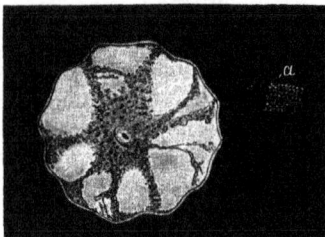

YOUNG ARCELLA. *a*, bit of shell

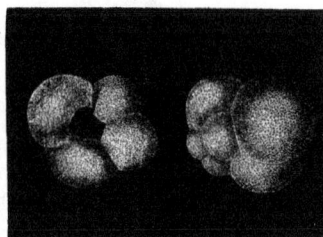

SHELLS OF GLOBIGERINA

the frontiers of China." The spindle-shaped *Fusulina*, which resembles a Nummulina drawn out at right angles to the plane in which its chambers are disposed, may be as much as half an inch in length. Thick limestones belonging to the Carboniferous period, and built up of its shells, are widely distributed in Europe, Asia, and North America.

SUN ANIMALCULES

Sun animalcules (order Heliozoa) are fresh-water forms with a rounded body and numerous stiff radiating pseudopods, which suggest the conventional figure of the sun, and hence gave the name to the group. They may be entirely without a firm skeleton, as in the common sun animalcule (*Adinophrys sol*), here figured, or there may be radiating siliceous spicules, as in the

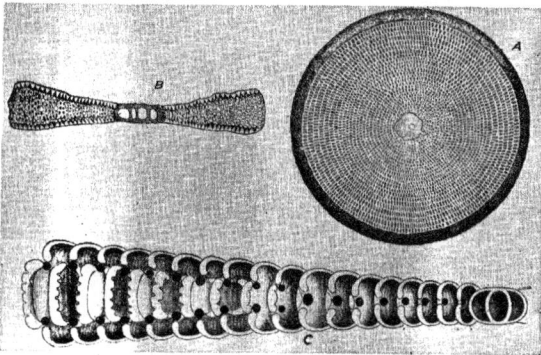

ORBITOLITES

A, from above; *B*, transverse section; *C*, diagram illustrating the transition from the simplest to the most complex type of chamber

green sun animalcule (*Acanthocystis chætophora*), which owes its colour to the presence of minute algæ within the protoplasm, or again, as in the lattice animalcule (*Clathrulina*, see figure), there may be a shell in the form of a hollow sphere with numerous rounded openings, through which the pseudopods are protruded. In the last case a stalk is present.

RAY ANIMALCULES

The ray animalcules (order Radiolaria) are among the most complicated cells to be found in the whole of the animal kingdom, one characteristic point being the presence of a membranous capsule within the body, internal to which is the nucleus. There is usually a flinty skeleton, which assumes the most complicated forms, as will be seen by reference to the illustration on page 1768, which typically includes some radiating spicules, whence the name of the order.

These animalcules live at various depths in the sea, and are of different colours—red,

blue, and yellow. The surface forms commonly contain "yellow cells," which are in reality minute algæ. The association is a case of mutualism, or symbiosis, where two forms of life are intimately associated to the benefit of both. Some of the deepest parts of the ocean are covered with "radiolarian ooze," mainly composed of the shells of these little creatures, which are able to resist the solvent action of sea-water under great pressure; this not being the case with the calcareous investments of foraminifera. Loose siliceous deposits of late Tertiary age, found in the Barbadoes ("Barbadoes earth"), Greece, Sicily, and the Nicobar Islands, are composed of radiolarian shells, and prove that parts of the floor of the very deep sea have been upheaved in comparatively recent times. Compact siliceous rocks ("radiolarian cherts"), which exist in some of the older geological formations, demonstrate the same thing for very remote times.

SOFT BODY OF POLYSTOMELLA
a, nucleus

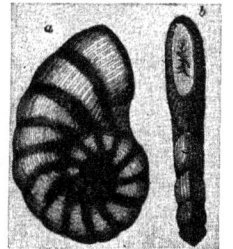

PENEROPLIS PERTUSUS
a, lateral, and *b*, front view

Propagation takes place by fission, while at times the body of a ray animalcule will divide into a number of motile helmet-shaped "spores" provided with flagella. These are sometimes all of the same size, or they may be of two sizes, large and small. It is probable that conjugation takes place between them. Some of the ray animalcules are colonial.

FUNGUS ANIMALCULES

The order Mycetozoa includes a number of problematic forms, considered by many botanists to be plants and not animals. They are commonly found as jelly-

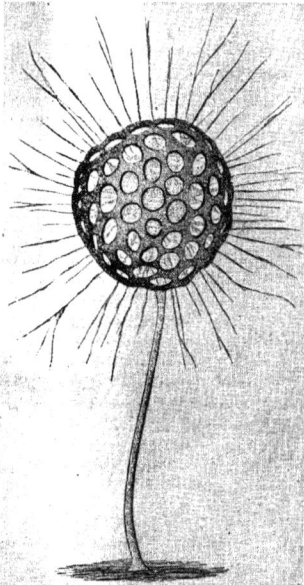

LATTICE ANIMALCULE

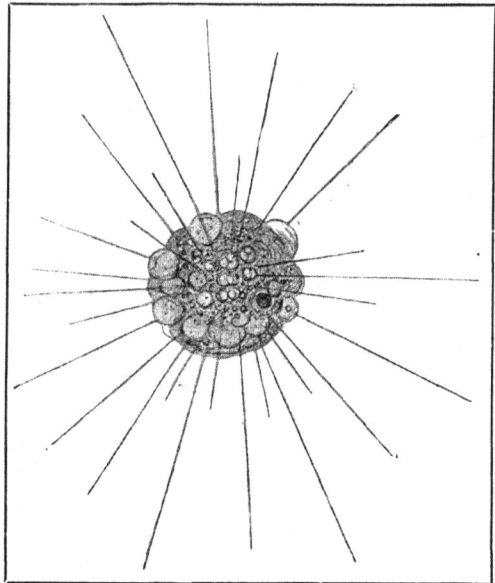

SUN ANIMALCULE

POLYSTOMELLA

like masses on bark, spent tan, and in similar places. Projecting from the mass are so-called "fruits" of various shape, the contents of which break up into a number of small rounded "spores," with firm coats. When such a spore germinates its coat splits open, and the protoplasmic contents, resembling at this time one of the flagellate animalcules with a single flagellum, but without any firm investment. After a time these little bodies draw in their flagella and assume the shape of minute amœbæ, large numbers of which fuse together to form

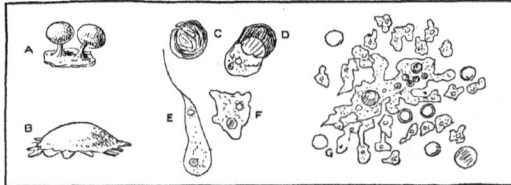

FUNGUS ANIMALCULES

A, B, fruits ; C, D, spores ; E, flagellate stage ; F, amœba stage ; G, plasmodium

a creeping "plasmodium," that at first moves about and feeds like a huge amœba, ultimately settling down and producing "fruits."

SPOROZOA

There is a large group of parasitic animalcules (Sporozoa) which reproduce by minute spores, and are responsible for many kinds of disease. It may be pointed out that malarial diseases are due to their presence, one stage of the life-history being passed through in a mosquito, and the other in the blood of a human being who has been bitten by that insect.

METAZOA

The animalcules, or Protozoa, which have just been described are distinguished from all other animals by the fact that their bodies are single cells, in other words, they are one-celled, or unicellular. It is convenient to lump together as Metazoa all the remaining groups of animals the bodies of which, being made of a number of cells, are multicellular. An individual animal of this kind may be looked upon as a colony of very closely associated units, divisible into tissues, or aggregates of cells specialised in different ways to suit them for the discharge of special functions. In higher animals, for instance, we have muscular tissue, made up of elongated cells concerned with movement ; nervous tissue, composed of star-shaped cells drawn out into slender conducting fibres ; and so forth. In the lower Metazoa, which will be first considered, the division of physiological labour is ill-marked, and therefore the tissues are not well defined.

In the foregoing account of Protozoa a number of colonial forms have been described or mentioned, but these differ from simple Metazoa in the fact that the individual cells of the colony are all alike. Epigrammatically expressed, such a colony is "morphologically multicellular, and physiologically unicellular." In such colonial Protozoa as *Volvox*, however, the cells are not all alike, and this perhaps gives a hint of the way in which Metazoa have descended from colonial Protozoa

1797

SUBKINGDOM PORIFERA—SPONGES

SPONGES are many-celled fixed animals living for the most part in the sea, although a small number are inhabitants of fresh water. They are undoubtedly to be considered as the lowest Metazoa, to the other sub-kingdoms of which they do not, however, seem to be related, having in all probability had an independent origin from one-celled animals akin to the Protozoa. Examination of the illustrations belonging to this chapter will show that these creatures exhibit a large variety in form, which is partly due to the fact that many of them are colonies in which the individual members are ill-defined. The familiar toilet article known as a " sponge " is the horny skeleton of such a colony, and though probably even yet regarded by many persons as of plant nature, it was recognised by Aristotle as belonging to the animal kingdom.

A simple sponge is shaped like a vase, with its walls perforated by numerous pores, and examination of living specimens shows that currents of sea-water continually flow through these pores into the central cavity, like the collar animalcules elsewhere described. Owing to this it has been conjectured that sponges have descended from colonial animalcules of this kind.

The more complex sponges, largely of colonial nature, differ from the simple form described in the reduction of the central cavity, and the thickening of the walls, which are traversed by a system of canals having a more or less elaborate arrangement, and in which the collar-cells are very generally restricted to localised rounded swellings of the canals.

We may take as an example the colonial form known as the crumb-of-bread sponge (*Halichondria panicea*) which is supported by siliceous spicules. This is common on the British coast, and may be found near low-tide mark as a green, brown, or orange crust spreading over rock surfaces. The surface of the colony is studded with conical projections, at the end of each of which an osculum is situated. Numerous pores can be detected between these. Probably each cone corresponds to an individual, but the boundaries between

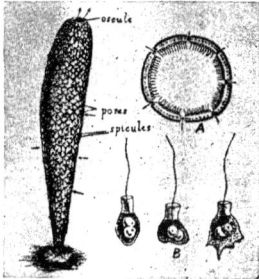

ASCON SPONGE
A, transverse section ; B, collar-cells

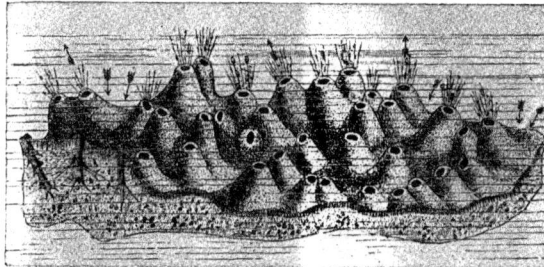

BREAD-CRUMB SPONGE
Arrows show currents entering surface and leaving by oscules

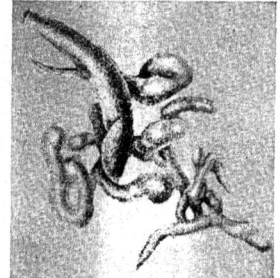

LEUCOSOLENIA
Calcareous ascon sponge

and make their exit by the main opening, or oscule, of the vase. In this way minute organisms and nutritive particles are brought to the sponge as food, together with the oxygen needed for breathing, while the various waste products are swept out of the central cavity.

The wall of the simple sponge is supported and strengthened by three-rayed spicules composed of carbonate of lime. Spicules of lime or silica, variously shaped and arranged, are found in the great majority of sponges, serving not merely as a support, but as a protection against predaceous fishes and other animals. The unpalatability of sponges is often increased by a deterrent odour, and the bright tints some of them possess may very possibly be considered as an example of warning coloration.

The constant flow of water-currents, upon which the life of a sponge depends, is brought about by the lashing of flagella, attached to collar-cells, very much

the members of the colony are not clearly defined. The accompanying figure represents a part of such a colony, and the directions of the water-currents are indicated by arrows.

Sponges are classified in accordance with the chemical nature and arrangement of the spicules, and the following scheme will serve as a guide :

CALCAREOUS SPONGES
Class CALCAREA—Spicules calcareous
GLASS-SPONGES
Class HEXACTINELLIDA—Spicules siliceous and six-rayed
COMMON SPONGES
Class DEMOSPONGIA—Spicules, when present, siliceous, and commonly supplemented by horny fibres
1. FOUR-RAYED SPONGES
Order TETRACTINELLIDA—Four-rayed spicules present
2. FLESHY SPONGES
Order CARNOSA—Outer part of the body firm, developed
3. SINGLE-RAYED SPONGES
Order MONAXONIDA—Most of the spicules simple needles or rods
4. HORNY SPONGES
Order CERATOSA—Skeleton chiefly composed of horny fibres

CLASS CALCAREA—CALCAREOUS SPONGES

THE calcareous spicules present in the class Calcarea may be simple needles, or else possess three or four rays. The simple sponge already described belongs to this group, and a common branching form of similar structure is *Leucosolenia*. Both belong to the " ascon " type, in which the central cavity is lined by collar-cells. The " sycons " are more complex, their walls consisting of a series of tubes, lined by collar-cells, and opening into the central chamber, from which such cells are absent.

A common British example (*Grantia compressa*) is a little flat, white bag, about an inch long, and found attached to rock or seaweed. The accompanying figure gives the early stages in the development of another sycon (*Sycandra raphanus*). The earliest stage shown (*a*) consists of large rounded granular cells at one end, and elongated flagellated ones at the other.

By means of the latter active swimming is effected, the flagella being in front. That the embryos or larvæ of fixed animals should possess active powers of locomotion is a general rule, and may be regarded as a provision for the distribution of the species. After a time the larva alters its shape (b), and then the flagellated half becomes tucked into the other, so that a double-walled cup results (c). After this the embryo becomes attached by the opening of the cup, pores and an osculum are formed, and numerous projecting spicules developed (d).

(oscule) is covered by a perforated plate. The cut represents a section through a small part of the wall, with six of the elongated chambers lined by collar-cells.

The Japanese glass-rope sponge (Hyalonema) is an extraordinary form in which the thick-walled cup constituting the body is anchored in the mud by a

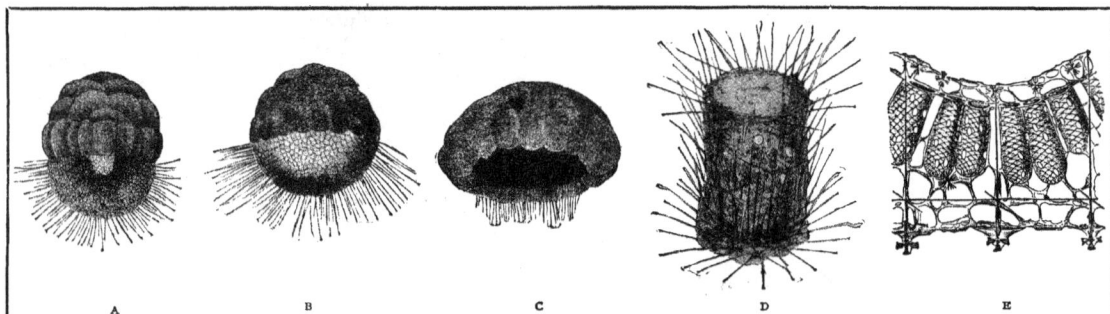

A-D, DEVELOPMENT OF SYCANDRA.   E, SECTION OF WALL OF VENUS FLOWER-BASKET, SHOWING SOFT PARTS

The canal system of the " leucon " sponges is still more complex, and the collar-cells are limited to special chambers connected by one set of canals with the exterior, and by another set with the central cavity.

Six-rayed spicules belonging to members of the class Hexactinellida are depicted in the accompanying figure. A great variety of form results from suppression or specialisation of one or more of the rays, and many of the spicules are interwoven or fused to constitute a continuous skeleton. The typical shape of the body is that of a vase. One of the simplest forms (Bathydorus) has been dredged from a depth of 2,900 fathoms in the North Pacific, and is in the form of a soft, thin-walled tube 7 inches long and 2 inches broad.

The beautiful sponge known as Venus flower-basket (Euplectella aspergillum), represented on pages 1767 and 1817, and a number of allied species are found up to depths of 2,550 fathoms in several parts of the world, the seas off Japan and the Philippines being particularly noted for them. The figured species is in the form of a curved tube, which may be as much as 18 inches long, and is supported by an elegant glassy skeleton, covered with brownish flesh in the living animal. The opening of the tube

" glass rope," consisting of a bundle of long, twisted spicules. An allied form (Semperella) is shown on the right of page 1767. Another very extraordinary type (Monorhaphis) does not possess a " glass rope," but is anchored by a single gigantic spicule, over 9 feet long and as thick as the little finger.

An interesting Philippine sponge (Polylophus), shown on the right front of page 1767, is a short, broad cup anchored by tufts of spicules, and remarkable for its bud-producing powers. A number of the buds are shown in the figure.

The Japanese genus Periphragella (4 on page 1767) is in the form of a curved funnel covered with a network of tubes, while in Farrea (3 on page 1767) there is a series of forked glassy tubes. Carpenter's glass-sponge (Pheronema) is another elegant type, of which an illustration is given.

The remains of six-rayed sponges abound in the Chalk, one of the most familiar being the elegant funnel-shaped Ventriculites. The numerous flints of the Upper Chalk are often deposited round such sponges, and the siliceous material of which they are composed has been largely derived from sponge spicules.

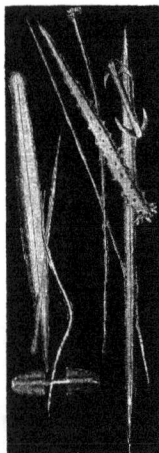

SILICEOUS SPICULES OF MONAXONID SPONGES

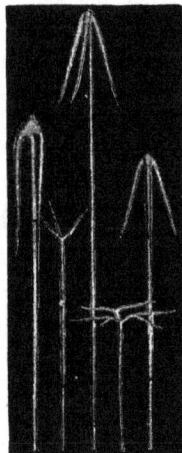

SILICEOUS SPICULES OF FOUR-RAYED SPONGES

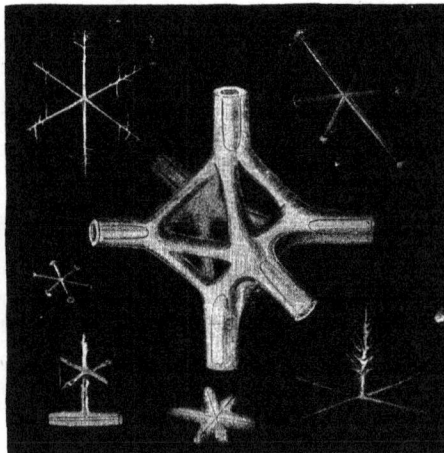

SILICEOUS SPICULES OF SIX-RAYED SPONGES

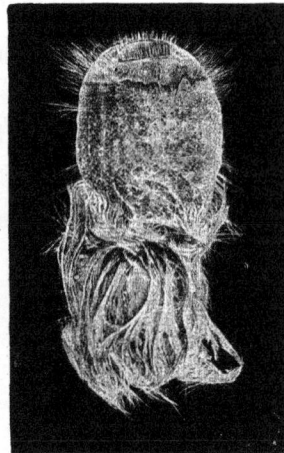

CARPENTER'S GLASS-SPONGE, PHERONEMA

COMMON SPONGES—CLASS DEMOSPONGIA

THE class Demospongia includes the four orders of four-rayed, fleshy, single-rayed, and horny sponges.

FOUR-RAYED SPONGES

Of these the four-rayed sponges (order Tetractinellida) of this group mostly live in shallow water, and the figure on page 1799 shows some of the typical four-rayed spicules.

FLESHY SPONGES

In the small group of fleshy sponges (order Carnosa) there is a firm bark-like outer layer, surrounding a softer portion in which are contained the chambers lined by collar-cells. The Mediterranean sea-kidney (*Chondrosia*) is here figured, and the large single oscule is clearly seen.

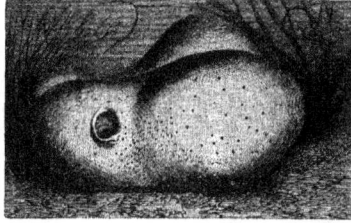

SEA-KIDNEY LEATHER-SPONGE

SINGLE-RAYED SPONGES

The order Monaxonida includes most of the British sponges, as well as the majority of shallow-water forms in all parts of the world. The firm skeleton consists of simple, needle-shaped spicules, commonly cemented together by horny material, and of smaller flesh spicules with thickened or anchor-like ends.

BREAD-SPONGE

The crumb-of-bread sponge (*Halichondria panicea*), to which allusion has already been made, is one of the commonest members of the order, and the large form known as Neptune's cup-sponge (*Poterium*), which is 3 or 4 feet in height, is also well known.

Oyster-shells and pieces of limestone riddled by numerous excavations may often be picked up on the shore. The tunnelling they exhibit is the work of a boring sponge (*Cliona*), but how it is effected is unknown. Such sponges are a notable pest in oyster fisheries, and on some limestone coasts they assist to a serious extent in the erosion of the land, for rocks which have been bored by them, it may be to a depth of several feet, disintegrate very rapidly.

NEPTUNE'S CUP-SPONGE

FRESH-WATER SPONGES

The order includes the fresh-water sponges, of which two common British species (*Euspongilla lacustris* and *Ephydatia fluviatilis*) encrust wooden piles and the stems of water-plants. The former species is often of a bright green colour, owing to the presence of granules containing chlorophyll in the surface layers. It is doubtful whether these granules are a part of the sponge itself or are to be regarded as algæ, but in any case the pigment present enables the associated protoplasm to build up organic matter from water and carbonic acid gas, thus aiding the nutrition of the sponge.

It commonly happens that fresh-water organisms die down during the winter, but have special ways of preventing the extinction of the species. This is the case with the fresh-water sponge, which produces innumerable buds, or gemmules, in the autumn. These are to remain dormant during the winter, growing into new sponges the following spring.

HORNY SPONGES

In the order Ceratosa the skeleton mainly consists of a complex network of horny fibres, the chemical composition of which resembles that of silk. Here belong the different kinds of toilet and bath sponges, which are of considerable economic importance, and for which no really efficient substitute has so far been invented.

The Turkey cup-sponges (*Euspongia officinalis*) are of the finest quality, the related zimocca, or hard, sponge (*E. zimocca*), largely imported from the West Indies, being harsher and more brittle. Its form is that of a rounded disc, with a flat upper surface and convex lower one. The larger and coarser bath-sponge, or horse-sponge, belongs to a different genus (*Hippospongia*). The finest toilet sponges come from the Eastern Mediterranean, the zimocca kind from the Mediterranean and West Indies, and bath-sponges from the West Indies and Florida. In the case of the Mediterranean sponge-fisheries, which date from very ancient times, divers and drag-nets are employed. The fisheries of the West Indies and Florida began about 1849, and the sponges are there secured by means of tridents attached to long wooden handles.

Sponges at some depth are difficult to discover when the surface is ruffled. This difficulty is surmounted by the use of a sponge-glass, which is simply a bucket or tube with a piece of glass at its lower end.

J. R. AINSWORTH-DAVIS

LIMESTONE BORED BY SPONGE

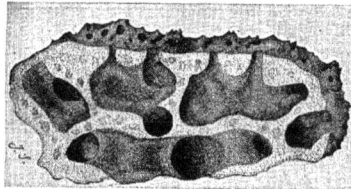

SECTION OF BATH-SPONGE

SUBKINGDOM CŒLENTERATA
JELLY-FISH, CORALS, AND SEA-ANEMONES

CORALS, jelly-fish, and sea-anemones constitute the great subkingdom known as Cœlenterata. The group comprises all those creatures in which the internal space corresponding with the alimentary canal of other animals is not a closed canal running through the body, but is commensurate with the whole cavity of the body. Consequently there are no spaces answering to the body-cavity of the vertebrates between the wall of the alimentary canal and the outer wall of the body.

A study of the earliest growth of the cœlenterates has shown that their internal cavities are nothing more than regular radiate outgrowths of the intestine, and, like the latter, come from the primitive intestine of the larva. The result of this development is a condition which does not occur elsewhere in the whole animal kingdom. We have no separate digestive canal, no closed blood vascular system, and no specialised respiratory apparatus. There is only a system of cavities, all in open communication with one another, occupying almost every corner of the body.

Again, the cœlenterates are radiate in structure—that is, when seen from above they are typically star-shaped ; and if a cœlenterate be cut across, every horizontal section shows a symmetrical arrangement of the parts around a centre. The shape, in fact, is precisely what might be expected in forms which are either fixed or swim without any side of the body taking the lead, for under such circumstances the surroundings are the same all round, and, the reaction of such surroundings being similar on every side, a radial form is assumed.

If the body of a swimming type steadily revolved, all points of the body would be exposed in turn to the same influences and the ideally symmetrical form—the sphere—would be attained. This, however, is not the case here. The side of a cœlenterate turned towards the light is always different in form and structure from the side that remains in comparative darkness. When cœlenterates form calcareous skeletal structures, the anterior end of the body, crowned with one or more circles of tentacles, remains soft and flower-like. The most highly developed of the free forms, however, such as the sea-anemones and the jelly-fish, have no hard skeleton at all, but are amongst the most delicate and beautiful objects in the realm of living Nature (page 1817).

Cœlenterates have failed to make any way in fresh water, not to speak of the land. A few free-swimming jelly-fish, and a minute attached polyp, are, indeed, found in fresh water, but these can hardly be looked upon as successes. It is very interesting to note that fresh-water jelly-fish are chiefly found in Lake Tanganyika, which there is reason to suppose was once an arm of the sea. The cœlenterates have a special interest, since they are considered to represent a stage in the development of animal life through which all the higher forms have passed.

GROUP CTENOPHORA—COMB-JELLIES

ALTHOUGH all are agreed that the comb-jellies are members of this subkingdom, their exact position is not clear. They are glassy, transparent creatures, either shaped like apples, melons, or Phrygian caps, or else forming bands, often a yard in length and thickened at the middle. Several types are shown in the illustration on page 1766. The marvellous transparency of all but one (Beroë) is specially remarkable. They inhabit the open sea, or are driven by currents and winds near the coast and into harbours. Their position in the water is usually more or less vertical, the mouth being turned downwards. The organs from which this group takes its name are the ribs, which either run from pole to pole, or else only for certain distances along the meridians, which are often symmetrically arranged.

These ribs consist of rows of short transverse combs, each being formed of a row of cilia. The cilia forming a comb are connected together at their bases, but are also capable of independent movement. As they wave to and fro, they constitute what is called a swimming or rowing plate. The activity of these rows of plates depends upon the will of the animal, which can move either the plates of a single rib or all the ribs together ; this latter movement resulting in slow locomotion in the direction of the apical pole—namely, the pole turned away from the mouth. The body is capable of various movements, for in addition to the rowing plates there are other structures, such as the oral umbrella, and the capturing filaments, or tentacles, with their hair-like branches. These tentacles, which are attached like arms at the sides, are capable of erection, or of withdrawal into pockets.

There is great variety in the development of these accessory organs of locomotion. For instance, the Cydippidæ have only arms, which, with their branches, serve for capturing food as well as for steering. In other groups, vertical, oar-like, dermal folds stand out from the body, by means of which the movements become more rapid and energetic. Some species of Eucharis, by suddenly shutting up the oral umbrella, can jerk themselves forward ; and when successive jerks of this sort cause the body to move with greater speed than usual, the arms are withdrawn into their pockets or stretched backward like a rudder. This power of free locomotion necessitates some regulating organ, so that the desired direction or position of the body may be maintained. Such an organ exists at the apical pole of the body, and may be described as consisting of a small weight borne on springs, by which the oscillations of the body or deviations from the line of movement can be instantly felt.

The ventrally-placed mouth is like a large slit between the folds of the umbrella, and leads into a stomach which is either tubular or flattened. The food

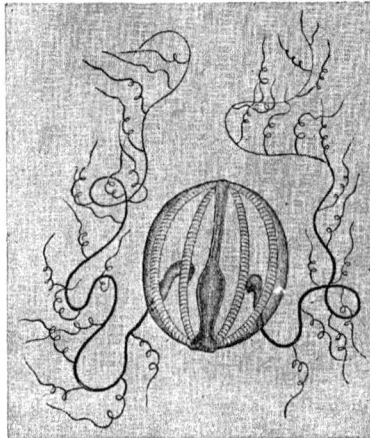

CYDIPPE

is digested in this stomach, the indigestible parts, mixed with mucus, being again ejected through the mouth. The upper end of the stomach is in direct communication with the funnel-shaped space of variable width. From this funnel-like cavity canals arise, which branch and run below the outer surface, following the lines of the ribs. This funnel further possesses an aperture of its own, opening on the exterior, in the region of the apical pole. Within the funnel is found a fluid substance containing particles of the food-pulp drawn in from the stomach, but consisting chiefly of water, taken in voluntarily ; this fluid being kept in motion by ciliary action through the canal system. Although water is also sometimes taken in through the proper apical aperture of the funnel, this aperture seems principally to serve for the ejection of the fluid when of no further use. It is then also mixed with waste matters from the body. Stinging-cells, such as occur in the next group, have as yet been found in only one species of ctenophore (*Haeckelia rubra*), and then only in small numbers.

Instead of stinging-cells, the ctenophores have adhesive cells, or small hemispherical knobs found on the tentacles, or capturing filaments ; these being provided with elastic, spirally-coiled stalks, but containing no poison. These knobs are beset with sticky globules, to which small animals, such as minute crustaceans, easily become attached. If the prey attempt to escape, the spiral thread by which the knobs are attached becomes stretched. When the thread is withdrawn, it more or less entangles the victim, and, being, like the knob, provided with a great number of sticky particles, renders escape impossible.

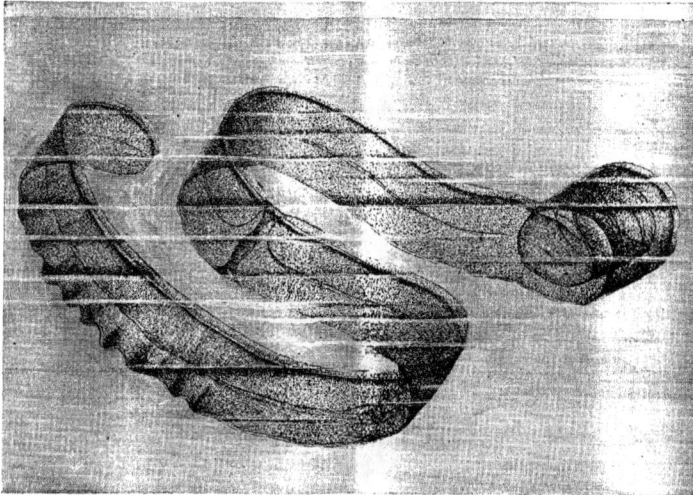

VENUS GIRDLE

Ctenophores feed upon all kinds of small pelagic animals, especially crustaceans, while they themselves fall a prey to the disc-shaped jelly-fish and sea-anemones. The largest specimens are, as a rule, found in waters sheltered from the wind. They are to be seen throughout the year, but are most plentiful during the spring months, and become rarer towards summer, when some species, such as Venus girdle, almost completely disappear. In the early autumn, however, great swarms appear, especially of *Cestus* and *Beroë*. Insignificant as these delicate creatures may appear, they delight the eye, both while living and after death, by their luminosity. This is principally displayed in the walls of the canals below the ribs.

The Ctenophora are hermaphrodite ; sexually mature animals of many species being found throughout the whole year, while others occur only in spring, summer, or winter. The young pass through a metamorphosis, or have larval stages which precede the definite form. In at least one species (*Eucharis multicornis*) sexually mature larvæ, or larvæ which are capable of reproduction as such, also occur ; these, when completely developed, become once more capable of reproduction as adults—a method of multiplication which has been called dissogony.

The most interesting, if not the most beautiful, of the Ctenophora are the *Beroidæ*—shown in the illustration on page 1766—which resemble Phrygian caps in shape. In section they are oval ; the mouth is wide, and they have no capturing filaments, or tentacles, and therefore no adhesive cells. They attain a size of 8 inches, and are of a delicate red colour, which appears marbled. This appearance is due to the branching of the eight principal canals above described, the ramifications forming a network. *Beroë forskalia*, shown on page 1766, is found in the Mediterranean. The *Beroidæ* are carnivorous, feeding on their own relations of other genera.

The *Cydippidæ* are conical, or barrel-shaped, with the ribs uniformly developed, and two opposite tentacles, as shown in the figure on page 1801. The species figured 4 on page 1766 is *Hormiphora plumosa* from the Mediterranean. The remarkable Venus girdle (*Cestus veneris*), shown in the annexed illustration, is so called because the body is lengthened out sideways like a ribbon, so that the mouth is found on the under edge of the ribbon half-way along it. The ribbon is edged with cilia, corresponding with the ciliated combs of the body proper. An additional charm is added to this beautiful form by its lively, graceful movements, the ribbon assuming all possible curves. If roughly touched, it rolls up spirally, beginning at one end. When undisturbed, its ribbon-like outgrowths are sometimes stretched out, sometimes more or less rolled up, or else the one is rolled up and the other extended. It can, like other Ctenophora, keep itself in motion by the mere play of its cilia, but it also uses the undulating movements of its ribbon-like body. The transformation of the larva after leaving the egg is complicated. The young larva is shaped liked a balloon, and possesses two principal tentacles provided with lateral filaments ; it has further, on each rib, four to five swimming-plates. At this stage this larva resembles the adults of some other species of Ctenophora, and only by degrees, after passing through many other stages, assumes the form of the girdle.

A few ctenophores (*Cœloplana* and *Ctenoplana*) have taken to creeping, and as a result of this mode of life they have become two-sided, though they are not bilaterally symmetrical—there being no distinction between head and tail ends. Probably, however, the bilateral symmetry characteristic of animals higher than cœlenterates resulted from the assumption of a creeping habit on the part of remote radially symmetrical animals, and it is interesting to notice that the creeping ctenophores resemble in many ways some of the lowest flat-worms.

GROUP CNIDARIA
CHARACTERISTICS OF THE STINGING CŒLENTERATES

THE Cnidaria, or stinging cœlenterates, which comprise the sea-anemones, corals, jelly-fish, and the little fresh-water polype or hydra of English ponds, receive their name from the so-called stinging-capsules found in their skin, which may be regarded as the homologues of the adhesive cells of the ctenophores. Before describing these offensive and defensive weapons, it is necessary to obtain some idea of the animals which use them, these cnidarians having departed less from the simple cœlenterate type than have the Ctenophora, in which this type is much disguised. Imagine, then, a long, footless stocking sewn up at each end. By thrusting one half of this stocking into the other half, there would be obtained a long bag with a double wall.

Suppose this bag fixed by its blind end to the ground, while the open mouth end stood up in the air, to catch anything that fell into it, and then suppose that, close round the mouth, the double wall grew out into arms, or tentacles, which could catch anything passing and draw it into the mouth, then we should have a structure somewhat resembling the fundamental form of the Cœlenterata. But it must further be supposed that the two woollen walls of the stocking are replaced by two layers of living cells, so that the outer one forms the skin (ectoderm), which is armed with the stinging-cells, while the cells of the inner layer (endoderm) are hungry creatures waiting to digest anything digestible which comes down into the bag. This is still not enough, as the whole animal must be able to move its tentacles, and to stretch or contract its body; so that between the layers there is a special gelatinous layer in which run muscle and nerve-fibres. Further, in order that the tentacles, when they seize a passing animal, may have no trouble with it, but may be able to bring it to the mouth as easily as possible, they are thickly covered with batteries of stinging-cells.

But how, it may be asked, can we get the beautiful bell-shaped jelly-fish from such a creature? The imaginary animal just described was fixed to the bottom of the sea, or to weeds and stones under water, and here it would grow. But there is a law of life that, after a certain size has been reached, further growth does not add to the animal's stature, but takes the form of buds, which may either be cast off as eggs to hatch and develop elsewhere, or may remain attached to and branching out from the parent animal. Both these processes take place in the simple cnidarians. Some branch and re-branch to form beautiful trees, or stocks, made up of living animals. Now, if all these animals were to drop eggs which fell to the ground to grow up around the parent stock, so fast would they grow that they would soon be killing one another through overcrowding.

Hence it has come to pass that in many forms only a certain number of the animals forming a stock produce eggs, and these are able to break away and swim off with their load of eggs, to drop them far away. In this way swimming-bells have been produced, originally only as carriers for scattering eggs broadcast, just as many trees have arrangements for

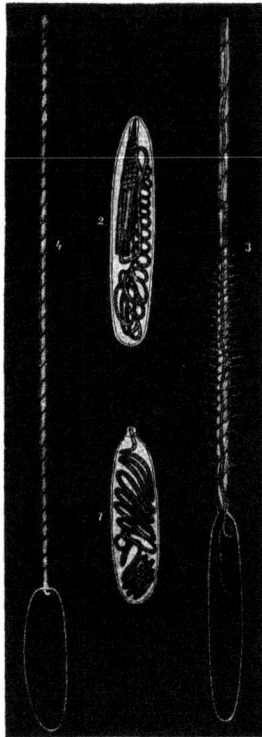

STINGING-CAPSULES
1 and 2, with retracted filament; 3, partly protruded; 4, fully protruded

scattering seeds as far as possible from the parent stem. From this beginning, all the race of jelly-fish appear to have sprung. The free-swimming life offered new fields for catching food. Myriads of small creatures swim near the surface of the water; the cnidarian fixed to the bottom of the sea may stretch its arms in vain for these, while the free-swimming bell can go amongst them and follow them along the surface currents, feeding as it goes. Hence, while the eggs of many jelly-fish when dropped develop first into fixed tree-like stocks, which, when grown, let loose another swarm of jelly-fish, the eggs of others, as if to save time, as it were, and impatient of the fixed tree-like stage, hatch out at once into young jelly-fish, which rise at one bound to all the free-swimming privileges of their immediate parents.

The former process is termed alternation of generations; the egg producing a stock, which is one generation, the stock producing a jelly-fish, which is a second, and these two alternating. In the latter case, when a jelly-fish produces a jelly-fish, one generation—the stock—has been suppressed. This is important, since there is evidence that this alternation of generations was, and, indeed, still is, widely spread in both the animal and the vegetable kingdoms. It is, however, as a rule, suppressed as animals rise in the scale of organisation.

One word as to the changes necessary to turn the simple cnidarian above described into a bell or umbrella-like jelly-fish. The principal change is in the gelatinous layer between the outer skin-wall and the inner stomach, or digesting layer. This middle layer develops into an enormous mass of glassy jelly of such a shape that, instead of the body being long from the mouth (oral pole) to the bottom of the sac (aboral pole), the animal is umbrella-like, the mouth being under the bell, while the top of the bell corresponds with the old base by which the parent polype was attached to the ground.

The stinging-cells, though all microscopic, vary considerably in size, without their structure being essentially affected. The surface protoplasm of the cell is modified into a tolerably firm shell, enclosing an oval or cylindrical vesicle. Closely associated with this structure is a pointed process, standing up far above the level of the skin, known as the trigger-hair.

Within the vesicle is found, either spirally rolled or in an irregular tangle, a long filament, or hollow tube, which is a prolongation of the vesicle, but turned outside in. This tube, which is more than twenty times as long as the cell, is pointed at the tip, and almost up to the tip beset with two rows of fine, spirally-arranged, barbed hooks. When the trigger-hair is touched or irritated, this filament is violently shot forth, being turned inside out like the finger of a glove. So long as the thread remains rolled up within the vesicle, the barbed hooks are, of course, in the tube; when it is shot out, they come on the outside. The rolled-up thread appears to be filled with some poisonous substance, which, when the tube is shot out, is ejected over the spot where the point strikes and wounds.

CLASS POLYPOMEDUSÆ—JELLY-FISH AND ALLIES

SIPHONOPHORA

We have already described the swimming-bells of the jelly-fish as the highest development of the stinging group. The Siphonophora, as represented by the Portuguese man-of-war, are, in their turn, the highest development of the swimming-bells. They are, in fact, colonies of bells, joined together in almost every possible way, and showing extraordinary modifications of individuals in the interests of a division of labour. For instance, some of the bells do nothing but row the colony along, others feed the colony, others are guards, and yet others produce the eggs.

As our first example of the group, we may take the creature known as *Physophora*, which consists of a long tube or central axis, surmounted by an individual which is nothing but an air-vesicle for holding the colony in an upright or a sloping position in the water. Below the air-vesicle come two rows of bells, which bring about by their contractions the movement of the whole colony. These rowing-bells force the water out of their cavities, and thus propel the colony. Below these, again, comes a circle of extremely mobile tentacles, which may, perhaps, be the tentacles of vanished bells. Among these tentacles are hollow structures, open at the end, which are the feeding-bells, now reduced to sucking-tubes, or stomachs, each of which endeavours to seize and digest for itself whatever in the shape of food (chiefly small crustaceans) is brought to it by the long capturing filaments and their branches, armed with stinging organs. The colourless nutritive fluid prepared by these two stomachs serves for the nourishment of the whole colony, and is carried to the various parts through the axial tube.

In the illustration, which has been chosen on account of its comparative simplicity, no reproductive or egg-bearing bells are shown. When present in the *Physophora*, these appear like clusters of grapes; in other genera they are capsules; in others, again, they may be actual swimming-bells, which become detached, and lead an independent life. This fact is of importance in helping us to understand this complicated organism. It shows that the *Physophora* is not a single animal, but a stock or colony. Of this there is evidence in the swimming-bells, as well as in the two, three, four, or more sucking-tubes, with distinct mouths and stomachs. And, lastly, we have the reproduction brought about, in some cases, by detached jellyfish-like individuals.

All the parts of the organism form a whole in a physiological sense; they belong to one life, and many are so modified as no longer to appear as individuals. But, on the other hand, some of them are fairly independent, and when they take the form of medusæ they are so highly developed that their individuality is at once manifest. We must therefore regard a siphonophore as a colony of highly-modified individuals, which, owing to the fact that these individuals differ greatly in form and function, constitute what is termed a "polymorphous colony."

One of the most beautiful and most dangerous of the Cœlenterata belongs to the Siphonophora. This is the Portuguese man-of-war (*Physalia*), several species of which are found in the southern seas. The air-bladder at the top of the stem is a large, oval vesicle, which projects above the surface, lying horizontally on the water. It is drawn out into two points at opposite

PHYSOPHORA
A compound jelly-fish

poles. A comb runs lengthwise and somewhat slantingly along the top of it. From its lower side, nutritive polypes, feelers on which the genital products develop, and very long tentacles hang down side by side below the surface of the water. Another strikingly beautiful species found in the Mediterranean is *P. pelagica*.

Lesson writes that these creatures "shimmer with the most splendid colouring. The air-bladder and its comb look like molten silver, adorned with light blue, violet, and purple. The small thickenings on the keel of the comb are of a vivid carmine, while the appendages are of an ultramarine blue." The tentacles of the *Physalia* are stiff with batteries of stinging-capsules, and those who are careless enough to touch them will repent.

During the Challenger expedition, deep-sea Siphonophora of a remarkable kind were brought to light. The most interesting belonged to a new family, the *Auronectidæ*. The colony, instead of being a long string of individuals, is here thickened, and shortened so as to be oval or round. It consists of a hard, cartilaginous mass, traversed by a close system of branching canals. The upper part of this mass is a large, round, hollow air-bladder (*p* in the figure on page 1805). This pneumatophore is surrounded by a circle of large, round swimming-bells (*n*), one of which (*l*) is modified in a remarkable way. It is not, like the rest, quite hollow, but it is traversed by a narrow canal attached to its walls by strands of gelatinous tissue. The free end of the canal opens outward through a short tube, while its attached end enters the great bladder of the pneumatophore. This specially modified swimming-bell has been called the aurophore, since it appears to regulate the quantity of air in the air-bladder. In order to sink to a greater depth, the *Stephalia* has only to contract its pneumatophore, discharging the air through the lateral canal. When the animal rises, the aurophore probably secretes a gas which fills the pneumatophore again. The lower end of the colony is occupied by a large feeding or nutritive polype, and at its sides there are several rows of smaller nutritive polypes (*s*), each of which, at its base, carries a capturing filament (*t*), and at its side grape-like clusters of reproductive bodies.

The siphonophora, as a rule, require frequent changes of depth. It does not appear that exclusively deep-sea forms are to be found in the Mediterranean, but that all siphonophora under certain circumstances and at certain seasons appear at the surface. Many pass through their larval development at a great depth, and the young *Physophora* larvæ found at the surface in the spring descend to greater depths at the beginning of summer, and only return, when their metamorphoses are complete, to develop into sexually mature animals. In the *Physophoridæ* we had the different individuals in a long series. In the *Auronectidæ* we found them arranged in a compact mass; and, lastly, in the *Velellidæ* the body is flattened out to a disc, which is traversed by a system of canals. On this disc lies the similarly-shaped pneumatophore, which is also traversed by concentrically arranged canals opening outwards.

The polypes hang on the lower side of the cartilaginous disc, a large nutritive polype occupying the centre, surrounded by concentric circles of smaller nutritive polypes. As in the *Auronectidæ*, these polypes carry at their bases genital clusters, but no capturing filaments. The tentacles are arranged round the margin of the disc, and are very short. The genus *Velella*, one species of which is frequently found in the Mediterranean, has an irregular oval disc, surmounted by a sloping comb, which acts as a sail. These jelly-fish, which are of a deep indigo colour, are often found in swarms.

HYDROMEDUSÆ

We now turn to the solitary swimming-bells, each one of which forms an individual competent to perform all the many functions required in its struggle for existence. There are hosts of these bells, of almost all sizes, some being large and beautiful, but dangerous to touch, while others are quite minute creatures, which have to be examined under the microscope. In regard to these swimming medusæ, it has been already mentioned that they were primitively individuals broken loose for a free-swimming life in the open sea from a stock attached to the ground at the bottom. The eggs of some of these forms have now given up passing through the attached stage, and hatch out at once as young medusæ.

Now, examination has shown that this host must be divided into two groups, having remarkable differences, the one being called the Hydromedusæ, and the other the Scyphomedusæ. The two came from two different kinds of attached stocks, and consequently, as free-swimming animals, in spite of their general resemblance to one another as jelly-fish, each has organs which the other wants.

Taking the Hydromedusæ first, as closest to the Siphonophora, we describe a few in detail, in order to give a clearer idea of the alternation of generations.

Among the Hydromedusæ there are the following different life-histories. Beginning with the highest, we have (1) jelly-fish alone, the eggs of which have given up forming stocks, but hatch out jelly-fish; (2) jelly-fish, the eggs of which still form stocks, some individuals of which swim away as jelly-fish; (3) stocks in which the sexual individuals do not swim away as jelly-fish. We need not here describe any of the medusæ in detail, since the much larger jelly-fish of the Scyphomedusæ will claim our attention presently, but two remarkable forms, which have taken to creeping on the ground, deserve attention. In Dalmatia, on seaweed, a delicate, pale object can often be discerned with a magnifying-glass creeping laboriously about on its long arms. If detached from the seaweed, it falls to the bottom, as it is unable to swim.

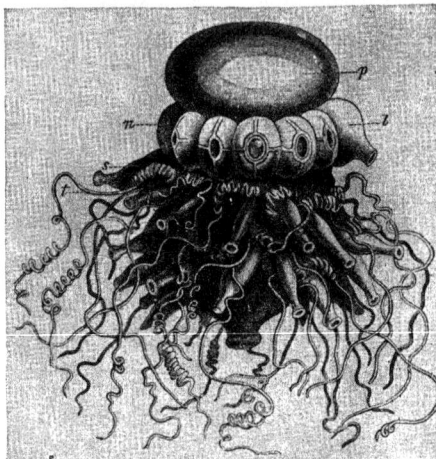

STEPHALIA—A JELLY-FISH
For explanation of lettering see previous page

In each point of its structure this animal is a medusa, related to the genus *Eleutheria*, or *Cladonema*, but still further removed from the ordinary medusa in one respect, since the *Cladonema* alternately swims and creeps. This creeping medusa (*Clavatella prolifera*) has six arms, the tips of which are provided with true suckers. On these it walks, as on stilts, while from each arm a short stalk rises, the swollen end of which is beset with stinging-capsules. The very extensile mouth-tube moves about tentatively, and easily seizes upon the small crustaceans to be found upon the seaweed. Just above the base of each arm lies a horseshoe-shaped eye-spot containing a well-developed lens. Somewhat higher up, between every two arms, a bud is to be found. None of the specimens of a certain size examined in May were without their six buds, these being at such different stages of development that their gradual growth could be clearly traced. On the riper buds the rudiments of a second generation of buds were to be seen. Multiplication by budding has been observed in other medusæ, and it is from such budding medusoid colonies that we may, perhaps, deduce the remarkable swimming colonies of the Siphonophora. As a rule, however, all medusæ multiply sexually by means of fertilised eggs; even the *Clavatella* at other seasons lays eggs.

Of the forms among which the reproductive individuals swim away as jelly-fish, we may take as an example *Corymorpha nutans*. Between the five individuals of this species grouped together in the illustration five small creatures, each provided with a filamentous appendage, are to be seen swimming, which are the medusæ belonging to this animal. Each egg of these minute medusæ, which are by no means large in size, develops into a ciliated larva, which, sinking to the bottom, grows into an attached *Corymorpha*. The illustration on the next page shows these animals, which in the polype form are always single.

Unlike most animals of this sort, they do not attach themselves to seaweed or stones, but live on fine sand, into which they sink the posterior end of the stem. Numerous thread-like appendages of this buried part penetrate the sand in all directions, thus firmly attaching the animal. The mouth at the anterior end is encircled by tentacles, a second circle of tentacles surrounding the widened part of the body which contains the stomach. Immediately above this latter circle, the buds stand in clusters; in summer they are found in all stages of development, and even while attached to their stalks assume the complete structure of a medusa. They move their umbrellas actively,

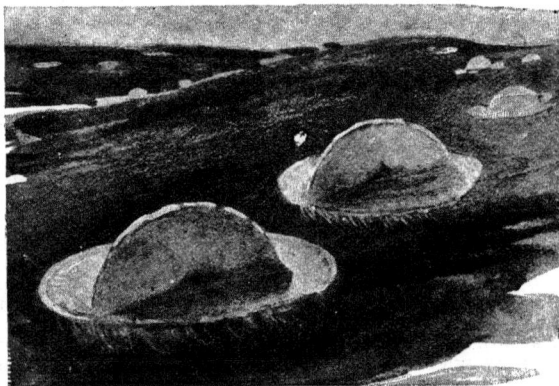

VELELLA SPIRANS

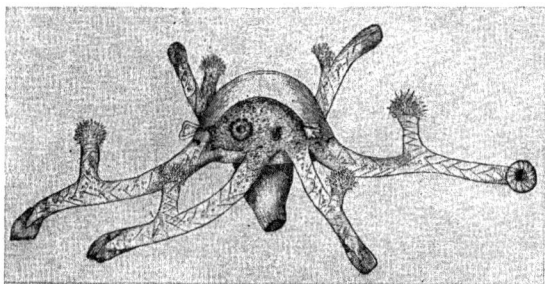

CLAVATELLA

break loose, and thus complete the circle of development or alternation of generations. *Bougainvillea ramosa* is another form in which, owing to the stock being branched, the division of labour is even more clearly seen ; some of the individuals are feeding and some are reproductive, these latter turning into swimming-bells and breaking loose. Both these forms are small, as indeed are the great majority of the hydroid stocks, but whole forests of hydroid-polype stocks may be seen on the reefs in the Pelew Islands almost as tall as a man, and with roots three or four inches in diameter. A bather entering such a forest is terribly stung, the pain lasting for hours. A solitary form (*Monocaulus imperator*)—the upper portion of which is here figured — nearly related to *Corymorpha*, and found in the Northern Pacific, attains still larger proportions. Specimens brought up during the Challenger expedition from a great depth were more than two yards in length, with a proportionate diameter.

As examples of stocks of which the reproductive individuals do not swim away as jelly-fish, we may select the pretty, feathered, plant-like creatures found along the seashore, which are often thought to be plants, but are really animal colonies, well-known types being *Sertularia* and *Plumularia*. In these cases, in addition to the nutritive individuals, there are the egg-bearing individuals, which never turn into free-swimming medusæ.

One small form which is not branched and feathered is *Hydractinia echinata*, found in the North Sea and on the English and Norwegian coasts, where it attaches itself to gastropod shells inhabited by hermit-crabs. The polype probably profits by changes of place for feeding, or else for some other reason adapts itself to the restless life of the crab, while the latter is protected by the stinging powers of its associates. The two organisms may be termed commensals, or messmates, and the term commensalism is applied to such cases. The part of the stock common to all the individuals is the skin-like portion which adheres to the surface of the shell or other object to which it is attached. This skin is raised up into spiny prominences, as shown in the figure on page 1807. A horny layer occurs in this integument, similar to that of which the single tubes consist. The nutritive canals running down the stems of the polypes are continued into this membrane, promoting its life and growth.

In such a stock there are never more than two kinds of individuals—namely, the nutritive individuals, distinguished by their long tentacles, mouths, and

digestive cavities ; and the reproductive individuals, male or female. These latter have no mouths, and are supplied with food through the system of canals running to them from the nutritive individuals. These reproductive individuals, instead of tentacles, carry at their tips a circle of stinging-batteries for the protection of their eggs, which are enclosed in capsules clustered together round the stalk a little way below the tips. The ciliated larvæ emerging from the eggs swim away, and eventually become attached and found new colonies. The egg-capsules in no way recall medusæ, but all medusæ which pass through a polype-like intermediate stage also pass through the simple capsule stage.

Two more of the hydroid stocks are worth mention, since they secrete masses of carbonate of lime, out of which the animals protrude like corals, which, indeed, they were thought to be. This error was made because only the massive skeletons of the Hydrocorallia—as they are called—and not the living animals, were known. Instead of the horny, often delicately branching integumentary skeleton usually found in the hydroid polype stock, that of the Hydrocorallia contains 97 per cent. of carbonate of lime, and forms rough, solid-looking masses, with lobed processes or bosses like those figured on page 1807, or else (*Stylasteridæ*) branches, like the precious coral of commerce.

The whole surface can be seen even with the naked eye, but still better with a lens, to be covered with small pore-like apertures. Closer examination shows that these are of two sizes, a larger central pore being surrounded by an irregular circle of from five to eight smaller ones. The mass of the colony is traversed by an irregular system of numerous branching canals of different sizes. In vertical section, indistinct layers can be seen running almost parallel with the outer surface. These form the floors of the polype-cavities. Only in the outermost layer of the stock is there life, the inner mass being composed of dead skeleton. In this living layer there is a close network of soft branching tubes, from which rise the small polypes, the bases of which are connected together by this network.

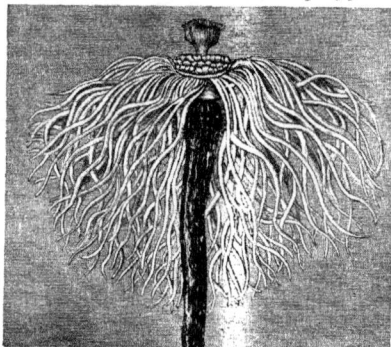

UPPER PORTION OF MONOCAULUS

CORYMORPHA, WITH DETACHED MEDUSÆ

The polypes lie in cup-like depressions, and, when undisturbed, project outward through the pores, retreating instantly at the slightest disturbance.

The polypes, like the pores, are of two sorts. Those inhabiting the larger pores are short and thick, with four short tentacles, resembling stalked globules, surrounding a comparatively spacious mouth. The polypes protruding from the more numerous smaller pores, which surround the large ones, are much longer and have no mouths. Each of these ends in a simple knob, below which, at intervals, and generally alternately on one side and the other, short, simple branches are given off. The central polype remains quite still, but those which surround it are constantly in undulating motion, often bending down to the mouth of the central polype, which they appear to be feeding. Here, again, there is division of labour in an animal colony, the larger central polype provided with a mouth being the feeding individual, while the mouthless nutritive individuals catch the prey. The smaller polypes also probably defend the colony, being far better armed with stinging-capsules than the larger polype. The knobs with which the tentacles end are stinging - batteries. They grow upon rocks or dead corals, often covering the skeletons of sea-fans (Gorgoniidæ), and are even found in the Bermudas on old bottles thrown into the sea. In the latter case the lower side of the stock is quite smooth as if polished, and reproduces exactly the surface of the glass with all its markings.

There are two families of these Hydrocorallia, as they are called—namely, the Milleporidæ and the Stylasteridæ. The members of the former family produce large numbers of little free-swimming jelly-fish as reproductive buds. These forms are of great interest as illustrating the marvellous adaptability of living forms. While the true corals, which are polype-colonies somewhat differently organised from these hydropolypes, secrete great masses of solid rock, we find two small families of minute hydropolypes also building up hard coral-stocks. This phenomenon is called convergence ; two different kinds of animals, starting from different points, become adapted to similar conditions of life, and eventually come superficially to resemble one another. Just as these hydropolypes forming coral were long

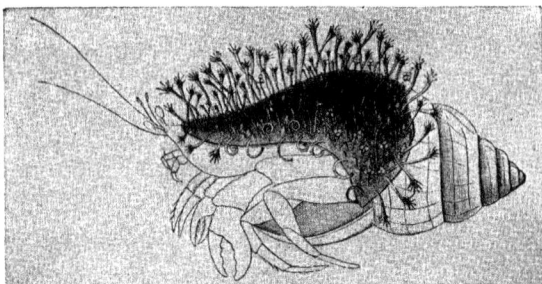

STOCK OF HYDRACTINIA ON WHELK-SHELL INHABITED BY A HERMIT-CRAB

thought to be true corals, so many other animals have, on account of their resemblance, been classed together which are now known to belong to different groups.

FRESH-WATER FORMS

Two other hydroid polypes which live in fresh water, while all the rest are marine, deserve mention. Of these, Cordylophora lacustris forms branched trees from one to three inches high, rising from a network of roots attached to stones, wood, mussel-shells, etc. The whole stock—except the club-shaped heads of the individual polypes, which are provided with proboscis-like mouths and irregularly-branched thread-like arms—is covered with a delicate horny envelope. In these stocks, which are of a red-grey colour, the sexes are separate. Until the middle of the nineteenth century, the Cordylophora had only been met with in brackish water on the coasts of Europe and of North America. After that it appeared from time to time in the lower courses of rivers, such as the Thames, the Elbe, etc., and now it has found its

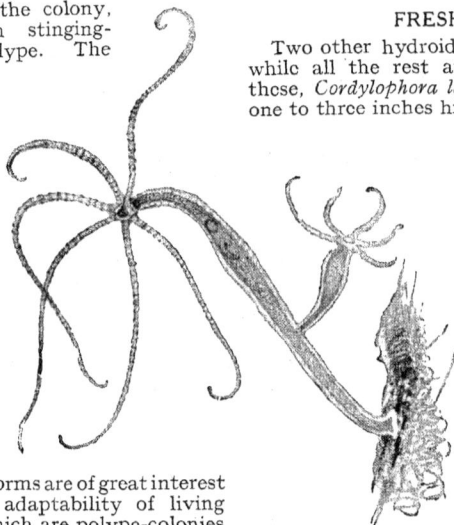

FRESH-WATER POLYPE

way far inland both in the Old and New Worlds. It occurs in the Saale, near Halle, and is specially plentiful in the slightly brackish lake of Eisleben. In Hamburg it has in some places invaded the water-pipes supplying the city, developing in them to such a degree as actually to stop them up.

This history of the migration of Cordylophora is instructive in helping us to understand the rise of at least a part of the fresh-water fauna. In this case, within our own experience, an animal inhabiting brackish water has in a few years become so adapted for living in fresh water as to be considered altogether a fresh-water form, without the least apparent change in its organisation. Whether a change in organisation would not gradually take place in the course of many years is, of course, another question, which for the present is unanswerable.

In the fresh-water polype, or Hydra, we have a hydropolype much better known and much more specially adapted to its habitat than the Cordylophora. These hydras, which are from one-eighth to one-third of an inch in length, form simple stocks of one or two branches, and as often as not are found single. They almost exactly resemble in form those polypes of the Hydractinia which are provided with a circle of tentacles.

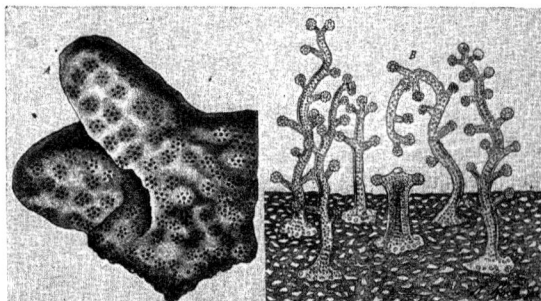

MILLEPORA

A, part of stock with polypes withdrawn ; B, five peripheral nutritive individuals round a central feeding individual

The water of stagnant pools or ponds in which water-plants are abundant will almost certainly yield one of the three species of the fresh-water hydra, if the water-plants be left undisturbed in a vessel. The little creatures often leave the weed and attach themselves to the sides of the vessel, where they can be examined with a lens. When undisturbed, the polypes begin to extend and spread out their six or eight tentacles like fine threads. Small creatures coming in contact with these tentacles remain attached to them, caught and held by the stinging-threads, whereupon the tentacle contracts, bringing the prey to the mouth, which is capable of great extension. The hydra usually multiplies by means of buds which grow out of the body. The offspring often remains attached to the mother until it, in its turn, has given rise to one or two buds. Single eggs, however, develop from time to time in the body-wall beneath capsule or wart-like prominences. The hydra is also remarkable on account of its capacity for regenerating lost parts of the body.

SCYPHOMEDUSÆ

In the Scyphomedusæ we again have free-swimming jelly-fish, stocks developing into jelly-fish, and persistent stocks which never form jelly-fish. Whereas in all the Hydromedusæ the mouth opens directly into the stomach, in the Scyphomedusæ, and their attached and related forms, the skin round the mouth has been drawn in to form a tube which opens some way down into the stomach; the drawing-in of this mouth-tube, or œsophagus, having led to the formation of ridges on the wall of the stomach, which hold the inner end of the tube in place, as shown in the illustration of *Monoxenia* on page 1809. Although this does not appear important, it indicates a higher specialisation.

Taking first the free-swimming jelly-fish, the larger and more characteristic forms are distinguished by their delicate colouring. The yellow and yellowish red *Chrysaora ocellata* are seen floating past in thousands off the southern coast of Norway. The western harbours of the Baltic Sea, after continuous northerly winds, are often filled with whole banks of the blue *Aurelia aurita*, and the splendid *Rhizostoma* are constantly to be met with in the Mediterranean and Adriatic Seas. On a fine spring day they are almost always to be found on the shore, where these large, reddish blue, living hemispheres are wrecked, and soon melt away.

In these medusæ—which are well known to all who dwell on the coast, and range from one to seven inches in diameter—we have the most highly-developed of the simple cœlenterates. Their body consists for the greater part of the circular umbrella, the margin of which is notched all round so as to hang down in small or large lobes. There are also, along the margin, from four to eight or more eye-like spots, and extensible filaments. At the centre of the lower side of the disc is the mouth, which in some forms lie at the end of a projecting stalk, and is almost always surrounded by several thicker folded processes for the capture of prey. In some cases the folded edges of these ribbon-like arms fuse together, leaving only sucker-like apertures.

Canals run from the sac-like cavity representing the stomach to the edge of the disc, where they enter a circular canal, often provided with apertures. The similarity between this apparatus of digestive canals and the arrangement obtaining in the Ctenophora is then evident. The reproductive organs lie either in special sacs round the stomach, or merely in widenings of the canals. The surface of the skin is provided with innumerable microscopically small stinging capsules, and, thus armed, these so-called discomedusæ float about in the water, their bodies being but little heavier than the water itself.

Although some of these splendid forms develop directly as jelly-fish from the egg, the great majority begin life as attached polypes, so that we have here again another instance of alternation of generations. The sexes are usually separate, and from the egg arises a ciliated larva, which is oval, hollow, and somewhat flattened, recalling the shape of a locket. This is the so-called *planula*, which for a time swims about, then attaches itself firmly by the end of its body and becomes pear-shaped, the stalk of the pear being represented by the attached end; a horny envelope is then secreted over the whole surface, the mouth breaks through the free end of the central cavity, four tentacles appear, and we have a four-armed polype, or *scyphistoma*.

The tentacles increase in number, and the *scyphistoma* can produce at its base a number of young polypes, which again can multiply by division. At a certain period this method of multiplication by budding of the polypes from the base ceases, and each *scyphistoma* divides up in quite a different fashion. The polype becomes horizontally constricted in several places, until it appears like a number of cups placed one inside the other; this is called a *strobila* (pinecone). When ready, the top cup breaks away, turns over, and swims off as a young form of medusa, called an *ephyra*, which gradually acquires the shape of the perfect discomedusa.

We thus have here an alternation of generations in which a sexual medusa-generation is succeeded by an asexually-reproducing polype-generation, this again being followed by another medusa-generation.

In relation to these, and constituting a kind of transition form connecting the discomedusæ and the polypes, are the Calycozoa, or cup-shaped medusæ, which either swim about freely or are attached by their apices, where the firm gelatinous disc attains its greatest thickness. At the margin of the disc, these forms carry eight to sixteen arm-like processes.

In the attached forms (*Lucernaria*) the ends of these processes are provided with short tentacles, occasionally broadened into discs and used for attachment, and also with stinging-capsules. The Calycozoa may leave their place of attachment and swim about for a time, with a rotatory motion, and then again settle down. *Lucernaria* has been found as deep as 3,300 feet, but appears to prefer to settle in shallower water.

The nearest relations of *Lucernaria* are the *Tesseridæ*. The species of *Tessera* are small, and swim freely, having an elegant, long, bell-like shape. The edge of the disc is drawn out into alternately longer and shorter arms, eight to sixteen in number.

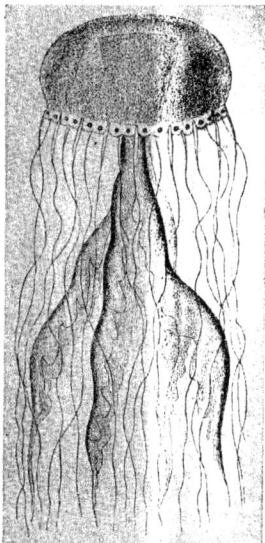

CHRYSAORA

CLASS ANTHOZOA—SEA-ANEMONES AND CORALS

WE turn from the free-swimming Scyphomedusæ to the permanently fixed polype forms—namely, the sea-anemones and corals. Although Aristotle and his contemporaries recognised the sea-anemones as animals, almost two thousand years elapsed before corals were considered to be related to them. In describing the development of a small coral discovered on the Arabian coast, and named *Monoxenia darwini*, Haeckel states that the polype, which is one-eighth of an inch long, is of strictly radiate structure, the mouth, which lies at the upper end of the cylindrical body, being surrounded by eight feathered tentacles. It is attached to some substratum by means of a flexible disc at the opposite end of the body to the mouth. It is clear that it has no hard skeleton, as the shape of its surface is changeable; and its internal structure must be shown by transverse and longitudinal sections.

The development of *Monoxenia* begins with the egg repeatedly dividing into many parts (*C, D, E*). This process, which is common throughout the animal kingdom, and is called egg-cleavage, in this case proceeds so simply and regularly that it ends in the production of a hollow sphere enclosed by a single layer of cells (*G*). Each cell sends out a long cilium or whip-like process (*F*), by means of which the larva turns about and swims in the body-fluid of the parent polype. One half of the sphere now becomes infolded into the other half (*H*), and forms what is called a gastrula (*I, K*). The term gastrula has taken a great place in zoology in recent years, since the Russian naturalist Kowalewsky found that many different classes of animals, in developing from the egg, passed through such a stage. Haeckel, generalising from these facts, invented his Gastrea theory, according to which all animals in which the gastrula stage occurs must have been descended from a common primitive form, Gastrea, which has, however, in its simplest form long been extinct, but of which the cœlenterates are the closest modern representatives.

The gastrula of *Monoxenia* is of the simplest kind, the infolding being complete, and the larva forming a sac, whose walls consist of two layers of cells, or germinal layers, an outer ectoderm and an inner endoderm (see section given in the illustration). The transition from the flat dish-shape (*H*) to the sac with a narrow mouth is at once clear, and the knowledge that all the cœlenterates proceed from a similar larva, and that

MONOXENIA DARWINI

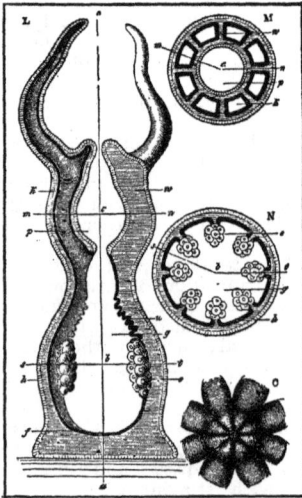

DIAGRAM EXPLAINING MONOXENIA

L., longitudinal section, on left through intersceptal cavity, on right through a partition wall; *M*, transverse section through line *m n*; *N*, transverse section through line *s b t*; *O*, eight-lipped mouth-aperture, with bases of arms; *a b c o*, principal axis; *p*, pharyngeal cavity; *g*, digestive cavity; *k*, divisions of digestive cavity; *w*, radial walls dividing up digestive cavity; *e*, masses of eggs; *u*, mesenterial filaments; *f*, masses of muscle and connective tissue

all the complications of their various systems are developed from such a simple gastrula, throws much light on their anatomy. During these transformations, the endoderm, whose cells multiply, continues as an uninterrupted lining to the stomach and its appendages, while the ectoderm yields the constituents of the skin. A third intermediate gelatinous layer, the mesoglœa, arises between the outer and inner layers; in this muscles and connective interstitial tissues appear. The chief part of the jelly forming the great umbrella of the Discomedusæ consists of this mesoglœa. In the mesoglœa of one division of the corals the calcifications take place. These internal calcifications play, however, but a very small part in the great rock-making activities of corals as a whole, the most important calcifications being external.

Returning to Haeckel's account of *Monoxenia*, although the transition from the gastrula larva to the adult animal has not been observed, there can be no doubt as to how it takes place; all the transformations having been watched in other species. The larva attaches itself with the end opposite to the mouth, the cilia disappear, and after the mouth-tube (*p*) has been formed by the folding-in of the anterior end along the longitudinal axis of the body (*L, o, a*), and has thus become marked off from the stomach (*g*), the eight hollow tentacles rise round the mouth as outgrowths of the body-cavity, or as direct continuations of the stomach.

Like all other corals, *Monoxenia* periodically multiplies by means of eggs, which arise either in the walls of the radiating stomach-partitions (or septa), or on their free edges, and have to be ejected through the mouth, as development does not in this case take place within the digestive cavity of the parent polype. As a rule, the polypes are either male or female, but in stock-forming species individuals of the two sexes may be mixed. Hermaphrodite individuals are less frequent.

Monoxenia may be taken as the simplest type of the regularly radiate polypes; in all radiate animals the different organs being repeated in regular rings round the central axis. *Monoxenia* and similar animals are considered simple, because the repeated organs develop similarly and simultaneously, and are comparatively few in number. The mouth, too, is circular. In many other polypes, however, the regularly radiate type is slightly

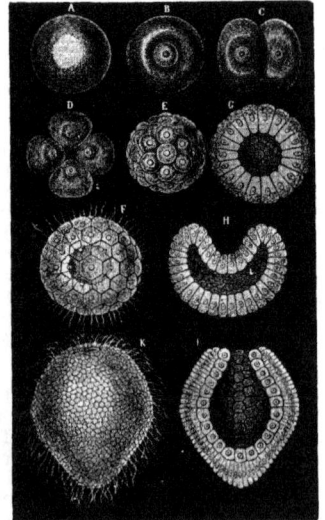

DEVELOPMENT OF MONOXENIA

The egg (A and B) divides into many parts (C, D, and E), producing finally a hollow sphere enclosed by a single layer of cells (G). Each cell sends out whip-like processes (F). One half becomes infolded in the other (H), forming a gastrula (I and K).

departed from ; the mouth, instead of being round, forming a long slit, while there is a tendency for the originally radiate animal to become bilateral.

From this account of a simple polype, it is easy to understand what kind of animal it is which makes coral ; and our readers, if they have not already done so, will give up speaking of "*insects* building up the coral-reefs." It is, however, by no means all such polypes that form coral, nor do those which form it produce it always in the same way Numbers of polypes, such as the beautiful sea-anemones, never produce any hard substance, but remain soft and delicate, though dangerous, at least to small animals, because of their stinging-cells. Many of these soft sea-anemones are highly specialised creatures, as may be seen from the picture on page 1817 ; but those which secrete coral are generally simpler, and smaller, and grow in vast colonies. It is the accumulation of all their little contributions of coral which, in the process of time, builds up islands or even continents.

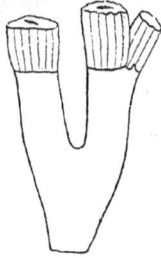

CAULASTRÆA
Outline showing living and dead portions

In regarding coral animals as reef-builders, we may leave out of account, as unimportant, those which form hard spicules within their bodies, and consider only those which perform most of the work. Imagine a crowd of small animals like sea-anemones fixed to a rock, each one secreting a layer of carbonate of lime between itself and the rock, and this layer becoming thicker and thicker till each polype rises on a little pedestal. There is probably a race between them, as there is among trees in a forest, which shall reach the highest to get most of the food as it passes by on the currents in the water. Now, it is obvious that a crowded colony of polypes like this would in a short time add a thick layer of solid carbonate of lime to the rock on which they first settled, and this is, in brief, the principle of reef-building. As a matter of fact, however, it is not quite so simple. The layer each polype secretes is not a smooth, flat disc, evenly secreted by the whole surface of its foot. Some parts of the foot secrete more than others, hence those parts rise up as spines, walls, and rings, which protrude into the body of the polype without, however, breaking through the skin. These probably help to fasten the polype to its pedestal, and prevent it from being swept off by strong currents. The figure of *Thecocyathus* on page 1812 is a good illustration of one of these plates.

Each genus of corals has a pattern of its own, each one perfect and beautiful in its way, and it is frequently a puzzle to discover how it is made. When crowds of polypes grow in contact, their pedestals will also grow in contact and form continuous masses ; this growing in contact being ensured by their ordinary method of multiplication. For a coral-polype does not have to wait until another takes up a position beside it, but as soon as it can feel freely it begins to bud or divide, producing a number of young polypes close around it. These also bud in their turn, and are soon surrounded by young polypes, and in this way such compact colonies are formed that it is a struggle among the inner ones to avoid being suffocated. We thus have densely-crowded colonies of polypes struggling upwards, each individual secreting a more or less beautiful and complicated pedestal. The pedestals are fused together in a hundred different ways, and from these different patterned pedestals, with their various ways of fusing together, are produced the almost countless different kinds of coral which together build up coral-reefs.

In a growing polype-stock the individuals usually remain in organic connection—that is to say, each first provides for itself, and then shares its superfluity with the others, sometimes by means of a continuous reticulated system of canals running from polype to polype, perforating the stony substance which often separates the members of the one stock from another. The whole stock may thus be physiologically one creature with many mouths. Where, however, the secretion of the pedestal is very rapid and the budding very slow the polypes may separate, each standing at the end of a branch, the illustration of *Caulastræa* showing an example of this. It will be understood from this description that only the layer of growing polypes with their intercommunications can be spoken of as living ; and as this layer rises higher and higher by secreting fresh layers of carbonate of lime, the living linings of the communicating canals are either withdrawn or die away, and all beneath the living layer is mere dead matter built up and left behind by the coral animals.

Before passing to our survey of the corals themselves, two other points deserve attention. Not all corals form stocks. Some remain single, like the mushroom-corals (*Fungidæ*), which grow to a very large size with a heavy solid skeleton ; and although these form new polypes by budding, the latter become detached and live as solitary individuals. Again, although coral-reefs are due to the great power of multiplying by division or budding, yet all corals, so far as is known, also at certain times produce eggs. The further development of these eggs gives rise ultimately to a small polype, which settles down and begins to secrete its pedestal and then to bud, thus starting a new coral-stock.

HEXACTINIA

The name of six-rayed polypes (order Hexactinia) assigned to these sea-anemones must not be taken too strictly. It is true that it was applied in good faith, because it was believed that this order always had exactly six or some multiple of six as the number of the tentacles ; but corals are tied by no such rigid rules, and all we can say is that the number of tentacles in this order generally approximates to some multiple of six. Among the Hexactinia the sea-anemones take the first place. They spread over all seas, being especially plentiful in the temperate zones, near the coast, at depths which bring them within the reach of every observer. They are distinguished by their solitary manner of life, their size, and their vivid and usually beautiful colouring. The skin is firm and leathery, and

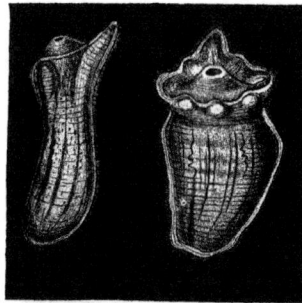

LARVÆ OF SEA-ANEMONE

often covered with warts. It does not secrete any calcareous skeleton either inside or outside, so that the animal is soft and capable of great contraction and changes of shape. Most sea-anemones use the basal disc for attachment, and can move from place to place by means of it, but a few species bore into sand with the posterior end of the body, or else secrete or build a sheath which they inhabit. In the coloured picture on page 1817 are depicted, in their natural brilliant colours, a number of sea-anemones of which the following are especially striking.

B, *Sagartia viduata*, the snake-locked anemone; B I, *Sagartia chrysosplenium*, the gold-spangled anemone; D, *Actino oba dianthus*, the plumose anemone; E, the scarlet-fringed *Sagartia miniata*; F, *Bunodes coronata*, the diadem pimplet anemone; G, G I, and G 2, varieties of *Actinia*; L, *Corynactis viridis*, the globehorn anemone; M, *Bunodes thallia*, the glaucous pimplet anemone; and R, *Stomphia churchiæ*, the beautiful gapelet anemone.

Other beautiful forms are found in the two striped anemones, *Ragactis pulchra* and *Cereactis aurantiaca*. The sun-anemone (*Heliactis bellis*), again, varies greatly in colour, but is always elegant, and the same may be said of the trumpet-anemone with its long tentacles (*Æptasia conchii*). In the foreground at the centre a hermit-crab is seen carrying with him on a whelk-shell his guest, *Sagartia parasitica*. A less conspicuous anemone (*Eloactis mazelii*) is provided with somewhat long cylindrical tentacles. The *Anemonia sulcata* lets its tentacles float gracefully, while the vestlet (*Cerianthus membranaceus*), of varying colour, hungrily stretches out its arms in all directions. *Cladactis costæ*, which is covered with warts, is no less voracious, but with apparent apathy allows its tentacles to droop around it.

Sea-anemones are extremely voracious, devouring large pieces of flesh, and sucking down mussels and oysters. Anemones only settle in places where the currents bring them the animal food they need; and are most plentiful where the current is strongest, as, for instance, at the entrance of harbours or on rocky coasts. Some species are in the habit of settling on other animals whose requirements make them frequenters of disturbed waters, hermit-crabs being especial favourites. Certain species, again, such as the large yellow-and-brown-striped *Actinia effæta* (see illustration above), are indeed always found fixed upon the shells inhabited by these crabs, the one mentioned being generally found with *Pagurus striatus*, a large Mediterranean hermit-crab which inhabits whelk-shells of suitable size. Two or three of these anemones often settle on one crab, which is protected by their stinging powers; while they in turn profit in the matter of food by the wanderings of their host. It is a good illustration of commensalism.

Prof. Alcock (in "A Naturalist in Indian Seas") speaks as follows of this kind of association: "Sea-anemones here — *i.e.*, on the Orissa coast— for the most part, were found attached to the shells of hermit-crabs, etc., a case of Hobson's choice sometimes, no doubt, but also sometimes illustrating that happy bond of commensalism, or Platonic union, which is one of the most valuable object-lessons for man's edification that marine zoology affords.

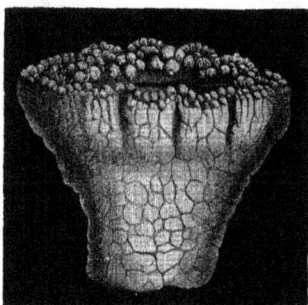
SHORT-TENTACLED ANEMONE, POLYSIPHONIA

When two animals of different grades in the zoological scale live together in such a fashion that each one assists the other in some definite way, while doing it no manner of harm they are termed commensals, or messmates. For instance, when a hermit-crab and a sea-anemone live together, the hermit-crab, being by nature a very ill-clad and vulnerable animal, acquires by the partnership a thick and easily adjustable greatcoat; while the sea-anemone, being by nature a hopeless lump of an animal, dependent on chance currents for its food and oxygen, acquires an engine and intelligent engine-driver all in one, which are always carrying it in the way of the necessaries of life; and yet with this mutual assistance there goes absolute independence in all other respects, such as mistresses and servants, who would both be none the worse for a little knowledge of the principles of zoology, never dreamt of."

SEA-ANEMONE, ACTINIA EFFŒTA

On account of the ease with which anemones are kept in captivity, their manner of reproduction has been well observed. With rare exceptions, they develop from eggs. Dalyell kept one for six years, and reared from it upwards of two hundred and seventy-six young ones. Two of these young lived for five years, producing eggs at ten or twelve months old, which hatched a couple of months later. He saw that the ciliated, infusorian-like larvæ settled down on the eighth day, losing their cilia, the first tentacles appearing during the process of attachment. Young anemones often pass through their whole development within the body-cavity of the parent.

Most anemones are provided with several circles of more or less similar cylindrical tentacles, but there are some specially beautiful species which, besides tentacles of the usual shape, have, either within or outside of the circle of ordinary tentacles, lobed or leaf-like tactile and seizing organs. These belong to the family of the *Crambactinidæ*. The beautiful *Crambactis* from the Red Sea, shown in the illustration, has, immediately round the mouth, several circles of delicate grasping tentacles, shaped like curly cabbage or endive leaves.

ENDIVE-ANEMONE, CRAMBACTIS

All the tentacles of the sea-anemones are hollow, with a fine aperture at the tip, through which, when the animal contracts, the water contained in the body-cavity can be expelled; but in the deep-sea forms these organs are very curiously modified. For instance, in the genus *Polysiphonia*, here illustrated, the tentacles are short and unsuited for catching and holding prey; but the aperture at the tip is large, and through it flows in water containing organic detritus which can be used as food. The allied *Sicyonis* has sixty-four wart-like tentacles with wide apertures standing in a double circle round the mouth, and in *Liponema* the body-wall is perforated by several hundred apertures leading into the digestive cavity and corresponding to the tentacles.

Although most members of the group arise as single individuals from eggs, some multiply by the detachment of small pieces from the pedal disc. Fischer observed this process in the translucent anemone (*Sagartia pellucida*) on the French coast. The pieces detached on August 23rd had, by September 7th, developed into small individuals with fifteen or sixteen tentacles. Multiplication by fission seems common in several species, such as *S. ignea*, and always ends in producing single individuals. Sea-anemones sometimes, however, form stocks, which are not very numerous, though some species can be found on European coasts. The genus *Zoanthus*, in which the separate individuals are united by a creeping branching root, is distinguished from *Epizoanthus*, in which the common stock resembles a root-like crust, on which the polypes form irregular groups of various sizes.

TRUE CORALS

From the foregoing observations it will be seen that in the soft division of the Hexactinia, or six-rayed ane-mones, there are both single individuals and colonies of individuals joined to-gether to form stocks; and there is also the same diversity in the skeleton-producing division—the corals proper, where we have both single individuals and stocks. Whereas, however, in the soft division, the simple individuals are the more numerous and the colonies comparatively rare, among the corals the opposite is the case, the colony-forming types presenting almost innumerable varieties. This is not difficult to under-stand, since the soft anemones cannot well form

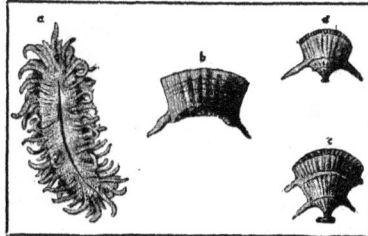
FLABELLUM
A scarlet crisp-coral

if the buds remain attached to their parents. When, however, solitary corals bud, the buds fall off, and lead solitary lives like their parents.

Most of the numerous species of the scarlet crisp-corals (*Flabellum*) are individuals, and are characterised by the slit-like form of the mouth. At *a* in the illustra-tion the living animal is seen from above, while *b* shows a side view of the skeleton, which is attached. It resembles a pair of fans fastened along their edges, and just inside the outer edges of the fans is the row of tentacles. The whole animal is as if the upper end of a circular polype had been squeezed, so that the mouth-area, instead of being round, formed a long oval (*a*). An interesting case of budding occurs in these corals, the buds falling off. In the illustration *c* shows the bud growing out at the top of an indi-vidual like *b*. In this budding con-dition the coral might pass for a different species of *Flabellum*. The bud, however, ultimately falls off (*d*), but, instead of becoming attached, is swept by the waves into some rocky fissure, where it spends the rest of its life. Besides the fact that it remains unattached, this bud differs from its attached parent in a far more important respect. It can produce

THECOCYATHUS
A simple coral

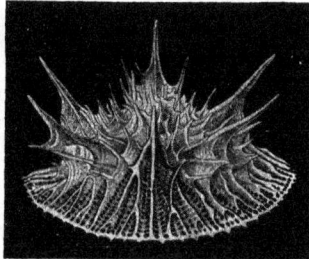
LEPTOPENUS DISCUS
A deep-sea coral

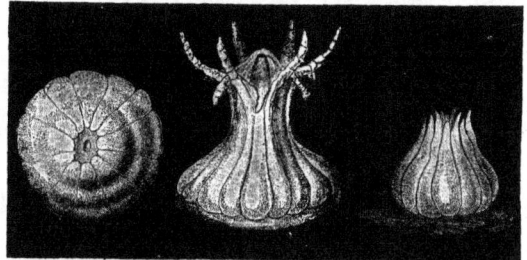
ASTROIDES CALYCULARIS
Showing developmental stages

complicated colonies, whereas the skeleton-forming polypes, by combining their skeletons, can build complicated structures, in order to raise themselves into more advantageous positions. We have first, then, to consider those corals which do not typically form stocks, but remain at the stage of a simple sea-anemone, only with a rigid, calcareous skeleton support-ing, and no doubt protecting, them in different ways.

All the corals found in British seas are (with the exception of the so-called tuft-coral) single, and generally very small. As an example of a regular, circular, solitary coral, we may take *Thecocyathus cylindra-ceus*, the skeleton of which is shown in the illustration above. The animal, when expanded, fills up the central depression, but when, on expelling the greater part of the watery contents of its cavity, it contracts, the whole body seems to sink into the hollow cup formed by its skeleton. In the illustration we see only the outer wall and the top of the ring of septa, which are solid vertical plates, rising up from the pedestal secreted by the foot and radiating outwards in all directions. Two other solitary corals are worth describing, as they show certain interesting specialisations. Both of them may increase by budding—that is, by the method which, in colony-forming corals, leads to the formation of stocks—

MUSHROOM-CORAL
Budding and forming small stock

eggs, which the fixed coral cannot do, so that we have here another case of alternation of generations. Out of the egg comes an attached form, which buds and produces the free unattached form, which again produces eggs, and so on.

The mushroom-corals (*Fungiidæ*) are another remark-able group of solitary forms, taking their name from their resemblance to the head of an expanded mushroom turned upside down. Turning to the figure of *Thecocyathus* (above), and imagining the circular wall pulled down all round, and drawing down the septa so that they radiate outwards, some idea of a *Fungia* may be obtained. Their skeletons are remarkable objects, which no one at first sight would in any way connect with a sea-anemone. Although the mushroom-corals are considered to be individuals, reproducing themselves by means of eggs, both budding and division into halves occur exceptionally; in the former case the buds sooner or later becoming detached. In some there is an alternation of generations, leading to the formation of compound stocks. In the illustrated form true mushroom-corals are produced at the ends of the branches; at *a* one has become detached, and the others are in different stages, the youngest being nearly cylindrical, like a typical polype, whereas the older ones spread out like a typical *Fungia*. When a bud has fallen off, the stem seems capable of developing another. This is the asexual generation, reproduction by eggs being the sexual generation.

Lastly, certain solitary corals have been discovered in the deep sea, where, on account of the presence of carbonic acid in sufficient quantities to make itself felt, there is little lime. On this account the calcareous skeleton is generally distinguished by great delicacy. A deep-sea coral with such a delicate skeleton (*Leptopenus*), found off the east coast of South America at a depth of over a mile, is shown in the illustration. Its pedestal is formed of a delicate network with fine rays or spokes, connected together in a regular manner by transverse supports.

Plentiful as are the solitary corals, they are surpassed in number by those which form compound stocks—that is to say, in which the buds do not fall off, but go on budding till coral-islands and barrier-reefs are built up. As it is impossible to give here more than a very few illustrations of the many different ways in which the coral-stocks grow, we can only select a few types. In *Dendrophyllia*, as shown in the illustration, we have a tree-like growth; each polype secreting a solid pedestal for itself, and living in a depression in

STAR-CORAL, ASTRÆA

bud is a living, feeding coral-animal, surrounded by its crown of tentacles. These madrepores play no small part in building up coral-reefs, and the many different elegant forms which they assume (while keeping to their method of budding) is astounding. Some corals, again, do not form true branches, but may cover the ground like a field of corn—a good example of this type being found in *Cladocora cæspitosa*, which inhabits the Mediterranean and Adriatic Seas. Here the single individuals form somewhat long tubes, and the buds arise laterally at the lower end, then bend upward and grow alongside of the parent, without any further connection or fusion. The spaces between the different rising polypes are not filled in with secreted hard matter, but the latter grow up side by side free. The stock, therefore, is easily broken. This coral flourishes extraordinarily in many places, covering areas of over one hundred square yards, with a growth of a foot in height.

The method of growth just described is shown also by another and quite different coral, *Astroides calycularis*. As in *Cladocora*, just described, the single polypes, with their calcareous tubes or pedestals, are not

BRANCHING CORAL, DENDROPHYLLIA
A, terminal branch of a stock ; *B*, section of a polype. *a*, depression ; *b*, septa ; *d*. columella

MASSIVE CORAL, ASTROIDES

the top. This is shown in the section *B*. Into this depression the soft animal can withdraw at the approach of danger, drawing all its tentacles (which also contract) down to *a*. The space occupied by the animal is not very roomy at the best, and it is further limited both by a great columella (*d*) rising up in its interior, and by the solid septa (*b*) projecting into it all round. It must not be forgotten that these parts are not in the animal, but outside of it, and as they are secreted they push the skin up, and never penetrate into the tissues themselves. These polypes bud at intervals, the apical polype most frequently; and the result is a simple branched stock, as seen.

A different kind of stock is developed when the polypes produce many buds, as in the madrepores. In these delicate stocks selected polypes spring up above the rest, and their sides become covered with small buds. Space would obviously not permit all these small buds to bud again in the same manner. A few favoured ones, however, which have sufficient room next spring out and become covered again with small buds. Each

A   BRAIN CORAL, MEANDRINA   B
A, stock with soft parts ; B, skeleton

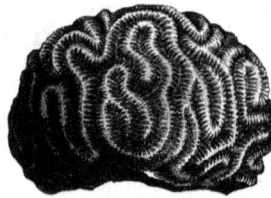

fused together by any cementing substance. The yellowish red polypes are seen standing out a great height above their cavities, much more so than is usual in corals. The larvæ of these corals leave the egg while still in the large chambered body-cavity of the parent, where they swim about for a time, till they escape through the mouth. They are long and worm-like, and slightly thicker at the posterior end, but may change considerably in shape. They swim about rapidly by means of their covering of cilia, the thicker end being foremost. The mouth appears at the thinner end soon after the larvæ leaves the body of the parent. Its free-swimming life has been known to last as long as two months, but under natural conditions it would probably be shorter. A strong sirocco had a marked effect upon the larvæ, which, after appearing exhausted, contracted, and became detached.

The transition from the worm-like larva to the polype takes place as in the anemones. The thicker end of the body is pressed against a hard rock, and the whole contracts into a thick, round disc ; while longitudinal

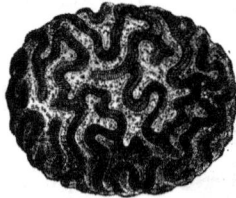

furrows become visible at the upper pole, where the mouth sinks deeper. At the ends of these furrows the twelve tentacles appear. The three illustrations on page 1812 show the stages which follow in rapid succession, resulting in a form almost exactly like a young sea-anemone. It has, however, already begun to secrete its calcareous skeleton. This is not formed as a connected whole, but begins as a number of separate centres of secretion between the polype and the substance to which it is fixed. These meet and fuse, till gradually the skeleton is produced. The polype begins to bud, and the buds develop their skeletons, the whole together forming a stock like that shown in the illustration.

ALCYONARIAN CORAL

The star-corals, which are some of the principal reef-builders, do not branch, but form great solid mounds; the polypes being all cemented together, and the budding so arranged that the whole colony forms a thin living layer or covering to the mass it and its parents have built up; all but this thin layer on the surface being dead coral. The illustration given on page 1813 is of *Astræa pallida*, a species which appears as a rounded mass, with flat base, and the individuals being quite distinct from one another, although their outer walls are in contact.

In the brain-corals, or *Meandrina*, we have animals budding, but not completely separating. No hard walls grow between the bud and its parent, although such separate the polypes less closely related. We thus get a system of valleys with rows of mouths, belonging to the polypes, which have budded off from one another.

HORNY CORALS

We have hitherto described skeletonless forms, and forms secreting solid, stony skeletons; the Antipatharia have horny skeletons, the method of secreting which will be described when we come to the horny skeletons of the next group. The polypes have only one instead of several rows of tentacles, and in most of them the tentacles are six in number. They form compound stocks, looking like delicate shrubs, with long branches, from which the polypes project, these branches being supported by a flexible horny axis.

ALCYONARIANS

Although the second order (*Octactinia*) of the corals contains a variety of forms, the appearance of the individual animals is more or less uniform, the number of tentacles being always eight. The tentacles are not hollow, but are usually somewhat flattened and notched round the edges like delicate leaves. These corals form stocks which are sometimes knobbed or lobed, and sometimes resemble a plant or tree with simple branches. The individuals of the stock are usually small, and rise like minute white blossoms above the soft fleshy surface

of the stock, which has a peculiar reddish yellow glistening appearance. The stock attaches itself by means of a stem, or else rests loosely in the sand, generally at a moderate depth. These corals secrete carbonate of lime, but in no case in the same way as do the true corals or the hydrocorals. In both these latter the solid skeleton is formed by the outer skin, while in the present group the secretion takes the form of minute calcareous particles of definite shape scattered about between the outer skin and the lining of the body-cavity.

The illustration above shows an alcyonarian, as these corals are called, with its lower part modified into a stem free from individual polypes; while the next figure exhibits a representative of another family, the sea-pens (*Pennatulidæ*), which also form stocks divided into a polype-bearing area and a stem resting on the sea-bottom. In one of the simplest forms of the sea-pens (*Veretillum*) the upper part is simply surrounded by polypes, and the lower a cylindrical stalk. A stock of this last-named type may lie for two or three consecutive weeks like a wrinkled turnip at the bottom of an aquarium, with all its activities suspended. After a time, however, the fine pores begin to take in water again, the surface becomes smoother, and gradually, as the individual polypes appear and stretch out their tentacles, the colouring of the whole stock becomes more vivid and more delicate. The stock lengthens and thickens, and the white crowns of tentacles stand out in dazzling contrast to the red of their bodies and of the common trunk. The foot swells out like an onion and becomes transparent, curves, and sinks into the stand; and the stock, which during its period of inactivity lay prone on the ground, assumes an erect position. In these sea-pens the parts of the stock may be compared with the parts of a feather, the whole being bilaterally symmetrical, and the single polypes being carried on the leaf-like lateral appendages of the stem. The sexual animals, which are provided with all the organs necessary for a polype, take in the food and reproduce themselves. The other less perfect brethren, called zooids, although more or less resembling these, have remained at a lower stage of development, are smaller, and have neither tentacles nor reproductive organs. These zooids appear to perform only one function—namely, to pump water through the body of the stock.

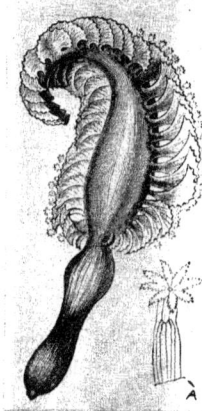

A SEA-PEN

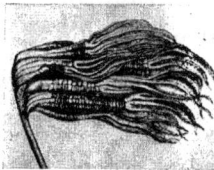

UMBELLULA THOMSONI

In addition to the small, isolated, calcareous particles already mentioned within the bodies of the individual polypes, sea-pens have a further support in the form of a calcified and often flexible axis, entirely concealed in the stock. The accompanying illustration represents *Pteroides spinosa*, in which the polype-bearing leaves are supported by a number of calcareous rays which project at the edges as spines.

The best known of the sea-pens is the phosphorescent *Pennatula phosphorea* of the Mediterranean and the Atlantic. In this form the capacity for giving light is not possessed by the whole surface of the stock, but only by eight band-like organs on the polypes themselves, the upper ends of which surround the mouth, while their lower parts run down over the stomach.

The higher forms of the sea-pens, or those which actually resemble feathers, are not found in deep water, none being recorded to exist below six hundred fathoms. Deep-sea forms have, however, been found; these being related to *Umbellula grænlandica*, which has long been well known. As early as the middle of the eighteenth century, when the presence of animal life at great depths was unknown, two specimens were brought up from a depth of two hundred and forty fathoms, sixty miles from the coast of Greenland. The polype-stocks consisted of a long, thin stem, ending in a bundle of polypes. The larger specimen was two yards long. These two specimens, soon after being described, were lost, but a very similar form (*U. thomsoni*) was obtained during the Challenger expedition; and other species have been discovered in various latitudes, at great depths. Two were found between Portugal and Madeira, at two thousand one hundred and twenty fathoms, while *U. leptocaulis* was taken in the Indian Ocean, some two thousand five hundred fathoms below the surface.

Another family of eight-rayed corals is that of the sea-fans, or *Gorgoniidæ*, of which the beautiful horny tree and bush-like growths give no idea of the living coral. As in the case of ordinary corals, the polypes secrete the horny branches beneath their bases, and on these they rise in gracefully branching colonies. All the sea-fans

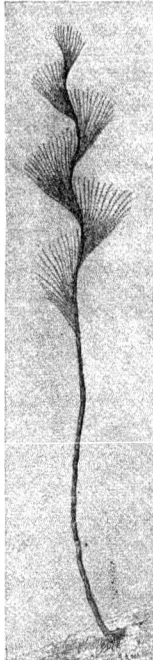

CORKSCREW
SEA-FAN

are attached, and branch in the most various ways, some in all directions, other only in one plane; in some cases simple branches run out at an angle or spirally, forming fans or nets, and so forth In most, the axis is horny and flexible, and they might be called horny corals, but single calcareous particles are enclosed in the axis, and its soft covering is crowded with them. The different kinds of these particles, found in different species, are of great importance in classification. One of the most common of the group is *Gorgonia verrucosa*, found in the Mediterranean and here illustrated. A shark's egg is shown in the illustration attached to the coral by means of its appendages.

A beautiful form is *Isidigorgia*, resembling a corkscrew with a long spiral. It sends off at right angles to the principal axis, and at short intervals, delicate branches, so that the whole structure looks like a spiral staircase constructed of fine cord. The illustration above shows an allied form (*Streptocaulus*) of the family *Chrysogorgoniidæ*, as yet found only in the Western Atlantic. The simple or branched colonies are as thin as horsehair, and the delicate axis has a golden sheen, with a beautiful display of colours.

The genus *Isis*, the stock of which is partly horny and partly purely calcareous, forms a transition between the above and the red coral (*Corallium rubrum*). In the latter the axis is calcareous, and built of numerous fine layers, the microscopic structure of which is so definite that a connoisseur can detect false from real coral. In the small illustration, a slightly magnified branch of a stock is represented, with several retracted polypes and two cut open. In the upper exposed calyx (*o*) eggs are seen, while the lower (*t*) contains a large

male vesicle, and at its side an egg (*o*). After hatching, the ciliated larvæ leave the egg while still within the chambered cavity of their parent (*B*). Two of the long worm-like larvæ (*f*, *g*) can be seen in the illustration through the delicate body-wall of a polype whose tentacles are retracted, and others are visible in a cell which has been cut open. In the uppermost cell a larva (*a*) is seen in the act of passing out through the mouth (*b*). The red coral is found only in the Mediterranean and Adriatic Seas; the most noted coral-fisheries being carried on off the Algerian and Tunisian coasts, at depths of forty to one hundred fathoms. The coral obtained in these fisheries varies greatly in value. The coral is made into articles of ornament both in Paris and Marseilles, but the chief industries are in Naples, Leghorn, and Genoa.

We conclude our description of these eight-rayed corals with the organ-pipe corals (*Tubiporidæ*), a family consisting of one genus (*Tubipora*), the members of which are neither numerous nor varied. The individuals resemble in form, in the number of their delicate tentacles, and in their soft anterior body, the other living members of the order. In the structure of their skeleton they are, however, unique among extant corals, and recall certain extinct forms. Each individual secretes a smooth-walled tube, without calcification of the vertical septa. These tubes, which, like the pipes of an organ, stand almost parallel, are united to form a stock by means of transverse platforms. These latter do not, however, correspond with the inner transverse partitions (*tabulæ*), by means of which the upper living part of the table is from time to time cut off from the dead part below.

The transverse platforms are neither regularly parallel nor continuous; nevertheless, they do indicate in a general way stages of growth. They are very richly provided with nutritive canals, and are of special importance for the whole stock, inasmuch as the young individuals bud out from their surfaces. As the longer tubes grow, the intervals between them increase, and as soon as there is room enough for a new polype, one buds out from the platform. Division of the individuals or formation of buds from the tubes themselves does not take place in this family.

A SEA-FAN, WITH SHARK'S EGG

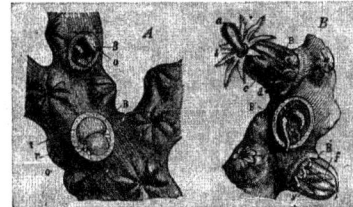

RED CORAL

A, part of a stock with retracted polypes, two have been opened (magnified); *B*, with polypes more or less extended; a larva (*a*) seen in the act of emerging through the mouth of the uppermost

CORAL REEFS AND ISLANDS

Having described a few typical corals, and explained their general structure and characters, some mention must be made of the importance of these creatures in the economy of Nature. Whereas most forms of animal life, in passing at death into their elementary constituents, leave no trace of their existence, the corals, or at least the numerous reef-making forms, build monuments which last for hundreds of thousands of years, and may be said to attain their greatest importance in the influence they exercise upon the life and development of the human race.

All reef-forming corals inhabit waters which in winter do not sink below a temperature of 68° F., the maximum summer heat in the Pacific Ocean being 86°. Two lines to the north and the south of the Equator, which would connect points where the winter temperature does not sink below 68°, waving in and out according to the currents, would enclose the zone of the reef-forming corals. Most of the stock-forming corals described above live exclusively within these limits of temperature, a fact that explains their rare occurrence in the Mediterranean, which is so favourable to other forms of animal life. The richest coral regions lie in the middle hottest zone—that is, between

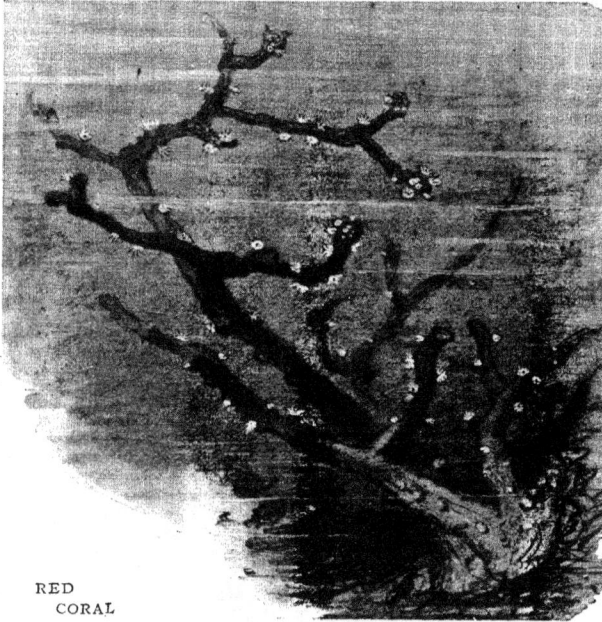

RED CORAL

15° and 18° north and south of the Equator, where the temperature does not fall below 72° F. The Fiji Islands fall within this region, and possess reefs extraordinarily rich in corals. The star-corals and brain-corals there reach their greatest development, while the madrepores are found as bushes, cups, or leaves, the latter often attaining a breadth of over six feet. In the Sandwich Islands, which lie outside of the hottest zone, the corals are less luxuriant and varied. The genera of corals found in the Indian Ocean and the Red Sea, as well as on the coast of Zanzibar, are essentially the same as those in the Pacific. The corals of the Gulf of Panama, although not in the hottest zone, have the character of the Pacific corals, and are different from those of the West Indies.

When the two Fosters and Cook discovered the coral islands of the South Sea, they were of opinion that the minute creatures to which these owe their origin began to build at unfathomable depths, gradually bringing their structures up to the surface of

ORGAN-PIPE CORAL

the water. Thus they thought that the same species were able to live at different depths. Recent researches have disproved this; and we now know that although many different animals live at enormous depths, all such are specially adapted to the conditions of life at those depths. Animals adapted to life at a great depth cannot exist at the surface. The number of deep-sea polypes is small, and among them there are no species forming reefs; and authorities are now agreed that reef-building corals can only live at moderate depths and within certain latitudes. One of the principal requirements is pure sea-water, some species flourishing in the canals between the reefs and in the shallower waters of the lagoons whereas others require the open sea.

Corals never flourish in impure water or on sandy or muddy coasts. They are not found at the mouths of rivers nor in excessively salt water; abnormal heating of the water of the lagoons also may cause their death. The dead portion of the stocks, and also, at times, that part which contains the living animals, suffer continually from the action of boring-worms and molluscs, while the boring-sponges cause still worse injuries. In addition to these foes, which tunnel through and help to disintegrate the skeleton of the coral-stocks, other enemies prey on the living polypes. These latter are richly provided with stinging-cells; and an unwary fish touching one with its lips is sure to be stung. Nevertheless, the greediest devourers of corals are certain fish which have acquired horny beaks like those of a parrot. These parrot-fish live by browsing on the living flowers of the coral-garden, having jaws which are untouched by their stings.

Coral-reefs are banks of coral-rock in the sea along the coasts of tropical countries. At high tide the reefs are usually under water, but at low tide are visible as wide, flat, naked expanses of rock, just above the level of the water, in marked contrast to the precipitous coasts of the islands they surround. At high tide the only sign of the presence of a reef is a line of breakers, which often extends for many miles at a distance from the land, a retreating wave only occasionally revealing a small portion of the rock. A small island may be surrounded by such a reef, the illustration on page 1821 showing a typical tropical island thus encircled. On the right side the reef forms a girdle stretching round the coast, and appears like a continuation of the land. This fringing-reef is also found on the left side, but beyond it, separated by a channel, is the barrier-reef. At one point the land is seen to dip down precipitously into deep water, and here, on account of the depth of water, the reef is wanting. The barrier-reef also is broken through at one point, forming the entrance to a harbour.

LIVING FLOWERS OF THE SEA

A. Aiptasia couchii—Trumpet anemone
A1. Echinus miliaris—Purple-tipped sea-urchin
B. Sagartia viduata—Snake-locked anemone
B1. Sagartia chrysosplenium—Gold-spangled anemone
C. Ophiocoma rosula—Brittle starfish
D. Actinoloba dianthus—Plumose anemone
E. Sagartia miniata—Scarlet-fringed anemone
F. Bunodes coronata—Diadem pimplet anemone
G. G1. & G2. Actinia mesembryanthemum—Beadlet anemone
H. Anthea cereus—Opelet anemone
J. Comatula rosacea—Pink feather-star on tube of sabella
K. Synapta digitata—A sea-cucumber
L. Corynactis viridis—Globehorn anemone
M. Bunodes thallia—Glaucous pimplet anemone

N. Tealia crassicornis—Dahlia wartlet anemone
O. Lophohelia prolifera—Tuft coral
P. Sagartia parasitica—Parasitic anemone on whelk-shell containing
Q. Euplectella aspergillum—Venus' flower-basket   [hermit crab
R. Stomphia churchiæ—Gapelet anemone
S. Serpula contortuplicata—Colony of tube-forming Annelid
T. T1. Peachia hastata—Arrow muzzlet.   Body in sand and exposed
U. Cucumaria hyndmanni—Sea-cucumber
V. Asterias rubens—Five-finger starfish
W. W1. & W2. Caryophyllia smithii—Devonshire cup-coral, open and
X. X1. Balanophyllia regia—Scarlet and gold star-coral, open and closed   [closed
Y. Sipunculus bernhardus—A siphon-worm in shell of turritella, on
      which is colony of balani
Z. Thalassema neptuni—A siphon-worm

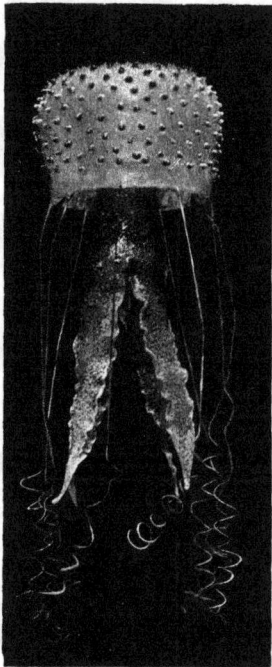

PELAGIA PANOPYRA HOLOTHURIA MACULATA PHYSOPHORA MYZONEMA

PHYSOPHORA HYDROSTATICA PENNATULA RUBRA STEPHANOMIA UVARIA

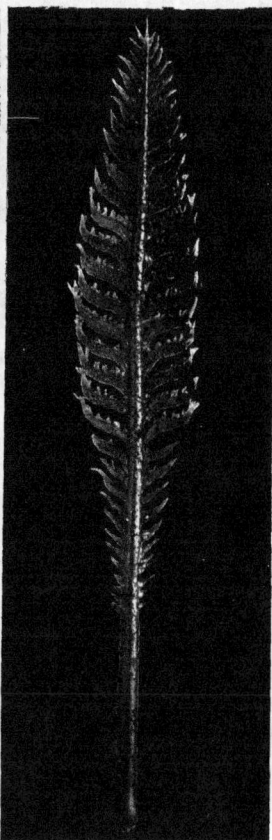

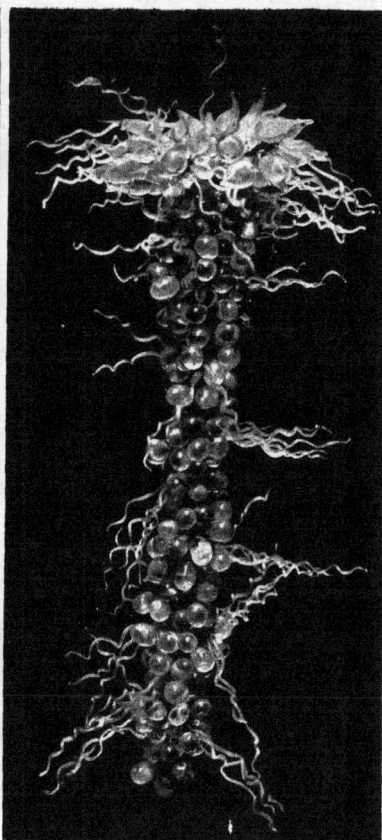

"The Siphonophora are the highest development of the swimming-bells. They are, in fact, colonies of bells, joined together in almost every possible way, and showing extraordinary modifications of individuals in the interests of a division of labour. For instance, some of the bells do nothing but row the colony along, others feed the colony, others are guards, and yet others produce the eggs."

PENNATULA GRISEA   SYNAPTA MAMILLOSA   PHYSALIA MEGALISTA

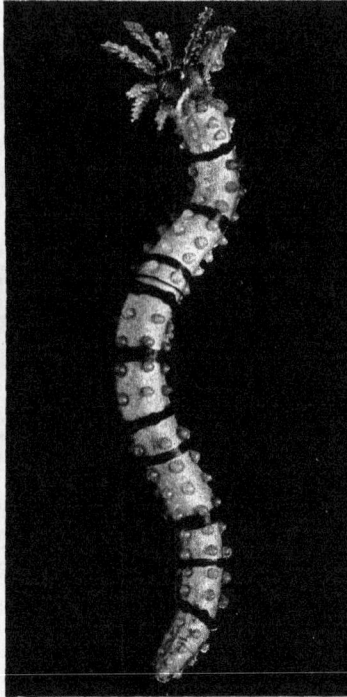

ACTINOLOBA DIANTHUS

"Among the Hexactinia the sea-anemones take the first place. They spread over all seas, being especially plentiful in the temperate zones, near the coast, at depths which bring them within the reach of every observer. They are distinguished by their solitary manner of life, their size, and their vivid and usually beautiful colouring. The skin is firm and leathery, and often covered with warts."

Many islands are bordered by a reef which protects the land from the sea like a mole. The barrier-reef may occasionally be ten or fifteen miles from land, and enclose several high islands. Various forms of reefs are found between the two extremes presented by such a barrier-reef and the simple fringing-reef.

The channel within the reef at low tide is sometimes hardly deep enough for navigation, or else it is blocked by masses of coral, which render its passage dangerous. At other times a reef encloses miles of open water, ten, twenty, or forty fathoms deep, but not free from hidden sources of danger; masses of living coral, from a few square feet to several square miles in extent, rising from the bottom. In the Fiji Islands all these kinds of channel formations occur.

The extent of the reefs, which include scattered banks and masses far below the surface of the water, varies greatly. On some coasts there are merely scattered groups or mounds of coral-rock, the tips of which project; while, on the other hand, to the west of the Fijis there is an area covered with reef of about three thousand square miles. Other reefs are one hundred or one hundred and fifty miles long, and the Australian barrier-reef attains a length of one thousand two hundred and fifty miles.

Passing from such a tropical island girt with coral-reefs, we come to what is more especially known as a coral-island proper, or atoll, which may be described as the encircling reef without any island to encircle. It surrounds a calm lake of blue water, in striking contrast to the restless ocean outside the solid circle. The ring of solid reef in this case is usually only one hundred to two hundred yards broad, and at some parts so low that the waves break over it into the lagoon. At other parts it is covered with tropical vegetation, but it rarely rises more than three to four yards above high-water mark.

Seen in the distance from a ship, a coral-island looks like a row of dark points, which are the tops of the coconut trees first seen above the horizon. On nearer approach, the lagoon with its green border is a wonderfully beautiful sight. Outside of the reef is the heavy surf, and within the white coral-strand is the thick band of verdure and the enclosed lake with its minute islands.

In colour, the water of the lagoon, where it is deep (ten to twelve fathoms), matches the blue of the surrounding ocean, but delicate apple-green and yellow tints mingle with the blue wherever the sand or coral-rocks approach the surface. Although the girdle of reef covered with vegetation occasionally surrounds the lagoon, it is more often broken up into a ring of separate islets of various sizes, between some of which navigable channels are found, giving admittance to the lagoon.

The submarine fields of living coral spread along the coasts of the islands and the mainland. Just as the accumulated remains of the primitive forests add layer after layer to the soil, so the coral-reefs are added to by

the breaking down of old corals, by the shells of bivalves and of other organisms. These fragments keep filling up the spaces between the separate living stocks, so that the level of the reef is constantly rising towards the surface. The currents and waves also take part in the building up of the reef. Masses of coral of all sizes, from great boulders to minute sand-grains, are broken up by the waves, and are cast upon the reefs, and then rolled about until quantities of fine detritus are produced, which, as calcareous mud, serves as a cement to bind the larger blocks together. A constant process of destruction goes on; some of the detritus being washed over the reef into the lagoon or canal, and some filling the spaces between the corals along the edge of the reef, while the rest remains upon the surface.

The layer of dead coral-rock forming the foundation of the reef is bordered by living coral. While this living coral is always extending the reef horizontally, the waves are piling up the dead masses vertically, till they rise above the surface of the water. Thus dry land begins to form, and by degrees islands arise well out of the reach of the waves. The ocean is thus the builder of the coral-island as it appears above the waves, the material having been supplied in the first place by the coral-animals. The moment the island is above water, plant-seeds reach it from distant lands, and ere long cover it with vegetation. The accompanying section of a coral-reef shows the slope of the reef, both towards the lagoon on the right and the open ocean on the left, the steep slope from shallow water to the land-level on the outer side, and the gradual slope on the inner side. The latter slope is then continued at almost the same angle, the quiet water not disturbing the slow accumulation and growth of the lagoon or canal shore. On the outer side of the reef, however, a broad terrace succeeds the steep slope, and surrounds the land which has risen above the sea, this terrace being exposed at low tide.

We have still to mention some of the causes of modification in the form and growth of the coral-reefs. The presence of harbours in reefs and atolls can, as a rule, be traced to the tides and to local ocean currents. There is generally an outflow through the canals and openings in the reefs. This is apparently due to the fact that water is constantly being thrown by the larger waves over the lower portions of the reef into the canal or lagoon, and seeks either to escape as an undercurrent in opposition to the flood-tide, or else strengthens the ebb-tide. These and other similar disturbances of the water in the canals bring with them much coral detritus, and render the bottom altogether unsuitable for the growth of corals. Where such currents are strong they keep the canals clean and open. The action of the oceanic currents is often increased by the fresh waters coming from the central islands, and harbours are therefore very often found at the mouths of valleys

ISLAND WITH FRINGING AND BARRIER REEFS

SECTION THROUGH A CORAL-REEF

CORAL-ISLAND, OR ATOLL

and of their small streams. The influence of the fresh water itself on the corals is not so great as is usually assumed, chiefly because it, being lighter than salt water, flows away on the surface of the latter and hardly touches the animals which grow below the surface.

The most important point which needs elucidation is why some reefs encircle islands as a fringe extending from the shore, while others run parallel with the land, no longer touching it ; others, again, forming circular lagoons with no island at all in the middle. This was the question which puzzled the first discoverers of reefs, and at one time it was supposed that instinct guided the animals in giving their structures the form best suited to withstand the force of the waves.

Darwin believed that all forms of reef arise by the gradual sinking of the land they surround. This

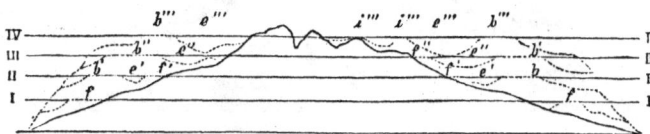

DIAGRAM EXPLAINING THEORY OF SUBSIDENCE OF CORAL-REEF
For explanation of numbers see text

theory has been confirmed in all essential points by Dana and Langenbeck. Other authorities have, however, differed.

It is a known fact that large countries, such as Sweden and Greenland, are in the act of sinking, and we also have direct proofs that reefs and their islands have subsided. The depth of a reef, although not directly measurable, can be approximately estimated, and must, in many cases, be at least three hundred yards. Since the living portion of a coral-reef cannot reach more than eighteen to twenty fathoms, such a depth of reef can only be explained by the sinking of the land on which it stands. If, instead of sinking, the land rises, the reef would be lifted out of water ; raised reefs three hundred feet high being known. This enormous thickness of reef can hardly be explained without a previous subsidence, inasmuch as such a height is greater than the known depth at which corals can live. The assumption that many reefs are the consequence of simple subsidence thus appears highly probable. The accompanying diagrammatic section through an island and its reef illustrates the action of gradual subsidence. The island at the water-line *I* has a simple fringing-reef (*f f*), a narrow rocky terrace at the level of the water, which first descends very gradually and then more steeply. Supposing the island to sink to the level *II*, what would happen ? While the land has sunk, the reef has risen, and there is a fringing-reef (*f'*) and a barrier-reef (*b*), with a narrow channel (*e'*) between them. A further subsidence to level *III* greatly increases the width of the channel (*e''*). On the one side (*f''*), the fringing-reef is retained, while on the other it has disappeared, a fact due to currents and other such agencies. Finally, when the

water is at the level *IV*, two rocky islands are visible in a large lagoon surrounded by the reef (*b''' b'''*), with two small reef-islands (*i''' i'''*) developed on mountain-peaks which have disappeared below the surface. The coral-rock has greatly increased in thickness, and almost entirely covers the former island.

The simultaneous occurrence of atolls, barrier-reefs, and fringing-reefs in neighbouring regions does not coincide with the theory, nor does the appearance of atolls and barrier-reefs in regions in which recent elevation of the land has been proved. The discovery of extensive submarine banks of sediment formed of the calcareous portions of foraminifera, deep-sea corals, molluscs, etc., makes it possible to explain the formation of atolls and barrier-reefs without the help of subsidence, this explanation being more probable than that involving the sinking of extensive areas of land. The formation of atolls can be explained by the better growth of the corals on the outer edges of the reef which are most exposed to the action of the surf, and the sweeping of the coral material out of the lagoon through the agency of oceanic currents, and the dissolving action of the carbonic acid contained in the sea-water. The deep canals which divide the barrier-reefs from the neighbouring mainland are formed in the same way. The enormous magnitude of the reefs which the theory of subsidence demands is nowhere realised.

Neither among modern reefs nor among geological formations do we find any traces of such gigantic masses of coral-rock. We are thus in face of a fascinating and important scientific problem, which still remains to be solved, a problem which was long thought to have found its solution. After Darwin's and Dana's subsidence theory had been generally for many years accepted as beautiful and completely satisfactory, it was urged that it is not always applicable, and that much simpler causes suffice to explain the phenomena. It is obvious, then, that we have an ample supply of possible explanations of coral-reefs, and it is most probable that among the many scattered reefs in the world, in one case one set of factors has played the chief part, in another case a slightly different set, and, further, a detailed and exhaustive study of any particular reef would probably reveal natural processes of no small importance which have not as yet been taken into account. A deep bore-hole put down in the island of Funafuti, one of the Ellice group, has proved that coral-rock descends to an enormous depth. So far as it goes, this experiment is, therefore, decidedly in favour of the subsidence theory.

CORAL-REEF AT LOW WATER

SUBKINGDOM PLATYHELMIA
FLAT-WORMS

THE Cœlentera, described in the last section, may be described as two-layered animals, because any one of them can be practically regarded as a double-walled bag, the cavity of which is a stomach. The flat-worms, and all the remaining subkingdoms to be considered, including the Vertebrata, possess a third layer (mesoderm) in which various organs are developed, between the two (ectoderm and endoderm) alone found in Cœlentera, and are therefore collectively known as three-layered animals.

Flat-worms are soft, usually flattened animals, and are mostly aquatic or parasitic, though some of them are found in damp places on land. The body is bilaterally symmetrical, exhibiting a distinction between head and tail ends and right and left sides, as well as between upper (dorsal) and lower (ventral)

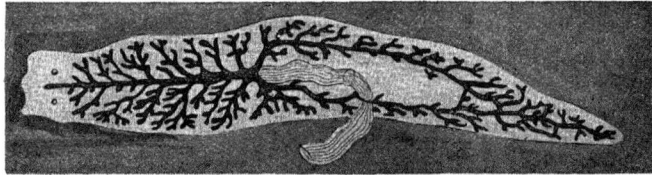

STRUCTURE OF A TRICLADE

surfaces. It is also unsegmented—all in one piece, and not made up of successive rings.

When a digestive cavity is present it opens to the exterior by a single aperture, which not only serves as a mouth, but also for the rejection of undigested parts of the food. The mesoderm is traversed by a system of fine canals which open to the exterior, and are concerned with getting rid of nitrogenous waste products. Both sexes are united in the same individual. There is a central nervous system, consisting of a pair of thickenings (ganglia), from which variously arranged cords are given off.

Flat-worms are divisible into three classes : (1) Whirl-worms (Turbellaria), (2) Flukes (Trematoda), (3) Tape-worms (Cestoda).

CLASS TURBELLARIA—WHIRL-WORMS

THE flat-worms included in the class Turbellaria glide over the surfaces of stones and water plants by the action of a uniform covering of cilia, and the little whirls or eddies set up have given the name of the group.

Whirl-worms are mostly found in salt and fresh water, but some of them live in damp places on land. Though the vast majority are of small size, some attain considerable dimensions. They are highly carnivorous. Three chief orders can be distinguished : (1) Polyclades, (2) Triclades, (3) Rhabdocœles.

Polyclades are marine whirl-worms which may be almost circular, oval, or elongated. The mouth is situated on the under surface of the body, and when it is in the centre, as in some of the rounded types, there is marked resemblance to the creeping ctenophores (Cœloplana and Ctenoplana),elsewhere mentioned (p. 1802). A muscular pharynx can be protruded from the mouth to secure prey, consisting of all sorts of small animals. The pharynx opens into an ovoid stomach from which branching tubes radiate in all directions. In the oval or elongated polyclades the mouth may be shifted either forward or backward, and this constitutes one of the points in which the various genera differ from one another.

The accompanying figure represents a polyclade gliding down the surface of a piece of seaweed, its front end being turned back. Two sensitive tentacle-like processes are clearly visible.

It may be added that polyclades, and the same is true for all the thin aquatic whirl-worms, is able to

A POLYCLADE

swim almost like a skate by undulations of the sides of the body. Creeping and swimming are mostly effected at night, the time when active feeding takes place.

The group Triclades includes marine, fresh-water, and also land forms, distinguished from the Polyclades by the presence of three main branching tubes opening into the stomach. This is clearly seen in the above figure of a fresh-water form (Planaria lactea) with protruded pharynx. The front end of the body, bearing two eye-spots, is to the left.

The terrestrial triclades are generally known as planarians, and a small species (Geodesmus bilineatus) is figured. They are of narrow, elongated form, and may be of large size. One such form (Bipalium kewense), first discovered in the greenhouses at Kew Gardens, may be as much as 18 inches long when fully extended. It is grey in colour, with three dark longitudinal stripes, and feeds on earthworms. Some of the tropical planarians are very beautifully coloured.

The Rhabdocœles are nearly all aquatic whirl-worms that live in salt or fresh water, and are distinguished by the simple tubular character of the digestive cavity. A typical fresh-water type (Mesostomum tetragonum) is figured, the head with two eye-spots being to the right. The mouth is here in the middle of the under side of the body. One very small marine form (Convoluta) is of particular interest because its tissues are crammed with minute algæ, enabling it to live like a green plant. Great numbers of these little creatures are found in tide-pools at Roscoff, on the coast of Brittany.

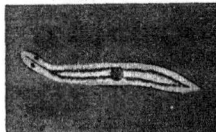

A PLANARIAN
Geodesmus bilineatus

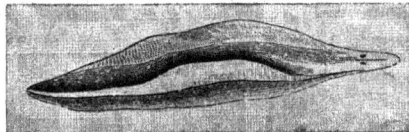

A RHABDOCŒLE
Mesostomum tetragonum

CLASS TREMATODA—FLUKES

THE class Trematoda includes flattened leaf-shaped parasites, provided with a varying number of adhesive suckers. The life-history is often complicated.

We may take as an example the well-known liver-fluke (*Fasciola hepatica*), which lives, when adult, in the liver of the sheep, and sets up " liver rot," which at various times has occasioned enormous losses to the farmer.

The leaf-shaped body is about an inch long, and there is a conical projection at its front end, terminating in a sucker, within which the mouth is situated. This leads into a muscular pharynx that acts like a suction pump, drawing blood and broken-down liver substance into the forked intestine, each half of which is beset with numerous branches (see figure). There is a second sucker placed far forward on the ventral surface. The body of the fluke is covered by a firm, horny cuticle, studded with innumerable minute spines, which prevent the parasite from slipping as it wriggles through the liver of its host.

A, LIVER-FLUKE LARVA ; B, LIVER-FLUKE ; C, DACTYLO-COTYLE ; D, ANTHOCOTYLE

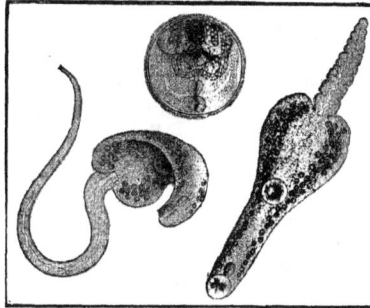

As in many parasitic animals—especially those which have a complicated life-history, spent in more than one kind of host—an enormous number of eggs are produced, the chances of survival being extremely small. The eggs pass down the bile-duct into the sheep's intestine, ultimately reaching the exterior of the body. Should a particular egg fall into water a ciliated embryo (A) hatches out from it, and swims actively about in search of the first host—a small water-snail (*Limnæa truncatula*).

If unsuccessful in its quest within about ten hours the embryo dies. But if a snail is found the embryo bores into its lung, assumes a rounded form, and loses the ciliated covering, becoming an inert sporocyst. Within this, as internal buds, are formed a number of the next, or redia, stage. This possesses a short, blindly-ending digestive tube, and a pair of muscular projections at the side of the body, which assist in locomotion. Liberated by rupture of the wall of its parent sporocyst, the redia makes its way to the liver of the snail, upon which it feeds. During the summer months several other generations of rediæ come into existence, being formed as buds within the bodies of their immediate predecessors.

On the approach of autumn the next, or cercaria, stage is produced, these being formed within the bodies of the last brood of rediæ. The cercaria resembles a minute tadpole, and is practically a young fluke provided with a tail. It works its way out of the snail, swims to the edge of the water, and fixes itself to a stem or leaf of grass, losing the tail and secreting a firm calcareous covering. Should one of these minute white cysts be swallowed

CERCARIA OF LIVER-FLUKE

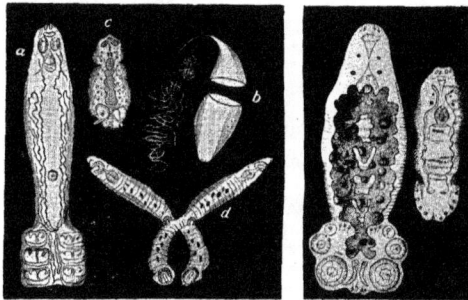

by a sheep its limy covering is dissolved by the gastric juice, and the young fluke, thus liberated, wriggles up the bile-duct into the liver of its host, and there becomes adult. From what has been said it will be realised that the original egg is able to give rise to a number of flukes, thus considerably increasing the chances of survival.

It is convenient to divide the flukes into two groups : (1) External Parasites, and (2) Internal Parasites.

EXTERNAL PARASITES

Before proceeding to describe the typical members of this group, mention may be made of a small ($\frac{1}{4}$ inch long) but particularly interesting fluke (*Temnocephala*) found attached to crayfishes and other fresh-water animals in the Southern Hemisphere. It looks something like a minute cuttlefish, and feeds on small crustaceans and microscopic plants. Its interest lies in the fact that in many ways *Temnocephala* resembles the whirl-worms, which suggests that flukes are a specialised branch from that stock. This particular fluke is not a parasite, and consequently has not undergone so much of modification as other members of its class.

Flukes possessing three or more suckers, and often hooks in addition, infest fishes, amphibia, and certain other animals. The figure represents two forms, *Dactylocotyle* and *Anthocotyle*, one (c) living attached to the gills of pollack, and the other (D) to those of whiting.

A very extraordinary life-history is exhibited by a form (*Diplozoon paradoxum*) which infests the minnow, and consists of two individuals (a in figure), permanently fused together. The eggs are laid on the gills of the host, and are drawn out into long anchoring threads (b). The survival of the larva (c) depends on its finding a minnow within a few hours. The larvæ attach themselves to the gills of their host, rapidly increase in size, acquire more suckers, and pass into what is known as the *diporpa* stage, a name given when they were supposed to be adult flukes. Each is provided with a minute knob on its dorsal surface, and a small cup on its ventral surface. After a time the diporpæ fuse in pairs to produce the mature form. One of the two first attaches its ventral cup to the dorsal knob of the other, which in turn twists round and accomplishes a similar feat. The two then become actually fused, and remain so for the rest of their lives.

A curious fluke (*Polystomum integerrimum*), represented in

LIFE-HISTORY OF DIPLO-ZOON PARADOXUM

a, perfect animal ; *b*, egg ; *c*, young ; *d*, diporpa

POLYSTOMUM AND LARVA

the accompanying figure, lives when adult in the urinary bladder of the frog. The hinder part of the body possesses an expansion provided with several suckers and a pair of hooks, by which attachment to the host is effected. In the spring the fluke

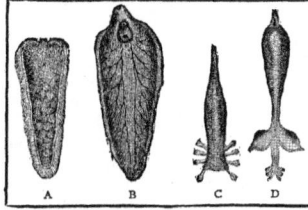

stretches part of its body out of the frog, and lays its eggs in the surrounding water. After several weeks a ciliated embryo devoid of suckers hatches out, and, if lucky, manages to attach itself to the gills of a passing tadpole, losing its cilia and gradually increasing in size. When the tadpole is becoming a frog, and losing its gills in consequence, the little fluke passes into the digestive tube of its host, ultimately finding its way into the bladder, where it becomes adult in five or six years.

INTERNAL PARASITES

The liver-fluke has numerous allies infesting a great variety of animals. They live in various parts of the body, and in many cases occupy three kinds of host in the course of their life-history. Perhaps the most interesting form (*Distomum macrostomum*) is one which, when adult, lives in the intestines of small birds. The eggs pass out of the body of the bird with the excrement, and if this happens to fall upon leaves they are liable to be swallowed by a small land-snail (*Succinea putris*).

Within the body of the snail a minute embryo hatches out, which bores through the wall of the digestive tube of its host, and becomes a sort of bag (sporocyst), that develops into a large branching structure.

Some of the branches are fertile, and large numbers of young flukes develop within them. Such branches push their way into the feelers of the snail, and acquire colours which give them a resemblance to caterpillars, each having a red spot at its tip and being marked with rings of white and green. They are distinctly visible from outside through the thin stretched walls of the tentacles, and they exhibit pulsations which increases the resemblance. Attracted by their appearance and movements, the small birds mentioned above peck off the tentacles of the snail, together with the fertile sporocyst branches and their contained young flukes.

Should these be swallowed by the adult bird, they are destroyed by the gastric juice, but if fed to nestlings their investments are dissolved, and they pass on into the intestine to become adults.

CLASS CESTODA—TAPE-WORMS

TAPE-WORMS are even more specialised than flukes, and just as these have in all probability descended from ancestors resembling the whirl-worms, so have the tape-worms taken origin from a fluke-like ancestral stock.

The common tape-worm (*Tænia solium*), when adult, inhabits the small intestine of human beings. It may be as much as 9 feet in length, and consists of a minute head (scolex) followed by a large number of flattened joints (pro-

dormant encysted, or bladder-worm, stage (*Cysticercus*), a tape-worm head being formed by an ingrowth of the original vesicle. No further development can take place in the body of the pig, but if an underdone piece of " measly " pork is eaten by a human being, the " measles," or cysts, resume active growth. The covering of the bladder-worm is dissolved by the gastric juice, the tape-worm head protrudes from the vesicle, and when the intestine is reached attaches itself by means of its hooks and suckers.

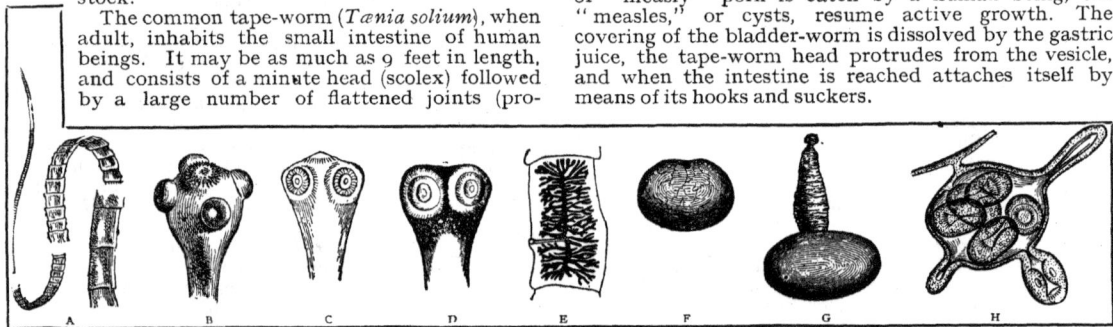

TAPE-WORMS

A, Tænia solium ; B, C, and D, head enlarged ; E, joint of Tænia saginata ; F, bladder-worm stage with head beginning to form ; G, with head protruded ; H, part of cyst of Tænia echinococcus

glottides). The head is provided with a circlet of hooks and four suckers, by means of which it is able to hold on to the lining of its host's intestine. There is no trace of a mouth or digestive cavity, nor are these necessary, for the parasite is bathed in a solution of pre-digested food.

The proglottides are most probably to be regarded as a long series of buds produced at the hinder end of the head. If this is a correct view, the tape-worm is not a single animal, but a chain of individuals. However that may be, each joint, or proglottis, contains a complete and complicated set of male and female reproductive organs, and there can be no doubt that self-fertilisation takes place.

The ripe joints at the end of the chain contain a very large number of eggs, within which embryos have already developed, and from time to time they become detached to pass out to the exterior.

Should such a joint, or the eggs liberated by its disintegration, be swallowed by a pig, the covering of each egg is dissolved by the gastric juice, and a little six-hooked embryo, shaped like a hollow sphere, emerges. By means of its hooks, the embryo bores into the wall of the stomach, gets into a blood-vessel, and is carried by the circulation to a muscle. Here it passes into a

The beef tape-worm (*Tænia saginata*) has a similar life-history, but in this case the ox is the intermediate host of which the muscles harbour the bladder-worm stage.

The disease known as " gid," or " staggers," in sheep is due to the presence of the bladder-worm stage of a tape-worm in the brain, in the form of a large cyst able to produce a number of tape-worm heads. In this case the adult worm (*Tænia cœnurus*) lives in the small intestine of the sheep-dog. A very small tape-worm (*Tænia echinococcus*), also found in the dog's intestine, passes its bladder-worm stage in the liver or abdominal cavity of various hoofed animals, and this stage is here a very large cyst, from the walls of which smaller cysts are produced internally. Each of the latter gives rise to a number of tape-worm heads. Human beings are liable to be infested by the cystic stage of this notorious parasite, and it is said that 2 or 3 per cent. of the population of Iceland suffer in this way.

It may be mentioned in conclusion that there are a few small tape-worms which consist of a head, or scolex, only, and do not give rise to a chain of proglottides. It is these which most nearly approach the flukes in structure, and come nearest the ancestral stock.

SUBKINGDOM NEMATHELMIA
THREAD-WORMS

THE subkingdom Nemathelmia includes an immense number of worms with cylindrical bodies tapering at the ends. There are no limbs, and the body is not made up of a series of rings or segments as in earthworms, etc., which belong to another subkingdom. There is usually a complete digestive tube, and the sexes are

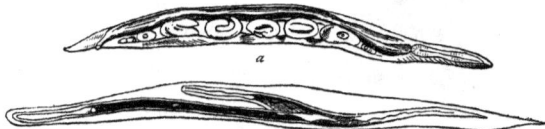

RHABDITIS
a, female form ; *b*, brood-pouch (enlarged)

generally distinct. Some of the thread-worms live in the sea, fresh water, or damp earth, but the great majority are parasites, not infrequently of dangerous character.

The subkingdom is divided into two classes : (1) Ordinary Round-worms or Thread-worms (Nematoda) and (2) Spiny-headed Worms (Acanthocephala). To these are commonly appended (3) Arrow-worms (Chætognatha), marine forms of doubtful affinity.

NEMATODA

A good example of a thread-worm which is not a parasite is the little vinegar-eel (*Anguillula aceti*). Common in sour paste or dirty vinegar, a female is represented in the annexed figure. The broad head is seen above, and the slender tail below. The digestive tube is shaded, and a number of eggs are to be distinguished on the right. Many of the near relatives of this form live in the soil, while others attack various cultivated plants, producing such diseases as " beet sickness."

Some of the largest members of the subkingdom are the ascarids, to which the name round-worm is more properly applied. One species (*Ascaris lumbricoides*), infests the small intestine of human beings, occasionally to the number of several hundreds. In this case the female is from 10 to 14 inches long, the much smaller male from 4 to 6 inches. The latter may be recognised by the sharply curled tail. A still larger species (*A. megalocephala*) is found in the horse and allied animals, and a smaller species (*A. mystax*) in dogs, cats, and other carnivorous mammals.

An extraordinary life-history is presented by a small form (*Rhabdonema nigrovenosum*) which in an early so-called rhabditis stage dwells in the excrement of frogs. Here pairing takes place, and the eggs hatch out within the body of the female, the internal organs of which are devoured by the larvæ. Having immolated their mother, these live for a time in water or mud,

and if swallowed by a frog make their way into its lungs, there growing to the length of about an inch. In this stage the sexes are united, the male organs

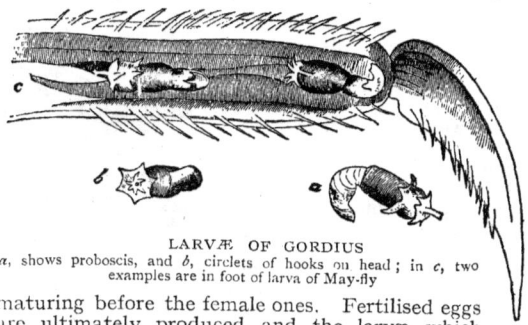

LARVÆ OF GORDIUS
a, shows proboscis, and *b*, circlets of hooks on head ; in *c*, two examples are in foot of larva of May-fly

maturing before the female ones. Fertilised eggs are ultimately produced, and the larvæ which hatch out of them pass into the digestive tube of the host, and thence to the exterior, when they pass into the rhabditis stage already mentioned.

The strongyles are a group of dangerous parasites, in which the mouth possesses spines or tooth-like projections, by which the tissues of the host are lacerated. One very objectionable species (*Dochmius duodenalis*) burrows in the wall of the small intestine in man, producing the often fatal disease known as miners' anæmia. They are swallowed with drinking water, and are often associated with imperfect sanitation. A related form, much dreaded by the poultry-keeper, is the gape-worm, or forked-worm (*Syngamus trachealis*), which infects the windpipe and other air-passages in fowls, causing the disease known as " gapes."

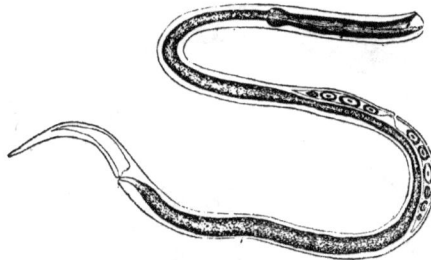

VINEGAR-EEL

The adult worm is about three-quarters of an inch long, and resembles a Y with a long stem. The slender branch of the Y is really the male, permanently attached to his partner. The embryos abound in the soil of neglected poultry-runs, and earthworms appear to have something to do with their distribution. It has been estimated by Crisp that this parasite causes the death of half a million pullets every year in England, a total loss of at least £25,000, a not inconsiderable sum.

We may probably regard a minute thread-worm known as *Trichina* as being the most dangerous parasite in the class. When adult it lives in the intestine of many mammals, including man, pig, and rat. The female is not more than one-sixth of an inch long, and the male (see figure) much smaller. The former produces living young, which bore into the wall of the intestine, and are mostly carried in the blood-stream to the muscles, where they become encysted. It is estimated that thirty or forty millions of them may be present in the same host, and the result of their presence is the disease called trichinosis, which is often fatal. Human

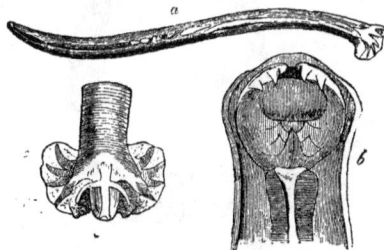

DOCHMIUS DUODENALIS
a, entire ; *b*, head, and *c*, tail (enlarged)

beings contract it by eating diseased pork which has not been sufficiently cooked, and the pig is infected by devouring diseased rats or the infected flesh of their fellows.

One of the tropical scourges to human beings is the guinea-worm, the female of which lives under the skin, and may attain a considerable length. It gives rise to painful swellings, and when these discharge the worm may gradually be removed by winding it round some narrow cylindrical body. If this is not done with extreme care the parasite is ruptured, and the numerous embryos escape, to develop into adult worms. Usually the embryos make their way into the bodies of minute fresh-water crustacea (*Cyclops*), and thus get taken up by human beings in drinking water.

The slender, elongated gordian worms are mostly found in fresh-water puddles, and are species of the genus *Gordius*. A number of individuals are often intertwined into a complicated tangle. The eggs are laid in strings on water plants, and when swallowed by insects undergo further development, ultimately making their way to the exterior. They were formerly believed by rustics to be horsehairs which had come to life.

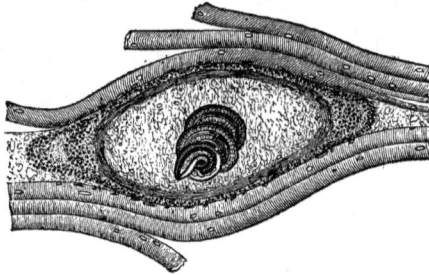

TRICHINA COILED UP IN HUMAN MUSCLE

boned animals. The front end of the body is provided with a kind of proboscis beset with spines, which enable attachment to the host. The spiny-headed thread-worm (*Gigantorhynchus gigas*), which infests the pig, is the largest member of the class, the female varying from 4 to 26 inches in length. The early part of life is passed within the grubs of various beetles, that of the cockchafer being a common host in Europe.

CHÆTOGNATHA

The arrow-worms (class Chætognatha) are marine forms that form part of the floating population (plankton) of the sea, and are of doubtful affinities. They are nearly transparent, and move with great rapidity by bending their bodies to either side alternately, fin-like expansions serving to balance them while swimming. The head is surrounded with a kind of hood, within which are a number of curved hooks that are employed in seizing the small organisms that serve as food. The complex internal organs include a well-developed nervous system; there are two eyes and other sense organs;

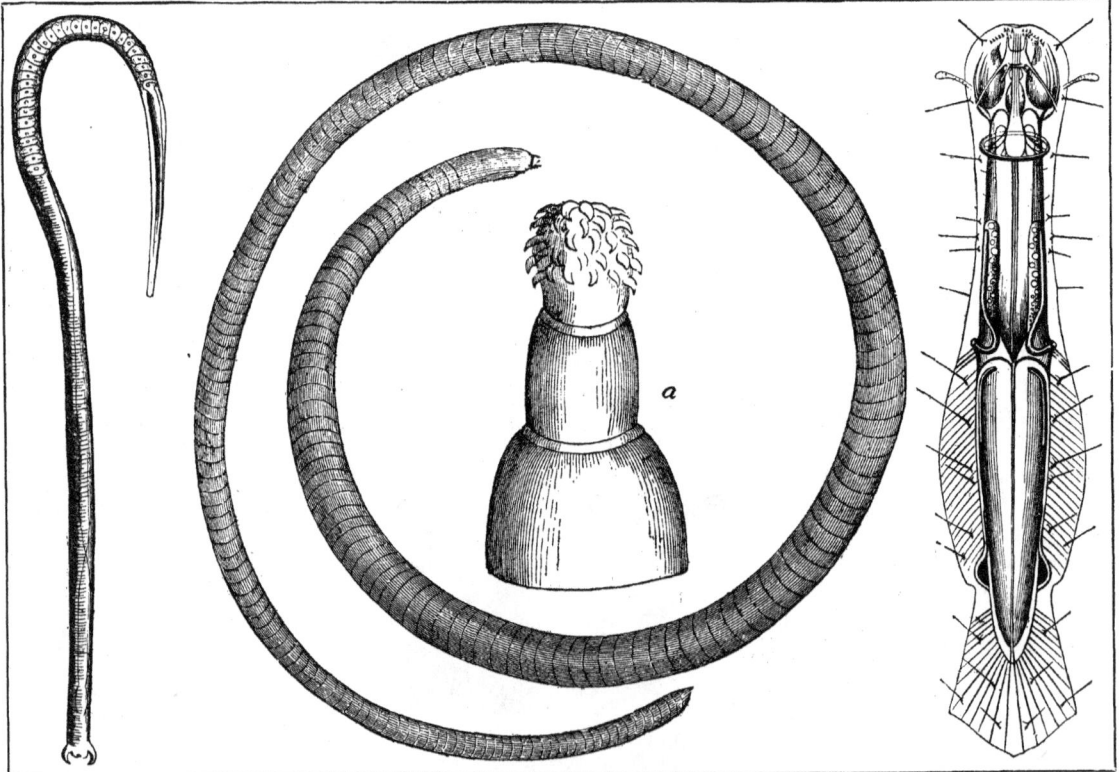

MALE TRICHINA

SPINY-HEADED THREAD-WORM
a, shows head enlarged

ARROW-WORM

ACANTHOCEPHALA

The members of the class Acanthocephala are remarkable parasites, devoid of a digestive tube, and living when adult in the intestine of back-

and the sexes are united in the same individual. Like other plankton animals, the arrow-worms form part of the food of various fishes. They occur in vast numbers, and are very widely distributed.   J. R. AINSWORTH-DAVIS

SUBKINGDOM ECHINODERMA

SEA-LILIES, STAR-FISHES, SEA-URCHINS, & SEA-CUCUMBERS

THE star-fish, the sea-urchin, the brittle-star, the feather-star, and the sea-cucumber—especially the three former—are well known to all frequenters of the seashore ; while various fossil forms are no less familiar to dwellers inland. They and their relatives are placed in one great group of animals, the subkingdom Echinoderma—a group corresponding in importance to the molluscs or the vertebrates.

If an echinoderm be touched with the finger, its skin will be found to have a rough surface, due to the presence of a crystalline deposit of carbonate of lime. In a sea-urchin, a brittle-star, or a feather-star, this deposit is in the form of little plates, which build up a more or less rigid test ; whereas in the star-fish it usually forms a kind of scaffolding, between which there stretches the more yielding, leathery skin. In the ordinary sea-cucumbers the deposit consists only of small spicules, easily seen under the microscope. It is this same deposit that forms the spines of a sea-urchin

SEA-PORCUPINE

and the stalked column of a crinoid ; and it is this which has enabled so many of the forms to be beautifully preserved as fossils. To this character is due the name of the group, derived from the Greek, *echinos*, a hedgehog, and *derma*, skin. The tendency of the echinoderms to deposit lime is not confined to the skin, the walls of the internal organs being often strengthed by similar material.

The next feature noticeable is the radiate symmetry, in many cases giving to the animal a star-shape, to which the common names star-fish, brittle-star, and the like are due. The ordinary red star-fish, or cross-fish, of the English coasts has five distinct rays, or arms; and this number five, to a greater or less extent, controls the arrangement of the organs in the majority of the echinoderms. The internal organs, as will be seen later on, are variously affected in the various classes of the echinoderms by this five-rayed symmetry.

Examining a star-fish or a sea-urchin, one sees, on the under surface of the rays in the former, and passing in five bands from top to bottom of the latter, a number of small cylindrical processes, which are usually gently waving about like trees in a wind. They lie in each band, or in each ray, along two rows, with a clear space between, like trees on either side of an avenue; hence the whole band of them in each ray is called an ambulacrum (garden-walk). Most of these little processes end in sucker-like discs, which the animal can stretch out and attach to smooth surrounding objects ; and it is thereby able either to hold itself firm against waves or currents, or to pull itself along. Hence these processes are usually called tube-feet ;

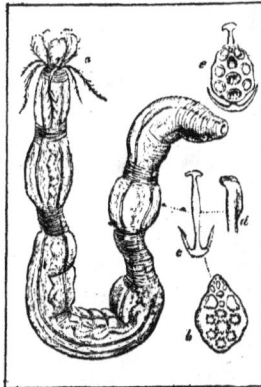

ANCHOR SEA-CUCUMBER

a, tentacles round mouth ; *e*, anchor and plate-shaped spicules ; *b*, *c*, and *d*, similar spicules of an allied form

but sometimes they end in a point, and cannot assist in locomotion, though they may help respiration, when they are sometimes called tentacles.

The movements of tube-feet are caused by the squeezing of a fluid into them ; for each foot is like an indiarubber tube closed at the end, and passing through the test (as the shell of the sea-urchin is termed) to join with one main tube, which runs along under the ambulacrum in a radial direction ; and before it joins this radial canal each tube-foot gives off a small swelling likewise filled with fluid, so that when this swelling is contracted all the fluid is squeezed up into the foot, and pushes it out like the finger of a glove when blown into. The radial canals pass along under the ambulacra till they join in a ring-canal surrounding the mouth.

Eventually this circular canal is connected with the surrounding water by a canal passing right across the body-cavity to the other side of the animal, near the vent, where it opens to the exterior through a plate pierced with a number of pores. This plate is called the madreporite, and the canal leading to it—owing to the limy deposits formed in its walls—the stone-canal. This whole system of fluid-filled canals is termed the water-vascular system. The foregoing description refers to its arrangement in a star-fish or regular sea-urchin ; but the system occurs, with various modifications, in all echinoderms, and is one of the features that separate the group from other animals.

The echinoderms are also peculiar in the possession of three, or perhaps four, different systems of nerves, of which three, or at least two, are present at the same time. One system supplies the skin, the tube-feet, and the intestine, its chief parts being a ring round the mouth and radial nerves radiating therefrom. The second system has a similar arrangement, but lies deeper, and supplies the internal muscles of the body-wall. The third system, which is most fully developed in crinoids, starts from the other side of the body, opposite to the mouth, and supplies the muscles that work the arms and stem. If the arm of a star-fish be opened from the back, there will be seen a pair of pleated extensions from the stomach. If these be removed, there will be exposed a pair of orange-coloured tubes, somewhat branched and knotty, which communicate with the exterior at the angles between the rays. These are the generative glands. In all echinoderms, except sea-cucumbers, these glands are affected by the radiate structure of the animals ; in crinoids the generative products are even produced in the extremities of the arms.

We may now shortly examine the main characters in which a sea-urchin, a star-fish, a crinoid, a brittle-star, and a sea-cucumber differ from one another. First may be noted obvious differences in form and in position in the living state. In an ordinary sea-cucumber the body is cucumber-shaped, with the mouth at one end and the vent at the other; between these run the five ambulacra, one or two of which are often more developed than the others, so that the animal crawls along on that side of its body, with its mouth foremost. A sea-cucumber has no arms or projecting rays, but its mouth is surrounded by a circlet of tentacles, often branched, which can be retracted at will.

A regular sea-urchin, such as the sea-egg (*Echinus*), resembles a sea-cucumber in being without projecting rays, but it is more spherical in shape, and moves with its mouth towards the sea-floor. On the other hand, in a heart-urchin (*Spatangus*), which moves through and swallows mud and sand, the body has become transversely elongate—that is to say, the long axis is at right angles to the position it occupies in a sea-cucumber; the mouth having moved a little forward, and the vent being transferred from the top of the body to its lower surface, so that both the mouth and vent lie on the under surface, at either end of the long axis.

In a star-fish, as in a regular sea-urchin, the mouth is in the centre of the under surface, while the vent is almost in the centre of the upper surface, although absent in a few forms. The body is either markedly pentagonal in outline, or more or less star-shaped. In the latter case it is said to consist of a central disc extended into arms, as in the illustration on page 1831. The number of these arms varies from five (*Asterias*) to over forty (*Heliaster*); but in each species with more than six arms the number may vary slightly, although constant during the life of the individual; in *Labidiaster*, however, fresh arms grow out even in the adult.

A brittle-star resembles a star-fish in which there is a sharp distinction between arms and disc; the mouth being on the under surface, but the vent wanting. And whereas the arms of a star-fish are simply extensions of the body, containing the generative glands and processes from the stomach, those of a brittle-star are mere appendages to the body, with a stout internal skeleton of separate ossicles, working on one another by well-developed muscles, and containing only blood-vessels, water-vessels, and nerves. The arms of the brittle-stars are nearly always five in number, though sometimes there may be from six to eight. As in

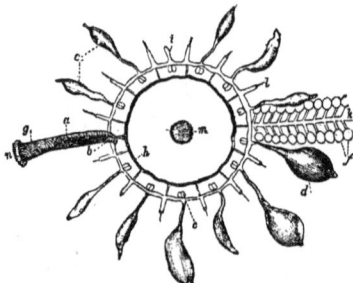

AMBULACRAL SYSTEM OF A STAR-FISH

f, small swellings connected with tube-feet; *k*, radial canal with which they unite; *e*, ring-canal into which radial canals open; *c*, *d*, membranous sacs that serve as reservoirs for water from radial canals; *a*, stone-canal, leading from ring-canal to madreporite, *n*; *m*, mouth.

the star-fish, the arms are unbranched, except in the family *Astrophytidæ*, where they fork ten or twelve times, and where the numerous branches interlace so as to form a kind of basket-work all round the disc, whence these animals are called basket-fish, or medusa-head star-fish.

A crinoid differs markedly from a sea-urchin, star-fish, or brittle-star in that the mouth faces upwards, the vent being also on the upper surface. This position is due to the fact that, so far as we know, all crinoids are at some time of their lives attached by a stalk to the sea-floor, or some other object, so that the mouth and vent naturally move up to that side of the body furthest from the stalk. This fixed state of existence has also caused the development of arms, five in number, but often forked many times, which arms stretch out from the body on all sides of the mouth, and contain extensions of the nervous, blood, vascular, water-vascular, and generative systems.

The representatives of the tube-feet are arranged along the sides of these arms, on their upper, or oral, surface, and between them is a groove, which is lined at the bottom with cilia, or extremely minute hair-like processes, that keep waving in the direction of the mouth, and so maintain a constant stream of water towards the latter; such water containing the minute animalculæ and fragments of decaying organic matter on which the crinoid feeds.

The extinct cystids and blastoids have their mouth in a similar position to that of the crinoids, and for a similar reason, but have not similarly branched arms. In the blastoids five grooves radiate down the body from the central mouth, and from the sides of these grooves there spring small, jointed, but unbranched processes, called pinnules. The stem of the blastoids is very short, so that when the pinnules have been lost, as is usually the case, the five-grooved body looks like a bud, whence the name of the class. It is difficult to describe a cystid as having any definite shape, for the various animals to which this name is applied differ greatly from one another in structure. Echinoderms are built upon one or other of the plans of structure just described. Echinoderms only live in the sea, where they find in solution the lime-salts from which their skeletons are built. None have become modified for a truly fresh-water existence, and in this respect they are peculiar among animals; a few

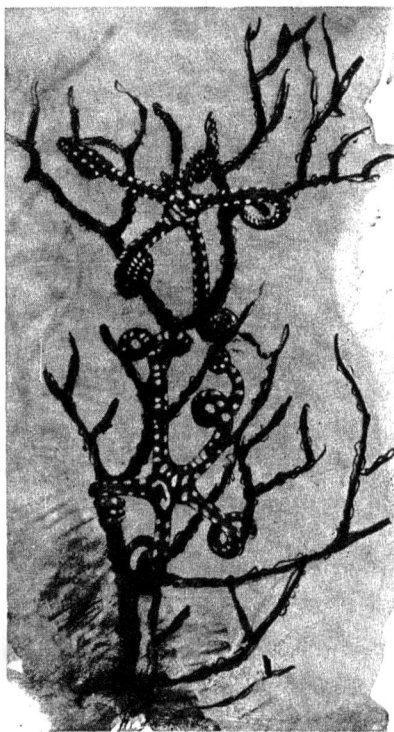

BRITTLE-STAR

holothurians, however, are found in the mud of some estuaries and brackish-water lagoons, while a star-fish (*Asteracanthium*) and a brittle-star (*Ophioglypha*) occur in the brackish waters of the Eastern Baltic.

CLASS CYSTIDEA—CYSTIDS

The Cystidea, a detailed account of which would be out of place here, have been extinct since the Carboniferous period. Not only are they among the oldest animals, but there is reason to suppose that they approach more nearly the primitive forms from which all the classes of the echinoderms were derived. Many have not that regularity of symmetry which characterises later echinoderms. Some forms are more round balls composed of a number of plates in which it is hard to see any arrangement. Some of them seem to have been unstalked, while in others the stalk is quite short. The arms are short, and vary in number, bearing but slight relation to the plates of the test.

Other cystids seem to be composed of an irregular number of plates; but they have become more definitely radiate in structure. Yet other cystids are definitely attached by well-developed stalks, and have their bodies enclosed by a limited number of plates arranged in regular order. Some of these present a six-rayed symmetry, while others are governed by a five-rayed symmetry.

Both of these groups have, as a rule, better developed arms, which sometimes branch, and are usually five or six in number, according to the symmetry of the cup. Hence these forms are much more like the crinoids than are the other cystids.

CLASS CRINOIDEA—SEA-LILIES

The crinoids differ from the more highly developed of the cystids in the greater regularity of their structure, the symmetry of which is nearly always governed by the number five, and in the greater development of the arms, which are often much branched. Slight consideration will show why the number five has been favoured by these particular animals. The body of a crinoid is encased by a limited number of relatively large plates, united together by the skin in which they are developed, and it is clear that the sutures between these plates are lines of weakness. Supposing that there were four plates in each circle, then the four sutures would be in opposite pairs, and the lines of weakness would run right across the body of the animal, which would easily be broken; and the same result would follow if there were six plates and three pairs of opposed sutures. Though the test might be more flexible, still there would be three lines of weakness in each circle instead of two. But when there are five plates, each suture lies opposite to the middle of a plate, and so the line of weakness does not run right across the body.

The alternation of the plates in a crinoid may be explained by similar mechanical considerations, for such an arrangement corresponds to the bonding of successive courses of bricks in a wall. There is reason to suppose that the ancestors of all crinoids, as well as most of the Palæozoic crinoids, were attached to the sea-floor or some other object throughout life by the stem.

Sea-lilies have been placed at a disadvantage in the struggle for existence by their inability to move about, and for long ages they have been declining in numbers, until they are now comparatively rare and only found in the deep sea. The only common members of the class at the present time are the feather-stars (*Comatulidæ*), of which the rosy feather-star (*Comatula rosacea*) occurs in British seas. These creatures have given up the stalk, although this is present in early life. The family dates from the Jurassic period, and includes not only numerous species of *Comatula*, but at least as many more of a closely allied genus, *Actinometra*, as well as three other less common genera.

Next to the stem, the most characteristic structures of a crinoid are its arms. Each arm starts from one of the five plates that form the uppermost circlet in the cup. The arms are said to be radial in position, and those plates from which they start are specially distinguished as the "radials."

The locomotion of the feather-stars is effected by the raising and depressing of alternate arms, and the movements of these arms are correlated by the peculiar nervous system that has its headquarters at the bottom of the cup. This swimming has been observed in both *Comatula* and *Actinometra* kept in an aquarium. As a rule, however, these animals remain attached by their cirri to rocks, to the bottom ooze, to seaweeds, or to other marine animals. In this position the arms are outspread, and the small branches, or pinnules, that line their sides are kept slightly waving. If the water be ruffled, the first impulse of the crinoid is to flatten its arms out suddenly and to hold on to the rock or other object with its pinnules.

The pinnules of a *Comatula* can be bent in any direction, those near the extremity of the arm being specially active. If its extremity be touched by any irritating substance, the arm is erected at right angles to the upper surface of the animal, and so removed

STALKED CRINOID, OR SEA-LILY

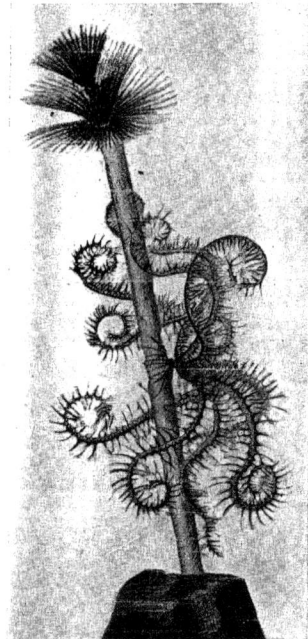

ROSY FEATHER-STAR CLINGING TO TUBE OF SABELLA-WORM

from the other arms ; while the pinnules move something like the legs of a fly that is cleaning itself. If, however, this proves ineffectual, the arm bends over to the one on the opposite side, the pinnules of which then assist in the operation. The pinnules move in this manner to disembarrass the arm of fragments of foreign matter that are too large ; but the hooks at the end of the pinnules can catch and retain minute fragments, which, as they decay, attract animalcules, and so furnish food for the animal.

At the present day crinoids live in all seas at depths between fifty and three thousand two hundred fathoms, but they prefer clear and undisturbed waters. The same has been the case in former geological periods, for while crinoids are abundant in limestones, of which their own remains form large masses, they are much rarer in sandstones and shales. Unstalked as well as

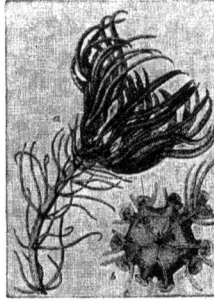

MEDUSA-HEAD
PENTACRINID

a, crown and part of stem; *b*, food-grooves passing to central mouth

stalked crinoids live chiefly in colonies, but this is due less to sociability than to limited powers of motion even in the larval state.

The food of crinoids consists chiefly of foraminifera, diatoms, and the adults of small and the larvæ of larger crustaceans. Crinoids themselves form food for fish, though nowadays their place seems to be taken by the brittle-stars and an occasional sea-urchin. As protection against such attacks, some crinoids have been provided with spines.

The stalk has no doubt been evolved to prevent these animals from being smothered by additions to the oozes in which they live. The stages in the evolution of a stalk may be gathered by studying the extinct cystids. It was at first simply a narrower region of the body, ultimately becoming an appendage of considerable length.

CLASS BLASTOIDEA—BLASTOIDS

THE Blastoidea constitute a compact group, pretty clearly marked off from both Cystidea and Crinoidea, which they resemble in the upward position of the mouth and the generally fixed habit. The chief character that separates blastoids from other echinoderms is the presence of an elongate plate, the lancet-plate, underlying the ambulacrum and pierced by a canal supposed to have contained the radial water-vessel. These five canals meet in a circular canal round the mouth, but there is no evidence that they were connected with tube-feet as in other echinoderms. Each side of each ambulacrum was lined by a row of delicate, unbranched arms ; and the food-grooves of these arms passed to a single groove running down the middle of the surface of the ambulacrum, and these five grooves then passed up to the mouth.

CLASS ASTEROIDEA—STAR-FISHES

WITH the Asteroidea we come to echinoderms that differ from those hitherto described, and resemble those to be dealt with, in that none of them are fixed, but all are free-moving, and in the fact that the mouth is not directed upwards. There is, however, reason to believe that these free forms are, like the free crinoids, descended from ancestors that were fixed. In a crinoid the mouth moves upwards to the surface opposed to this organ of attachment, and there becomes surrounded by arms, which similarly face upwards ; but in an asteroid the mouth and its surrounding arms are bent downwards so as to face the sea-floor, and the animal, instead of collecting its food from the water above, extracts it from the mud below. Correlated with this mode of life, the vent and madreporite are on the upper side of the body.

The so-called pedicellariæ are interesting and characteristic structures. These are small pincer-like organs that occur in star-fishes and sea-urchins, on the surface of the test, as shown in the illustration. The movable spines covering the surface of these animals, and varying in size from minute, delicate, bristle-like structures to long rods, which may be thin and pointed, or thick, or even globular, are familiar to all. The pedicellariæ are prob-

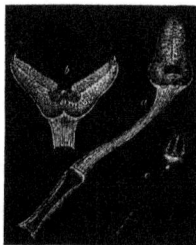

PEDICELLARIÆ
a, two-jawed, closed ; *b*, open ; *c*, three-jawed

ably derived from the smaller spines ; two of these united at the base by a muscle, and slightly curved so as to approach one another at the ends, form the simplest kind of pedicellaria ; and by gradual modifications of this type all the varieties may be derived.

Many uses have been ascribed to the pedicellariæ, such as holding pieces of food, or removing dirt from the surface of the test. In some sea-urchins they are provided with poison-glands, which seems to show that they serve as weapons of offence in those cases. It has been considered that in sea-urchins their chief use is to catch hold of fronds of seaweed and keep them steady until the spines and tube-feet can be brought into action. The inner surfaces of the forceps in the pedicellariæ are remarkably sensitive, and the blades close on any minute object immediately their inner surfaces are touched by it.

Besides spines and pedicellariæ, star-fish also have on the surface of the skin small tubular processes, containing an extension of the body-cavity. These have very thin, contractile walls, and doubtless serve to assist respiration. All star-fishes have tube-feet, but in some these have no suckers at the end, and in all cases those which are at the ends of the rays are used only as feelers, and are

A STAR-FISH

1831

stretched in the direction in which the animal is moving. At the extremity of each arm is a single tube-foot, which is the first to be formed, and is known as the unpaired tentacle, this being always stretched straight out. Immediately above this tentacle is a small eye, coloured by red pigment, and protected by small spines.

Star-fish are sluggish animals, rarely moving of themselves, and staying for days in the same position. Those kept in tanks or in glass vessels prefer to cling to the side instead of lying on the bottom. When disturbed, however, a star-fish can travel at a considerable pace. Those star-fish that have suckers crawl by means of their tube-feet, while those that have no suckers still use their tube-feet slightly, but also progress by the muscular movements of the rays.

The short-armed *Asterina* and *Astropecten* can right themselves in less than a minute, and accomplish the act by raising themselves on the tips of four rays, and then turning a somersault by throwing over the fifth ray. *Asterias* takes rather longer, and effects its purpose by first twisting over one or two of the rays and catching hold of the ground by the suckers. It then gradually turns over the rest of the body. *Cribrella* rights itself in the same way as *Asterias*, but, apparently because of the stiffness of its skeleton, takes much longer over the process. Star-fish, like other echinoderms, are a sociable class. Even the deep-sea forms sometimes live in swarms. Many shallow-water forms also are gregarious, and some species have been observed to pair at the breeding season.

CLASS OPHIUROIDEA—BRITTLE-STARS

THE name Ophiuroidea, given to the brittle-stars, refers to their long, serpent-like arms, which are attached to a relatively small and usually rounded body or disc. The digestive and generative systems do not extend into the arms, but are confined to the body ; so that the arms are appendages to the body rather than portions of it. They are cylindrical, and have no groove on the under side, such as exists in star-fish, but have little openings through which the tube-feet pass. In this class it is the arms themselves and not the tube-feet that are used for locomotion. The tube-feet accordingly have no terminal suckers, but are very sensitive to touch, and probably assist respiration. The greater part of each arm is formed by a central axis of successive calcareous segments, not unlike the vertebræ of a backbone. Each arm-ossicle or vertebra is, however, composed of two parts, one on either side, and united in the middle line ; the successive ossicles being connected by pairs of strong muscular bundles, and articulating with one another by tenon-and-mortice joints.

The spines, which are clearly shown in the annexed figure of *Ophiothrix*, are borne on the side-plates of the arm, and aid the animal in locomotion. The integument of the disc also bears plates or scales of various sizes, often more or less covered with granules and minute spines. The precise arrangement of the plates on the top of the disc varies in different species, but five pairs of plates, known as the radial shields, are always present at the base of the arms, and are shown in the figure. On either side of the arms where they join the disc, there is seen on the under surface a slit-like opening. These openings, known as the genital slits or clefts, are usually single but sometimes double ; they lead into thin-walled pouches or bursæ at the sides of the rays.

In a living ophiurid the disc alternately expands and contracts, and thus water is pumped into and out of the pouches, through the slits. The entering water brings oxygen, which it exchanges, through the thin walls of the pouch, for the carbonic acid contained in the water of the body-cavity, and then goes out again by the return current. Hence the pouches are called respiratory bursæ. But they have another function, since the ovaries enter into them, and the ripe ova may either be carried out by the current through the slits, or they may remain and undergo direct development in the pouches themselves.

Around the mouth are a number of short, flat processes, or papillæ, serving as strainers, and keeping foreign bodies that are not wanted for food from entering the stomach. Round the mouth are also twenty tentacles, which are really the modified tube-feet of the two first arm-segments of each arm. They are in a state of continual movement, assisting the food to enter, and clearing away the undigested residue which is ejected from the mouth.

The branched ophiurids, or *Clad-ophiuræ*, are sedentary, attaching themselves by coiling their branching arms around corals and such-like animals, but can move when they please. The same mode of life is also affected by a few of the simpler forms ; but, as a rule, ophiurids have considerable powers of locomotion. Most walk rather than creep, raising themselves on their five arms as upon legs, stretching out one or two arms in front, and drawing the rest of their body in the same direction. In other forms, however, the rays of the body undulate laterally, and produce a creeping serpentine movement.

The little *Amphiura*, which lives under stones, among the roots of seaweed, can turn its arms very quickly around its disc, and so form itself into a little ball ; thus, if it is disturbed, it can roll and sink quickly into deeper parts of the water. Sometimes ophiurids are seen to progress by a kind of rowing motion of the arms.

COMMON BRITTLE-STAR

CLASS ECHINOIDEA—SEA-URCHINS

THE sea-urchins are the best known, as they are the most numerous of all echinoderms. The illustration on page 1834 shows the test, or shell, of the egg-urchin, with the spines on the right side, but scraped away from the left. The plates of the test are seen to be covered with rounded tubercles of various sizes, and it is to these that the spines are attached by a ball-and-socket joint, surrounded by muscles that can move the spines in any direction. The tubercles do not, however, cover the whole test indiscriminately, but are disposed chiefly in five broad zones, extending from one pole to the other. Alternating with these are five narrower zones, bearing smaller and fewer tubercles, and pierced by small holes arranged in regular rows. Through these holes pass the tube-feet, which are all provided with suckers at the end. These latter zones

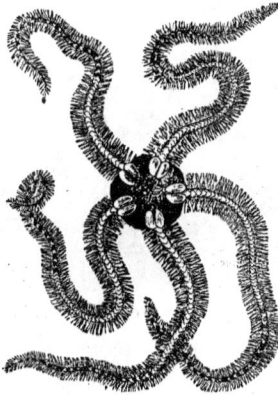

are, therefore, the ambulacral zones, one of them being seen in the middle of the illustration. The other zones are called interambulacral.

All the zones converge towards the summit of the test, where the vent is situated in a circular space covered with membrane. This membrane contains a few irregular granules, and is surrounded by five large interradially placed plates, pierced by the ducts of the generative glands. One is also pierced by a large number of small water-pores, and is called the madreporite. Outside these five plates, and alternating with them, are five other plates, each situated at the top of an ambulacral zone, and pierced by the unpaired tentacles, which terminate the water-canals, and represent the unpaired tentacles near the eye at the ends of the arms of a star-fish.

At the other pole of the body is another membrane, surrounding the mouth-opening, through which may be seen five pointed teeth. They belong to a very elaborate masticating apparatus, shown in the illustration, and found in all the regular urchins, as also in the Clypeastrida among the irregular urchins. This consists of twenty principal pieces arranged into a five-sided conical mass, compared by Aristotle to a lantern (*a*). In the centre of the whole are five teeth (*b*, *c*), working in body-sockets, or pyramids, connected by muscles with one another, with the interior of the test, and with the arched processes, known as auricles (*d*), that surround the mouth-opening.

Such a sea-urchin as that described preserves as much as any echinoderm the five-rayed symmetry of the group; but in many forms the five-rayed type is not so obvious, for the animal has become elongated along one of the axes, so as to have a superficial two-sided symmetry. This is naturally connected with constant movement in one direction, as though the animal had a head and tail; and such modification is found among those urchins that live on muddy bottoms, and especially in those from considerable depths. Not only is the test elongated, but the mouth moves forward to the front margin, and the vent downwards to the hinder margin, so as eventually to lie on the under instead of on the upper surface of the test.

An earlier stage in this modification is shown in the illustration of the shield-urchin (*Echinarachnius*), and a fully developed one in the heart-urchin (*Brissopsis*), with its long tube-feet extended in the act of walking towards the left. These heart-urchins, as they move along through the sand and mud, scoop it up into their mouths, and pass it through the intestine, extracting on its passage such nutriment as the minute organisms it contains can afford. To enable them to scoop it up in this way, the hinder margin of the mouth is produced forwards in a kind of shovel shape, as is shown in the illustration of a *Pourtalesia* test

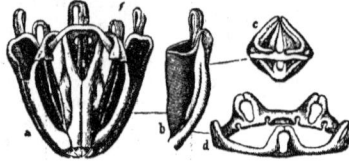

JAWS OF STONE-URCHIN
a, lantern; *b*, *c*, teeth; *d*, auricles

from which the spines have been removed. These animals live at very great depths in the sea, and are the urchins most modified in this particular direction.

Urchins of the heart-shaped type have short, delicate spines, and move almost entirely by their long tube-feet, in the manner described; but the greater number of the regular urchins progress chiefly by the aid of their spines, which are much stouter, while the tube-feet often have the suckers very imperfectly developed. The spines of sea-urchins also serve as organs of protection; but their efficacy varies much in different forms. For instance, *Diadema* and *Cidaris* have sharp spines, 8 or 10 inches long, which prick one almost before one can see them, and can pierce the stoutest boot; their danger being increased by the gregarious habit of the animals. Some sea-urchins have poison-glands attached to their spines. It is the smaller spines that are protective, and they are placed for this purpose near the main openings and organs of the body, such as the vent, genital pores, and eyes; they also protect the ambulacra, and bases of the larger spines.

A *Porocidaris* feeling itself free from danger, in well-aerated water, walks from one side to the other, doubtless in search of food; its ambulacral tentacles being stretched out as feelers, and its long spines moving as described. The smaller spines are depressed to permit of the free movement of the larger ones, and those of the ambulacra raised to permit the extension of the tube-feet. If one slightly wounds the animal when thus expanded, the larger spines immediately stiffen on their tubercles, while all the smaller spines depress themselves, each over the organ that it is destined to protect. Though the tube-feet may not be used for locomotion, they are put to another useful purpose. If a *Strongylocentrotus* be placed in a tank with some dead shells or similar objects, it will raise them on to its back, and hold them there by means of the tube-feet, as a kind of concealment. Some sea-urchins cover themselves all over in this way with bits of seaweed, shell, and small pebbles, and so move about unobserved.

Other sea-urchins do not move from place to place, but always stay in one spot, where they are generally found living in a hole. Sometimes the hole may have been there before the sea-urchins; sometimes it may have been formed by the growth of calcareous algæ around the sea-urchin; but sometimes the urchin itself has bored the hole. This is accomplished not by any acid secretion—for on the west coast of Africa an *Echinometra* has been found boring into a solid lava—but by the continuous movement of the teeth and spines. The common *Strongylocentrotus* is a well-known example of a boring sea-urchin. When the waves wash up against the urchin it sets its spines rigidly against the sides of its hole, and so holds fast,

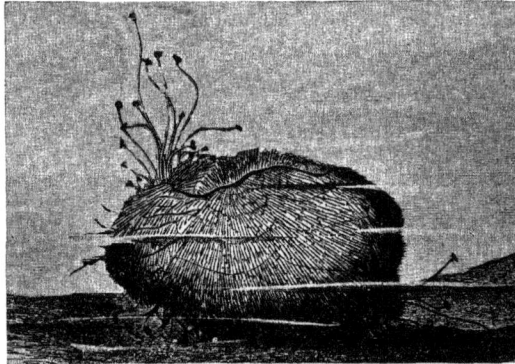

HEART-URCHIN, MOVING TOWARDS THE LEFT

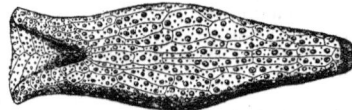

TEST OF PHIAL-SHAPED POURTALESIA, WITH SPINES REMOVED

Although most of the sea-urchins have a rigid test, yet there are some in which the plates are only loosely joined together, so that the test is flexible. This is the case in an *Astropyga ;* but is still more pronounced in cavity protruding between the membrane round the mouth and the plates of the test. In the irregular urchins some tube-feet are modified for respiration, becoming broad, flat, and somewhat lobed ; the hinder

TEST OF EGG-URCHIN   SHIELD-URCHIN   LEATHER-URCHIN

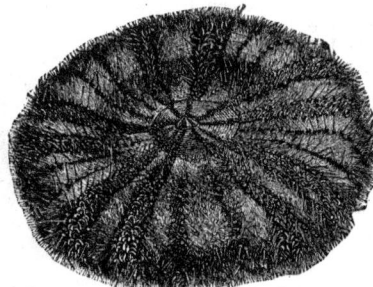

The spines are removed from the left half of the egg-urchin, and the shield-urchin is viewed from above

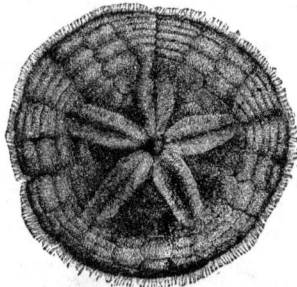

the leather-urchin (*Asthenosoma*), and other members of the family *Echinothuridæ.* Respiration is effected in the regular sea-urchins by ten gills near the mouth. These are thin-walled ciliated extensions of the body-end of the intestine seems to be respiratory in function. Some sea-urchins possess eyes. These creatures are both animal and vegetable feeders, and are even cannibals when opportunity offers.

CLASS HOLOTHUROIDEA—SEA-CUCUMBERS

SEA-CUCUMBERS (class Holothuroidea) are, as we have seen, elongated and worm-like creatures, with a mouth at one end and a vent at the other. The skin is leathery, and contains a comparatively small amount of calcareous matter. Usually this occurs in small spicules, which assume very definite shapes, such as anchors or wheels ; but in some forms the spicules increase in size, so as to form a plated integument. There may also often be a ring of calcareous plates round the gullet, five of which plates have the same relation to the radial water-vessels as the auricles around the jaws of a sea-urchin, and they likewise serve for the attachment of muscles. In such a common form as *Cucumaria* there are five rows of tube-feet passing from mouth to vent. The five-rayed symmetry is not obscured, and is traceable in the arrangement of nerves and muscles, although it does not affect any portion of the digestive or generative systems.

Around the mouth in *Cucumaria* is a fringe of branched tentacles, connected with the water-vascular ring. In most other echinoderms, it will be remembered, a canal passes from this ring and opens to the exterior by a madreporite ; and in a few holothurians of primitive structure this is similarly

SEA-CUCUMBER

the case. But in *Cucumaria*, as in most, the connection with the exterior is lost, and the canal, with its madreporite, hangs down into the body-cavity. In *Cucumaria* the tentacles are used like a net to intercept floating organisms in the surrounding water. Many holothurians swallow a great deal of sand, and the intestines of those that live near coral-reefs generally contain fragments of coral. They usually attach themselves by their tube-feet to rocks or seaweed, and wave the tentacles around. *Holothuria atra*, which lives on the Great Australian Barrier Reef, inserts its hinder extremity within a crevice of the rock, into which, on being disturbed, it speedily retreats.

Some curious modifications of form have taken place among the holothurians. In the plated sea-cucumbers (*Psolus*), of which a specimen is illustrated on page 1835, the animal has become flattened, and the tube-feet restricted to three out of the five ambulacra, and by these three the animal creeps about, or holds itself fixed to the rock. A similar modification is carried to excess in the deep-sea holothurians known as *Elasipoda.* Here, as in the illustrated *Scotoplana*, there are a couple of rows of thick tube-feet, forming little stumps, with which the animal moves, as a centipede moves by its legs. In front there is a sort of funnel or scoop formed by the short tentacles, while a few of the tube-feet form long horns or feelers on the upper side. In the deep-sea *Psychropotes*, on the other hand, mouth, vent, and tube-feet are confined to a flat sole ; while the posterior part of the body is extended in a long tail.

Some holothurians live in mud ; and by reason of constantly keeping both mouth and vent above the surface their bodies have become curved in U-fashion, as seen in the U-shaped *Ypsilothuria*. This is carried still further in the club-like *Rhopalodina*, a form shaped like a cherry, with a thick stalk ; the openings of both mouth and vent being at the top of this stalk. A yet stranger modification is the holothurian described

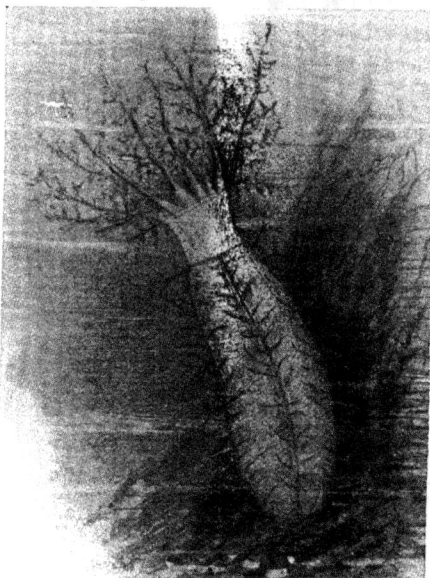

under the name *Pelagothuria*, which lives in the East Pacific, on the surface of the ocean. It has no calcareous spicules, the longitudinal muscles being mostly changed into a jelly tissue. Around the mouth the body is extended into a kind of disc, prolonged into thirteen to sixteen feelers. The animal swims by the movements of this disc.

Holothurians have no means of offence, but protect themselves for the most part by assuming the colour of their surroundings. The huge *Synapta besseli*, which reaches a length of 6 feet, has a habit, when taken in the hand, of squeezing the fluid contents of its body towards the portion that is grasped, till it becomes too big to hold. Some, when much irritated, seem to fade away and dissolve by breaking up their tissues; while others have an objectionable habit of shooting out

DEEP-SEA HOLOTHURIAN

a part of their intestines in long, viscous strings; and it is owing to this that a common British form has gained the name cotton-spinner. It has been suggested that this habit has been acquired in connection with the presence in the intestine of small parasites, and these, in their turn, have their habits affected according as they live in holothurians that are or are not cotton-spinners. Among the parasites of holothurians should specially be mentioned a little fish of the genus *Fierasfer*, that inhabits the intestine of some species, and has its food provided for it by the holothurian; this fish is described on page 1773.

Holothurians are of interest, as furnishing a food known as trepang, which ranks with edible birds' nests among the delicacies of a Chinese table. The fishing for trepang, or *bêche-de-mer*, as the holothurians are called by the Portuguese, takes place very largely in Oriental countries, and is being extended to the Barrier Reef of Australia. All kinds are not equally esteemed, for some have too much calcareous deposit in their skin, and others get rid of their insides, and so become too lean.

DEVELOPMENT OF ECHINODERMS

Few things about echinoderms are more remarkable than their modes of reproduction, which include both a sexual method, from the fertilised egg, and also one by budding or splitting of a single individual into two. Many echinoderms, as we have already seen, have the curious power of breaking off portions of themselves, as the brittle-star and crinoid can break off their arms. Also they are able to eject the whole or a part of their viscera, a faculty which has been specially developed in some of the holothurians. It is still more remarkable that the portions so broken off can be grown again, and that they themselves can, in many cases, grow fresh bodies, and become complete individuals. A star-fish of the genus *Linckia* commonly avails itself of this faculty;

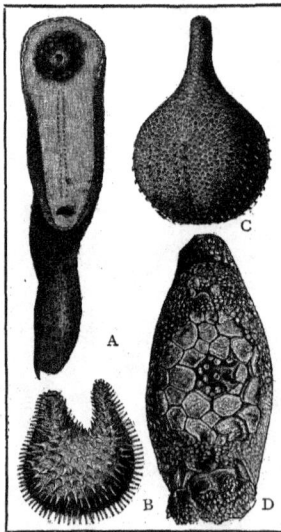

A. DEEP-SEA HOLOTHURIAN
B. U-SHAPED SEA-CUCUMBER
C. CLUB-LIKE SEA-CUCUMBER
D. PLATED HOLOTHURIAN

In D some plates above brood-pouch are removed

and it is by no means rare to find big arms with a small body at one end, and four little arms growing out of it; these are known as comet-forms.

This power of reproduction is probably due to the extension of all the systems of the body into the arms; the arms of brittle-stars, which do not contain all the systems of the body, have not been known to reproduce individuals. In some cases echinoderms have been seen definitely to reproduce themselves by fission, or splitting in half. Such division is well known to take place among the sea-cucumbers; and it is believed by some to take place even in brittle-stars. Just as a butterfly does not develop directly from the egg, but passes through the intermediate larval stage of the caterpillar, out of whose chrysalis the butterfly springs, so the sea-urchin or the star-fish egg gives rise to a larval form, in whose body, as it were, the mature form is developed. The particular shape of the larva varies in the different classes of echinoderms; but the differences are not essential, and it is clear that all the larval forms are modifications from one primitive type. The changes passed through in the development of the sea-urchin known as *Strongylocentrotus drœbachiensis* are depicted in the illustration on page 1836, in which the drawings are very greatly magnified.

The fertilised egg divides and subdivides until a round ball of cells is formed. This is then pushed in at one end, as one might push in a soft india-rubber ball, so that there is formed a little sac with a double wall to it (stages 1, 2). Stage 3 shows this in outline as though transparent; and one sees the opening (*u*) turned downwards, and the canal (*d*), which foreshadows the intestine. At the upper pole of the embryo, near the number 3, is a small tuft of cilia, by the motion of which the embryo swims about. In stage 4 this ciliated area is seen to have extended downwards to the letter *v*. The intestine now develops in such a way that the original opening (*a*) remains as the vent, the middle part (*d*) widens into a stomach, and a fresh mouth-opening (*n*) is pierced through at its upper end. This is seen from the side in stage 4, and from the front in stage 5.

But before the mouth is formed two ear-like processes (*w*) show themselves, which are important as being the beginnings of the ambulacral and water-vascular systems. There also appear a few delicate symmetrically laced rods of carbonate of lime, which by-and-by grow into the skeleton of the larva, in shape something like an inverted easel. The two lower ciliated bands now grow towards one another, so that the vent comes to lie beneath them (stages 7 and 8). They also join themselves to the two upper bands, so

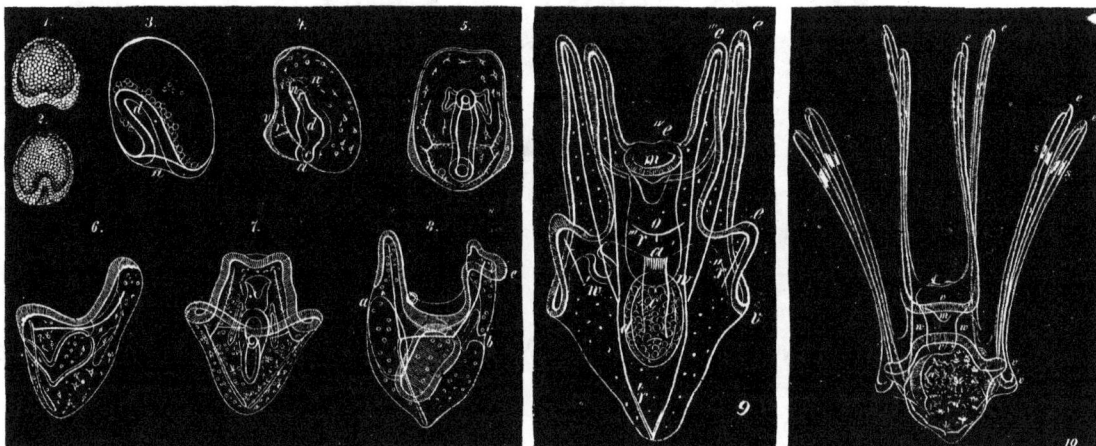

STAGES IN THE DEVELOPMENT OF A SEA-URCHIN

Stages 1—8 are in proportionate size, 9 is reduced slightly, and 10 more so. For explanation of letters, see text

that there is formed a single zone of cilia, which persists to the end of the larva's life.

Already can be distinguished the beginnings of the apex and of the processes (*e*), which finally lengthen into the arms that give such a strange appearance to the larva, in the sea-urchins and also in the star-fish and brittle-stars. In stage 8 can be seen, at *b*, the pore that admits water to the water-vascular system ; and at this point will lie the madreporite of the future sea-urchins. The next illustrations shows all these parts in a rather more advanced stage : *a* is the vent ; *c*, the hinder intestine ; *d*, the stomach, around which a deposit of spicules indicates the first beginnings of the body of the sea-urchin ; *o*, the gullet ; *m*, the mouth ; *e*, the arms of the larva ; *r*, the calcareous rods that support them ; *v*, two more strongly developed and slightly projecting portions of the ciliated band ; and *w*, the water-vessels. The larva in its full development is called a *pluteus*, on account of its frequent changes of shape, as it swims about with its arms constantly moving. It will be noticed that through the whole of its development it retains a two-sided symmetry, such that, if cut down the middle, it would be divided into two precisely similar halves. This is very different from the five-rayed symmetry of the sea-urchin, and difficulties arise both in this class and in the others when we try to discover how the five-rayed form was produced from the two-sided one.

From this larva only the stomach and the water-vascular system are continued in the sea-urchin, whose prickly body is now being formed around the stomach of the larva ; and it is in just those two systems, especially in the madreporite and in the intestine, that we note in the adult the traces of the primitive bilateral symmetry. When the little body of the sea-urchin, which at first is like a flat box, has become provided with a mouth of its own, and with a circlet of comparatively large spines, then the parts not necessary to the new structure disappear. The calcareous

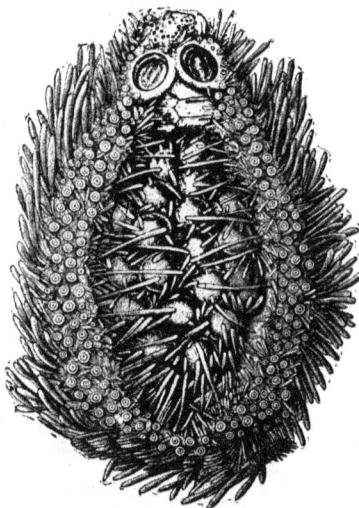

BROOD-POUCH OF A SEA-URCHIN

skeleton of the larva is absorbed, and the lime salts thus set free help to build up the test of the sea-urchin. The arms sink in, and at last the outer larva remains as nothing more than a skin over the test of the urchin.

The mode of life of the little sea-urchin, about one millimetre in diameter, is now completely altered. It is no longer carried about through the water, but crawls by means of its tube-feet and its spines. We cannot here follow the further changes that it undergoes. It is not, however, all echinoderms that pass through these curious larval stages, for in many species the young are developed in the shelter of the mother. We have already seen this to be the case with many brittle-stars, which are protected in the so-called genital bursæ.

In one sea-urchin (*Hemiaster philippi*) there are depressions between the ambulacra, which are called brood-pouches, for in these the young develop from the egg, covered over by the spines of the parent, as in the annexed illustration. In some holothurians the young are attached to the body of the parent ; but in others, as in *Psolus ephippifer*, they live on the back of the mother under some large mushroom-like plates. Some star-fish, too, such as *Pararchaster*, have a kind of tent of plates in the middle of the disc, where the young grow up as in a nursery.

The direct development from the egg to the adult in these protected forms seems to show that the elaborate shapes of the various larvæ have been developed secondarily for the special purpose of transporting the young and aiding in the dispersal of the species, and, therefore, that they are not relics of any ancestral forms. There can, however, be little doubt that the echinoderms were originally derived from some form or forms with a two-sided symmetry ; and it is certainly curious what a close resemblance their assumed primitive larval form presents to the larva of *Balanoglossus*, the worm-like chordate described on page 1787.

SUBKINGDOM ANNELIDA
SEGMENTED WORMS

THE invertebrates with which we have so far been concerned are all unsegmented—namely, the body is all in one piece. But the members of the present subkingdom, to which such familiar forms as earthworms and leeches belong, possess bodies divided into a series of rings or segments, and are therefore said to be segmented. Hence the name of the group, Annelida, derived from the Latin *annulus*, a ring. The symmetry is bilateral, there being a clear distinction between the head and tail ends, while there is typically a spacious cavity (cœlom) between the wall of the body and that of the digestive tube, and this communicates with the exterior by pairs of small tubes, which discharge nitrogenous waste products. There is a well-developed central nervous system, consisting of a ring encircling the front part of the digestive tube and a double longitudinal cord running backwards from this near the under or ventral side of the body. The upper side of the ring is thickened into a pair of ganglia, which may be regarded as a rudimentary brain, and the ventral cord swells into a pair of ganglia in each segment, with few exceptions.

The subkingdom is divided into three classes: (1) Primitive segmented worms (Archiannelida), (2) Bristle-worms (Chætopoda), and (3) Leeches (Discophora).

CLASS ARCHIANNELIDA—PRIMITIVE SEGMENTED WORMS

THE class Archiannelida is a very small group of marine worms of simple structure, though this must not be regarded as a positive proof of their primitive nature, for some comparatively simple animals have, no doubt, descended from more complex ancestors, as appears to be the case with the fixed sea-squirts, which possess a notochord and well-developed nervous system when young, but lose the former and degenerate the latter before assuming the adult characters.

The form belonging to this group which has the most claim to be considered primitive is a minute creature (*Dinophilus*) that abounds in rock-pools at Plymouth and elsewhere, and is represented in the figure. The projection in front of the ventral mouth is termed the prostomium, and it bears a pair of eyes. It is followed by five or six segments, behind which is a short tail-piece. Swimming is effected by rings of cilia. In certain

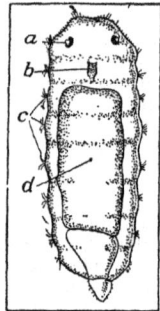

DINOPHILUS

a, eye; *b*, mouth; *c*, ciliated rings; *d*, stomach

points of structure *Dinophilus* resembles the whirl-worms, and it may possibly give us an idea of the characters of the ancestral stock from which annelids have taken their origin.

The only other member of the class requiring notice is *Polygordius*, a slender, reddish form, a little over an inch long, which lives in sand on the shallower parts of the sea-floor in the Mediterranean and elsewhere. The head bears a pair of small feelers, or tentacles. The food of *Polygordius* consists of the small organisms and nutritive particles contained in sand, which it constantly swallows, and the minute organisms and nutritive particles contained in which constitute its food. The eggs are fertilised, and undergo their development in the surrounding sea-water, and very characteristic larvæ, known as trochospheres, hatch out from them. The trochosphere, of which the form will be gathered from the figure, swims by means of a double circlet of cilia. It is destined to become the head of the adult, the body being developed as a narrow outgrowth, which becomes divided into a series of segments, from its hinder end. The head is at first large, but soon becomes reduced in size.

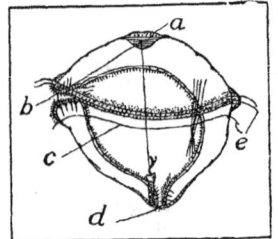

TROCHOSPHERE

a, brain; *b*, mouth; *c*, stomach; *d*, vent; *e*, circlet of cilia

CLASS CHÆTOPODA—BRISTLE-WORMS

IN the class Chætopoda is included a great multitude of marine, fresh-water, and terrestrial worms, which are far more complex in structure than the members of the preceding class, and all agree in possessing bristles, or chætæ, projecting from the sides of the segments and primarily serving as an aid to crawling, though they may also discharge other functions.

The class may conveniently be divided into two orders: (1) Many-bristled worms (Polychæta), and (2) Few-bristled worms (Oligochæta), although there is not a sharp boundary line between the two groups.

MANY-BRISTLED WORMS

The vast majority of the members of the order Polychæta are marine, and the sexes are distinct. While some creep, burrow, or swim freely, others inhabit tubes and are much modified in structure in accordance with their mode of life. A distinct head

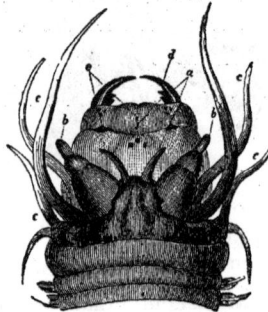

HEAD OF NEREIS

a, *b*, *c*, feelers; *d*, jaws; *e*, teeth

is typically present, bearing feelers and eyes, and obvious bundles of bristles project from the sides of the segments. As a rule there is a trochosphere larva in the life-history, essentially resembling the one described for *Polygordius*, but presenting modifications in detail, especially as to the arrangement of the bands of cilia present.

The average external characters of polychætes may be gathered by examination of the sea-centipede (*Nereis*), often to be found creeping over seaweeds and other objects near low-tide mark. The head, with its feelers, is shown in the accompanying figure, and there are also four simple eyes, which are not represented. The projection at the upper end of the illustration is the protruded pharynx, provided with powerful curved jaws for securing prey. Under ordinary circumstances it is not visible from the exterior. The long body is divided into a considerable number of segments, of which the first two can be seen

I Q

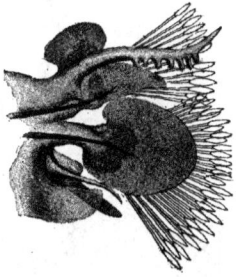

FOOT-STUMP & BRISTLES
OF HETERONEREIS

in the figure. There are hollow unjointed projections on their sides known as foot-stumps (parapods), and it is in these that the bristles are imbedded.

A nereis can crawl with considerable rapidity, its body undulating from side to side, and enabling the parapods to serve as feet. Under these circumstances there is a superficial resemblance to a centipede, hence the common name. It may be added that a parapod consists of a dorsal and a ventral moiety, each being supported by a strong axial bristle (aciculum) and bearing numerous smaller ones.

During the breeding season nereis undergoes some remarkable changes, in which it is not typical. These involve, among other things, enlargement and specialisation of the parapods in the hinder, or reproductive, section of the body. The worm is so much altered in appearance at this time that it was formerly supposed to be a distinct form, and received the name of *Heteronereis*. The male and female differ somewhat in appearance, and a male in the *Heteronereis* stage is represented in the illustration. A detached parapod is represented above, and this shows very clearly the two strong axial bristles, and groups of delicate spear-shaped ones.

Pretty closely related to nereis is an interesting form, of which Dr. Benham says: " A peculiar worm—*Palolo viridis* —is used as food by the natives of Samoa and Fiji. The worm . . . lives in fissures among corals on the reefs, at a depth of about two fathoms. At certain days in October and November they leave the reefs and swim to the shores of the above islands, probably to spawn ; and this occurs on two days in each of the above months—the day on which the moon is in her last quarter, and the day before. The natives, who call the worm " Mbalolo," give the name " Mbalolo laili " (little) to October, and " Mbalolo levu " (large) to November, thereby indicating the relative abundance of the worms in these two months. The natives eat them either alive or baked, tied up in leaves."

The so-called *Palolo viridis* is imperfect and headless, and the matter has now been fully elucidated. The complete annelid—*Eunice viridens*—burrows into the reef-rock of Samoa, the reef, when prised open with a crowbar, proving shortly before the swarming season to be absolutely alive with palolo. Owing to the great length of the entire worm, its fragile structure, and its intricate association with the honeycombed reef, the extraction of complete specimens is a matter of considerable difficulty, demanding very delicate manipulation.

The complete annelid consists of two distinct parts, a broad anterior " atokal " portion, sharply marked off from a slender and much longer " epitokal " portion, which at the swarming season becomes detached and constitutes the free-swimming palolo. It is com-

parable to the modified hinder part of a sea-centipede in the heteronereis stage. The total length averages 16 inches, of which about the first fourth is formed by the thick atokal portion. From 250 to 430 is the approximate number of segments in the atokal region, the smaller number occurring in a female and the larger in a male. In the males the colour is reddish brown, and in the females bluish green. These sexual colours are most strongly marked in the epitokal region, where they are due to the sperm and ova, the collapsed integument being quite colourless after the discharge of those elements.

Palolo, as above mentioned, are by no means confined to Samoa and Fiji. " A similar swarming of marine annelids," writes Mr. Woodworth, " and at corresponding seasons is known for other islands of the Pacific, though the worms have not everywhere been identified. Powell speaks of them in the Gilbert Islands, where they are known to the natives as *te nmatamata*, and Codrington gives a detailed account for Mota in the Banks Islands, where they are known as *un*. Brown mentions the annual occurrence of a palolo on the east coast of New Zealand, and the *wawo* of Rumphius, which occurs at Amboyna, in the Moluccas, is doubtless the same. Seeman mentions the occurrence of the worm in the New Hebrides. It is reasonable to suppose that a systematic search would show the palolo, or some allied form, to have a wider distribution in the coral-reefs of the Pacific than has been as yet recorded. That the annelid is best known from Samoa and Fiji is accounted for by these groups of islands having been most visited and longest inhabited by whites."

Many of the free-living polychaetes are able to swim more or less, and in those which are distinguished by this habit the parapods are broadened into paddles, as in *Phyllodoce*. One interesting little worm, *Tomopteris*, belongs to the floating population of the sea. The parapods are elongated, and devoid of bristles; each contains a star-shaped light-producing organ. The head is shaped like a hammer, and possesses two short and two long tentacles, supported by axial bristles. The latter appear to serve as balancing organs, and are probably modified parapods.

Among the objects commonly cast up on the shore are short, broad worms popularly called sea-mice, of which an example is figured. They are abundantly

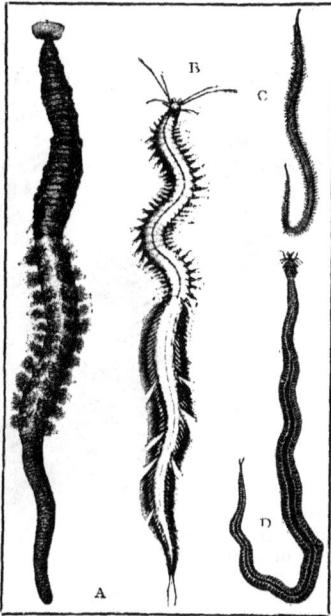

A, ARENICOLA PISCATORUM ; B, HETERONEREIS ; C, GLYCERA ; D, PHYLLODOCE LAMINOSA

SEA-MOUSE

provided with bristles, but this does not deter some fishes from swallowing them. The species (*Aphrodite aculeata*) to which the name sea-mouse is most commonly applied, and which burrows in the sand, is distinguished by the rainbow colours displayed by its long bristles, and by a feltwork of small detached bristles that cover its back. Another kind is figured.

The small-headed *Glycera* is a burrowing form which leads on to the tube-dwellers, of which a common type, the lug-worm (*Arenicola piscatorum*), much used for bait, is represented in the same illustration. The head and parapods are greatly reduced, and several pairs of branching gills are present in the middle section of the body. It inhabits U-shaped burrows, the walls of which are to some extent consolidated by a secretion of the skin to form a temporary tube. Professor J. A. Thomson says that this form ". . . burrows like the earthworm, not only forcing the anterior part of its body onwards, but eating the sand for the sake of the organic particles and small organisms which it contains. The sandy castings, which pass from the end of the food canal, and are got rid of at the mouth of the tube, fall into spiral coils.

"It has been calculated that in a year the average volume of sand per acre thus brought up in castings is about 1,900 tons, representing a layer of 13 inches spread out over the surface. This work, comparable to that of the earthworms, tends to cleanse the sand and to reduce it to a fine powder. When getting rid of the casting, the worm lies with its tail upwards and its head downwards, or with its body bent like a bow ; when the tide comes in, the mouth may protrude at the other end of the U-shaped tube."

In the extraordinary form *Chœtopterus*, here figured, there is a U-shaped tube, as in the lug-worm, but it is of a more permanent character, and consists of a parchment-like substance formed by the hardening of a secretion of the skin.

In cases where worms of this order live in a narrow, firm tube the head bears tufts of filaments serving for breathing, and being also clothed with cilia, the action of which sets up food-bearing currents, as in some of the fixed Protozoa and Echinoderma, elsewhere described. As the food consists of minute organisms and the like, a powerful, jawed pharynx, capable of protrusion, as, *e.g.*, in nereis (page 1837), is unnecessary and is not present. One such form, *Terebella*, is figured. It possesses stout branching gills and numerous long tentacles. The latter collect the sand-grains of which the tube is composed.

TEREBELLA

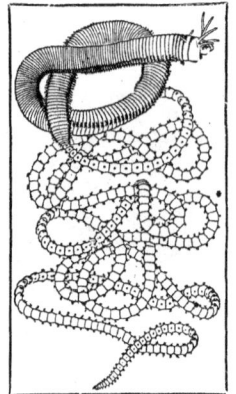

CHÆTOPTERUS

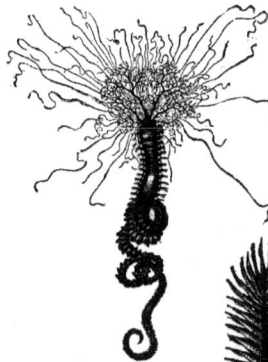

COMMON EARTHWORM

The empty shells of molluscs, loose stones, etc., may often be found more or less covered with wavy calcareous tubes. These are the habitations of *Serpula* and its allies. The head bears two beautiful coloured plumes, which serve the purposes already mentioned. One tentacle is converted into a conical stopper, or operculum, which closes the mouth of the tube when the animal draws itself in on the approach of danger. As is usual in such cases, the plumes serve to some extent as eyes, and by their means the worm can detect sudden variations in the amount of light. This is a matter of practical importance, for a passing shadow often means the approach of some rapacious fish. In examining aquarium specimens protruding from their tubes, a lightning-like withdrawal can generally be brought about by suddenly intercepting the light. It may be added that the head-plumes of some tube-worms are studded with innumerable eyes, which doubtless enhance the power of distinguishing variations in the illumination.

Some of the brown seaweeds between tide-marks will often be seen to have numerous little white spiral tubes adhering to them. These belong to the tubeworm, *Spirorbis*, and in this case the stopper closing the tube is hollowed out to serve as a brood-pouch, in which the eggs undergo their early development.

As a postscript to the polychætes, we may take the members of a small family (*Myzostomatidœ*) most probably related to them. They are small, round, or oval flattened creatures, found as external parasites on crinoids and star-fishes. The underside of the body bears a small number of parapods provided with bristles.

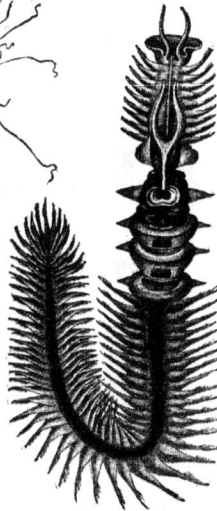

EUNICE VIRIDIS

FEW-BRISTLED WORMS

The members of the order Oligochæta, distinguished by the possession of comparatively few bristles, are nearly all inhabitants of the soil or of fresh water. There are no parapods, though these were doubtless present in ancestral forms—which almost certainly were an offshoot from the polychætes—but a varying number of bristles are present in the body-wall, serving as aids to locomotion. The head is devoid of tentacles, and usually of eyes, exhibiting the same kind of reduction as in burrowing polychætes, such as the lug-worm, and outgrowths of the body serving as gills are only present in a few aquatic types.

The earthworms, of which the commonest British species mostly belong to two genera (*Lumbricus* and *Allolobophora*), are of extreme interest and importance, and have attracted much attention since the publication of Charles Darwin's classic work, "The Formation of Vegetable Mould Through the Action of Worms," and which embodied the results of observations and experiments made at intervals during a period of over forty years.

The cylindrical body of an earthworm is clearly divided into a large number of segments: the head end is pointed, and the tail somewhat broad, and flattened from above downwards. The form and structure of the animal afford an excellent illustration of adaptation to a burrowing life. As everyone knows, these creatures tunnel in the soil, and their operations are usually limited to a depth of 12 or 18 inches. In extremely dry or cold weather, however, they descend to a much greater distance, and pass into a torpid condition until the external conditions become more favourable.

A worm in burrowing forces its narrow head between the particles of soil, and may be said to eat its way through the ground, as it constantly swallows earth for the sake of the contained organic matter, after the fashion of some marine forms. It comes to the surface from time to time for the purpose of voiding the earth which has passed through its body, and these worm "castings" may often be seen on the surface of lawns. The proper aeration and drainage of the soil largely depend upon the burrows of these creatures, and as the earth which has passed through their bodies is in a very fine state of division, and they constantly bring up material from below to the surface, thereby effecting a thorough mixing, they do the same kind of work that farmers effect by means of the plough and other implements of tillage.

Worms also feed on bits of leaves and other vegetable substances, and devour fragments of animal origin. There are no jaws nor teeth, and the food is taken in by the action of a muscular pharynx, which acts like a suction pump. Thence it passes through a narrow gullet into a dilated crop, that serves for temporary storage, and from this into a muscular, rounded gizzard, the mill-like action of which is enhanced by small, hard fragments of stone and the like which are swallowed from time to time. It is much the same sort of arrangement that is to be found in a bird.

Darwin says : " In many parts of England a weight of more than ten tons annually passes through their bodies, and is brought to the surface on each acre of land ; so that the whole superficial bed of vegetable mould passes through their bodies in the course of every few years." Various objects on the surface of the ground are gradually buried by the constant deposition of castings, and in this way many objects of antiquity have been preserved that otherwise would have perished.

Worms protect their burrows from being flooded, and from the entry of enemies, by blocking the openings with bits of leaf, stalk, or similar objects, and display much discrimination in laying hold of the narrow ends of things employed for this purpose.

A worm crawls, burrows, and even climbs by alternate elongation and contraction of its body, the bristles serving as holdfasts during locomotion. These may easily be detected by drawing a worm the wrong way between the finger and thumb.

Although worms are entirely devoid of eyes and ears, their skins are provided with numerous sense-cells: they possess an acute sense of touch, and can both smell and taste. They are extremely sensitive to vibrations in the ground, and the front parts of their bodies react to light, as may be seen by suddenly turning the glare of a lantern on worms stretching out of their burrows at night, when they retreat with extreme rapidity. Under such circumstances the broad posterior end of the body gives a purchase on the wall of the burrow, without which such sudden withdrawal would be impossible. In spite of the fact that the sexes are united, worms pair in order that the eggs may be fertilised. Spindle-shaped cases, or cocoons, are constructed for the reception of the eggs, and these are formed by the hardening of a secretion poured out from a band-like part (clitellum) of the skin. This is not far from the front end of the body, and is popularly, but erroneously, supposed to be due to injury by the spade.

Should a worm be accidentally cut in half, the severed portions are often able to renew the missing parts, the head end growing a tail, and the tail end a head. Considering the numerous enemies possessed by these unfortunate creatures, such as birds and moles, it is clear that this regenerative power assists in the preservation of the species. Most persons, for instance, have seen a thrush extracting worms from a lawn. Sometimes only half the victim is secured, in which case the part that escapes has a fair chance of repairing its injuries.

BEAKED NAIS

Some species of earthworms native to Africa, the East Indies, and Australia are veritable giants, attaining the length of several feet and the thickness of a man's finger.

Among fresh-water forms may be mentioned the little red river-worm (*Tubifex rivulorum*), large numbers of which can often be seen in the mud of ponds, the projecting bodies of numerous individuals looking like a red patch, which quickly disappears when the water is disturbed. The beaked nais (*Nais proboscidea*), represented in the figure, is another common form, in which the head is drawn out into a sensitive proboscis, and a pair of eyes are present. The bristles are extremely long, as is commonly the case in fresh-water types. This is one of the few-bristled worms able to propagate regularly by a process of transverse division, or fission.

A few members of the order have become external parasites. Having lost their bristles, and acquired a sucker at each end, and also horny jaws, they present a superficial resemblance to leeches, for which they were long mistaken. The best-known form (*Branchiobdella*) lives on the gills of the fresh-water crayfish.

CLASS DISCOPHORA—LEECHES

The leeches (class Discophora) are broad, flattened worms, inhabiting the sea, fresh water, or damp places on land, and externally resembling the last-

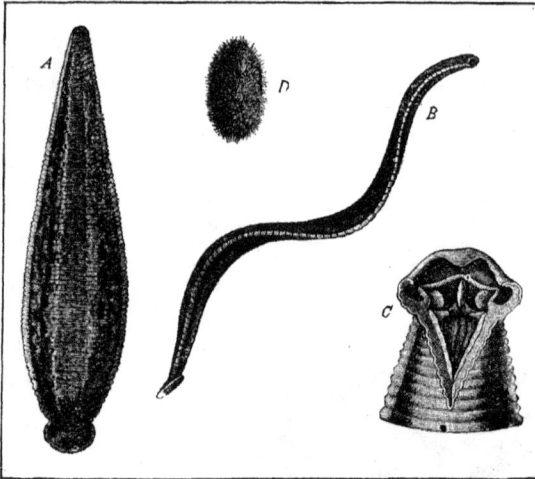

MEDICINAL LEECH

A, top view; B, side view, swimming; C, head cut open to show mouth; D, egg-cocoon

mentioned members of the few-bristled worms, possessing, as they do, a sucker at each end of the body, and being nearly always devoid of bristles. They are probably an offshoot from the last order, although they differ from them in internal structure. Their peculiarities are largely due to the parasitic mode of life which is affected by most of them.

The medicinal leech (*Hirudo medicinalis*) was formerly largely used for blood-letting, at one time a very favourite operation, so that the name "leech" used to be commonly applied to a physician. It is a flat, lancet-shaped creature, inhabiting fresh water, and handsomely marked with black and olive green on the upper surface. At the pointed front end of the body is a sucker for effecting attachment to the body of the victim, which is commonly fish, frog, or mammal, and the mouth is situated in the middle of this organ. There is a second large round sucker at the hinder end. By the alternate attachment of the suckers a leech is able to crawl in a looping fashion, and it can also swim with facility by means of undulations of the body.

A large number of transverse grooves appear to suggest a corresponding number of segments, but as a matter of fact several of these annulations are contained in each segment. There are ten black eye-spots on the upper side of the head region, and numerous peculiar sense organs arranged in transverse rows, one across the front of each segment. There are no special organs of respiration, but breathing is effected by the general surface of the skin. In this, the union of the sexes in the same individual,

STRUCTURE OF LEECH

A, jaw; B, anterior end, showing eyes; C, alimentary canal

a, œsophagus; *b*, saccular crop; *c*, last pair of pouches

and the deposition of the eggs in cocoons, there is a resemblance to the earthworms and their allies.

A leech pierces the skin of its victim by means of three saw-edged jaws, and the flaps of the three-rayed incision they make are easily raised by the sucking action of the pharynx. The blood is prevented from coagulating by a fluid secretion poured out on to the wound, and this accounts for the difficulty experienced in stopping the flow of blood when a leech is removed from a human patient.

A full meal can only be obtained now and then, and the digestive tube of the parasite mostly consists of a large pouched crop, in which a considerable amount of blood can be stored up to be gradually digested afterwards.

The group of jaw-possessing leeches with red blood to which the medicinal form belongs also includes the land-leeches, which are troublesome pests in some tropical countries, and the horse-leech (*Hæmopis gulo*). The last is not the greedy bloodsucker traditionally supposed, for it mostly preys on earthworms and other invertebrates. Another common fresh-water species (*Nephelis vulgaris*) feeds on the juice of water-snails, and its egg-capsules are deposited on water-plants in proximity to the eggs of these molluscs, so that parasites and prey hatch out side by side.

A second group of leeches is distinguished by the absence of jaws and the colourless nature of the blood. The muscular pharynx can be protruded like a proboscis. A well-known fresh-water form (*Clepsine complanata*), which preys on water-snails, displays a certain amount of parental solicitude, for the hinder part of the body is curled up into a sort of pouch, in which the eggs develop, and even after the young leeches are hatched out they attach themselves for a time to the under surface of the parent. There are other jawless leeches which are parasitic on fishes, an example of these being the rock-leech (*Pontobdella*), which infests sharks and rays, and is remarkable for the warty nature of its skin.

ROCK LEECHES

SUBKINGDOM GEPHYREA
SIPHON-WORMS

THE subkingdom Gephyrea includes a number of more or less cylindrical marine worms, which do not present any obvious traces of segmentation, and in which the sexes are distinct, the male being in some cases small and degenerate. There are commonly a certain number of horny spines or hooks, in some cases possibly comparable to the bristles of the Chætopoda, with which the group probably has affinities, although it is a rather artificial assemblage of forms, and some authorities break it up into three distinct subkingdoms. These are here termed classes—(1) Echiurids (Echiuroidea). (2) Sipunculids (Sipunculoidea), and (3) Priapulids (Priapuloidea).

ECHIURIDS

The worms of the class Echiuroidea possess an anterior projection or proboscis, at the base of which the mouth is situated, while the vent is at the hinder end of the body. There are two hooks on the ventral surface.

The spine-tailed worm (*Echiurus*) is so named because the hinder end of the body is surrounded by a single or double circlet of perianal bristles. Of its habits, Dr. Shipley says : "*Echiurus* is almost always found in U-shaped tubes or passages in the sand, which it digs out for itself by the rapid contractions of its body-wall, aided by its bristles. It does not long remain in the same hole, but frequently changes its home. As a rule, the *Echiurus* sits near the mouth of its tube, which is often a foot or even two in depth, and sends out its proboscis in every direction ; at the least sign of disturbance it withdraws into the deeper recesses. The walls of the tube are kept from falling in by a layer of mucus, which makes a smooth lining to the passage. The perianal bristles, which can be withdrawn or protruded at will, enable the animal to fix itself at any level in the tube."

Bonellia is a shorter, plumper form, in which the degenerate male resembles a whirl-worm. The female, represented in the figure, possesses a pair of ventral hooks, but no posterior bristles, and her proboscis is forked at the end. The habits of the green *Bonellia viridis* were observed by Eisig in the aquaria of the Naples Zoological Station. It hides under stones or in clefts in the rock, and stretches out its proboscis in search of food, part of which was observed to consist of sea-squirts. Under these circumstances, the proboscis was extended to a length of over one and a half yards, though but a few inches long in the unextended condition. Its forked tip is used as a seizing organ.

SIPUNCULIDS

Among the Sipunculoidea are included the large majority of the Gephyrea, and they are distinguished by the fact that the front part of the body, or introvert, can be drawn in by special muscles. This region of the body may be studded with hooks, and when fully extended is terminated by a crown of tentacles, within which the mouth is situated. The members of the class burrow in the sand, which they swallow for the sake of the nutritious particles contained in it, and the vent is shifted far forwards on the dorsal surface.

The common siphon-worms (*Sipunculus*) include the largest members of the class, and some of them may be as much as a foot long. There are no hooks upon the introvert. The somewhat similar *Phascolosoma* includes smaller forms, in which the introvert, which may bear hooks, is sharply demarcated from the rest of the body. The species of *Phymosoma* are mostly tropical, and live in holes in coral-reefs. Hooks are generally present on the introvert. Some species of *Aspidosiphon* live in association with certain corals.

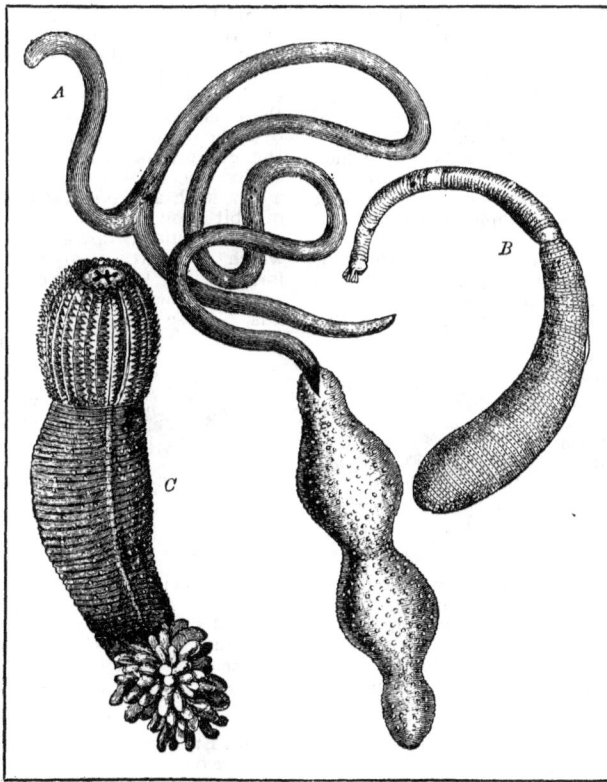

GEPHYREAN WORMS
A, Female Bonellia ; B, Phascolosoma ; C, Priapulus

PRIAPULIDS

The Priapuloidea form a small group of Gephyrea, in which there is a short introvert studded by circlets of hooks. The vent is situated at the extremity of the body, and not shifted forwards as in the members of the preceding class.

One of the most typical species (*Priapulus caudatus*) is represented in the figure, and is distinguished by the fact that its tail end is provided with numerous thin-walled outgrowths, which presumably have a respiratory function. This conclusion is strengthened by the fact that when the worm burrows their tips remain at the surface, and they may be pulled out to a length which may be greater than that of the worm itself.

The species of the only other genus (*Halicryptus*) in the family are studded with numerous small spines and devoid of breathing appendages.

SUBKINGDOM POLYZOA
MOSS-POLYPES

THE animals known as moss-polypes and lamp-shells are practically unknown to any except professional zoologists and geologists, or the members of natural history clubs. They are often associated together in one subkingdom as Molluscoida, which once was held to include the sea-squirts as well, and the meaning of the word—"mollusc-like"—serves to remind us that they were once included with the true molluscs, or shellfish. Although the affinities of the two groups are still obscure, it is best, for the present, to regard them as distinct subkingdoms.

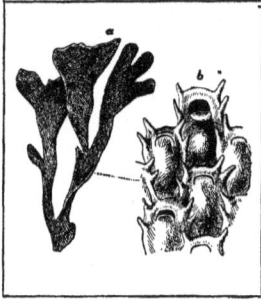

SEA-MAT
a, stock; *b*, five cells (magnified)

Moss-polypes (subkingdom Polyzoa) are small generally colonial animals, the skeletons of which assume a great variety of forms. These are among the common objects of the shore, and are often mistaken for seaweeds or for zoophytes belonging to the subkingdom Cœlentera. Some moss-polypes inhabit fresh water, and the accompanying figure will give an idea of the structure of an individual belonging to one such form (*Paludicella*). The colony is of branching form, and is supported by a horny skeleton divided into a number of compartments or cells, each of which accommodates a member of the colony, and is drawn out into a sort of cup. From this, in the figure, the polype is seen projecting, its free end being surmounted by a lophophore or crown of tentacles (*a*), which are clothed with cilia. These, by their constant movement, set up currents in the surrounding water, promoting respiration and bringing minute organisms and nutritious particles that serve as food.

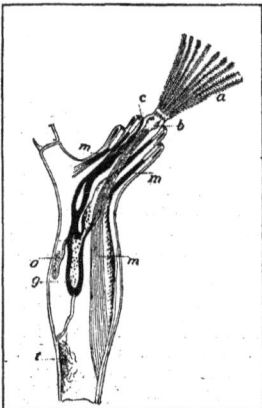

The arrangement is very much like what is present in some of the tube-worms, such as *Serpula*. The mouth (*b*), placed within the circlet of tentacles, leads into a gullet that opens into a pouch-like stomach (*g*), and from this the intestine runs back to open by the vent (*c*), placed outside the tentacles. On the approach of danger the polype can be drawn back into its cell by means of muscular bands (*m, m*). The sexes are united, and the male and female glands are shown at *t* and *o*. The former is close to a cord that connects the stomach with the wall of the cell.

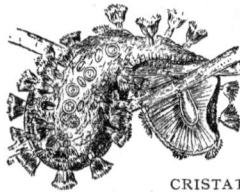

SECTION OF INDIVIDUAL OF PALUDICELLA
See text for description of lettering

Two classes are recognised in the Polyzoa: (1) Ectoprocta, and (2) Entoprocta.

ECTOPROCTA

In the class Ectoprocta, which includes the vast majority of species, the vent is outside the crown of tentacles, as in the example described. There are two orders, (1) Gymnolæmata, and (2) Phylactolæmata, to which, however, it will only be possible to make brief reference.

In the order Gymnolæmata, which includes most Polyzoa, the lophopore is in the form of a circlet of tentacles, and there is no lip projecting over the mouth. Although the great bulk of the included species are marine, a few, such as *Paludicella*, inhabit fresh water.

The sea-mat (*Flustra*), of which a figure is given, forms a flat, branching colony, and when the polypes are withdrawn into their cells these are closed by small lids. These are absent in the round-mouthed types, as may be seen in the illustration of *Tubulipora*.

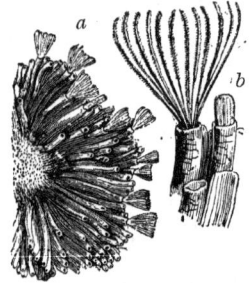

TUBULIPORA
a, part of stock; *b*, a few cells more highly magnified

The order Phylactolæmata includes a number of fresh-water forms, distinguished by a horseshoe-shaped lophophore, and the presence of a lip (epistome) overhanging the mouth. The form selected for illustration (*Cristatella*) is somewhat exceptional, for, instead of being fixed, it is

CRISTATELLA
a, statoblasts, with three young animals

able to creep about from place to place. In common with other fresh-water Polyzoa (including *Paludicella*) internal buds (statoblasts) are produced, which remain dormant during the winter, giving rise to fresh colonies the following spring, when the original ones have succumbed to the cold.

ENTOPROCTA

The Entoprocta represent a small group of marine Polyzoa, which are regarded as the most primitive members of the subkingdom. The species figured belongs to a genus (*Loxosoma*) which is unique among the Polyzoa in that its members are not colonial, for though it produces buds, these do not remain attached to the individual producing them. These forms are not attached to stones or sea-weeds, but to other kinds of animals, these being sponges in the species illustrated.

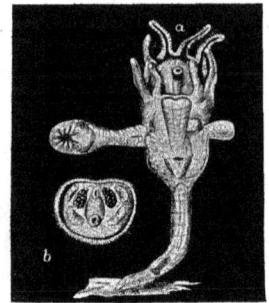

LOXOSOMA, WITH LATERAL BUDS
a, stock; *b*, swarm-larva

SUBKINGDOM BRACHIOPODA
LAMP-SHELLS

THE lamp-shells are animals enclosed in a horny or calcareous shell composed of two pieces, on which account they were for a long time included with the bivalve molluscs, such as cockles, mussels, and the like. Not only, however, do these shells differ entirely in structure from those of the molluscs mentioned, but they are differently situated, being dorsal and ventral, instead of right and left. The dorsal valve is usually smaller than the ventral, while both are bilaterally symmetrical, which is not the case in such animals as cockles, where, too, the valves are of equal size. As a rule the ventral valve is drawn out at the back into a sort of perforated beak, through a hole in which protrudes a fleshy stalk, by which the lamp-shell is usually attached to some firm object.

On opening the shell we find that two tentacle-bearing prolongations of the body project from the sides of the mouth. These have received the mis-leading name of "arms," hence the name of the subkingdom (Gk. *brachus*, an arm; *pous*, *podos*, a foot). They are clothed with cilia, by which food and oxygen bearing currents are set up, and in this respect are comparable to the lophophore of a moss-polype, though this must not be taken as an indication of affinity.

These arms vary in complexity, and are in many cases supported by projections from the dorsal shell, these often being loop-like in form. When a lamp-shell is feeding, the two valves are slightly separated and the arms may protrude between them. Each valve is lined by a soft membrane, the edges of which are usually fringed with bristles comparable to those of the chætopod worms, and which probably help to prevent the entry of parasites or foreign particles. The shell is opened as well as closed by the action of special muscles.

Lamp-shells were once a very dominant marine group, and their shells are amongst the earliest fossils known, but they have been for many ages a declining race, and at the present time are far from common, though they occur in large numbers in some localities. Their affinities are very uncertain, but they are not improbably distant relatives of the chætopod worms. They may, broadly, be divided into two classes, (1) Hingeless Lamp-shells (Inarticulata), and (2) Hinged Lamp-shells (Articulata), of which the former are more primitive.

UPPER VALVE AND ANIMAL OF CRANIA

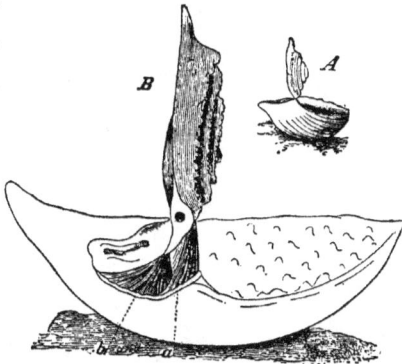

THECIDIUM MEDITERRANEUM
A, natural size; *B*, section through shell (magnified)
a, closing muscles; *b*, opening muscles

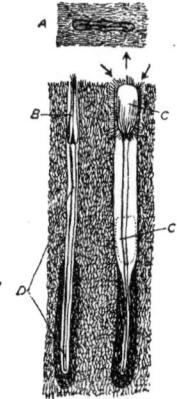

TONGUE-SHELL IN BURROW
A, opening of burrow; *B* and *C*, side views; *D*, stalk

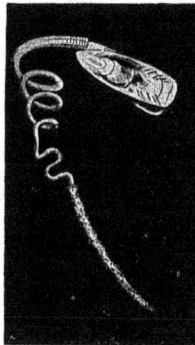

LINGULA PYRAMIDATA   TEREBRATULA VITREA
Photograph of Terebratula by Prof. B. H. Bentley

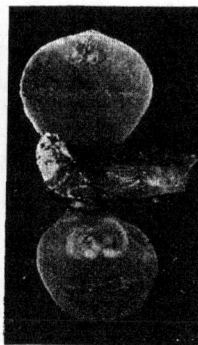

HINGELESS LAMP-SHELLS

In the Inarticulata the two valves are not provided with a hinge. They include the horny tongue-shells, and the figure given of a North American species (*Lingula pyramidata*) shows the long stalk with its end surrounded by a sheath of sand-grains, and also the bristles guarding the edges of the shell. The absence of a hinge enables the valves, which are almost the same size, to be shifted to some extent laterally, and there are special muscles by which such movements are effected. The habits of a species (*L. anatina*) common in the Philippines, New Hebrides, and neighbouring islands have been carefully studied, and are illustrated in the second figure.

It will be seen that the creature lives in a narrow vertical burrow in the sand, and when the stalk is fully extended (*D*) the front edge of the shell is just at the surface (*B* and *C*). The direction of the currents setting into and out of the shell are indicated by arrows. Should danger threaten, the stalk rapidly contracts, pulling the shell to the bottom of the burrow (*C*), the upper part of which collapses.

Another widely-distributed hingeless form (*Crania*) is of rounded form, and the ventral valve is firmly attached by its substance to some submarine object, there being in this case no stalk. The upper valve is conical in shape, and it is drawn from below in the figure, the body of the animal with its two spiral arms being also shown.

HINGED LAMP-SHELLS

The presence of a distinct hinge in members of the class Articulata prevents the lateral shifting movements possible in *Lingula*. The annexed figure shows the action of the hinge in a small stalkless recent type (*Thecidium mediterraneum*), which is represented with the lid-like upper valve wide open. Two opposite sets of muscles (*a* and *b*) open and close the shell respectively.

The remaining illustration is of a beautiful form (*Terebratula vitrea*) with translucent shell. Two individuals are shown, attached by their stalks to a piece of stone.

In another living type (*Rhynchonella*) the shell is beautifully ribbed, and its margin presents a well-marked fold.

J. R. AINSWORTH-DAVIS

SUBKINGDOM ROTIFERA
WHEEL ANIMALCULES

THE wheel animalcules are an attractive group of minute translucent aquatic creatures, present almost everywhere. Many of them creep or swim freely, while others are fixed and inhabit some kind of case, sometimes built up from foreign particles.

The main features in the structure of a free form are indicated in the annexed diagram. At the front end of the body is a thin disc (A) fringed with cilia, which set up currents bringing food and oxygen, and also enable the animalcule to swim. This is the so-called "wheel organ," a name originally given because the successive movement of the marginal cilia gives rise to an optical illusion, the disc appearing to rotate like a wheel. This organ is of the most varied shapes in different genera, as may be gathered from the figures.

The food particles, after entering the mouth (E), are conducted into a muscular gizzard (F), provided with complex horny thickenings that serve to break them up. Thence the crushed food passes on to stomach and intestine (G), where the process of digestion is completed. Two forked glands (see figure) pour a digestive juice into the stomach.

At the hinder end of the body is the so-called "foot," terminating in a pair of pincers. By means of this organ creeping movements can be effected, and it also enables the animal to fix itself temporarily to some firm object. There is a simple nerve-thickening (D) serving as a brain, and a pair of delicate feelers are placed not far from this (B).

Two interesting fixed forms inhabiting fresh water are also represented. One is the flower animalcule (*Floscularia*), where the ciliated disc is provided with long tufts of bristles that serve to entangle food. The narrow end of the animalcule is sheltered within a gelatinous tube, not shown in the figure.

The other type represented (*Melicerta ringens*) is even more interesting, for it possesses an apparatus by means of which mud can be moulded into little pellets and built up into a protective tube.

Wheel animalcules are notable for being able to endure a large amount of drying without losing their vitality, and, being blown about with dust in this condition, are distributed almost everywhere. They may be kept for years in a dried condition, reviving when placed in water. Baker, writing about the middle of the eighteenth century, gives the following quaint account of a form (*Philodina roseola*) which abounds in roof-gutters and similar places: "I call it a 'water animal' because its appearance as a living creature is only in that element. I give it also, for distinction sake, the name of 'Wheeler,' 'Wheel Insect,' or 'Animal,' from its being furnished with a pair of instruments which in figure and motion appear much to resemble wheels. It can, however, continue many months out of water, and dry as dust, in which condition its shape is globular, its bigness exceeds not a grain of sand, and no signs of life appear. Notwithstanding, being put into water, in the space of half an hour a languid motion begins, the globule turns itself about, lengthens by slow degrees, becomes in the form of a lively maggot, and most commonly in a few minutes afterwards puts out its wheels, and swims vigorously through the water in search of food; or else, fixing by its tail, works them in such a manner as to bring its food to it. But sometimes it will remain a long while in the maggot form, and not show its wheels at all . . .

"If the water standing in gutters of lead, or the slimy sediment it leaves behind, has anything of a red colour, one may be almost certain of finding them therein, and if in summer, when all the water is dried away, and nothing but dust remains, that dust appears red, or of a dark brown, one shall seldom fail, on putting it into water, to discover multitudes of minute reddish globules, which are, indeed, the animals, and will soon change their appearance, in the manner just now mentioned. . . .

"A couple of circular bodies, armed with small teeth like those of the balance-wheel of a watch, appear projecting forwards beyond the head, and extending sideways somewhat wider than the diameter thereof.

"They have very much the similitude of wheels, and seem to turn round with a considerable degree of velocity, by which means a pretty rapid current of water is brought from a great distance to the very mouth of the creature, who is thereby supplied with many little animalcules and various particles of matter that the waters are furnished with.

"As these wheels (for so, from their appearance, I shall beg leave to call them) are everywhere excessively transparent, except about their circular rim, or edge, on which the cogs, or teeth, appear, it is very difficult to determine by what contrivance they are turned about, or what their real figure is, though they seem exactly to resemble wheels moving round upon an axis. . . .

"As the animal is capable of thrusting these parts out, or drawing them in, somewhat in the way that snails do their horns, the figure of them is different in their several degrees of extension and contraction."

The affinities of the wheel animalcules are very uncertain, though many naturalists have thought them to be related to the bristle-worms. Prof. Hartog has pointed out that they in some ways resemble a particular kind of larva found in certain nemertine worms.

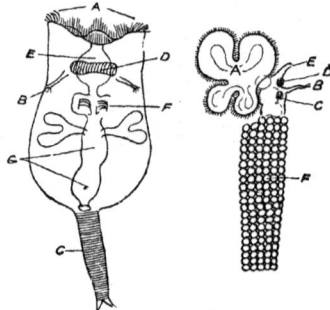

WHEEL ANIMALCULE

A, ciliated disc; B, feeler; C, foot; D, brain; E, cavity of mouth; F, gizzard; G, rest of digestive tube

MELICERTA

A, ciliated disc; B, feeler; C, cement gland; D, ciliated cup and pellet; E, flap overhanging same; F, dwelling of pellets

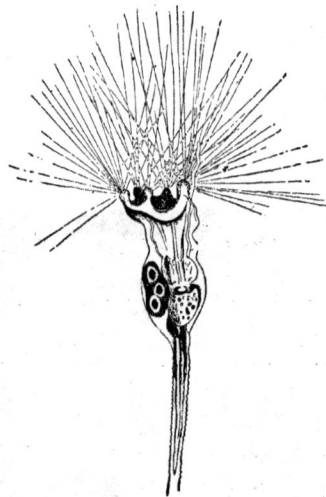

FLOWER ANIMALCULE

SUBKINGDOM NEMERTEA
NEMERTINE WORMS

THE subkingdom Nemertea includes a number of soft, elongated forms, with rounded, somewhat flattened bodies covered by a ciliated skin and not divided into segments. Most of them are marine, but a few live in fresh water or on land.

Nemertine worms are highly carnivorous, feeding more especially on various kinds of bristle-worms. Their prey is secured by means of a tubular proboscis, which when not in use is contained in a sort of sheath, situated above the digestive tube, and opening by a pore at the front of the body, usually above the mouth. It can be shot out with great rapidity, and again withdrawn into its sheath by means of a muscular band attached to the body-wall.

(*Polia crucigera*) selected for illustration has a green body marked with white streaks. This species is about 16 inches long, and the proboscis can be protruded for as much as 6 inches.

The nemertines are not, as a rule, distinguished for powers of swimming, an exception being afforded by a flattened form (*Pelagonemertes*) native to the Indian Ocean. The accompanying figure represents an individual with proboscis partly protruded and the nervous system clearly indicated.

A nemertine is hatched out of the egg as larva, one kind of which is shaped like a helmet, and known as a *Pilidium*.

The affinities of the subkingdom are very doubtful.

END OF PROBOSCIS OF
TETRASTEMMA

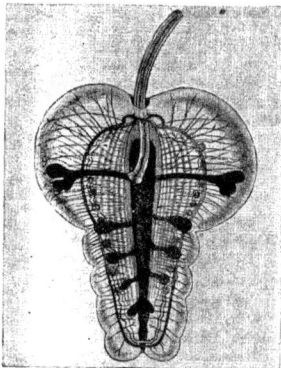

PELAGONEMERTES
Proboscis partly protruded

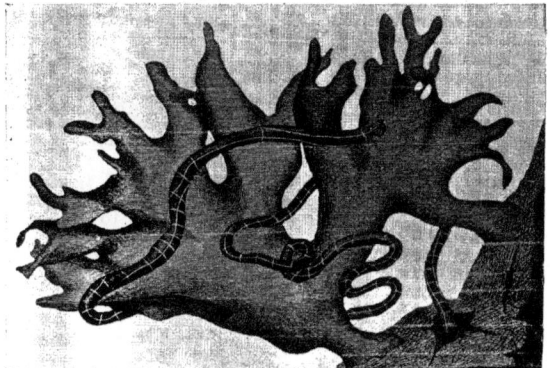

POLIA CRUCIGERA
Cross-bearing nemertine on a coral

Tetrastemma is one of the so-called " armed " nemertines, because the tip of the protruded proboscis is provided with a sharp spine, near the base of which poison-glands open. It is shown in the annexed figure. The front part of the central nervous system consists, as in annelids, of a ring, but this, instead of encircling the digestive tube, surrounds the proboscis sheath. From the ring a couple of nerve-cords run back, one on either side, and these, with the ring, are indicated in the larger figure of *Tetrastemma*, which also represents the four eyes present in this species.

Many of the nemertines are " unarmed," the proboscis being devoid of a spine. One common kind

They were considered for a long time as forming a group of the turbellarian worms, and were afterwards promoted to the rank of a distinct but allied group of flat-worms. It is by no means unlikely that they are distantly related to the ancestral stock from which backboned animals have taken origin, and Hubrecht, who has elaborated this view, gives plausible reasons for supposing that the proboscis sheath is comparable to a notochord, and a median dorsal nerve—often present in addition to the lateral nerve-cords—the equivalent of a spinal cord. However that may be, there is no doubt that the distinctive characters of the group entitle it to rank as a special primary branch of the animal kingdom.

SUBKINGDOM MOLLUSCA—SHELLFISH

CUTTLE-FISHES and the pearly nautilus; snails and slugs; cockles, mussels, and oysters; together with certain other less familiar forms, make up the great subkingdom of Mollusca, the species of which abound in the sea and fresh water and on the land.

The name Mollusca (Lat. *mollis*, soft) is rather unfortunate, and, as a matter of fact, was first applied to the cuttle-fishes only, afterwards being extended to embrace all the other forms now included in the subkingdom.

Some of the main characters of molluscs can be ascertained by examination of a garden snail in the act of crawling. It will be seen that the body is not divided into rings or segments—that is to say, is *unsegmented*—that there is a firm protective calcareous

shell, and that there is nothing by way of limbs, the animal creeping by means of a muscular thickening of the under side of the body, technically known as the " foot." The least constant of these characters is the shell, but when absent this is often due to a process of reduction—namely, forms which are now devoid of this protective covering are descended from forms in which it was present. The common black slug, for instance, has no doubt originated from shell-bearing snail-like ancestors, and some of the existing land-slugs still possess a small shell.

Shellfish are divided into five classes : (1) Primitive Shellfish (Protomollusca), (2) Bivalves (Lamellibranchia), (3) Tusk-shells (Scaphopoda), (4) Snails and Slugs (Gastropoda), (5) Cuttle-fishes and their allies (Cephalopoda).

CLASS PROTOMOLLUSCA—PRIMITIVE SHELLFISH

O N turning over stones between tide-marks on the British coast, we shall often find below them small oval flattened creatures, with a broad muscular foot on the ventral side, and a series of eight overlapping shelly plates on the dorsal side. These are mail-shells (*Chiton*), which are of particular interest, because they are in many respects extremely primitive, and give us some idea of the remote ancestors from which all shellfish have descended. The word "remote" is used advisedly, for some of the fossils found in the Cambrian rocks, which are the oldest yielding undoubted remains of animals, belong to at least three classes of shellfish, and prove that the subkingdom must first have taken origin at a time inconceivably remote.

The possession of several overlapping shelly pieces on the upper side of the body cannot, however, be regarded as a primitive character, and is peculiar to the chitons. It is undoubtedly a protective device, and when the animal rolls up, as these overlapping pieces of armour enable it to do, some at least of its foes are kept at bay. One such enemy is the common shore-crab, and it has been observed in an aquarium to seize an unrolled chiton in one of its pincers and pick away the soft body of its victim with the other.

The chiton possesses an ill-marked head, and is able to protrude from its mouth a complex rasp-ing organ, by means of which it scrapes off bits of vegetable matter to serve as food. This organ, which will be described elsewhere, is possessed by all shellfish except those belonging to the class of bivalves.

The chiton is also interesting in respect of its breathing organs. When an aquatic animal is covered entirely or largely by soft skin, special structures of the kind are unnecessary. But here the development of shell-plates on the one side and a fleshy foot on the other render some special provision necessary, and we find a series of delicate outgrowths, or gills, on each side of the body, lodged in grooves to right and left of the foot. Even in some allied forms these organs are reduced in number, and concentrated towards the posterior end of the body, while in other shellfish they are reduced to two or one, and never exceed four in number.

Some of the foreign relatives of the British chitons are distinguished by the possession of numerous eyes on the dorsal side of the body, and these probably give warning of shadows that may mean danger, and give the signal for curling up.

Allied to the chitons are a number of little-known forms (*Neomenia, Chætoderma*, etc.), some of which are worm-like in form, and are only known to the professional zoologist.

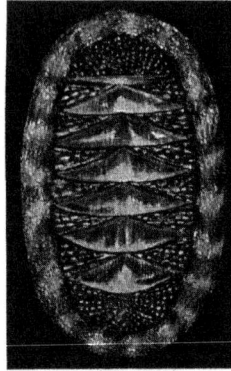 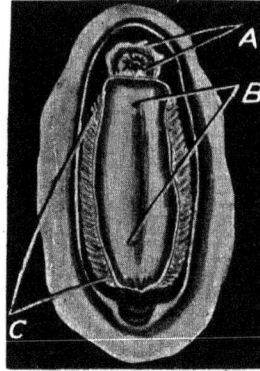

Upper side   Lower side

MAIL-SHELL, CHITON
A, head ; *B*, foot ; *C*, gills

CLASS LAMELLIBRANCHIA—BIVALVES

M ANY familiar shellfish, such as cockles, mussels, and oysters, are covered by a shell consisting of two pieces, or valves, and hence receive the name of "bivalves," thus contrasting with such molluscs as snails, in which the shell is in one piece only, and therefore known as "univalve."

Examination of the swan-mussel (*Ano-donta cygnæa*), a common fresh-water bivalve, will illustrate some of the leading characters of the class. This species is of oval form, from 4 to 6 inches long, and about half as much in breadth. The valves are right and left, and are not bilaterally symmetrical like the valves of a lamp-shell.

They are united together dorsally, being there connected by an elastic band known as the ligament. This is on the stretch when the shell is closed, and thus is able to pull them apart when the muscular bands that effect closure cease to contract.

Examination of the outer surface of a valve will reveal the presence of numerous curving lines of growth that follow the outline of the shell and represent successive additions to its substance. They are concentric to a point

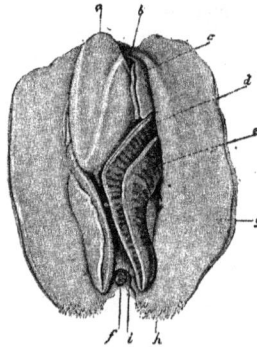

SOFT PARTS OF SWAN-MUSSEL

a, foot ;   *b,* mouth ; *c,* labial palps ; *d, e,* gills ; *f,* vent ; *g,* lobes of mantle ; *h,* tentacles

on the dorsal edge, which is known as the beak or umbo, and is the oldest part of the shell. It receives its name from the fact that in many bivalves it forms a well-marked projection.

The shell is made up of three layers : (1) an external greenish horny coat, (2) a firm white layer composed of prisms of calcareous substance, and therefore called the prismatic layer, and (3) a calcareous lining made up of alternating films of different kind, the edges of which form a series of microscopic ridges that break up light like a prism, producing rainbow colours. This layer is the source of "mother o' pearl" obtained from various molluscs, and actual pearls are composed of similar material.

The figure represents a mussel taken out of its shell, and looked at from below. To right and left are seen two soft flaps (*g*) with thickened pigmented edges. These line the valves of the shell, and are known as the lobes of the mantle. They secrete the shell, but only their thickened edges are concerned with making its two outer layers. Between them the projecting body is drawn out into a muscular orange-coloured organ (*a*), flattened from side to side. This is the foot, corresponding to the similar but differently shaped thickening seen in a creeping snail. In front of the foot a mouth (b) can be seen, but there is no distinct head, on which account the name Acephala (Gk. a, without ; cephalon, a head) is sometimes given to the class.

There are no jaws, nor is there a rasping organ within the mouth as in a chiton or snail. On either side of the mouth are two sensitive flaps, the labial palps (c), with a groove between them. Further back may be seen the gills, consisting of an inner and an outer plate (d and e) on each side. The form of the gills in this and many other members of the class has suggested the name Lamellibranchia (Lat. lamella, a plate ; Gk. branchia, gills). Close to their hinder end is the vent (f). In this particular bivalve the mantle-lobes are not united at their edges, but when these are brought together two apertures are left at the back, the lower of which is fringed with sensitive tentacles (h). It may be added that the shell is closed by the contraction of two strong bands of muscle, which run across from one valve to the other. These are the anterior and posterior adductors.

By studying the habits of the swan-mussel we are able to understand the meaning of some of the characters just described. The creature burrows in the mud, to which the compressed, sharp-edged form is an obvious adaptation, for it offers a minimum resistance to progression, friction being also reduced by the smooth surface of the shell. Locomotion is effected by the foot, which, like the body, is flattened from side to side, resembling in shape the head of an axe ; hence the name Pelecypoda often applied to the class (Gk. pelekus, an axe ; pous, podos, a foot). It is expanded and protruded by being pumped full of blood, after which it is shortened by contraction of its muscular wall, the body being thus pulled forward. The rate of movement is slow, about fifteen feet a day, and the animal appears to move little except at night.

When a mussel is actively feeding, its hinder end projects from the mud, the shell opens slightly, and two openings between the mantle-lobes can be seen, an upper and a lower, the edges of the latter being fringed with delicate tentacles. A continuous stream of water flows in through this latter "inhalent" aperture, carrying with it minute organisms and organic particles to serve as food and the dissolved oxygen required for breathing. The various waste products of the body are carried away by a counter-current flowing out through the upper, or "exhalent," aperture.

As in several kinds of animal previously described, such as tube-worms and lamp-shells, these currents are the result of ciliary action. The labial palps, inner surface of the mantle-lobes, and the gills are, in fact, clothed with enormous numbers of cilia,

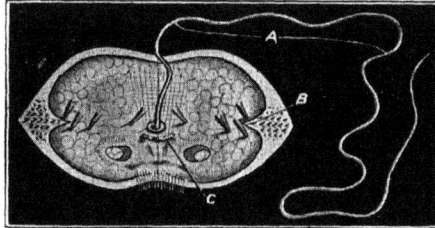

GLOCHIDIUM

A, byssus ; B, hook ; C, mouth

LARGE SWAN-MUSSEL

which work together in an orderly fashion to produce a regular flow of water. The food-particles are conducted on each side along a groove between the labial palps into the mouth.

It is clear that the food and way of feeding do not necessitate a rasping organ, the absence of which is probably due to a process of reduction, for it was most likely present in ancestral forms. These, no doubt, possessed a distinct head, but this has gradually dwindled as a result of the mode of life, for creatures burrowing in mud, having the body enclosed in a shell, and feeding by ciliary action, do not require a well-developed head provided with various sense organs and containing a specialised brain. It is, in fact, the same kind of reduction found in an earth-worm as compared with a free - living bristle - worm (see page 1839), and is explained on similar lines.

The life-history of the swan-mussel is by no means typical, but presents somewhat peculiar adaptations to fresh-water life, which may be briefly described.

It should first be mentioned, however, that in a marine bivalve—e.g., the cockle—a larva hatches from the egg which is not unlike the trochosphere characteristic of marine bristle-worms (see page 1837), being provided with a circlet of cilia serving as a swimming organ. The head end soon develops a velum, or ciliated flap, which performs the same function, and the larva is now called a veliger. This is typical not only of marine bivalves, but also of marine snails and slugs. Were such a minute fragile larva produced by a mollusc inhabiting fresh water, it would be liable to get washed away by the current, ultimately perishing in the sea, if it reached so far. The life-history of the swan-mussel is modified so as to prevent this calamity.

The eggs, instead of passing out into the surrounding water to take their chance, enter the tubular cavities of the outer gill-plates, which serve as brood-pouches.

In this sheltered position they develop into a curious larval form, known as a glochidium, this name having been given before the life-history of the mussel was known. As will be seen from the figure, the glochidium is covered by a triangular shell quite unlike that of the adult, and provided with a strong, curved hook at the tip of each valve. The foot and gills are at first undeveloped, and a long, sticky thread, the byssus, projects from the shell. After the larvae have been hatched they are ejected into the water, and settle down in clumps on the mud, their threads waving above them.

Should a stickleback, minnow, or other small fish come in contact with one of these threads it will adhere to the part touched, and the glochidium to which it belongs will be carried away, rapidly opening and shutting its shell meanwhile. In this way it swims up to its bearer, and by means of the sharp hooks on the shell grips the fish's skin. The irritation set up causes the growth of a sort of gall, within which the further development of the glochidium takes place. After about three months it has assumed the form of a young mussel, and is liberated by the rupture of the gall, falling down into the mud, where it ultimately grows into an adult.

The bivalve shellfish are divided into five orders, differing in the structure of the gills and other characters, namely: (1) Protobranchia, (2) Filibranchia, (3) Pseudolamellibranchia, (4) Eulamellibranchia, and (5) Septibranchia.

PROTOBRANCHIA

The most primitive bivalves are included in the order Protobranchia, the widely distributed *Nucula* being one of the best-known types. Each of the simple gills is made up of a double series of straight leaflets, or gill filaments, and the foot has a flattened under surface like that of a snail, and can be used for creeping.

FILIBRANCHIA

The bivalves of the order Filibranchia possess gills of more complex structure than those of the Protobranchia. Each of them is made up of a double series of filaments, which are turned up at the ends and united together by long interlocking cilia. The gills, in fact, are becoming effective organs for setting up currents by ciliary action.

Among the most primitive members of the order are the ark-shells (*Arcidæ*), which are very widely distributed, as is commonly the case with ancient forms, and are distinguished by the possession of a rectangular shell. Another ancient family (*Trigoniidæ*), in which the shell is roughly triangular, are now represented only by some small species in Australian seas, but were at one time very dominant and widely distributed. Their remains help to build up some of the Jurassic limestones, as at Weymouth, where they make up a layer of calcareous rock several feet thick. That these forms should still survive in the Australian region is a fact of considerable interest, for, as elsewhere shown, the land animals of that part of the world are also to a great extent archaic, and the same thing is true for the land plants.

Probably the most familiar form included in the order is the edible sea-mussel (*Mytilus edulis*), distinguished by its bluish wedge-shaped shell. The figure shows (above) the narrow foot, and (below) the inhalent and exhalent apertures, the former being large and fringed, the latter small and rounded. Notice also the bundle of strong, silky fibres (*byssus*), by which the mollusc is attached to some firm object. The attachment is not, however, permanent, for if conditions become unfavourable the mussel can cast off the byssus and creep to some more desirable place, there to re-attach itself. The sea-mussel is of considerable

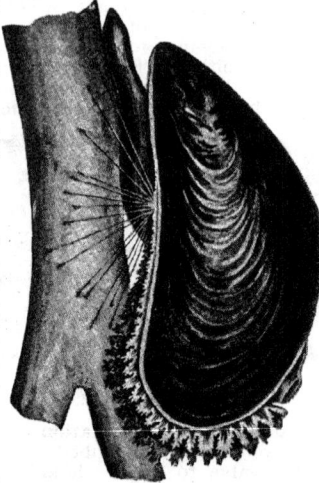

EDIBLE SEA-MUSSEL CLOSED AND ATTACHED BY THE BYSSUS

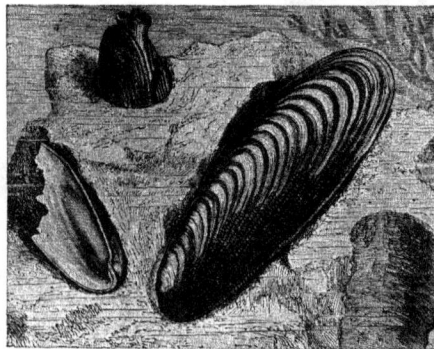

DATE-SHELLS IN ROCK

economic importance, for it is largely used as bait, and also for human food, though it sometimes proves poisonous, bringing out a rash on the skin, or even causing fatal illness. The mussel fisheries of this and some continental countries yield a considerable revenue. Special methods of culture are sometimes employed, as at Esnaudes, on the west coast of France, where the annual yield is worth as much as £52,000.

DATE-SHELLS

In date-shells (*Lithodomus*) the foot is used as a boring organ, and the figure represents a piece of limestone that has been tunnelled by these bivalves, which belong to the same family as the sea-mussel. The hinder end of one individual is seen projecting from its burrow, and it will be observed that the mantle-lobes are drawn out into two tubes, or siphons, at the end of which the inhalent and exhalent apertures are placed. Such siphons, sometimes of considerable length, are found in a great many bivalves, and enable their possessor to feed and breathe when sheltered more or less deeply in burrows, either in firm substances or in the sand or mud of the sea-floor.

PSEUDOLAMELLIBRANCHIA

The gills in the order Pseudolamellibranchia, though not so complex as those of the swan-mussel, present an advance on those described for the last order in that the constituent filaments are actually united together instead of being merely attached to one another by interlocking cilia.

The wing-shells (*Aviculidæ*), which in the type genus *Avicula* are drawn out into wing-like projections, are of special interest because they include the pearl-oysters, or, more correctly, pearl-mussels (*Margaritifera*), of which the most notable are those producing the " orient " pearls of Ceylon. Until comparatively recently it was supposed that pearls were generally due to the formation of calcareous shelly layers round foreign particles, such as grains of sand accidentally introduced into the shell of a bivalve and setting up irritation. Small pearls ("seed" pearls), it is true, may be thus formed, but large ones are usually, if not always, deposited round the dead bodies of some stage in the life-history of an internal parasite, which may be either of the fluke or tapeworm kind. They belong to the latter in the case of *Margaritifera*, and the life-history has been worked out by Professors Herdman and Hornell.

The adult tape-worm (a species of *Tetrarhynchus*) lives in the intestines of a large Indian ray (*Trygon*), and its embryos, passing out into the water, are ultimately carried in large numbers into the inhalent openings of pearl-mussels. Entering the bodies of these bivalves, they develop into a further stage. The mussels are devoured by trigger-fishes (*Balistes*), which crunch them up with their strong teeth. Within the stomach of the trigger-fish the embryo tape-worm is liberated from its investment by the action of the gastric juice, bores through the wall of the stomach, and passes into a dormant encysted stage. No further development is possible unless the trigger-fish falls a victim to the voracity of one of the sting-rays already mentioned. Should this happen, the parasite becomes an adult tape-worm.

A certain number of the minute stage infesting the pearl-oyster dies within their hosts, and being covered over by successive layers of calcareous deposit, similar to that which lines the shell, pearls are the result. It is a whimsical reflection that the proud wearer of a historic pearl necklace handed down as a much-prized family heirloom, is in reality adorned with the costly sepulchres enclosing the mummified bodies of a series of tape-worms that died in infancy, and if, indeed, there be any truth in the familiar story, the proud Cleopatra would have hesitated before drinking the pearl dissolved in vinegar had her zoological knowledge been as complete as our own.

The importance of scientific research as applied to practical matters is well illustrated by what has been said about the pearl tape-worms. When the Ceylon pearl fisheries (after a record yield in 1891) failed for some ten years, it was suggested that the trigger-fish, well known to devour the oysters, was responsible for the failure, and should be exterminated. Had this been done the tape-worm would also have been rooted out, and there would have been no more pearls.

OYSTERS

The oyster family (*Ostreidæ*) is of economic as well as scientific interest, and we may take the well-known "native oyster" (*Ostrea edulis*) as an example. As compared with an ordinary free-moving bivalve, it is of a rounded shape, and during most of its life is fixed to some firm object by the substance of its left valve, which is deeply convex, while the right valve is flattened, and serves as a lid. The anterior adductor muscle has disappeared, and the posterior one has shifted to the middle of the shell, the most convenient position for it. The foot has also disappeared. The shell is opened by means of what is termed an "internal" ligament, consisting of an elastic pad, lodged in a pit at the hinge of the shell (above in the figure). When the shell is closed by the action of the adductor muscle, this is compressed, and when the muscle slackens its elasticity comes into play and separates the valves. In this species of oyster the sexes are united, but they are distinct in the American oyster (*O. virginiana*) and the Portuguese oyster (*O. angulata*). An enormous number of eggs are produced and hatch out into free-swimming embryos, which secure the distribution of the species, and compensate for the fixation of the adults. They are popularly known as "oyster spat."

Oysters formed part of the food of prehistoric man, and their shells help to make up the rubbish-heaps ("kitchen middens") by which the wanderings of primitive communities are marked on the coasts of many parts of both the Old and New Worlds. Oyster-culture of a more or less elaborate kind has been

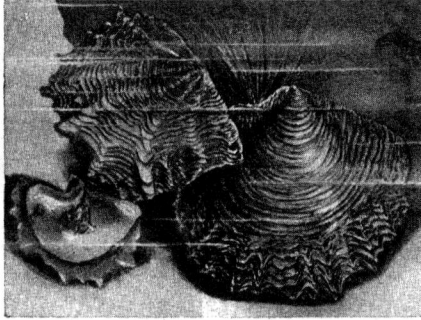
PEARL-OYSTERS

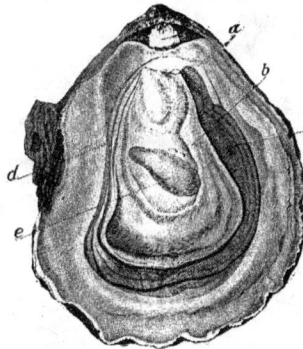
ONE VALVE AND SOFT PARTS OF NATIVE OYSTER
a, mouth; *b*, mantle; *c*, adductor; *d*, junction of mantle-lobes; *e*, gills

practised by many civilised nations from the time of the Romans to the present day, and we know that in the luxurious days of Rome the flavour of the British "natives" was appreciated.

Mr. A. H. Cooke makes the following remarks about Roman oyster-culture: "The first artificial oyster-cultivator on a large scale appears to have been a certain Roman named Sergius Orata, who lived about a century B.C. His object, according to Pliny the Elder, was not to please his own appetite so much as to make money by ministering to the appetites of others. His *vivaria* were situated on the Lucrine Lake, near Baiæ, and the Lucrine oysters obtained under his cultivation a notoriety which they never entirely lost, although British oysters eventually came to be more highly esteemed. He must have been a great enthusiast in his trade, for on one occasion when he became involved in a law-suit with one of the riparian proprietors, his counsel declared that Orata's opponent made a great mistake if he expected to damp his ardour by expelling him from the lake, for, sooner than not grow oysters at all, he would grow them upon the roof of his house. Orata's successors in the business seem to have understood the secret of planting young oysters in new beds, for we are told that specimens brought from Brundusium, and even from Britain, were placed for a while in the Lucrine Lake to fatten after their long journey, and also to acquire the esteemed 'Lucrine flavour.'"

That oysters are still appreciated may be judged from the fact that the Whitstable Company alone sells from 10 to 12 million annually, about two-thirds of these being natives, while the annual yield of all the British fisheries is worth about £155,000, and that of the world over £3,728,000, the oyster fisheries of the United States coming first with a value of about two and a half million pounds sterling.

It would take too much space to describe the details of oyster-culture, but probably the most elaborate methods are practised in France, where crates of tiles, covered with lime, are placed on the shore in order to attract the "spat" when these pass into the fixed stage. The young oysters are subsequently flaked off the tiles, and placed for a time in shallow trays covered with wire gauze. From these they are transferred to properly prepared and protected oyster-beds and in some cases fattening and other processes constitute a final stage.

SCALLOPS

The pectens, or scallops (*Pecten*), are typical representatives of another family (*Pectinidæ*), which presents specialisation comparable in some ways to that of oysters, but not carried so far. The fan-shaped shell, with two ear-like projections near the umbo, is often beautifully marked and coloured. A small foot is present. The large edible scallop is sedentary, like the oyster, but the shell is not fixed. In this case it is the right valve which is undermost and of convex shape. In most species, however, the valves are of similar size and shape. A few scallops are fixed by a byssus, while many are able to swim by opening and shutting the valves of the shell. The edges of the mantle in these shellfish are fringed with numerous sensitive tentacles, and usually studded with numerous eyes of complicated structure and red or green in colour. One kind of scallop (*Pecten jacobæus*) was used in old times as a badge by pilgrims returning from the Holy Land.

FILE-SHELLS

Except for their white colour, the file-shells (*Limidæ*) are not unlike the pectens in structure and habits. The edges of the mantle are fringed by long tentacles, but do not possess eyes. Some species can swim freely by flapping the valves, others are attached by a byssus, and these may construct a sort of shelter or nest by fixing various hard bodies together with their byssal threads.

EULAMELLIBRANCHIA

The Eulamellibranchia form the largest subdivision of bivalves, and contain a great number of families, only a few of which can receive mention here. The gills are plate-like, and of more complex structure than in any of the other orders. Seven suborders are recognised : (1) Submytilacea, (2) Tellinacea, (3) Veneracea, (4) Cardiacea, (5) Myacea, (6) Pholadacea, and (7) Anatinacea.

FRESH-WATER MUSSELS

In the suborder Submytilacea, the swan-mussel, already described, is a good example of the fresh-water mussels (*Unionidæ*), which have an almost cosmopolitan distribution. A peculiarity of the swan-mussel is the absence of interlocking projections and pits (teeth and sockets) along the hinge-line of the shell, hence the name of the genus, *Anodonta* (Gk. *an*, without; *odous*, *odontos*, a tooth). These, however, are present in the allied genus *Unio*, the shell of which is lined with a thick layer of mother-o'-pearl, furnishing the material from which pearl buttons are manufactured in Iowa and Illinois. River pearls are also chiefly derived from species of *Unio* (and *Anodonta*), the most notable being *Unio margaritiferus*, which was the chief object of the once renowned pearl fisheries of Scotland and North Wales, the rivers Spey and Conway being particularly notable. This species and some of the pearls, are represented in the figure.

We have seen (page 1850) that orient pearls are formed round tapeworm embryos, and river pearls are also due to the presence of parasites, in this case flukes. British pearls were famous so far back as Roman times, and a Conway pearl is said to be one of the jewels in the British crown. So late as 1865 the annual yield of the Scottish pearl fisheries was worth about £12,000, and at the present time the river pearls of Bavaria, Saxony, and the United States are of considerable importance.

A second fresh-water family (*Dreissenidæ*) is abundantly represented in our canals and rivers by a species

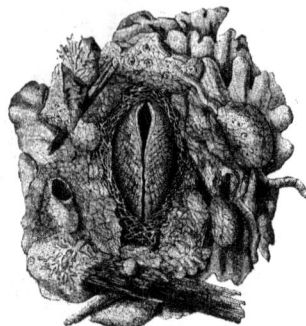

FILE-SHELL IN NEST

(*Dreissena polymorpha*) which resembles the sea mussel in shape, and attaches itself to all sorts of firm objects by means of a byssus. It affords an interesting case of rapid dispersal, being native originally to the rivers flowing into the Caspian and Sea of Aral. It was first noticed in England in 1824, in the Surrey Docks, where it was probably introduced with Russian timber, and since then has quickly spread through the fresh waters of this country, France, and Belgium.

The family *Cyrenidæ* includes both estuarine and fresh-water species, *Cyrena* being the type genus for the former, while a small British river bivalve (*Cyclas coenea*) will serve to represent the latter. If some specimens of cyclas are placed in a glass vessel full of water, each of them will soon protrude two delicate translucent tubes from the back of the shell, these affording an excellent illustration of the siphons present in many bivalves. By adding a little carmine to the water the direction of the water-currents may be easily observed, and it will be seen that the lower siphon serves for taking in water, and the upper one for the passage of the outward current, which takes away the various waste products. The foot of cyclas can be used not only for burrowing, but for climbing, and one of these molluscs can easily crawl up the vertical glass front of an aquarium, using the tip and hinder region of its foot as suckers. As in fresh-water mussels, the eggs develop within the gills, but in this case it is the inner plates of these organs that are used for this purpose.

The fresh-water oysters of North Africa constitute another family (*Ætheridæ*), the members of which crawl about when young, but ultimately settle down and become fixed.

One of the best-known marine families (*Cyprinidæ*) includes a large thick-shelled rounded form (*Cyprina islandica*), not uncommon on the Scottish coast. It is an essentially northern species, ranging south from Greenland on either side the Atlantic.

TELLENS

The suborder Tellinacea includes several families, distinguished by an elongated flattened foot and two distinct siphons. They are mostly marine.

The large and widely-distributed tellens (*Tellinidæ*) are represented by nine British species, the small flat shells of which are common seashore objects, and are often very beautiful, being delicate, translucent, and tinted with pink or yellow. Another family (*Scrobiculariidæ*) includes one common British form (*Scrobicularia*

PEARL-MUSSELS, SHOWING PEARL WITHIN SHELL AND DETACHED PEARLS (1-8)

piperata), which lives in the mud of estuaries, and can stretch out its siphons to four or five times the length of the shell, thus being able to feed and breathe with comfort when its body is deeply buried in the mud. This arrangement, affording protection against seabirds and predatory fish, is very common among bivalves ; and it may be mentioned that the siphons are very sensitive to changes in illumination, being rapidly drawn in if a shadow passes over them. This appears to be associated with the presence of pigment, and it is a case of " seeing without eyes," of which examples are also found among the tube-worms (see page 1839).

Another family (*Donacidæ*) is represented on British coasts by an elegant little bivalve (*Donax politus*), somewhat elongated in form, purplish in colour, and marked with radiating pale streaks. It is not unlike a tellen, but the shell is much stronger. The trough-shells (*Mactridæ*) are squarish forms, widely distributed in shallow seas where the bottom is sandy, and represented by six native species.

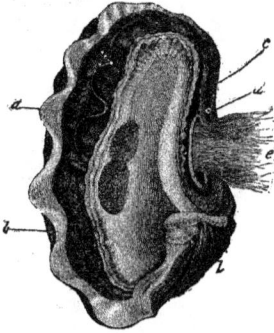

VALVE AND SOFT PARTS OF TRIDACNA

a, inhalent ; *b*, exhalent ; and *d*, foot openings ; *e*, byssus ; *c*, adductor muscle ; *l*, shell

VENUS SHELLS

In the members of the suborder Veneracea the siphons are usually short. The most important family is that of the Venus shells (*Veneridæ*), named in reference to their elegant sculpture and attractive markings. They are marine forms, and there are several British species, of which perhaps the most familiar are the carpet shells (*Tapes*). One of the " clams " (*Venus mercenaria*) largely eaten on the Atlantic coasts of the United States is a member of this family.

COCKLES

The cockle family (*Cardiidæ*), the most important in the suborder Cardiacea, includes bivalves with a long, sharply bent foot and short siphons. The common cockle (*Cardium edule*) possesses a strongly ribbed shell of rounded outline, and with strongly convex valves. It burrows in the sand, and is particularly abundant in estuaries. It is a popular article of food, and the British cockle fisheries are of considerable importance. By suddenly bending and strengthening the foot, cockles are able to execute springing movements. The common cockle ranges into the Baltic, and is also found in the Caspian and Black Seas. Another and larger British species is the red-nosed cockle (*C. rusticum*), so called from the scarlet colour of its foot.

TRUE CLAMS

The true clams (*Tridacnidæ*) include tropical and subtropical forms, with a short foot, and attached by a byssus. The mantle is brilliantly coloured. The shells of the type genus (*Tridacna*) are broadly oval in form, extremely massive, and strongly ribbed. The edges of the valves are deeply fluted. There are six or seven species, which are often to be found living in the fissures of coral-reefs with the under side facing upwards, so that they are a source of danger to the natives.

The shells of *Tridacna* are sometimes employed for religious purposes, as in St. Sulpice, Paris, where two valves over 500 pounds in weight are used as *bénitiers*.

The thorny oysters (*Chamidæ*), as their name indicates, are fixed forms. They are tropical or subtropical, and the often brightly coloured shells are covered with ridges or spines.

GAPERS

In the suborder Myacea are included a number of families in which the shell is more or less elongated in adaptation to the burrowing, or it may be boring, habit. The edges of the mantle-lobes are fused together, by which entry of sand and the like is to a great extent prevented. A short slit is left, however, for the protrusion of the foot, which is generally rather small, and the hinder part of the mantle is drawn out into siphons. These are sometimes very large, and cannot be entirely drawn back into the shell, which, when empty, is consequently seen to " gape " at the back.

The gapers (*Myidæ*) inhabit deep burrows in sand or mud, and the common sand-gaper (*Mya arenaria*) is a common British form, and also abounds on the Atlantic coast of North America, where it is known as the long-neck or soft-shell clam, and is largely eaten. The name " long-neck " has reference to the large and elongated siphons, which are fused together and covered with horny, wrinkled skin. Like marine bivalves generally, the gaper hatches out as a minute free-swimming larva. After a time this anchors itself to some firm object by means of a byssus thread. After attaining a length of about one-fifth of an inch, the young gaper throws off the byssus and begins to burrow.

RAZOR-SHELLS

The razor-shells (*Solenidæ*) are more perfectly adapted to burrowing than any other bivalves, and this is associated with the elongated narrow shell, to the shape of which they owe their name. The foot is exceedingly large, and directed straight forwards, this being the most favourable position for the exercise of its locomotor function. Razor-shells, used in some localities for food, burrow with such extreme rapidity that they are very difficult to capture. There are two common British species, in one of which the shell is straight (*Solen siliqua*), while in the other (*S. ensis*) it is curved.

Another family (*Saxicavidæ*) includes forms able to bore into the softer rocks, though in many cases they are found attached to various objects by means of a small byssus.

Gastrochæna, the type of a distinct family, lives in the sand, and is remarkable for secreting a flask-shaped tube, covered with adherent particles.

PIDDOCKS

The suborder Pholadacea contains boring forms with long united siphons and a short foot with a fl..tened end.

The piddocks (*Pholas*) are elongated bivalves enclosed in thin white shells,

PIDDOCK IN BURROW

the front ends of which are studded with hard projections, and serve as a file, by means of which burrows are excavated in all sorts of substances, including compact siliceous rocks. The foot is often considered to be the chief boring organ, but this is more than doubtful. Its sucker-like tip enables the animal to grip firmly the substance being bored, the actual boring being effected by a twisting movement that brings the rough front of the shell to bear.

SHIP-WORMS

Ship-worms (*Teredinidæ*) are chiefly wood-borers, notorious for the damage they inflict on the wooden timbers of ships, the piles used in the formation of dams for exclusion of the sea, and so forth. In the former case the introduction of copper sheathing was in the first instance a device to limit their ravages. The great dam-break in Holland at the beginning of the eighteenth century was due to their attacks.

As will be seen from the figure, the body is long (6 inches to a foot) and worm-like, the small valves of the shell (seen at the top) being used for boring, while the hinder end of the body is drawn out into long siphons. At the base of these are a pair of shelly plates (pallets), probably used to close the burrow when the siphons are drawn back. The tubular dwelling is lined with a shelly layer secreted by the surface of the body.

PANDORAS

The bivalves included in the suborder Anatinacea are exceptional in the fact that the sexes are united in the same individual.

One of the families (*Pandoridæ*) is represented on the British coast by the genus *Pandora*, so called with reference to the magic box in which that goddess kept her gifts. It is somewhat narrow at the hinder end, and has a beautiful pearly lining. The right valve is flat and lid-like, the left valve convex.

Another family (*Clavagellidæ*) includes the remarkable watering-pot shell (*Brechites*), that lives in sand or mud on the shores of the Indian and Pacific Oceans. It is enclosed in a shelly tube, one end of which is closed by a perforated plate surrounded by a frill, as shown in the figure. This is the front end of the tube, and is directed downwards, while the other end contains the united siphons, the apertures of which are close to the top of the burrow. The foot is much reduced, and does not possess a byssus. At first sight this mollusc would hardly be recognised as a member of the order, but close inspection will show a pair of minute valves imbedded in the side of the tube, near the perforated plate.

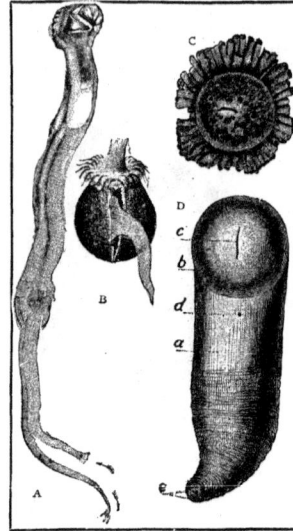

A, shipworm ; and B, its larva. c, animal of watering-pot shell ; and D, anterior end of shell. *a*, mantle ; *b*, front end ; *c*, mantle ; *d*, foot ; and *e*, siphonal openings

SEPTIBRANCHIA

The group formed by the order Septibranchia includes a small number of unfamiliar bivalves mostly living in deep water, and distinguished by the fact that the gills are represented by a perforated partition, or septum, stretching horizontally between the mantle-lobes.

CLASS SCAPHOPODA—TUSK-SHELLS

SHELL - COLLECTORS on the northern coast of Britain not infrequently pick up small curved tubes, which may be described as slender, hollow cones, open at each end. These belong to members of the present class (Scaphopoda). One species, the elephant tusk-shell (*Dentalium entalis*), is white in colour ; while a second, the grooved tusk-shell (*D. tarentinum*), is delicately grooved at the

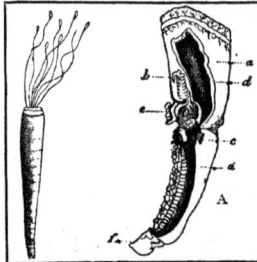

COMMON TUSK-SHELL

A, in section ; *a*, front mantle-cavity ; *d*, hinder mantle-cavity ; *b*, mouth process ; *c*, anal opening ; *d*, foot-cavity ; *e*, tentacle supports ; *f*, posterior end

broad end, and of a pinkish tint at the other. The enclosed shellfish burrows with facility by means of a trilobed foot that can be protruded from the broad or front end of the shell, and resembles the corresponding organ of a bivalve. In side view it has an obscure resemblance to the prow of a ship, whence the name of the class (Gk. *scaphe*, a skiff ; *pous, podos*, a foot).

The food consists of minute animals and plants, which are seized by the swollen, sticky ends of a bundle of slender thread (*captacula*), which can be thrust out from the shell, as shown in the figure. They probably are also useful as breathing organs. The head is ill-developed, though larger than in bivalves, and the mouth is situated at the end of a sort of snout. A rasping organ is present, as in snails and slugs.

J. R. AINSWORTH-DAVIS

CLASS GASTROPODA—SNAILS AND SLUGS

THE name of the class Gastropoda (Gk. *gaster*, belly ; *pous, podos*, foot) expresses one of the most obvious characters present in an average example—namely, the possession of a muscular expansion on the under side of the body, known as the foot, and usually serving for crawling over some firm substance. It is also seen in mail-shells (page 1847), but in these is not so sharply marked off from the rest of the body. Examination of a creeping snail, periwinkle, or pond-snail will demonstrate its use, though the exact way in which it is employed is not by any means easy to understand. Under such conditions it will be seen that the front end of the body possesses a fairly distinct head, provided with feelers or tentacles (the so-called " horns " of a common garden snail).

Another well-marked feature of an average gastropod, such as any one of those above mentioned, is the elevation of the dorsal side of the body into a projecting hump, covered by a spiral shell, which, being in one piece, is said to be univalve. The hump is due to the fact that most of the internal organs have been collected together into a compact mass, which is clearly marked off from the foot, so as not to hinder locomotion. It is twisted into a spiral form for the sake of compactness, again in the interests of locomotion, and the formation of a shell of corresponding form is evidently a protective measure. In fact, a great many snails are able to entirely draw back the body into the shell, and not infrequently there is a horny or shelly plate, which serves under such circumstances to give further security by acting as an operculum to close the mouth of the shell. The periwinkle-eater is well aware of its presence, for it gives a firm hold to the pin by which the animal is plucked from the shell to please the palate. When the periwinkle or other operculate snail is crawling, this plate is borne on the upper side of the back part of the foot.

It is a matter of common observation that a garden snail is able to gnaw holes in cabbage leaves, while many marine forms may be observed to cut pieces out of seaweeds. This is the work of the rasping organ (*odontophore*), already mentioned as being present in mail-shells and their allies, as well as in tusk-shells. A few words as to its nature will not be out of place here, especially as we shall find later on that a similar organ is possessed by the cuttle-fish group. In fact, the bivalves are the only molluscs in which this structure is entirely absent.

By making a vertical longitudinal cut through the head of a snail, the main features of the rasping organ may be observed.

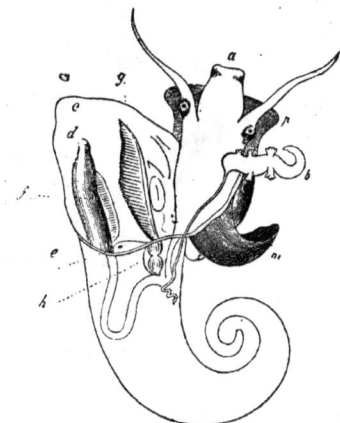

PERIWINKLE
Male, with shell removed

a, mouth; *b*, sexual organ; *c*, reflexed mantle; *d*, vent; *e*, kidney; *f*, slime-gland; *g*, gill; *h*, heart; *m*, shell-muscle; *p*, foot

The mouth leads into a muscular oval pharynx, often known as the buccal mass, and from the back end of which the gullet passes backwards. Rising up on the floor of the pharynx is a swelling, supported by gristly pieces, and provided with a complicated set of muscles. Stretching over this swelling, from front to back, is a horny ribbon (*radula*) studded with numerous sharp projections, or teeth.

By means of appropriate muscles this apparatus can be protruded from the mouth, when other muscles cause the radula to execute scraping or licking movements, by which particles are detached from the substance used as food. By one or more firm horny jaws this substance is firmly held while it is being rasped. The radula passes at the back into a radula pouch, or sac, from which it grows forward as worn away.

A few other features in the structure of a typical marine snail are shown in the accompanying figure of a periwinkle removed from its shell and partly dissected. Opening just above the neck of such a creature is a wide opening leading into a fairly large gill-cavity, the thin roof of which is termed the mantle, comparable to the mantle-lobes of a bivalve.

In the specimen figured, the roof of the gill-cavity has been cut through along the right side, and the mantle has been turned over to the left. The last part of the intestine opening by the vent is seen on the left of the figure (*d*). It was originally on the right, but has been displaced. To its right is a comb-like gill (*g*), and behind this the heart (*h*), consisting of two chambers, a thin-walled auricle in front, and behind this a thick-walled ventricle. The auricle receives purified blood from the gill, and squeezes it into the ventricle, by which it is pumped to the body.

Such a heart, containing nothing but pure blood, is said to be systemic, as in all molluscs. It contrasts markedly with the heart of fishes (lung-fishes alone excepted), which contains nothing but impure blood, received from the body and pumped to the gills for oxygenation. The latter type of heart is known as branchial. On the right of the gill is indicated a sensory structure, the water-testing organ (*osphradium*), which tests the quality of the water entering the gill-cavity.

A marine snail or slug hatches out as a little free-swimming veliger larva, resembling that of a bivalve, except in the character of a shell.

Snails and slugs are divisible into three orders : (1) Fore-gilled Gastropods (Prosobranchia), (2) Hind-gilled Gastropods (Opisthobranchia), and (3) Lung-Snails and Slugs (Pulmonata).

FORE-GILLED GASTROPODS

The periwinkle, mentioned above, is a good example of the order Prosobranchia, the position of the gill in front of the heart giving the name to the group. That the intestine should open just over the neck seems a curiously insanitary arrangement, and it is quite unlike what is found in shells and bivalves, where the vent is at the posterior end of the body. The forward position has resulted from the twisting of the visceral hump into a spiral, which has also brought the gill in front of the heart. The remote ancestors of snails and slugs were no doubt entirely bilaterally symmetrical.

The order is divided into two suborders, and these again into families. The former are : (1) Shield-gilled Gastropods (Scutibranchia), and (2) Comb-gilled Gastropods (Pectinibranchia).

SIDE-SLIT SNAILS

The suborder Scutibranchia derives its name from the shape of the gill (or it may be gills), which is of a pointed oval shape, and consists of a central axis, with a series of leaflets on each side.

One of the most interesting families (*Pleurotomariidæ*) contains the side-slit snail (*Pleurotomaria*), now very rare and represented by only six species, but once very abundant, including more than a thousand fossil species. Two gills are present, and a slit on the outer side of the mouth of the shell serves for the passage of water from the gill-cavity. The presence of a single gill in periwinkles and the great majority of marine forms is undoubtedly due to reduction.

SEA-EARS

The sea-ears, or ormers (*Haliotidæ*), are also two-gilled forms, and one species (*Haliotis tuberculata*) is well known in the Channel Islands, where it is esteemed as a delicacy. The visceral hump has been here flattened out, and the spiral mostly obliterated. As a result of this the shell, which is beautifully coloured, is an oval moderately convex plate ; and there is a row of holes over the gill-cavity, corresponding to the slit in a side-slit shell, and serving the same purpose. The oval foot is extremely muscular and powerful, enabling the former to hold with great tenacity to rocks or stones.

SLIT AND KEYHOLE LIMPETS

Still another family (*Fissurellidæ*) includes two-gilled forms with a conical visceral hump and shell of corresponding shape. The slit-limpets (*Emarginula*) possess a slit on the front margin of the shell serving the same function as in *Pleurotomaria*, while the apex of the shell is drawn out into a backwardly bent beak. In *Rimula* the slit is replaced by a small hole near the front margin of the shell, while in the keyhole-limpets (*Fissurella*) there is an opening at the top of the cone. In the course of its development one of the latter passes through stages in which there is first a slit and then a hole, placed as in *Rimula*.

DOLPHIN-SNAIL

TOP-SNAILS AND ALLIES

The remaining shield-gilled snails never possess more than one gill, and are therefore more specialised than the forms already mentioned. The well-known top-snails (*Turbinidæ*) and a closely allied family (*Trochidæ*) include a great variety of very attractive forms, including a number of British species. The operculum is horny in the former family and shelly in the latter. The members of both families are distinguished by the beautiful pearly appearance of their shells when the outer layer has been removed, and these are therefore much employed for ornamental purposes, especially in the construction of boxes and other fancy articles often seen for sale in seaside resorts.

DOLPHIN-SNAILS AND NERITES

The dolphin-snails (*Delphinulidæ*) are tropical forms, of which an example is figured. Nerites (*Neritidæ*) are somewhat unattractive snails, of which a typical marine genus (*Nerita*) is distinguished by the possession of a stout spheroidal shell, able to resist the buffeting of the waves, while a fresh-water relative (*Neritina*) has a thinner shell.

TORTOISESHELL LIMPETS

The tortoiseshell limpets (*Acmæidæ*) are so named from the markings on the shell, which is devoid of slit or opening. One species (*Acmæa testudinata*) is fairly common in parts of Britain, and is often known in Scotland as John Knox's limpet, on account of a mark inside the shell which bears a faint resemblance to portraits of that divine.

TRUE LIMPETS

The true limpets (*Patellidæ*) have a stout conical shell, which in the common species (*Patella vulgata*), and still more markedly so in the horse-limpet (*P. athletica*) is ribbed from apex to margin. Limpets are employed as bait, and to some extent as human food. A large proportion of their food consists of minute algæ encrusting the rocks between tide-marks, and the radula, which is subjected to unusually rapid wear on this account, is of relatively great length, being, in fact, longer than the shell. Of course, only a small part of this is in actual use, the remainder being a reserve contained within the radular sac.

The rocks upon which these creatures live are generally so studded with acorn barnacles and other organisms that it is difficult for the foot to obtain a firm hold, and a limpet makes its home on a definite spot, which is thus kept smooth, and gradually deepened into a seal-like depression, or " scar," by the attrition of the edge of the shell. Limpets feed when the tide is down, often wandering several feet from their homes, but returning as the tide comes up, and holding on to

ANIMAL OF MUREX
b, gill ; *b'*, water-testing organ ; *p*, purple gland

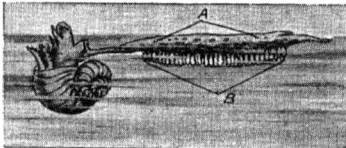

the rock with very great force. When well covered with water, feeding is often resumed. The " homing " faculty of limpets is difficult to explain, for the sense-organs are ill-developed, and it is not a matter of sight or smell.

An interesting peculiarity of limpets is the absence of organs corresponding to the gills of forms so far mentioned. The gill-cavity, however, is present in the usual place, and two little orange-coloured projections on its floor probably mark the position of the vanished gills. They are replaced functionally by a large number of delicate plates, lodged in a groove above the foot, and shown in the accompanying figure.

VIOLET SNAILS

In the suborder Pectinibranchia there is never more than one gill, and this has lost one set of its leaflets, the axis with the remaining set looking something like a comb, hence the name of the suborder. The very numerous families are grouped into tribes, largely according to the nature of the radula ;

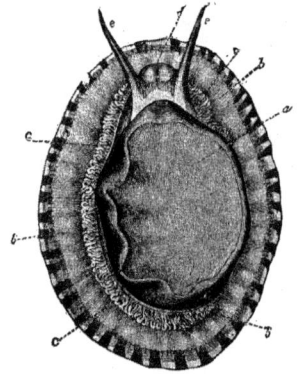

UNDER SURFACE OF LIMPET
a, foot ; *b*, mantle ; *c*, gills ; *d*, mouth ; *e*, tentacles

but this is too technical a matter for discussion here, and it must suffice to give some account of well-known or remarkable types.

The beautiful, thin-shelled violet snails (*Ianthinidæ*) swim in the open sea, and are of carnivorous habit, feeding chiefly upon jelly-fish. The type genus (*Ianthina*) is figured to show the remarkable egg-raft which these creatures construct of hardened mucus, rendered buoyant by the addition of air-bubbles.

This is the work of the curved foot, which is kept upwards. The raft serves not only for the protection of the egg-capsules, which are attached to its under surface, but also as a float for the benefit of the adult, enabling it to drift here and there with the currents and obtain an abundant supply of food.

WENTLE-TRAPS AND MURICES

The wentle-traps (*Scalariidæ*) possess elegant ribbed shells of elongated spiral form. There are four British species, one of these, the common wentle-trap (*Scalaria communis*) being extremely beautiful.

Murices (*Muricidæ*) constitute a large family of marine snails, in which the shell is often handsome. In the type genus (*Murex*) it is elongated and commonly spiny, and the opening is drawn out into a tube for lodging the siphon, a fleshy half-tube by which water is introduced into the gill-cavity. Such an arrangement is typical of carnivorous forms, and is perhaps related to their active habit, which necessitates vigorous breathing. It follows, as a general but not invariable rule, that the nature of the food of a snail can be determined by inspection of the empty shell. Vegetarian forms, such as periwinkles, are devoid of a siphon, and consequently the mouth of the shell is entire (holostomatous). The shell of a carnivorous type, on the other hand, has a mouth which presents an indentation or tube (*i.e.*, is siphonostomatous) for the lodgment of the siphon.

The figure shows a partly dissected Murex, the gill-cavity having been laid open. In front will be noticed the tapering tentacles, with eyes at their bases. Projecting forwards from between them is the protrusible

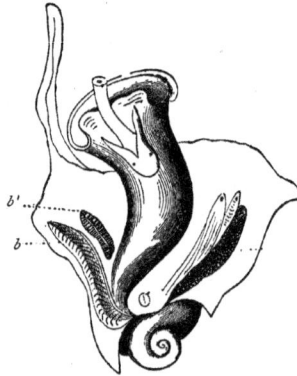

proboscis, and to the left is the long siphon. Further back, on the same side, is the gill (*b*), and by its side the water-testing organ (*b'*). Below, to the right, is the last part of the intestine, below and attached to which (darkly shaded) is the purple gland (*p*).

This purple gland is of particular interest because its secretion in species of murex and some of the allied forms was the source of the famous Tyrian purple. Heaps of the shells of *M. trunculus* can still be seen on the Tyrian shores; while on the coast of Greece the same thing is true for another species (*M. branderi*),

RHIZOCHILUS ANTIPATHUM

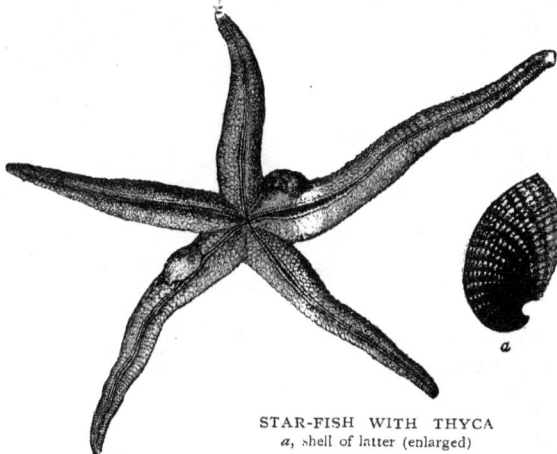

which was doubtless used for the same purpose. It would indeed appear that the ancient Phœnicians, in pushing their way from east to west along the Mediterranean, were partly impelled by the desire of finding fresh supplies of these valuable dye-producing snails.

Murices attack other molluscs, especially bivalves, perforating the shells of their prey by means of their powerful rasping organs, probably aided by an acid secretion. Our only British species, the sting-winkle (*M. erinaceus*) is a notorious pest to oyster fisheries.

The purple-snail (*Purpura lapillus*), closely allied to the murex, is exceedingly common on rocky coasts round Britain. As the name implies, it also yields a purple secretion. The thick white shell enables its possessor to resist the buffeting of the waves between tide-marks. The purple-snail may often be noticed boring holes in the strong calcareous investments of acorn barnacles, which commonly encrust the rocks upon which it lives. The eggs are contained in elegant little flask-shaped capsules.

A small but remarkable family (*Coralliophilidæ*) includes a number of snails devoid of a radula, or rasping ribbon, and associated with various corals. In the figured species (*Rhizochilus antipathum*) the adult is fixed by out-growths from the mouth of the shell to a branching coral (*Antipathes*), and the food probably consists of minute organisms taken in through the siphon with the water used for breathing. The dog-whelks (*Nassidæ*) include a common British species (*Nassa reticulata*) which rivals the sting-winkle in its depredations on oyster fisheries.

WHELKS

The whelks (*Buccinidæ*) are a large family best known from the common species (*Buccinum undatum*), which is of considerable value as bait, and has a wide distribution round the shores of the North Sea and the Atlantic coast of North America. A section through the shell is represented in the accompanying figure.

STAR-FISH WITH THYCA
a, shell of latter (enlarged)

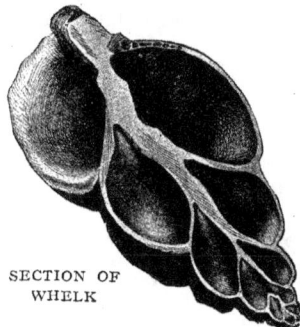

SECTION OF WHELK

This species is somewhat variable, not only as to the shape and nature of the shell, but also as to the operculum, which is sometimes duplicated, or in rare cases triplicated.

Among the most attractive members of the whelk family are the ivory-snails (*Eburna*), the species of which range from the Red Sea to the Cape and Australia. The smooth white shell is spotted with dark red, and the animal is also spotted.

TULIP-SNAILS

The tulip-snails (*Fasciolaridæ*) are spindle-shaped forms, some of which attain a large size. One species, for example (*Fasciolaria gigantea*), native to the shores of South Carolina, is enclosed in a shell which may be 2 feet long.

CHANK-SNAILS

The chank-snails (*Turbinellidæ*) deserve notice on account of the fictitious value attached to left-handed sinistral specimens. On this point A. H. Cooke speaks as follows: " The chank-shell (*Turbinella rapa*) is of especial interest from its connection with the religion of the Hindoos. The god Vishnu is represented as holding this shell in his hand, and the sinistral form of it, which is excessively rare, is regarded with extraordinary veneration. The chank appears as a symbol on the coins of some of the ancient Indian Empires, and is still retained on the coinage of the Rajah of Travancore.

" The chief fishery of the chank-shell is at Tuticorin, on the Gulf of Manaar, and is conducted during the north-east monsoon, October to May. In 1885-86 as many as 332,000 specimens were obtained, the net amount realised being nearly Rs. 24,000. In former days the trade was much more lucrative, four or five millions of specimens being frequently shipped. The Government of Ceylon used to receive £4,000 a year for licences to fish, but now the trade is free. The shells are brought up by divers from two or three fathoms of water. In 1887 a sinistral specimen was found at Jaffna, which sold for Rs. 700. Nearly all the shells are sent to Dacca, where they are sliced into bangles and anklets to be worn by the Hindoo women."

MITRE-SNAILS

Mitre-snails (*Mitridæ*), so named from their shape, are abundant on the shores of the islands in the warmer parts of the Indian and Pacific Oceans, and include a number of attractive forms, often of moderately large size. One of the best-known species (*Mitra episcopalis*) possesses a strong shell handsomely spotted with reddish orange.

VOLUTES

The volutes (*Volutidæ*) include a great variety of elegant shells, often beautifully coloured and marked. The last turn of the spiral is large, and the mouth of corresponding size. The music-shell (*Voluta musica*) of the West Indies, which will serve as an example, is marked with streaks and dots that have suggested its name.

OLIVES

The olives (*Olividæ*) possess highly polished shells, the shape and general appearance of which gives them their appellation. They are mostly tropical forms, and highly carnivorous, ploughing through the sand in pursuit of their prey, which consists of bivalves. The front part of the foot (to right in the figure) is sharply marked off from the rest, and protects the head region, at the same time enabling easy progress through the sand, to which the smooth shell offers but little friction. The eyes are placed about half-way up the tentacles.

HARP-SNAILS

The harp-snails (*Harpidæ*) include but a single genus (*Harpa*), the species of which are native to the Indian and Pacific Oceans. They possess large, brightly coloured shells, and an unusually large foot. Harp-shells are of interest because they are among the small number of snails possessing the power of self-mutilation. When suddenly disturbed they can cut off the hinder part of the foot by pressing the shell against it. In certain other cases such mutilation appears to be resorted to in order to distract the attention of enemies, and possibly this may be the case here.

CONES

The cones (*Conidæ*) make up a large family, of which most members are tropical. They are famous for the beauty of their shells, the general shape of which can be seen in the accompanying figure of the textile cone (*Conus textilis*). The large siphon is represented as fully extended, and the eyes are visible on the outer sides of the tentacles. Some, at least, of the cones can inflict a poisoned bite by means of the powerful teeth on the radula, which are provided with poison-glands.

High prices have been given by collectors for the shell of certain species, the rarest being, of course, the most valuable. Of one species (*Conus gloria-maris*) from the Philippines and Moluccas only some dozen examples are known, and one of these was sold by auction in 1854 for the sum of £43 1s.

NATICAS

The naticas (*Naticidæ*) possess a globular, smooth shell, and are modified, like the olives, for burrowing in the sand in pursuit of bivalves. The eyes are reduced or absent. Some species of the type genus (*Natica*) are British. The spawn is in the form of a broad sandy band rolled up in a spiral.

CARRIER-SNAILS

The curious carrier-snails (*Xenophoridæ*), which progress by a series of leaps, attach all sorts of foreign bodies to their conical shells, a habit which no doubt helps to make them inconspicuous, increasing their chance of escaping from enemies.

CAP-SHELLS

The cap-shells (*Capulidæ*) are attached forms like limpets, and modified in a somewhat similar fashion. The Hungarian cap-shell (*Capulus hungaricus*), a British species, is covered by a conical shell, the tip of which is rolled back into a short spiral, like a cap of liberty.

In another native form (*Calyptræa chinensis*) there is a curious projection into the shell, suggesting a cup in appearance, hence the popular name, "cup-and-saucer limpet." An unusual mode of life is adopted by a member of the family (*Thyca crystallina*) native to the Indian Ocean. As shown in the figure on page 1857, it is parasitic on star-fish, and its elongated snout is sunk within the body of its host.

TEXTILE CONE

PERIWINKLES

The periwinkles (*Littorinidæ*) are familiar vegetarian forms, mostly living between tide-marks, and very widely distributed. The edible species (*Littorina littorea*), eaten in incredible numbers in the larger centres of population, is our largest native species. A small and very familiar British form (*L. obtusata*) is enclosed in a small rounded shell devoid of a spire, and very variable in colour, exhibiting a range of tints from olive-green to bright orange-yellow. It abounds on brown sea-weeds between tide-marks. The intertidal zone is of particular interest, because it presents conditions favourable to the evolution of terrestrial forms from aquatic ones, by a modification in the way of breathing.

In this connection some of the periwinkles are worthy of special study. One of our native species (*L. rudis*), with a short-spired shell, lives at or near high-water mark, and in this form the gill is of relatively small size, while the roof of the gill-cavity is beginning to perform the functions of a lung. The process is carried still further in some of the tropical American periwinkles, which invade the land, and are even found climbing on trees. The gelatinous spawn of periwinkles may be found adhering to various objects.

Another member of the family is the horned winkle (*Placuna divaricata*), with a handsome banded shell, extremely long tentacles, and a pair of feeler-like organs borne on the back part of the foot, as shown in the illustration on this page.

What has just been said about periwinkles prepares us for the fact that there are two allied families (*Cyclophoridæ* and *Cyclostomatidæ*), of which the members are purely terrestrial, having entirely lost their gills, and converted their gill-cavities into lungs. They must not be confounded with the true lung-snails, including ordinary land-snails and slugs, and which will be considered in the sequel. Most of the forms in question are tropical, but some small species, such as *Cyclostoma elegans*, are British. In these and other members of their family (*Cyclostomatidæ*) the foot is

HORNED WINKLE

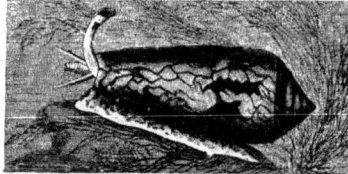

BLACK OLIVE

VIVIPAROUS RIVER-SNAILS
Male on left, female on right, and young in front

traversed by a deep longitudinal groove, and its right and left halves are moved alternately in creeping.

SPIRE-SNAILS

The spire-snails (*Rissoidæ*) are small marine and brackish-water forms with elongated, elegant shells, presenting a rib-like sculpture. There are a number of British species. A closely related family (*Hydrobiidæ*) includes species inhabiting fresh and brackish water. Regarding one of them, A. H. Cooke remarks : " Specimens of *Hydrobia ulvæ*, taken on the wet sands at the mouth of the Dee, are found to have several little rounded excrescences scattered over the surface of the shell. These, on examination, are found to be little masses of sand-grains, in the centre of which is a clear jelly containing segmenting ova or young embryos. In all probability the shell is the only comparatively stable object, in the expanse of shifting sands, on which the eggs can be laid."

RIVER-SNAILS AND APPLE-SNAILS

The river-snails (*Viviparidæ*) are not unlike large thin-shelled peri-winkles, as may be seen from the figure. The eggs are hatched internally, and the newly-born young have bristly shells, the bristles being ultimately lost.

The apple-snails (*Ampullariidæ*) inhabit tropical marshes, and are remarkable from the character of their breathing organs. The gill-cavity is divided into two parts, one of which contains the gill, while the other serves as a lung. Semper says that they " breathe not only with both gills and lungs, but they do so in regular alternation ; for a certain time they inhale air at the surface of the water, forming a hollow, elongated tube by incurving the margin of the mantle, so that the hollow surface is closed against the water and open only at the top. When they have thus sucked in a sufficient quantity of air, they reverse the margin of the mantle, opening the tube, into which the water streams." During the dry season these snails remain imbedded in the mud in a torpid state.

TURRET-SHELLS AND WORM-SHELLS

The marine turret-shells (*Turritellidæ*) merit a word in passing, because the much-elongated, gently tapering shell of a common species (*Turritella communis*) is one of the most familiar objects of the shore. The little blind-shell (*Cæcum trachea*) belongs to the same family, and its investment is in the form of a slightly curved tube.

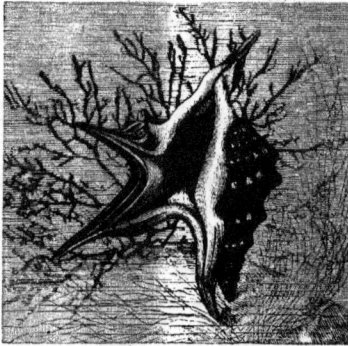

PELICAN'S FOOT

Much more remarkable are the worm-shells (*Vermet-idæ*), for, as will be seen from the figured species (*Vermetus lumbricalis*), the last few turns of the shell are not in contact. When young these snails creep about in the usual way, but ultimately become attached to various firm objects.

WING-SNAILS

The wing-snails (*Strombidæ*) are of large size and mostly tropical, with thick, heavy shells, in which the outer edge of the mouth is drawn out into a wing-like expansion. They move by a succession of leaps, effected by suddenly straightening the bent, narrow foot, seen on the upper side of the accompanying figure of a stromb (*Strombus lentiginosus*). The eyes are large and complex, and borne at the end of unusually thick tentacles.

An allied species (*S. gigas*) is the fountain-snail, or conch, of the West Indies, the pink shell of which may weigh as much as 4 or 5 pounds. It is largely imported and ground up to make fine porcelain, besides which it is one of the molluscs used for cameos, as the shell is made up of layers of different colour. It is also noted as the source of pink pearls. In the allied scorpion-snails (*Pteroceras*) the expanded mouth of the shell is drawn out into a number of claw-like processes.

A related family (*Chenopodidæ*) includes the pelican's foot-snail (*Chenopus pes-pelecani*), of which an illustration is given. It is often cast up on our shores, or seen among the contents of a dredge.

COWRIES AND TUN-SNAILS

Cowries (*Cypræidæ*) are for the most part tropical forms, and are remarkable for the character of their shells, the spiral form of which is only revealed by sections, since the turn last formed covers and conceals the others. One beautiful little ridged species (*Cypræa europæa*) is not infrequent on our coasts, while some of the larger kinds are very ornamental. The money-cowry (*C. moneta*) has played an important part for currency purposes in India and Africa. It was formerly imported into this country in large quantities for use in the West African trade.

The general appearance of the mainly tropical tun-snails (*Doliidæ*) will be gathered from the figure on page 1860 of a typical species (*Dolium perdix*). The widely expanded foot and exceedingly long proboscis are represented, as is also the siphon, bent back on the upper surface of the shell. The glands connected with the mouth secrete a fluid containing an appreciable amount of free sulphuric acid, and this helps the

WORM-SHELL   STROMB

tun-snail to perforate the shells of other molluscs upon which it feeds.

HELMET-SNAILS AND TRITONS

Helmet-snails (*Cassididæ*), of which one species (*Cassis glauca*) is figured, possess thick, heavy shells, much used for cameo carving. In one of the kinds (*C. madagascariensis*) most esteemed for this purpose the outer layer of the shell is white, under which is a dark brown portion, that makes an effective background to the finished cameo.

The tritons (*Tritonidæ*) are large marine forms, with strong, elongated shells. In a well-known Pacific species (*Triton tritonis*) this may be a foot or more in length, and is used as a war-trumpet by the South Sea Islanders, the apex of the shell being broken off and serving as a mouthpiece. A Mediterranean species (*T. nodiferus*) was similarly used as a trumpet by the ancient Romans.

HETEROPODS

A very interesting group of comb-gilled snails, containing several families, is that of the Heteropoda, including a number of translucent creatures living in the open sea in almost all parts of the world. Part of the foot is modified into a flattened fin, by the movements of which swimming is effected, the upper side of the animal being directed downwards. Heteropods are highly carnivorous, and the mouth is placed at the end of a long, blunt snout. The eyes are extremely well developed, as also are the so-called organs of hearing, which have to do with the balance of the body.

One of the least modified heteropods is *Atlanta*. In this case there is a spiral shell, into which the animal can be entirely retracted, the opening being then closed by an operculum. In most other heteropods the shell is reduced, and in most specialised species, such as those of *Pterotrachea*, has altogether disappeared.

HIND-GILLED SNAILS

In the members of the last order we have seen that the dorsal part of the body, in typical cases, is rolled up into a spiral mass, covered by a shell of corresponding shape, into which the animal can withdraw itself for safety. The result of the spiral twisting, which has been in a direction contrary to that in which the hands of a clock move, has been to bring the gill-cavity to the front, and the contained gills or gill are anterior to the heart, hence

HELMET-SNAIL

MONEY-COWRIES

the term "fore-gilled." This spiral arrangement is in many respects inconvenient, and, as a matter of fact, has been done away with in limpets and certain other forms.

The same thing applies to the members of the order Opisthobranchia, in which the spiral part of the body has more or less unrolled, bringing back the heart and gill to the right side of the body, the latter being commonly a little behind the former, whence the term "hind-gilled." This will be understood by reference to the figure on page 1860, which represents the blood system of a typical form. The spiral region has not merely unrolled, but, as it were, been absorbed into the rest of the body, which has become more or less slug-like. At the same time the shell has gradually been reduced, and in the sea-slugs, which are the most specialised members of the order, has disappeared altogether, together with the gill-cavity and gill, breathing being effected in other ways. There are numerous families, grouped into two suborders—(1) Tectibranchia, and (2) Nudibranchia.

TECTIBRANCHIA

In the suborder Tectibranchia are included a number of forms, in which a shell is present as a rule, with a widely open gill-cavity, containing a single gill on the right side of the body.

The members of the least specialised family (*Actæonidæ*) possess a spiral shell with projecting spire, and form a link with the fore-gilled types. Bubble-snails (*Bullidæ*), so named on account of the appearance of the thin shell, are rather more modified, for the shell has lost its spire, as shown in the figure of a typical species (*Acera bullata*), which abounds on the mud-flats of many estuaries.

In a third family (*Scaphandridæ*) the shell is still further reduced, and the body cannot be withdrawn into it. The food consists of small molluscs, which are swallowed whole, being ground up by the action of a powerful gizzard.

Another family (*Philinidæ*) includes a small, very abundant British species (*Philine aperta*), which is white and slug-like, and burrows in the sand and mud of the shallower parts of the sea. It is devoid of tentacles and eyes, and the small plate-like shell is imbedded in the body.

The sea-hares (*Aplysiidæ*) are so named from a faint

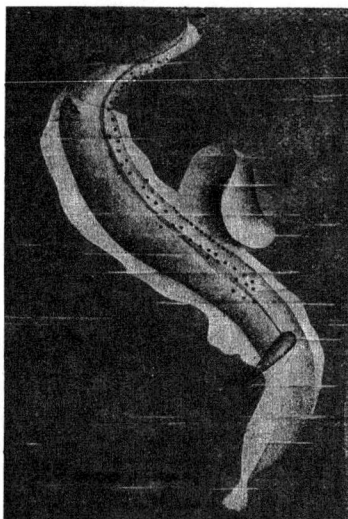

PTEROTRACHEA

resemblance in form to a crouching hare, as will be seen from the illustration of a typical species (*Aplysia depilans*), which ranges from the French coast into the Mediterranean. The shell is a thin, curved plate, almost entirely enclosed in a pouch on the upper surface of the body. These creatures can not only creep, but also swim, there being a muscular expansion on either side of the body, which can be flapped up and down. In the figure these expansions are folded up over the back. Although apparently helpless, the sea-hares are protected from their enemies by the possession of an unpleasant odour, and they are able to discharge a purple fluid that probably serves as a defence.

Belonging to this suborder are several families of small, translucent, pelagic snails, popularly known as sea-butterflies, and which are found in vast shoals at the surface of the open sea. They were formerly

TUN-SHELL

obvious and conspicuous object, it exemplifies warning coloration, apprising fishes and other enemies of unpleasant properties that render it undesirable as food.

Among the largest and most widely distributed sea-slugs are the sea-lemons (*Dorididæ*), of which a species (*Doris pilosa*) is figured. The peculiar ridged tentacles are seen on the left, and the delicate branching leaf-like gills on the right. The latter are in the form of a circlet round the vent, and can be protected by being drawn back into pouches, as also can the tentacles.

It may be noted that in the sea-lemons and other nudibranchs the spawn is in the form of a jelly-like band, coiled up into a spiral, and with one edge attached to some firm object. The larva that hatches from the egg possesses a spiral shell, and also an operculum, which is one reason for believing that these creatures have taken origin from

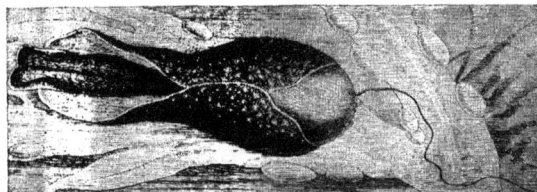

BUBBLE-SNAIL

CAVOLINIA

SEA-LEMON

assigned to a special order of wing-footed snails (Pteropoda). Some of them are enclosed in elegant shells, of varied form, but not usually spiral. The figured species (*Cavolinia tridentata*) possesses a brown shell, drawn out into spines, and the large fin-like expansions that serve as swimming organs are clearly depicted. In many kinds of sea-butterfly the shell is absent.

NUDIBRANCHIA

The most specialised of the hind-gilled forms are included in the suborder Nudibranchia. Shell, gill-cavity, and gill have disappeared, and the body has become almost or entirely bilaterally symmetrical. On account of their general appearance they are popularly known as sea-slugs. Breathing is in some cases effected by the general surface of the body, and in others by special outgrowths, which may be called gills, though they are not comparable to the gills which are typical for average molluscs.

The colour and shape of sea-slugs may be either such as to render them inconspicuous in their natural surroundings, or the opposite. In the former case the arrangement comes under the heading of protective coloration and form, and some, at least, of such creatures are eaten readily by fishes. When, on the other hand, a sea-slug is an

CIRCULATION IN SEA-SLUG

shell-bearing ancestors. In a related family (*Polyceridæ*) the gills are not retractile. The little species represented (*Ancula cristata*) is white in colour, with orange tips to the projections from the body; and as it is certainly unpalatable to fishes, this must be regarded as a case of warning coloration.

In the members of another family (*Tethyidæ*) the head is provided with a large hood-like flap, as shown in the species figured (*Tethys leporina*). This is a Mediterranean form, which swims actively, and as it possesses no radula, the food probably consists of minute organisms. There are two rows of red-tipped processes along the upper side, which contrast with the translucent body. The animal gives out a bright phosphorescent light, especially when irritated.

A protectively coloured family (*Dendronotidæ*) is distinguished by the branching processes from the body, which probably help to some extent in breathing, though their chief use is most likely to bring about a resemblance to seaweed. The figured species (*Dendronotis arborescens*) is coloured with shades of brown spotted with white, and as a result of its shape

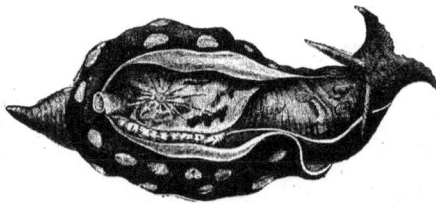

SEA-HARE

and colours it harmonises with surrounding objects so as to be inconspicuous. This must help it to escape the attention of fishes by which it is known to be relished. The species is not infrequent between tide marks on our coasts.

ÆOLIDIA PAPILLOSA

TETHYS LEPORINA

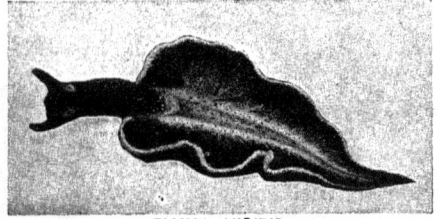

ELYSIA VIRIDIS

One of the most remarkable families (*Phyllirhoidæ*) includes little transparent fish-shaped creatures, devoid of foot, and met with in the open sea. Their general appearance will be gathered from the figured species (*Phyllirhoë bucephala*), which, like so many transparent pelagic animals, is highy phosphorescent.

The æolids (*Æoliaiidæ*) are creeping slug-like forms, narrowing greatly at the hinder end, and with numerous tapering processes on the back. The largest British species (*Æolidia papillosa*) is represented in the figure. The animal is coloured so as to be conspicuous, and is not palatable to fishes, partly, no doubt, because the tips of the body outgrowths are provided with stinging-cells, closely resembling those possessed by jelly-fishes and sea-anemones. This is therefore undoubtedly a case of warning coloration. The creature itself is highly carnivorous, and particularly fond of sea-anemones, the stinging properties of which protect them from the attacks of most other animals.

The last family to be noticed (*Elysiidæ*) includes small forms, of which the only British species (*Elysia viridis*) is figured. Here the sides of the body are drawn out into flat expansions that are supposed to serve as breathing organs. Elysia exemplifies protective coloration, harmonising with the seaweeds upon which it lives, and being able to change its colour according to the hues of its temporary home. When found on green seaweed it is green, while on brown seaweed it is brown.

LUNG-SNAILS AND SLUGS

The true land snails and slugs, together with certain fresh-water and brackish-water forms, make up the cosmopolitan order Pulmonata. In nearly all cases they are devoid of an operculum, and the equivalent

DENDRONOTIS ARBORESCENS

of the gill-cavity of an ordinary marine snail does not contain a gill, but is converted into a lung. Its opening is reduced to a small hole on the right side, an arrangement by which desiccation is prevented. The sexes are united. The relationship of the order to those already described has yet to be determined, but its remote ancestors were undoubtedly marine, and probably the invasion of the land was by way of estuaries and rivers.

Two suborders may be distinguished (1) Sessile-eyed Pulmonates (Basommatophora), (2) Stalk-eyed Pulmonates (Stylommatophora).

EARLET SNAILS AND POND-SNAILS

The suborder Basommatophora embraces forms in which an external shell is always present, and there is only one pair of tentacles, which cannot be retracted and bear the eyes at their bases. Most of them inhabit fresh water.

The earlet snails (*Auriculidæ*) are land forms living in the neighbourhood of the sea. The members of the type genus (*Auricula*) often have large, strong shells, that of one species (*A. judæ*) being figured. This and its allies are chiefly found in the swamps of tropical islands.

The pond-snails (*Limnæidæ*) are fresh-water forms found in most parts of the world, and represented in Britain by a number of species. In one of the most familiar genera (*Limnæa*) the shell is thin, horny, and generally sharply pointed, as will be seen from the figure of our largest native species (*L. stagnalis*). Like other aquatic pulmonates they breathe air, and are therefore obliged to come at intervals to the surface. In one of the individuals represented a bubble of air is being expelled from the lung-cavity.

These snails may often be seen creeping in an inverted

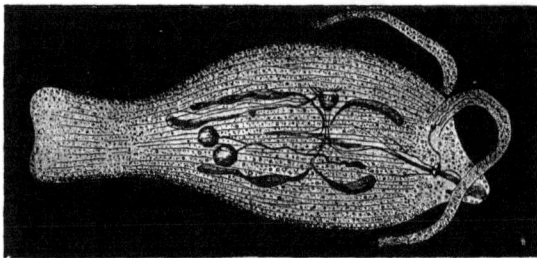

PHYLLIRHOË BUCEPHALA

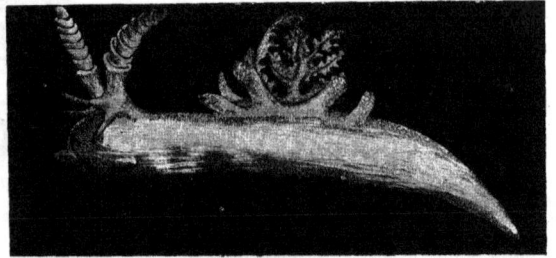

ANCULA CRISTATA

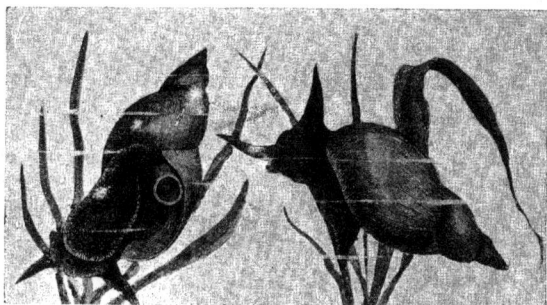

POND-SNAILS

position just below the surface of the water. The eggs are deposited in jelly-like clumps on water plants or other submerged objects. One small species (*L. truncatula*) has been mentioned elsewhere in connection with the liver-fluke, to which it serves as an intermediate host.

A related genus (*Physa*) includes small forms in which the shell is left-handed. A common British species (*P. hypnorum*) is fond of crawling just below the surface of the water, like some species of the preceding genus. Like many of its allies, it is able to spin a thread of hardened slime, sometimes over a foot long, by which it hangs from the surface. The thread is expanded into a little float above. The snail climbs up this thread when a fresh supply of air is required.

Trumpet-snails (*Planorbis*) are easily recognised by the flat spiral shell. The largest British species (*P. corneus*) is figured. The little fresh-water limpet (*Ancylus fluviatilis*) is found attached to stones at the bottom of streams, and possesses a thin conical shell. Unlike other members of the family, it does not come to the surface at intervals in order to take in air, and probably breathes the air dissolved in water.

AMBER-SNAILS AND LAND-SNAILS

In the suborder Stylommatophora there are two pairs of tentacles, and the eyes are borne at the tips of the posterior longer pair. The tentacles are hollow, and can be drawn by special muscles into the interior of the body, as may be seen by touching the extended " horns " of a creeping garden-snail. They include the typical land snails and slugs, and are found in all parts of the world.

The amber-snails (*Succineidæ*) are small forms which live in water and damp places on the land. The delicate shell is yellowish or reddish in colour, and its shape will be gathered from the figure given of a common species (*Succinea putris*).

The typical land-snails (*Helicidæ*) include a vast number of species from all parts of the world, and living in almost every conceivable situation. The type genus (*Helix*) embraces over 1,600 species, the shells of which are of a great variety of shapes. The most familiar species is the common garden-snail (*H. aspersa*), well known for its depredations on cultivated plants.

Like many land-snails this form passes through the unfavourable season of the year in a torpid condition, withdrawing into its shell, and closing the opening with a plate of hardened slime, in the centre of which a three-rayed slit is left for the passage of air.

Garden-snails, in common with many of their kind, possess a strong homing faculty, and are known to return time after time to some sheltered corner they have selected as a refuge. Their rate of crawling, which is proverbially slow, has been carefully investigated, and has been found on the average to amount to a mile in sixteen days fourteen hours. Their muscular strength is very great, for a vigorous snail can lift nine times its own weight vertically, and drag fifty times its own weight along a smooth horizontal surface. An athlete weighing 12 stone, who wished to emulate the former feat, would have to climb a vertical ladder with a weight of $13\frac{1}{2}$ cwt. hanging from his waist. The eggs are laid in a small hole in the ground, made by the parent for the purpose.

A much larger species, with a thick, yellowish white shell, is the Roman snail (*H. pomatia*), so named because it is supposed to have been introduced by the Romans. However that may be, they, and other species, were certainly esteemed as food by epicures of that nationality, and were carefully tended in " cochlearia," or snail-gardens. Such gardens (escargotières) are now to be found in many parts of France and certain other countries.

AMBER-SNAIL

The much smaller field-snail (*H. nemoralis*) is remarkable for the variable colour and markings of the shell. This may be entirely yellow or brownish pink, or may be marked in addition with a number of dark bands. Something like ninety different colour varieties of the kind have been described. A closely related species (*H. hortensis*), represented on page 1820, is also very variable.

The little glass - snails (*Vitrina*), of which a typical British species (*V. pellucida*) is figured, possess a thin, horny shell, into which the whole of the body cannot be withdrawn. By suddenly jerking the hinder part of its foot it is able to spring off a stone or twig on which it may happen to be crawling, and this habit diminishes the chance of capture by predatory birds. The same part of the foot can also be detached voluntarily, a defensive

TRUMPET-SNAIL

ROMAN SNAIL

BLACK SLUG

whole body. Prof. Semper has described Philippine species as practising self-mutilation in an extraordinary manner, the hinder part of the foot being convulsively moved up and down until it falls off.

The numerous species of the type genus (*Limax*) possess a reduced internal shell. The largest British species (*Limax maximus*), represented in the coloured plate, may be 6 inches or more in length. The field-slug (*L. agrestis*) is a serious pest to crops and garden plants (see coloured picture on page 1820).

The cellar-slug (*L. flavus*) is one of the forms possessing a strong locality sense, for it has been observed to move to and fro along a regular track. It has an objectionable habit of getting into flour-jars and pans of cream.

SHELL-SLUGS

The shell-slugs (*Testacellidæ*) differ from the forms so far described in being highly carnivorous. The extent to which the shell is developed varies greatly. In the species figured (*Testacella haliotidea*) it is in the form of a cap on the tail. This species, which is found in Britain, lives underground, and devours earthworms.

measure which is here and there illustrated in various groups of animals, especially lizards.

A South American genus (*Strophocheilus*) contains some large and handsome forms, one of which (*S. maximus*) bears a shell sometimes six inches in length. A smaller species belonging to another genus (*Buliminus montanus*) is shown in the coloured plate on page 1820.

BLACK SLUGS

The familiar black slugs (*Arionidæ*) are best known in Britain by a species (*Arion empiricorum*) represented in the figure, which shows the round opening into the lung. This species presents many variations in colour, for though typically black, it may be brown, red, yellow, greenish, or white. Two varieties are shown in the coloured plate on page 1820. The sea-slugs have undoubtedly descended from shell-bearing ancestors, and the same thing is true for land-slugs. In this case all traces of the shell have disappeared, except some calcareous granules imbedded in the skin of the back.

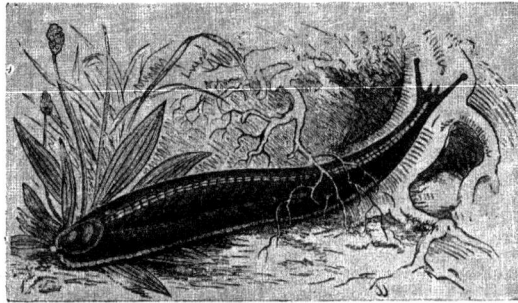

SHELL-SLUG

WARTY SLUGS

A peculiar family of warty-looking slugs (*Onchidiidæ*), which are mostly tropical or subtropical, includes one species (*Onchidium celticum*) found in South Devon and Cornwall. These creatures live on sandy shores near high-tide mark, and swallow sand for the sake of the contained organic matter. The little shore-haunting fishes known as mud-skippers (*Periophthalmus*), elsewhere mentioned as natives to the shores of the Indian Ocean and part of the Pacific, prey upon slugs of this kind, and it is remarkable that throughout the range of mud-skippers the species of *Onchidium* possess very numerous dorsal eyes of complex structure. They defend themselves against these enemies by contracting their skins, by which the secretion of certain glands is forced out suddenly like a charge of small shot. It has been suggested by Prof. Semper that the dorsal eyes enable the slugs to perceive the shadows that mark the approach of the mud-skippers, but this has not been definitely proved.

J. R. Ainsworth-Davis

CHRYSALIS SNAILS

The chrysalis snails (*Pupidæ*) includes small land forms with elongated left-handed shell, the opening of which is narrowed by projections. One species is represented in the coloured plate, and a related genus is *Clausilia*.

GLASS-SNAIL

AGATE-SNAILS

Among the agate-snails (*Stenogyridæ*) the genus *Achatina* includes species, of which one (*A. fulica*) from East Africa is figured, with large handsome shells. A West African form (*A. variegata*) is the largest existing land-snail, its shell being sometimes over 7½ inches in length.

TRUE SLUGS

The true slugs (*Limacidæ*) are for the most part devoid of a shell, or possess but a small one. An exception is afforded to this by species of the genus *Zonites*, which possess thin, somewhat flattened spiral shells, into which they can entirely withdraw themselves. A smaller but still sufficiently obvious shell is present in members of a genus (*Helicarion*) very like that containing the glass-snails (*Vitrina*), elsewhere described. The shell in this case is not large enough to contain the

AGATE-SNAIL

CLASS CEPHALOPODA—CUTTLE-FISHES AND ALLIES

THE cuttle-fishes and their kind are undoubtedly the most highly organised and the most predaceous of all molluscs. They are exclusively marine, and possess a powerful rasping organ, together with horny jaws, not unlike those of a parrot, as shown in the figure. When a shell is present at all, it is in the form of an internal calcareous or horny plate. A large part of what has been said, however, does not apply to one existing form, the pearly nautilus, which is a sluggish creeping animal invested with a spiral calcareous shell, and agreeing in this respect with the great majority of extinct types.

Alike in pearly nautilus and cuttle-fishes, the body is bilaterally symmetrical, and there is a marked peculiarity in the head region. In the pearly nautilus there are a number of tentacle-bearing lobes round the mouth, and in cuttle-fishes, etc., either eight or ten sucker-bearing processes in the same region. It is generally held that these structures are in reality the front part of the foot, which has wrapped round and fused with the head proper.

We have seen in typical snails that the dorsal region of the body is drawn out in a twisted hump covered by a shell, but that the ancestors of such forms were probably in many ways similar to the existing primitive creatures known as mail-shells, or chitons.

Creatures of much the same kind were probably ancestral to the members of the present class, but as the remains of these are among the most ancient known fossils, it is only possible to speculate in a tentative fashion upon the course of evolution. We may, however, suppose that the remote creeping ancestors of cephalopods developed a dorsal hump, as in the case of snails, differing, however, in being bilaterally symmetrical. It was probably covered and protected by a cap-like shell. The gill-cavity was at the back of this hump, and contained at least one pair of gills. The arrangement was essentially an adaptation to swimming, the creeping habit being gradually abandoned, and the foot shortened up in consequence, as represented in the figure, which is an attempted restoration. If we may judge from existing types, the method of swimming was somewhat peculiar, the motive power being supplied by waste water squirted out of the gill-cavity, so as to propel the animal backward.

Recent cephalopods, to which our attention must mainly be directed, are divided into two orders (1) Four-gilled Cephalopods (Tetrabranchiata), and (2) Two-gilled Cephalopods (Dibranchiata).

FOUR-GILLED CEPHALOPODS

Among the Tetrabranchiata, the pearly nautilus (*Nautilus pompilius*) has a rounded body contained in the last compartment of a spiral-chambered shell. A sort of cord (*siphuncle*) runs from the body through the remaining chambers, which are separated from one another by curved partitions, or septa, with wavy edges. The parrot-like beak is surrounded by lobes fringed with sticky tentacles, that

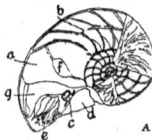

ANCESTRAL CEPHALOPOD

a, visceral hump ; *b*, shell ; *c*, gill cavity ; *d*, gill ; *e*, head ; *f*, foot ; *g*, side-flaps (epipods)

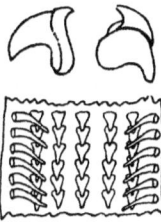

FEEDING ORGANS OF CUTTLE-FISH

a, beaks ; *b*, piece of rasping ribbon

probably help to secure the food, consisting in this case of bivalve molluscs.

On either side of the head is a large eye of remarkable structure. It essentially consists of a pigmented cup, with a minute opening to the exterior that freely admits sea-water, and it appears to act on the principle of the pinhole camera. At the back of the body is a capacious gill-cavity containing four feathery gills, similar to those of sea-snails in structure. Water is admitted into the cavity by a wide slit, and expelled again through a muscular tube, the funnel consisting of two halves rolled together.

There has been much discussion as to the nature of the funnel, which some think represents part, or even all, the foot of other molluscs; but it is most probably comparable to the muscular side-flaps (*epipods*) by which swimming is effected in sea-hares and sea-butterflies, as previously described.

The nautilus lives on the sea-floor in the South-West Pacific, especially round New Guinea and the Philippines, at a depth ranging from 325 to 2,300 feet deep. It creeps by means of its tentacle-fringed lobes, and it has been suggested that the funnel can be unrolled and used for the same purpose. The chambers of the shell, marking regions the animal has outgrown and partitioned off, are said to contain a certain amount of gas similar in composition to air, and secreted by the siphuncle. In this way a heavy encumbrance is buoyed up and prevented from impeding locomotion. The nautilus is able to swim by ejecting water through the funnel.

The chambered shells of extinct cephalopods are very abundant as fossils, and we can trace back the completely spiral shell of nautilus through various transitionary forms to one that was perfectly straight (*Orthoceras*). This seems, in fact, to be the primitive kind of chambered shell, and rolling up gradually took place as a matter of convenience.

TWO-GILLED CEPHALOPODS

The order Dibranchiata includes cuttle-fishes, squids, and octopi, all of which agree in the possession of two gills only, while the body is never enclosed in a chambered external shell, like that just described for nautilus. Attached to the last part of the intestine is a pouch-like gland, the ink-bag, from which a dark fluid can be ejected, so as to cloud the surrounding water. This is done as a defensive measure, often enabling these animals to elude their enemies, these being predaceous fishes, or, in some cases, cetaceans.

These creatures are also able to undergo rapid colour-changes, by which they harmonise with their surroundings and are rendered inconspicuous. This has a double meaning, for on the one hand it no doubt helps them to escape their foes, and on the other to approach their own prey, largely consisting of fishes, without giving a premature alarm. The change of colour is effected by the agency of little pigment bodies (*chromatophores*) in the skin. These can contract to a very small size, rendering the body pale, or they may be pulled out to considerable dimensions

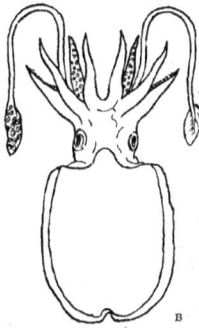

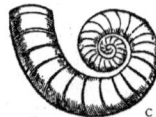

A, PEARLY NAUTILUS ; B, SEPIA ; C, SPIRULA

a, body ; *b*, siphuncle ; *c*, eye ; *d*, hood ; *e*, tentacles ; *f*, shell-muscle ; *g*, funnel

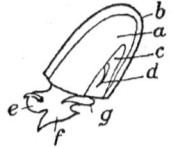

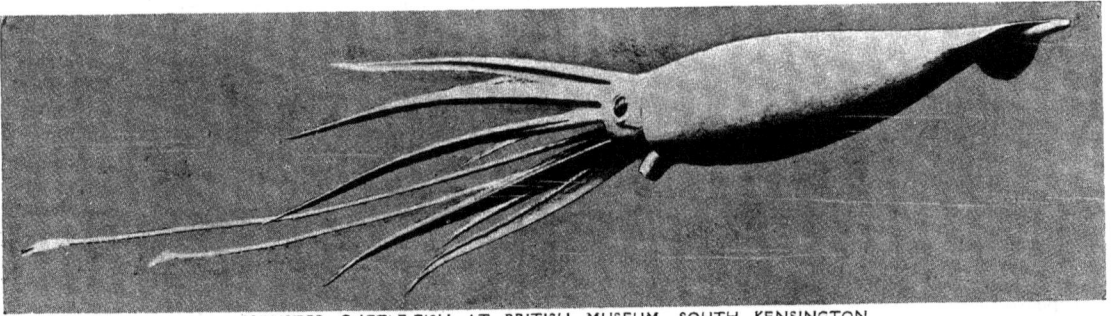

MONSTER CUTTLE-FISH AT BRITISH MUSEUM. SOUTH KENSINGTON

by the action of attached muscle-fibres, in which case the animal assumes a reddish hue.

It may further be added that the funnel is a complete tube, the two curved folds which are distinct in nautilus having here become fused. The eyes are more complex than those of nautilus, and possess a spheroidal lens.

There are two suborders : (1) Ten-armed Dibranchs, (Decapoda), and (2) Eight-armed Dibranchs (Octopoda).

CUTTLE-FISHES & ALLIED FAMILIES

The cuttle-fishes and squids, which constitute the suborder Decapoda, are provided with ten sucker-studded prolongations, or arms, around the mouth, eight of these being of equal length, while the other two (conveniently called tentacles) are relatively long, the suckers being also limited to a terminal expansion.

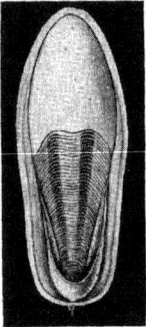

CUTTLE-BONE

PEN OF SQUID

The most primitive family (Spirulidæ) is represented by an existing type (Spirula), of which but few specimens have been secured, though their shells abound on some tropical beaches. One of these is represented in the figure, from which it will be seen to be a chambered spiral, the coils of which are not in contact. It is imbedded in the end of the body, but partly projects to the exterior, and is of small size compared to the animal. Undoubtedly we have here the dwindled remnant of what was once an external chambered shell, like that of nautilus. Several comparable cases of shell reduction were described in dealing with the last class of molluscs.

The cuttle-fishes (Sepiidæ) are broad, flat forms with a narrow lateral fin, and long tentacles, which can be drawn back into pouches. There is an internal shell in the form of a thick, laminated, calcareous plate, commonly known as a "cuttle-bone," and enclosed in a sort of pouch of skin on the front side of the body. It is provided with a little spine—probably

equivalent to the last vestige of a chambered shell—at the end farthest from the head. The shell serves to maintain the firmness of the body during swimming and gives attachment to special muscles.

The common cuttle-fish (Sepia officinalis), its shell, and grape-like egg-capsules, are represented in the accompanying figures. The members of the family are very widely distributed, but do not attain very large dimensions. They are used as food in India and the Far East, being sold in the dried form. The ink-bags were the original source of the brown pigment sepia. The powdered cuttle-bones are still used in dental preparations, and under the name of "pounce" served the purpose of blotting-paper before that useful material was invented.

EGG-CAPSULES OF CUTTLE-FISH

In all the remaining families of decapods the shell is reduced to a narrow, horny "pen," so called because it is not unlike a quill pen in general outline, as will be seen from the figure. The forms with shells of the kind are sometimes called calamaries, a name which has reference to the same circumstance (Latin calamus, a quill pen).

The least specialised of these families (Sepiolidæ), includes the little species here figured (Sepiola rondeletii), which abounds in the Mediterranean, and is not infrequent in British seas. Buried up to the eyes in the sea-floor, it patiently waits for prey, and the little pit in which it is ensconced is partly excavated by jets of water from the funnel.

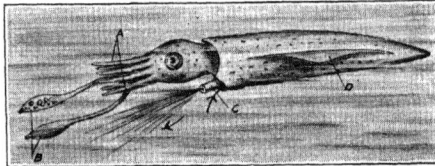

SWIMMING SQUID
A, arms ; B, tentacles ; C, funnel ; D, fin

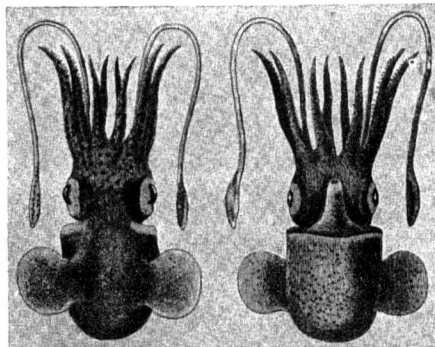

UPPER AND LOWER VIEWS OF SEPIOLA

SQUIDS AND ALLIED FAMILIES

The squids (Loliginidæ) are elongated and conical, with wide triangular fins. They are cosmopolitan, and the common squid (Loligo vulgaris), a well-known British form, is figured, with its pen. It is much esteemed as bait. The numerous eggs are laid in a number of radiating gelatinous tubes.

THE OCTOPUS

function. It is passed into the gill-cavity of the female, and becomes detached, surviving for some time afterwards. Being formerly mistaken for a parasitic worm, it was given the name *Hectocotylus*, hence the word hectocotylised.

An allied form (*Eledone moschatus*) is common in the Mediterranean and on British shores. The arms have one row of suckers, while there are two rows in *Octopus*.

In two families of octopods (*Cirrhoteuthidæ* and *Amphitretidæ*) the arms are united together into a sort of umbrella by intervening membranes, thus forming an apparatus for entangling prey, and which can also aid swimming by opening and shutting.

ARGONAUTS

Lastly may be mentioned a very interesting family (*Argonautidæ*), which includes the argonaut, or paper-nautilus, about which all sorts of extraordinary stories were once current. Two of the arms of the female terminate in large, flat expansions, by which an elegant ribbed shell is secreted, this being shown in the figure. It is quite unlike any other kind of cephalopod shell, and its chief use is to serve as a receptacle in which the eggs are developed. Swimming is effected in the usual way, by expulsion of water through the funnel. The male is very small, and one of his arms is hectocotylised in a marked way, as shown in the accompanying illustration.

In conclusion, it may be pointed out that fishes and members of the present class have for long ages contested the mastery of the sea. The course of evolution in either group may be said to have been determined largely by that of the other.

J. R. AINSWORTH-DAVIS

In an allied family (*Ommatostrephidæ*) the body is proportionately longer and the pen narrower than in the ordinary squids. Shoals of these sea-arrows, or flying-squids (species of *Ommatostrephes*), are common in the open sea, and are preyed upon by dolphins and sperm-whales. The popular names have reference to their power of projecting themselves for considerable distances out of the water, not infrequently landing on the decks of ships. Here, too, belong squids of gigantic size referable to another genus (*Architeuthis*).

From various reliable descriptions we may conclude that the bodies of such creatures may attain a length of 10 feet or more, with tentacles some 30 feet in length, and a weight of perhaps half a ton. Some of the accounts of a supposed great sea-serpent are no doubt founded upon gigantic squids of the kind.

SHELL OF FEMALE ARGONAUT

OCTOPI AND ALLIED FAMILIES

In the members of the suborder Octopoda there are eight sucker-bearing arms, but the long tentacles of cuttle-fishes and squids are absent. The internal shell is either entirely absent or only represented by vestiges.

The typical octopi (*Octopodidæ*), of which the common form (*Octopus vulgaris*), is figured, possess a rounded body devoid of fins, and though able to swim with ease and rapidity, spend a good deal of their time in creeping. They lurk in rock-fissures and similar places, and are particularly destructive to crustacea. In some localities, such as the island of Sark, they attain very considerable dimensions, and are known to attack human beings.

The male exhibits a peculiar modification common in octopods, one arm becoming specially modified, or " hectocotylised," in connection with the reproductive

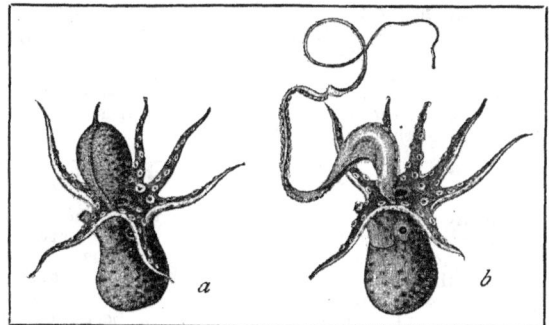

ARGONAUT

a, male with hectocotylus enclosed in sac ; *b*, male with hectocotylus further developed

SUBKINGDOM ARTHROPODA
JOINTED-LIMBED ANIMALS

WE now come to the last and most highly special-ised subkingdom of invertebrate animals—the Arthropoda, or jointed-limbed forms. In several important respects they agree with the bristle-worms, and are probably descended from the same ancestral stock as these. They are, for example, divided into rings, or segments, and their central nervous system consists of a ring round the front part of the digestive tube, thickened into a brain above, and continued behind into a ventral cord, which swells into a varying number of ganglia.

Instead, however, of possessing paired hollow foot-stumps at the sides of the segments, they are provided with pairs of solid jointed limbs, of which at least one pair are modified into jaws, while the excretory organs are for the most part of different nature. Cilia, which play so important a part in the invertebrate sub-kingdom, so far considered, are entirely absent.

Seven existing classes, divided into two groups, are recognised, as follows: A. Aquatic Arthropoda (Branchiata): (1) Sea-spiders (Pycnogonida), (2) King-crabs, etc. (Gigantostraca), (3) Crustaceans (Crustacea), including lobsters, shrimps, crabs, etc. B. Air-breathing Arthropoda (Tracheata): (1) Scorpions, Spiders, and Mites (Arachnida), (2) Primitive Tracheates (Prototracheata), (3) Centipedes and Millipedes (Myriapoda), (4) Insects (Insecta).

CLASS PYCNOGONIDA—SEA-SPIDERS

THE animals belonging to the class Pycnogonida present such a marked general resemblance to the true spiders that they have been included in the same class. On the other hand, from their marine mode of life, some writers have come to the conclusion that their affinities are rather with the crustaceans. As a matter of fact, it appears impossible to affiliate them with either of these groups, and the general opinion is that they are entitled to form a class by themselves. In all these creatures the adult is provided with four pairs of well-developed legs, composed of a large and varying number of segments, and each tipped with a single claw.

FEMALE OF SLENDER SEA-SPIDER, WITH EGGS

These limbs, which are often exceedingly long and slender, radiate from the sides of the cephalothorax, which is produced into stalks for their support. In front of them, and attached to the headpiece, are sometimes three additional pairs of appendages. Hence the full complement of limbs is seven, and not six pairs as in the true spiders. The first pair of appendages, forming the mandibles, are short and often pincer-like; the second pair, or palpi, being also short; while the third pair, which are only developed in the females, are shorter than the true legs, and, from their function, are termed the egg-bearing legs.

SHORE-SPIDER

In some cases, however, these three pairs of appendages have entirely disappeared, as in the shore-spider (*Pycnogonum littorale*). Projecting forwards from the front end of the body is a long, rigid beak, or proboscis, at the tip of which the mouth is situated. This beak is not formed by the fusion of limbs, like that of the ticks, but results from the great development of the area immediately around the mouth. The cephalothorax is divided into four distinct segments, of which the first, or head, supports the first four pairs of appendages, and has on its summit a pair of eyes, while the rest bear the three posterior pairs of limbs. Attached to the rest of these segments, and projecting backwards between the last pair of legs, is the abdomen, which is reduced to a mere tubercle or rod-like process. The greater part of the body-cavity is occupied by the stomach, which sends prolongations almost down to the extremities of the four pairs of walking legs. No breathing organs are known. The sea-spiders are exclusively marine, and range from shallow water to depths of sixteen hundred fathoms or more. The conditions of life in the deep sea have by no means a dwarfing effect upon them, since the species living in the abysses of the ocean attain a size never equalled by those frequenting the coast. Some of the former are of a very large size, *Colossendeis gigas*, for instance, covering a span of nearly 2 feet from toe to toe. None are able to swim, but all crawl slowly amongst the branches of seaweed. The embryo emerges from the egg as a larva, provided with a beak and three pairs of appendages, representing the short anterior three pairs of the adult; the four pairs of great locomotor limbs being subsequently produced by outgrowths from a posterior elongation of the body.

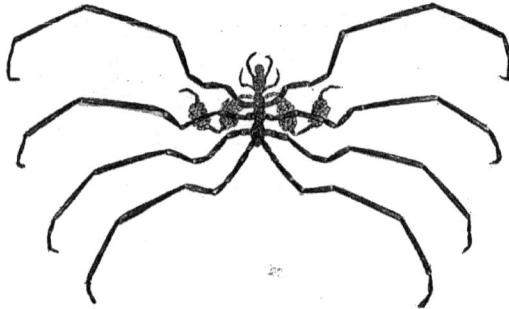

CLASS GIGANTOSTRACA—KING-CRABS

IN many respects the representatives of the class Gigantostraca occupy a position intermediate between the scorpions and spiders and the crustaceans. From the fact that they are marine and breathe by means of gills, they were formerly always classified with the crustaceans, but a large amount of evidence has been brought forward to show that, whereas the earliest kinds are related to the primitive crustaceans, the more specialised kinds are strikingly like some of the scorpions. The class contains three orders, named Xiphosura, Merostomata, and Trilobita. The last two of these are now entirely extinct, and the first-named nearly so, since it is represented at the present day by only a single genus, the king-crabs, or horseshoe-crabs (*Limulus*).

XIPHOSURA

In the existing group, forming the order Xiphosura, the body is armed behind with a long spike-like tail, movably articulated to the middle of the hinder border of the abdomen. The abdomen consists of a large unsegmented pentagonal plate, armed on each side with six movable spines, and hollowed out below to receive six pairs of large flattened limbs, attached to the anterior part of its lower surface. With the exception of the first, each limb supports on its hinder surface a bunch of fine branchial plates, arranged one above another like the leaves of a book. In front of the abdomen comes the cephalothorax, which is covered above with an enormous carapace, having its border semicircular and its hinder angles acutely produced.

The carapace is furnished above with four eyes, two being small and simple ocelli, situated close together some little distance behind the front border, while the others are large kidney-shaped compound eyes, placed at a corresponding distance from the lateral margin. The great size of the carapace is due to the prolongation of its edges into a wide, sloping, shelf-like expansion, concealing the walking limbs. Of the six pairs of the latter, the first are placed in front of the mouth, and are short, three-jointed nippers; while the rest are longer, generally six-jointed, and all but the last nipper-like, the last or sixth ending in a number of flattened plates. The basal segments of the second, third, fourth, and fifth limbs are furnished with large processes, projecting into the mouth and studded with numbers of slender softish spines. The mouth is thus situated between the bases of these limbs, near the middle of the lower surface of the cephalothorax.

CHINESE KING-CRAB

The males differ from the females in having the second or second and third pairs of limbs thickened and otherwise modified. In the male of the round-tailed king-crab (*Limulus rotundicauda*) the second and third pairs are considerably swollen, and the two fingers of the nippers cross each other when closed; whereas in the Moluccan king-crab (*L. moluccanus*) the immovable fingers of these limbs are reduced to short processes.

In distribution king-crabs are limited to the east coast of the United States, to the shores of China and Japan, and of the Indo-Pacific Islands, ranging from the Moluccas to Singapore and Java. In the last-named area two species, *L. moluccanus* and *L. rotundicauda*, occur. The Chinese species is known as *L. longispinus*, on account of the long and strong spines projecting from the carapace and abdomen; while the North American species is *L. polyphemus*.

The habits of the last-named species are tolerably well known. It spends the greater part of the year in water from two to six fathoms deep, and, being unable to swim, creeps about the bottom of the sea in search of food, or even lives buried in mud, into which it scoops its way. This it effects by thrusting the front edge of the carapace forwards and downwards into the mud, the tail behind being used as a prop while the legs are engaged in raking up the mud and pushing it out sideways. The tail is also of service in helping the animal to regain its proper position if turned upside down. Digging the tip of the organ into the soil, the crab raises its body, and, after a few efforts, succeeds in struggling over. In fact, were it not for the possession of a long tail, the king-crab would be as helpless on its back as a tortoise in the same position.

King-crabs feed almost exclusively upon soft marine worms and bivalve molluscs. The food is seized and tucked into the mouth by means of the legs, where the spines on the basal segments of these appendages crush and tear it to pieces. In May, June, and July, large numbers of king-crabs approach the coast in couples to spawn. Choosing spring-tides, they advance along the bottom until the water is shallow enough to allow the carapace to project above the surface. The female then scrapes a hollow in the mud, lays her eggs, and hurries back with her mate into deep water. By the action of the waves the eggs are soon covered with a layer of sand, and at ebb-tide are exposed to the warmth of the sun.

When first it emerges from the egg, the young king-crab is a minute, nearly spherical creature, with a fringe of stiff bristles running round the body, and differs from the parent in having no tail. Subsequently it undergoes a succession of moults, during which the form of the adult is gradually acquired, the tail appearing at the second change. The casting of the skin is effected by the splitting of the integument of the cephalothorax all round, immediately beneath the margin of the carapace. Through the aperture thus made the creature struggles forth, leaving its old shell behind.

Before the growth of the tail the young king-crab is in a helpless state, the slightest obstacle turning it upside down. In this emergency it starts a vigorous flapping of its gill-plates, which causes it to rise in the water. Then, ceasing the agitation, it at once descends, with a chance of alighting right side up.

The existing king-crabs are the typical representatives of the family *Limulidæ*, and fossil remains of *Limulus* occur in the Tertiary rocks as well as in the Cretaceous, Jurassic, and Triassic beds of the Secondary epoch. In the Palæozoic strata the class is represented by a number of forms, such as *Belinurus* from the Carboniferous, *Protolimulus* from the Devonian, and *Hemiaspis* from the Silurian, which resemble *Limulus* in most of their characters, but differ in having the abdomen composed of at least nine distinct segments. On this account they are referred to a distinct family, *Hemiaspididæ.*

It is, however, interesting to note that in the young king-crab the abdomen is also composed of nine segments, so that, just as in the life-history of each individual king-crab the final and adult stage with a solid abdomen is preceded by a transitory stage in which the abdomen is jointed, in the history of the class the existing and final stage, represented by the adult king-crab of our own day, was preceded by a transitory stage, which, in the segmentation of the abdomen, was on a level with the young king-crab.

1870

MEROSTOMATA

The seas in which these fossil forms lived were also inhabited by some nearly allied types, differing from the king-crabs both in habits and some important points of structure. The carapace, for instance, was much smaller and did not conceal the legs, the last pair of which were generally thickened and flattened, and transformed, as in *Eurypterus*, into powerful short paddles. In one form, however, named *Slimonia*, the legs of the last two pairs were enormously elongated, evidently to serve the purpose of oars. The abdomen was used as a propeller, and it was long and divided into twelve flexible segments, the last of which bore the tail-plate, or telson. As in the king-crab, the bases of most of the cephalo-thoracic limbs were armed with teeth and acted as jaws, but those of the anterior pair formed either short tactile organs or long and powerful nippers, as in *Pterygotus*.

The Merostomata, as these animals are termed, appear to have lived both in fresh and salt water, and their organisation seems to show that they were powerful swimmers ; considering, too, the large size which some of the species attained, examples of *Pterygotus* reaching a length of from 4 to 6 feet, there is little doubt that these monstrous creatures were the masters of the ocean in Palæozoic times.

TRILOBITES

A third order is represented by the extinct trilobites, or Trilobita, which swarmed in the seas of the Palæozoic epoch, and are among the earliest of known fossils. The name trilobite, or three-lobed, is given to them because in the best-known and typical members of the group the body is divisible into three distinct parts—an anterior cephalic shield corresponding to the head of the Crustacea and to the cephalothorax of *Limulus*, and formed, as in crustaceans, of five fused segments ; a median thoracic portion, composed of a variable number of freely movable segments ; and the

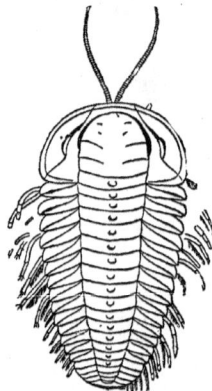

TRILOBITE

pygidium, also composed of a variable number of segments, but usually fused to form a great caudal shield. The lateral portions of the segments are produced sideways into great pleural plates, which mostly conceal the limbs, and the hinder angles of the cephalic shield are frequently prolonged into sharp spiniform processes, sometimes so long that they project backwards beyond the hinder end of the body.

On the upper side of the cephalic shield there is a pair of large kidney-shaped compound eyes, but no sign of the simple eyes present in the Xiphosura and Merostomata has been discovered. For many years no trace of limbs could be detected, but it is now known that a pair of limbs was attached to the lower surface of each of the segments of the head and body ; though, instead of there being two pairs situated in front of the mouth, as in crustaceans, there was only one, as in the Xiphosura, Merostomata, and Arachnida. These, however, take the form of long filiform antennæ, and are placed on each side of a large upper lip, or *labrum*, behind which comes the mouth.

The rest of the appendages of the head, as well as those of the thorax, are alike, consisting of a large basal segment, from which spring two branches, an inner, which was used for crawling, and an outer, many-jointed and fringed with bristles, which was perhaps used for swimming. The basal segments of these limbs in the head region were utilised as jaws, and in the pygidium the inner branches, or *endopodites*, were flattened and more or less leaf-like, as in the lower crustaceans, such as *Apus*.

There is little doubt that trilobites, instead of swimming in the open sea and leading an active predatory life, spent their time crawling or swimming slowly along the bottom, feeding upon worms, burrowing in the mud, and, in case of danger, rolling up tightly into a ball, like wood-lice. Many specimens are found fossilised in this condition, with the lower surface of the pygidium pressed against the head.

CLASS CRUSTACEA—CRABS, LOBSTERS, CRAYFISH, & ALLIES

THE crustaceans comprise a large assemblage of arthropods, presenting great diversity of structure. Some of the parasitic species have become so simplified in organisation that they appear to present no relationship with the higher members of the class, such as crabs, lobsters, and wood-lice. Yet it is certain that all the species, whether terrestrial or aquatic, free-living, sessile, or parasitic, belong to the same stock, and may be derived from the same fundamental plan of structure.

Essentially the body consists of a large number of segments, to each of which is attached a pair of two-branched appendages, the external branch being called the exopodite, and the internal the endopodite. Five segments at the front end of the body unite to form a head ; the appendages of the first two of these segments being situated in front of the mouth, and performing the office of feelers, or antennæ, while those of the remaining three segments are transformed into jaws, the first pair of jaws being the mandibles and the following two pairs the maxillæ. The rest of the appendages are variously modified, and to some are attached respiratory organs in the form of gills.

According to this definition, crustaceans may be distinguished from the centipedes, millipedes, insects, etc., by the presence of two pairs instead of one pair of antennæ, and by possessing branchial and not tubular (tracheal) respiratory organs. The Arachnida may be

separated from crustaceans by having in front of the mouth only one pair of appendages, acting as jaws and not as antennæ, while respiration is effected by means of saccular or tubular ingrowths of the integument.

Nor can there be any confusion between crustaceans and the sea-spiders, since the latter have no antennæ and all their appendages are placed behind the mouth, which is situated at the extremity of a tubular proboscis. But when we come to the Xiphosura it is not so simple to point out the differential characters of the crustaceans. It is true that the king-crabs are easily distinguishable, and appear to be more nearly related to the Arachnida, yet the trilobites, which seem to be ancestral forms of the king-crabs, show marked affinities to the primitive crustaceans.

In a few crustaceans, especially those leading a terrestrial life or inhabiting fresh water, the young is very similar to the adult, and gradually attains maturity without going through any marked change of form ; but in the majority the young upon leaving the egg is unlike the parent, and only acquires its definite form after undergoing a series of moults. The earliest stage, which has been called the *Nauplius*, is a minute oval body, showing no trace of segmentation, and provided with a single median eye, and three pairs of swimming appendages, which become the two pairs of antennæ and the mandibles of the adult.

This stage, however, is by no means of invariable occurrence, but is chiefly characteristic of the lowest members—the Entomostraca—and is rare in the higher Malacostraca. In some members of the latter group, nevertheless, it does occur, as in one of the shrimps (*Penæus*). In this, the nauplius passes into a stage called the *zoæa*, in which four pairs of appendages, representing the maxillæ and the two following pairs of limbs of the adult, have appeared, and the abdominal region increased in length, although, like the greater part of the thorax, it is still limbless. A pair of compound eyes is present on the sides of the head.

After this so-called copepod stage, the large eyes become stalked, the abdomen continues to increase in length, and takes on the function of swimming, which was before performed by the antennæ, and the remainder of the thoracic and abdominal limbs appear. Since the thoracic limbs are provided with a distinct exopodite, as well as the principal branch, or endopodite, as in the cleft-footed shrimps (Schizopoda), this larva is known as the schizopod stage.

Lastly, the median eye and the exopodites of the locomotor thoracic limbs disappear, and the adult form of the *Penæus* is attained. It is, however, exceptional among the higher forms for the young to be set free in the nauplius stage. The young of the lobster, for instance, hatches in the schizopod condition; while that of the common crab appears in the zoæa form, although characterised by the presence of a long dorsal spine, and a sharp beak on the carapace. Moreover, the two pairs of antennæ, the mandibles and maxillæ, are of small size, while the following two pairs of limbs are relatively large, and forked. By means of these, the minute transparent creature swims, and, after undergoing several moults, passes into a stage termed the *megalopa*, which is much like the adult, but has enormously large eyes, and swims by flapping its long, jointed abdomen like a shrimp.

SUB-CLASS MALACOSTRACA

Much difference of opinion still obtains as regards the classification of crustaceans, which are here divided into two main subclasses. In the subclass Malacostraca,

NAUPLIUS LARVA OF BARNACLE

ZOÆA STAGE OF CRAB

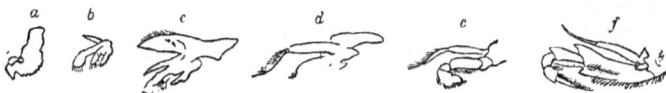

JAWS OF CRAYFISH

a, mandible; *b, c*, maxillæ; *d, e, f*, first, second, and third maxillipedes

comprising the largest and most familiar forms, the number of segments in the body is very generally nineteen (but never more), and each has a pair of appendages. The first five segments compose the head, which, except in some blind species, bears a pair of compound eyes, two pairs of antennæ, and three pairs of jaws—namely, a pair of mandibles in front and two pairs of maxillæ behind. The eight segments behind the head, which constitute the thorax, may be united with the head, as in crabs, when the whole region is termed the *cephalothorax*, and the shield that covers it the *carapace*.

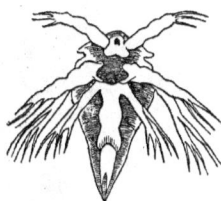

Sometimes, too, as in the cray-fish, the anterior three pairs of thoracic appendages are transformed into jaws, and on this account are called the *maxillipedes*, or foot-jaws; and in such cases only the remaining five pairs, called the trunk-limbs, are large, and used for locomotion or seizing prey. In less highly organised forms, all the maxillipedes may be free and foot-like, as in the mantis-shrimps, or only the anterior pair, as in sand-hoppers, may act as jaws.

The remaining six segments, forming the abdomen, are usually provided with six pairs of small two-branched limbs, and to the last of these segments there is articulated a single plate, or telson, while the limbs, or uropods, are generally of large size, and form with the telson the tail-fin.

The Malacostraca are divisible into two series, the Podophthalmata, containing those in which the eyes are perched on movable stalks, and the Edriophthalmata, containing those in which they are sessile, or, if raised upon stalks, not movable. The former are further distinguished by having the fore part of the body generally covered by a carapace; in the latter some of the thoracic segments are movable, and there is generally no carapace.

DECAPODS

The first order (Decapoda) of the stalk-eyed series is characterised by having the posterior five pairs of thoracic limbs strongly developed, and forming either walking or swimming legs, or prehensile pincers. The three pairs of maxillipedes are generally transformed into jaws; but in some of the lowest forms, as shrimps, the third, or

EDIBLE CRAB

last, pair are long and limb-like, so that in reality there are six pairs of large thoracic limbs.

The gills, which are attached to the sides of the cephalothorax and to the basal segments of its limbs, are concealed in a gill-chamber, formed by the lateral portions of the carapace. Each gill may be compared to a plume consisting of a central stem, to which are attached a number of delicate processes in the form of flattened plates or of filaments. The front aperture of the gill-chamber is closed by a movable plate called the *scaphognathite*, and attached to the second maxilla. During life this plate is in constant motion, baling out the impure water through the anterior opening, and thus compelling a flow of fresh fluid into the chamber through the openings at the hinder end of the carapace above the bases of the limbs.

CRAES

Decapods are divisible into two suborders, the Brachyura, or short-tailed, and Macrura, or long-tailed group. The first-named suborder contains those members of the order which may be called crabs. Here the abdomen, or so-called tail, is small, and shorter than the cephalothorax, against the lower surface of which it is usually tucked away. In the males it is generally narrow, and bears only one or two pairs of appendages, but in the female it is broader and is furnished with four pairs of limbs. In neither sex is its last segment furnished with a pair of uropods forming the tail-fin. The lower surface of the cephalothorax is generally broad and triangular, and the third pair of maxillipedes are short and flattened, and form, when in contact, a plate completely covering the rest of the mouth-organs.

The group is divisible into five tribes, the first of these being the Cyclometopa, or those with rounded foreheads. It includes most of the commoner species, such as the edible crab (*Cancer pagurus*) and shore-crab (*Carcinus mænas*). The former belongs to the family *Cancridæ*, characterised by having the carapace much wider than long. As an article of food the male is more esteemed than the female, being larger and having larger claws. The two sexes, as in all crabs, may be distinguished by the size of the tails, this organ in the male being

SWIMMING-CRAB

narrower, more pointed, and having fewer and smaller appendages than in the female. The family *Cancridæ* is represented in tropical seas by a large number of species and genera, some of which, such as *Actæa*, have the carapace covered with granules and ornamented with a network of deep grooves.

The members of the family *Portunidæ* may be recognised by a modification of the last pair of legs. In the great majority of crabs these legs are like the rest, ending with a long, slender-pointed foot, which bears evidence to its being an organ for running, climbing, or crawling ; but in the *Portunidæ* these legs are much flattened, the last segment in particular being dilated into an oval plate. The creatures are thus equipped with a pair of oars, by means of which they swim. Several species of the typical genus *Portunus* are found in British waters, and many of them are handsomely coloured, although none are such expert swimmers as the tropical species, especially those inhabiting the gulf-weed of the Atlantic. The peculiar motion of the oar-like feet has given rise to the name of fiddler-crabs, so often applied to the group. The figured species (*Thalamita natator*) is tropical.

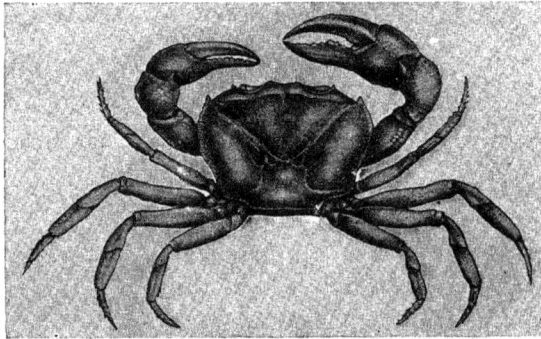

INDIAN LAND-CRAB

The common British shore, or green, crab (*Carcinus mænas*), which is referred to this family, differs from the rest in having the last pair of legs adapted for walking, being armed with a claw, and not flattened into a paddle. Connecting the present with the following section is the family *Thelphusidæ*, which contains a number of genera and species found in fresh-water streams or on land, and sometimes ascending mountains, in temperate and tropical regions. One of the best-known species is the South European *Thelphusa fluviatilis*, which swarms on the muddy banks of the Lake of Albano, and is also abundant in the neighbourhood of Rome, where it is captured for sale. Another well-known form is the Indian land-crab (*T. indica*), to which the species here figured is nearly allied.

The second tribe (Catometopa) is characterised by the broad and squared frontal region of the carapace being bent downwards. It is typically represented by the family *Gecarcinus*, which has representatives in both

SWIFT LAND-CRAB

THORNBACK CRAB

frequent the same spot. Each has a burrow to itself, and if one of them tries to enter by mistake the burrow of another, the rightful occupant makes a loud scraping noise to warn the intruder of its error ; whereupon the latter retreats in search of its own abode. So strong is this instinct against trespassing that a crab will always risk the chance of a fresh run for safety rather than persevere in seeking concealment in the home of another.

Nearly allied to the foregoing are the calling-crabs (*Gelasimus*), represented by a number of species from the warmer parts of the world. The carapace is broad and squared in front, and the long, slender eye-stalks lie, when at rest, along its front border, sunk in the orbits. But the most remarkable characteristic is the enormous size of one of the pincers in the male ; and it is from the habit of brandishing this claw, as if beckoning, that the name of calling-crab is derived.

So abundant are these crabs that they may frequently be seen by the thousand either running over the sand or peering out of their holes. These holes, which are thickly scattered over wide areas, lead into burrows frequently a foot or more in depth. The crabs scrape up a heap of sand, and, grasping the pellet with three of the legs of one side, carry it to some distance before letting it drop, then, raising their eyes and peering round, dart back to the burrow, scrape together another heap, and persevere in the same manœuvre till the burrow is of the required depth.

It was long supposed that the pincers of the male were used as weapons of attack and defence ; but, in addition to its size, this limb is noticeable for its

the Eastern and Western Hemispheres. Two inhabit the West Indian Islands, and of one of these (*G. ruricola*) from Jamaica a full account has been given by Mr. Browne.

These crabs are generally found at a distance of from two to three miles from the sea, where they spend the day under stones, or in other sheltered situations. Pairing takes place in the spring ; and shortly afterwards the whole population makes a move for the sea, in which the females lay their eggs. When seized with this migratory instinct, nothing can turn them from their course. Issuing from hollow trees, from under rocks, and out of innumerable holes, they muster in a host so fast that they thickly cover an area more than a mile long and upwards of forty yards wide. The males lead the way, and the band proceeds in a straight line to its destination, climbing over everything that comes in its road, be it hedges, houses, churches, hills, or cliffs, and rather clamber up at the peril of their lives than make a circuit. Having reached the sea, the females lay their eggs, and the young hatch out as miniature copies of their parents. At the time of moulting, which takes place late in the summer, the crabs retire to their burrows, close up the apertures, and remain there out of harm's way until the old shell is cast and the new integument hardened. It is while still in the soft state that these crabs, which are eaten by the natives, are considered most palatable.

The second family (*Ocypodidæ*) is typically represented by the swift land-crabs (*Ocypoda*), which appear to be less strictly terrestrial than the last, although unable to endure a long sojourn in the sea. Indeed, from the adoption of a land life their breathing-organs have become so modified that these crabs may be drowned by an immersion of twenty-four hours. They frequent sandy beaches, and, when chased, run with such speed as to make their capture a matter of difficulty. They burrow deep perpendicular holes in the sand. In these they stay when the tide is up, but at low water they wander over the beach in search of food, which consists of sand-hoppers or any offal cast up by the waves.

These crabs are gregarious, in the sense that numbers

CALLING-CRAB

bright colours, and Mr. Alcock, who observed a number of males of an Indian species (*G. annulipes*) waving their large claws in the presence of a female, has suggested that their object in so doing is to make a display of their gaudy ornamentation and thus influence her choice of a mate.

The third family (*Grapsidæ*) contains species which for the most part are shallow-water forms. They are widely distributed, and attract the attention of travellers both on account of their bright colours and their extraordinary activity. Possessing long and powerful legs, tipped with sharp, strong claws, they are able to dart amongst the rocks on the coast with amazing speed, while by means of their flattened carapace and limbs they can slip away into the narrowest clefts and chinks.

In the next tribe (Oxyrhyncha) the carapace is generally narrowed in front and wide behind, and furnished between the eyes with a distinct beak, which is sometimes double and of great length. On its dorsal surface the carapace is usually roughened with spines or tubercles. and frequently furnished with hooked hairs. These crabs frequent deep water, and, at least on the English coast, are regarded by fishermen as spiders.

The characteristics of this group are shown in the

DROMIA CRAB

figures of two British species, the thornback crab (*Maia squinado*) and the long-beaked spider-crab (*Macropodia longirostris*). Also belonging to this tribe is *Macrochira kæmpferi*, which is not only the largest crab, but the largest crustacean known. It inhabits the seas of Japan, and is said to be able to span eleven feet with its outstretched pincers.

A peculiarity of many of this group is their extreme untidiness, owing to the quantities of scaweeds, zoophytes, and other marine objects affixed to the carapace and limbs ; and it has been ascertained that the presence of these extraneous bodies is not the result of chance, but that they are placed there, presumably for the purpose of concealment, by the crabs themselves. This feat they are enabled to perform owing to the flexibility of their pincers, and to the hooked hairs and spines with which the carapace is studded. Some examples of *Hyas*, deprived of their covering of foreign bodies, were placed under observation in an aquarium of which the bottom was covered with a layer of sponge.

Contrary to their habitual sluggishness of manner, the crabs appeared much perturbed, running first to one side, then to the other in the aquarium. Soon, however, by means of their pincers they tore off small fragments of the sponges, and, after first putting them to their mouths, placed them finally upon the dorsal surface of the carapace or limbs, sticking them there with a rubbing movement. Sometimes, after several vain efforts, the crab brought the fragment afresh to its maxillipedes, and then repeated its efforts to make it adhere. The crab ultimately succeeded in completely changing its appearance, and in rendering itself indistinguishable amongst the objects that surrounded it. The crab proceeded in exactly the same fashion when the bottom of the aquarium was strewn with seaweeds or any kind of zoophytes. Moreover, it was observed that some specimens, clothed with seaweed, which were

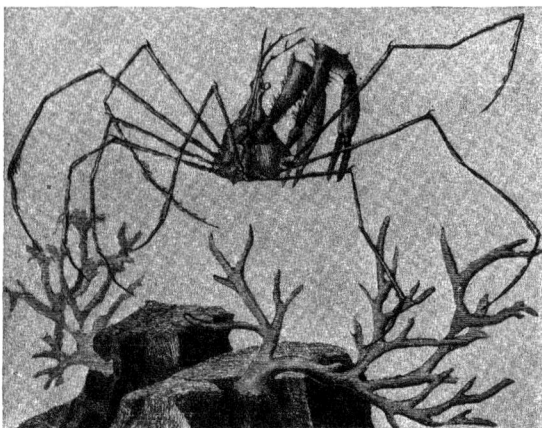

LONG-BEAKED SPIDER-CRAB

left in an aquarium of which the bottom was covered with sponges did not hesitate to take off their old clothing and put on a new one of sponges.

The present tribe also contains the family *Parthenopidæ*, the species of which, although not armed with spines and hooked hairs for holding foreign objects, are yet protected in the midst of their surroundings, by having the carapace covered with pits and variously shaped depressions, giving it a roughened corroded appearance, and consequently imparting a resemblance to pieces of rock or fragments of dead coral.

The sharp-nosed crabs, forming the tribe Oxystomata, are so called because the carapace is produced in front into a short beak-like prominence, while the external maxillipedes which cover the mouth are narrowed and pointed at the apex. The families belonging to this group present great diversity of structure and habits, the *Matutidæ* being active swimmers and resembling the *Portunidæ* in having their posterior legs transformed into flattened paddles, while the *Calappidæ*, which live a sluggish life on the floor of the sea, have the sides of the carapace produced into shelf-like plates covering the legs, and the upper edges of the pincers supplied with large crests. The latter shut in the cephalothorax in front, so that when at rest the whole animal is enclosed in a casing of shell, and resembles a smooth pebble on the sea-bottom.

Some of the species of the family *Leucosiidæ* are remarkable for the porcelain-like appearance and texture of the carapace, while in others, as in *Ebalia*, this plate is granular and corroded. Three or four species of the latter genus occur in British waters, but the majority of the Oxystomata are inhabitants of the tropics. In the genus *Dorippe*, belonging to the family *Dorippidæ*, the last two pairs of legs are short and raised on the upper surface of the body behind the carapace. In this position and structure they are adapted for carrying foreign bodies to serve as a protection.

The aberrant forms constituting the tribe Anomala differ from the other members of the suborder in having sometimes as many as fourteen pairs of gills, and also in that the apertures of the oviducts are situated upon the basis of the third pair of legs and not upon the breast-plate of the cephalothorax. Moreover, as in the *Dorippidæ* of the preceding tribe, the last or last two pairs of legs are shorter than the rest, and dorsally placed, as shown in the illustration of the common *Dromia vulgaris*. The crab uses these limbs to hold foreign bodies like sponges and shells, beneath which it thus lies concealed.

HERMIT-CRABS

HERMIT-CRABS

The long-tailed group (suborder Macrura), comprising the lobsters, hermit-crabs, prawns, and shrimps, is distinguished by having the abdomen, or tail, usually of large size, and constituting a powerful flapper for swimming, in which function it is assisted by the enlargement of the appendages of its last segment to form with the telson a powerful tail-fin. The external maxillipedes are slender and leg-like, and the antennæ usually longer than the body. The first tribe, Anomura, contains forms which typically have a symmetrical tail. With these were originally classified the anomalous crabs, and there is no doubt that some of the species bear a striking resemblance to the latter. This is shown in the illustration of the broad-clawed porcelain crab (*Porcellana platycheles*), which frequents rocks and seaweed at low water. It may be distinguished from the true crabs by its long antennæ, the

BROAD-CLAWED PORCE-
LAIN CRAB

so much as sniff at it if protected by an anemone.

One of the commonest deep-water British hermit-crabs, *Eupagurus prideauxi*, is invariably found associated with an anemone, but the latter adheres to the lower surface of the shell in such a manner that its mouth and tentacles are situated immediately below the forepart of the crab's body. It is thus able to share in the meals that the crab procures for itself, and the companionship is consequently mutually beneficial to the two. An advantage conferred upon the crab by the presence of the anemone results from the fact that the latter gradually absorbs the shell in which the former is lodged, so that there is no occasion for it to change its abode with growth, the soft tissues of the polype offering no resistance to the crab's increase in size.

Certain hermit-crabs have forsaken the sea as a permanent abode, and spend the greater part of their lives on land. For instance, the genus *Cenobita*, which

COCO-NUT CRAB

CRAYFISH

presence of a tail-fin, and the slender unflattened external maxillipedes.

The most familiar members are the hermit-crabs, which abound in all seas, and are represented by several British species. In the typical forms, the integument of the abdomen is soft, and, aware of its defencelessness, the hermit-crab invariably thrusts itself for protection into some empty shell, which it subsequently never willingly quits, save for the purpose of changing its abode for a larger one, when compelled by the exigencies of growth. It is not an uncommon thing to find shells containing a hermit-crab surmounted by a large anemone. The advantage to the crab of this association is considerable, for anemones are so distasteful that no fish will bite at them twice, and, consequently, a fish that would, under ordinary circumstances, greedily swallow a hermit-crab, shell and all, will not

occurs both in the West Indies and India, may be met with in forests far from the coast. The best known of these terrestrial forms is the great coco-nut crab (*Birgus latro*), here figured, found in the islands of the Indo-Pacific seas, and remarkable not only for its great size and habits, but also for having the abdomen symmetrical and covered above with a series of horny plates.

These animals inhabit deep burrows, which they hollow out beneath the roots of trees, and carpet with fibres stripped from coco-nuts. Periodically, however, they are compelled to visit the sea to moisten their gills ; and here they lay their eggs, the young being hatched and living for some time on the coast. They live principally upon coco-nuts which fall from the palms, but they do not climb the trees after the fruit. To get at the contents of the nut, the crab first tears away the fibre

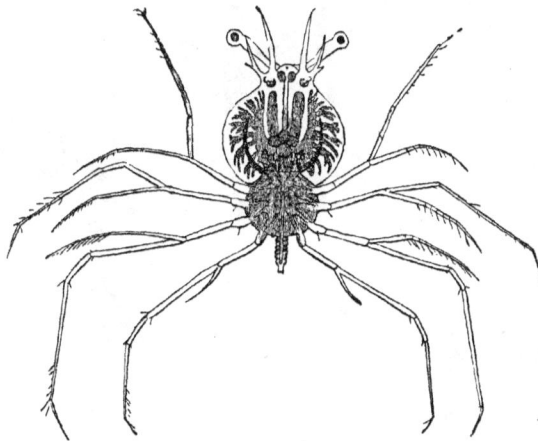

A GLASS-CRAB

overlying the three " eyes," and then hammers away with its claws at the latter until a hole is made, when it extracts the kernel by means of its smaller pincers. Some observers state that, after drilling through the perforated eye, the crab grasps the nut in its claws and breaks it against a stone.

In the next tribe, or Thalassinidea, the carapace is much compressed and has a small rostrum, but the abdomen is well developed and often wider in the middle than in front. As in the short-tailed group, the fourth pair of thoracic limbs is enlarged and generally completely chelate, while the four succeeding pairs, of which the last is smaller than the rest, usually terminate in simple claws.

All the members of this group are exclusively marine, living at the bottom of the sea buried a foot or more in the mud. The species *Thaumastocheles zeleuca* obtained at a depth of four hundred fathoms in the West Indies is characterised by the extraordinary development of the pincers of the right claw, which are not only very long and slender, but beset with spine-like teeth. The creature is totally blind, having lost both eyes and eye-stalks.

ROCK-LOBSTERS AND GLASS-CRABS

In the tribe of the Scyllaridea none of the limbs of the thorax are truly chelate, and the antennæ are not furnished with an external scale-like basal piece. The best-known members of this group are the *Palinuridæ*, or rock-lobsters, one member of which the crawfish (*Palinurus vulgaris*), may be seen for sale in England. It is larger than the lobster, and has enormously long, stout antennæ, and a spiny carapace, but no claws. This species is figured on the left side of the coloured plate on page 1870.

The second family, *Scyllaridæ*, contains a considerable number of genera (*Scyllarus*, *Ibacus*, etc.), mostly from tropical seas, remarkable for having the carapace broad and flattened, with the eyes enclosed in complete orbits on its upper surface, and the antennæ short and scale-like. In this tribe the larvæ are unlike those of crabs or lobsters. On account of their transparency and delicacy they are called glass-crabs. The body is formed of three distinct parts, a large round-sided head, a smaller but also round-sided thorax, and a minute jointed limbless abdomen, which projects from the hinder end of the thorax.

ONE-CLAWED LOBSTER

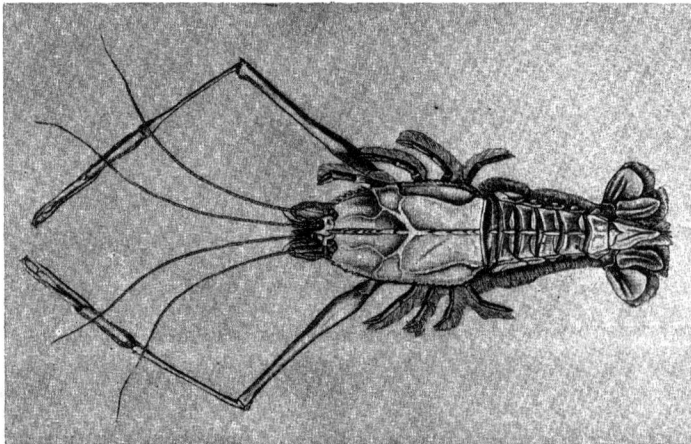

SLENDER-CLAWED CRAYFISH

LOBSTERS AND CRAYFISH

The lobsters and crayfish (Astacidea) have at least three pairs of the large thoracic limbs pincer-like, the first being much larger than the others. The antennæ are furnished with a distinct basal scale-like plate. The first family (*Eryontidæ*) contains several genera found in deep water in various parts of the world, the slender-clawed *Willemæsia leptodactyla* occurring in both the Pacific and Atlantic Oceans, at depths varying from thirteen hundred to over two thousand fathoms. As in many deep-water species, the eye-stalks are rudimentary. The five posterior pairs of thoracic limbs are chelate in both sexes, and the first pair of antennæ have their inner branches long, while the carapace is flattish with a small rostrum.

The remaining three families— namely, the *Nephropsidæ*, or lobsters, the *Potamobiidæ*, and *Parastacidæ*, or true crayfish—are nearly allied. Among the former, the Norway lobster (*Nephrops*) is smaller than the common lobster, and has the pincers long, slender, and covered with scale-like tubercles. The common lobster (*Astacus gammarus*), from a commercial point of view, is one of the most important crustaceans. The crayfish (*Potamobiidæ*), which live exclusively in fresh water, are very like small lobsters ; one species (*Potamobia fluviatilis*) being British.

Throughout the day crayfish usually lurk under stones or the edge of banks, and creep out in the evening in search of food, which consists of worms, water-insects, small frogs or fish, and plants and roots of many kinds. During the winter they seek the shelter of crevices or excavate deep burrows in the banks. In these they lie, with their antennæ stretched forward, and their claws ready to seize any passing object that may serve for prey. Pairing takes place in the autumn, and the female retires to her winter quarters to deposit her eggs, which vary in number from one to two hundred. After being laid, the eggs are attached to the abdominal limbs of the mother. During the winter they develop slowly, and are not ready to hatch until late spring or early summer. The young, which resemble the parent, adhere to their mother's limbs, and do not leave her until able to shift for themselves. Growth, however, although fast at first, is a slow process, the crayfish not

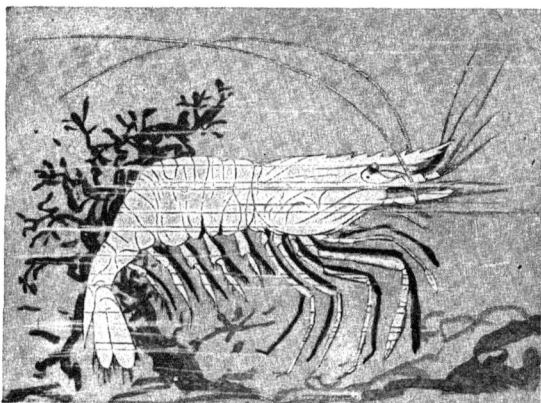

COMMON PRAWN

reaching maturity until about five years after birth. They probably live, under favourable conditions, for about fifteen or twenty years.

SHRIMPS AND PRAWNS

The next tribe, Caridea, embraces the shrimps and prawns, in which the last three pairs of thoracic limbs are never chelate, although the two pairs in front of them are frequently so. The tribe is divided into three sections. The first of these, or Crangoninea, contains the family *Crangonidæ*, or shrimps, characterised by having the first pair of trunk-limbs subchelate— that is, with the terminal segment capable of being folded back upon the penultimate.

The common shrimp (*Crangon vulgaris*) occurs in shallow water on sandy coasts of temperate countries of the Northern Hemisphere. Its colour is a speckled grey, corresponding closely with that of the sandy sea-bottom upon which it lives, and in which it buries itself when threatened with danger. To escape the vigilance of fish, shrimps resolutely keep themselves hidden during the day, but come forth at night to hunt for food. The presence of this they perceive by means of scent, since a blind shrimp will find food as quickly as an uninjured one.

A second British species is Allman's shrimp (*Crangon allmani*), abundant in deep water in the Irish Sea and on the west coast of Scotland. It may be distinguished at once by the presence of two fine keels on the upper side of the sixth segment of the abdomen. Both have a short rostrum and no spines on the carapace; but some of the other members of the family have crests of spines on the carapace, and sometimes a largish rostrum, as in the Arctic *Sclerocrangon boreas*. In *Rhynchocinetes typus*, from the South Pacific, this rostrum is not only large, but movably jointed to the carapace.

The section Monocarpinea differs from

the last in having the first and second trunk-limbs completely chelate, and the second pair larger than the first. To this section belong a number of fresh and salt water forms, and among them the *Palæmonidæ*, or prawns. The general form of the body is shown in the figure of the common prawn (*Leander serratus*). In its native haunts the prawn is nearly invisible, being almost colourless, translucent, and marked merely with streaks of various tints. In the rivers of tropical countries occur prawns (*Palæmon*) rivalling lobsters in size, and remarkable for the length of their pincers, which may exceed that of the body. Among the largest are *P. jamaicensis* from the West Indies and Central America, and the Indian *P. lar*, so much esteemed when cooked as a curry.

Also belonging to the same section is the family *Atyidæ*, containing a few genera, such as *Atya* and *Caridina*, found in both Eastern and Western Hemispheres in fresh-water streams and lakes. In *Atya* the trunk-limbs are curiously constructed, the first two pairs being short and subequal, with the two fingers of the pincers tipped with a long tuft of hairs. The remaining three pairs, of which the first is much the stoutest, end in simple claws, and are studded with scale-like or spiny tubercles. It feeds on the organic matter contained in the mud which it gathers up in its nippers, compresses into pellets, and transfers to its mouth.

The Polycarpinea contain those species in which the wrist of the second pair of trunk-legs is divided into several secondary segments. In other respects they are nearly allied to the last group. A common British representative is the red shrimp (*Pandalus montagui*), which gives its name to the family *Pandalidæ*, and is abundant upon many parts of the British coasts. This tribe is abundantly represented in tropical seas by the hooded shrimps (*Alpheidæ*), remarkable for the concealment of the eyes beneath the edge of the carapace, and for the enormous size, bright colours, and peculiar shape of the right or left pincers. With this instrument the hooded shrimps, which frequent holes and crevices in the coral-reefs, are able to produce a clicking sound when angry or alarmed by the approach of danger.

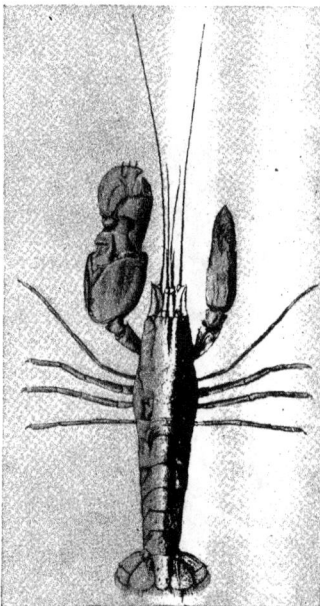

HOODED SHRIMP

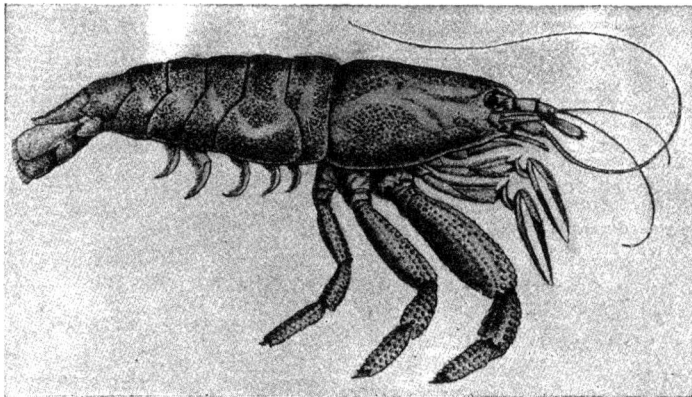

WEST INDIAN PRAWN

The last tribe of the sub-order, known as Penæidea, appear similar to the Monocarpinea, but may be distinguished by the circumstance that the first three pairs of trunk-limbs are chelate, so that only the posterior two terminate in simple claws. Some of the species of the genus *Penæus*, belonging to the family *Penæidæ*, attain a large size in tropical seas, and form an important article of commerce. Nearly allied is the little *Spongicola venusta*, which makes its home in glass-sponges. In this neighbourhood may be placed the anomalous family, *Sergestidæ*, in which the gills are impoverished or lost, while the first pair of trunk-legs, and sometimes the second, are simple, the chelæ of the third are minute, and the fourth and fifth pairs are feeble, rudimentary, or absent. To this family belongs the genus *Leucifer*, remarkable for having the eyes and antennæ supported at the end of a long neck which extends in advance of the mouth. The gills are absent, respiration being effected by means of the general integument, which is so thin that the internal organs can be seen. In the figure the dark line (*n*) is the ventral nerve-cord, which throws out finer branches from ganglionic swellings in each of the segments; (*h*) is the heart, while immediately below the latter is the stomach, passing forwards into the gullet and backwards into the intestine.

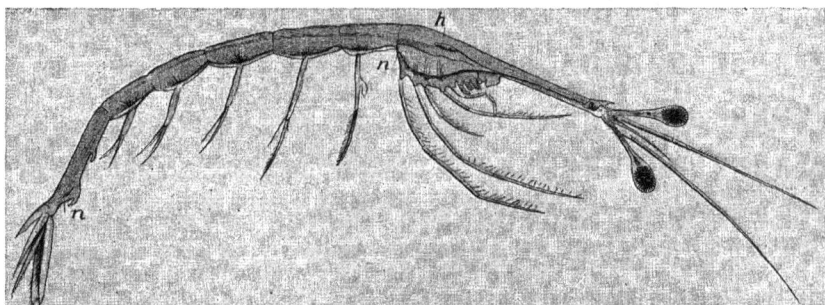

LONG-NECKED SHRIMP

n, ventral nerve-cord ; *h*, heart

SCHIZOPODS

The name of Schizopoda is applied to a group nearly allied to the long-tailed decapods, the chief difference between them being in the fact that in the present order the eight thoracic limbs are similar in structure, each being pediform and provided with a distinct exopodite on the second segment. The gills, which are attached either to the thoracic or abdominal appendages, generally

SEROLIS BROMLEYANA

project into the water, and are but rarely concealed in a chamber. The eggs are carried by the female beneath the trunk, and are frequently protected by the development of a pouch. The order contains several families embracing a large number of mostly marine forms, some of which occur at great depths.

Of the British species, the finest is *Nyctiphanes norvegica*, which forms an important part of the food of herrings. It has luminous organs on the thorax and abdomen, and when swimming in a glass vessel, in a darkened room, appears like a flash of light. The young, as in all the members of the family *Euphausiidæ*, are hatched in the nauplius stage. Most schizopods are small, but species belonging to the genera *Lophogaster* and *Gnathophausia* measuring as much as six inches in length have been obtained. To the family *Mysidæ* belongs the genus *Mysis*, or opossum-shrimps, among which is *M. veheta* from certain lakes in Northern Europe. Into these lakes the species is presumed to have entered while they were connected with the sea, a supposition borne out by the fact that it is nearly related to *M. oculata*, now living in the Arctic Ocean.

STOMATOPODS

The mantis-shrimps (*Squillidæ*) of the order Stomatopoda, which owe their name to the resemblance that their seizing limbs bear to those of the insect mantis, are abundant in tropical seas, where they sometimes reach a large size. Although bearing a general likeness to the long-tailed decapods, they may be recognised by certain prominent characters. A glance at the accompanying figure will show that the carapace is so short as to leave the hinder segments of the thorax uncovered, and, since the gills are attached to the abdominal limbs, it forms no branchial chamber. Only three pairs of limbs are modified into

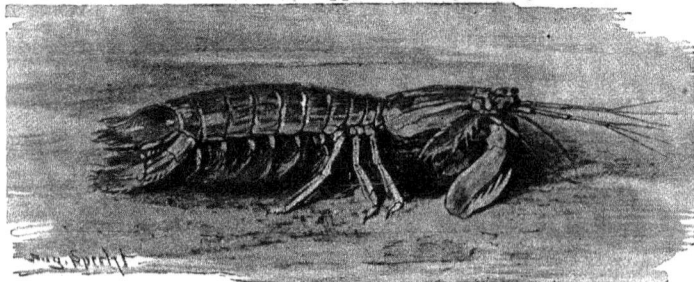

MANTIS-SHRIMP

COMMON AND PILL WOOD-LICE

jaws, these being the mandibles and two pairs of maxillæ. The remaining eight pairs of thoracic limbs are foot-like, the large prehensorial pair corresponding to the second maxillipedes of a decapod.

Two kinds of mantis-shrimp are occasionally met with in the English Channel, namely, *Squilla desmaresti* and *S. mantis*. The former is not uncommon along the shallower parts of the shores of Jersey, but as it lives in deep burrows among the roots of sea-grass, in a zone never uncovered by the tide, its appearance is infrequent.

CUMACEA

We now come to the Edriopthalmata, the second great series of the Malacostraca, in which the compound eyes are generally sessile, and never mounted on movable stalks. As a rule, the last seven segments of the thorax are not covered by the carapace, and the last four are always free. The first order, Cumacea, is in many respects intermediate between the typical members of the last and present series, the thorax being larger and broader than the abdomen, while the carapace covers all but the last five segments. The front angles of the carapace are produced to meet in a kind of beak in front of the head, and the eyes are generally united in a single cluster of ocelli. None of the thoracic limbs are prehensile or chelate. The first five segments of the abdomen have no appendages in the female, although such limbs are present in the male. In the sixth segment appendages are present in both sexes, and form a fork-like termination to the body.

Two of the best-known genera of the group are *Cuma* and *Diastylis*. The order, however, is relatively a small one, containing only a little over a hundred species. It has, nevertheless, a wide distribution, forms being met with in shallow and deep water in all seas, although the Arctic Ocean produces individuals of the largest size and in the greatest abundance.

ISOPODS

Unlike the last, the second order of the series (Isopoda) exhibits great diversity of structure. As a rule, the posterior seven segments of the thorax are free, and at least the first three, and generally the first five, segments of the abdomen are short and sometimes fused together, while the sixth is the largest, and bears the telson and a pair of appendages. The other abdominal appendages usually overlap, and are modified to act as gills. The seven thoracic limbs are generally large, and perform the function of walking or swimming organs, while the posterior five pairs at least have no exopodites.

There are four pairs of jaws—namely, the mandibles, two pairs of maxillæ, and one pair of maxillipedes.

Although there are many exceptions, it may be said that, as a rule, the body is broad, short, and flattened. Corresponding to the structural variations, the isopods exhibit great diversity of habits and habitat. Most are marine, occurring in shallow waters or at great depths. Some live in fresh water, others on land, while others have taken to a parasitic life, and have thus to a great extent lost the characters of the order.

Of the five tribes, the Valvifera have the posterior pair of abdominal appendages, or uropods, transformed into valves or flaps, covering part of the lower surface, and constituting a chamber for the gills. The tribe contains two families, the *Arcturidæ* and *Idoteidæ*. The former are distinguished by their slender cylindrical shape, long lower antennæ, and the length of the fourth thoracic segment, which separates the posterior three pairs of thoracic legs from the anterior pairs by a wide space. The anterior thoracic feet are slender and hairy on the inner side, while the posterior feet are strong and prehensile, and enable the animal to fix itself to the branches of corallines.

In the *Idoteidæ* the body is longish and narrow, the thoracic segments being all of about the same size and shape, and their appendages short, stout, and used for walking. The anterior segments of the abdomen are short, and the posterior fused into a caudal shield. The species of the genus *Idotea* live in shallow water, and frequent places where there is an abundance of decomposing seaweed. They are essentially carnivorous, feeding on dead fish, worms, and molluscs.

The tribe Flabellifera contains a part of marine species, in which the abdomen terminates in a tail-fin, formed as in the macrurous decapods from the telson and the limbs of the last segment. There are too many families to mention, but some of the characteristic forms are shown in the accompanying illustrations. In the genus *Serolis*, which alone represents the family *Serolidæ*, the body is depressed and broad, the segments of the thorax being furnished with long pointed sideplates, which impart to the animal a superficial resemblance to a trilobite. The legs and two pairs of antennæ are long.

It is stated that the *Serolidæ* " live by preference on sandy ground, into which they burrow with their flat bodies up to the caudal plate. Their nourishment appears chiefly to consist of the organic materials distributed in the fine sand, diatomacea and organic detritus. Their locomotion is carried on less by swimming than by backward movements on the sandy ground, wherein the widely separated feet are used as a point of support." The species figured (*S. bromleyana*) is the largest, and has been taken at a depth of nineteen hundred and seventy-five fathoms. In the *Sphæromidæ*

MALE GNATHIA   FEMALE GNATHIA

the convex body is capable of being rolled into a ball. Several species of *Sphæroma* occur on the coasts of Britain, and may be found, sometimes in numbers, sometimes isolated, beneath stones or amongst seaweed at low water.

FISH-LICE AND ALLIES

The next family (*Gnathiidæ*) contains the genus *Gnathia*, in which the males and females are so dissimilar that they were referred to two families. In the adult male the mandibles are powerful and prominent, and the head is large, squared, and at least as wide as the thorax. In the adult female, on the contrary, the head is small and triangular, without visible mandibles, and the thorax is much dilated. Many species are known from the European coasts, and one has been obtained at a depth of nine hundred fathoms. Belonging to this tribe, but representing a family by itself, is *Limnoria lignorum*, known to fishermen as the gribble, which is a persistent destroyer of submerged wood. The creatures are about one-sixth of an inch long, and of an ashy grey colour; and the destruction they bring about is due to their habit of boring into timber below water-mark. They are vegetarians, and feed on the wood which they excavate.

The members of the group known as fish-lice are mostly of large size, the body being longish and oval, and the antennæ fixed on the front of the head, which bears in addition two large eyes. The anterior three pairs of thoracic limbs are stout and prehensile, terminating in strong curved claws, while the posterior four pairs are longer and thinner, and adapted for crawling. By means of their powerful fore feet the *Cymothoidæ* attach themselves to both marine and fresh-water fish, and have a liking for the inside of the mouth of their hosts.

Another tribe is the Epicaridea, the members of which live parasitically upon other crustaceans. The form of the body in the female is, as a rule, distorted and unsymmetrical; but the smaller males are symmetrical, and are usually found adhering to the females. No group of crustaceans seems exempt from the attacks of these parasites, but it is said that each species has its peculiar kind.

The best-known example of the tribe Asellota is *Asellus aquaticus*, distributed in fresh-water ponds and ditches all over Europe. The creature is of a greyish colour, mottled with paler markings; and the male, which is longer than the female, measures about half an inch long. The body is long, narrow in front, with a small head, and the antennæ of the second pair are about as long as the body and head taken

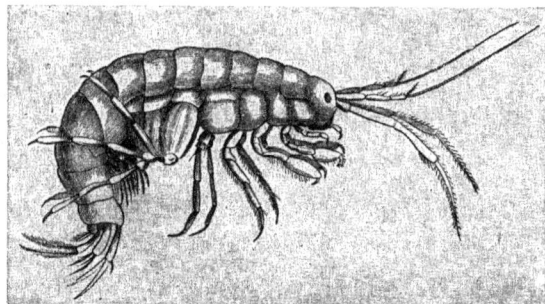

FRESH-WATER SHRIMP

together. The seven segments of the thorax are free and of large size, but those of the abdomen have coalesced into a plate, from the end of which the long, slender, forked uropods project. The seven thoracic limbs are long, slender, and increase in length from the first to the seventh.

WOOD-LICE AND ALLIES

The tribe Oniscoidea contains the wood-lice, in which the abdominal appendages are modified for breathing air. Like all crustaceans that have adopted a terrestrial life, they seem able to live only in air saturated with moisture. The body is usually broadly oval, convex above, and flat or hollow beneath, widest in the middle, and gradually narrowing towards the head and tail. The head is small, but the thorax large and seven-jointed, the abdomen being short. Representatives of this tribe are found in all quarters of the globe. A familiar British species is the sea-slater (*Ligia oceanica*), a large species living amongst the stones and rocks upon the coast above high water. The creature is nocturnal, and unless disturbed is not often seen during the day, but issues from the cracks and clefts of rocks in numbers at night.

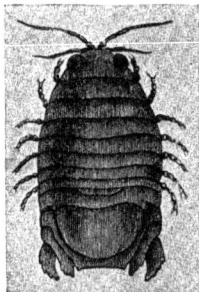

SPHÆROMA

More obtrusive are the common wood-lice *Porcellio scaber* and *Oniscus asellus*, the former distinguishable from the latter by its duller colour and the granules upon its segment; *Oniscus* being smooth and more or less variegated. Both these are rather flat, and incapable of rolling up into a ball; but the pill wood-louse (*Armadillidium vulgare*) has the dorsal surface more convex, and when handled rolls up into a ball.

In addition to its rounder shape, the pill wood-louse may be recognised by the fact that the appendages of the last abdominal segment (the uropods) do not project like a couple of small tails from the hinder end of the body. Members of this group, differing but little from the species described, are widely distributed in all temperate and tropical parts of the world.

The tribe Phreatoicidea contains the genus *Phreatoicus*, of which two species—both inhabiting fresh water—are known, one from New Zealand, and the other from Australia. The body is long and laterally compressed, the seven thoracic appendages are well developed, and the first is sub-chelate, as in many Amphipoda, and the abdomen consists of six distinct segments, with the gills attached to its appendages.

The last tribe, Chelifera, containing the genus *Tanais* and others, approaches the next order, and is distinguished by having the first pair of appendages following the jaws—that is, the second maxillipedes—pincer-like. It further differs in that the abdominal limbs are used rather for swimming than for respiration, and the breathing-chamber is situated in the posterior portion of the thorax.

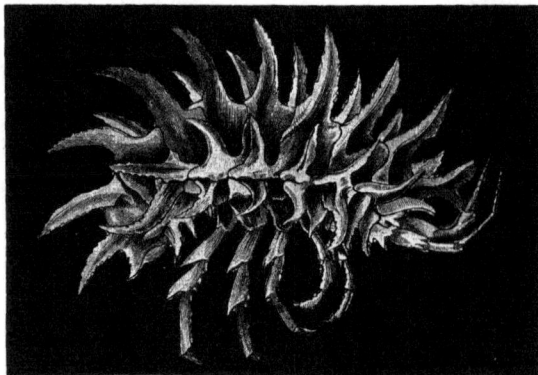

SPINY SHRIMP

AMPHIPODS

The amphipods (order Amphipoda) are allied to the isopods, but the majority are recognisable by having the body narrow and flattened from side to side, instead of broad and flattened below. Moreover, the gills are attached to the thoracic feet, and the latter, instead of being broad, leaf-like, and overlapping, are foot-like, elongate, and used for leaping or swimming.

In the suborder Gammaridea the eyes are mostly of small size, and seldom prominent. The head does not coalesce with the first segment of the thorax, and the maxillipedes have a distinct palp ; the abdomen being well developed and bearing appendages. The form of the body is shown in the illustrations of the fresh-water shrimp (*Gammarus pulex*) and the sand-hopper (*Talitrus locusta*). The latter lives near the edge of the sea, beneath seaweed or other substances which prevent the evaporation of the moisture from the sand. Sand-hoppers usually progress on land by leaps ; and although some nearly

SKELETON-SHRIMP

allied forms are found far from the sea, the majority of the Gammaridea are marine, swimming by means of the constant play of their abdominal appendages, and, when thrown on the land, wriggling helplessly along on their sides. The fresh-water shrimp is common in the streams and ditches of Europe. During the cold months of the year they bury themselves in the mud, but emerge from

PHRONIMA

SAND-HOPPER

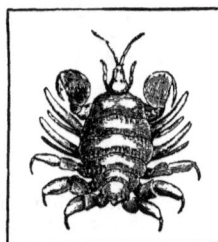

WHALE-LOUSE

their winter-quarters on the first warm days of spring.

Among deep-water forms, perhaps the most noteworthy are *Acanthechinus tricarinatus* and *Andania gigantea*. The former has developed a spiny process at almost every possible point, each of the principal segments bearing three large and pointed spines, which have their edges armed like the blades of a saw. Very different is *Andania*, which is one of the largest amphipods, reaching a length of 2 inches. Many members of the group construct tubular dwelling-places, in which they take shelter, and lay their eggs. For instance, the British *Amphitha rubricata*, which is a brilliant crimson colour, builds a nest of particles of seaweed cemented together with threads ; while another species of the same genus (*A. lit'orina*) makes a tube by cementing together the edges of a leaf of growing weed, so as to make a tube open at both ends.

In the tribe Caprellina the head has coalesced with the first segment of the thorax, and the abdomen is reduced in size, with most of its appendages wanting. The two principal families are the *Caprellidæ*, or skeleton-shrimps, and the *Cyamidæ*, or whale-lice. In

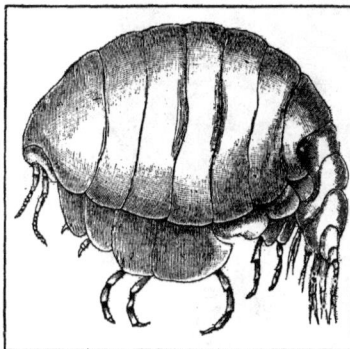

GIGANTIC ANDANIA

the former the thorax is cylindrical, and the abdomen, with its limbs, rudimentary. In the typical *Caprella* the third and fourth thoracic segments are without legs, but bear a pair of branchial vesicles ; the appendages of the second pair are developed into clusters, and those of the three last pairs are of the ambulatory type. These shrimps seldom swim, but climb amongst the branches of seaweeds and zoophytes. When at rest, they grasp the stems of the weeds with their hind limbs, and, holding the body in an erect position wave their long antennæ in search of prey. In the whale-lice, which live parasitically upon cetaceans, the short and conical head is united to the first segment of the thorax, which consists of six free, flattened segments. As in *Caprella*, the third and fourth segment of the body bear no limbs, but are furnished with a pair of gills, usually turned over the animal's back. In the female these segments carry beneath them plates, forming sacs for the eggs. The second, fifth, sixth, and seventh segments are provided with short limbs, terminating in a sharp, pointed segment, which closes against the enlarged penultimate segment as the blade of a pocket-knife closes on its handle. By means of these chelate appendages the animals fasten themselves to the skin of cetaceans, thrusting their sharp claws into the epidermis, and adhering so firmly as to be able to withstand the dash of the waves.

The members of the tribe Hyperina differ from the last by the larger size of the head, the more prominent eyes, and the absence of a palp on the maxillipedes. To this tribe belongs the large *Cystosoma*, a pelagic animal, probably retiring during the day to a considerable depth, but occasionally coming to the surface. The animal is colourless and transparent, so that by transmitted light the internal organs can be seen. The head is large and inflated, with its upper surface occupied by two enormous eyes. The genus *Phronima* also contains species attaining a considerable size, examples of the European *P. sedentaria* exceeding an inch in length. As in *Cystosoma*, the second pair of antennæ is obsolete ; the head is large, with the eyes placed upon its summit. There are seven pairs of large thoracic appendages, the third from the end forming a large and strong pincer.

The species are widely distributed, although most abundant in the tropics. Like many pelagic animals, they are translucent, and mostly live in the mantle-cavity of the ascidians *Pyrosoma* and *Doliolum*, where the eggs are laid and the young hatched.

LEPTOSTRACA

To a certain extent connecting the Malacostraca with the Entomostraca is a group of crustaceans known as the Leptostraca, and containing the three recent genera *Nebalia*, *Nebaliopsis*, and *Paranebalia*, and a number of fossil forms. The affinities of the group seem to lie with the phyllopods on the one hand, and the schizopods on the other. The body is laterally compressed, and the whole of the cephalothorax and the first four segments of the abdomen enveloped in a carapace, which springs from the head, and is formed of two movable valves, closed by a muscle. Although the eight thoracic segments are overlapped by the carapace, they are distinct and movable. The abdomen consists of eight movable segments, or two in excess of the normal number; but there are only nineteen pairs of appendages. The head bears a small, movable rostrum and a pair of stalked eyes. The two pairs of antennæ are well developed, and there are three pairs of jaws. The appendages of the thorax are foliaceous. The members of this group are marine, and widely distributed, being found in cold and warm latitudes. The female carries the eggs attached to her thoracic feet.

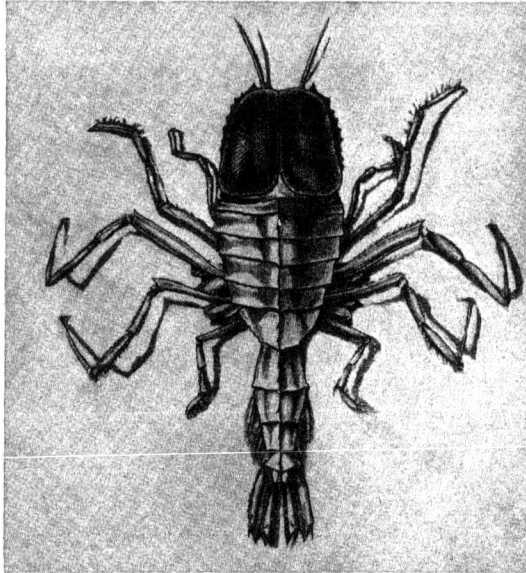

TRANSPARENT OCEAN-SHRIMP

ENTOMOSTRACA

The crustaceans of the subclass Entomostraca are small, and vary much more than the Malacostraca, from which they differ in the following features. The number of body-segments is not constant, but either greater or less than nineteen, and, as a rule, there are no appendages to the abdomen. In the majority of cases the young are hatched as a nauplius.

CIRRIPEDES

The adult members of the order Cirripedia are so unlike typical crustaceans that it can hardly be a reproach to the older naturalists that they failed to discover their affinity. Two well-known members of the order are the barnacles so frequently attached to the bottoms of ships or floating timber, and the acorn-barnacles covering the rocks on the coast. The barnacle (*Lepas*) consists of a tough longer or shorter stalk, one end of which adheres tightly by means of a cement to the timber or ship, while to the other is attached an oval compressed body encased in pieces of shell, through two of the valves of which can be protruded six pairs of slender, bristly, two-branched, filamentous limbs. These limbs, being the appendages of the thorax, keep up a constant sweeping motion, whereby particles of food are washed into the mouth that lies below them. The abdomen is undeveloped; but the rest of the body is enveloped in a fold, or mantle, supporting the outer shelly skeleton. The jaws consist of two pairs of maxillæ and a pair of mandibles, and the lower part of the head is inferiorly continued into the stalk, which contains the gland secreting the cement. If a barnacle be carefully removed from its point of attachment, the remains of the first pair of antennæ may be observed on the adhesive surface.

BARNACLES ATTACHED TO PUMICE

STALKLESS BARNACLE

When first hatched, the young are in the nauplius stage, being furnished with a median eye, and three pairs of appendages, of which the posterior two are branched. After swimming for a while by means of these appendages, the larva moults several times, and passes into a second stage, in which, with its two eyes and compressed carapace, it resembles a *Daphnia*. The rudiments of the six pairs of thoracic legs appear behind the mouth, and the first pair of swimming appendages become antenniform, each being provided with a sucker. By means of these suckers the larva fixes itself to its permanent resting-place, and the cement-gland, pouring out its secretion, glues the creature firmly to its point of attachment. Hence it follows that the fixed end of the stalk is the front extremity of the body. In the allied stalkless barnacle (*Megalasma*) the shell is attached directly to the support.

We are thus led on to the acorn-barnacles (*Balanus*), in which the entire animal is enclosed in a shell formed originally of six pieces, which grow into a tube of variable length. Some of the latter group (*Balanidæ*)—namely, the genus *Coronula*, or coronet-barnacles—attach themselves to the skin of whales. The burrowing barnacle (*Tubicinella*) has the same instinct. When adult it is long and cylindrical, consisting of a stout stony rod, marked with a series of annular ridges. This is buried deeply in the skin of whales, sometimes penetrating so far as the blubber.

These cirripedes are not true parasites, inasmuch as they do not extract nourishment from the animal to which they are attached; but many members of the group live exclusively upon other living beings, and nourish themselves at their expense. One form, for instance, *Proteolepas*, is, in the adult condition, a maggot-shaped, limbless, shell-less body, living within the mantle-chamber of other members of the same order; while the root-headed cirripedes (*Rhizocephala*) live parasitically upon the higher crustaceans. They are degenerate forms, possessing neither appendages nor segments, the body being a mere sac, devoid of alimentary canal, and absorbing nutriment by means of the root-like processes branching throughout the body of the host.

OSTRACODS

The order Ostracoda is a small assemblage, characterised by the possession of a bivalved shell, formed from the right and left halves of the carapace, and furnished with an elastic hinge to separate the valves and a muscle to keep them shut. The shell encloses the body,

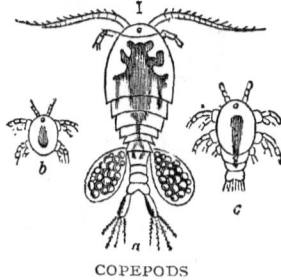

COPEPODS

a, female cyclops with egg-sacs; *b*, *c*, nauplius and later larvæ of same

which is unsegmented, has a rudimentary abdomen, and bears seven pairs of appendages—namely, two pairs of antennæ, three pairs of jaws each belonging to the head, and two of limbs attached to the thorax. These limbs, however, are stout and narrow, and, as a rule, there are no special respiratory organs.

Ostracods occur both in fresh water and the sea, the best-known forms being *Cypris* and *Cythere*. The former contains species found in ditches and ponds in England. When the waters in which they live dry up, the species of *Cypris* bury themselves in the mud until rain falls, the eggs, which are spherical, being attached to aquatic plants. The species of *Cythere* are mostly marine, haunting rocky pools on the coast, and crawling among the seaweed. In *Cypridina*, on the contrary, which is also marine, the animals dart about with velocity, the females carrying their eggs between the valves of the shell attached to their feet.

COPEPODS

In the free-living members of the order Copepoda the body is elongate and segmented, the thorax bears four or five two-branched swimming-feet, and the abdomen is without appendages. A common fresh-water form is *Cyclops*, the structure of which serves as a type of that of the order. The body is broad in front and tapering behind, being thus pear-shaped in outline. The normal five pairs of head-appendages are well developed, the first pair of antennæ being long and acting as oars. The dorsal elements of the head are fused to form a carapace, which bears a single eye in front and is behind united to the first thoracic segment, the remaining five of this region being free. The abdomen consists of four narrow cylindrical, limbless segments; but the last bears a pair of processes severally tipped with a tuft of four long bristles.

The eggs are carried by the mother in a couple of oval sacs attached to the last segment of the thorax, and so prolific are these creatures that a female, it has been calculated, will in a year produce over four thousand million young. The young, when hatched, is an oval *nauplius* (*b* in the figure), which gradually acquires the characters of the adult.

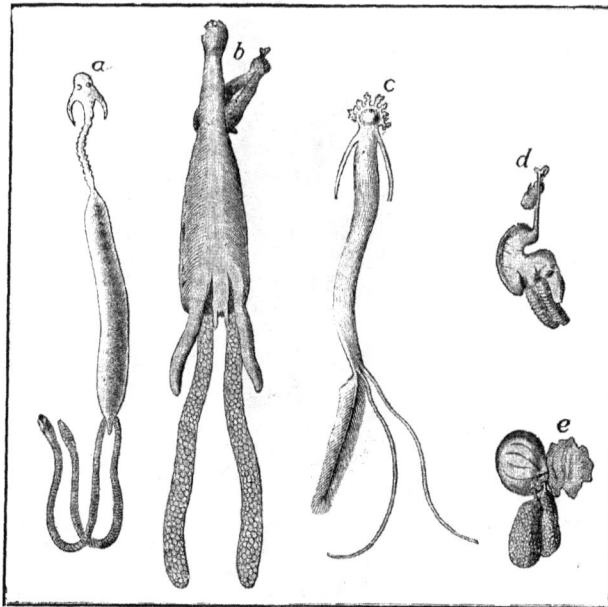

FISH-LICE

a, lernæonema; *b*, brachiella; *c*, pennella; *d*, hæmobaphes; *e*, caligus

Closely resembling the preceding is the marine *Cetochilus*, which is devoured in large quantities by whalebone whales. These crustaceans are of a bright red colour, and when seen in myriads give the sea the appearance of being stained with blood.

The copepods hitherto noticed are spoken of as the Eucopepoda, but we now come to a number of genera which have taken to a parasitic life, these Epizoa, or

Parasitica, being strangely unlike the higher forms. As one of the least modified types, may be mentioned the carp-louse (*Argulus*). Of the more degenerate types, the structure is exemplified in the annexed figure. In these the body may be broad and flat, as *Caligus* (*e*),

are passed into a brood-pouch, separating the upper surface of the thorax from the backward extension of the carapace. Here the summer eggs hatch, but the winter set are enclosed in a kind of capsule developed from part of the carapace. This capsule, called the

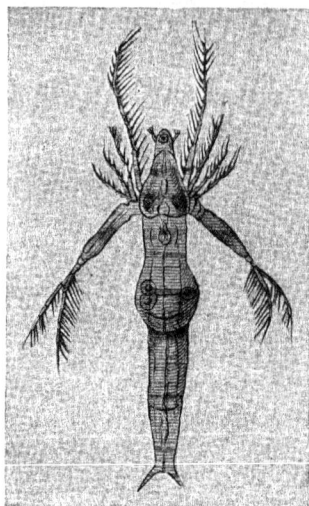
GLASSY LEPTODORA

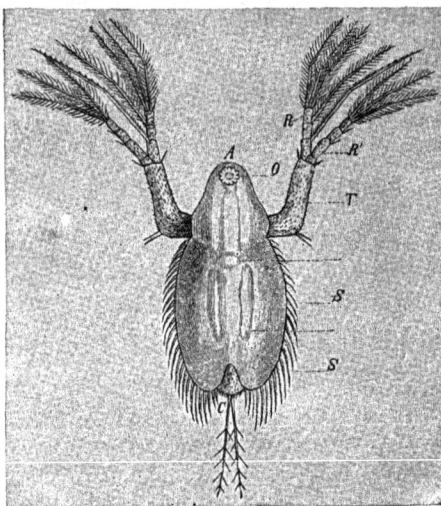
SPINY-TAILED WATER-FLEA
A, head; *C*, abdomen; *O*, compound eye; *R, T*, antennæ; *S*, carapace

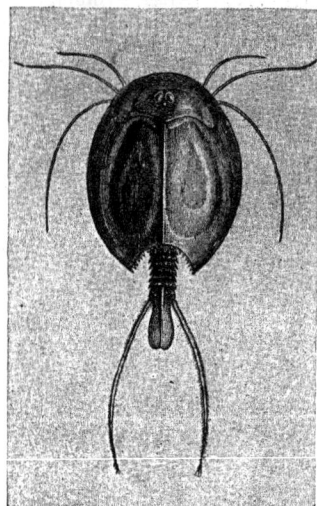
SCALE-TAILED APUS

frequently found upon codfish and brill, or long and worm-like, as in *Lernæonema* and *Pennella* (*a* and *c*), the former being a common parasite on herring and sprat, while in *Lernæa* (the gill-sucker) to which *Hæmobaphes* (*d*) is allied, the body is swollen and twisted. The two long processes represented in the figures projecting from the posterior end of the body are the egg-sacs. The appendages of the head and thorax are either absent or converted into adhesive organs.

CLADOCERA

The members of the order Cladocera take their names from the large and branched antennæ, which serve as swimming organs. They are all small, and the carapace forms a bivalve shell enclosing the greater part of the body, this carapace being an extension of the dorsal surface of the head-segments. An example of the order is the water-flea (*Daphnia pulex*), of which *Acanthocercus*, represented in the middle figure above, is nearly allied. Here the body is narrowed in front, and at the posterior end, where the carapace (*S*) is deeply notched, is the tip of the abdomen (*C*), bearing the pair of rigid barbed setæ, from which the genus takes its name. At the front of the head (*A*) is a large compound eye (*O*), and the branched and plumed appendages projecting from beneath the sides of the head are antennæ (*R, T*). The first pair of antennæ are small and simple. The jaws consist of the mandibles and the first pair of maxillæ, the second pair of maxillæ being obsolete in the adult. The thorax comprises five segments, each bearing a pair of leaf-like swimming-limbs. The abdomen consists of three segments, and is limbless.

The males of *Acanthocercus* are smaller than the females and much rarer, being generally met with in the autumn. Eggs are laid both in summer and winter, and

ephippium, is cast off with the next moult of the mother's integument. Another type is the glassy *Leptodora hyalina*, so called on account of its semi-transparency, which inhabits the open water or fresh-water lakes. The shell is so much reduced as scarcely to envelop the animal.

PHYLLOPODS

Some of the members of the order Phyllopoda are relatively large, the body being long and composed of a great number of segments, of which the thoracic, and sometimes the abdominal, are furnished with leaf-like gill-bearing appendages.

In the family *Apodidæ*, containing the genera *Apus* and *Lepidurus*, the anterior end of the body is covered with a carapace, projecting from the head over the free segments of the thorax. The hinder border of this carapace is deeply cut out, and near its front end there is a pair of contiguous compound eyes. The mouth is bounded in front by a large upper lip and behind by a deeply cleft *metastoma*, or lower lip. Both pairs of antennæ are short. The jaws consist of a pair of mandibles and two pairs of maxillæ; these are followed by eleven pairs of thoracic limbs, and there are appendages on the abdomen, sometimes numbering as many as fifty-two pairs. The last segment of the abdomen bears a pair of long filaments, and sometimes, as in *Lepidurus*, a distinct caudal plate. These crustaceans occur in the fresh waters of most countries. They swim on their backs, using the legs as paddles; and the eggs are capable of surviving long periods of drought.

a, male of Branchipus grubei; *b*, female of B. grubei; *c*, male of Artemia salina

In the second family—the *Branchipodidæ*—the body is also elongate, but there are no appendages to the abdomen, which consists of nine segments, while there are eleven pairs of thoracic appendages. These forms likewise swim upside down. Some (*Branchipus*) occur in fresh waters, but others (*Artemia*) prefer briny pools.

CLASS ARACHNIDA—SCORPIONS, SPIDERS, AND MITES

THE rest of this work will be occupied with a review of the air-breathing arthropods, or Tracheata, which are adapted, for the most part, to a life on land. The present class, though not the least specialised, is first considered because in all probability it is most closely allied to the gill-bearing arthropods already described, especially to the king-crabs and some of their extinct allies, which are often included in the Arachnida.

Indeed, it seems probable that the existing king-crabs are in many ways similar to the ancient aquatic stock from which scorpions and the like have taken origin. The whole question is an extremely difficult one, and involves elaborate anatomical details, the discussion of which would be out of place in the present work.

It will be convenient, in order to give a preliminary idea of the nature of arachnids, to describe briefly the chief characters of a scorpion. As will be seen from the figure, the body is obviously made up of rings or segments, those towards the front being closely united together. We can distinguish between three regions of the body: (1) the head; (2) the leg-bearing thorax; and (3) the extensive abdomen, devoid of obvious limbs. The first two are intimately united into a region known as the cephalothorax.

The head bears two pairs of appendages, there being a small pincer (*chelicera*) on each side of the minute ventral mouth, and behind this a pair of large pincer-claws (*pedipalps*), the bases of which work against each other, and serve as jaws. The hinder part of the cephalothorax bears four pairs of walking legs, as against the three pairs of an insect. This part of the body is provided with a pair of large eyes near the middle line, and a number of smaller ones near the front margin. Though fairly complex in structure, the eyes are not very efficient, which may be correlated with the fact that scorpions are nocturnal animals.

The abdomen consists of an anterior broader part, made up of seven segments, and a narrow, tail-like posterior part composed of five. To the tip of the latter is attached a swollen, sharply pointed sting, on the end of which opens a poison-gland, secreting a fluid that easily kills small vertebrates, insects, centipedes, and spiders, upon which the animal chiefly feeds, not chewing them, but sucking their juices. This is effected by the alternate expansion and contraction of a muscular pharynx serving as a suction-pump. To the under side of the first abdominal segment is attached a sort of flap known as the *operculum*, which is not infrequently double, and represents a pair of appendages fused together. Behind it, on the second abdominal segment, are a pair of curious comb-like structures (*pectines*), which may be regarded as organs of touch.

The entire body of a scorpion is invested in a horny coating, comparable to that of a higher crustacean, except that it is not impregnated with salts of lime. Its firm nature necessitates special breathing organs, and these consist of four pairs of lung-books, opening by narrow slits on segments three to six of the abdomen.

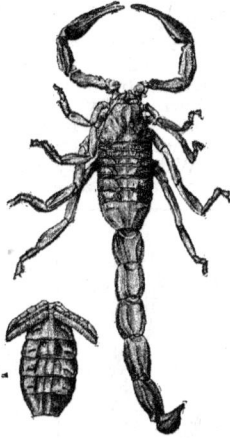

SPANISH YELLOW SCORPION

a, lower surface of abdomen, showing combs and stigmata

A section through one of them shows the presence of a number of thin, flat projections resembling the leaves of a book, and the blood circulating in these is purified by the surrounding air. The floor of the lung-book has been compared to an appendage that has fused with the body, and the whole structure may perhaps be considered equivalent to the gill-bearing appendage of a king-crab, sunk in the body, and adapted to air-breathing.

Arachnids are divided into the following seven orders : (1) Scorpions, Scorpiones ; (2) Whip-scorpions, Pedipalpi ; (3) Spiders, Araneæ ; (4) False Spiders, Solifugæ ; (5) False Scorpions, Pseudo-scorpiones ; (6) Harvestmen, Opiliones ; (7) Mites and Ticks, Acari.

SCORPIONS

Scorpions have a very wide distribution south of 40 deg. N. lat. in the Old World, and 45 deg. N. lat. in the New. They are absent from South Patagonia, New Zealand, and the Antarctic region. They differ greatly in size, 8 inches in length being the maximum. The largest are native to South India and Tropical Africa, and one of the latter (*Scorpio viatoris*) is figured. Scorpions lurk during the day in all sorts of dark crevices under stones, or it may be bark, and some of them dig small pits in the ground as hiding-places.

Certain scorpions are able to make a rasping noise by rubbing together the roughened bases of some of the limbs, these being the pedipalps and first legs in the case of the large black rock-scorpions.

WHIP-SCORPIONS

The members of the order Pedipalpi are widely distributed in South Asia, Africa, and the tropical parts of America. Although some of them are somewhat scorpion-like in appearance, they differ in many ways from true scorpions. The cheliceræ do not end in pincers, the pedipalps are strong seizing organs of varying character, and the first pair of legs are converted into feelers, which may be of considerable length.

In the whip-scorpions proper, of which a typical species is figured (*Thelyphonus caudatus*), and which are native to South Asia and the East Indies, the tail is drawn out into a slender whip-like thread, the pedipalps end in pincers, and the first pair of legs are not greatly modified as feelers.

The order also includes forms devoid of a whip-like appendage, as seen in the species represented (*Phrynus reniformis*), which is native to Tropical America. The

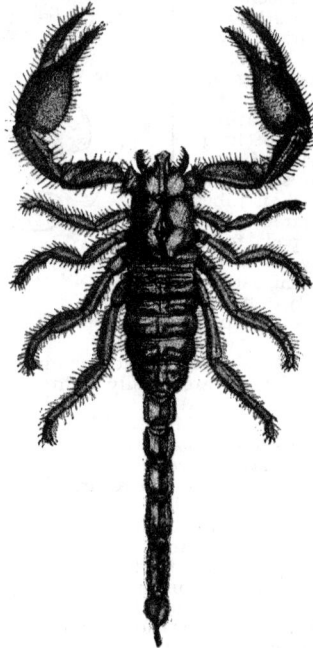

AFRICAN ROCK-SCORPION

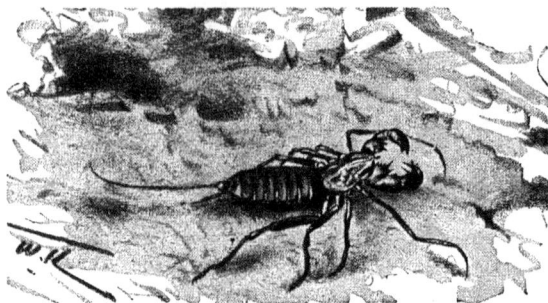

WHIP-SCORPION

greatly elongated pedipalps are not provided with pincers, but the last joint is a curved spine which can be turned down like the blade of a pocket-knife. The first pair of legs are transformed into very long, slender feelers.

TRUE OR WEB SPIDERS

Spiders (order Araneæ) constitute a large, clearly defined group, of wide distribution. They possess a rounded abdomen, which is generally soft, and marked off by a well-marked narrowing from the cephalothorax. Its segments are, as a rule, so intimately fused together as to show no traces of segmentation. The cheliceræ are short, and, like those of *Phrynus*, devoid of pincers, but with a curved end-joint that can be folded back, and upon which a poison-gland opens. The pedipalps are in the form of short feelers. Breathing is effected by two pairs of lung-books, or in the majority of cases by one pair of these structures supplemented by special air-tubes (*tracheæ*), such as are characteristic of the orders of Arthropoda to be described subsequently. On the under side of the abdomen, usually far back, are from two to six small, blunt projections, the spinnerets, from which exudes a sticky

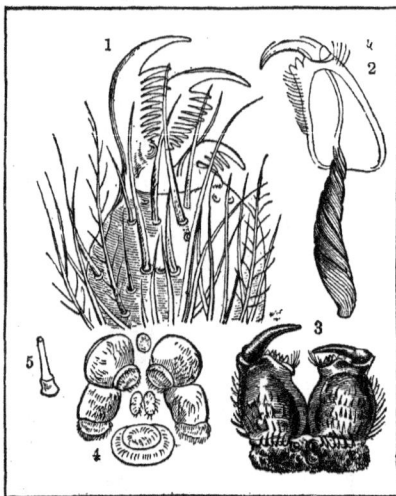

ANATOMY OF COMMON CROSS-SPIDER

1, foot with claws and hairs ; 2, mandible with poison-gland and duct ; 3, face and jaws ; 4, spinning mammillæ : 5, one of the spinning papillæ

fluid that hardens into the slender threads commonly associated into webs. The eyes are usually eight or six in number.

Of the members of this order Mr. Cecil Warburton remarks : " It is perhaps unfortunate that the obtrusiveness of particularly unattractive specimens of the race has always caused spiders to be regarded with more or less aversion. This prejudice can hardly fail to be modified by a wider acquaintance with these animals. There are certainly few groups which present points of greater interest in respect to their adaptation to special modes of life and the ingenuity displayed in the construction of their nests and the ensnaring of their prey. Spiders are wingless, yet they may often be observed travelling through the air. They are air-breathing, yet many are amphibious in their habits, and one species, at least, spends the greater part of its existence beneath the surface of the water. On land they may be found in all imaginable localities which admit of the existence of that insect life on which they depend for food."

The original use to which the power of spinning was put appears to have been the construction of a cocoon,

in which the eggs are sheltered ; while the lining with silk of a burrow, as in the trap-door spiders, and the construction of webs or snares were probably further stages in adaptation. In cases where the female has no settled home, constructing no web, but wandering about in search of prey, the cocoon is often carried about for safety, as in the figured specimen of a wolf-spider (*Parmosa amentata*).

Perhaps the most interesting use to which the spinning power is put is in the formation of " gossamer " threads by young spiders of various kinds. Probably the word gossamer was originally " goose-summer," in allusion to the fact that these floating threads are most abundant at Michaelmas, when geese are largely slaughtered. A young spider desirous of aerial transit raises its abdomen into the air, and spins a line, which is blown out by the wind until long enough to bear the spinner away. Regarding this performance Professor Krieghoff speaks as follows :

" Webs not only enable spiders to secure their prey, but also serve as a means of locomotion. In autumn young spiders begin to disperse themselves, casting out in any desired direction threads of such wondrous fineness as only to be visible in sunlight. By clinging to these with legs drawn in, these daring aeronauts are able to traverse considerable distances. The object of this procedure is probably to reach suitable winter quarters, as their weakness unfits them for the struggle for existence with older and stronger relatives in the original home. And by this instinct they unconsciously increase the area of distribution of their species, and contribute to its preservation.

" On fine autumn days we may see thousands upon thousands of these wind-borne gossamer threads, often compacted into thick, web-like masses, some sweeping through the air, some hanging from flowers or other objects, others spread out over meadows and stubble-fields, where, shining and sparkling in the sun like silver and diamonds, they produce what is known as ' wives' summer ' or ' old wives' summer.' As soon as spiders emerge from their winter quarters in spring-time the same appearance repeats itself, though in much less degree, and is

FEMALE OF WOLF-SPIDER CARRYING COCOON

then known as 'maidens' summer.' These shimmering meteors in the air, woven of dew and silver, are described by folk-lore as the cunning handiwork of elves, taught by Freya and Holda, the renowned spinners of Asgaard."

Male spiders are in many cases smaller than the females, and often brightly coloured. In some species they perform elaborate courtship dances, when these colours are displayed to the best possible advantage. But the love antics of a small male must be rather trying to the nerves, for if he fails to find favour it is by no means unlikely that the object of his affections may spring upon him and add him to her dietary.

Two suborders of spiders are recognised: (1) Segmented Spiders, Mesothelæ; (2) Unsegmented Spiders, Opisthothelæ.

SEGMENTED SPIDERS

The suborder Mesothelæ includes a few large spiders native to South-East Asia. The upper side of the abdomen is clearly segmented, and presents a series of nine hard plates. The spinnerets are eight in number and placed in the middle of the under surface of the abdomen, and breathing is effected by two pairs of lung-books.

UNSEGMENTED SPIDERS

The great majority of spiders are included in the suborder Opisthothelæ, the group being distinguished by the absence of obvious segmentation in the abdomen, to the hinder end of which the spinnerets—never exceeding six in number—are shifted.

The bird-eating and trap-door spiders (Aviculariidæ,

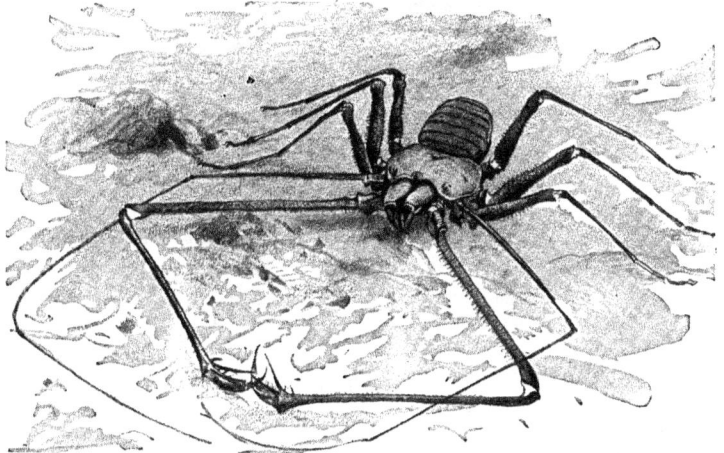

PHRYNUS RENIFORMIS

FEMALE AND MALE FIELD-SPIDERS

or Mygalidæ) live in the warmer parts of the world, and are not represented in Britain. They possess four lung-books and four spinnerets.

The bird-eating spiders are tropical and subtropical forms, of which the largest are found in the northern parts of South America, and are said to be sometimes nearly as large as rats. One species (Avicularia vestiaria) is figured. Large forms of the kind were formerly included in one genus, Mygale.

Speaking of such creatures in Equatorial Brazil, Mr. H. W. Bates says: "Many species of Mygale, those monstrous hairy spiders, half a foot in expanse, which attract the attention so much in museums, are found in sandy places at Nazareth. The different kinds have the most diversified habits. Some construct, among the tiles or thatch of houses, dens of closely woven web, which, in texture, very much resembles fine muslin. These are often seen crawling over the walls of apartments. Others build similar nests in trees, and are known to attack birds. One very robust fellow, the Mygale blondii, burrows into the earth, forming a broad, slanting gallery, about two feet long, the sides of which he lines beautifully with silk. He is nocturnal in his habits. Just before sunset he may be seen keeping watch within the mouth of his tunnel, disappearing suddenly when he hears a heavy foot-tread near his hiding-place."

And again: "At Cameta I chanced to verify a fact relating to the habits of a large hairy spider of the genus Mygale in a manner worth recording. The species was M. avicularia, or one very closely allied to it; the individual was nearly 2 inches in length of body, but the legs expanded 7 inches, and the entire body and legs were covered with coarse grey and reddish hairs. I was attracted by a movement of the monster on a tree-trunk; it was close beneath a deep crevice in the tree, across which was stretched a dense white

BIRD-EATING SPIDER

MALE AND FEMALE HOUSE-SPIDERS

web. The lower part of the web was broken, and two small birds, finches, were entangled in the pieces; they were about the size of the English siskin, and I judged the two to be male and female. One of them was quite dead, the other lay under the body of the spider not quite dead, and was smeared with the filthy liquor or saliva exuded by the monster. I drove away the spider and took the birds, but the second one soon died."

The trap-door spiders are widely distributed, and include a number of South European species. The one figured (species of *Pachylomerus*) is native to Jamaica. These forms dig burrows in the ground and line them with silk, the opening of the burrow being closed by an earthen lid attached by a silk hinge. Lurking within the burrow the spider clings tightly to the lid, preventing it from being raised without the exercise of considerable force. The burrows of some trap-door spiders are provided with two openings to the exterior, serving as front and back doors. Some species make their nests in trees, and one such form (a species of *Pseudidiops*) from Bahia, in South America, is represented in the illustration. In this case a furrow is excavated in the bark of a palm-tree, a silken tube woven in the depression, and the exposed surface of the tube covered with bits of wood and lichen to make it harmonise with the surroundings.

In the rest of the spiders to be here noticed there is only one pair of lung-books, the second pair being replaced by air-tubes. We may first mention a small kind of field-spider (*Segestria sexoculata*), which is dark in colour, and possesses six eyes. The different appearance of the male and female is shown in the figure. This spider constructs a small tubular nest in the crevice of a

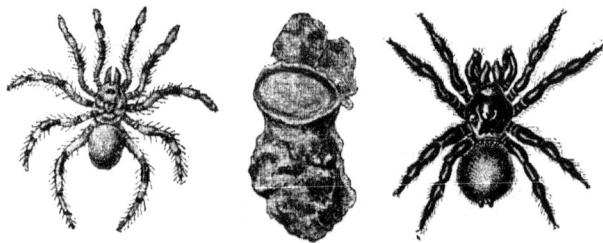

A, PALM TRAP-DOOR SPIDER, AND, B, ITS NEST; C, JAMAICAN
TRAP-DOOR SPIDER

FEMALE AND MALE CRAB-SPIDERS

wall, and threads radiating from the opening of the tube to surrounding objects serve to ensnare prey.

HOUSE-SPIDER

The common house-spider (*Tegenaria domestica*), which is also figured, has an evil reputation on account of the cob-webs it fabricates. Each of these is a sheet-like horizontal snare, behind which is a chamber in which the spider lurks. This possesses a back door, enabling hasty exit should the occupant be molested. Grains of sand or other heavy particles are attached to the under side of the lurking chamber to give it stability. This kind of spider was no doubt the commonest victim of mediæval superstition, which attributed remarkable curative powers to such creatures. An example is given by Mr. F. E. Hulme: "When a child has whooping-cough, one of the parents should catch a spider and hold it over the head of the patient, repeating three times, 'Spider, as you waste away, whooping-cough no longer stay.' The spider must then be hung up in a bag over the mantelpiece, and when it has dried up the cough will have disappeared." In times when blood-letting, was a frequent practice cobwebs were used for arresting the flow of blood, and within recent times they were esteemed as a styptic by old-fashioned persons, as the present writer can vouch from early recollections.

A larger species of house-spider (*T. parietina*), an object of peculiar aversion, is sometimes called the cardinal spider, from an obscure connection with the name of Cardinal Wolsey.

WATER-SPIDER

The water-spider (*Argyroneta aquatica*), represented in the accompanying figure, is a particularly interesting

WATER-SPIDERS AND THEIR NESTS

native species, which lives for most of its time under water, being enabled to breathe by a film of air that adheres to its hairy abdomen. It weaves a thimble-shaped water-tight nest, moored to water-plants, and filled with air brought down in successive bubbles. Within this the spider lays her eggs, and remains during winter in a dormant state. Young water-spiders often shirk the task of making a nest, using the empty shell of a water-snail for the purpose, as shown on the left of the figure.

HEDGE-SPIDER

Closely woven sheet-like webs are often to be found in hedges and on bushes in this country, and at the back will be found a tube-shaped refuge, with a back door of escape, much as in a house-spider. These are the work of the hedge-spider (*Agalena labyrinthica*), which ranges through Europe and North and Central Asia. Among its allies are certain marine spiders (species of *Desis*), which are stated to live on coral-reefs between tide-marks, and to protect themselves from the sea-water by weaving water-tight screens in front of their lurking-holes. They are said to feed upon small crustaceans.

GARDEN-SPIDER

The orb-spinners (*Epeiridæ*) are so called on account of the circular wheel-like snares they construct. Of these the most familiar is the garden-spider (*Epeira diademata*), remarkable for the size and beauty of the webs it makes in late summer and early autumn. On account of a peculiar white mark on the back, well shown in the illustration, it is known as the cross-spider in Germany.

The method of construction of the web in spiders of this and allied species is a highly interesting process. The individual threads are not, as often supposed,

ORB-SPINNER AT REST IN ITS SNARE

threads are made must be very considerable, for Mr. Cecil Warburton has succeeded in drawing out 100 yards of silk from a single individual, which finally put an end to the experiment by breaking the thread.

In making its web a spider first constructs a rough four-sided frame. Dropping from the middle of the top foundation line, she then makes a vertical diameter, climbs up to the middle of this, where a new line is fixed, and then, spinning as she goes, attaches this above to make a radial thread. By continually working from the centre, which is constantly strengthened, the spider completes the radii, or spokes of the wheel, so to speak. Beginning once more at the centre, a short spiral, the so-called "notched zone," is attached, and then a little way outside this a scaffold spiral is woven. The last operation of primary importance consists in the formation of a close spiral from without inwards, and this, being made of a sticky silk, serves to entangle the prey. The web-maker either waits for her catch in the notched zone, which is not sticky, or lurks some little distance away at the end of a signal line, the vibrations of which apprise her of captures. Small victims are bound with silk and carried bodily from the web, while larger ones are closely invested with sticky threads before being attacked.

In laying the foundation lines of a web the aid of the wind may be taken advantage of, and in this way streams and other barriers can be effectively bridged. Mr. Warburton has measured bridge-lines of 11 feet long in the common garden-spider, while tropical species are able to construct their snares across streams several yards broad.

Another species of orb-spinner figured above (*Tetragonnatha extensa*) is distinguished by its long, narrow

GARDEN OR CROSS SPIDER

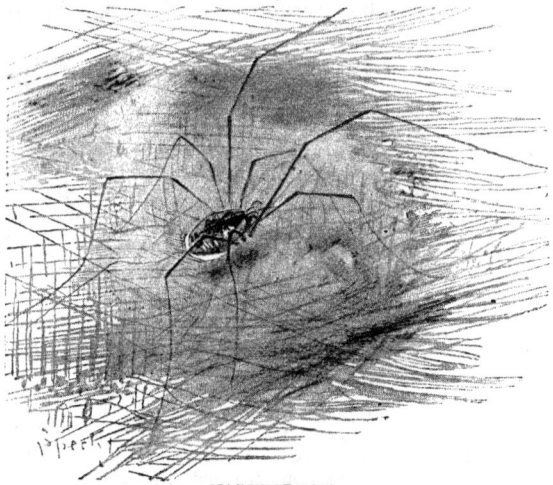

HARVESTMAN

woven from many finer ones, although such a thread is fixed by a root-like bunch of excessively fine filaments—a fact that has given rise to the misconception. The supply of the sticky material from which the

abdomen, the great size of its cheliceræ, and the fore-and-aft way in which the legs are disposed when at rest. It is said that the webs of some tropical orb-spinners are strong enough to catch small birds.

CRAB-SPIDERS

The little crab-spiders (*Thomisidæ*), so called from their sidling mode of walking, lurk in flowers for the purpose of seizing insect visitors, and are charac-

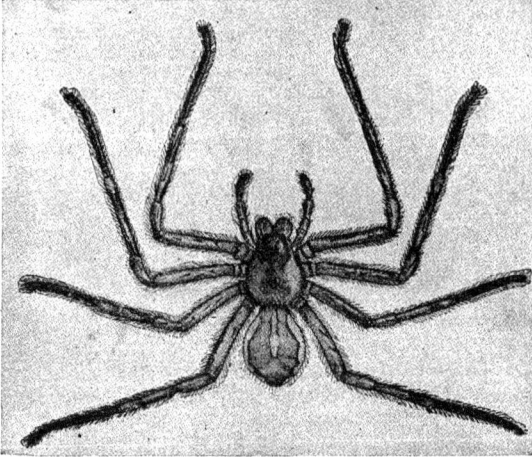

SOUTH AFRICAN CRAB-SPIDER

terised by the shortness of the two hinder pairs of legs.

In a related family (*Heteropodidæ*) similar characters are found, and the accompanying figure of a South African form (*Palystes*) will give an idea of the general form of the larger members of the family.

WOLF-SPIDERS

The wolf-spiders (*Lycosidæ*) are rapidly-moving forms, of which small species are found in this country. They hunt down their prey, and the female carries about her egg-bag with her. The newly-hatched young anchor themselves to their mother until able to fend for themselves.

A very interesting form (*Lycosa tarantula*), here figured, is one of those to which the name of "tarantula" is given in South Europe, and which were believed to be exceedingly poisonous. This species constructs a silk-lined burrow in which she lurks for prey. In the Middle Ages hysterical complaints were supposed in Italy to be the result of tarantula bites, and violent dancing was a favourite remedy. Regarding this, R. Radau states: "The sufferers . . . dance and leap until they fall exhausted. They are cured by a strongly-marked music, which excites them more and more, till it brings them to a crisis. When the disease was raging in Italy, musicians roamed the country to offer their assistance. The rapid dance they played was known by the name of 'Tarantella,' a name that reminds us that the malady was supposed to be caused by the bite of the 'tarantula,' a large and venomous spider."

RAFT-SPIDERS

The raft-spider (*Dolomedes fimbriatus*) is a large and rare South English species, representative of a distinct family (*Pisauridæ*), and named from its habit of

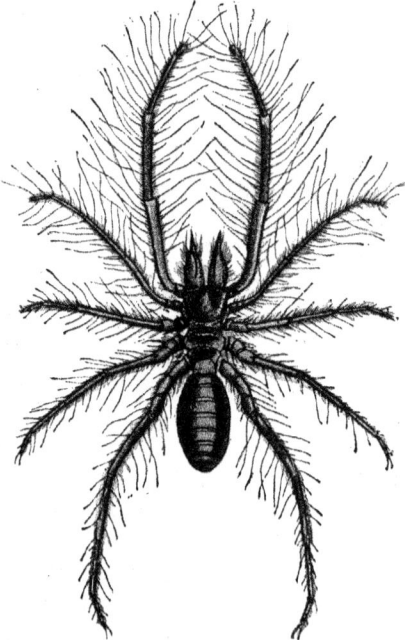

ITALIAN TARANTULA SPIDER

fastening leaves together into a raft, from which it dives in pursuit of aquatic prey. It is also able to run along the surface of the water, and the female carries her egg-bag in her cheliceræ.

JUMPING SPIDERS

Jumping spiders (*Attidæ*) constitute the largest family of the order, about four thousand species being known. Though widely distributed, their headquarters are in the tropics. Some of them are brilliantly coloured, especially the males, which often display their charms in courtship dances, elsewhere mentioned. One common British species (*Epiblemum scenicum*) is handsomely striped with black and white. Some species closely resemble ants in form, and are found associated with these insects, imitating their mode of walking, and even holding their front pair of legs in the air, which makes them look as if, like their associates, they possessed antennæ. The Australian parachuting spider (*Attus volans*), already mentioned, belongs to the same family.

FALSE SPIDERS

The false spiders (order Solifugæ) are large, hairy creatures, native to hot regions, and superficially resembling spiders, as will be seen from the figure given of a Persian form belonging to the genus

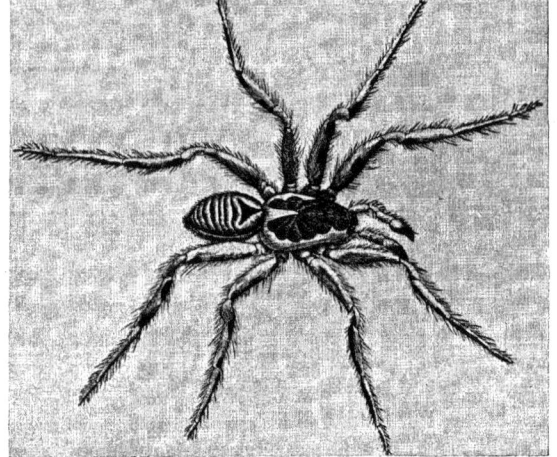

MALE OF PERSIAN FALSE SPIDER

Galeodes. They differ from spiders, however, in many ways, as, for example, in the obvious segmentation of the body, suggesting insects; in the pincer-bearing cheliceræ, and the absence of spinning organs. Breathing is entirely effected by air-tubes, there being no lung-books; and the pedipalps are very long. They are erroneously believed to be extremely poisonous.

FALSE SCORPIONS

The minute false scorpions (order Pseudoscorpiones), though widely-distributed, are seldom seen, owing to their habit of lurking in all sorts of dark corners. The most familiar species is probably the book-scorpion (*Chelifer cancroides*), common in library and museum collections, and which, as will be seen from the figure, looks something like a very small tailless scorpion, though its internal structure is very different. Spinning glands are present, but these are not comparable to those of spiders. They open on the cheliceræ, and their use appears to be to construct a sort of refuge.

HARVESTMEN

The harvestmen (order Opiliones) form a widely-distributed group of spider-like creatures, with compact bodies and very long legs. They can, however, be easily distinguished from spiders by the fact that the abdomen is fused with the rest of the body; while the cheliceræ terminate in large pincers, and there are no spinning organs. Many of them run with great swiftness, while others escape observation when danger threatens by remaining perfectly still, their colour harmonising with the surroundings.

MITES AND TICKS

The large order of mites and ticks (Acari) includes a host of forms, some fairly large, but many extremely small. They are of very considerable economic importance, as many of them attack cultivated plants, or are parasitic upon domesticated animals; while some of them harbour the germs of serious stock-diseases, such as Texas fever and louping ill.

The cephalothorax and abdomen of a mite are completely fused together, as in harvestmen, although a groove marks the boundary between them, but are still more specialised, for the abdomen has lost all trace of segmentation. In most cases the typical four pairs of legs are present, though they are occasionally reduced to two pairs.

The cheliceræ sometimes terminate in pincers, and are sometimes transformed into piercing stylets; while the pedipalps may partly fuse together into a kind of lower lip.

Mites either have no special respiratory organs or breathe by means of air-tubes. They hatch out as six-legged larvæ, a fourth pair of legs being subsequently developed. Two suborders are recognised : (1) Typical Mites and Ticks (Acarina), and (2) Worm-like Mites (Vermiformia).

TYPICAL MITES AND TICKS

The downy mites (*Trombidiidæ*) include a great variety of hair-clad, usually

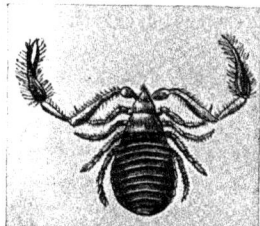

BOOK-SCORPION

active forms feeding upon the juices of plants or animals. The little harvest-mite (*Trombidium holosericeum*) is clothed with scarlet hairs, and has an evil reputation from the fact that its six-legged larva, the so-called "harvest-bug," attacks human beings, causing an intolerable itching. Related tropical species may be half an inch in length. The spinning mites belong to the same family, and one of them (*Tetranychus telarius*), the "red spider," is a serious pest on hops and some other crops.

Beetle-mites (*Gamasidæ*) are so named because some of them are parasitic upon beetles, as shown in the figure of a typical species (*Gamasus coleoptratorum*). Some troublesome pests are included in this family, as, for example, the little red fowl-mite (*Dermanyssus avium*), which abounds in dirty fowl-houses, and also attacks small cage-birds.

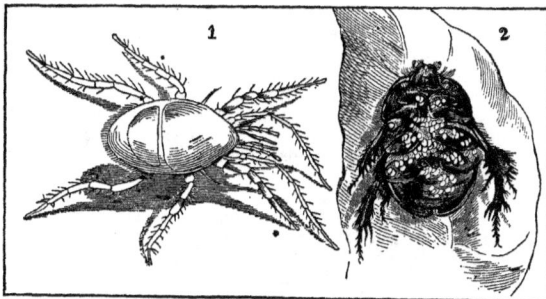

1, BEETLE-MITE ; 2, DOR-BEETLE INFESTED WITH GAMASUS

Ticks (*Ixodidæ*) are comparatively large forms, of which the females are greedy blood-suckers, attacking mammals, birds, and even reptiles, for which they lie in wait on bushes or among grass. The ordinary sheep-tick (*Ixodes ricinus*) is illustrated.

Another family (*Sarcoptidæ*) contains a number of notorious parasites responsible for mange and itch. Some of these simply prey upon the epidermal appendages, while others, like the itch-mite (*Sarcoptes scabiei*), burrow in the skin.

DEMODEX FOLLICULORUM

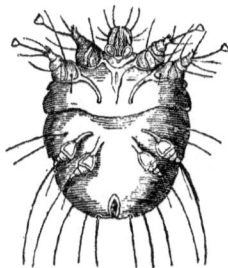

ITCH-MITE
Lower view of female

WORM-LIKE MITES

Worm-like mites (suborder Vermiformia) are so called on account of their worm-like elongated bodies. One family (*Demodicidæ*) includes a minute parasite (*Demodex folliculorum*) which lives in the hair-sheaths of the human skin. The gall-mites (*Phytoptidæ*) are also very minute, and their two last pairs of legs have disappeared. They infest various plants, and one of them, the black-currant gall-mite (*Eriophyes ribis*), is a serious pest, producing the condition known as "big bud." The mites live in the young buds, which swell up as a result of the irritation produced. Such buds either never open, or give rise to crippled shoots.

BEAR-ANIMALCULES

It is usual to describe as a sort of appendix to the Arachnida two small groups of animals of which the affinities are very doubtful. These are the Bear-animalcules (Tradigrada) and Tongue-worms (Linguatulina, or Pentastomida).

Bear-animalcules are microscopic creatures living in damp places. They have been compared to unlicked bear cubs in appearance. There are four pairs of

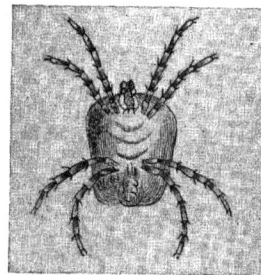

HARVEST-MITE

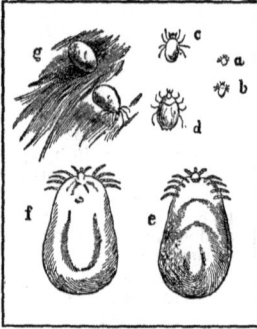

ENGLISH SHEEP-TICK

a, six-legged young; *b*, eight-legged young; *c*, male; *d*, female not distended; *e*, female distended with blood, from below; *f*, same from above; *g*, specimen clinging to the hairy integument of a mammal.

stumpy legs provided with strong claws, and a protruding snout, with a pair of horny teeth. Dr. Shipley has devised an interesting way of finding these and other small creatures that live among plants. "In searching amongst the heather of the Scotch moors for the ova and embryos of the nematodes, which infest the alimentary canal of the grouse, I have recently adopted a method not, as far as I am aware, in use before, and one which in every case has yielded a good supply of tardigrades, otherwise so difficult to find. The method is to soak the heather in water for some hours, and then thoroughly shake it, or to shake it gently in a rocking machine for some hours. The sediment is allowed to settle, and is then removed with a pipette, and placed in a centrifugaliser. A few turns of the handle are sufficient to concentrate at the bottom of the tube a perfectly amazing amount of cryptozoic animal life, and amongst other forms I have never failed to find Tardigrada."

TONGUE-WORMS

The creatures included in the group Linguatulina, or Pentastomida, are elongated worm-like parasitic forms, with two pairs of horny hooks near the mouth, and devoid of limbs, except in the embryo, where two pairs are present. One of the most familiar forms (*Pentastomum tænioidea*) lives in the nose of dogs and allied forms. The eggs are sneezed out to the exterior, and, if swallowed by a hare or rabbit, undergo further development, ultimately settling down in the liver. If an infected hare or rabbit is devoured by a dog, the adult condition is assumed.

J R AINSWORTH-DAVIS

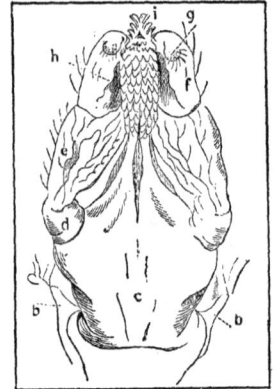

MOUTH-ORGANS OF SHEEP-TICK

c, snout; *d*, *e*, *f*, *g*, segments of palp; *i*, spiny beak, formed by fused mandibles

CLASS PROTOTRACHEATA—PRIMITIVE TRACHEATES

THE remaining groups of arthropods embrace forms which breathe by air-tubes, never by lung-books, as in some of the Arachnida, and possess a pair of feelers, or antennæ, on the head, on which account they are sometimes distinguished as Antennata.

The class Prototracheata only includes a number of widely distributed caterpillar-like animals conveniently included in the genus *Peripatus*, although it has been proposed to break this up into several others. The general appearance of these creatures may be gathered from the figure given of a Venezuelan species (*Peripatus edwardsii*). The distributional range includes the following: Africa (Cape Colony, Natal, and the Gabun) the Malay Peninsula (and possibly Sumatra), Australia, Tasmania, New Britain, South and Central America, and the West Indies. This exceedingly wide distribution of animals without marked facilities for dispersal is an indication of great antiquity, and this conclusion is fully borne out by the primitive anatomical characters of the group.

The rounded body represents no obvious traces of segmentation, except that it bears a varying

PERIPATUS EDWARDSII

number of stumpy legs, provided with pairs of claws at their tips. These are not divided into joints, as in more typical arthropods, and suggest the foot-stumps of chætopods or bristle-worms, a resemblance which is intensified by the fact that they are hollow. *Peripatus* may, in fact, be regarded as representing a transition stage between segmented worms and arthropods. In front of the first pair of legs are a pair of blunt projections, the oral papillæ, upon which open peculiar slime-glands. The mouth is surrounded by a swollen circular lip, inside which are two jaws armed with cutting blades, and representing, like the oral papillæ, a pair of modified limbs. The possession of only a single pair of jaws is peculiar to *Peripatus* among Arthropoda. The head bears a pair of blunt antennæ, and there is a fairly complex eye near the base of each of these.

The coloration of the body varies with the species, and is more or less attractive. Professor A. Sedgwick speaks of these animals as follows: "*Peripatus*, though a lowly organised animal, and of remarkable sluggishness, with but slight development of the higher organs of sense, with eyes the only function of which is to enable it to avoid the light—though related to those animals most repulsive to the æsthetic sense of man, animals which crawl upon their bellies and spit at, or poison, their prey—is yet, strange to say, an animal of striking beauty. The exquisite sensitiveness and constantly changing form of the antennæ, the well-rounded, plump body, the eyes set like small diamonds on the side of the head, the delicate feet, and, above all, the rich colouring and velvety texture of the skin, all combine to give these animals an aspect of quite exceptional beauty.

"Of all the species which I have seen alive, the most beautiful are the dark-green individuals of *Capensis*, and the species which I have called *balfouri*. These animals, so far as skin is concerned, are not surpassed in the animal kingdom. I shall never forget my astonishment and delight when, on tearing away the bark of a rotten tree-stump in the forest on Table Mountain, I first came upon one of these animals in its natural haunts, or when Mr. Trimen showed me in confinement, at the South African Museum, a fine fat, full-grown female, accompanied by her large family of thirty or more just-born but pretty young, some of which were luxuriously creeping about on the beautiful skin of their mother's back."

Peripatus breathes by means of minute air-tubes, bundles of which open to the exterior by scattered pores. In the more highly organised tracheates to be subsequently considered the air-tubes are much better developed, and open by a comparatively small number of pores, which have a regular paired

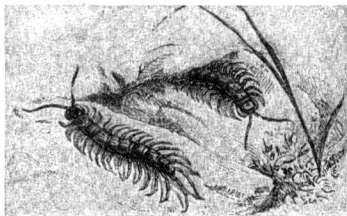

ENGLISH CENTIPEDES

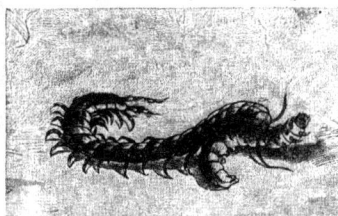

SCOLOPENDRA

arrangement. How these organs arose in the worm-like ancestors of tracheates as an adaptation to breathing ordinary air is quite unknown, but it is possible that they developed from glands in the skin.

As already indicated, *Peripatus* lives in all sorts of dark places. Its chief food would appear to consist of insects, and slime can be forcibly ejected from the oral papillæ to some distance, probably serving to entangle the prey, besides which it is most likely a means of defence.

CLASS MYRIAPODA—CENTIPEDES & MILLIPEDES

THE arthropods associated in the class Myriapoda usually possess elongated bodies, and numerous or very numerous pairs of jointed legs. The head bears a pair of antennæ, and more than one pair of jaws, are present. The air-tubes are much better developed than in *Peripatus*, and open by regularly arranged stigmata on the sides of the body.

There are two orders : (1) Centipedes, Chilopoda ; and (2) Millipedes, Diplopoda. Although the old class Myriapoda is retained here as a matter of convenience, some recent authorities consider the two orders mentioned to be entitled to the rank of distinct classes.

CENTIPEDES

The body of a centipede consists of a head, followed by a trunk consisting of from 15 to 100 segments, which for the most part are very similar. Upon the head are two long antennæ, and commonly two lateral groups of small eyes. Outside the mouth are four pairs of jaws, but only the first two pairs belong to the head, the others springing from the modified front end of the trunk. We can, in fact, arrange *Peripatus*, centipedes, and insects, in a series, marked by increasing specialisation of the body from before backwards. The jaws are : (1) biting mandibles ; (2) soft, flat, forked maxillæ, fused together at their bases ; (3) leg-like first maxillipedes ; and (4) second maxillipedes, which terminate in strong curved poison-claws.

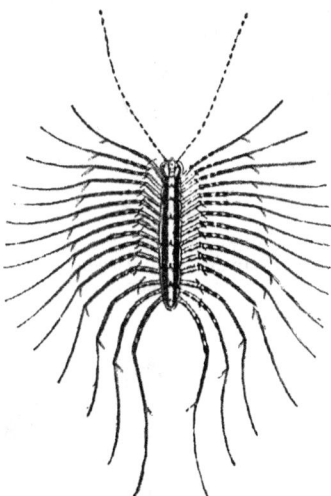

SCUTIGERA

The trunk is flattened from above downwards, and each segment is covered by horny dorsal and ventral plates, connected by a soft area of skin on either side. The well-developed legs are attached at the side to this softer region, where also will be found, with rare exceptions, the openings (stigmata) of the air-tubes. Each segment bears one pair of legs, and every other segment possesses a pair of stigmata. The total number of pairs of legs is always odd.

The first family (*Scutigeridæ*) to be mentioned includes the genus *Scutigera*, of which one species is figured. These forms are widely distributed in the warmer parts of the globe, and are exceedingly active, this being associated with unusually long legs. The antennæ are also of great length. Most centipedes avoid the light, but those belonging to *Scutigera* may often be seen hunting their prey in the sunshine, and, in accordance with this habit, possess a pair of large compound eyes. The breathing organs differ from those of other centipedes, and there is only a single series of stigmata, which open along the middle of the back.

A more familiar family (*Lithobiidæ*) includes several British species, the largest of which (*Lithobius forficatus*) is figured. It is an active, chestnut-coloured form with fifteen pairs of legs, and may be found lurking under stones. The eggs are laid singly, and rendered inconspicuous by an earthy coating.

GEOPHILUS GRAPPLING WITH AN EARTHWORM

Many of the large tropical centipedes belong to a family (*Scolopendridæ*) of which the members may be a foot in length, and are much feared on account of the severity of their poisoned bite. One of them (*Scolopendra morsitans*) is represented on a reduced scale in the figure, in the act of attacking a beetle grub.

One widely distributed family (*Geophilidæ*) includes slender forms which live underground, and feed chiefly upon earthworms, as indicated in the illustration. These centipedes are often phosphorescent, one British species (*Linotænia crassipes*) being especially notable in this respect.

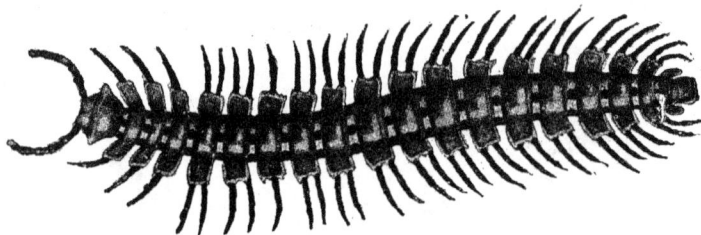

SUMATRAN MILLIPEDE

Passing mention must be made of one group of Chilopoda (*Symphyla*), because it includes short centipedes, all belonging to one genus (*Scolopendrella*), which in many respects come near the lowest insects, and probably give some idea of the ancestral stock from which insects have been derived. They are minute creatures native to North America, India, and Sumatra, and are represented in Britain.

MILLIPEDES

The sluggish vegetarian forms known as millipedes (order Diplopoda) differ in many ways from centipedes, which they only resemble superficially. The antennæ are short, and each of them usually consists of seven joints only. There are generally two groups of eyes, and two pairs of jaws are present: (1) mandibles, and (2) maxillæ, which are united together into a flattened lower lip. The body is usually rounded, and the legs are generally feeble and take origin near the middle of the under surface. Most of the segments bear two pairs, as can be seen in the figured Sumatran species (*Platyrrhachus mirandus*), and this indicates the fusion of pairs of segments, a conclusion strengthened by the fact that each apparent segment possesses a pair of stigmata. In many cases there are stink-glands, ejecting a malodorous fluid as a protective measure.

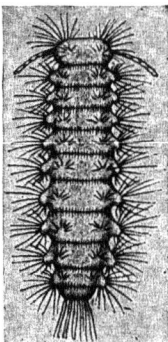

POLYXENUS

ENGLISH PILL-MILLIPEDES

One rather exceptional family (*Polyxenidæ*) includes small bristly forms which live under stones or bark, and the general appearance of which may be gathered from the illustration of Polyxenus.

The short pill millipedes (*Glomeridæ*) are able to roll up as a means of protection, as shown in the figure of an English form. The most familiar millipedes (*Julidæ*) are chiefly found in Europe and the United States. One species (*Julus sabulosus*) is figured at the bottom of this page.

Some small millipedes do considerable damage to the underground parts of certain crops, and are popularly known as "false wireworms," the true wireworms being the larvæ of small beetles.

Millipedes are an extremely ancient group, dating back to an early geological period. By studying fossil forms, and investigating the development of recent ones, it has proved possible to explain some of their peculiarities. The rounded shape, for instance, has been brought about by excessive development of the upper or dorsal surface, while the under part of the body between the attachment of the limbs, has remained comparatively undeveloped. Each apparent segment, bearing two pairs of legs, exhibits a doubleness of its internal organs, which negatives the idea of its being either a single segment or two segments intimately fused. We may regard this apparent segment therefore as a pair of segments, covered above by a single hard plate, which has resulted from the union of the dorsal plates of two original segments.

The study of extinct myriapods also demonstrates that the millipedes and centipedes have sprung from a common stock, which probably resembled in many ways *Peripatus*, that lowly organised and remarkably sluggish animal, whose characteristics are dealt with above.   J. R. AINSWORTH-DAVIS

JULUS SABULOSUS

CLASS INSECTA—INSECTS

THE term insect, although originally and, according to the meaning of the word, correctly employed in a wide sense to embrace all those animals in which the body is externally divided into a number of segments, including, of course, butterflies, beetles, bugs, spiders, scorpions, centipedes, millipedes, not to mention crabs and shrimps, is now, by common consent, used in a much more restricted sense to apply solely to such members of the Arthropoda as have only six walking legs.

In allusion to this feature the class is nowadays often called the Hexapoda, the term being much more precise and applicable than that of Insecta. In addition, however, to the possession of six legs, insects are characterised by certain other well-marked features, serving to distinguish them from all other arthropods. The body is divided into three distinct regions, arranged in a longitudinal series, and named respectively, from before backwards, the head, thorax, and abdomen.

The head, which varies much in size and shape in different groups, bears the eyes, the antennæ, and the jaws. The eyes are of two kinds, simple and compound. The latter, of which there is a single pair, situated one on each side of the head, and often so large as to occupy the greater part of its right and left half, consist externally of a multitude of lenses, often exceeding many thousands in number. The simple eyes, or ocelli, on the other hand, are fewer in number —usually only two or three—and placed upon the fore part of the head.

The antennæ are movably articulated by means of a special socket to the front of the head, usually below or near the inner edge of the compound eyes. They vary much in structure and length, being sometimes long and pliable, and composed of a large number of segments, as in the cockroach, and at other times short, like those of the house-fly, and consisting of a few segments only. There is no doubt that the antennæ contain highly important organs of sense, the bristles with which they are studded being probably tactile, and some of the other organs possibly olfactory in function.

The front edge of the head, or its lower edge when carried vertically, is often movably jointed to the rest of it, and constitutes an upper lip, or labrum. In the formation of the jaws, which are attached to the lower surface of the head, three pairs of appendages, respectively named the mandibles, the maxillæ, and the labium, are involved. But these parts are susceptible of an extreme amount of variation in structure and function, being sometimes formed for mastication, as in the mandibulate forms, such as the cockroaches and beetles, and sometimes for piercing or sucking, or both combined, as in the so-called sucking forms like the flies, butterflies, and bugs. There is no doubt that the mandibulate type of mouth, in which the gnathites, or jaws, are more foot-like in structure, is the most primitive of all. In this case the mandibles usually consist of a stout pair of one-jointed skeletal pieces, the inner edge of which is furnished with biting teeth. Sometimes, as in the males of stag-beetles, the mandibles are enormously large, and simulate horns.

The maxillæ are much more complicated in structure; each consists of a basal piece, composed of two segments—the cardo and stipes—from which spring two branches, an outer or palp, which has the appearance of a dwarfed limb, and an inner, which is, in its turn, double, the inner blade being called the lacinia and the outer the galea. The jaws of the third pair,

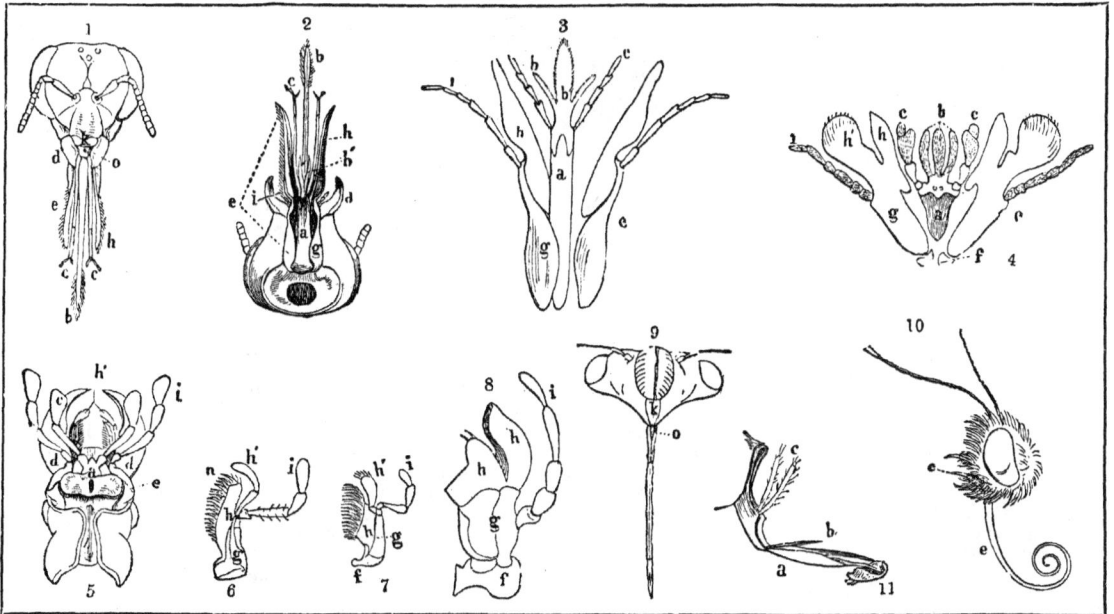

MOUTH ORGANS OF INSECTS

1, head of honey-bee, from the front; 2, head of humble-bee, from below; 3, maxillæ and labrum of a bee (Andrena); 4, maxillæ and labrum of saw-fly (Cimbex) 5, lower view of head of carnivorous beetle (Procrustes); 6, maxilla of carnivorous beetle (Cicindela); 7, maxilla of cock-tail beetle (Staphylinus); 8, maxilla of locust; 9, head of bug (Cicada), from the front; 10, head of butterfly; 11, proboscis of horse-fly (Tachina). a, mentum; b, ligula; b', paraglossæ; c, palp; d, mandible. Maxilla (e), with (f), cardo; g, stipes; h, lacinia; h', galea; i, palp; k, head-plate; n, teeth on lacinia

constituting the so-called labium, or lower lip, are constructed upon the same principle as the maxillæ, but the parts usually considered to correspond to the cardo are united to form a plate—the mentum— which is articulated by its hinder part to a sternal plate of the head, called the submentum. In front of the mentum there are externally the jointed palpi, resembling those of the maxillæ, and between these there is a median, sometimes bilobed, piece, called the ligula, and a pair of pieces termed the paraglossæ.

The degree of development of the several parts varies greatly in different orders, and it is often a matter of considerable diffi- culty to determine the exact corres- pondence that exists between them in two insects be- longing to different orders. This is especially the case when the jaws have been modified to form the different organs of suction that are met with. The structure of these will be des- cribed in detail when the species that possess them are discussed. Another organ to be mentioned in connection with the jaws is a mem- branous lobe, called the hypopharynx, or tongue, projecting into the interior of the mouth from the floor of the labium.

The thorax, or median part of the insect's body, is formed of three segments, called the prothorax, meso- thorax, and metathorax, each of which is composed of several distinct pieces. The dorsal areas of the three segments are termed the pronotum, mesonotum, and metanotum ; the lateral regions the pleura ; and the inferior regions the sterna. To the pleura are articulated the three pairs of legs, each of which con- sists primarily of five segments, named respectively, from the base to the apex, coxa, trochanter, femur, tibia, and tarsus ; the last, which constitutes the foot, being generally tipped with two claws, and subdivided into several—often as many as five—smaller segments.

To the sides of the upper surface of both the meso- thorax and metathorax are usually attached a pair of wings, which are very characteristic organs of all the higher insects, although absent in the lowest forms, and in many species degenerate through parasitic habits. The wings differ much in structure, thickness, clothing, and so on, in different orders of insects, but in all cases they seem to consist of an upper and a lower membranous layer, traversed by narrow bands of thicker material, the nervures.

The abdomen in insects is marked off from the thorax by the absence of true appendages. It may consist of as many as ten distinct segments, but never of more, and generally of fewer. Each segment is protected above by a dorsal plate, or tergum, and below by a ventral plate, or sternum, the two being connected laterally by membrane. The last segment is often provided with a pair of appendicular struc- tures, which may be long, many-jointed, and anten- niform, or short and one-jointed, like the pincers of an earwig. And, in addition to these, certain other struc- tures, such as the stings of bees and wasps, and the ovi- positors of locusts and ichneumon flies, are frequently con- nected with the hinder segments of the abdomen.

The only other external structures that need be men- tioned here are the stigmata, or aper- tures, of the respira- tory organs. These pierce the lateral surfaces of the thoracic and abdo- minal segments, and vary much in num- ber, size, and form, being generally far more plainly seen in the larvæ than in the adults. There may be as many as eleven pairs, but usually the number falls short of this.

In exceptional cases, as in the plant-lice (Aphidæ) belonging to the order Hemiptera,

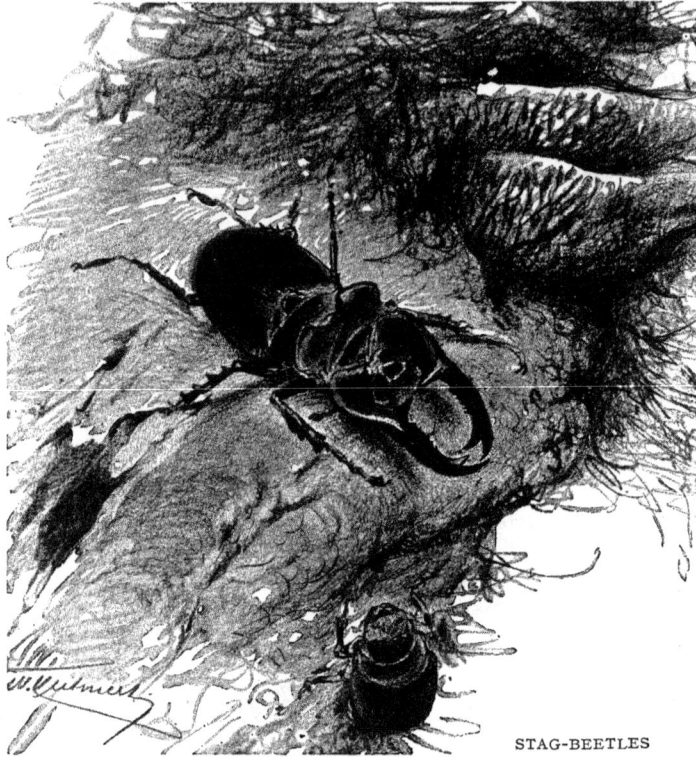

STAG-BEETLES

and in certain parasitic flies of the group Pupipara, the young are born in an advanced stage of development, the eggs developing within the body of the parent without being first deposited. But in the vast majority of species the young make their first appearance in the world in the egg-stage.

Between the time of its escape from the egg-shell and the attainment of maturity, the young undergoes a succession of moults, or castings of the skin. In some cases the change of structure that an insect presents during the course of its growth is, com- paratively speaking, trifling, the young being hatched in a condition in which in outward form it substantially resembles the parent in everything but size, and, in the case of species that bear wings in the adult, in the entire absence of these organs. A familiar instance of this method of growth is found in the cockroaches and grasshoppers, in which the young emerge from the egg as miniature and wingless copies of their parents.

In other cases, however, as in the flies (Diptera) and butterflies (Lepidoptera), an extraordinary change of form takes place during growth, the young upon hatching being so totally unlike the adult that no one unacquainted with the facts of insect development would suppose the two to belong to the same category of animals. In these two orders, as well as in some others, the new-born young has the appearance of a fleshy grub ; and the grub-like condition is retained unchanged, except in size, until the time for the last moult approaches. It then undergoes a startling change of condition, and, losing its organs of sense and ceasing to feed, passes into a state of quiescence, during which the final changes in its organisation are more or less rapidly passed through, and the final moult sets free the mature insect, perfect in all its structural details.

The immature stages of insects that present a complicated development of this kind are variously spoken of as grubs, maggots, caterpillars, or, more comprehensively, larvæ ; while the quiescent stage is termed the chrysalis or pupa, and the final sexually mature stage the imago or perfect insect. Moreover, such species are said to undergo a complete metamorphosis, or to be holometabolous, as opposed to those like the cockroach, whose growth is accompanied by but little change of form, and are said to present an incomplete metamorphosis, or to be ametabolous.

It must not, however, be supposed that all insects are either completely or incompletely metamorphic in their development. The familiar types that we have mentioned exhibit almost, although not quite, the extremes of change that are offered in the class ; but between these occur other types which show developmental phenomena more or less intermediate in their nature, being less complicated than those of the blow-fly and more complicated than those of the cockroach. An account of these various methods of development will be given under each order as it is described.

Like the Crustacea, Arachnida, Millipedes, and all the main divisions of the Arthropoda, with the exception of the Prototracheata (*Peripatus*), and possibly the Centipedes, insects are an exceedingly ancient group, having left their remains in strata of Silurian age. The exact nature and affinities of these primeval remains has not, however, yet been satisfactorily determined, and some authors, indeed, seem to doubt whether they are rightly referred to insects.

Still, there is no question that species of this group flourished in abundance during the Carboniferous period ; but the conclusion that all the known fossil insects from these strata form a natural order, distinct from all the existing groups of this rank, can hardly be regarded as finally established, seeing that, in the opinion of some authors, they are assignable to places in our classification of existing species, and are nearly related to the orders Orthoptera (cockroaches, grasshoppers, and dragon-flies) and Hemiptera (bugs and plant-lice). In the Secondary rocks insect remains, considering the small chances of the preservation of such creatures in stratified deposits, are fairly abundant ; and none of the species present ordinal differences from those which now exist. So, too, the hosts of species that have been discovered in Tertiary deposits, in the amber-beds and elsewhere, are referable to existing orders.

It has been estimated that in numbers of species insects excel all other land animals of the world taken together, and a recent computation has put the total of described forms at 250,000, and yet, according to Lord Walsingham, only about ten per cent. of existing species have hitherto been discovered. But this is not the only respect in which the animals of this class are in advance of all other groups. In brightness of colour, beauty of pattern, and gracefulness of form some of the species can hardly be equalled even by the most gorgeous birds, while in mechanical perfection of structure, as testified by activity and strength, others of the group are unsurpassed in the animal kingdom.

It has been stated that if a man could leap in proportion to his stature as far as a flea can hop, he could clear at a bound a wall over one hundred feet high, and if he could sing as loudly as the cicada, his voice could be heard for a distance of many miles. Indeed, even in matters about which man is wont to especially pride himself, such as those touching social organisation, he might with advantage go to the ant to learn wisdom, since many of the problems of modern civilisation, involved in the questions concerned in the regulation of increase of population, the proper division of labour, and the support of useless individuals, have been satisfactorily solved by many of the species of insects that live habitually in communities.

Speaking in a general way, insects may be said to be terrestrial animals, since all the species are fitted more or less completely for atmospheric respiration and for progression on the land ; many of them, in addition, being furnished with wings, which propel them through the air with amazing velocity. In many of the orders, however, as, for instance, in the beetles and bugs, there are species that have adopted an aquatic mode of life, and spend their days in

GREEN SAW-FLIES

fresh-water ponds and streams in various quarters of the globe. Others, again, like some of the gnats and dragon-flies, live in fresh water during the larval stages of their existence, but quit it on attaining maturity. Insects, too, are sometimes found on the coast beneath stones and seaweed at low water, but there are only a few species of insect that can strictly be called marine ; these are bugs such as *Halobates*, sometimes met with in numbers on the surface of the ocean thousands of miles from land.

The phenomena known as mimicry and protective resemblance are strikingly exemplified in insect life. The term mimicry is usually applied to cases where a species, otherwise unprotected, lives unmolested owing to its resemblance to another which is gifted with defensive weapons in the form of poison-glands, or with a nauseating flavour that renders it distasteful. Such species as these are usually rendered conspicuous by contrasting patches of bright colour. It is noticeable, for instance, that the patterns of bees and wasps are strikingly diversified, in order that the insects may be readily recognised and not slain by mistake for other species. Bees and wasps, then, being species that enjoy immunity from attack, are often imitated, or mimicked, by perfectly harmless flies and moths, and some beetles and animals allied to crickets similarly mimic ants. But the phenomenon of protective resemblance—or the mimicry of inanimate objects—by which a species is rendered practically invisible amongst its surroundings on account of its resemblance to a leaf, stone, twig, or bird-dropping, is of far commoner occurrence. In the coloured picture on page 106, a key to which is given on page 120, a few instances of this kind of adaptation to surroundings are portrayed. In 12, 13, and 18 are the larvæ, or caterpillars, of different species of Lepidoptera, the first two in colour and shape simulating branches, and the last a snail-shell; 1, 2, 9, and 14 are leaf-like pupæ, or chrysalids, of other kinds of Lepidoptera; while 3, 5, 7, 11, 15, 23, and 24 are the adult stages of members of the same order under different disguises. The most noticeable of these is 11, representing a large and handsome butterfly, which, when at rest with its wings folded back, exactly resembles a dead leaf, even to the midrib and stem; while 23 and 24 exhibit two small moths, which might be readily mistaken for bird-dung.

In the Orthoptera, as the insects allied to the cockroaches and grasshoppers are called, the phenomenon is carried to an extent elsewhere unsurpassed in the animal kingdom. This is well shown in the case of the leaf-insect (4), the stick-insect (8), and the leaf-like locust (10). Most of the other figures in the picture are of less importance. Attention, however, may be drawn to the water-bug (16), the young dragon-fly (6), the beetle (19), the curious bugs (20) which in attitude and colour closely approximate to the stems or bark to which they cling; 25 and 26 show two beetles resembling sheep's droppings; 17 exhibits one of the May-flies like a dead leaf, and 21 two plant-bugs which secrete threads of white wax and appear as tufts of woollen matter.

HYMENOPTERA

The general characters of the order Hymenoptera will be more or less familiar to most readers from their acquaintance with the well-known members of the wasp, bee, and ant tribes. The group includes the saw-flies, wood-borers, gall and parasitic wasps, ichneumons, ants, spider-killing wasps, solitary and social wasps, and solitary and social bees. The number of species known is from 30,000 to 40,000, though, from our knowledge of the proportion which they bear to other orders, it is computed that there may be upwards of 150,000 species yet to be discovered.

In specialisation of structure they undoubtedly rank amongst the most highly developed of the Insecta. The neat, agile frame, hard, shining integuments, stout mandibles, strong, light wings, and movable abdomen, bearing, in the case of the female, at its apex an ovipositor of great power and precision of application, or modified into an instrument for sawing and boring in some species, and in several families becoming a sting. All these features combine with a temperament of extreme nervous energy to give them a character for general intelligence, and a power of adapting means to ends such as are manifested in no other allied

GIANT DRAGON-FLY
From the restoration in Herr Carl Hagenbeck's Park, Stellingen

order. The web-making spiders alone resemble them in this respect, and we are able to find few analogies nearer than the intelligent action, individual or concerted, of man himself.

The social Hymenoptera, such as ants, bees, and wasps, have solved, on their own life-plane, industrial difficulties and social problems pressing for solution in the various societies of men. Doubtless this has been accomplished to a certain extent only at the cost of a loss of individuality such as civilised man would not tolerate for a moment. When we find that the worker ants, bees, and wasps have, during their specialisation as workers pure and simple, lost their sexual faculties, that the members of a species of Amazon ant, during their specialisation as warriors, have lost the power of even feeding themselves, being entirely dependent on slaves for their food, we may well pause before concluding that such solutions of important problems are in the end for the best, at any rate so far as concerns the human race.

The actions of insects also differ from those of human beings in being mainly instinctive, and not intelligent. An action is instinctive when performed without previous training, as the result of a given stimulus, even when it is unnecessary. Some nest-making bees go through a series of building operations of which the order cannot be varied. If for a half-built cell one of full height be substituted, the bee goes on raising it to an unnecessary height, not perceiving that part of her work has been done. Intelligence involves the power of profiting by experience.

Without entering into the more minute details of structure, the general characters by which the order may be distinguished are as follows: The possession of four transparent wings, a head, thorax, and abdomen distinct from each other, the latter joined to the thorax by a narrow stalk, or, in the case of the *Tenthredinidæ*, by a broad uniting joint. The integuments are strong, hard, shiny, and often hairy. The mandibles are well developed for biting purposes, while the subordinate mouth parts are, in the case of the honey-bees, modified to form a long, tongue-like proboscis for extracting nectar from flowers. The head is more or less globular, bearing compound eyes and several ocelli on the crown between and just behind the antennæ. Besides the mastication of food, the mandibles are used for digging holes in the ground or for gnawing timber, and various other purposes.

In some ants the soldiers have the head enormously developed, as are also the mandibles; their function being to protect the society from enemies, and also to carry on war against neighbouring communities. The antennæ are in most cases long, jointed, and filiform, constituting sensitive organs of touch and recognition. The thorax is composed of the usual three pieces, prothorax, mesothorax, and metathorax. It bears the wings, four in number, above, and the legs, six in number, beneath, the latter being modified in many species for special purposes, such as, in the bees, for gathering pollen from the blossoms of the plants visited for the sake of honey. Often the legs are armed with long spines, which in the sand-wasps materially assist in the excavation of the pits in which these insects bury their victims and deposit their eggs.

The wings are ample, strong, and light, formed of a transparent membrane strengthened with fine nervures, or veins. The arrangement of these nervures varies much in different groups, and is of importance in the classification of members of the order. The relative importance of this character is, however, not the same in every family, being in the saw-flies, perhaps, of the greatest value. Species which are wingless in one or both sexes are found in many of the families; while in the genus *Oxyura* of the family *Proctotrypidæ* the wings consist merely of a fine central stalk with a battledore-shaped plumose tip. The abdomen is united to the metathorax either throughout its whole width, as in the *Tenthredinidæ*, or, as in most of the other families, by a narrow stalk, or petiole. These two characters serve for the division of the order into the groups of Sessiliventres and Petiolata.

The organs of reproduction are situated at the apex of the abdomen; while in the female the instrument for depositing the eggs has become in the section Aculeata developed into a sting; in the *Ichneumonidæ* it is sometimes enormously long, and used for piercing the larvæ in which they lay their eggs. In the case of the large wood-borers (*Sirex*) it is used as a boring instrument, while in the saw-flies it is serrated on the edges and employed to wound the tender shoots in which the eggs are deposited. Amongst the *Pompilidæ* and some other families, the sting is used to paralyse the victim in which the insects lay their eggs, or leave in the cell to feed the larvæ as they hatch. Probably no pain is given to the victim, and, even in the case of those grubs that feed internally upon the tissues of caterpillars, in all probability less inconvenience is caused than we suppose.

In all cases the metamorphosis is complete. The egg may be laid in a cell prepared either by the female or the workers for that purpose, and the grub is fed by the attendants on a preparation of pollen or other foods specially prepared. In other cases the eggs may be laid on the foliage of trees and plants on which the larvæ feed, or they may be deposited upon or in the bodies of living or paralysed caterpillars, grubs of other species, or spiders, locusts, and the like. The *Cynipidæ* with the poison from their sting, and other causes combined, produce a large gall upon the leaves of trees, especially oaks; and on the fleshy cell-structure of these galls the grubs feed when they emerge.

Larvæ of two different kinds are met with in the order. Thus, whereas those of the saw-flies have legs, sometimes even more in number than those of the Lepidoptera, the grubs of the majority lack functional legs. The former live a life of greater liberty, feeding on the foliage of trees; the latter are free, so far as they are not confined within an egg-membrane, but being internal feeders, whether in foliage, larvæ, wood, or shut up as solitary hermits, each in its several cell passes a larval period of limited freedom. It is a curious fact that the legs of some larvæ are more evident in the early than in the latter stages, thus proving that the habit of cell-life is a comparatively recent departure from a former habit, when in all probability the larval life was passed in greater freedom.

The phenomenon of parthenogenesis is one which crops up in various orders of insects, being simply the production by the female of eggs or young without the fertilisation of the egg-germs within the female, by the stimulative elements necessary to the production of young in the higher animals. It is not, however, a chance phenomenon, appearing as a race-preserving expedient, on the sudden failure of male forms, but one of Nature's resources for preserving the continuity of species. It is constant in many species of the Hymenoptera, in the form of what is known as the alternation of generations; in some species, however, it is supposed to be the sole form of reproduction, for the males of these species have never yet been discovered. Whether we regard the fertilisation of the female egg-germs by the male elements as dynamic or stimulative, or as merely a matter of the interchange of character determinants between the two sexes, it appears to be beyond a doubt that a continuous succession of virgin-reproductions must inevitably tend to the degeneration and ultimate extinction of the race.

Parthenogenesis, or virgin-reproduction, may be of three kinds. First, resulting in the production of the male sex only; second, of the female alone; and, thirdly, in cases when the young are produced not as eggs in the first instance, but alive, as in the case of the plant-lice, or *Aphidæ*. It seems that parthenogenesis does not favour the production of one sex more than another. We should therefore be cautious how we accept too hastily the commonly received belief that male bees are necessarily the offspring of non-fertilised eggs. It by no means follows that because an egg was not fertilised that therefore the sex produced in it is the direct result of non-fertilisation. The question, however, is still a matter of controversy, and more evidence is needed before final conclusions can be reached.

COMMON TAILED-WASP AND CORN SAW-FLIES

1, Sirex juvencus, with larva and pupa; 2, Cephus pygmæus and larvæ in rye-stalks; 3, Pachymerus calcitrator, a wasp parasitic on Cephus; 4, enlarged larva and pupa of Cephus

That the members of this order are, on the whole, useful to man cannot be doubted—more useful, perhaps, than the majority of insect forms — whether as bees, with their honey - storing instincts, or as the ichneumon tribes dealing destruction to thousands of the larvæ—those insect pests which would otherwise work terrible havoc with our corn crops and garden produce. On the other hand, it must be confessed that the larvæ of the saw-flies often work damage to the foliage of forest trees, while in many tropical climates ants are a devouring scourge to all that belongs to man, and may greatly annoy man himself.

1, PINE SAW-FLY, WITH LARVÆ AND OPEN AND SHUT PUPA-CASES; 2 BROAD-BODIED SAW-FLY, WITH LARVÆ ON NEST

We must now leave these introductory lines, but, before passing on to a more or less detailed description of certain species and their peculiar characteristics of structure and of habit, the subjoined outline of classification of the various families of the order will give a general idea of the different groups, which are more obviously separated by certain broad distinguishing characters.

Order HYMENOPTERA

Suborder SESSILIVENTRES
    1. Family TENTHREDINIDÆ—Saw-Flies
    2.   ,,   SIRICIDÆ—Wood-Borers
Suborder PETIOLATA
    Section PARASITICA
    1. Family CYNIPIDÆ—Gall-Wasps
    2.   ,,   PROCTOTRYPIDÆ—Egg-Wasps
    3.   ,,   CHALCIDIDÆ—Parasitic Gall-Wasps
    4.   ,,   ICHNEUMONIDÆ—Large Larvæ-Wasps
    5.   ,,   BRACONIDÆ—Small Larvæ-Wasps
    6.   ,,   EVANIIDÆ—Hymenoptera Parasites
    7.   ,,   CHRYSIDIDÆ—Burnished Wasps
    Section ACULEATA
    1. Family FORMICIDÆ—Social Ants
    2.   ,,   MUTILLIDÆ—Parasitic Ants
    3.   ,,   THYNNIDÆ—   ,,   ,,
    4.   ,,   SCOLIIDÆ—   ,,   ,,
    5.   ,,   SAPYGIDÆ—   ,,   ,,
    6.   ,,   BEMBICIDÆ—   ,,   ,,
    7.   ,,   POMPILIDÆ—Spider-Wasps
    8.   ,,   SPHEGIDÆ—Locust-Wasps
    9.   ,,   LARRIDÆ
    10.   ,,   NYSSONIDÆ
    11.   ,,   CRABRONIDÆ—Fly and Aphid Wasps
    12.   ,,   PHILANTHIDÆ—Andrena Parasites
    13.   ,,   MASARIDÆ—Solitary Wasps
    14.   ,,   EUMENIDÆ—Mud-Wasps
    15.   ,,   VESPIDÆ—Paper-Wasps
    16.   ,,   ANDRENIDÆ—Solitary Bees
    17.   ,,   APIDÆ—Honey and Humble Bees

THE SAW-FLY GROUP

The saw-fly group (suborder Sessiliventres) contains the various species of saw-flies, and may be subdivided into the saw-flies proper (*Tenthredinidæ*) and the wood-borers, or tailed-wasps (*Siricidæ*), although it also comprises the little pith-boring *Cephidæ* and the rare and little-known species of *Oryssidæ*. The food of the larvæ of these insects consists entirely of vegetable matter. In the case of the first-named family, the leaves of trees and shrubs; in that of the second, the solid wood of various trees; and in the case of the third, the tender pith of the stalks of rye and also the shoots of pear and other trees.

Such grubs as are internal feeders are either limbless, or have at most six more or less rudimentary thoracic legs. Those, on the other hand, which live a free life and feed on foliage are very similar in general appearance to lepidopterous larvæ, from which they may be distinguished by the greater number of their legs; these varying from twenty to twenty-two, whereas those of the Lepidoptera have but sixteen at most.

They also differ by the shining and almost naked skin, and the curious habit possessed by many of curling inwards the posterior segments, raising them at the same time and depressing them with a rhythmic movement. This action, which may be for the purpose of frightening away foes, coupled with the melancholy-looking eyes, gives them a grotesque appearance, not observable in the caterpillars of the Lepidoptera, save in a few instances. When full-grown, the majority of the larvæ leave the food-plant and spin in or on the surface of the ground, or under dry leaves and moss, a barrel-shaped cocoon in which they pass the winter, turning to a chrysalis only a short time before the perfect insect emerges. At least a thousand species are known, though this is probably but a small moiety of those that exist.

STEM SAW-FLIES

The larvæ of the slender, delicate, armoured-stem saw-flies (family *Cephidæ*) pass the lives in the stems of plants or young shoots of trees; and the adults are characterised by the saw of the female being partially concealed by two integumental flaps. As an example of the typical genus, we may take the corn saw-fly (*Cephus pygmæus*), of which the perfect insect flies actively in the sunshine, flitting from blossom to blossom among buttercups in May, and thence onwards through the summer.

The larvæ cause serious damage on the Continent to rye-crops, and more rarely in wheat-fields, where they crawl up and down within the stems, feeding on the delicate tissues. When full fed, they construct a transparent cocoon in which to pass the winter, becoming pupæ, and a little later in May emerging as full-grown saw-flies. The parasitic insect (*Pachymerus calcitrator*), figured in the illustration on page 1900, is one of the petiolate Hymenoptera which seems to be exclusively parasitic on the present species.

TAILED-WASPS

In the family *Siricidæ* the female is furnished with a long, boring ovipositor for piercing the bark of trees; the eggs being laid in the orifice thus formed, and the larvæ feeding on the wood. In the accompanying illustration of the boring apparatus of one species, *c, c, a*, shows the whole of the muscular structure with which the boring is carried out. The perfect insects are usually of large size and conspicuously coloured. Among the typical forms the common tailed-wasp (*Sirex juvencus*) is a very rare species in England, although more plentiful on the Continent. The females, which are sometimes surprised in the act of depositing their eggs on pine-trees, may be easily caught, as the ovipositor can only be withdrawn with considerable difficulty. Indeed, the abdomen breaks in half if the insect be roughly grasped.

The much larger giant tailed-wasp (*S. gigas*) is far commoner among pine-trees, and is distinguished by its bands of black and yellow. Although it does considerable damage, it does not attack a perfectly healthy tree, unless recently felled. How long the larvæ may live in the interior of the tree, and how long it is before the perfect insect appears, is not known, but cases are often quoted of this insect appearing in houses soon after their completion, having evidently emerged from the wood of the joists and beams.

Another genus is well represented by the broad-bodied saw-fly (*Lyda campestris*). In this species the grubs feed on the young shoots of the Scotch fir, in which the eggs are laid. When hatched, the larvæ spin a slight web, in which they remain concealed, protruding the fore part of the body when feeding on the pine-needles. When all the needles in the neighbourhood have been devoured, the web is extended, so that a greater number of young shoots may be embraced and destroyed. The perfect insect is shining blue-black, with some of the abdominal segments reddish yellow.

TRUE SAW-FLIES

In the exceedingly numerous and widely distributed group of true saw-flies (family *Tenthredinidæ*), a well-known example is the pine saw-fly (*Lophyrus pini*), of which the larvæ are sometimes found in such numbers in pine-woods, where they feed upon the needles, that the trunks are often coloured yellow and the branches

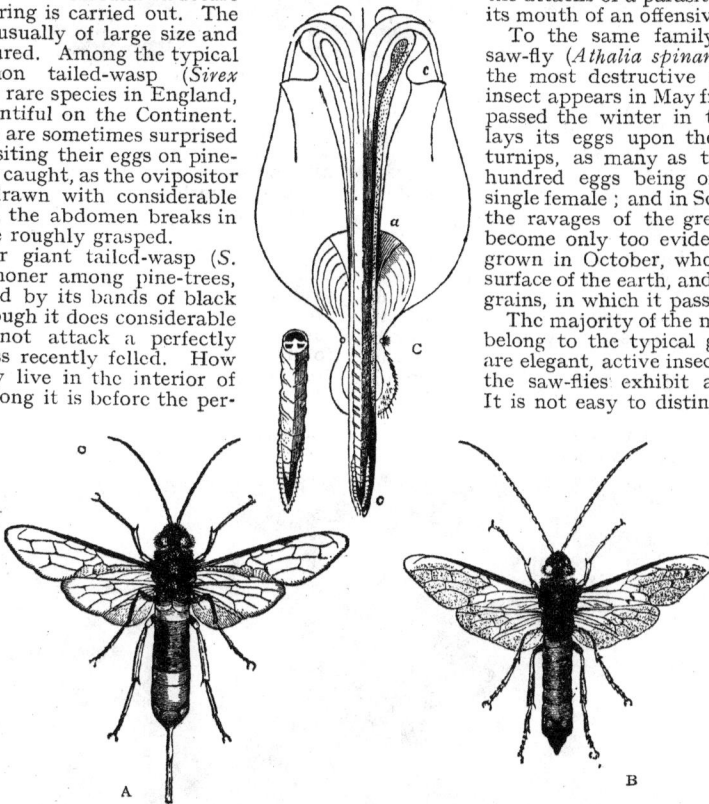

A, FEMALE; B, MALE; AND C, ENLARGED BORING APPARATUS OF GIANT TAILED-WASP

weighed down. Towards the end of July, the perfect insect emerges by gnawing off the cap of the barrel-shaped pupa-case. The eggs are laid in incisions made in the needles, these wounds being subsequently closed with a viscid secretion which protects the eggs. As many as twenty eggs may thus be deposited in a single needle. When young, and also just before turning into pupæ, the grubs are very susceptible to sudden cold or heavy rain, which will kill off thousands.

In addition to these destructive agencies, nearly forty different kinds of parasites infest the grubs, while mice devour numbers of the pupæ. The illustration shows all the stages of development, one of the grubs being drawn in the act of endeavouring to ward off the attacks of a parasite by the ejection from its mouth of an offensive fluid.

To the same family belongs the turnip saw-fly (*Athalia spinarum*), which is one of the most destructive species. The perfect insect appears in May from larvæ which have passed the winter in their pupa-cases, and lays its eggs upon the leaves of rape and turnips, as many as two hundred or three hundred eggs being often deposited by a single female; and in September and October the ravages of the green and black larvæ become only too evident. The grub is full grown in October, when it descends to the surface of the earth, and forms a cell of earth-grains, in which it passes the winter.

The majority of the members of the family belong to the typical genus *Tenthredo*, and are elegant, active insects, which alone of all the saw-flies exhibit a carnivorous habit. It is not easy to distinguish the males from the females, though the difference in the colour is of some assistance. It has been noticed, for instance, that in cases where the abdomen of the female is entirely black, that of the male is black and red. Of the green saw-fly (*T. scalaris*), the larva is common on the willow, and is pale green with black spots on the back, sometimes blending to form a central band. The pretty brush-horned rose saw-fly (*Hylotoma rosæ*), which in size and colour closely resembles the turnip saw-fly, extends throughout Europe; the larva being found from July to October on the wild and cultivated roses.

Regarding the action of the saws, Dr. Sharp says: "There is no difficulty in observing the operation; indeed old Réaumur, when speaking of the placid disposition of the saw-flies, suggests that it was given them that we may easily observe their charming operations. We cannot but regret we are unable to take so complacent a view of the arrangements of nature."

When turning to a pupa, it spins an outer meshed envelope, and a more densely woven inner one; early larvæ pupating at once, and emerging as perfect insects early in August. The later broods, however, pass the winter in the pupa case, and appear

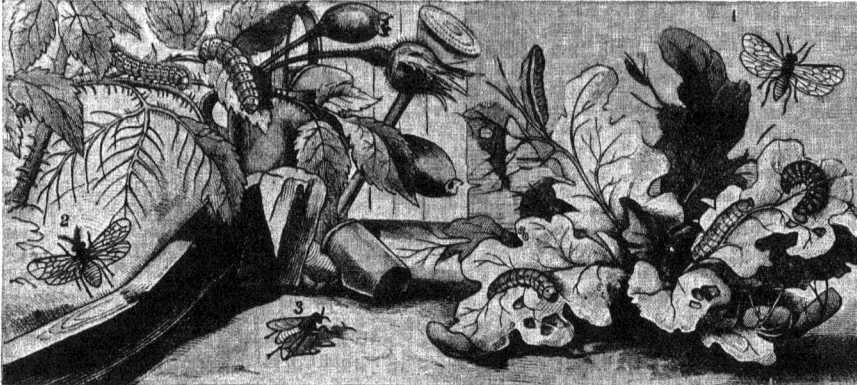

1, TURNIP SAW-FLY AND LARVÆ ; 2, 3, MALE AND FEMALE ROSE SAW-FLIES WITH LARVÆ

As to what exactly gives rise to the resultant gall, which follows sooner or later upon the wounded plant, is not known with any certainty. It has hitherto been supposed that the fly injects an irritating fluid into the wound, but recent researches tend to show that this serves rather as an adhesive security to retain the egg on the selected spot. It is probable that the different stimulative irritants offered, first by the inflicted wound, next by the presence of the eggs, and thirdly by the movements of the larva after it is hatched, together with the action of a fluid exuded by the grub itself, all tend to produce the strange modifications of cell-structure which manifest themselves in the forms of the various galls.

The larvæ of the *Cynipidæ* almost entirely feed internally upon galls produced on oak-leaves and the oak-blossoms. These galls are entirely closed, and the grub dwells within a hard cell, called the larval chamber. In some cases there may be several such chambers, as, for instance, in the bedeguar gall on the wild rose-tree formed by *Rhodites rosæ*. We have said that each species confines itself to one portion of the plant, and the form of the gall is the same ; but an exception is furnished by the galls of *Spathegaster baccarum*, which occur upon the leaves as well as on the flower-tassels of the oak.

The phenomenon known as the alternation of

in the following spring. The female makes an incision on the twigs of rose-bushes, in which she lays her eggs, after which the twig withers away.

PETIOLATA

The insects belonging to the second subdivision of the order are distinguishable from the last by the petiole, or short stalk joining the abdomen to the thorax. Sometimes this stalk is so short that the abdomen and thorax are closely united, while in others it is longer, and thus these characters form a fairly natural subdivision of the sub-order Petiolata into the pseudosessile and pedicellate forms. For general purposes they may, however be divided into Parasitica, or those in which the females are furnished with an ovipositor, and Aculeata, or those in which the ovipositor has become modified into a retractile sting.

GALL-WASPS

Of the former, or parasitic, section of the suborder, our first representatives are the gall-wasps (*Cynipidæ*), all of which are small and inconspicuous insects, varying in colour from black to brown and brownish red. The wings are furnished with few nervures, and the dark stigma on the anterior margin is absent ; while in some species the females have the wings either rudimentary or altogether wanting. Of the galls so common on the foliage of trees and other plants, some are produced by beetles, aphides, flies (gall-midges), and others by the members of the present family and some of the *Tenthredinidæ*.

In the gall-wasps each species selects some special portion of the plant for its attack, which it pierces with its ovipositor, and lays an egg in the wound.

GALL-WASPS WITH GALLS AND PARASITES

1, sponge gall-wasp, with an old gall—beneath is a new gall, whence the wasps have not yet made their exit ; 2, oak-root gall-wasp, with its gall ; 3, bramble gall-wasp, with its gall ; 4, a gall of the same slit in half ; 5, Synergus facialis, and 6, Figites scutellaris, parasites ; 7, Ibalia cultellator, parasitic on Sirex juvencus

generations—that is to say, where sexually and asexually produced generations alternate with each other—has been clearly shown to exist amongst the *Cynipidæ*. It is a remarkable fact, too, that the galls produced by a parthenogenetic female are different in form from those produced by a female originating from the normal sexual process. The insects produced by these different galls were for many years looked upon as distinct species. It is, of course, on the cell-tissues of the gall that the larvæ of the *Cynipidæ* feed and thrive ; they themselves, however, in their turn being subject to the attacks of numerous hymenopterous parasites of various kinds.

Of the typical genus, we may take the common oak gall-wasp (*Cynips folii*) as a familiar example. It is a glistening black insect, which forms an oak-gall on the under side of oak-leaves. A parasite (*Torymus regius*) lays its own egg upon the larva of the *Cynips* lying within the gall, when the latter is about half grown. Another species (*Cynips gemmæ*) is produced from conical scale-covered galls, sprouting from the young shoots of the oak, in the interior of which the grubs feed. The

stems, from which in due course emerge the perfect insects. In the figure on page 1903 is shown the oak-root gall-wasp (*Bioriza aptera*). In this form the female is wingless, but the male is unknown. The galls are formed on the rootlets of the oak-trees beneath the surface of the ground.

In the common rose gall-wasp (*Rhodites rosæ*), which produces the so-called bedeguar gall on roses, the larvæ are full-fed in autumn, although the perfect insect does not appear till the following spring. Their beautiful, mossy, pink-tinted galls furnish a home for many other insects, such as various species of *Synergus*, but especially parasites belonging to the families *Pteromalidæ* and *Braconidæ*. *Synergus facialis*, of which a figure is given in the lower illustration on page 1903, is parasitic on the gall-wasps.

OAK GALL-WASPS

1, Cynips folii on its gall ; 2, Torymus regius, its parasite ; 3, gall of Cynips gemmæ

illustration on this page shows the gall produced by insects of this species.

To the same family belongs the sponge gall-wasp (*Teras terminalis*), which emerges from many-chambered spongy galls. In spring these galls are light coloured ; but later on, when the insect has made its escape, become brown. The female insects may or may not possess wings, whereas the males are always provided with these appendages. Upwards of forty parasites have been reared from the galls of this species. Yet another familiar type is the bramble gall-wasp (*Diastrophus rubi*), which in spring produces hard and often twisted swellings on bramble-

So, too, is *Figites scutellaris*, shown in 6 of the same illustration. These are gall-wasps, so far as structure is concerned ; but as regards their habits they are in no way different from ichneumons, living in the larval state in the bodies of various insects. *Figites scutellaris*, as well as most other members of the group, is parasitic on the larvæ of the flies ; while *Ibalia cultellator* is parasitic on the larvæ of the giant saw-flies.

EGG-WASPS

The members of the obscure family *Proctotrypidæ* are minute insects, with scarcely a trace of nervures in the wings in some species, and the ovipositor protrusible and withdrawable at pleasure. Though some of the species are wholly unlike the Aculeata, yet others approach them so nearly in general characters that the present classification must be regarded as tentative. The habits of these minute insects are imperfectly known, though some are parasitic on the eggs of insects and spiders. The perfect

ROSE GALL-WASP AND ITS GALL

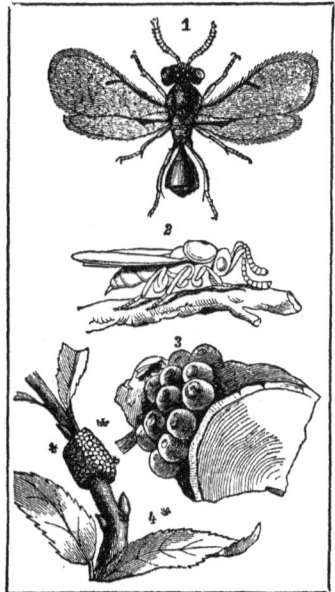

EGG-WASPS

1, Teleas læviusculus ; 2, Teleas terebrans ; 3, eggs of a moth with a Teleas upon them about to pierce and lay its eggs within ; 4, eggs

insects, small and black, with variously-shaped plumose wings, seem to prefer damp, dark localities, such as are furnished beneath fallen leaves and débris of hedges.

Here also may be placed the two species of egg-wasps (*Teleas læviusculus* and *T. terebrans*), which are both of shining black and very minute insects. They are shown in the accompanying illustration buzzing round the eggs of a moth, into which they are ready to insert their own. The females usually deposit their eggs in those of the family *Bombycidæ*, as, for instance, those of the common lackey.

PARASITIC GALL-WASPS

The family *Chalcididæ* includes a large number of small, brightly-coloured insects with metallic lustre, nearly three thousand European species being known, while the tropics have not yet furnished their contingent of species. The antennæ are always elbowed, and the wings broad, with few nervures. Some of the larvæ live in galls, devouring the grubs of the gall-wasp or those of the other inhabitants of the galls. The members of the present order, scale-insects, and plant-lice, are alike subject to the attacks of the species of this family.

One species (*Leucopsis gigas*), found in Southern Europe, lays its eggs in the larvæ of a mason-bee, which makes a cell of hard cement to protect its grub. The attacker has a boring apparatus, and it is a problem how it ascertains the whereabouts of a grub, bores through the hard masonry, and lays eggs in the inmate. The cells are not distinct, but the whole number, which are made in a sort of colony, are covered with cement, so that the task is doubly difficult. With the divining powers apparently situate in the antennæ, a suitable spot is chosen, and after, it may be, an hour or so of continuous boring, the succulent morsel is reached and the egg laid. How the wasp ascertains where the grub lies is unknown. It seems to have the power, if not of seeing, at any rate of feeling literally through a brick wall.

One of the largest members of the family is the gouty-legged wasp (*Smicra clavipes*), the egg of which is laid in the larvæ of certain water-insects. The wasp is glistening black with reddish legs, and has the wings better furnished with nervures than other members of the family. In the chrysalis-stinger (*Pteromalus puparum*) the egg is laid in the chrysalis of several common butterflies during summer, while the larvæ remain in their host all through the winter sometimes to the number of fifty.

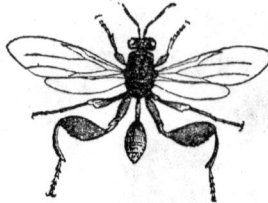

GOUTY-LEGGED WASP

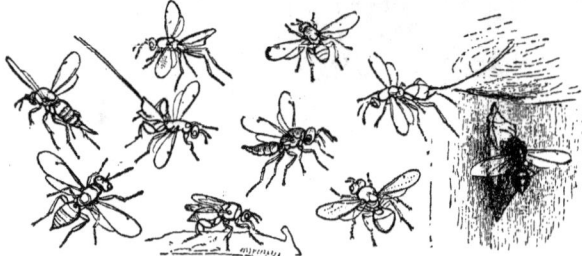

VARIOUS CHALCIDIDÆ AND THE CHRYSALIS STINGER

ICHNEUMON-WASPS

The species included in the vast family *Ichneumonidæ* number upwards of six thousand, and doubtless more remain to be discovered. The majority are parasitic on the larvæ of Lepidoptera, rendering good service to the agriculturist and gardener by holding in check the enormous quantities of larvæ hatched every year. Some, however, attack other insects, as well as spiders. The importance of these insects in forestry is also fully recognised. During pest-years they rapidly increase in numbers until what may be called the equilibrium of Nature has been regained. It has even been proposed to breed them artificially, but this appears too difficult and too expensive.

The family is distinguished by the structure of the wings, though these appendages vary too slightly to be of much value for generic or specific purposes. The antennæ are of uniform thickness, many-jointed, and, as a rule, filiform, though in some exceptional cases club-shaped. The ichneumon-wasps do not hum, either when quiescent or on the wing, and are thus enabled to approach the victim within whose body they wish to lay their eggs with a greater chance of success.

Having selected a suitable caterpillar, the female deposits an egg with her ovipositor either on or beneath its skin. The egg soon hatches, and the grub feeds upon the tissues of the larvæ until full-fed, when they pupate in or around the now almost empty skin of the caterpillar.

The family has been divided into five groups, sufficiently distinguished from each other in their typical forms, but merging into one another through transitional species. Our first example is the ichneumon (*Exenterus marginatorius*) figured in the illustration on this page, which belongs to the subfamily *Tryphoninæ*, and is found chiefly in pine-woods, where it is parasitic on *Lophyrus pini*, described on page 1901. The female attaches an egg by means of a hooklet to the skin of the green larvæ, when nearly full grown. When the insect forms its barrel-shaped pupa, in which to pass the winter, the parasite remains attached to the skin of the larva, whose tissues it gradually absorbs.

The perfect insect makes a small hole in the pupa-case when it emerges, and does not, as does *L. pini*, bite off a little cap at the top. Another type is *Bassus albosignatus*, which frequents the honey-dew dropped by aphid colonies. It lays its eggs on various larvæ which

ICHNEUMON-WASPS

1, Exenterus marginatorius, about to sting larva of Lophyrus pini ; 2, pupa-case of the latter with parasite emerged ; 3, with the proper saw-fly emerged ; 4, Bassus albosignatus, about to attack a Syrphus-larva ; 6, Banchus falcator ; 7, pupa of ichneumon

feed upon the aphides. In the allied genus *Banchus* the species are parasitic on caterpillars, especially those of the hawk-moths. The affected larvæ do not even reach the pupal state, but shrivel away, while the parasites form pupæ within the empty skin.

The members of the typical genus and subfamily, such as *Ichneumon pisorius*, are among the largest and most brightly coloured of the group ; their colours, which are white, black, red, and yellow, occurring in great variety of combination. The females are usually more brightly coloured than the males. The former sex is easily distinguished by the filiform antennæ, which are sometimes knotted, and may be observed to coil after the insect is dead. Many fine species may be taken from moss in the spring, where they hibernate, though the great majority appear in the summer, and do not live through the winter.

The European species named is one of the largest, and may be regarded as typical of the general appearance of members of the family. It is found from June onwards in pine-woods, where it attacks the larvæ of the pine hawk-moth, depositing a single egg in each victim. The caterpillar maintains its general health, and passes into the chrysalis state as though nothing were amiss ; the only difference being that a large ichneumon fly emerges instead of the expected moth. An illustration of the parasite is given in the illustration below, together with a pupa-case, with the cap removed, whence the fly has escaped.

Of the other forms here figured, the male of *Cryptus tarsoleucus* gives a good idea of the general appearance of the males of the ichneumons, with their narrow, elongate abdomen. All the species of *Cryptus* are parasitic on the larvæ of the saw-flies, and the *Bombycidæ*, the female laying several eggs in each larva. A fine handsome form is the one known as *Mesostenus gladiator*, on account of its long, needle-like ovipositor. It flies in June, and may be found in the vicinity of old crumbling walls, where bees of various kinds make their nest in the holes and crevices.

In the same illustration is figured *Ephialtes manifestator*, representing the subfamily *Pimplariinæ*. In some members of this group the ovipositor issues from a ventral cleft in the abdomen, and in others from the tip itself, the instrument being sometimes three times the length of the entire body. All the species of the genus are much alike in general appearance, the smaller kinds being parasitic on small larvæ, and the larger on

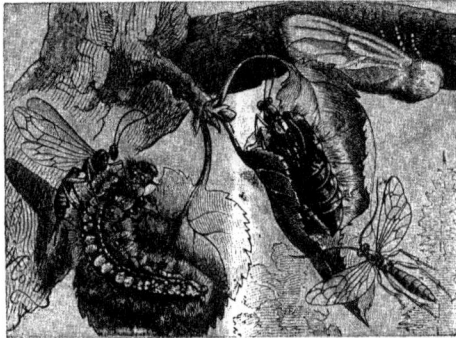

PIMPLA INSTIGATOR

Female to the left, stinging the larva of the satin-moth. To the right is the moth, beneath it the pupa, from which emerges the adult, while the male of the parasite is seen below

those of superior size. They may be seen flying about in woods in summer, in search of the wood-boring larvæ in whose bodies they lay their eggs.

With intelligent agility the female hurries over the trunk, but by what sense she ultimately detects the presence of a larva within, and directs the ovipositor straight down to the spot, it is impossible to say ; sight can be of no assistance, nor, one would judge, can touch. Can the antennæ be used, as the divining-rod is supposed to be used in the search for water, when commonsense methods have failed ? Possibly, however, the sense of smell assists, and thus the seemingly miraculous becomes once more a commonplace. The females apparently follow the borings of the larvæ, for it would be next to impossible for them to penetrate the hard fibres of the timber in which their victims burrow.

One of the commonest members of the family, and one of the largest English forms, is *Pimpla instigator*, which preys upon many species of larvæ, especially those so destructive both in gardens and the forests. The perfect insect may be seen on tree-trunks, in woods and hedgerows, searching for larvæ. The illustration represents this species attacking the larvæ of the satin-moth.

OTHER ICHNEUMON-WASPS

The members of the family *Braconidæ* are very similar in general appearance to those of the last, though the differences in the number and form of the cells enclosed by the wing-nervures forms an easy distinction. In habits, the *Braconidæ* are similar to the *Ichneumonidæ*, attacking, as a rule, the larvæ of Lepidoptera, although they are found as well in those of other insects. Upwards of a thousand parasitic grubs of the genus *Microgaster* have been taken from a single caterpillar. It must be remembered that the grubs are not in reality gnawing at the vitals, but are nourished by the fluids circulating through the system.

ICHNEUMON-WASPS

1, Ichneumon pisorius, male, and empty pupa of pine hawk-moth, whence the parasite has emerged ; 2, Cryptus tarsoleucus, male ; 3, Mesostenus gladiator, female ; 4, Ephialtes manifestator, male and female, the latter laying her eggs

As an example of the family, we may take the genus *Microgaster*, which comprises many of the commonest species. The females of all, except two, which are parasitic on *Aphides* and the eggs of spiders, attack the larvæ of Lepidoptera, especially those clothed with hair. They are themselves the victims of the attacks of a species of *Pteromalus*—a genus of Hymenoptera briefly noted above.

JAVELIN-WASPS

In the family *Evaniidæ* the abdomen is attached above the middle of the metanotum, not to its lower margin. Among these is the javelin-wasp (*Fœnus jaculator*), a species parasitical on Hymenoptera, which breed in old walls. In the typical genus *Evania* the species are believed to be parasitic on the cockroach, depositing their eggs in the egg-capsules, and this habit will account for the presence of a certain species on board ships, where cockroaches abound.

BURNISHED WASPS

The members of the family *Chrysididæ* are not easily mistaken for those of any other, being of moderate size, and distinguished by the brilliancy of their colour, not only in the tropics, but even in temperate climates. The integuments are more or less coarsely punctured, and the whole body glistens with metallic lustre, golden yellow, fiery red, blue, and green, all these being, as a rule, in combination. The perfect insects are most numerous in the summer months, and may be observed among flowers, on decaying timber and old walls. The females lay their eggs in the nest of the various burrowing Hymenoptera. It is probable that the grub devours the store of food garnered for its own progeny by the careful mother. Possibly it makes little distinction between the food supply and the tissues of the organism nourished by them. The common ruby-tailed wasps belong to this family.

The golden burnished wasp (*Stilbum splendidum*) is entirely steel-blue or golden green. It occurs on the shores of the Mediterranean, and is also found in Asia. It is one of the largest of the European forms. Among these, the burnished blue wasp (*Chrysis cyanea*) is universally distributed throughout the whole of Europe. The females lay their eggs in the larvæ of those species of Hymenoptera which make their nests in bramble stems.

The common golden wasp (*C. ignita*) may be seen flying in search of the larvæ of Hymenoptera whose burrows are made in old posts, walls, sand-pits, and other such places. Of the royal gold-wasp (*Hedychrum lucidulum*), another of the commoner and more beautiful species, a figure appears in the illustration above. In the same illustration is also shown the brazen-tailed wasp (*Elampus æneus*), of which the female deposits her eggs in the grub of a small species of the *Sphegidæ*.

SOCIAL ANTS

The ants bring us to the section Aculeata, the members of which differ from the preceding section in that the females are furnished with a retractile sting in place of an ovipositor. As a family (*Formicidæ*) ants are characterised by having the first segment of the abdomen, and sometimes also the second, reduced in size to form a stalk for the rest of the abdomen. The workers, moreover, are without wings. On account of their remarkable habits and intelligence, these insects demand a fuller notice than is accorded to other groups. As regards their visual powers, ants are very sensitive. While disliking any strong light suddenly thrown into their nests, they prefer rays transmitted through a red medium, but object more to those coming through green and yellow, while those through a violet medium they abhor.

Though sight is well developed, hearing seems much less so, vibrations of the air produced by tuning-forks, violin-strings, or whistling being little heeded. Neither has any sound emitted by the ants themselves been detected, even with the most sensitive instruments.

The sense of smell is evidently keen, for brushes dipped in scent arouse distinct curiosity. When the scent left in its track by an ant is obliterated, the ants next following are baffled, like hounds at fault, until, after a little casting about, they pick it up on the other side. In seeking for an object of whose existence and position they are aware, ants make use of both sight and smell; but it is in the latter that they place most confidence, for if the object be removed only the space of an inch from its position the ant in search of it will make a number of cross journeys over the old resting-place before it is successful. The scent, too, seems to be rather that left by former footsteps proceeding from the object itself. This sense of smell, and perhaps touch combined, is obviously manifested in the caressing or recognition of friends with the delicate antennæ.

The mysterious sense of direction is, after all, but sensitiveness to the direction in which the rays of light fall from a luminous object, and, as such, is but a form of sight. This is proved as follows. Ants made to cross a wooden bridge would, in most cases, instantly turn round, if their heads were turned in an opposite direction, by the bridge being made to rotate on a point. And they would at once lose the sense of direction if the light was shut out from the artificial track prepared for them, while if the candle were moved round in the same direction as the bridge over which they travelled, though the direction be changed, the ant does not become aware of it, because

BURNISHED AND GOLD WASPS

1, golden burnished wasp; 2, burnished blue wasp; 3, common gold wasp; 4, royal gold wasp, female; 5, brazen-tailed wasp

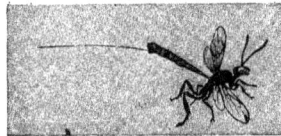

JAVELIN-WASP

MICROGASTER NEMORUM
Her larvæ are shown feeding on a large caterpillar

the rays of light fall from the same point. Nevertheless, the sense of smell is evidently the stronger, for ants carrying larvæ from a cup to the nest still continue their course, although the board on which they are travelling be turned right round. They follow the scent of former tracks rather than take notice of the direction in which the light falls.

It is obvious that without some faculty representing, at any rate, the rudiments of memory, ants would not be able to recognise even the scent left by comrades on the ground, nor would they persistently seek for an object which had been removed. They exhibit, however, all the phenomena of true memory. A fact, by repetition, becomes more firmly fixed as a sense-impression on their brains. It fades away if not refreshed. Evidence in favour of a highly developed sense of memory is furnished by the fact that ants from a certain nest were in the habit of journeying year by year, during the season of activity, to a chemist's shop, not six hundred yards distant, to a syrup-jar. It is scarcely likely that the jar was found every year by fresh ants, so that memory alone will account for the circumstance.

It is, perhaps, in the recognition of friends, however, that ants manifest the most extraordinary powers of memory. They invariably recognise a friend, while a stranger is almost instantly slain. Ants held captive for months, and returned to the nest, are recognised as lost friends, and caressed with the antennæ. This recognition might be merely a matter of the well-known odour of a friend ; but even then it must be a national smell, for it is scarcely possible that each can recognise the personal scent of every individual. Not only do they recognise the perfect ants, but even the offspring, or eggs, removed or hatched in other nests, and returned home full grown, are recognised as kith and kin, while their foster-mothers are slain. One can hardly suppose that the scent, unless such be inherited, would account for such recognition.

Whereas ants show evidence of such feelings as rage and combativeness, the emotion of sympathy is by no means as constant or intense as might be supposed from their general intelligence and power of recognising friends. Mutilated ants, and those in difficulties, are passed by on the other side ; but an intoxicated ant staggering in its tracks does not fail to excite astonishment, and is carried off as a sort of curiosity to the nest. Chloroformed ants, however, are dropped into the water, where they are of course, motionless.

That ants have the power of communicating intelligence admits of no doubt. Two ants were introduced, the one to three hundred or six hundred larvæ in one glass, the other to two or three in another glass. Each took a larva and returned to the nest. A larva was added to the second glass every time one was taken. In forty-seven and a half hours the ant which was introduced to the six hundred larvæ had brought two hundred and fifty-seven friends to help, while the other in fifty-three hours had brought but eighty-two.

The swarms of ants which in spring rise in clouds are males and females. This is their nuptial dance, and for hours they circle and sport in the sunshine. The males fall and die, or are destroyed by numerous foes. Nor is any assistance offered them by the workers, who well know that their vocation in life has been fulfilled, and they themselves are no longer of any use. The females, having divested themselves of their wings, with claws and legs, set about founding new colonies. The eggs, however, must be nursed if they are to hatch, and are subjected to much licking by the nurses. Then the larvæ must be fed ; next, they are carefully cleansed and carried for their daily walk through the lanes of the nest.

Not even after the grub has become a pupa is the ant allowed to emerge without assistance. Büchner writes that " the little creature when freed from its chrysalis is still covered with a thin skin, like a little shirt, which has to be pulled off. When we see how neatly and gently this is done, and how the tiny creature is then washed, brushed, and fed, we are involuntarily reminded of the nursing of human babies." Next, they are taught their domestic duties, and to distinguish between friend and foe. If the nest is attacked, the older and more experienced fight, while the younger members remove the pupæ to a place of safety.

Ants not only feed upon the honey-dew dropped by plant-lice upon leaves, but also rear aphide eggs, and feed the insect for the sake of their secretion. Tunnels, or covered ways, are made by some ants up the branches of the trees where the aphides live, so that the insects are enclosed and kept prisoners. Certain portions of the tunnels are enlarged to form stables, where the aphides are penned, the doors being large enough for the narrow ants to enter and leave, but not for the rotund plant-lice to escape. The "cows" are induced to part with a drop of honey-dew by a gentle stroking with the antennæ, and general encouragement of other kinds. Ants are far in advance of human dairymaids in the matter of tact in dealing with their cows. Colonies of aphides have been carried by ants to fresh pastures.

It is no long step from cow-keeping to slave-making. At least three species of ants indulge in this reprehensible practice. A raid is organised against a neighbouring nest—warriors and workers are slain and the pupæ carried off, hatched, and reared, soon to work and fight for their masters in the land of their captivity. In some cases the slaves are kept for indoor occupation, and are carried off, as part of their goods and chattels, by their masters, when they migrate into new quarters. Another species does not work at all, neither males nor females ; the workers—sterile females—capture slaves, but do no more. They neither feed their young nor make their nests—a city-state entirely dependent on slave-labour.

Not only, however, do slave-making ants engage in expeditions against other communities for the purpose of securing servants, but even many ants whose energies are confined to agriculture not infrequently

RED WOOD-ANTS

wage war for the sake of plunder on others whose habits of life are similar. An expedition of the former tribes usually consists of a general attack upon the nest of a species which they are in the habit of enslaving. Single scouts are sent out to reconnoitre, whose business it is to investigate the position of the nest and the whereabouts of the entrances. Having satisfied themselves of the feasibility of an attack, they return to their own nest, and summon forth the hosts of ferocious warriors. These, encouraging one another with taps of the antennæ, march on the unhappy colony whose baby inhabitants they propose to enslave.

Of all the warriors the most warlike are the Amazons (*Formica rufescens*), robber-ants of great size, strength, and courage. A column is formed, and, guided by the scent of their prey, as they come within the radius of their victims' pathways to and from their city, in hundreds they rush onwards. An hour, it may be, after the start, the nest is reached and entered, and soon the struggle becomes a furious battle, on the one hand to save, on the other to carry off the larvæ. Up the neighbouring trees the owners fly with their precious burdens, a harbour of refuge, secure from danger, for here the Amazons cannot follow—specialised to kill but not to climb. Others hang on the flanks of the retreating columns and harass the thieves bearing off the tender pupæ. A nurse seizes one end of her nursling, the Amazon has the other; imperceptibly the jaws of the latter steal up, still holding on, towards the far end, till the nurse's head is pierced. Sometimes the Amazon lets go, and the nurse is gone in a trice, and the pupa with her.

The slaves left behind in the city are ready to receive the plunder; and soon more slaves are hatched, whose prison is now their home, for they have never been conscious of another. But success does not always smile upon their expeditions; an entire army may lose the way, courage may fail the leaders, disputes may arise, and general unaccountable want of *esprit de corps* breaks their resolution, and the attack is abandoned. Many a warrior loses its way, emerging from the ravaged nest by passages which open to the thicket far from it they entered by.

Another robber-ant (*Formica sanguinea*), not so well furnished with offensive weapons, but larger and more intelligent than the former, also sallies forth in search of slaves. Both may meet in combat on the march, and the dead and dying, mangled remains, and heads and legs nipped off, bear witness to the consequences. These robber-ants do not attack a nest with a rush, as do the Amazons. They lay deliberate siege to it, surround it, securing the entrances and exits.

Of the other inmates of ants' nests, such as beetles, crickets, spiders, wood-lice, and the like, want of space forbids mention, and, indeed, the reason of their presence is not obvious. The supposition that they are kept as pets possibly derives support merely from the analogy drawn from similar whims amongst human beings. That ants sleep is an undoubted fact, and so, too, that they bestow much care upon their toilet, assisting each other in this respect. Bates writes that "here and there an ant was seen stretching forth first one leg and then another, to be brushed and washed by one or more of its comrades, who performed the task by passing the limb between the jaws and tongue, finishing off by giving the antennæ a friendly wipe."

Recreations, too, are not unknown to them; running after each other in hide-and-seek, followed either by a rough-and-tumble game. Stranger still, they hide away the dead bodies of their friends in chinks and crevices far

I, HONEYPOT-ANTS; 2, PARASOL-ANTS; 3, DWELLINGS OF HUSBANDMEN-ANTS

from the nest, and thus perform a sort of burial. That the habit is more than the desire to be rid of what is useless, or may be injurious, seems doubtful. The idea that ants grew corn—the so-called "ant-rice"—is a mistake.

Of the British species, the largest is the red wood-ant (*F. rufa*), represented on page 1908. It abounds in fir plantations in the southern counties of England, and the huge heaps of pine-needles it gathers over its nest are familiar objects to frequenters of the forests; while the size, ferocity, and numbers of the ants themselves become a nuisance even before their ways have ceased to be amusing. If the nest is disturbed, the fumes of formic acid burst out full in the face of the intruder, while the jaws of the enraged inhabitants render further operations impossible. Numbers of nests, however, are annually ransacked of their pupæ for young pheasants, which often seem surprised by the flavour of the ants, which they pick up with the pupæ. Highways cross the paths in every direction around the nest, and the ants may be seen coming and going continuously throughout the day, bringing in twigs, caterpillars, and fragments of all kinds of insects, to be safely stored away in the nest.

Still larger is the Hercules ant (*Camponotus herculeanus*), which inhabits wooded highlands in continental Europe, and constructs its nest in decayed tree-trunks. The female measures more than half an inch in length; and the insects, when swarming, gather in a cloud around the base of some tree. In colour the body is glistening grey, while the tips of the wings are yellow.

The honeypot-ant (*Myrmecocystus mexicanus*), of which the habits are alluded to above, inhabits the highlands of Mexico and South Colorado. The nest is constructed in the ground, usually beneath hillocks, in a gravelly soil, and contains passages and chambers arranged in different storeys, some for food, others for the larvæ, and the third for the honey-pots. The inhabitants condemned to servitude in the honey-secreting department of this community are never allowed out. An allied species is found in Australia.

Still more curious is the South American saüba, or parasol-ant (*Œcodoma cephalotes*), dreaded on account of the havoc it works amongst the foliage of plantations. Agriculture, too, becomes next to impossible where these destructive insects abound. They are not without their uses, however, for the Indians regard the females when full of eggs as a delicacy. Seizing the insects by the thorax, they nip off the luscious morsel with their teeth, much as we may see monkeys behave towards a fly. The nests of this species are prodigious. Mr. Bates speaks of hills forty yards in circumference, or about twelve yards across, while others are of even larger size. This hill, huge as it is, is merely the outer covering of a network of galleries extending deep and far into the ground, with many outlets into the surrounding country, usually carefully secured.

The workers, of which there are two forms, look after the progeny and gather food; while the soldiers, with broad heads and terrible jaws, sally forth if danger threatens their citadel. The stronger workers march in daily procession to the plantations in search of leaves, and return, each with a piece securely held in its jaws. The more slightly built remain at home, engaged in the less arduous operations of domestic economy, and rarely venture far from their nest.

These leaf-cutting expeditions are directed chiefly against coffee and orange plantations, and the ants, accompanied by a detachment of soldiers, partly, no doubt, to keep order, and more especially to guard the caravan against freebooters, march in large columns to the groves, climb the trees, and begin to reap their daily harvest. Each ant, having cut with its toothed mandibles a piece of leaf half an inch in diameter, descends the tree, holds its booty high in the air, edge upwards, and so homewards. The leaf-discs thus held above their heads have earned for these insects the name of "parasol-ants." The path they travel on is soon beaten down with footsteps, and worn till it becomes a deep groove; but even height does not end their activity and mischief, for they make raids on the houses of the planters in search of groceries and sweet-

to fifteen hundred. The females of members of the first two are wingless, while those of all three families possess a formidable poison-sting. Of the European *Mutilla europæa*, the males may be seen, though not commonly, among flowers, and frequenting foliage infested with aphides. The wingless female may, however, often be met with on sandy commons in summer. The larvæ are found in the nests of humble-bees, where they feed upon the grubs. All species of

I, FEMALE, AND 2, MALE MUTILLA EUROPÆA; 3, MALE, AND 4, FEMALE SCOLIA HEMORRHOIDALIS

the family, however, are not parasitic on humble-bees, for in South America, where the tribes of the former are scantily represented, those of the latter are numerous. Of the third family, we take as example the formidable *Scolia hemorrhoidalis*, which is found in Turkey, Hungary, Greece, and Southern Russia. Not very much is known of its habits and life-history, but such as is points to a larval life parasitic on various beetles; while other members of the family have been taken from nests of the parasol-ant. In the *Scoliidæ* the wings are present in both sexes. Figures of the male and female are given in the illustration above.

The members of the family *Bembicidæ* are distinguished from the under-mentioned *Sphegidæ* by the formation of the labrum, which is much produced. In general appearance they resemble the hornets and larger wasps. *Bembex rostrata*, figured on page 1911, is found not uncommonly throughout Europe, but becomes

I, POMPILUS NATALENSIS; 2, POMPILUS TRIVIALIS; 3, LARVA OF LATTER ON GARDEN-SPIDER; 4, PRIOCNEMIS VARIEGATUS; 5, AGENIA PUNCTUM, WITH ITS TWO CELLS

stuffs, appearing often in swarms. There are several species of this genus with similar habits, and all are known by the natives of Brazil under the single name saüba. When the leaves are stored in the nest they are overgrown with a minute fungus, upon which the ants feed.

SOLITARY WASPS

The species included in the families *Mutillidæ*, *Thynnidæ*, and *Scoliidæ* number from twelve hundred

more local in the northern countries. The insects fly in circles, with a loud hum of their powerful wings, round and round the burrows which the female makes in the loose sand or earth. Here are stowed away the bodies of large flies, reduced by an application of the sting to a state of unconsciousness; and in each nest a single egg is laid, the grub when hatched feeding upon the food which it finds placed within its reach.

RUNNERS

In the family *Pompilidæ* the males are characterised by their slender form and small size ; and both sexes may be recognised by their energetic hurrying to and fro with quivering wings and antennæ, moving rapidly on all sides, as they search sandy commons for a suitable spot to burrow in, as well as for the spiders which they numb with a sting and store up for the larvæ. The members of the family are universally distributed, but are larger and more brilliant in tropical countries. Some make their nests in the beetle-borings of old trees and posts, and prey upon all kinds of insects and their larvæ ; others prey exclusively on spiders, and confine their burrowing operations to sandy soils.

Not only do spiders of the family *Lycosidæ*, which run freely on the surface of the ground, but make no nest, fall victims to the *Pompilus*, but the *Epeiridæ* are snatched from the very centre of their maze and carried off, their powers of resistance rendered futile by one paralysing stroke of the poisonous sting. Well are these spiders aware of the danger, for they drop instantly from their webs into the herbage when the hum of wings warns them of the near presence of a wasp. Others, however, whose staple food consists of bees and wasps, are not so easily alarmed, and learn to distinguish between friends and foes.

The figured *Pompilus natalensis* is of considerable service in Natal, since its habit is to search every nook and cranny for house-frequenting spiders. Up and down the windows, in and out amongst the rafters, the female passes to and fro in search of the large spiders which lodge in their webs hung up amongst the woodwork. The victims, when captured, are buried with the egg in a hole in some suitable corner within or without the house. A large species of this genus attacks spiders of the genus *Lycosa* on English commons, and buries them in a somewhat similar fashion. The second species figured in the illustration (*P. trivialis*) also attacks spiders, especially *Lycosa inquilina*.

PERFECT-STINGERS

Many of the handsome insects belonging to the family *Sphegidæ* are uniformly black, black and red, or yellow and black. The majority, however, are black with brilliant yellow or white markings, and shine with the lustre of burnished metal. These markings are very variable even in the same species, rendering their identification difficult for the student, though on account of that contrast of colour and the activity of their movements the members of this family are amongst the most attractive of all hymenopterous insects. Some species prey upon lepidopterous larvæ, others on grasshoppers, while another provisions its nest with three or four crickets. These latter, however, are not captured without a severe tussle. The *Sphex* leaps upon the cricket's back, delivers a couple of stings, and all is over.

The numerous members of the family *Crabronidæ* are usually black with yellow markings. Their nests are formed either in the ground or in decaying timber ; the tunnels of wood-boring beetles being utilised in the latter case. While the smaller species feed chiefly on aphides, the larger kinds are more partial to flies. Figures of three species—viz., *Crossocerus scutatus*, *C. elongatulus* and *Crabro patellatus*—are given in the illustration below. Another form is *Mellinus arvensis*, usually met with in pine-woods, where it may be seen searching about on the sandy soil, and is particularly fond of the honey-dew deposited by aphides. A smaller form (*M. sabulosus*) is likewise shown in the illustration.

The same illustration also shows *Trypoxylon figulus*, a black insect, which may be observed throughout the summer flying busily to and fro among posts and decaying trees. A variation in the mode of making its cell will be noticeable. Selecting a long tunnel, the female brings in aphides or small spiders, lays an egg, deposits a suitable supply of food, and fits on the top a wad of mud ; above this again another cell is constructed, similarly capped with mud, and so on till the tunnel is full.

ANDRENA PARASITES

As an example of the family *Philanthidæ* may be taken *Philanthus triangulum*, the larva of which feeds upon the honey-bee and other members of the same group. In the accompanying illustration a figure of this species is given. Since at least five bees are provided for each larva, the havoc caused in hives where these insects abound must be considerable. A separate nest, in some warm, sunny slope, is made for each egg. Another form is *Oxybelus uniglumis*, figured in the

I, MALE, AND. 2, FEMALE OF MELLINUS ARVENSIS ; 3, M. SABULOSUS ; 4, BEMBEX ROSTRATA ; 5, PHILANTHUS TRIANGULUM ; 6, MALE, AND, 7, FEMALE OF CERCERIS ARCUARIA ; 8, TRYPOXYLON FIGULUS ; 9, FEMALE, AND, 10, MALE OF CRABRO PATELLATUS ; 11, MALE CROSSOCERUS SCUTATUS ; 12, C. ELONGATULUS ; 13, OXYBELUS UNIGLUMIS

illustration. In this species the female excavates tunnels in sandy ground, to which the sunshine has free access, and flies are mainly used to provision the nest, as a rule, one only to each cell. The fly is attacked from above, knocked down, stung in the neck, and carried off to the nest. A third form (*Cerceris arcuaria*), shown in the same illustration, is a black insect with yellow bands on the abdomen, as are most of its kindred.

WASPS AND BEES

Before taking into consideration the families into which wasps and bees are divided, it is advisable to give an account of some points connected with their habits, as well as a notice of their special senses. As regards sight, the large size of their compound eyes, in addition to the presence of ocelli, indicates their high degree of visual power. In respect of perception of colour, experiments have shown that if honey be placed on cards of different colour, bees show a decided pre-

though it flew many times straight to a certain conspicuous leaf close above the booty, doubtless a landmark, yet it could not for a long while—and after repeated pounces in the wrong direction, and more it seemed by good luck at last—succeed in finding it.

No one who has heard the cry of an angry wasp, and experienced the pain which has followed, will doubt that anger and malice have their places in the wasp's nature. Often do these insects seem to make straight for an innocent bystander, and sting from pure

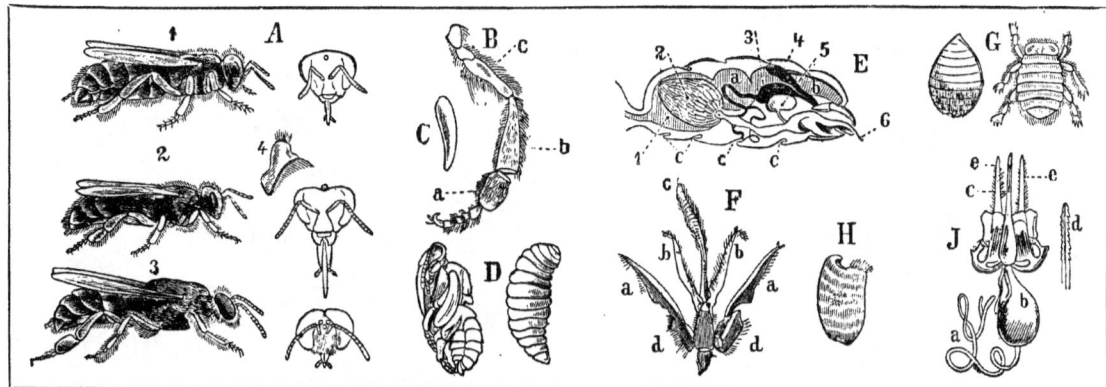

INMATES OF A HIVE

A—1, queen ; 2, worker (non-fertile female) ; 3, drone or male ; 4, mandible from outside ; B, hind-leg of worker ; c, thigh ; b, shank ; a, first tarsal joint. C, egg. D, larva and pupa. E, longitudinal section of abdomen of a worker ; 1, honey-crop ; 2, egg-sac ; 3, poison-sac ; 4, oil-gland ; 5, semen-sac or spermatheca ; 5, sting ; c, segmental interstices, whence the wax issues. F, mouth-parts ; a, maxillæ ; a, basal joint of same ; b, labial palpi ; c, tongue. G, bee-louse and its pupa. H, brush. J, poison-apparatus ; a, poison-gland ; b, poison-vesicle ; c, sting-groove ; d, sting ; e, sting-sheath

ference for special tints, blue and purple being the prime favourites. Similarly, no doubt, the colours of flowers have a greater or smaller degree of attraction for these insects. Indeed, it is beyond question that the fertilisation of flowers by the visitation of bees has tended to the development of the special colours patronised by the insects, while blossoms which were of less favourite hues have gradually disappeared. Black, white, and green flowers are not so common as yellow, orange, blue, or red ; and black is less prevalent than either of the others.

Although experiments to prove or disprove the sensibility of bees to sound have so far been negative, yet from the fact that they are exceedingly sensitive to a certain peculiar cry occasionally emitted by the queen, which acts like an electric shock, it would appear that hearing is likewise well developed. That bees and wasps are able to find their way, and to fly off apparently without hesitation straight for home, needs no proof. But this power does not necessarily indicate some mysterious sense of direction, enabling them to perceive their bearings by occult means. Rather may it be looked upon as due to the ordinary observance of conspicuous landmarks, such as are utilised for guidance even by man himself. Bees, for instance, have been taught the way to a store of honey by the repetition of single experiences, proving that they pass from the unaccustomed to the well-known little by little.

Naturally, the direction of a point to which whithersoever they may wander out, they must invariably return many times a day, soon passes from the sphere of calculation and enters the region of simple intuition ; so rapid and unconscious are the various acts of perception involved. That these insects do thus take note of landmarks has been shown by Bates, who describes how a sand-wasp carefully marked the spot where half of a larva had been left by circling round and alighting in the vicinity. And even then, when it returned,

spitefulness. Sympathy for the ailing and wounded, as among the ants, so among the bees, seems to be more noticeable than it is towards those actually in distress—though uninjured. It has been doubted, indeed, whether bees show any affection for one another ; the caressing antennæ, as well as the personal attention to each other so noticeable in the case of ants, are certainly lacking. As in ants, however, the antennæ seem to be the chief organs of communication.

As regards habits, there are two chief operations in which bees and wasps engage—namely, the procuring of food and the rearing of a progeny. This food is of two kinds—honey gathered from the nectaries of flowers, and bee-bread, or flower-pollen moistened with honey, kneaded by the workers, and stored away, for feeding the larvæ. The workers, or honey-gatherers, do not bring in more than one sort of pollen at the same time ; and when the nurses, or domestic bees, receive the pollen from the honey-gatherers they keep it carefully separate. One sort of pollen is more nutritious than another, and a female larva fed on the more nutritious bee-bread will become a queen, or fertile female, and one hive cannot afford more than a few of such luxuries. Those fed on the less nutritious bread turn out workers, or non-fertile females. For the males special conditions are arranged by the queen when laying the eggs. Royal cells, framed for the feeding of queens, are much larger than those for workers.

In secreting wax for the cells, bees, having eaten as much honey as they can conveniently carry, hang in a cluster from the top of the hive. Soon the wax begins to burst from glands beneath the edges of the segments of the body, and is rubbed off with the legs. Cell-construction now begins, and in addition to the wax, a sort of resinous cement, drawn from the sap of conifers, is used to strengthen the walls at their angles, and also to cover the inside of the hive. The six-sided form of the cells of the honey-bee appears to have been evolved after ages of gradual modification from the simple cylinder which would be formed by a cylindrical body—as that of a bee—moulding wax around itself ; this form alone admitting of the greatest number of cells being placed side by side, and tier by tier, without leaving waste, vacant spaces between. The greater the number of the cells, the stronger the colony ; the stronger the colony, the more numerous the swarms, and the greater the chance of the perpetuation of the race.

The intermediate form between the cylinder and the regular hexagon is found in the comb of the Melipona bee, which forms cylindrical cells, but so close together that the partition wall becomes a flat plate, since it is impossible for a thin sheet to be concave on both sides at once. Modifications from this form, combined with modified instincts, would eventually produce a regular hexagon. It is to be borne in mind, however, that this form arises not because the bees are aware that a regular hexagon is the most economic form of cell they can adopt, but simply because, when a group of bees stand close to each other, and form cells of pliant wax—whose walls break through at all points on account of their proximity, rendering it necessary to build up a flat wall between—they cannot fashion it in any other way. For at all points of a single cell, six bees at the sides and six bees below are constantly encroaching and fitting in the sides and corners of their own cells, around that of each single bee.

Bees have proved in practice what to the mathematicians is inevitable in theory. Nevertheless, bees are not compelled to form their combs in this or that way without any power of adaptation to special circumstances. They construct their comb, and hang their connections wherever the holding seems likely to be most secure, and thus, on a less complicated plane of intelligence, carry out precisely what human beings accomplish under more complex conditions—namely, they adapt means to ends. The difference is one of degree, not of kind.

The fact that eggs are laid by a single female of unusual size is noteworthy. Bee colonies, however, unlike those of the social wasps, are permanent, hibernating during the winter. Each wasp colony or

above, larger cells are set apart for the queens, the difference between these and the non-fertile females being entirely brought about by the difference in food. This, however, is not the case with the males, and it is a disputed point whether the queen can control the sex of any particular egg, or whether she can select a male egg as she proceeds with the laying.

Certain it is, at any rate, that when she reaches a drone, or male, cell, which is larger, she deposits an

HORNETS

egg which will become a male. It has usually been asserted that unfertilised eggs become males, while those which are fertilised turn out females. This may be the case, and certainly would tend to bear out the general truth that absence of nutrition tends to give the male element greater preponderance in the progeny, though the immediate physical conditions on which the sex of the offspring depends are imperfectly known.

A superabundance of males is, as a rule, associated with failing provisions and loss of bodily energy, and this is born out by the fact that when the queen is old she is apt to lay too many drone eggs. This, however, is a failing which the community cannot put up with, and if the queen be unable to produce profitable offspring she is put to death. Still, both bees and queens well know that the one

MUD-WASPS

1, Odynerus parietum, female and nest ; 2, Chrysis ignita ; 3, Polistes gallica, female and nest

nest originates from a single female, which survives through the winter, and by herself lays the foundation of a new colony. Among bees a certain number of workers, or non-fertile females, are set apart as maids-in-waiting, who attend to the queen's wants in the matter of food, which are considerable during the period of laying.

A single egg is laid in each cell, and, as mentioned

supreme calamity which can befall the bee community is to be left without a queen, not because they need her rule, but because on her alone rests the future of the colony. And it has been asserted that if two queens only remain, and are contesting for the mastery, and each should simultaneously have the chance to deliver a sting which might prove fatal to both, each releases the other, dreading to leave the hive queenless.

Inseparable from these phenomena is that of swarming, or the budding-off of new colonies from the mother-hive. Owing to the instinct of the workers, who can arrest or accelerate development by regulating the food supply, a new queen is always ready when a swarm of bees is prepared to leave the overcrowded hive. This queen is, however, not permitted to leave her cell till the actual moment of flight, and all along has to be protected from the reigning queen, by whom, if opportunity were afforded, she would be killed. Indeed, when the swarming season is over, the actual sovereign is permitted to make short work of all her rivals.

The function of the nurses, as their name implies, is to rear the young, and, if necessary, preserve the queens. After the males, or drones, have fulfilled their duties, they are massacred in thousands by the workers, even the young grubs and pupæ being dragged from their cells and killed. In many wasp societies these matters are, however, more leniently arranged, since the males usually assist in the general duties of the colony. Still, even these exhibit an unaccountable habit, all the grubs and pupæ being dragged out and slain as winter approaches. Whether the wasps themselves begin to experience the pinch of hunger, and wish to close mouths which must otherwise starve, or what may be the motive for such action, is beyond our ability to guess. Since every wasp, save here and there a large female, or queen, perishes at the approach of winter, the massacre cannot be justified on the score of prudential social policy.

HERMIT WASPS AND MUD-WASPS

The true wasps may be conveniently divided into hermit and social wasps, although there is more or less complete transition between the two. Of the typical hermit wasps (*Masaridæ*), which are mostly tropical forms, and constitute a link between the parasitic wasps described above and the *Vespidæ*, but little is known. Some kinds are, however, parasitic, and possibly many may be so. On the other hand, the *Eumenidæ* are hermit wasps which make their nests chiefly in mud walls or sandstone cliffs, some constructing a series of mud cells in the hollow stems of plants, and supplying their grubs with caterpillars for food. A well-known European example is the figured *Odynerus parietum*, a variable insect, making its appearance in May and June. The nests are made in holes of old mud walls, or the banks of clay-pits, and are filled with grubs of beetles belonging to the family *Chrysomelidæ*, or with the caterpillars of small moths.

SOCIAL WASPS

The members of the family *Vespidæ* form a link between the foregoing and the true bees, since each species includes a fertile female, or queen, infertile females, or workers, and males, or drones. The nests are formed of a kind of paper manufactured from the dry parings of old posts and trees. Since we have already dealt briefly with the general habits of the *Vespidæ*, further reference to them, save as occasion for their mention arises in the course of subsequent description of species, will be unnecessary. The members of the family may themselves be distinguished at once from all other Hymenoptera by the peculiar arrangement of the wings when folded at rest. The fore wings partly enclose the hind wings, both pairs lying along the sides of the abdomen, not concealing it from above. The food of wasps consists of the saccharine matter derived from various vegetable products, and also from animal matter. As regards the distribution of species—apart from the usual increase in size and beauty of colouring—it may

BROOD-CELLS OF HORNET

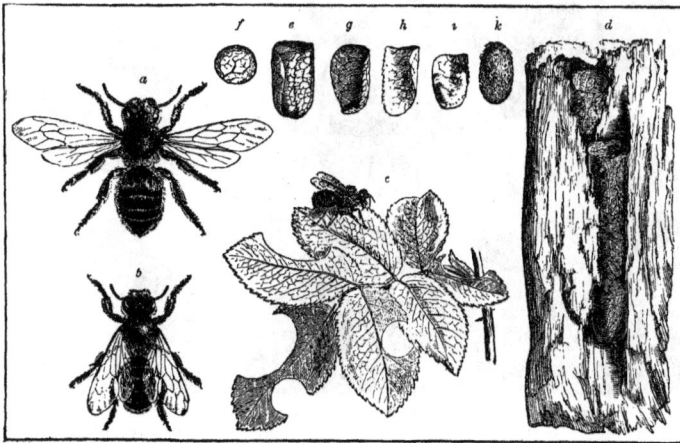

LEAF-CUTTER BEES
a, b, female and male (enlarged) ; *c*, rose-leaves with several pieces clipped out and a bee at work ; *d*, nest in a willow-stem ; *e*, single cell ; *f*, lid of same ; *g, h*, pieces of leaf ; *i, k*, side-pieces

be remarked that the closer the equatorial regions are approached, the more numerous do members of this group become.

Of the better-known forms the common hornet (*Vespa crabro*) is readily distinguished from other species of wasps by its large size and the prevailing red tint on the anterior portions of the body. It is universally distributed throughout Europe, and occurs as far north as Lapland. The solitary female, after her hibernation, begins to build the first foundation of her nest in May on some convenient beam in a loft or outhouse, or frequently in the holes made in the eaves of thatched cottages by sparrows. The food of the grubs consists of the bodies of insects, such as bees, which the workers chew up for their benefit. On the approach of autumn the remaining larvæ, which have not yet been hatched out, are torn from their cells and left to perish.

Under the title of common wasps no less than five species may be included, although *V. vulgaris* is the

SOUTH AFRICAN WASP AND ITS NEST

common wasp *par excellence*. *V. germanica* may be recognised by the three black spots on a yellow clypeus. *V. vulgaris* presents a longitudinal black line dilated at the extremity. *V. rufa* is rare in Northern Europe. *V. media* has the yellow markings of the abdomen darker than in the other species. The wood-wasp, *V. silvestris*, hangs its nest on the bough of a tree or shrub. About 150 species of the genus are known to science. Among other forms, space only admits mention of the South African wasp (*Belonogaster*), of which the comb is shown in the figure on page 1914. Common in houses at the Cape, this insect is much dreaded on account of the severity of its sting. This wasp, like many others of the group, is remarkable for the great length and slenderness of the part connecting the thorax with the abdomen.

MASON-BEES

1, nest with bee emerging, and larva in an open cell ; 2, male ; 3, females fighting

SOLITARY BEES

The solitary bees (family *Andrenidæ*) may be recognised by the fact that the pollen - collecting organs are situated on the femora and coxæ of the hind legs and the neighbouring sides of the thorax. The genera *Andrena* and *Hylæus* comprise the greater part of all the wild bees of Central and Northern Europe. The perfect insects appear in the early spring, making their nests in sandy soil. In the first genus three species (*A. schencki*, *A. cineraria*, and *A. fulvicrus*) are figured in the accompanying illustration. In the second, *H. grandis*, shown in the same illustration,

GROUP OF SOLITARY BEES

1, 2, hairy-legged bee (Dasypoda hirtipes) ; 3, 4, Shenck's earth-bee (Andrena schencki) ; 5, 6, grey-haired earth-bee (Andrena cineraria) ; 7, 8, brown earth-bee (Andrena fulvicrus) ; 9, 10, large burrowing bee (Hylæus grandis). A male and female of each is figured

flies in July and August, and forms a large number of holes—a kind of colony—in some sunny slope. The species of both *Andrena* and the allied *Halictus* are parasitic, and display a very curious habit. When retiring to rest, they fasten upon a twig or the edge of a leaf with their mandibles, fold their wings, draw up their legs, lay the antennæ neatly along their backs, and, having induced a temporary lock-jaw, hang securely until the morning, when they loose their hold and hurry off once again to play the parasite on their relatives.

Another species figured in the same illustration is the hairy-legged bee (*Dasypoda hirtipes*), which appears on the wing in July, and constructs a nest of about six cells in sandy ground. The burrow runs obliquely at first, afterwards descending perpendicularly. Another well-known type of the family is exemplified by the

mason-bees, of which one species (*Chalicodoma muraria*) is represented in the annexed illustration. These insects make their appearance in Europe during May, when the female forthwith sets about constructing her nest.

This includes not more than ten simple cells, and is attached to old walls or houses ; the cells being formed of grains of sand glued together with the saliva of the builder. In 1886 some bees of an allied genus (*Osmia*) constructed their nests in the locks of a door at Deptford. The cells had completely choked the works of the locks, and in one case a portion of the nest was forced out by the insertion of the key without driving away the bees. As the locks were in pretty constant use, it would appear that all the nests must have been built within a few days.

The leaf-cutter bees, of which an example (*Megachile centuncularis*) is figured on the illustration on page 1914, take their name from lining their nests with cells made from fragments of leaves nipped out by the strong jaws of the insects. These cells may be placed either in the holes of trees, in clefts and crannies of old walls, or in specially constructed burrows in the ground. Among the leaves most generally employed are those of the poplar, hornbeam, privet, poppy, and rose. The mode in which these insects work, and the structure of their cells and burrows, are exhibited in the illustration.

Yet another type of building is exemplified by the carpenter-bees (*Xylocopa*), which are among the finest members of the entire family. Their cells are built in rows in the solid wood of trees, and the method of procedure will be observed in the illustration exhibiting the violet carpenter-bee (*X. violacea*). This species, which is rare in Northern Europe, forms a series of cells, in each of which lies a larva, and since the lower ones are obviously the oldest, it is somewhat difficult to understand how the newly-emerged perfect bee escapes into the upper air from the lower cell. At present it is not altogether clear what course it takes ; whether it gnaws its way through the chambers where brothers and sisters are peacefully awaiting future developments—at the imminent risk of arresting all chances of such by thus breaking into their

VIOLET-WINGED CARPENTER-BEE

With cells cut out in a tree-stem

FLOWER-BEES AND LONG-HORNED BEES
1, female, and 2, male hairy-legged flower-bees; 3, female, and 4, male tufted flower-bees; 5, female wall-nesting flower-bee; 6, female, and 7, male long-horned bees

bedrooms, to the detriment of nervous systems not yet hardened to bear the strain—or whether it gnaws its way straight out at the side, seems a matter of doubt. Some authorities state that the female has already foreseen and guarded against such undesirable contingencies, by preparing a door of escape at the bottom of the lower cell. And they record as a remarkable fact that the bees, each in turn, gnaw through the floor of its cell, and of course find their elder brother or sister already flown from the cell next below. They never go in the opposite direction through the roof.

Our next examples of this family are the flower-bees (*Anthophora*), of which three species are shown in the illustration above. In general appearance these insects closely approximate to humble-bees. They build their nests in burrows in the ground, in holes of trees, or clefts and cracks in walls; the cells being separated by partitions, and made of the ruins of the burrow or cleft. Generally the whole nest has the form of a twisted tube. Like their allies, these bees are solitary, and, like humble-bees, are much infested by parasites. Finally, we have the long-horned bees, of which one species (*Eucera longicornis*) is shown in the same illustration. These bees construct smooth tunnels in the earth, divided as usual into sections, each of which contains one egg, together with a supply of pollen and honey for the future larva.

TRUE BEES

Among the true bees (family *Apidæ*) are included not only the various kinds of honey-bees, but likewise their more clumsy cousins the humble-bees. Such a well-known insect as the common honey-bee (*Apis mellifica*), of which the habits have been already referred to, requires no special notice; but it is important to observe that the honey-bees of the equatorial zone differ somewhat from those inhabiting more temperate regions, in consequence of which they are assigned to distinct genera, such as *Melipona*, *Trigona*, and *Tetrasoma*. All these are rather small and stingless bees, making up for the absence of a special weapon of offence by a free use of their jaws. Their brood-cells and combs resemble those of the common wasp, each forming but a single layer, and clay and resinous substances being chiefly used for closing the entrance of the cavities in which the nest are placed. The characteristic transitional features in the shape of the cells, intermediate between the simple cylindrical and the perfect hexagonal forms, have already been noticed in the short introductory remarks. *Melipona*

and its allies form the connecting link between the solitary and the hive bees.

As in the wasps, each family in the humble-bees owes its origin to a single female which has hibernated — usually in some hole in the ground which it excavates for the purpose. The hive-bees, on the contrary, swarm—that is, they send off a full-grown population under a queen ready to enter upon the organised life of an industrial community at once. The different forms of humble-bees are much the same as those of the hive-bees—namely, large females; workers, or undeveloped females; small females which are similar to the large (or queens) in structure; and males. One very strange habit has been recorded and confirmed by subsequent observations. A small female is set apart for the duty of awakening the nest every morning with her piercing note, and has been called the "trumpeter." It seems that only those nests which are large and have plenty of spare hands can afford this luxury.

Humble-bees, both as regards appearance and habits, are too well known to need description. Of the two species figured in the illustration on page 1917, the common humble-bee (*Bombus terrestris*) forms small rounded nests of carded moss. On the other hand, the stone humble-bee (*B. lapidarius*) makes its habitation in cavities among stones, where it forms an oval nest, of which only the sides are covered with moss and grass. Although the fact that bees are essential to the fertilisation of most flowers is familiar, it is less well known that in many cases the work is done by smaller insects. In the case, for instance, of the blue flag, or iris, it has been shown that, in addition to the bees by which they are fertilised, the flowers of this plant are visited by a number of insects of other kinds. The visits of these latter appear for the most part to have been hitherto unnoticed; and, as most of these illicit visitors are of no use for the purpose of fertilisation, the ill-adapted ones are habitually deceived by the flower itself as to its proper entrance. Of the various visitors, two small bees of the genera *Clisodon* and *Osmia* were thoroughly at home in the flower, alighting at the entrance and passing immediately down the narrow passage leading into the nectary, and as quickly emerging and flying off. On the other hand, numerous kinds of syrphid flies spent a much longer time on the flower, which many of them visited only for pollen. Other visitors were certain small flower-beetles and weevils, which never by any chance succeeded in reaching the nectary.

ORDER DIPTERA

As implied by their scientific name, the typical members of the order Diptera, now claiming attention, are distinguished from all other insects by the possession of but a single pair of wings. In this case one pair of these organs has disappeared, and examination will reveal the fact that it is the front pair that is retained in full functional importance, while the hinder pair has become reduced to a couple of short, slender, club-like organs, known as *halteres*, or balancers. From their small size it might be supposed that these balancers were organs of but little physiological importance, but the experiment of removing them will show that this is not the case, for an insect thus mutilated is thereby entirely deprived of the power of maintaining its equilibrium and of directing its course in the air. Hence the name balancers that has been assigned to these vestigial wings.

The mouth parts, instead of being of the primitive mandibulate type, are formed for purposes of piercing or sucking. In the former kind of structure, as represented, for instance, in *Pangonia longirostris*, one of the horse-flies (*Tabanidæ*), these organs are composed of seven pieces, which have been interpreted by Mr. Waterhouse as follows. The uppermost is a long, pointed instrument, the labrum. Immediately below this, and more or less concealed by it, is an almost equally long and slender piece, which is probably the hypopharynx. The mandibles are modified into a pair of sharp lancets, and below them are two extremely slender instruments, which, from the presence of palpi, are recognisable as parts of the maxilla. All these pieces lie concealed in the basal half of the proboscis, which, for part of its length, is gutter-shaped, but afterwards assumes the form of a tube, and is

I, COMMON HUMBLE-BEE WITH NEST; 2, STONE HUMBLE-BEE

HONEY-BEES

believed to be comparable to the labium. In the gnats the mouth is formed upon the same plan, but the lancets are all more slender. In piercing the skin the lancets only are used, the labium or proboscis serving merely as a guide.

In the flies that use the mouth for sucking—as, for instance, in the blow-flies and drone-flies—the jaws are still more modified, so that the identity of the separate pieces is difficult to establish. The most prominent part is the proboscis, the expanding terminal lobes of which are the *paraglossæ* of the labium. The maxillæ are represented by two scales or short stylets closely adherent to the sides of the proboscis, and of two club-like palpi; but the mandibles seem to have disappeared.

The only character that need be specially noticed in the wings is that they are usually naked—being but rarely furnished with short hairs—and that the veins are almost all longitudinal—that is, they run from the base or point of attachment of the wing to its free margin. These veins are represented in the figures on the next page by the letters a, b, c, d, e, f, g. The transverse veins x, y, on the contrary, are always few in number. The shape and size of the spaces (indicated by the numbers 1, 2, 3, etc.) circumscribed by these veins form valuable systematic characters for distinguishing the species and genera of this order. The balancers may be entirely exposed, as in the common daddy-longlegs, but are sometimes concealed by a scale-like membrane, as in the blue-bottle fly.

In connection with the wings may be noticed the buzzing of flies. This appears to be the result of two distinct sounds, one produced by the rapid vibration of the wings, and the other by the vibration of the thorax. The latter movement is the more rapid of the two, and gives rise to the shrill note heard the moment a blow-fly is seized; while the former is the ordinary buzzing produced when the insect is in flight. According to recent calculations, the thoracic vibrations in the case of one of the humble-bee flies (*Volucella*) amounted to thirteen hundred per second, while those of the wings were just one-half this number—namely, six hundred and fifty per second.

The legs possess the normal five segments; the tarsi, or feet, which are also divided into five segments, being armed with two claws, and, in addition, often supplied with adhesive pads, by means of which the insects are enabled to ascend perfectly smooth surfaces. These pads are composed of a multitude of funnel-shaped hairs, each supposed to act as a minute sucker. Some authors assert, however, that they secrete a sticky fluid, and that the insect maintains its hold by this means.

The antennæ vary considerably in structure. In their least modified form, as presented by the gnats and their allies, they are simple and thread-like organs, consisting of a series of subequal segments, often modified by the presence of long symmetrically arranged bristles, which impart to them a feather-like aspect. In most of the members of the order the antennæ are, however, curiously constructed. The three basal segments are stout, the third being especially large and produced into a great lobe-like plate, sometimes projecting as far as the extremity of the terminal part of the organ, which frequently has the form of a plume-like whip, the *flagellum*, although sometimes reduced to a bristle.

Not infrequently the antennæ differ greatly in structure according to sex. In the males of gnats, for example, they are large and feathery, while in the females they are only furnished with short hairs. The males and females of most of the common flies, on the contrary, may be recognised by the development of the compound eyes. In the former sex these organs are almost in contact on the summit of the head, while in the latter there is a widish space between them. Rarely the sexual characters are much more pronounced, as, for instance, in the stag-horned flies, in which the head of the male is furnished with large branching processes, and the stalk-eyed flies, in which the eyes in this sex are supported upon long, horizontal, immovable stalks.

Like the other higher orders of insects, flies, in the course of their development, go through a complete metamorphosis; the larvæ—of which, perhaps, the commonest are maggots and cheese-hoppers—being worm-like, and passing into a partially or wholly quiescent pupal stage before attaining maturity. These larvæ differ much in structure in some of the families; those of the gnats having a well-developed head, with the antennæ, mandibles, maxillæ, and labium always recognisable; whereas in the maggots of the blow-fly the head is narrow and pointed, without antennæ, and with the mouth parts reduced to a pair of retractile hooks, the opposite extremity of the body being broad and square-cut.

It must not be supposed, however, that the larvæ of all the members of this order are of one or other of these two types. On the contrary, the structure varies according to habitat, and almost every gradation is found linking the two together. Some species live in fresh-water ponds and streams, others in the earth amongst roots of grass, others in rotting animal or vegetable matter, and others, like the maggots of the warble-fly, in the stomachs of the hosts they infest. Thus the nature of their food and surroundings is extremely varied, and that the larvæ are likewise so may be seen by a glance at the illustrations in the following pages.

Upon reaching its full size the larva passes into the pupal stage. The pupa, however, exists under two conditions. In one case, as in the gnats, it emerges from the skin of the larva and leads an independent life of longer or shorter duration, until the attainment of maturity; in the others, as in the fly called *Stratiomys*, it remains within the larval skin, which becomes thickened and constitutes a protective covering for it.

Again, the rupture of the larval skin for setting free the pupa is effected in one of two ways. In the first case the opening is T-shaped, consisting of a longi-tudinal split on the back behind the head, or rarely of a transverse split between the seventh and eighth segments of the body; in the second case a circular split occurs behind the head, which is pushed off like a kind of cap.

These two methods of splitting of the larval skin have been used as characters for dividing the Diptera into two suborders, those in which the pupa escapes in the former way being termed straight-seamed flies, or Orthorrhapha, and those in which the pupa escapes in the latter way circular-seamed flies, or Cyclorrhapha.

For the rupture of the larval skin, the pupæ of the Cyclorrhapha are furnished with a bladder-shaped excrescence on the front of the head. In the vast majority of flies the young make their first appearance in the form of eggs. In some few cases, however, as in the genera *Sarcophaga* and *Mesembrina*, belonging to the family *Muscidæ*, the young are born as active maggots; while in the forest-flies and their allies only one young one matures at a time, and this is retained by the mother and nourished at her expense until it has passed into the pupal stage. The most anomalous method of reproduction occurs in one of the gall-midges, where the larvæ themselves produce other grubs by a process of internal budding.

That flies were abundant in early Tertiary times, when they were not very different from those that now exist, is shown by the abundance of their remains preserved in the amber-beds of the Baltic. Strata of the same age at Florissant, Colorado, have also yielded fossil flies. A few have been obtained from Secondary rocks.

STRAIGHT-SEAMED FLIES

The first section of the suborder Orthorrhapha contains the gnats and mosquitoes (*Culicidæ*), daddy-longlegs (*Tipulidæ*), true midges (*Chironomidæ*), and fungus-midges (*Mycetophilidæ*).

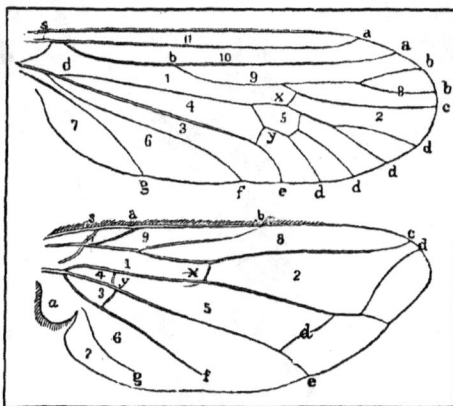

WINGS OF DADDY-LONGLEGS AND OF BLOW-FLY
See text on page 1917 for explanation of lettering

These families are sometimes spoken of collectively as the Nematocera, or flies with thread-like antennæ, on account of the length and thinness of those organs, which usually consist of as many as ten or more segments. The maxillary palpi also are elongate, and the body and limbs present, as a rule, the type with which we are familiar in the gnats and daddy-longlegs.

MOSQUITOES AND GNATS

The mosquitoes and gnats (*Culicidæ*), although often regarded as distinct, are in reality identical. They abound in all lands, and may be met with in cold barren countries like Iceland and Lapland, as well as in the dense forests of tropical climes, everywhere being the plague of travellers on account of their insatiable thirst for blood and the intense irritation caused by their bite. It is, however, only the females that bite and suck blood, and in this connection it may be pointed out that no members of the Diptera sting in the sense in which the word is used with regard to ants and wasps—that is to say, the wound, although giving rise to a sharp stinging sensation, is inflicted by jaws, and not, as in the case of ants, by the organ especially designed for the purpose placed at the hinder extremity of the abdomen.

GIANT SWIFT MOTHS

"The splendid giant swift moth of Australia is one of the finest of all moths. The caterpillar is long, cylindrical, and fleshy, usually making its burrows in the wood of the grey gum-tree. The larva changes into the chrysalis in December, and the moths appear early in March."

A GROUP OF INSECTS

A. Arm-cheeked beetle
A¹. Perforated tortoise-beetle
B. Plant bug (Cicada fraxina)
B¹. Lace-wing fly
C. Eyed leaf-insect
C¹. Tsetse-fly
D. Giant wood-wasp
D¹. Death's-head moth
E. Irish burnet moth

E¹. Lantern fly
F. Cockchafer
F¹. Chrysochroa
G. Praying mantis
G¹. Demoiselle dragon-fly
H. Mole-cricket
H¹. Yellow longhorn beetle
I. Winged stick-insect
J. Chinese tortoise-beetle

J¹. Australian longhorn beetle
K. Carabus uleus
L. Lamprima aurata
M. Scarlet tiger-moth
N. Great green grasshopper
O. Countess beetle
P. Anthicus pedestris
Q. Rose-chafer
R. Great dragon-fly

S. Greenbottle fly
T. Hornet
U. Ichneumon fly
V. Stag-beetle
W. Humble-bee
X. Cerocoma schæfferi
Y. Great water-beetle
Z. Serricornia sternicornis

The figure representing the banded gnat (*Culex annulatus*), a species sometimes found in houses, and noticeable for being the largest British form, is selected to illustrate the mode of life characteristic of the members of this family. The long, slender eggs, amounting to some three hundred or more, laid by the mother in batches on the surface of a pond or ditch, give rise to worm-like larvæ furnished with a distinct head, a large, somewhat squared thorax, and a tapering jointed abdomen. Along each side of the body there is a row of bristle-tufts, one for each segment, and the last segment is, in addition, produced into a couple of tubular tails, at the extremity of which open the tracheæ, or breathing-tubes. Thus equipped, the young gnat hangs suspended in the water, its heavy head directed downwards, and the tip of its forked tail just projecting above the surface, so that the apertures of its breathing apparatus are in communication. Occasionally when the surface of the water is disturbed, or from any other reason causing alarm, the larva wriggles to the bottom of the pond, soon, however, to return to its accustomed place at the surface.

During growth, the larva undergoes a series of three moults before reaching its full size, the newly-clothed insect escaping from the old skin through a longitudinal slit behind the head. At the fourth moult emerges the pupa, which is a very different-looking creature from the larva, showing the cases for the antennæ, wings, and legs of the adult, while from the sides of its thorax project a pair of tubes, analogous to those of the larval tail, and like these carrying the apertures of the tracheæ. By means of its jointed abdomen the pupa jerks itself about in the water in company with others of its kind. At the appointed time a longitudinal split occurs on the back behind the head, and, extricating itself from its pupal-case, the adult gnat appears on the surface of the water, where, under favourable conditions, its skin hardens and its wings unfold, while it floats upon the water, using its discarded clothing as a raft. If this time of danger be successfully overcome, the insect takes wing and joins its companions in their mazy dance; but, before acquiring strength to do so, it is at the mercy of every wave or gust of wind, and if once swept back into the water its chances of survival are small. The above-mentioned banded gnat may be distinguished from other British species by its large size, its spotted wings, and striped legs and abdomen. The common gnat (*C. pipiens*), which is often abundant in houses in the autumn, is much smaller and without the ornaments characteristic of its larger ally.

Most of the species of the genus *Culex* have the wings uniformly coloured, but in all the members of the allied *Anopheles*, as typified by *A. maculipennis*, they are spotted.

In regard to the torments of mosquitoes in and near the tropics, the following passage from Mr. Bates's work on the Amazons may be quoted: "At night it was quite impossible to sleep for mosquitoes; they fell upon us by myriads, and without much piping came straight at our faces as thick as raindrops in a shower. The men crowded into the cabins, and then tried to expel the pests by the smoke from burnt rags, but it

was of little avail, although we were half suffocated by the operation."

But it is not solely as mere temporary torments that mosquitoes are to be dreaded. They are the carriers of both malaria and yellow fever. Malaria, or ague, is a disease peculiar to man. It is caused by extremely minute parasites which live in the red corpuscles of the blood. Formerly malaria was believed to be contracted by merely breathing the air of marshy districts, but it is now known that it is due to parasites transmitted from man to man by the "bite," or, rather, "stab," of a mosquito or gnat.

The common mosquito, or stabbing gnat (*Culex pipiens*) does not transmit the malaria parasite; the spot-winged mosquitoes, of the genus *Anopheles*, some of which are abundant in England, and others in nearly all parts of the world, being the carriers of the parasite. This parasite multiplies not only in the human blood, but in the stomach and tissues of the gnat; but to describe the mode in which it develops in both situations would occupy too much space.

Yellow fever, on the other hand, is propagated by a species of the genus *Stegomyia* —namely, the banded mosquito, *S. fasciata*. Although in temperate regions the mortality from this disease is not high, in one year in the United States the deaths due to malarial fever were 3,976 per 100,000, and in a later year 2,673 per 100,000. In Italy the average death-rate from this cause alone is 15,000 annually, while in India five million deaths were ascribed to "fever" in 1892; and in Italy 2,000,000 persons suffer annually in one way or another from malaria. An interesting question arises in connection with mosquitoes as to the nature of the food of the vast hordes of them that frequent the tropics.

It is true that the females alone bite; but the proboscis is a highly perfected organ for piercing and sucking, and it might be supposed that it is extensively used for the purpose. Yet it has been pointed out that the vast majority of mosquitoes can never taste mammalian blood. In various places, such as parts of India, for example, mosquitoes are found in swarms in spots never visited by human beings, and in which there are no large mammals. It has been suggested that, failing to obtain blood, mosquitoes support themselves on the juices of plants, but no observations in support of this have been recorded.

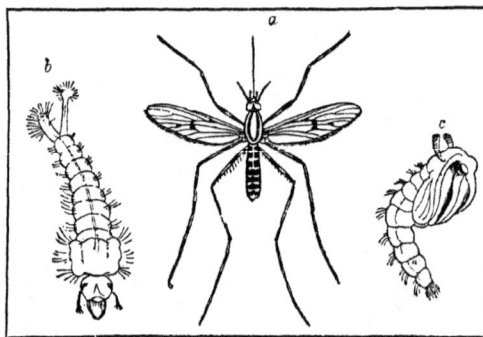

BANDED GNAT
a, female; *b*, larva; *c*, pupa

DADDY-LONGLEGS

The daddy-longlegs (*Tipulidæ*) contain the finest species of this division of the order; the largest European form being the giant daddy-longlegs (*Tipula gigantea*), which has its wings clouded with brown, and measures about 1¼ inches in length. Much larger kinds are, however, met with in Burma and China. The short and fleshy proboscis is not adapted for piercing, but merely for absorbing fluids; and the antennæ are not feathery, as is so often the case in the gnats and midges, although in the species of the genus *Ctenophora* —which are of stouter build, and often brightly coloured black and yellow, thereby resembling some of the saw-flies—these appendages are comb-like in the male.

In this family the eggs are laid and the larvæ undergo

their growth and change of form either in water or earth. The females of two of the commonest British species (*T. oleracea* and *T. paludosa*) may be seen in summer and autumn flying about meadows and depositing their eggs here and there in the soil. When hatched, the larvæ start feeding upon the roots of grass and corn, thereby doing considerable damage .to farmers and gardeners, to whom they are known by the name "leather-jackets."

TRUE MIDGES

The true midges belonging to the family *Chironomidæ* are nearly allied to the gnats, with which they are often confounded ; but the mouth parts are rarely adapted for piercing, the proboscis being short and soft. In the genus *Ceratopogon* the jaws of the females are, however, lancet-shaped, and capable of drawing blood. The little black midge that in the summer settles upon the hands and face and inflicts a sharp prick belongs to this genus ; but the best-known member of the family is the plumed midge (*Chironomus plumosus*), which on summer evenings may be seen dancing in swarms along roads and lanes. Its name has been given to this species on account of the beautiful feathery-like antennæ of the male.

In connection with this species a case of luminosity has recently been recorded. An observer in Russian Asia found on the shores of Lake Issykkul a number of examples of this midge, and of an allied form belonging to the genus *Corethra* emitting a phosphorescent light. Failing to discover any luminous organ, he came to the conclusion that the light was due to the presence in the insect of multitudes of parasitic bacteria, an opinion strengthened by the observation that the shining individuals were sluggish and never seen on the wing.

FUNGUS-MIDGES

The fungus-midges (*Mycetophilidæ*) take their name from the fact that the larvæ of most species feed upon fungi of various kinds. The perfect insects, which frequent damp situations, are all of small size, and mostly pale in colour. To this family belongs the so-called yellow-fever fly, a species of the genus *Sciara*, which in North America is said to appear when yellow fever is prevalent.

But perhaps the most notorious species is the so-called army-worm fly (*Sciara militaris*), which has long attracted attention on account of the peculiar habits

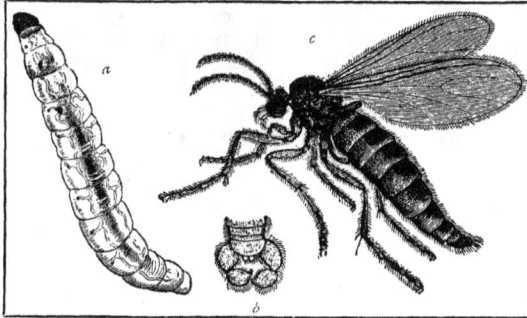

ARMY-WORM FLY
a. larva ; *b*, end of abdomen of male ; *c*, female

A COLUMN OF THE ARMY-WORM

of the larvæ (*a* in figure, enlarged). This fly is completely black, with the exception of its legs, which are brownish. The female, which is represented, greatly enlarged, at *c* in the accompanying illustration is larger than the male, and has the abdomen terminating in a pointed stylet.

In the male, on the contrary, there is at the apex of the abdomen a pair of thick two-jointed claspers, and between these a couple of small adjacent processes, as shown at *b*. The extremely small eggs are laid by the mother, to the number of about one hundred, upon soil among fallen leaves, on which the larvæ feed. On attaining maturity, these larvæ measure nearly a quarter of an inch long, and have the form represented at *a*. The black head is distinct, and furnished with eyes, and the semi-transparent body consists of thirteen segments, some of which are furnished with lateral black stigmata.

In many countries of Europe where this insect is met with, vast hosts of these maggots, forming a compact mass, sometimes several feet long and an inch or two broad, have been seen at times creeping along at a slow pace through the woods like a greyish serpent. The maggots crawl along, not only side by side, but also one over the other, all adhering together by their sticky surfaces, and continually changing their position in the column. At the close of their march, when fatigue or want of nourishment causes them to rest for a time, the larvæ composing a single train collect into a ball, which gradually diminishes in size, and finally disappears by the burrowing into the mould of those that are lowest in the mass. For a long while the reason for this peculiar habit remained wrapped in obscurity, and perhaps even yet we do not understand its full significance. It has been suggested, however, that when the supply of food for the multitude runs short, the whole army is moved by a sudden impulse to start in search of fresh supplies.

It is almost superfluous to add that the peasantry of the countries where this strange phenomenon is observable, failing to understand its true significance, have from time immemorial regarded it as something supernatural, and as foretelling various events in the future, some looking upon it as a sign of the imminence of war, others of the destruction of their crops, and so forth. The pupa of the army-worm is shorter but considerably thicker than the larva. This stage lasts from eight to twelve days; but the perfect insect is short-lived, the female surviving apparently only long enough to pair with the male and lay her eggs.

When writing of one of the true midges, reference was made to a pathological case of phosphorescence, but in the present family there are two instances known of the normal occurrence of this phenomenon—not, however, in the adult insect, but in the larval or pupal stages. The first instance is furnished by *Ceroplatus sesioides*, a midge which, although not yet known to occur in England, has been met with in several of the countries of Europe. Here the luminosity is said to resemble that of the glow-worm, but proceeds from the entire animal, and from members of both sexes.

The larvæ, which are found in small colonies on the under side of a fungus, exhibit, when crawling in the dark, a moving streak of light, less bright than that emitted by the pupæ. The insect also shines when lying in the cocoon, so long as its abdominal rings are still transparent and have not attained their complete colouring. The cocoons themselves are not luminous, but allow the light to be transmitted as through a paper lantern; and since, as a rule, several of them are situated together, a more extensive glow is displayed, whereby both the cocoons themselves and the surrounding objects are illuminated. When the insect is about to emerge from the cocoon, the luminosity gradually diminishes, and ultimately ceases altogether.

The second instance is presented by a New Zealand midge, called *Boletophila luminosa*, the larva of which is known as the "glow-worm." Here the female is luminous in all three stages of its existence; but in the male the luminosity disappears two or three days before the emergence of the perfect insect. The luminous organ, which is situated in the posterior part of the body of the larva, consists of a gelatinous, semi-transparent structure, capable of extension, contraction, and other changes of form, and, like its luminosity, is completely under the animal's control. As to the part played by this organ in the midge's economy, authors are at variance, one believing that the light serves to attract small creatures, so that they become entangled in a web of mucus, which the larva suspends in some niche in the soil.

COLUM-
BATSCH
FLY

GALL-MIDGES

The gall-midges (*Cecidomyidæ*) are minute, fragile insects, in which the wings are furnished with few veins, are often hairy, and always fringed on the edges.

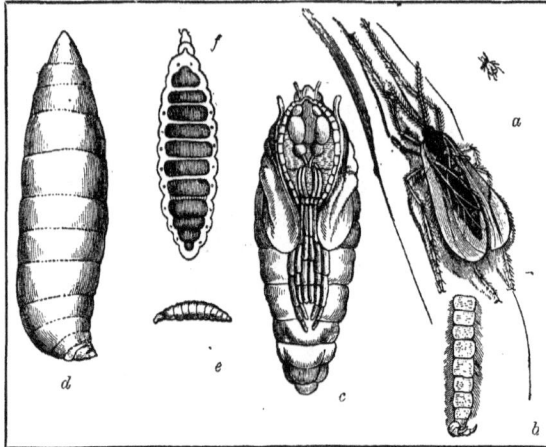
DEVELOPMENT OF HESSIAN FLY
a, female enlarged and of natural size; *b*, abdomen of male; *c*, pupa; *d*, skin of larva forming pupa-case; *e*, larva seen in profile; *f*, larva from above

From an agricultural point of view, these insects are the most important of all the gnat-like flies, since much damage is at times done to crops by their larvæ. The most notorious is the Hessian fly (*Cecidomyia destructor*), represented in its various stages in the accompanying illustration. This insect was believed to have been introduced into North America by the Hessian troops at the time of the War of Independence, whence the inhabitants of the United States gave it the name by which it is now commonly known. The adult female, which measures rather less than a tenth of an inch, is mostly of a velvety black colour, varied with blood-red, especially on the abdomen; while the rather larger male is browner, with the red clearer.

These flies may be observed on the wing during the second half of April. They live, however, only for a few days, and perish soon after laying their eggs, which amount to about eighty or a hundred. These are placed separately or in pairs upon the leaves of the wheat-plant, and in a short time hatch, when the larvæ crawl down the leaf, reach the stalk, and burrow in it to take up their abode and feed upon its tissues. This does not immediately, nor in a direct manner, cause the death of the plant, but, weakening its stem, renders it liable to be beaten down by wind or rain, and causes it to bear inferior corn if it reach maturity. Towards the end of July the larvæ are full grown, and pass into the pupal stage; while at the end of August or the beginning of September the adults again appear and lay their eggs on winter wheat, the larvæ that are hatched from these passing the winter in the pupal state and completing their development in the spring.

Nearly allied is the wheat-midge (*C. tritici*), which, as its name indicates, also attacks the wheat-plant, to which it at times does great damage. The female lays her eggs—often in numbers—not on the leaves or stems, but in the heart of the blossom, and their presence either entirely prevents the formation of any seed, or renders that produced of a poor kind.

There are many other species of gall-midges which attack different kinds of plants, such as the willow, hawthorn, etc., but lack of space forbids further reference to them. One only (*Miastor metroloas*) demands attention on account of the remarkable powers of reproduction of its larvæ. In the majority of cases insects are only able to reproduce their kind after attaining the adult state, the larvæ being merely the young modified for a free and active existence; but the larvæ of this midge, which are found under the bark of

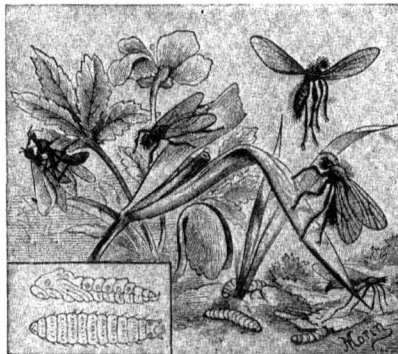
ST. MARK'S FLY
The larva and pupa are shown in left corner

trees, possess the remarkable power of producing living young within their bodies. These grow to a certain size at the expense of their parent, whose vitals they devour, then rupture the empty skin and start life on their own account. The second larvæ repeat the same process of reproduction, and so the phenomenon continues through the colder months.

SAND-FLIES

The two families now to be mentioned have been termed the anomalous, or fly-like, Nematocera, since, although their antennæ are many-jointed, they are shorter than in the foregoing families, and their limbs and bodies, instead of presenting the aspect of those of the gnats and midges, are shorter, thicker, and closely approach in this respect those of ordinary flies. To the family *Simuliidæ* belong the minute "sandflies" of the tropics, which surpass even the mosquitoes in their venomous bite, and on account of their minute size are far more difficult to cope with. In these insects the mouth parts are adapted for piercing; and the early stages of life are passed in water.

The best-known European example is the Columbatsch fly (*Simulia columbatzensis*), taking its name from a village in Servia, where it is a great pest. In fact, in all the countries irrigated by the lower waters of the Danube, this fly, hardly larger than a flea, abounds; and it is said that in Hungary cattle and sheep have been destroyed by hundreds owing to the tortures they have suffered from these insects.

The second family of the group (*Bibionidæ*) contains the well-known St. Mark's fly (*Bibio marci*), a large, black, hairy, slow-flying insect, common in spring, and taking its name from its being frequently seen in numbers on or about St. Mark's Day. The two sexes differ greatly in many respects, the male having the wings clear, whereas those of the female are dusky again the eyes in the male are so large that the entire head seems to be composed of them, but in the female these organs are small and wide apart.

The eggs—in number amounting to about one hundred and fifty—are laid on the ground among vegetable or animal débris, on which the larvæ subsequently feed. After passing the winter in the soil in an immature state, the larvæ ascend to the surface in the spring, and enter the pupal stage, from which, after about a fortnight's time, the perfect insects emerge.

HORSE-FLIES

Although related to the gnats and midges by the nature of the slit through which the pupa makes its escape from the larval skin, and consequently referred to the section Orthorrhapha, the flies of this family approach those of the second section in the shortness of their antennæ, and since all the Diptera with short antennæ were formerly termed the Brachycera—as opposed to Nematocera—these and the remaining families of the suborder are often grouped together as Orthorrhapha Brachycera. Although the horse-flies (*Tabanidæ*) are often termed gad-flies, the latter name is proved by Anglo-Saxon literature to have been originally applied to the *Œstrus* group of the *Muscidæ*.

Horse-flies are distributed all over the world, and vary but little in outward form, usually having large, flat bodies, and being generally of a dull reddish brown colour. They are all blood-suckers, and the mouth parts attain a high degree of perfection as piercing instruments. A common representative of the family in England is the so-called clegg (*Hæmatopota pluvialis*), a greyish insect which has a habit of pitching quietly upon the hands or face, and inflicting a sharp prick almost before the victim is aware of its presence. Fortunately, however, it is easily killed, for, instead of taking flight, it generally stays where it has settled, and allows itself to be crushed.

A larger, though scarcer, British species is the great horse-fly (*Tabanus bovinus*), the female of which

GREAT HORSE-FLIES

sucks the blood of large mammals, such as horses, asses, and cattle. The males always frequent flowers; and the larvæ in form and habits show considerable resemblance to those of the daddy-longlegs, living in the soil and feeding upon the roots of grasses. In this way they spend the winter, reaching maturity in May, when they pass into the pupa stage, the fully-formed insect appearing in June.

"The 'coachman-fly' (doubtless one of the family *Asilidæ*) is said to feed on the horse-flies; and will sit through a whole drive on the collar or some other part of the harness, or even on the steed itself, in order to pounce on the insects as they settle. The curious thing is that the horses seem to know the difference, for directly a horse-fly comes, even if it does not sting, they become restless, tossing their heads, and lashing with their tails, but the 'coachman' may rest on any part of them for any length of time, and never be interfered with or driven off."

ROBBER-FLIES

The flies of the family *Asilidæ* are generally of a somewhat slender build, the body being long and parallel-sided, while the legs and wings are long and strong. All are provided wth a short, powerful, piercing proboscis, and prey upon insects of various kinds, often seizing and carrying off butterflies much larger than themselves. The general form of the members of this family is shown in 1 of the illustration, representing *Dioctria oelandica*, a species from the island of Oeland, off the coast of Sweden, with a shining black body and wings of the same colour.

Many species of the genus *Asilus* are found in Britain, but the largest and handsomest of all is the hornet robber-fly (*A. crabroniformis*), measuring upwards of an inch in length and of a yellowish colour variegated with black, there being four stripes of the latter colour upon the thorax, and a broad transverse band across the base of the abdomen.

Some of the tropical members of the family are far larger, those belonging to the genus *Mydas*, from South America, being scarcely surpassed in dimensions by any member of the order.

The fly represented in 2 of the illustration is the tesselated empis (*Empis tesselata*), belonging to the family *Empidæ*, the species of which are predaceous, like the *Asilidæ*, and resemble them in form, but differ in certain structural details which need not be dwelt upon. The tesselated empis—the largest member of the group found in Britain—is ashy grey in colour, and has its abdomen ornamented with a chess-board pattern. As Mr. Dallas expressed it, "when paired, the females of this and of many other of the larger species of the family are always found to be busily engaged in sucking out the juices of some other insect. It seems probable that the male seizes the opportunity of his intended partner being thus occupied to make his advances ; if her mouth were free he would in all likelihood himself fall a sacrifice to her voracity."

BEE-FLIES

The families of short-horned, straight-seamed flies hitherto considered resemble each other in the fact that the larvæ live in the earth, and feed upon the roots of grass or other vegetable matter, while the adults prey upon other animals, whose blood they suck. But in the bee-flies (*Bombyliidæ*) —so called from the likeness in hairiness and shape they present to humble-bees—the larvæ, so far as known, live parasitically on other insects, attacking grasshoppers, caterpillars, etc., while the adults suck the juices of flowers. The genus *Bombylius* is represented in England by a small number of species, although in the tropics there are large numbers of forms. In all the thick, fat body is covered with long, yellow hairs. The wings are powerful, and the head is furnished with a long proboscis, which is thrust into blossoms while the insect (8 in figure on page 1927) stays poised in mid-air, like a hawk-moth when similarly occupied.

The black-and-white bee-fly (*Anthrax semiatra*) is mostly of a black tint, and clothed with hair of the same colour ; but the hairs on the front part of the thorax and abdomen take a yellowish tinge, the wings, as shown in the illustration, being black in the basal half but clear elsewhere. These insects may be seen on the wing in dry, sunny spots, stopping from time to time to suck a flower or rest upon a stone, and seeking for the cells of solitary bees wherein to deposit their eggs. The left-hand figure shows the cocoon of one of these bees, with the pupa-case, from which the fly on the right has just emerged, protruding from it.

For the last family of this section (*Stratiomyidæ*) the common *Stratiomys chamæleon* may be taken as the type. This is a rather large insect, with a short, broad abdomen, variegated at the sides with pale spots ; the sides of the face and posterior parts of the upper surface of the thorax being also yellow. The antennæ are longish, and the hinder part of the thorax is armed with a pair of spines. The females, which may be seen on the wing in the neighbourhood of marshes, ponds, and ditches, lay their eggs on the leaves of water-plants, and the larvæ spend their time wriggling about in a helpless way.

In these larvæ the body consists of twelve segments, is somewhat depressed, pointed at each end—though more so towards the tail than the head—and covered with a tough blackish brown skin. The head is small and pointed, and the retractile tail-segments are furnished at the tip with a breathing orifice surrounded by a circlet of barbed hairs. By means of these the larva is enabled to suspend itself from the surface of the water, hanging vertically downwards with the orifice just above the water's level, and is also able by the folding in of the hairs to take a bubble of air below the surface when it sinks to the bottom. The larvæ feed on such particles of matter as they find in the water, and when ready to pass into the pupal stage creep to the land, and take refuge beneath a stone or in some other place of safety. The development of the pupa and perfect insect takes place only in the front part of the larval skin.

A curious choice of habitat for the young on the part of some flies belonging to this family has been recorded from Wyoming. These larvæ were found in a cup-shaped depression at the top of a cone about twenty inches high situated a few feet from a large sulphur mound, under which the boiling water could be heard. Through small apertures in the bottom the hot water rose and filled the cup. It was in this that the larvæ were found ; and it is estimated that the temperature of the water was only twenty or thirty degrees below boiling point.

They are not the only animal forms which have become adapted to a life in very hot water, and a parallel may be found among some lower plants.

ROBBER-FLIES
1, Dioctria oelandica ; 2, Empis tesselata

FEMALE OF STRATIOMYS CHAMÆLEON

BLACK-AND-WHITE BEE-FLIES, WITH PUPA SKIN PROTRUDING FROM COCOON OF BEE

CIRCULAR-SEAMED FLIES

The suborder Cyclorrhapha, which is characterised by the circumstance that the pupa escapes from the larval skin through a circular aperture formed by the pushing off of the head-end, contains the majority of ordinary flies. It is divisible into two sections, the first of which includes those that present the normal method of development, the young being hatched from eggs laid by the mother, although very rarely the eggs hatch immediately before being laid. The second embraces those in which the young are retained within the parent's body, and nourished at its expense until the pupa stage is reached. The flies of the last category are for this reason generally called pupipara.

HOVER-FLIES

The family *Syrphidæ* includes a number of species which, although differing considerably in external form, may be distinguished from other members of the suborder by the presence of the so-called spurious vein in the wing—a vein lying between the third and fourth longitudinal veins, and crossing the short transverse vein (marked *y* in the figure on page 1918) which unites them. They also bear considerable superficial resemblance, both in colour and shape, to various bees and wasps. The best known types are the hover-flies (*Syrphus*), drone-flies (*Eristalis*), and humble-bee flies (*Volucella*).

The hover-flies of the genus *Syrphus*, which with their black and yellow bands mimic wasps, are so named on account of their habit of hovering in flower gardens in summer, darting from blossom to blossom, and often sustaining themselves poised in mid-air, after the manner of a hawk. The females lay their eggs singly on leaves and stems infested with plant-lice; and the larvæ devour numbers of these pests, seizing them in a most voracious manner, sucking them dry, and rejecting the empty skins.

DRONE-FLIES

Like the hover-flies, drone-flies (*Eristalis*) frequent flower gardens, where they may be seen in numbers on various blossoms. As their name indicates, these flies resemble honey-bees, the likeness being so close that it is difficult to persuade an uninitiated person that they may be handled with impunity. The resemblance, which is enhanced by the ceaseless twitching of the abdomen, appears indeed to be more deeply seated than might at first be supposed, for spiders, which recognise their prey by touch and not by sight, treat the drone-flies with caution. Thus, a blue-bottle fly placed in a web of the field-spider was immediately and without hesitation seized and devoured, although a humble-bee was avoided by the spider, which—evidently fearing to come to close quarters—let out a thread, and rushing round and round its victim at a

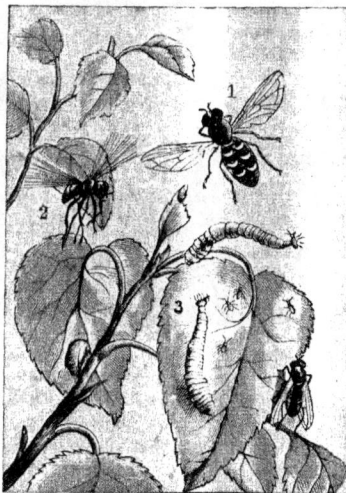

HOVER-FLIES
1, fly; 2, fly hovering; 3, larvæ devouring plant lice on leaf

distance, succeeded in winding it up, and then approaching, inflicted a bite which soon put an end to the insect's struggles. When a drone-fly was thrown into the web, the spider darted at it as before, but as soon as it touched the fly with its fore-legs, recoiled, as if in alarm, then returning to the attack, dealt with the harmless victim just as it had acted with the humble-bee.

The larvæ of the drone-flies live mainly in ditches and feed upon decaying organic matter, and are commonly known as rat-tailed maggots, on account of the long tail-like appendages at the hinder end of the body. With this flexible and telescopic tail, traversed by tracheal tubes opening at its tip, the maggot is able to breathe while below the water, by keeping the tip of its tail above the surface, where it is supported by the rosette of hairs round the extremity. The eggs of drone-flies are also laid in dead carcases and other refuse, and it is now believed that the legend of the ox-born bees is traceable to this habit of the fly, in conjunction with its striking resemblance to the honey-bee.

The belief that honey-bees are produced by spontaneous generation from carcases of dead animals has prevailed for more than two thousand years, but, according to Osten Sacken, " the original cause of this delusion lies in the fact that a drone-fly (*Eristalis tenax*) lays its eggs upon the carcases of animals, that its larvæ develop within the putrescent mass, and finally change into a swarm of flies, which in their shape, hairy clothing, and colour look exactly like bees, although they belong to a totally different order of insects."

HUMBLE-BEE FLIES

Scarcely less interesting than the drone-flies are the species of *Volucella*. These large flies (page 1927) mimic humble-bees in colour and form; and it was long supposed that the females were thus enabled with impunity to enter the nests of humble-bees and lay their eggs amongst those of the proper owners. But although it is true that the eggs of the *Volucella* are laid and the larvæ reared inside the nests of various Hymenoptera, it has been ascertained that the species which resemble humble-bees visit for the same purpose the nests of wasps, to which the flies bear no particular resemblance. And it is hardly credible that the wasps give access to the flies under the delusion that they are members of the community, as was conceivable in the case of the bees. We are compelled, therefore, to conclude that the flies are allowed by the bees and wasps to come and go without interference for some reason apart from the resemblance that exists between the two sets of insects.

It is, of course, possible that the similarity offered by the flies to bees and wasps is more deeply seated than was supposed, and affects such senses as touch or smell, or some other unknown sense, but there seems no evidence to justify this supposition; and if the maggots of the flies feed on the larvæ of the bees or wasps, we are not yet in a position to offer an explanation of the phenomenon. If they play the part of scavengers, clearing the hive of waste matter, the reason for the admittance of the flies becomes clear.

WASP-FLIES

Closely resembling many of the *Syrphidæ* in their banded coloration, which imparts to them a wasp-like aspect, the members of the family *Conopidæ* may be recognised by the absence of the spurious vein in the wings, and also by their broad heads, of which the fore part is produced into a conspicuous prominence bearing the long antennæ. Like the horse-flies, the *Conopidæ* in the adult stage frequent flowers, but they lay their eggs in the bodies of various Hymenoptera, like bees and wasps, and also in crickets and other Orthoptera. Here the eggs hatch and the larvæ feed upon the living tissues of their prey, and here they undergo their metamorphosis, although they do not invariably quit the place of their development upon the death of the victimised host. Taschenberg, for example, found the pupa of *Conops vittatus* emerging from the abdomen of a humble-bee which had been for six months in his collection.

The *Conopidæ* are widely distributed, and especially abundant in the tropics. Mr. Bates gives an account of the habits of a species which he noticed hovering over the armies of foraging ants. These ants, he says, " are accompanied by small swarms of a kind of two-winged fly, the females of which have a very long ovipositor, and which belongs to the genus *Stylogaster*. These s w a r m s hover with rapidly-vibrating wings, at a height of a foot or less from the soil over which the ants are moving, and occasionally one of the flies darts with great quickness towards the ground. I found that they were not occupied in transfixing ants . . . but most probably in depositing their eggs in the soft bodies of insects which the ants were driving away from their hiding-places. These eggs would hatch after the ants had placed their booty in their hive as food for their young."

TYPICAL FLIES

The family *Muscidæ* embraces a large and varied assortment of species, of which house-flies and blow-flies are well-known examples. The characteristic structure of the wings may be seen by referring to the figure on page 1918. The proboscis is adapted for sucking, and usually ends with two fleshy lobes. The flagellum of the antennæ is generally plumed with hairs on both sides, though sometimes, as in the tsetses, the hairs are restricted to one side, while in the spiny-flies it may be naked. The relative size of the three basal segments of the antennæ varies in different genera, but usually, as in the blow-flies, the house-flies, and the tsetses, the third segment is at least three times the length of the second (see *b* in figure on page 1929, and 10 in that on this page). It may also be mentioned that the upper surface of the thorax is marked with a transverse suture, and that the feet are furnished with a pair of adhesive pads (as in the upper figure on page 1928).

The family is divided into several subfamilies, and these may be grouped in two sections, based upon the presence or absence behind the wings of a membranous scale which, when present, covers the *halteres* or balancers. The subfamilies that possess this scale are termed the calypterate *Muscidæ* ; while those that are without it are, in contrast, called the acalypterate *Muscidæ*. Taking the calypterate *Muscidæ*, we begin

INSECT LIFE IN SUMMER

1, common wasp ; 2, 3, honey-bees ; 4, hairy-legged bee (Dasypoda) ; 5, solitary wasp (Pompilus) ; 6, stor humble-bee ; 7, common humble-bee ; 8, bee-fly (Bombylius) ; 9, humble-bee fly (Volucella) ; 10, spiny fly (Tachina) ; 11, noctuid moth (Anarta) ; 12, 13, field tiger-beetle, crawling and flying ; 14, wood tiger-beetle ; 15, rose beetle (Cetonia) ; 16, dung-beetle (Typhœus) ; 17, field-cricket ; 18, grasshopper (Stenobothrus).

with the subfamily *Muscinæ*, of which the house-fly (*Musca domestica*) is the typical representative.

This species may be found during summer in numbers in every house, crawling up the window-panes, flying in companies about the middle of the room, or creeping about the table in search of food. It is the unwelcome companion of man in every country, following him in his travels, taking up its residence with him wherever he may choose to settle, and resisting equally well the cold of northern latitudes and the heat of tropical climes. For the most part, the eggs are laid and the larvæ undergo their development in excrement ; but the choice of the female does not seem to be always restricted to matter of this sort, since she sometimes selects meal, bread, or fruit, for the purpose. These flies are liable to the attacks of a parasitic fungus (*Empusa muscæ*) which causes their death, and in autumn it is not uncommon to find their bodies killed by this means, with the abdomen much distended, and showing the soft membrane between the segments.

The common blue-bottle, or blow-fly (*Calliphora erythrocephala*), is too well known to need description. One of the most noteworthy features connected with this fly is the extraordinary keenness of the sense—perhaps smell, which enables it to discover the whereabouts of carcases, however small, or of particles of meat. In these it hastens to lay its eggs ; and in a longer or shorter time, according to temperature, the eggs hatch, and the larvæ, feeding upon the meat, rapidly grow until they reach maturity and pass into the pupa stage.

Many persons believe that blue-bottles are full-grown examples of the house-fly, and when informed that such is not the case, and that these insects after reaching the winged stage are incapable of growth, point out that blue-bottles vary greatly in size, and ask what may be the explanation of the difference. The answer is, that the size of the blue-bottle in its final stage depends upon the size of the maggot before pupating, and the size of the maggot upon the amount of nourishment it is able to obtain before its supply of food was exhausted. In any given case, when the supply is limited, the maggots that are the first to hatch will get more food than those that appear later, and, in consequence, when the whole of it is exhausted, will have attained a greater length and fatness than the others, and thus become converted into larger flies. Or, again, if three or four hundred eggs be laid in a dead mouse and the same number in a dead rabbit, it is clear that in the former case the supply of food will be smaller for each larva, and will sooner come to an end than in the latter.

The grey flesh-fly (*Sarcophaga carnaria*) is a handsome species, measuring in the female half an inch in length. Seldom entering houses, it is not uncommon in the open country, where it may be seen basking in the hot sun upon stones or walls. Its prevailing colour is pale slate-grey, variegated on the thorax with black bands, and the abdomen with square black spots, set corner to corner like the squares of a chess-board. A noteworthy fact connected with this species is that the eggs hatch within the parent before being laid, so that the young are born alive ; they feed upon decaying animal and vegetable matter.

The blow-flies belonging to the genera *Calliphora* and *Lucilia*, respectively known as the blue-bottle and green-bottle flies, as a general rule deposit their eggs upon dead animal matter. This, however, is by no means always the case, there being many instances on record of their laying and hatching of the eggs upon

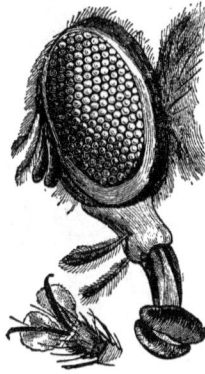

HEAD OF HOUSE-FLY AND FOOT OF GREY FLESH-FLY

living animals. Thus it is by no means uncommon for sheep to be attacked in this way by a green-bottle fly (*L. silvarum*).

Toads and frogs also seem to be frequently selected as objects of attack on the part of these flies. In one case the eggs of a green-bottle fly were laid on a toad's back, and the larvæ upon hatching migrated into its eyes. In other cases the laying of the eggs and the migration of the larvæ have not been actually observed, but toads have been found with their nostrils infested with maggots ; and it is possible that the latter may have effected an entry from the outside, as described above.

Cases are also on record of the death of lizards from maggots of blow-flies, which testify to the extraordinary vitality of the latter. In one instance a gecko fed on blue-bottles was found to have the whole abdominal region greatly distended. It soon afterwards died, and on dissection its intestines, lungs, and liver were found to be almost entirely destroyed by maggots, whose presence was naturally attributed to eggs from gravid female blue-bottles, which had been swallowed as food. In another case, some lizards fed on the living maggots of the blue-bottle died in consequence of the attacks on their internal organs by their intended food. Far more important are the cases of infection of human beings ; the resulting sickness, which often entails great suffering, and may end in death, being known as *myiasis*.

The sharp-mouthed fly (*Stomoxys calcitrans*), represented in 9 of the figure on this page, closely resembles the house-fly in size, shape, and colouring, but may be recognised by its sharp, horizontally projecting proboscis, and also by the flagellum of the antennæ being hairy upon one side only. By means of its proboscis this fly pierces the skin of cattle and horses, or even of man

Resembling *Stomoxys* in habits and in the structure of the antennæ and mouth - parts, the tsetse flies (*Glossina*) of Tropical Africa, although barely equalling a blow-fly in size, are some of the greatest pests to domestic cattle, as the following accounts amply testify. As shown in the illustration on opposite page, the proboscides of the flies are long and prominent, and the antennæ (*b*) are peculiar in that the third segment is very long and produced almost as far as the apex of the flagellum, which is furnished with barbed hairs along its outer surface only.

Writing of the South African species, Livingstone says : " This insect is certainly not very much larger than the common house-fly, and is nearly of the same brown colour as the honey-bee. The after part of the body has three or four yellow bars across it. It is remarkably alert, and evades

GROUP OF FLIES AND THEIR GRUBS

1, blow-fly ; 2, eggs ; 3, larvæ ; 4, pupa ; 5, newly-born larva of grey flesh-fly ; 6, grey flesh-fly ; 7, adult larva of the same ; 8, house-fly and larva ; 9, sharp-mouthed fly ; 10, carcase of house-fly killed by fungus growth

dextcrously all attempts to capture it with the hand at common temperatures. In the cool of the mornings and evenings it is less agile. Its peculiar buzz when once heard can never be forgotten by the travellers whose means of locomotion are domestic animals, for its bite is death to the ox, horse, and dog. In this journey, though we watched the animals carefully, and I believe that not a score of flies were ever upon them, they destroyed forty-three fine oxen."

A most remarkable feature is the perfect harmlessness of the bite to man and wild animals, and even calves so long as they continue to suck the cows, though it is no protection to the dog to feed him on milk. The poison does not seem to be injected by a sting, or by ova placed beneath the skin, for, when the insect is allowed to feed freely on the hand, it inserts the middle prong of the three portions into which the proboscis divides somewhat deeply into the true skin. It then draws the prong out a little way, and it assumes a crimson colour as the mandibles come into brisk operation. The previously shrunken belly swells out, and, if left undisturbed, the fly quietly departs when it is full. A slight itching irritation follows the bite.

In the ox the immediate effects are no greater than in man ; but a few days afterwards the eyes and nose begin to run, the coat stares, a swelling appears under the jaw, and sometimes at the navel ; and though the poor creature continues to graze, emaciation begins, accompanied with a peculiar flaccidity of the muscles. This proceeds unchecked until, perhaps months afterwards, purging comes on, and the victim dies in a state of extreme exhaustion.

With the gradual spread of civilisation, it might be supposed that the ravages of this pest would become lessened, but this does not appear by any means to be the case. Writing in 1881, Mr. Selous remarks that " nowhere does this virulent insect exist in such numbers as to the westward of the Victoria Falls, along the southern bank of the Zambesi and Chobi. It is usually found in great numbers near the rivers, becoming scarcer and scarcer as one advances inland, till at a distance of a few miles it disappears, except in some particular patches of forest. Along the water's edge they are an incredible pest, attacking one in a perfect swarm, from daylight till sunset ; and without a buffalo or giraffe tail to swish them off, life would be unendurable. . . . About one in every ten bites (that perhaps touches a nerve) closely resembles the sting of a wasp or bee, as it will cause one, when seated to spring up as if pricked with a needle.

I think that this plague of the tsetse flies along the Chobi and Zambesi is due to the enormous numbers of buffaloes that frequent their banks, as they always seem very partial to these animals. The bite of this remarkable insect, as is well known, though fatal to all kinds of domestic animals, is innocuous to every species of game and to man. A general belief exists that among domestic animals, the donkey, dog, and goat are exceptions to this rule ; but this is a mistake, for I have seen all three die from the effect of its bites."

The genus to which the common tsetse belongs is represented in Tropical Africa by several species, all of which seem to be similar in habits. It ranges from Somaliland in the east and the Congo in the west, southwards as far as the Limpopo. Tsetses are by no means universally distributed throughout the country, being somewhat local in their distribution, and inhabiting definite tracts of land, corresponding with the beds of rivers, from which they do not appear to spread to any great distance.

From all other flies, tsetses are readily distinguished when at rest by the complete closure of the wings, which shut like the blades of a pair of scissors, instead of having their tips divergent. The shortness and the stoutness of the proboscis is likewise a characteristic.

As already mentioned, the mouth-parts of these blood-sucking flies are adapted to pierce the thick skins of large mammals, such as zebras and antelopes. The blood of some of the animals thus attacked is infested with a host of somewhat eel-like parasites, known as *Trypanosoma brucei*. These are carried adhering to the proboscis of the tsetse, and are thus introduced into the blood of domesticated animals, like the horse and the ox, thereby producing the tsetse-disease, or nagana. The flies act simply as carriers of the blood-parasites, which undergo no development in the bodies of the former ; the process is therefore unlike malaria contagion, in which the parasites undergo a part of their development within the body of the mosquito. Tsetses are also carriers of the parasite which produces the dreaded " sleeping sickness " in the human subject.

Another group of flies constitutes the subfamily *Tachininæ*, of which the best known examples are the spiny-flies (*Tachina*), so called on account of the thickness of the bristles with which their bodies are clothed. Of stout and robust build, these flies present a great resemblance to blow-flies and their allies, but have the bristles of the antennæ naked, or feathery only at the base, and the scales covering the balancers of larger size. The larvæ, like those of the *Conopidæ*, live parasitically upon other insects, such as beetles, grasshoppers, and caterpillars.

The great spiny-fly (*Echinomyia grossa*), rather a local species, is the largest representative of the family found in Britain. It is about two-thirds of an inch long, with a short, broad, oval abdomen ; the shining black of its body being relieved by the reddish yellow colour of the head and the base of the wings. The allied species (*E. ferox*) represented in the illustration, is brownish, with the abdomen tinted with red at the sides. Belonging to the same subfamily is the Australian fly *Rutilia*, remarkable among the order for being ornamented with bright metallic green spots.

By reason of their external form and general colouring, the flies of the subfamily *Anthomyinæ* appear to the casual observer to be nothing but ordinary house-

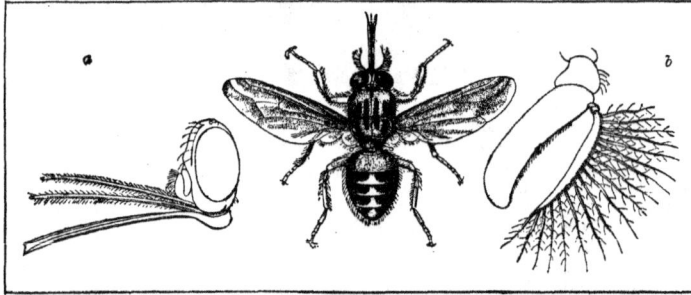

SOUTH AFRICAN TSETSE FLY
a, side view of head ; *b*, antenna

SPINY FLY

flies, but they may be distinguished from the latter by the absence of the apical transverse vein on the wing (marked *d* on the figure of the fly's wing on page 1918). The scales, moreover, which cover the halteres are very small, and lead up to the condition found in those flies in which they are absent. The larvæ, which differ from those of the house and blow flies in being covered with spines, live on plants of various kinds, those that have attracted most attention being the species that attack cultivated vegetables, such as onions, cabbages, lettuces, radishes, and the like.

MALE AND FEMALE ASPARAGUS FLIES

a, front view of head

Those members of the family having no scales covering the balancers and assigned to the subfamily *Trypetinæ* are generally of small size, many being very obnoxious on account of the damage inflicted by their larvæ on various marketable vegetables. Of the numerous species, it is only possible to notice a few. The first is the painted-winged asparagus-fly (*Platyparea pœciloptera*), which, as its name indicates, has variegated wings, and attacks asparagus. The male (left of figure) is smaller than the female, and the latter sex may be recognised by the possession of a long ovipositor, by means of which she deposits her eggs between the scales of the head of the asparagus. The laying takes place about the beginning of May, and in two or three weeks, according to the season, the eggs hatch, and the larvæ burrow into the stalk of the plant. In a fortnight or so the latter reach maturity, and, after passing through the pupa stage, develop into flies towards the end of June.

Many more or less nearly allied species are found in England and other countries, but it will suffice to indicate a few of the more important. Of these the cherry-fly (*Spilographia cerasi*) and the olive-fly (*Dacus oleæ*) devour in their larval stages the fruits after which they are named ; while the various species of the genus *Ceratitis* similarly attack the orange. Recently *C. capitata* was very destructive to the mandarin oranges in Malta, and seems to have been first introduced into the island about thirty years ago.

This fly is lively and hardy, as shown by the fact that a specimen kept under a glass shade without food, maintained its activity for twelve days. When egg-laying, the female chooses the side of the fruit exposed to the sun, where it perforates the rind so that the larvæ, upon hatching, start at once to devour the nutritious food. The infected fruit drops to the ground, and the larvæ, when mature, pass out to become pupæ beneath the earth. Besides oranges and other acid fruits, peaches and melons are attacked by this fly.

The figure represents another of these injurious little insects (*Chlorops tæniopus*), a shining yellow fly variegated with black bands. This species and its allies, which are most destructive in the larval stage to cereals and grass, much resemble in the cycle of their development the above-mentioned Hessian fly. Allied to the preceding in structure and habits are the members of the subfamily *Ortalinæ*, containing the genus *Ortalis* and others. A curious representative from the Malay Archipelago, known as the staghorn-fly

(*Elaphomyia*), takes its name from the development of the sides of the head into large branching horns. This, however, is only a sexual character, and confined to the male. Finally, the small black fly (*Piophila casei*) known in the grub-stage as the cheese-hopper, belongs to that group of *Muscidæ* in which there are no scales to cover the balancers.

GAD-FLIES AND BOT-FLIES

The flies of the family *Œstridæ* are mostly of large size, and many present superficial resemblance to various kinds of bees. In structural characters they are nearly allied to house-flies, but the head is larger and broader, and the mouth-parts are reduced. In the larval stage gad-flies infest, either as internal or external parasites, various mammals, but since those that attack domestic cattle have been more thoroughly studied than the others, attention will mainly be directed to three of the best known forms, namely, those that infest respectively horses, oxen, and sheep.

The horse bot-fly (*Gastrophilus equi*), which resembles the honey-bee in size, colour, and form, lays its eggs on the skin of horses, asses, and mules, which seem to have an instinctive dread of the insect. It has been noticed, moreover, that the gad-fly instinctively selects for the purpose a spot that is well within reach of the quadruped's mouth. The reason for this, although not at first very obvious, becomes clear when it is understood that the larval fly can only obtain its proper nourishment in the alimentary canal of its host.

CHLOROPS TÆNIOPUS

As soon as the maggot emerges from the egg, it starts to irritate the horse's skin. Thereupon the horse, to remove the irritation, licks the infested spot and swallows the maggots, which then attach themselves by means of their hook-like mandibles to the inner wall of the stomach, or œsophagus, making little excavations, and nourishing themselves by sucking up the secreted mucus. Here in perfect security they live and grow for about a year, after which, when nearly full grown, they enter the intestine and pass out of the body with the excrement. Falling to the ground, the maggots bury themselves in the soil and enter upon the pupal stage. In favourable weather, the perfect insect is produced from the pupa in about six weeks.

The ox-bot, or ox-warble (*Hypoderma bovis*) deposits

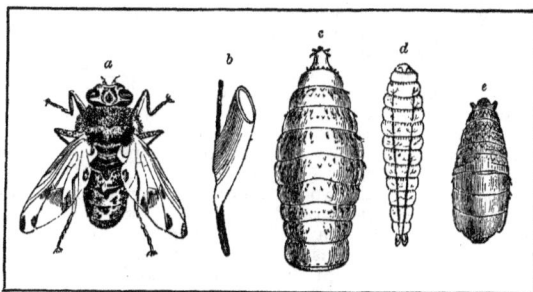

DEVELOPMENT OF HORSE BOT-FLY

a, adult fly ; *b*, egg attached to a hair ; *c*, mature larva ; *d*, newly-hatched larva ; *e*, pupa

its eggs in the hair of the skin of cattle, and the maggots, after hatching, burrow through the skin and take up their lodging in the tissues beneath, where, in course of development, they give rise to the large tumours known as warbles, each of which opens to the exterior by means of a small aperture. In these tumours the maggots remain for ten or eleven months until practically full-grown, when, quitting their host, they fall to the ground, bury themselves, and in the course of a month or six weeks emerge from the pupa stage as fully developed flies. The species most commonly met with in England is not *H. bovis*, but *H. lineatum*.

It can be easily understood from the fact that since no fewer than four hundred maggots, each growing to an inch in length, have been known to infest a single beast, the loss occasioned by the attacks of

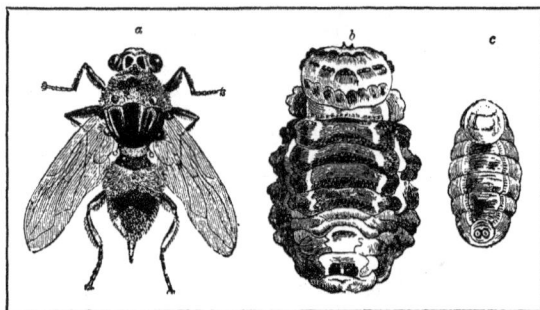

OX-WARBLE FLY

a, adult fly ; *b*, larva ; *c*, lower side of pupa

this fly is considerable, and it has been estimated, indeed, by Stratton, that in the United Kingdom alone a loss of something like £8,000,000 per annum is sustained. The mischief begins in the summer, when the cattle gallop about in terror in their vain efforts to escape the flies seeking to deposit their eggs upon them. This causes waste of milk, and damage to health. Then there is the damage to the meat by the destruction to the tissue just under the hide, resulting in what butchers call licked meat or jelly. And, lastly, there is the evidence of tanners as to the damage to hides ; one estimate given by a firm putting the loss on hides sold at two markets in Birmingham during seven weeks at £545 ; while a Nottingham authority reckons the loss in that town at £1,500 to £2,000 per annum.

The sheep bot-fly (*Œstrus ovis*) lays its eggs in the nostrils of sheep, and the maggots, after being hatched, pass up the nasal passages and enter the chamber in the bones of the forehead, where they nourish themselves on the mucus to which the irritation of their presence gives rise. The presence of these parasites, which are seldom fewer than seven or eight at a time, is most injurious to the infested animal, and gives rise to a sickness of a very serious nature. At the end of about nine months the larvæ reach maturity, and, making their way again into the nostrils, are expelled by the sneezing of their host, and reaching the ground bury themselves, and remained concealed until they emerge as perfect insects from the pupal stage.

The three species above mentioned serve as types of the life-histories of the entire family, which contains, in addition, a large number of genera and species infesting various kinds of animals. Even man himself is not exempt from their attacks, and all kinds of domestic cattle and beasts of burden, such as reindeer, camels, and elephants, are liable to be infested with them.

Two notices of the occurrence of larvæ in human beings were published by John Howship in 1833. In both cases the larvæ, named *Œstrus humanus*, were extracted from tumours, the sufferer in one case being a soldier in Surinam, and in the other a carpenter

in Columbia. In addition to the mammals mentioned, others, such as hares, rabbits, mice, and voles, often suffer from these parasites. Their larvæ have also been met with in birds and frogs. Schneider, for instance, states that two larvæ much resembling those of *Hypoderma* were obtained from under the skin of the head of a young sparrow, where they had produced two large hard tumours, and Krefft has given descriptions of specimens belonging to the genus *Batrachomyia* that were found living parasitically upon Australian frogs. The larvæ were situated between the skin and the flesh behind the drum of the ear, and could be squeezed out through apertures in the skin.

FOREST-FLIES

The family *Hippoboscidæ* brings us to the second section (Pupipara) of the Cyclorrhapha, all the members of which are no less remarkable among flies for the strangeness of their appearance than for their method of development. They are all short and flat, with longish and powerful legs, which enable them to run with great speed, some of them being entirely wingless, with the mouth parts much reduced ; but in the mode of their development they are absolutely unique in the entire order. In the first place only a single young one at a time is produced, and this, instead of being laid in the egg stage, remains within the mother, nourished at her expense by means analogous to those which obtain in the higher mammals. When born, the young is either actually a pupa or immediately assumes the pupa state, being motionless, without segmentation, and entirely protected by a horny shell, which imparts to it the appearance of the seed of a vetch.

The members of this section, which are mostly parasitic on birds or mammals, are referable to three families. Of these, the forest-flies are represented by several genera, all the members of which are parasitic upon mammals or birds, and are frequently spoken of as ticks. The species known from its abundance

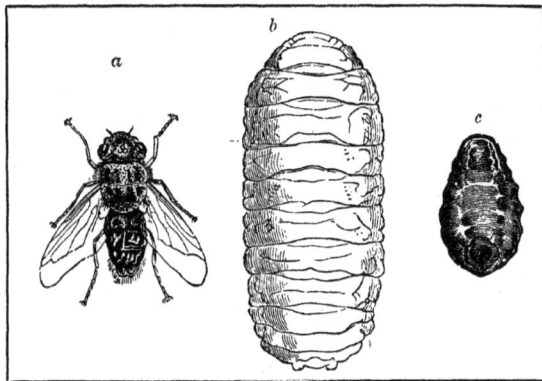

SHEEP BOT-FLY

a, adult fly ; *b*, upper side of larva ; *c*, under side of pupa

in the New Forest as the forest-fly (*Hippobosca equina*) has the wings well developed. It infests horses and oxen, usually attaching itself to those parts of the body where the covering of hair is scanty. A second kind, known as *Ornithomyia avicularia*, occurring, as its name indicates, on birds of almost all kinds, also possesses a pair of fully developed wings ; but in another species, *Stenopteryx hirundinis*, which is found on swallows and about their nests, the wings are narrow and sickle-like, and scarcely fitted for flight.

A fourth species, the so-called deer-tick (*Lipoptena cervi*), is provided with wings upon issuing from the pupa case ; but, after flying about for a time, the insects settle upon deer, and drop their wings by fracturing them at the base. The last member of the family to be mentioned, the so-called sheep-tick—which must not be confounded with the mite of that name—is entirely wingless from its birth. We thus get in this family a series of forms starting with the fully-winged forest-fly and leading through the swallow-tick with its wings reduced in size, and the deer-tick which can cast its wings, to the sheep-tick which has entirely lost these organs.

The second family of the group, *Nycteribiidæ*, contains the single genus *Nycteribia*, the species of which live parasitically upon bats. All are wingless and have lost their compound eyes, but possess the balancers. The legs are long, powerful, and furnished with strong hooked claws, by means of which they cling to the hosts they infest. The bee-louse (*Braula cæca ;* G on page 1912), the type of the family *Braulidæ*, is a minute, blind, and wingless insect infesting honey-bees, being found upon the workers as well as upon the drones and queen, but seeming to have a preference for the two latter as hosts.

COMMON FOREST-FLY

FLEAS

The fleas, which by some are regarded as an order (Aphaniptera), may be considered to be aberrant flies, their mouth organs, which are adapted for piercing and sucking, being modified upon the same principles as obtain in the flies. They further resemble that group in undergoing a complete metamorphosis, but differ from the majority of flies in being destitute of wings. The group is divisible into two families.

In the true fleas, or *Puli-cidæ*, the body of the adult is strongly flattened from side to side, and this, in conjunction with the smooth, hard, and nearly naked integument, enables the insect to swiftly traverse the hairy coating of its host. Some of the segments, however, are usually armed with strong b a c k w a r d l y - projecting spines. There are no compound eyes, but each side of the head is furnished with a simple eye, the legs being able, strong, and fitted for leaping. The eggs are laid about the floors of houses, kennels, etc., and the larvæ, which are slender, worm-like creatures, devoid of legs, but furnished with a biting mouth, live on particles of decaying organic matter found in the dust of the places they infest. When adult the larva, or maggot, is said to spin a cocoon within which the pupa state is passed.

In addition to mankind, fleas (*Pulex*) live parasitically upon other animals, such as dogs, cats, badgers, pigeons, fowls, moles, hedgehogs, squirrels, etc. They are, moreover, even more abundant in tropical than in temperate countries.

To the family *Sarcopsyllidæ* belongs the dreaded chigoe, or jigger (*Sarcopsyllus penetrans*), of tropical

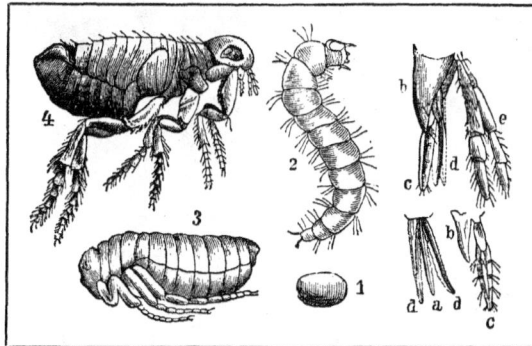

COMMON FLEA AND ITS STRUCTURE

1, egg ; 2, larva ; 3, pupa ; 4, perfect insect ; *a*, labrum ; *b*, labium ; *c*, labial palpi ; *d*, mandible ; *e*, maxillary palpi

countries. The adult female burrows beneath the skin of the foot, and shortly after effecting an entrance her body becomes swollen up with eggs, and grows to the size of a pea. At this stage she may be easily extracted, and as the young are not parasitic it is seldom that serious results ensue. According to Mr. W. H. Blandford, " the recorded distribution of the chigoe extends over Tropical America and the Antilles, from 30° N. to 30° S., and in late years it has been exported in ballast to Africa, and has established itself in Angola, Loango, and the Congo." It also occurs in British Central Africa, where quite recently it occasioned much suffering among the natives, and by laming the native postmen caused delay in the transmission of mails.

As in the case of the *Pulicidæ*, the fleas of this family do not confine their attentions to mankind. There is, for example, a genus known as *Vermipsylla*, which occurs in Turkestan, and is said to be very injurious to cattle. Speaking of the occurrence of the jigger in Florida, a correspondent writes that " the wooden houses are built on piles, and under them the sand is infested with jigger-fleas. All dogs are attacked by them, and fowls and puppies frequently killed ; in fact, sitting hens must regularly have their combs covered with lard and brimstone, and insect powder dusted over their wings, to keep them alive.

LEPIDOPTERA

The beautiful insects comprehended in the order to which the name Lepidoptera, or scale-wings, has been given are familiar to the majority of readers without any lengthy introductory description. The butterflies, or Rhopalocera, and the moths, or Heterocera, though they form two distinct sections of the order, cannot be divided by any hard-and-fast lines. They may generally be distinguished from one another by the manner of the folding of the wings at rest, or more precisely by the different character of the antennæ. The wings of the moths, too, are locked together by a tiny hook on the inner margin of one wing fitting into an eye on the inner margin of the other. The butterflies never possess this curious structure.

The Lepidoptera are easily distinguishable from other orders of insects by the four ample wings, with more or less regular veins or nervures, clothed with the minutest, exquisitely chiselled scales, of many shapes, and great variety of external chasing. These scales are but modified forms of hairs, broadened out, flattened and fashioned to cover the delicate membrane of the wing with an overlapping armament of beauty. And it is to this wondrous sculptured dust, breaking up the rays of sunlight as it plays upon the surface of their wings, that butterflies and moths owe their tender shades, brilliant colours, and metallic lustres. A few butterflies are clear-winged, with scarcely any scales, such as the *Ithomia* of Brazil, while the *Sesiidæ* represent the clear-wings amongst the moths. Some orders of insects, such as the Hymenoptera, have four membranous wings like the Lepidoptera, but these are transparent and not clothed with scales. Others, such as the beetles, have the upper pair horny and useless for purposes of flight, the second pair being membranous but not scaly.

The mandibles, or jaws, found in most other four-winged insects except the Hemiptera, or bugs, are replaced in the Lepidoptera by a long tubular proboscis, or suctorial apparatus, used for exhausting the contents of honey-bearing flowers, or drawing in nutriment from less tasteful sources. In common with all other insects, the Lepidoptera have the body divided into three separate sections—the head, bearing the eyes, proboscis, and antennæ; the thorax, whence originate the legs below and the wings above; and, lastly, the abdomen, bearing along the sides the spiracles for breathing, and the generative organs at the apex. The abdomen is never attached by a narrow stalk, or pedicle, as in the Hymenoptera. So close may their general resemblance be to other insects that, as is the case with the hornet clear-wing moth, none but a naturalist could distinguish it from the common hornet. A general resemblance of body-plan may coexist in individuals of two widely separate orders, together with a habit of life and temperament, and likewise essential characters, wholly distinct and different.

The Lepidoptera also resemble the insects of most other orders in passing through several sharply-defined phases before the last and perfect stage is attained. All first appear in the form of an egg laid by the mother on some food-plant or tree. On hatching, the eggs give rise to a free-walking, feeding, sleeping, and breathing larva or caterpillar; thence, after successive changes of the skin, this passes into the quiescent, trance-like state, called the pupa or chrysalis stage; from this it at last emerges, at a suitable season of the year, as the fully-formed butterfly or moth.

At the beginning of life the butterfly or moth is a thing of beauty even in the egg state. Butterflies' eggs, though falling into distinct groups of resemblance, on which even systems of classification have been based, are as various as they are beautiful. Globular, oval, flat, barrel-shaped, bottle-shaped, green, white, or brown, the egg is usually of a hue which renders it not easily visible on the leaf where it has been deposited. After a time the shell bursts, the tiny larva creeps forth, and begins feeding either on the egg-shell or on the food lying in abundance near at hand.

The larvæ are long, cylindrical, creeping, worm-like objects, with short legs, and a more or less hairy or

quite naked body. The greater number feed upon the leaves of trees, shrubs, plants, and grasses; while many are internal feeders, burrowing deep into the decaying hearts of various trees. Others mine in the pith of thistles; while many more burrow at the roots of grass or devour turnip-roots, to the detriment of the crops. The larvæ of the mining moths (*Tineæ*) make sinuous channels between the upper and lower skin of various leaves. These in the perfect form are amongst the smallest and most lovely of all the Lepidoptera. Others, again, feed on clothing and other woollen stuffs, gnawing ragged holes, and when the imago, or perfect insect, appears the mischief has been done. So voracious are larvæ that huge oak forests may be in a few days swept bare of almost every vestige of foliage.

The body of the larva consists of a head with mandibles for nipping off the edges of leaves or gnawing amongst decaying timber; a pair of small, short antennæ form appendages on either side; and just behind three simple eyes, or ocelli, on either side, very different from the large compound eyes of the perfect insect. Behind the head lie eleven segments, or movable rings. Three of these, close behind the head, correspond to the thorax of the adult, and bear the three pairs of thoracic legs, short and horny, exactly corresponding to the three pairs of legs of the butterfly or moth. The other segments carry the pro-legs, or claspers—varying in number from one to five pairs—used for clinging to leaves and other surfaces. In some of the moths the last pair are obsolete as legs, and are developed into a pair of horns, supposed to be for protective purposes. as, for instance, in the puss-moth. A caterpillar may thus possess sixteen legs, though often there is not the full number.

A very curious form of larva is that producing the insects known as the geometers, so called because of the peculiar gait of the caterpillar, which measures out the surface over which it passes with a regular series of equal strides or loops. Their body is long, and, since there are only four pro-legs, they cannot crawl, but by bringing up the hinder legs, advancing the head, and again bringing forward the tail, the caterpillar spans the space to be traversed by a series of looping strides. Hence the Americans call them " span-worms." These larvæ, too, are remarkable for their resemblance—when the head is stretched outwards—to a broken twig, a likeness which undoubtedly secures them from many dangerous foes.

Many larvæ are protected by their similarity in colour to the surrounding foliage, and it has been supposed that the pigment from the leaves which the caterpillar eats lends its characteristic hue to its

KALLIMA, OR DEAD-LEAF BUTTERFLIES, AT REST
Photograph by A. Ullyett

devourer. From the moment of hatching until the final moult, when the caterpillar enters the pupa state, it undergoes a series of from eight to ten changes of the skin. These changes form crises in the lives of larvæ, which, at any rate in captivity, sometimes die during the process.

The stage immediately preceding that of the perfect form is usually called, when reference is made to the butterflies, the chrysalis state ; but in the case of the moths, the pupa state, though there is no essential difference between the two. In this strange, quiescent state the wings, legs, antennæ, and proboscis of the future insect can be seen fully formed and folded tightly within the outer covering. The only signs of vitality are given by wriggling movements of the segmented abdomen, when the pupa is irritated. The hard external covering is useful for resisting the attacks of predaceous insects, though, of course, not securing them immunity from mice, birds, or moles, which devour them with avidity.

The chrysalis of a butterfly is usually angular and gilded. Some are suspended simply by the tail, others have a silken girdle round the middle to keep them fast, while some spin a very slight cocoon. The pupæ of the moths, on the other hand, are dull red, usually smooth mummy-like objects, to which likeness the word pupa or " pupet " doubtless refers. The greater part of these lie simply in the earth, beneath moss or bark, wherever the larva has crawled to effect the change, without any additional covering. Others form a hard cocoon, of the grains of mould, to which consistency is given by means of a gummy secretion furnished by the larva. Many form with this secretion a hard case, the outer side covered with chips of the surrounding bar, which, owing to their similarity to the surroundings, serve as a protection from observation. Others spin a silken egg-shaped cocoon, sometimes flocculent and broken, sometimes formed of yards and yards of silken thread, emitted from the mouth and passed over and under, across and round, until the cell is complete.

Among the more interesting of these cocoons is that of the emperor-moth, which forms a short tubular exit closed against the entrance of earwigs and other insects by a circular series of fine bristles directed outwards and converging to a point. The principle of contrivance is the same as that employed in the manufacture of lobster-pots ; but here the process is reversed, for in this case it admits of a ready egress, but prevents any entrance. Moths whose pupa stage is passed within an external cocoon have a double task before them when the time is ripe for an emergence. The pupa itself—as does also the chrysalis of butterflies—splits at the dorsal suture above the thorax, and the moth emerges, ferreting a way through one end of the cocoon, which seems to be softened by moisture from within, and thus escapes. The imago, or perfect insect, having now emerged, climbs to some point of vantage, where the wings, still very small, though completely formed, are allowed to hang downwards, expand, and harden in the air. After a few hours they are stiff and ready for use.

At no stage in their lives are lepidopterous insects free from the attacks of enemies. In the egg state they fall a prey to beetles and small birds, and as larvæ they are extremely liable to receive a deadly thrust with the ovipositor (or sting) of an ichneumon. As the ichneumon grubs grow at the expense of their host, scarcely a tissue in the whole body may remain,

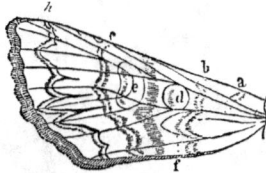

LEFT FORE WING OF NOCTUA MOTH

a, b, c, h, indicate normal position of the transverse bands. *a,* costal margin ; *f,* inner margin ; *d,* discoidal spot ; *e,* reniform spot

save those needful for the carrying out of life-supporting functions. And at last, when the grubs are themselves ready to pupate, and have no further need of their host, they finish up the rest and the larva dies—chiefly because there is nothing left to live. The enemies of the imago, whether butterflies or moths, are numerous. Birds, bats, dragon-flies, and the like, pursue and harass them whenever they happen to meet with them. The marvel is that any remain alive to lay eggs and perpetuate the species.

In the struggle to escape detection and capture, all unconscious though it may be, arises the phenomenon alluded to above, and known as protective mimicry. The kindred phenomenon of protective colouring, when the moth or butterfly merely resembles in hue the bark, leaf, or twig on which it rests ; also protective resemblance, simply when insects take the form of objects, such as twigs, dead leaves, bits of decayed wood, flakes of white bird-droppings—these are all well known. But protective mimicry means more—it implies the actual mimicking either the form, colour, or habits of some other insect which is either too savage or unpleasant to make it a desirable object of food ; as, for instance, the clear-wing moths mimic gnats, bees, wasps, ichneumons, and so forth.

Perhaps the most curious instance is that of the *Kallima,* or dead-leaf butterflies, of Northern India, whose upper sides are richly coloured, while the under sides are dull brown mottled and veined with darker colours. So conspicuous a butterfly would not fail to fall a ready prey to foes. If it but settle for an instant, however, the sharpest eye will not detect it. The secret lies in the colour and veining of the under side. The fly settles, clings to a twig, presses the tails of the under wings—now folded together against it—and nothing but an old withered leaf remains where but just now was a gaudy butterfly.

A species of the genus *Heliconius,* an insect avoided by birds on account of its bitter flavour, is closely mimicked by another butterfly of the genus *Mechanitis.* Though very sweet-flavoured, it escapes unmolested amongst its less agreeable companions. The mimicry involved in the feigning of death by many species of moths is, of course, protective. It has even been asserted that a specimen of the magpie-moth continued to feign death three hours after its head had been severed from the thorax.

If all the dangers noted above have been passed through with impunity, in due time, at various seasons of the year, the perfect insects—butterfly, or moth, as the case may be—will emerge. These vary in size from 12 inches or more in the expanse of the upper wings to a quarter of an inch—the latter being among the smallest moths, or Microlepidoptera.

We have remarked that the body is divided into three distinct divisions—head, thorax, and abdomen ; we must now shortly notice the various structures peculiar to each division.

The first division of a lepidopterous body is itself divided into four main divisions—the occiput, next to the thorax ; the epicranium, bearing the antennæ ; and, in some moths, the ocelli, or simple eyes ; the clypeus, lying in front of the epicranium, just on the mouth parts, which latter themselves fall into at least five or six distinct structures ; the proboscis, long, and capable of being rolled up beneath the labrum when at rest. The labrum, lying at the base of the proboscis, above ; the maxillary palpi (absent or vestigial in the butterflies); the labial palpi, and vestigial mandibles, aborted in many cases, complete the mouth structures.

It is in the structure of the mouth parts, perhaps, that the butterflies and moths differ most from other insects, and more especially from the fact that the mandibles of the insects have in the Lepidoptera become modified into a long, spirally curled, retractile proboscis, composed of three distinct hollow tubes, soldered to each other along their inner margins. Indeed, it has much the appearance of a double-barrelled gun, with a third tube lying below beneath the suture of the upper and larger pair. But it is with this latter alone that nutrition is imbibed, and it is supposed that the other pair may furnish air in addition to that obtained through the spiracular orifices along the abdomen.

The ocelli, or simple eyes, resembling those of the larvæ, the small eyes on the upper part of the head of bees and other Hymenoptera, as well as those of other arthropods, such as we find to the number of from four to eight in the spiders, are not discoverable in the butterflies, but are present in the moths alone. The large compound eye, composed of numerous facets, is, however, present in both sections of the order, lying on either side of the epicranium, just below the point of insertion of the antennæ. Whether they see Nature with these " as through a veil," or appreciate every detail as we do ourselves, is a matter of speculation, but not easy of solution.

The pair of thread-like, many-jointed organs, which take their origin from the side of the epicranium, just above and within the compound eyes, are among the butterflies, with the exception of the family *Hesperiidæ*, thread-like, and abruptly clubbed at the apex. In the latter family they are gradually thickened towards the end, which often terminates in a hook-like point.

The moths, however, as their name Heterocera implies, furnish us with far greater variety in the form of the antennæ, quite apart from the fact that they differ in both sexes of the same species. Thread-like, for instance, in the female, pectinate in the male sex, we find at least ten different forms of antennæ among members of this section : *Filiform*, or thread-like, gradually tapering to a point ; *fusiform*, broadened from the base onwards to the tip, which is again narrowed ; *dilate*, narrow from the base to about one-third its length, then rather suddenly enlarged, and again narrowed at the tip ; *ciliate*, clothed with the finest hairs ; *setigerous*, each joint furnished with a bristle on either side ; *setigerous* and *ciliate*, furnished with both fine hairs and paired bristles ; *fasciculate*, each joint furnished with a group or tuft of short bristles, like a small brush ; *dentate*, or toothed, each joint produced into a sharp tooth-like process at the side ; *lamellate*, where each joint is produced at the margin into a small plate-like prominence ; *serrate*, saw-like, each joint produced into a short, sharp point at the side, giving the whole antennæ the appearance of a saw, with the teeth not so long as in the dentate antennæ ; *pectinate*, each joint furnished with long plume-like hairs, or a pair of such on either side.

The second division of the body, like that of the Hymenoptera, is composed of three closely united rings, each bearing beneath a pair of legs, while the posterior pair carry also on their upper or outer sides each a pair of well-developed wings. The *prothorax* bears the fore legs ; the *mesothorax* the mid-legs and fore wings ; and the *metathorax* the hind legs and hind wings. The legs are not used normally for walking, but are chiefly serviceable for clinging to objects while settling or at rest. They do not call for any special mention, and are not of great account for purposes of classification, except in the butterflies, where in the case of the males of the three families *Nymphalidæ*, *Erycinidæ*, and *Lycænidæ*, the fore pair are much reduced in size, being in some cases almost rudimentary.

By far the most important structure in the eyes of the general naturalist, though not necessarily so in the opinion of the expert, are the beautiful membranous, scale-clad pinions. These give the distinctive character to lepidopterous insects, and render them so fascinating to the lover of Nature. Broad and ample on the whole among the butterflies, more narrowed as a rule amongst the moths—the hawk-moths, for instance —they are formed of the finest transparent membrane stretched out between the stiff rib-like nervures, or, more properly speaking, veins, which carry the nutritive fluids from the central system to all parts of the structure.

The nerves, as custom will persist in terming them, in the butterflies take a bow-like or ellipsoidal sweep from the base of the wing, forming what is called the *discoidal cell*, whence there branch off to the edges a series of horizontal, almost parallel, slightly divergent nervures. On the position of these the identification of species is most securely based.

SMALL TORTOISESHELL BUTTERFLY

In the moths, on the other hand, the *discoidal cell* is less conspicuous, though nervures branch off divergently from the base of the wing in a somewhat similar manner to those of the butterflies. One of the most remarkable features in the wings of the Heterocera, as distinguished from those of the Rhopalocera, is the existence of the frenulum and retinaculum, briefly referred to above as the hook-and-eye arrangement.

As already said, the scales are modified hairs, which take a more and more perfect scale-like form towards the centre of the wing. They lie in regularly arranged rows, overlapping each other, attached by a short stalk to a small fovea, or pit, in the membrane, to the number of many hundreds of thousands on each insect. Of different shapes and sizes, they are themselves, owing to their exquisitely sculptured surface, objects of extreme beauty. And it is to these alone that butterflies and moths owe their manifold tints, from the sombrest browns to the most resplendent metallic greens, golds, and purples.

The third division of the body is composed of a series of nine rings, or segments, sometimes, as in the case of many of the moths, tufted along the dorsal line, and also at the extremity. The spiracles, through which the air passes to the tracheal system, lie along the sides of the abdomen, while the organs of reproduction are placed at the extremity in both sexes.

BUTTERFLIES

As distinguished from the moths, the butterflies (suborder Rhopalocera) may be recognised as a general rule by their antennæ, which, as suggested by their name, are slender and abruptly clubbed at the extremity. In some cases, however, in the family of the skippers these organs are gradually enlarged towards the tip, which is itself often slightly hooked. Butterflies have not, in any case, the hook-and-eye arrangement by which the upper and under wings are interlocked in the moths. The fore legs are not always well developed, and this is particularly noticeable in members of the male sex, forming a trustworthy character in the broad subdivision of the Rhopalocera into families.

Butterflies are mostly diurnal, although a few take wing only towards evening. Their eggs and larvæ differ in many respects from those of the moths, while the chrysalis is seldom enclosed in even the finest network of silk, and in no case is wrapped in a distinct cocoon, nor even buried beneath the earth, very rarely even close to the surface. Usually the chrysalis is angular and blotched and speckled, with gold and silver ornamentation ; sometimes it is suspended to a branch or twig by the tail, and sometimes, while fastened by the tail, is also engirdled with a line of silk around the middle, thus lying in a position horizontal to the plane to which the larva has attached itself.

The following broad subdivisions of butterflies may be made : Firstly, those which have four perfect legs .y in both sexes, the fore pair being rudimentary or undeveloped ; while the chrysalis is suspended by the tail without any girdle. These include the family *Nymphalidæ*. Secondly, those having four

GROUP OF BUTTERFLIES

1, peacock butterfly ; 2, the same just emerged ; 3, its caterpillar ; 4, its chrysalis ; 5, meadow-brown ; 6, its caterpillar

perfect legs in the male and six in the female, while the feet of the former have no claws at their extremity ; the chrysalis being raised, resting on a leaf or suspended. The *Erycinidæ* represent this group. Thirdly, we have the family of the blues (*Lycænidæ*), in which there are six perfect legs in the female, and the chrysalis is suspended. The fourth group is that of the swallow-tails (*Papilionidæ*), in which both sexes possess six perfect legs, while the chrysalis is attached by the tail and girdled by a silken thread. Lastly, the *Hesperiidæ* agree with the preceding as regards the legs, but the chrysalis is either attached by threads or enclosed in a loose cocoon. As a rule, mountainous regions are those which abound most in butterflies, although there is a marked exception in the case of the valleys of Tropical America.

FRITILLARY GROUP

The family *Nymphalidæ* includes an extensive assemblage of butterflies, among which are the fritillaries, peacocks, painted ladies, tortoiseshells, and admirals. Here also come the leaf-butterflies, purple emperors, white admirals, Camberwell beauty, and the large high-flying blue *Morphos*. We have also the subfamily *Satyrinæ*, which includes the ringlets, marbled whites, meadowbrowns, and graylings, besides many others too numerous to mention. First we may notice, as an example of the subfamily *Danainæ*, the butterfly shown on the lower right-hand corner of the coloured picture on page 1969, which is known as *Euplœa harrisii*. In common with several other species, it belongs to a genus of large blue and brown winged tropical butterflies, in which the upper surface of the wings is usually spotted with white. At the top left-hand corner of the same picture is figured the male of the orange scallop-wing (*Cethosia biblis*), which may be taken as a representative of the subfamily *Nymphalinæ*. It is an inhabitant of North-Eastern India. Its black and spiny larvæ have the body banded with red and yellow, and the head surmounted with a pair of horn-like processes.

The better-known fritillaries (*Argynnis*), are mostly confined to north temperate regions. In this genus, the British silver-washed fritillary (*A. paphia*) is among the finest representatives of a large number of orange-red or fulvous insects whose hind wings on the under side are spotted, spangled, or slashed with silver upon a dusted green ground. Not uncommon throughout England, it occurs in abundance in the glades of the New Forest, where the larva feeds on the dog-violet or wild raspberry. The dark green fritillary (*A. aglaia*), a near relative, frequents the southern grassy downs along the margins of the cliffs, or sports in the fern-embroidered dells of the Lake District valleys. The high brown fritillary (*A. adippe*) is a rather smaller form, whose hind wings, like those of the last-named species, are spotted with silver discs, while those of the silver-washed are slashed obliquely towards the lower angle. The Queen of Spain (*A. lathonia*) is a much rarer insect, while the two elegant little pearl-bordered fritillaries (*A. euphrosyne* and *A. selene*) are also British.

The greasy fritillary (*Melitæa aurinia*) brings us to another genus, the members of which closely resemble those of the former but are as a rule smaller. So many figures of all the British species have been published that detailed description is superfluous. The greasy fritillary inhabits low-lying marshy meadows in various localities in England, where the larvæ feed on the plantain.

The heath fritillary (*M. athalia*) is a very similar though very local species ; while the glanville (*M. cinxia*) is rare in Britain, where it is confined to the Isle of Wight. Many handsome species of this genus are found in all the more northern regions of the world, but undoubtedly the most numerous occur in the South-Western United States. The magnificent fritillary *A. childreni*, which measures nearly 5 inches from wing-tip to wing-tip, is indigenous to the Himalaya.

Closely allied to the fritillaries is the map-butterfly (*Araschnia levana*) of Central Europe. It presents two very distinct forms, one of which (*A. levana*) appears in the spring, the other (*A. prorsa*) later on in the summer, while an intermediate form (*A. porima*) is also recognised.

The form known as the spring brood is fulvous red with scattered black spots, presenting also three white spots near the tip of the wing. The summer brood has black wings with a red marginal line, having besides a broad broken white bar across the wings and some white spots near the margin. The larvæ feed on the nettle in June and September. The insect, though common on the Continent, has not been recorded from England.

The curiously-shaped butterfly known as the comma (*Polygonia c-album*) was formerly much more common in England than it is at present. The wings are rufous with black spots, and very strongly emarginate along the edges, and angular. The white c-shaped spots on either hind wing beneath render it not easily mistaken for any other British species.

The handsome butterflies designated tortoiseshells (*Vanessa*) are among the most widely distributed of the family, though confined to the Northern Hemisphere. Most inhabit the more temperate regions of Europe, Asia, and America, although a few now occur in India, Ceylon, the Malay Peninsula, and Mexico. The caterpillars feed on plants and trees, and are usually dark and spinous. The chrysalis, angular and distinguished by its brilliant lustre, is suspended by the tail, and forms a beautiful object. The large tortoiseshell (*V. polychloros*), so common in woods in England, is usually found settling upon the trunks of trees in summer and autumn. The wings are rich fulvous red, blotched and margined with black, and having a narrow broken vein of blue just before the outer fringe. The larvæ feed on the leaves of various trees, and the chrysalis is pale pink relieved with golden blotches. The small tortoiseshell (*V. urticæ*), whose jet-black spiny larva feeds on the nettle, is among the commonest British butterflies.

The peacock butterfly (*V. io*) is well known on account of the large eye-like blotches on the upper and under wings. The larvæ also feed upon the nettle ; and the insect is found throughout Europe and Northern Asia as far as Japan, but not in Northern Africa. One of the handsomest, and at the same time one of the rarest, of British butterflies is the Camberwell beauty (*V. antiopa*). Its large angular wings are rich brown above, with a broad yellow border, enclosing on its inner margin a row of blue spots.

In the tropics the place of the preceding genus is taken by *Junonia*, the members of which are not perhaps so richly coloured as the tortoiseshells. They occur all over Eastern and Southern Asia, and are also found in North and South America, the Oriental countries, and Africa. The caterpillars are spinous, as are those of the two tortoiseshells. A figure of the beautiful although dark-coloured Swinhoe's tortoiseshell (*J. swinhoei*) is given at the lower left-hand corner of the coloured plate on page 1969 (9).

As an example of the genus *Pyrameis*, we may take the red admiral (*P. atalanta*), which is a well-known and richly-coloured British butterfly, appearing in the autumn in woods, and also in orchards, where it feeds upon the juices of decaying apples. The large black wings, with a scarlet band across the upper and a margin of the same colour around the lower, together with the group of pure white blotches towards the tip of the former, render it a very conspicuous insect. When, however, the wings are closed, the mottled black and brown render it almost invisible. The larvæ are black and spinous, and feed upon the common nettle ; and the species is found all over Europe and North Africa, North and West Asia, and North and Central America. In many other regions its place is taken by some very closely allied forms.

In the painted lady (*P. cardui*), of which the caterpillars feed upon the thistle, the wings are orange-red, black-spotted, and black-tipped, the latter area bearing a group of white spots. It is abundant in almost every country of the world, except the Arctic regions and South America.

Nearly allied are the porcelains (*Cyrestis*), which measure from 2 to 3 inches across the wings, and are found in India, the Malay Archipelago, and a few in West Africa and Madagascar. The sooty-veined porcelain (*C. thyodamas*), represented in the coloured illustration on page 1969 (6), is an inhabitant of Madagascar.

Of the genus *Limenitis*, the large white admiral (*L. populi*) occurs in Central Europe, South Scandinavia, and Finland, but has not been met with in the British Isles or in Holland. It is nearly twice the size of the English white admiral (*L. camilla*), its wings being brown with a row of lunate orange marks near the hinder margin of the lower pair. The arrangement of the white bars on the upper wings is the same as

WALL-BROWN

RESPLENDENT PTOLEMY

2 A

that of the British form, but these are almost obliterated in the male sex. The under side is of a beautiful orange-yellow colour, broken with white, and elsewhere suffused with various shades of purplish and bluish grey.

Closely allied to the admirals are the mango-butterflies (*Euthalia*), which are almost entirely confined to India, the Malay Peninsula, and the adjacent islands. They measure from 2 to 4 inches across the wings, and the larvæ feed on the leaves of the mango. A coloured illustration of the black mango-butterfly (*E. lubentina*) will be found on page 1969 (3).

The emperors (*Apatura*) are widely distributed over the world, except in Africa. Two species alone are found in Europe, and these are much more brilliant insects than the majority of the temperate species. The caterpillars are not hairy, but smooth, and bear a pair of horns on the head, as also does the chrysalis. In Britain the purple emperor (*A. iris*) is confined to the southern counties of England. Its strong purpleshot, white-banded wings, 3 inches in expanse, carry it with a grand sweeping flight far above the highest oak-trees, whence it descends — alas for imperial predilection!—to a savoury banquet of putrid flesh, set out in some suitable locality. The caterpillar feeds upon the sallow, and the perfect insect appears in July.

Passing over many genera, containing some of the loveliest foreign forms, we reach the subfamily *Morphinæ*, in which the caterpillars are remarkable for their bifurcate tail and notched or bifid head.

The species of the typical genus are giant butterflies of almost every hue, the most conspicuous being of a dazzling metallic sky-blue. Their long, satiny wings bear them aloft far out of the reach of the collector's net. In the illustration on page 1938 is figured, from the under side, the resplendent ptolemy (*Morpho neoptolemus*). The upper side is rich black brown, with broad transverse blue bands, shot with delicate lilac across both wings. A pair of white spots is conspicuous on the top of the fore wing.

We have now to notice briefly a number of much less brightly-coloured butterflies, many of which will be familiar to most readers, forming the subfamily *Satyrinæ*. They include the ringlets (*Erebia*), speckled-woods (*Pararge*), marbled whites (*Melanargia*), meadow-browns and heaths (*Epinephele* and *Cænonympha*), wall-browns (*Satyrus*), graylings and common wood-ringlet (*Hipparchia*), and many others. The caterpillars are mostly smooth, fusiform, and green, having two horns on the head and a bifurcate tail. They feed on grasses. These butterflies fly somewhat feebly over meadows, downs, highlands, and heath districts.

As an example of the typical genus *Satyrus* may be taken the common British wall-brown (*S. megæra*). Here the wings are rufous brown, spotted, speckled, and streaked with black, having also a single eye-like spot on the upper wing at the tip, and three on each lower wing, near the margin. As a rarity, collectors prize a specimen in which the fore-wing spots are bipupilled—having twin pale centres.

Of the graylings (*Hipparchia*), the British *H. semele* is abundant in the heath and mountainous districts of England. Owing to its beautifully grey-mottled under side, it is absolutely invisible when settled upon rocks or among the grey stones of the moorlands. The nearly allied meadow-browns and heaths (*Epinephele*), which do not present a very great number of species, are most abundant in the Mediterranean region and Western Asia. They fall into two groups, of the first of which *E. janira* is a good example; while *E. tithonus*, the large heath or gatekeeper, illustrates the other. The former, which is the commonest of British butterflies, abounds in fields and meadows in the summer, ceasing to fly the moment the sunbeams are obscured by a passing cloud. Specimens with pale patches on the wings are valued by lovers of varieties. The upper figures on page 1936 represent the adult and caterpillar.

ERYCINIDÆ

The small family *Erycinidæ*, of which the characters are given on page 1936, includes species chiefly found in the tropics. *Erycina aulestes* of Brazil is peculiar in having the

GROUP OF BRITISH BUTTERFLIES

1, large white admiral ; 2, female, and 3, male of golden-rod copper ; 4, small copper ˙ 5, azure or Clifden blue ; 6, silver-studded skipper ; 7, Duke of Burgundy

mid-wings produced into a tail-like projection. As an example of the family we may take the Duke of Burgundy butterfly (*Nemeobius lucina*), an illustration of which is given in the coloured picture on page 1969 (8). Its brown, yellow-spangled wings once earned for it a place amongst the fritillaries. It is, however, the sole British representative of a family whose members are so abundant in Brazil.

BLUES AND COPPERS

The large family *Lycænidæ*, represented by many small brightly-coloured insects, includes the blues, coppers, hairstreaks, and many others. Of the hairstreaks (*Thecla*) the purple hairstreak (*T. quercus*) is a familiar example. This butterfly has the wings brown-black, shot with purple, and abounds all through Europe wherever oak forests exist. It flits round the foliage, laying its eggs, and resting on the leaves, and is a common British butterfly. The green hairstreak (*T. rubi*) is a smaller species than the rest, with a bright green under side, and is not uncommon in some districts flying around bramble bushes in summer.

In the allied genus *Polyommatus*, we mention the large copper (*P. dispar*) as one would speak of a departed friend, for, although formerly abundant in the fens of Cambridgeshire and other counties, it has not been seen alive for over half a century in Britain. The small copper (*P. phleas*) is, however, very abundant both in England and on the Continent. It is shown in 4 of the illustration on page 1938. Of the golden-rod copper (*P. virgaureæ*) figures are given in 2 and 3 of the same illustration. This species is abundant on the Continent, though unknown in Britain. It flies in July and August, and the larva feeds on the golden-rod.

The elegant little butterflies known as blues (*Lycæna*) have the upper side of the wings in the male sex of various shades of blue; those of the female, on the other hand, being usually brown, shot with a bluish or purple tinge. The larvæ are wood-louse shaped, and feed mainly on grasses of various kinds. The common blue (*L. alexis*) is one of the most abundant of British butterflies, whose white-fringed, pale blue upper side and speckled under side in the male are familiar to everyone.

Of the many blues found in England, such as the silver-stud, the chalk-hill, the holly blue, and the little or Bedford blue, the Clifden blue (*L. adonis*)—the azure blue of many authors—is the most beautiful. It occurs not infrequently, though locally, upon the Chalk downs of the southern coasts, and in some other localities. A figure of the male is given in the illustration on page 1938. The wings are of a much brighter blue than those of *L. alexis*.

SWALLOW-TAILED GROUP

The immense family *Papilionidæ* includes the giant *Ornithoptera*, or bird-winged butterflies of the tropics, the swallow-tails, Apollo butterflies, whites, brimstones, and many others. As mentioned above, this family and the next are characterised by the possession of six perfect legs in both sexes. The chrysalids of the present family are suspended by the tail and girdled with a thread of silk. The largest of the butterflies (*Ornithoptera*) belonging to this family measure nearly a foot across the expanded wings. A species (*O. paradisea*) is figured on page 1940. The typical members of the family are the swallow-tails (*Papilioninæ*), which are large butterflies characterised generally by the presence of a long tail-like process to the hind wings. Occasionally, however, as in the female of *Papilio merope*, these appendages are wanting.

The two uppermost figures of the illustration on this page exhibit the scarce swallow-tail (*P. podalirius*), which is a large, strong insect with triangular front wings, and a long tail at the lower angle of the hinder pair. In colour the wings are pale yellow, with oblique transverse black bars. This splendid butterfly, although common in Southern Europe, North Africa, West Asia, and Persia, is only very rarely taken in England. The larvæ feed on leaves of the sloe, apple, plum, and other

GROUP OF TROPICAL BUTTERFLIES

1, scarce swallow-tail, with larva and chrysalis; 2, map-butterfly, spring brood; 3, larvæ; 4, summer brood; 5, chrysalis of same

orchard trees. The common swallow-tail (*P. machaon*) was formerly very abundant in the fen districts of England, but since these have been drained it has become scarcer. The wings are sulphur yellow, black at their base, with black veins, and the hind pair of the same colour, with a band of blue towards the margin, and a red spot on the inner angle, close to where the tail springs. This larva feeds on the common carrot. This species has a very wide range, occurring in the Kashmir Himalaya. Of the royal swallow-tail (*Tinopalpus imperialis*), from Sikkim, a figure is given in 5 on page 1969. The females are less brilliantly coloured than the males, and have a pair of tails to each hind wing.

The whites, clouded yellows, orange tips, brimstones, and the like, represent the second subfamily (*Pierinæ*) of this assemblage, in which there are no tails to the hind wings. One of the rarest British butterflies is the black-veined white (*Aporia cratægi*), shown in all stages of development in the illustration on this page. Its caterpillar feeds on the leaves of the blackthorn and other bushes. Of a foreign representative of the group, the black-tailed sulphur (*Dercas verhuelli*), an illustration is given in 7 in the coloured picture on page 1969. It is nearly allied to the common brimstone butterfly (*Rhodocera rhamni*), so abundant in spring in English lanes and hedgerows.

SKIPPERS

The skippers (family *Hesperiidæ*) differ from all the others in the broad, thick head ; the hind tibia (with some few exceptions) being armed with two pairs of spurs. There are hundreds of species belonging to this interesting family, the majority being indigenous to South America. Many are distinguished by their powerful build, brilliant colours, and long-tailed hind

MALE AND FEMALE OF ORNITHOPTERA PARADISEA

wings. The European species are all small and more or less sombre-coloured, averaging about an inch across the wings.

In the puss-tailed skipper (*Goniurus catillus*) of Brazil the front wings are brown on the upper side, with five or six pale yellow spots ; the hind wing also being brown, and ending in long, broad flat tails, quite as long as the hind wing itself. The antennæ are strongly hooked at their apex.

Telegonus alardus, from Venezuela, has large wings, 2 inches across, brown, shot at their base with blue and green, but only

BLACK-VEINED WHITE, WITH LARVÆ AND CHRYSALIS

very slight tail-like prominences on the hinder wings.

To *Pamphila* and the following genera belong all the small, quick-flying butterflies, known as the skippers, properly so called. When at rest many of these insects raise the upper wings, leaving the lower ones horizontal, a habit not unknown among butterflies of other families. The Lulworth skipper (*P. actæon*) is a rare, or rather local, small brown skipper, confined in England to a few spots along the south coast.

Among others are *P. silvanus*, the large skipper, *P. linea*, the small skipper, and *P. lineola*, the scarce small skipper lately added to the British list. The dingy skipper belongs to another genus (*Nisoniades*), as does the chequered skipper (*Cyclopides*). The grizzled skipper (*Hesperia malvæ*) is a black or brown butterfly, with white spots on the upper side, common in England in summer. The silver-studded skipper (*H. comma*) is confined to some of the midland and southern counties of England, though abundant on the Continent. Figures of this butterfly will be found in the illustration on page 1938, and on the coloured plate on page 1969 (4).

MOTHS

Since limitations of space will only admit mention of a few of the genera and species of butterflies, we pass on to the moths, in which the antennæ are of many different forms, but never distinctly clubbed (suborder Heterocea). Moths are vastly more numerous—both in genera and species—than butterflies ; and, as already observed, are for the most part nocturnal insects. The other distinctive features having been already mentioned, we proceed to the first family of the group.

EMPEROR-MOTHS

The splendid moths included in the family *Saturniidæ* are probably among the most beautiful, as they certainly are among the largest, of all known Lepidoptera, ranging in size from the atlas moth (*Attacus atlas*), which measures a foot at least in expanse of wing, down to the English emperor-moth, of two or at most three inches in diameter. They do not, however, vary so very much in the comparative beauty of their richly-coloured ocellated wings. The larvæ, too, are not only of remarkable beauty, but have great commercial value; for it is from members of this family that China and Japan obtain vast quantities of a strong, though less expensive, silk than that produced by the ordinary silk-worm. These species are the oak silk-moth of China (*Saturnia pernyi*), and its near relative, *Antheræa yama-mai* of Japan.

In all their stages these lovely insects are remarkable, differing widely in their general characters from the majority of moths. The larvæ, with their clear rich green velvet bodies, deeply cleft into separate, well-marked segments, their rounded warts, golden, rose-coloured, and sky-blue, emitting long, sinuous hairs, the latter, sometimes enlarged at the extremity, cannot fail to attract attention both for their unusual aspect and their beauty.

When this stage is past, and the insect reposes in the large, leathery sombre brown cocoon, there is no lack of interest. The mouths of these cocoons, as noted at the beginning of the chapter, are fashioned for the better security of the slumbering pupa. No earwigs, beetles, or other prowling enemy can find its way into the cocoon to destroy the inmate, though the moth can readily emerge as soon as the outer shell of the enclosed pupa has been burst. For with a subtle ingenuity, no less wonderful because instinctive, the larva has carefully provided against these contingencies. It has arranged stiff, springy bristles round the orifice, each pointing outwards,

ATLAS MOTH

gathered in at their tips, so that unwelcome visitors cannot gain an entrance.

But beyond all these interesting features, the perfect insects are themselves sufficient to enlist our admiration. The enormous strong fore wings with prominent anterior angles; the rich browns, purples, and greys in every shade and gradation; the large crescent-shaped or eye-like blotch on both fore and hind wing render the members of this family not easily to be mistaken for any other lepidopterous insects. True, the eye-like blotches recall to mind those of the peacock-butterfly, but the stout, woolly bodies, the plumose antennæ, and the feathered legs of the emperor-moths will show clearly enough that the resemblance is but superficial, and that there is no close relationship between them. The males fly swiftly, with a somewhat erratic flight in the broad daylight; and if the female, held captive in some receptacle, be placed in the open woods, many of the former sex will eagerly gather round the cage, and thus themselves fall victims to the net of the naturalist.

There are many species included in the family *Saturniidæ*, though mention can be made of only a few. The common emperor-moth (*Saturnia carpini*), one of the dwarfs of the family, is abundant in England, where, in the heather-districts, the beautiful emerald larva, studded with rose or golden-yellow warts, may often be discovered wandering over some open sandy space or footpath. It is, however, at times scarcely distinguishable as it nestles among the heather-stems, since the rosy warts on the back and sides assimilating closely with the pink heather-blossoms secure it from observation. The moth itself—smaller and darker in the male sex—is of a deep purple brown. The fore wings, richly variegated with greys, are bordered with a snow-white fringe, while the hind pair are orange margined with brown.

DEATH'S-HEAD MOTH

Both fore and hind wings bear a black eye-like blotch, ringed with a narrow line of blue in the centre. The tough and dry empty cocoon may often be seen spun up amongst the heather-stems. The common emperor is found all through Europe and in Northern and Western

SPURGE HAWK-MOTH AND CATERPILLAR BEING ATTACKED BY ICHNEUMON-FLY

Asia, while a much larger form, the peacock-moth (*S. pyri*), is not uncommon in Southern Europe, and has been caught so far north as Paris.

Passing on to the Chinese oak silk-moth (*S. pernyi*), we find that its chief interest lies in the fact of the commercial value of its cocoon ; a value which has not been fully recognised for more than thirty or forty years. The Abbé Perny, from whom it derives its scientific name, was the first to introduce it to the notice of European silk-merchants, and from him we have a description of the method adopted by the Chinese in breeding and rearing the larvæ and winding off the silken treasure. Coppices of dwarf oak-tree are cultivated, the earth is smoothed and cleansed with great care beneath the trees, while attendants are always at hand to shift the larvæ from one bush to another, or restore them to the foliage when they have fallen to ground. The best of the cocoons from last year's cultivation are placed in a carefully regulated temperature, and the moths are hatched off exactly at the season when the oak-leaves are beginning to be ready for the larvæ. This will be about the month of April, when the females are laid in wicker trays where they may deposit their eggs. Soon, within ten days, the tiny larvæ creep forth and mount the oak-twigs laid in the trays for their reception.

Carried forth to the tender oak-foliage, they quickly begin to feed, while the keepers are always on the watch to protect them from insect-vermin, birds, and so forth, which, if permitted, would soon clear off the whole plantation. Forty-five days at the outside, and the larvæ are full-fed ; they then spin their cocoons, pass into the pupa state, and the winding off of the silken harvest begins. The largest cocoons are selected and set aside for the breeding of larvæ for another year. The rest are exposed to a high temperature, which destroys the pupæ within. Boiling water—in which the earthy salts of buckwheat ashes, cleaned for this purpose, have been dissolved—renders the cocoon fit for being unwound.

The silk is wound off in strands—five, six, or eight in number—a single strand from each cocoon, according to the strength of thread required. The silk thus prepared is much stronger than that from the silk-worm moth, though it is neither so fine in texture nor so valuable. The Japanese oak silk-moth (*Antheræa yama-mai*) is closely allied to the above, and the process of cultivation of the insect much the same.

SILK-SPINNERS

The only species belonging to the family *Bombycidæ*, known in Europe, is the one mentioned above as the true silk-worm moth (*Bombyx mori*). This insect has become acclimatised in many parts of Southern Europe, where, as in China, it is cultivated for its silken produce. The larva is itself not remarkable, save perhaps for its resemblance to the caterpillars of the hawk-moths, with its smooth naked skin, and short erect tail. It is, however, by far the most valuable caterpillar yet discovered. Ages ago, from two to three thousand years before the Christian era—if Chinese records be reliable—this larva was well known in the far East, and already silk-culture was a well-established element in the national industry.

History relates how the eggs were first brought to Europe, in the reign of the Emperor Justinian, by Persian monks, concealed in their hollow bamboo staves ; and from these silk-culture in Europe took its origin. It was, at any rate, carried on at Constantinople in A.D. 520. The Arabs introduced the industry into Spain, whence it spread in the twelfth century to Sicily, and thence to Italy and all the south of Europe. So far as England is concerned, both James I. and George I. endeavoured to introduce the cultivation of the silkworm for commercial purposes, but without success. The actual mode of cultivation and preparation of the cocoon differs in no very essential feature from that of the oak silk-moth, save that it is usually conducted under cover in well-ventilated rooms ; the wicker trays of silk-worms being

EYED AND HUMMING-BIRD HAWK-MOTHS, WITH THEIR LARVÆ

arranged in rows one above the other on light bamboo racks.

HAWK-MOTHS

The large moths included in the family *Sphingidæ* are either diurnal or subnocturnal in their habits, flying powerfully both in the daytime or just before nightfall. Among other characteristics, the antennæ are gradually thickened towards the tip, which terminates in a hook. The fore wings are elongate, narrow, and usually pointed towards the apex; while the hind wings are comparatively of small size. The larvæ are smooth, generally with a horn on the last segment of the abdomen. They make no cocoon, but the pupa lies in the earth, into which the larva burrows before the transformation takes place. As is the case with almost all, they are protected by their colouring, which assimilates to that of the food-plant. These fine insects are divided into several subfamilies and many genera.

As the type of the subfamily *Acherontinæ* may be taken the well-known death's-head moth (*Acherontia atropos*), which is by far the largest of British moths. It is a very stout, bulky insect, with strong, broad wings; its thorax having on the upper side a pale mark, which bears some small resemblance to a human skull, whence it derives its scientific and trivial names. The fore wings are dark plum colour, lined and spotted with yellow; the hind wings yellow, with two sinuous transverse bars of black; and the body dark plum colour, with black transverse lines, and a yellow patch at the side of each segment.

The most remarkable fact about the moth is that it is capable of producing an audible squeak. Whether this is produced, as was formerly supposed, by the friction of the palpi against the coiled proboscis, or by the sudden passage of air—previously drawn into a cavity in the stomach—through the œsophageal orifice and the proboscis, acting upon a cleft at the extremity of the latter, is not certain. If, as has been asserted, the squeak does not abate even on the decapitation of the moth, the air-passage theory suffers a shock, and evidently does not entirely account for the noise. The cleft at the end of the proboscis would perform a somewhat similar function to that of the tongue in a penny trumpet, the reed in certain wind instruments, or the orifice in a whistle-pipe.

The handsome larva (green, with large, pale yellow, swollen anterior segments, and yellow, black-speckled oblique stripes across the sides), with its spinous tail, may be sometimes discovered on the jasmine and in potato fields. Not infrequently, the large pupa tumbles from its friable earthen case when the potato crop is dug. The moth flies strongly at night, feasting usually upon the sap oozing from the trees. It does not, however, hesitate to rob the hive of the honey-bee, and apparently without molestation.

To the typical genus of the second subfamily, *Smerinthinæ*, belong several well-known British species, among which the eyed hawk-moth (*Smerinthus ocellatus*) is figured on page 1942 as an example. This moth is characterised by its angular, slightly scalloped fore wings and rose-coloured hind wings, each bearing an eye-like black spot, ringed with blue, near the inner angle. The larva is delicate green, its skin rough with minute warty points, with a series of oblique white stripes across the segments at the sides, and a short, sharp tail. It feeds on the willow and other trees, assimilating well in colour with the leaves and their oblique veins;

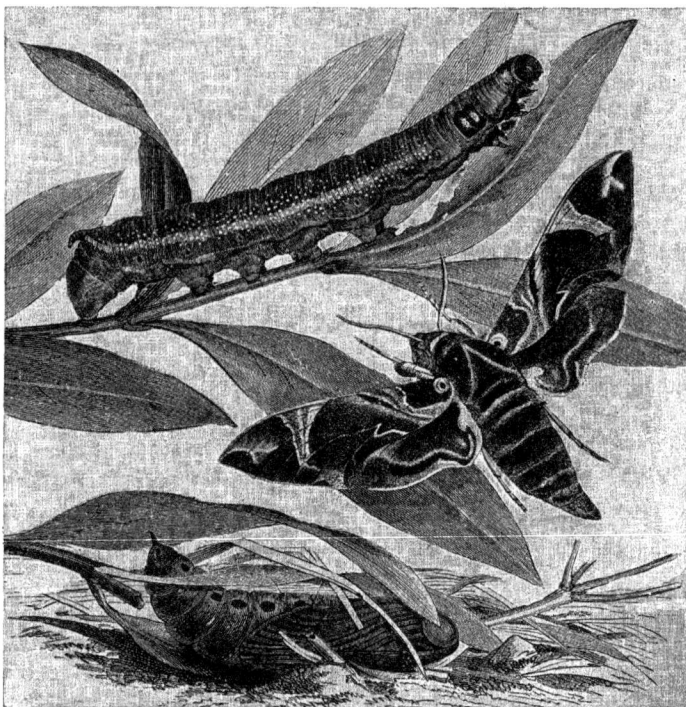

OLEANDER HAWK-MOTH, WITH LARVA AND PUPA

while the moth, hanging with half-closed wings, closely resembles a half-detached withered leaf. The insect is found throughout Europe and Northern Asia.

One of the largest and most beautiful of the tribe is the oleander hawk-moth (*Daphnis nerii*). In this species the fore wings are rich green, veined with white, having towards their base a triple, transverse rose-coloured bar, whose posterior arm runs along the hind margin of the wing to the thorax. The hind wings, thorax, and abdomen are green. The larva is green, with a pale band and numerous white speckles on the sides. The first three segments are suffused with yellow, and the third bears a large bilobate blue spot, outlined with black, on either side. The moth occurs throughout Europe, Africa, and Southern Asia; but neither larva nor perfect insect are often taken in England. The caterpillar feeds on the oleander and periwinkle in summer.

The elephant hawk-moth (*Chærocampa elpenor*) typifies a third subfamily (*Chærocampinæ*). In this species the front wings are green, margined and veined with delicate rose-colour; the hind wings black, with rose-coloured borders; the thorax and abdomen of the same tint of green, with a central rose-coloured band along the back, another at the sides, while the two last segments of the abdomen are rose-coloured. The larva is black, with three eye-like spots at the sides of segments three, four, and five, which are much enlarged, having also a rose-coloured band along the sides. It feeds on fuchsia, bed-straw, willow-herb, and the like, and is common in Europe and Northern and Western Asia in June.

To the same subfamily belong the members of the genus *Deilephila*, which have a world-wide distribution, although specially common in Southern Europe; among these, one of the commonest on the Continent is the spurge hawk-moth (*D. euphorbiæ*). Although the adult is rare in England, the caterpillar has been observed in some numbers in Devonshire, feeding on the sea-spurge. The fore wings are grey and rose-colour in blended tints, with a large dull green spot at their base, and an oblique submarginal band of the same colour, besides two smaller crescent-shaped spots towards the tip, the hind wings delicate rose, with black base, a deep crimson transverse bar, followed by a narrower black one a little beyond the middle ; and the thorax and abdomen green, the latter with white sides.

The caterpillar is black, speckled with yellow, having a dorsal rose-coloured central line, a row of yellow spots along either side, and another below of red and yellow spots blended. It feeds on the sea-spurge from July to September. In the figure on page 1942 the larva is repelling the attack of an ichneumon by ejecting noxious fluid into its face.

The pine hawk-moth (*Sphinx pinastri*) belongs to the typical subfamily (*Sphinginæ*), and is a dull grey species, scarcely to be discerned as it rests on the similarly tinted bark of the pine-trees on which the larva feeds. The moth lays her pale green eggs upon the pine-needles, and in about a fortnight the larvæ emerge, and at once attack the needles. They have occurred in such abundance on the Continent as to ruin whole forests of pine-trees, to the extent of many thousand acres.

Although the moth is common throughout Europe, and several specimens have been taken in England, it is very doubtful whether a genuine British-bred specimen has ever occurred. The larva, which changes to a pupa beneath the earth, is green, with narrow longitudinal bands of red and white ; these lines being naturally a great protection amidst the longitudinal lights and shades of the pine-needles.

Yet another subfamily (*Macroglossinæ*) is represented by the humming-bird hawk-moth (*Macroglossa stellatarum*), shown in the figure on page 1942. This small and swift species, which hovers with a darting, fluttering course over flower-beds in the sunshine, is double-brooded, and occurs almost all the year round. It has often been mistaken for a humming-bird, whose flight it closely resembles, while travellers familiar with the latter mistake the long proboscis from which the moth derives its generic name for the slender bill of the humming-bird.

The fore wings are dark black-brown, and the hind wings pale copper-red. The sides of the abdomen are blotched with white, its extremity being thickly tufted. The larva is green or pinkish brown, with a pale stripe along its sides, and feeds on the lady's bed-straw. The autumn brood of larvæ hibernate in the pupa state, the perfect insects emerging in the spring.

PROMINENTS

The prominents (family *Notodontidæ*), which are of moderate size, with stout, hairy bodies, long, ample wings, sometimes with a tooth-like tuft of scales on the inner margin, are very similar in general appearance to members of the family of owl-moths (*Noctuidæ*). The antennæ are usually pectinate in the male, and simple in the female, but in some genera comb-like in both sexes.

The larvæ, which in many species assume strange abnormal shapes and attitudes, are smooth and shiny, and without the last pair of claspers. In some cases the terminal segment bears a pair of tail-like processes, which can be raised or depressed, spread widely apart, or closed at pleasure. When full fed, the larva forms a tough cocoon, covered with chips of wood or other debris, in which it turns to a pupa. The

PINE HAWK-MOTH WITH LARVÆ AND PUPA

perfect insects fly at night, and may sometimes be found during the day resting on the trunks of trees, palings, or other suitably coloured objects.

A common British representative is the buff-tip (*Phalera bucephala*), although it is more often met with in the larval state than adult. Yellow-and-black-spotted, the young larvæ may be found together, feeding gregariously upon elms and other trees. The silver-grey wings, streaked and barred with rich browns, their tips painted with a patch of pale yellow, appear when closed, as the moth rests on the grey bark of a tree, exactly like a short grey stick with the top bevelled off on either side, and partially decayed.

The puss-moth (*Dicranura vinula*) is another common British species often found on poplar-trees in the larval state, though the perfect insect is seldom met with. The latter has white fore wings, tinged and marked with grey, the thorax being spotted with black. The compressed, globular, dull red egg is laid in the summer months on the leaves of the poplar or sallow, and the tiny caterpillars are at first quite black, but become greener as they grow older. When full grown, they assume, at rest, the characteristic position represented in the illustration on opposite page, whence they derive their name of puss-moths, from some fancied resemblance to a cat. The bifurcate tail emits thin red filaments from the apex of each branch when the larva is irritated ; the colour being then bright green, with a red-brown or chocolate-pink patch margined with white behind the head, narrowed and then broadened at the sixth segment, and narrowing again to the tail.

The cocoon is very tough, formed in some crevice of the bark gnawed into a convenient cup by the strong jaws of the larva. On the top are glued the chips thus obtained, and, with bits of lichen added, it almost defies detection amongst the surrounding knobs and rounded bits of bark. The species is common throughout Europe and Asia.

The caterpillar of the lobster-moth (*Stauropus fagi*) resembles nothing to be found in Nature save those of the closely allied species, as may be seen from the illustration. The moth is found, but not commonly, throughout Europe, and the larva feeds in July upon the oak, birch, and other trees. It is supposed that the extraordinary attitude, with head and tail erect, has proved beneficial in warning off noxious enemies.

Another type is represented by the figure-of-eight moth (*Diloba ceruleocephala*), in which the fore wings are lead-colour, with a pair of white spots which sometimes bear a very close resemblance to figures of eight. The larva is blue-green, with a central yellow stripe along the back, another below the spinners, while each segment bears a number of black warts, each with a black hair springing from the top. Illustrations of the moth and larva are to be found on page 1950.

Of other forms, the dromedary prominent (*Notodonta dromedarius*), the zigzag (*N. ziczac*), the kitten-moth (*Cerura bifida*), and the swallow prominent (*Pheosia dictea*) are among the more remarkable of the *Notodontidæ* indigenous to England. But we must leave this interesting group, and pass over the family *Cymatophoridæ*, including the peach-blossom (*Thyatira batis*), frosted green (*Polyphloca ridens*), buff-arches (*Habrosyne derasa*), and others.

curious habit, which increases the deception, and renders its likeness to some hostile wasp still more striking. If surprised sitting in the sunshine upon a poplar-trunk, the abdomen will be arched upwards, and the tail tapped against the bark with a veritable—to all appearances—stinging movement. The larva burrows in the wood of the poplar, and the pupa-skin may be found half out of one of the galleries when the moth

1, PUSS-MOTH, WITH 2, CATERPILLARS, AND 3, COCOON; 4, CATERPILLAR OF LOBSTER-MOTH

has emerged. The insect is common all through Europe and Northern and Western Asia.

CLEAR-WINGS

The elegant insects the clear-wings (family *Sesiidæ*)—whose transparent wings, attenuated bodies banded with yellow and red, dilate and hooked antennæ, give them no small resemblance to members of the Hymenoptera—are diurnal in their habits, flying swiftly to and fro in the bright sunshine. The larvæ are what are called internal feeders, burrowing in the trunks of various trees or in the pith of shrubs. The pupæ are armed with little hooks, which enable them to move up and down their tunnelled galleries.

There are many species even in England, one of the

SYNTOMIDÆ

One small family, the *Tiniageriidæ*, must be passed over, and a brief reference made to the moths of the family *Syntomidæ*, which introduces us to the well-known burnets. The *Syntomidæ* include small moths with broad, triangular-spotted wings, and body extended behind the hind wings. The members of this family are very similar in general appearance to the burnets, but differ in the absence of the ocelli. They are widely extended, and take the place of the burnets in the tropics of the Eastern Hemisphere. Among them, the spangled white (*Syntomis phegea*) is a common moth in some localities on the continent of Europe, with blue-black wings spotted with white, as represented in the illustration on page 1950. The larva is black, thickly clothed with hair, and feeds on the dandelion, while the perfect insect flies, somewhat like the burnets, in the sunshine, and settles upon flower-heads. It is not found in England, though extending through Europe to Northern and Western Asia.

HORNET CLEAR-WING AND GOAT-MOTH WITH LARVÆ AND PUPÆ

largest being the hornet clear-wing; and so closely do this moth (*Trochilium apiforme*) and its near relative (*T. bembiciforme*) resemble the common hornet, or perhaps more closely the female of one of the smaller wasps, that only a practised naturalist would be able to tell the difference, and then only on a close examination. The wings are transparent, and the body is black, striped and spotted with yellow. The moth has a

We may also notice the handmaid moth (*Naclia ancilla*), a very rare species in England, but not uncommon in the woods of Southern and Central Europe in June and July. Its larva is black, with yellow lines on the back and sides, and it feeds on tree and rock lichens in spring.

BURNETS

The burnets (family *Zygænidæ*) are for the most part small moths, with long, rather narrow fore wings, and stout bodies extending beyond the hind wings. Their usual colour is black, green, or dark blue, spotted with red, white, or yellow. The hind wings are grey, red, or similar in colour to the fore wings, with a narrow black margin, and the antennæ are somewhat abruptly narrowed towards the extremity. The burnets are local, though, from their gregarious habits, abundant where they occur. The larvæ are rather compressed, tapering at both ends ; and the cocoon is long, spindle-shaped, yellow or white, of fine shiny silk, and attached longitudinally to grass stems.

Of the six-spotted burnet (*Zygæna filipendulæ*) the caterpillar feeds late in the autumn, and hibernates until the following spring. It is short, stout, slightly hairy, dull yellow, with two rows of black spots along the back, and feeds on grasses of various kinds.

The moth flies heavily in broad daylight and may often be seen, two or three together, hanging upon flower-heads in chalk pits and on downs by the sea. Its fore wings are black, with metallic green lustre, having six bright red spots in three pairs ; and the hind wings are bright crimson, with a narrow black border. The species, which is abundant in certain parts of England, as well as on the Continent, is shown in various stages of development in the illustration on page 1950, and the adult is figured here.

CASE-WEAVERS

An interesting group of moths, although not noticeable either for size or coloration, is that of the case-weavers (family *Psychidæ*). Their chief claim to notice is from the curious habits of the larvæ, which form from vegetable débris, twigs, chips, etc., a case in which they dwell, protruding merely the thoracic segments, with the three pairs of legs belonging to them. Some other moths, as for instance the genus *Coleophora*, also construct a tough case of a somewhat similar nature, but manufactured entirely of silk. Among other insects the same habit of the larvæ is found among the caddis-flies, which creep on river-beds protected by a case of encrusted shell, pebbles, twigs, etc.

In the moths of the present family the males alone possess well-developed wings, the females being worm-like, and often without antennæ, legs, or wings. The phenomenon known as parthenogenesis has been observed amongst members of this family. The moths are mostly dull brown insects, and the various species are better distinguished by a comparison of the larval-cases than of the insects themselves.

Of the many species embraced in this family, only one can be described, and this but briefly. This species (*Psyche unicolor*) is a dull brown little moth, common in Central and Eastern Europe but not found in England. The larva of the male moth makes a larger and more conspicuous case than does the grub which will produce the wingless female. The larvæ hibernate securely enclosed in their cases, which are spun on a tree-trunk or other convenient object. In the spring the silken attachments are severed, and the larva

SIX-SPOTTED BURNET

continues to feed until the time of pupation has arrived, when it again spins up the mouth of the case to a tree or post, and changes within it to the pupa. The male then emerges as a perfect moth, but the female, which is devoid of eyes, ovipositor, or any appendages worthy of being styled antennæ or legs, remains in the larval-case even after it has emerged from the pupa. The organs for the production of eggs are, however, complete, and parthenogenesis must, as in many other cases, be looked upon as exceptional.

COSSIDÆ

The moths belonging to the family *Cossidæ*, like those of several others, do not possess any proboscis, the antennæ being pectinate in both sexes. The larvæ are smooth, and feed sometimes for several years before pupating in the centre of tree-trunks of various kinds ; a cocoon being formed of chips of wood within which the pupa awaits its final development. The family is typified by the goat-moth (*Cossus ligniperda*), in which the front wings are of a rich brown, streaked and mottled with darker tints, while the hind pair are dull brown.

The larva—often known as the auger-worm—is exceedingly destructive to forest trees, the holes which it bores in its ravages being often half an inch, and even more, across. Its odour recalls that of a goat, hence the name given to the moth. A large, long, flat, broad larva, flesh-coloured, with short hairs scattered over the body, it is seldom met with, though it sometimes may be found as it crosses a road or footpath when seeking for a suitable place in which to spin its cocoon. The caterpillar lives for over three years in the larval state, and makes a very tough cocoon from wood chips, glued together with a gum which it secretes. The species, which is a native of Europe and Western Asia, generally appearing in June and July, is figured on page 1945.

GHOST-MOTHS AND ALLIES

The next family (*Arbelidæ*) must be dismissed without further remark. The *Hepialidæ* include the insects known as ghost-moths, one of which, the largest British species (*Hepialus lupulinus*) has the wings white above and brown below, so that when it flies in the dusk of the evening it appears and disappears in rapid sequence owing to the practical invisibility of the dull colour of the under side, in sharp contrast to the vivid white of the upper side. A near ally of the ghost-moth, likewise referable to the family *Hepialidæ*, is the splendid giant swift moth (*Zelotypia stacyi*) of Australia, which is illustrated in the coloured plate on page 1919 as being one of the finest of all moths. As the coloration and characters of this magnificent insect are sufficiently indicated in the illustration, it will only be necessary to give some account of its habits. Originally described from imperfect specimens found at the Manning River and in the neighbourhood of Newcastle, this moth was subsequently obtained in some numbers by the miners of the latter district. Mr. A. S. Oliff writes that " as the insect is rarely found in the perfect, or imago, condition, the larva has to be sought for and reared—a matter of no little difficulty, as it lives, like those of the allied genus *Charagia*, in cylindrical burrows, which it makes in the interior of the stems or branches of trees, sometimes near the surface of the ground, and sometimes at a height of fifty or a hundred feet. By searching for these burrows, and rearing the larvæ, or pupæ, when found, a considerable number of specimens have been obtained by the miners ; but I am informed that the supply is by no means equal to the demand."

The caterpillar is long, cylindrical, and fleshy. Above its general colour is pale yellow, with the divisions between the segments inclining to reddish brown. The first three segments are rather bright red ; and the following segments, with the exception of the two last, are marked with three pale spots in the middle, and two on each side. The finely rugose head is black, as are the claws of the short legs. In the long and cylindrical pupa each of the abdominal segments beyond the extremities of the wing-covers is provided with a transverse serrated horny ridge near the front margin ; the seventh to the tenth segments bearing similar but less prominent ridges ; while the hinder extremity is armed with small sharp spines.

Usually the caterpillar makes its burrows in the wood of the grey gum-tree ; but there is some doubt as to whether it does not occasionally resort to another species of gum. Regarding the habits of the larva and pupa, Mr. Froggart writes that the former " changes into the chrysalis in December, after having eaten off the web in front of the bore, and placed a thick felty wad, or button, just inside the opening of the bore ; but as soon as the chrysalis skin has become hard and firm, it pushes the wad away, and moves freely up and down the bore, which varies in depth from ten to twelve inches. It can move up and down the passage very rapidly, the curious file-like rings on the lower edge of the abdominal segments being evidently adapted to helping its locomotion. When nearly mature it has the habit, particularly in the afternoons, of resting in the bore, with the top of its head just level with the floor of the cross-bore, and plainly visible from the outside. The moths appear early in March. It has been found that they never come out after three o'clock in the afternoon ; and chrysalids under observation, if not out at that hour, can be safely left until the next day."

The next family (*Callidulidæ*) must be omitted ; but the *Drepanulidæ* may be briefly referred to as containing the British species *Cilix spinula*, and the common hooktip (*Drepana falcataria*), and allied forms. Of the *Thyrididæ* there is but one European genus (*Thyris*), and no British species of this.

The *Limacodidæ* are small, stout moths of wide distribution, represented in Britain by two species. The larva of the North American hag-moth (*Phobetron pithecium*) possesses elongated appendages covered with down, and its habits are very interesting. It lives on orange-trees, and in many cases secures a suitable place for turning into a pupa by killing some of the leaves. These then assume a light brown colour, with which the cocoon assimilates, for it is covered with the down of the larva, and this is of similar hue. The cocoon is complex and provided with a very perfect lid.

LIFE-HISTORY OF PINE-LAPPET MOTH

a, male ; *b*, female ; *c*, eggs ; *d*, larva ; *e*, cocoon ; *f*, a beetle (Calosoma) attacking larva ; *g*, larva of Calosoma ; *h*, ichneumon laying its eggs in the pupa ; *i*, small parasites emerging from their cocoons on the remains of the larva which they have devoured

LASIOCAMPIDÆ

The lappets, drinkers, and eggars are well-known species included in the large family *Lasiocampidæ*. These moths are large, for the most part, 2 inches to 2½ across the expanded fore wings, others being smaller, about 1 inch only in expanse of wing, with stout, hairy bodies and strong wings. They fly rapidly in broad daylight or at night. The larvæ are clothed with soft hair, that on the sides being often directed downwards in a tufted form. To the genus *Gastropacha* belong the lappet (*G. quercifolia*) and the oak-eggar (*G. quercus*) ; the common drinker pertaining to another genus (*Odonestis*), with the specific name *potatoria*.

As examples of the former genus we select for description the pine-lappet and the procession moth, both abundant on the Continent, but not occurring in England. The larvæ of both these moths spin silken cocoons.

MIGRATION OF LARVÆ OF PROCESSION-MOTH
1, moth ; 2, highly magnified single hair of larva ; 3, segment of larva ; 4, pupa ; 5, cocoons of several larvæ spun up together

Having the front wings grey, tinted with different shades of brown, the pine-lappet (*Gastropacha pini*) is a large moth measuring from 2½ inches across the wings. The larvæ are ashen grey, with a dorsal row of dark blotches, a lateral brown stripe, and a pair of blue transverse bands on the third and fourth segments. This handsome larva is often very destructive to the pine-forests, where it feeds upon the needles of the trees, and sometimes appears in overwhelming numbers.

In coping with the enormous quantity of caterpillars of this moth which devastate the district on these occasions, man is materially assisted by other creatures. Thus, a tree-frog feeds upon the larvæ ; ichneumons of different species sting, and thus destroy, thousands ; an internal fungus establishes itself in the caterpillar, with the same result ; and, lastly, a beetle and its larvæ, which are represented in the illustration, render no small assistance in clearing off the pest.

The caterpillars are hatched in the autumn and hibernate, remaining throughout the winter in the moss at the foot of the trees. In this state, coiled round in a spiral form, they may be frozen quite stiff, yet on the return of spring they regain vitality, and climb the trees in search of their usual provender. The red-brown cocoon is spun sometimes between the needles of the tree, as represented in the illustration, or else beneath some semi-detached piece of bark.

In the procession-moth (*Gastropacha processionea*) the fore wings are yellow-grey, with a glossy sheen, and dark indistinct oblique transverse bars. The larvæ are hairy with a blue-black back, pale sides, and red or grey warts on each segment. At night the caterpillars march out to feed in a regular orderly procession, as represented in the illustration. One, the leader, marches at the head, followed by two, three, and so on, forming a wedge-shaped column. They ascend the oak-trees and return again in the same manner to their resting-place. They also spin

their cocoons together, as in 5 of the illustration. The species is common throughout Central and Southern Europe in August and September.

As our last representative of the family we take the lackey-moth (*Clisiocampa neustria*), which is common in England and all through Europe and North and Western Asia during July and August. The fore wings are dull ochre-brown, with two transverse brown bars. The eggs are laid by the female in the late summer in a firmly attached ring round some small twigs, as shown in the illustration. The larvæ hatch in the spring following, and are brown with blue, white, red, and yellow longitudinal stripes ; all feed on the leaves of the pear and other fruit trees, and spin a long sulphurous yellow cocoon among the leaves.

LYMANTRIIDÆ

The family *Lymantriidæ* includes a number of moths in which the males have the antennæ strongly pectinated, while in the case of the genus *Orgyia* the female is wingless. None possesses a proboscis. The larvæ are hairy, and clothed with long, thick tufts, springing in some places from wart-like prominences. The hairs of the larvæ are woven into the cocoon, and if they come in contact with the skin cause great irritation. In this family are included some well-known British moths, such as the vapourer (*Orgyia antiqua*), the pale tussock (*Dasychira pudibunda*), the black arches (*Lymantria monacha*), the gold-tail and brown-tail, the satin-moth, and many others.

In the gipsy-moth (*Ocneria dispar*) the wings of the male are smoky black, while those of the female are grey ; the appearance of the two sexes being very different indeed. The larvæ feed on various trees, and, though very rare in England, are sometimes so abundant on the Continent as to prove very destructive to all kinds of trees and herbage, stripping even maize and millet fields, orchard, and vegetable produce. The cocoon is formed in a few folded leaves spun together with silk, or in a crevice in the bark. Its

LACKEY-MOTH
The eggs, larvæ, and cocoon are shown

wings, antennæ, and the dark half of the thorax and abdomen on the left side are of the colouring and form peculiar to the male, while those on the right resemble the form peculiar to the female. The lower figure on this page illustrates the stages in the development of the black-arches moth, which is not altogether abundant in England but much more commonly met with on the Continent.

Indeed, so abundant is it at times that it causes great injury to forest trees. In Prussia, Lithuania, and Poland the havoc has been particularly severe. In 1863 the moth appeared in countless thousands, driven up as a regular insect storm by the south wind. Within a few hours the moths spread over the whole countryside, buildings were completely covered by them, and the very surf of the lake assumed a more snowy whiteness, due to the colour of the hosts of moths drowned in the waters. The woods seemed as though visited by a violent snow-storm, so thickly were the insects massed in the foliage. In 1852 whole forests were felled, in order if possible to be rid of the pest. The trunks were searched for eggs, and every tree-trunk in an area of fourteen thousand acres was examined. Often an ounce of eggs would be taken from a single tree, and, at the computation of thirty thousand to the ounce, we get, at one hundred trees per acre, upwards of thirty hundred million larvæ at work upon the trees in that area when the eggs hatched.

Spotted woodpeckers, finches of all kinds, the larva of a longicorn beetle, *Clerus*, all assisted in the work of destruction. Yet, in spite of all this, it needed a hundred labourers with twenty foremen to carry out the destruction of the young larvæ hatched from eggs which were overlooked in a single acre of forest. The ground too, after the season was over, was white with the cocoons of countless thousands of *Ichneumonidæ*, so that millions of the larvæ can never, from the attacks of these alone, have reached maturity.

The pale tussock-moth (*Dasychira pudibunda*) derives its trivial name from the tufts or tussocks of hair, so noticeable a feature in the hairy clothing of the

GIPSY-MOTHS
1, male ; 2, female ; 3, pupa ; 4, larvæ in different stages

larvæ. The fore wings are grey with a smoky transverse bar. The larva is green with a transverse bar of velvet black between the segments from five to eight. Each of these segments bears a thick, squarely truncated tuft of upright yellow hairs, and the last carries a long tail or brush of hair. The species is abundant in England and all Europe.

In the brown-tail moth (*Porthesia chrysorrhœa*) the wings are snowy white, while the body is white with a brown tufted tail in the male, which in the female is much larger. The hairs of the tuft are deposited upon the eggs as a covering when laid by the female. The larva is short, thick, and black, with four rows of spiny tubercles along the sides. It is common in Great Britain and also on the Continent.

Very similar to the last is the gold-tail (*Porthesia auriflua*), but the front wings are dotted with three or more black spots, while the tuft at the extremity of the abdomen is formed of golden hairs instead of brown. The larva has rows of tubercles along the sides, whence issue numerous hair-like bristles. Each of the tubercles of the second row bears tufts of white hair. The third row is bright red. A bright vermilion double stripe runs along the back, while between the tenth and eleventh segments is a cup-like scarlet protuberance.

The satin-moth (*Porthesia salicis*) is another well-known member of the family, taking its name from the white satiny wings, the antennæ and thorax being also white, and the body black, clothed with white hairs. The larva feeds on the poplar, and is abundant in England and throughout Europe.

TIGER-MOTHS

Two families, including many tropical species, come between the *Lymantriidæ* and the *Arctiidæ*, namely, the *Pterothysanidæ* and the *Hypsidæ*. The forms included under the name *Arctiidæ*, embracing a number of beautiful moths, such as the tigers and ermines, are usually divided into

BLACK-ARCHES MOTHS
1, 2, males ; 3, 4, 5, females ; 6, young larvæ ; 7, full-grown larvæ ; 8, pupa

1, 2, 3, TIGER-MOTHS, AND, 4, LARVA; 5, SIX-SPOT BURNET, AND, 6, LARVA; 7, SPANGLED WHITE

When young, they spin together the needles of the pines, and often drop themselves by a thread to various points, whither they may feel inclined to descend. The pupa may be found in plenty amongst the moss which so often carpets the ground in pine-woods. The moth itself is cinnamon-red, with white blotches and spots. It is common in England and on the Continent.

The merveil du jour (*Dipthera orion*), indicates another subfamily (*Acontinæ*). It has the fore wings of a pale green, with longitudinal white stripes, and three broken transverse black bars, the fringe being spotted with black and white. The larva is black, with large primrose yellow spots on the back of the third, fifth, and eighth segments. It feeds in September upon the oak and birch, and the pupa is enclosed in a cocoon of bark chips, or fragments of decayed wood.

In the same group, the caterpillar of the white-spotted pinion (*Cosmia diffinis*), as well as that of the closely-allied *C. trapezina*, are remarkable for their habit of preying upon their fellow-caterpillars if confined together, otherwise their food consists of the leaves of various trees. The moth of the species figured in the illustration is very beautiful, being of a satiny chestnut, suffused with reddish grey, and having two somewhat transverse slashes from the margin of the wing. The crimson underwings (*Catocala*), which indicate another subfamily (*Guadrifinæ*), and are known in the New Forest as the crimsons, are rich chocolate-brown of various hues, with deep crimson under wings, marked with a pair of transverse black bands. They are common in some parts of England. The finest of these

four subfamilies, the *Arctiinæ*, represented by the tigers, properly so called, the *Lithosiinæ* including the footmen, the *Nolinæ*, and the *Nycteolinæ*.

Of the first subfamily, the most familiar member is the common tiger-moth (*Arctia caja*), which in summer comes freely to light. The fore wings are rich chocolate brown with cream-coloured markings; and the hind wings crimson with black blotches. Two very beautiful varieties of this exceedingly variable moth are figured in the accompanying illustration. The larva is the well-known woolly bear, a large, swiftly-moving caterpillar, clothed with long, bristling, black hairs, red at their base, which spins a loose web, and turns to a naked pupa.

WHITE-SPOTTED PINION-MOTH AND PINE-MOTH, WITH THEIR LARVÆ

OWL-MOTHS

Passing over the family *Agaristidæ*, we reach the true night-flying moths, now included in the family *Noctuidæ*. This enormous group has been subdivided into no less than ten subfamilies. Of the first subfamily (*Trifeinæ*) the rustic shoulder-knot (*Hadena basilinea*) is a well-known example. In this moth the fore wings are grey-brown, with a central transverse darker band, and a distinct dark streak at the base of the wing. The larva is grey-brown, with three white lines along the back. It feeds on various kinds of grass, and often on wheat grains.

As its scientific name implies, the pine-moth (*Trachea piniperda*) is in the larval state very destructive to pine-trees in seasons favourable to a great increase in their number.

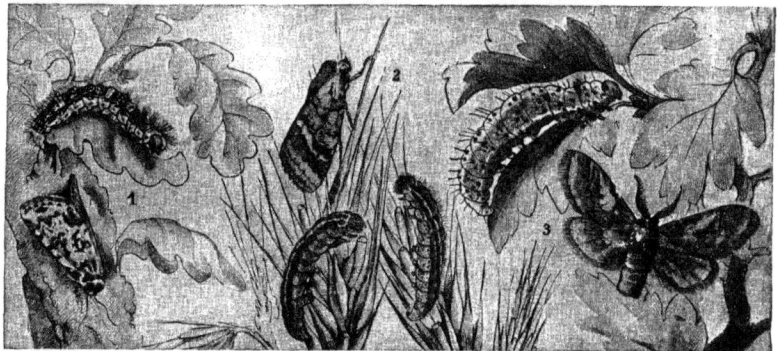

I, MERVEIL DU JOUR; 2, RUSTIC SHOULDER-KNOT, AND, 3, FIGURE-OF-EIGHT MOTH, WITH THEIR LARVÆ

beautiful insects is the Clifden nonpareil (*Catocala fraxini*), very rare in England, but more abundant on the Continent. Scarcely less striking is the red under-wing (*C. nupta*), in which the grey wings are mottled with darker shades, rendering it difficult to detect when resting on the grey bark of some forest-tree. The hind wings are pale crimson, with a central curving transverse black bar, and another broad black band along the margin. The caterpillar is grey, with darker brown markings, bearing a pale yellow prominence on the ninth segment. It feeds on a species of willow, *Salix fragilis*, and the adult appears on the wing in August and September; being not uncommon in England, but found more abundantly on the Continent.

In the angle-shades (*Brotolomia meticulosa*), which is one of the most beautiful, as it is one of the commonest of British moths, the larva is delicate green, smooth, and velvety, thickly speckled with minute white spots. It feeds on groundsel. The perfect insect, which appears on the wing in May and June, with a second brood in September, is common throughout Europe.

In the prettily-marked species known as the feathered gothic (*Neuronia popularis*) the fore wings are dark brown, with white nervures. The orbicular and vermiform spots are of the same colour. The antennæ are pectinate in the males, and simple in the female; while the hind wings are dull white, with darker margins. The larva is brown, streaked and spotted with black and rosy brown, with a pale stripe along the sides, and four others, more interrupted, along the back. It feeds on the various kinds of grasses in April and May, while the perfect insect appears on the wing in the early part of September.

LOOPERS

1, male, 2, female, and 3, larva of mottled umber; 4, male, and 5, female of scarce umber; 6, male, 7, female, and 8, larva of winter-moth

The next form for notice is the so-called antler-moth (*Charœas graminis*), which is probably one of the most destructive species in Britain, when, under the influence of a favourable season, the larvæ appear in very great numbers. The larvæ feed upon the roots of grasses, and it is no uncommon thing for whole districts of pasture-land to become brown and withered, owing to their attacks. The perfect insect appears on the wing in August and September.

1, FEATHERED GOTHIC, WITH LARVA; 2, ANGLE-SHADES; 3, ANTLER-MOTH

LOOPERS

The moths belonging to the family *Geometridæ* resemble in many respects the butterflies, having large, ample wings, a small head, and a narrow elongated body. The antennæ are not, however, clubbed; those of many of the males being pectinated. The palpi protrude only slightly, the proboscis is present in different degrees of development, while the head bears no ocelli on the top. When at rest, the majority of these moths carry their delicate wings slightly expanded, or closed over their bodies, like the roof of a house, sloping from the centre on either side. They are semi-nocturnal in their habits, appearing at dusk, and lying concealed during the day in bushes, trees, and herbage, whence they may be easily driven by beating the foliage.

The larvæ differ very decidedly from those of the other families, several pairs of the pro-legs being wanting, so that locomotion is possible only by alternately advancing the front and hinder segments, the central portion of the body being thus raised in the form of a loop. The pupæ are sometimes, as in the butterflies, encircled with a silken thread, but the majority spin together

RED UNDER-WING, WITH LARVA

MALE AND FEMALE BORDERED-WHITE   ARGENT-AND-SABLE   MAGPIE-MOTH, PUPA, AND LARVA

a few leaves, and change within the receptacle thus formed, or burrow into the earth among dead leaves and moss.

Of the first subfamily (*Boarmiinæ*) we select as a representative the handsome pepper-moth (*Biston betularia*), which is one of the largest of the European geometers, and resembles members of the family *Bombycidæ* in the possession of a stout abdomen. The form of the larva, however, is quite distinct, and closely resembles that of a dead twig. Doubtless such a likeness saves it somewhat from the attacks of birds and ichneumon-wasps. When fully extended, and clinging only by its hindmost claspers, the caterpillar assimilates so marvellously with the brown and olive tints of the boughs among which it takes up its station that it is almost indistinguishable from its surroundings.

Another handsome member of the same group is the mottled umber (*Hibernia defoliaria*), which appears very late in the season, long after the majority of the members of the order have completed the term of their existence. By night the male circles around the trunks of trees in search of his wingless partner. In the former sex the large wings are pale in colour, with a darker wavy transverse bar. The female, on the other hand, is variegated black and ochreous yellow, and looks like a spider. The larvæ feed on the buds of various trees, and descend into the earth to change into the pupa; the latter being dark mahogany, with a sharp spine at the tail. The species is not rare in England and on the Continent.

PEPPER-MOTH, WITH LARVA AND PUPA

The scarce umber (*H. aurantiaria*), which is figured in the same illustration, is less common than the last, but appears at the same season. Nearly allied is the winter-moth (*Cheimatobia brumata*), which in mode of life is somewhat similar to the mottled umber, but, as indicated by its scientific name, flies still later in the year. The larva lives partially secluded amongst the leaves which it draws together with silk. When occurring in great numbers, these caterpillars do serious damage to forest trees and orchards. The male is of a dusky grey colour, with three darker bands across the upper wings; while the female is wingless. In order to prevent the females from ascending the trees and laying their eggs on the foliage, it is the custom to ring the trunk with a narrow band of some sticky substance.

The bordered-white (*Bupalus piniarius*) is another well-known member of the group. In this species the males are very abundant, flying amongst fir-plantations in England and on the Continent. The females are no less common, but do not take wing so readily. The larva is pale green, with whitish stripes, and pale yellow spiracles, and feeds during the months of August and September on the spines of the Scotch fir.

One of the most familiar of the British loopers is the magpie-moth (*Abraxas grossulariata*), which at times makes its appearance in great numbers. The perfect insect is prettily mottled with white and black, and on this account is called in Germany the harlequin-moth. Another species, the scarce or

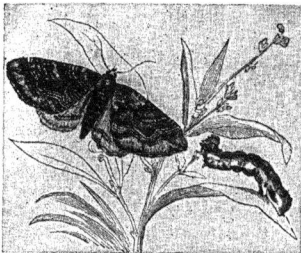

DARK SPINACH-MOTH AND LARVA

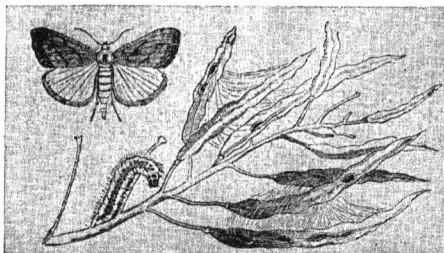

MOTHER-OF-PEARL MOTH AND LARVA

PURPLE-BARRED YELLOW AND LIME-SPECK

clouded magpie (*A. ulmata*), is more abundant in the midland counties of England than the common magpie, though less so in the south. Of the common species the larva feeds on the gooseberry and black currant, doing considerable damage at times. It is one of the most strikingly marked of the geometric larvæ, and turns to a yellow-banded pupa within a slightly woven web.

The little moth commonly known as the dark spinach (*Larentia chenopodiata*) may be taken to represent the subfamily *Larentiinæ*. Appearing in July and August, it is a common species on the Continent, and is specially abundant in gardens and shrubberies, where it may be found nesting either on the bark of trees or the walls of buildings. The caterpillar is greyish brown in colour, and feeds on the goose-foot. The group to which this species belongs are often termed carpet-moths.

In another genus, known as pugs (*Eupithecia*), the lime-speck moth (*Eusignata*) may be mentioned. The ground colour of the wings is milk-white, with grey blotches and specks, and a broad red grey band on the margin. These moths fly commonly at night in England and on the Continent, while the larva, which is very variable in colour—bluish green, yellow green, or pinkish white—feeds in August and September on various annuals, such as golden-rod, ragwort, and so forth.

By no means a common species in England, although found occasionally in districts where birch-trees abound, the argent-and-sable (*Melanippe hastata*) appears in May, flying round trees. The larva may be found amongst the birch foliage, in a receptacle formed of several leaves drawn together with silken threads. The pupal state is passed in the ground.

The purple-barred yellow (*Lythria purpuraria*), figured in the annexed illustration, is a not uncommon species on commons, pasture-lands, and stubble-fields in England and the Continent. The ground colour of the wings is pale olive yellow, the upper pair banded with two or three pale vinous purple bars. The larva, which is brownish yellow with a pale longitudinal dorsal stripe, feeds on sorrel and docks.

SNOUT-MOTHS

The snout-moths (family *Hypenidæ*) are intermediate between the *Geometridæ* and *Pyralidæ*, bearing

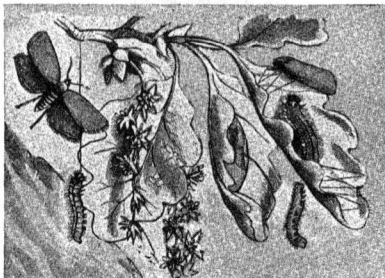

OAK-TORTRIX IN VARIOUS STAGES

characters which ally them to both families and yet exclude them from either. The common snout (*Hypena proboscidalis*) is a pale brownish yellow moth, transversely marked with rusty brown; and is abundant throughout England and the Continent from June to September. *H. obsitalis* has only once been taken in England.

MICROLEPIDOPTERA

The whole of the remaining members of the order are of minute size, and are hence generally indicated by the name Microlepidoptera, although it must be understood that many of them are closely allied to some of the foregoing. They are divided into a large number of families—with their subfamilies and genera—of which only a very few can be even mentioned here.

Among these the pearls (*Pyralidæ*) are represented by the mother-of-pearl moth (*Botys margaritalis*) which in June or July may be seen in Britain hovering over the fields in the dusk of the evening, where the female lays her eggs on the seed-pods of the flax and other plants. When the caterpillars emerges it spins a few threads between the pods, and bores through their outer shell in order to feed upon the seeds. The moth itself is of a dull, sulphur-yellow, with two transverse rusty yellow bands, intersected by a rusty brown stripe running obliquely from the tip of the wing. It is common in June and July on the Continent.

OAK-GALL AND LARCH-TORTRIX MOTHS
1, oak-gall tortrix; 2, pupa appearing from the resin-gall; 3, Glypta resinanæ (ichneumon); 4, the larch tortrix; 4a, pupa; 5, larva in a larch-bud; 6, pupa appearing from gall

To the same family belongs the meal-moth (*Asopia farinalis*), found in summer wherever corn, meal, or grains are stored in quantities. It rests on the rafters and walls in the daytime, flying at nightfall. The larva feeds on corn, meal, grain, bran, etc., and passes its life in concealment in a silken tube, of which the outer side is encrusted with particles of the foodstuffs on which the larva feeds. The larval state lasts for nearly two years.

The wax-moth (*Galleria mellonella*) may be taken to illustrate another family—the *Tortricidæ*. This remarkable moth is double-brooded, appearing on the wing in the springtime, and again in July and onwards. The larva feeds in the hives of honey-bees, and, according to some, in the nests of wild bees as well. The wax, however—not the honey—forms its food-stuff, and through the combs it eats long tunnels, which it lines

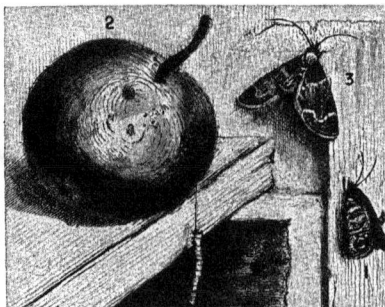

1, CODLIN-MOTH; 2, ITS CATERPILLAR; 3, MEAL-MOTH

with silk as it goes. It does not seem particularly choice in the matter of diet, and has been successfully reared on heather, woollen stuffs, dry leaves, paper, and so forth. In the case of the wax-eaters, the second brood nourishes itself upon the excrement of the first brood, which seems to differ in no way from the original wax itself. The moth appears on the wing in May.

Another member of the same family is the oak-tortrix (*Tortrix viridana*), figured on the previous page. This beautiful little moth, bright green with shining grey hind wings, may be found flying about in June in swarms in woods where oak-trees abound. The larvæ, which feed on the leaves, and roll themselves carefully within the folded leaves, are sometimes so numerous as to become a perfect pest. Acres and acres of oak-plantation may be seen completely stripped of the foliage, while the green moths flutter about in countless thousands. The pupal state is passed in a folded leaf, or in the chinks of the bark or other suitable crevice.

The larch-tortrix (*Retina buolinana*) is a bright, foxy-red moth, with habits very similar to those of the last-named species. The moth may be seen in July flying among the trees in young planta-tions, and laying its eggs amongst the buds at the tip of the shoots. The larvæ are hatched in the autumn, and immediately begin to gnaw the buds, giving rise to the exudation of resin.

In the allied pine-gall tortrix (*R. resinella*) the adult has dark fore wings, streaked and mottled with transverse silvery bars and blotches. The larvæ feed within the stem of the buds of the pine-needles, their ravages causing a drop of resin to exude from the twig, which grows larger as the activities of the internal bur-rower increase. If the drop of resin be examined a small pas-sage at the base will be found passing into the pith of the pine-twig, and here the larva may be found. This lump of sticky gum, which attains the size of a filbert, and in which the larva passes the pupal state, has been misnamed a gall; but a

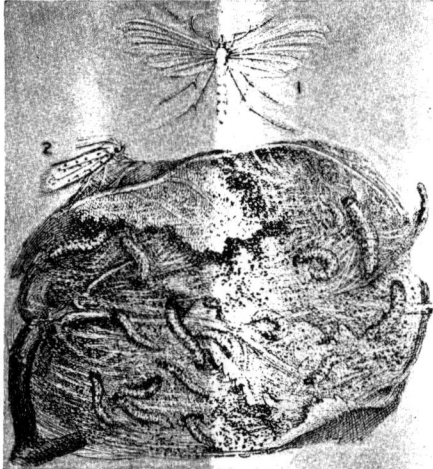

I, PLUME-MOTH; 2, MOTH; AND 3, NEST OF HYPONOMEUTA MALINELLA

gall is not an exuding juice or gum—it is a distinct outgrowth of the cellular structure of the plant. The figure on page 1953 gives illustration of the moth, the resin-drop, and the pupa. A figure is also given of the ichneumon-fly, which seeks the larva with its long needle-like ovipositor, and from its eggs emerge the grubs which will in due course devour their nest.

An especial interest attaches to the pea-moth (*Grapholitha dorsana*), whose larva is the so-called maggot which attacks green peas. When full fed it seeks the earth, and constructs a cell in which to pass the pupal stage. These larvæ also are not averse to a provender of dry peas, to which it often causes considerable destruction. The moth appears on the wing in May. The well-known codlin-moth (*G. pomonella*) takes its name from the circumstance that the larva feeds within growing apples, eating, however, not so much the flesh as boring into the heart and feasting upon the pips. It is rosy red, paler beneath, with grey tubercles, each bearing a long bristle. This moth flies in June, and conceals itself in the daytime in a crevice in the bark, with whose tints its grey mottled wings readily assimilate.

The family of the clothes-moths (*Tineidæ*) is typically represented by the lesser clothes-moth (*Tinea pellionella*), although it must be borne in mind that there is not one particular moth which destroys cloth-ing, but that the larvæ of several species are equally destructive. *T. pellionella* is one of the smaller of these, whose larvæ, of a silky yellow colour, attack all kinds of clothing, as well as the upholstery of our furni-ture. *T. tapetzella*, a larger species, attacks more exclusively furs, skin-rugs, and so forth. In the allied corn-moth (*T. granella*) the caterpillar is very destruc-tive to corn in granaries, feeding indiscriminately upon various kinds of grain. The female lays one or two eggs on a single corn-grain, and after the deposition of all the eggs, the bodies of the adults may be found in numbers in spider-webs in places which they frequent. The presence of the caterpillar may be known by the "frass" or excrement on the grains. Several grains may be spun together, the larva feeding within the shelter of the receptacle thus formed.

Of certain allied species there are no English names, so that they must be mentioned by their scientific titles. Among these, *Depressaria nervosa* appears on the wing from June to September, and has reddish grey fore wings mottled and streaked with black dots. The female lays her eggs upon cumin, and the larvæ soon after they emerge spin together the flower-heads, feeding on the seeds and blossoms. When about to enter the pupal state, the larva bores its way into the centre of the food-plant, gnaws out a suitable chamber, closes the entrance with a little door of silk, and remains safe from the attacks of insidious insect foes.

A familiar moth during June and July in English apple-orchards is *Hyponomeuta malinella*. The satiny white fore wings, with three longi-tudinal rows of black dots, render it a beautiful and conspicuous object as it rests on the apple-tree by day, or flies to and fro beneath the trees as the evening draws on. The female lays her eggs in an elongated cluster on an apple-twig, and the presence of the larvæ first becomes apparent owing to the silky gauze net with which the tiny larvæ spin the leaves together, enlarging their domicile as occasion requires. When full fed, they pupate also in the web, so that numbers of tiny pupæ nestle side by side where the larvæ were wont to feed.

Another family is typified by the genus *Coleophora*, which embraces about seventy species of small moths, characterised by their long, narrow wings, margined with long delicate fringes, the first joint of the antennæ often bearing a tuft of hair. The larvæ live in little cases, in which they pass the winter, turning to the pupa in the spring. As an example of the genus, we mention the larch-mining moth (*C. larcinella*), which is a dull-coloured moth, whose larvæ eat their way into the needles at the tip of young larch-trees. The cater-pillar is full fed towards the end of May, when it spins its little case fast to a larch-needle, and turns to a pupa within. A few weeks later the moth emerges.

Finally, we have the beautiful plume-moths (*Ptero-phoridæ*), in which the larvæ are hairy, and, when full fed, suspend themselves by their anal claspers, turning to pupæ without any covering. The pupæ themselves are often hairy also, though many of them are quite smooth. The plume-moths, as a family, may be recognised by their feathery wings, slender bodies, and long spinous legs.

COLEOPTERA

The beetles are in general easily distinguished from all other insects, and though they seem almost endless in their variety, and comprise an immense number of distinct specific forms, constitute a very well defined order (Coleoptera). The chief characters that serve to distinguish them are briefly as follows. They undergo a complete metamorphosis. Their mouth—which is fitted for taking in solid food—is furnished with biting jaws (mandibles), a pair of maxillæ with palpi, and an undivided or very slightly divided lower lip (labium), which also bears palpi. The antennæ are extremely variable in form, but seldom possess more than eleven joints. The prothorax is usually large and is freely articulated with the following segment (mesothorax), over which it fits behind in such a manner as almost to completely cover it on the upper side.

The fore wings are converted into a pair of stiff, horny structures called elytra, which, in a state of rest, usually meet by their edges in a straight line along the middle of the back, and serve to protect the hind wings and the soft hind parts of the body. The hind wings are in beetles the only true organs of flight; these are membranous and transparent, provided with few nervures, and when not in active use are generally folded transversely beneath the elytra. Many beetles are without hind wings, and are said to be apterous, but it is to be remembered that very few beetles, except in the larval state, are completely apterous in the sense of being without both hind wings and elytra.

In the wingless species, the elytra are generally well-developed, and frequently fastened together along the suture where they meet. The presence of elytra, though not exclusively peculiar to beetles, is still one of their most characteristic features, and affords in most cases a ready means of recognising them. Elytra very similar to those of some Coleoptera are, however, met with among earwigs, and the elytra of beetles do not invariably meet in a straight suture. Thus, in the oil-beetles (*Meloë*) one elytron folds partly over the other; while in certain other groups— the *Rhipiphoridæ*, for example—the elytra are of such a form that they either do not meet at all, or only just touch at the base, and are sometimes so small and so little like the ordinary elytra of beetles that their true nature is not at first sight very apparent.

We have alluded above to the great variety that is to be met with among beetles. No insects exhibit greater extremes of size, and we find on the one hand beetles so small that a pin's head is large in comparison, while on the other we get those giants of their race, the elephant and goliath beetles, which are nearly as

big as a man's fist, and the still larger titan from South America, which is sometimes quite half a foot long, and scarcely less broad in proportion. Even within the limits of a single species beetles are not always of a nearly uniform size, and it is not uncommon to find that in certain species some individuals may be very much larger than others, frequently two or three times as large, and occasionally even as much as five times.

In their external form beetles also afford the most striking contrasts, and the differences of form are not confined to the general shape, but extend to nearly all parts of the body. The head, especially, varies to a great extent both in its shape and the direction which it takes. It is somewhat ring-like behind, where it fits more or less deeply into the cavity of the prothorax. The part between the eyes and the prothorax may be as wide as or even wider than the rest of the head, or may be abruptly or gradually narrowed behind to form a sort of neck.

In most beetles this part of the head is rather short, but its length varies, and there is one remarkable species from the Philippines which presents a most comical appearance owing to the extraordinary length of its neck. This species belongs to a group of leaf-rolling beetles, and doubtless finds its long neck extremely useful.

HONEYCOMB-MOTH

1, larva ; 2, wax-moth ; 3, pupæ ; 4, honeycomb

The fore part of the head is most variable in shape, and though generally short, is in some beetles quite out of all proportion in its length. In the weevils it is prolonged in the form of a rostrum or snout, which is sometimes much longer than all the rest of the body. What is called the "front" of the head frequently faces upwards, being on the same plane, or nearly so, with the occiput or posterior part of the upper surface.

But in many beetles the fore part of the head is bent down, so that the front looks forward, and sometimes even to such an extent that the mouth is drawn back against the prothorax, and the front of the head looks downward. The lower or anterior part of the front of the head is called the clypeus, and to this—usually by the intervention of a short, flexible piece known as the epistome—the upper lip (labrum) is attached.

Running along the middle of the under side of the head there is a piece, generally marked off by a line on each side, which in its posterior part is named the gula, and in front the submentum. The submentum —sometimes prolonged beyond the margin of the head in the form of a peduncle—gives attachment to the lower lip (labium), which consists of a basal

LARCH-MINING MOTH

piece of variable size and form called the mentum, and a terminal part, the ligula. The latter usually bears two lobes (the paraglossæ) at its extremity, while, from its base, known as the hypoglottis, the labial palpi arise.

Between the labrum and labium lie the mandibles and maxillæ. The mandibles are strong biting jaws, and are attached to the sides of the head by pivot-like joints, which permit only of lateral movements. They are often much larger in the males than in the females, and in the males of some forms, such as stag-beetles, attain monstrous proportions. Each of the maxillæ consists typically of a stem, composed of two pieces—cardo and stipes—with a four-jointed palp attached to the outer and two lobes to the inner side of the free end of the stipes.

Except in the larval state, beetles rarely possess those eyes with a single lens which are known as ocelli. The compound eyes, on the other hand, are generally large and well developed, but vary considerably in form, and in the size and number of their facets. They are often simple in outline, sometimes slightly notched in front and reniform, or the notch may extend more deeply and divide the eye into two distinct lobes. Each eye may even be completely divided into two parts, more or less widely separated from one another, so that some beetles appear to have four eyes instead of two. This appearance is very strongly marked in certain water-beetles, in which one part of each eye is on the upper and the other on the under side of the head.

The eyes of some beetles look coarse and granular, while in others they appear quite smooth and glassy-looking owing to the small size and slight convexity of their facets. Among the longicorn beetles it is generally found that in the nocturnal species the eyes are coarser and more granular than in those species which fly during the day, so that the size of the facets seems to have some relation with the conditions of light depending on the habits of the insects. But this curious fact does not, so far as we know, apply to any other family of beetles. Exceptionally also it is found among beetles that the facets in the upper part of the eye are different in size from those on the lower part.

The antennæ of beetles are scarcely less important in their functions than the eyes. They are in most cases sensitive to touch, and there is reason to believe that these organs are also the chief seat of the senses of smell and hearing. They appear under a variety of different forms, some of which, while subject to minor modification, are pretty constant throughout certain large groups of beetles, and thus account for the names Clavicornia, Lamellicornia, and so forth, given these groups.

As a rule, the antennæ, no matter what their length, are made up of eleven joints or segments; but this number may be increased, in some cases to thirty or forty (as in *Rhipicera*), and even to as many as fifty (in the longicorn genus *Polyarthron*), or it may be reduced even to so low a number as two (in *Platyrho-palus*). When the joints are more or less cylindrical in form, the antennæ may be either filiform, if of nearly uniform thickness throughout, setaceous if they taper towards the extremity, or moniliform if each of the joints is short and bead-like. The antennæ are said to be clavate when thickened at the extremity in the form of a knob or club; lamellate when three or more of the terminal joints spread out in broad processes which lie flat upon one another; serrate when the joints have on one side short angular processes like the teeth of a saw; pectinate, or comb-like, when the processes are fairly long and stand out nearly at right angles; or flabellate if the processes are proportionately very long. These are some of the chief types of antennæ met with in the Coleoptera; others of less frequent occurrence will be mentioned when we come to treat of the different families.

The sense of smell is undoubtedly very acute in a great many beetles, as anyone acquainted with their habits could easily testify; and it is considered probable that certain minute pits scattered over the surface of the antennæ, or crowded together on special areas, are in some way connected with this sense.

Though it is not so easy to prove that beetles can hear, it seems hardly open to doubt that in some cases at least they possess this faculty. Everyone has heard of the death-watch beetle (*Anobium*), which lives in old furniture and woodwork of houses, and makes a noise like the ticking of a watch. This little beetle produces the noise by hammering against the wood with its head, and apparently does so for the purpose of attracting its mate, who replies by making a similar tapping sound. It is easy by imitating their sounds to get the beetles to answer back; so that here at least there is some evidence that these insects are endowed with the faculty of hearing.

Many other beetles are able to make sounds, which, though not nearly so intense as the chirping of the crickets and grasshoppers, and not usually confined to one sex, are produced somewhat after the same manner by the friction of one part of the body over another. In beetles the sound sometimes arises from the rubbing of the hind legs against the edge of the elytra, but in most cases it results from the rubbing of an edge over an adjacent area which is crossed like a file by a number of fine parallel ridges.

This stridulating area is in some beetles placed on the upper side of the back part of the head, or on the gular surface underneath, so that when the head moves in its socket the upper or lower edge of the prothorax, as the case may be, scrapes along the file and thus gives rise to the sound. The prothorax of beetles is, as we have already stated, freely articulated with the mesothorax. Its dorsal arch or pronotum ordinarily covers over the whole of the mesonotum, with the exception of the small piece known as the scutellum; but when the prothorax is bent down a considerable part of the mesonotum in front of the scutellum comes into view. It is on this part that the stridulating area of most of the longicorns and of some phytophagous beetles (*Megalopinæ*) is situated.

These insects make a sort of squeaking noise—which is sometimes fairly loud—by rapidly bending the prothorax up and down, and so causing its hind edge to move backwards and forwards over the ribbed surface of the mesonotum. In other beetles the stridulating area may be either on the upper surface of one of the hinder segments of the abdomen or on the sides of one of the anterior

ZABRUS GIBBUS AND ITS LARVÆ

segments, the sound being produced in the one case by the friction of the area against the edge of the elytra, in the other by that of the posterior thighs against the sides of the abdomen.

Beetles are among the most active of insects when on the ground, and, in accordance with their running powers, we find that their legs, though generally slender, are strong and well developed. But in certain groups, where the habits and environment of the insects require it, the legs are adapted to various other purposes. Beetles that jump usually owe their leaping powers to the greatly thickened femora and straight and relatively long tibiæ of the hind legs. It would, however, be a mistake to suppose that when a beetle has thickened and strongly developed hind legs it must consequently be able to jump. Some burrowing species, and others that are not very active in their movements, have very thick hind legs; though, as a rule, it is the front pair of legs which is thickened and otherwise modified to serve as digging organs in those beetles that burrow underground.

In aquatic beetles the swimming legs are disposed like oars, and have all their parts broad and flat, while their breadth is further increased by rows of bristles. Either the hind legs only, which is the rule, or the middle pair also, as in the whirligig beetles (*Gyrinidæ*), may be thus transformed into swimming organs. The coxæ, or basal joints of the legs, vary much in shape and in the mode in which they are inserted in their sockets on the under side of the thorax. Those of each pair are sometimes close together, sometimes widely separated from another, while a longer or shorter distance may intervene between the coxæ of the different pairs of legs, and especially between those of the two hinder pairs.

Considerable importance attaches to the number of joints in the tarsi or feet. In classifying beetles this number is one of the first things to be noticed. If a beetle has five joints in each of its tarsi, it is placed in that section of the order which is known as the Pentamera; if it appears to have only four joints in each foot, it belongs to the Tetramera; and if but three, to the Trimera. When there are five joints in each of the four anterior feet, and only four in the hind feet, the beetle may be regarded as one of the Heteromera.

To these general rules there are a few exceptions, which need not be discussed here; but we must point out that, although in the Tetramera the tarsi appear to be four-jointed, and in the Trimera three-jointed, they are really composed of five joints and four respectively. The fourth joint in the one case, and the third in the other, are, however, usually so small as not to be noticed except upon very close examination.

The abdomen is never stalked in beetles, but attached to the thorax by a broad base, which is applied against the posterior coxæ; exceptionally, however, as in certain mimicking species, its base may be more or less narrowed. It is generally somewhat flattened in shape, and on the upper side eight segments are usually distinguishable, which, so far as protected by the elytra, have a soft and but slightly horny integument. Five or six segments are generally visible on the ventral side, but in certain cases the number may be reduced. The terminal segments are usually retracted within the abdomen, and completely hidden from view; but in the females of many species they can be exserted in the form of a tubular ovipositor, which enables the insect to lay its eggs deep in the crevices of bark.

Although beetles do not always exhibit differences in external form by which the sexes may be distinguished, such differences frequently exist, and are sometimes of the most pronounced character. As a rule, the male is more slenderly built than the female, and has longer and more fully developed antennæ; his eyes also are often larger, and in the length and shape of the legs, and in the width and structure of the tarsi, differences in the two sexes are frequently to be noticed. When the male is fully equipped for flying, the female may be without wings, or even, as in the case of a glow-worm, without elytra; and whenever there is any decided difference in coloration, it is almost invariably the male which displays the brightest and most conspicuous colours.

The great projecting horns and processes on the head or prothorax, which give so grotesque an appearance to many beetles, are generally wanting or only feebly developed in the females; and these and other differences are sometimes so strongly marked that it is difficult to recognise in the two sexes individuals of one and the same species.

The larvæ of beetles do not in outward appearance exhibit anything approaching the great diversity seen in the perfect insects. They seldom display conspicuous markings, and are mostly of dingy white, brownish, or black colours. The external structure and form vary sufficiently to make it possible to tell to what family, or

CARNIVOROUS BEETLES AND THEIR PREY
1, Carabus nemoralis; 2, Calosoma sycophanta; 3, Carabus auratus, and 4, its larva

division of a family, a larva belongs; but, so far as species are concerned, our knowledge of the larvæ is extremely limited, and applies to a relatively very small proportion of the whole number of known species of Coleoptera.

In the weevils, and some other beetles, the larvæ are soft white grubs with scarcely any trace of legs, but in most of the other larvæ the legs are fairly well developed, though not so completely as in the perfect insects. The head is always horny, and furnished with jaws for biting and grinding solid food. Exceptionally, as in the carnivorous larvæ of some water-beetles, the mandibles are adapted for sucking up the juices of the animals on which these larvæ prey. The antennæ are short and few-jointed, and in some cases quite inconspicuous. Eyes, when present, are always in the form of ocelli, which are grouped together in varying number on each side of the head.

The head is followed by a series of rings or segments, of which the first three—scarcely different in form from the rest—constitute the thorax, and give attachment to the legs. A pair of pro-legs is sometimes present on the last segment, but in beetle-larvæ the intermediate segments never carry those false legs, which are so often found in the caterpillars of Lepidoptera and Hymenoptera. The spiracles—which are mostly hidden by the elytra in the perfect insects—are generally quite conspicuous in the larvæ, and appear as a row on each side of the body. Their number varies; and in those aquatic larvæ which breathe by means of tracheal gills they are altogether wanting. When about to pupate, some larvæ construct cocoons of earth, or, in the case of wood-boring species, they may make a shell out of fine chips and dust glued together with a sticky secretion. The pupæ, whether enclosed in a cocoon or not, are inactive, and show all their appendages lying freely against the body, with each appendage wrapped round by its own special covering of integument.

The larval existence of beetles varies from five or six weeks in some groups to almost as many years in others ; and when conditions arise to interfere with the proper nourishment of the larvæ, the period may be unduly prolonged. Some of the wood-boring larvæ seem to

TIGER-BEETLES
Cicindela hybrida, with larva and pupa ; Collyris longicollis

live an exceptionally long time. There was, some time ago, in the Natural History Museum in London a block of wood containing a living longicorn larva, which for about five or six years had been feeding and burrowing in the wood. The larva was brought to the museum in a boot-tree, which its owner previously had in constant use for over fourteen years. Other cases are on record in which beetles have been seen to emerge from furniture in houses, after having apparently passed an even more prolonged larval existence.

Beetles, whether from the extent of their numbers or the variety of their shapes and instincts, are well qualified to play an important part in the economy of Nature. Their chief function is that of universal scavengers. Not only do they dispose of the smaller quantities of dead and decaying animal and vegetable matter passed over by larger animals, but, by their own peculiar methods, they are enabled to attack and clear away even the carcases of quadrupeds of large size, and the dead trunks of the largest trees. Owing to the compactness of their shape and the solidity of their outer covering, they are adapted for a much greater diversity in modes of life than is possible for insects of other orders.

Besides groups fitted to act as scavengers, we find further series of forms that live in, and prey upon, all kinds of plant life. There are groups, again, either of terrestrial, arboreal, or aquatic habits, which seek for, and prey upon, living animals of the smaller kinds. Some beetles live within the depths of the darkest caverns ; and in such cases, having no use for eyes, they are generally blind. Others are to be found dwelling as " guests " in the homes of the ants and termites.

Although the beetles cannot boast of such a long line of ancestry as the cockroaches and other Orthoptera, yet their records go back to an early period in geological history. There is no certain evidence that they existed in Palæozoic times, and their first appearance has not been traced farther back than the beginning of the Secondary epoch. The earliest undoubted fossil remains of Coleoptera occur in the Swiss Trias, and from this period onwards fossil beetles are to be met with in greater or less abundance in rocks of different ages. They are especially well preserved in amber ; and from the Tertiary amber-beds on the Baltic

thousands of specimens have been collected. Of the beetles now existing, over one hundred and thirty thousand different species have been described, and as new species are being yearly added to the list at a rapid rate, it is probable that the number of named species will eventually reach 150,000 or even 200,000.

PENTAMERA

Beetles in which all the tarsi are five-jointed are classified in the section Pentamera. In this group there comes first a great tribe of beetles, which, on account of their carnivorous tastes and predaceous habits, are known as the Adephaga. Their whole organisation seems well adapted to enable them to capture and devour their prey, and it is in the modifications directed to this end that some of the chief distinguishing characters of the tribe are to be found. Their legs are fitted for speedy locomotion, and their jaws for the cutting and tearing operations to which they are usually applied. The mandibles are acutely pointed, and have sharp cutting edges ; and the inner lobes of the maxillæ are hard and hooked at the end. The outer lobes of the maxillæ are two-jointed and slender, and resemble palpi, which explains the fact that these beetles are often described as having three pairs of palpi. The antennæ are usually simple, and never clubbed. The tribe is divided into the Geodephaga and Hydradephaga, one subtribe containing terrestrial, the other aquatic forms.

TIGER-BEETLES

The *Cicindelidæ* consist of about one thousand known species, which are distributed throughout the world, but are much more abundant in tropical than in temperate or cold countries. In Europe two genera only are represented—*Tetracha*, which comprises nocturnal and twilight-loving species, and *Cicindela*, whose species are found in the hottest and sunniest places. The tiger-beetles are extremely pretty insects of remarkably active habits, and exhibit the predaceous type of structure

ELAPHRUS RIPARIUS

to perfection. Besides possessing great speed of foot, most of them make ready use of their wings, and they are further characterised by large and prominent eyes, and mouths well adapted for seizing and holding their prey, the mandibles being long and provided with a number of sharp teeth, while the inner lobe of the maxillæ is furnished with a movable claw or

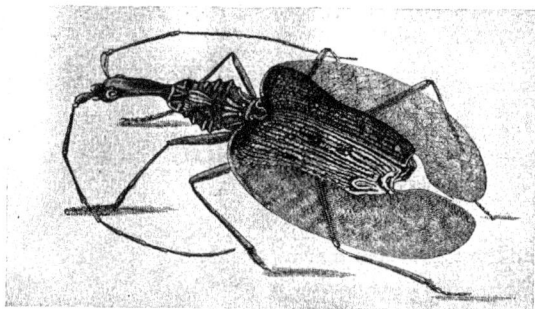

MORMOLYCE PHYLLODES, OR VIOLIN-BEETLE

hook at the tip. The fact that this hook is movable, and not firmly fixed to the blade of the maxillæ, affords a means of distinguishing the tiger-beetles from all the other beetles of the tribe Adephaga.

More than half of all the known species of the family

belong to the single genus *Cicindela*, and this is the only genus which is cosmopolitan. With the exception of a few species of an almost entirely ivory-white colour, the *Cicindelidæ* exhibit greenish, bronzy, or darker metallic tints, frequently varied with white or pale yellow spots and bands, which in the case of a great many species run together to form more or less intricate and pretty patterns.

While their shape is usually such as is shown in the figure of *C. hybrida*, we get, on the other hand, remarkable exotic forms, in which the body is narrow and elongated, and broadest towards the hinder end. *Collyris* and other genera of the various Oriental countries—where the species are found pursuing their prey on the trees in the forests—afford examples of this type. From its great resemblance in colour and form to *Collybris*, a rare and curious longicorn beetle, found in the same localities, has been named *Collyrodes ;* and beetles of the family *Cicindelidæ* are amongst those most frequently mimicked by other beetles.

CARNIVOROUS WATER-BEETLES
Dytiscus marginalis—1, male ; 2, female ; 3, eggs ; 4, pupa ; 5, larva attacking a tadpole ; 6, Hydrocharis caraboides ; 7, its larva ; 8. Acilius sulcatus. female

GROUND-BEETLES

In external structure the carnivorous ground-beetles (*Carabidæ*) approach the *Cicindelidæ*, from which they may in most cases be distinguished by their general shape, as well as by the fact that they never exhibit the coloration and markings characteristic of that family. Other points of difference may be seen in their less prominent eyes, in the absence of an articulation in the hook of the maxillæ, and in the shape of the mandibles, which, though occasionally long, do not exhibit the slender curved form and sharp dentition met with in the tiger-beetles.

The number of species of *Carabidæ* at present known can scarcely be less than eleven thousand. This family seems better represented in temperate and colder regions than within the tropics, though species, in more or less abundance, are to be found in every country and island of the world. Whilst the species are almost all predaceous in their habits, we find them under a variety of different forms and with several distinct peculiarities of structure, many of which are to be regarded as special adaptations.

The *Carabidæ*, like all other beetles, have their enemies, but we never find mimetic and protective disguises ; and to escape from their enemies the ground-beetles have mostly to rely upon their speed of foot, or the readiness with which they can take flight or disappear amongst the herbage. Many species are, however, provided with anal glands that secrete an acrid or stinging liquid, which is sometimes ejected with considerable force when the insect is handled.

SCARITES GIGAS

In the "bombardier beetle" (*Brachinus crepitans*) and others of the same group, the secretion is volatilised on emission, and issues as a little cloud of smoke, which is accompanied at each discharge by a slight sound ; and when the insect is irritated it repeats the discharge several times in succession, but each time with diminished force. The "bombardier" is a rusty-red species, with dull blue-black elytra, and a narrow head and prothorax, and is pretty common, especially on chalk, in different parts of the south and south-east coasts of England.

Among those species of the family that in habits and general appearance most closely resemble the *Cicindelidæ* are the little beetles of the genus *Elaphrus*. . These love to run about in the rays of the sun, not so much in dry places as on the muddy banks of rivers, on the sands of the seashore, and in other damp situations. They have large prominent eyes, a narrow prothorax, slender legs, and curiously marked elytra. This genus is confined to the Northern Hemisphere. The species which is figured (*Elaphrus riparius*), like some other beetles of the family, is able to produce a stridulating noise by rubbing the back of its abdomen against a projecting nervure on the under side of the elytra. Those tiny little beetles of a glistening bronzy-black appearance, and with beautifully sculptured elytra, which are to be seen on almost any bright day in the spring or summer, running quickly over garden beds or paths, belong to the genus *Notiophilus*.

The genus *Carabus*, after which the family is named, contains over three hundred species, and is somewhat remarkable in its distribution ; for, with the exception of a small group of species found in Southern Chili, it is restricted in its range to the North Temperate zone. Six or seven species are found in Britain ; *Carabus violaceus* and *C. nemoralis* are perhaps the two most frequently met with. The first is nearly smooth, of a dull blue-black colour, with purplish borders to the thorax and elytra, and is of about the same size as *C. nemoralis* (represented in the figure on page 1957). The latter has a purplish thorax and bronzy elytra, marked with a few rows of conspicuous

BRITISH ROVE-BEETLES

1, the devil's coach-horse (Ocypus olens); 2, Staphylinus pubescens; 3, Philonthus æneus; 4, Oxyperus rufus; 5, Pæderus riparius; 6, Staphylinus cæsareus

predaceous insects, some at least of the species of *Zabrus* and a few others are largely, though probably not wholly, addicted to a vegetable diet. The species (*Zabrus gibbus*) figured on page 1956 lives in cornfields, and has at different times committed great havoc among crops—wheat, barley—rye, in various parts of Germany and Italy.

CARNIVOROUS WATER-BEETLES

The *Dytiscidæ*, or carnivorous water-beetles, resemble the *Carabidæ* in many of their structural features, and differ chiefly in the modifications undergone to fit them to an aquatic mode of life. Thus we find, as in the latter family, the mentum is usually broad and deeply emarginate in front, the outer lobe of the maxillæ is two-jointed and palpiform, the antennæ are moderately long and slender, and the trochanters of the hind legs are prominent.

punctures. Another species which is figured on page 1957, *C. auratus*, is very rare in England and doubtfully indigenous, but in France it is common and does much service by destroying the cockchafers and their grubs.

The genus *Calosoma* approaches *Carabus* in many of its characters, but may be easily distinguished by its shorter, broader, and more rounded prothorax, and the greater relative width of the elytra. *Calosoma inquisitor*, though rare and found only in parts of England, may be regarded as a true British species; but the species figured on page 1957 (*C. sycophanta*) is only an occasional visitant to this country.

The *Carabidæ* as a whole, though sufficiently varied in their external structure, do not exhibit any very unusual or striking peculiarities of form, and the species already considered, with a few more presently to follow, may be taken as typical of the commoner forms met with throughout the family. In the genus *Mormolyce* we have, however, a remarkable exception.

WHIRLIGIG-BEETLE

On the other hand, the antennæ are always smooth; the head is broad and fits deeply into the prothorax, while the latter is applied by a broad base against the elytra, so that the outline of the body is continuous and the general shape more or less oval; the hind legs, which, with their tibiæ and tarsi flattened and furnished with rows of bristles, are adapted to serve as oars in swimming, are somewhat longer than the other legs, and come off from the body at a considerable distance behind them, while their coxæ appear as broad, flat plates firmly joined to the metasternum, for parts of which they might at first sight be very readily mistaken.

The species of this strange genus—three in number, and all very much alike—have been found in Java, Sumatra, and other East Indian islands. They are of pitchy-brown colour, and have the body much flattened and the head slightly elongated, while their antennæ are also very long; but, as will be seen from the figure, the chief peculiarity in the appearance of these extraordinary insects is due to the great lateral expansions of the borders of the elytra, and the curious manner in which these expansions are prolonged behind. *M. phyllodes*, the best-known species, occurs in Java, Borneo, and the Malay Peninsula; and the people of Java, struck, no doubt, by its peculiar shape, call it "the violin." Some of the largest individuals of the species are nearly three and a half inches long, and measure more than an inch and a half across the broadest part of the elytra.

We have alluded in the introduction to the burrowing habits of some of the *Carabidæ*. The *Scaritinæ* are a group that possess such habits, and the figure on page 1959 of *Scarites gigas* will give an idea of the general form characteristic of nearly all the species of the group The genus *Scarites* comprises a large number of species, all of a uniform black colour, and most of them of a moderate size. They make their burrows in the banks of streams, the seashore, or other suitable places, and rarely leave them during the day, lying in wait for their victims at the mouth of the holes.

The genus *Zabrus*, which we have next to notice, forms, so far as its habits are concerned, one of those exceptions that go to prove the rule. For, while it is true that almost all the *Carabidæ* are carnivorous and

The males may be distinguished from the females by the shape of their fore tarsi, in which the first three joints are strongly dilated, and furnished underneath with sucker-like hairs; while in this sex also the back is generally smooth and glossy, the elytra of the females frequently have a ribbed or corrugated surface.

The *Dytiscidæ* seem especially fond of stagnant waters, and some of the species are common objects in our ponds and ditches. They come to the surface when it is necessary to take in a fresh supply of air beneath the elytra. These organs fit very closely against the sides of the body, and so prevent the air from escaping while the beetle is swimming about under the water; but the air meanwhile is being used up in breathing by means of the thoracic and abdominal spiracles. The beetles fly strongly, and on fine summer

CLAVIGER TESTACEUS, CARESSED BY ANTS

evenings may sometimes be seen winging their way to new quarters, a change which is often necessitated by the drying up of the pools in which they had previously been living. The large brown water-beetle *Dytiscus marginalis* is one of the commonest British species Another common species *Acilius sulcatus*, is also represented in the figure on page 1959.

WHIRLIGIG-BEETLES

The *Gyrinidæ*, or whirligig beetles, are a small but very well-defined group, and in many points of structure are sharply distinguished from the other families of the tribe Adephaga. In their oval shapes they resemble the *Dytiscidæ*, though they are usually somewhat flatter below and a little more convex on the upper side. But in the relative proportions of the three pairs of legs they are entirely different. The fore legs are long and slender, and when stretched out look like arms, whereas the two hinder pairs are short and broad being modified for use as paddles in swimming.

Another very distinctive feature is presented by the eyes, each of which is divided by a ridge on the side of the head. These beetles appear, in consequence, to have four eyes ; one pair, as it is said, though there is no proof of the fact, for espying objects above them, the other for looking at things in the water below. From the *Dytiscidæ* and *Carabidæ* they differ further in having their antennæ shorter than the head, and the outer lobe of the maxillæ either completely atrophied or else in the form of a slender spine.

The *Gyrinidæ*, though widely distributed and represented in almost all parts of the world, include altogether rather less than three hundred known species. The genera are few in number, and two only occur in Europe. Some of the British species, such as *Gyrinus natator*, are commonly to be seen in ponds and canals or " holes " in reedy, sluggish streams, where the shiny little beetles attract attention by the ease and rapidity. of their movements as they skim about on the surface

BURYING BEETLES

SILPHA ATRATA AND LARVA

of the water, performing a variety of intricate evolutions, some sweeping along in graceful curves, others going round in circles or spiral tracks, now all collecting together in groups, and then, if startled, suddenly darting off with amazing speed in every direction.

ROVE-BEETLES

The next beetles we have to consider are those which, on account of their abbreviated wing-cases, are known as the Brachyelytra. This tribe, to which, however, not all beetles with short elytra belong, contains a single very large family—the *Staphylinidæ*. Owing to the shortness of their elytra, and the usually narrow and elongated form of their bodies, the rove-beetles have an easily recognised and characteristic appearance. The head is generally large and flat, with a narrow neck behind where it fits into the prothorax. The antennæ—composed of eleven or occasionally twelve joints—are usually filiform, but are often slightly thickened towards the extremity, and in some cases end in a distinct club.

Though prominent and conspicuous in a few genera, the eyes are, as a rule, raised but very little above the general surface of the head. It is interesting to note that ocelli, which are of such rare occurrence in adult beetles, are to be found in certain groups of this family ; two ocelli being present in *Homalium* and its allies, and a single ocellus in the genus *Phlœobium*. The mandibles vary in form according to the habits of the species ; they are usually strong, often sharply curved and pointed at the end, and of a distinctly carnivorous type. Attached to the base and running a little way alongside the inner margin of each mandible, there is to be seen in many species a narrow flexible plate fringed, or not, with hairs at the end. This piece, first made known by Kirby, who called it the *prostheca*, is rarely met with except in the *Staphylinidæ*. The ligula is narrow, and bears distinctive paraglossæ ; and the outer lobe of the maxilla is never palpiform.

The rove-beetles are for the most part carnivorous, and prey upon all kinds of larvæ and other insects, as

HISTER FIMETARIUS AND LARVA

well as upon slugs, snails, and worms, but they feed largely on carrion, and to some degree on vegetable matter. Several species live in fungi, some in flowers, others under bark and in rotten wood, while in the case of certain genera, such as *Lomechusa* and *Atemeles*, the species are to be sought for in or about ants' nests. Some of these latter species are welcome guests, since, like the aphides, they secrete a liquid which is eagerly swallowed by the ants ; others may possibly act as scavengers. Among the species of the genera *Spirachtha* and *Corotoca*, which live with the termites in South America, some are very remarkable from the fact that the females give birth to living young.

Many of the British species of beetles belong to this family. Everyone has seen the devil's coach-horse, that long, black, ugly-looking, but useful, insect which is to be found under stones and earth, or roving about in gardens, and which, when you attempt to stay its progress by pointing with a stick or finger, stands with threatening jaws and upturned tail as if ready to accept the challenge. This species, which, with a few others, is figured on page 1960, is scientifically known as *Ocypus olens*, and is one of the largest rove-beetles.

CLUB-HORNED BEETLES

We now come to a series of small families, forming the group known as the Clavicornia, or Necrophaga. This group, however, rests on no true scientific basis, and is more or less artificial in its character. Most of the species included in the group feed upon decaying animal or vegetable matter, hence the name Necrophaga. The antennæ exhibit in general a tendency to be thickened towards the tip, and in many cases the last three joints form a distinct club ; but in some of the families antennæ of quite another shape are to be found. Though usually five-jointed, the tarsi display much variation in the number of their joints.

The family of the *Paussidæ* includes probably less than two hundred known species, the majority of which have been discovered in the tropics of Asia and Africa, though one species (*Paussus favieri*) occurs in the south-west of Europe. They are mostly reddish-brown insects of rather small size, oblong form, and in general appearance little attractive, were it not for the extraordinary shapes of their antennæ. These organs are generally very broad and flat, in some species resembling a paper-knife in shape ; the number of joints varies from ten to two, and the last joint frequently has a bulbous or discoidal form. So far as at present known, all the species live in ants' nests, and, unless sought for in these situations, they are rarely seen except at night, when they occasionally fly into rooms.

The tiny beetles belonging to the *Pselaphidæ* resemble the *Paussidæ* in exhibiting certain anomalies in their structure, and their lives are passed in similar obscure situations. But while the *Paussidæ* may possibly be related to the *Carabidæ*, the very short elytra of the *Pselaphidæ* and the entirely horny nature of the dorsal plates of the abdomen seem to indicate an affinity with the *Staphylinidæ*. In other points of structure, however, these two families are different. In the *Pselaphidæ* the lobes of the maxillæ are soft and membranous ; and the abdomen, which in one group (the *Clavigerinæ*) is composed of five segments, with the basal rings fused together, is quite incapable of the movements so characteristic of the rove-beetles. The joints of the antennæ vary in number from eleven to six, or even two, and are in most cases clubbed at the end. While in one division of the family the palpi are usually composed of three or four joints, and are long and conspicuous, in the other they are one-jointed and scarcely visible. The tarsi are three-jointed, the first and second joints often very short, while the third is long and often bears only a single claw.

The *Pselaphidæ* are distributed throughout most parts of the world. They are to be found under stones, moss, dead leaves, and other vegetable refuse, as well as under the bark of trees and in damp marshy situations ; but the most interesting species are those which live in ants' nests. They are all of small size. The genus *Claviger*, comprising about eighteen European and one or two Asiatic species, has six-jointed antennæ, and is further remarkable for the fact that the long, cylindrical head is entirely devoid of eyes. The best-known species, *C. testaceus*, is in Britain met with chiefly in the nests of the common yellow ant (*Lasius flavus*), though on the Continent it is found also in the nests of other species. It is about a tenth of an inch long, yellowish brown in colour, wingless, with the elytra fused together, and with a deep impression on the base of the abdomen.

The relation between the ants and their guests is of a most interesting character. Whenever an ant meets one of these guests in a gallery of the nest, it gently touches and caresses it with its antennæ, and while the beetle responds in a similar manner, the ant sucks at the tufts of hair near the end of the beetle's elytra, and then licks the whole anterior surface of the back of its abdomen. The ants feed the beetles in very much the same way as they feed their larvæ. When the beetle is hungry it expresses its desire to be fed by licking an ant near the mouth, and occasionally stroking the sides of its head with gentle movements of its antennæ. The attention bestowed by the ants on the beetles is as great as that which they give to their own larvæ, and they frequently feed the hungry ones among them before looking after their own brood.

MELIGETHES ÆNEUS

BURYING-BEETLES & ALLIES

The orange-banded bury-ing-beetles of the genus *Necrophorus* are probably the best-known members of the *Silphidæ*, though they are not to be considered the most representative, either in habits, size, or general appearance. The many genera of which the family is composed differ greatly in size and outward form, while the burying instinct is almost entirely confined to the genus *Necrophorus*. In nearly all cases, however, the antennæ, consisting usually of eleven joints, are thickened towards the tip or furnished with a dis-tinct club ; the prothorax is usually broad and flat, with sharply defined lateral margins, while the elytra frequently do not reach the tip of the abdomen ; the coxæ of the four anterior legs are large, prominent, and conical in shape ; and the tarsi are usually five-jointed, though occasionally with a less number of joints.

The carrion-beetles are widely distributed, though chiefly characteristic of the colder and temperate zones. In the genus *Necrophorus* the antennæ terminate in an almost globular, four-jointed mass ; the body is broadest across the ends of the elytra, which are abruptly truncated, leaving the tip of the abdomen exposed. The species of this genus are black in colour, but in most of them the elytra are crossed by two broad orange bands. They feed upon dead animals of all kinds, and their habit of burying the smaller carcases, such as those of mice, moles, small birds, etc., has gained for them the name of " sexton." or " bury-ing " beetles.

Their mode of operation is to creep underneath and dig the earth away until they have made a hole big enough to receive the dead body ; as the latter sinks the loose soil closes over it, and in time completely hides it from view. The females then lay their eggs in the carcase, which subsequently serves as food for the larvæ. These insects must have a very acute sense of smell, for in a very short time after a mole has been killed some of them may be seen hovering over the body, although not previously observed anywhere in the vicinity.

Out of about a dozen species of *Necrophorus* occurring in Europe, seven are found in Britain, *N. vespillo* being perhaps the one which is most widely distributed. Most of the species of the genus *Silpha*—from which the family name is derived—are dark, sombre-looking insects, somewhat ovate in shape, the prothorax being broad and closely applied to the base of the elytra, while the elytra usually extend to the tip of the abdomen. The head is small, and when turned down is hidden under the pronotum.

The beetles themselves are generally met with in or about dead animals, but some of the species display a partiality for a vegetable diet ; thus in France the adult *Silpha reticulata* has been found to attack wheat, while *S. nigrita* devours strawberries in the Alps and Pyrenees. The larvæ of most of the

GREAT BLACK WATER-BEETLES
1, larva ; 2, male ; 3, female with egg-cocoon

species are somewhat like wood-lice in shape, with the posterior angles of the abdominal segments sharply produced. Those of *S. opaca* and *S. atrata* are some-times very destructive to the leaves of sugar-beet and mangold-wurzel.

HAIRY-WINGED BEETLES

The *Trichopterygidæ*, or hairy-winged beetles, are exceedingly minute insects, the smallest, in fact, of all the beetles, many of the species being less than the fiftieth part of an inch in length. They are further remarkable on account of the structure of their wings. These organs are very long and narrow, each consisting of a strip of membrane attached to a horny stalk, and fringed on each side with long and closely-set hairs.

HISTERIDS

The *Histeridæ* form a well-defined family, widely distributed, and numbering considerably more than twelve hundred species. In colour they offer little variety, being mostly either black, dark blue, or green, the elytra being occasionally spotted with red or yellow. They are compactly oval or oblong-oval in form, and nearly always present a highly polished appearance. The antennæ are short, with a long basal joint and a very distinct terminal club, and as a rule are capable of being turned back into grooves beneath the thorax. The elytra are truncate at the tips, leaving the last two segments of the abdomen exposed ; they are generally marked with a series of finely impressed longitudinal lines, the number and disposition of which afford useful characters in dis-tinguishing between the different species of a genus. In the division of the family to which *Hister* belongs, the prosternum is produced in front, forming a promi-nent " chin-piece " which serves to protect the lower part of the head when the latter is retracted. In *Saprinus* the " chin-piece " is wanting.

NITIDULIDÆ

The *Nitidulidæ* have some resemblance in external form to the *Histeridæ*, though they are generally of smaller size, with their integuments less hard, and their colours a little more varied. The elytra are slightly truncate behind, leaving a variable number of the segments of the abdomen exposed. The antennæ are eleven-jointed or, exceptionally, ten-jointed, with the last two or three joints forming a knob; the maxillæ have, as a rule, but a single lobe, and the tarsi are five-jointed, though in a few genera the males, at least, have only four joints in the posterior tarsi. Many of the species are found feeding and breeding in decaying vegetable or animal substances, such as rotten wood, bark, fungi, and in carcases or bones; some frequent the exuding sap of trees; while a very large number are to be seen on flowers, among which are the brightly coloured little beetles of the genus *Meligethes*.

SCARABÆUS VARIOLOSUS

The species figured (*M. æneus*) is one of the commonest, and met with chiefly on the flowers or leaves of cruciferous plants. In Germany these little beetles are well known, on account of the depredations they commit on crops of rape. A few days after emerging from their winter sleep, the beetles lay their eggs in the buds. In about a fortnight the larvæ are hatched, and proceed to feed on the undeveloped or full-blown flowers; while later on they attack the young pods, to which they do more damage than the beetles themselves.

RASPBERRY BEETLE

The small family *Byturidæ* may also be mentioned here. The genus *Byturus* contains only four or five known species, which are confined to Europe and North America, and one of which is familiar to gardeners and others as the "raspberry beetle." This

SCARABÆUS SACER

species (*B. tomentosus*) is somewhat oblong in form, from an eighth to a sixth of an inch in length, of a dirty yellowish colour, and covered with a yellow down. Though found on flowers of many different kinds, it is especially common on raspberry blossoms, and the cylindrical brownish larvæ sometimes do much damage to the flowers and fruit.

MUSEUM-BEETLE & ALLIES

The *Dermestidæ* have a special interest, owing to the destructive habits of many of the species. The beetles themselves are small in size, oblong or oval in shape, sometimes nearly round, and usually clothed with fine, closely lying hairs or scales, which frequently give rise to greyish or yellowish spots or bands on the elytra. The front of the head, except in the genus *Dermestes*, bears a single ocellus; the short antennæ, consisting usually of eleven joints, are clubbed at the end; the abdomen is entirely covered over above by the elytra; and the tarsi are always five-jointed.

While certain species are met with only on flowers, the

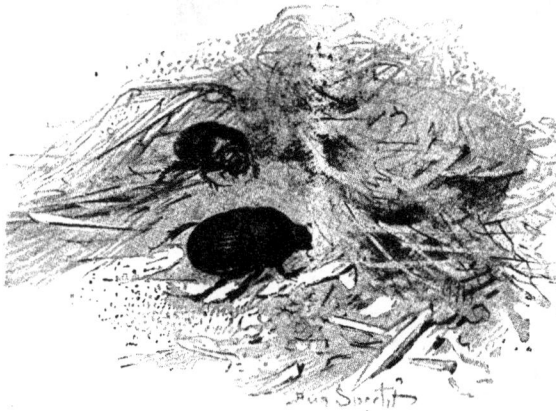

DUMBLE-DORS, OR SHARD-BORN BEETLES

majority live in dried animal matter—furs, skins, and the like, as well as articles of food, such as bacon and cheese. The perfect insects do comparatively little damage, the real depredators being the larvæ, including those of many species which in the adult state frequent flowers. The larvæ are little hairy creatures of a dark colour, looking like small caterpillars, with the hairs sticking out straight and arranged more or less in tufts or bundles.

The larvæ of *Anthrenus musæorum*, the so-called museum-beetle, have to be carefully guarded against in museums, as they are very destructive to zoological collections, and more especially to those of dried insects. *Attagenus pellio* is another very common species of this family, usually found in houses, and well known on account of the ravages of its larva in natural history collections, furs, hair-stuffed couches, and so forth. The larva is of a brown or red-brown colour above, and covered with long hairs pointing backwards; it is broader in front, and tapers towards the hinder end, where it carries a tail-tuft of very long hairs.

WATER-BEETLES

In the *Hydrophilidæ* the antennæ are short, and composed of from six to nine joints, of which the first is relatively long, and the last three or so thickened in the form of a club; the mentum is a large shield-like plate without a notch in front; the lobes of the maxillæ are not toothed, and the palpi are long and slender, frequently much longer and more conspicuous than the antennæ. These characters afford a ready means of distinguishing these herbivorous water-beetles from the carnivorous water-beetles, to which in general shape many of them bear a close resemblance. The great length of the maxillary palpi has given rise to the name *Palpicornes*, by which the family was formerly known.

In the perfect state, all the members of the family feed upon vegetable matter; but it is only those of the subfamily *Hydrophilinæ* — of which the great water-beetle, *Hydrophilus piceus*, may be taken as the type— that are truly aquatic in their habits; the second subfamily, the *Sphæridiinæ*, though including certain marsh-frequenting species, is composed mainly of land-insects which are found chiefly in vegetable refuse or in the droppings of herbivorous mammals. Of the *Hydrophilinæ* some are found in stagnant, others in running water, but they are nearly all poor swimmers, while a large number progress by simply crawling along the surface film upside down; in their slow movements they present a marked contrast to the active predatory *Dytiscidæ*.

STAG-BEETLE AND ALLIES

We now pass to the Pectinicornia, a small tribe containing only two families, one of which has no European representative, while both are somewhat limited in the number of their species. In the *Lucanidæ* the antennæ are ten-jointed, with the first joint long and set at an angle with the rest of the antennæ, of which from three to seven of the last joints are furnished with rigid tooth-like processes on one side. The outer lobe of the maxillæ ends in a pencil of hairs, while the inner lobe has very often the form of a claw; the ligula is membranous or leathery in texture, and is attached to the inner face of the mentum; the elytra cover over the abdomen, which on the ventral side shows five or, in the male, six segments; and the tarsi are five-jointed, with a long, slender spur projecting between the claws of the terminal joint, and carrying at the end two long bristles. The male insects are remarkable for the massive development of their jaws, which in many cases are forked and branched

COCKCHAFERS

The common stag-beetle (*Lucanus cervus*), may, in full-sized males, attain a length of over 2 inches, or, if the mandibles be included, more than 3 inches. It is most abundant in the neighbourhood of oak-woods, and in England is not uncommon in the southern counties, where the males may be often observed on the wing on fine summer evenings, flying with a loud hum.

The *Passalidæ* are a small family of about two hundred known species, which are almost entirely restricted to the warmer parts of the world, the greater proportion being found in America. In the form of the antennæ and in some other respects they show an affinity with the *Lucanidæ*, though easily distinguished by the character of the mouth parts. The ligula is horny, and lies in a deep quadrangular emargination in the mentum; the lobes of the maxillæ both resemble claws; and the mandibles offer a peculiarity of structure met with in no other family, each being provided with a movably articulated tooth placed close to the molar surface.

BURROWING-BEETLES

The Lamellicornia—comprising the burrowing-beetles, cockchafers, and a host of other forms, differing both in habits and external structure—are represented in all parts of the world. We

SUMMER-CHAFERS

have only to mention the goliath-beetles of West Africa, and the elephant and hercules beetles of Tropical America, to indicate the great size attained by some of the species; while as regards beauty and brilliancy of coloration no beetles can rival many of those belonging to the two subfamilies *Cetoniinæ* and *Rutelinæ*. The male stag-beetles, as we have just seen, are distinguished by their large heads and monstrous jaws, but in the males of the present family it is, as a rule, the prothorax which is greatly enlarged or otherwise modified in form, and often furnished, like the head, with processes of various kinds, sometimes short, in others taking the shape of huge curved or branching horns.

The family admits of two principal divisions. In the first division the ligula of the lower lip is more or less membranous and distinct from the mentum, and the spiracles of the abdomen are all situated in the connecting membrane between the dorsal and ventral plates. Among these we may mention the genus *Scarabæus*, over sixty species of which are known, most of them African, some occurring in Asia, and a few, including *S. sacer*, one of the sacred beetles of the Egyptians, found also in South Europe. Among the coprophagous species met with in Great Britain, those of the genus *Aphodius*, which represents a second subfamily, are the most numerous. They are somewhat oblong in form, as in *Aphodius fossor*, one of the largest and best-known species, and are usually shining black, though in many the elytra are of a reddish or yellow colour, in some cases spotted with black.

A type of another subfamily is found in the genus *Geotrupes*, of which we have in this country several species, including the well-known "dumble-dor," or "shard-born" beetle (*G. stercorarius*). The species almost all exhibit dark blue or black colours, and in most cases the sexes differ little in external form.

The plant-feeding, or phytophagous, subfamilies belong to the second division of the *Scarabæidæ*. In these the ligula is consolidated with the mentum, and the abdominal spiracles are placed, some in the connecting membrane between the dorsal and ventral plates, the others on the sides of the ventral plates. One of our most familiar insects, the common cockchafer (*Melolontha vulgaris*), gives a good idea of the general form and style of coloration prevailing in the subfamily *Melolonthinæ*. Examples of some of the other *Melolonthinæ* are *Polyphylla fullo*, one of the finest European species, which, though not indigenous to Britain, has occasionally been found on the south coast, and *Rhizotrogus solstitialis*, a common British insect, generally known as the summer-chafer.

The *Rutelinæ* have some resemblance in external form to the *Melolonthinæ*, but can in general be easily recognised owing to the difference in length between the two claws of each of their tarsi. The *Dynastinæ* are mostly confined to the warmer parts of the world, and chiefly remarkable on account of the great sexual differences exhibited by the species. In the hercules-beetles (*Dynastes hercules*), of the West Indies and Tropical America, the male is sometimes over 5 inches long. The elephant-beetle is a more massive insect, though, having relatively much shorter horns, its total length is not so great. The rhinoceros-beetles, of which the male of the European species (*Oryctes nasicornis*) is figured below, are of smaller size. The next subfamily, the Cetoniinæ, includes the rose-chafers, as typified by the British *Cetonia aurata*, with its burnished golden-green armour. The goliath-beetles belong to this sub-family. In some of the genera, such as *Ceratorrhina* and *Goliathus*, the males may be recognised by the shape of the head, which is often excavated above, and furnished with hooks or horns, as shown below.

AGRIOTES LINEATUS AND ITS LARVA, THE WIRE-WORM

BUPRESTIDÆ

The *Buprestidæ*, together with the click-beetles

distributed, and include many of the finest species of the family, such as the *Euchroma gigantea* of South America, and the species of *Catoxantha* found in the East Indies. *Chalcophora mariana*, occurs in many pine-forests of the Continent, and is one of the largest European species. The *Buprestinæ* are more numerous than the other two groups, and are found in all parts of the world.

CLICK-BEETLES

The click-beetles are, as a rule, narrower and more elongated than the *Buprestidæ*, and differ also in having the posterior angles of the pronotum sharply produced behind, and the prosternal process laterally compressed and slightly curved, with its point resting in a deep cavity in the mesosternum. Their antennæ—consisting of eleven, or rarely twelve, joints—are usually serrate, though in many cases, especially in the males, they are either pectinate or flabellate. These beetles owe their name of skip-jacks to the power they have, when fallen on the back, of springing into the air and alighting on

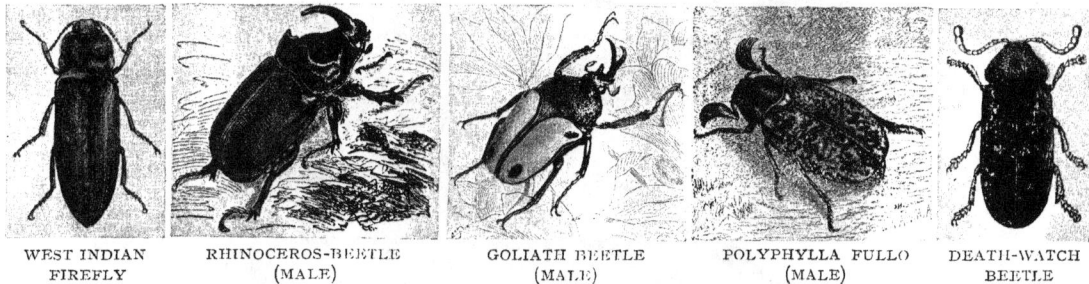

| WEST INDIAN FIREFLY | RHINOCEROS-BEETLE (MALE) | GOLIATH BEETLE (MALE) | POLYPHYLLA FULLO (MALE) | DEATH-WATCH BEETLE |

(*Elateridæ*), and a few smaller families, constitute the tribe Serricornia. Distinguished chiefly by their serrated or flabellated antennæ, the beetles of this tribe agree also in having the tarsi five-jointed, and the prosternum prolonged behind and fitting into a cavity of the mesosternum. They are generally of an elongated form, with the elytra narrowed from the base to the tip and completely covering the abdomen. The *Buprestidæ* have short, serrated antennæ, composed of eleven joints, which, with the exception of three or four nearest the base, are covered on special areas with very minute pits supposed to be of an olfactory nature ; these areas may be spread over nearly the whole of each joint, or confined to one side or the end of the joint, and their position affords one of the most important characters used in the classification of the family.

The family is divided into three principal groups— the *Julodinæ*, *Chalcophorinæ*, and *Buprestinæ*. The first group is chiefly restricted to Africa and the East Indies. The *Chalcophorinæ* are more widely

their legs again. The larvæ of some species eat into soft succulent roots and tubers, and in this way prove destructive to many of our cultivated plants. These pests are well-known to farmers under the name of wire-worms. The larva of *Agriotes lineatus* is one of the worst, being destructive not only in the fields but also in the kitchen-garden. It is of a pale yellowish brown colour, differing little in general appearance from the larvæ of other species, and lives for probably four or five years, passing then into a pupa, which remains concealed in the ground for a few weeks before changing into the perfect insect. Among the exotic members of this family, the most remarkable are the fire-flies, found in the West Indies and America. There are several species of these beetles, all belonging to the genus *Pyrophorus*, one of which, *P. noctilucus*, is illustrated on this page. They have a dark brown or reddish brown colour, obscured by a covering of short grey hairs, and may be easily recognised by the two slightly raised yellow spots placed near the hind angles of the prothorax. In the living insect these spots glow

GLOW-WORMS

with a rich yellowish green light. A stronger, but more diffused light, of a reddish colour, is given off from the abdomen when the beetles are flying.

GLOW-WORMS

The remaining families of the section Pentamera are included in the tribe Malacodermata. The beetles of this tribe are distinguished by having the elytra less solid and compact, and the body in general softer and more flexible than is usual in other groups. The *Lycidæ* are deserving of notice, inasmuch as they form one of those groups of insects which are most frequently mimicked by species of other families. They have a characteristic appearance, owing to the small size of the head and prothorax, as compared with the greatly expanded elytra.

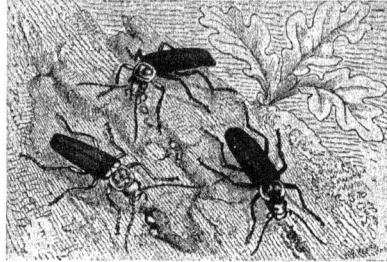

SOLDIER-BEETLES

To their unusual shape these beetles generally add a conspicuous coloration ; tawny yellow and red, varied in many cases with black spots and bands, being the predominant colours throughout the family. They are found on the flowers and leaves of trees, and are sometimes seen in great abundance ; and it is said that they secrete a nauseous liquid, which gives them immunity from the attacks of insectivorous animals.

The *Lampyridæ* are remarkable on account of the luminous properties possessed by nearly all the species. In these insects the head is small and, being retracted under the pronotum, generally invisible from above ; the eyes are large, especially in the males, the mandibles small but sharply pointed, and the antennæ come off close together from the front of the head. The phosphorescent organs are situated in the abdomen, their position being shown in most of the species by pale yellowish or whitish areas on the ventral surface of certain of the segments.

These beetles are found in nearly all parts of the world, though most numerous in Tropical America. In *Lampyris* and certain other genera the females are frequently apterous. The female of *Lampyris noctiluca*—the common glow-worm —is not only without wings, but has even no trace of elytra, so that in appearance it is not unlike the larva of the same species, though it

CLERUS FORMICARIUS

may be distinguished by its broad semicircular prothorax, its more fully developed legs, and much greater luminosity. In the genus *Lucicola*—which is represented by two or three species in South Europe—

both sexes are winged, and the males are even more luminous than the females.

SOLDIER-BEETLES

The *Telephoridæ* are distinguished from the two preceding families by having the head more exposed, the bases of the antennæ more widely separated from one another, the pronotum somewhat square in shape, the maxillary palpi ending in a hatchet-shaped joint, and the mandibles longer and often bifid at the end, or toothed on the inner side. Some of them are among the commonest and most familiar of our insects—being known to schoolboys as " soldiers " and " sailors "—and few can fail to recognise the species figured. This species (*Telephorus fuscus*), and a few others of the same genus—some of which are of an almost entirely yellowish red colour —are very plentiful on flowers at certain times of the year.

CLERIDÆ

The *Cleridæ* are generally brightly coloured, of cylindrical form, with the prothorax narrower than the elytra, the eyes notched in front, the antennæ either serrate, pectinate, or clavate, and the tarsi furnished underneath with membranous lobes. *Clerus formicarius* is very abundant in pine-forests, where it plays a useful part in hunting for and devouring wood-boring beetles ; while the larva is still more active in following under the bark the larvæ of various kinds which are there to be met with. Another species (*Trichodes apiarius*) hunts for its prey on flowers, especially those of the *Umbelliferæ*, and the larvæ are found in beehives, where they devour many of the young brood.

DEATH-WATCH BEETLE AND ALLIES

The *Ptinidæ* are all small insects, usually of a somewhat cylindrical form, rounded at each end, and with the head retracted under a hood-like covering, formed by the prothorax. They are obscurely coloured and chiefly interesting on account of their mischievous propensities. In the larval state *Ptinus fur* is very destructive in herbaria, and natural history collections generally.

The best known of the *Ptinidæ* are the death-watch beetles of the genus *Anobium*, to which we have already referred at the beginning of this chapter. These beetles seldom show themselves openly, so that to most people

HERCULES-BEETLES

they are only known by the sounds they produce or the holes with which the larvæ riddle furniture and the woodwork of houses. The holes with which old books are sometimes seen to be perforated are also made by the larvæ of a species of *Anobium*, which for this reason are known as bookworms.

HETEROMERA

The Heteromera are those beetles in which the tarsi of the fore and middle legs are five-jointed, those of the hind legs being four-jointed. The *Tenebrionidæ* exceed in number of species the rest of the Heteromera together. The antennæ are inserted under a projecting angle or ridge on each side of the head, and composed of eleven or, exceptionally, ten joints, of which the third is generally the longest; the coxæ of the front legs are usually rounded, with their sockets separated by a fairly broad prosternal process, and completely closed in behind; and the claws of all the tarsi are simple. Many of the obscurely coloured species are without wings, and frequently have the elytra fused together. The churchyard-beetles (*Blaps*) and the meal-beetle (*Tenebrio*) are probably the best known members of the family. *B. mucronata* is the commonest species in England; it differs from *B. mortisæa*, which also occurs, though, rarely, in that country, in having shorter points to the elytra.

Of the genus *Tenebrio*, two species occur in Britain, one of which (*T. molitor*) is almost cosmopolitan in its range, having been carried in flour to nearly every part of the world. The larvæ, known as meal-worms, are long and narrow, of a light yellowish red colour, with the integument hard, and the last segment conical in shape and ending in two processes, armed each with a small black spine.

OIL-BEETLES

The *Meloidæ* are chiefly distinguished from the other Heteromera by having the head abruptly constricted behind in the form of a short neck, the coxæ of the anterior and middle legs long and prominent, and placed close to one another in the middle line, and the claws of the tarsi accompanied each by a slender hook, so that they appear double. Many of the species possess vesicating or blistering properties.

The larvæ are interesting on account of their habits and the changes of form they undergo in the course of their development. These changes are well illustrated in the case of the oil-beetles (*Meloe*). The larvæ of these when first hatched from the egg are active little creatures furnished with six legs. They climb on to flowers, and wait in readiness to fasten themselves to the hairs of bees coming to gather the honey.

In this way they get 'carried to the nest, where they devour the eggs of the bee. They now cast their skin, appear as little, maggot-like grubs, with much reduced legs, and feed on the honey intended by the bee for its own young. After a time they change to the form of a pupa, from which, instead of the perfect insect, a third form of larva, somewhat similar to the second, emerges, while a further change is still required before the true pupal stage is reached.

Seven species of the *Meloe* occur in Great Britain, but, with the exception of one or two, are very rare. When handled or irritated they exude an acrid, oily-looking liquid of a yellow colour from certain of their joints.

PARASITIC BEETLES

The *Stylopidæ* are remarkable little insects, which live parasitically in the bodies of wasps, bees, and bugs, and present a type of structure distinct from that of all other beetles. The male is a winged insect, with coarsely-faceted prominent eyes, large fan-shaped wings, and extremely small inconspicuous elytra; the first two thoracic rings are very short, while the metathorax is greatly elongated and covers the base of the abdomen; the hind legs are placed a long way behind the middle pair, and the tarsi of all the legs are membranous underneath, and without claws at the end. The female, on the other hand, is a grub-like creature, without legs, wings, or eyes. She never leaves the body of her host, and from her eggs active little six-legged larvæ develop, which make their way out and get carried into the nests of bees and wasps, where they bore into the bodies of the grubs. The *Stylopidæ* are very rarely seen, and the number of species known is small. They have been arranged in four or five genera, based upon slight differences in the structure of the males, all of which have the general appearance shown in the figure of *Xenos pecki*.

TETRAMERA

The *Curculionidæ*, or weevils, are distinguished from all other beetles by well-marked characters. The head is produced in front in the form of a short snout or a more or less elongated and narrow beak, which carries the mouth at its extremity; the prothorax rarely has sharp lateral edges, and the coxal cavities on the under side of that segment are closed in behind by the extension inwards of the epimera to meet in the middle line; and the antennæ are elbowed, with the first joint as a rule long, and some of the joints at the end forming a club.

CHURCHYARD-BEETLE

OIL-BEETLES AND LARVÆ

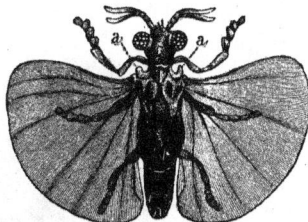

XENOS PECKI—MALE
a indicates rudimentary elytra

A GROUP OF BUTTERFLIES

1. Orange scallop-wing; 2, Common blue; 3, Mango admiral; 4, Silver-studded skipper; 5, Royal swallow-tail; 6, Sooty-veined porcelain.
7, Black-tipped sulphur; 8, Duke of Burgundy; 9, Swinhoe's tortoiseshell; 10, Harris's snowflake.

2 C

The weevils have been arranged in a number of subfamilies, but it is impossible in a limited space to describe the various modifications of structure on which these divisions are based, and we must content ourselves here with a brief reference to some of the typical and more interesting forms. In the genus *Sitones* we have examples of those weevils in which the snout is short and comparatively broad. *S. lineatus* is a well-known species which lives on papilionaceous plants, and frequently does much mischief by devouring the young leaves of peas and beans. It is a little yellowish grey or drab-coloured beetle with three pale lines along the thorax, and a number of rows of punctures along the elytra. Its colour is due to a thick covering of scales, some of which, when looked at closely, are seen to have a golden tint.

Weevils are, as a rule, most destructive during the larval state, the adult insects doing a comparatively small amount of injury to vegetation ; but as regards *Hylobius abietis*, known as " the large pine-weevil," one of the worst enemies of young conifers, the injury done to the trees is altogether the work of the beetles, while the grubs are quite harmless. The genus *Apion* comprises a large number of little, long-snouted weevils, having in general the form shown in the figure of *A. apricans*.

Though the British species are numerous and some of them common everywhere on clover, trefoil, and other leguminous plants, they are seldom noticed owing to their small size. In *Apoderus, Attelabus,* and *Rhynchites* we have a group of genera which are interesting on account of the leaf-rolling habits of the females, and remarkable also, in the case of the first genus, for the great length of neck displayed by some of the species. The females deposit a single egg, or in some cases two or even three eggs, in each of the little rolled-up leaf packages, which serve afterwards both as a shelter and food-supply for the larvae. Three or four species of these leaf-rolling weevils are found in Britain.

The illustration of *Apoderus longicollis*, a Javan species, shows what an extraordinary length the neck may attain in the males of some of the tropical representatives of the genus, although in this species it is not nearly so long in proportion as in an allied form (*A. tenuissimus*) found in the Philippine Islands.

The nut-weevil (*Balaninus nucum*) affords a strong contrast in the shape of its head to the species just mentioned. It will be noticed that in this weevil the head is very short behind the eyes, whereas the beak is greatly elongated, with the antennae inserted near the middle of its length. The female lays her eggs in hazel-nuts while the latter are still in a half-developed condition ; she first pierces a hole in the soft shell of the nut, and then depositing an egg in the opening pushes it in with her beak. The grub feeds inside the nut, remaining in it until autumn, when it bores a round aperture in the shell, and, escaping from the nut, makes its way into the soil, where it surrounds itself with a cocoon formed of fragments of earth.

The " apple-blossom weevil " (*Anthonomus pomorum*) is another species which, on account of its injurious habits, deserves some notice. It is about a quarter of an inch long, of a greyish brown colour, with an oblique white band on the elytra, and three whitish lines on the thorax. The female deposits her eggs in the unopened flower-buds of the apple, and the larva by feeding on the stamens and pistil causes the bud to wither and die. In about fifteen days the larva attains its full size, changing then to a pupa within the bud, and the beetle appears about eight days later and escapes through an opening which it makes in the side. A closely allied species (*A. pyri*) proves injurious in the same way to pear blossoms.

The cabbage-gall weevil (*Ceutorhynchus sulcicollis*) and certain species of *Baridius* attack cruciferous plants; the larvae of the former live inside galls which they raise on the roots of cabbages and turnips, while those of *Baridius* may be found living in the lower part of the stem.

The grain-weevils, which are most numerous in tropical countries, are represented in Britain by two almost cosmopolitan species, the corn-weevil (*Sitophilus granarius*), and the rice-weevil (*S. oryzae*). These are both small species, but belong to a subfamily (*Calandrinae*) which includes a number of the largest tropical weevils, such as the palm-weevil (*Rhynchophorus palmarum*).

The *Scolytidae* and two other small families, the *Brenthidae* and *Anthribidae*, are associated with the weevils in the tribe Rhynchophora. The *Scolytidae* are little beetles which live under bark, and often prove very injurious to trees. They have four-jointed tarsi, clubbed antennae, and the head produced in front into a short snout. The females lay their eggs along the sides of galleries which they

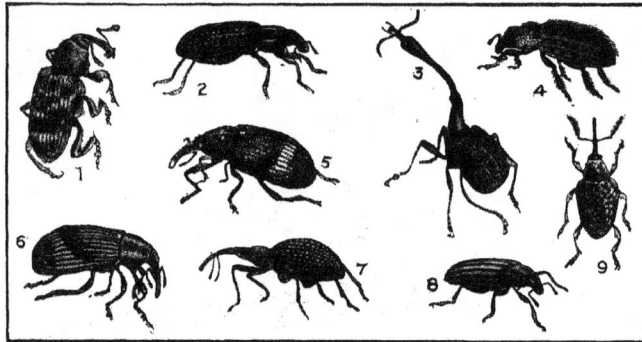

A GROUP OF WEEVILS
1, pine-weevil ; 2, 4, and 8, Sitones lineatus and allies ; 3, Apoderus longicollis ; 5, apple-blossom weevil ; 6, pear-blossom weevil ; 7, Apion apricans ; 9, nut-weevil

burrow out under the bark ; the larvae, when hatched, make tracks at right angles to the mother-galleries, and thus form curious and characteristic patterns.

LONGICORNS

The *Cerambycidae*, or longicorns, have in most cases a characteristic appearance by which they may be easily recognised, though, owing to a great variety in their form and structure, the family as a whole is not easily defined. Thus the great length of the antennae to which these beetles owe their name is not always a distinguishing feature, for in many genera the antennae are much shorter than the body. The longicorns resemble the Rhynchophora in having the first three joints of the tarsi furnished underneath with a brush-like covering of hairs, and the fourth joint very small and hidden between the lobes of the third ; but they are distinguished from that tribe by the fact that the epimera of the prothorax do not meet, while the head, though sometimes produced into a short muzzle, is never prolonged in the form of a beak.

The larvae all have a strong family likeness, and are quite unlike those of the *Chrysomelidae*. They are of a dirty-white or pale yellow colour, with a rather soft skin, and in general form most resemble the larvae of

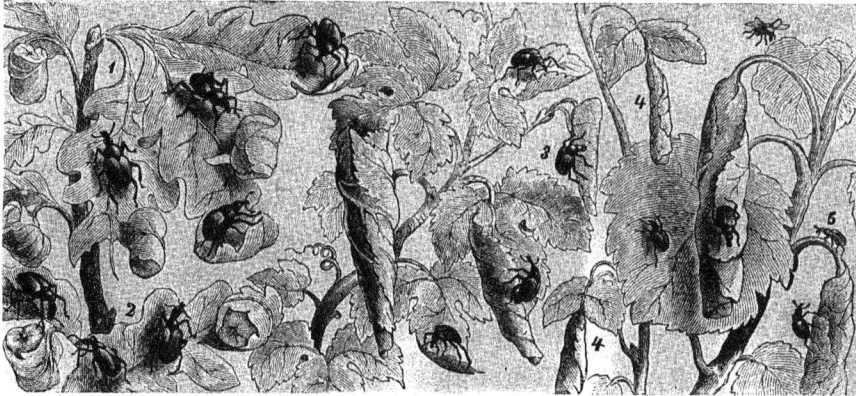

LEAF-ROLLING WEEVILS
1, Attelabus curculionoides ; 2, Apoderus coryli ; 3, Rhynchites betuleti ; 4, R. populi ; 5, R. betulæ

Buprestidæ. These larvæ all live in the interior of plants ; some feeding just under the bark, while the great majority bore tunnels in the woody tissue, or live exclusively in the pith. The males have, as a rule, longer antennæ than the females, and may often be distinguished by the larger size of the eyes, jaws, or prothorax, or the greater length of the legs. The females are provided with a flexible ovipositor, which can be protruded some distance beyond the end of the body.

In the subfamily *Prioninæ* the anterior coxæ are strongly transverse, and their sockets widely open behind, the sides of the prothorax are sharply edged, the palpi are never pointed at the end, and the front tibiæ are without a groove underneath. This subfamily is the least numerous in species, though many of these are distinguished for their great size. *Titanus giganteus,* a Brazilian species, sometimes measures over half a foot long, and is the largest of all known beetles, while the sawyer-beetle (*Macrodontia cervicornis*) and other species occurring in Tropical America are not much smaller.

Most of the *Prioninæ* are found in the warmer parts of the world. They are represented in Europe by *Prionus coriarius, Ergates faber,* and a few other forms. *P. coriarius* is the only species which occurs in England, and is not very common, being met with chiefly in oak-woods, where the larvæ live in the rotten trunks of trees.

The *Cerambycinæ* are widely distributed, and include a very large number of species. They are, as a rule, narrower than the *Prioninæ,* and without sharp lateral edges to the prothorax, while the sockets of their front legs are seldom strongly transverse and are often rounded and completely closed behind. Most of the species have a stridulating area on the meso-notum, and by this means are enabled to produce

PALM-WEEVIL

sounds. This sub-family includes a number of very re-markable mimetic forms, some with broadly expanded elytra, and black and tawny colours resem-bling *Lycidæ,* others with the elytra greatly reduced in size, and the shape of the body modified in imitation of various Hymenoptera.

The metallic-coloured beetles of the group *Callichro-mides,* most of which are found in tropical countries, exhale a scent resembling otto of roses. In England this group is represented by the musk-beetle (*Aromia moschata*), figured on page 1974, a handsome insect of a golden-green or bronzy colour, which is met with on willow-trees. Among the European species of *Ceram-bycinæ,* the groups *Lepturides, Molorchides,* and *Clytides* are best represented. Some of the *Lepturides,* such as *Strangalia* and *Toxotus,* are flower-fre-quenting insects, others, like *Rhagium,* are found on the trunks of pine-trees. In the *Molorchides,* the elytra are usually short or very narrow, and the abdomen slender and constricted at the base, so that many of the species have a resem-blance to Hymenoptera.

The European *Necydalis major* looks like a hornet, but in many of the tropical forms these resemblances are more pronounced. The *Clytides* are found on flowers, chiefly of the umbelliferous kind, and two or three species are among the prettiest of British beetles. Some of the *Clytides* and species of *Hylotrupes* and *Callidium* are occasionally met with in houses, being introduced in the woods in which the larvæ feed.

The *Lamiinæ* are more numerous than the other longicorns, and distinguished by having an oblique groove on the lower side of the front tibiæ, the last joint of the palpi usually pointed at the end, and the front of the head in most cases turned down vertically, or sometimes even inclined backwards, bringing the mouth close to the prosternum. The species of the

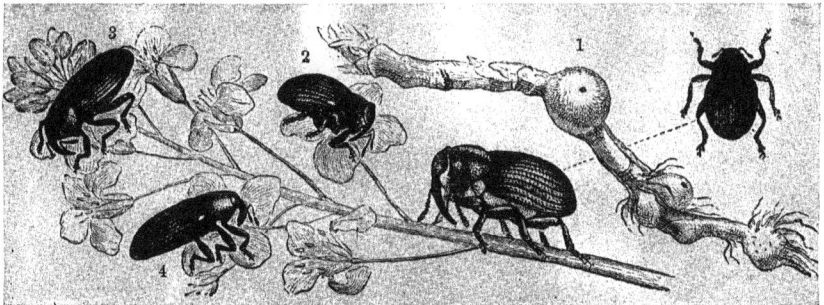

CABBAGE-GALL WEEVIL AND ALLIED BEETLES
1, Ceutorhynchus sulcicollis ; 2, C. assimilis ; 3, Baridius chloris ; 4, B. cuprirostris

HYLOTRUPES BAJULUS, WITH LARVÆ

genus *Lamia* are few in number, and by no means typical of the subfamily ; they are clumsy-looking, dull black insects, one of which (*Lamia textor*) is found on willow-trees and in osier-beds in some parts of Britain. In the genus *Acanthocinus* the antennæ attain their greatest length, being four times as long as the body in the male. *A. ædilis* is found in pine-woods in Scotland, and is met with occasionally in other parts of Great Britain and even in London, where it is sometimes introduced in timber. Among the exotic species of this subfamily, the harlequin-beetle (*Acrocinus longimanus*) is one of the most remarkable, being distinguished, not only by its curiously variegated colours, but also by the extraordinary length of the front legs in the male.

PEA AND BEAN BRUCHUS

The *Bruchidæ* are a small but widely-spread family of little beetles which are found chiefly on leguminous plants. The larvæ live in the seeds, eating up all the internal parts and changing to pupæ within the outer shell. These beetles were at one time classed with the weevils, but are now generally recognised as being more nearly allied to the next family. They are illustrated on page 1975 : 1 is the pea-bruchus (*Bruchus pisi*) ; 2, the bean-bruchus (*B. rufimanus*) ; and 3, *B. granarius* and larva ; all being enlarged.

PLANT-EATING BEETLES

The *Chrysomelidæ*, more commonly known as the Phytophaga — though this name is equally applicable to many other beetles — all live upon plants, feeding chiefly upon the foliage, while some also attack the flowers. They are almost as numerous as the weevils, and in their own way quite as destructive to vegetation. The family is divided into four sections. The Eupoda include those forms which most resemble the longicorns. Many of the

STRANGALIA ARMATA

beetles belonging to this section have thickened hind legs, but instead of being active jumpers, as might be suspected, they are really very slow and awkward in their movements. In the males of the genus *Sagra*, the hind legs are enormously developed, the species of this genus being for that reason sometimes known as kangaroo-beetles.

The *Donacinæ* live upon aquatic plants of various kinds ; they have a bright metallic coloration, which in many species is veiled by a delicate covering of silky hairs, and their elytra are marked with rows of deep punctures. The larvæ feed under water upon the roots of the plants, and change to pupæ which are enclosed in oval cocoons.

In the beetles of the section Camptosomata the body is short, the head vertical and deeply sunk in the pro-thorax, and the abdomen slightly curved, with its middle segments contracted ; the antennæ are short and serrate or pectinate in the first subfamily, while in the second they are rather long and filiform. The larvæ move about surrounded by a sack-like case, from which the head and anterior part of the body are free. They retract themselves completely within the case and close up the opening when about to undergo their metamorphosis.

FLEA-BEETLES AND ALLIES

The Cyclica comprise four subfamilies, of which the first, the *Eumolpinæ*, is almost entirely composed of exotic species, though one of the few species found in Europe (*Bromius vitis*) is only too well-known on account of the damage it inflicts on the leaves of the vine. With the *Chrysomelinæ* we come to the most typical forms belonging to the family. These beetles are distinguished by their oval and convex shapes, having in many cases a great resemblance to lady-birds (*Coccinellidæ*), while their colours are nearly always brightly metallic or otherwise conspicuous. Some of the species are very gregarious in their habits. *Lina tremulæ* is often found in large numbers in all its stages on the leaves of aspen ; the larvæ are somewhat like those of lady-birds, and have the habit of exuding a strong-smelling yellow liquid from the mouth and other parts of the body. The Colorado potato-beetle (*Leptinotarsa decemlineata*) is very destructive to the potato crops in North America. The *Galerucinæ* are poorly represented

PRIONUS CORIARIUS (FEMALE) AND ERGATES FABER (MALE)

in Great Britain, but of the flea-beetles (*Halticinæ*) there are a large number of species, of which the best known are the turnip-flea (*Phyllotreta nemorum*), and other little jumping beetles which attack cruciferous plants. The larvæ of the *Halticinæ* usually mine in the tissues of the leaf underneath the epidermis; in this respect differing from the larvæ of most of the other *Chrysomelidæ*.

TORTOISE-BEETLES

The Cryptostomata are distinguished by having the front of the head inclined backwards, so that the mouth is almost completely hidden. Two subfamilies are included in this section. The *Hispinæ* are remarkable for the sharp projecting spines with which many of the species are armed, while the *Cassidinæ* have the characteristic form to which they owe their name of tortoise-beetles. In Great Britain, the tortoise-beetles are represented by half a dozen or more species of the genus *Cassida*, one of which (*C. nebulosa*) in its different stages is illustrated on page 1976. A Brazilian species (*Desmonota variolosa*), remarkable for its deeply sculptured elytra and bright golden-green colour, is also shown in the same figure.

LADY-BIRDS

Two families, are included in the Trimera. The *Coccinellidæ*, or lady-birds, are familiar to everyone. These charming little insects have always been held in much respect, as the different names

MUSK-BEETLE

given to them testify, and it is well that it should be so. For while the species of a few genera (*Epilachna, Lasia*) are herbivorous in their habits, the great majority live—especially in the larval state—upon green-fly and plant-lice, and, by keeping these noxious insects in check, perform a useful service to man.

The lady-birds are found in nearly all parts of the world, and over a thousand different species are known. Among several species occurring in Britain the two commonest are, perhaps, the large seven-spotted *Coccinella septempuncta* and the small two-spotted *C. bipunctata*. The latter varies in colour to a great extent, so that between the typical form with red elytra marked with two black spots, and others in which the elytra are entirely black, one meets with almost every intermediate condition.

The larvæ of these species may be recognised by their slate-blue colour, marked with some yellow dots, and by the greed with which they devour aphides. The larvæ, when about five or six weeks old, are ready to pupate. Fixing themselves by the tail-end to a leaf, they cast their skin, and the pupæ, resting upon the cast-off larval skin, remain attached to the leaf. The beetles emerge about eight days later.

NEUROPTERA

The Neuroptera form the last order of insects which undergo a complete metamorphosis in the course of their development. In this order it was formerly

NECYDALIS MAJOR

1, RHAGIUM INDAGATOR; 2, R. BIFASCIATUM

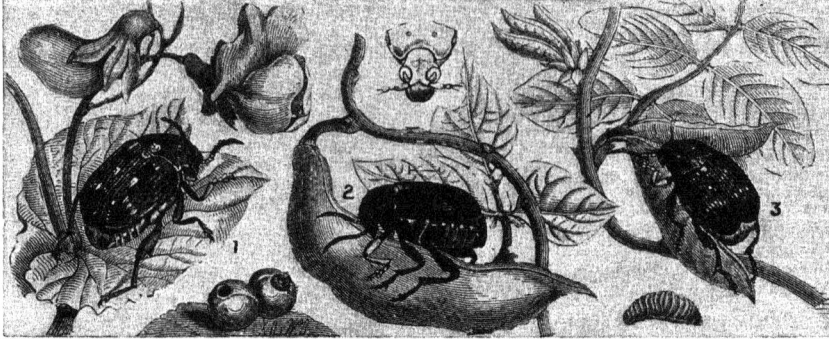

1, PEA-BRUCHUS; 2, BEAN-BRUCHUS; AND 3, CORN-BRUCHUS

usual to include certain groups of insects, such as the dragon-flies, May-flies, white ants, etc., none of which pass through a period of prolonged inactivity, or pupal stage, before reaching the perfect condition. But although it is largely a matter of convenience whether these groups be placed, as they are in this work, in the order Orthoptera, or arranged in a series of separate orders, no one, taking into consideration the great difference in their mode of development, would now think of associating them in the same order with the true Neuroptera.

The adult insects of the present order have their mouth organs, when fully developed, adapted to biting and grinding, and never formed for piercing or sucking; in which respect they differ from three of the other orders of the meta-bolous insects—namely, the Hymenoptera, Diptera, and Lepidoptera. From the Coleoptera they are easily distinguished by the structure of their fore wings, which are never hard and horny like the wing-cases, or elytra, of the latter. Both pairs of wings are membranous, and usually traversed by numerous more or less closely reticulating veins, whence the name of Neuroptera given to the order. The hind wings are often very similar to the fore wings, but sometimes differ considerably in size and shape. In one section—the caddis-flies—they are capable of being folded like a fan, but in the other section (Planipennia) they always remain flat, and are spread horizontally or obliquely in repose.

CADDIS-FLIES

The caddis-flies (sub-order Trichoptera),

MALE AND FEMALE OF TOXOTUS MERIDIANUS

forming the first of the two great divisions of the order, are in general appearance rather like some of the smaller kinds of moths; and since they differ a good deal from the typical Neuroptera, they are often treated as a distinct order. In their adult state they have two pairs of wings, in which the neuration is comparatively simple, with few transverse nervures. The wings are generally clothed with hairs, and the hind wings usually shorter, broader, and less hairy than the front pair. When at rest, the hind wings are folded fan-wise, with the fore wings covering them over like a roof.

The caddis-flies have a rather small head, which bears two long, tapering, and many-jointed antennæ. They have round and prominent eyes, and usually also three ocelli, placed on the forehead. With the exception of the palpi, their mouth organs are feebly developed. Their legs are long, and possess five-jointed tarsi; and the tibiæ are generally furnished with spurs, whose number and disposition are of considerable value in distinguishing the genera.

These insects fly chiefly in the evening or at night, and, attracted by the light, frequently enter houses; some of the smaller species flying in swarms over water. The larvæ, with few exceptions, are aquatic in their habits; some being carnivorous, although most feed on vegetable matter. Found in streams, lakes, and ponds, or any piece of water in which plants grow, caddis-worms, as the larvæ are called, are well known to

1, LAMIA TEXTOR; 2, FEMALE, AND 3, MALE OF ACATHOCINUS ÆDILIS; 4, SAPERDA CARCHARIAS; 5, SAPERDA POPULNEA

TORTOISE-BEETLES

1, Cassida nebulosa ; 2, its larva ; 3, Desmonota variolosa ; 4, portion of its elytron ; 5, its leg

anglers, by whom they are frequently used as bait. The eggs from which they are hatched are laid sometimes in the water, or on aquatic plants or trees overhanging water. Females have occasionally been captured with a coating of dry mud on their abdomen, showing that they had gone to some muddy pool to lay their eggs.

The cases, made out of all sorts of materials with which many of the larvæ surround their bodies, have long been objects of interest to the naturalist. Some larvæ pick up bits of sticks and leaves, grains of sand, and fragments of shells, or whatever else comes handiest, and fasten them together in a rough sort of fashion ; but many exercise a choice in the selection of materials, and exhibit great dexterity and neatness in piecing them together. The shape of its dwelling and the nature of the materials used are often characteristic of the family, sometimes of the genus or species, to which a larva belongs. In the family Phryganeidæ, for example, the larvæ construct their cases with bits of leaves or twigs, cut into suitable lengths, and arranged side by side in such a manner as to form a spiral band passing many times around the case (see 7 in figure).

The species of Limnophilus fashion their cases in various styles ; the larvæ of L. pellucidus using entire leaves, so that the case may have a flattened form, wide in proportion to its depth. The case made by L. rhombicus consists of bits of sticks or fibres placed transversely, with shells sometimes added ; while those of L. flavicornis are often built almost entirely of the shells of different

small molluscs, more especially those of Planorbis.

What is still more remarkable about these cases is the fact that the case-worms do not necessarily select empty shells, but take those with living occupants as well, and fasten them all together around their backs. Grains of sand, of finer or coarser kind, are used by many larvæ in the construction of their cases ; and the latter may be either cylindrical in form or slightly curved, or, as in the exotic genus Helicopsyche, they may, like snail-shells, have a distinct spiral curvature.

The grubs of other species arrange bits of sticks transversely in four different directions, using longer pieces as they progress, so that the complete case is four-sided, with the sides gradually widening from one end to the other ; and there is a type in which the four sides, instead of being straight, are carried round in a gentle spiral curve.

The interior of each larval case is a tubular chamber, lined with silk, open at each end, and about wide enough to enable the larva to turn inside. At the fore end, which is generally a little wider, the head, thorax, and the six legs of the larva may be seen projecting ; whereas the hinder end is usually closed by a silken partition pierced with holes. The body of the naked larva is made up of a number of segments, of which the first three—carrying the legs—are, like the head, hard and of a brownish colour ; while those that follow, about nine in number, are soft, white, and partly transparent. On the last segment are a pair of horny hooks, which enable the larva to grip tightly to its case.

On the first abdominal segment three fleshy protuberances are often seen—a longer one above and a shorter one on each side—which appear to be used in enabling the larva to steady its body in the case, and to regulate its position with regard to the sides, so that the water necessary for breathing may pass freely in and out. The larva breathes by means of rows or

LIFE-HISTORY OF DONACIA CLAVIPES

LADY-BIRDS

1, Micraspis duodecimpunctata ; 2, Coccinella septempunctata ; 3, larva (enlarged) ; 4, C. impustulata ; 5, C. bipunctata and dark variety ; 6, Chilocorus bipustulatus

tufts of soft white filaments—the tracheal gills—attached to the sides of all the abdominal segments except the first and last, and differing in arrangement in different species.

Previously to entering the pupal stage, the larvæ of many species provide for their protection during that inactive and helpless period of their existence. They shut themselves up in their cases, some by closing the openings at each end with sieve-like plates of silk, which, while allowing free access to the water necessary for breathing, may serve to keep out their enemies; others by placing stones loosely over the openings, and so accomplishing the same purpose. There are a few larvæ, moreover, which, in their earlier days, make cases out of leaves, but add stones as they grow older, until just before pupation begins the case is entirely made of stones. Before the pupa is transformed into the perfect insect, it extricates itself from its case, and leads an active life, swimming and running with agility. It then climbs up the stem of a plant to undergo its transformation. In some of the smaller species the pupa does not leave the water, but rises to the surface, and the fly emerges from the floating pupal skin.

Caddis-flies are divided into seven families, arranged in two groups chiefly distinguished by the number of joints in the maxillary palpi of the male insect. In the first section—Inæquipalpia—the maxillary palpi of the male are composed of two, three, or four joints, never five; thus differing from those of the female, in which the number of joints is always five. This section contains four families; the life-history of a species of the typical genus (*Limnophilus*) being depicted in our illustration. The second section—Æquipalpia—is characterised by the fact that the maxillary palpi of the male are five-jointed like those of the female; it includes three families.

VARIOUS FORMS OF CASES MADE BY CADDIS-FLY LARVÆ
1-5, cases composed of sand and pebbles; 6, a case made of small snail-shells; 7-10, cases made of parts of plant

LIFE-HISTORY OF COLORADO BEETLE

FLAT-WINGED GROUP

The members of the suborder Planipennia are distinguished from the last by having both pairs of wings formed nearly alike, and usually provided with a closely reticulated system of nervures, with numerous transverse branches. The wings—which are incapable of being folded up—are for the most part naked, and, when at rest, are turned back in a slanting position against the sides of the body. The mouth organs are well developed.

SCORPION-FLIES

The scorpion-flies (*Panorpidæ*), have a slender body, and the head turned downwards and prolonged in the form of a beak. The mandibles are rather short and narrow; the maxillæ, which are fused with the mentum, have five-jointed palpi; and the narrower lower lip is bifid at the extremity, with three-jointed palpi. The antennæ are setiform, and inserted between the rather prominent eyes, and below the ocelli, which are usually distinct. The prothorax is short and collar-like, and the wings of these insects are but little reticulated.

LIFE-HISTORY OF CADDIS-FLY
1, larva; 2, pupa; 3, larva in its case; 4, perfect insect

The common scorpion-fly (*Panorpa communis*), which may be taken as the type of the family, is a shiny black insect about half an inch or more in length, with long, transparent, spotted wings, and a yellow beak and legs. The three last body-segments of the male are narrower, and can be curved like a tail, and have a reddish colour; and the last carries a pair of pincer-like claws. It is from this circumstance that the insect has received its name, though it must be remembered that it does not possess a sting like a scorpion.

strong legs and can walk well, breathes by means of tracheal gills, having the form of jointed appendages attached in pairs to the sides of the first seven abdominal segments. When the time for pupation arrives, generally about May or June, the larva leaves the water and seeks a place to bury itself in the earth. Having excavated a little cell, it throws off the larva skin and becomes a pupa, which has the legs and wings free from the body, but enclosed in special sheaths. After a few weeks longer it is transformed into the perfect insect.

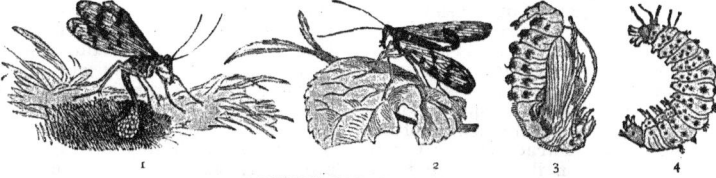

COMMON SCORPION-FLY

1, female depositing her eggs; 2, male; 3, larva, 4, pupa

SNAKE-FLIES AND ALDER-FLIES

In the snake-flies and alder-flies (*Sialidæ*), forming the second family, the head is comparatively large, and often inclined in front, but never elongated in the form of a beak. The antennæ are bristle-like, and not so long as the body; the prothorax being strongly developed.

The camel or snake flies (*Rhaphidia*) have the head long and narrow behind, and freely articulated with the long and narrow prothorax. The latter can also move freely at its articulation with the segment which follows, and this explains how the prothorax is raised, and the head bent forward in the characteristic attitude these insects always adopt when about to seize their prey, which consists usually of various small insects.

The European alder-fly (*Sialis lutaria*) is at first sight rather like a caddis-fly, but has a stouter body, and may be distinguished by its more completely developed mouth organs, as well as by the different structure of its wings. It emerges from the pupa about May or June. The winged insects fly slowly and heavily, and are to be met with about trees and shrubs, or walls and palings, at no great distance from water. The female, which is somewhat larger than the male, lays her eggs in patches on a plant or other object in the vicinity of water. There may be several hundred eggs packed closely together in a single cluster; they stand upright, being cylindrical in form, with rounded ends, and each terminating above in a little white projection.

The larvæ hatch in a few weeks, and then find their way into water, where they creep on the mud in search of the aquatic creatures on which they feed. When full grown, they are about an inch long, with a body tapering slightly towards the head, and, more gradually, towards the long and narrow tail. The head and three thoracic rings are horny, the rest of the body having a softer integument. The larva, which has

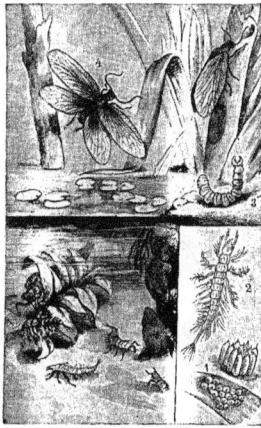

LIFE-HISTORY OF ALDER-FLY

1, eggs; 2, larva; 3, pupa; 4, imago

MANTIS-FLIES

The lace-wing flies, ant-lions, mantis-flies, and some other families, have been associated in a third group of Planipennia, to which the name Megaloptera is given. In all, the wings are relatively large and closely reticulated; the prothorax being variable in size and form, and the joints of the tarsi not dilated. The mantis-flies (*Mantispidæ*) take their name from the shape of the forelegs, and their position near the front end of the long prothorax, in which respect they resemble the mantis, or praying, insect. One species is common in South Europe. The larvæ live parasitically in the nests of spiders and tree-wasps, and, while they are at first free and active, they afterwards become almost legless, like those of certain beetles. The allied family *Nemopteridæ* is mainly characteristic of the countries around the Mediterranean Sea. These insects have elongated and narrower, or almost linear hind wings, often widened out a little before the tip.

ANT-LIONS

The ant-lions (*Myrmeleontidæ*) may be recognised by their clubbed antennæ, and their long and closely reticulated wings, rounded off to an obtuse point at the extremity.

Of the European species the common ant-lion (*Myrmeleon formicarius*) is one of the best known. It lives in pine-woods. The winged insect, which may be seen in July and September, rests during the day clinging to a plant, with its wings spread like a roof over the hind part of its body. At sunset it becomes active, and executes a slow flight in its search after food or a mate. The larva, to which the name ant-lion properly belongs, has the habit of making pitfalls to entrap its prey. It is somewhat oval in the shape of its hind body, and has a narrow prothorax resembling a neck, and a rather big head, provided with a pair of

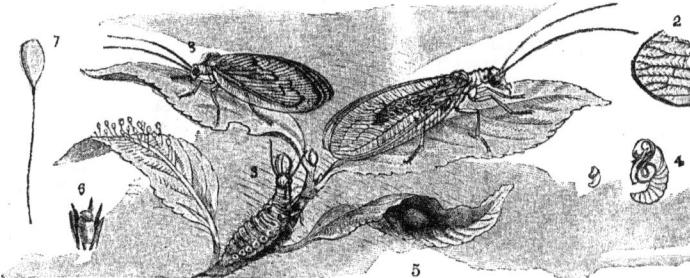

LACE-WING FLIES

1, Chrysopa vulgaris; 2, the tip of its wing; 3, larva; 4, pupa; 5, 6, cocoon; 7, egg; 8, Hemerobius hirtus

long, curved, and sharply pointed mandibles, each of which has three teeth on the inner side. Its body is arched up in the middle, and has wart-like protuberances thickly covered with hairs at the sides.

When about to make a pit, the ant-lion selects a dry and sandy spot, and begins by tracing out a circular furrow to mark its outer limit. Placing itself inside the circle, it buries its abdomen in the sand, and then proceeds with the work of excavation. With one of its fore legs it shovels the sand on to its large, flat head, to which it then gives a sudden jerk, and sends the sand out over the border. It repeats this process, walking backwards and maintaining a spiral course all the while, until finally it reaches the centre of the cavity. Sometimes, however, instead of continuing to work altogether in one direction, it turns round and works the opposite way, thus giving relief to the leg which had previously been employed. And, as the sand is always taken from the inner side, it is the leg on that side which is always used as a shovel. The pit, when completed, is shaped like the mouth of a funnel, being wide above and gradually narrowed to the bottom. Its size is adapted to the size of the larva, which when full grown makes a pit about two inches deep, and three inches wide at the top.

Buried in the sand at the bottom, with only its antennæ and the tips of its mandibles projecting, the ant-lion waits until an ant or some other creature falls down the loose sides of the pit, when it is immediately seized with the pincer-like jaws, and retained until all the juices of its body have been sucked out, and nothing left but the dry and shrivelled skin. The latter is cast outside the pit, and the larva again lies in wait. If by chance the victim should escape the first onslaught, and endeavour to scramble up the sides of the pit, its attempt is soon frustrated, for the ant-lion throws up sand with its head, causing the victim to tumble once more to the bottom.

LACE-WING FLIES

The lace-wing flies (*Hemerobiidæ* and *Chrysopidæ*) are smaller and more delicate insects than the ant-lions, and have setiform antennæ. The golden-eyed fly(*Chrysopa vulgaris*),figured on page 1978, may be taken as a typical species. It is slender, with long and richly veined wings of a tender green colour, as is also the body. Its antennæ are long and tapering, and its prominent eyes shine like hemispheres of gold. The larvæ of the lace-wings are not unlike the ant-lion, although somewhat longer and narrower in proportion to the size of their bodies, and less hairy. Their mandibles, moreover, have no teeth on the inner side. In their carnivorous habits they resemble ant-lions, but, instead of making pits and remaining stationary, they rove about in search of their prey, which consists of the different kinds of green-fly and plant-lice.

ORTHOPTERA

This order being taken to include, not only the true Orthoptera, but various other groups formerly placed in the Neuroptera, and hence known as Pseudoneuroptera, it is necessary in defining the group to mention only such characters as are common to the whole of these insects. None of the members of the group undergo a distinct metamorphosis ; the development from the larval to the adult condition taking place by a succession of changes, and the perfect insects being distinguishable from advanced larvæ by little more than the possession of complete wings. The wings are, however, in some cases confined to one sex, while in others they are altogether wanting in both sexes. The mouth organs, when not reduced to a functionless condition, are adapted to biting ; the lower lip (labium) is nearly always divided in the middle at its free end, and each of the two halves often subdivided into a pair of lobes. On the floor of the mouth, concealed by the labium, there is, as a rule, a membranous or more or less horny structure, known as the tongue (lingua), which is free from the labium in its anterior part.

Though poor in the number of species, as compared with some other orders, the Orthoptera contain many of the most interesting forms of insect life ; some, like the leaf and stick insects, remarkable for their size and the variety of their protective disguises, others, as the white ants, for the wonderful development of their social habits. The day-flies are noted for their short lives, the dragon-flies for their beauty, while many other forms are well known from some particular feature or habit.

Larva   Imago   Pupa
LIFE-HISTORY OF THE ANT-LION

In past epochs of the earth's history Orthoptera were well represented, their remains being found in rocks of various ages extending back to Palæozoic times. The oldest reputed insect is known by the impression of an orthopterous wing (*Palæoblattina*), from the Silurian sandstone of Calvados in France. There is some doubt as to which group of the order the insect belonged, and even as to whether the impression owed its origin to an insect at all.

However this may be, traces of undoubted Orthoptera, as well as of Neuroptera, are met with in rocks of Devonian and Carboniferous ages. The Orthoptera of the latter period included numerous cockroaches (*Blattidæ*), together with stick-insects, ephemerids, and dragon-flies, some of which greatly exceed in size any existing form. One of the dragon-flies (*Meganeura monyi*) was 13 inches in the length of its body, and each of its wings was quite a foot long.

PSEUDONEUROPTERA

The Pseudoneuroptera are distinguished from the Neuroptera by the absence of a pupal stage. While agreeing in this respect with the typical Orthoptera, these insects differ by certain characters not generally considered of the first importance. Both pairs of wings in this group are thin and membranous, resembling one another in structure, and the hind wings do not fold up ; whereas in the true Orthoptera the fore wings are usually thicker and harder than the hind wings, and the latter are capable of being folded like a fan.

DRAGON-FLIES

The first group is that of the dragon-flies (Odonata), the general appearance of which is too well known to need description. All have a large head, the sides of which are covered almost entirely by the two big, glassy-looking, compound eyes, while on its crown are two or three small simple ocelli. Each of the short and bristle-like antennæ has a stouter basal portion by which it is inserted on the forehead. The mouth faces downwards, and has a large semicircular lip (labrum) in front, the jaws being strong, horny, and well provided with teeth. The maxillæ are without palpi, but their narrow and palp-like outer lobe is often regarded as the real palp.

Succeeding the jaws behind is the lower lip (labium), which at its free end is usually slightly cleft in the middle, while its palpi take the form of two dilated and often two-jointed lateral lobes, these lobes sometimes overlapping one another in front to hide the free end of the lip. The thick and cylindrical thorax is followed by a long slender abdomen, which usually carries at the end two leaf-like or pincer-like appendages. When looked at from the side the two hinder segments of the thorax appear oblique, with the wings set rather far back above, and the legs pushed forward below. The wings are long, transparent, and traversed by a rich network of veins. The legs are often spiny, and their tarsi are always three-jointed. The position of the accessory organs of the male on the under side of the second abdominal ring is a feature distinguishing dragon-flies from other insects.

The female dragon-fly deposits her eggs in such a position that the larvæ, when hatched, find themselves either in their natural element, the water, or very close to it. In some species the female, accompanied by the male, goes under the water to lay her eggs ; others drop them into the water ; while in many species the female makes incisions in some aquatic plant and there deposits her eggs. The larvæ are even more fiercely carnivorous than the adult, and are distinguished from all other aquatic larvæ by the possession of a peculiar structure fixed under the head, known as the mask.

In their mode of respiration dragon-fly larvæ are also peculiar, some being provided with external tracheal gills—in the form of three leaflets placed near the tail end—which serve also to assist in locomotion, while others breathe by means of gills of an exceptional character. The latter are situated in the hinder part of the intestine, and consist of six longitudinal bands in its walls, crossed by several transverse folds, supplied with numerous fine branches from the tracheal trunks. Water can be sucked in at the opening, guarded by five valves at the hind end of the body, and when it becomes vitiated can be squirted out again either gently or with considerable force. When it is suddenly and violently

THE MAY-FLY

The upper picture shows an adult male, the lower a May-fly imago escaping from the skin of the sub-imago ; the larva is in the water below

expelled it serves to propel the insect forwards at a rapid rate.

The larvæ live about ten or twelve months, during which time they undergo several moults, rudiments of wings appearing some time before the final transformation. When this is about to take place the larva leaves the water by climbing the stem of a plant, or to some other dry spot. As the time approaches, its eyes, which were before dull and opaque, become bright and transparent. Its skin dries up, and soon begins to crack along the middle of the thorax ; the thorax appears through the cleft, and swelling up causes it to extend ; the head is next disengaged, and the legs are then drawn out of their sheath. The insect now throws its head farther and farther back, and by this means gradually frees the hinder part of its body with the exception of the last few segments, which still remain enclosed in the larval skin. After a while it suddenly bends its body forwards, grasps the sides of the sheath with its legs, and, doubling up its abdomen, finally extricates the rest of its body.

Dragon-flies are divided into three families, of which the first two have more in common with one another than with the third. The *Libellulidæ* are distinguished by their comparatively stout bodies ; by the size of their eyes, which cover almost all the sides of the head, and very nearly meet on its crown ; and by the structure of their lower lip, in which the median terminal piece is short and slightly divided at the end, while the very broad palps spread out and overlap it in front. The last character is useful in distinguishing the *Libellulidæ* from the next family, which in many respects they resemble. Their larvæ breathe by means of internal gills, and have a mask which is hollowed out on the inner side, and somewhat resembles a helmet. Members of this family are found in most parts of the world, and about twenty species occur in Europe.

The *Æschnidæ* have eyes even larger than those of the *Libellulidæ*. The end piece (ligula) of their lower lip is not divided in front, and not exceeded in length by the palpi, while each of the latter is armed with a strong tooth or spine. The abdomen is long, narrow, and cylindrical. Their larvæ are more elongate, and have bigger eyes than those of the last family. The flat mask has the palpi narrow, and armed with a movable hook at the tip. Like the larvæ of the *Libellulidæ*, they are provided with intestinal gills. Some of the biggest dragon-flies belong to this family.

The *Agrionidæ* form a family of slender-bodied dragon-flies, which have both pairs of wings shaped nearly alike. They are further distinguished from the other two families by the shape of the head, the smaller size of the eyes, and the structure of the lower lip. The head has a projection at each side, at the end of which is placed one of the two hemispherical eyes, and on the wide space lying between the compound eyes there are three ocelli arranged in a triangle. The lower lip consists at its free end of three parts of nearly equal length, the median piece (ligula) being notched in the middle, while the two palpi consist of

two joints, of which the first is large and terminates in an inwardly curved spine, whereas the second is small and articulated with the first outside the base of its spine.

The larvæ may be known by the three leaf-like tracheal gills at the end of their bodies, which are wanting or inconspicuous in those of other families. This family contains many of the most brilliantly and variously coloured dragon-flies, the sexes of the same species often differing in coloration. Some of the exotic species attain a great length, but this is brought about by the elongation of their slender abdomen without a corresponding increase in the proportions of the other parts of the body.

MAY-FLIES

The day-flies, or May-flies (*Ephemeridæ*), constituting the second group of the Pseudo-neuroptera, are comprised in a single family. They have soft and fragile bodies, with a long ten-jointed abdomen, bearing at the extremity two or three long, bristle-like, and many-jointed tails. The hind wings are sometimes wanting, and, when present, are always much smaller than the front pair, the latter being usually three-sided, with the corners rounded off. Three ocelli, in addition to the two large compound eyes, are borne upon the head, and the antennæ are short, and composed of two stout basal joints, followed by a slender, many-jointed bristle. In the adult the mouth organs are never well developed, but remain small and soft. The jaws have no function to perform, as the perfect insects do not eat.

The common notion that the life of the May-flies in the winged state lasts but a single day is sometimes, but not generally, true, many being able to live several days, provided the atmosphere be not too dry. There are some, however, which do not live for even the proverbial day, but emerge one evening, only to perish before the sun again appears. There is less truth in the supposition that these insects appear only in May; May-flies of one species or another being seen on fine days throughout the summer and autumn. They are to be found in the neighbourhood of rivers and lakes, some flying only by night, and others during the cooler hours of sunlight, or on favourable evenings until a little after sunset. During the heat of the day they seek repose, with their wings raised vertically. If the day be cold and raw, they seldom fly, but remain under shelter. In fine weather, however, they may sometimes be seen assembled together in swarms about sundown, till some time after sunset.

The peculiar up-and-down movement which marks the flight of some species has been often observed; and the mazy dance of the May-flies has been described by more than one author. In these dancing assemblies the male insects always greatly outnumber those of the other sex. The larvæ of the *Ephemeridæ* live in water; a few kinds are carnivorous, but most feed upon the minute vegetation scattered through the mud or covering stones, and the larger aquatic plants. Many

remain concealed in the banks or under stones, while others rove among water-reeds, and swim with celerity.

The larvæ of some genera are found only in large rivers. The eggs are, in some cases, deposited at the surface of the water, and then sink to the bottom; but in others the female creeps into the water to lay her eggs in patches on the under side of stones. The eggs are exceedingly numerous, and vary in shape according

LIFE-HISTORY OF DRAGON-FLIES
1, larval skin of a dragon-fly; 2, larva with mask exserted; 3, Libellula depressa; 4, advanced larva of Libellula; 5, the same about to undergo its final transformation

to the genus. The larvæ cast their skin several times; they are at first without special organs of respiration, but when they are about eight or ten days old tracheal gills begin to appear and ultimately develop into forms which vary somewhat in different genera. The gills are attached in pairs to the sides of some, or all, of the first seven segments of the abdomen, in some species standing out straight from the sides, and in others turned over the back. The mouth organs of the larvæ are better developed than in the adult, the mandibles being nearly always strong and toothed, and sometimes giving off a tusk-like process.

At their transformation most May-flies do not change directly from the larval form into the imago, but first pass through a stage, known as the subimago, in which they have their wings expanded, and breathe through the spiracles like the perfect insect. In this form they are distinguished by the dulness of their integument, the shortness of the fore legs and tail-bristles, and the less prominent and duller eyes. The subimago emerges from the larval skin at the surface of the water, and, after standing awhile upon the water, flies to a more convenient resting-place. At the next moult, which soon follows, the perfect insect makes its appearance. The emergence of May-flies takes place at different periods during summer and autumn, and that of any one species may last for several days in succession. At this time they sometimes appear in countless numbers, as thick in the air as snowflakes, and at the end of their brief existence leave their dead bodies to cover land or stream.

Nearly fifty species of *Ephemeridæ* are found in the British Isles. Two of the commonest (*Ephemera vulgata* and *E. danica*) are, in the subimago stage, known to anglers as " green drake " and " grey drake." They are four-winged species, with a body furnished at the end with three very long tails. The fore legs are extremely long, especially in the males, which sex is distinguished also by the much larger size of its eyes. The larvæ of *E. vulgata* burrow in the mud, or hide under stones, in ponds and sluggish streams. They have rather long antennæ, and the tusks of their mandibles project a good way, and cross one another in front of the head. They have six pairs of tracheal gills, which are turned up over the back, each gill consisting of two narrow blades, united at the base, and fringed with hairs along each side. The final transformations of the larvæ occur about the end of May or early in June, at which time, on a fine evening, the winged insects may sometimes be seen in hundreds, dancing in the air.

STONE-FLIES

The stone-flies (*Perlidæ*), forming the last group of Pseudoneuroptera with aquatic larvæ, are narrow, elongated insects of a flattened form, with a good-sized head, rather long, many-jointed antennæ, and four not very closely reticulated wings, which shut horizontally over the body when at rest. The abdomen usually carries two long, multiarticulated styles at the extremity The mouth organs are weakly developed in the adult insects; the mandibles and maxillæ are membranous; the maxillary palpi long, with slender terminal joints, and the labial palpi three-jointed. The thorax is square or oblong, with its three segments almost equally developed. The tarsi are three-jointed, and have their claws separated by a bilobed pad.

The species of this family are not numerous, though some are almost world-wide in their distribution. The adults appear about the same time as dragon-flies and alder-flies, and frequent nearly the same places. Though they have large enough wings, they fly heavily, and not for any considerable distance at a stretch, and are generally most active in the evening. The female fastens her eggs loosely together, and drops them in masses as she flies over water. The larvæ are mostly found in rapid streams, where they keep under stones or among broken pieces of wood, and live by preying actively upon the weaker creatures inhabiting the same waters. They have strongly developed jaws and rather long palpi. They breathe by means of tracheal gills, in the form of tufts of filaments, attached to the bases of the legs and the sides of the integument which joins the three thoracic and the first abdominal rings to one another. The two filamentous tails may have a pair of tracheal tufts at their base.

In later stages of their life the larvæ exhibit rudiments of wings.

When the time for its transformation arrives, the full-grown larva, or nymph, leaves the water by climbing the stem of a plant, or crawling some distance up the bank until it finds a dry stone on which to stand, when the emergence of the imago takes place in the usual way, preceded first by a splitting of the larval skin along the middle of the thorax. When the insect is free, its wings dry rapidly, and it is soon ready to fly.

A fact of importance, first noticed in the *Perlidæ*, though it occurs in some other groups, is that the tracheal gills are retained by the perfect insects, where they are attached in the same places as in the larva, but much reduced in size, and probably, in most cases, functionless. As an example of the *Perlidæ*, one of the best-known British species, *Perla bicaudata*, is figured on this page.

COMMON STONE-FLY
1, larva; 2, fly escaping from larval skin; 3, perfect insect

TERMITES, OR WHITE ANTS

The termites, or white ants (*Termitidæ*), differ considerably in one respect from all the other groups of Pseudoneuroptera. They live in societies which are of a highly organised and complex nature and most resemble those met with among insects of the highest type, such as bees and ants. This is, however, the only direction in which the termites diverge to any extent from the rest of the Orthoptera; for, like all these, they pass from the larval to the adult state by a series of gradual changes; while in the structure of their bodies they show an affinity with some of the lowest groups of the order. In the termites the head is free and distinct, with the antennæ composed of a number of small bead-like joints, and rather short.

The perfect insects have compound eyes, and, as a rule, two ocelli; but the wingless individuals are generally without eyes of any kind. The mouth parts, which are constructed on a clearly orthopterous plan, are not very unlike those of a cockroach, and consist of a distinct upper lip (labrum), two strong horny mandibles, a pair of two-lobed maxillæ with five-jointed palpi, and a lower lip (labium), divided at the end into four lobes, and bearing three-jointed palpi. In the thorax the first segment is well developed, and its dorsal plate, or pronotum, is rather broad and flat; the other two segments being less strongly developed, though in the winged insects attaining a fair size. Both pairs of wings are much alike; they are long, narrow, not very closely veined, each wing being marked by a transverse suture at a short distance from the base; and in a state of rest they are laid flat over the back. The legs are slender, and well fitted for running, and their tarsi are four-jointed. The abdomen has a slightly elongated or oval form, and carries two very short appendages—the cerci—near its extremity.

The common habitation of a society of white ants is known as a nest; and in each nest, which is divided into a number of cells, or chambers, communicating with one another, there may be found several different kinds of individuals in addition to the larvæ. Some are provided with wings, or with the rudiments thereof, and are distinguished also by having eyes. These are the sexually developed males and females, which are capable of reproducing their kind; though this function is, as a rule, carried on by a single couple in each nest. The king and queen—as this couple are named—are lodged in a large cell near the middle of the nest, and may be recognised by their large size, and the fact that they retain but small stumps of the wings which they once possessed. The royal cell is larger than the others, and has thicker walls; while the passages leading into it are too small to afford the occupants a means of escape, though large enough to admit the workers, which come and go, some to bring food to the royal pair, others to carry away the eggs laid by the queen. At this time the abdomen of the queen, owing to the number of eggs it contains, is swollen to an enormous size. " She lies there," writes Drummond, in reference to one of the African species, " a large, loathsome, cylindrical package, 2 or 3 inches long, in shape like a sausage, and as white as a bolster." Her eggs are discharged at a rapid rate, amounting in a single day to several thousands, and the process is continued with the same activity for months in succession.

Both workers and soldiers are wingless members of the community, and, in the majority of species, have no eyes. The workers have small and rounded heads, with short mandible and well-developed maxillæ and palpi; whereas the soldiers are easily recognised by their big, square, or oblong heads, and long mandibles. The workers are the most numerous class, and have many duties to perform in the way of building, tunnelling, and providing food for the young larvæ and for the king and queen. The soldiers look after the protection of the workers, and act generally in defence of the community. In one genus there are no true mandibulate soldiers; but there is, instead, a class of individuals known as "nasuti," from the fact that their pear-shaped heads are prolonged in front in the form of a beak. The exact part which the nasuti play is not yet clearly known; but, like the soldiers of other species, these individuals appear at the first sign of danger, and shake their heads and palpi in a most menacing way. The eggs of the queen termite are, as mentioned, carried away by bodies of workers, and placed in special chambers, or nurseries.

When the young larvæ are hatched, they are at first indistinguishable from one another, and are little blind creatures, with soft and pale integument; and it is only after the first or second moult that they begin to show those differences which subsequently distinguish the larvæ of the various classes. They are fed with a special kind of food, consisting of comminuted dead wood, mixed with saliva, which certain of the workers prepare for them. By varying the quantity and quality of the food supplied, the termites appear able to arrest or deviate the development of larvæ that would, in the ordinary course, become perfect insects, or, in other words, they can produce workers and soldiers from larvæ which, if fed upon a different diet, might develop into winged insects fitted to become kings and queens. And it has been shown that neither the soldiers nor the workers of the termites belong to one particular sex only, as is the case with the neuters of bees and ants, but that individuals of both sexes, in an imperfect sexual condition, are comprised in each class.

The winged insects into which many of the larvæ develop are most abundant at certain periods of the year, especially after rains; they do not remain long in the nest, but, after a few days at the most, make their way out, or are led out by the workers, and shortly afterwards take flight. They may often be seen flying in swarms, and at night sometimes enter houses, being attracted by the light. Many are devoured by birds, which seize them as they leave the nest. When they have finished their flight, and alight on the ground, they shed their wings, which easily snap off at the line of suture near the base. If a couple, chancing to be near a termite burrow, are found by some workers, they are brought in, a royal cell is prepared for them, and, as king and queen, they become the parents of a new colony.

Some larvæ develop into individuals which, although fitted to perform the functions of perfect insects, never possess complete wings, but are provided at most with wing-pads, or rudiments of wings. These individuals, which somewhat resemble the nymphs of the perfect insects, are known as substitution kings and queens, and take the place of true royal couples, when from any cause the latter are not to be found.

The food of white ants consists ordinarily of decaying wood, or similar vegetable matter, which, when it has passed in a half-digested state through their bodies, is eaten again. These insects have also the habit of devouring their dead, which makes it possible to destroy a whole colony by placing a little arsenic or mercuric chloride in their food; for the few that die through first partaking of the poison are eaten by others, which in their turn are also devoured, and so the poison is spread through the entire population.

About two hundred species of termites have been described; and these inhabit chiefly the tropical

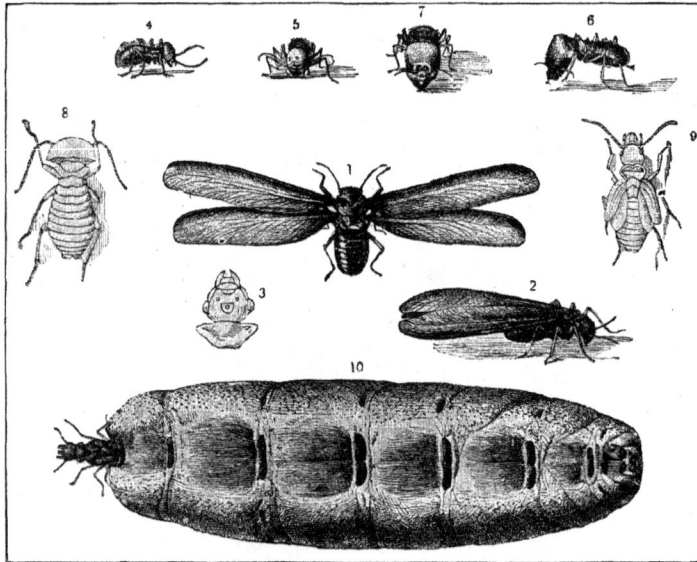

TERMITE, OR WHITE ANT

1, male of Termes dirus; 2, the same, seen from the side; 3, head, enlarged; 4, worker; 5, the same, front view; 6, 7, soldier, side and front view; 8, worker (much enlarged); 9, nymph; 10, queen

and subtropical parts of the world, although two small species are found in the south of Europe. Some species live in hollows they have eaten out in the interior of the trunks and branches of trees, or in timber. They line the galleries they make, which are often so close together as to be separated only by a thin wall, the wood in the interior being almost all eaten away. A few make openings to the exterior, and form nests around the branches of trees; these nests being sometimes as large as a sugar-barrel, though the size varies considerably. The nests of most species are usually placed entirely below the level of the ground, and often lie beneath mounds of earth raised above the surface.

Some of the larger African species, such as *Termes bellicosus*, build mounds of earth, frequently reaching a height of 12 or 14 feet. These mounds, which may stand singly or in groups of varying size, are divided inside into chambers and galleries communicating with one another and with the nests and galleries underground. The nests of this kind, which consist

HOUSE-CRICKET AND FIELD-CRICKETS

1, male, 2, female, with, 3, young, and, 4, old larvæ of field-cricket ; 5, male of house-cricket

and carrying prominent eyes, three ocelli, and bristle-like antennæ. Their mandibles are horny at the tip, but the other parts of the mouth are usually soft a n d membranous ; t h e maxillæ being bilobed, with four-jointed palpi, and the palps of the bifid labium rudimentary. The middle segment of the thorax is the largest, and the pro-thorax is usually very short and narrow.

The wings, which are wanting in some species, are slanting in repose, like the sides of a roof and cover over the abdomen ; they are of an almost glassy transparency, and have, as a rule, an open system of neuration. The tarsi are composed of two or three joints. Most species of *Psocidæ* live in the open air, and feed on fungi, lichens, and the fragments of other plants ; the largest European species (*Psocus lineatus*) being scarcely more than a quarter of an inch long.

almost entirely of clay, become in time quite hard and solid, and are much more durable than those which are composed of particles of dead wood pasted together with sticky saliva or with excrementitious matter. From the central nests termites construct underground galleries or tunnels leading in different directions, and sometimes reaching hundreds of feet in length. When it is necessary for the workers to go above ground in search of food, they protect themselves by building covered ways leading to the object they desire.

Their tunnels sometimes lead to the interior of houses, and when once ter-mites gain admittance in this way there is scarcely any limit to the mis-chief which may result from their operations. The wooden pillars that support the roof, the woodwork of the roof itself, and even articles of furniture, may be destroyed before the inhabitants become aware of what is taking place. For in tunnelling through wood termites take care to leave the outer shell intact ; and what appears on the outside to be a solid piece of wood may consist in the interior of nothing but a series of galleries lined with white-ant mortar.

These insects easily make their way into wooden boxes, and quickly destroy the books, papers, clothing, or whatever else they contain. The rapidity with which they work is remarkable, and in a single night they have been known to burrow up through the leg of a table, then across the table, stopping on the way to devour the articles lying on it, and down through another leg into the floor again. Forest trees, also, are often ruined by the action of termites which, in order to get at the dead branches, will sometimes bore their way up through the trunk, and thus bring about its premature decay.

BOOK-LICE

The book-lice and the other insects classed with them in the family *Psocidæ* form another small group of Pseudoneuroptera. They are mostly very small in-sects, with a proportionately big head, swollen in front,

PSOCUS LINEATUS

BIRD-LICE

The Mallophaga, commonly known as bird-lice, are small, wingless in-sects, resembling ordinary lice to some extent, but differing from them in many characters, and especially in the structure of the mouth, which is fitted for taking food by biting instead of sucking. They form a distinct group, now generally placed in the Pseudoneuroptera, though some ento-mologists assign it a position near the Pediculina, or true lice.

The bird-lice are flat-bodied insects, with a broad head, varying a good deal in form, and a thorax which usually appears to consist of only two segments. Their antennæ are short, and composed of three, four, or five joints ; and their eyes, when present, are simple. The mandibles appear as short hooks, sometimes toothed on the inner side ; the maxillæ are short, and said to be always palpless ; while the lower lip is distinct, and often bears palpi. The legs are short and stout, and have two-jointed tarsi, each of which carries at the end either one or two claws. As bird-lice are

MOLE-CRICKET, WITH EGGS AND LARVÆ

found on mammals as well as on birds, their name is to some extent misleading.

TRUE ORTHOPTERA

The insects of the suborder Orthoptera genuina differ chiefly from those of the last group in the characters of their wings, which are usually stiff and tough, and in some cases horny, serve as wing-covers, and are generally spoken of as elytra ; whereas the hind pair are membranous, and capable of being folded longitudinally, or both longitudinally and transversely. The division of the ligula, or terminal piece of the lower lip, into two or four lobes is usually more complete than in the Pseudoneuroptera.

It is usual to divide the true Orthoptera into two series or tribes—the Saltatoria, with strongly developed hind legs, adapted for leaping, and the Cursoria, in which the hind legs are not thus developed, but are better fitted for use in running and walking. The Saltatoria, or jumpers, are sometimes spoken of as . the musical Orthoptera, since the males of nearly all these insects, such as crickets, grasshoppers, and locusts, produce loud, chirping sounds.

The females are supposed to be attracted by the chirping of the males ; they seldom emit any sound themselves, and when they do it is generally of a very feeble character. It is probable that most insects can hear, but by what means they do so is, in the majority of cases, to a great extent a matter of conjecture. The saltatorial Orthoptera are, however, remarkable in possessing very definite organs of hearing.

CRICKETS

The crickets (*Gryllidæ*), which form the first family of the suborder, have a somewhat rounded head, supporting long, whip-like antennæ. Their mandibles are strong and toothed ; the inner lobe of the maxillæ being devoid of teeth, and the outer one long and slender. The fore wings, or elytra, do not differ from one another in structure, and, when at rest, are closely applied to the somewhat thick and massive hind body. The hind wings are folded many times, and may generally be seen projecting beyond the tips of the elytra. The hind legs are generally used in jumping, while the other two pairs are better adapted for walking, although in the mole-crickets the fore legs are thickened and otherwise modified for use in burrowing. The tarsi of all the legs are composed of either two or three joints. The abdomen bears two flexible, velvety appendages, and in females it usually carries also a long exserted ovipositor.

GREAT GREEN GRASSHOPPER

The chirping of crickets is produced by rubbing the base of one elytron over the other ; in which respect these insects differ from most grasshoppers and locusts, and resemble only those grasshoppers with long antennæ which belong to the family *Locustidæ*. They resemble the latter also in having their organs of hearing placed on the fore legs. These organs are lodged in the upper part of the tibia, a little below their articulation with the femora, and consist externally of two small depressions, or pits, on opposite sides of each tibia, with a thin membrane stretched across the bottom of each depression. Inside the leg a tracheal vessel widens out between the two tympanic membranes to form a vesicular expansion.

Crickets are found all over the world, but only four species are British. Of these, one (*Nemobius sylvestris*) may be recognised by its small size, being little more than a third of an inch long. It is usually found among the dead leaves in woods, and appears to be restricted in its range to the southern counties. The field-cricket (*Gryllus campestris*), which sometimes measures an inch in length, is generally of a black colour, and lives in dry fields, where it is often heard, though seldom seen on account of its retiring habits. The house-cricket (*G. domesticus*) has a reddish brown colour, and is somewhat smaller than the field-cricket. It has well-developed wings, and the female has a long ovipositor.

The mole-crickets, of which there is one British species (*Gryllotalpa vulgaris*), have such a peculiar structure that they are easily distinguished from all other insects. They have a long, smooth, shiny prothorax, rather short, close-fitting elytra, and under wings which, when rolled up, look like a tail curving down over the tip of the abdomen. The abdomen itself carries two long, flexible tails. The fore legs are thicker, but shorter, than the hind legs; they have very short tibiæ, each ending below in four strong claws spread out like the fingers of a hand.

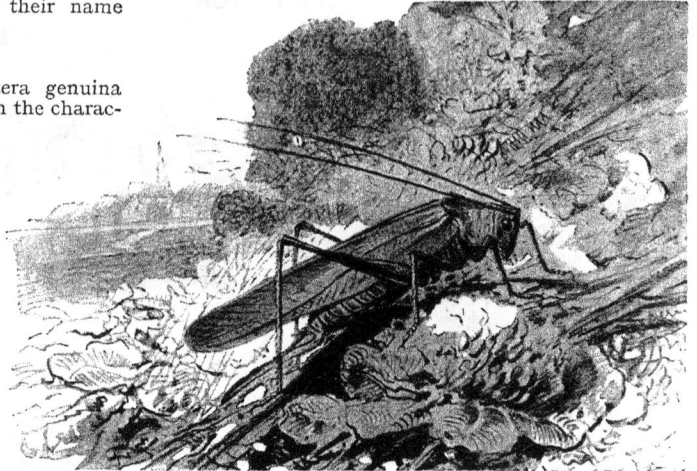

I, FEMALE OF HETRODES SPINULOSUS ; 2 MALE AND FEMALE OF MECONEMA VARIUM

LONG-HORNED GRASSHOPPERS

Although named *Locustidæ*, this family does not comprise the locusts, but includes only those grasshoppers in which the antennæ are long and tapering, and the tarsi are four-jointed ; while the female is provided with a long ovipositor. Besides these characters, there are some others which help to distinguish the *Locustidæ* from the members of the next family. In the present group the organs of hearing are placed, as in the crickets, in the tibiæ of the fore legs, and the chirping of the males is produced by the friction of the wing-covers over one another. The wing-covers, instead of being both alike, as in crickets, exhibit a certain amount of difference in structure.

Taking the male of the large British green grasshopper as an example, it will be seen that on the portion of the right elytron which folds horizontally over the trunk there is, near the base, a somewhat irregularly circular area, which has a glistening appearance, like a piece of talc. This area is bordered by a strong, prominent vein. In a corresponding position on the left elytron, which, when closed, overlaps the right, there are also some thick, transverse veins, but the cells enclosed by these veins have a similar texture to the rest of the membrane. When the insect rubs its left elytron rapidly over the right, the veins projecting on the under side scrape on the margin of the mirror and set the latter in vibration, thus giving rise to the well-known sound. The chirping of the *Locustidæ* is generally louder and more prolonged than in the other grasshoppers. In certain North American species known as katydids, the song seems

MIGRATORY LOCUST OF SOUTH-EASTERN EUROPE AND ITS LARVÆ

to consist of these words repeated again and again, with a slight variation.

The life-history of the *Locustidæ*, so far as it is known, does not differ in any essential respect from that of the *Acridiidæ*. It is probable that in most cases the female uses her long ovipositor to lay her eggs at some depth in the ground. These grasshoppers are less herbivorous in their habits than those belonging to the next family. The *Locustidæ* are most numerous in species in America and Asia, there being not many more than two hundred species in Europe, of which about ten are British.

In the large green grasshopper (*Locusta viridissima*), which is nearly an inch and a half long, and is easily distinguished by its size from all the other British species, the male makes a harsh and strident noise, by which attention is attracted, when otherwise, owing to its green colour, it might altogether escape notice. Green is the prevailing tint in very many species of this family. In some species the elytra have the most exquisite resemblance in colour and venation to green leaves · while in others they look more like withered leaves. Nowhere is this style of protective coloration better displayed than in the exotic genera *Cycloptera* and *Pterochroza*, one of the species of which is figured in colours on page 106. The shape, colour, and venation of the wings are not only exactly like those of leaves, but there may be seen, here and there, little glistening, transparent patches of cuticle, which reveal, as it were, the work of an insect grub. In others, fungi seem to grow on the leaves.

Among the species of the family remarkable in other respects we have space to mention only a few. In the genus *Hetrodes* the adult insects of both sexes are without wings ; the prothorax is very large, and is armed above with a number of spines. An idea of the general appearance of the adult insect may be gathered from the figure of *Hetrodes spinulosus*. This species is found in Arabia and Syria.

For the sake of contrast the male and female of a small British grasshopper (*Meconema varium*) are figured beside it. The latter is winged in both sexes ; it is found in oak-trees, and belongs to a subfamily which is peculiar from the fact that the elytra of the male have no stridulating organs.

LOCUST TRIBE

The locusts and short-horned grasshoppers (*Acridiidæ*) are distinguished by easily recognised characters from the other two families of the suborder. The antennæ are short, seldom attaining more than half the length of the body, the tarsi are three-jointed, and the female always has a very short ovipositor. They differ also in the position of the auditory organs, and in the mode by which the males produce the chirping. In these insects the organs of hearing appear externally as two pits, somewhat crescentic or semilunar in shape, placed one on each side of the first abdominal segment, immediately behind the thorax. At the bottom of each pit there is a tense membrane, which on its inner side is brought into relation with the terminal rods and fibres of a nerve which arises from the last thoracic ganglion.

It was thought that these pits were in some way concerned in the production of sound, but it is evident from their structure that this is not the case, while they really seem capable of serving no other function than that of ears. Moreover, it is now known that the chirping of these insects is produced by rubbing the hind legs up and down against some of the projecting nervures in the sides of the closed elytra. When the insect is stridulating it keeps the tibia of the leg folded up against the femur. In some species the sound is heard at both the upward and downward stroke of the legs, in others at the downward stroke only. The sound varies in intensity in different species, and for this reason some of the commoner species may be recognised even before they are seen.

In most of these insects the front of the head is vertical or slightly inclined backwards, but in some (*Tryxalinæ*) it is much inclined backwards, and the whole head seems prolonged in a way that makes it look like a cone or wedge, with the antennæ and eyes near the apex, and the mouth placed below under its base. The *Acridiidæ* are usually provided with three ocelli in addition to the compound eyes, the ocelli being, as a rule, more distinct than in the *Locustidæ*. The mouth organs are well developed, consisting of a large upper lip; strong, toothed masticatory jaws; five-jointed maxillary palpi; and a lower lip, divided at the end into two or four lobes, and bearing three-jointed palpi. The prothorax is generally large, much longer above than below, and often carrying a prominent crest along the middle. Wings are usually present, but the hind pair are wanting in the females, or even in both sexes of some species.

In their general life-history the *Acridiidæ* are probably much alike. The female lays her eggs at a short depth below the surface of the ground, or attaches them to the stalks of grasses, and usually surrounds them, in mass, with some sort of protective covering. Later on in the same year, or in the spring of the year following, the larvæ are exuded. They soon become active, and—except that they are without wings, have shorter antennæ, and are of smaller size and no definite colour—do not differ much in appearance from the perfect insects. After undergoing, as a rule, about six moults, the larvæ which are hatched in the spring become adult late in the summer. It is generally in the days immediately following their entry into the perfect state that the male insects are loudest and most persistent in their song.

Few of the British *Acridiidæ*, of which there are about a dozen, are remarkable for the brightness of their colours; nor do any cause trouble by a great excess of numbers. But among the exotic species there are many exhibiting vivid tints of colour, and some which are capable of multiplying to such an extent as to become a serious source of mischief in the places where they abound. It is to the species accustomed to assemble together, and migrate from place to place, in vast swarms that the name of locusts is more especially applied; this habit really constituting almost the only difference between the locusts and many of the other grasshoppers of this family.

Grasshoppers feed chiefly on the grasses of different kinds, including most of the cultivated grains; but locusts leave scarcely anything in the nature of vegetation untouched, when, as often happens, they invade a district where the ordinary herbs and grasses are insufficient to support their vast numbers. Trees and shrubs are then stripped bare of their leaves, and the bark and wood even are not spared. Pressed by hunger, locusts do not refrain from attacking plants which at ordinary times they seem to avoid. They frequently devour their own dead, and even carry their cannibalism so far as to kill and eat the newly-moulted and soft-skinned larvæ.

Different species of these destructive insects are found in all the great regions of the world; though North Africa is, perhaps, the one which suffers most from their ravages. The locusts referred to in Scripture belonged, in all probability, for the most part to the species known as *Acridium peregrinum*, which has its chief home in the Sahara and surrounding districts.

Several other species are found in North Africa, and in South Africa *Pachytylus migratorioides* is one of the most widely distributed. Great swarms of locusts of this species have been seen at different times in recent years; one which passed over Pretoria in 1891 was estimated to be twenty-five miles long, one and a half broad, and half a mile in depth. It was probably to this species also those locusts belonged of which Barrow, giving an account of their ravages in the year 1797, states that the whole surface of the ground over an area of about two thousand square miles was literally covered with them; and that when driven into the sea by a north-west wind they formed a bank on the shore three or four feet high and fifty miles long. Among European locusts, the best known is *P. migratorius*, which occurs chiefly in the south-east, and is found also in Egypt and in West and Central Asia. The locust of North-West India is *Acridium peregrinum*.

Passing from the locusts, we may briefly notice a few of the other insects of the family. The *Tryxalinæ* are remarkable on account of the peculiar shape of their head, to which we have already alluded. No species

TETTIX SUBULATA

LEAF-INSECT

of this subfamily is found in Britain. In the allied *Tettiginæ* the pronotum is produced behind into a long process, which in some of the species reaches beyond the tip of the abdomen. Two of the smallest species of grasshoppers found in Great Britain belong to the genus *Tettix*—the typical genus of this subfamily. The genus *Pneumora*, which is represented only in South Africa, is characterised by the bladder-like dilatation of the abdomen in one of the sexes. The hind legs in this genus are rather short, and are scarcely adapted for leaping.

STICK-INSECT AND LARVA

STICK AND LEAF INSECTS

The stick and leaf insects (*Phasmatidæ*) are chiefly interesting on account of their resemblance to the objects after which they are named. They form one of the families of cursorial Orthoptera, and, in addition to the easily recognised shape of their bodies, are distinguished by the following characters. The head is distinctly visible from above, and is set somewhat obliquely, with the mouth placed well forwards on the under side. The short prothorax is much shorter, as a rule, than the next segment, or mesothorax. The legs, which in shape usually harmonise with the shape of the body, are inserted somewhat close to the sides of the thorax, those of each pair being separated from one another by a rather broad sternal plate; the tarsi are five-jointed, and exhibit a pad-like lobe between the claws of the terminal joint.

In the stick-insects the trunk is long, narrow, and cylindrical; the legs are generally long, and when stretched out unsymmetrically from the body, as they habitually are in the resting posture, look like smaller branches coming off from a thicker jointed stem. Many stick-insects have no wings at any stage of their life, and it is difficult in such cases to distinguish the adult insects from some of the older larvæ. In the winged species the fore wings are usually very short, and often cover only a small part of the hind pair; the latter exhibit a division into two distinct areas—one more membranous and transparent, and often brightly coloured; the other, which is narrower, and placed next the anterior border, being coloured like the elytra. When the wings are at rest, the brightly coloured portion is folded beneath the other part, which alone is then exposed to view, so that there is nothing to detract from the general stick-like appearance of the body.

These insects are usually found amongst underwood, or on shrubs and the stems of long grasses. They are mostly inactive during the day, and are not easily seen, owing to the way in which their form and colours harmonise with their surroundings. They roam about

PRAYING INSECTS

at night, and feed upon leaves. Many inhabit tropical and subtropical countries, and amongst them are some of the largest insects known, more than one measuring over 13 inches in length. Two species are found in South Europe, belonging to the genus *Bacillus*, and are both wingless forms of rather large size. One of these is figured in the illustration; and, as examples of some of the more finely coloured tropical forms, two species from the island of Borneo are represented on the coloured plate of Orthoptera.

The leaf-insects, though belonging to the same family, exhibit a marked contrast to stick-insects in the shape of the body, which, instead of being narrow and cylindrical, is broad and flat. The male is narrower than the female, and distinguished also by having moderately long antennæ, well-developed hind wings, and short fore wings. In the female the antennæ are very short

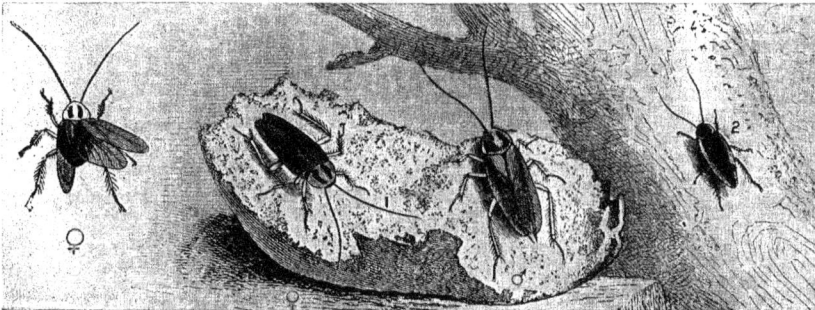

COCKROACHES

1, male and female of Phyllodromia germanica ; 2, Ectobia lapponica the hind wings are rudimentary, and the elytra are fairly large, leaf-like structures, which in some species almost entirely cover the broad, flattened abdomen. The legs have broad, leaf-like expansions on both the femora and tibiæ, contributing to the general leaf-like appearance.

It is remarkable that the colour of these insects, which is either the green of a living leaf, or some shade of yellow or brown, like that of a withered leaf, is due to a substance similar in its nature to chlorophyll, or the green colouring matter of plants ; and it is stated that the internal structure of the elytra bears a striking resemblance to that of a plant. All these curious insects belong to the single genus *Phyllium*, and are found in the Oriental countries, and in some islands of the Indian Ocean.

PRAYING INSECTS

The praying insects, or *Mantidæ*, constituting the next family of the suborder, have the head turned down, with the face inclined backwards, so that the vertex projects in front, while the mouth lies close to the lower edge of the prothorax. They have many-jointed, bristle-like, or comb-like antennæ. The prothorax is generally much longer than the other two segments of the thorax taken together. Whereas the two hinder pairs of legs are long, and resemble one another, the fore legs—which are inserted close to the front and wider end of the prothorax—exhibit a

1, MALE OF THE LARGE EARWIG ; 2, COMMON EARWIGS

peculiar form and structure, their coxæ being long and three-cornered, and often spined on the angles, and the femora broad, flattened, and grooved below to receive the tibiæ, which can be folded back upon them like the blade of a knife. The tarsi of all the legs are five-jointed. These insects usually have two pairs of wings, of which the fore wings, or elytra, are ordinarily of the length of the abdomen.

The characteristic posture which these insects assume when resting on a tree or shrub, with their prothorax raised and the fore legs doubled up in front of them, accounts for their common names of soothsayers and praying insects. They are amongst the most predaceous and bloodthirsty of creatures, living on flies and other insects, which they seize with their raptorial fore legs, in the manner shown in the illustration. *Mantidæ* are chiefly found in the warmer parts of the world, but a few species occur in South Europe. The best known of these is the figured *Mantis religiosa*. Some species, such as the African *Harpax ocellata*, are curiously marked, while others are prettily coloured. The colours are sometimes so disposed that the insect in its resting attitude resembles a flower, and thus draws towards it other insects, which, when they have approached near enough, are suddenly caught.

COCKROACHES

The cockroaches (*Blattidæ*) constitute one of those families in which the legs are more specially fitted for

1, SYROMASTES MARGINATUS ; 2, ITS LARVÆ ; 3, NEIDES TIPULARIUS

running. They have a rather short head, with a large, flat face, looking slightly downwards, and the mouth brought close to the prosternum. The eyes are large and compound, and in the place usually occupied by the lateral ocelli there are often to be seen two pale soft spots in the integument. The long and tapering antennæ are inserted close to the eyes, and composed of a stouter basal joint, followed by a number of short joints.

The strong and horny jaws are toothed or spined on the inner side, and thus well adapted to biting ; and the head is scarcely visible from above, being overlapped by the large, shield-like plate of the prothorax. The legs are long, with spiny tibiæ, and end in five-jointed tarsi. The pulvillus, which projects between the tarsal claws of these and many other insects, constitutes a sixth joint, although not usually reckoned as such. Cockroaches are generally provided with two pairs of wings, the front pair being stiff and horny, while the hind pair are of a more membranous texture, and in a state of rest are folded longitudinally, and almost entirely covered by the elytra. The abdomen is broad and flat, and carries two jointed appendages —the cerci—near its extremity.

About six species are found in Britain, of which three only are really indigenous, the others having

SHIELD-BUGS

1, Pentatoma rufipes ; 2, Acanthosoma dentatum ; 3, Eurydema oleraceum ; 4, Ælia acuminata

been imported. The common cockroach (*Periplaneta orientalis*) is believed to have belonged originally to the East. These insects are commonly spoken of as "blackbeetles," though not beetles, and not black, but having a reddish brown colour. The male is easily recognised by the wings scarcely reaching beyond the middle of the abdomen.

The female is broader in the body, and has very short rudimentary fore wings and no hind wings. Her eggs are arranged in a horny case, opening at the top, and shaped like a purse, which she carries about with her for some time, protruding from the end of her abdomen. She finally deposits the egg-capsule in a crevice in the walls or below the floor, and after some interval the young larvæ are extruded. During growth they shed their skin several times. The new skin is at first soft and of a pale or nearly white colour, but gradually hardens and gets darker.

The American cockroach (*P. americana*) which is such a pest on many ships, and is found about the docks and warehouses of seaport towns, is larger than the common species. It has two pale bands on the prothorax, and is winged in both sexes.

The German cockroach (*Phyllodromia germanica*) is another imported species, said to have first arrived with the soldiers returning from the Crimean War, but now plentiful in some houses, especially in bakeries and restaurants. It may be distinguished by its smaller size, and pale yellow-brown colour, with two dark brown bands along the pronotum. Both sexes have wings. In some parts of Central Europe they live in woods, resembling in this respect many other species, including three, belonging to the genus *Ectobia*, found in woods in England.

EARWIGS

The earwigs (*Forficulidæ*), which form the last family of cursorial Orthoptera, possess distinct characters, and are sometimes treated as a separate order, under the name of Dermaptera. Easily recognised by the narrow body, short, squarely-cut, horny elytra, and the pincer-like appendages of their abdomen, these insects are further distinguished by the intricate folding of the hind wings. The elytra, or fore wings, do not overlap one another as in most Orthoptera, but, like those of beetles, simply meet by their edges along the middle line. The hind wings, which are thin and membranous throughout most of their extent, are folded, partly like a fan, by means of folds radiating from near the middle of the anterior margin, and also transversely. In this way they occupy a small space, and are almost completely covered by the elytra, a tiny piece only being left projecting behind. When fully expanded, each wing is somewhat elliptic in outline, with a straighter anterior and more rounded posterior margin. To these characters it is only necessary to add that the tarsi are three-jointed, and the ligula deeply two-lobed.

This family is represented in almost all parts of the world, but not more than two or three species are commonly met with in Britain. The species are distinguished chiefly by the size and shape of their forceps, the length and number of joints of the antennæ, the state of development of the wings, and so forth.

The common earwig (*Forficula auricularia*), found all over Europe, is the best-known species. The female is usually smaller than the male, and her forceps are shorter, and without teeth at the base. Her eggs are laid under stones, moss, or in other such places, and she watches over them with care. It was long ago observed that the female earwig sits over her eggs like a hen in a nest, and if they happen to get scattered, gathers them all together again. The young larvæ when hatched keep close to her, clustering under her body, and sometimes climbing on to her back. They are not very unlike their mother in appearance, but are without wings, and of much smaller size.

RHYNCHOTA

The numerous insects included in the order Rhynchota exhibit great differences in their external form, and while some, such as the *Flatinæ*, rival the butterflies and moths in the beauty and delicacy of their colours, others are amongst the most loathsome of creatures. But whatever be their form or colour, all agree in two

HOTTENTOT-BUGS

essential characters, the first consisting in the fact that their development takes place without a complete metamorphosis; and the second that all have the mouth taking the form of a beak, or rostrum, adapted for piercing and sucking. The beak consists chiefly of the lower lip (labium), which is long and narrow, composed of three or four joints, and grooved along the whole length of its upper or anterior surface. This groove forms a sort of sheath, in which are lodged four long, slender blades, corresponding to the mandibles and maxillæ of other insects, but here transformed into piercing organs. All these parts are covered at the base in front by the narrow and slightly elongated upper lip (labrum). From the structure of their mouth, which is fitted only for the reception of liquid nutriment, it is easy to infer that these insects live by piercing tissues of plants and animals, and extracting the juices.

The larvæ differ little from the adults except in size, the absence of wings, and their usually shorter and more slender antennæ. In many, however, the females are without wings at all stages; and in some cases both sexes are thus unprovided. When wings are present, they may be all of similar texture, or the front pair may be somewhat stiffer and less

RED BUGS

membranous than the hinder. Wings of both these kinds are found in the section Homoptera. In other cases, while the hind wings are entirely membranous, the front pair are stiff and horny for some distance from their base, and thin and membranous towards the extremities. Such wings, which characterise the section Heteroptera are known as hemi-elytra.

All the Heteroptera, no matter how different they may be in external form or mode of life, are termed bugs, although this name was originally applied only to the bed-bug and a few closely allied species. Most are winged insects, in which the fore wings, known as hemi-elytra, or simply as elytra, always have the form described above. Their antennæ are either short and inconspicuous, as in the water-bugs, or distinctly visible, as in the land-bugs, and are generally composed of a small number of joints. As a rule they have two compound eyes, and often two or three ocelli. The first segment of the thorax is usually large, with the head sunk deeply into it. The abdomen generally has an oval, flattened form, and the legs are mostly slender. With few exceptions, bugs are characterised by a peculiar and somewhat unpleasant odour, which arise from a liquid secreted by special glands placed in front of the abdomen, and opening to the exterior by means of two small apertures on the ventral surface of the metathorax.

CALOCORIS STRIATELLUS

LAND-BUGS

Bugs are divided into two tribes, based upon their mode of existence, and the fact that in one tribe—the land-bugs, or Geocorisa—the antennæ project, and are distinctly visible, while in the other—the water-bugs, or Hydrocorisa—they are very short, and hidden below the eyes.

SHIELD-BUGS

The shield - bugs (*Pentatomatidæ*), which constitute one of the largest families of the Geocorisa, are so called on account of their large scutellum, which reaches at least to the middle of the abdomen, and sometimes quite to its extremity, covering it over completely. The fore wings are some-times chitinised only near the basal margin, especially in those species with a very large scutellum. The body has in general an elliptical outline, or is shaped like a scutcheon, owing to the projecting lateral angles of the somewhat hexagonal pronotum.

These bugs are mostly found on low plants, some in conceal-ment, many showing themselves openly, and often attracting observation by their striking colours. The adults pass the winter sheltered under bark or dried leaves. In early spring the females lay their eggs on the foliage of low plants, shrubs, and pine-trees. The oval or spherical eggs are provided with an oper-culum, or lid, and disposed in patches resembling honeycomb. The larvæ moult several times in the course of their growth, and thus gradually effect a change in their form and coloration. They feed on the juices of plants, or, in some cases, of animals, and attain their full size towards the end of summer.

The European species are rather limited in number, but many forms are found in other parts of the world. The Hottentot-bug (*Eurygaster maurus*) is the name

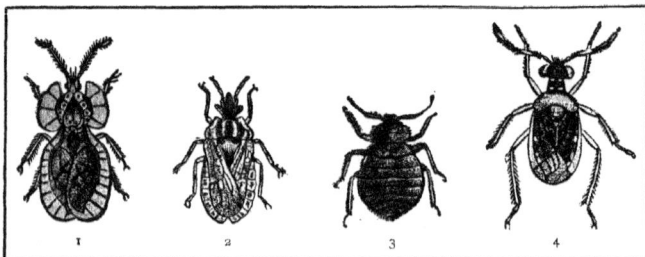

1, TINGIS AFFINIS ; 2, ARADUS CORTICALIS ; 3, CIMEX LECTULARIUS ; 4, SALDA ELEGANTULA

REDUVIUS PERSONATUS AND ITS LARVA

given to a species with a very large scutellum, found throughout nearly all Europe. It is of a yellow, dark brown, or black colour, with two clear spots on each side of the base of the scutellum. Some rather pretty bugs of the genus *Scutellera*, belonging to the same subfamily, and characterised by a similar large scutellum, are found in Australia and the Eastern Archipelago. They are of a short, broad, and convex form, and have a very fine metallic blue coloration, often spotted with bright yellow.

FOREST-BUGS

The forest-bugs (*Pentatoma*) have strongly projecting angles to the pro-thorax, and also a long triangular scutel-lum. The species figured (*P. rufipes*) is common throughout Europe, on birch and other trees, and renders service by destroying certain caterpillars. On page 1989 are figured three other species of this family—*Acanthosoma dentatum*, which is common on willows ; *Eurydema oleraceum* a bluish green or metallic green species, with red or white markings, which in some places is injurious to plants of the cabbage tribe, but also lives on other plants, and has often been seen to prey upon insects ; and another common species met with near the outskirts of woods and in fields and meadows.

COREIDÆ

The family *Coreidæ* includes a number of land-bugs, which vary a good deal in form, but which possess in common the following characters : antennæ four-jointed, set rather high up on the head ; two ocelli generally present ; scutellum short and triangular ; elytral membrane strongly and thickly veined. These bugs mostly inhabit the warmer parts of the world, not more than about sixty species being found in Europe. Their habits are not very well known. Some of the European species live during the winter under leaves, and when disturbed in their retreat make by their movements a peculiar rustling sound. In summer they are to be found among herbs and shrubs seeking their food, or they may sometimes be seen flying actively in the sunshine. The figures—one representing a stout, strongly - built insect (*Syromastes marginatus*), the other a species (*Neides tipularius*) with a body as slender almost as that of a daddy-longlegs— illustrate what considerable differences of form are met with in this family, even among the common European species.

LYGÆIDÆ

The *Lygæidæ*, the next family of land-bugs, may be characterised as follows : antennæ four-jointed, arising from below an imaginary line drawn from the middle of the eye to the base of the rostrum ; two ocelli usually present, and placed close to the compound eyes ; sheath of rostrum composed of four nearly equal joints ; scutellum short and triangular ; membrane of elytra traversed by four or five longitudinal veins. They live for the most part under stones, dead leaves, or moss at the foot of trees, where they are often found together in large numbers ; and it is from their love of such obscure places that the name *Lygœus* has been given to the typical genus. They feed on the juices of plants or the dead bodies of other insects. A few species only show themselves in broad daylight.

The species of the genus *Pyrrhocoris*, and others associated in the same subfamily, are distinguished by the fact that they have no ocelli. *P. apterus* is a common and widely-spread European species, occasionally met with in Britain, which may be known by its red and black colours, and the want of hind wings, as well as of a membranous part of its elytra.

PLANT-BUGS

The plant-bugs (*Phytocoridæ*) have the following characters. Head triangular in shape, tricarinate above and without ocelli ; antennæ long, four-jointed, with the second joint longest, and the last two very slender ; rostrum four-jointed, resting against the under side of the thorax, and almost reaching to the end of it ; tarsi three-jointed ; elytra with an appendix, or smaller angular piece, divided off by a transverse suture from the rest of the coriaceous part of the elytra, and coming between it and the membrane. This family is well represented in temperate regions, and about three hundred European species are known. They are mostly soft-bodied, fragile bugs, presenting a considerable variety of colour, of which green is in many cases the predominant tint. They live principally on honey, and are to be found on flowers and in meadows. *Phytocoris tripustulatus*, a species with black elytra, marked with three orange spots on the outer margin, is common on nettles. Another species (*Calocoris striatellus*), widely

1, CICADA SEPTEMDECIM AND LARVÆ ; 2, MALE AND FEMALE CICADA ORNA

distributed throughout Europe, and met with chiefly on umbelliferous plants, is figured on page 1991.

FLOWER-BUGS

The *Acanthiidæ* form a family of mostly very small bugs, which are usually without ocelli, and have a three-jointed beak—lodged in a groove under the side of the head—and two-jointed tarsi. These bugs frequently have a somewhat peculiar appearance, owing

to the membranous or vesicular lobes with which the thorax, abdomen, and elytra are often furnished. For this reason they are sometimes known as membranaceous bugs. The species of the genus *Tingis* are seldom more than one-sixth of an inch long, and distinguished by the knob-like ends to their antennæ, as well as by the foliaceous expansion of their thorax and the extension of the latter behind to cover the scutellum. The common *T. affinis* may be recognised by the brown colour of its body, its transparent borders, with transverse brown nervures, and the X-shaped spot on the middle of each elytron. This species may be found on sandy soil among the roots of grasses, or under plants, such as wormwood, belonging to the genus *Artemisia*.

Another species (*T. pyri*) is noted for the injury it does to pear-trees, by pricking holes in hundreds on the under side of the leaves and extracting the sap. It is of a brown colour, with pale yellow or white elytra, marked with a brown spot at the base, and another at the extremity. *Aradus corticalis* is a common species, found under bark, which is figured on page 1991 to give an idea of the flattened form and membranous appearance of the bugs of the subfamily *Aradinæ*. These bugs have a longer rostrum and more cylindrical antennæ than those of the *Tingitinæ*.

The bed-bug (*Cimex lectularius*), which also belongs to this family, is a wingless species, with four-jointed antennæ, and a beak composed also of four joints, which can be turned back to lie in a groove under the throat. The shape of the insect may be seen from the figure, as well as the two lobes lying at the sides of the scutellum, which are all it has in the way of elytra. Closely allied species are found in dovecots, and in the nests of martins and swifts.

REDUVIIDÆ

The *Reduviidæ* are predaceous bugs, in which the head, narrowed behind in the form of a neck, carries two ocelli in addition to the compound eyes. Their antennæ are composed of four joints, though these are often subdivided in such a way that the number may appear much greater. The rostrum is short and strong, and three-jointed ; their legs are long, and have three-jointed tarsi ; and the fore legs often serve as prehensile organs, their tarsi being specially adapted for that purpose. *Reduvius personatus*, the largest British species, is three-quarters of an inch long, of a black-brown colour, with red legs, which, as well as the prothorax and antennæ, are somewhat hairy.

The *Saldidæ*, which, on account of their large projecting eyes, are sometimes known as Oculati, form, with the next family, a sort of transitional group between the land-bugs and water-bugs. They live in the neighbourhood of water, either by the sea-shore or along the sandy banks of inland waters ; and not only run with great rapidity, but often advance with leaps and bounds, their long, spiny hind legs being well fitted for this mode of locomotion. One of the species of the typical genus *Salda* is represented in the illustration.

POND-SKATERS

The pond-skaters (*Hydrometridæ*) have moderately long, conspicuous antennæ, and present other points of structure showing that they are nearly related to the true land-bugs. In some species wings, and in others elytra also, may be wanting. These insects may be seen walking or gliding about on the sunny surface of stagnant or slow-flowing waters ; and those of one genus (*Halobates*) are found on the surface of the sea, sometimes right out in mid-ocean. The true pond-skaters (*Gerris*), move about very quickly on the surface of the water, and use their fore legs in seizing their prey. *Limnobates stagnorum* is a more sluggish insect, walking slowly on the surface of the water or on the grassy banks; and is remarkable for its elongated, slender body, whence its name of needle-bug or water-gnat. This species is figured on this page, together with *Gerris paludum* and the larva of *Velia currens*.

WATER-BUGS

The water-bugs, Hydrocorisa, are distinguished from the land-bugs, not only by their mode of life, but also by their short, inconspicuous antennæ, and are mostly dull and uniformly coloured insects, frequenting stagnant waters, where they swim, some on their back, others with the back uppermost.

WATER-SCORPIONS

Of the two families, the water-scorpions (*Nepidæ*) have a small, narrow head, and their fore legs are specially modified to serve as prehensile organs. Whereas some swim actively, others drag themselves slowly along the bottom of the ponds in which they live. They are furnished with an appendage looking like a long tail, but consisting of two separate pieces, grooved on their inner surface, and capable of being locked closely together to form a tube, which leads to the two spiracles placed at the hind end of the body. When the insects come up to breathe, the tip of this breathing-tube may be seen emerging just at the surface of the water. The form of the body is in some (*Nepa*) broad and flat; in others (*Ranatra*) elongated. The female of *Nepa* lays her eggs in chains on aquatic plants, and each egg has seven short processes radiating from one end. The eggs of the *Ranatra* are laid one by one in notches, which the female makes in the stems of the plant. Certain exotic species of this family are remarkable for their great size, attaining in the genus *Belostoma* a length of over four inches.

WATER-BOATMEN

The water-boatmen (*Notonectidæ*) may be recognised by the large, broad head without ocelli, and the short, thick rostrum. They have long hind legs fringed with hairs on one side, which they use like oars in swimming. When the insect comes to the surface to breathe, it rests with these long legs, stretched out like a boatman leaning on his sculls. Though the name *Notonectidæ* has reference to their mode of swimming on the back, this habit is not characteristic of all the species. All are predaceous bugs, like all the rest of the same tribe, and are found abundantly in stagnant waters. Two of the common species, *Notonecta glauca* and *Corixa geoffroyi* are figured above.

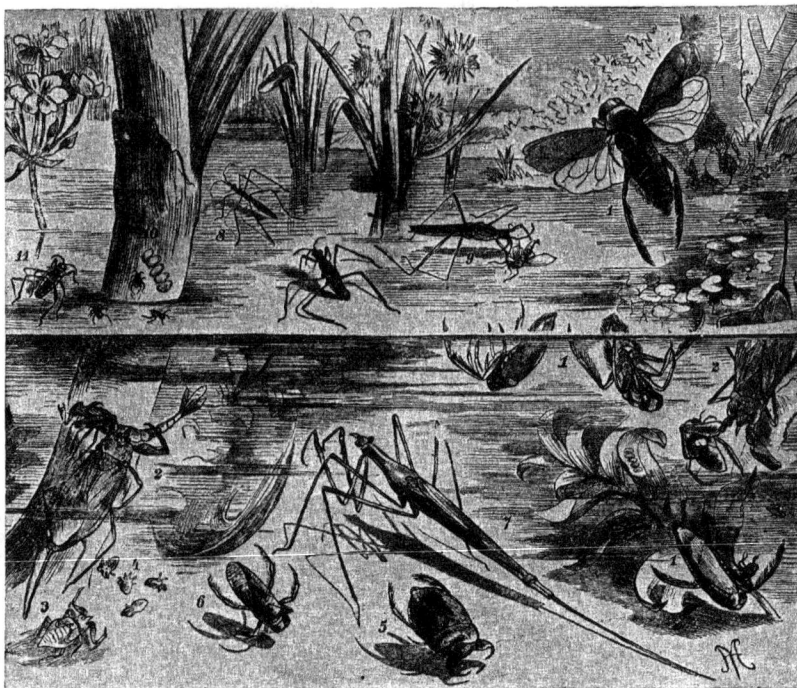

BRITISH WATER-BUGS

1, water boatman (Notonecta glauca); 2, water scorpion (Nepa cinerea), with, 3, its larva and, 4, its eggs; 5, Naucoris cimicoides; 6, Corixa geoffroyi; 7, Ranatra linearis: 8, Limnobates stagnorum; 9, pond-skater (Gerris paludum), with, 10, its eggs and larvæ; 11, larva of Velia currens

HOMOPTERA

The Homoptera present much greater variety in external form than the insects of the preceding group, from which they differ in the following characters. The beak arises from the lower and hinder part of the head, and is, therefore, almost completely hidden from view. The fore wings are, when present, of the same texture throughout the whole of their extent, and in many cases placed slanting, like the sides of a roof, when at rest. All the members of the section live by sucking the juices of plants; the females being often provided with a horny ovipositor—generally composed of three toothed plates, sheathed by two valves—for the purpose of making incisions in plants where the eggs are deposited. Unlike most bugs, they are not odoriferous insects, although many have special glands for the secretion of a kind of white waxy substance, often seen covering part of their body.

CICADAS

The cicadas (*Cicadidæ*) are stout-bodied insects, with a short, broad head, bearing prominent lateral eyes, and three distinct ocelli, which are often brightly coloured and resemble tiny jewels set near the middle of the forehead. The short antennæ are like small bristles inserted on the sides of the head just below the front margin of the eyes. The prothorax is short and broad, and the mesothorax also broad, on the upper side stretching back some distance behind to form a kind of shield. The fore wings are longer than the hind pair, both being often glossy and transparent, but sometimes finely coloured and more or less opaque.

Cicadas remain for a long period in the larval state, in many cases for several years; a North American species, *Cicada septemdecim*, being known as the seventeen-year locust, since that period is the interval between one generation of winged insects and the next. They inhabit chiefly the warmer regions of the earth, of the four or five hundred species known not more than eighteen being found in Europe, and these mainly in its southern parts. The song of the cicadas, which has been celebrated from very early times, is only produced by the male insects. " Happy " writes a Greek poet, " are the cicadas' lives, for they all have voiceless wives." The females are necessarily silent, since they are without the special apparatus for producing sound distinctive of the males. The two scaly plates which in the latter cover the under side of the base of the abdomen are not, as sometimes supposed, the sound-producing organs. But if one of them be stripped off, there will be disclosed a cavity, divided by an oblique horny ridge into two portions, the inner one somewhat irregular in shape, and exhibiting tense glistening membranes in its walls, while the outer portion is narrow, and opens by a narrow mouth towards the side. Hidden in the wall of the latter chamber lies the membrane which is the chief organ concerned in the production of sound. These membranes are set in vibration by the contraction and relaxation of a pair of strong muscles attached to their inner faces and lying inside the body. The other membranes in their neighbourhood seem to serve the purpose only of modulating the sound. The cicadas figured are two of the commoner species from South Europe. Both live on ash-trees, although *Cicada orni* selects by preference the manna ash. The specimen with its under side exposed may be easily recognised as a male, on account of the two plates, or opercula, covering the cavities in which the sound-apparatus is lodged.

1, 2, LEDRA AURITA ; 3, APHROPHORA SPUMARIA, AND 4, ITS LARVA

LANTERN-FLIES AND ALLIES

The lantern-flies and other insects included in the family *Fulgoridæ* are characterised by never having more than two ocelli, these being placed, one on each side, near the inner margin of the compound eyes. The latter are not large, and below them are inserted the short and inconspicuous antennæ. The front, vertex, and sides of the head are usually separated from one another by sharp crests, and the head itself is in some cases greatly prolonged in front. The fore wings are either similar in texture to the hind pair, or else somewhat harder and more leathery.

The Chinese lantern-fly (*Hotinus candelarius*), so widely distributed in Asia, is one of the best known ;

the common names said to be given to it in China being very suggestive of its luminosity, although so far there is no trustworthy evidence to show that it possesses any such property. A figure of the great lantern-fly (*Fulgora laternaria*) is given on page 1995.

Lantern-flies are nearly all prettily coloured ; and of the other insects belonging to the same family there are some, like those of the genus *Flata*, rivalling in the delicacy of their colours the most beautiful butterflies or flowers ; while others, as in the genus *Flatoides*, exhibit that curious mixture of grey and black which, in combination with the flattened form of their bodies, gives them the most astonishing resemblance to lichen-covered bark. The species of *Flata* and other genera are remarkable also for their white tufted tails of wax, which are found more especially in the larvæ, but are often present also in the winged insects. These insects do not stir far from their food-plant, on which they may be seen both in the larval and adult state, clustered together in large numbers, somewhat after the manner of plant-lice.

The European species of *Fulgoridæ* are not remarkable for their size or the brilliancy of their colours. *Issus coleoptratus* is perhaps the largest British species, and we figure *Cixius nervosus*, another widely distributed British and European species, together with *Pseudophana europæa*, the sole representative in Europe of its genus, and sometimes spoken of as the European lantern-fly.

FROG-HOPPERS

The frog-hoppers (*Cercopidæ*) are mostly small insects with a short, broad head and stiff, opaque elytra. They usually have two ocelli placed on the vertex of the head between the compound eyes, and their antennæ are inserted, not below the eyes, as in the *Fulgoridæ*, but between and a little way in front of them. These insects can give most vigorous leaps, and their hind legs are generally thickened or otherwise adapted for that purpose. They feed on various plants, and in the summer the frothy masses in which their larvæ lie concealed may be seen in numbers. It is from this habit the larvæ have of surrounding themselves in a mass of froth, known as cuckoo-spit, that the name *Aphrophora* (froth-bearing) has been given to one of the principal genera. A species of that genus and another form (*Ledra aurita*)—remarkable for an ear-like lobe on each side of the prothorax—

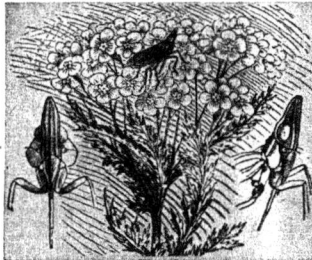

PSEUDOPHANA EUROPÆA

CENTROTUS CORNUTUS

are represented in the upper figures on this page.

MEMBRACIDAE

The family *Membracidæ* includes mostly exotic insects, which have in many cases an extraordinary appearance, owing to the shape of the prothorax, or the curious way in which it is armed with spines or knobs, or with both combined. In these insects the head is somewhat vertical, and usually placed rather

GREAT LANTERN-FLIES

soft, pulpy little creatures, with rather long antennæ and conspicuous round eyes, so commonly seen crowded together on the under side of leaves, in buds and flowers, in clefts in the bark of trees, and sometimes even on the roots. The antennæ are composed of from three to seven joints, on some of which are a number of curious rounded pits, probably of a sensory nature. The eyes are placed on the sides of the head, and each has often a sort of supplementary eye attached to its hind border; while in the winged aphides there are three

low down; it carries very short antennæ inserted near the front margin, and there are two ocelli between the compound eyes. The family is best represented in Tropical America, very few species being found in Europe, and two only in Britain. *Centrotus cornutus*, one of the two latter, may be recognised by the form of its prothorax, which carries on each side a horny spine, and is prolonged behind in another horny process, reaching almost to the end of the body.

LEAF-FLEAS

The leaf-fleas (*Psyllidæ*)—included with the next two families in that section of the order to which the name Phytophthires has been given—are little jumping insects, winged in both sexes, and using their wings not so much for the purpose of flying as to assist in their leaps. They have moderately long antennæ, consisting of eight or ten joints, and are thus easily distinguished from the *Cercopidæ*. The head is provided with three ocelli, in addition to the compound eyes, and the tarsi are two-jointed. Owing to their method of loco-

CIXIUS NERVOSUS

motion, these insects are not liable to be mistaken for plant-lice, although, like these insects, they infest the leaves and buds of plants. They prick the leaves to feed on the sap, their puncture being often followed by the formation of gall-like swellings. *Psylla genistæ* feeds on the broom, and other species are found on apple and pear trees.

PLANT-LICE

The plant-lice (*Aphidæ*) are small insects, which make up in numbers what they lack in size, and, owing to the injury they inflict on plants, must be ranked amongst the greatest pests with which the gardener and horticulturist have to contend. They are those

occlli on the crown of the head. The beak is composed of three joints, and the tarsi are two-jointed and terminate in two claws. Wings, as a rule, are found only in the adult males and in some of those generations of asexual individuals to be mentioned presently. The fore wings are longer than the hind pair, and placed in repose like a roof over the hind part of the body. Both pairs have a scanty venation, consisting in each wing of a single longitudinal vein, and of some simple or forked branches given off obliquely from it.

The number of species is considerable, and there is scarcely a single kind of plant that does not suffer as the special host of some one or more. Many are green, whence the name of green-fly by which they are commonly known; others are black, red, or some other colour. They are usually named after the plants on which they more particularly live, though each species is not necessarily confined to one kind of plant. Thus we have the plant-louse of the rose (*Aphis rosæ*), the green aphis of the apple (*A. mali*), which is found also on the pear and sloe tree, the cherry aphis (*A. cerasi*), and a host of others named in the same manner.

The life-history of plant-lice is very complicated, and, although differing somewhat in different species, is always characterised by what is known as an alternation of generations. There are several broods or generations of these insects in the course of a year, but it is only in the last autumn brood that true sexual individuals are found. The males are generally provided with wings, but the females are larger and wingless; they lay fertilised eggs, from which, in the following spring, the first brood of the year is produced. The insects of this brood are usually wingless, and give birth to living young, or, as in the genus *Phylloxera*, lay eggs from which the young subsequently develop. The new brood, thus produced parthenogenetically, resembles the one from which it has sprung, and gives rise to a fresh brood in a similar manner. As many as nine or ten generations may succeed one another in this way during the course of the season, before the appearance in the autumn of the last sexual generation. The brood preceding and giving rise to the latter often consists of winged individuals, which leave the plant on which they were born and fly to some other.

In the genus *Phylloxera* the males are wingless, and each of the sexual females lays but a single egg, known as the winter egg; but in other forms the number is often much greater. Each of the parthenogenetic females of *Phylloxera* may in the course of its life lay as many as two hundred eggs, and each of the viviparous females of other species may give birth before they die to forty or fifty young. When we consider that there are several generations every year, it can be easily understood how it is that these insects spread with such rapidity; and a sum in geometrical progression would show that the individuals which might arise in the course of a year from a single winter egg of *Phylloxera* are not to be counted by hundreds or thousands, but by millions. Other species are capable of multiplying as rapidly. Fortunately, plant-lice have many enemies, such as the larvæ of lady-bird beetles, of lace-wing flies, and of the flies of the family *Syrphidæ*. These larvæ devour great numbers, and ichneumon flies also help to keep them in check.

Plant-lice are divided into a number of subfamilies, of which the first is represented by the genus *Aphis*. In this genus the antennæ are seven-jointed and about as long as the body; the two horny tubes called *cornicles*, which project from the back of the abdomen, are also characteristic. Through these tubes the lice secrete a sweet kind of liquid much sought after by ants, who, in an affectionate way, come and caress the aphides in order to obtain it. The sticky substance known as honey-dew, which is often spread in a shiny layer over the upper surface of leaves, is, in most cases, nothing but the liquid dropped by the crowds of plant-lice living above on the under side of other leaves.

The members of the allied subfamily *Lachninæ* have six-jointed antennæ, and instead of cornicles possess prominent glandular structures placed on the back of the abdomen. The figured *Lachnus punctatus* is found on the willow. The apple-blight insect (*Schizoneura lanigera*), which may be recognised by the white fluff covering in the wingless individuals the back of the abdomen, belongs to another subfamily. The winged individuals of this species are black, whereas those devoid of wings are of a yellowish or reddish brown colour, and live in the crevices of bark. The species is supposed to have been introduced from America, hence its name of American blight.

In the genus *Phylloxera*—distinguished among other characters by the three-jointed antennæ—one species lives on the leaves of the oak-tree, while a second (*P. vastatrix*) is the dreaded insect so destructive to the leaves and roots which they attack. These, like many other species of the family, cause the formation of galls on the leaves and roots which they attack. The curious galls with the appearance of small fir cones, to often seen on young shoots of the spruce-fir, are caused by a species (*Chermes abietis*) remarkable for its complicated life-history.

SCALE-INSECTS

The scale-insects (*Coccidæ*), which owe their name to the fact that the larvæ and females of many species look like oval or rounded scales attached to the bark and leaves of plants, are very dissimilar in the two sexes. The adult males are provided with one pair of wings; the hind wings being rudimentary or altogether absent; they have rather long antennæ, distinct eyes, and, in some cases, are furnished with two long bristle-like tails. These winged males are very rarely seen, which is accounted for by the fact that their mouth-parts are atrophied, so that they are incapable of taking nourishment, and live only a short time. The females are always wingless, and usually remain fixed to one spot, with their beak buried in the tissues of the plant, and their back often spread out in the form of a shield covering the head and body. The beak is generally three-jointed, the antennæ are short, and in the tarsi, which appear at first sight to consist of but one joint, two or three joints may on close examination be distinguished, the last ending, as a rule, in a single claw. In many species the female dies shortly after laying eggs beneath her, when her body dries up and remains as a protective cover for them.

When the larvæ are hatched they soon leave this shelter, and rove about the food-plant in search of a suitable place in which to insert their beaks and begin the operation of pumping up the sap. They cast their skin several times in the course of their growth; and those which become adult females undergo no great change in appearance, beyond an increase in their size, a gradual lengthening of the antennæ, and a partial or almost complete obliteration of the segmentation of their bodies. With the male larvæ the case is different; these, unlike all others belonging to the

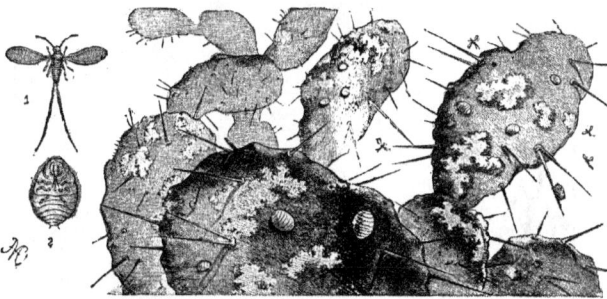

COCHINEAL INSECTS
The enlarged figures represent, 1, the male, and, 2, the female

order, undergoing a true metamorphosis before reaching the perfect state. Each prepares for itself a sort of cocoon, and it becomes transformed into a quiescent pupa, from which, after a certain lapse of time, the winged insect emerges.

In *Orthezia* and other genera the female, instead of keeping to one spot on the food-plant, moves about and taps it at different points in order to extract the sap. When the eggs are laid, she envelops them in a kind of white cottony secretion and leaves them. Some species penetrate beneath the epidermis of their food-plant, and often cause the formation of galls, which, growing up around them, sometimes take the most extraordinary shapes.

Scale-insects are probably more numerous within the tropics than in more temperate regions, although comparatively few of these tropical species have been described. These insects are found on the bark and leaves, and sometimes even on the roots of several different kinds of plants. They multiply rapidly, and often prove as injurious as the most noxious plant-lice. The orange, apricot, olive, peach, fig, and other fruit trees, as well as ornamental shrubs like the rose, have each their own species, from which they sometimes suffer severely. Some years ago the orange plantations of California were threatened with ruin owing to the ravages of *Icerya purchasi*, which had been accidentally imported from Australia, and had spread with great rapidity. Experts were sent to Australia to try and discover the natural enemies of the insect in its native country; it was found that the scale-insect was there kept in check by dipterous and hymenopterous parasites, but chiefly by the larvæ of a lady-bird beetle. A number of these beetles and parasitic insects were brought to America, and set to prey upon the *Coccidæ*. When they had multiplied sufficiently, they were distributed amongst several orange plantations, with the satisfactory result that many were soon almost entirely cleared of the scaly-bug.

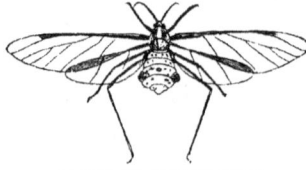

LACHNUS PUNCTATUS

Though many species of *Coccidæ* have to be combated because of their injuries, there are a few which are cultivated on account of the useful products they yield. Among these, the cochineal insect (*Coccus cacti*) is a native of Mexico and other parts of Central America, where it feeds on a species of cactus; but it has been introduced into Spain, Algeria, and a few other countries. The male is of a dark red colour, with pale wings; the female has a reddish brown colour, but her body, which shows a distinct segmentation until the time of laying, is covered with a white powder. About seventy thousand dried bodies of these insects, chiefly females, are said to be contained in a single pound of cochineal.

Long before the introduction of cochineal into Europe, two native species of *Coccidæ* had been used for similar purposes. The dye with which the ancients produced their deep red or crimson colours was obtained from *Chermes vermilio*, known to the Greeks as kokkos and to the Arabians and Persians as kermes or alkermes.

Another species (*Porphyrophora polonica*), formerly known as the scarlet grain of Poland, is found in many parts of Central Europe, and was at one time extensively collected for the sake of the red dye it afforded. The lac-insect (*Carteria lacca*) of the Oriental countries, not only furnishes the colouring matter called lac-dye, but causes also an exudation of a resinous substance, gum-lac, from the bark of the trees on which it lives. Stick-lac is the name given to this substance in its native state while still adhering to the twigs of the tree; when separated, pounded, and freed by washing from its colouring matter, it is known as seed-lac, which, after further preparation, becomes lump-lac, or shellac.

The Pediculina, or true lice, as distinguished from the bird-lice of the order Orthoptera, are provided with piercing and suctorial mouth parts, and live on the blood of animals, to which by this means they are enabled to gain access. Though they are without wings, and were at one time associated with other wingless insects in a separate order, lice are now generally regarded as degraded forms of Rhynchota, in which the wingless condition has been brought about as an adaptation to their parasitic life.

In these insects the head is set horizontally, and carries short, cylindrical, and usually five-jointed antennæ; the eyes are small and simple; and the mouth consists externally of a soft, retractile beak, somewhat conical in shape, and furnished below with a row of hooks for attachment. Within the fleshy beak there are four grooved pieces, forming by their juxtaposition an inner membranous tube, which can be extended beyond its sheath, and acts both as a piercing organ and as a conduit for the passage of the blood which is sucked up by the insect.

The thorax is small and not distinctly divided into segments, while the abdomen is relatively large, generally somewhat elliptical in outline, and exhibits seven or eight clearly marked segments. The tarsi are two-jointed, with the second joint in the form of a claw, which can be turned back towards the first. Lice multiply rapidly, one generation succeeding another in a short space of time. Their pear-shaped eggs are generally found attached to the bases of the hairs; the young, which are hatched after about eight days, undergo no metamorphosis, and, in some cases, require only about eighteen days before becoming adult.

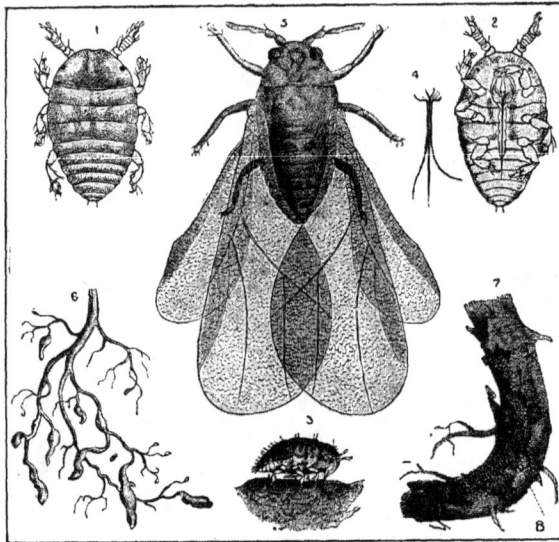

VINE-PHYLLOXERA

1 and 2, wingless form found on the root, seen from above and below; 3, the same from the side; 4, piercing organs; 5, winged individual; 6, rootlets of vine, with swellings caused by Phylloxera; 7, old root stock; (8) hibernating individuals

THYSANOPTERA

The insects comprised in the order Thysanoptera—some of them familiar enough to gardeners and others, by whom they are known as thrips—are all small. A few species only exceed four or five lines in length, while the great majority are less than a tenth of an inch long. They are distinguished from all other insects by certain peculiarities in the structure of their mouth and of their wings and tarsi. The mouth lies far back on the under side of the head; its mandibles are transformed into a pair of piercing setæ, while the upper lip, maxillæ, and labium—the two latter provided with short palpi—are united together to form a short suctorial tube. The wings are small and narrow, contain few nervures, and are thickly fringed all round with long hairs. Two

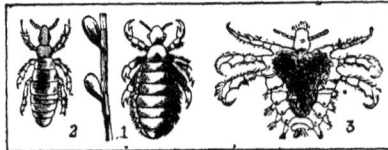

1, HEAD-LOUSE AND EGGS; 2, BODY-LOUSE; 3, CRAB-LOUSE

pairs of such wings are generally present, but in some cases they may be rudimentary or altogether wanting. The tarsi, which consist of one, two, or three joints, are without claws at the end, but are furnished instead with small vesicular lobes, by means of which they adhere to the surface on which they rest.

To these characters of the order we may add that the body is narrow and cylindrical; the thorax is formed of three, and the abdomen of ten segments; there are only three or four pairs of spiracular openings—two on the abdomen and one or two on the thorax; three ocelli are generally present on the head in addition to the fairly large faceted eyes; and the antennæ are composed of from seven to nine joints. The larvæ have a general resemblance to the adult insects, and in their last stage they remain inactive and take no nourishment.

Less than a hundred species of Thysanoptera, belonging mostly to the European fauna, have been described. These little insects are frequently to be seen on flowers, and on other parts of plants. They feed upon the juices, and when present in large numbers are capable of doing an appreciable amount of injury. Some destroy the pollen grains, and so prevent the fertilisation of the flowers. The corn-thrips (*Thrips cerealium*) sucks the young grains on the ears of corn, and stops their further growth. *Heliothrips hæmorrhoïdalis*, another species, is common in hothouses, where it may be found on the young buds of several kinds of plants.

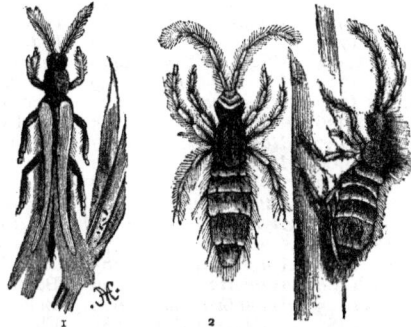

1, THRIPS CEREALIUM; 2, PODURA VILLOSA

THYSANURA

The Thysanura are active little insects, which live generally in obscure places and are mostly of too small a size to attract much attention. They never exhibit any trace of wings, undergo no metamorphosis, and have a distinctly segmented body, which is usually covered with hairs or scales and furnished behind either with a forked tail, used as a springing apparatus, or with two or three long, jointed appendages, which sometimes serve a similar purpose. Characterised on the whole by a somewhat primitive type of structure, and in general appearance resembling the larvæ rather than the adult forms of other insects, the Thysanura are in some cases distinguished by special features.

SPRING-TAIL OF THE GLACIERS

The spring-tails, forming the suborder Collembola, are all furnished on the under side of the first abdominal segment with a curious tube or sucker, from the mouth of which a glandular process, secreting a viscid matter, can be protruded; they are remarkable also from the fact that in most of them no trace of a tracheal system has yet been discovered. In the Collembola the eyes,

when present, are in the form of simple or grouped ocelli; the antennæ number not more than six joints, and the abdomen has at most but six segments and very often only three. The forked tail, which is attached to one of the hinder segments, is usually turned forwards and held in position under the body; when released, it springs back, striking the surface of support, and causes the insect to bound up into the air.

These little insects are to be found commonly enough under flower-pots, leaves, and stones, or under the bark of trees and in other such situations. They may sometimes be seen collected together in great numbers, and spread over the surface of the ground like a layer of powder. Some species, such as *Podura aquatica*, may frequently be seen floating in patches on pools of water, and by striking their tails against the surface of the water they can spring up into the air just as readily as others do from the ground. The glacier flea (*Desoria glacialis*) is an interesting species, found in Alpine regions, where it is often to be met with on the surface of the ice.

The bristle-tails (Thysanura proper) form but a small number of genera, some of which are very remarkable in having a series of small rudimentary legs on each side of the abdomen in addition to the ordinary six legs borne by the thorax. In all the genera the antennæ are formed of a large number of joints; and the abdomen shows ten distinct segments, and, except in the genus *Japyx*, carries at the end two or three long jointed tails. *Japyx* has instead a pair of short pincers like an earwig.

The little silver-fish (*Lepisma saccharina*) is one of the best-known insects belonging to this suborder. Found very often in damp corners in houses among old books or papers, it may be recognised by the silvery scales covering its body, and by its three bristle-like tails, of which the middle one is the longest. It feeds on the paste in the binding of books, and on sugary and starchy substances generally, though it is credited also with eating paper and linen.

Two species of *Machilis* are found in Great Britain; one being common about rocks at the seaside, while the other is to be met with under stones in different parts of the country. *Campodea staphylinus*, the last insect we have to mention, is a pale, soft-bodied little creature, which is common almost everywhere under stones and in loose garden soil. It runs actively, and has two very long tails, which it sticks up in the air or turns forward over its body. It has no eyes; the antennæ are shorter than the tails and of equal thickness throughout; and the abdomen has seven pairs of rudimentary appendages.

INDEX TO HARMSWORTH NATURAL HISTORY

INDEX

INDEX

END OF THIRD VOLUME